35th Air Infiltration and Ventilation Centre Conference (AIVC 2014), 4th TightVent Conference and 2nd Venticool Conference 2014

Ventilation and Airtightness in Transforming the Building Stock to High Performance

Poznan, Poland
24 – 25 September 2014

ISBN: 978-1-5108-3763-8

Printed from e-media with permission by:

Curran Associates, Inc.
57 Morehouse Lane
Red Hook, NY 12571

Some format issues inherent in the e-media version may also appear in this print version.

Printed by Curran Associates, Inc. (2017)

For permission requests, please contact International Network for Information on Ventilation and Energy Performance (INIVE) at the address below.

International Network for Information on Ventilation and Energy Performance (INIVE)
Lozenberg 7
B-1932 Sint-Stevens-Woluwe
Belgium

Fax: +32 (0)2 529 81 10

info@inive.org

Additional copies of this publication are available from:

Curran Associates, Inc.
57 Morehouse Lane
Red Hook, NY 12571 USA
Phone: 845-758-0400
Fax: 845-758-2633
Email: curran@proceedings.com
Web: www.proceedings.com

35th Air Infiltration and Ventilation Centre Conference (AIVC 2014), 4th TightVent Conference and 2nd Venticool Conference 2014

Ventilation and Airtightness in Transforming the Building Stock to High Performance

Poznan, Poland
24 – 25 September 2014

INIVE

Table of Contents

Poznań, Poland
Andersia Hotel
24–25 September 2014

35th AIVC Conference
4th TightVent Conference
2nd venticool Conference

Ventilation and airtightness in transforming the building stock to high performance

PROCEEDINGS

In cooperation with:

Acknowledgments

The conference organisers gratefully acknowledge the support from:

AIVC with its member countries : Belgium, Czech Republic, Denmark, Finland, France, Germany, Greece, Italy, Japan, the Netherlands, New Zealand, Norway, Poland, Republic of Korea, Sweden and USA.

Since 1980, the annual AIVC conferences have been the meeting point for presenting and discussing major developments and results regarding infiltration and ventilation in buildings. AIVC contributes to the programme development, selection of speakers and dissemination of the results. pdf files of the papers of older conferences can be found in AIRBASE. See www.aivc.org.

TightVent Europe

The TightVent Europe 'Building and Ductwork Airtightness Platform' was launched in January 2011.

It aims at facilitating exchanges and progress on building and ductwork airtightness issues, including the production and dissemination of policy oriented reference documents and the organization of conferences, workshops, webinars, etc. The platform has been initiated by INIVE EEIG (International Network for Information on Ventilation and Energy Performance) and receives active support from the following organisations: Aeroseal, Buildings Performance Institute Europe, BlowerDoor GmbH, European Climate Foundation, Eurima, Lindab, Retrotec, Soudal, Tremco illbruck, and Wienerberger. More information can be found on www.tightvent.eu.

venticool

The international ventilative cooling platform, venticool (venticool.eu) was launched in October 2012 to accelerate the uptake of ventilative cooling by raising awareness, sharing experience and steering research and development efforts in the field of ventilative cooling. The platform supports better guidance for the appropriate implementation of ventilative cooling strategies as well as adequate credit for such strategies in building regulations. The platform philosophy is pull resources together and to avoid duplicating efforts to maximize the impact of existing and new initiatives. venticool will join forces with organizations with significant experience and/or well identified in the field of ventilation and thermal comfort like AIVC (www.aivc.org) and REHVA (www.rehva.eu). Venticool has been initiated by INIVE EEIG (International Network for Information on Ventilation and Energy Performance) with the financial and/or technical support of the following partners: Agoria-NAVENTA, ES-SO, Velux, Wienerberger and WindowMaster

Poznań University of Technology

Poznań University of Technology (PUT, www.put.edu.pl) was established in 1919 and currently, with over 20 thousand students and more than 1 000 academic staff, is one of the leading technical universities in Poland.

10 Faculties cover wide range of technical knowledge: Architecture, Chemical Technology, Civil and Environmental Engineering, Computing, Engineering Management, Electrical Engineering, Electronics and Telecommunications, Mechanical Engineering and Management, Technical Physics, Machines and Transportation. PUT offers Bachelor, Master and Doctorate courses in Polish and English. European Credit Transfer System (ECTS) is implemented, many foreign students study in the Erasmus program. PUT participates in Polish and international scientific projects and is involved in cooperation with industry.

INIVE EEIG (International Network for Information on Ventilation and Energy performance)

INIVE was founded in 2001.

INIVE is a registered European Economic Interest Grouping (EEIG), whereby from a legal viewpoint its full members act together as a single organisation and bring together the best available knowledge from its member organisations. The present full members are all leading organisations in the building sector, with expertise in building technology, human sciences and dissemination/publishing of information. They also actively conduct research in this field - the development of new knowledge will always be important for INIVE members.

INIVE has multiple aims, including the collection and efficient storage of relevant information, providing guidance and identifying major trends, developing intelligent systems to provide the world of construction with useful knowledge in the area of energy efficiency, indoor climate and ventilation. Building energy-performance regulations are another major area of interest for the INIVE members, especially the implementation of the European Energy Performance of Buildings Directive.

With respect to the dissemination of information, INIVE EEIG aims for the widest possible distribution of information.

The following organisations are members of INIVE EEIG (www.inive.org):

BBRI - Belgian Building Research Institute - Belgium
CETIAT - Centre Technique des Industries Aérauliques et Thermiques - France
CSTB - Centre Scientifique et Technique du Bâtiment - France
IBP - Fraunhofer Institute for Building Physics - Germany

SINTEF - SINTEF Building and Infrastructure - Norway
NKUA - National & Kapodistrian University of Athens - Greece
TNO - TNO Built Environment and Geosciences, business unit Building and Construction - Netherlands

The following organisations are associated members.

CIMNE - International Center for Numerical Methods in Engineering, Barcelona, Spain
eERG - End-use Efficiency Research Group, Politecnico di Milano, Italy
ENTPE - Ecole Nationale des Travaux Publics de l'Etat, Vaulx en Velin, France
TMT US - Grupo Termotecnia, Universidad de Sevilla, Spain

INIVE

PROPOSED CHANGE IN SPANISH REGULATIONS RELATING TO INDOOR AIR QUALITY WITH THE AIM OF REDUCING ENERGY CONSUMPTION OF VENTILATION SYSTEMS

Linares, Pilar*[1], García, Sonia[1], Sotorrío, Guillermo[1], Tenorio, José A[1]

1 Eduardo Torroja Institute for construction sciences-CSIC
4, Serrano Galvache St.
Madrid, Spain

**Corresponding author: plinares@ietcc.csic.es*

ABSTRACT

The ventilation required in order to maintain acceptable indoor hygiene standards results in a significant consumption of energy. Currently the Spanish regulations on indoor air quality (IAQ) require minimum rates for delivery-to and extraction-from the habitable rooms of residential buildings. These rates are not adjustable, so ventilation systems based on variable ventilation rates, are not normally deemed acceptable unless a comprehensive statement of compliance is provided, justifying the proposed ventilation solution. However the use of variable ventilation systems is desirable, as it would almost certainly produce a reduction of the overall ventilation rate and, consequently, a reduction in the heating and cooling energy demand while maintaining a good level of air quality.

This paper presents part of the ongoing research towards the modification of the Spanish regulations in order to adapt required ventilation rates to real needs. This would mean allowing reduction in ventilation rates and energy demand but without any impact on indoor air quality.

The objective behind this research is to propose to the Spanish Government the substitution of the current required constant ventilation rates by maximum values of CO_2 concentration as an indicator of air quality. By establishing maximum values, the implementation of ventilation systems based on variable ventilation rates will be enabled because the justification will be more easily provided.

KEYWORDS

Ventilation, IAQ, regulations, energy efficiency

1 INTRODUCTION

The current Spanish Building Code was enforced in 2006 including IAQ provisions for dwellings which represented a big regulatory step. However the provisions are not as performance-oriented as was initially anticipated, requiring minimum rates for delivery-to and extraction-from the habitable rooms. These rates are not adjustable, so ventilation systems based on variable ventilation rates are not normally deemed acceptable, unless a comprehensive statement of compliance is provided, justifying the proposed ventilation solution.

In 2010 with the adoption of the recast EPBD, EU Member States faced new challenges. The new goal is to increase the level of performance towards nearly-zero energy buildings by 2020. In order to achieve this goal in Spain, a deep review of the energy requirements has been made in 2013 increasing the energy efficiency of buildings. However, this is not enough, energy efficiency in buildings is affected as well by ventilation systems.

Therefore the use of variable ventilation systems is desirable, as it would almost certainly produce a reduction of the global ventilation rate and, consequently, a reduction of the heating and cooling energy demand, while maintaining a good air quality level.
As a consequence research is in progress to modify the Spanish regulations to allow the use of more efficient systems, adapting required ventilation rates to real needs. This would mean a reduction in ventilation rates and energy demand but without impact on indoor air quality.
The goal of this research is to update the regulations which should require an IAQ level that shall be provided. Equally, the goal is to provide a simplified verification method that facilitates the fulfilment of this IAQ level.

2 CURRENT IAQ REQUIREMENT

The current IAQ requirement establishes minimum ventilation rates (see Table 1) for delivery-to and extraction-from the habitable rooms. These rates have to be provided in a continuous way.

Table 1. Minimum ventilation rates

Rooms	**Per person**	**Per usable floor area m^2**	**Per room**
Bedrooms	5 l/s		
Living and dining rooms	3 l/s		
WC and bathrooms			15 l/s
Kitchens		2 l/s	

3 PROPOSED IAQ REQUIREMENT

IAQ level is usually characterized by a maximum level of pollutants that may affect people´s health and comfort and that could be achieved by different ventilation systems.
However, common pollutants are not easy to assess, so generally an indicator is used to model the state of the rest of the pollutants. Among the possible pollutants that are commonly produced indoors, CO_2 is the most commonplace and closest related to human activity. In spite not supposing a health risk in the usual concentrations that is found in dwellings, CO_2 is a very good indicator of the ventilation rate. This way is the most common indicator used in regulations and guides.
The required CO_2 concentration is limited in two ways:

- 900 ppm maximum yearly average;
- 500.000 ppm per hour maximum yearly accumulated above 1.600 ppm. This parameter shows the relationship between the CO_2 concentrations reached above a limit value and their duration over a year. It can be calculated as the sum of the areas (in ppm•h) within the representation of the CO_2 concentration as a time function and the limit value (See Fig. 1).

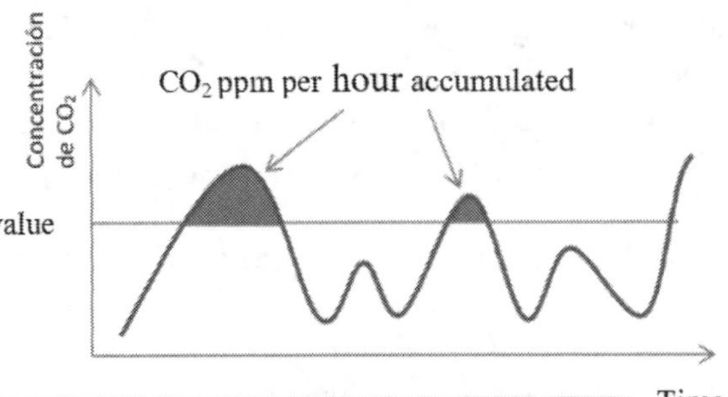

Fig. 1. CO_2 concentration over time

These required concentration levels shall be achieved under certain design conditions (such as occupancy scenarios, CO_2 production rate, yearly average outdoor CO_2 concentration, etc.)

that should be set in the regulation. This means it is a design performance because it would only be measurable *in situ* under these conditions.
These values have been chosen based on the specified values in the RITE taking into account an outdoor CO_2 concentration of 400 ppm: the value corresponding to IAQ 2 for the maximum yearly average concentration and the IAQ 4 value for the base over which to calculate the maximum yearly accumulated. (See Table 2)

Table 2. IAQ classes according to RITE

IAQ Classification		**CO2 concentration (ppm)**
1	Best quality indoor air	750
2	Good quality indoor air	900
3	Medium quality indoor air	1.200
4	Low quality indoor air	1.600

4 PROPOSED VERIFICATION METHOD

The fulfilment of the requirement can be achieved by the use of expertise methods (like specialized software), but it is convenient as well that at least a simplified verification method is provided by the regulations for designers to use. This simplified method shall be easy to use by non-expert practitioners and will consist of different ventilation rates (continuous and variable) that will provide fulfilment of the requirement.
These ventilation rates are obtained from the results of an analysis of the IAQ of different dwelling types (case studies) assessed with pollutants distribution software CONTAM. The analysis consists of simulating these dwellings (with an occupancy scenario) with different ventilation rates in order to optimize them achieving the required IAQ.
By using CONTAM we have obtained for each dwelling CO_2 concentrations for certain ventilation flows in each room. (See Fig. 2).

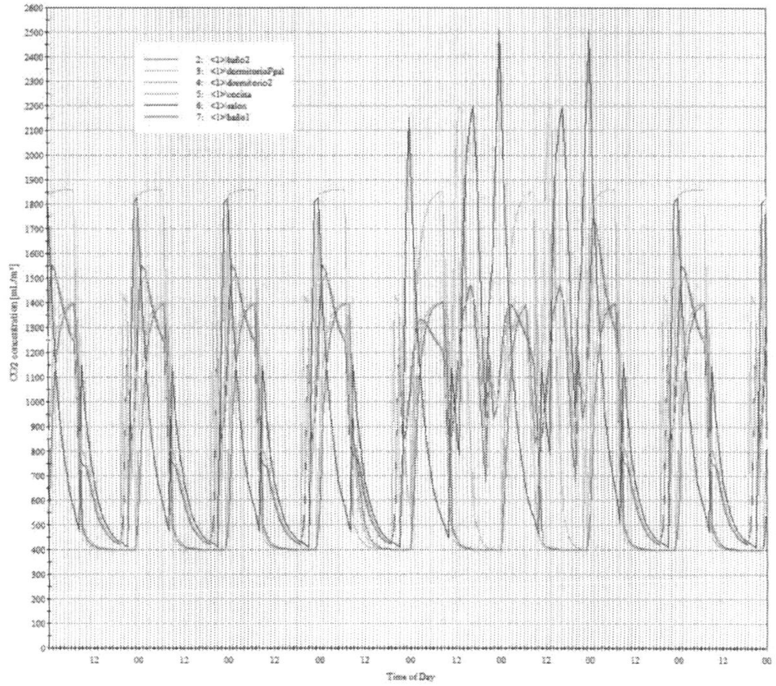

Fig. 2. Results for a week for dwelling 2 with variable ventilation flow:5-14 l/s. CONTAM.

From these data we can derive the yearly average and the yearly accumulated over 1.600 ppm, and compare them with the IAQ requirements. These ventilation flows are optimized, choosing the minimum ones that allow fulfilling the IAQ requirement.
Several dwelling types have been chosen for the analysis taking into account their bedroom and bathroom counts (See Table 3). They are real dwellings representative of the dwellings that have been built recently (the Spanish population and dwelling census has been used). The results for each of these dwellings can be extended to the rest of cases of the same type.

Table 3. Dwelling types

Kind and composition of dwelling	**Type**
Flat: Living/Kitchen+1 Bedroom+1 Bathroom	1
Flat: Living+Kitchen+2 Bedrooms+2 Bathrooms	2
Flat: Living+Kitchen+3 Bedrooms+2 Bathrooms	3
Flat: Living+Kitchen+4 Bedrooms+2 Bathrooms	4
Terrace house: Living+Kitchen+4 Bedrooms+2 Bathrooms	5

The occupancy scenario allows setting the CO_2 production for each room at any time. The number of occupants in each dwelling has been determined depending on the number of bedrooms. (See Table 4) based on the Spanish population and dwellings census.

Table 4. Dwellings occupancy

Bedroom number	**Occupants**
≤ 1	2
2	3
≥ 3	4

The scenario shall specify for each occupant: sleeping hours, number and duration of each stay in every room and times when the occupants are out from home. (See Table 5).

Table 5. Occupancy scenario for one person

Room	**Monday to Friday**	**Room**	**Saturday and Sunday**
Bedroom 1	0:00-8:00	Bedroom 1	0:00-8:00
Kitchen	8:00-8:30	Kitchen	8:00-8:30
Bathroom 1	8:30-9:00	Bathroom 1	8:30-9:00
	----------------	Living room	9:00-10:00
Living room	17:00-18:00		----------------
Bathroom 1	18:00-18:05	Living room	12:00-13:00
Living room	18:05-20:00	Kitchen	13:00-14:00
Kitchen	20:00-21:00	Living room	14:00-15:30
Bathroom 1	21:00-21:05	Bathroom 1	15:30-15:35
Living room	21:05-00:00	Living room	15:35-18:30

		Kitchen	20:30-21:30
		Bathroom 1	21:30-21:35
		Living room	21:35-00:00

5 RESULTS

Table 6 shows the continuous flow values for the different dwelling types.

Table 6. Results with continuous flow

Dwelling type	Continous flow [1] (l/s)	Total continous flow (l/s)	Yearly average CO_2 concentration (ppm)	Yearly accumulated over 1.600 ppm (ppm·h)
1	6	12	898	11.700
2	7	21	875	417.040
3	12	36	828	437.840
4	13	39	870	497.380
5	9	27	844	441.220

(1) In kitchen and each bathroom

Table 7 shows the possible variable flow values for the different dwelling types.

Table 7. Results with variable flow

Dwelling type	Variable flow (1)		Yearly average CO_2 concentration (ppm)	Yearly accumulated over 1.600 ppm (ppm·h)
	MAX (during occupation) (l/s)	MIN (during no occupation) (l/s)		
1	7	5	901	0
	10	4	883	0
2	11	6	862	261.560
	14	5	892	244.660
	21	4	898	106.340
3	14	11	844	421.720
	18	9	880	380.900
	33	6	897	185.640
4	22	12	874	264.420
	30	10	884	235.820
	47	8	897	127.920
5	14	7	864	398.840
	34	4	878	307.320

(1) In kitchen and each bathroom

5 CONCLUSIONS

Results show how it should be possible to achieve target IAQ requirements based on variable ventilation rates and lower continuous ventilation rates than the ones that are currently required, thus saving energy for heating and cooling without impacting air quality.
Ongoing research will quantify the saved energy in each case study.

6 REFERENCES

*Código Técnico de la Edificación (*Building Code*)*. RD 314/2006.

Documento Básico sobre Ahorro energético (Basic procedure for Building Energy Efficiency). RD 235/2013

RITE. Reglamento de Instalaciones Térmicas en los Edificios. RD 238/2013.

Gavira, M, Linares, P et al (2005). *Comportamiento higrotérmico de la envolvente del edificio según el CTE. ... Soluciones alternativas: sistemas de ventilación por caudal variable (*Hygrothermal behaviour of the building envelope according to CTE...Alternative solutions: variable flow ventilation systems*). I Jornadas de investigación, 2,* 739-756.
EPBD. Directive 2010/31/EU of the European parliament and of the council on the energy performance of buildings (recast).
Spengler, J., Samet, J.M., Mc Carthy, J. *Indoor Air Quality Handbook.* McGraw Hill. 2000.
UNE-CEN/TR 14788:2007 In *Ventilation for buildings - Design and dimensioning of residential ventilation systems.*
CONTAM *Multizone Airflow and Contaminant Transport Analysis Software*. National Institute of Standards and Technology (NIST)
Censo de Población y Viviendas (Population and dwellings census) 2001, INE
•*Réglementation aération des logements.* Arrêtés du 24.03.82 et 28.10.83.

DURABLE AIRTIGHTNESS IN SINGLE-FAMILY DWELLINGS: FIELD MEASUREMENTS AND ANALYSIS

Wanyu Rengie Chan* and Max H. Sherman

Lawrence Berkeley National Laboratory
1 Cyclotron Road, Mail Stop 90R3058
Berkeley, CA 94720, USA
**Corresponding author: wrchan@lbl.gov.*

ABSTRACT

This study presents a comparison of air leakage measurements collected recently (November 2013 to March 2014) with two sets of prior data collected between 2001-2003 from 17 new homes located near Atlanta, GA, and 17 homes near Boise, ID that were weatherized in 2007-2008. The purpose of the comparison is to determine if there are changes to the airtightness of building envelopes over time. Durability of building envelope is important to new homes that are increasingly built with improved levels of airtightness. It is also important to weatherized homes such that energy savings from retrofit measures, such as air sealing, are persistent. Analysis of the multi-point depressurization data shows that the blower tests characterized the air leakage at 50 Pa pressure difference well. This is shown by good agreement between air changes per hour at 50 Pa (ACH50 or n_{50}) as measured, and as estimated from the fitted values of leakage coefficients and pressure exponents to the multi-point depressurization data. We used Student's t-test to compare the current two sets of air leakage measurements with their respective prior data. Results suggest that the mean of 6.5 ACH50 measured recently from the new homes was higher than the mean of 5.6 measured previously in 2001-2003. Calculations of the percentage change with respect to the prior ACH50 show that all but one new home show increases in ACH50. The median percentage increase in ACH50 is about 20% for new homes, but it is nearly zero for the weatherized homes. We performed a regression analysis to describe the relationship between prior and current measurements of ACH50. For the new homes, best estimate of the slope factor is approximately 1.15, meaning that the regression model predicts a 15% increase in ACH50 over ten years. On the other hand, analysis of the weatherized homes suggests no significant increase (slope factor near 1). Further analysis of the data is underway that will characterize the potential increase in air leakage among new homes using data from ResDB (LBNL's Residential Diagnostic Database). More understanding of the factors associated with building envelope durability will eventually lead to improvements in building materials and practices that are better at sustaining airtightness in the long run.

KEYWORDS

Blower door, fan pressurization measurements, air leakage, new construction, weatherization

1 INTRODUCTION

The building industry has made great progress over the past 30 years in building homes with improved airtightness. Most homes have demonstrated improved levels of airtightness through testing shortly after construction, however, little is known about how the airtightness changes with time as houses age. This is also a concern in retrofitted homes, where the energy savings from air sealing might be short-lived if the airtightness improvements are not durable.

Analysis of the LBNL Residential Diagnostic Database (ResDB, Chan et al. 2012) suggests that the air leakage of new US single-family detached homes improves at a rate of roughly 1% per year, such that the airtightness testing results when the homes are new shows that recently constructed homes are about 10% tighter than homes built ten years ago. The database has limited test results available from homes that were built in the same year, but tested at different times after construction. Analysis of these tests also showed that for homes all built within a given year, there is an increase in air leakage at about 1% per year with respect to the age of the home when the blower door test was performed. However, this result is uncertain because there are many external factors that the regression analysis cannot account for.

To better address the question of air leakage changes with time, this study performed air leakage testing in homes where a blower door test was performed approximately five to ten years ago. This study targeted two types of homes. The first category of homes were built between 2001 and 2003, and with the blower door test performed prior to occupancy. These data will reflect a potential change in air leakage after approximately ten years. Homes to be recruited in this category had not had had any major renovations. The second category of homes had undergone retrofits, with the air-sealing work and blower door test performed between 2007 and 2008.

2 METHODS

We collaborated with two subcontractors to collect air leakage measurements of single-family homes on this project: Southface Energy Institute in Atlanta, GA, and Community Action Partnership of Idaho (CAPAI) in Boise. These organizations were selected because they had access to homes in the above two categories, i.e., (i) homes built between 2001 and 2003 that were tested for air leakage when new; (ii) homes that were weatherized between 2007 and 2008. Both Southface and CAPAI were very knowledgeable about the characteristics of the homes in their area because they worked closely with builders and homeowners in their communities. Their field technicians routinely conduct blower door tests, and are comfortable with performing variations of the blower door test besides the typical single-point measurement at 50 Pa depressurization. In this study, both pressurization and depressurization tests were performed at multiple pressure points as a way to evaluate the extent to which testing conditions may influence the air leakage results.

Southface tested 17 homes that were built between 2001 and 2003 from Atlanta and its surrounding neighborhoods of Alpharetta, Cumming, and Decatur. CAPAI also tested 17 homes that participated in low-income weatherization program between 2007 and 2008 from Boise, Caldwell, Nampa, and Notus. Southface and CAPAI reached out to potential homeowners by phone and by using mailing materials. The recruitment materials and phone scripts were prepared by LBNL and approved by LBNL's Institution Review Board (IRB) for protection of human subjects. Each participant signed a consent form, and received a small financial incentive for completing the blower door test. Personal identifiable information, such as homeowner names, full street address, and phone number, were treated as secured data by Southface and CAPAI. This information is not shared with LBNL or included in any of our reporting or analyses.

Southface and CAPAI recruited homes and conducted blower door tests between November 2013 and March 2014. In addition to the blower door test, other basic information about the homes was also collected, including floor area, number of stories, number of bedrooms, year built, foundation type, presence of an attached garage, and the type of heating and cooling equipment. General descriptions about the air barrier if presence, caulking, weatherstripping, use of spray foam and mastic at the different building components were also noted in some of the homes.

3 RESULTS

3.1 Descriptions of Sampled Homes

All the 2001-2003 new homes belonged to an energy efficiency program. Table 1 shows the basic characteristics of the 17 homes recruited for this study. They are typically 275 m^2 (3000 ft^2) in floor area, ranging between 170 m^2 and 400 m^2. Most of the homes (14 of 17) are two-stories. Many of them (9 of 17) have a basement. The number of homes with finished (5 homes) and unfinished basement (4 homes) is about equal. The remaining homes are either built on slab (4 homes) or have a crawlspace (4 homes). Most of the homes (15 of 17) are heated by forced-air furnace and cooled by centralized air-conditioning. Only two homes are the exceptions, where heat pumps are used instead for heating and cooling. Due to the large size of these homes, many of them (9 of 17) have two heating and cooling systems, where one of them is in the attic, and the other is in the basement.

Table 1: House characteristics of new homes built between 2001 and 2003.

ID	City	Year Built	Floor Area (m^2)	Ceiling Height (m)	Stories	N Bed-room	Foundation	Heating/Cooling (x2 = two systems)
N1	Cumming	2001	256	3.7	2.5	4	Crawlspace (unvent)	Heat pump
N2	Cumming	2001	243	2.7	2	4	Crawlspace (unvent)	Furnace/AC
N3	Cumming	2003	305	3.0	2.5	5	Basement (cond)	Furnace/AC
N4	Cumming	2003	191	3.0	2.5	5	Slab	Furnace/AC (x2)
N5	Alpharetta	2002	287	3.0	2	4	Basement (cond)	Furnace/AC (x2)
N6	Alpharetta	2003	305	3.7	2.5	2	Basement (cond)	Furnace/AC (x2)
N7	Cumming	2001	277	3.0	2.5	4	Basement (cond)	Furnace/AC (x2)
N8	Cumming	2001	336	3.0	2	5	Basement (cond)	Furnace/AC (x2)
N9	Cumming	2002	203	3.2	2.5	5	Basement (uncond)	Furnace/AC (x2)
N10	Cumming	2003	281	3.0	2.5	3	Slab	Furnace/AC
N11	Alpharetta	2001	330	3.0	2	3	Basement (uncond)	Furnace/AC (x2)
N12	Atlanta	2001	405	3.7	1	2	Slab	Heat pump (x2)
N13	Decatur	2002	170	2.6	1	2	Crawlspace (vent)	Furnace/AC
N14	Decatur	2002	202	3.0	1	3	Slab	Furnace/AC
N15	Decatur	2002	289	2.6	2	3	Crawlspace (vent)	Furnace/AC (x2)
N16	Cumming	2002	296	3.0	2	3	Basement (uncond)	Furnace/AC
N17	Cumming	2002	281	3.0	2	4	Basement (uncond)	Furnace/AC

Table 2 shows the characteristics of the homes sampled by CAPAI. The homes were smaller in size, with a mean floor area of 130 m^2 (about 1400 ft^2), and all of them are single-story. The common foundation types are crawlspace (11 homes) and basement (6 homes). Most crawlspaces are vented (10 of 11). There are homes with conditioned basement (4 homes) and unconditioned basement (2 homes) in the dataset. Most of the houses (13 of 17) are heated by a forced-air furnace. Three of the homes use electric baseboard as the main heating equipment, and one uses a wood pellet stove. A wide range of cooling equipment was used in these homes, including centralized AC, wall AC, window AC, or an evaporative cooler (swamp cooler). There were also three homes that currently do not have a cooling system.

It was noted by the field technician that weatherization work in these homes typically include doors/windows upgrade (14 homes received doors/windows replacement or air sealing around them), insulation of floor (11 homes) or ceiling (7 homes), and duct sealing and/or insulation (9 homes).

Table 2: House characteristics of homes weatherized between 2006 and 2008.

ID	City	Year Built	Floor Area (m^2)	Ceiling Height (m)	Stories	N Bed-room	Foundation	Heating/Cooling
W1	Caldwell	1959	87	2.4	1	3	Crawlspace (vent)	Furnace/Wall AC
W2	Boise	1978	103	2.3	1	2	Crawlspace (vent)	Furnace/Evap Cool
W3	Boise	1930s	151	2.7	1	3	Crawlspace (unvent)	Furnace/AC
W4	Boise	1977	94	2.3	1	3	Crawlspace (vent)	Furnace/AC
W5	Caldwell	1960s	101	2.4	1.5	3	Crawlspace (vent)	Furnace/AC
W6	Caldwell	1951	234	2.4	1	4	Basement (cond)	Furnace/Evap Cool
W7	Caldwell	1970s	102	2.4	1	3	Crawlspace (vent)	Furnace/Widw AC
W8	Boise	1970s	99	2.3	1	2	Crawlspace (vent)	Furnace/(none)
W9	Caldwell	1967	89	2.3	1	2	Crawlspace (vent)	Elec. /(none)
W10	Caldwell	1979	116	2.3	1	2	Crawlspace (vent)	Furnace/AC
W11	Nampa	1948	114	2.2	1	2	Basement (uncond)	Furnace/Evap Cool
W12	Notus	1974	125	2.3	1	3	Basement (cond)	Elec. /Evap Cool
W13	Nampa	1927	204	2.4	1	4	Basement (cond)	Furnace/AC
W14	Boise	1900s	117	2.7	1	2	Basement (uncond)	Elec./Wall AC
W15	Boise	1968	188	2.4	1	4	Basement (cond)	Furnace/Evap Cool
W16	Boise	1960s	122	2.5	1	3	Crawlspace (vent)	Furnace/(none)
W17	Nampa	1967	181	2.3	1	4	Crawlspace (vent)	Wood/Evap Cool

3.2 Air Leakage Measurements

All prior measurements of air leakage were collected from single-point depressurization test at a pressure difference of 50 Pa. The new homes were tested following RESNET test protocol, and the weatherized homes were tested post-weatherization following testing procedure specified by the Weatherization Assistance Program. The air change rates at 50 Pa, n_{50} orACH50 (h^{-1}), were computed from the reported airflow rates at 50 Pa divided by the house volume (Equation (1)). House volume, V (m^3), is estimated by multiplying the floor area by the ceiling height.

$$ACH50 = Q_{50Pa}/V \quad (1)$$

Both depressurization and pressurization measurements were collected from the two sets of homes, with differential pressures ranging between ±30 to ±60 Pa. The leakage coefficient, C ($m^3/s\text{-}Pa^n$) and pressure exponent, n (-), were fitted from the depressurization test results using Equation (2).

$$Q = C \times \Delta P^n \quad (2)$$

where Q (m^3/s) is the air flow through the blower door at a differential pressure ΔP (Pa).

Estimated C for new homes has a mean of 0.10 $m^3/s\text{-}Pa^n$ (std. dev. = 0.041). For weatherized homes, C has a mean of 0.065 $m^3/s\text{-}Pa^n$ (std. dev. = 0.019). Estimates of n for the new homes and weatherized homes have a similar mean values of 0.68 (std. dev. = 0.074) and 0.66 (std. dev. = 0.043), respectively.

The new homes have larger leakage coefficients partly because they are larger in size than the weatherized homes. Using the fitted values of C and n, estimates of ACH50 for new homes has a mean of 6.5 (std. dev. = 2.4). For weatherized homes, the estimated mean of ACH50 is 10.2 (std. dev. = 3.6). ACH50 estimated from the fitted values of C and n are essentially the same as the ACH50 calculated from the single point measurement at 50 Pa, as shown in

Figure 1. The remaindering of this analysis will use the ACH50 measured at a single point of 50 Pa depressurization, which is closer to the test protocol used to collect the prior measurements.

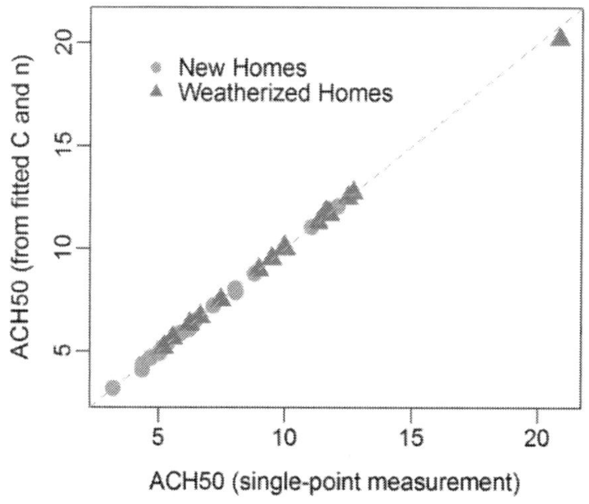

Figure 1: Comparison of the air changes per hour at 50 Pa (ACH50) in two groups of homes, calculated using a single-point measurement at 50 Pa depressurization (x-axis), and by using fitted values of C and n (y-axis).

3.3 Changes in Air Leakage

Figure 2 compares the ACH50 measured previously with the current measurements. Results from the t-test (Table 3) suggest that there is a change in ACH50 among the 2001-2003 at a 75% confidence interval (from p-value = 0.248). The confidence level improves when the ACH50 data is log-transformed that there is an increase in ACH50. On the other hand, the t-test results show no change in mean ACH50 for the weatherized homes in 2007-2008. The log-transformation has little impact on the analysis for this group of homes.

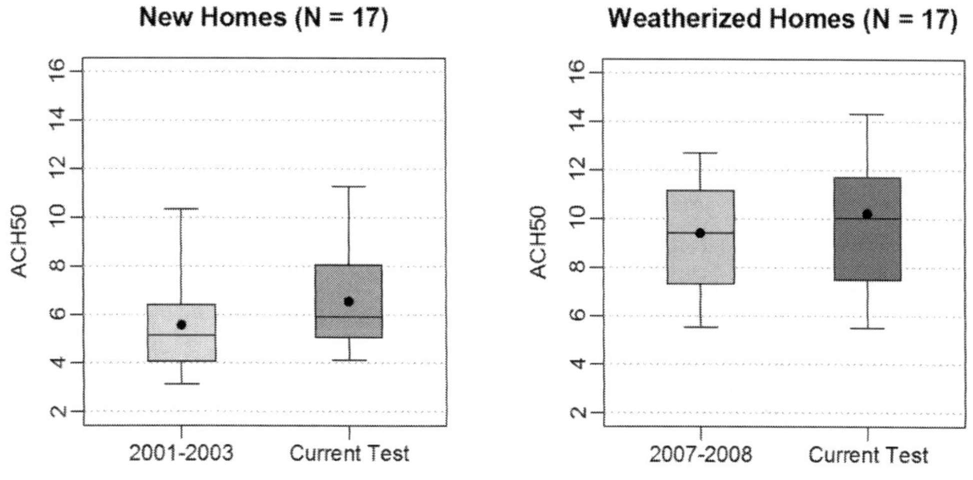

Figure 2: Comparison of the air changes per hour at 50 Pa (ACH50) in two groups of homes, where air leakage were measured previously and were repeated again recently. The boxplot shows interquartile range (25th to 75th percentile), and the median. The whiskers extend to 5th and 95th percentiles. The solid triangle shows the mean ACH50.

Table 3: Summary statistics of Student's t-test results

Dataset	Parameter	Prior Test Mean ACH50	Current Test Mean ACH50	t-Test: Difference b/w Prior and Current Mean Values	
				p-value	95% Conf. Interval
New Homes	ACH50	5.56	6.54	0.248	-0.71 to 2.66
New Homes	log (ACH50)	1.62	1.82	0.180	-0.10 to 0.49
Weatherized Homes	ACH50	10.21	9.40	0.460	-1.41 to 3.03
Weatherized Homes	log (ACH50)	2.27	2.20	0.577	-0.16 to 0.29

Figure 3 shows the change in ACH50 calculated from the current tests with respect to the prior tests measured in 2001-2003 for the new homes, and 2007-2008 for the weatherized homes. All but one of the new homes show an increase in ACH50. The median change is about 20%. On the other hand, there are roughly equal numbers of weatherized homes that show positive and negative changes in ACH50. The median change is nearly zero.

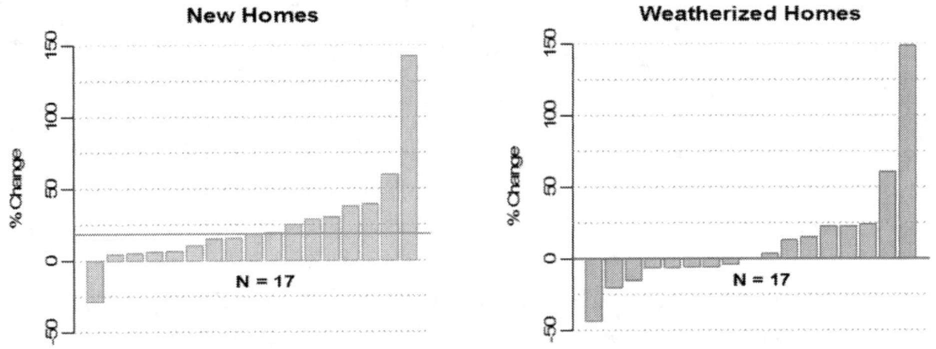

Figure 3: Percentage change in ACH50 calculated with respect to the prior values measured in 2001-2003 for the new homes, and 2007-2008 for the weatherized homes. The horizontal line indicates the median % change.

3.4 Regression Analysis

A regression analysis was preformed to describe the relationship between the ACH50 measured from the prior air leakage tests and now. The intercept was set to zero in the linear regression, as follows:

$$ACH50_{current} = a \times ACH50_{prior} \tag{3}$$

Figure 2 compares the ACH50 measured with results from the prior tests plotted on the x-axis, and results from the current tests on the y-axis. Overlaid on these figures are the results from the linear regression. Solid blue line shows the model-fit from least square estimate. The dotted blue lines shows the 95% confidence interval from the predictions. The regression results are summarized in Table 4.

For the 2001-2003 new homes, three of homes (N9, N12, and N14) indicated in Figure 4 by the "X" symbol appear to be outliners, suggesting possible abnormality with the blower door tests. If these three homes were excluded, the slope of the regression line would increase but only by a small amount. This is illustrated by the solid orange line in Figure 4. Field technicians reported problems with keeping the attic hatch door closed during the blower door test in one of these three homes (N14). N9 and N12 had the highest and lowest ACH50 estimated for this group of homes tested when new. This is perhaps an indication that their construction was unique in some ways from the rest of the group, or perhaps the data is inaccurate for some reason. Unfortunately, the prior test reports on N9 and N12 did not record any detailed information that could explain their extreme values. Excluding these three homes resulted in a slightly higher slope estimate of 1.18, instead of 1.12 (see Table 4). Based on these two estimates of the slope factor, the increase in ACH50 is about 1.15, i.e. 15% increase over ten years for these new homes.

For the weatherized homes, there was one home that showed substantial increase (W13) in ACH50, and another home that showed substantial decrease (W12) in ACH50. Table 4 shows that these two data points have negligible effect on the slope estimate. The slope estimates do not preclude zero at the 95% confidence interval. Based on this analysis, which is in agreement with the percentage change calculations presented above, we concluded that there is no significant change in the air leakage of homes that were weatherized in 2007-2008.

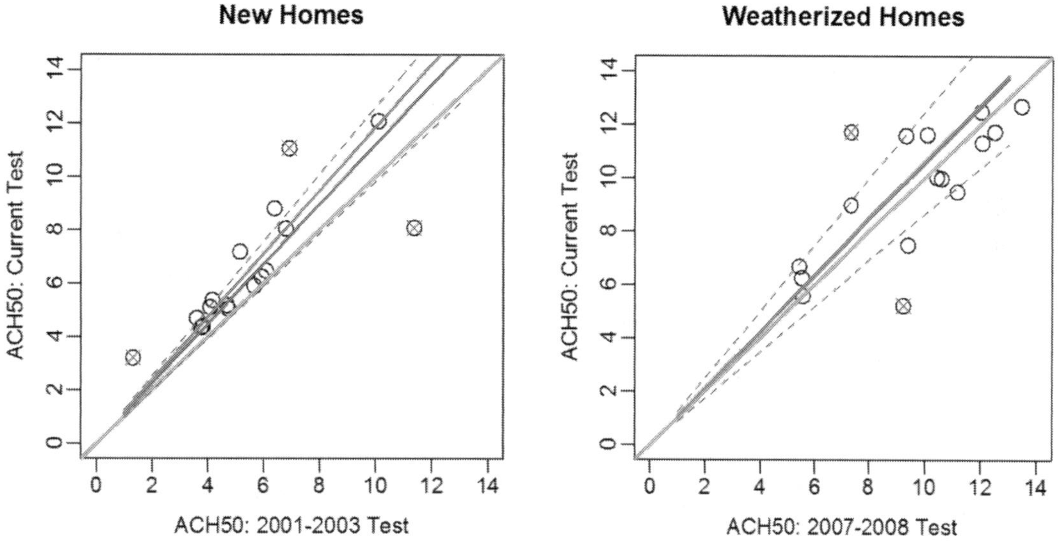

Figure 4: Comparison of the air changes per hour at 50 Pa (ACH50) in two groups of homes, where air leakage were measured previously and were repeated again recently. The boxplot shows interquartile range (25th to 75th percentile), and the median. The whiskers extend to 5th and 95th percentiles. The solid triangle shows the mean ACH50.

Table 4: Results of linear regressison.

Dataset (see Figure 4 for data excluded)	Estimate of *a*	Std. Error	p-value	95% Conf. Interval	R^2
New Homes	1.12	0.066	1.17e-11	0.98 to 1.26	0.944
New Homes (excl. 3 data points)	1.18	0.031	9.60e-15	1.12 to 1.25	0.991
Weatherized Homes	1.05	0.090	2.84e-9	0.86 to 1.24	0.889
Weatherized Homes (excl. 2 data points)	1.06	0.091	1.36e-8	0.87 to 1.26	0.900

4 CONCLUSIONS

Blower door measurements in homes built between 2001 and 2003 show an increase in air leakage of about 15% in ten years. The rate of increase in air leakage observed in this study is about the same as previously analyzed in ResDB (approximately 10%), thus confirming our hypothesis that aging of the building envelope is the cause. On the other hand, effectively no increase in air leakage was observed in homes that were weatherized about six or seven years ago between 2007 and 2008. This suggests that the improvements made from weatherization were still effective. Moreover, aging appears not to occur in this later group of homes that were all built prior to 1970s.

The vastly different results from these two groups of homes suggest that the leakage sites may be different. For the weatherized homes, the joints between building components that were sealed, such as by caulking and weatherstripping around windows and doors, do not change their leakage characteristics with time. On the other hand, in new homes, the new and moist wood materials can shrink over the first several years, potentially causing leaks in the building envelope. The effect of this drying process may be similar to the relationship between air leakage and indoor humidity observed by Kim and Shaw (1986). In addition, past work by Proskiw (1998) measured the airtightness of 17 Canadian homes over an 11-year period and found leakage occurring at the floor drains, around duct penetrations and windows, even though the air barrier remained effective. The drying of wood frames leading to shrinkage and therefore gaps between building components may be one reason that could explain the leaks that were found.

Since the finding of this study is based on a small set of data from two groups of homes located near Atlanta, GA and Boise, ID. More data is needed to determine if aging of the building envelope leading to increase in air leakage is a widespread issue in the US housing stock. We plan to incorporate the measurements collected from this study with the air leakage model developed using data from ResDB. If further analysis also suggests that there is an increase in air leakage over time, then there is an opportunity for improvements in building materials and practices that can better sustain airtightness and realize the energy savings. While the change in air leakage is only about 15% over ten years based on this study, some homes experienced rather substantial increases in ACH50 (>30%). Future work to identify factors that are associated with durability issues would provide valuable information on how to improve airtightness not just test-when-new, but also in the long run.

5 ACKNOWLEDGEMENTS

Support for this work was provided by the U.S. Dept. of Energy Building America Program, Office of Energy Efficiency and Renewable Energy under DOE Contract DE-AC02-05CH11231; by the U.S. Dept. of Housing and Urban Development, Office of Healthy Homes and Lead Hazard Control through Interagency Agreement I-PHI-01070; by the U.S. Environmental Protection Agency Indoor Environments Division through Interagency Agreement DW-89-92322201-0; and by the California Energy Commission through Contract 500-09-042.

We want to thank all the homeowners who participated in this study for their time and cooperation. We also want to acknowledge the field team who performed this work, led by Eyu-Jin Kim at Southface Energy Institute, and Hans Berg at Community Action Partnership of Idaho (CAPAI).

6 REFERENCES

ASTM (2010). E779-10 Standard Test Method for Determining Air Leakage Rate by Fan Pressurization. ASTM International, Conshohocken, PA.

Chan W.R., Joh J., and Sherman M.H. (2012) Analysis of Air Leakage Measurements from Residential Diagnostics Database. LBNL Report 5967E. Lawrence Berkeley National Laboratory, Berkeley, CA.

Kim A.K. and Shaw C.Y. (1986) Seasonal variation in airtightness of two detached houses. Measured Air Leakage of Buildings: A Symposium, Issue 904. American Society for Testing and Materials, p.17-32.

Proskiw G. (1998) The variation of airtightness of wood-frame houses over an 11-year period. Thermal Performance of The Exterior Envelopes of Buildings VII Conference Proceedings. December 6-10, Clearwater Beach, FL.

IMPACT OF A PHOTOCATALYTIC OXIDATION LAYER COVERING THE INTERIOR SURFACES OF A REAL TEST ROOM: VOLATILE ORGANIC COMPOUND MINERALISATION, RISK ASSESSMENT OF BY-PRODUCT AND NANOPARTICLE EMISSIONS.

Franck Alessi*[1], Didier Therme[1]

1 CEA, INES, SBST, LGEB,
50 avenue du lac Léman (bâtiment HELIOS)
F-73377 Le Bourget-du-lac, France.
franck.alessi@cea.fr

ABSTRACT

Many studies about photocatalytic oxidation (PCO) have been carried out in laboratories. They use an inert test chamber with ideal indoor conditions: a low volume, a controlled temperature and humidity, and a constant injection of one to five specific gases. The principal aim of this study was to implement, in a real test room (TR) of an experimental house, a titanium dioxide (TiO_2) layer to quantify its efficiency. This layer, directly in contact with the indoor air (IA), was one of the four layers embedded in a passive system (PS) specifically designed to improve the indoor air quality (IAQ) and the thermal comfort. A specific monitoring in the TR assessed the removal rate of the volatile organic compounds (VOCs), as well as formaldehyde (HCHO) as a possible intermediate, and the nanoparticle (NP) emissions. In addition, a comparison was made with a reference room (RR) which was not equipped with the PS.

KEYWORDS

Photocatalytic oxidation, Indoor air quality, Volatile organic compounds, Nanoparticles, Real test room.

1 INTRODUCTION

The photocatalytic oxidation may be divided into six elemental mass transfer processes occurring in series (Zhong et al., 2010): 1) Advection (pollutants are carried by airflows), 2) External diffusion of reagent species through the boundary layer (BL) surrounding the catalyst or catalyst pellet, 3) Adsorption onto the catalyst surface, 4) Chemical reaction at the catalyst surface, 5) Desorption of reaction product(s), 6) Boundary layer diffusion of product(s) to the main airflow.
One of the most common choices of photocatalyst is titanium dioxide. This TiO_2 has two crystal forms: Anatase and Rutile with respectively an energy band-gap of 3.23eV and 3.02eV. When the TiO_2 semiconductor is illuminated by photons, an electron in an electron-filled valence band (VB) is excited by photoirradiation (the energy hν must be equal or greater than the band-gap energy) toward a vacant conduction band (CB), leaving a positive hole in the VB (Mo et al., 2009). The key step in photocatalysis is the formation of hole-electron pairs on irradiation with UV-light. The energy of UVA [$320 \leq \lambda(nm) \leq 400$], UVB [$280 \leq \lambda(nm) \leq 320$], UVC [$100 \leq \lambda(nm) \leq 280$], are widely used because it is equal or greater than the 3.2eV band-gap energy of TiO_2 (Zhong et al., 2010).

These electrons and positive holes drive reduction and oxidation, respectively, of compounds adsorbed on the surface of a photocatalyst such as oxygen 'O_2', water vapor 'H_2O', 'VOCs' (Ginestet et al., 2005; Auvinen et al., 2008; Zhong et al., 2010; Mo et al., 2009).
The activation equation can be written as:

$$TiO_2 + h\nu \rightarrow h^+ + e^- \quad (1)$$

In this reaction, h^+ and e^- are powerful oxidizing and reducing agents, respectively. The oxidation (2) and reduction (3) reactions can be expressed as:

$$h^+ + OH^- \rightarrow OH^\bullet \quad (2)$$

$$e^- + O_{2ads} \rightarrow O^-_{2ads} \quad (3)$$

Where O_2^- is a superoxide anion. When organic compounds are chemically transformed by a PCO, it is the hydroxyl radical ($OH^\bullet$), derived from the oxidation of adsorbed water or adsorbed hydroxide ion (OH^-), that is the dominant strong oxidant. Its net reaction with a VOC can be expressed as (Mo et al., 2009):

$$OH^\bullet + VOC + O_2 \rightarrow nCO_2 + mH_20 \quad (4)$$

Some reactants will generate partial oxidation products (intermediates) which are relatively more harmful to people's health (Mo J, Zhang Y et al., 2009). The oxidation process sometimes stops along the way yielding aldehydes, ketones and organics acids (Mo et al., 2009). For example, formaldehyde is frequently quoted as one of the main intermediates (Ginestet et al., 2005; Auvinen J et al., 2008; Kolarik J et al., 2010).
Furthermore, the photocatalyst may potentially emit some nanoparticles of TiO_2 in the indoor air. The TiO_2 dust, when inhaled, has been classified as possibly carcinogenic to humans (Group 2B) WHO (2010). The size of these nanoparticles may vary from 10 to 50nm INRS (2012).
A lot of studies use the PCO in a laboratory test chamber. Their volumes are very small (often less than $1m^3$) and most of the time the tests are carried out at constant air temperature (e.g. 20°C) and relative humidity (e.g. 50%), coupled with a constant injection of one to five different gases able to represent, partially, the indoor air.
In this study, a PS was developed embedding 4 layers with a total thickness of 5.5cm: an adhesive layer (0.3cm), a thermal insulation layer (3cm), a thermal storage layer (2cm), and a photocatalytic layer (0.2cm, directly in contact with the IA). This photocatalyst contained 5% of TiO_2 doped with carbon. Consequently, only 2.32 eV has to be transferred into this layer instead of 3.23 eV for the pure anatase. This study focuses only on the PCO impact on the indoor air quality of a room equipped with the PS (TR) and without the PS (RR).

2 METHODOLOGIES

2.1 Location of the tests

The tests were carried out in two rooms (the test room -TR- and the reference room -RR-) of an experimental test house located on the INCAS platform of the French National Institute of

Solar Energy (INES). To compare the TR and the RR, each room was covered with an identical sarcophagus (SG) on the walls and the ceiling. This SG was made of plasterboard and was covered with 2 layers of white paint. After the SG installation, the TR and the RR had the same volume 3.62 x 3.10 x 2.34 m^3 and identical building materials, insulation, and paint. The rooms were south facing and they were juxtaposed on the first floor as shown in Figure 1.

To understand the PCO impact on the indoor air, the TR was equipped with the passive system (PS) and the RR was without the PS. No furniture was added in the TR and the RR. There was no air exchange rate to avoid new pollutants entering from the outside, and the doors and the windows remained closed. The sources of pollution mainly came from the PS in the TR and from the SG in the RR. The floors of the TR and the RR were covered with identical linoleum 4 years ago.

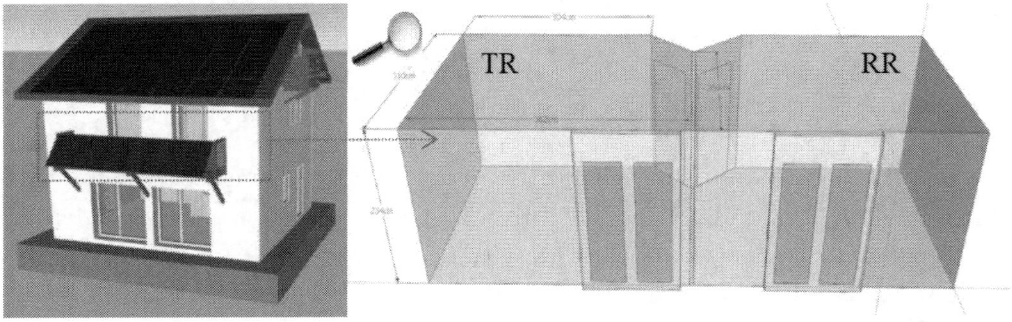

Figure 1 From the left to the right: illustration of the experimental test house, and localization of the test room (TR) and the reference room (RR)

2.2 Measurement campaigns

Four measurement campaigns were carried out to quantify the impact of the PS on the IA.

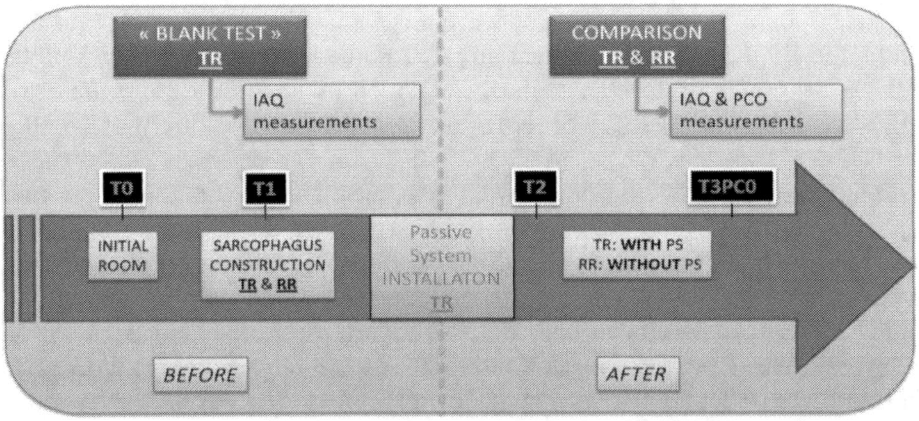

Figure 2 Four different steps, from January to May, to quantify the impact of the PS on the indoor air

The steps T0 and T1 were an "IAQ blank test" focused only on the TR. T0 corresponded to the room in its original state whereas T1 was equipped with the SG, see Figure 2. The parameters monitored and their positioning are indicated in the Table 1.

The step T1 monitored, directly after the installation of the SG, the same parameters as the previous step T0. In addition, the HCHO concentration was monitored to assess the SG emission and to compare this result with the HCHO concentration coming from the PS in the next step T2.

The step T2 was done right after the PS installation both in the TR and the RR. To highlight the PCO efficiency (T2) and to compare the results with the RR and the previous steps (T0 and T1), the tests were carried out without interference: no occupants, no intrusions, no furniture, doors and windows closed, mechanical ventilation system (MVS) was off, and the roller blind remained open.

In the final step T3PCO, the difference with T2 came from the use of specific actuators to modify the indoor environment modifying the behaviour of the PCO. The TR and the RR were equipped with identical resistances to make a ramp of temperature, visible lamps and UVA lamps (only in the TR) to activate the PCO, fans to mix the IA and improve the contact time with the TiO_2, roller blinds opened or closed to show the impact of the natural light on the PCO, and finally the MVS turned 'on' coupled with the opening of the door or the window to compare the PCO efficiency and the ventilation on the IAQ.

2.3 Parameters monitored

For the main parameters (Table 5), the monitoring was principally focused on the TR where the passive system was installed. To highlight the PCO effects, the measurements were made before and after the PS installation, and in comparison with the reference room (RR) when possible. The air temperature (Ta) and the relative humidity (RH) were monitored at each step in the TR, then in the two rooms after the PS installation. These parameters are some key factors for the PCO efficiency as well as the pollutant concentration, the type of photocatalyst and its quantity, the type of lighting and so on (Mo et al., 2009; Juan et al., 2003). As the photocatalyst layer should remove the VOCs and give off some carbon dioxide (CO_2), the total volatile organic compounds (TVOC) and the CO_2 were monitored. The formaldehyde (HCHO) was also measured as one possible intermediate as well as the number and the diameter of the nanoparticles (NP) to highlight a possible release of TiO_2 in the indoor air. The TVOC, CO_2, HCHO, NP were monitored before and after the PS installation and compared with the RR when it was possible (for more details, cf. Table 1).

Table 1. Information about the sensors used

Parameters	Sensor model	Type of sampling	Rooms monitored	Steps	Probe positioning	Duration (Days)	Time step
CO_2 (ppm)	VAISALA MI70 & GMP70	Continuous	TR	T0/T1 T2/T3PCO	CR H: 125cm	≥ 7	5 min
NP (p/l & Ø:nm)	GRIMM NanoCheck model 1.365	Continuous	TR	T0/T1 T2/T3PCO	CR H: 125cm	≥ 7	10 s
HCHO (ppb)	ETHERA Profil'air® Dynamic Kit	Integrated measurement (every 2h from 9am to 5pm)	TR	T1/T2 T3PCO	CR H: 125cm	≥ 5	4 S/D
TVOC (mg/m^3)	INNOVA 1412i & 1303	Continuous	TR & RR	T0/T1 T2/T3PCO	CR H: 125cm	≥ 7	≤ 5 min
Ta (°C) RH (%)	VAISALA HMP 110	Continuous	TR → TR & RR →	T0/T1 T2/T3PCO	CR H: 125cm	≥ 7	1 min

CR: Centre of the room; S/D: sample per day; p/l: Number of nanoparticles per litre; Ø: diameter of nanoparticles (nm).

3 RESULTS AND DISCUSSION

The Table 2 shows a normal carbon dioxide concentration which fluctuated, for T0 and T1, around the ambient CO_2 concentration of 400ppm ASHRAE (2007). For T2 and T3PCO, the CO_2 concentration was very low ($^{+}/_{-}$40ppm) and ≈10 times less than the ambient CO_2

concentration (Cf. Table 2). These results didn't come from a measurement mistake or a CO_2 stratification because additional measurements were made to invalidate these hypotheses. This low concentration was due to a sink effect between the CO_2 and the calcium hydroxide $Ca(OH)_2$ embedded in the PS to form calcium carbonate ($CaCO_3$) and water. The equation can be written as:

$$Ca(OH)_2 + CO_2 \rightarrow CaCO_3 + H_2O \tag{5}$$

After the PS installation there was not a significant difference? of air temperature between the TR and the RR. The main difference concerned the relative humidity parameter with a higher level in the TR (almost twice as high) than in the RR especially during the step T2. This high concentration in the TR came from the PS containing 80% of water in its structure.
For the step T1, the sarcophagus multiplied the TVOC concentration by 7.5, inside the TR in comparison with T0 (cf. Table 2).

Table 2. Average values of all parameters monitored from the steps T0 to T3PCO

Steps	T0		SG C*	T1		PS C*	T2		T3PCO	
Date (year 2013)	28/01→04/02		14/02	18/02→25/02		06/03	18/03→26/03		19/04→02/05	
Rooms	TR	RR	SG in TR and RR	TR	RR	PS only in TR	TR	RR	TR	RR
CO_2 (ppm)	392	/		440	/		36	/	53	/
TVOC (mg/m^3)	4.23	/		30.6	/		212.9	15.5	53.2	20.7
HCHO (ppb)	/	/		175.3	/		22.9	/	137.6	/
NP p/l	2.04x10^6	/		9.61x10^5	/		3.66x10^6	/	7.2x10^7	/
NP Ø (nm)	47	/		67	/		45	/	25	/
Ta (°C)	16.0	/		20.6	/		14.2	14.4	27.7	25.2
RH (%)	36.6	/		40.7	/		87.3	52.2	64.1	43.4

SG: Sarcophagus; PS: Passive system; C*: Construction; p/l: Number of nanoparticles per litre;

Compared to the step T1, the HCHO concentration decreased and was reduced by a factor of 9 after the installation of the PS (step T2), whereas at the same time, the TVOC concentration increased strongly by a factor of 7.
The decrease of the HCHO concentration is likely due to the high RH inside the TR (around 87%, Table 6) because the formaldehyde has a good solubility in the water INRS (2011). The HCHO was probably dissolved in the adsorbed water or condensed water (Pei et al., 2011) on the windows. Furthermore, a high water vapour concentration saturates the photocatalyst surface (Mo et al., 2009; Wang et al., 2007; Juan et al., 2003; Gaya et al., 2008). An excessive water vapour competes with pollutants such as VOCs for an adsorption site on the photocatalyst, thus reducing the pollutant removal rate. This is the "competitive adsorption" between the water vapour and the pollutants (Zhong et al., 2010; Mo et al., 2009; Wang et al., 2007). This phenomenon could explain the high TVOC concentration for the step T2.
Compared to T2, the step T3PCO had a TVOC concentration divided by 4 whereas the HCHO concentration increased and was multiplied by 6. Even if the TVOC concentration decreased, the value was still high (>50mg/m^3) compared with the RR (Table 2). The decrease of the TVOC concentration can be explained by a long period between T2 and T3PCO (23 days), a natural decrease of the VOCs emissions in the time coming from the PS, and a lower RH (64%) in the TR limiting the "competitive adsorption" phenomenon.

3.1 Impact of the actuators on the PCO efficiency

At the beginning of T3PCO, the **Error! Reference source not found.** shows an increase of the TVOC concentration in the TR and the RR. The doors and the windows were just closed and the MVS was turned off. The previous days before T3PCO, the doors were opened and the MVS was turned on to dry out the room and reduce the RH.

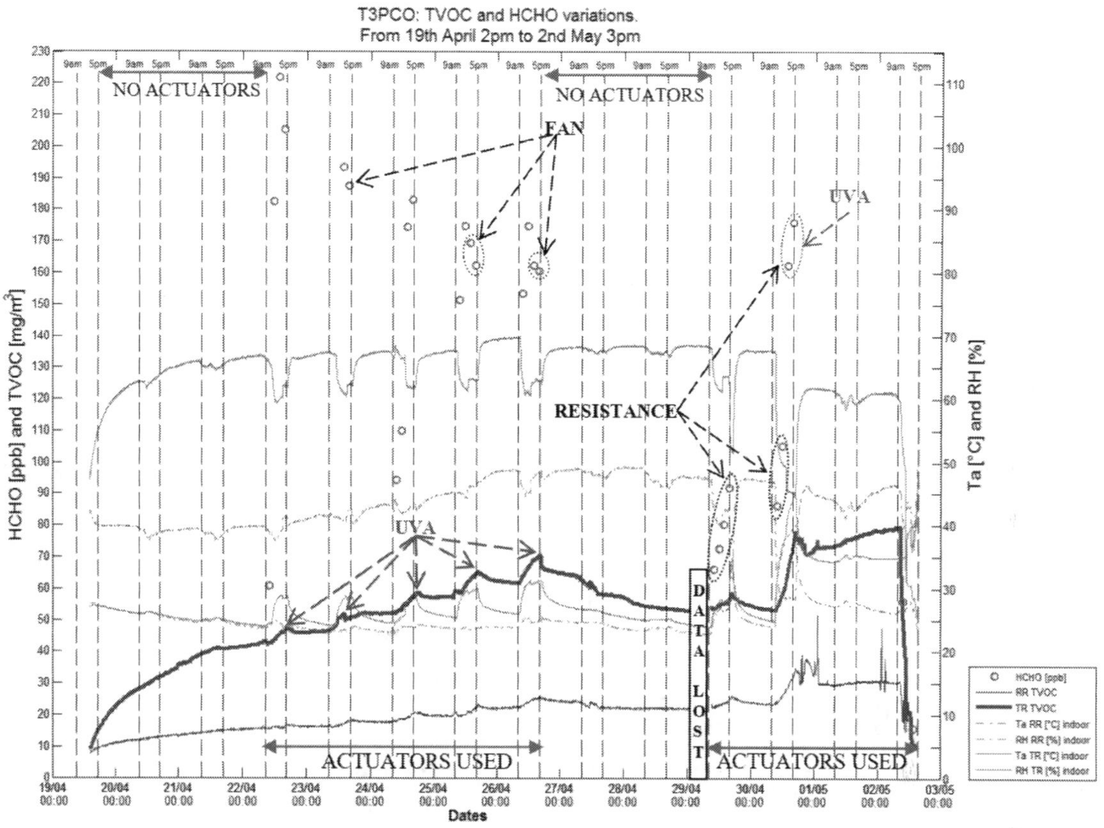

Figure 3. HCHO and TVOC concentrations for the step T3PCO

Every night, the Figure 3 shows a decrease or a stabilization of the TVOC and HCHO concentrations. Every night there was no heating system (Figure 3). Consequently the air temperature naturally decreased and an adsorption phenomenon of the pollutants occurred (Zhong et al., 2010).

Figure 3 shows the ramp of temperature from 9am to 1pm on the 30th April. It raised the TVOC concentration by +8mg/m^3 and the HCHO concentration by +19.1ppb. In the afternoon the UVA lamps were added to the heating system. The impact on the TVOC concentration is significant (+16.9 mg/m^3) as well as for the HCHO concentration (+89.9ppb). When the UVA lamps were turned on the TVOC increased as well as the HCHO (Auvinen et al., 2008; ADEME, 2013). The UVA activates the HCHO emission as an intermediate.

From 30th April 5:00pm to 2nd May 9:00am, a long test period was made with only the UVA lamps turned on to test the impact on the PCO. The results showed a 'linear' increase of TVOC. There was no PCO effect.

The fans coupled with the UVA lamps seem to stabilize or slightly reduce the HCHO emissions (Cf. 23rd, 25th, 26th April of the Figure 3). This is probably due to better air recirculation on the photocatalyst surface as well as a better contact time ADEME (2013).

The last day (2nd May at 9am), the MVS was turned on for 2 hours in the TR and the RR, then the door was also added for 2 hours, as well as the windows for the following 2 hours. The TVOC and the HCHO concentrations were almost divided by 4 in 6 hours. At the end, the TVOC concentrations were equivalent and slightly inferior to 10mg/m^3 for the TR and the RR.

3.2 Nanoparticles results and discussion

The bar graph shows (Figure 4.) a lower number of NP coupled with a higher diameter than before the installation of the PS (steps T0 and T1). After the PS installation, the Figure 4. shows an increase of the number of NP coupled with a decrease of their diameter (Table 2). Under UVA radiation, the bar graph shows a huge increase of TiO_2 nanoparticles (the number of NP per litre was multiplied by 19.5 for T3PCO in comparison to T2).The reference OQAI (2012) speaks about an increase of TiO_2 nanoparticles <200nm under a UV radiation. To finish, the diameter of the NP measured in the step T3PCO (Ø: 25nm) tended toward the crystalline size of the TiO_2 used in the photocatalyst layer (Ø: 15nm for the step T3PCO).

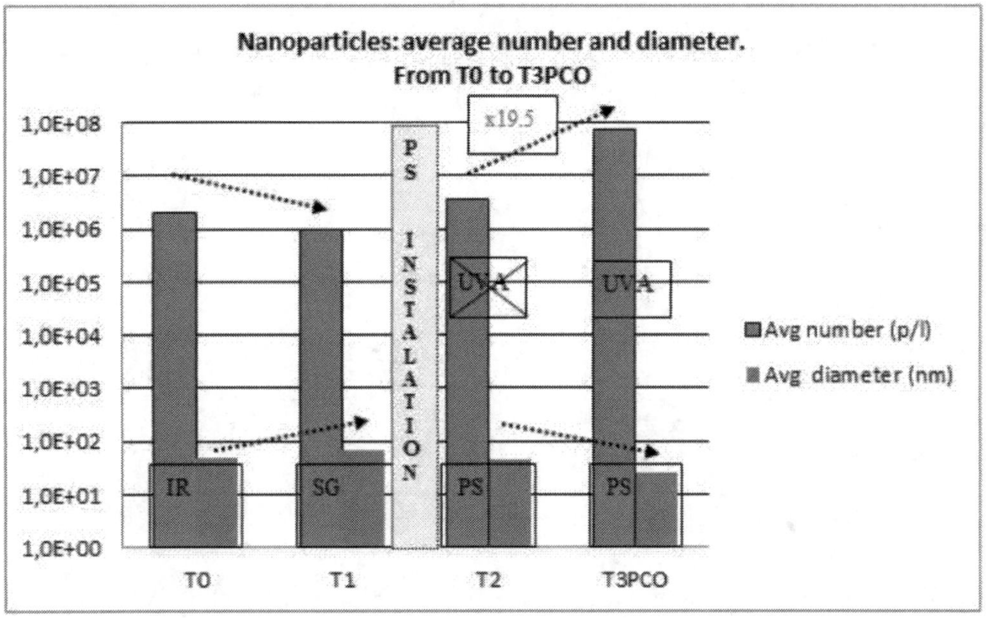

Figure 4.Nanoparticles measured in the TR, from the steps T0 to T3PCO
(IR: Initial Room; SG: Sarcophagus; PS: Passive system)

4 CONCLUSIONS

In this study, the efficiency of the PS generated some disappointing results in a real indoor environment. This PS was embedded with 3 different layers plus a specific TiO_2 layer to improve the IAQ. The PS was applied onto the walls and the ceiling of the TR and was compared to a RR without anything added. There are several key factors to enhance the PCO efficiency: the temperature (Zhong et al., 2010; Mo et al., 2009), the water vapour concentration (Whang et al., 2007), the contaminant mixture (Ao et al., 2004), the quantity of photocatalyst substrate OQAI (2012), the amount of UVA lamps (Juan et al., 2003), etc.
For the step T2, immediately after the installation of the passive system, there was a high emission of VOCs in comparison to the RR and to the previous steps T1 and T0.
The high amount of water embedded in the PS (80%) increased the RH up to 90% in the TR, for the step T2. A high RH is responsible for a competitive adsorption between the water

vapour and the pollutants onto the photocatalyst surface thus reducing the PCO process. This could explain the high TVOC concentration for the step T2, as well as the low HCHO concentration compared to T1. The HCHO is extremely soluble in water and there was probably a dissolution of this pollutant in the water vapour which was preferentially adsorbed by the photocatalyst surface rather than the TVOC pollutants.
For the step T3PCO, the results were not better with the use of actuators, especially the UVA-lamps. In the TR and the RR, there was a low UV light intensity because the window filtered the UV wavelengths. This lack of UV was compensated by the UVA lamps. When they were turned on for a long time (≈48h), the UVA lamps did not enhance the PCO process but increased the TVOC and the HCHO emissions (intermediate).
Moreover, the results showed an increase of the air temperature due to the heating system and the UVA lamps. These 2 actuators were responsible for raising the TVOC and HCHO concentration. But the UVA raised the TVOC and the HCHO concentration more than the heating system. The combination of the opening of the door, window, and MVS showed a good improvement of the IAQ during the last day of the step T3PCO.
Finally, immediately after the PS installation (T2), there was an increase of NP and especially during the step T3PCO under UVA radiations. Without laboratory analysis, it is difficult to conclude formally about their origins even if their diameter (Ø: 25nm for T3PCO) tended toward the crystalline size of the TiO_2 nanoparticles used in the PS (Ø: 15nm).
The next step of further studies will consist of improving the efficiency of the PCO layer by dispersion of TiO_2 on a specific perlite substrate. This new PCO layer will be tested alone then with the PS.

ACKNOWLEDGEMENTS

The work described in this paper was supported by an FP7 research project funded by the European Commission: CETIEB (FP7 Grant Agreement No. 285623). I wish to thank my Work Package 5 partners and Jay Stuart.

REFERENCES

Zhong L et al (2010) Modelling and physical interpretation of photocatalytic oxidation efficiency in indoor air applications. *Building and environment*, **45**, 2689–2697.

Mo J et al (2009) Photocatalytic purification of volatile organic compounds in indoor air : A literature review. *Atmospheric Environment*, **43**, 2229–2246.

Ginestet et al (2005) Development of a new photocatalytic oxidation air filter for aircraft cabin. *Indoor Air*, **15**, 326–334.

Auvinen J et al (2008) The influence of photocatalytic interior paints on indoor air quality. *Atmospheric Environment*, **42**, 4101–4112.

Mo J, Zhang Y et al (2009) Determination and risk assessment of by-products resulting from photocatalytic oxidation of toluene. *Applied Catalysis B: environmental*, **89**, 570–576.

INRS (2012) Préconisation en matière de caractérisation des potentiels d'émission et d'exposition professionnelle aux aérosols lors d'opérations mettant en œuvre des nanomatériaux. *Institut National de Recherche et de Sécurité*, (HST ND2355-226-12).

WHO (2010) World Health Organization, International Agency for Research on Cancer, Monographs on the Evaluation of carcinogenic Risks to Humans, Volume 93, Carbon black, Titanium Dioxide, and talc, Lyon France.

Kolarik J, Wargocki P (2010) Can a photocatalytic air purifier be used to improve the perceived air quality indoors. *Indoor Air*, **20**, 255-262.

Juan Z, Xudong Y (2003) Photocatalytic oxidation for indoor air purification: a literature review. *Building and Environment*, **38**, 645–654.

ASHRAE (2007) Ventilation for acceptable indoor air quality, ASHRAE standard 62.1-2007, American Society of Heating, Refrigerating and Air–Conditioning Engineers, Inc., Atlanta, GA.

Wang S et al (2007) Volatile organic compounds in indoor environment and photocatalytic oxidation: State of the art. *Environment International*, **33**, 694–705.

Gaya UI, Abdullah AH (2008) Heterogeneous photocatalytic degradation of organic contaminants over titanium dioxide: A review of fundamentals, progress and problems. *Journal of Photochemistry and Photobiology C: Photochemistry Reviews*, **9**, 1–12.

INRS (2011) Fiche toxicologique aldehydes formiques et solutions aqueuses, Institut National de Recherche et de Sécurité, FT7.

Pei J, Jianshun S et al (2011) An experimental study of relative humidity effect on VOCs' effective diffusion coefficient and partition coefficient in a porous medium. *Chemical Engineering* Journal, **167**, 59–66.

ADEME (2013) Fiche technique de l'ADEME, Epuration de l'air par photocatalyse. Agence de l'Environnement et de la Maitrise de l'Energie.

OQAI (2012) Journée technique OQAI, RECUEIL DES RESUMES, L'épuration de l'air intérieur par les procédés photocatalytiques : efficacité et innocuité, Observatoire de la Qualité de l'Air Intérieur, Paris, France.

Ao CH, Lee SC (2004) Photodegradation of formaldehyde by photocatalyst TiO_2: effects on the presences of NO, SO_2 and VOCs, *Applied Catalysis B: Environmental*, **54**, 41–50.

THE ENERGY IMPACT OF ENVELOPE LEAKAGE. THE CHILEAN CASE

Ariel Bobadilla[1,3], Felipe Ossio[*2,3], Rodrigo Figueroa[1], Alex González[1], Muriel Díaz[1], and Roberto Arriagada[1]

1 CITEC UBB, Universidad del Bío Bío.
Av. Collao 1202, Concepción, Chile

2 Escuela de Construcción Civil, Pontificia Universidad Católica de Chile.
Av. Vicuña Mackenna 4860, Santiago, Chile
**Corresponding author: felipe.ossio@uclouvain.be*

3 Architecture et Climat, Université catholique de Louvain
Place du Levant 1, Louvain-la-Neuve, Belgium

ABSTRACT

Improving the airtightness of housing is an issue that concerns the Chilean state. Building ordinances do not currently include any requirement to limit infiltration and its associated energy loads. This situation affects the energy and environmental performance of housing, and has economic and social consequences of great importance for inhabitants and the State. This text presents part of a research project commissioned by the Chilean Ministry of Housing and Public Works, with the aim of defining acceptable airtightness standards for buildings by territorial zone in Chile. The results are presented of a baseline study intended to measure the impact of infiltration on the energy demand for the thermal conditioning of dwellings in Chile.

Using pressurization techniques, air tightness was measured in a sample of 185 dwellings built in 2007 and 2010, which is representative of more than 95% of the construction carried out in those years in the 54 provincial capitals of Chile. The LBL model was used to link airtightness properties of dwellings with climate variables in each of the provinces to obtain standardized coefficients of infiltration. Finally, with the dynamic simulation software TAS, energy loads due to infiltration and total energy demand were measured by building type and province.

It was concluded that the energy loads caused by infiltration in Chile vary significantly and have different effects depending on the type of energy demand. In particular, in the sectors with predominant heating needs, air infiltration generates demands of between 23 and 203 kWh/m².year, and can represent between 20 and 50% of the total energy demand for thermal conditioning. Also, it was confirmed that in the country's current state of development, an important step in the energy improvement of the building stock necessarily lies in the improvement of its airtightness characteristics.

KEYWORDS

Air infiltration; Airtightness; Energy efficiency of buildings

1 INTRODUCTION

Through an agreement signed by the National Commission for Scientific and Technological Research CONICYT, the University of the Bío-Bío and the Pontifical Catholic University of

Chile, it was agreed to carry out FONDEF project D10R1025, "Establishment of acceptable kinds of building infiltration for Chile", with an implementation deadline of 30 months starting in December 2011. This work was commissioned by the Chilean Ministries of Housing and Urban Development and Public Works.

The project was submitted to the XVIII FONDEF Projects Competition by a technology consortium, under the mandate of the Chilean Ministries of Housing and Urban Development, and Public Works. The technology consortium in charge of its execution and transfer is comprised of: the Technical Division of Housing Development and Study (DITEC) at the Ministry of Housing and Urban Development; the Directorate of Architecture at the Ministry of Public Works; the Chilean Energy Efficiency Agency at the Ministry of Energy; the School of Civil Construction at the Pontifical Catholic University of Chile and its Directorate of Outreach in Construction (DECON UC); the Center for Research in Construction Technologies at the University of the Bío-Bío (CITEC UBB); the Catholic University of Louvain in Belgium; the Technology Resource Center in France (NOBATEK); and the Pocuro S.A., Venteko S.A., Indalum S.A., Alcorp S.A. and Wintec S.A. construction companies.

The Fondef Project proposed the development of standards for airtightness and acceptable kinds of building infiltration by territorial zone in Chile, with the aim of reducing to acceptable limits the impact of air infiltration on energy demand and consumption in the building industry. In addition, it involved the development of knowledge and technological solutions to support the design, implementation and quality control of projects, in order to obtain buildings with airtightness levels adjusted to minimum optimum energy use needs throughout the nation (CITEC UBB & DECON UC, 2010).

This report, part of Fondef Project D10I1025, presents the results of a baseline study designed to measure the current impact of air infiltration on the energy demand for the thermal conditioning of dwellings in each of the 54 provinces of Chile. This knowledge is necessary to define criteria to establish kinds of airtightness and to understand gaps in quality by building type and Chilean province.

2 DESCRIPTION OF THE PROBLEM

Air infiltration, is defined as the passage of uncontrolled air through hidden cracks and unplanned openings in the envelope. It generates thermal loads, due to cold or heat depending on the season, which affect a building's energy performance, and also transport noise and air pollutants that affect environmental comfort (Liddament, 1996). It is produced by pressure differences that induce flow through the envelope. These are caused by wind action, indoor-outdoor temperature differences, or by the operation of mechanical ventilation equipment.

Chile is located between 17°30' south and the South Pole, primarily in the temperate and polar zones of the southern hemisphere. Because of its size, location and geographical configuration, it has a great variety of climates and micro-climates. More than 80% of the population is located in the central and southern parts of the country where the energy demands for the thermal conditioning of buildings are mainly for heating.
Thermal regulations, which are intended to establish energy savings objectives and requirements for residential buildings in Chile, are under construction. Newly created in the year 2000, the regulations establish seven thermal zones differentiated by heating degree days, and for each, set only thermal insulation requirements. At present, different activities

are being carried out including the Fondef project mentioned above, which aim to define the requirements necessary to regulate other important matters, such as airtightness and ventilation, with the purpose of improving the construction ordinance in the areas of energy and environment. This is nowadays esteemed to be essential in reducing the energy demand of buildings in Chile to acceptable limits (Bobadilla, 2014).

In the last decade, regulations have been able to limit transmission losses through building envelopes, but have not produced significant results in the energy improvement of buildings, as demonstrated by several recent studies (Damico *et al.*, 2012; Escorcia *et al.*, 2012). The most consistent explanation attributes this fact to the reduced ability of buildings in Chile to resist air infiltration.

A baseline study of air infiltration, also carried out in the context of Fondef Project D10I1025, found that the average airtightness of the new housing stock in Chile, its n50 value, is located at the level 12.9 1/h (Bobadilla *et al.*, 2014; Figueroa et al., 2013). Average values for stock are expected to be in the range of between 11.1 and 14.7, at 95% confidence level. These indicators are enlightening in regards to the degree of airtightness in buildings in Chile and the current state of development of design and building techniques. However, they are also indicators of the energy efficiency of the stock. Less than 5% of stock meets the 3.0 1/h standard reference limit of airtightness that is in effect in several European countries. Particularly critical is the standard in timber construction, which presently exhibits airtightness indices around 24.6 1/h. All of these indicate the absence or deficient use of sealing techniques, as well as poor implementation quality, which should be improved to ensure the acceptable energy performance of the building stock in Chile.

Alternately, case studies showed that air infiltration can come to represent 60% of energy demand in some areas (Ossio *et al.*, 2012). This situation plays a key role in the energy and environmental performance of dwellings, in the quality of life of the population and, to a significant degree, in energy costs, which have economic and social consequences of great importance for the population, industry and State of Chile.

The acceptable type of airtightness defines in practice what society accepts as the maximum permissible heat loss in an area or country due to air infiltration. In accordance with international practice, the standard is set by attempting to successfully combine at least two criteria: that the type is such that at least a percentage of the building stock complies and; that energy demand attributed to infiltration does not exceed a certain value (ASHRAE, 2004). It is this context that justified this investigation, which aimed to establish the energy impact of air infiltration on new building stock in Chile. The results are presented in this report.

3 METHODOLOGY

For research purposes, the target population was defined as all dwellings with building permits requested in 2011, distributed throughout the 54 provinces of Chile. The universe included 151,071 housing units, which have a combined floor area of 10,431,888 m^2. From this universe, a sample of 185 dwellings was selected. This was representative of more than 95% of the construction undertaken that year in Chile and had a sampling error of 10%. The sample was defined considering features of interest, such as the predominant material in walls (concrete; brickwork; brickwork plus lightweight structure; timber; other materials) and the type of grouping (detached, semi-detached and row houses). In this way, 20 distinct building types were obtained in the 54 provinces, which are detailed in Table 1.

The investigation involved the following activities:

- Measurement of the sample
- Dynamic Simulation

Table: 1 Housing types that make up the sample. Source: CITEC UBB & DECON UC, 2013.

Code	Description	Surface (m^2)
T1	House - brickwork - one floor - detached	67
T2	House - brickwork - one floor - row houses	67
T3	House - brickwork - two floors - detached	75
T4	House - brickwork - two floors - row houses	75
T5	House - concrete - one floor - detached	66
T6	House - concrete - two floors - detached	66
T7	Apartment building - brickwork and/or concrete	46
T8	House - concrete blocks - one floor - detached	50
T9	House - concrete blocks - two floors - detached	75
T10	House - concrete blocks - two floors - row houses	75
T11	House - timber - one floor - detached	50
T12	House - timber - one floor - row houses	50
T13	House - timber - two floors - detached	80
T14	House - timber - two floors - row houses	80
T15	House - SIP - one floor - detached	66
T16	House - other materials - one floor - detached	66
T17	House - other materials - two floors - detached	75
T18	House - other materials - two floors - row houses	75
T19	House - brickwork plus lightweight structure - two floors - detached	75
T20	House - brickwork plus lightweight structure - two floors - row houses	75

3.1 Measurement of the sample

The Blower Door pressurization technique was used as described in the standard UNE -IN 13829 "Thermal performance of buildings -- Determination of air permeability of building -- Fan pressurization method", to measure air permeability, and with it the airtightness of the housing sample. This physical property was synthesized with the indicator air changes per hour, 1/h, measured at 50 pascals of pressure difference.

Table 2 shows the results obtained. A considerable difference was observed in airtightness properties when the different materials categories were compared; the average n50 value of the sample was determined to be 12.9 1/h, with minimum values of 2.6 1/h and a maximum of 49.8 1/h.

Table: 2. Summary of Study Results, Method A. Source: Figueroa *et ál*, 2013.

Predominant Material in Walls	Minimum n50 value (1/H)	Maximum n50 value (1/h)	Average n50 value (1/h)
Concrete	2.6	28.6	9.0
Brickwork	4.3	19.6	11.8
Brickwork plus Lightweight Structure	2.3	49.2	15.0
Timber	4.5	49.8	24.6
Other materials	3.3	15.7	10.2

3.2 Dynamic Simulation

The energy modeling software *TAS* version 9.2.1.4 was used to determine energy loads from infiltration and the total demand of dwellings, including the distinct building types present in the 54 provinces of Chile.

It was necessary to create new meteorological databases to cover the entire nation and make improvements to the existing databases. The new databases include time information for weather variables in the 54 provincial capitals of Chile and were obtained through a process of adaptation and validation of the databases that the software *Meteonorm* 6 generates. The adjustment method used at the University of Southampton (Hampton *et al.*, 2013), originally applied to correct the effects of global warming on weather databases, was employed in the correction and adaptation of data. This methodology made it possible to modify the time data of the *Meteonorm* databases, so as to coincide with official monthly information and other specific local sources (Gonzalez *et al.*, 2013).

The thermal properties of the dwellings were deduced from the plans and technical specifications for the housing projects. To consider air infiltration, standard coefficients of infiltration were determined for the distinct dwellings in each of the provinces using the LBL model developed at the Lawrence Berkeley Laboratory (Sherman & Modera, 1986). These coefficients of infiltration linked the typical airtightness properties of the envelope of each building type (determined in 3.1) with local climate variables. This was used to measure the energy load from air infiltration with the energy simulation program TAS.

4 RESULTS

4.1 . Standardized coefficients of infiltration

In 14 of the 54 provinces, the standardized coefficients of infiltration are less than or equal to 1 1/h, within the range 0.7 -1.0 1/h. They are associated with inland locations, mainly in the north central area of the country, some of which have micro-climates, average wind speeds under 2.5 m/s and several months officially designated as calm weather (less than 2.0 m/s). These localities are also compatible, in some cases, with relatively high average exterior temperatures, which minimize the effects of thermal draft. The greatest weighted provincial values are in the range of 1 and 2.9 1/h and are associated with coastal localities, mainly in the southern and austral areas, with annual average speeds greater than 2.5 m/s and/or low average temperatures. Variations in standardized infiltration by building type and province are quite a bit larger. They range from 0.4 1/h for a detached, 1-floor, block concrete dwelling in the city of Vallenar (annual average wind speeds less than 2.0 m/s and an average temperature of 14.9 °C), to 4.0 1/h for a detached, 1-floor, wood construction in Punta Arenas (mean annual wind speed of 8.2 m/s and an average temperature of 5.9 °C).

4.2 Energy demand by building type and province

Table 3 shows energy demands by building type present in the provinces of Antofagasta, Santiago, Concepción and Punta Arenas. Heating, cooling, and total demands are given considering the load associated with the standard infiltration of each building type in the province, which are also listed. The table also includes the weighted demands by province (Depp), which are an indicator that takes into consideration the participation, by built surface area, of the different building types in the housing stock.

Table: 3 Energy demand of residential buildings in the provinces of Antofagasta, Santiago, Concepción and Punta Arenas. Source: Bobadilla, 2014.

	PROVINCES - BUILDING TYPE																	
Parám.	**Antofag.**		**Santiago**			**Concepción**							**Punta Arenas**					
	T6	**T7**	**T3**	**T6**	**T7**	**T3**	**T6**	**T7**	**T11**	**T13**	**T17**	**T19**	**T11**	**T13**	**T12**	**T14**	**T17**	**T18**
In	1.3	1.0	0.9	0.9	1.6	1.4	1.5	2.6	2.3	2.7	1.7	1.4	2.9	3.3	4.0	3.3	2.1	1.0
Cei	24	22	37	31	63	61	66	112	106	117	75	61	238	258	336	256	169	81
Dec	86	60	97	104	60	114	94	74	125	93	149	175	273	197	248	183	237	226
Der	9	8	11	12	9	0.0	0.0	0.0	0.0	0.0	0.0	0.0	0.0	0.0	0.0	0.0	0.0	0.0
Dett	120	89	145	147	131	175	160	186	231	210	225	236	511	455	585	439	406	307
Depp	92		134			207							449					

In : Standardized infiltration coefficient (1/h)
Cei : Energy load from infiltration (kWh/m^2.year)
Dec : Energy demand for heating (kWh/m^2.year)
Der : Energy demand for cooling (kWh/m^2.year)
Dett : Total energy demand by building type (kWh/m^2.year)
Depp : Total weighted provincial energy demand (kWh/m ^{2}.year)

Tables 4 and 5 present the weighted provincial energy demands by built surface area, of the main building types present in the various provinces. Demands were measured with infiltration coefficients of 0 (1/h) and the corresponding standardized coefficient Ln (1/h).

An example of the results is shown in Figure 1, where 7 provinces are represented and the contribution of air infiltration to energy demand is noteworthy.

Figure: 1 Weighted energy demands by provincial capital with energy load from infiltration.
Source: Bobadilla, 2014.

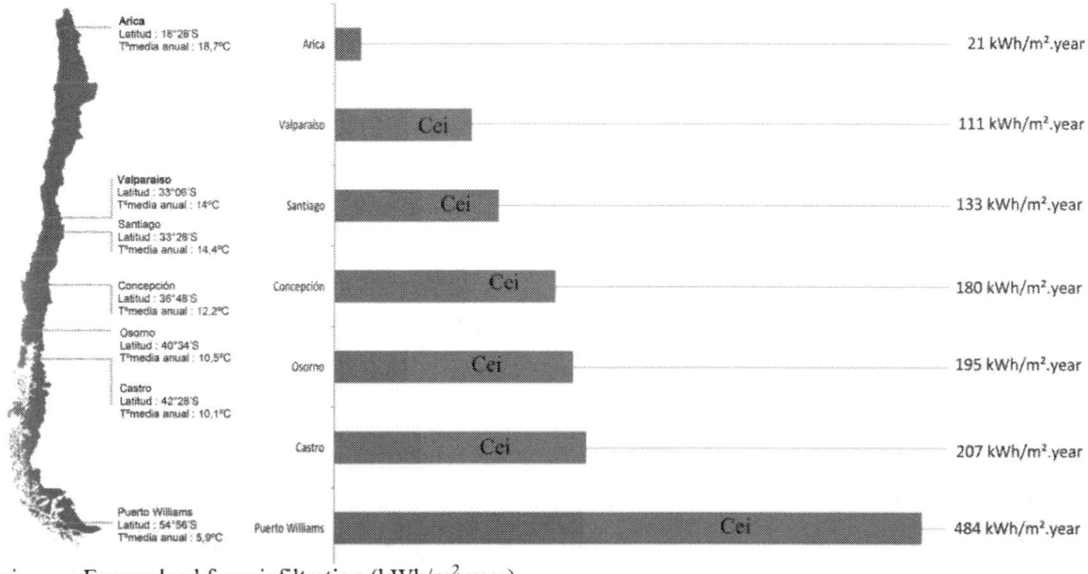

Cei : Energy load from infiltration (kWh/m^2.year)

These tables were used to create the provincial map of energy demands for the thermal conditioning of housing stock with 2011 building permits in Chile. At this level, the predominance of heating needs in residential buildings in Chile can be observed. In 21 of the 54 provinces only heating needs are reported. 31 provinces present combined heating and cooling demands. In 18 of these, cooling demands do not exceed 15 kWh/m^2. year; in 2 they do not exceed 28 kWh/m^2.year; and in 2 there are only cooling requirements. These are mostly concentrated in the provinces in the north and some in the center of the country, including Santiago.

Heating requirements are present in virtually the entire nation, mainly in the Andean area and center south of the country. In 45 of the 54 provinces, the demand exceeds 100 kWh/m^2.year; in 15 provinces it exceeds 200 kWh/m^2.year; and in 4 provinces it exceeds 400 kWh/m^2.year. These are all indicators that reflect a low level of energy quality in the new building stock in Chile.

The energy load due to air infiltration varies significantly and has a different impact depending on the type of energy demand. In the provinces with predominant heating needs (Only heating is required or cooling demand does not exceed 10% of heating.), infiltration means demands of between 23 and 203 kWh/m^2.year, which may represent between 20 and 50% of the total energy demand for thermal conditioning.

Table: 4 Weighted energy demands by provincial capital - North and Center. Source: Bobadilla, 2014.

Zone	Province	Det_0	In	Detp	Decp	Derp	Cei
17°35'S	**Arica**	19	1,3	21	0	21	-3
	Putre	135	1.5	209	209	0	74
	Iquique	21	1.2	28	20	9	7
	Pozo Almonte	297	1.5	378	367	10	81
N	**Tocopilla**	12	0.9	11	0	11	-1
O	**Calama**	137	2.8	269	259	9	132
R	**Antofagasta**	70	1.0	93	79	14	23
T	**Chañaral**	54	1.0	60	49	11	6
H	**Copiapó**	75	1.7	104	76	28	29
	Vallenar	74	0.7	92	81	11	18
	Coquimbo	91	1.4	133	122	11	42
	Ovalle	79	1.5	113	93	20	34
31°46'S	**Illapel**	127	0.7	168	143	26	41
32°15'S	**La Ligua**	90	1.1	124	100	23	34
	Los Andes	74	1.9	138	116	22	64
	San Felipe	109	1.3	152	119	33	43
	Quillota	113	0.9	151	133	18	38
	Valparaíso	53	2.2	111	111	0	58
	San Antonio	73	2.1	119	119	0	46
	Hanga Roa	34	1.0	35	1	33	1
	Quilpué	51	1.5	110	110	0	59
	Colina	106	0.9	140	124	16	34
	Santiago	77	1.5	133	119	14	56
C	**Puente Alto**	103	0.8	134	117	17	31
E	**San Bernardo**	98	0.9	148	134	14	50
N	**Melipilla**	109	0.9	130	109	21	21
T	**Talagante**	114	1.1	160	149	10	46
E	**Rancagua**	92	0.9	113	103	10	80
R	**San Fernando**	109	0.9	137	126	11	94
	Pichilemu	78	1.0	96	96	0	23
	Curicó	106	1.0	141	132	9	33
	Talca	87	1.1	129	112	17	131
	Linares	111	0.7	133	123	9	22
	Cauquenes	99	1.6	159	143	16	60
	Chillán	89	1.3	142	133	9	53
	Los Ángeles	91	1.2	150	136	14	59
	Concepción	73	1.7	180	180	0	107
38°20'S	**Lebu**	79	2.3	197	197	0	118

Det_0 : Total provincial energy demand without infiltration (kWh/m^2.year)
In : Standardized infiltration coefficient (1/h)
Detp : Total energy demand with standardized provincial infiltration (kWh/m^2.year)
Decp : Provincial energy demand for heating (kWh/m^2.year)
Derp : Provincial energy demand for cooling (kWh/m^2.year)
Cei : Energy load from Infiltration (kWh/m^2.year)

Table: 5 Table: 6 Weighted energy demands by provincial capital - South and Austral. Source: Bobadilla, 2014.

Zone	Province	Det_0	In	Detp	Decp	Derp	Cei
37°40'S	**Angol**	75	1.2	110	104	6	35
	Temuco	155	0.8	202	202	0	47
S	**Valdivia**	117	1.0	171	170	0	54
O	**La Unión**	108	1.0	154	154	0	46
U	**Osorno**	137	1.1	195	195	0	58
T	**Puerto Montt**	137	1.3	192	192	0	55
H	**Castro**	145	1.2	207	199	7	62
43°37'S	**Chaitén**	155	1.0	201	201	0	46
44°45'S	**Coyhaique**	181	1.5	264	264	0	83
A	**Puerto Aysén**	161	1.6	260	260	0	99
U	**Chile Chico**	187	2.9	364	364	0	177
S	**Cochrane**	117	2.7	227	227	0	110
T	**Puerto Natales**	221	1.7	325	325	0	104
R	**Punta Arenas**	213	2.8	446	446	0	233
A L	**Porvenir**	199	1.2	277	277	0	78
54°56'S	**Puerto Williams**	281	2.1	484	484	0	203

Det_0 : Total provincial energy demand without infiltration (kWh/m^2.year)
In : Standardized infiltration coefficient (1/h)
Detp : Total energy demand with standardized provincial infiltration (kWh/m^2.year)
Decp : Provincial energy demand for heating (kWh/m^2.year)
Derp : Provincial energy demand for cooling (kWh/m^2.year)
Cei : Energy load from infiltration (kWh/m^2.year)

An analysis at the level of building type by province shows more significant differences, given that the average values hide the marked dispersions observed in those provinces where construction systems with very different airtightness properties coexist.

The localities that demand the most energy are: Puerto Williams, Punta Arenas, Puerto Natales, Chile Chico and Pozo Almonte, all of which have weighted provincial demands greater than 325 kWh/m^2.year. In contrast, the localities with the least energy demand are: Arica, Iquique and Tocopilla, with values below 28 kWh/m^2.year, corresponding to only cooling demands. In the provinces where cooling needs predominate (Only cooling is required or heating demand does not exceed 10% of cooling.), in 4 of the 54 provinces of Chile, infiltration helps to reduce energy use.

Dwellings located in localities in the Andean zone, which runs from the north to the south of the country, present demands above 200 kWh/m^2.year. These are high but expected considering the harsh climate of this area and the typically low levels of thermal protection in the buildings in this territory. From north to south in inland areas, progressive demands can be observed in general terms, ranging from a bit more than 10 KWh/m^2.year in Arica and Tocopilla (cooling), up to more than 400 KWh/m^2.year in Punta Arenas and Puerto Williams (heating), which are explained in the same way.

5 CONCLUSIONS

Improving the airtightness of housing is an issue that concerns the Chilean state. Building ordinances do not currently include any requirement to limit infiltration and its associated energy loads. This situation affects the energy and environmental performance of housing, and has economic and social consequences of great importance for inhabitants and the State.

In the current state of development, an important step in the energy improvement of the building stock in Chile is, necessarily, to improve its airtightness characteristics. The baseline of airtightness in buildings completed between the years 2007 and 2010 is 12.9 1/h, with expected values between 11.1 and 14.7 1/h with 95% confidence. This is an overall measure of the current techniques used and quality associated with sealing and airtightness of residential buildings in Chile. Especially critical is the standard of airtightness in timber construction, which has indices around 24.6 1/h. All of these indicate the absence or deficient use of sealing techniques, as well as poor implementation quality, which should be improved in Chile to ensure the acceptable energy performance of all building types constructed in Chile.

The standardized coefficients of infiltration range from 0.7 to 2.9 1/h throughout the nation. The lowest values are associated with inland locations, mainly in the north central area of the country, some of which have micro-climates, that are compatible with low average wind speeds and/or relatively high average exterior temperatures. Meanwhile, the highest standardized coefficients of infiltration are associated with coastal localities, mainly in the southern and austral areas, with wind speeds over 2.5 m/s and/or low average exterior temperatures.

The results of the dynamic simulations show that energy loads due to infiltration in Chile vary significantly and impact in different ways depending on the type of energy demand. In particular, in the sectors with predominant heating needs, air infiltration generates demands of between 23 and 203 kWh/m^2.year, and can represent between 20 and 50% of the total energy demand for heating and cooling.

All regulation that seeks to organize and regulate construction should be based on actual physical data so that its dictates are effectively feasible. For these purposes, this research delivers important references in order to understand and reduce gaps in quality in terms of sealing and airtightness of buildings in Chile, and criteria for the establishment of airtightness requirements, an undertaking that is currently in progress in Chile.

6 ACKNOWLEDGMENTS

The authors wish to thank CONICYT for funding this initiative, and the entities DITEC, the Department of Architecture at the Chilean Ministry of Public Works, the Chilean Energy Efficiency Agency at the Ministry of Energy; and the companies Pocuro S.A., Venteko S.A., Indalum S.A., Alcorp S.A., and Wintec S.A. for their support and cooperation.

7 REFERENCES

CITEC UBB & DECON UC (2010). FONDEF D10I1025 *Establecimiento de clases de infiltración aceptable de edificios para Chile. XVIII Concurso de Proyectos de Investigación y Desarrollo*. Santiago, Chile.

Liddament, M. (1996). *A Guide to Energy Efficient Ventilation*. Brussels, Belgium: Air Infiltration and Ventilation Center.

Bobadilla, A. (2014). *Energy and Indoor Environmental Quality of Residential Construction in Chile. Analysis of the trends, achievements and pending challenges*. Université catholique de Louvain, Louvain-la-Neuve, Belgium.

Damico, F. C., Alvarado, R. G., Bruscato, U., Kelly, M. T., Oyola, O. E., & Diaz, M. (2012). *Análisis energético de las viviendas del centro-sur de Chile*. Arquitetura Revista, 8(1), 62–75. doi:10.4013/arq.2012.81.07.

Escorcia, O., García, R., Trebilcock, M., Celis, F., & Miotto, U. (2012). *Mejoramientos de envolvente para la eficiencia energética de viviendas en el centro-sur de Chile*. Informes de la Construcción, 64(000). doi:10.3989/ic.11.143.

Bobadilla, A., Figueroa, R., Díaz, M., Vargas, G., Arriagada, R., Espinoza, R. (in press) (2014). *Establecimiento de línea base de infiltraciones de aire para la limitación de la demanda energética en edificios residenciales existentes. El caso de Chile*. Informes de la Construcción.

Figueroa, R., Bobadilla, A., Besser, D., Días, M., Arriegada, R., & Espinoza, R. (2013) *Air infiltration in Chilean housing: A baseline determination*. 29th Conference, Sustainable Architecture for a Renewable Future PLEA2013, Munich, Germany.

Ossio, F, De Herde, A, & Veas, L. (2012). *Exigencias europeas para infiltraciones de aire: Lecciones para Chile*. Revista de la Construcción, 11(1), 54-63.

ASHRAE. (2004). *Air Leakage performance for Detached Single-Family Residential Buildings*. ANSI/ASHRAE Standard 119-1988 (RA 2004). Atlanta, USA.

CITEC UBB & DECON UC. (2013). *Informe Hito 3: Clases de infiltraciones aceptables. FONDEF D10I1025 "Establecimiento de clases de infiltración aceptable de edificios para Chile"*. CONICYT, Chile

Comité Europeo de Normalización, (2010). UNE-EN 13829: *Aislamiento térmico. Determinación de la estanquidad al aire en edificios. Método de presurización por medio de ventilador*. Madrid, España.

Hampton Jentsch M.F., James P.A.B., Bourikas L. and Bahaj A.S. (2013) *Transforming existing weather data for worldwide locations to enable energy and building performance simulation under future climates*, Renewable Energy, Volume 55, pp 514-524.

González, A., & Díaz, M. (2013). *Función e impacto del archivo climático sobre las simulaciones de demanda energética*. Hábitat Sustentable, 3(2), 75–85.

Sherman, M., & Modera, M. (1986). *Comparison of measured and predicted infiltration using the lbl infiltration model. Measured Air Leakage of Buildings, ASTM STP 904 (págs. 325-347)*. Philadelphia: American Society for Testing and Materials.

AIRTIGHTNESS IMPROVEMENT OF STRUCTURES TO IMPROVE INDOOR AIR QUALITY

Katariina Laine *2

2 Vahanen Oy
Building Physics Specialist Services
Linnoitustie 5
02600 Espoo, Finland
**Corresponding author: katariina.laine@vahanen.com*

ABSTRACT

The aim of improving air tightness of structures is to prevent the uncontrolled air leakages through structures. Built environments contain microbes, particulate and gaseous impurities but removing them is not always necessary. For example, an ageing building envelope commonly contains microbial impurities even when there is no obvious moisture damage. Air leaks convey impurities to indoors where they can lead to poor indoor air quality and associated health problems. Air leaks have also negative impact to energy efficiency and living comfort.

An air tight building envelope prevents air leaks through the envelope structure. Common air leakage places are the joints of structures, i.e. the joint of ground slab and external wall and untight inlets of the envelope. The joints of the structures are typically not air tight without detailed planning. In Finland an air tight building envelope is taken into account in the building regulations and therefore airtightness is relatively good in the new construction. In the existing buildings improving airtightness of structures is a relatively new concept. It can be used in a combination with other renovation methods to solve indoor air quality problems in buildings. In Finland several techniques have been implemented in order to prevent uncontrolled air flux through the structures since 1980's. In practise the air tightness renovation method has been successful in indoor air quality problems in buildings.

The main principle of moisture or microbial damage renovation is that before installing new building materials all moisture or microbially damaged materials are removed and the source of the problem is identified and fixed. However, in practise it is often impossible to remove all impurities and therefore improving airtightness of the new structure has an important role in completing a successful renovation. Improving of airtightness must be applied on to the whole building envelope. Air tightening renovation is never a single renovation action but always a part of other action. The air tightening material must be elastic, safe and long lasting and tested for its purpose. Improving airtightness requires planning, accurate execution on the construction site and supervision. Quality control measurements are in a key role to success. Quality control measurements are done visually and with the trace-gas leak test. It is very important always adjust the ventilation system to match the changed pressure difference.

In Finland there are no guidelines for planning, execution or supervising air tightness renovation method. It is planned to start certification training and publish first guideline in 2014.

KEYWORDS

Improving air tightness, air leakages, microbial damages, indoor air quality, repair method

1 INTRODUCTION

Poor indoor air quality associated problems can have a major impact on occupant health and comfort. Indoor air quality problems in western countries in homes, schools and offices are often caused by complex, inter-related issues. The problems can be caused i.e. by moisture damages, microbial damages, harmful substances, radon or other impurities. In external envelope structures and in the soil under building's ground slab there are almost always some microbial based impurities and radon which can affect negatively to the indoor air quality when transported with the air flux into the room.

Structure's joints are not automatically completely tight and therefore their implementation requires detailed planning. Common air leakage places are the joints of structures, i.e. the joint of ground slab and external wall and untight inlets of the envelope. There is usually air flow from outside to inside the building caused by the building's pressure ratio. The building's pressure ratio is determined by interaction of wind, chimney effect and ventilation as well as the usage of the premises. The pressure ratio will vary and can change very fast and strongly. As a result the air flows from one room to another, between floors or through envelope structures. The air flow conveys heat, moisture and impurities such as particles, mineral wool fibres, microbial and gaseous impurities, radon and other harmful substances.

The aim to improve air tightness of structures is to prevent the uncontrolled air leakages through structures. Methods used to improve the air tightness in structural renovation are called air tightness renovation. Using air tightness renovation energy consumption can be decreased, moisture convection can be prevented and uncontrolled air leakages through structures, junctions and inlets are cut off. The poor indoor air quality associated problems can be decreased and prevented with airtight structures. In Finland, improving airtightness of structures has long been a stabilized renovation method for prevention of radon. In practise air tightness renovation method as part of a larger entity has successfully solved the indoor air quality problems in buildings.

Air tightness factor has been taken into consideration in the National Building Code of Finland and in the regulations of new building. In the existing buildings improving airtightness of structures is a relatively new concept. Practice has shown that the air tightness renovation as its own is not sufficient but requires other actions to solve the problems. An air tightness renovation can fail due inadequate planning, careless implementation or lack of quality control at the construction site. Additionally some structures might have been air tightened where it is not a suitable method. Lack of understanding structures building physical functionality has even caused further damages. In Finland the experts working with the indoor air quality problems have different opinions regarding using the air tightness renovation method, its effect in the indoor air quality, its durability and its effect to the life cycle costs for buildings. Therefore several projects have been started to develop common procedures and guidelines this year.

Demolishing and building new is one option to solve an indoor air quality problem, when the structure is at the end of its lifecycle. In some cases demolishing and building new might be economically better option than renovating old. Demolishing and building new is not always possible for economic reasons. In old protected buildings demolishing is often not permitted. In some cases to remove a small quantity of impurities would require unreasonable demolishing actions. In old buildings it is not possible to remove all risk structures or contaminated materials. It is a fact, that built environments contain microbes, particulate and gaseous impurities but removing them is not always necessary. For example, an ageing building envelope commonly contains microbial impurities even when there is no obvious moisture damage. The impurities are simply conveyed from outdoor and from soil under the building. As it is not always possible to demolish and build new structures, optional renovation methods are needed, i.e. air tightness renovation. There is also need for more solutions to solve indoor air quality problems.

2 AIR TIGHTNESS OF STRUCTURES IN RADON PREVENTION

Radon is odourless, colourless, harmful and dangerous gas. Harmful quantities should not be released into living spaces. In the National Building Code of Finland it is stated that tight uniform slab solution is safe in respect of radon. Air tightening of the joint between a ground-supported slab and the enclosure wall is necessary, when the enclosure walls are constructed separately. Additionally tightening of lead-ins is extremely important in all structural solutions. A tight concrete structure not susceptible to cracking prevents the penetration of radon through the structure. (Ministry of the Environment, 2004)

Radon prevention methods are e.g. air tightening of the ground slab and enclosure wall and under pressure solutions of soil under the ground slab. In Finland, the air tightening method to prevent radon has been used for decades and it is an established renovation method. The oldest radon renovations with air tightness solutions have been done in the 1980's. Radon is a good indicator for successful air tightness renovation i.e. how much soil-borne air containing radon leaks into the room spaces. Therefore radon measurements before and after the renovation can be used in evaluating the success of the air tightness renovation in radon areas. (Kettunen et. al, 1991)

The essential requirement for preventing radon by air tightening structures is that the air flow paths are cut off totally not only partly. The implementation of air tightening has to be done carefully. When the renovation has been planned, the work has been performed carefully with quality controlling, can the radon containing air leaks into indoor air cut off totally. When the air leakages are cut off, the only way radon can be conveyed through structure is by diffusion. The impact of diffusion is irrelevant for indoor air quality. (Kettunen et. al, 1991)

2.1 Air tightness of external envelope structures in new buildings

The National Building Code of Finland for new buildings requires air tightness for external envelope structures. The air tightness is usually done with vapour barrier which functions at the same time as an air barrier. A massive concrete or brick structure can also function as a vapour and air barrier structure. In an air tight building the supply air is not taken through air leakage but controlled through supply air valves or through mechanical ventilation. (Ympäristöministeriö, 1998, 2010, 2012a, 2012b)

According the National Building Code of Finland, in respect of functioning of the ventilation system, moisture technical functionality, good indoor air quality and energy efficiency reasons the air leakage value q_{50} for envelopes in new buildings should be not more than 1 $m^3/(h \cdot m^2)$. However, it is allowed to be up to 4 $m^3/(h \cdot m^2)$, in some special cases even higher. However a low air tightness value measured from the building does not guarantee completely air tightness of the envelope structures. There can still be locally significant air leakage points. Therefore careful tightening of joints and inlets is very important. (Ympäristöministeriö, 2012b; RIL 107-2012)

Structure's joints are not automatically completely tight and therefore the implementation requires detailed planning. E.g. for ground floor concrete slab it is typical that after casting, cracks will occur at the edges of the new concrete slab due shrinkage which is associated with drying of concrete. Also cracks can occur at the corners and around pillars due to drying shrinkage. Therefore in concrete structures' joints there are without exception discontinuity points for air leakage if the air tightness is not specifically planned. There is an example of air tightening principle for the ground floor concrete slab in figure 1. The same principal is also used in preventing radon.

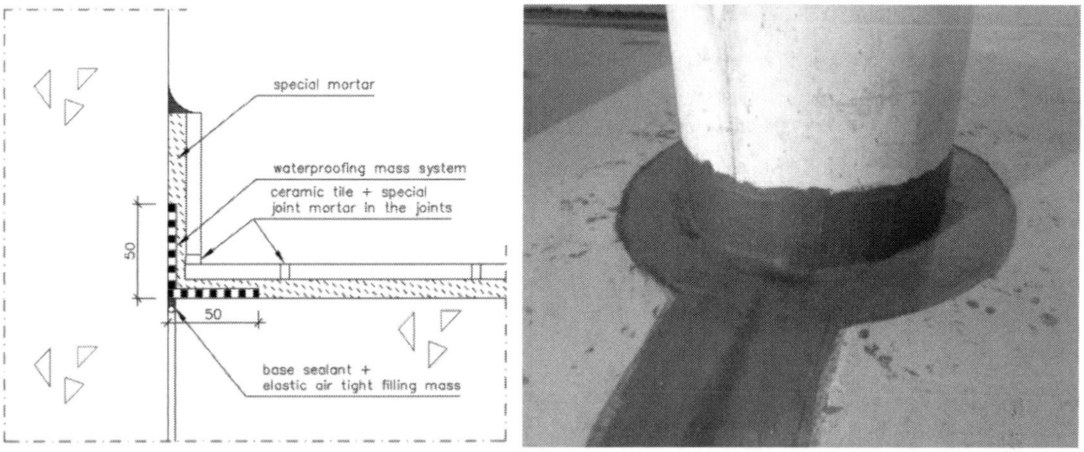

Figure 1: Principal solutions for ground floor concrete slab and concrete wall or pillar joint air tightness renovation with waterproofing mass system. Extract from a plan and photos from impletion at the site.

3 AIM FOR AIR TIGHTNESS OF STRUCTURES

3.1 Effects of air tightness on structure's functionality

Improving air tightness of envelope structure has a positive effect on the following factors:
1. Reduction of energy consumption
2. Prevention of moisture convection
3. Prevention of impurities entering through air leakage flows

Additionally an air tight envelope is a fundamental condition for a reliable and energy efficient mechanical ventilation system. By improving air tightness the living comfort can be improved as the feeling of draft is decreased and the temperatures of inner surface will increase. (Vinha, 2014)

Improving air tightness of envelope structure has a negative effect on the following factors:
1. Air flows drying the envelope structure will decrease
2. The natural moisture of new construction materials e.g. concrete dries more slowly.

However not all air flows through structures have a drying effect. The relative moisture content of soil's porosity is almost 100 % RH and therefore the air flows through ground slab do not dry the structures. In general, the structures should not be tightened without survey, planning and supervision. The building physical behaviour of the tightened structure must always be evaluated prior to the actual renovation.

New low energy construction concepts are challenging especially regarding thick insulation layers and drying of structures. The significance of water vapour and air tightness of inner surface of external envelope will increase. The moisture obtained during construction time dries out more slowly in thicker insulation layers in the absence of air flows through structures. This sets tighter requirements for the site moisture management and weather protection. In some cases the structures need to be dried during installation.

Reduction of energy consumption. The actual effect of air leaks on energy consumption depends on e.g. air leak distribution and pressure differential over the envelope structure. Air flow through structures reduces thermal insulation and cools the structures. Improving air tightness with current heat insulation material thicknesses is a one of the most significant and inexpensive means to reduce energy consumption in modern, already energy efficient buildings. However, the usage of buildings and the habits of the residents have an effect on energy consumption. E.g. window ventilation can increase significantly the need for heating energy in otherwise air tight building. (Vinha, 2014)

Prevention of moisture convection. Air leakages may carry significant amount of water vapour into the building envelope. Moisture damages can occur, when the moisture condensates on the cool surfaces. Moisture provides favourable condition for local microbial growth inside the structures. Harmful moisture convection will decrease when the inner layer of the envelope structures are air tight. Also cooling of the surfaces due to cold air leaks will cause condensation risk which can be controlled by preventing air flows through structures.

Prevention of flow of impurities and its effect on the indoor air quality. There are several factors influencing the indoor air quality. Physical factors, chemical and microbiological impurities, concentration of particles and fibres can be examined using different measurement methods and results can be verified against guidelines in the literature in order to get objective results. Indoor air quality is experienced by each individual subjectively and is affected by individual differences and physiological factors. (Sosiaali- ja terveysministeriö, 2003)

The impurities have a negative effect on the indoor air quality. With air flow through leakages microbial impurities, particles, radon and chemical impurities or other harmful substances can be conveyed into indoor air. Also noise and smell from smoke or food can enter indoor. By air tightness renovation, the inner surface of the structures is made airtight so that the air leakages are cut off. The air tightness of structures will also improve fire safety by slowing the spreading of flue gases. In general, air tightness of inner shell structure has a positive effect to indoor air quality. (Sosiaali- ja terveysministeriö, 2003)

Studies related to air tightness renovation has been done relatively few. In general it is known that with air tightness renovation the impurities entering through air leakages can be prevented and the flow of gaseous impurities will be slowed down. It is not known, what kind of concentration of impurities can be conveyed with air flow. The speed and significance of microbial transportation through materials with diffusion is also unknown. There are no health based limits to evaluate the significance of microbial based exposure. Generally it can be estimated that air leakages convey more significant concentration of impurities than diffusion. The importance of the air tightness of the structures of the indoor air quality is not specific determined. There is a need for some research.

4 PLANNING OF AIR TIGHTNESS RENOVATION

4.1 Air tightness testing methods in existing buildings

Before planning for air tightness renovation and implementation on site, the required actual data has to be collected. The locations of the air leaks in structures can be clarified accurately with tracer-gas leak testing method. In the test the tracer gas is led into the tested structure from where gas leaks along with the air flux into the direction of under pressurized room. In the room the trace gas is detected and the leakage points are located with tracer gas analyser.

The air tightness of building envelope structures can be evaluated from architectural and structural drawings showing the implementation method and materials used. The discontinuity points in structure such as gaps and cracks can be found by visual inspection during site inspection. by opening structures e.g. by removing skirting and window trims, the implementation method of joints can be evaluated. Air movement near joints can be observed by sensory smoke gas. Thermal leakages and some air leakages can be detected with an infra-red camera.

4.2 Principles in planning improving air tightness renovation

When renovating an existing building the aim is principally to achieve the same air tightness as it should have had originally. Renovations are always compromises between technical and building history point of view and e.g. the protection degree of the building may have an effect on the renovation method. When planning an air tightness renovation, the air tightness of the whole envelope structure should be taken into consideration. Air tightness renovations implemented on individual structures or premises do neither significantly improve the air tightness of the building nor decrease the impurities in the indoor air.

Air tightness renovations are not suitable for all structures. In general, the structures should not be tightened without survey, planning, supervision of renovation and control. The building physical behaviour of the tightened structure must always be evaluated prior to the actual renovation. Structures with microbial or moisture damage should not be tightened. Damage is removed from the structure and the cause of the damage is fixed.

Improving air tightness is not a sufficient method for blocking all harmful substances from entering indoor air. E.g. for petroleum hydrocarbons other actions need to be used. When removing of harmful substances is not possible, encapsulation renovation can be used. Encapsulation is a renovation method that aims at preventing the entrance of harmful substances or other impurities from entering into the indoor air both with convection and diffusion. Products used in encapsulation of harmful substances prevent or slow the diffusion of gases.

Air tightness renovation requires structural engineering designing. The designing of air tightness renovation is always done case by case and should not be applied directly on other sites. The renovation designing requires expertise and knowledge of material and methods used. Building's thermal and moisture technical functionality, which can change after implementation of renovation, has to be taken into consideration. If there is a danger that after tightening the moisture content might increase detrimentally in moisture sensitive structures, tightening should not be done. The drying ability of the structure should be taken into consideration in renovation especially for external wall structures so that there is a functional ventilation gap.

Air tightening renovation should never be the only renovation action but always a part of a larger entity. When improving air tightness, at least the ventilation system should be inspected and when required adjusted to correspond the changed pressure conditions. The optimal pressure difference over the external envelope is after the indoor air renovation close to 0 Pa. Controlled supply air should be taken into consideration. In moisture and microbe damage renovation, the damaged materials are removed, problem is identified and fixed. The structures are renovated so that they building physics properties stay correct. Implementation should be long-lasting and reliable. (Ympäristöministeriö, 1997)

The methods and materials used in air tightness renovation are determined by structural designer. The materials or methods determined in the designs cannot be changed without prior approval from the designer. The expected life expectation of the structure is stated in the designs. The structural engineering designs consist of work plans and required drawing with necessary details of implementation method of the renovation. The extension of dismantling, its implementation and preparation of the surface is determined in the designs. Preparation of the surface requires typically the removal of the surface material because old finishes seldom provide a solid enough surface for a good attachment of the sealing material. The surface should be even enough, strong and clean from any dirt or dust or any other factors reducing adhering.

5 MATERIALS USED IN AIR TIGHTNESS RENOVATION

In Finland, there are presently no requirements by the authorities for the materials used in air tightness renovation. The materials used should however, fill several requirements for the purpose. The products should have sufficient adhesive and elasticity, long-term durability and airtightness for the purposed use. When large surfaces are treated, the designer should pay special attention on material emissions and on their effect on indoor air quality. In Finland, building materials with air permeability max $1\cdot10^{-6}$ $m^3/(m^2\cdot s\cdot Pa)$ are considered as air barriers. E.g. sealing strip, sealant, polyurethane foams, butyl or other special tape or special adhesive tape can be used for improving the air tightness of structures. When using a sealant, polyurethane foam or special tape it should be verified that the tightening is made using a broad enough adhesive area. The air tightness of the structure can also be improved with material combinations. A common method is to use a liquid spreadable waterproof mass system. Liquid waterproofing barriers have an estimated life span of 40 years. An air tight system installed onto brickwork consists of tight filler, glass fibre mesh and a surface treatment e.g. special paint. Air tight building materials include also several coatings and epoxy primers. Also tight building boards such as gypsum board with carefully sealed seams can function as a part of air barrier in the structure.

6 IMPLEMENTATION OF AIR TIGHTNESS RENOVATION ON SITE

Careful implementation of the work is detrimental for the success of renovation. The technical implementation of the air tightness renovation is confirmed on site with appropriate quality control methods. Quality control methods for the assessment of the technical implementation of renovation are determined in the design documents. Control methods include supervision, implementation and evaluation of models and measuring the air tightness with tracer-gas leak test. Model works of different structures to be tightened are important for the success of the implementation. The quality control of air tightness renovation is done on site before installing the surface structures. The quality control investigation is documented. At the moment there are no official qualification requirements for the workers or for the quality controlling. Certification training in planned to for people doing renovation work and supervision in autumn 2014.

6.1 Quality control with tracer-gas leak testing method

In the trace-gas leak testing method the tracer gas is led into the tested structure from where gas leaks along with the air flux into the direction of under pressurized room. In the room the trace gas is detected and the leaking points are located with tracer gas analyser. A constant negative pressure of 10…15 Pa over the examined structure should be maintained (in the testing room) to obtain comparable results. The negative pressure condition is implemented with a mechanical exhaust ventilator machine. Prior to testing the pressure differences of structure to the room should be verified by measurement. The stability of the prevailing under pressure is recommended to be verified by repeated measurement during the tracer gas testing.

Leakage points are detected with tracer gas analyser and the results are interpreted qualitatively. The conclusion is simple: there is ''either leakage'' or ''no leakage''. The concentration of tracer gas in the structure is not known. Therefore the qualification of the investigator is very important. The detected leak points are marked and the tightness is improved with a suitable method. The tracer-gas leak test should be repeated until all leak points have been eliminated. A quality control report should be issued where the detected leak points are marked e.g. to the layout. Figure 2 explains the quality control assessment of air tightness renovation using tracer-gas leak test.

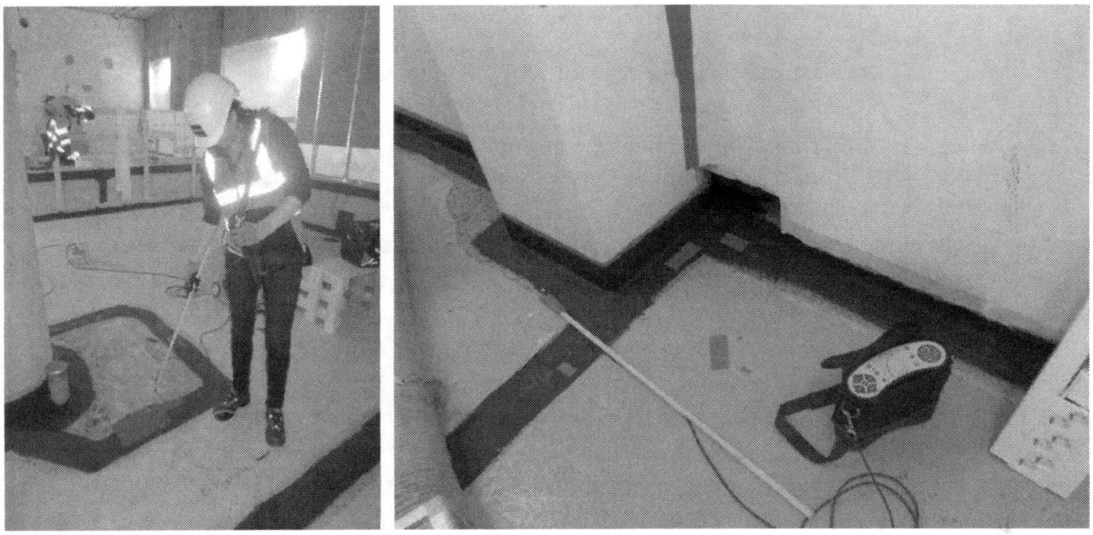

Figure 2: Air tightness renovation of concrete ground slab is done with waterproofing mass system. The tracer-gas leak test is done as part of quality control. In the picture on the right hand, in the test found leakages are marked with an orange tape, so they can be fixed.

7 SUCCESS OF AIR TIGHTNESS RENOVATION AND ITS DURABILITY

7.1 Evaluating the success of the (air tightness) renovation

The evaluation of the success of the renovation is based on user's subjective experience of condition in the premises and objective measurements to evaluate the technical success of renovation. The success is evaluated by visual inspection, different types of measurements as well as with user questionnaires. The process is defined in the quality control plan. The measurements are verified against guidelines in the literature in order to get objective results.

Because air tightness renovation should never be the only renovation action but always a part of other action, the success of the whole renovation should be evaluated. For example, if the air tightness is done well but the final cleaning has been neglected or the ventilation system does not work as planned, it is obvious that the user experiences that indoor air quality problem is not solved. Feedback is often a good measure for evaluating the success of renovation.

The most important evaluation methods are:

- feedback from users and user questionnaires prior and after renovation
- control measurements of structure's air tightness (trace-gas leak test)
- inspection of ventilation system's functionality
- inspection of cleaning level
- measurement of indoor air factors (temperature, humidity, carbon dioxide, pressure)
- measurements of indoor air's chemical and microbiology concentration, if needed.

In practise evaluations after renovation have been done in individual buildings and the structures have been found as air tight as during quality control on-site. However, practise has also shown, that even the most carefully executed renovation and mould dust cleaning is not enough for the most sensitive people and their symptoms return after returning to the premises.

8 CONCLUSIONS

The National Building Code of Finland for new buildings requires air tightness for external envelope structures. The air tightness is done with vapour barrier or with an air barrier. Structure's joints are not automatically completely tight and therefore the implementation requires detailed planning. In external envelope structures and in the soil under building's ground slab there are almost always some microbial based impurities and radon which can affect negatively to the indoor air quality when transported with the air flux into the room. The impurities can lead to poor indoor air quality associated problems. The aim to improve air tightness of structures is to prevent the uncontrolled air leakages through structures. It is a relatively new renovation method and therefore renovation designing requires expertise and knowledge of material and methods used. Research studies related to air tightness renovation has been done relatively few. Some research still needs to be done.

Air tightness renovation is not suitable for all structures. In general, structures should never be tightened without inspection, planning and supervision. Structures with microbial or moisture damage should not be tightened. Air tightening renovation should never be the only renovation action but always a part of a lager entity. When improving air tightness, at least the ventilation system should be inspected and when required adjusted to correspond the changed pressure conditions. Careful implementation of the work is detrimental to the success of renovation. The essential requirement for preventing air leakages is that the air flow paths are cut off totally not only partly. The technical implementation of the air tightness renovation is confirmed on site with appropriate quality control methods i.e. addition to visual inspection the tightness is tested with tracer-gas leak testing method.

In practise air tightness renovation as part of a larger entity has successfully solved the indoor air quality problems in buildings. In Finland the air tightness renovation method to prevent radon is an established renovation method and it has been used for decades. Air tightness renovation has been used to solve microbial based indoor air quality problems about 15 years. The materials used should fill several requirements for the purpose. In practise evaluations after renovation have been done in individual buildings and the structures have been found as air tight as during quality control on-site. The air tightness renovation can be one solution to solve poor indoor air quality problems.

9 REFERENCES

Kettunen, A-V., Slunga E., Viljanen, M. (1991). *Radonin merkitys talonrakennustekniikassa. Radonsuunnitteluohje normaalin radonluokan alueille. Julkaisu/Report 114.* Teknillinen korkeakoulu, Rakennetekniikan laitos. 1991. Espoo.

Ministry of the Environment (2004). *National Building Code of Finland. B3 Foundations. Regulations and Guidelines 2004.* Helsinki.

Rakennusinsinöörien Liitto RIL ry. (2012). *RIL 107-2012 Rakennusten veden- ja kosteudeneristysohjeet.* Helsinki.

Sosiaali- ja terveysministeriö. (2003). *Asumisterveysohje 2003:1. Asuntojen ja muiden oleskelutilojen fysikaaliset, kemialliset ja mikrobiologiset tekijät. Sosiaali- ja terveysministeriön oppaita.*

Vinha, J. (2014). *Rakennusten rakennusfysikaalisen suunnittelun ja toteutuksen periaatteet.* Tampereen Teknillinen yliopisto, Rakennustekniikan laitos. Rakentajain kalenteri 2014. s.351-373.

Ympäristöministeriö. (1997). Ympäristöopas 29. *Kosteus- ja homevaurioituneen rakennuksen korjaus.* Tampere.

Ympäristöministeriö. (1998). *Suomen Rakentamismääräyskokoelma, osa C2 Kosteus, määräykset ja ohjeet 1998.* Helsinki.

Ympäristöministeriö. (2010). *Suomen Rakentamismääräyskokoelma, osa C3, Rakennusten lämmöneristys 2010, Määräykset.* Helsinki.

Ympäristöministeriö. (2012a). *Suomen Rakentamismääräyskokoelma, osa D2, Rakennusten sisäilmasto ja ilmanvaihto. Määräykset ja ohjeet 2012.* Helsinki.

Ympäristöministeriö. (2012b). *Suomen Rakentamismääräyskokoelma, osa D3, Rakennusten energiatehokkuus. Määräykset ja ohjeet 2012.*Helsinki.

DEMAND CONTROLLED VENTILATION IN RENOVATED BUILDINGS WITH REUSE OF EXISTING DUCTWORK

Mads Mysen[*1], Kari Sørnes[1]

1 SINTEF Building and Infrastructure
Forskningsveien 3B
NO-0314, Oslo, NORWAY
**Corresponding author: mads.mysen@sintef.no*

ABSTRACT

Most existing non-residential buildings have Constant Air Volume (CAV) ventilation leading to over-ventilation in periods with low or no occupancy. Demand controlled ventilation (DCV) can considerably reduce the ventilation airflow rate and energy use for fans, heating and cooling compared to constant air volume (CAV) ventilation. There is a potentially enormous upcoming marked for converting from CAV to efficient DCV in existing commercial buildings.

Conversion from CAV to DCV with reuse of existing ductworks, was one of several energy measures carried out in Solbraaveien 23, a Norwegian office building. The building was originally built in the early eighties and is considered to be representative for a large number of buildings in need for an upgrade. Total delivered energy use was reduced from 250 kWh/m^2 to 80 kWh/m^2, and the indoor environment was improved.

Reuse of existing ductworks was very profitable. The ductwork cost in Solbraaveien 23 was roughly cut in half compared to the alternative which was demolition and new ductwork installation.

Based on the experiences from Solbraaveien 23, it is specified a step by step procedure for reuse of existing ductwork that can be used in all projects where such reuse is considered.

The following success criteria are identified for the successful conversion from CAV to DCV with reuse of existing ductwork:

- Can the original system partition be reused?
- Do shafts have sufficient capacity and availability?
- Does the ductwork have sufficient access and quality?
- Are there any visible corrosion?
- Are there risks for any duct parts with asbestos?
- Is the ductwork sufficiently airtight?
- Are the drawings up to date and easily accessible?

KEYWORDS

CAV, DCV, Reuse, Ductwork, Energy

1 INTRODUCTION

IEA reports that CO_2 emissions from buildings must be reduced from 50 $MtCO_2$ in 2010 to 5 MtCO2 in 2050 in the Nordic countries to avoid severe problems of global warming (IEA & Nordic energy research, 2013). A consequence for the building sector is that a widespread conversion of buildings to very low energy consumption and even zero energy buildings is necessary.

Demand controlled ventilation (DCV) can considerably reduce the ventilation airflow rate and energy use for fans, heating and cooling compared to constant air volume (CAV) ventilation (Maripuu, 2009) due to relatively low simultaneous occupancy in office buildings (Halvarsson, 2012). The central ductwork-components will probably have more capacity in a DCV-system than in a CAV-system because of limited simultaneous use. Hence, conversion from CAV to DCV with reuse of existing ductwork could increase the system capacity and actually solve a former capacity problem. There is a potentially huge upcoming marked for converting from CAV to efficient DCV in existing commercial buildings (Mysen et. al., 2011).

A Norwegian office building has been upgraded from CAV to DCV with reuse of existing ductwork. This paper presents this upgrading to modern DCV (Mysen et. al., 2014)

2 DESCRIPTION OF SOLBRAAVEIEN 23

Solbraaveien (Figure 1) is an office building built early in the eighties. It is situated in the municipality of Asker, close to Oslo.

Figure 1. Solbraaveien 23 before and after refurbishment.

It was originally built with CAV-ventilation with reports of annoying noise from the ventilation system. The air inlet was below the windows, blowing upwards with room air induction. Such air inlets are space consuming (Figure 2).

Figure 2. Left: Before retrofitting, the old air-inlets were space consuming. Right: After retrofitting.

The following main retrofit measures were carried out:

- New air-handling-units
- Conversion from CAV to DCV
- The windows were changed, new U-value of 0,8 W/m^2K
- Additional insulation on walls and roof
- Reduced leakage
- Air-water heat pump

Total delivered energy use was reduced from 250 kWh/m^2 before retrofitting to 80 kWh/m^2 after retrofitting and the indoor environment was improved (Mysen et. al., 2014).

3 RESULTS

3.1 Premises and procedure for re-use of existing ductwork

The procedure for re-use is developed by the entrepreneur (GK AS) and SINTEF in the R&D-project UPGRADE Solutions (Mysen et. al., 2014). It is existing ductwork at the "user-side" of the air-handling-unit that is of interest to re-use in upgraded DCV.

A stepwise procedure is developed and shown in figure 3.

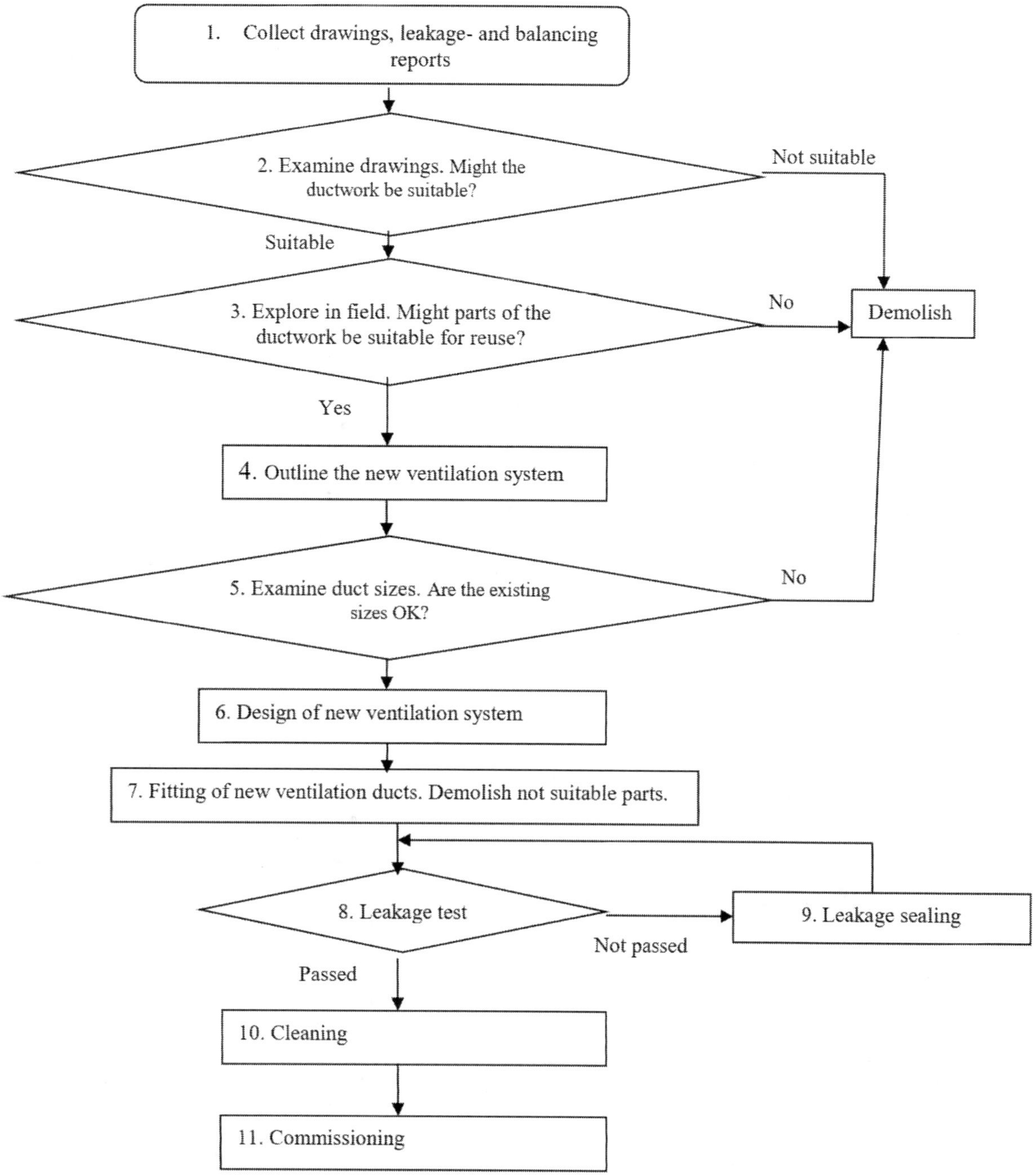

Figure 3. Stepwise procedure for re-use of duct-work.

3.2 DCV system solution

The ventilation system is upgraded with the use of variable supply air diffusers (VSAD). The DCV-units (same as VAV-damper) are integrated in the air diffusers, making it espcially suitable for upgrading to DCV. Figure 4 shows a schematic diagram where variable supply air diffusers are regulated by a controller, and communication is performed via bus.

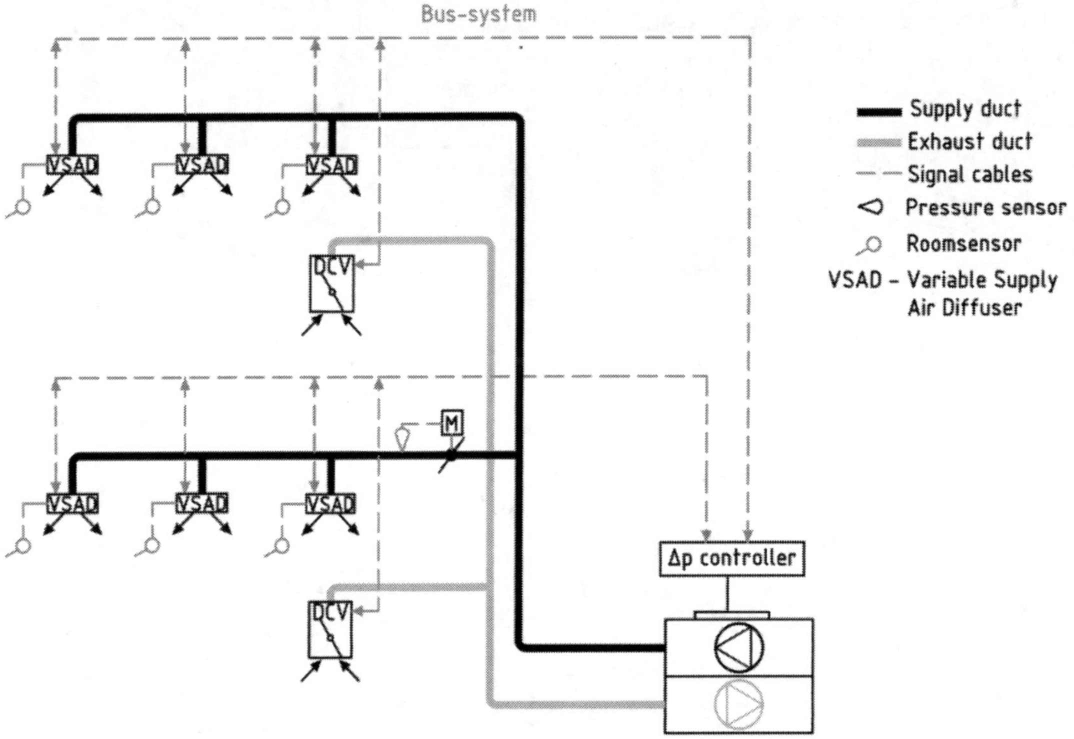

Figure 4.Schematic diagram with VASD regulated by a main controller.

The controller records the required airflow rate, the supplied airflow rate and the damper angle for all the DVC-dampers, and regulates the fan speed such that one of the VSAD is in a maximum open position opened on the supply side, and such that one of the DVC-damper is in a maximum open position on the exhaust side. The integrated motor-driven damper makes sure that the pressure remains in the working range of the VSADs. This damper should normally remain in a maximum open position and only throttle if the pressure in the duct becomes too high relatively to the working range of the VSADs. Such a situation can happen in the branches closest to the fan in large ventilation systems.

VSAD is combined with overflowing arrangement from the offices to corridors and outlets controlled by traditional VAV-dampers.

Table 1. New and re-used parts of the ventilation system after retrofitting.

Ductwork	90-95% is re-used
Air inlets	New VSAD
Air outlets	New, controlled by new VAV-dampers
Air-handling-units (AHU)	New
All ventilation-parts between outside air and AHU (main building air-intake and air exhaust)	New

3.3 Investment costs for re-use versus new ductwork

Table 2 shows the estimated costs in Solbraaveien 23 compared with a new ductwork-solution. Additional costs related to demolishing or fitting of new duct-work is roughly estimated based on Norwegian prices and experiences from Solbraaveien which has a total net area of 10.000 m^2.

Table 2. Costs with reuse of ductwork compared with new ductwork.

Activity	*Total cost in Solbraaveien (with reuse) [Euro/10.000 m²]*	*Total costs with new ducts (traditional solution) [Euro/10.000 m²]*
1. Collect drawings, leakage- and balancing reports	1 250,-	
2. Examine drawing. Might the ductwork be suitable?	1 250,-	
3. Explore in field. Might parts of the ductwork be suitable for reuse?	1 250-13 000,-	
4. Outline the new ventilation system	No difference	
5. Examine duct sizes. Are the existing sizes OK?	2.500 – 6.000	
6. Design of new ventilation system	0	
7a. Demolish not suitable parts.	19 000,-	150 000 – 200 000,-
7b. Fitting of new ventilation ducts	50.000-62.500,-	400 000 - 500 000,-
8. Leakage test	No difference	
9. Leakage sealing	6.250	
10. Cleaning	112.500 -225.000	
11. Commissioning	0	
12. Unforeseen costs		
SUM	194 000 – 328 000	550 000 – 700 000

This rough estimate shows that reuse was a very profitable alternative to new ventilation ductwork in Solbraaveien 23. Maximum additional cost for reuse was estimated to 40 Euro/m^2, while the minimum alternative cost for demolishing and installation of new ductwork was estimated to 70 Euro/m^2. Reduction of the demolishing costs is an important cause of the profitability.

4 DISCUSSION AND CONCLUSIONS

Conversion from CAV to DCV was one of several energy measures carried out in Solbraaveien 23. Total delivered energy use was reduced from 250 kWh/m^2 to 80 kWh/m^2, and the indoor environment was improved.

Reuse of existing ductwork might require some compromises when it comes to normal requirements for specific fan power, maximum air velocity, noise generation and leakage. Before considering ductwork reuse, one has to clarify that the builder owner has a pragmatic attitude towards such normal requirements.

Furthermore, one must clarify if the ductwork is suitable for reuse as early as possible in the process. Based on the experiences from Solbraaveien 23, it is specified a step by step procedure for reuse of existing ductwork that can be used in all projects where such reuse is considered (Figure 3).

The following success criteria are identified for the successful conversion from CAV to DCV with reuse of existing ductwork:

- Can the original system partition be reused?
- Do shafts have sufficient capacity and availability?
- Does the ductwork have sufficient access and quality?
- Are there any visible corrosion?
- Are there risks for any duct parts with asbestos?
- Is the ductwork sufficiently airtight?
- Are the drawings up to date and easily accessible?

Reuse of existing ductworks was very profitable in Solbraaveien 23. The ductwork cost was roughly cut in half compared to the alternative which was demolition and new ductwork installation. Reuse of the existing ductwork can potentially reduce the refurbishment period and therefore reduce loss of rental income. This is not included in the economical consideration.

5 ACKNOWLEDGEMENTS

This paper is funded by contributions from industry partners and public funding from the Norwegian Research Council as part of the project "UPGRADE Solutions".

6 REFERENCES

IEA & Nordic energy research (2013). *Nordic Energy Technology Perspective*. Paris/France, OECD/IEA. www.IEA.org.

Maripuu, M.-L. (2009). *Demand controlled Ventilation (DCV) systems in commercial buildings: functional requirements on systems and components*. Goteborg/Sweden. School of Electrical and Computer Engineering. Chalmers tekniska högskola.

Halvarsson, J. (2012). *Occupancy Pattern in Office Buildings - Consequences for HVAC system design and operation.* Trondheim/Norway. Doctoral Thesis, NTNU.

Mysen, M. Schild, P. Drangsholt, F. Larsen, B.T. (2011) Convertion from CAV to VAV - a key to upgrade ventilation and reach energy targets in the existing building Stock. *Proceedings Roomvent 2011*, Paper 142, ISBN: 978-82-519-2812-0, June 2011, Trondheim/NORWAY.

Mysen, M. Aronsen, E. Johansen, B.S. (2014) *Conversion to DCV with reuse of ductwork (In Norwegian).* Oslo/Norway. SINTEF Fag 15. SINTEF akademiske forlag. ISBN: 978-82-536-1377-2.

REQUIREMENTS AND HAND-OVER DOCUMENTATION FOR ENERGY-OPTIMAL DEMAND-CONTROLLED VENTILATION

Mads Mysen*[1], Axel Cablé[1], Peter G. Schild[1], and Kari Thunshelle[1]

1 SINTEF Building and Infrastructure
Forskningsveien 3B
NO-0314, Oslo, NORWAY
*Corresponding author: **mads.mysen@sintef.no**

ABSTRACT

Demand controlled ventilation (DCV) considerably reduce the ventilation airflow rates and energy use compared to Constant Air Volume (CAV) systems. DCV in commercial buildings is probably a prerequisite to achieve ambitious energy-goal. However, evaluation of real energy use demonstrates that the energy saving potential is seldom met. DCV-based ventilation systems must become more reliable to close the gap between theoretical and real energy-performance. These unfortunate experiences with DCV have many causes, including: inadequate specifications and hand-over documentation, balancing report not suitable for DCV, communication errors and lack of knowledge about DCV-systems among decision makers.

There is also a significant difference in performance between DCV-systems and simpler systems that, for example, vary the airflow rate with pre-set air damper positions, or that use a single sensor for several rooms. In order to verify that a DCV system fulfils the expectations in terms of indoor climate and energy use, one must specify measureable objectives of performance.

In this paper, the most important control points during commissioning of a DCV system are described. Measureable objectives of performance should be specified. In particular, recommendations are given in terms of:

- control and measurement of Specific Fan Power
- balancing procedures and control of airflow rates
- requirements for sensors and dampers
- hand-over documentation
- deviations during commissioning and corrective procedures.

KEYWORDS

Energy use, Demand controlled ventilation, Specific fan power

1 INTRODUCTION

Demand controlled ventilation (DCV) considerably reduces the ventilation airflow rates and energy use compared to Constant Air Volume (CAV) systems. DCV in commercial buildings is probably a prerequisite to achieve Nordic and European energy-goal (IEA and Nordic Energy Research, 2014).

When correctly implemented, DCV can reduce the energy consumption of ventilation by more than 50 % (Maripuu, 2009). However, evaluation of real energy use demonstrates that the energy saving potential is seldom met (Mysen et al., 2010). DCV-based ventilation systems must become more reliable to close the gap between theoretical and real energy-performance. These unfortunate experiences with DCV have many causes, including: inadequate specifications and hand-over documentation, balancing report not suitable for DCV, communication errors and lack of knowledge about DCV-systems among decision makers. Based on this experience from case-studies, an expert group has developed new requirements and hand-over documentation (Mysen and Schild, 2013). The work is carried out in the Norwegian R&D-project reDuCeVentilation (http://www.sintef.no/Projectweb/reDuCeVentilation/).

DCV systems are ventilation systems in which the air flow is controlled automatically according to a measured demand at room level. This means that DCV-systems must have a sensor in the room giving a continuous measurement of the indoor air quality. This signal is then used to control the airflow rate according to the desired indoor air quality level.

VAV stands for Variable Air Volume. It is a broader term than DCV, as it encompasses all systems with variable airflow rate, irrespective of type of control. Only VAV-systems that control the airflow rate according to a measured demand in the room, and not according to a preset value, are considered as DCV in this paper. Normal VAV-dampers in DCV systems are denoted DCV-damper in this paper.

The commissioning, balancing, and control of a DCV-system consist of the following work steps:

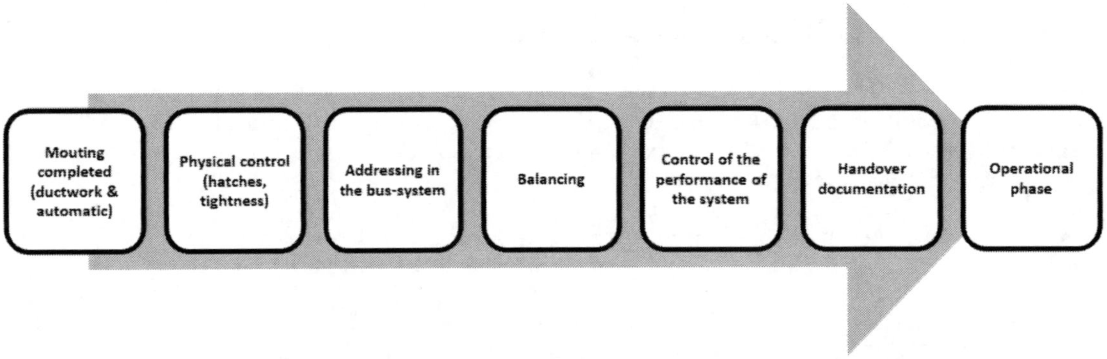

Figure 1. Recommended work steps subsequent to the mounting of the ventilation system.

This paper addresses only requirements and the balancing, control- and hand-over steps.

2 DCV- PRINCIPLES AND BALANCING PROCEDURE

2.1 General

There are several DCV-systems, but they can roughly be divided into the following principles:

"Pressure Controlled DCV", "Static Pressure Reset DCV", "Damper optimized DCV" and "Variable Air Supply Diffusor DCV". These principles are defined and described by Mysen (Mysen and Schild, 2011). Some of the DCV-principles require special balancing procedure and this paper address "Pressure Controlled DCV" and "Damper optimized DCV". Balancing procedure for "Static Pressure Reset DCV" is similar to "Pressure Controlled DCV". Balancing procedure for "Variable Air Supply Diffusor DCV" is similar to "Damper optimized DCV".

2.2 Pressure controlled DCV

Pressure Controlled DCV (PC-DCV) is the traditional DCV systems (Figure 2). The purpose of static pressure control is to indirectly control the airflows by controlling the pressure in a strategic duct position. PC-DCV requires installation of active DCV-units, or known as VAV-units, controlling supply and exhaust air flows to each DCV-room/zone. Controlling fan speed to maintain a constant static fan pressure rise, will result in unnecessary throttling along the critical path during most of the air-handling-units (ahu's) operating time, and therefore unnecessary fan energy use. The duct path with the greatest flow resistance from the ahu to any terminal, is called the 'critical path' dimensioning necessary fan pressure rise.
One unfortunate experience of pressure controlled DCV system is that minor changes in room demand just redistribute airflow in the duct system with the airflow in the AHU being more or less constant. The consequence is that no energy saving is achieved, or the supply air is insufficient. This is normally caused by inadequate precision or wrong placement of the pressure sensor.

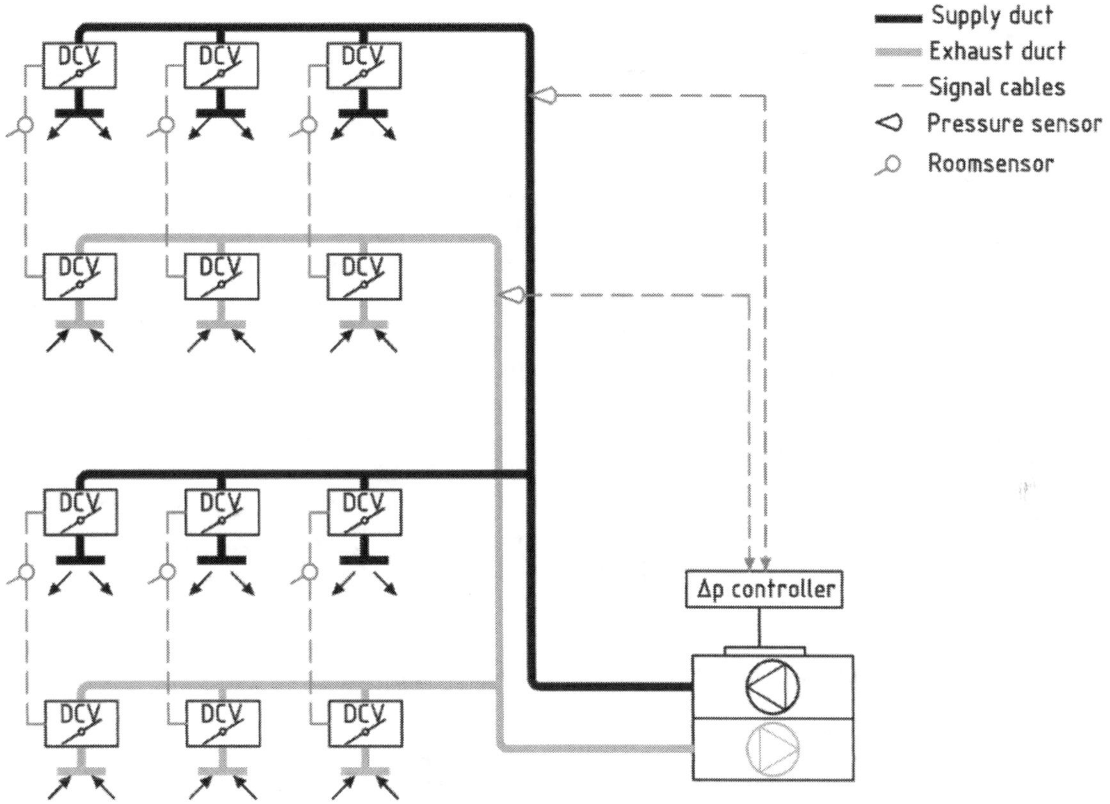

Figure 2. Pressure controlled DCV system. The fan speed is controlled so as to keep constant static pressure in the main ventilation duct, at the location of the pressure tap.

The main purposes of balancing a pressure controlled DCV system are:

- controlling the placement of the pressure sensor
- setting the right pressure set point

In addition, the balancing will reveal connection and communication errors.

Balancing of a pressure controlled DCV-system consists of the following steps:

- Control that all the DCV-units have supply voltage and no polarity error
- Control that the pressure sensor is mounted on a location with stable static pressure or uniform velocity profile, by performing measurements over the duct cross section with a Pitot-tube or a hot-wire anemometer.
- Select a pressure set point which is slightly higher than necessary. This can be deduced from pressure drop calculations or empirically.
- Program the actual maximum and minimum airflow rates values (V_{max} and V_{min}) on each DCV-damper and set the dampers to automatic mode. Control that all the DCV-dampers get the maximum airflow rate, and read the degree of opening. Find the index damper, which is the damper with the highest degree of opening.
- Adjust the pressure set point until the index DCV-damper gets the maximum airflow rate without throttling (about 80 % degree of opening). You have then found the energy optimal pressure set point, which is the lowest pressure set point which provides the right airflow rates according to the designed values.
- Complete the VAV control form. The completed control form should be included in the documentation of the ventilation system.

2.3 Damper optimized DCV

Damper optimized DCV consists in controlling the airflow rate in the main duct according to the position of the dampers, such that at least one of the dampers is in a maximum open position (Figure 3). The purpose is to ensure minimum fan energy consumption by looking for a minimum pressure rise over the fan. This is achieved if one duct path (critical path) is always open. With damper-optimized DCV, the required airflow rate, the supplied airflow rate as well as the damper angle are recorded for all the DCV-dampers. This information is sent to a controller which regulates the fan speed.

Balancing of DCV-units in damper-optimized systems is very simple, and consists in specifying minimum and maximum design airflow rate for each DCV-unit. This can be done either through the bus-system or by connecting a programming device directly on the DCV-units. Various programming devices are used by the different suppliers.

Balancing of a Damper optimized DCV -system consists of the following steps:

- Control that all the DCV-units, room sensors/ room regulators etc. have supply voltage and no polarity error
- Program the actual maximum and minimum airflow rates values (Vmax and Vmin) on each DCV-damper and set the dampers to automatic mode.
- If the DCV-units do not give the expected response, check the polarity on the supply voltage.
- Complete the VAV control form. The completed control form should be included in the documentation of the ventilation system.

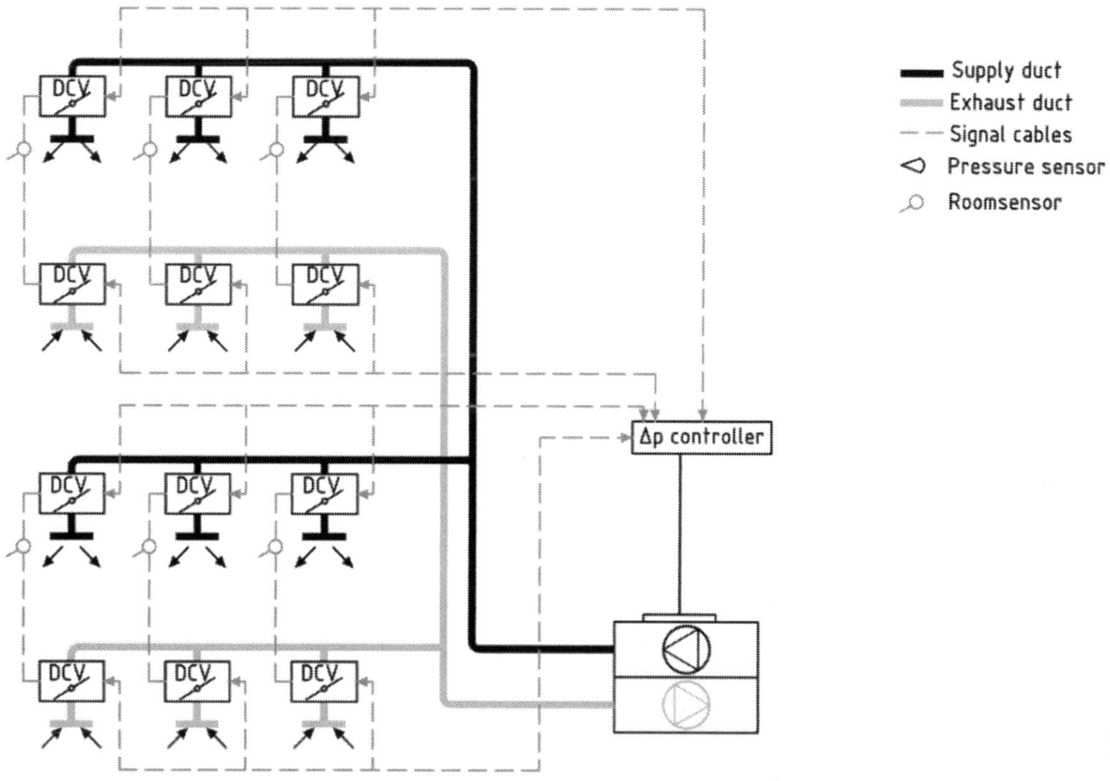

Figure 3. Damper-optimized DCV-system.

3 REQUIREMENTS

It must be possible to control the specified requirements. Specific Fan Power (SFP) is normally required and controlled at maximum air flow. However a DCV system will typically have air flow rates between 30 and 80% of maximum air flow, depending mainly on diversity factor for dimensioning and base ventilation level. At maximum airflow, there are only small differences between the system's SFP (Figure 4, r=1), but at lower airflow rates there are major differences depending on the control strategy. It is important to require maximum SFP-value for two operating scenarios, maximum airflow and reduced airflow rate, to ensure an energy efficient control strategy (Figure 4).

Fitting a DCV-system, typically involves several contracts including BMS (Building Management System), Ventilation system and Electrical Equipment. However, the overall responsibility for the system functionality should be clearly defined and placed in one contract.

Adequate specifications, hand-over documentation and balancing report suitable for DCV-systems must be used.

Critical components, such as sensors, must have proper functionality and acceptable measurement uncertainty throughout their predicted life expectancy, for instance:

- CO_2-sensors +/- 50 ppm
- Temperature sensors +/- 0.5 °C

Some of the critical components like CO_2-sensors should be controlled at site. One control point for CO_2-sensors is to check if they all give the same ppm-results when the building is empty during evening/night.

An airflow change in any room should give approximately the same change of the total airflow through the air-handling-unit (ahu). This test will reveal pressure controlled DCV system that redistributes parto of the airflow caused by inadequate precision or wrong placement of the pressure sensor.

A DCV is a dynamic system and should be tested and tuned in for both summer and winter conditions. There should be an inspection, function test and review of the DCV- system after a period of normal operation, e.g. 1 year.

The most important control points are presented in figure 4.

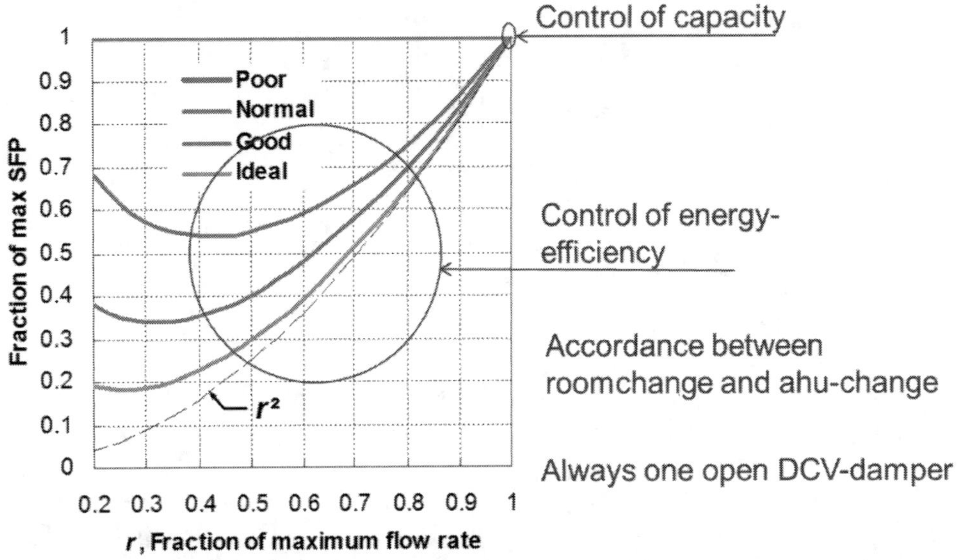

Figure 4. The most important control points. Measurement of SFP with partial load, control of the compliance between airflow rate at room level and total airflow rate, and control that there is always a DCV-damper in max open position with the help of the BMS (Schild&Mysen, 2009).

SFP shall be measured such that power losses in Variable Speed Drives are included, preferably using a suitable 3-phase energy analyzer, or read directly on the AHU.

Deviations during commissioning are normal and should be expected. Therefore, it is important to either forecast time to improve the system, or to create a model for economic compensation to take into account the deviations from the requirements which affect the energy consumption.

Furthermore, new discrepancies will occur during the operational life. It is essential that the automatic controls and the Building Management System (BMS) make it easy to detect faults. It is also important that the control components are accessible for inspection, service and replacement. DCV- dampers in exhaust ducts are especially exposed for dust and must be accessible for inspection and cleaning.

4 HAND-OVER DOCUMENTATION

4.1 VAV-control form

Problems during operation occur most often at maximum or minimum airflow rates. Tests should therefore be carried out for these two operating situations. For each of these situations, it is necessary to consider each DCV-unit and override the control signal from the roomsensor (eg. temperature) in order to force the DCV-unit to respectively max and min airflow rate, and document both airflow rate and degree of opening. The degree of opening tells whether the DCV-units work in a favorable range (40 to 80%) and whether the pressure set point is balanced.

This requires four control measurements per DCV-unit. Such a control procedure is particularly relevant for DCV-systems with pressure control and limited control possibilities from the BMS.

A special VAV-control form is designed for this purpose (Figure 5).

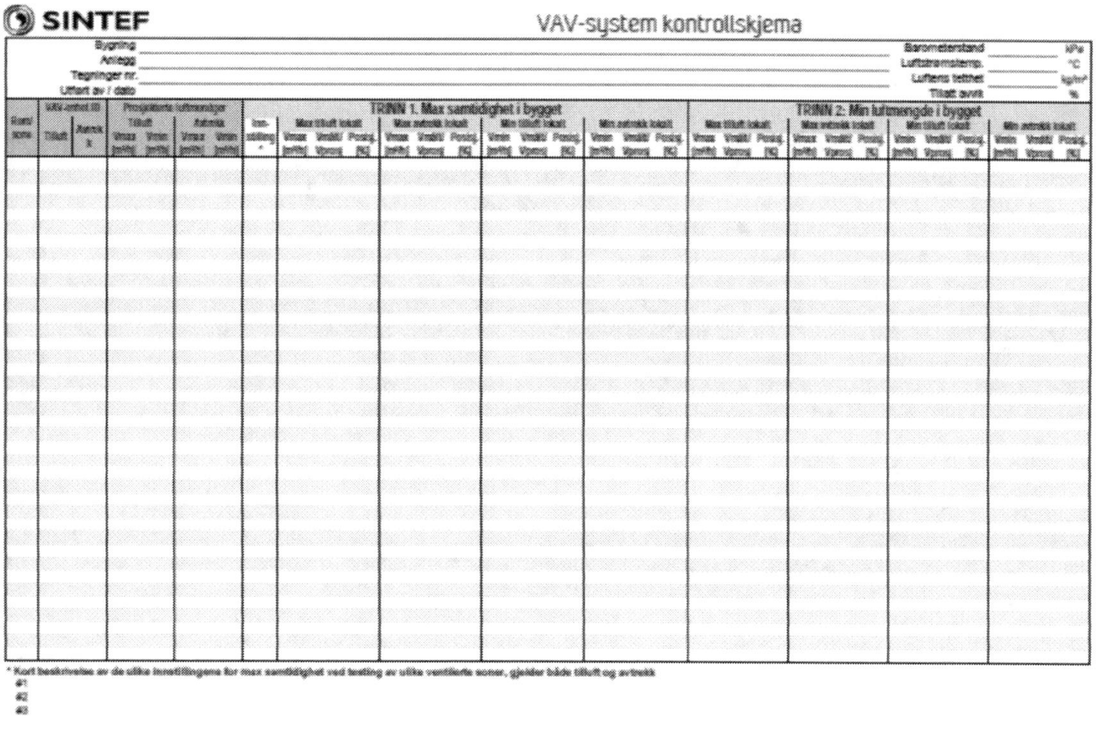

Figure 5. Recommended VAV-form (http://www.sintef.no/Projectweb/reDuCeVentilation/)

There are developed procedures for load tests at maximum and minimum loads/airflow rates (Mysen and Schild, 2013).

4.2 Automated load test

Manual load test are time consuming, and it has proven to be very difficult to override DCV-units in a load test. One should therefore strive to automate the load test completely by programming it in the control panel or in the BMS. This has several advantages: it can be a complete test (not spot-checking) with all combinations of overriding, it reduces costs significantly, and can be repeated as often as needed (one time per year during normal operation, or after changes in the system).

5 REFERENCE

IEA & NORDIC ENERGY RESEARCH (2014). Nordic Energy Technology Perspective, Pathways to a Carbon Neutral Energy Future. OECD/IEA France.

MARIPUU, M.-L. (2009). *Demand controlled Ventilation (DCV) systems in commercial buildings: functional requirements on systems and components,* Göteborg, School of Electrical and Computer Engineering, Chalmers tekniska högskola.

MYSEN, M. & SCHILD, P. G. (2013). DCV- Requirements and hand-over documentation (In Norwegian). *SINTEF Fag 11.* ISBN: 978-82-536-1369-7 (71 s)

MYSEN, M. & SCHILD, P. G. 2011. Requirements for well functioning Demand Controlled Ventilation. *REHVA European HVAC Journal,* 48, 14-19.

MYSEN, M., SCHILD, P. G. & DRANGSHOLT, F. 2010. Robustness and True Performance of Demand Controlled Ventilation in Educational Buildings – Review and Needs for Future Development. *31st Air Infiltration and Ventilation Centre Conference 2010 : (AIVC 2010) : low energy and sustainable ventilation technologies for green buildings : Seoul, South Korea, 26-28 October 2010.* Seoul: International Network for Information on Ventilation and Energy Performance.

MONITORING RESULTS AND OPTIMIZATION OF A FAÇADE INTEGRATED VENTILATION CONCEPT FOR BUILDING RETROFIT

Fabien Coydon, Maxime Duran, Arnulf Dinkel, Sebastian Herkel

Fraunhofer Institute for Solar Energy Systems ISE
Heidenhofstrasse 2
79110 Freiburg, Germany
fabien.coydon@ise.fraunhofer.de

ABSTRACT

An office building of the Fraunhofer Institute for Solar Energy systems (Fraunhofer ISE) in Freiburg was retrofitted in 2012 with an innovative concept based on technology integration in the façade. Prefabricated window modules integrating air inlets and outlets, façade integrated air ducts and a heat and moisture recovery ventilation device were implemented. A long term monitoring was set up including energy, temperature, CO_2 and humidity measurements. The results of this monitoring and an analysis of the system are presented, focussing on the influence of façade integrated air ducts on the energy performance of the whole system and on the thermal comfort.

A specific definition of the global energy efficiency of façade integrated ventilation systems is proposed and used to evaluate the performance of the system.

On the basis of the monitoring results, a simulation model could be defined and used to optimize the integration concept by testing different geometries and materials. The last developments of the integration system are presented such as the next step of this work: the implementation of the system on the façade of a multifamily home in Frankfurt.

KEYWORDS

Central ventilation, Façade integration, Heat recovery, Building retrofit, Energy efficiency

1 INNOVATIVE FAÇADE RETROFIT WITH INTEGRATED AIR DUCTS

1.1 Advantages and risks of façade integration

In order to reach the goals for the reduction of CO2 emissions set by the European Union, the rate of energy retrofit has to be multiplied by 2 or 3. The main difficulties are the costs which are often high because of the multiplicity of tasks to be realized. Walls, roof and floor must be insulated, windows and heating device must be replaced and a ventilation system must be installed. The dwellings are mostly unfit for habitation during a few weeks, a few months or sometimes for the whole duration of the retrofit. In some European countries, the installation of ventilation is mandatory in retrofitted buildings, as for example in Germany where the DIN norm 1946-6 (DIN 2009) imposes since 2009 to install a ventilation system as soon as a third of the windows are replaced.

Integrating the ventilation system in the façade can represent a good solution (Coydon, Dinkel et al. 2013) since it avoids cumbersome air ducts inside the dwellings, simplifies the retrofit work by avoiding core holes and allows to let the tenants inside the dwellings during the work.

The main risk of integrating air ducts into an external wall is the heat loss leading to a reduction of the heat recovery potential which was already investigated by the Fraunhofer IBP for a similar integration concept (Ziegler, Krause et al. 2012). This study presents an analysis of this risk.

1.2 Demonstration building and its retrofit concept

A new façade concept was tested in 2012 for the energy retrofit of an office building belonging to the Fraunhofer ISE (Figure 1):

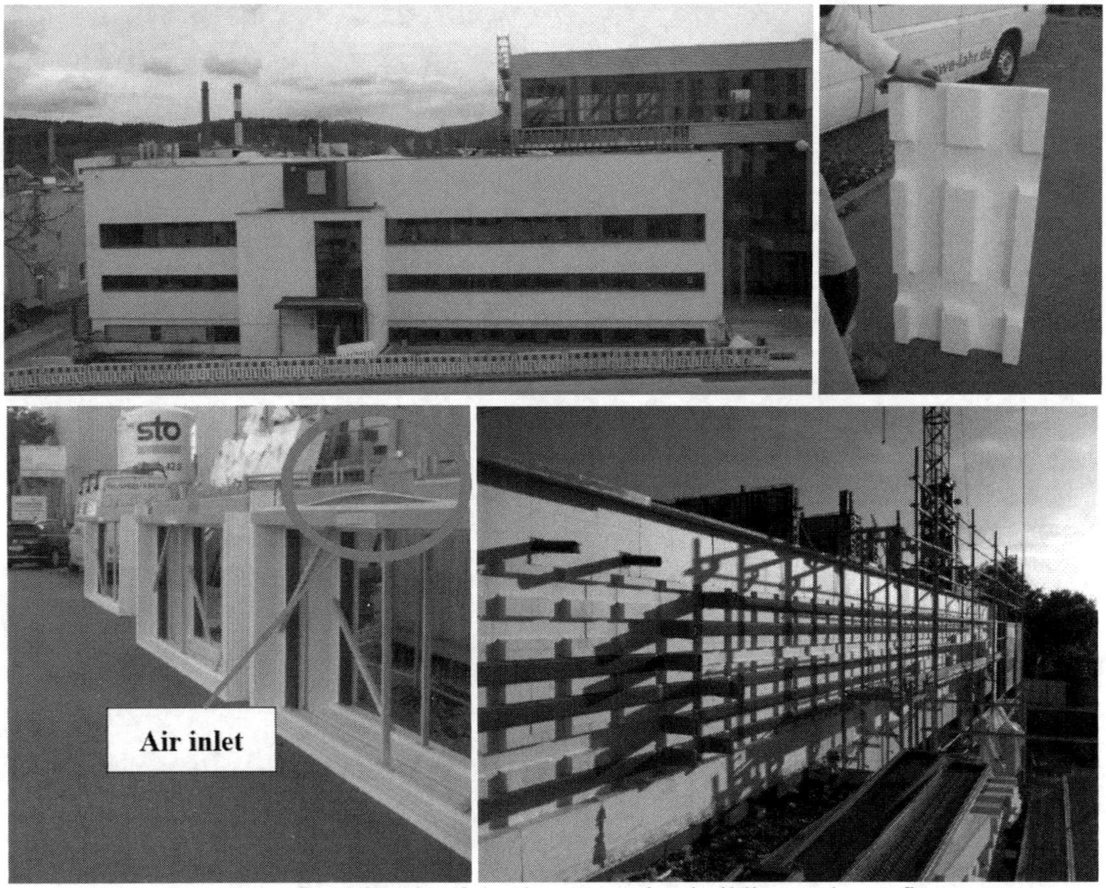

Figure 1 Retrofitted façade of the demonstration building and retrofit concept

This retrofit concept is based on insulation boards (standard EPS material) in which air ducts can easily be clipped-in. These insulation boards are fixed on the façade after what the air ducts can be installed and covered by a second layer of insulation boards. Thanks to these systems, air ducts are integrated in a structured manner allowing fast and replicable work. This layer allows clipping-in the pipes horizontally or vertically into the prepared channels.

Prefabricated window modules are including roller shutters and air inlets so that the windows openings can be directly used for penetrating into the building and no core hole is necessary.

The panels cover the wall of 6 offices situated on the second floor (see Figure 2). In that case, the air handling unit is placed outside on a separate roof in a container.

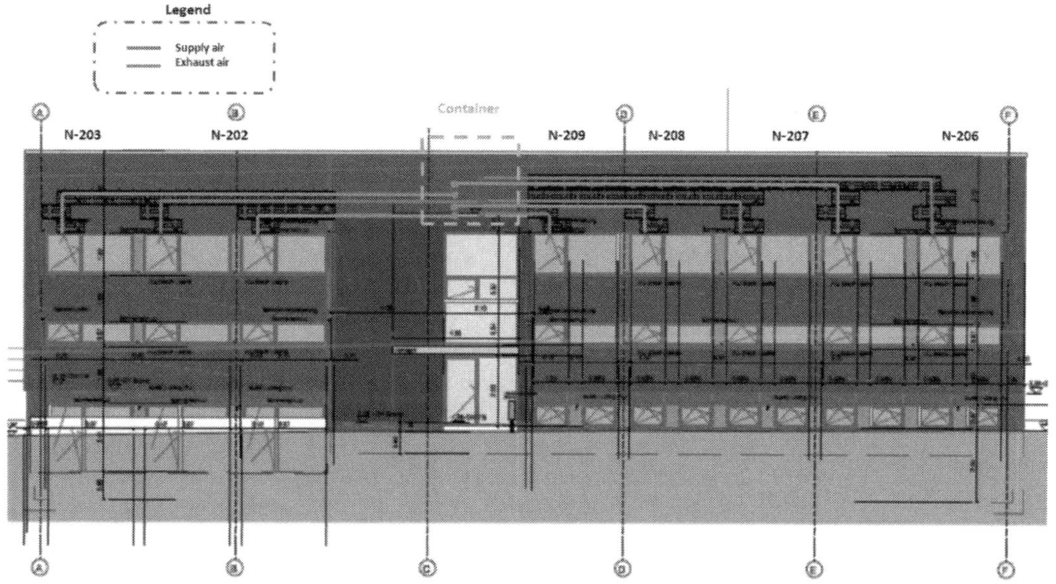

Figure 2 Renovated façade with air ducts paths

2 ANALYSIS OF THE SYSTEM

2.1 Monitoring concept

The measurements include temperature and humidity sensors at each strategic point of the system (inlets and outlets of the ventilation device and of each air duct) but also sensors for airflows and electrical consumptions (Figure 3).

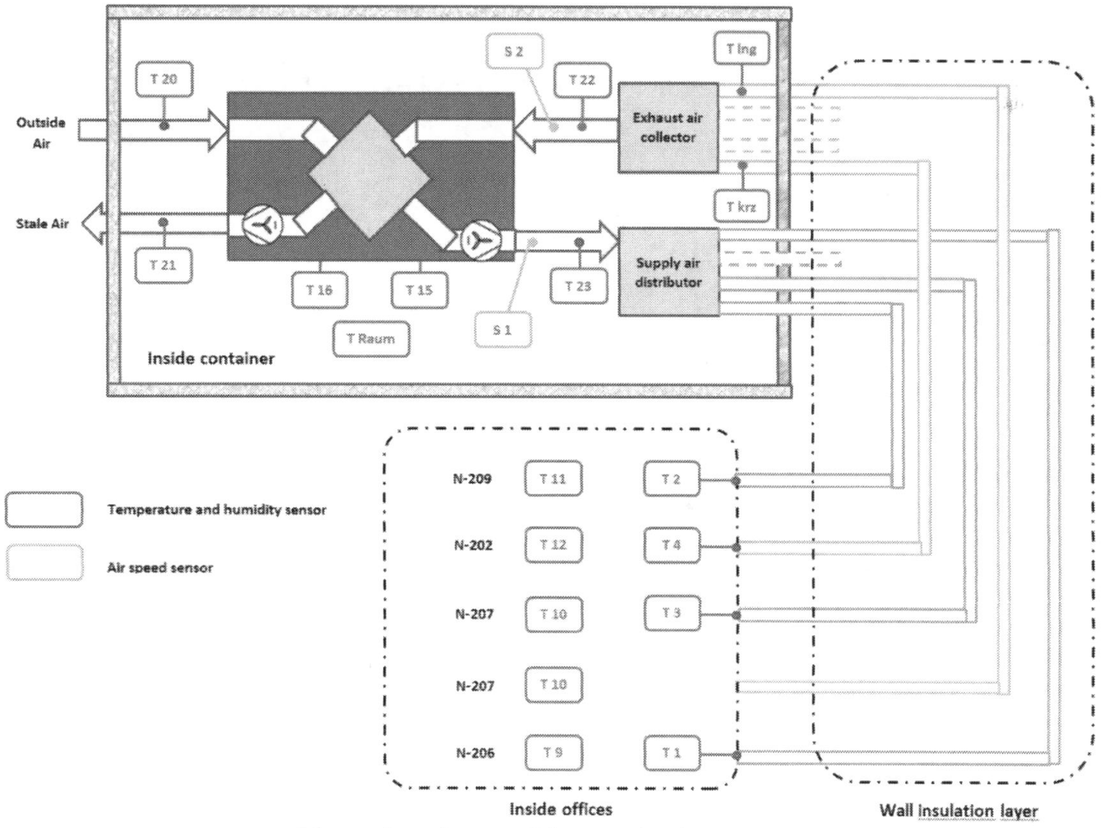

Figure 3 Setup of the monitoring concept

2.2 Energy efficiency of the ventilation system: definitions

The norm DIN EN308 (DIN 1997) defines the heat recovery rate of a ventilation system with a simple ratio of temperature differences. This definition is based on the comparison between a ventilation device and a theoretical reference case which is a ventilation system without heat recovery providing the same airflow. The temperatures taken into account are directly measured at the inlets and outlets of the ventilation device:

$$\eta_{EN308} = \frac{T_{supply-1} - T_{outside}}{T_{exhaust-1} - T_{outside}} \quad (1)$$

In order to describe the energy performance of the system, the electrical power of the fans must be taken into account. For the theoretical reference case, the electrical power is supposed to be equal to zero:

$$\eta_{device} = \frac{\dot{m}.Cp.(T_{supply-1} - T_{outside}) - P_{elec}}{\dot{m}.Cp.\ (T_{exhaust-1} - T_{outside})} \quad (2)$$

The main issue of façade integrated air ducts is their heat losses. Therefore, we used the supply and exhaust temperatures inside the rooms instead of the temperatures at the interface between façade integrated ducts and device:

$$\eta_{with\ ducts} = \frac{\dot{m}.Cp.(T_{supply-2} - T_{outside}) - P_{elec}}{\dot{m}.Cp.\ (T_{exhaust-2} - T_{outside})} \quad (3)$$

In order to respect the heat balance with the new boundaries, we had to take the influence of the air ducts on the building envelope insulation into account:

$$\eta_{with\ wall} = \frac{\dot{m}.Cp.(T_{supply-2} - T_{outside}) - P_{elec} + HL - HL'}{\dot{m}.Cp.\ (T_{exhaust-2} - T_{outside})} \quad (4)$$

With:

HL: Heat Losses through the wall of the reference case

HL': Heat Loss through the wall with façade integrated air ducts

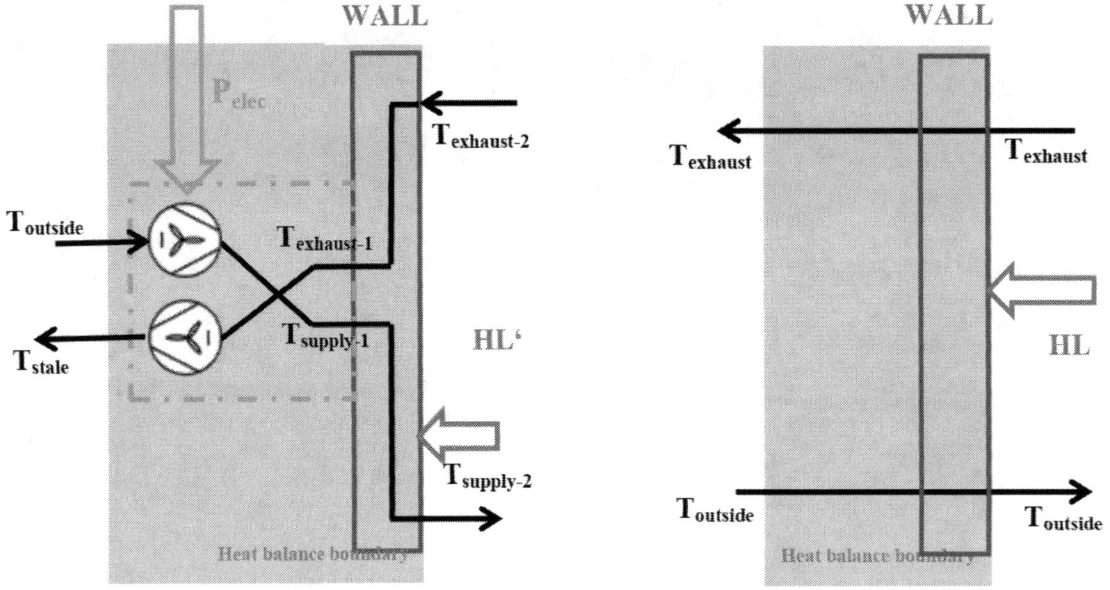

Figure 4 Thermal balances of the analysed system (left) and reference case (right)

2.3 Energy efficiency of the ventilation system: results

In order to distinguish the influence of heat losses of the air ducts and of the external wall, we compared the results obtained by equations (2), (3) and (4) and represented the results on Figure 5.

As the boundary of the heat balance is set on the inner surface of the wall, the presence of air ducts in the façade is reducing the heat losses coming from the inside air. In the relevant domain (T_{ext} < 14°C), the efficiency of the device itself, taking the electrical power into account is around 85 %.

This efficiency is reduced to around 65 % by the heat losses in the air ducts but a part of these heat losses is regained by the reduction of losses through the wall. Thus the global efficiency of the system is around 75 %.

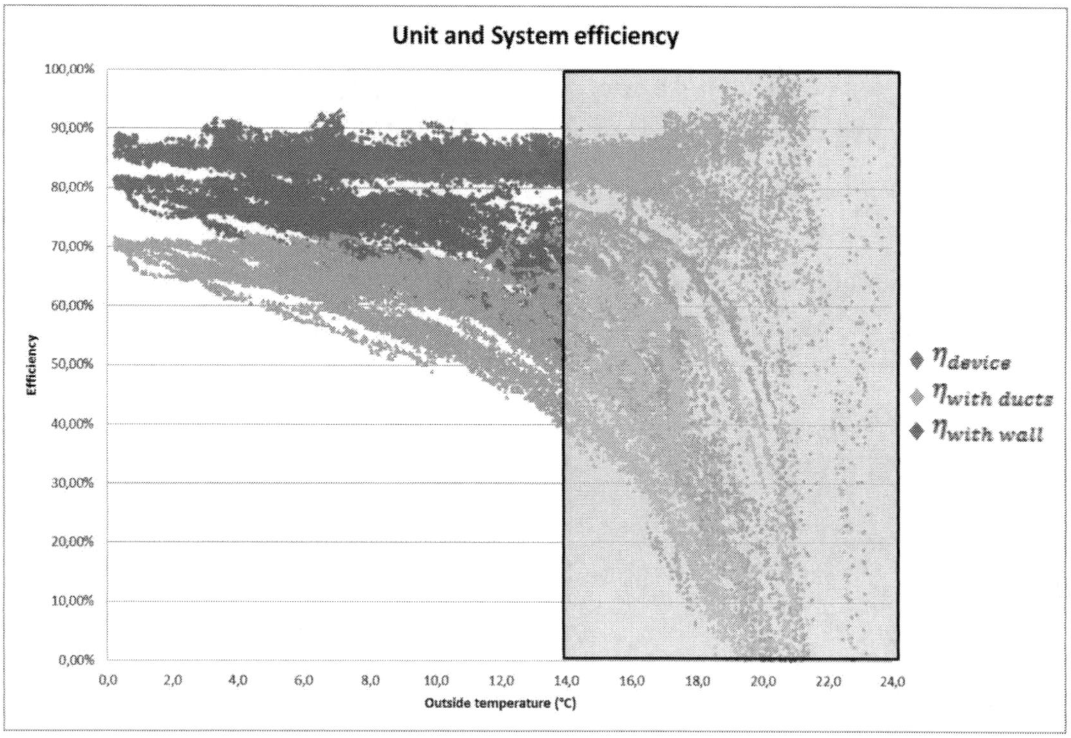

Figure 5 Energy efficiency of the ventilation system according to 3 different definitions

In order to improve this result and reduce the influence of the heat losses in the ducts, a numerical model was developed at the Fraunhofer ISE, so that modifications in the system could be simulated.

3 NUMERICAL MODEL

3.1 Description of the model

The model aims to optimize the panels by calculating the heat losses of the air ducts and through the wall according to several contexts.

The mesh size for the model is 5 mm x 5mm in a vertical section of the wall and around 1 m for the dimension parallel to the air ducts. The model allows a combination of exhaust and supply airs and can only analyse steady states.

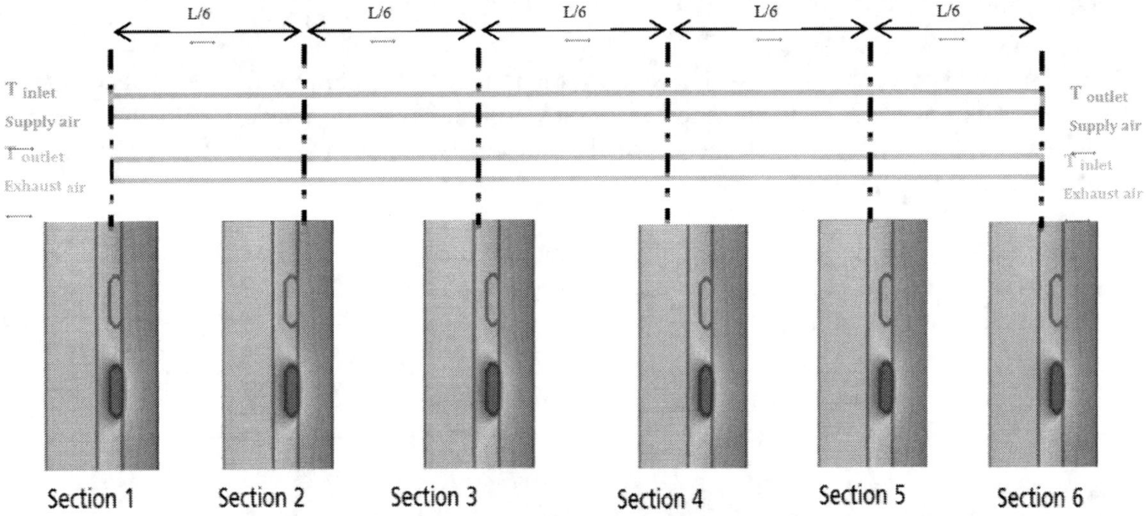

Figure 6 Description of the numerical model

1.1 Validation of the model

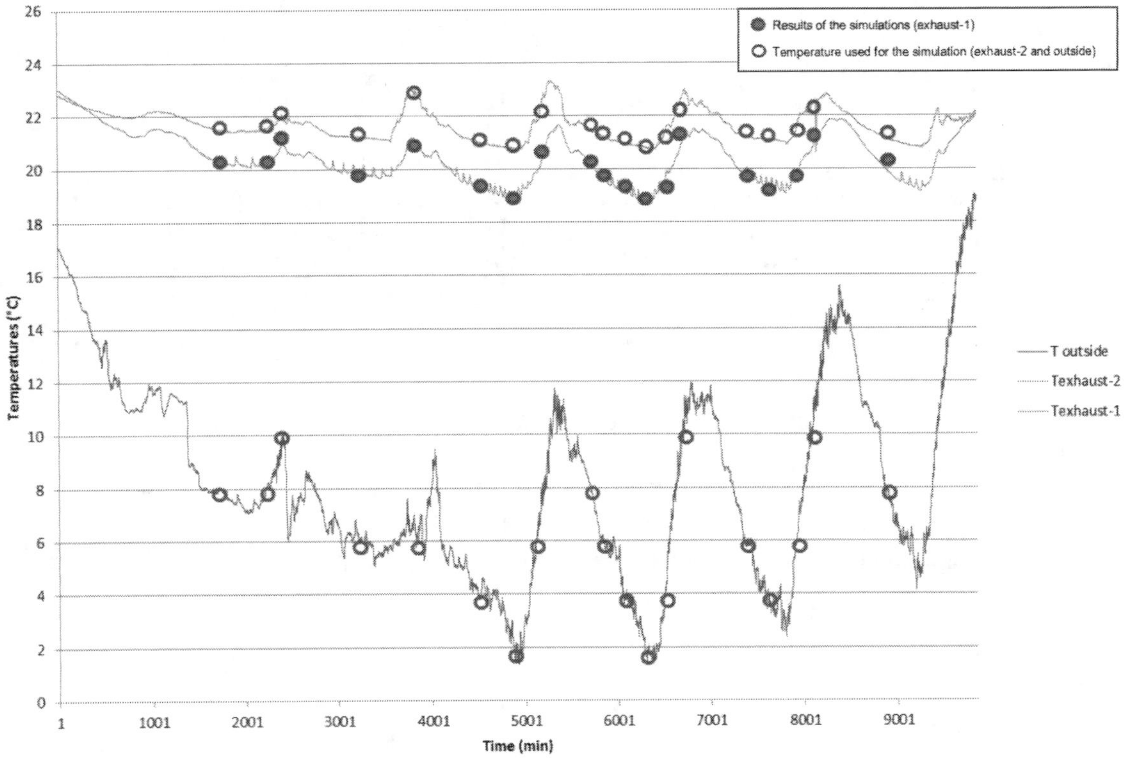

Figure 7 Results difference between the model and measurements

Figure 7 presents the results of the validation tests for the longest exhaust air duct (duct where the heat losses are the highest). Only the heat losses of the air ducts could be validated whereas the model is also calculating the heat losses of the wall but this point seems to be less critical.

The same validation tests were done for the other air ducts and for different airflows and gave all acceptable results.

4 OPTIMISATION OF THE INTEGRATION SYSTEM

4.1 Context of the work

Within a European project (Retrokit) focusing on prefabrication for building retrofit, a demonstration building will be renovated in 2015 in Frankfurt (Germany) with façade integrated HVAC networks. This demonstration building is a multifamily home (20 dwellings) where the external walls will be insulated, the windows and the heating system will be replaced and a new façade integrated ventilation system will be installed.

Figure 8 shows the design phase on one façade of the building with air ducts and heating pipes. Figure 9 shows the foreseen integration concept.

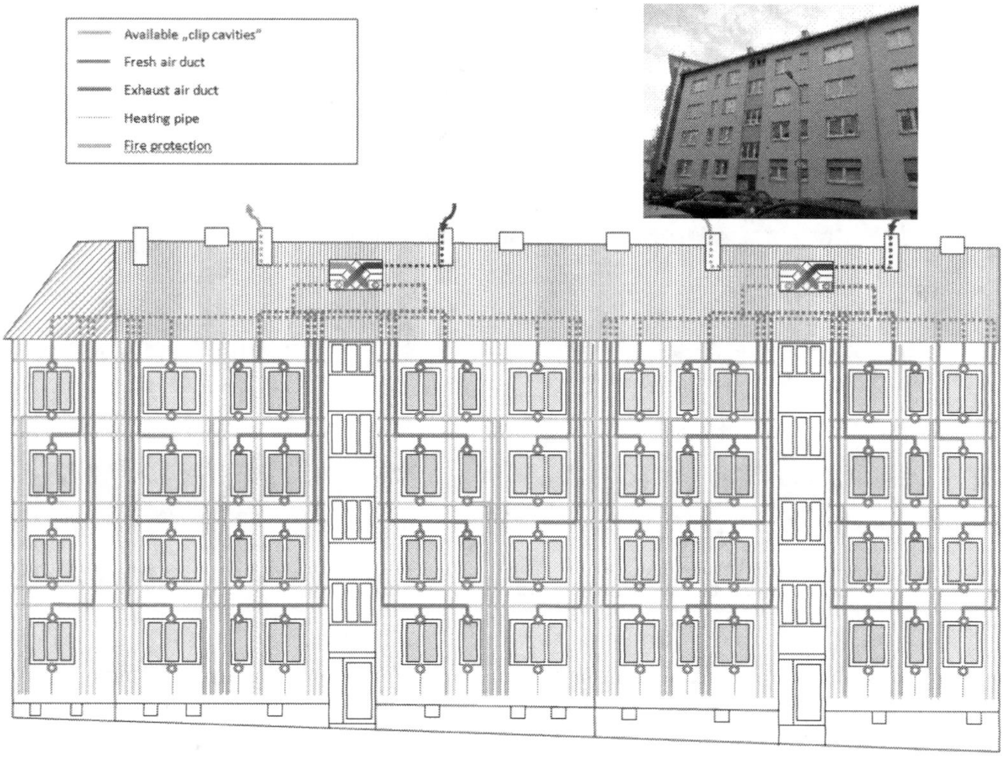

Figure 8 Networks layout on one of the façades

As the new developments are done for a multifamily home, a special focus had to be set on the fire protection. This issue had a strong influence on the choice of the materials:

- the insulation boards will be made of mineral wool,
- the air ducts will be either PVC or Aluminium ducts.

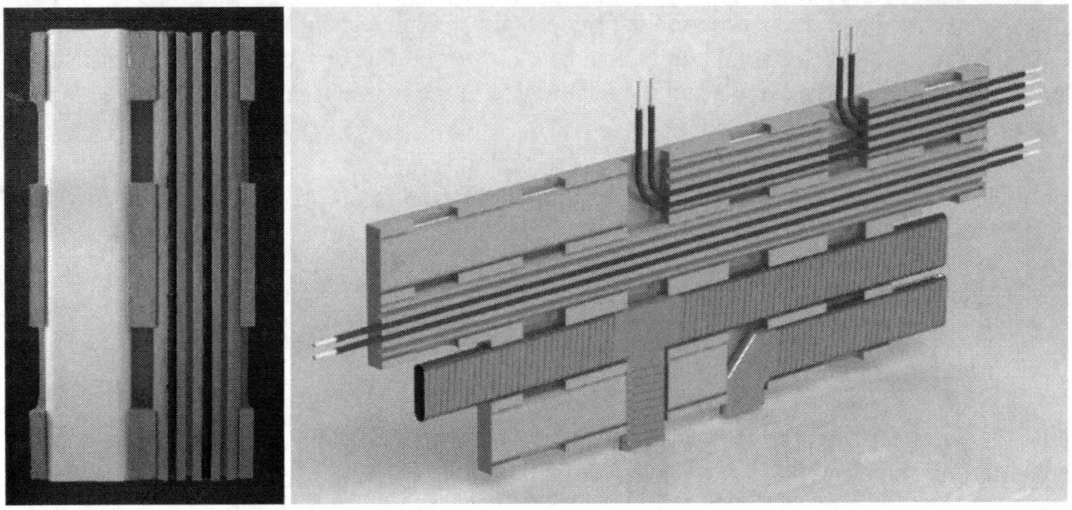

Figure 9 Picture and 3D view of the new foreseen integration concept

4.2 Optimisation

The main parameter used for the optimization was the thickness of both insulation layers (Figure 10). The total thickness of the insulation is 160 mm (fixed by the building owner). The energy point of view is playing a key role in the choice of these thicknesses but the stability of the insulation boards (which has to resist to the working conditions of construction sites) is also an important point. Different geometries of air ducts were also investigated regarding their heat losses but also other issues like their simplicity to be cleaned.

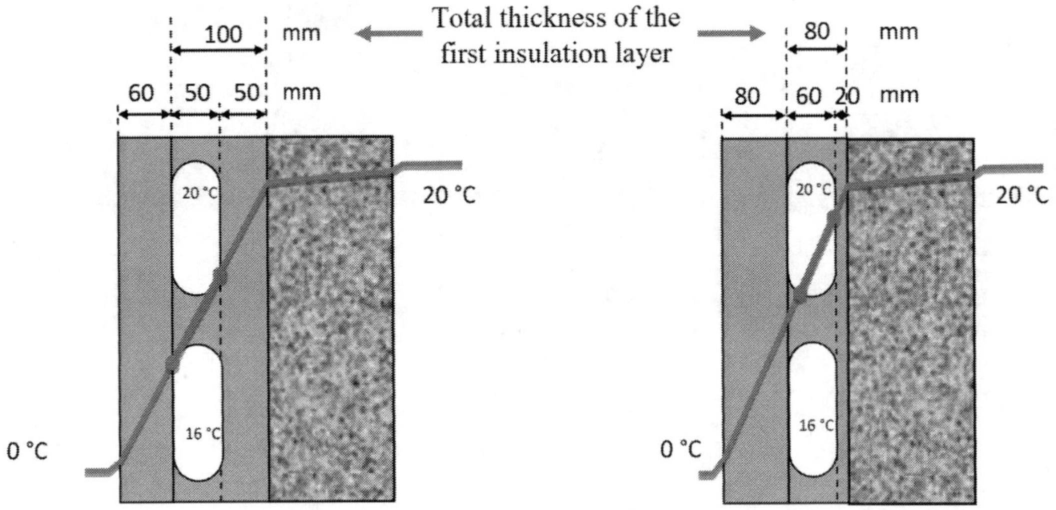

Figure 10 Two different configurations of the insulation layers thickness

Figure 11 is showing the temperature difference between inlet and outlet of the longest air duct planned for the demonstration building presented in 4.1.

The results presented here are corresponding to the following conditions:

- External dimensions of Tube 1: 205 mm x 55 mm
- External dimensions of Tube 2: 205 mm x 40 mm
- Inside temperature (= inlet of the exhaust pipe): 22°C
- Outside temperature: -20°C
- Temperature at the inlet of the supply air duct: 17°C
- Airflow set on its minimal value: 23 m³/h

The thickness indicated on Figure 11 for each case is the thickness of the first insulation layer.

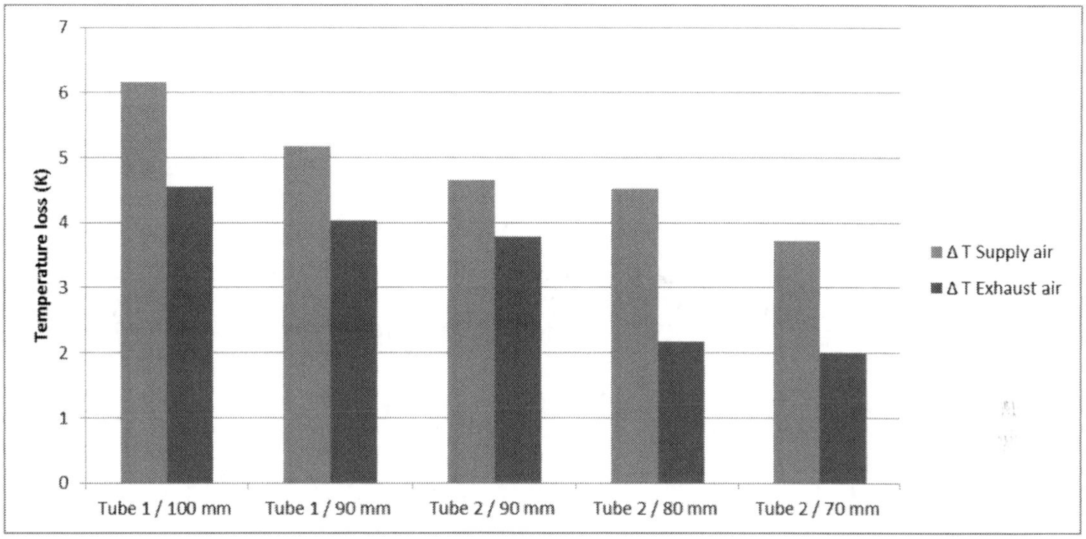

Figure 11 Temperature differences between inlet and outlet simulated for a 15m long duct

Thanks to the reduction of the thickness of the first insulation layer, the air ducts are positioned in a warmer domain and the heat losses of the air ducts can be reduced.

As presented in 2.3, the former integration system has led to a reduction of the energy efficiency of 10 %. With the last developments, we are aiming at keeping this reduction under 4 %.

5 CONCLUSION

Integrating ventilation ductworks in the façade of retrofitted building is a promising solution to make renovations easier, cheaper and more comfortable for the building tenants. It is possible to install air ducts in the insulation layer of an external wall without risking high heat losses with a standard insulation thickness of 16 cm.

The next steps of this study will be to confirm the simulation results by comparing them to the measurements which will be implemented after retrofit on the demonstration building in Frankfurt.

The integration of heating pipes is currently investigated and will be the subject of future publications.

6 ACKNOWLEDGEMENTS

Part of the work presented was funded by the German Ministry of Economics and Labour BMWi under the reference 03ET1035A and 0327400 P. Other parts were funded by the European Commission under the contract number 314229 – RETROKIT in the FP7 Programme.

7 REFERENCES

Coydon, F., A. Dinkel, et al. (2013). Rapid Retrofit Solutions Integrating HVAC into Prefabricated EWIS Systems. CLIMA 2013: 11th REHVA World Congress & 8th International Conference on IAQVEC. Prague.

DIN (1997). DIN EN 308. Wärmeaustauscher - Prüfverfahren zur Bestimmung der Leistungskriterien von Luft/Luft- und Luft/Abgas-Wärmerückgewinnungsanlagen. Berlin, Deutsches Institut für Normung e.V. **DIN EN 308**.

DIN (2009). DIN 1946-6. Raumlufttechnik - Teil 6: Lüftung von Wohnungen - Allgemeine Anforderungen, Anforderungen zur Bemessung, Ausführung und Kennzeichnung, Übergabe/Übernahme (Abnahme) und Instandhaltung
Germany, Deutsches Institut für Normung e.V. **DIN 1946-6:**.

Ziegler, M., M. Krause, et al. (2012). Thermische und strömungstechnische Simulation einer Fassadendäammung mit integrierter Luftführung für die Bestandssanierung. BauSIM 2012. IBPSA. Berlin, IBPSA.

USE OF DCV FOR HEATING AND THE INFLUENCE ON IAQ IN PASSIVE HOUSE BUILDINGS

Vegard Aslaksen[1]

1 Norconsult AS
Vestfjordgaten 4
1338 Sandvika,Norway

ABSTRACT

Measurements were performed in a test room at SINTEF building and infrastructure, Oslo. The test room is 16 m^2 and built according to NS-EN 442-2. Measurements of various air flow rates (9 l/s, 18 l/s, 34 l/s and 50 l/s) and different supply air temperatures (2, 4, 6 and 10 degrees over room temperature) were performed. Tracer gas (SF_6) measurements were performed to evaluate ventilation effectiveness and age of air in occupied zone.
These tests were performed with 2 different exhaust positions (down by the floor and up under the ceiling) to check if the position of the exhaust influence the ventilation efficiency and age of air in the occupied zone.
The test room used an Active supply air diffuser from Lindinvent, (http://www.lindinvent.com/products/air-diffusers/ttc/) designed to keep a constant air velocity with various air flow rate.
The results show variable local air change index in the breathing zone based on air flow rate and the difference between room temperature and supply air temperature. The heating demand that's needed can be delivered without influencing the IAQ in passive house buildings.
Results show the different local air change index in the breathing zone for the different exhaust positions.

There are preformed measurements with heat sources, and they appear to increase mixing, and measurements done without heat sources are a worst case scenario.

KEYWORDS

Air heating, indoor climate, Tracer gas, passive housing, cold climate

1 INTRODUCTION

The Norwegian government are implementing new and stricter regulations for new buildings. The new regulations in TEK15 mean that new buildings after 2015 will have to meet passive house requirements. New requirements for isolation and airtightness will reduce the heating demand. And due to energy regulations, new buildings often use ventilations systems with variable air volume (VAV) as opposed to constant air volume (CAV).
This project is part of the research project "For Klima". "For Klima" is a research project which is researching the possibility to simplify the solutions for heating, while maintaining a good indoor climate.

This project has been focused on the possibility of using the ventilation system to supply heating. If the ventilation system can be used, this will reduce both investment- and operating costs.

The heating demand in passive house buildings has been calculated, and the data has been used to estimate how frequent heating is used. There is also been built a test room to measure the local air change index with different heating demands covered by using heated supply air and with varying air flows.

2 METHOD

2.1 Heating demand and occurrence

The testing plan for the tracer gas measurements are based on the calculated heating demand. The calculated heat deamand is based on heat loss and internal loads from "Prosjektrapport 42" (Mads Mysen 2009) from SINTEF building and infrastructure.

Table 1: Test plan for tracer gas measurements. It will be done with two different placements of the exhaust.

Test plan					
Air Flow: 50 l/s					
ΔT	2	4	6	10	Isothermal
Air Flow: 34 l/s					
ΔT	2	4	6	10	Isothermal
Air Flow: 18 l/s					
ΔT	2	4	6	10	Isothermal
Air Flow: 9 l/s					
ΔT	2	4	6	10	Isothermal

The test plan consists of four different airflows, where 18 l/s is air flow equivalent to requirements set by TEK10(DIBK). The test plan is carried out twice, with exhaust located both in the ceiling and by the floor. This is done to examine how exhaust location affect local air change index.

Heating demand is used in combination with metrological data to estimate the occurrence of outdoor temperatures that leads to a heating demand in passive house buildings.

2.2 Tracer gas measurements

The test room used in tracer gas measurements is 16 m^2 room built according to NS-EN 442-2(Norge 1996) and is used to simulate an office. To simulate a heating demand in the room, it's possible to regulate the temperatures of the walls. This is done by pumping water trough pipes in the walls. Only one wall was cooled down during testing, this was done to simulate an office with one outer wall. The test room is built with a separate air supply connected to a active supply air diffuser from Lindinvent (http://www.lindinvent.com/products/air-diffusers/ttc/) for easy regulation of the air flow. Heating of the supply air is done with a heat coil. Temperature measurements during testing includes: Surface temperatures in the room, supply air temperature and room temperatures. These measurements are used to make sure there are stable conditions during the tracer gas measurements.

The tracer gas measurements are analysed using both a step up and step down analysis to determine local air change efficiency in the breathing zone. The analysis is done by using a analysis-tool developed by Peter Schild, SINTEF Building and infrastructure / HiOA.

3 RESULTS

3.1 Heating demand

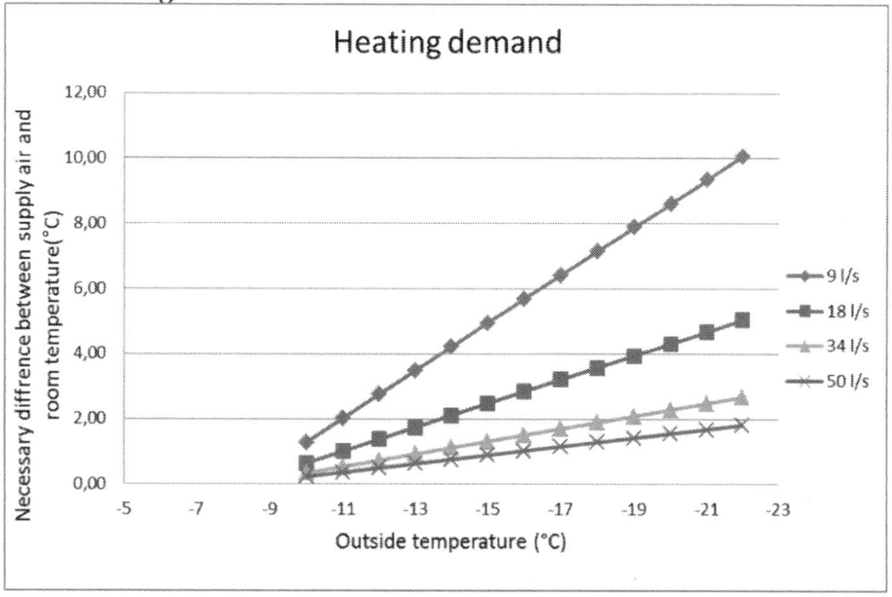

Figur 1: The temperature difference between supply air and room temperature necessary to cover the heating demand at different outside temperatures with different air flows.

Figure 1 shows the increasing difference between the supply air temperature and the room temperature necessary to cover the heating demand at different outside temperatures. The result shows that passive house buildings don't have a heating demand until the outside temperatures reach as low as -8 °C, and that the air flow has a significant impact on the necessary difference between supply and room temperatures.

The necessary difference between air supply and room temperature at an outside temperature of -22 °C varies between 2 °C and 10 °C based on the size of the air flow.

Air flow according to TEK10 (18 l/s) gives a necessary difference between supply air and room temperature at 5 °C at an outside temperature at -22 °C.

Table 2: The occurrence of days with temperatures below a set point. Data from the last 10 years collected at Blindern, Oslo. From eklima, Metrologiske institutt.

Measuring station Oslo-Blindern							
Results from 2004-2013 (10 years)							
	Jan	Feb	Mar	Nov	Des	Tot	Snitt pr. år
Days with -8 °C	85	78	45	8	74	290	29
Days with -14 °C	18	16	6	0	17	57	5,7
Days with -20 °C	2	0	0	0	0	2	0,2

Table 2 shows the number of days with occurrence of temperatures below -8, -14 and -20 ° C. -8 ° C is the highest outside temperature resulting in a heating demand in passive house buildings. While -14 ° C is the outdoor temperature that will require a difference between the supply and room temperature of 2 ° C when using air flow based on the requirements in

TEK10. These figures will vary depending on your location and altitude, these figures represent the situation in Oslo, Norway.
The figures indicate the occurrence of the given temperature or below during a 24 hour period, so the occurrence during working hours is lower than the number given, since the coldest periods of the day often occurs outside working hours.
The incidence of days that requires heating is low and very rarely is there a need that exceeds 2 ° C difference between the air supply and room temperature, when using air flow based on requirements in TEK10.

3.2 Tracer gas measurements

Tracer gas measurements to find the local air change index in the breathing zone were conducted at isothermal conditions. Measurements were then performed according to the test plan, with the corresponding variations in air flow, temperature difference between supply air and room temperature and exhaust position. The results are then presented by comparing local air change index in the breathing zone with the corresponding result from the measurements done under isothermal conditions. The results therefore shows how the temperature difference between supply air and room temperature affects local air change index in the breathing zone compared to isothermal conditions.

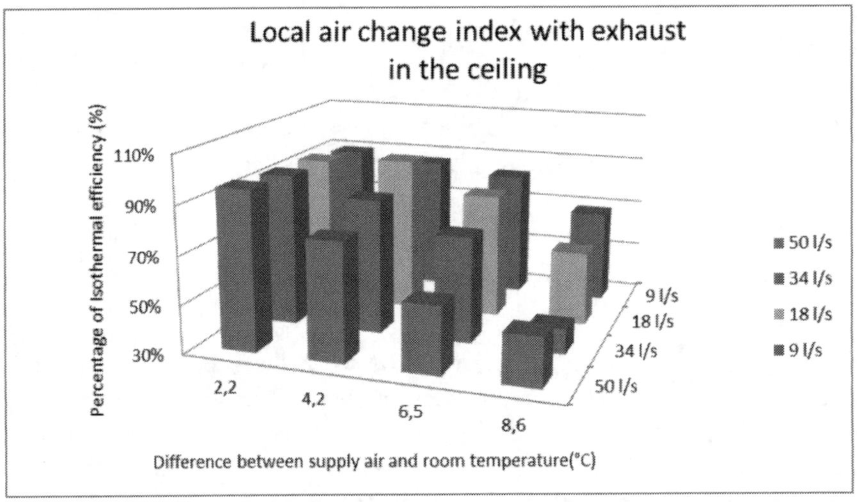

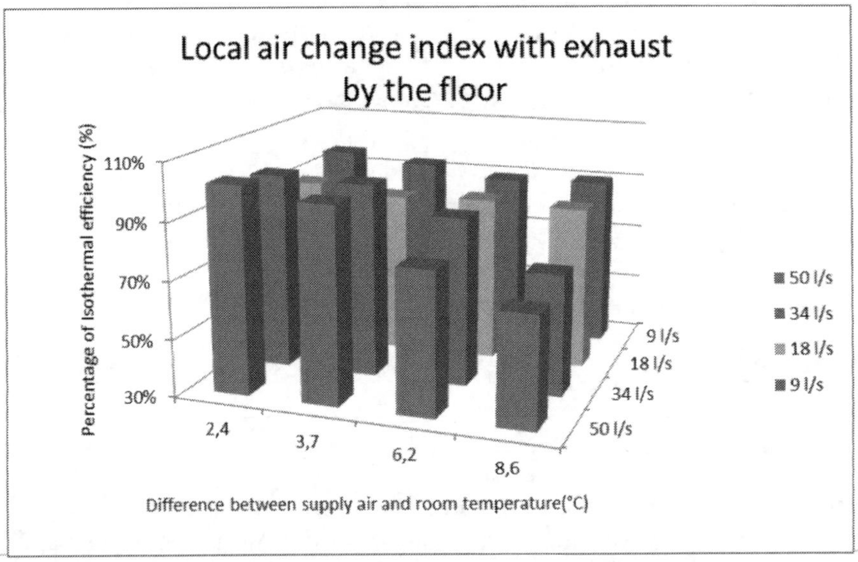

Figure 2: Development of local air change index for different air flows with increasing temperature difference between supply air and room temperature. Graphs for both exhaust positions.

Figure 2 shows the development of local air change index in the breathing zone with increasing temperature difference between the supply air and the room temperature with exhaust placed both in the ceiling and down by the floor and with different air flows.

Local air change index in the breathing zone is affected by both the exhaust location and the difference between the supply air and room temperature. The results shows that placing the exhaust down by the floor, provides a consistently higher local air exchange index in the breathing zone than the exhaust positioned at the ceiling. Local air change index will also decrease with increasing difference between supply air and room temperature. While placing the exhaust down by the floor, gives the same trend, the local air change index decreases slower and at lower air flow rates (9 l / s and 18 l / s) the curve is flat, and you keep a constant local air change index in the breathing zone.

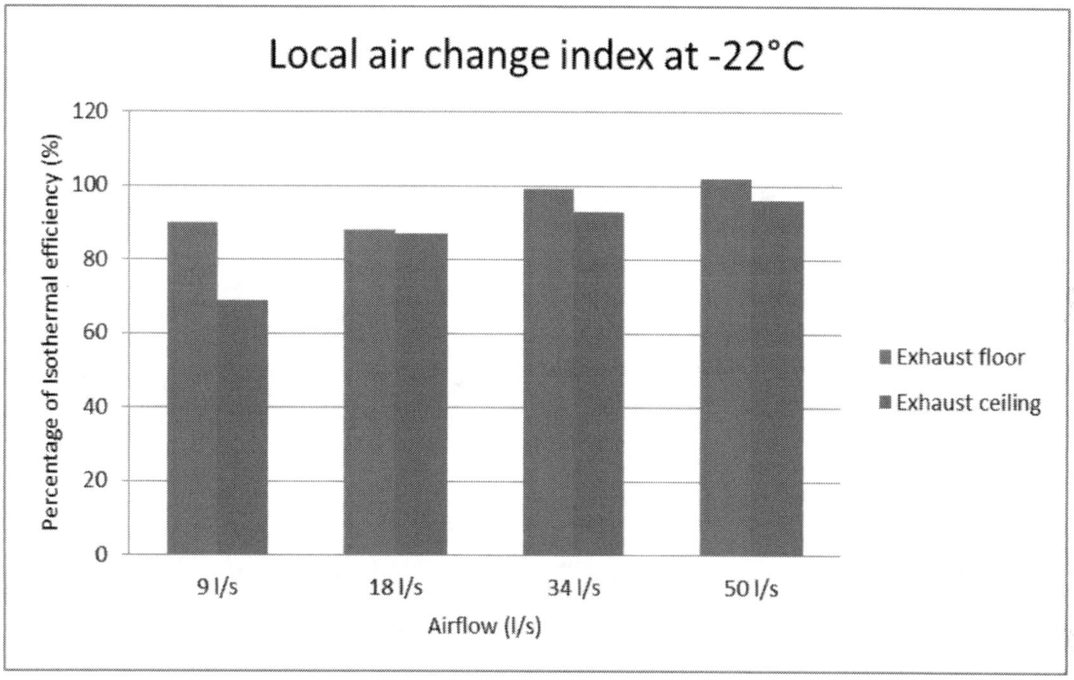

Figur 3Local air change index with a temperature difference between supply air and room temperature that covers the heating demand at an outside temperature at -22 °C. Compares different air flows and exhaust position

Figure 3 shows the local air change index in the breathing zone for different airflows with a difference in supply air and room temperature that corresponds to a heating demand with an outside temperature at -22 °C. It also shows how the location of the exhaust affects the results under these circumstances.

The figure shows that the exhaust location is less important at the above temperatures, when you use an airflow corresponding to the requirements in Norway or higher (18 l/s or higher). Exhaust placement is of greater importance when the difference between supply air and room temperature are higher, which corresponds to greater heating demands than passive house buildings, or a lower airflow.

3.3 Heat sources

Trace gas measurements conducted and presented in the results are done without heat sources in the room. A room without heat sources is thought to be a worst case scenario, and introducing heat sources in the room will increase mixing and increase the local air change index in the breathing zone. To test this assumption, two tests were conducted with heat sources in the room and compared to the same test done without heat sources. The heat sources that were introduced were a dummy to simulate a human working by the desk and a lamp placed at the desk.

Table 3: The effect of heats sources in the room.

Effect of heat sources				
Air flow: 34 l/s				
	Exhaust by the floor		Exhaust in the ceiling	
	Temperature difference (ΔT)	Local air change index(%)	Temperature difference (ΔT)	Local air change index(%)
Without heat sources	9,9	72	4,1	86
With heat sources	10,2	90	5,1	97

Table 3 shows the results of the tests. In both cases, the temperature difference is slightly higher with the introduction of heat sources than without. Introduction of heat sources increased local air change index by 18% with exhaust located by the floor, and 11% with the exhaust located in the ceiling. This is a good indication that the assumption was correct, and heat sources increases the mixing and therefore the local air change index in the breathing zone.

4 CONCLUSION

The results for the calculation of heating demands and use of meteorological data shows that buildings on passive house level, rarely needs a temperature difference between supply air and room temperature that exceeds 2 °C when using air flows according to TEK10. At this temperature difference the exhaust position has a negligible effect and the results for local air change index in the breathing zone shows that it's possible to maintain a good IAQ.

The results show that even on the coldest days of winter (-22 °C) it's possible to maintain a high local air change index and the position of exhaust is negligible when using air flow according to TEK10. This means that you can maintain a good IAQ even on the coldest days of winter.

This means that the use of the ventilation system for heating buildings at passive house level can be performed while maintaining good air quality in the breathing zone on even the coldest days of the year.

5 REFERENCES

DIBK. "TEK10." from http://www.dibk.no/no/BYGGEREGLER/TEK10-Forskrift-om-tekniske-krav-til-byggverk-med-veiledning/.

Mads Mysen, M. K., Matthias Haase, Tor Helge Dokka (2009). "Kriterier for passivhus- og lavenergibygg- Yrkesbygg, Prosjektrapport 42."
Norge, S. (1996). NS-EN 442-2 Radiatorer og konvektorer. Standard.no, Standard Norge.

THE EFFECT OF ENTHALPY RECOVERY VENTILATION ON THE RESIDENTIAL INDOOR CLIMATE

Bart Cremers

Zehnder Group Nederland
Lingenstraat 2
8028 PM Zwolle, The Netherlands
bart.cremers@zehndergroup.com

ABSTRACT

The indoor climate in residential buildings is affected by the people that live in the house and their activities. One of the goals of a ventilation system is to prevent excess humidity in the house by removing part of the moisture. The moisture balance can however be distorted in winter with a low humidity in the house as a result.

An enthalpy exchanger can be used in a recovery ventilation system to reduce the risk of very low indoor humidity levels in winter. While heat recovery ventilation (HRV) is recovering energy in terms of temperature, enthalpy recovery ventilation (ERV) is recovering energy in terms of both temperature and moisture.

In a house in Rotterdam, The Netherlands, the temperatures and humidities of the air streams in a recovery ventilation system were monitored. Eight day periods with HRV and ERV are compared with each other during mild outdoor conditions.

In the monitored house, the measured recovered amount of absolute humidity with ERV is about 1-2 g/kg with an average humidity recovery efficiency of 65%. In the monitored house, ERV brings the indoor relative humidity up to 10 percentage point higher than HRV.

The measured average thermal efficiency and humidity efficiency correspond well with the laboratory values from the specifications.

KEYWORDS

Enthalpy recovery, residential ventilation, indoor air quality, moisture balance

1 INTRODUCTION

The indoor climate in residential buildings is affected by the people that live in the house. The presence of people and their activities (like cooking and showering) and the presence of plants make the indoor climate rise in humidity. One of the goals of a ventilation system is to prevent excess humidity in the house by removing part of the moisture. The indoor air is replaced by winter outdoor air, which usually has a lower moisture content than the indoor air.

The moisture balance can however be distorted in winter (e.g. with high ventilation rate and very low outdoor air humidity) with a low humidity in the house as a result. An indoor relative humidity lower than 30% can affect our well-being as it causes dry throat or nose, and it can affect construction of the house or the (wooden) carpets or furniture.

An enthalpy exchanger can be used in a recovery ventilation system to reduce the risk of very low indoor humidity levels in winter. While heat recovery ventilation (HRV) is recovering energy in terms of temperature, enthalpy recovery ventilation (ERV) is recovering energy in terms of both temperature and moisture.

The effect of ERV can only be seen in houses that are air tight, and when there are moisture sources (cooking, showering, plants) in the house. Without any moisture sources there is nothing to recover, and there is no difference between ERV and HRV.

2 THE MONITORING SET-UP

Figure 1: The monitored house in Rotterdam

A house (fig. 1) in Rotterdam, The Netherlands, is ventilated with a balanced ventilation system. The air distribution system is a flexible circular duct system with separate channels leading to the rooms in the house. The ventilation unit is controlled by a main CO_2 control in the living room. The air flow rate is set to a minimum of 210 m^3/h and is increased automatically when the CO_2 level in the living room rises or as a result of a humidity sensor in the bath room.

In the winter of 2012 the ventilation unit was equipped alternately with a heat exchanger and an enthalpy exchanger. Unfortunately, the winter was a very short one with only two week period of cold sub-zero temperatures. In the cold period, the heat exchanger was used, but after placement of the enthalpy exchanger, the very low outside temperatures have not returned!

In order to compare the indoor climate for similar outdoor conditions, two periods of eight days were compared. These days had similar outdoor conditions with temperature between 5 and 10 °C and absolute humidity between 4 and 6 g/kg. During these periods, the air temperatures and humidities were monitored for each of the four air streams: outdoor air, supply air, extract air and exhaust air.

3 ENTHALPY RECOVERY

A heat exchanger is transferring heat between two air streams and therefore in winter the supply air stream is increased in temperature. An enthalpy exchanger is transferring heat between two air streams and also transferring moisture between two air streams. Therefore the supply air stream in winter is not only increased in temperature, but also increased in humidity.

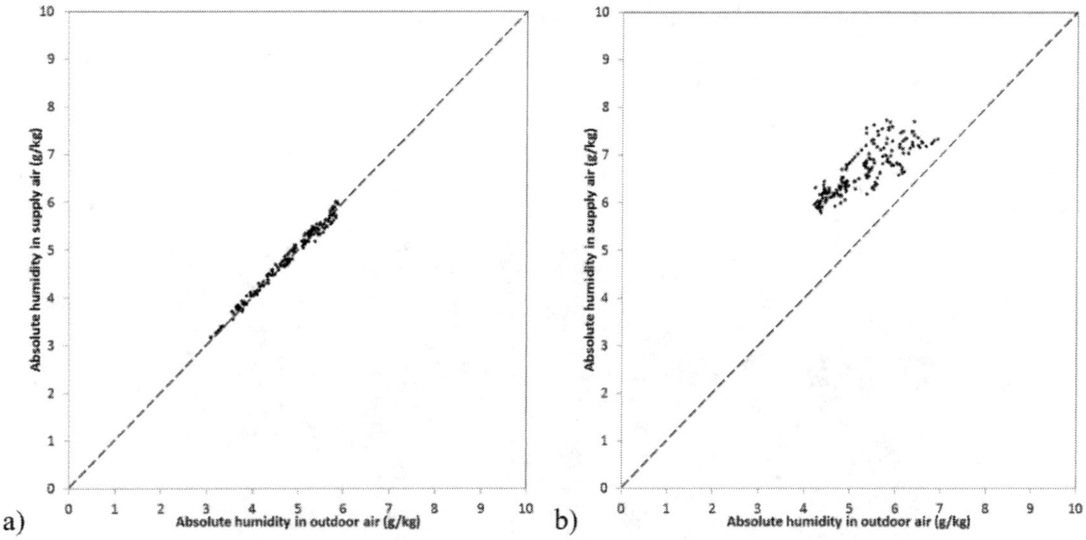

Figure 2: The effect of heat recovery (a) and enthalpy recovery (b) on the humidity of the supply air.

In fig. 2 the effect of heat exchanger (HRV) and enthalpy exchanger (ERV) on the supply air humidity is shown by hourly averaged values in the 8-day period. The absolute humidity of the outdoor air is shown on the horizontal axis and the absolute humidity of the supply air is shown on the vertical axis. HRV (fig. 2a) has no effect on the humidity of the supply air, but for ERV (fig. 2b) it is apparent that the supply air is increased in humidity. This humidity is recovered from the extract air, and thus from the humidity in the house.

The average increase of absolute humidity in the supply air is about 1 – 2 g/kg. For a typical ventilation air flow of 150 m^3/h this means a recovery of 4 – 8 liter of water per day.

4 THE INDOOR CLIMATE

In fig. 3 the hourly values of temperature and humidity levels are shown in Mollier diagrams for the 8-day periods with heat exchanger (fig. 3a) and with enthalpy exchanger (fig. 3b).

Focusing first on the supply air stream, the air coming from outside is expressed with green dots and the supply air just after the exchanger is expressed with red dots. Fig. 3 shows that for HRV the temperature is increased by heat recovery to a level of 20 °C, almost the indoor temperature. For ERV, the same is happening for the temperature, but also the moisture level is increased as indicated by the black arrow that is slanted to the right.

In fig. 3 the extract air from the house is expressed with yellow dots, and it is shown that for the same outdoor conditions, the indoor humidity is higher when the enthalpy exchanger is used. For HRV, the indoor relative humidity varies in this (relatively mild) period between 35% and 45%, while for ERV the indoor relative humidity varies between 45% and 50%.

From this we can conclude that the use of ERV increases the indoor relative humidity in winter up to 10 percentage point, with HRV as reference.

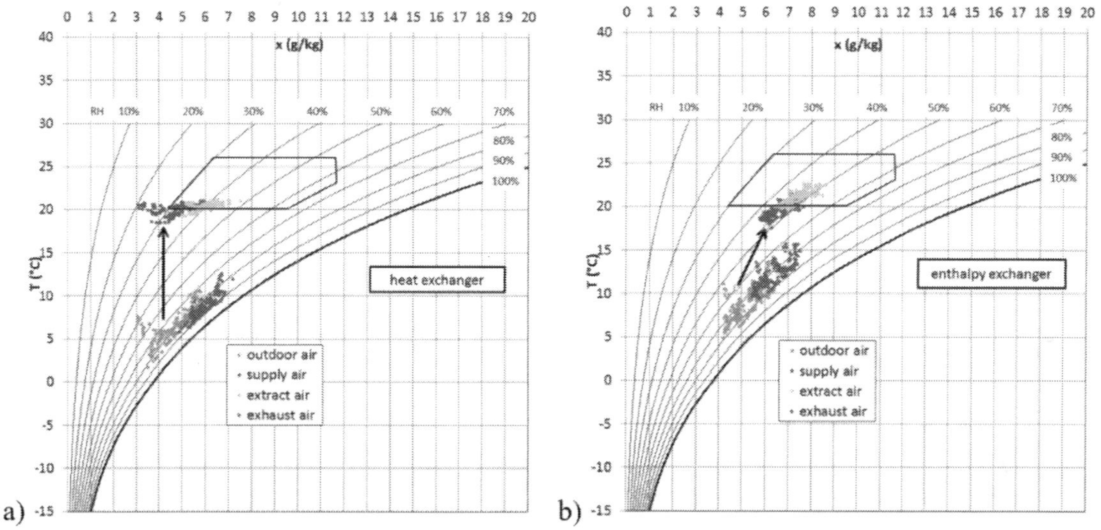

Figure 3: The effect of a heat exchanger (a) and an enthalpy exchanger (b) on the supply air and the indoor air.

5 RECOVERY EFFICIENCY

For the whole monitored winter the efficiencies are summarized in table 1. The thermal efficiency η_{th} can be calculated from the temperatures T and the humidity efficiency (or latent efficiency) η_{hum} can be calculated from the absolute humidities x according to the following formulas.

$$\eta_{th} = \frac{(T_{supply} - T_{outdoor})}{(T_{indoor} - T_{outdoor})} \tag{4}$$

$$\eta_{hum} = \frac{(x_{supply} - x_{outdoor})}{(x_{indoor} - x_{outdoor})} \tag{5}$$

For the heat exchanger, the average thermal efficiency is 89% for mild outdoor temperatures. With cold outdoor air, condensation in the return air of the exchanger may be formed and the flow of return air maybe partly blocked, with the effect that the thermal efficiency drops to 70% in condensing states. As the heat exchanger does not transfer moisture, the humidity efficiency is zero.

For the enthalpy exchanger, the measured thermal efficiency has an average value of 88%, while the humidity efficiency is averaged to 65%. These values correspond well with the values in the specifications.

Table 11: Thermal efficiency and humidity efficiency of the two exchanger types.

Exchanger type	Thermal efficiency	Humidity efficiency
Heat exchanger	89% (non-condensing) 70% (condensing)	0%
Enthalpy exchanger	88%	65%

6 CONCLUSIONS

Ventilation removes stale air, and therefore also humidity from the indoor climate of a house. In winter, the moisture balance may be distorted causing relatively dry indoor climate, because dry outdoor air is entering the indoor climate. An enthalpy exchanger recovers a part of the humidity from the extract air to the supply air, helping to maintain the moisture balance.

In the monitored house, the recovered amount of absolute humidity is about 1-2 g/kg. Therefore, the use of enthalpy recovery brings the indoor relative humidity up to 10 percentage point higher than the use of heat recovery (for instance from 35% to 45%). In the monitored house, the thermal efficiency and the humidity efficiency correspond well with the values from the specifications.

AIR RENEWAL EFFECTIVENESS OF DECENTRALIZED VENTILATION DEVICES WITH HEAT RECOVERY

Fabien Coydon[1], Jens Pfafferott[2]

1 Fraunhofer Institute for Solar Energy Systems ISE
Heidenhofstrasse 2
79110 Freiburg, Germany
fabien.coydon@ise.fraunhofer.de

2 University of Applied Sciences Offenburg
Badstraße 24a
77652 Offenburg, Germany
jens.pfafferott@hs-offenburg.de

ABSTRACT

Central ventilation systems with heat recovery have shown their limits especially within the context of building energy retrofit. The difficulties to install these systems in existing buildings, to find available space for devices, air ducts, silencers and fire dampers and to independently control the air flow in each room according to the real ventilation needs have led to an increasing market for decentralized ventilation devices.

A frequent criticism of decentralized devices is the poor ventilation effectiveness due to a high short-circuiting risk. Therefore, the University of Applied Sciences of Offenburg and the Frauhofer Institute for Solar Energy Systems are evaluating together the inside air quality reached with decentralized ventilation systems.

The tests described in this paper were performed in a test lab representing a single room were a decentralized ventilation system with heat recovery, heating and cooling functions was implemented. In a first part of the evaluation process, tests are realized to characterize the air distribution provided by the decentralized device by smoke visualization. The results are showing interesting differences between isothermal, heating and cooling cases.

The characterization of air distribution in the volume of the room is not enough to evaluate the capacity of a ventilation system to provide a good inside air quality. Therefore, a second series of tests was performed with CO_2 as a tracer gas, enabling a more precise quantification of the ventilation effectiveness. The results show the CO_2 concentrations over time in zones with different air renewal levels for each of the 3 modes.

KEYWORDS

Decentralized ventilation, Ventilation effectiveness, Heat recovery, Inside air quality

1 INTRODUCTION

The work presented in this paper is included in a research program aiming at evaluating ventilation concepts for the energy retrofit of residential buildings. Central and decentralized ventilation are compared on the basis of a full list of criteria including their energy performances, comfort, acceptance and life cycle costs.

As a part of this work, an analysis of the inside air quality provided by decentralized ventilation systems will be detailed here.

As no ductwork is necessary, decentralized ventilation systems are a very interesting solution to provide fresh air in retrofitted buildings where place is rarely available to install voluminous equipment. A frequent criticism against decentralized devices is their low ventilation effectiveness since the air inlets and outlets are often very close to each other risking a short-circuit inside the ventilated room as well as outside.

Decentralized ventilation systems have already been evaluated (Mahler, Himmler et al. 2008; Haase and Gruner 2013) but only a few studies have a specific focus on the ventilation effectiveness of these systems (Ender, Gritzki et al. 2004; Ajaji and André 2013). At the Wuppertal University, the microclimatic influence of the external side of the façade was investigated (Voß and Voss 2012) but not the short-circuiting risk.

At last, the combination of heating by warm air and ventilation was investigated by the Technical University of Denmark and the Slovak University of Technology but in a context of central ventilation (Krajčík, Simone et al. 2012).

The main goal of the study presented here is to evaluate a decentralized ventilation system including heating and cooling functions in different running modes.

In order to describe the capacity of decentralized ventilation to provide a good inside air quality, one parameter, the ventilation effectiveness, is often used with different definitions. A first possibility, related to smoke visualisation of the ventilation air movements, is based on time measurements describing how long the fresh air needs to reach each point of a room. A second definition, related to tracer gas evaluations, is based on concentration measurements in different points of the room and in the fresh and exhaust air (Mats 1981; Recknagel, Sprenger et al. 2007/2008).

2 DESCRIPTION OF THE TESTS AND EVALUATION METHODS

2.1 Ventilation device

The decentralized ventilation system tested here and shown on Figure 1 is foreseen to provide fresh air, to heat and to cool office rooms until 6m deepness. It includes 2 fans, 1 heat exchanger, 1 heating/cooling coil, 2 filters and can run with 3 different airflows.

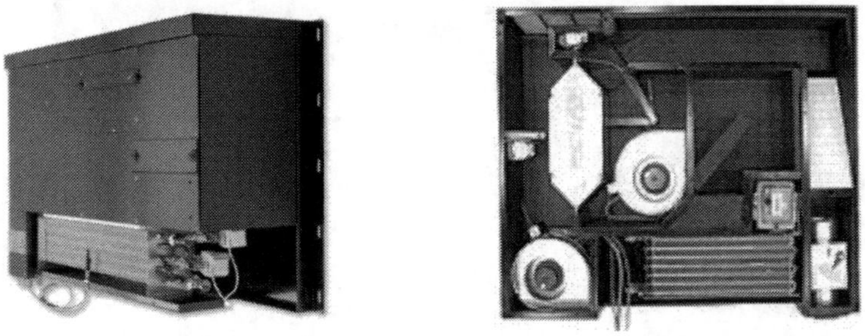

Figure 1 Tested decentralized ventilation device

2.2 Test room

The decentralized ventilation device was evaluated in the test facilities of the University of Applied Sciences Offenburg. The room dimensions are 5,2 x 3,8 x 3 m^3. Two walls are producing heating or cooling loads.

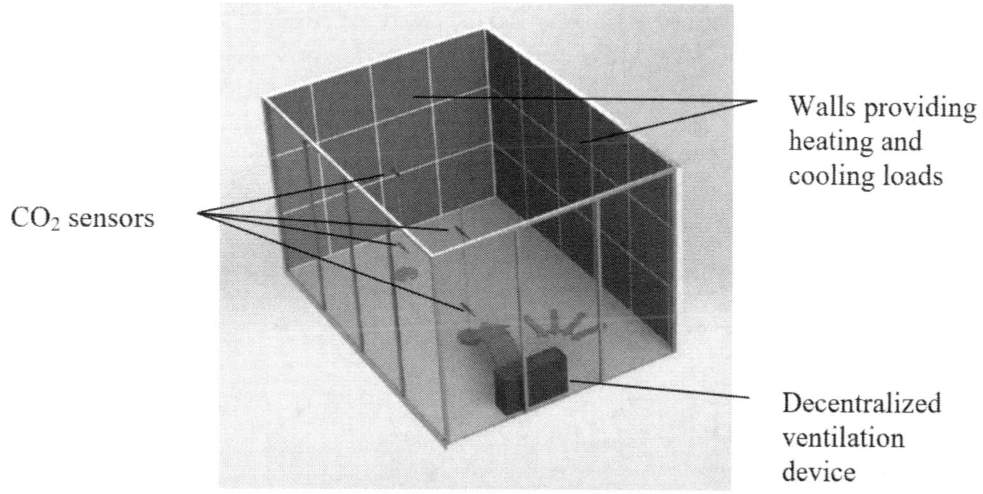

Figure 2 3D view of the test room and positions of the CO_2 sensors

2.3 Smoke visualisation

The first series of tests was done with smoke visualisation. We focussed on two dimensions in the area directly placed in front of the device. The first dimension was the deepness of the room so that we could verify that the device was able to treat a 5m deep room. The second dimension was the height which allows describing an eventual stratification of the fresh air provided by the ventilation device.

Figure 3 View of a smoke visualization test

In many test cases, the smoke was not visible any more before it filled the entirety of the room volume and in particular before it reached all of the four points were CO_2 sensors were placed. This phenomenon made it impossible to determine a ventilation effectiveness based on time measurements and could lead to the conclusion that the ventilation device is not able to treat the air in the whole room volume.

2.4 CO_2 measurements

The second series of tests was done with a tracer gas. As the main pollutants removed by ventilation systems are humidity and CO_2, we choosed CO_2 as a tracer gas. The choice of another tracer gas could have led to errors in the evaluation due to different behaviors of gases regarding the diffusion and mixing phenomenons in the air. To compare the results with the smoke visualisation, we used 4 CO_2 sensors placed in front of the ventilation device at horizontal distances of 1,5 m and 4 m and at 2 different heights ("Down": 0,5 m and "Up": 1,8m).

The room was filled with CO_2 and the analysis began after homogenisation of the 4 measured CO_2 concentrations as shown on Figure 4.

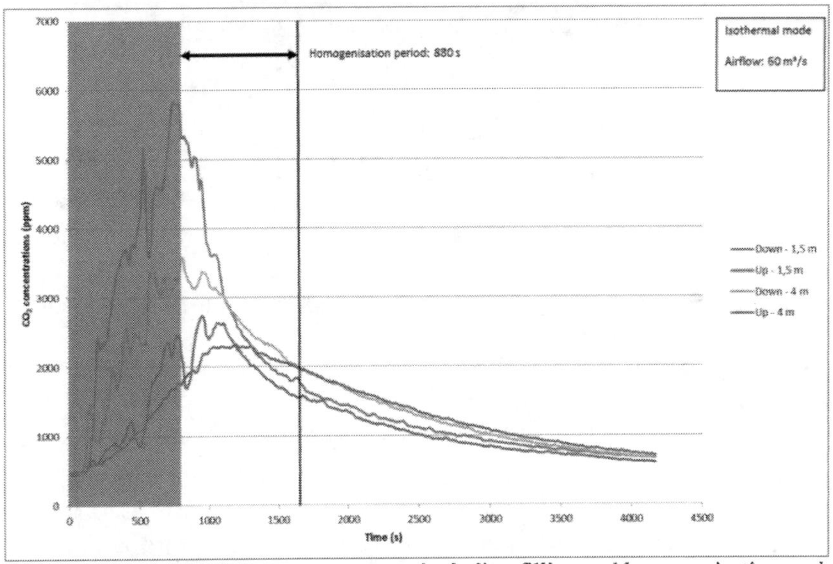

Figure 4 Example of CO_2 measurements including filling and homogenisation periods Isothermal mode (60 m³/h)

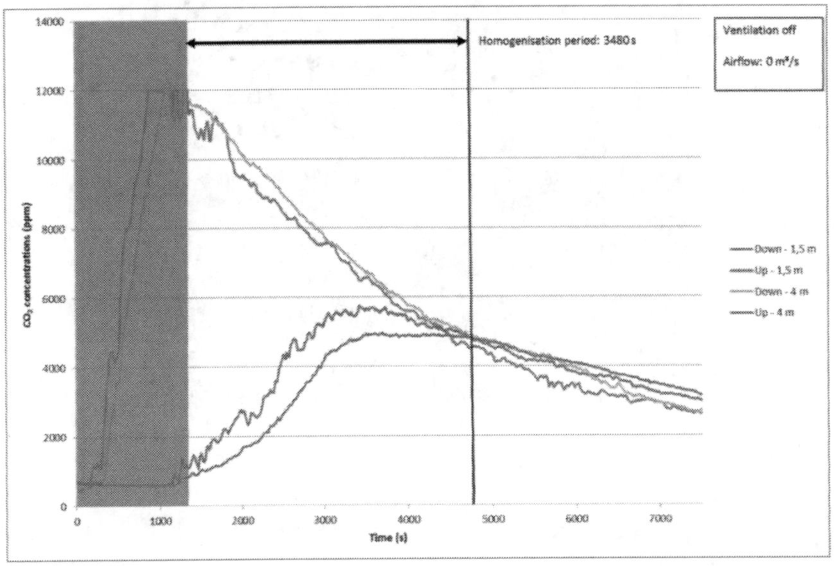

Figure 5 CO_2 homogenisation period without ventilation

A first conclusion can be drawn from the measurement of the homogenisation period which is the period where the CO_2 concentrations dynamyc is still due to the initial concentration differences. In order to define this homogenisation period, we used the theoretical evolution of each CO_2 concentration, having an exponential form ($c_{CO2} = c_0 + A.e^{-b.t}$). After the homogenisetion period, the measured concentration is actually close to the theoretical one.

We considered that the end of the homogenisation period occurred as the difference between theoretical and measured CO_2 concentrations were lower than 5% of the initial difference.

We compared 2 different cases:

- ventilation off – corresponding to Figure 5
- ventilation on – minimal airflow (around 60 m³/h) – corresponding to Figure 4

The first one lasted around 3480 s, whereas the second one lasted only 880 s. This means that even set on its minimal airflow, the device is able to homogenise pollutant concentrations in the room which reduces short-circuiting risks.

In order to compared the diferent cases, we used the definition of the ventilation effectiveness proposed by Mats and Recknagel (Mats 1981; Recknagel, Sprenger et al. 2007/2008):

$$\eta = \frac{c_{exhaust} - c_{fresh}}{c_{average} - c_{fresh}} \quad (1)$$

Where:

- η: ventilation effectiveness
- $c_{exhaust}$: CO_2 concentration in the exhaust air
- c_{fresh}: CO_2 concentration in the fresh air
- $c_{average}$: average CO_2 concentration of the 4 measured points

In our case, this effectiveness does not describe the capacity of the device to ventilate the whole volume of the room, but only the vertical plane situated in front of the ventilation device.

3 ANALYSIS OF ISOTHERMAL, HEATING AND COOLING MODES

An important part of the work is to compare isothermal, heating and cooling modes with both evaluation methods. The results presented here were obtained with a middle airflow (around 80 m³/h) with following temperatures:

- isothermal mode: $T_{supply\ air} = 22$ °C $T_{room\ inside} = 22$ °C
- heating mode: $T_{supply\ air} = 40$ °C $T_{room\ inside} = 23$ °C
- cooling mode: $T_{supply\ air} = 12$ °C $T_{room\ inside} = 24$ °C

3.1 Smoke visualisation

Figure 6 is presenting the results of the smoke visualisation tests. As mentioned in 2.3, the measurement of the time needed by the fresh air to reach each of the 4 points corresponding to the CO_2 sensors was not possible especially for the cooling mode. These pictures are showing important differences between each case. The repartition in the room of the fresh air provided by the ventilation device is good in the isothermal mode but in the cooling mode, the fresh air is only filling the lower part of the room and in the heating mode, only the upper part.

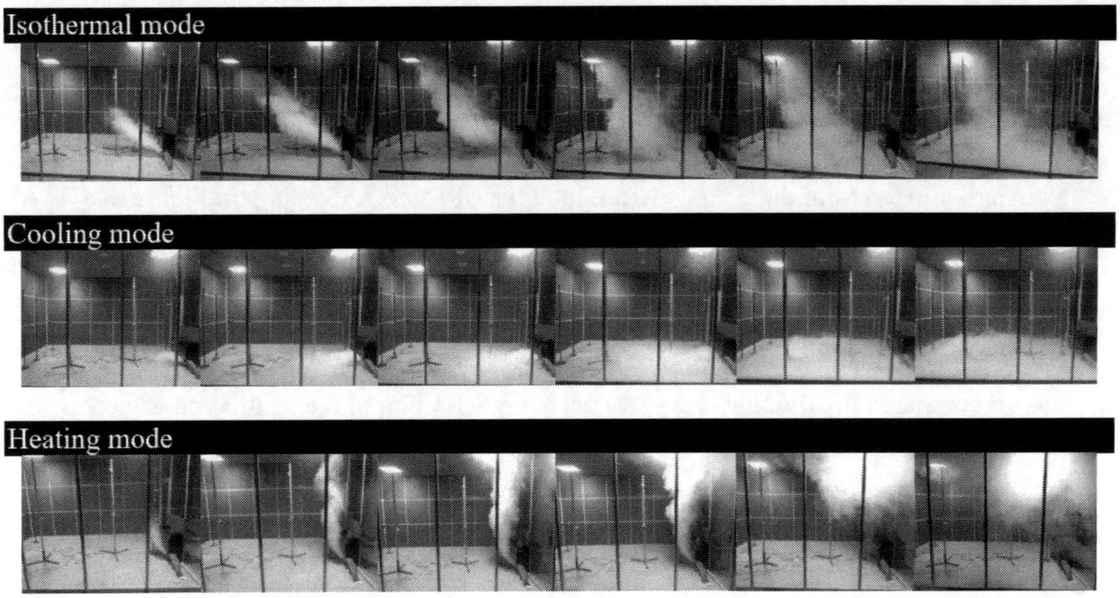

Figure 6 Smoke visualisations after 5, 10, 15, 20, 30 and 50 s for the 3 different modes

A purely qualitative analysis of these pictures leads to the conclusion that only one of these cases presents an acceptable air distribution.

3.2 CO_2 measurements

In order to quantitatively verify these results, 0, 0 and 0 are presenting the CO_2 measurements obtained for the same cases.

A first observation can be done regarding the differences between the 4 sensors:

- Isothermal mode:
  The lowest CO_2 concentrations after homogenisation were obtained at the 2 points situated near the device. No stratification of the CO_2 concentration is appearing for this case (no difference between the upper and the lower sensors for both locations).
  The differences between the 4 measured concentrations and their average value remains under 15 %.

- Heating mode:
  The lowest CO_2 concentration after homogenisation was obtained at the upper point situated near the device. A vertical stratification of the CO_2 concentration only appears at the nearest location.
  The differences between the 4 measured concentrations and their average value also remains under 15 %.

- Cooling mode:
  The lowest CO_2 concentration after homogenisation was obtained at the lower point situated at 4 m from the device. A vertical stratification of the CO_2 concentration only appears at this location but not at 1,5 m from the device.

The differences between the 4 measured concentrations and their average value are reaching 40 %.

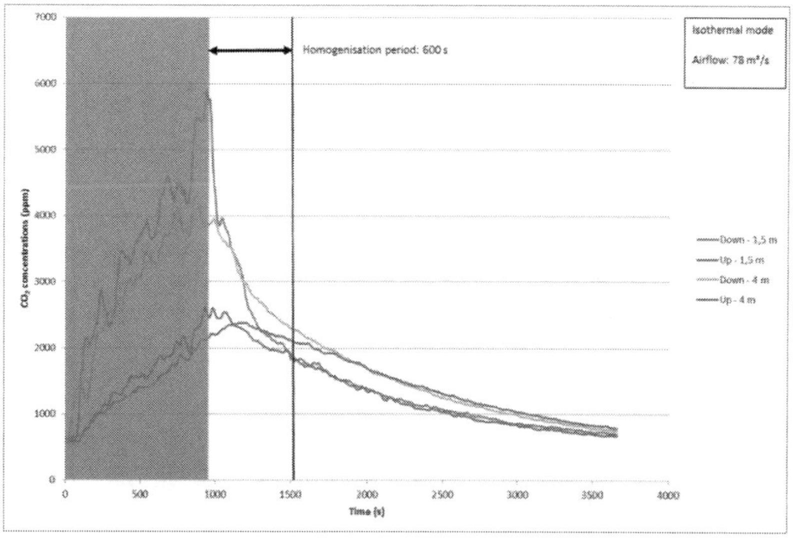

Figure 7 CO_2 measurements in isothermal mode (80 m³/h)

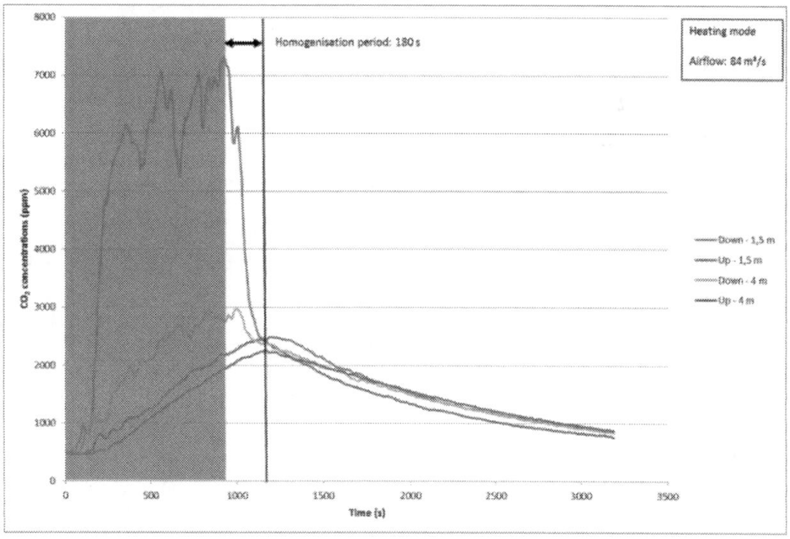

Figure 8 CO_2 measurements in heating mode (80 m³/h)

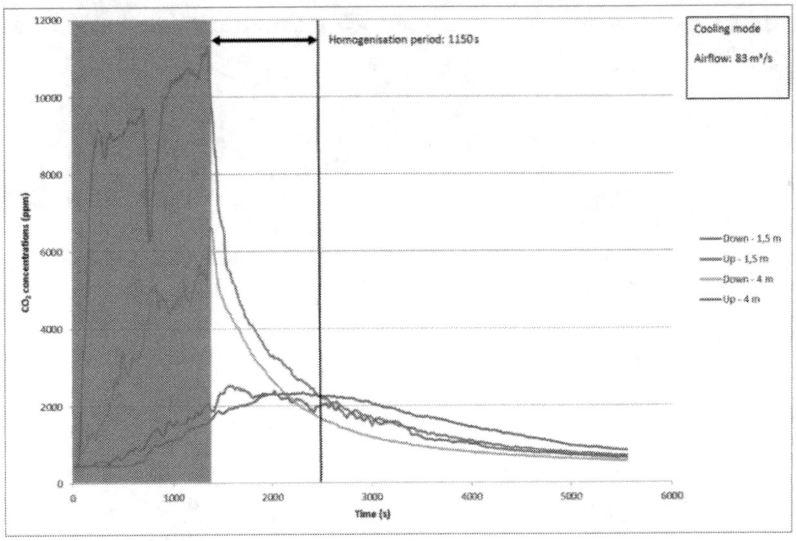

Figure 9 CO_2 measurements in cooling mode (80 m³/h)

The ventilation effectiveness was calculated for each measurement and is represented on Figure 10. On this figure we can see that the values obtained during the homogenisation period are not relevant.

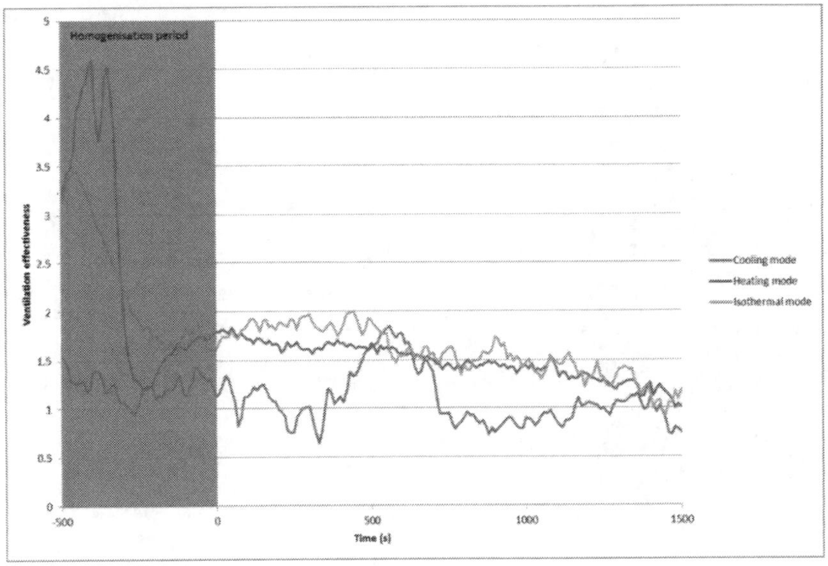

Figure 10 Comparison of ventilation effectiveness in isothermal, heating and cooling modes

For all 3 modes, values above 1 can be found on the chart. This is due to the position of the CO_2 sensors used for the determination of the average concentration in the room. As they are positioned in the vertical plane directly aligned with the fresh air direction, they are in the best ventilated part of the room. This means that no conclusion can be drawn from the absolute values but only comparisons between the 3 cases.

The average values of the ventilation effectiveness are:

- isothermal mode: 1,6
- heating mode: 1,5
- cooling mode: 1,1

These results confirm that the best air distribution is obtained for the isothermal mode and the worst for the cooling mode. They show that the air distribution obtained for the heating mode is quite similar to the isothermal mode.

4 INFLUENCE OF THE AIR FLOW RATE

The same investigations were made to analyse the influence of the airflow rate on the ventilation effectiveness. These investigations are presented here for the isothermal mode.

4.1 Smoke visualisation

Figure 11 presents the results of the smoke visualisations for the 3 different airflows. On these pictures, we can see that the air speed allows a faster penetration of the smoke in the room after 5 seconds, but no real difference can be seen in the repartition of the smoke in the room volume between the 3 cases after 20, 30 or 50 seconds.

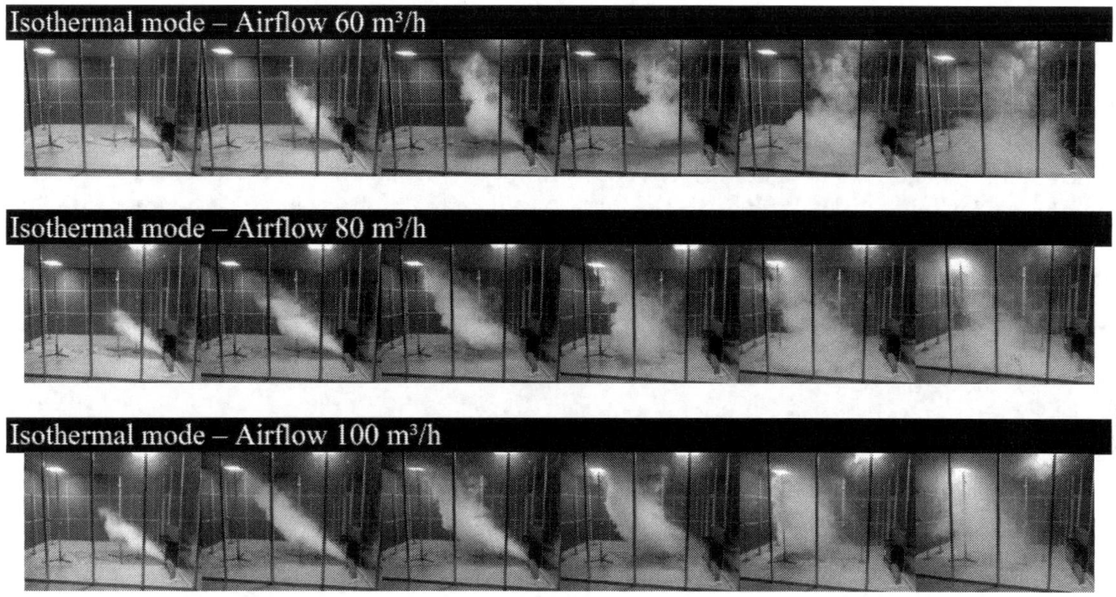

Figure 11 Smoke visualisations after 5, 10, 15, 20, 30 and 50 s for the 3 different airflows

4.2 CO_2 measurements

The CO_2 measurements can be seen on Figure 4 (60 m^3/h), Figure 7 (80 m^3/h) and Figure 12 (100 m^3/h). The repartition is by each case identical: no vertical stratification of the CO_2 concentrations and a notable difference between both locations (1,5 m and 4 m).

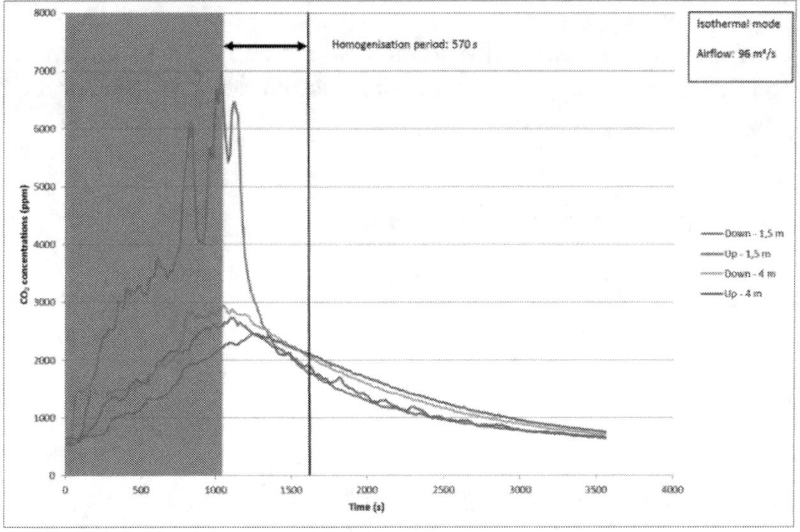

Figure 12 CO_2 measurements in isothermal mode (100 m³/h)

The average values of the ventilation effectiveness obtained with each airflow are:

- 60 m³/h: 2,0
- 80 m³/h: 1,6
- 100 m³/h: 1,3

The results are showing a higher effectiveness for the lowest airflow. We suppose that there is an optimal airflow for which this effectiveness would be the highest. It would be interesting to test airflows lower as 60 m³/h in order to verify that this optimum exists. We also suppose that the results would be different if the effectiveness had been determine using sensors in the whole volume of the room.

5 CONCLUSIONS

The main goal of a ventilation system is to remove pollutants like CO_2 from a room. In order to quantify its capacity to reach this goal, smoke visualisation is not the right tool. Indeed, we have seen for the heating mode that the smoke seems to stay in the highest layers of the air volume but the CO_2 concentrations are still reduced at each measured point with a similar speed.

One important conclusion is about the cooling mode for which the ventilation effectiveness was the lowest obtained in our tests. It might be interesting to equip the air inlet of such a reversible device with a system adapting the speed and direction of the airflow to the inlet temperature of the air.

6 ACKNOWLEDGEMENTS

Part of the work presented was funded by the German Ministry of Economics and Labour BMWi under the reference 03ET1035A and 0327400 P. Other parts were funded by the European Commission under the contract number 314229 – RETROKIT in the FP7 Programme.

7 REFERENCES

Ajaji, Y. and P. André (2013). A Decentralized Mechanical Ventilation System with Heat Recovery: Experimental Assessment of Thermal Comfort and Pollutant Removal Ability. Clima 2013. Prague.

Ender, T., R. Gritzki, et al. (2004). Bewertung von dezentralen, raumweisen Lüftungsgeräten für Wohngebäude sowie Bestimmung von Aufwandszahlen für die Wärmeübergabe im Raum infolge Sanierungsmaßnahmen. [Abschlußbericht]. Stuttgart, Fraunhofer-IRB-Verl.

Haase, M. and M. Gruner (2013). Evaluation of Decentralized Façade-Integrated Ventilation Systems in Nordic Climate. Clima 2013. Prague.

Krajčík, M., A. Simone, et al. (2012). "Air distribution and ventilation effectiveness in an occupied room heated by warm air." Energy and Buildings **55**: 94-101.

Mahler, B., R. Himmler, et al. (2008). DeAL - Evaluierung dezentraler außenwandintegrierter Lüftungssysteme - Abschlußbericht. Stuttgart, Steinbeis-Transferzentrum und Steinbeis Forschungszentrum Energie-, Gebäude- und Solartechnik.

Mats, S. (1981). "What is Ventilation Efficiency." Building and Environment **16**(2): 123-135.

Recknagel, H., E. Sprenger, et al. (2007/2008). Taschenbuch für Heizung und Klimatechnik, Oldenbourg Industrieverlag, München.

Voß, T. and K. Voss (2012). Mikroklimatische und baukonstruktive Einflüsse bei dezentraler Lüftung. BauSIM 2012. IBPSA. Berlin, IBPSA.

PERCEPTION OF A COOLING JET FROM CEILING - A LABORATORY STUDY

Henna Maula*, Hannu Koskela, Annu Haapakangas and Valtteri Hongisto

Finnish Institute of Occupational Health
Lemminkäisenkatu 14-18 B
20520 Turku, Finland
**Corresponding author: henna.maula@ttl.fi*

ABSTRACT

The effect of a cooling jet from ceiling on thermal comfort, perception and subjective performance in warm office environment (29.5 °C) was studied. Altogether, 29 participants (13 male and 16 female) participated. All participants were tested in both thermal conditions and the order of the thermal conditions was counterbalanced between the participants. During the experiment, participants filled questionnaires and performed computerised tasks. Using the cooling jet significantly improved the whole body and local thermal comfort. It also improved the perception of the whole working environment. The indoor air was perceived fresher when the cooling jet was provided. Fatigue increased in time significantly only when the cooling jet was not provided. Eye symptoms increased in time only with the cooling jet. The cooling jet increased the sensation of draught, but still the warm room air temperature was perceived to impair the performance more than draught. The room air temperature was perceived to be better for working effectively for a long period of time with the cooling jet.

KEYWORDS

Thermal comfort, cooling jet, perception, symptoms, self-rated performance

1 INTRODUCTION

Room temperature is an important part of the indoor environment. Too warm room air affects negatively occupant's thermal comfort and performance (Häggblom et al. 2011). Local cooling can be produced by increasing the air movement (Melikov et al. 2013).

Table fans are commonly used in offices during warm summer periods. Huang et al. (2013) studied the user requirements for air movement with an online survey in China and a series of climate chamber experiments. They concluded that the flow generated by electric fans can be used as an effective cooling method to maintain a comfortable environment at 28 °C-32 °C. However, in open-plan offices with high density of workstations the fans also increase substantially the thermal load of the office increasing the need of cooling and the energy consumption. Supplying an air jet from the ceiling can be integrated to the air conditioning system thus removing the need for a local fan.

Toftum (2014) summarized the factors influencing the perception of air movement. According to the summary, occupants feeling warmer than neutral do not feel draught at air velocities up to 0.4 m/s. In the high temperature range up to 30 °C, very high air velocities, such as 1.6 m/s, were found to be acceptable. However, they concluded that very high air movement should be under the control of the exposed occupant.

ASHRAE 55(2010) and ISO 7730 (2005) standards allow elevated air speed to be used to increase the maximum operative temperature for acceptability under certain conditions. According to the ASHRAE standard, if sedentary office occupants don't have control over the local air speed in their space, the upper limit to air speed should be 0.8 m/s for operative temperatures above 25.5 °C.

The aim of this study was to investigate how the cooling jet from ceiling is perceived and how it affects the thermal comfort and subjective work performance.

2 METHODS

The study was carried out in the open-plan office laboratory (82 m^2) of FIOH in spring 2013. The room had twelve workstations and was designed to resemble a neutral clerical office. Four workstations out of the twelve were used in this study (Figure 1). The HVAC of the office laboratory was adjusted to room air temperature 29.5 °C, relative humidity 20 % and supply air flow rate 280 l/s. Supply air was blown with five terminal units and with one cooling jet. The cooling jet was produced with seven adjustable nozzles installed symmetrically into the end of one of the inlet ducts (Figure 3). Noise level and lighting conditions met current recommendations.

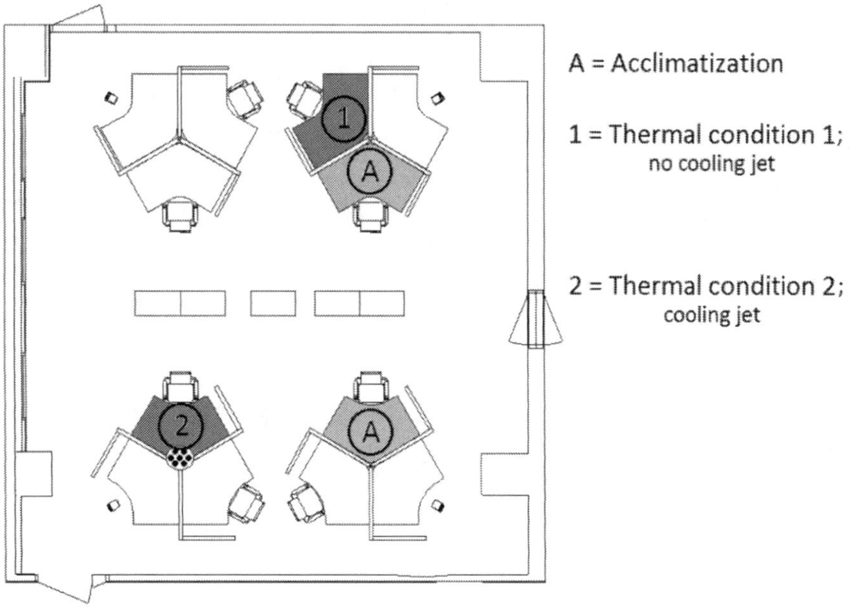

Figure 1: Layout of the open-plan office laboratory.

Figure 2: Seven nozzles installed symmetrically into the end of one of the inlet ducts to produce the cooling jet.

Two thermal conditions were tested; (1) no cooling jet and (2) cooling jet with fresh supply air (ΔT = -3.5 °C and target velocity 0.8 m/s). Air velocity was measured in the occupied zone with hot-sphere anemometers while the dummy subjects (heat load 90 W/dummy) were present to simulate a sitting person (Zukowska-Tejsen et al., 2012). The cooling jet in thermal condition 2 was located in front of the occupant in the suspended ceiling (height 2.55 m). The angle of the cooling jet was 45° from the vertical axis. Target velocity of the cooling jet was set to upper limit of acceptable air speed without personal control according to ASHRAE standard 55 (2010) and ISO standard 7730 (2005). Both thermal conditions were tested simultaneously in the open-plan office laboratory.

Altogether, 29 participants (13 male and 16 female) were recruited. The participants were native Finnish speakers and aged between 20 and 33 years (average = 24). The study had a repeated measures design. All participants were tested in both thermal conditions two at a time. The order of thermal conditions was counterbalanced between participants. Clothing insulation and activity level were controlled. The participants were advised to wear trousers, short sleeve shirt, socks and ankle-length shoes. The estimated clothing insulation including office chair (0.1 clo) was 0.71 clo. The main activity of the participants during the study was typing. The estimated activity level was 1.1 met.

The experiment day started with 30 minutes acclimatization (Figure 3). Workstations meant for acclimatization didn't differ from the thermal condition without the cooling jet. The purpose of the acclimatization was to secure that each participant had similar acclimatization despite of in which thermal condition they took part in first. During acclimatization the participants filled questionnaires and computerised performance tasks were introduced to them. After the acclimatization, participants moved to the experimental workstation and the first session started with questionnaires. After the questionnaires the participants performed computerised tasks followed by questionnaires at the end of the session. The session in both thermal conditions lasted for 40 minutes. After the first session the participants changed places with each other and the second session in different thermal condition but with an identical procedure started immediately. There were no breaks in between the two sessions.

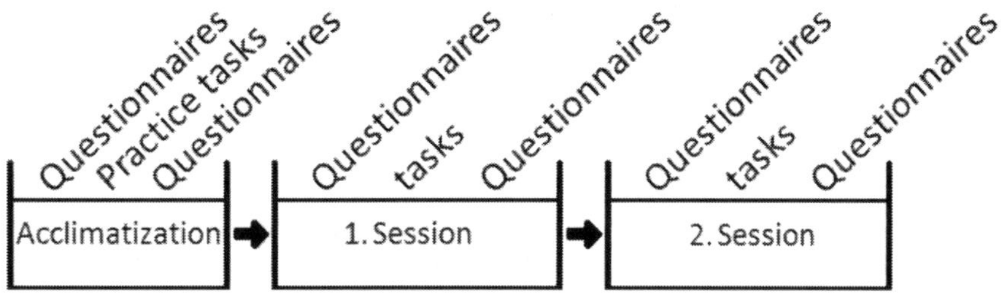

Figure 3: Test procedure

Whole body thermal comfort, local thermal comfort, symptoms, subjective performance ratings and overall experience of the environment were assessed with questionnaires, which were repeated throughout the session. Thermal comfort was evaluated using a seven-point thermal sensation scale. Fatigue was measured with a modified version of the Swedish Occupational Fatigue Inventory (SOFI; Åhsberg et al., 1995, Åhsberg et al., 1998) including three factors which were sum variables of three items (tiredness: sleepy, yawning, drowsy; lack of energy: worn out, exhausted, drained; lack of motivation: uninterested, indifferent, passive).

Statistical analyses were conducted with SPSS Statistics 20 with a confidence interval of 95 %. The normality of the data was tested with Shapiro-Wilk. A repeated measures ANOVA was used when data was normally distributed or when distributions were similarly skewed. Whenever parametric methods were used, the results were also checked with the nonparametric Wilcoxon test. When an interaction effect was found, paired comparisons between conditions were performed using t-test or Wilcoxon. Benjamini-Hochberg procedure was used in paired comparisons.

3 RESULTS AND DISCUSSION

Table 1 shows the percentage of participants experiencing the cooling jet to be pleasant or unpleasant in certain body parts. The perception of the cooling jet was contradictory. Over 50 % of the participants reported the cooling jet to be pleasant in the arms, face or at the top of the head. Respectively the face and the top of the head received most votes of the airflow being unpleasant. On the open questionnaires some participants reported the airflow to be unpleasant in the eyes but to be pleasant everywhere else in the face. Those answers are included in both pleasantness categories. Also the intensity of the cooling jet distributed the opinions. Some of the participants would have preferred more air flow and some less.

Table 1: The percentage of participants reporting airflow pleasant or unpleasant in certain body parts.

	Top of the head	Face	Neck	Front neck	Torso	Arms	Hands
Pleasant	52	66	7	38	45	72	31
Unpleasant	21	41	10	10	0	10	0

The differences in thermal comfort were statistically significant ($p<.01$, Figure 4). 67 % of the participants considered the thermal conditions to be acceptable with the cooling jet, while only 28 % considered them to be acceptable without the cooling jet. No gender differences were found. The improvement of thermal comfort under the influence of the cooling jet was statistically significant in each body part except the thighs. However, a trend of the cooling jet improving thermal comfort in thighs was found (p=.068). Thermal comfort was significantly better with the cooling jet even in legs and feet, although they were not under direct influence of the airflow.

The occurrence of fatigue was minor. Fatigue increased in one measured factor, tiredness, toward the end in both thermal conditions, but the increment of fatigue was significant only in the condition without the cooling jet (p<.01; Figure 5). Eye symptoms increased in time only in the condition with the cooling jet (p<.001), but the intensity of eye symptoms was minor even at the end of the session.

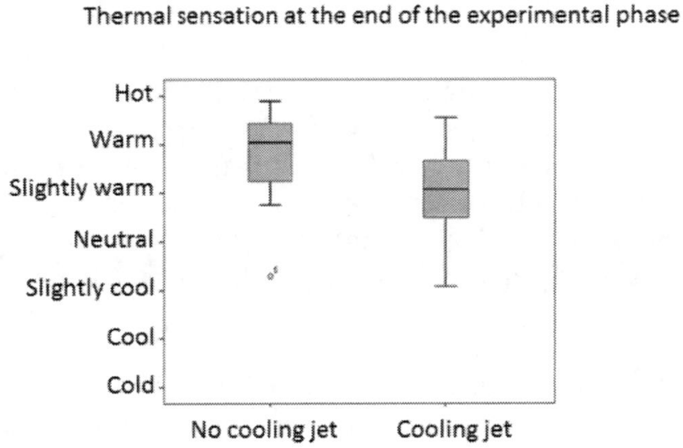

Figure 4: Thermal sensation votes

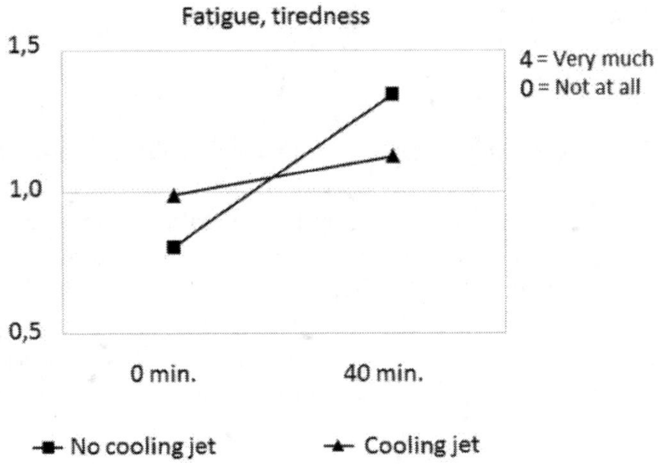

Figure 5: Fatigue, factor tiredness

The indoor air was perceived fresher when the cooling jet was provided (p<.01). Under the cooling jet the air temperature was perceived better (p<.001) but there was more sensation of draught (p<.001). The whole working environment was perceived better in the thermal condition with the cooling jet (p<.05).
There was a trend that the subjective performance was better when working under the influence of the cooling jet, but the difference between the conditions was not statistically significant (p=.064). However, the participants were more frustrated during the most difficult task in the condition with no cooling jet (p<.05). The stuffiness of the room air was perceived

to impair the performance more without the cooling jet than with cooling jet ($p<.05$). Draught disturbed the performance more in the condition with the cooling jet ($p<.01$), but in that thermal condition heat was perceived to be more disturbing than draught ($p<.01$). The room air temperature in the condition with the cooling jet was perceived to be better for working effectively for a long period of time ($p<.05$).

4 CONCLUSIONS

The results show that providing local cooling with a supply air jet improves the thermal comfort and the perception of the working environment in a warm office. The study was carried out with a cooling jet with air speed in the upper acceptable limit when local control over the cooling jet is not provided. The perception of the cooling jet was contradictory. Results showed the need for individual control over the airflow already at the air velocity of 0.8 m/s. In the future, the effect of individual adjustment in similar conditions on the thermal comfort, perception and subjective performance should be studied.

5 ACKNOWLEDGEMENTS

This work was carried out within the RYM Indoor Environment research program which was financially supported by Tekes and the participating companies.

6 REFERENCES

ASHRAE Standard 55-2010. Thermal environmental conditions for human occupancy. Atlanta: American Society of heating, refrigerating, and air-conditioning engineers. 2010.

Huang L.,Ouyang Q., Zhu Y., Jiang L. (2013). A study about the demand for air movement in warm environment. Building and Environment 61 (2013) 27-33.

Häggblom H., Hongisto V., Haapakangas A. and Koskela H. (2011). The Effect of Temperature on Work Performance and Thermal Comfort – Laboratory Experiment. Proceedings of Indoor Air 2011, The 12th International Conference on Indoor Air Quality and Climate, Austin, Texas, June 5-10

ISO 7730 2005. Ergonomics of the thermal environment. Analytical determination and interpretation of thermal comfort using calculation of the PMV and PPD indices and local thermal comfort criteria. International Organization for Standardization, Genève.

Toftum J. (2004). Air movement - good or bad? Indoor Air 2004; 14 (suppl 7): 40-45.

Zukowska-Tejsen D., Melikov A. and Popiolek Z. 2012. Impact of geometry of sedentary occupant simulator on the generated thermal plume: Experimental investigation. HVAC&R Reseach, Vol. 18, No.4, p.795-811

Åhlsberg E., Gamberale F., Gustafsson K. (1995). Upplevd trötthetskvalitet vid olika arbetsuppgifter. Utveckling av mätinstrument. Arbete och Hälsa, 1995:20

Åhlsberg E., Gamberale F., Gustafsson K. (1998). Upplevd trötthet efter mentalt arbete. En experimentell utvärdering av ett mätinstrument. Arbete och Hälsa, 1998:8

PROMEVENT: IMPROVEMENT OF PROTOCOLS MEASUREMENTS USED TO CHARACTERIZE VENTILATION SYSTEMS PERFORMANCE

Adeline Bailly*[1], Cedric Lentillon[1]

1 Cerema
Direction Territoriale Centre-Est
46 rue Saint-Théobald
38081 l'Isle d'Abeau, FRANCE
**Corresponding author: adeline.bailly@cerema.fr*

ABSTRACT

For the coming energy-efficient buildings, the guarantee of energy performance becomes a major challenge. It is therefore crucial to implement accurate and reliable measurements, in order to ensure this performance. The in-force French EP-regulation RT2012 already imposes compulsory justification of envelope airtightness. Moreover, the Effinergie+ label requires ventilation systems control and ductwork airleakage performance. These requirements, ventilation controls for IAQ concern and regulatory compulsory controls of buildings need reliable diagnostic protocols.

In 2014, several French partners, led by the Cerema[1], proposed a new project, PROMEVENT, which main objective is to improve ventilation systems measurements protocols. It aims to deal with different phases of a diagnostic:

1) Visual control
2) Airflow measurements and pressure differences measurements at air vents
3) Ventilation ductwork airleakage measurements.

Several points will be tested through repeatability and reproducibility assessments. Then measurements will be performed on laboratory benches, and on ventilation systems of single-family and multi-family dwellings. By the end of the PROMEVENT project, recommendations and a first version of a protocol for the measurement of residential buildings ventilation systems will be proposed. Moreover, one of the main objective is to produce a more reliable and optimised protocol which will be written as a proposed draft standard.

This paper presents the context in which the PROMEVENT project has been defined, and describes its different steps.

KEYWORDS

Ventilation – Measurements – Airtightness – Airflow - Improvement

1 CONTEXT AND OBJECTIVES

The recent French energy performance regulations and labels have resulted in a new buildings generation. Since 2000, airtightness requirements have been gradually implemented in French

[1] A new public scientific organism born from the merging of 11 scientific institutes (including the CETE de Lyon) of the French ministry for ecology, sustainable development and energy (MEDDE).

regulations, leading to a reinforcement of air renewal systems and a need to ensure their reliability. First labels and mostly Effinergie-BBC label have imposed a requirement on building envelope airtightness for residential building. Since January, 2013, the in-force EP-regulation RT2012 imposes airtightness requirements for all new residential buildings.

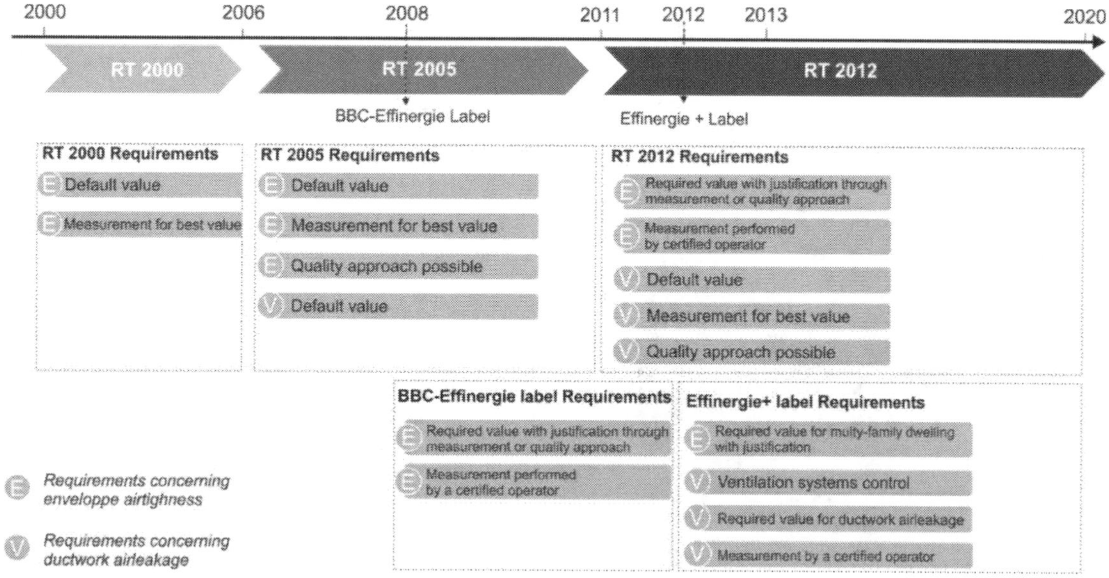

Figure 1: Evolution of French Thermal Regulations

These highly airtight buildings create an issue for both comfort and indoor air quality. Indeed, in these dwellings, air change rates are provided by ventilation system, which have to be efficient to ensure good indoor air quality, while limiting heat losses due to air change. Several recent studies illustrate this concern. The OQAI (French indoor air quality observatory) performed a national IAQ campaign from 2003 to 2005 (Kirchner, 2008). 567 dwellings (chosen in order to represent the national housing stock) have been investigated through ventilation systems diagnostics and indoor air quality measurements. This national study has concluded that the air change rate and the duration windows are opened are the most important factors of the indoor air quality. Moreover, calculations analysis which have been performed during the project QUAD-BBC have shown some typical evolution of pollutants in highly airtight low consumption buildings (Boulanger, 2012).

The French in-force regulation concerning ventilation requires a general and permanent ventilation for residential buildings. It also imposes minimal airflow of exhaust air. So as to meet those two seemingly divergent objectives, technically advanced mechanical ventilation systems have been developed. Nevertheless, high quality and technical skills are required during design phase, implementation and maintenance, which are often neglected. Ventilation systems have an influence on the sanitary aspects of the supplied and indoor air, through moisture development for example (Van Herreweghe, 2013). Moreover, inhabitants may have not understood the functioning of new mechanical systems, especially for balanced ventilation, and might decide to take it down. The OQAI has recently carried out a field survey in seven new built energy-efficient houses in France (Derbez, 2014). All inhabitants have experienced some difficulties with their Mechanical Ventilation with Heat-Recovery systems, because they are difficult to use, the user's manual is complex, high noise levels can be produced or they cause a lack of comfort. But if MVHR systems are turned out or voluntary degraded (airvents closed for example), indoor air quality can become poor and present a risk to human health.

Therefore, in many countries, several studies have been launched to realize a state of the art of ventilation systems in dwellings. In France, a survey (Jobert, 2013) has been carried out through control reports performed between 2008 and 2011 concerning 1287 dwellings (88% are multi-family dwellings). Almost all buildings are equipped by simple exhaust ventilation systems.

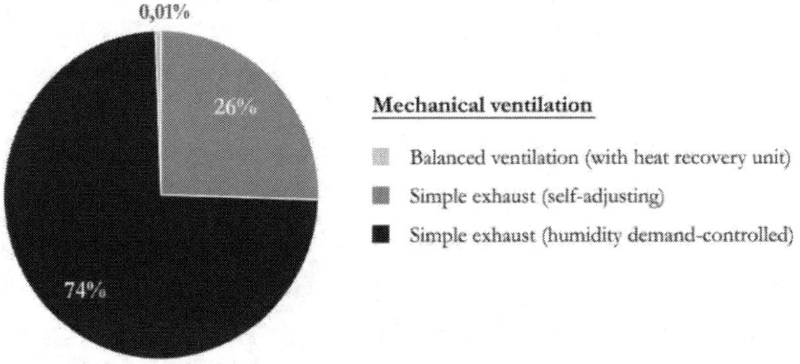

Figure 1: Ventilation system repartition in the analysed sample (Jobert, 2013)

47% of the sample do not comply with the airing regulation, which means that they present at least one non-compliance remark (68% for single-family dwellings, and 44% for multi-family dwellings). Those non-compliances are generally related to design errors, poor implementation and lack of maintenance. Same issues have been observed in many European and American countries (Van Den Bossche, 2013). Two practices could improve the quality of ventilation systems. On one hand, a quality approach could be set up. In France, such approaches have been successfully adopted for building envelope airtightness (Charrier, 2013). Moreover, the VIA-Qualité project is testing the feasibility of such approach for ventilation systems and IAQ (Jobert, 2013). In a second hand, as the French thermal regulation imposed the airtightness level justification, the ventilation system performance could also be controlled, what is already compulsory with a recent label, "Effinergie +". Those controls are in accordance with new approaches, which impose that standards and regulations compliance is ensured by on-site verifications. Those approaches might lead to financial or organizational consequences. Currently, ventilation systems controls are generally performed in France in three cases:

- For buildings applying for new Effinergie label (Effinergie +)
- During regulatory compulsory control (by the technical civil servants network of the Ministry in charge of the Construction's sector)
- When IAQ issues have been set out for a building.

Therefore, control protocols have to be unquestionable. In these cases, several diagnostic protocols are used: either labels reference documents or good practice guide, such as the Effinergie protocol, the DIAGVENT method and the European standard EN12599. The reliability of these protocols may be not sufficient. The PROMEVENT project proposes to test repeatability, reproducibility and feasibility of these protocols in order to define a more reliable protocol for ventilation system controls.

This project may have to deal with such issues as: how can it be representative of all different situations? How can it characterize the equipment use impact? How will it overcome airflow measurement difficulties? The PROMEVENT established program is described in the second part of this paper.

2 PROMEVENT PROGRAM

PROMEVENT is an 8 organizations project whose main objective is to improve the quality of ventilation system controls. More precisely, it aims to ensure the reliability of ventilation measurement protocols concerning pressure and airflow differences at air vents, and ductwork airleakage. This program will focus on residential buildings. According to current practices in new constructions, two "dwelling/ventilation systems" associations will be studied:

- Single-family dwellings equipped with balanced ventilation
- Multi-family dwellings equipped with single humidity demand-controlled ventilation.

This project includes three workpackages. The first one consists in a characterization of existing protocols in order to evaluate their reliability, through experimental campaigns. The second one is the elaboration of a new protocol which must ensure controls reliability. The last one will lead to a verification of the new protocol feasibility, reliability and relevancy, through on-site tests.

The PROMEVENT project has been proposed to a call for proposals launched by ADEME within the subject "toward responsible buildings in 2020". The Consortium is constituted of 8 French partners, both private and public sectors: a public institution (Cerema[2] - ex CETE de Lyon), a laboratory (CETIAT), 5 consultancies (ALLIE'AIR, ICEE, PLEIAQ, CETii, PBC) and an association (Effinergie). They are quite complementary on scientific and operational dimensions. The Consortium, thus formed should be able to defend the importance of a unique and standardized protocol which will have to take into account uncertainty requirements and implementation concerns. It should also convey its discourse to all relevant professionals.

2.1 Existing protocols reliability

Several protocols are used in France and abroad to control ventilation systems performance, including visual diagnostic, proper functioning at air vents control and ductwork airtightness measurement. There are described in label reference documents, campaign protocols, standards or good practice guides. EN12599, DIAGVENT method, Effinergie protocol or OQAI protocol are some of them. In a first step, an inventory is being taken to compare the different protocols considering risks and sources of uncertainty, investigated components and technical feasibility. Moreover, different technologies and equipment used to perform measurements (concerning airflow, pressure difference and ductwork airtightness) are being listed and studied to describe their field of application and product liability according to their manufacturers. This review is also including a state of the art of international projects and studies dealings with this topic, in order to be able to integrate current evolutions.

This first step is preparing the second one: experimental campaigns. A methodology will be established to evaluate feasibility, relevancy and reliability of existing protocols. Two

[2] Centre for expertise and engineering on risks, environment, mobility, urban and country planning

campaigns will be performed in order to conduct repeatability and reproducibility evaluations from 2014 to 2015.
One campaign will be carried out in laboratory. It will study measurement equipment performance and calibration in different configurations: proper functioning without, then with components which might disrupt the flow. Thanks to a ventilation system assembled in laboratory, the impact of some specific configurations will be evaluated without risking a building deterioration.
A second campaign will be performed in situ on:

- 10 single-family dwellings equipped with balanced ventilation;
- 2 multi-family dwellings equipped with single humidity demand-controlled ventilation.

4 operating-teams will contribute to this campaign. Roles are defined in the following table.

Table 1: First in-situ campaign program

		Team 1	Team 2	Team 3	Team 4
2 multi-family dwellings	Building 1: *airflow, pressure difference at airvents and ductwork airtightness*	B1	B1		
	Building 2: *airflow and pressure difference at airvents*			B2	B2
10 single-family dwellings	House 1: *airflow, pressure difference at airvents, and ductwork airtightness*	H1	H1	H1	
	House 2: *airflow and pressure difference at airvents*			H2	H2
	House 3 to 10: *airflow and pressure difference at airvents*	H3 ; H4	H5 ; H6	H7 ; H8	H9 ; H10

This dwellings repartition has been defined to optimise repeatability and reproducibility evaluation regarding to geographical, financial and temporal limits. Several measurements (as much as possible) will be performed in each dwellings (for repeatability and equipment reproducibility evaluation). Ventilation systems of 4 buildings (B1, B2, H1 and H2) will be controlled by 2 or 3 different teams (reproducibility evaluation).

2.2 New protocol formulation

The results of the two campaigns and the review will be analysed to determine which points of currently used protocols are unreliable, to what extent and why.
The laboratory campaign analysis will help to determine measurement uncertainty of each equipment depending on the application, and the need of calibration. Impact of air vents type or non-alignment measurement will be assessed. It is also expected to detect the impact of some specific issues related to ventilation ductwork through tests performed on the ventilation system assembled in laboratory.
The in-situ campaign results will be analysed through several aspects:

- Repeatability of tested protocols
- Reproducibility concerning equipment impact
- Reproducibility concerning operator impact
- "Out of hand" in-situ conditions impact.

Main objectives are to evaluate uncertainty of the tested protocols and to specify uncertainties sources.

According to those results, the Consortium will formulate a new protocol. At this step, a group of relevant professionals will be consulted. It could be composed of ventilation experts and industrials, controllers, control equipment manufacturers, architects, certification organisms. Discussions on new protocols will focus on several points:

- Ensure sufficient reliability
- Ensure technical and financial feasibility
- Define self-checking equipment conditions
- Define needs and organisation of operators training, qualification and control.

2.3 New protocol validation

The last milestone of this project consists on the new protocol validation. It will be based on a second in-situ campaign results. According to geographical and financial limits and buildings availability, fewer buildings will be tested. Each team will perform several measurements (as many as possible), with different equipment. The following table presents the second in-situ campaign program.

Table 2: Second in-situ campaign program

		Team 5	**Team 6**	**Team 7**	**Team 8**
1 multi-family dwellings	Building 3: *airflow, pressure difference at airvents and ductwork airtightness*	B3	B3	B3	
4 single-family dwellings	House 11: *airflow, pressure difference at airvents and ductwork airtightness*			H11	H11
	House 12: *airflow, pressure difference at airvents and ductwork airtightness*	H12	H12		
	House 13 & 14: *airflow, pressure difference at airvents and ductwork airtightness*	H13 ; H14			

In order to evaluate the technical feasibility and the relevancy of the protocol, new teams will be composed with new operators: they will not have been part of the project before this step. Each team will be accompanied by an operator of team 1 to 4 (first campaign and all project) whose role will be to assist, to guide and to audit new operators. It will lead to evaluate of the understanding of the protocol. It will also ensure that diagnostics have been performed properly to evaluate repeatability and reproducibility of the new protocol.

Experimental results will be analysed to evaluate:

- New protocol relevancy: improvement of reliability
- Implantation ease, understanding and appropriation by new operator
- Financial feasibility: difference between costs of current used protocols and new protocol
- The self-checking relevancy.

Those evaluations will lead to a final protocol that will be drawn up as a proposed draft standard. It will also lead to a guide intended to ventilation operators. This guide will explain and illustrate each step of this ventilation system diagnostic protocol. It will be available and free for download.

3. CONCLUSIONS

The PROMEVENT project objective is to define a new protocol for controls of ventilation systems performance. This protocol will be established based on many existing protocols currently used. Several laboratory and in-situ campaigns will be carried in order to test repeatability, reproducibility and feasibility of those protocols, including the equipment choice impact. It is expected that conclusions of this project will lead to a new standard which will be imposed in new buildings regulation, in order to impose compulsory check of ventilation systems performance. At the end of this project, training and recommendations will be provided to operators through a practical guide, which may be useful for measurements performed for label.

4. ACKNOWLEDGEMENTS

The "PROMEVENT" project receives funding from the French Environment and Energy Management Agency. The contribution of Cerema is funded by the French ministry for ecology, sustainable development, and energy (MEDDE).
The sole responsibility for the content of this publication lies with the authors.

5 REFERENCES

S. Kirchner, M. Derbez, C. Duboudin, P. Elias, A. Gregoire, J-P. Lucas, N. Pasquier, O. Ramalho, N. Weiss. *Indoor air quality in French dwellings*. Indoor Air 2008. Copenhagen, Denmark - Paper ID: 574. 2008

X. Boulanger, L. Mouradian, C. Pele, Y-M. Pamart, A-M. Bernard. *Lessons learned on ventilation systems from the IAQ calculations on tight energy performant buildings*. 33rd AIVC Conference. Copenhagen, Denmark. 2012.

J. Van Herreweghe, S. Caillou, M. Roger, K. Dinne. *Sanitary aspects of domestic ventilation systems: an in situ study*. 34th AIVC Conference. Athens, Greece. 2013.

Jobert, R., Guyot, G. *Detailed analysis of regulatory compliance controls of 1287 dwellings ventilation systems*. 34th AIVC Conference. Athens, Greece. 2013.

M. Derbez, B. Berthineau, V. Cochet, M. Lethrosne, C. Pignon, J. Riberon, S. Kirchner. *Indoor air quality and comfort in seven newly built, energy-efficient houses in France*. Building and Environment. N°72, p. 173-187. 2014.

P. Van Den Bossche, A. Janssens, D. Saelens. *Securing the quality of ventilation systems in residential buildings: existing approaches in various countries*. 34th AIVC Conference. Athens, Greece. 2013.

S. Charrier, J. Ponthieux, A. Huet. *Airtightness quality management scheme in France: assessment after 5 years operation*. 34th AIVC Conference. Athens, Greece. 2013.

IMPACT OF A POOR QUALITY OF VENTILATION SYSTEMS ON THE ENERGY EFFICIENCY FOR ENERGY-EFFICIENT HOUSES

Adeline Bailly*[1], Franck Duboscq[1], Romuald Jobert[1]

1 CEREMA
Direction Territoriale Centre-Est
46 rue Saint-Théobald
38081 l'Isle d'Abeau, France
Corresponding author: adeline.bailly@cerema.fr

ABSTRACT

The "VIA-Qualité" project (2013-2016) focuses on low energy, single-family dwellings. It proposes the development of quality management approaches (ISO 9001) which aim to increase both on-site ventilation and indoor air quality. One of the main benefits of those approaches is the improvement of ventilation system performance, especially thanks to a rigorous follow-up from design to installation. Efficient ventilation system performance is rewarded in the French EP-calculation, through a primary energy consumption estimation. In order to evaluate the energy impact of the proposed quality approaches, some sensitivity studies have been carried out for single humidity-controlled ventilation system. The primary energy consumption of typical single-family dwellings has been estimated along various parameters, such as:

- Ductwork airleakage
- Exhaust and incoming airflow
- Electrical fan power
- System localization (to take into account leakages in and out of heated volume).

This paper presents variations of the estimated dwellings energy consumptions as an indicator of the impact of several typical dysfunctions of ventilation systems which have been observed during different campaigns and controls.

KEYWORDS

Ventilations systems – Dysfunctions – Energy impact – Single-family house

1 INTRODUCTION

In France, the energy performance regulation (RT2012) generalizes requirements of the BBC-Effinergie label, in particular regarding the envelope airtightness. For a single family dwelling, the requirement is $Q_{4Pa\text{-}Surf}$ [1] $\leq 0.6\ m^3.h^{-1}.m^{-2}$ (around $n_{50} \leq 2.3\ h^{-1}$). In those airtight dwellings, the air change rates have to be adequate to insure a good indoor air quality, and at the same time they have to induce low thermal losses in order to comply with requirements of the regulation. Nevertheless, the RT2012 does not include any new requirement on ventilation rates, which are provided by another 30-years-old regulation (JO, 1983). Therefore, without

[1] Air permeability at 4 PA divided by the loss surfaces area excluding basement floor

compulsory check of the proper functioning of ventilation systems, how inhabitants can be sure that the air renewal of their houses is adequate to ensure a good indoor air quality? In addition, in the current energy context, mechanical ventilation systems are spreading, including high technical systems such as single humidity-controlled ventilation systems and balanced ventilation systems. These systems require knowledge and skills from each of the three main actors: designers, installers and inhabitants, who in most cases are not even aware of the ventilation principles. Indeed, various campaigns (Paul Van Den Bossche, 2013) in different countries have brought forward the poor quality of ventilation systems, mainly related to a poor design or installation and a lack of maintenance. In France, a study (Romuald Jobert, 2013) has been carried out from regulatory compliance control reports of almost 1300 dwellings. This survey confirms that ventilation system dysfunctions are very frequently observed in dwellings, and it gives clear information about their localization and qualification.

Therefore, we need solutions to prevent those dysfunctions and ensure a good indoor air quality. Various countries (Paul Van Den Bossche, 2013) are developing different schemes to secure the quality of ventilation systems through actions during all steps of the process of a building construction. Such a system is already in place since 5 years in France regarding the envelope airtightness. The good experience of this quality approach (Sandrine Charrier, 2013) and the current ventilation systems quality assessment have motivated the "VIA-Qualité" project. Started in 2013, this 3-years French project proposes to develop quality management (QM) approaches (ISO 9001) with the goal of increasing both on-site ventilation and indoor air quality (IAQ). It focuses on low energy, single-family dwellings, which mainly concerns in France the individual homebuilders sector. The benefits would be to: 1- Improve ventilation system performance, especially thanks to rigorous monitoring from design to installation; 2- Limit indoor internal pollution sources, monitoring materials selection (Wargocki P, 2012); 3- Increase final users' awareness and understanding. An assess of the effectiveness of these QM approaches will be carried out, through ventilation measurements and IAQ measurements in 8 test houses. Moreover, in order to ensure the reproducibility of this kind of operation, the economic and energetic interests of these QM approaches have to be evaluated. This analysis will be carried out in two steps: 1- evaluation of the impact of ventilation system performance on energy consumptions; 2- crossed analysis between the energy-savings and the additional costs due to materials, studies, controls... This paper presents an analysis of impacts of ventilation system dysfunctions on the regulatory energy performance calculations (RT2012) for the three first single-family houses of the VIA-Qualité project.

2 VENTILATION PARAMETERS IN THE FRENCH EP-CALCULATION

The in-force French EP-regulation (RT2012) is mainly based on 3 kind of performance requirements: 1- energy efficiency (independent of systems); 2- primary energy consumption [Cpe[2]] and 3- summer comfort (for buildings without air-conditioning). In this study, the variations of the energy consumptions have been calculated as an indicator of the energy cost of several ventilation systems dysfunctions.

[2] In France, it is noted Cep

The EP-calculation is run with XML data which define the values of various input parameters. The following part briefly explains how ventilation is taken into account in the EP-calculation for a classic single-family house, with a single humidity-controlled ventilation system. Concerning ventilation system, the input parameters used in this study are listed in table 1.

Table 1: Income parameters concerning ventilation systems in the French EP-calculation

Income parameter	Definition	Possible values
Air inlet module ($m^3.h^{-1}$)	Coefficient which is used to estimate the total airflow incoming through all air inlets	From 0 to ∞
Q_{ext} ($m^3.h^{-1}$)	Total extraction airflow	From 0 to ∞
Type_air_intake (-)	Two types of air intake are identified in the EP-calculation: humidity & pressure controlled air intake or pressure controlled air intake	0: humidity & pressure controlled air intake 1: pressure controlled air intake
Engine_Power (W)	Engine power of the fan	From 0 to ∞
C_{dep} (-)	Coefficient of design quality	From 1 to ∞
Ratfuitevc (-)	Ratio of leaks to heated volume	From 0 to 1
Cletres	Airtightness class of the ductwork (A,B,C,D): defined the value of the ductwork leakage coefficient K_{res}	C: K_{res} = 0,003.10^{-3} $m^3.s^{-1}.m^{-2}$ B: K_{res} = 0,009.10^{-3} $m^3.s^{-1}.m^{-2}$ A: K_{res} = 0,027.10^{-3} $m^3.s^{-1}.m^{-2}$ Default (2.5A): K_{res} = 0,0675.10^{-3} $m^3.s^{-1}.m^{-2}$

2.1 Some equations of the EP-calculation concerning ventilation systems

The following equations and fixed values of some coefficients are defined in the EP-calculation method.

Exhaust airflow model

For a single-family house, the regulated exhaust airflow is defined with two indicators: $Q_{ext,\,min}$ which represents the base flow, and $Q_{ext,\,max}$ which represents the flow at full load. The values of these two airflows have to be consistent with the French ventilation regulation (JO, 1983). The average flow $Q_{ext,\,regul}$ is calculated according to equation 1:

$$Q_{ext,regul} = \frac{Q_{ext,max} * Dugd + Q_{ext,min} * (168 - Dugd)}{168} \quad (1)$$

Dugd [hour/week] is the duration of use at full load expressed in h/week. For a single-family house, Dugd = 7 hours per week.

Then, the EP-calculation introduces the coefficient C_{dep} in order to take into account some dysfunctions of the system due to design. Therefore, the extraction flow becomes $Q_{ext,\,dep}$:

$$Q_{ext,dep} = C_{dep} * Q_{ext,regul} \quad (2)$$

The default value of C_{dep} is 1.25, and in most cases C_{dep}=1.1 (justification with a certified document).

The following equation defines the leakage rate through all duct leaks $Q_{ext, leaks}$ for a pressure difference ΔP:

$$Q_{ext,leaks} = 3600 * K_{res} * \Delta P^{0,667} * A_{duct,ext} \quad (3)$$

K_{res} [$m^3.s^{-1}.m^2$] depends on the leaktightness class of the network
$A_{duct, ext}$ [m^2] is the surface of the air duct, which can be estimated as a percentage of the floor area (the uncertainty of this parameter value is not analysed in this study).

The final extraction flow is calculated according to equation 4:

$$Q_{ext} = Q_{ext,dep} + Ratfuitevc * Q_{ext,leaks} \quad (4)$$

K_{hv} [-] describes the part of the exhaust duct in the heated volume (from 0 to 1).

Intake airflow model
In a single humidity-controlled ventilation system, the characteristic curve which represents the air inlets performance has a straight line format. The following equation corresponds with the principal part, which is used when the pressure difference at the air inlet is under 20 Pa.

$$Q_{AI}\,(\Delta P) = C_d * \left(\frac{2}{\rho_{ref}}\right)^{0.5} * 10^{-4} * M * \left(\frac{10}{|\Delta P_{ref}|}\right)^{0.5} \quad (5)$$

Q_{AI} [$m^3.h^{-1}$] is the airflow which enters through the air inlets.
C_d [-] is the coefficient of discharge. Its value is 0.68.
ρ_{ref} [$kg.m^{-3}$] is the air density at 19°C. Its value is 1.2 $kg.m^{-3}$.
M is the sum of the air inlet modules.
ΔP_{ref} [Pa] is the pressure difference for which the module is defined. Its value is 20 Pa.

The two airflows Q_{ext} and Q_{AI} are then used to estimate the indoor temperature for each time step. The difference between the indoor temperature and the setpoint temperature is an input for evaluating the heating needs. With those needs and the performance of the house systems, the EP-calculation establishes the regulatory primary energy consumptions for 1 year and for 1 square meter of a specific floor area named $SHON_{RT}$ (without some parts like non-heated places, balcony...): C_{pe} [$kWhpe.m^{-2}.year^{-1}$]. In order to respect the EP-regulation (R2012), the C_{pe} of a house have to be lower a limit value $C_{pe\ max}$, which depends on various parameters including climate zone and altitude. The average limit value is 50 $kWhpe.m^{-2}.year^{-1}$.

2.2 Definition of dysfunction scenarios

A list of dysfunctions of ventilation systems has been established based on the analysis of Jobert, 2013, and supplemented with results of a 20 houses campaign realised during the first step of the VIA-Qualité project. In this list, dysfunctions which could have an impact on energy consumption of the house have been identified. For each of them, related parameters of the EP- calculation and the appropriate variation range have been identified. Those criteria have been introduced into six scenarios which correspond to different common situations in low-energy houses. Table 2 presents those scenarios.

Table 2: Various scenarios representing common ventilation system dysfunctions

Scenario	Characteristic	Concerned parameters
1: Lack of humidity-control	In this scenario, the installer has put in place the wrong air inlets: they are not humidity-controlled, but just pressure-controlled	• Type_air_intake • Air inlet module [74.9 $m^3.h^{-1}$; • 186 $m^3.h^{-1}$]
2: Excessive number of air inlets	This scenario corresponds with a common situation where 1 additional air inlet has been installed in a room	• Air inlet module [74.9 $m^3.h^{-1}$; • 115 $m^3.h^{-1}$]
3: Low battery in toilets air outlets	Most of the air outlets which are installed in the toilet fitted out with a presence sensor. In most cases, the sensor runs with batteries. This scenario represents the impact of low batteries, which impact on the exhaust airflow.	• Q_{ext} [59 $m^3.h^{-1}$; 110.1 $m^3.h^{-1}$]
4: Duct leakage	In this scenario, various configurations are texted, which correspond with different airtightness classes of the ductwork and different duct positions in the heated volume	• Airtightness class of the ductwork [2.5A ; C] • K_{hv} [0.25 ; 1]
5: Over ventilation	In some cases, fans are wrongly adjusted, which induces an over ventilation of the house. In this scenario, various level of over ventilation are simulated.	• Q_{ext} [59 $m^3.h^{-1}$; 100 $m^3.h^{-1}$]
2: Fan performance	In this scenario, various fan with different performance are tested.	• Engine_Power [8W ; 45 W]

2.3 Houses presentation

Those scenarios have been modelled with the EP-calculation for three low-energy houses. Each of them are equipped with a single humidity-controlled ventilation system, which equips most of new single-family houses. Table 3 describes some characteristics of those houses. A first simulation in normal situation (no modification of the project input data) has been performed for each house. The calculated value of the C_{pe} [Consumption calculated in primary energy[3]] is then used as the reference value for the six previously described scenarios.

Table 3: Houses presentation

House	Type	Heating system	C_{pe} (total) / $C_{pe\ max}$ [kWhpe.m^{-2}.$year^{-1}$]	C_{pe} (heating consumption) [kWhpe.m^{-2}.$year^{-1}$]
House 1	Two-storey house Floor area: 97 m²	Heat pomp	**51.3** / 60	24.4
House 2	Single-storey house Floor area: 84 m²	Heat pomp	**55.2** / 64.6	26.9
House 3	Single-storey house Floor area: 90 m²	Condensing boiler	**57.2** / 73.2	39.9

[3] Consumptions are presented in energy primary: for electrical source, a multiplier factor equal to 2.58 is applied.

3 RESULTS

Each of the 6 scenarios has been simulated for the three test houses with the EP-calculation. In this study, every consumption is expressed in primary energy: electrical source is penalised (1 kWh of final energy = 2.58 kWh en primary energy[4]). Houses 1 and 2 use only electrical source for all energy consumers (heating, cooling, lighting, domestic hot water, ventilation fans and distribution systems). House 3 uses gas source for heating and a part of the domestic hot water.

Except for scenario 6, the analysed dysfunctions only impact on the heating consumptions. The following part presents those impacts and gives some explanations of those results. Results of scenario 6 are presented in a second time, as the impact of the studied dysfunctions does not concern the heating consumption.

3.1 Impact of several ventilation dysfunctions on regulatory heating consumption

As the relative impacts on the heating consumption of each dysfunctions are almost the same for the three houses, figure 1 introduces the average impact for each scenario (except scenario 6). Table 4 gives results for each house.

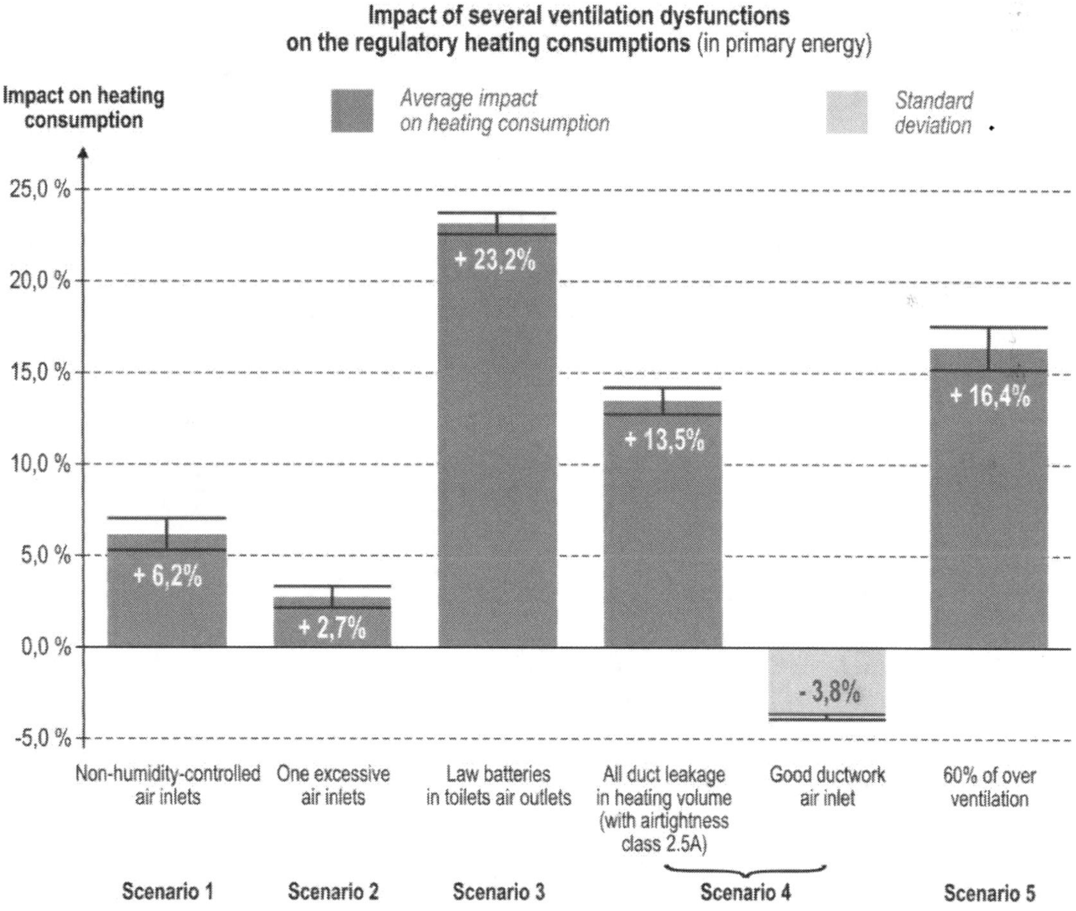

Figure 1: Impact of several ventilation dysfunctions on the regulatory heating consumptions *[in primary energy]*

[4] This coefficient is defined in the French regulation

Whereas different kinds of heating systems with a 2.58-coefficient for electrical source, the standard deviations of the impacts of these dysfunctions are small. Indeed, house 3 (which uses gas-source) has been built in a colder climate, so even with a 1-coefficient for primary energy consumption, the part of heating consumption is almost the same than for the two other houses. It will be interesting to perform the same study with a gas-source house built in a "hot" climate, for which the part of heating consumption should be significantly lower.

Table 4: Impact of several ventilation dysfunctions on the regulatory heating consumption depending on the test house (in primary energy)

	Relative impact on the regulatory heating consumption (primary energy)					
	Scenario 1: Non-humidity-controlled air inlets	Scenario 2: One excessive air inlet	Scenario 3: Low batteries in toilets air outlets	Scenario4: All duct leakage in heating volume	Good ductwork airtightness	Scenario 6 : 60% of over ventilation
House 1	+7.4%	+2.9%	+22.5%	+13.9%	-3.7%	+16%
House 2	+5.6%	+3.3%	+23.4%	+12.6%	-3.7%	+16%
House 3	+5.5%	+2.0%	+23.6%	+14.0%	-4.0%	+18%
Average	**+6.2%**	**+2.7%**	**+23.2%**	**+13.5%**	**-3.8%**	**+16%**
Standard deviation	*0.9%*	*0.6%*	*0.5%*	*0.6%*	*0.1%*	*1.2%*

According to (Jobert, 2013), among the key elements of a ventilation system, air inlet is one of the worse installed. Within those dysfunctions, 18% concern the non-compliance with prescribed rules and regulations and 18% concern the presence of an additional air inlet.

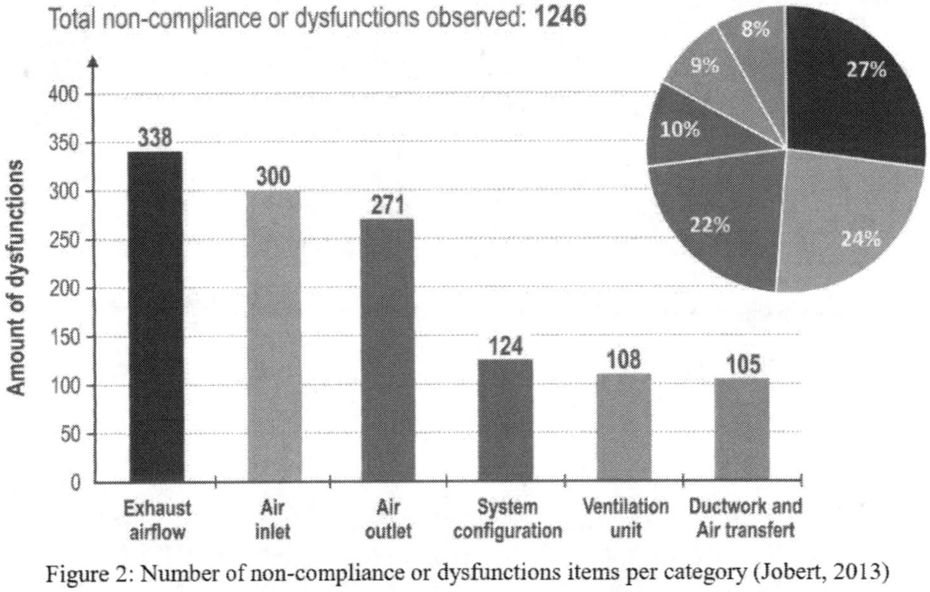

Figure 2: Number of non-compliance or dysfunctions items per category (Jobert, 2013)

Scenario 1 corresponds with an installation of wrong air inlets. Indeed, in France, there are two types of single-flow humidity-controlled ventilation systems called A and B. For the A system, air outlets are humidity and pressure controlled, but air inlets are pressure-controlled only. On the contrary, for the B system, outlets and inlets are humidity and pressure controlled. In most houses, the B system is installed, which induces a lower total incoming airflow than the A systems. Figure 1 shows that this error could increase the heating consumption of about 6%.

Scenario 2 also corresponds with an air inlet dysfunction: it models the implementation of one excess air inlet somewhere in the house. The model of the EP-calculation does not let to precise in which room (utility room or main room) the air inlet is installed. The only effect considered in this study is the increase of the total airflow incoming. With this hypothesis, the excess air inlet induces an overconsumption of almost 3%. This result is not really significant. Nevertheless, depending of the localisation of the excess air inlet, it could induce a short circuit of the house sweep and therefore be responsible of a poor air renewal in some rooms. This energy impact should be supplemented with sanitary impact.

Air outlets are also often affected by dysfunctions (22% of cases). Indeed, in a single humidity- controlled ventilation system, outlet in the toilet are equipped with a system which momentarily increases airflow when someone uses the toilet. In most new houses, the system is a presence sensor running on batteries. The lack of battery is a common dysfunction: it represents 15% of dysfunctions concerning outlets (people often did not know they have to replace them). In this case, the outlet stays in the position where it was when the batteries have stopped: it could be either the lower airflow or the big airflow. Scenario 3 corresponds with the "worst" situation: the batteries of the two toilets outlets are fallen out of order when the outlets were in the big airflow position (it could happen!). In this case, the total exhaust airflow is more important than the predicted one. The impact of this dysfunction is significant: about 23% of over consumption for heating. It could induce an additional cost of more than 60 € each year, while batteries for 2 outlets cost about 10€. This situation is the "worst", but in the other one, when the airflow is blocked in the lower position, the air renewal in the toilets will be insufficient. So, a lack of batteries will either induce an important additional cost each year, or lead to a moisture development in toilets. Therefore, it should probably be more relevant to pay 10€ and change the batteries.

The previous dysfunctions concern an over energy cost when the exhaust airflow is blocked at a high value. Nevertheless, is some cases, the poor quality of the ductwork prevents the airflow from reaching this value. The EP-calculation takes into account this quality with 4 different ductwork airtightness classes. Without any measurement, the worse class has to be used in the calculation. Scenario 4 includes several configuration, with classes varying for different positions of the ductwork leaks (in or out the heated volume). Figure 1 shows that a good ductwork airtightness should induce almost 4% of energy consumption gain. This gain is estimated with the hypothesis of a fixed exhaust airflow at the outlet (independent of the airtightness class). Therefore, with more leaks, there is an additional exhaust airflow through those leaks. With this hypothesis, the impact of the localisation of the leaks compared to the heated volume is significant: for a bad airtightness (class 2.5A) and a duct entirely in the heated volume, the overheating energy consumption could raise more than 13%. The EP-calculation do not estimate the over consumption of the fan needed to secure the airflow at the outlet, so this dysfunction impact might be higher. In practice, the exhaust airflow at the outlets is often not secured, and in this case, the impact of this dysfunction is a sanitary impact with a possible very low exhaust flow. Therefore, a bad ductwork could induce either an important energy consumption or a sanitary issue, or both.

Then, scenario 5, frequent for balanced ventilation system, is less frequent for single-flow ventilation systems: it deals with over ventilation. This dysfunction happens if the fan has not be well set. Various total exhaust airflow have been tested, the most important is a 60% of over ventilation. For this value, the ventilation may be responsible of a 16% increase of the energy consumption for heating, hence the importance of the fan setting.

3.2 How initial financial savings could induce significant losses

Some dysfunctions are due to design and installation, others to maintenance. The following part concerns dysfunctions due to initial financial savings. Indeed, fans and ventilation terminals are available in various models, at various prices. Scenario 6 is related to fan. For each of the three test houses, the fan which is foretold in their thermal study is a high-performance fan: the nominal power is 8 W. Some other lower performance fans exist for this type of ventilation system, for the same delivered airflow. For financial reasons (up to a 100 €-difference between fan prices for an initial 200€-price), it is possible than one of this fan is finally installed instead of the 8 W fan. Table 5 presents the impact of the fan choice on the total energy consumption C_{pe} compared with the initial C_{pe} (with an 8W-fan).

Table 5: Impact of a bad performance-fan on the total energy consumption (primary energy)

Fan Nominal power	**Impact on the C_{pe} (heating consumption)** [kWhpe.m^{-2}.year^{-1}]		
	House 1	**House 2**	**House 3**
12 W	1%	2%	2%
30 W	8%	9%	8%
45 W	13%	15%	14%

Those results prove the importance of the fan choice. Indeed, with a low-performance fan (45 W), the regulatory energy consumption can significantly increase (until 15% for the house 2). In practice, the impact on the energy bill is less important: until 11% (above 45€). Nevertheless, this additional cost for a 8W-fan is low enough to be quickly paid off: 2 years and 3 months only!

Moreover, the impact on the C_{pe} could be critical: for house 2, with a 45W-fan, the recalculated C_{pe} is almost equal to the limit value $C_{pe\ max}$. If an other dysfunction exists, the $C_{pe\ max}$ could therefore be overtaken. Then, the house would not respect the RT2012.

On this point, air inlet are also affected by financial choice. For example, scenario 2 corresponds with an initial 30% saving. In order to evaluate the energy and final economic impact of a first price installation, a combination of scenario 1 and 2 have be modeled (non-humidity controlled air inlet and a 45W-fan). Moreover, a dysfunctions combination is a current situation: according to Jobert, among all controlled dwellings, 29% get two non-compliances or more. Figure 3 presents results of this particular combination.

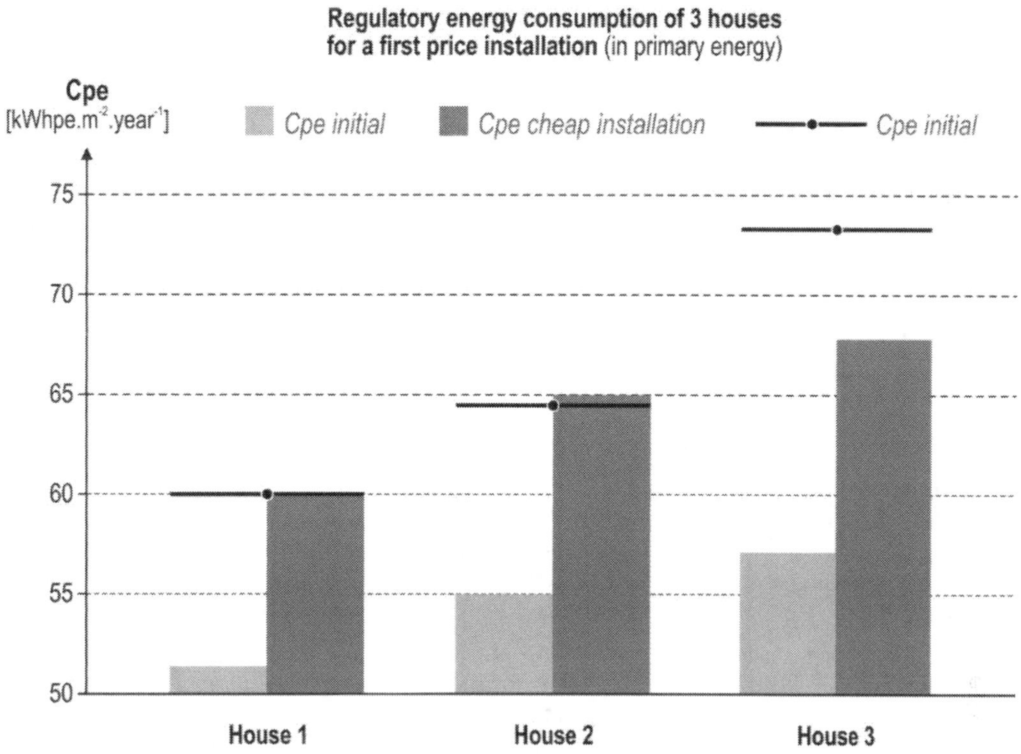

Figure 3: Regulatory energy consumption (primary energy) of 3 houses depending on the ventilation fan performances

In this probable situation, the regulatory C_{pe} recalculated with the real equipment will exceed the limit value C_{pe} max for two houses (1 and 2). In a first hand, energy consumptions would be significantly higher than the predicted ones, and in the other hand, the house would not respect the EP-regulation RT2012! Therefore, make some initial financial savings should rapidly induce many issues, including energy losses and poor indoor air quality.

4 CONCLUSIONS

Ventilation has to be understood as a principle and not just a system. A dysfunction of one element impact the whole buildings, and may increase the total primary energy consumption [Cep] by about 16% (22% of the heating consumption). Not only the energy bill may be significantly increased, but also the house might not respect the EP-regulation requirements. As ventilation systems dysfunctions are a main issue for single-family houses, a scheme as a quality management approach may increase ventilation quality. To that end, the VIA-Qualité project develop tools for each of the three main actors (designer, installers and inhabitants) in order raise awareness among them about the ductwork, products and maintenance quality.

5 ACKNOWLEDGEMENTS

The "VIA-Qualité" project has received funding from the French Environment and Energy Management Agency under the contract ADEME/1304CO014 and from the Rhône-Alpes region. The contribution of CEREMA is funded by the French ministry for ecology, sustainable development, and energy (MEDDE).

The sole responsibility for the content of this publication lies with the authors.

6 REFERENCES

JO. (1983). Arrêté du 24 mars 1982 consolidé relatif à l'aération des logements : aération générale ou permanente. *JO du 15 novembre 1983*.

Paul Van Den Bossche, A. J. (2013, Septembre 25-26). Securing the quality of ventilation systems in residential buildings: existing approaches in various countries. *34th AIVC Conference, Athens, Greece*, pp. 70-72.

Romuald Jobert, G. G. (2013, September 25-26). Detailes analysis of regulatory compliance controls of 1287 dwellings ventilation systems. *34th AIVC Conference, Athens, Greece*, pp. 74-76.

Sandrine Charrier, J. P. (2013, September 25-26). Airtightness quality management scheme in France: assessment after 5 years operation. *34th AIVC Conference, Athens, Greece*, pp. 42-44.

Wargocki P, H. T. (2012, October 11-12). Health-Based ventilation Guidelines for Europe - HealthVent project. *33th AIVC Conference, Copenhagen, Denmark.*

DEMAND-CONTROLLED VENTILATION

20 YEARS OF IN-SITU MONITORING IN THE RESIDENTIAL FIELD

Jean-Luc Savin[1], Stéphane Berthin[1] and Marc Jardinier[1]

1 Aereco
62 rue de Lamirault
Collégien
77615 Marne la Vallée cedex3
France

ABSTRACT

Is Demand-controlled ventilation a relevant answer to face the new challenges of the Building sector, which requires everyday higher energy efficiency and better indoor air quality? Can Demand-controlled ventilation be considered as an alternative to heat recovery ventilation, through an affordable and low maintenance solution? Since the take off of the DCV in the early 80's, these questions have been considered many times. Following the expansion of the technology over the world (more than 4.5 million dwellings are already equipped today, only considering humidity controlled ventilation), numerous in-situ monitoring have been conducted to try to better understand the behaviour and the performance of the DCV. Precursor in the field as inventor of humidity controlled ventilation in 1983, Aereco has carried out numbers of monitoring on different ventilation types, in various countries and on specific buildings, measuring in total hundreds of dwellings. With a first large-scale study based on natural ventilation in multi-storey buildings realised in 1993, the research has been completed with experiments on hybrid ventilation and on mechanical exhaust ventilation, individually or collectively managed, to recently lead to the follow-up of on an innovative full room-by-room DCV balanced system with heat recovery.

This proceeding will specially focus on four major monitoring: demonstration project "EE/166/87" (natural ventilation, France, Belgium and Netherlands, 1991-1993), "HR-VENT" (hybrid ventilation, 55 dwellings, France, 2004-2005), "Performance" (collective MEV, 29 dwellings, France, 2008-2009) and the last monitoring that concerns several houses in Germany and France equipped with a full DCV heat recovery (started in 2013).

In parallel to the ventilation systems evolution, the one of the measurements methods and instruments has been remarkable since the first monitoring. From tracer gas to collect the airflows at the first monitoring, we have reached today a level of technology that enables more reliable and more precise multi-parameters measurements through electronic sensors, sending data via internet.

Whatever the ventilation technique it is associated with, the automatic control of the airflows according to the demand has demonstrated benefits at various levels. The first monitoring highlighted the correlation between CO2 and humidity, even in the technical rooms, conferring relevancy to humidity controlled at the exhaust. On all monitoring, the seasonal behaviour of the system when humidity controlled has been found out: following the lower absolute humidity level in winter, the airflows are reduced when the dwellings are unoccupied, leading to energy savings on both the heating and on the fan electrical consumption. On the indoor air quality side, we have checked that the automatic control to the demand enables to optimise it: the system improves the repartition of the airflows among the rooms according to their specific needs, increasing the air renewal in the most occupied and polluted rooms. In addition, DCV monitoring have shown specific phenomena such as the ability to optimise the available pressure through a time-repartition of the exhaust demand dwelling-to-dwelling, when connected to a collective duct. Besides parameters measurements and data collection we brought specific care on occupant's acceptance, making surveys every time it was possible. The results have shown that the system was in general very well accepted, highlighting the relevancy of non intrusive and low maintenance systems.

KEYWORDS

Monitoring, Demand-controlled Ventilation, Humidity controlled, Indoor Air Quality, Energy Savings.

1. INTRODUCTION

Since the take off of the Demand-controlled Ventilation in the early 80's, the question of its performance and of its ability to position as an alternative to the heat recovery ventilation has been considered many times. Following the expansion of the technology over the world -more than 7 million dwellings are already equipped today only considering humidity controlled ventilation-, numerous in-situ monitoring have been conducted to better understand the behaviour and the performance of the DCV. Precursor in the field as inventor of humidity controlled ventilation in 1983, Aereco has carried out numbers of monitoring in residential buildings on different ventilation types, in various countries, measuring in total hundreds of dwellings. This paper aims at giving a global overview of the major monitoring that have been hold on Demand-controlled ventilation systems in residential application, dealing with different topics such as indoor air quality, energy efficiency and evolution of measurement instruments

2. 20 YEARS OF MONITORING

Monitoring one or several dwellings to assess the performance of a specific building equipment always raises the question of the significance and of the relevancy of the study: how to be sure that these dwellings are representative, so that we can approach the statistical and average behaviour of a system through the follow up of a limited set of dwellings? The multiplication of monitoring campaigns as well as the quantity of monitored dwellings is the only way to evaluate as best the performance of a ventilation system. As application of this principle, Aereco has conducted numerous in-situ experiments on various places testing its Demand-controlled ventilation systems in the residential field, as presented on Figure 1.

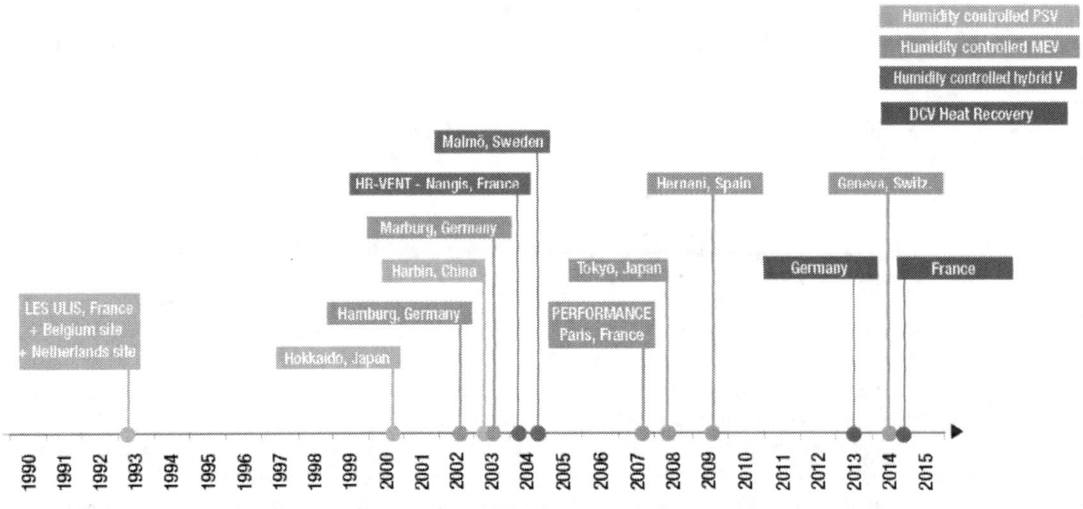

Figure 1: Historic presentation of Aereco major monitoring campaigns on ventilation (Starting dates).

Figure 2: One of the five buildings monitored in "HR-VENT" project (left); One of the two buildings monitored in "Performance" project (right)

3. EVOLUTION OF MEASUREMENT METHODS AND MEANS

The precision and the quality of results of in-situ experiments are very dependent from the sensors and the methodology used. Since our first monitoring of DCV realised in 1993 up to now in 2014, the evolution of the means has been continuous and strong, improving the accuracy of the measurements and enlarging the spectrum of measurable parameters.
From tracer gases used in 1993 in "Les Ulis" monitoring giving punctual measurements up to electronic sensors generalised in the latest monitoring, the airflow measurements have become much more precise and reliable, offering possibilities of a permanent follow up that can be applied to a large set of dwellings.
The data collection has also been considerably eased with the development of WiFi and internet: a recent in-situ experiment applied on a new Demand-controlled heat recovery ventilation system now includes climatic sensors that are directly connected to the local WiFi, sending the data through internet.

4. DEMAND-CONTROLLED VENTILATION SYSTEMS

The typical system measured during the various monitoring we are talking about in this paper is an unbalanced system with humidity controlled air inlets (Figure 3) in the bedroom and living room, humidity and / or presence detection exhaust units (Figure 4) in the wet rooms, and a centralised fan connected to these units. These systems are presented on the following schemes for hybrid ventilation (Figure 5) or for mechanical exhaust ventilation (Figure 6). In these systems, the fan has the singularity to provide a constant pressure for a varying airflow.

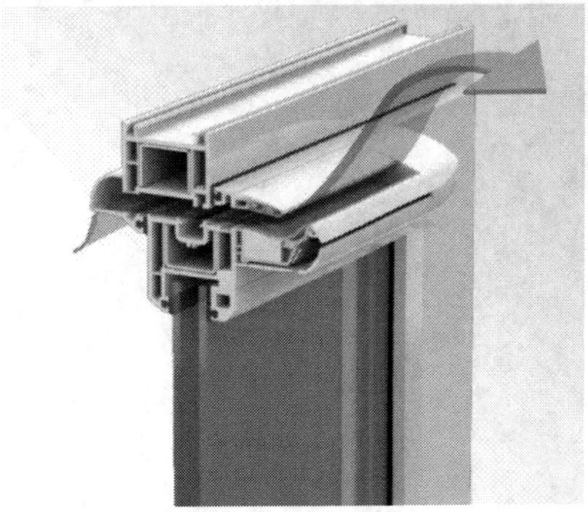

Figure 3: Humidity controlled air inlet

Figure 4: Demand-controlled exhaust unit (MEV)

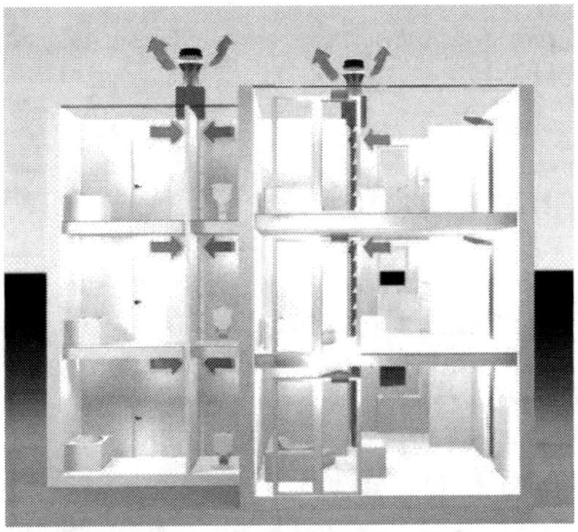

Figure 5: System scheme in hybrid ventilation

Figure 6: System scheme in mechanical exhaust ventilation (MEV)

5. IN-SITU WORKING OF DEMAND-CONTROLLED VENTILATION SYSTEMS

5.1 Instantaneous behavior

One of the first objectives of a monitoring on ventilation is to check that the system in-situ working conforms to the laboratory measurements. The dynamic working of the Demand-controlled ventilation, with its airflows constantly moving according to the humidity or to the presence[1] has been measured in most of the experiments as a key result.

[1]for the majority of the products in Aereco's range

HR-VENT monitoring (humidity controlled system with hybrid ventilation) has given the opportunity to check the working of the extract grilles. Figure 7 presents the day variation of the aperture according to the relative humidity for a humidity sensitive extract grille located in a bathroom. We can see that the variations of aperture –so the airflow for a given pressure- follow well the evolution of the relative humidity, with a high reactivity.

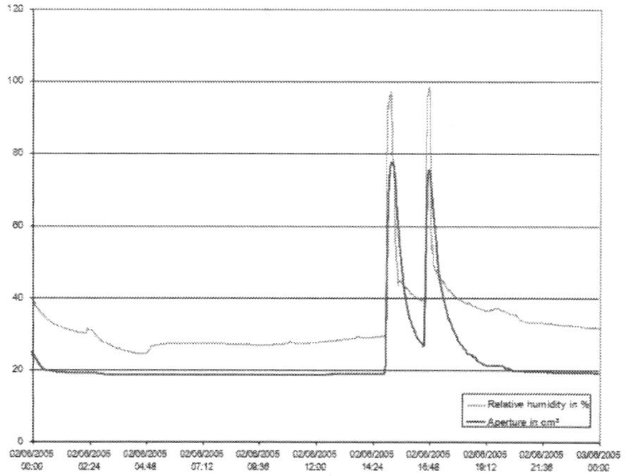

Figure 7 : Aperture vs. relative humidity of a humidity sensitive extract grille in the bathroom. "HR-VENT" project.

The working of the system has also been observed at the level of the air inlets during the "Performance" monitoring (humidity controlled system in mechanical exhaust ventilation). The scheme Figure 8 shows that the airflow at 10 Pa (noted "Qea 10 Pa") of the humidity controlled air inlet follows quite well the variations of the relative humidity (RH) in a bedroom.

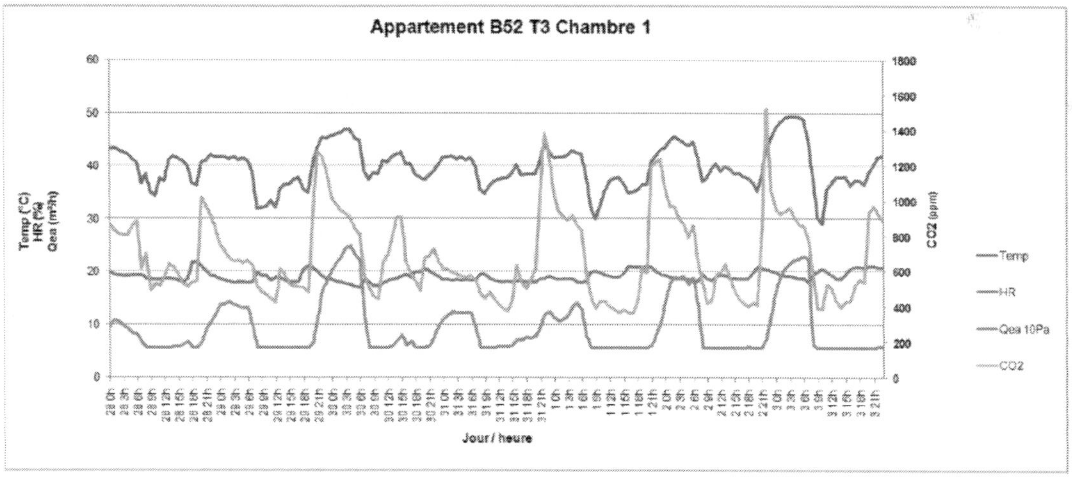

Figure 8 : Airflow @ 10 Pa ("Qea 10 Pa"), Relative humidity ("HR"), CO2 and Temperature in a bedroom equipped with a humidity controlled air inlet during one week. "Performance" project.

Microscopic and macroscopic behavior

Long term monitoring campaigns enable to highlight a major phenomenon inherent to the humidity controlled ventilation: its statistical seasonal behavior. On the chart Figure 9 we can compare a microscopic view and a macroscopic view of the aperture of the humidity controlled extract grille for different climatic seasons. Due to the evolution of the average indoor relative humidity along the year, the grille is mainly closed in cold season; in hot season the grille aperture opens wider. In parallel to this statistical behavior, the charts on the left column remain that the humidity controlled extract grille always reacts to the humidity giving punctually a high airflow when needed, in cold as in hot season. Only the average aperture varies according to the thermal season during the year. This lower airflow level during the cold season allows to save energy on thermal losses due to ventilation.

Instantaneous / punctual behavior (on day)	Statistical behavior (two months)

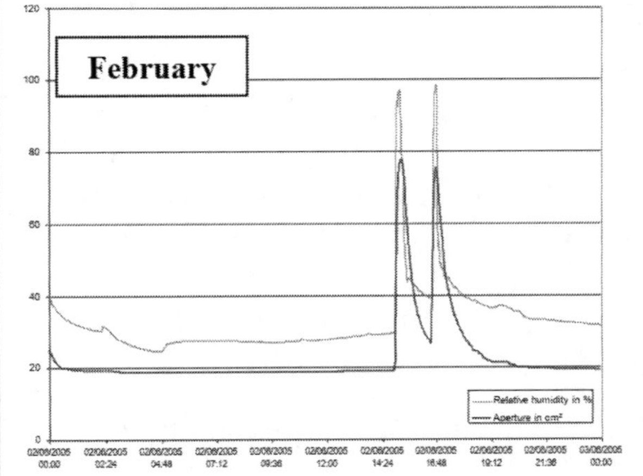

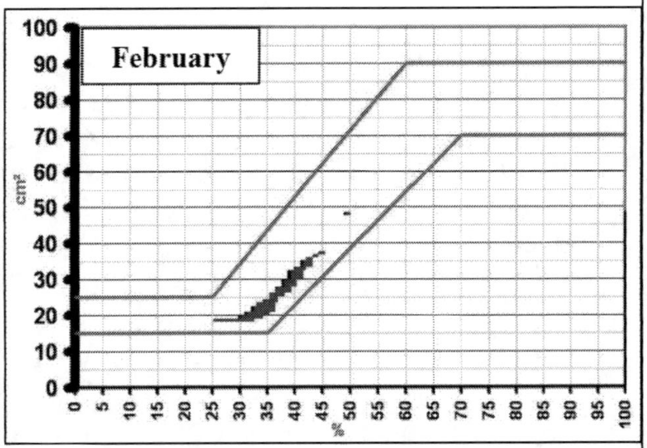

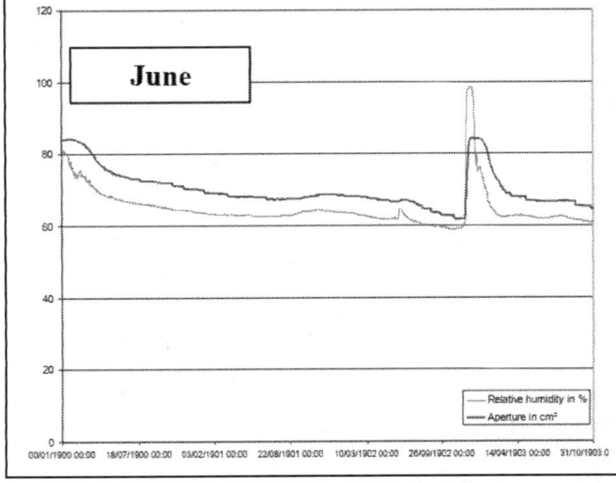

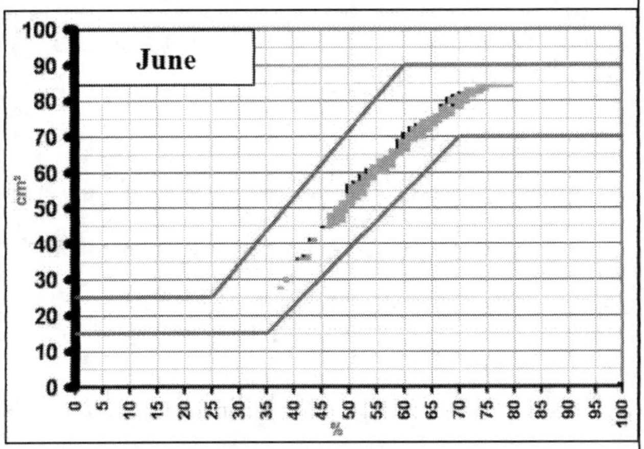

Figure 9 : Microscopic day-view and macroscopic monthly view of the variation of Aperture Vs Relative humidity of a humidity sensitive extract grille in a bathroom. HR-VENT project.

6. INDOOR AIR QUALITY

6.1 IAQ in the technical rooms

From 1993, a first monitoring ("Les Ulis", humidity controlled passive stack ventilation) gave the evidence that CO_2 elevations are correlated with H_2O ones in the technical rooms. Either humidity and CO_2 evolve through a linear –or close to- scheme (case of the toilets – see Figure 10), or H_2O increases more than CO_2 (in the bathroom or in the kitchen – see Figure 11 and Figure 12). Considering this phenomenon, driving the exhausted airflow according to humidity offers a guarantee of well reacting to potential elevations of CO_2, whatever the technical room, giving full consistency to the humidity controlled extract grilles.

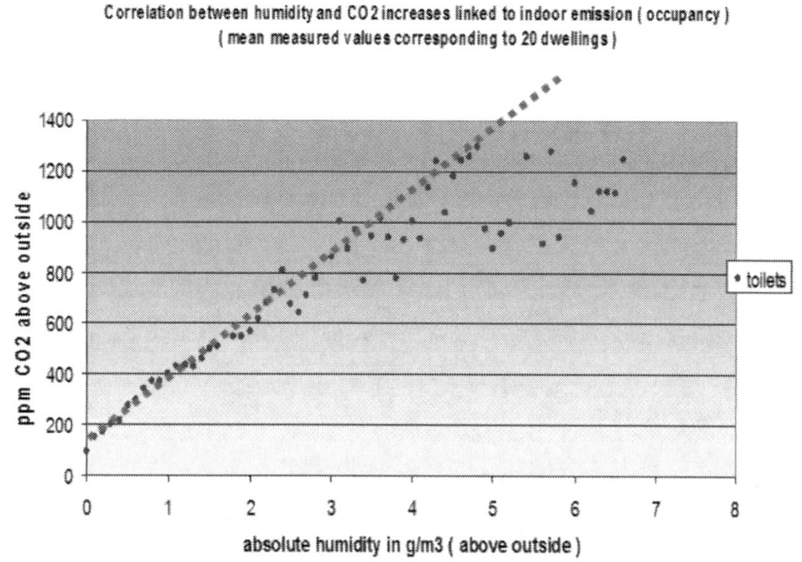

Figure 10 : Correlation between elevations of CO_2 and absolute humidity in the toilets

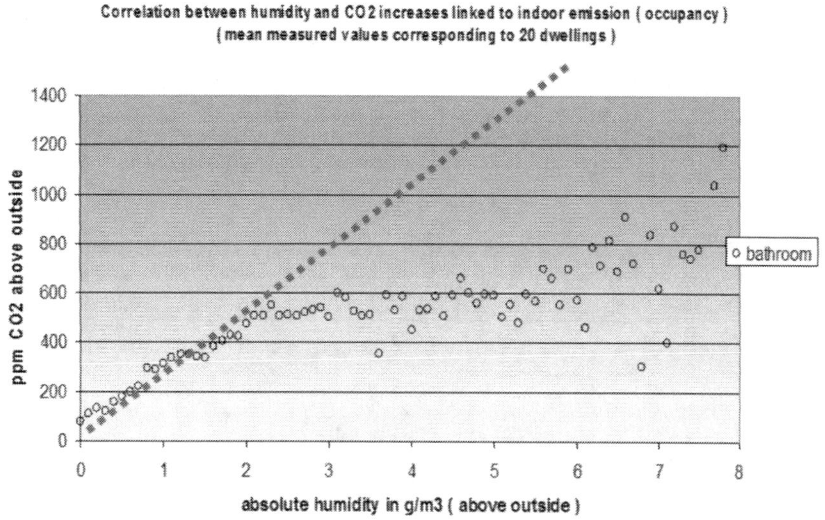

Figure 11 : Correlation between elevations of CO_2 and absolute humidity in the bathroom

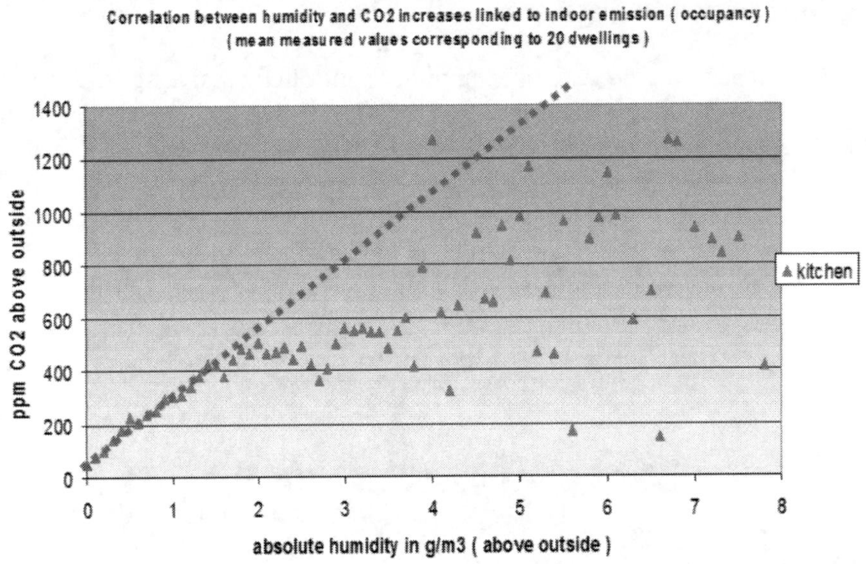

Figure 12 : Correlation between elevations of CO_2 and absolute humidity in the kitchen

6.2 IAQ in the main rooms

"Performance" project has given complementary results at the level of the humidity controlled air inlets. The measurements of CO_2 in two different bedrooms equipped with humidity controlled air inlets during one year show that the indoor air quality can be ensured in a low occupied bedroom (with one adult) as well as in a high occupied one (with four adults) as presented Figure 13. The peak of CO_2 concentration has shifted from 700 ppm in the low occupied bedroom to 950 ppm in the highly occupied one, but even in that case, the 1500 ppm level is not exceeded more than a very few hours.

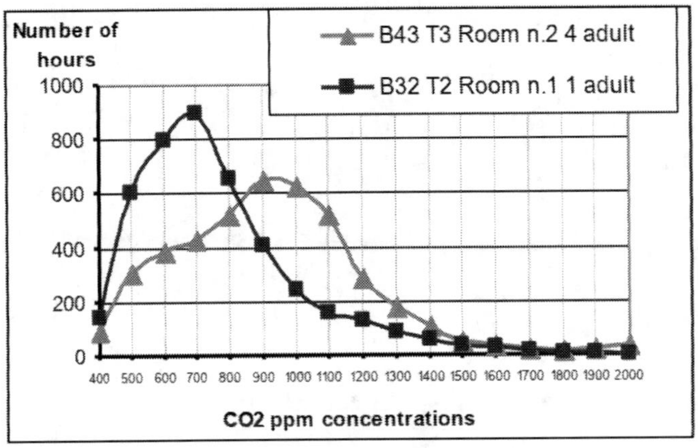

Figure 13 : CO_2 concentrations (absolute) in two bedrooms with different occupations (blue - square: 1 adult, red triangle: 4 adults).

7. ENERGY PERFORMANCE

7.1 Energy losses due to the air renewal

The Figure 14 presents the average measured equivalent airflow for energy[1] per dwelling during a complete heating season.

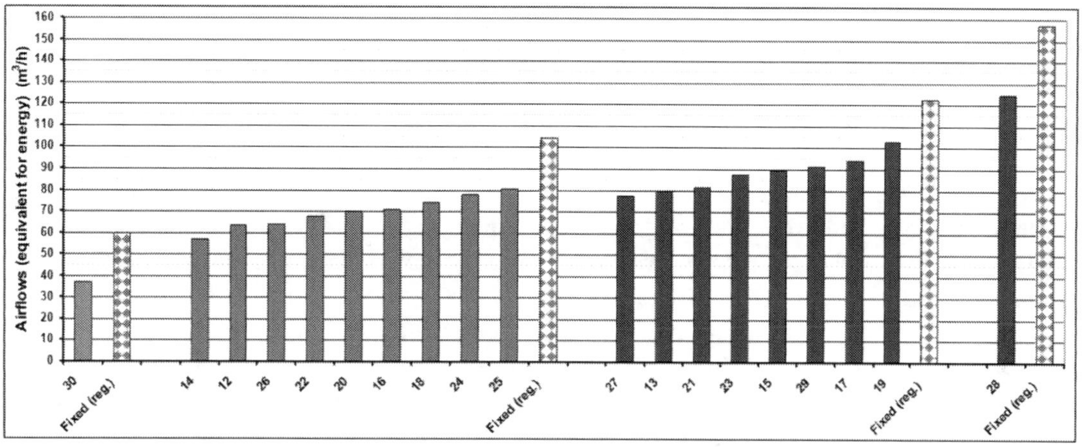

Figure 14 : Statistical equivalent airflows for energy per dwelling on Paris site. Rated by dwelling types. Comparison with French regulatory constant airflow (grey squares). 2007-2008 heating period.

The disparity of the measured equivalent airflows results from the adaptation of the ventilation system to various occupations, activities, occupant behaviours and dwelling sizes. The comparison with the French regulatory reference (fixed airflow, grey triangles bar) shows the statistical airflow reduction –thus the heating energy savings- given by the DCV system. The measured savings on the airflow for this project have been evaluated at 30% in average (55% if extrapolating to the statistical average occupancy in France). It is important to understand that this statistical airflow reduction does not affect the IAQ as the system still reacts to punctual high needs with high airflows, as discussed before in the paper.

7.2 Fan consumption

An additional advantage of the Demand-controlled ventilation is to reduce the average power consumption of the fans by reducing the average global exhaust airflow. Through "Performance" project, the measurements have shown that the energy consumption of the fan has been reduced between 35% to 50% in comparison with the French reference for a constant airflow ventilation system.

1 The equivalent airflow for energy corresponds to the fix equivalent airflow in terms of heat losses through ventilation. It takes into account the indoor-outdoor temperature difference.

8 CONCLUSIONS

As a specialist of demand-controlled ventilation, Aereco has carried out numerous in-situ monitoring in the residential field to evaluate the performance and the behaviour of its ventilation systems since 1993. The evolution of the technology for measurements devices has offered new opportunities in the experiment field for larger, more accurate and multi-parameters monitoring. The working of the demand-controlled ventilation systems has been validated in real occupancy conditions, notably the one of the humidity controlled ventilation, at the level of the air inlets as well as for the exhaust units.

Long term monitoring campaigns enable to observe a major phenomenon inherent to the humidity controlled ventilation: its statistical seasonal behavior. Giving naturally an average airflow lower in winter than in summer, the humidity controlled ventilation saves energy on heat losses, from 30% to 55% in the measured dwellings. The monitoring campaigns have also demonstrated that the instantaneous airflow can be very high when needed with this system, even in winter where the average airflow is low. The indoor air quality is then optimized at every moment. A complementary advantage of the demand-controlled ventilation is to reduce the average fan power consumption, as a consequence of the low statistical airflow induced by the system. The very positive results from the numerous monitoring conducted allow to consider the demand-controlled ventilation as a real alternative to the heat recovery ventilation systems in the residential field.

In-situ monitoring are and will always be the most effective way to assess and to validate the real performance of ventilation systems, moreover when they are innovative: the occupants behavior and the dwelling configuration can hardly be strictly repeated in laboratory.

9 REFERENCES

Air.H. (2009). *"Performance de la ventilation et du bâti" Final report (2009).* Ademe convention n.0504C0114.

Jardinier, M. (2006). *"Demand controlled ventilation : conciliating IAQ and energy savings."* Moscow, Cold Climate 2006 conference.

Savin J.L., Berthin S. and Jardinier M. (2005). *"Assessment of improvements brought by humidity sensitive and hybrid ventilation/ HR-Vent project"*. 26th AIVC conference, Brussels.

Savin, J.L., Jardinier, M. and Siret, F. (2004). *"In-situ performances measurement of an innovative hybrid ventilation system in collective social housing retrofitting"*. 25th AIVC conference.

Savin J.L and Bernard A.M., 2009. *"Performance" project: Improvement of the ventilation and building air tightness performance by implementing a quality approach when building and monitoring of results in occupied dwellings in France.* AIVC Conference 2009.

Siret, F., Savin, J.L., Jardinier, M. and Berthin, S. (2004). *" Monitoring on hybrid ventilation project - first results."* 25th AIVC conference.

Siret, F. and Jardinier, M. (2004). *"High accuracy manometer for very low pressure."* 25th AIVC conference.

Wouters, P., Geerinckx, B, L'Heureux, D., Simonnot, J. and Phaff, J. (1993). *"Passive humidity controlled ventilation for existing dwellings"* - Demonstration project EE/166/87. December 1993

SUMMER PERFORMANCE OF RESIDENTIAL HEAT RECOVERY VENTILATION WITH AN AIR-TO-AIR HEAT PUMP COOLING SYSTEM

Bart Cremers

Zehnder Group Nederland
Lingenstraat 2
8028 PM Zwolle, The Netherlands
bart.cremers@zehndergroup.com

ABSTRACT

Increasing airtightness and isolation of residential buildings in today's climates cause challenging situations for the summer indoor climate. In combination with ventilation for fresh air, it calls for intelligent control of passive cooling when available, and active cooling when needed.

The combination of heat recovery ventilation and an air-to-air heat pump cooling system is a solution to these challenging situations. With the exhaust air heat pump cooling system, heat is transferred from the supply air (which is getting colder) to the exhaust air (which is getting warmer).

Such a ventilation system is monitored throughout the summer for an actual installation in The Netherlands. The control of such a system depending on the actual indoor and outdoor conditions is explained in a control diagram.

Correlation diagrams show how the ventilation supply air temperature and humidity varies with outdoor temperature in accordance with the passive or active cooling mechanism. It has been shown that the sensible cooling can be doubled with active cooling when compared to passive cooling. The total (sensible + latent) cooling of the air-to-air heat pump amounts up to 1100 W with 150 m^3/h and up to 1700 W with 270 m^3/h.

Limitations of the technology are explained with diagrams for conditions where the condenser gets too hot. For these situations the ventilation air flow rate is first automatically increased to allow the condenser to cool. If necessary, the heat pump is shut off intermittently to prevent damage to the condenser.

KEYWORDS

Residential ventilation, indoor air quality, ventilative cooling, air-to-air heat pump

1. INTRODUCTION

Increasing airtightness and isolation of residential buildings in today's climates cause challenging situations for the summer indoor climate. In combination with ventilation for fresh air, it calls for intelligent control of passive cooling when available, and active cooling when needed.

The combination of heat recovery ventilation and an air-to-air heat pump cooling system is a comprehensive solution for bringing fresh air into a residential building for a good indoor air quality. More than only bringing fresh air, the heat recovery saves energy for heating the ventilation demand in the cold season. In the warm season, the addition of the air-to-air heat pump gives top cooling and dehumidification via the supplied fresh air. The working principle and the practical performance are reported in this article.

2. HEAT RECOVERY WITH AIR-TO-AIR HEAT PUMP COOLING

The working principle of heat recovery ventilation with an air-to-air heat pump cooling system is explained using fig. 1. A fan is extracting air from the building (ETA) through a heat exchanger that is transferring the heat (in winter) or the cold (in summer) to the supplied fresh air. Under favorable conditions, the extracted air is bypassing the exchanger when heat recovery is not needed. The extracted air is leaving the building as exhaust air (EHA).

Another fan is bringing fresh outdoor air (ODA) into the building through a heat exchanger where it is transferring heat (in winter) or cold (in summer). Before this so-called pre-supply air (P-SUP) is supplied in to the rooms, it can be cooled by the air-to-air heat pump. This heat pump brings energy from the supply air (which gets colder) to the exhaust air (which gets warmer). Eventually the fresh air is supplied to the rooms (SUP) in a comfortable way.

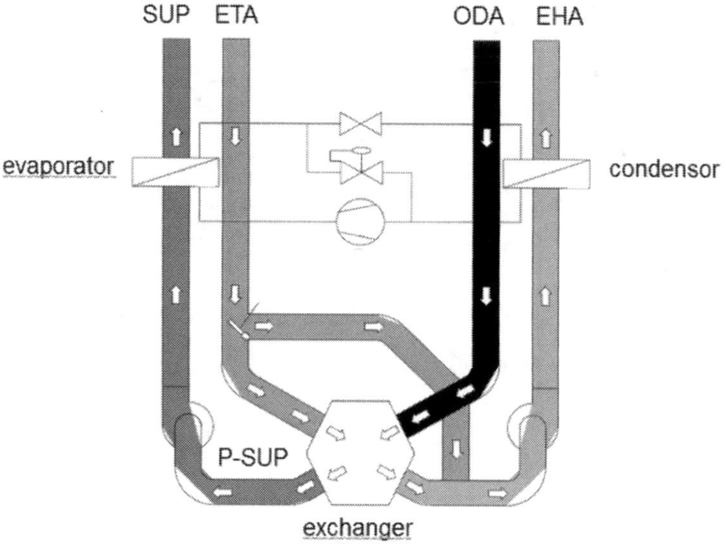

Figure 1: Schematic working principle with outdoor air (ODA), pre-supply air (P-SUP), supply air (SUP, extract air (ETA) and exhaust air (EHA).

The various states that this ventilation system can take are explained using the state control diagram in fig. 2. Depending on the outdoor temperature and indoor temperature (as measured in the outdoor air ODA and the extracted air ETA, respectively) and the setting of the comfort temperature, a specific state (expressed in colors) is entered. Example values of temperatures are given in italics for better understanding. The bypassing of the heat exchanger (to switch off heat recovery) is expressed by BP and the air-to-air heat pump is expressed by CC.

With low outdoor temperature (typically below 13 °C), the bypass is always closed, even when the indoor temperature is higher than the comfort temperature. This is to prevent that very cold air flows through the supply air ducts and produces condensation on the outside of the ducts in the house[1]. Above 13 °C outdoor temperature, and when the indoor is above the

[1]Although ventilation systems with air-to-air heat pump cooling have insulated supply air ducts, this is the standard algorithm for products without heat pump cooling

setting of the comfort temperature, first the bypass is opened to allow the cool fresh air to enter the building. When the outdoor air is not sufficiently cool anymore with respect to the comfort temperature, the air-to-air heat pump is automatically switched on to provide cooling of the fresh air.

When the outdoor temperature is higher than the indoor temperature, the bypass is always closed to take benefit of the lower indoor air. Therefore, heat recovery (or in this case: cold recovery) precools the incoming fresh air to a level close to the indoor air temperature. When the indoor air temperature gets too warm, i.e. above the setting of the comfort temperature, the air-to-air heat pump is automatically switched on to further cool the supply of fresh air.

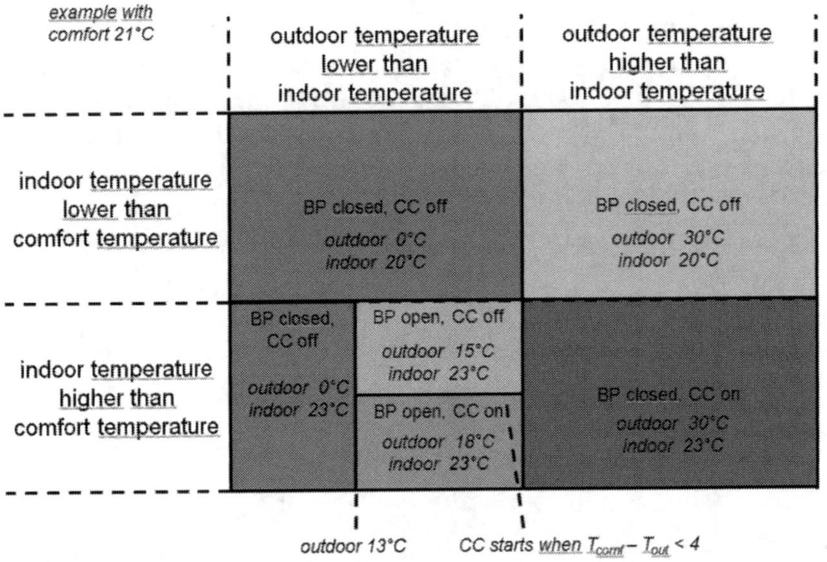

Figure 2: State control diagram

3. MONITORING OF SUPPLY AIR

During the summer of 2012, from the end of July until the beginning of October, a monitoring campaign was held in a house in Emmeloord, The Netherlands. The heat recovery ventilation system combined with an air-to-air heat pump cooling system was mostly used in middle fan speed or high fan speed, with fresh air flow rates 150 m^3/h and 270 m^3/h respectively. The setting of the comfort temperature was initially set to 23.5 °C, but the residents varied the setting during the monitored period between 23 and 24 °C according to their own needs.

The temperatures of the air streams ODA, P-SUP, SUP, ETA and EHA are measured internally in the ventilation unit. Additional Sensirion humidity sensors were added in the ducts to record the humidity of the air streams ODA, P-SUP, SUP and ETA. Fan percentages, the setting of the comfort temperature and the state of the ventilation system was also recorded internally on the PCB of the unit. The recordings were saved at an interval of 5 min and afterwards transformed into 1 hour average values for further analyzing.

Figure 3 shows a correlation diagram between the temperature of the supply air SUP and the temperature of the outdoor air ODA. The hour average values appear in the chart as groups of points that are indicated in accordance with the various states of the ventilation system from the state control diagram.

With closed bypass and outdoor temperatures below indoor temperature, the supply air temperature is at a comfortable level between 16 °C and 23 °C, in relation to the actual indoor air temperature because of the heat recovery. The average practical heat recovery efficiency based on the supply air temperature during the monitoring period is 87%. The heat recovery in this period saves heating energy for the central heating system of the house, indicated as avoided heating.

With closed bypass and outdoor temperature above indoor temperature, the supply air temperatures are also held close to the indoor air temperature because of the heat (cold) recovery. This gives a comfortable supply of fresh air and reduces the heat load for the building. This state is however not occurring often as the indoor air temperature rises to a level above the comfort temperature where the air-to-air heat pump is started.

When the indoor air temperature rises above the comfort temperature (and outdoor air temperature is above 13 °C), cooling is requested. With enough cooling capacity with the outdoor air, passive cooling is started (bypass open, heat pump off), resulting in supply air temperature 1 to 2 °C above the outdoor air temperature. The small rise in temperature is the effect of heat gains in outdoor air ducts and heat dissipation by the supply fan.

If the outdoor air itself had not enough cooling capacity, the air-to-air heat pump is started with open bypass to give a supply air temperature ranging between 10 °C for relatively low outdoor temperature to about 16 °C for outdoor temperature close to indoor air temperature. In this state, the ventilation system is giving passive cooling and active cooling at the same time, resulting in comfortable cooling throughout the house with the minimum of cooling energy.

When the outdoor air temperature is above the indoor air temperature, the bypass closes so that the incoming fresh air is first reduced in temperature by heat (cold) recovery and afterwards by the air-to-air heat pump. This results in supply air temperatures ranging from 12 °C to 22 °C depending on outdoor air temperature.

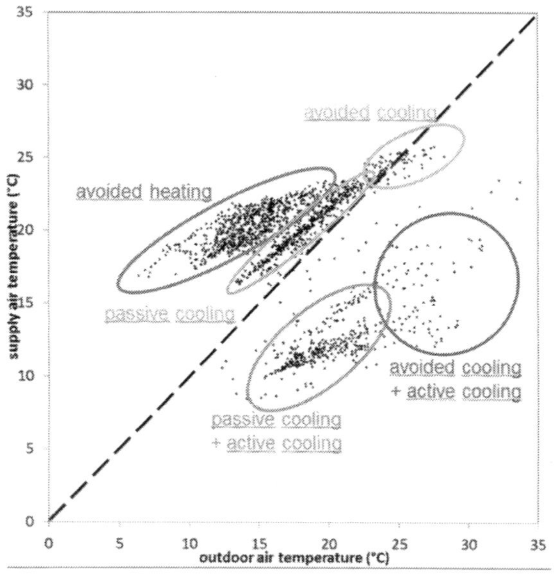

Figure 3: Monitoring results of supply air vs. outdoor air.

Figure 4 shows the correlation between the supply air and the indoor air, both in terms of temperature and absolute humidity. The sensors for this graph are positioned at another place as the internal sensors, so the values differ slightly. Without any cooling, the supply air is close to the indoor air, thanks to heat recovery. With passive cooling (heat pump off), the fresh air is supplied with a temperature 1 to 6 °C lower than the indoor air temperature. With active cooling (heat pump on), the fresh air is supplied with a temperature 6 to 12 °C lower than the indoor air temperature. Therefore, one could say that the sensible cooling power for the indoor air with active cooling can be doubled when compared to passive cooling.

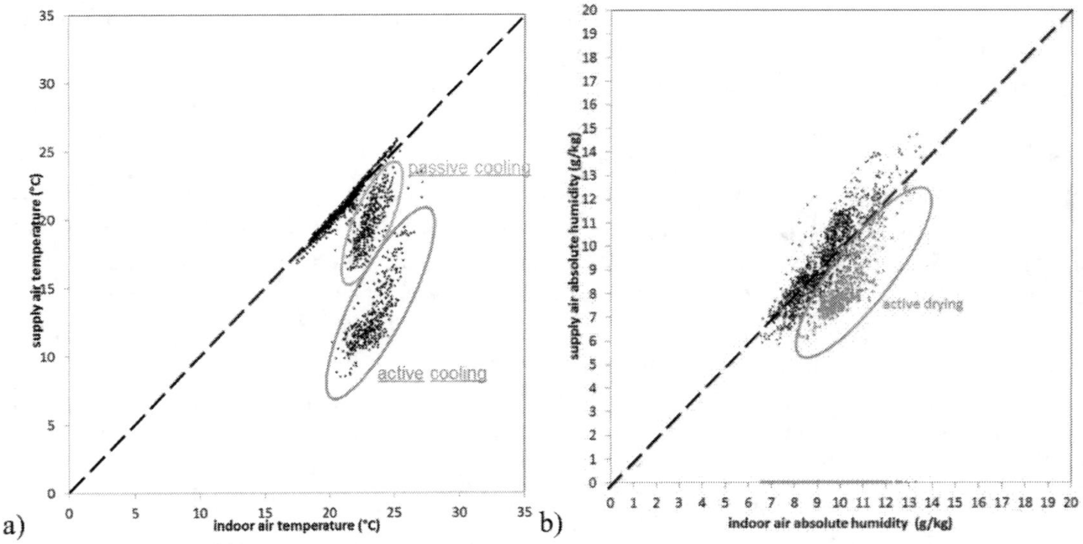

Figure 4: Monitoring results of supply air vs. indoor air: a) temperature, b) absolute humidity.

Figure 4b shows the absolute humidity of the indoor air and the supply air. When the heat pump is on, the dots are colored blue. It is obvious that the air-to-air heat pump is also supplying the fresh air with lower humidity than the indoor air (difference up to 3.5 g/kg), maintaining comfortable summer humidity levels in the house.

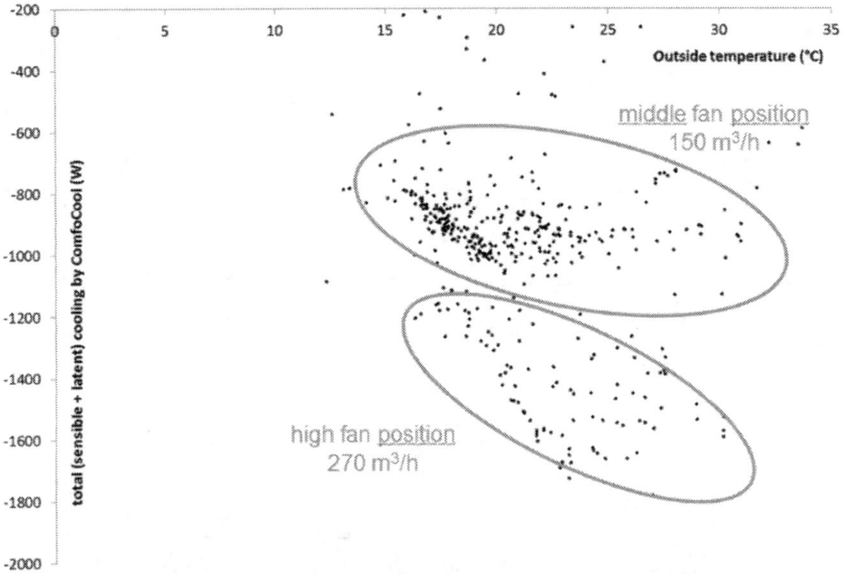

Figure 5: Total cooling power

The total cooling power (sensible + latent) of the heat pump has been calculated for the air-to-air heat pump by the difference in conditions in front of the evaporator (P-SUP) and after the evaporator (SUP) of the heat pump. Figure 5 shows the total cooling power in relation to the outdoor air temperature. Depending on actual temperature and humidity conditions, the total cooling power ranges between 700 and 1100 W for middle fan position (150 m^3/h) and ranges between 1200 and 1700 W for high fan position (270 m^3/h). The power consumption of the system was not monitored in this project. According to the specifications of the product, the heat pump has a power consumption of 800 W when active.

With the help of this cooling system, comfortable fresh air is supplied into the rooms, keeping the indoor temperatures throughout the whole house at a maximum of 26 °C as indicated by figure 4a.

4. WORKING AREA

A technical limitation of the air-to-air heat pump is to prevent damage to the condenser when it gets too hot. If the condenser temperature is reaching a safety limit, the fresh air flow rate is first increased to the maximum. If the safety limit is reached with maximum air flow rate, the heat pump is shut off intermittently to allow temporary cooling down of the condenser.

This effect is shown in figure 6 where the condenser temperature is shown in relation to the outdoor air temperature. With the heat pump shut off, the condenser is close to the outdoor air temperature. When the heat pump is on, the condenser takes temperatures roughly between 35 °C and 57 °C for outdoor air temperatures ranging from 15 °C to 30 °C. For outdoor temperature above 30 °C, the hourly values of the condenser temperatures start decreasing because of intermittent switching on and off of the heat pump.

Although the monitoring period did not show very warm periods, extrapolation of the monitoring data indicates that for high outdoor temperatures the air-to-air heat pump will not be active anymore. The combination of heat recovery ventilation and air-to-air heat pump cooling system is therefore intended for climates where the summer outdoor air temperatures are not often above 32 °C. This limitation holds only for the selected refrigerant and heat pump parts in the product used in this monitoring study.

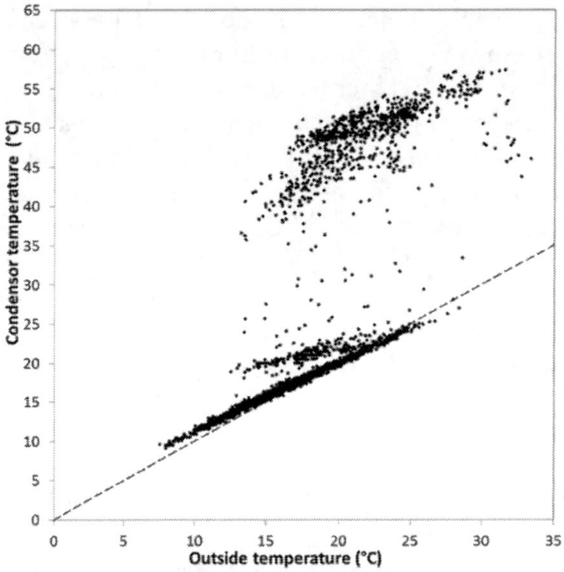

Figure 6: Condenser temperature

5. CONCLUSIONS

The heat recovery ventilation system in combination with an air-to-air heat pump cooling system is investigated by a summer monitoring period in a house in Emmeloord, The Netherlands. The ventilation system can act in various states that are explained by a state control diagram.

The monitoring results show that the various states produce a comfortable supply of fresh air throughout the whole building. When cooling is requested, the system automatically enters the optimal cooling strategy: passive cooling when possible, and active cooling when necessary.

The active cooling strategy can double the sensible cooling power for the house when compared to the passive cooling strategy, and it can be used in a wider range of conditions. The total cooling power of the air-to-air heat pump cooling system can reach up to 1100 W at 150 m^3/h and up to 1700 W for 270 m^3/h.

The total ventilation and cooling system has proven to provide a comfortable summer indoor climate with indoor air temperatures at maximum 26 °C for outdoor air temperatures up to 30 °C.

Opposite to circulation air-conditioning units, this ventilation/cooling combination not only brings fresh air to the house, but also provides comfortable indoor climate in all of the ventilated rooms in the house.

HEATING "PASSIVE HOUSE" OFFICES IN COLD CLIMATE USING ONLY THE VENTILATION SYSTEM – COMPARISON OF TWO VENTILATION STRATEGIES

Hugo Lewi Hammer[1*], Mads Mysen[2], Axel Cablé[2], Kari Thunshelle[2]

1 Oslo and Akershus University
College of Applied Sciences
Pilestredet 35
N-0166 Oslo
Norway
**Corresponding author: hugo.hammer@hioa.no*

2 SINTEF Building and Infrastructure
Forskningsveien 3 b
N-0314 OSLO
NORWAY
Address, Country

ABSTRACT

In this article we compare to ventilation strategies to heat a "passive house" office building using only the ventilation system. Two ventilation strategies with supply air temperature above and below the current room temperature were compared through a cross over experiment. A questionnaire was used to measure the perceived health and well being. Both strategies documented very good indoor climate with highly positive scores on the questionnaire. The strategy with supply air temperature above the room temperature resulted in a little better perceived health and well being compared to the other strategy.

KEYWORDS

cold climate, heating, questionnaire, passive house, ventilation strategies

1. INTRODUCTION

GK environmental house is the first office building in Norway satisfying the passive house standard (NS 3701, 2012). The building is located in Oslo and has been operational since August 2012. The building is mostly heated by supplying warm air into the rooms through the ventilation system. In addition, active air supply diffusers with integrated presence and temperature sensors are used in each room to control the ventilation rate according to room demand. In this paper, we investigate the potential effects on health and well-being when the heating demand is covered entirely with the ventilation system. The concern is whether air mixing is poor when over-tempered air is supplied into the rooms. In this context, we assessed the perceived indoor climate resulting from two different ventilation control strategies. The first strategy (named strategy 1) resulted in a supply temperature somewhat higher than room temperature, while the second strategy (strategy 2) resulted in a supply temperature slightly below room temperature. Perceived health and well being is measured using a questionnaire.

2. MATERIAL AND METHODS

Both control strategies used consisted in having an initial room temperature of around 21°C when the users arrived in the building. The supplied air temperature at the exit of the Air Handling Unit (AHU) was then controlled depending on the average temperature measured in all the rooms. The curve used to control the supply temperature was different for both strategies and is presented on Figure 1 and Figure 2.

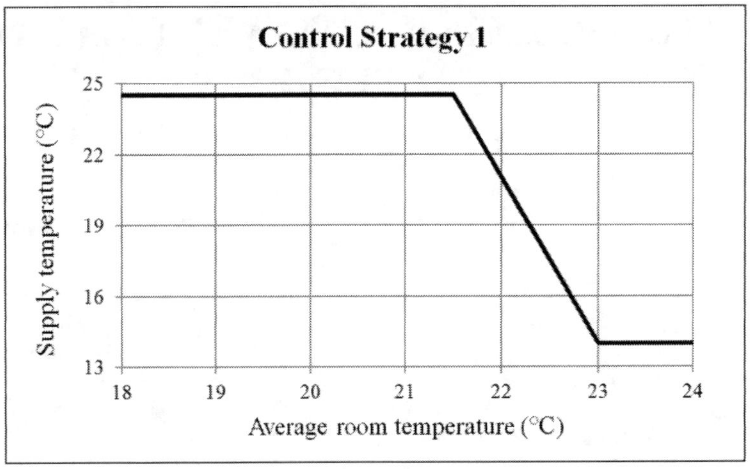

Figure 1: Control of the supply temperature at the exit of the AHU according to average room temperature in the building: strategy 1.

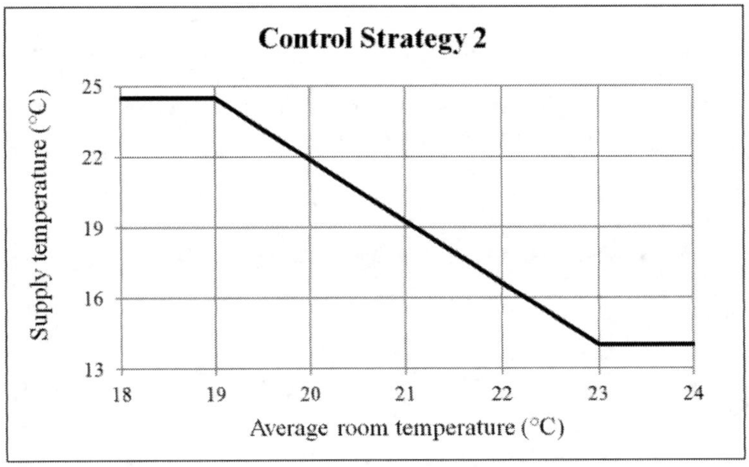

Figure 2: Control of the supply temperature at the exit of the AHU according to average room temperature in the building: strategy 2.

Globally, strategy 1 results in a higher supply temperature than strategy 2.

The two ventilation strategies described above were compared using a cross over design, see Table 1.

Table 1: The plan for the cross-over experiment.

	19th Nov.	13th January		14th January	
Part of building	Both	South	North	South	North
Ventilation strategy	Normal	1	2	2	1
Ventilation temperature control	Normal	Fast (Figure 1)	Slow (Figure 2)	Slow (Figure 2)	Fast (Figure 1)
Ventilation rate [l/s]	Varying	Varying	Varying	Varying	Varying

On the 13th of January the users in the south and north part of building were exposed to strategy 1 and strategy 2, respectively and on the 14th of January the exposing were switched (Table 1). With such a cross-over design, each user in the building was exposed to both ventilation strategies. On the 19th of November the ventilation were run as normal and the indoor climate were expected to be good. This case was used as control.

A questionnaire was used to measure the users' perceived health and well-being under the different cases in Table 1. The questions are shown in Table 3. On each question the users gave a value between one and ten ranging from very uncomfortable and very comfortable. The sum, *S*, of all the scores for a given user, represents the overall health and well-being for this user. The questionnaire also included health question like whether a used suffered from asthma or cold.

Data from the questionnaires were analyzed using a random effect linear regression model with *S* as the dependent variable. The three cases 19th November, strategy 1 and strategy 2 represent the main independent variable, but also office temperature is included. The office temperature is varying with date and place in the building and including this as an independent variable, we were able to study the effect of the ventilation strategies independent of the room temperature. We also include as independent variables gender and whether the participants suffered from cold or asthma such that we are able to compare the ventilation strategies independent of these factors. We expect that repeated measures from the same user is correlated, e.g. some users are always too cold and some users are always tired, and this is taken into account by including a random effect on the user level in the regression model. Statistical analyses were performed using the program R (R development team, 2013) and the R package lme4 (Bates et al., 2013).

3. RESULTS

Table 2 shows some technical measurements from the experiment.

Table 2: Technical measures.

Date	19.November	13 th January		14 th January	
Part of building	Both	South	North	South	North
Ventilation strategy	Normal	1	2	2	1
Initial supply air temperature (7 am)[°C]	North: 15.9 South: 17.8	24.3	17.3	15.8	24.4
Average room temperature (6 am – 2 pm)	North: 22.1 South: 22.0	21.7	20.7	21.5	21.4
Average supply air temperature (6 am – 2 pm)	North: 15.6 South:15.9	22.8	19.3	21.0	23.4
Outdoor temperature [°C]	4.1	– 10.9		– 5.9	
Outdoor condition	Partly cloudy	Cloudy		Cloudy	

The average supply temperature during the length of the questionnaire (6 hours) as well as the average room temperature are presented in Table 2. We can see that strategy 1 resulted in a supply temperature above room temperature, while strategy 2 resulted in a supply temperature slightly below room temperature . Furthermore, the supply temperature was relatively low on November 19 (15.6-15.9°C). This results from the fact that the outside temperature was relatively high (4.1°C) compared to the tests performed on January 13 and January 14 (– 10.9°C and – 5.5°C, respectively). It was therefore necessary to compensate for the internal heat gains by supplying cool air into the rooms.

For each of the three dates in the experiment, 19th November, 13th and 14th January, the questionnaire was sent out to 133 users and we received 46, 35 and 36 answers, respectively. 18 users answered all the three questionnaires. Table 3 shows average score on each question in the questionnaire under the two ventilation strategies and 19th November.

Table 3: Questionnaire with average score under the two ventilation strategies.

Questions	19th November	Vent. strat. 1	Vent. strat. 2
Are you tired?	6.43	7.97	7.46
Does your head feel heavy?	6.83	8.70	7.68
Do you have a headace	8.07	8.77	8.32
Do you feel faint or dizzy?	8.26	9.17	8.49
Do you have problems concentrating?	6.87	8.00	7.61
Do you feel itching or burning in your eyes?	8.39	8.43	8.32
Do you feel hoarse or dry throat?	8.17	8.77	8.39
Do you feel itching or burning in your face or on your hands?	9.20	8.90	8.85
Do you feel nauseous or otherwise unwell?	9.37	9.47	9.58
Is it too warm?	7.76	8.77	8.29
Is there bothersome warmth because of sunshine?	8.26	9.60	8.68
Is it too cold?	8.00	7.57	7.17
Do you feel a draught around your feet or your neck?	9.37	8.93	8.73
Does the temperature in the room vary?	8.85	8.90	8.43
Does the air feel heavy?	7.80	9.00	8.26
Does the air feel dry?	8.02	8.57	8.14
Is there any unpleasant smell?	9.48	9.70	9.39
Do you have a stuffy or runny nose?	9.30	8.50	8.41
Do you cough?	8.96	8.70	8.53

We see that overall the users feel quite well with almost all average scores above eight. We also see that the scores under strategy 1 are higher (more comfortable) than under strategy 2.

Table 4 shows the results from the regression analysis. All answered questionnaires (46+35+36=117) were included in the analysis.

Table 4: Results from linear regression.

Parameter	Estimate	St. err.	df	t value	p value
Vent. strat. 1 (reference: vent. strat. 2)	6.6	2.9	42.5	2.26	0.029 *
19th November (reference: vent. strat. 2)	2.8	2.9	44.9	0.98	0.332
Office temperature	5.1	6.0	46.4	0.84	0.405
Gender male (reference: female)	20.8	6.9	67.9	3.01	0.004 **
Asthma (reference: No asthma)	– 31.1	10.9	69.6	2.83	0.006 **
Cold (reference: No cold)	– 5.3	4.1	49.4	1.28	0.205

Signif. codes: p-value < 0.01:**, p-value < 0.05:*

We see that the perceived health and well-being is significantly better under strategy 1 compared to strategy 2 and that strategy 2 is not significantly poorer than the control autumn (19th November). We also observe that the perceived health and well-being is better for male users and users with no asthma or cold.

4. DISCUSSION

For the winter conditions considered (13 and 14 January), a better perceived indoor climate was obtained for a supply temperature higher than room temperature (strategy 1) than for a supply temperature slightly lower than room temperature (strategy 2).

Furthermore, no significant discomfort regarding perceived indoor air quality was obtained compared to the control case (19 th November). This is an indication that the active air supply diffuser employed display good mixing properties for a broad range of supplying conditions, and that the short-circuiting of the fresh and warm ventilation air is reduced to a great extent.

Therefore, supplying warm ventilation air to cover the heating demand appears to be a relevant solution for office buildings with passive house standard.

5. REFERENCES

Standard Norge NS 3701 (2012) *Criteria for passive houses and low energy buildings*.

Douglas Bates, Martin Maechler, Ben Bolker and Steven Walker (2013) *lme4: Linear mixed-effects models using Eigen and S4*, R package version 1.0-5. Available at http://CRAN.R-project.org/package=lme4

R Development Core Team (2013) *R: A Language and Environment for Statistical Computing*, R Foundation for Statistical Computing, Vienna, Austria. Available at http://www.R-project.org/.

MONITORING OF AN INNOVATIVE ROOM-BY-ROOM DEMAND CONTROLLED HEAT RECOVERY SYSTEM ON FOUR LOCATIONS

Colette Madeline, Stéphane Berthin, Pierre Kraus

Aereco SA
62, rue de Lamirault
77615 Marne la vallée, France

ABSTRACT

Demand controlled heat recovery ventilation systems, which combines heat recovery (HRV) and demand controlled (DCV) is growing fast among ventilation manufacturers.

Several categories can be identified, from global dwelling regulation, to fine room-by-room regulation of the airflow rate. Simulations show that room-by-room demand controlled heat recovery ventilation is the best compromise to optimize at the same time indoor air quality, comfort, and energy savings.

To reinforce this assessment, four room-by-room demand controlled HRV have been monitored since May 2013. The system is an innovative ventilation for individual treatment, which modulates automatically the supply airflow according to CO_2 concentration in the main rooms, and exhaust airflow according to humidity concentration in the wet rooms. The balance of the airflow is made using two specific compensation valves at the exhaust and at supply.

Three projects are located in Germany: Nuremberg, Dortmund and Frankfurt, and one takes place in France, near Paris. The dwellings have from three to five main rooms, and a real occupancy of two to five people. They are all existing dwellings. The comparison was made between three different types of installation: two in the heated space, one in the attic (slightly insulated), and one with the heat recovery unit in the heated space and the ductwork in the attic (non insulated).

The results have shown the good functioning of the system in terms of balancing, adaptation of the airflow rates to the needs, and filtration. The relevance of using CO_2 concentration to modulate the airflow rate at supply has been verified by presence sensors in the main rooms. Temperature sensors revealed the necessity to carry on the installation in the heated space to optimize energy savings. Besides, the collected data gave the opportunity to validate a simulation model of the system using CONTAM (multizone indoor air quality and ventilation analysis program developed by NIST).

The monitoring also demonstrated the good performance of this type of heat recovery, in terms of indoor air quality, comfort and energy savings. The CO_2 and the humidity level stayed in the comfortable range thanks to demand controlled ventilation. The supply temperature, when the HRV is located in the heated space, is mostly between 18 and 20°C, thanks to heat recovery (the outdoor temperature being in the range of 0 to 14°C). In-situ measures and surveys occupants supplied complementary information regarding the acoustic and thermal comfort.

1. KEYWORDS

Heat recovery ventilation; demand controlled ventilation; indoor air quality; energy savings

2. INTRODUCTION

The need to reduce more and more the energy consumption in residential buildings has lead several manufacturers of ventilation systems to associate heat recovery ventilation (HRV) with demand controlled ventilation (DCV). Some systems control the supply or the exhaust only, and some other control both, room by room or in a global way. Simulations made on CONTAM (a multizone indoor air quality and ventilation analysis program) show that room-by-room demand controlled heat recovery ventilation is the best compromise to optimize at the same time indoor air quality (IAQ), thermal comfort, and energy savings. A monitoring of this last type of HRV was carried out in four dwellings. This study shows the results of this monitoring campaign and some of the results of the simulations.

3 SIMULATIONS

3.1 Energy efficiency / Indoor air quality

The simulations compare five types of HRV systems on the energy and IAQ aspects in a three bedroom house occupied by three people (Table 1). As CO_2 concentration results are very low in the dwelling (no exposure over 2 000 ppm for none of the HRV systems), the CO_2 comparison level has been set to 1 200 ppm.

Table 1: Simulated HRV systems

System	Exhaust	Supply	Balance
Fixed airflow rate	Fixed airflow in the wet rooms according to French regulation	$Q_{supply\_room}=Q_{extr}/4$	Fixed balanced airflow rates
Preset speed level	Humidity sensor in the wet rooms	CO_2 sensor in the parents' bedroom	$Q_{tot} = f(Max(RH; CO_2))$ Instantaneous distribution of the total airflow rate between the rooms
Exhaust DCV	Humidity controlled exhaust units in the wet rooms	$Q_{supply\_room}=Q_{extr}/4$	Instantaneous distribution of the total airflow rate in the main rooms
Supply DCV	$Q_{extr.\_room}=Q_{supply}/3$	CO_2 controlled supply units in the main rooms	Instantaneous distribution of the total airflow rate in the wet rooms
Exhaust and supply DCV	Humidity controlled exhaust units in the wet rooms	CO_2 controlled supply units in the main rooms	Supply and exhaust compensation valves

Looking at the results below (Figure 1, Figure 2, Figure 3), the supply DCV is the best system regarding energy and CO_2 aspect (very few heat losses and acceptable CO_2 concentration), but the RH (relative humidity) in the bathroom and in the kitchen is very high. The exhaust DCV has a good RH level in the wet rooms but it has more hours of exposure over 1200 ppm (which is easily satisfactory), and the fixed airflow and the preset speed level systems have too many heat losses. According to the simulation results, the exhaust and supply DCV seems to be the best compromise to optimise at the same time the indoor air quality and the energy consumption.

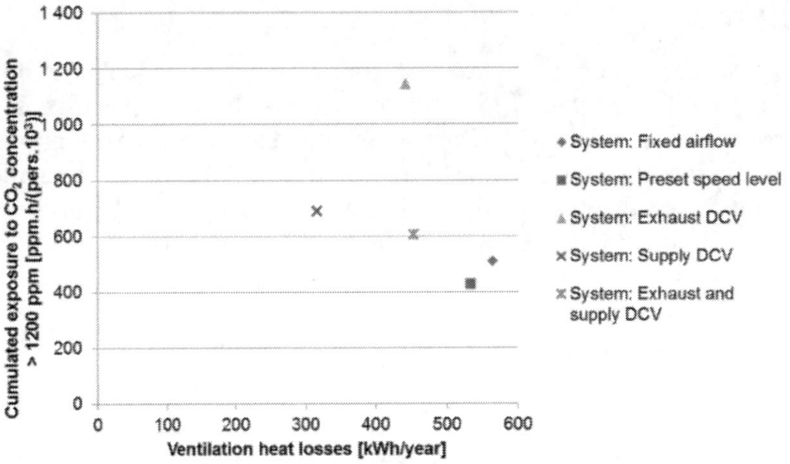

Figure 1: HRV energy / CO_2 exposure over the heating season (CONTAM simulation)

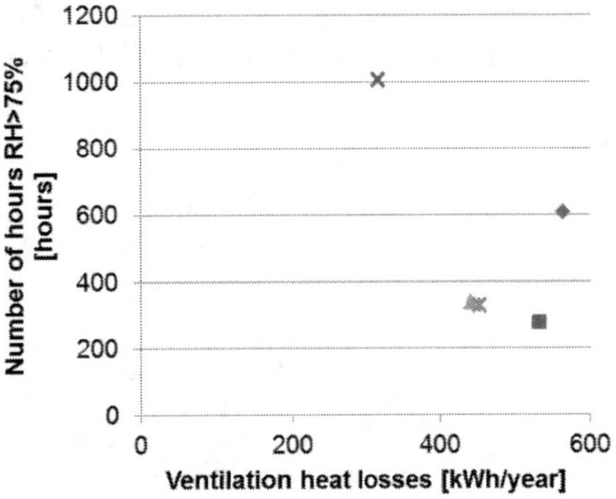

Figure 2: HRV energy / RH in the bathroom (CONTAM simulation)

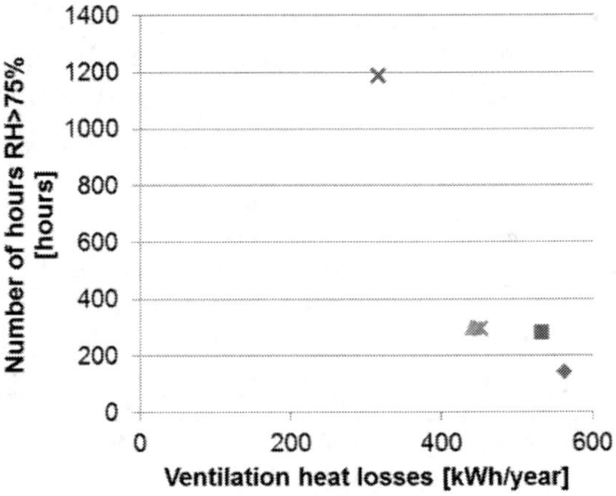

Figure 3 : HRV energy / RH in the kitchen (CONTAM simulation)

3.2 Filter clogging

The monitored system tests the total pressure drop of the HRV every two months in order to give a signal to change the filters if necessary. This filter test has been simulated on 5 HRV systems using G4 + F7 filters at supply. The results are showed for filter F7 only (Figure 4).

During the four first months of the simulation, the pressure drop is more or less the same for all the systems. The dust mass of the filter increases a lot after eight months of operation in the system with fixed airflow and in the preset speed level system. DCV systems have almost the same dust mass increase, which is the slowest of the five simulated systems, thanks to the modulation of the airflow rates.

In Dortmund, after eleven months of functioning, filter F7 has a pressure drop of 21.5 Pa, which is even less than the result showed by the calculation. In Nuremberg, the filter F7 has a pressure drop of 26.4 Pa after 9 months of functioning.

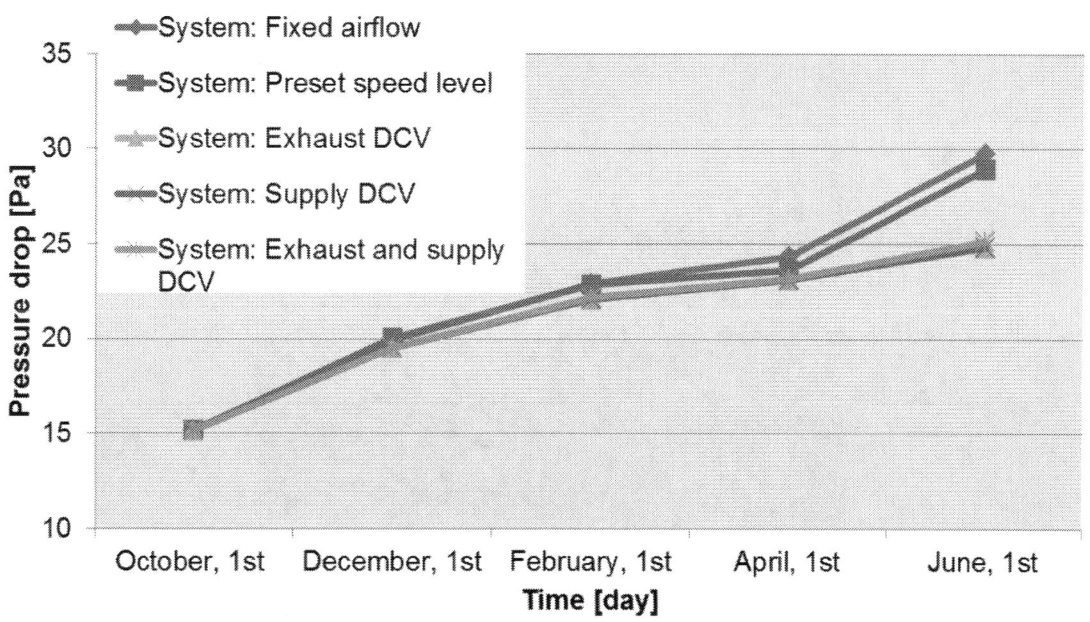

Figure 4: F7 filter tested at 170 m^3/h every 2 months (CONTAM simulation)

4 MONITORING

4.1 Monitored room-by-room demand controlled HRV

Figure 5: Monitored room-by-room demand controlled HRV

The ventilation is a heat recovery system involving a heat recovery unit connected to exhaust units and a distribution box that modulates the supplied airflows (Figure 5).

The airflows are automatically modulated according to the needs of each room in the dwelling: the air supplied to the bedrooms and living room and the air exhausted from the kitchen, the bathroom, and the WC. Each supply unit is directly connected to the distribution unit, which adjusts the supplied airflow in the main rooms according to the level of CO_2. On the exhaust side, the units adjust the airflow, according to parameters that depend on the room concerned: humidity in the bathroom, occupancy in the WCs, humidity and switch for the boost airflow in the kitchen.

At all times, the total supplied and exhausted airflows are measured in the heat recovery unit and balanced by two controlled compensation valves that are located in the living room, the kitchen, or a corridor. For example, when the need increases in the kitchen as a meal is being prepared, and is not accompanied by high demand in a bedroom or in the living room, the exhaust airflow can be balanced by opening the compensation valve of the supply ductwork.

4.2 On-site installations and monitored values

Figure 6: Monitored dwelling in Dortmund

Three projects are located in Germany: Nuremberg, Dortmund (Figure 6) and Frankfurt, and one takes place in France, near Paris. The dwellings have from three to five main rooms, and a real occupancy of two to five people. They are all existing dwellings. The comparison was made between three different types of installation: two in the heated space (Figure 8), one in the attic (slightly insulated, Figure 7), and one with the heat recovery unit in the heated space and the ductwork in the attic (non insulated).[1]

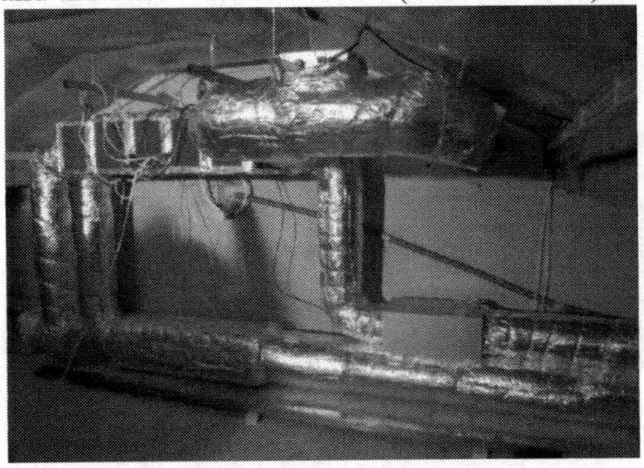

Figure 7: Installation in the attic

Figure 8: Installation in the false ceiling

Every parameter of the system has been monitored every 1:30 minute in Nuremberg and in Dortmund, and every minute in Frankfurt and in Paris (Table 2). The first projects in Nuremberg and Dortmund have presence sensors in every main room (presence detection measure every 1:30 minute). The last projects in Frankfurt and Paris have temperature and relative humidity sensors in every main room (measure every minute).

[1]As the last installation in Paris was installed in the beginning of May 2014, there is not enough data to include its results in this study for the moment.

Table 2: Monitored parameters

Parameter	Location	Monitoring project	Parameter	Location	Monitoring project
CO_2 sensors	In every main room	All	Exhaust airflow rate	At the main unit	All
Opening of the damper at supply	At the supply dispatching box	All	Supply airflow rate	At the main unit	All
Fresh air temperature	At the main unit	All	Pressure at the exhaust	At the main unit	All
Supply air temperature	At the main unit	All	Pressure at supply	At the main unit	All
Exhaust air temperature	At the main unit	All	Speed rotation of the exhaust fan	At the exhaust fan	Nuremberg, Dortmund
Used air temperature	At the main unit	All	Speed rotation of the supply fan	At the supply fan	Nuremberg, Dortmund
Opening of the exhaust compensation valve	At the exhaust compensation valve	All	Defrosting state	At the main unit	All
Opening of the supply compensation valve	At the supply dispatching box	All	By-pass state	At the main unit	All
Exhaust fan setting	At the exhaust fan	All	Supply fan setting	At the supply fan	All
Presence sensors	In every main room	Nuremberg, Dortmund	Temperature and humidity sensor	In every room	Frankfurt, Paris

4.3 Check of the working of the HRV

The graph presenting the total supply airflow rate as a function of the total exhaust airflow rate shows a positive and linear correlation between supply and exhaust airflow rate (Figure 34). These results allowed to highlight a program default for airflow rates under 40 m^3/h, which was corrected on the standard mass-production product.

Considering the total airflow rate over 40 m^3/h only, the correlation coefficient is 0.86. Moreover, the peak of correlation happens with no interval, so the balance of the airflow is instantaneous. The instantaneous difference between supplied and exhaust airflow rate is mainly around zero, and is 97.98 % of the time included in the interval [-10; 10[m^3/h (Figure 10).

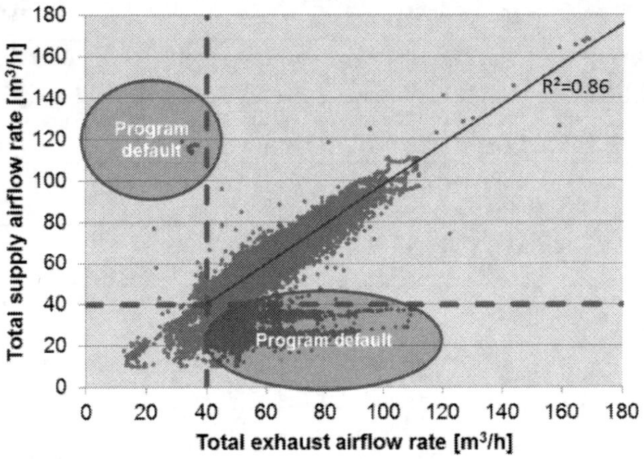

Figure 9: Relationship between total exhaust airflow rate and total supply airflow rate.
(Frankfurt, from 2014/01/03 to 2014/05/31)

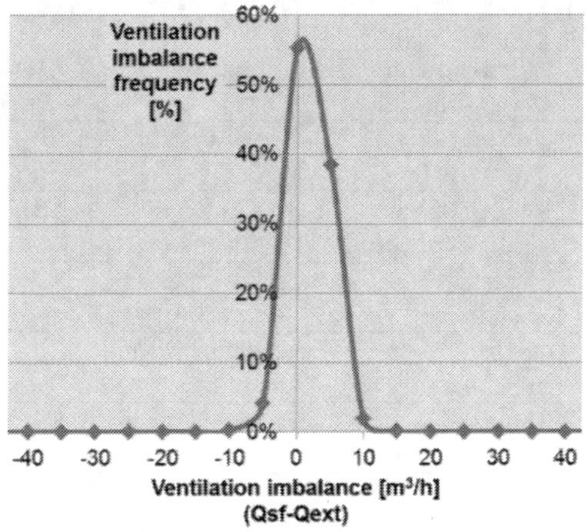

Figure 10: Balance of exhaust and supply airflow rate.
(Frankfurt, from 2014/01/03 to 2014/05/31; above 40 m^3/h).

Four preset supply airflow rate are defined, according to the concentration of CO_2 in each main room. The results show that the supply airflow is well distributed among the 4 preset rates, depending on the occupancy of the room (Figure 11, Figure 12).

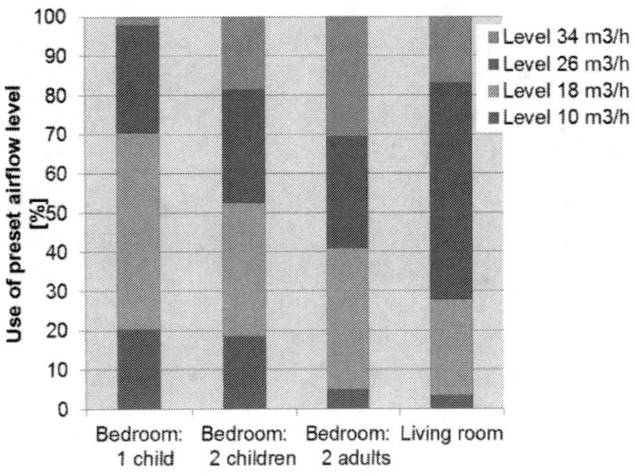

Figure 11: Use of preset supply airflow rate
(Frankfurt, from 2013/11/15 to 2014/05/31)

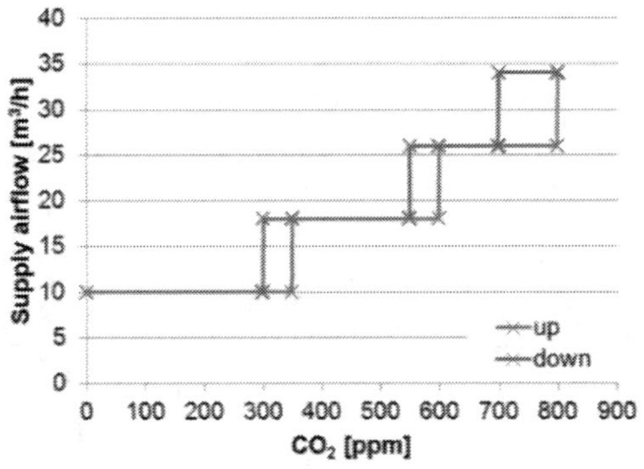

Figure 12: Preset CO_2 level
(Above ambient concentration)

5 PERFORMANCE OF THE ROOM-BY-ROOM DCV HRV

5.1 Indoor air quality

The RH distribution in the dwelling shows that the concentration is between 30 and 60 % for the considered period (from 2014/02/13 to 2014/05/31). This range of RH concentration is comfortable according to ASHRAE comfort zones (Figure 13).

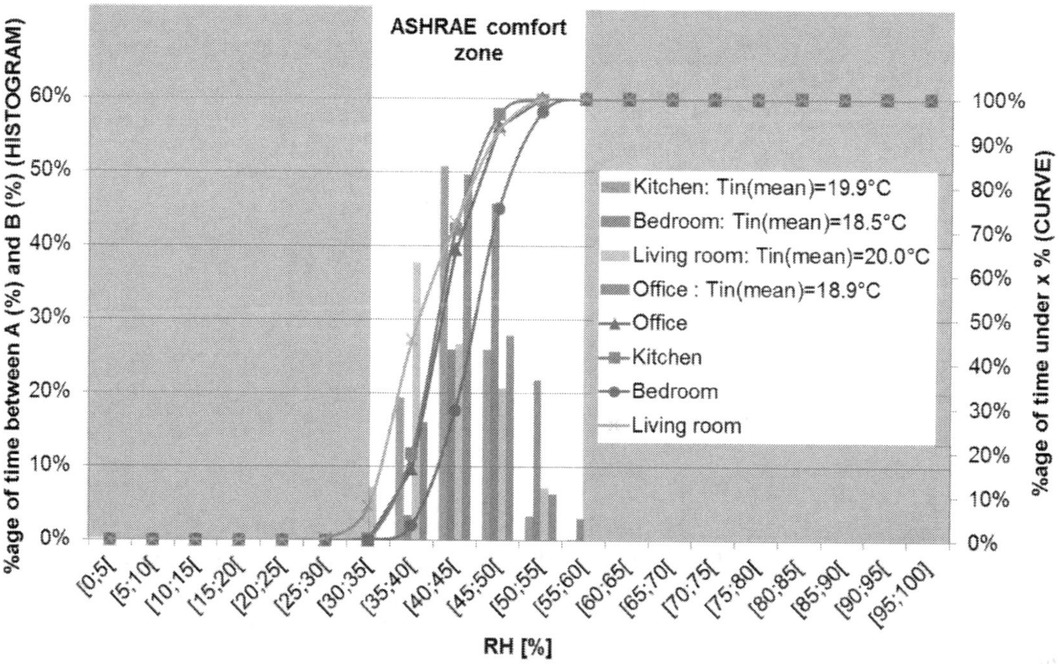

Figure 13: RH distribution (Frankfurt, from 2014/02/13 to 2014/05/31)

The risk of condensation in the dwelling is almost null if we consider that such an efficient HRV system should be installed in a dwelling having an energy efficient envelop (Table 3). The worst case would occur in the kitchen, with a maximum of 2 hours of condensation risk on a PVC double glazing with an air space of 12 mm (note that the bathroom has not been monitored). All the main rooms have nearly zero risk of condensation, independently of the external wall composition.

Table 3: Risks of condensation in the dwelling (Frankfurt, from 2014/02/13 to 2014/05/31)

Room (Mean dew point temperature [°C])	Mean internal wall temperature [°C] (March 2014)				Risk of condensation [number of minutes] (From 2014/02/13 to 2014/05/31)			
	Double glazing U=1.2 W/(m².K)	Double glazing U=2.3 W/(m².K)	Insulated ext. wall U=0.51 W/(m².K)	Non insulated ext. wall U=2.25 W/(m².K)	Double glazing U=1.2 W(/m².K)	Double glazing U=2.3 W/(m².K)	Insulated ext. wall U=0.51 W/(m².K)	Non insulated ext. wall U=2.25 W/(m².K)
Kitchen (6.5)	16.2	12.6	18.5	12.8	26	116	3	112
Bedroom (6.2)	15.1	11.9	17.1	12.1	0	2	0	2
Living room (5.7)	16.4	12.7	18.7	12.9	0	0	0	0
Office (5.5)	15.4	12.1	17.4	12.3	0	0	0	0

CO_2 distribution in the bedrooms has been observed thanks to the sensors located in each main room, which control the supply airflow rate. In the bedrooms with an occupancy of one person, the CO_2 concentration is less than 1 150 ppm 80 % of the time. In the bedrooms with an occupancy of two people, the CO_2 concentration doesn't exceed 1 250 ppm 80 % of the time. In any case, CO_2 concentration won't exceed 1 650 ppm, whatever the occupancy (Table 4).

Table 4: CO_2 distribution: main results in bedrooms

Monitoring ref.	Max distributed area (peak %)	100 % of time under X ppm	90 % of time under X ppm	80 % of time under X ppm
Nuremberg: 1 child	950-1000 ppm 26.7 %	1 400	1 150	1 100
Nuremberg: 2 children	1100-1150 ppm 25.5 %	1 600	1 300	1 200
Nuremberg: 2 adults	1150-1200 ppm 15.0 %	1 650	1 350	1 250
Dortmund: 1 adult	1000-1050 ppm 19.8 %	1 250	1 150	1 100
Dortmund: 1 child	1100-1150 ppm 23.1 %	1 550	1 250	1 150
Frankfurt: 2 adults	950-1000 ppm 10.9 %	1 600	1 200	1 050

The CO_2 concentration distribution shows a fast reduction on the right side of its peak, which implies a good management of the air renewal in case of high CO_2 concentrations (Figure 14).

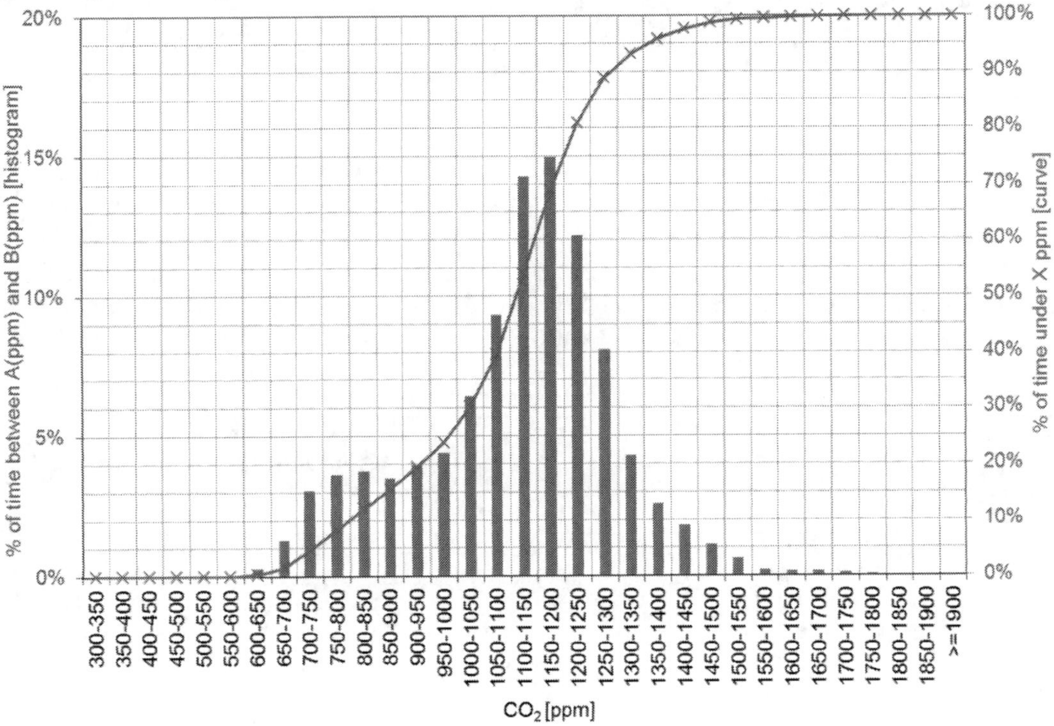

Figure 14: CO_2 distribution in a bedroom occupied by 2 adults (Nuremberg, from 2013/11/15 to 2014/05/31)

5.2 Thermal comfort

The survey fulfilled by the occupant in February showed no annoyance due to the supply air in the main rooms in Nuremberg and in Dortmund. If we look at the supply temperature distribution, we can observe that during the heating period the supply temperature is between 16 and 26°C 99.3 % of the time in Nuremberg, 95.6 % of the time in Dortmund, and 39.5 % of the time in Frankfurt.

These results show the huge impact of the location of the installation. In Frankfurt, the ductwork is insulated with 2*30 mm of glass wool, and the main unit is also insulated with blocks of polyurethane foam. The thermal efficiency is decreased by more than 50 % for an insulated installation located in the attic compared with an installation located in the heated space (Figure 15).

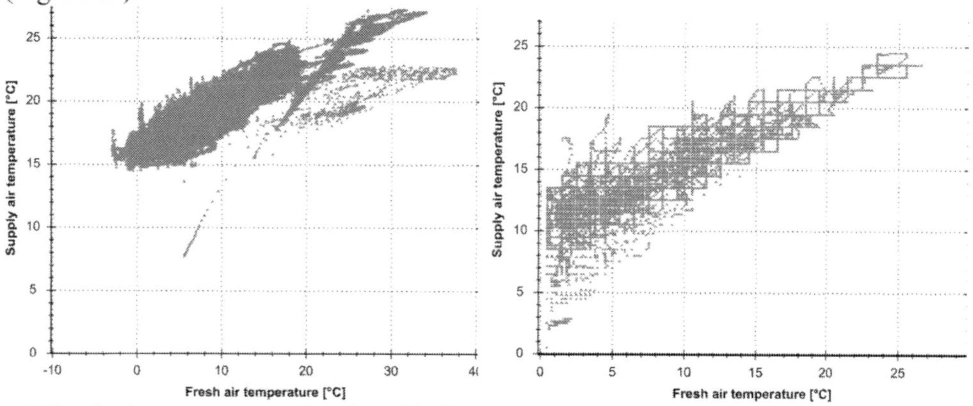

Figure 15: Supply air temperature as a function of fresh air temperature (From left to right: Dortmund and Frankfurt)

The discomfort due to draught can be expressed as the percentage of people predicated to be bothered by draught. This criterion is called draught rate (DR) and is calculated through empirical values. According to the draught rate and some other parameters, it is possible to define the comfort level category (ISO 7730:2005). An estimation of this criterion has been calculated at 2 meters from the unit, using the instantaneous airflow rate, the local air temperature being the supply and the extract air temperature average at the main unit. The DR is less than 15 % almost 100 % of the time in Nuremberg and in Dortmund, as recommended by the scientific literature. In case of an installation in the attic (as showed in Frankfurt), the DR which depends on the supply temperature is higher than for an installation in the heated space (Figure 16).

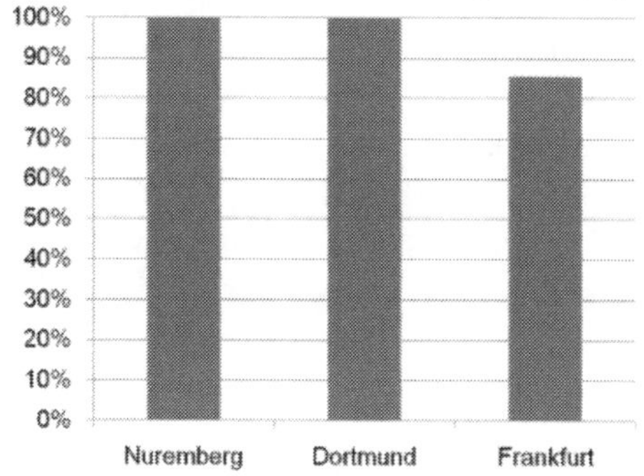

Figure 16: DR < 15% in parents' bedroom (at 2 meters)

5.3 Energy efficiency

An estimation of heat losses due to the ventilation system has been calculated, using the temperatures and the instantaneous total airflow rates at the main unit. The calculation is not made for fresh air temperature over 15°C, and the extract air temperature is taken as an estimation of the inside temperature. The monitoring in Dortmund has very low heat losses in comparison with the monitoring in Nuremberg, because the extract air temperature is about 3°C more in Nuremberg than in Dortmund[2]. The electrical consumption of the fans has also been estimated, using the instantaneous speed level of the fans and the total airflow rates (Table 5).

[2]The monitoring in Frankfurt is not presented because its data doesn't include the whole heating period (beginning of the results in January).

Table 5: Mean airflow rates, heating needs, and electrical consumption of the fans over the considered period

Monitoring ref. (Surface m²)	Considered period	Mean airflow rate [m³/h]	Mean extract /supply temperature [°C]	Heat losses (final energy) [kWh/(m².an)]	Mean fan power (exhaust + supply) [W]
Nuremberg (106 m²)	2013/11/15 to 2014/05/31	101.6	23.6/19.4	6.5	21.9
Dortmund (111 m²)	2013/11/15 to 2014/05/31	116.9	20.3/19.1	2.2	30.7

Considering a gas appliance for heating, with a mean cost of 6.89 c€(incl tax)/kWh, and a cost of 29.21 c€(incl tax)/kWh for electricity, the mean monthly cost due to the monitored HRV would be 10.13 €(incl. Taxes)/month in Nuremberg, and about 6.91 €(incl tax)/month in Dortmund during heating season. Prices of S2-2013 in Germany (Eurostat European comission, 2014).

6 ADDITIONAL SENSORS

The two first monitoring (Nuremberg and Dortmund) have been equipped with presence sensors in the main rooms. From these data, two analyses can be made: Is CO_2 a good indicator of human presence? Is it possible to use presence sensors instead of CO_2 sensors to control the supply airflow rate?

In addition, relative humidity and temperature sensors have been added in monitoring Frankfurt since the middle of February, and in Paris since May. There are located in the main rooms (bedrooms, living room, office) and in the wet rooms. These sensors have been added in order to study the possibility of using humidity sensors instead of CO_2 to control the supply airflow rates.

The results confirm the fact that CO_2 is a good indicator for presence detection, the CO_2 concentration increasing in the dwelling when a detection of presence occurs. Presence sensors, as an alternative to CO_2, could be relevant using more accurate sensors. In the same way, relative humidity sensors could be used instead of CO_2, but they would need to be corrected from outside humidity variation, like the humidity controlled air inlets currently used in MEV (mechanical exhaust ventilation) systems for instance.

7 CONCLUSIONS

This monitoring campaign has shown the good working of the tested demand controlled heat recovery ventilation system. It has also helped to improve the management of the program for specific points. Regarding the efficiency of the system, the indoor air quality is very good: no hours over 2 000 ppm of CO_2, and a relative humidity between 30 and 60 % in the main rooms. For the installations in the heated space, the supply temperature is more than 95 % of the time between 16 and 26°C. Looking at the results for the installation in the attic, the study confirms that it is necessary to put the HRV in the heating space, in order to use the entire potential of the HRV system.

In terms of energy consumption, the results are very positive, with a maximum of 6.5 kWh/(m².year) for the ventilation heating needs, and a mean electrical power of 30.7 W in the worst case.

This monitoring is going to continue until spring 2015, in order to be able to study the summer functioning and to generalize the efficiency results of the HRV during winter.

8 ACKNOWLEDGEMENTS

This project would not have been possible without the agreement of the owners of the dwellings, that is why we would like to thanks a lot the occupants of the four monitored dwellings.

9 REFERENCES

Eurostat European comission. (2014, 05 20). *Eurostat Database*. Retrieved 06 24, 2014, from http://epp.eurostat.ec.europa.eu/portal/page/portal/statistics/search_database

ISO 7730:2005. *Ergonomics of the thermal environment - Analytical determination and interpretation of the thermal comfort using calculation of the PMV and PPD indicies and local thermal comfort criteria.*

CLEANLINESS OF AIR FILTERS IN THE EXPERIMENTAL PASSIVE HOUSE

Michal Michalkiewicz [1], Malgorzata Basinska *[2]

[1] *Poznan University of Technology*
Pl. Sklodowskiej – Curie 5
Poznan, Poland

[2] *Poznan University of Technology*
Pl. Sklodowskiej – Curie 5
Poznan, Poland
**Corresponding author:*
malgorzata.basinska@put.poznan.pl

ABSTRACT

An inherent element of the passive house is the system of exhaust ventilation in air supply. According to their class, air filters used in ventilation systems stop the contamination, but may also be the main source of secondary indoor contamination during long-term use.

The aim of the study was to determine the contamination of different air filters after several months of their continuous work. The weight of the filter and its pollution with dust and microbiological contamination were determined. Research was carried out for four sets of filters working in the experimental passive house in the Poznan University of Technology – DoPas. The measurements were focused on the quality of the microbiological contamination of air filters. During microbiological examination of filters, the following microorganisms were determined: the general count of mesophilic bacteria, the general count of psychrophilic bacteria, the count of Staphylococcus (*Staphylococcus*) mannitol positive (type α) and mannitol negative (type β), the count of *Pseudomonas fluorescens* bacteria, actinomycetes (*Actinobacteria*) as well as the general count of microscopic fungi. Performed measurements of studied microorganisms count revealed that psychrophilic bacteria, microscopic fungi and mesophilic bacteria were most numerous. Their count was about ten times higher in the intake duct filter than in the exhaust duct filter. After disinfecting air ducts their cleanliness level dropped from moderate contamination to low.

KEYWORDS

Passive house, office building, air filters, microbiological contamination, ventilation installations

1. INTRODUCTION

Recent years have witnessed a growing focus on the quality of air in buildings, since people spend over 70% of their time there. More and more buildings (especially office complexes) are equipped with mechanical ventilation systems, which is connected, among others, with limiting the amount of energy supplied to the building through air flow control. Moreover, the purposed of ventilation systems is to eliminate contamination generated indoors or dilute it till it reaches an acceptable level. Despite vast interest in indoor air cleanness, some parameters that contribute to the quality of air have not been fully known. One of these elements is an air filter whose purpose is to stop physical as well as microbiological contamination of air. Depending on the quality of filtering material and the type of filter, it stops:

– insects, dust, plant spores, bacteria – on coarse filters and fine filters,
– microorganisms, viruses, particles floating in the air – on HEPA, ULPA and special filters.

The quality of filtering depends not only on proper class of filter used for various expected results but also on the filter's own cleanness which deteriorates while the filter is in service. Figure 1 presents a picture of both a clean filter and a contaminated one, magnified 160 times.

a) b)

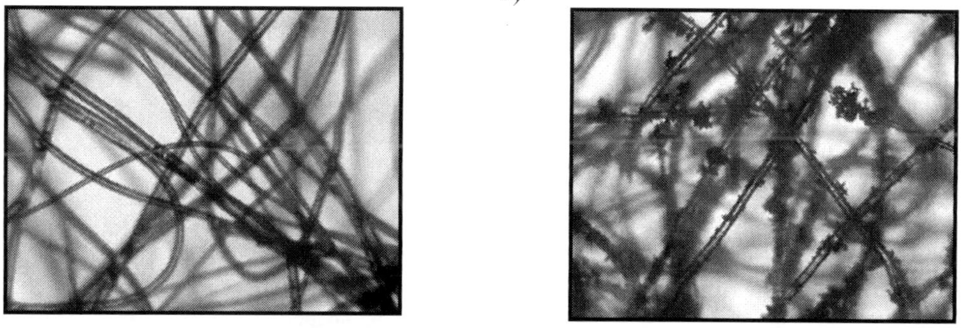

Figure 1: Air filter cleanness level, magnification 160 times, a) clean filter, b) contaminated filter

Chin (Chin 1999) writes that a hygroscopic filtering material can absorb moisture from the air and when moisture level on dust particles stopped by the filter is high enough, a process of sprouting and growth of a fungus colony may start. The author claims that the primary bio-contamination is fungi and bacteria, the next is mites, insects and nematodes, whose count is related to the growth of fungi. The dampness in a ventilation system also leads to an increase in the count of bacteria in the system, which was proven a study by Ahearn (Ahearn et al., 1997) who revealed that an air conditioning system can contribute to an increase in fungi contamination indoors.

In order to evaluate the cleanness of filters after an eighteen-month use period, a microbiological contamination assessment of the filters was performed. The measurements were taken in an experimental passive house in the Poznan University of Technology – DoPas, modernized in 2007 to meet the standard of passive construction. Having a light frame structure, the building is a detached house, with single floor, fully cellared. The cellar has mass structure, with walls made of concrete. Total area of the building is 143,7 m^2, with cubage of 620 m^3. As a heat source, the building utilizes a central AHU 1 (Air Handling Unit) with two EU5 type filters mounted on intake and exhaust lines. As an independent source (an option to diversify heat sources) a compact heating and cooling unit can be installed, connected with a ground heat exchanger and a solar collector (AHU 2). External air, having been initially cooled in the exchanger, can be a bottom heat source for the heat pump in the unit. One of the elements of the passive house ventilation system is a GWC ground exchanger with a pipe that is 38 m long with a diameter of □=200 mm, placed in the ground at a depth of 1.8 to 2.2 meters.

2. MATERIALS AND METHODS

The operation of replacing filters in the ventilation system in the DoPas passive house in Poznan University of Technology allowed to perform a microbiological examination of contamination in air filters. Replacement of filters after an eighteen-month use period, performed by a specialized company, was completed with cleaning and disinfecting the ventilation system. During the operation, the ventilation system, the AHU 1 ventilation unit, as well as the intake and exhaust were mechanically and pneumatically cleaned and disinfected. Mechanical cleaning was done with power brushes driven externally and with compressed air. During the process of duct cleaning, the cleaning machinery collected impurities do level I and II bags. For the purpose of disinfection, a biocide was used, registered as medical product, with CE conformity marking and approval from the National Institute of Hygiene.

Microbiological examination employed:

- EU5 type filter from the AHU 1 ventilation unit, mounted on air intake – **Filter 1**,
- EU5 type filter from the AHU 1 ventilation unit, mounted on air exhaust – **Filter 2**,
- level I bag type EU 5 from cleaning machine, used for mechanical and pneumatic duct cleaning – **Filter 3**,
- level II bag type EU 5 from cleaning machine, used for mechanical and pneumatic duct cleaning – **Filter 4**.

Table 1 presents the parameters of aforementioned filters and bags.

Table 1: Characteristics of examined filters

Filter	Filter pack weight [g]	Filter type	Filter parameters
Filter 1	230,0	EU5	Ventilation unit, intake line,
Filter 2	1100,0	EU5	Ventilation unit, exhaust line,
Filter 3	320,0	EU5	Bag from ventilator of level I cleaning machine (acc. to PN-EN 779)
Filter 4	190,0	EU5	Bag from ventilator of level II cleaning machine

Filters and bags were dismounted very gently and packed in sterile gauze. Complete filters with their casing were weighed in the laboratory, then filter packs were removed and also weighed. Segments were cut out from the removed filter pack and weighed on sterile Petri dishes. After weighing they were placed in glass-stoppered flasks containing 100ml of sterile water. Next, flasks with water and filter pack segments were shaken and the obtained water-based suspension with microorganisms was used for microbiological examination.

Water samples with microbe suspension were used to make deep inoculations on Petri dishes, flooded with proper nutrients. Various kinds of dilutions were used while making inoculations and only those dishes that yielded from a dozen to 100 microorganism colonies were selected for calculations. Knowing the volume of the inoculated fluid, the value of dilution, the count of microbes grown and the weight of the shaken segment of filter pack it was possible to recalculate the results from the culture into 1g of dry filtering material and the entire weight of the filter pack.

While conducting microbiological examination of filters and bags, the following microorganisms were determined: the general count of mesophilic bacteria, the general count of psychrophilic bacteria, the count of Staphylococcus (*Staphylococcus*) mannitol positive (type α) and mannitol negative (type β), the count of *Pseudomonas fluorescens* bacteria, actinomycetes (*Actinobacteria*) as well as the general count of microscopic fungi. Examination results were presented as number of colonies (colony forming unit [CFU]) which had grown on the Petri dish. Microbiological examination utilized nutrients and culture conditions recommended for the analysis of microbiological contamination of air according to Polish Standards PN-89/Z-04111.02 and PN-89/Z-04111.03 (table 2).

Möritz and Martiny (Möritz, Martiny, 1997) used similar methodology of isolating bacteria and microscopic fungi from filters used in HVAC systems.

Table 2: Culture conditions of selected microorganisms according to Polish Standards (PN-89/Z-04111.02, PN-89/Z-04111.03)

Microorganisms	Type of medium	Temperature incubation [°C]	Time incubation [h]
Mesophilic bacteria	Nutrient agar culture	37	48
Psychrophilic bacteria *	Nutrient agar culture	22	72
Staphylococcus	Chapman medium	37	48
Actinobacteria	Pochon medium	26	120
Pseudomonas fluorescens	King B medium and	26	120
	identification of colonies in UV rays	4	168
Microscopic fungi	Waksman medium	26	168
	Czapek-Dox medium	26	168

* not included in the Polish Standards

After culture period was finished, microorganism colonies that had grown on Petri dishes were counted. The obtained results were recalculated into 1g of filtering material and the entire weight of the filter.

3. EXAMINATION RESULTS

Tables 3, 4, 5 and 6 compile the examination results for microorganisms taken from the filtering material of the dismounted filters and bags from the cleaning machine. Microbes obtained from the culture were recalculated into 1g of filtering material and the weight of the entire filter or bag.

Table 3: Average count of mesophilic and psychrophilic bacteria (CFU) abd standard deviation [σ] in the examined filtering material

Filter	Average bacteria count			
	Mesophilic		Psychrophilic	
	CFU/1g of filter	CFU/ entire filter	CFU/1g of filter	CFU/ entire filter
Filter 1 – intake	14 700 σ=1152,9875	3 381 000	25 640 σ=3041,7810	5 897 200
Filter 2 – exhaust	1 235 σ=91,9967	1 358 500	2 980 σ=284,5182	3 278 000
Filter 3 – level I cleaning machine	1 990 σ=636,2199	636 800	19 250 σ=1341,4436	6 160 000
Filter 4 – level II cleaning machine	3 910 σ=150,8235	742 900	15 325 σ=2002,9593	2 911 750

Table 4: Average count of staphylococcus (CFU) and standard deviation [σ] in the examined filtering material

Filter	Average count of staphylococcus			
	Mannitol positive		Mannitol negative	
	CFU/1g of filter	CFU/ entire filter	CFU/1g of filter	CFU/ entire filter
Filter 1 – intake	1 370 σ=184,9711	315 100	170 σ=25,0233	39 100
Filter 2 – exhaust	300 σ=50,8299	330 000	128 σ=9,2171	140 800
Filter 3 – level I cleaning machine	186 σ=58,3811	59 520	62 σ=13,3936	19 840
Filter 4 – level II cleaning machine	80 σ=8,6410	15 200	40 σ=8,6410	7 600

Table 5: Average count of microscopic fungi (CFU) and standard deviation [σ] in the examined filtering material

Filter	**Average count of microscopic fungi**			
	Waksman medium		Czapek-Dox medium	
	CFU/1g of filter	CFU/ entire filter	CFU/1g of filter	CFU/ entire filter
Filter 1 – intake	17 100 σ=1847,0111	3 933 000	20 500 σ=5447,0569	4 715 000
Filter 2 – exhaust	2 550 σ=261,9876	2 805 000	1 280 σ=92,1705	1 408 000
Filter 3 – level I cleaning machine	12 420 σ=1105,0776	3 974 400	16 150 σ=3550,1899	5 168 000
Filter 4 – level II cleaning machine	2 420 σ=183,3578	459 800	7 650 σ=1150,6156	1 453 500

Table 6: Average count of actinomycetes and *Pseudomonas fluorescens* (CFU) and standard deviation [σ] in the examined filtering material

Filter	**Average bacteria count**			
	Actinomycetes		*Pseudomonas fluorescens*	
	CFU/1g of filter	CFU/ entire filter	CFU/1g of filter	CFU/ entire filter
Filter 1 – intake	855 σ=298,2304	196 650	0	0
Filter 2 – exhaust	850 σ=91,8105	935 000	0	0
Filter 3 – level I cleaning machine	2 485 σ=182,8910	795 200	0	0
Filter 4 – level II cleaning machine	1 610 σ=150,6017	305 900	0	0

4. EXAMINATION CONCLUSIONS

Analyzing the count of the examined microorganisms present in 1g of filtering material it is concluded that most numerous microbes in Filter 1 (intake line) were psychrophilic bacteria (25 640 CFU/1g), microscopic fungi (17 100 – 20 500 CFU/1g) and mesophilic bacteria (14 700 CFU/1g). Among staphylococcus, mannitol positive staphylococcus (type α) prevailed over mannitol negative staphylococcus (type β), with eight times the count than the latter. Actinomycetes were found at relatively low numbers – 855 CFU/1g of filtering pack weight (Fig. 2-4). Most numerous microbes in Filter 2 (exhaust line) were also psychrophilic bacteria, microscopic fungi and mesophilic bacteria, however their count was 10 times less than that of Filter 1 in some cases. The count of staphylococcus was also lower, especially mannitol-positive staphylococcus, while actinomycetes indicated similar strength to the ones in Filter 1. In level I cleaning machine bag (Filter 3) the prevailing microbes were psychrophilic bacteria (19 250 CFU/1g) and microscopic fungi (12 420 – 16 150 CFU/1g) in 1g of filtering material, while in level II cleaning machine bag (Filter 4) the count of psychrophilic bacteria and microscopic fungi was lower than in Filter 3, however the count mesophilic bacteria was much higher (by 100%) reaching 3 910 CFU/1g. In both bags, staphylococcus count was low while the count of actinomycetes was higher than in Filter 1 and Filter 2. In all examined filters, no presence of *Pseudomonas fluorescens* was recorded. Figures 2-4 compile average count of microorganisms from all examined filters, recalculated into 1g of filtering material.

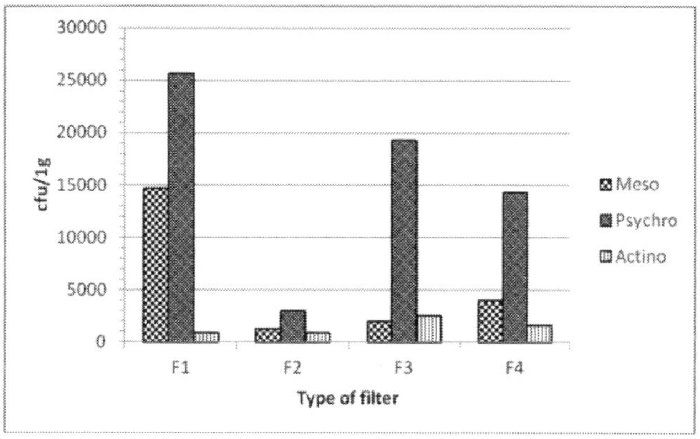

Figure 2: Count of mesophilic bacteria (Meso), psychrophilic bacteria (Psychro) and actinomycetes (Actino), recalculated into 1g of filtering material

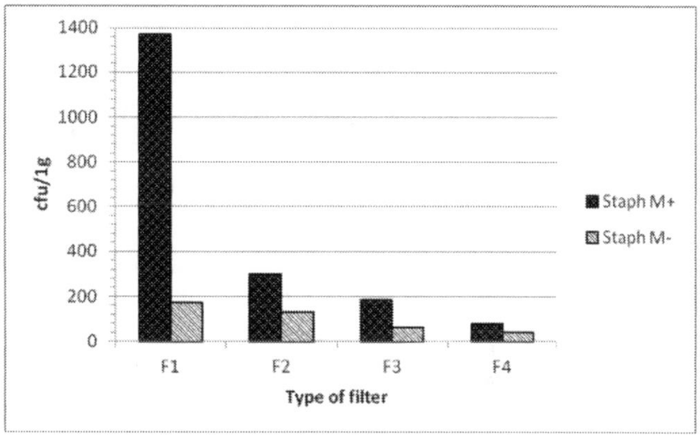

Figure 3: Count of staphylococcus mannitol positive (Staph M+) and mannitol negative (Staph M-) recalculated into 1g of filtering material

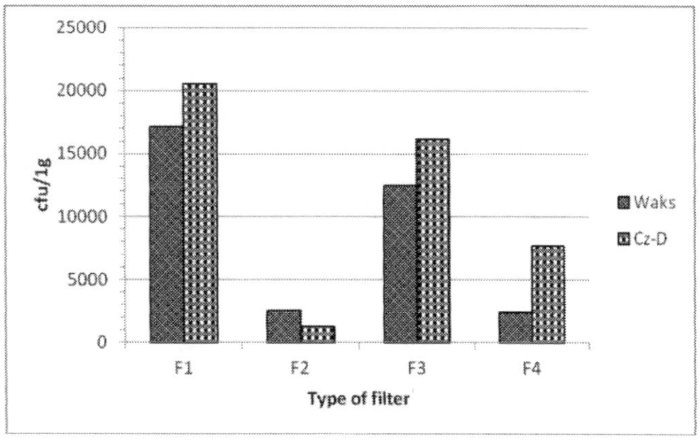

Figure 4: Count of microscopic fungi grown on Waksman medium (Waks) and Czapek-Dox medium (Cz-D) recalculated into 1g of filtering material

Because of different weight of the examined filter packs, other values were obtained after recalculating the count of microorganisms on the weight if the entire filters (table 3-6). Each filter pack had the following weight: Filter 1 = 230 g, Filter 2 = 1100 g, Filter 3 = 320 g,

Filter 4 = 190 g. Considering these values, the highest count of mesophilic bacteria was recorded in Filter 1 (over 3.3 million CFU/230g), the highest count of actinomycetes and staphylococcus mannitol positive and mannitol negative was recorded in Filter 2 (over 935 thousand CFU/1100g, 330 thousand CFU/1100g and over 140 thousand CFU/1100g respectively) and the highest count of psychrophilic bacteria and microscopic fungi was recorded in Filter 3 (over 6.1 million CFU/320g and over 5.1 million CFU/320g respectively). Filter 4 normally yielded the lowest count of the examined microbe groups. One might understand that if Filter 1 and Filter 2 functioned as air cleaners for external air inflowing from outside environment and for air contained indoors, with users inside the building, these filters stopped mostly human-related microbes (mesophilic bacteria and staphylococcus) as well as microscopic fungi that penetrated indoors from outside environment. While the mechanical cleaning of air intake ducts was in progress, numerous impurities found there were released from the surface where they settled and penetrated the bags of the cleaning machine. Thus, especially in level 1 bag (Filter 3), the strength of numerous microbes naturally present in the environment was high (psychrophilic bacteria, actinomycetes, microscopic fungi), while the count of human-related microbes (mesophilic bacteria, staphylococcus) in the bags of the cleaning machine was lower. In vast majority of the examined microorganisms, their count was higher in level 1 cleaning machine bag than in level II bag. This proves, that the first stage of duct cleaning was already very effective in terms of eliminating bacteria and fungi that settled in the lines supplying air indoors to the rooms of the passive house.
In order to assess the influence of filter replacement and air duct cleaning on the quality of indoor air in the passive house, the level of microbiological contamination in the air was analyzed before and after filter replacement. Air examination, performed with MAS-100 Eco microbiological air sampler and using impact method, was conducted by the same company that performed the ventilation system cleaning and disinfecting process. The examination revealed that, after a complex procedure, the microbiological quality of air supplied to the rooms of the passive house was good. At the same time, it was revealed that after cleaning and disinfecting the system, a decrease in the general count of microbes was noted, compared to the quality of air before replacing the filters and cleaning the ducts (table 7). During identification process of the detected microorganisms, microbes were grown that form saprophytic and opportunistic flora represented by *Micrococcus spp.*, mould fungi and sporulating bacilli.

Table 7: Microbiological cleanness of air from intake duct before and after the ventilation system cleaning and disinfecting process

Determination	Before cleaning [CFU/m^3]	After cleaning [CFU/m^3]
General microorganism count	107	40
Air pollution level [4]	Moderate	Very low

Air quality assessment was also performed in the experimental house, according to the guidelines of Polish Standards (PN-89/Z-04111.02, PN-89/Z-04111.03). The examination confirmed an improvement in the quality of air and a decrease in the general count of mesophilic bacteria, psychrophilic bacteria, staphylococcus bacteria (*Staphylococcus*) mannitol positive (type α) and mannitol negative (type β), *Pseudomonas fluorescens*, actinomycetes (*Actinobacteria*) and the general count of microscopic fungi.

5. CONCLUSIONS

Usage period of filters and the condition of the ventilation system in the building has a very big influence on the quality of indoor air. Polluted filters in the passive house (DoPas) caused a drop in the effectiveness of air filtration. The impurities gathered and the growth of microorganisms in filters could influence their filtrating capabilities making the filters contribute to a deterioration of the quality of indoor air. Over many years, ventilation ducts had gathered numerous physical and gas impurities. Apart from blowing new impurities from external air, various substances concentrated in ventilation ducts could also penetrate indoors. They could be dust and solids dated back to the construction of the system, soot, grease particles, gases and bioaerosol-forming biological impurities, including flower pollen, viruses, bacteria and fungi. They can impair the well-being of people in the building. It depends on the count of substances blown indoors and the type of negative effect they have on humans. Ventilation system contamination is a problem connected both with the health of residents in the ventilated rooms and with the technical condition of the system which for instance may be subject to biological corrosion due to large numbers of microbes. Such situation contributed to the deterioration the quality of indoor air, one symptom of which being the so-called sick building syndrome (SBS). The occurrence of ailments related to SBS is mainly connected with insufficient amount of supplied clean air and its bad quality. The sources of indoor air contamination may be construction materials, furnishing elements, also residents themselves, microbe metabolites (endotoxins, enterotoxins, mycotoxins, glucans), contaminated ventilation and air conditioning systems and low quality of outdoor air. Symptoms of SBS include headaches, dizziness, fainting, malaise, fatigue, allergic reactions and other ailments described by researchers such as (Charkowska, 2003; Joshi, 2008; Gąska-Jędruch, Dudzińska, 2009; Joe et al., 2013; Miaśkiewicz-Pęska, Łebkowska, 2011).

With favorable temperature, humidity and access to nutrients, air filters can become a habitat for numerous microbes. Examining the filtering material can yield the presence of both harmless and pathogenic microorganisms, among them dangerous species of bacteria, e.g. *Pseudomonas aeruginosa* and *Legionella pneumophila*, as well as fungi, e.g. *Alternaria spp.*, *Aspergillus spp.*, *Fusarium spp.*, *Mucor spp.*, or *Rhizopus spp.* In order to limit the growth and colonization of microbes in filters and air ventilation and conditioning systems, it is necessary to perform periodical maintenance and cleaning of the systems and use antibacterial agents (e.g. $AgNO_3$), to cover the filtering material (Charkowska, 2000; Miaśkiewicz-Pęska, Łebkowska, 2011).

Organic particles stopped in filters are a source of carbon for microbes and moisture contained in the flowing air is their source of water. Such conditions not only help microbes survive but also help them grow in significant numbers. As a result, it leads to an emission of microorganisms from filters and secondary contamination of the treated air. An examination of used filters conducted by Miaśkiewicz-Pęska (Miaśkiewicz-Pęska et al., 2007) revealed very high concentration of living microorganisms, among which most numerous were moulds (10 000 CFU/cm^3 of filter), and the number of bacteria reached 6000 CFU/cm^3 (Miaśkiewicz-Pęska et al., 2007).

Therefore, it is necessary to perform periodical inspection, cleaning and disinfection of ventilation systems as well as regular filter replacement (according to manufacturer recommendation) to ensure proper operation of the system and to create air comfort in rooms (Fanger et al., 2003; Gołofit-Szymczak, Skowroń, 2005; Charkowska, 2000; Miaśkiewicz-Pęska et al., 2007).

6. REFERENCES

Chin S. Yang. (1999). *Biological contamination in the HVAC system.* www. invironment.com.

Ahearn, D.G.; Crow, S.A.; Simmons, R.B.; Price, D.L.; Mishra, S.K.; Pierson, D.L. (1977). *Fungal colonization of air filters and insulation in a multi-story office building.* Production of volatile organics. Curr. Microbiol., 35, 305-308.

PN-89/Z-04111.02. (Polish Standard). *Air purity protection. Microbiological testings. Determination number of bacteria in the atmospheric air (imision) with sampling by aspiration and sedimentation method (Badania mikrobiologiczne. Oznaczenie liczby bakterii w powietrzu atmosferycznym (imisja) przy pobieraniu próbek metodą aspiracyjną i sedymentacyjną)* [In Polish].

PN-89/Z-04111.03. (Polish Standard). *Air purity protection. Microbiological testings. Determination number of the fungi in the atmospheric air (imision) with sampling by aspiration and sedimentation method (Badania mikrobiologiczne. Oznaczenie liczby grzybów mikroskopowych w powietrzu atmosferycznym (imisja) przy pobieraniu próbek metodą aspiracyjną i sedymentacyjną)* [In Polish].

Möritz, M.; Martiny, H. (1997). *Fungi and bacteria on air filters from heating, ventilation, and air-conditioning systems: A method for the determination of fungi and bacteria on air filters.* Zentralbl Hyg Umweltmed., 199(6), 513-526.

Charkowska, A. (2003). *Contamination of air conditioning systems and methods to eliminate it (Zanieczyszczenia w instalacjach klimatyzacyjnych i metody ich usuwania).* IPPU MASTA Sp. z o.o. [In Polish].

Joshi, S.M. (2008). *The Sick building syndrome.* Indian J Occup Environ Med., 12(2), 61-64.

Gąska-Jędruch, U.; Dudzińska, M.R. (2009). *Microbiological pollution in indoor air (Zanieczyszczenia mikrobiologiczne w powietrzu wewnętrznym).* Polska Inżynieria Środowiska pięć lat po wstąpieniu do Unii Europejskiej, Monografie Komitetu Inżynierii Środowiska, T 2, 59, 31-40 [In Polish].

Joe, Y.H.; Ju, W.; An, J.H.; Hwang, J. (2013). *A quantitative determination of the antibacterial efficiency of fibrous air filters based on the disc diffusion methods.* Aerosol and Air Quality Research, 14, 928-933.

Miaśkiewicz-Pęska, E.; Łebkowska, M. (2011). *Effect of Antimicrobial Air Filter Treatment on Bacterial Survival.* FIBRES & TEXTILES in Eastern Europe, 19, 1 (84), 73-77.

Charkowska, A. (2000). *Contamination of air conditioning systems and its elimination (Zanieczyszczenia instalacji klimatyzacyjnych i ich usuwanie).* Problemy jakości powietrza wewnętrznego w Polsce'99. Materiały konferencyjne, Wydawnictwo Instytutu Ogrzewnictwa i Wentylacji Politechniki Warszawskiej, Warszawa, 19-37 [In Polish].

Miaśkiewicz-Pęska, E.; Karwowska, E.; Grzybowski, P. (2007). *Microorganisms in office buildings. Biological purity of ventilation system elements (Mikroorganizmy w budynkach biurowych. Biologiczna czystość elementów instalacji wentylacyjnej).* Chłodnictwo i Klimatyzacja, 8, 60-65 [In Polish].

Fanger, P.O.; Popiołek, Z.; Wargowski P. (2003). *Indoor environment. Its influence on health, comfort and productivity (Środowisko wewnętrzne. Wpływ na zdrowie, komfort i wydajność pracy).* Gliwice [In Polish].

Gołofit-Szymczak, M.; Skowroń, J. (2005). *Microbiological threats in office rooms (Zagrożenia mikrobiologiczne w pomieszczeniach biurowych).* Bezpieczeństwo pracy, 3, 29-31 [In Polish].

COMPARISON OF TWO VENTILATION CONTROL STRATEGIES IN THE FIRST NORWEGIAN SCHOOL WITH PASSIVE HOUSE STANDARD

Axel Cablé[*1], Hugo Lewi Hammer[2], and Mads Mysen[1,2]

1 SINTEF
Forskningsveien 3B
N-0314, Oslo, Norway
**Corresponding author: axel.cable@sintef.no*

2 Oslo and Akershus University College of Applied Sciences
Pilestredet 35
N-0166, Oslo, Norway

ABSTRACT

The Marienlyst School is the first educational building in Norway built according to the passive house standard. This building benefits from a super-insulated and airtight envelope. While this reduces the heating demand largely, it also enhances the risk for poor indoor air quality and overheating compared to conventional buildings. It is therefore particularly important to implement an efficient ventilation strategy in order to avoid adverse effects on the health, well-being and productivity of the pupils.

In this context, the perceived indoor climate resulting from two different ventilation control strategies was evaluated in one classroom of the building. Both strategies consisted in varying the ventilation rate according to room demand, *ie.* Demand Controlled Ventilation (DCV). The existing strategy consisted in varying the ventilation rate in order to maintain a constant carbon dioxide concentration of 800 ppm in the classroom. A new strategy was implemented which consisted in a combined CO_2 and temperature DCV, *ie.* to control towards a proportionally lower CO_2 concentration when the indoor temperature increased. The aim with this strategy was to address both overheating and the fact that perceived indoor air quality decreases when temperature rises.

Indoor climate measurements, as well as questionnaires on the perceived indoor air quality and thermal comfort filled up by the pupils were used to compare both strategies. The data from the questionnaires were then analyzed using a random effect linear regression model. The regression analysis revealed that the initial ventilation strategy was responsible for discomfort resulting from too high variations in the indoor temperature. The new combined CO_2 and temperature DCV strategy provided a perceived indoor climate which was significantly better than the existing strategy. Therefore, the developed ventilation strategy appears to be a relevant solution in order to address the problem of overheating and perceived indoor air quality in educational buildings with passive house standard.

KEYWORDS

Demand-Controlled Ventilation, passive house, questionnaires, indoor climate, school

1. INTRODUCTION

The Marienlyst School is the first educational building in Norway built according to the passive house standard (NS 3701, 2012); see Fig.1. It has been operational since 2010, with around 500 pupils of age 13 to 15. This building benefits from a super-insulated and airtight envelope (U-values ranging from 0.05 to 0.12 W/m²K for the envelope, and air leakage inferior to 0.6 vol/h under 50 Pa). While this reduces the heating demand largely, it also enhances the risk for poor indoor air quality and overheating compared to conventional buildings. It is therefore particularly important to implement an efficient ventilation control strategy in order to avoid adverse effects on the health, well-being and productivity of the pupils and school staff (Wargocki, 2000) (Sundell, 2011).

Figure 1: a) Marienlyst school, Oslo. b) View of a classroom.

In educational buildings, large variations of occupancy occur between periods with empty and occupied classrooms. This results in large variations of the heat and pollutant loads in the rooms, which is especially challenging in terms of control of the ventilation system.

Demand Controlled Ventilation systems (DCV) consist in varying the ventilation rate according to a demand measured at room level, and seems therefore particularly fitted to educational buildings. In fact, previous studies have reported that DCV systems can reduce the energy use due to ventilation in the average classroom up to 51% compared to a system operating with full airflow from 7:00 am to 5:00 pm (Mysen, 2005). It is however necessary to evaluate how DCV strategies perform in practice, particularly in the new context of educational buildings with low energy demand.

In this prospect, a new combined CO_2 and temperature DCV strategy was implemented in a classroom of the building. The objective of this study was to assess whether the new strategy provided a better perceived indoor climate than the existing ventilation control strategy, using both indoor climate measurements and questionnaires filled in by the pupils.

2. METHODOLOGIES

2.1 Ventilation control strategies

The existing control strategy in the building consisted in varying the ventilation rate in order to maintain a constant carbon dioxide concentration of 800 ppm in the classroom (constant CO_2 control). In fact, the indoor CO_2 concentration can be used as a proxy for the bio-effluents concentration in the room. Staying under a given value therefore guarantees a certain indoor air quality level.

In addition, an "overheating mode" was included in the existing strategy. It consisted in forcing the opening rate of the Variable Air Volume ventilation dampers to 100% when the indoor temperature reached 23.8°C and until the temperature got down to 23°C, independently of the CO_2 concentration in the room. This resulted in a larger amount of outdoor air being supplied through the ventilation system, which aimed to prevent overheating in the classroom.

A new strategy was implemented which consisted in a combined CO_2 and temperature control. It consisted in controlling the ventilation rate towards a proportionally lower CO_2 concentration when the indoor temperature increased over 22.5°C; see Figure 2. The aim with this strategy was to address both overheating and the fact that perceived indoor air quality decreases when indoor temperature rises. In fact, previous studies have revealed that a lower concentration of bio-effluents is perceived as acceptable when the indoor temperature is high (Mysen, 2005). As a consequence, the ventilation rate should be increased for higher indoor temperatures in order to maintain an acceptable perceived indoor air quality in the classrooms.

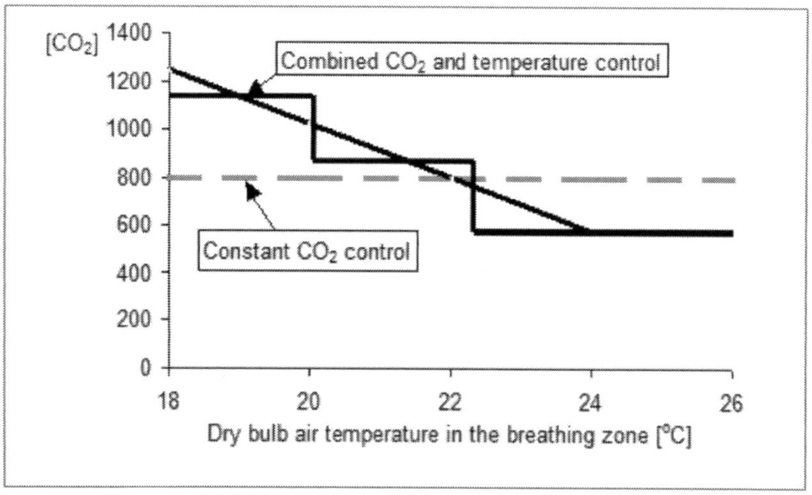

Figure 2: Combined CO_2 and temperature control (Mysen, 2005).

2.2 Questionnaires and regression analysis

In order to compare the performance of the two ventilation strategies described above, a case cross over study was carried out in one classroom of the building during February, March, and May 2012, see Table 1.

Table 1: Date and ventilation control for the interventions

Date	20 February 2012	14 March 2012	30 May 2012
Ventilation control	Constant CO_2 with "overheating mode"	Combined CO_2 and temperature	Combined CO_2 and temperature

At the end of each class session, the pupils were given a questionnaire with 19 questions relating to Sick Building Syndrome-symptoms and perceived indoor climate. On each question, the pupil gave a value between 0 and 1, ranging from very comfortable to very uncomfortable. The questionnaire also included health questions, such as whether a pupil suffered from asthma or cold.

Data from the questionnaires were analyzed using a random effect linear regression model with the score for each question as the dependent variable. The two ventilation control strategies: "Constant CO_2 control" and "Combined CO_2 and temperature control" represent the main independent variable. We also include gender and whether the pupils suffered from cold or asthma as independent variables, in order to be able to compare the ventilation control strategies independently of these factors. Statistical analyses were performed using the program R (R development team, 2013) and the R package lme4 (Bates, 2013).

3 RESULTS

3.1 Actual conditions during the interventions

The CO_2 concentration inside of the room, as well as indoor temperature were measured every 2 minutes.

The measurements during the intervention on February 20 with constant CO_2 control are plotted on Figure 3. The setpoint for the indoor CO_2 concentration derived from the ventilation control strategy is also indicated on Figure 3, as well as the periods where the "overheating mode" was used (override of the constant CO_2 control and full opening of the ventilation dampers).

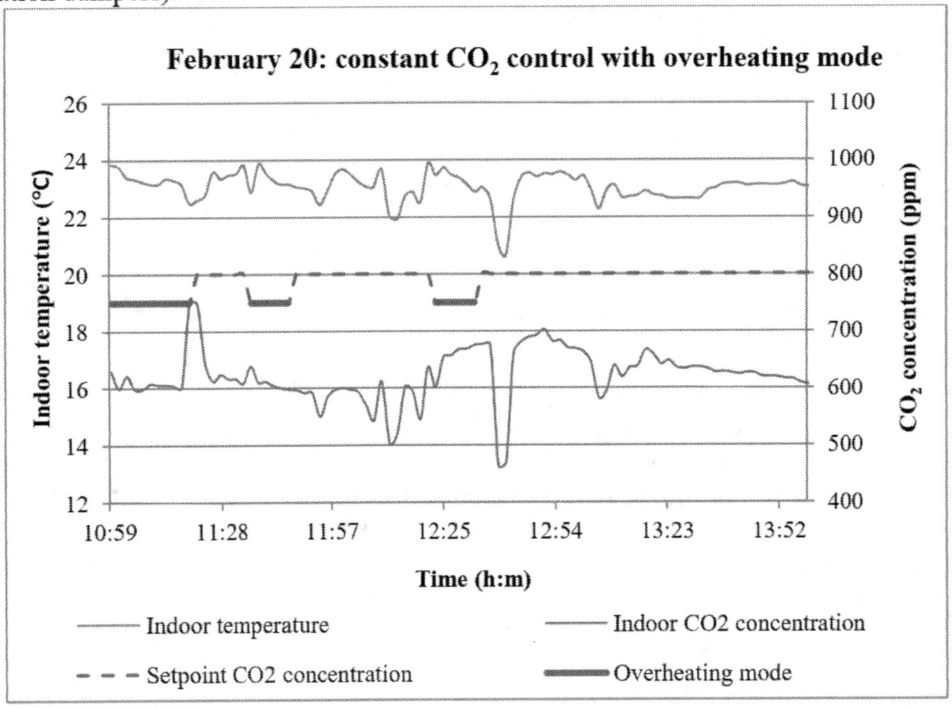

Figure 3: Measured indoor climate parameters on February 20 (constant CO_2 control with "overheating mode")

The measurements during the interventions on March 14 and May 30 with combined CO_2 and temperature control are plotted on Figure 4 and Figure 5, respectively. Similarly, the setpoint for the indoor CO_2 concentration derived from the curve presented on Figure 2 is also indicated.

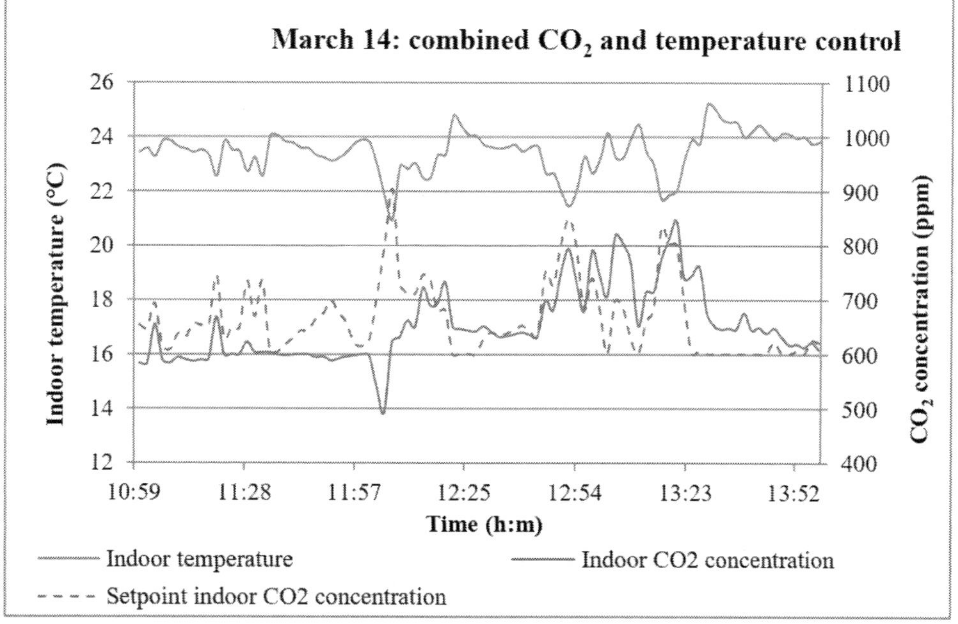

Figure 4: Measured indoor climate parameters on March 14 (combined CO_2 and temperature control)

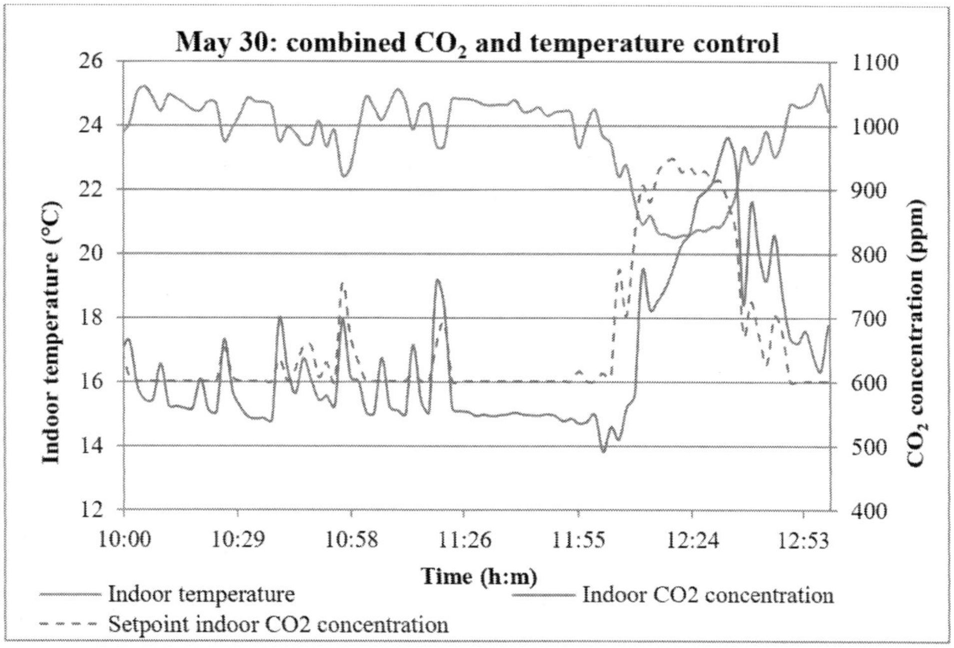

Figure 5: Measured indoor climate parameters on May 30 (combined CO_2 and temperature control)

The following observations were made:

- On February 20, the indoor temperature reached 23.8°C on three occasions, causing the "overheating mode" to override the constant CO_2 control of the ventilation rate, and supplying an increased amount of fresh air into the room. As a consequence, the indoor CO_2 concentration was lower than 800 ppm. A clear drop of the indoor temperature and indoor CO_2 concentration can be noticed subsequently to each of period of "overheating mode".

- On March 14 and May 30, the indoor temperature exceeded 22.5°C. As a consequence, the combined CO_2 and temperature control strategy controlled towards a lower concentration of CO_2 (see the red dotted line on Figure 4), down to 600 ppm. The actual CO_2 concentration in the room agreed fairly well with the CO_2 concentration setpoint on both days.
- On May 30, the classroom was empty from 12:00 to 12:30. As a consequence, the indoor temperature dropped down to 20.5°C, which caused the combined CO_2 and temperature control strategy to control towards a higher concentration of CO_2. This illustrates the ability of this ventilation control strategy to help saving energy by reducing the ventilation rate when the room is empty.
- The temperature in the classroom was relatively high for all cases, ranging from 23.1°C to 23.7°C on average. This is most certainly due to the high efficiency of the building envelope. Moreover, the temperature was relatively similar during all three interventions; see Table 2 for average values during the interventions.
- Likewise, the indoor CO_2 concentration was similar for all cases and relatively low, ranging from 622 ppm to 652 ppm, see Table 2.

Table 2: Average indoor temperature and CO_2 concentrations during the interventions

	Average indoor temperature (°C)	**Average indoor CO_2 concentration (ppm)**
20 February (constant CO2 control with overheating mode)	23,1	622
14 March (combined CO_2 and temperature control)	23,5	652
30 May (combined CO_2 and temperature control)	23,7	635

3.2 Regression analysis

39 pupils answered to the questionnaires. Among them, 19 answered to all three questionnaires and were considered in the regression analysis.

The regression strategy revealed that there is a significant relationship between the ventilation control strategy and to what extent the pupils are bothered by temperature variations in the classroom (p-value = 0.018); see Table 5 for the results from the linear regression. The new combined CO_2 and temperature strategy reduced the discomfort by variations of the indoor temperature significantly compared to the existing strategy (constant CO_2 control with "overheating mode"). The fact that the pupil had a cold, or the gender of the pupil did not have any significant impact on the results.

The other questions did not reveal any significant difference between both control strategies.

Table 5: Results from the linear regression

Variables	Estimate	Std. Error	Df	T-value	P-value
Combined CO_2 and temperature control (reference: constant CO_2 control)	–0.28	0.11	26.24	–2.53	0.018 *
Gender male (reference: female)	–0.22	0.14	19.33	–1.62	0.121
Has a cold (reference: Does not have a cold)	0.15	0.12	34.45	1.21	0.238

*Signif. codes: p-value < 0.01:**, p-value < 0.05:**

4 DISCUSSION

The regression analysis revealed that the initial ventilation strategy was responsible for discomfort resulting from too high variations of the indoor temperature. The hypothesis is that the rough control of the ventilation rate for this control strategy was accountable for this.

In fact, right after a period under "overheating mode" with high ventilation rate, the indoor CO_2 concentration was much lower than 800 ppm; see Figure 3. As a consequence, as soon as the constant CO_2 control mode was active again (*ie.* indoor temperature 23°C), it controlled towards the lowest ventilation rate. This resulted in fluctuations between maximum and minimum ventilation rate, which may be responsible for the obtained results.

The combined CO_2 and temperature control strategy allowed to have more gradual variations of the ventilation rate according to room temperature, which is likely the reason why it allowed to significantly reduce the perceived discomfort by temperature variations.

Moreover, the results from the measurements revealed that the indoor temperature was somewhat high in the classroom, which is certainly accountable to the efficient building envelope. This underlines the importance of having a ventilation control strategy capable of providing a higher ventilation rate when this occurs, in order to maintain an acceptable perceived indoor air quality.

5 CONCLUSIONS

A combined CO_2 and temperature ventilation control strategy was efficiently implemented in a classroom of the first school in Norway with passive house standard.

Questionnaires concerning the perceived indoor comfort and indoor air quality were carried out. The regression analysis revealed that the existing ventilation strategy was responsible for discomfort resulting from too high variations in the indoor temperature.

The new combined CO_2 and temperature strategy provided a perceived indoor climate which was significantly better than the existing strategy. Therefore, this ventilation strategy appears to be a relevant solution in order to address the problem of overheating and perceived indoor climate in educational buildings with passive house standard.

6 ACKNOWLEDGEMENTS

This paper is funded by contributions from the industry partners VKE, Undervisningbygg Oslo KF, Skanska, Optosense AS, MicroMatic AS, Swegon AS and TROX Auranor AS, and public funding from the Norwegian Research Council as part of R&D project "reDuCeVentilation".

7 REFERENCES

Standard Norge NS 3701 (2012) Criteria for passive houses and low energy buildings.

Wargocki P, et al. (2000) The effects of outdoor air supply rate in an office on perceived air quality, sick building syndrome (SBS) symptoms and productivity. Indoor Air, 10, 4: 222-236.

Sundell, J., Levin, H., Nazaroff, W. W., Cain, W. S., Fisk, W. J., Grimsrud, D. T., ... & Weschler, C. J. (2011). Ventilation rates and health: multidisciplinary review of the scientific literature. Indoor air, 21(3), 191-204.

Mysen, M. (2005). Ventilation systems and their impact on indoor climate and energy use in schools: studies of air filters and ventilation control.

Bates, D., Maechler M., Bolker B. and Walker S. (2013) lme4: Linear mixed-effects models using Eigen and S4, R package version 1.0-5. Available at http://CRAN.R-project.org/package=lme4

R Development Core Team (2013) R: A Language and Environment for Statistical Computing, R Foundation for Statistical Computing, Vienna, Austria. Available at http://www.R-project.org/.

SELF-EVALUATED THERMAL COMFORT COMPARED TO MEASURED TEMPERATURES DURING SUMMER IN THREE ACTIVE HOUSES WHERE VENTILATIVE COOLING IS APPLIED

Peter Foldbjerg *[1], Thorbjørn Færing Asmussen[1], Moritz Fedkenheuer[2], and Peter Holzer[3]

1 VELUX A/S
Daylight Energy and Indoor Climate
Ådalsvej 99
2970 Hørsholm, Denmark
**Corresponding author: peter.foldbjerg@velux.com*

2 Humboldt-Universität zu Berlin
Unter den Linden 6
10099 Berlin, Germany

3 Institute for Building research & Innovation
Luftbadgasse 3/8
A - 1060 Wien, Austria

ABSTRACT

The thermal comfort of the residential buildings Sunlighthouse in Austria and LichtAktiv Haus in Germany are investigated with a particular focus on the summer situation and the role of solar shading and natural ventilation. The houses have generous daylight conditions, and are designed to be CO_2 neutral with a good indoor environment. The thermal environment is evaluated according to the Active House specification (based on the adaptive method of EN 15251), and it is found that the houses achieve category 1 for the summer situation. It is found that ventilative cooling through window openings play a particularly important role in maintaining thermal comfort in the houses and that both window openings and external solar shading is used frequently. The occupants have reported their thermal sensation daily during most of a one-year period. The thermal sensation votes show good correlation with the thermal comfort category.

KEYWORDS

Thermal comfort; ventilative cooling; residential buildings; solar shading, Post Occupancy Evaluation

1 INTRODUCTION

Five single-family houses in five European countries were built between 2009 and 2011 as a result of the Model Home 2020 project. Sunlighthouse (SLH) in Austria and LichtAktiv Haus (LAH) in Germany were both completed in 2011. The houses were occupied by test families in a one-year period (SLH) and a two-year period (LAH), and measurements were made during the period in both houses (Foldbjerg, 2013). The present paper focuses on the occupant's evaluation of the thermal comfort.

The houses follow the Active House principles (Eriksen, 2011), which mean that a balanced priority of energy use, indoor environment and connection to the external environment must be made. The design has particularly focused on excellent indoor environment and a very low use of energy. There is a particular focus on good daylight conditions and fresh air from natural ventilation.

Measurements of IEQ include light, thermal conditions, indoor air quality, occupant presence and all occupant interactions with the building installations, including all operations of windows and solar shading. Use of natural ventilation for summer comfort is based on ventilative cooling principles (Venticool, 2014).

The presented results focus on thermal conditions, and the occupant's evaluation of the thermal environment. Some demonstration houses in Scandinavia have experienced problems with overheating, often due to insufficient solar shading and use of natural ventilation (Isaksson, 2006 and Larsen, 2012). Two British government reports similarly find that both new and refurbished low energy residential buildings have an increased risk of overheating (AECOM, 2012 and Carmichael, 2011).

The houses use natural ventilation in the warm part of the year to prevent overheating. Additionally there is external automatic solar shading on all windows towards south, east and west.

Figure 1. Photos of Sunlighthouse (left) and LichtAktiv Haus (right) by Adam Mørk.

Each room is an individual zone in the control system, and each room is controlled individually. There are sensors for humidity, temperature, CO_2 and presence in each room. The building occupants can override the automatic controls, including ventilation and solar shading at any time. Override buttons are installed in each room, and no restrictions have been given to the occupants. As house owners they have reported a motivation to minimise energy use on an overall level, and to maximise IEQ on a day-to-day basis.

The recorded temperature data is evaluated according to the Active House specification (Eriksen, 2011), which is based on the adaptive approach of EN 15251 (CEN, 2007). The results presented here are based on the measurements and analyses for the period in which test family have occupied the house, i.e. from March 1, 2012 to February 28, 2013.

The occupants responded to a questionnaire every day, where they reported their thermal sensation on a simplified 5-step version of the ISO 7730 (ISO, 2005) thermal sensation scale (as used in the present paper: hot, warm, neutral, cool, cold). They were asked to provide a response that represented the entire day.

2 RESULTS

Figure 2 shows thermal comfort categories for the two houses. Sunlighthouse experiences temperatures in category 1 for 85% of the year or more for most main rooms. The temperatures outside category 1 are mainly in category 2 (low), i.e. between 20°C and 21°C. Temperatures above category 1 are very limited, and all main rooms achieve category 1 when temperatures below category 1 are disregarded.

LichtAktiv Haus shows very similar performance for the summer situation, with very few hours above category 1 for the main rooms. The "Hall" and "Technique" rooms are secondary rooms and not used for occupancy.

Given the large glazed areas of both houses, the practically non-existing overheating in both houses is remarkable.

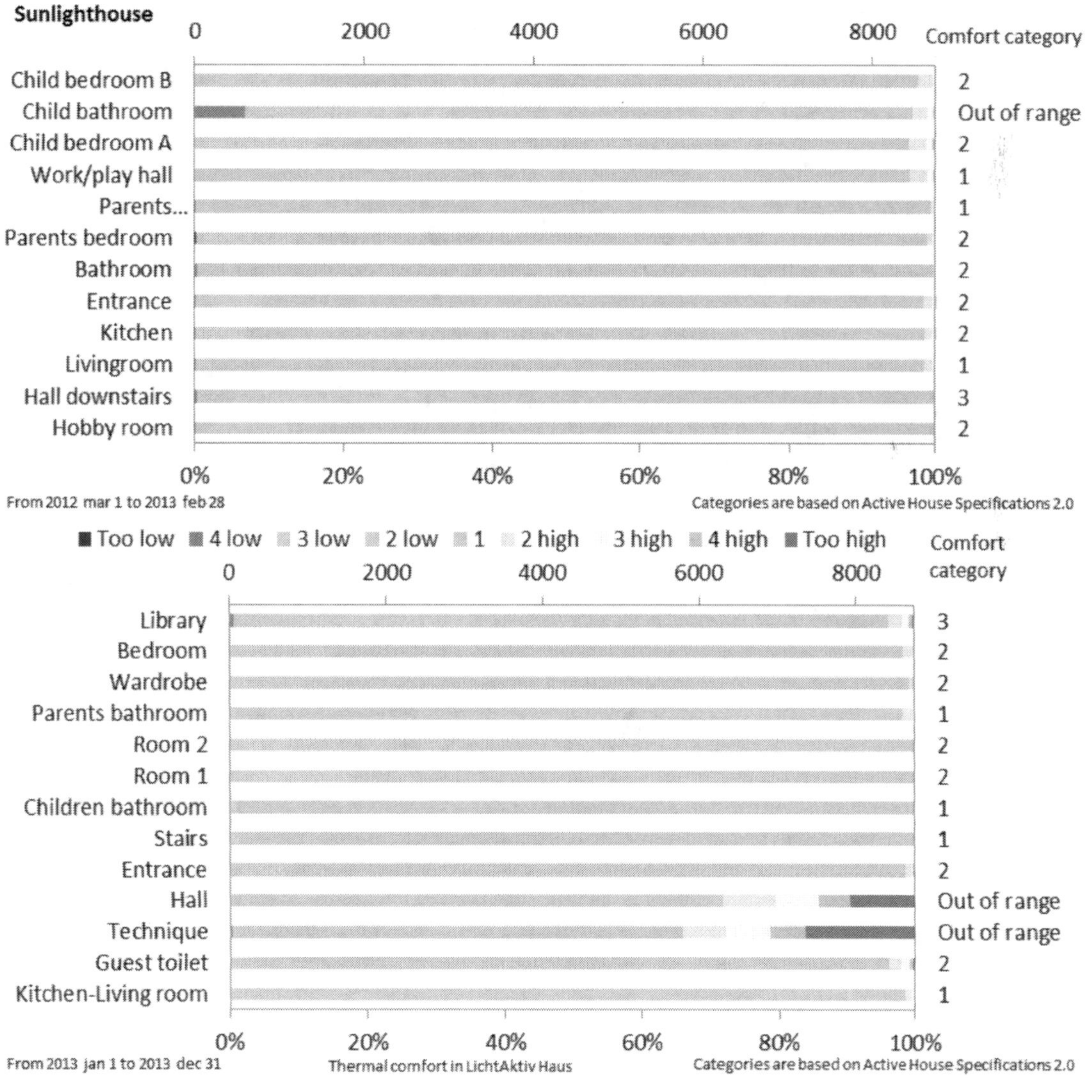

Figure 2. Sunlighthouse (above) and LichtAktiv Haus (below). Thermal comfort for each of the rooms evaluated according to Active House specification (based on adaptive method of EN 15251). Criteria are differentiated between high and low temperatures.

The results in Figure 2 sum up the rooms' performance as regards thermal summer comfort over the stretch of one year. The variation over time-of-day and time-of-year is further investigated in

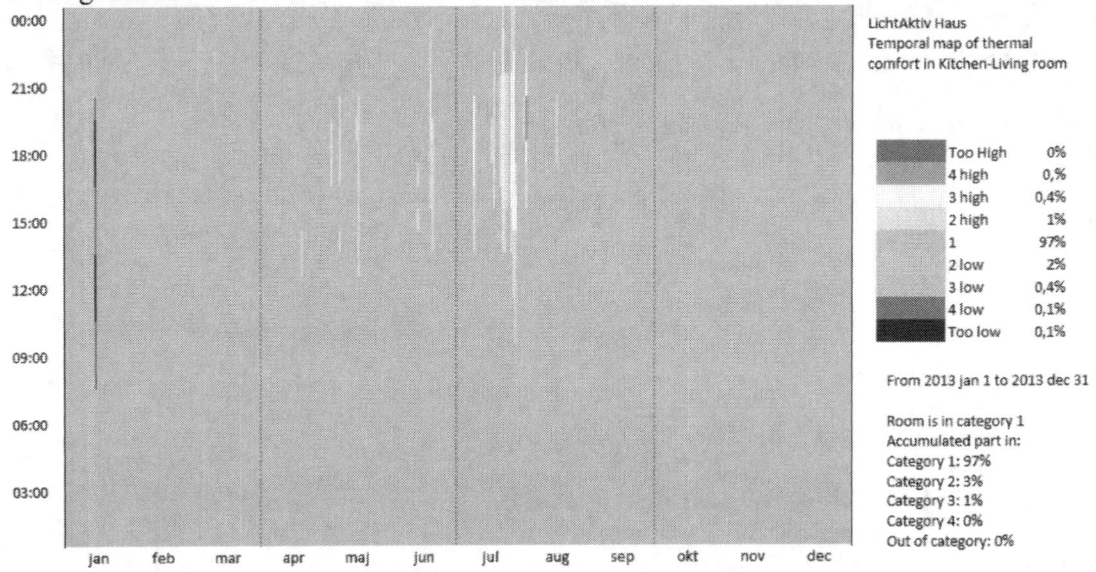

Figure 3, which is using temporal maps (carpet plots), indicating each hour of the year according to its position in the day-of-year (horizontal axis) and time-of-day (vertical axis). The figure focuses on the kitchen-living room in each house, as both of these rooms have particularly large glazed areas and therefore an increased risk of overheating.

For Sunlighthouse, three to five episodes with temperatures below category 1 are seen, each lasting a day or two. In June, a few episodes with temperatures that reach category 2 are observed between 16:00 and 23:00. These episodes last for 2-3 days. LichtAktiv Haus experience similar results.

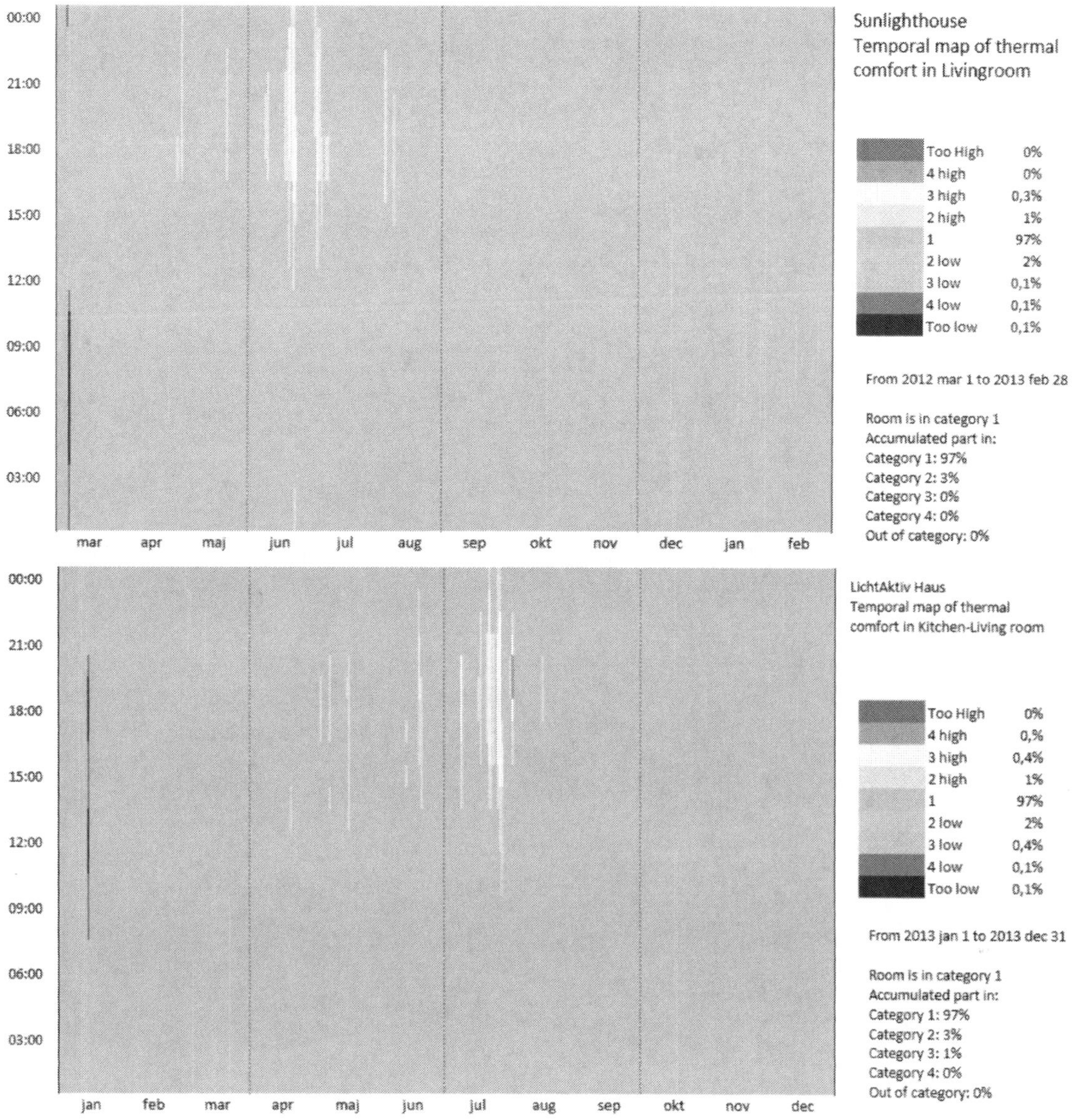

Figure 3. The comfort category for the kitchen-living room of each hour of the year is plotted as a temporal map. Sunlighthouse (above) and LichtAktiv Haus (below). Evaluated according to Active House specification (based on adaptive method of EN 15251). Criteria are differentiated between high and low temperatures.

The important role of window openings and solar shading in maintaining thermal comfort has been reported previously (Foldbjerg, 2013).

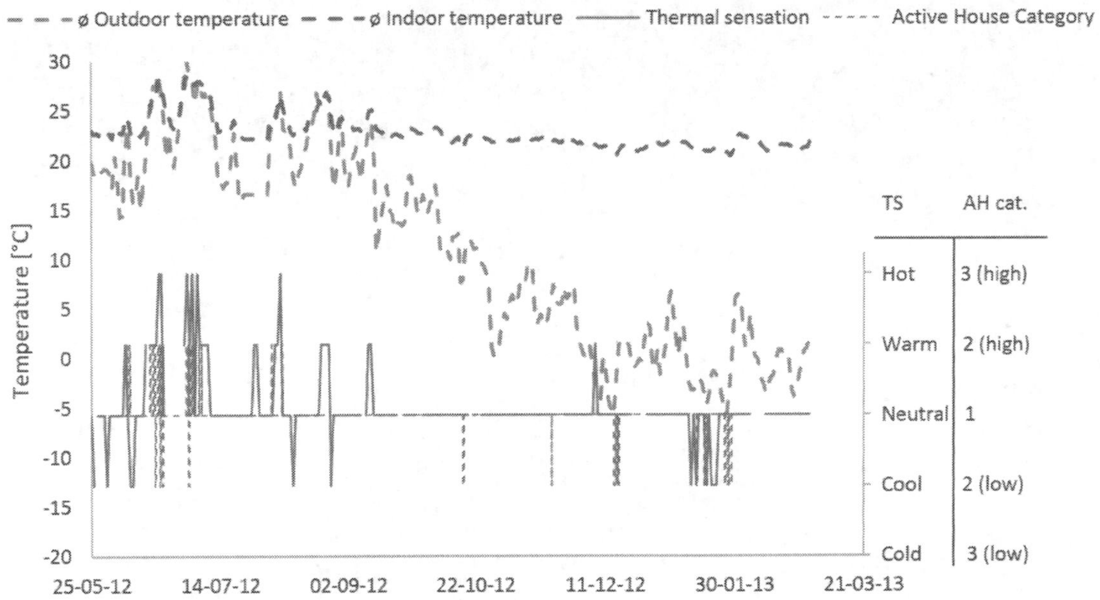

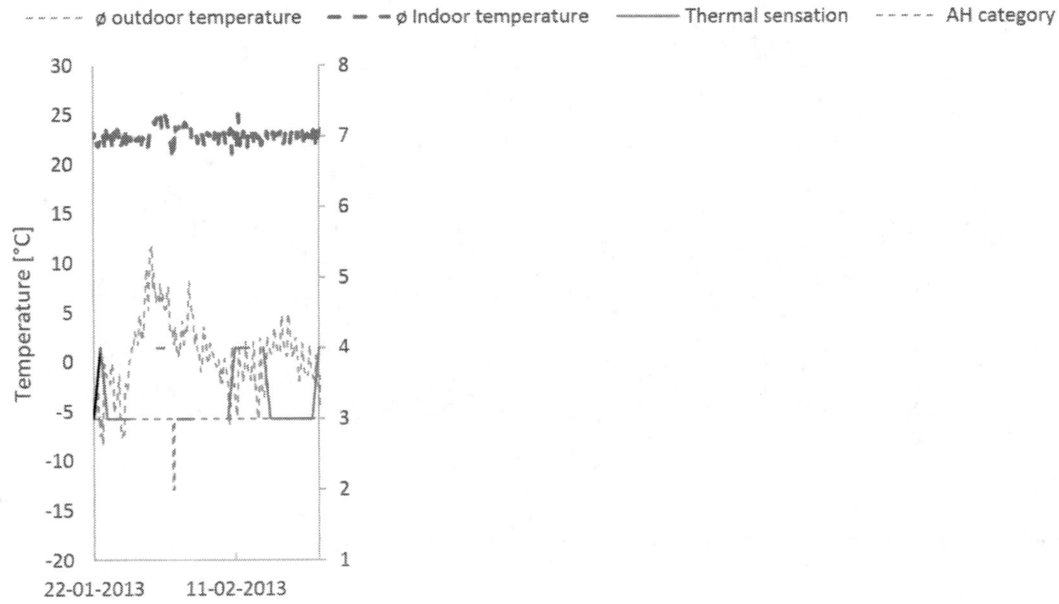

Figure 4 shows measured temperatures, thermal comfort category, and occupant-reported thermal sensation in the two houses. During the very warm period in Austria at the end of June and beginning of July, the outdoor temperature reached 30°C. The peak indoor temperatures in Sunlighthouse are 1 to 2°C lower during these warm periods, which means that the cooling potential of the outdoor air is used to a high degree. The indoor temperature does rise above category 1, but not higher than category 2.

In the summer period the occupants of SLH rate their thermal sensation as "warm" or "hot". The same correlation is seen in the beginning of August. During the last weeks of January 2013, the indoor temperature drops slightly below 20°C, which corresponds to category 2 (low). On these days the occupants reported their thermal sensation as cool.

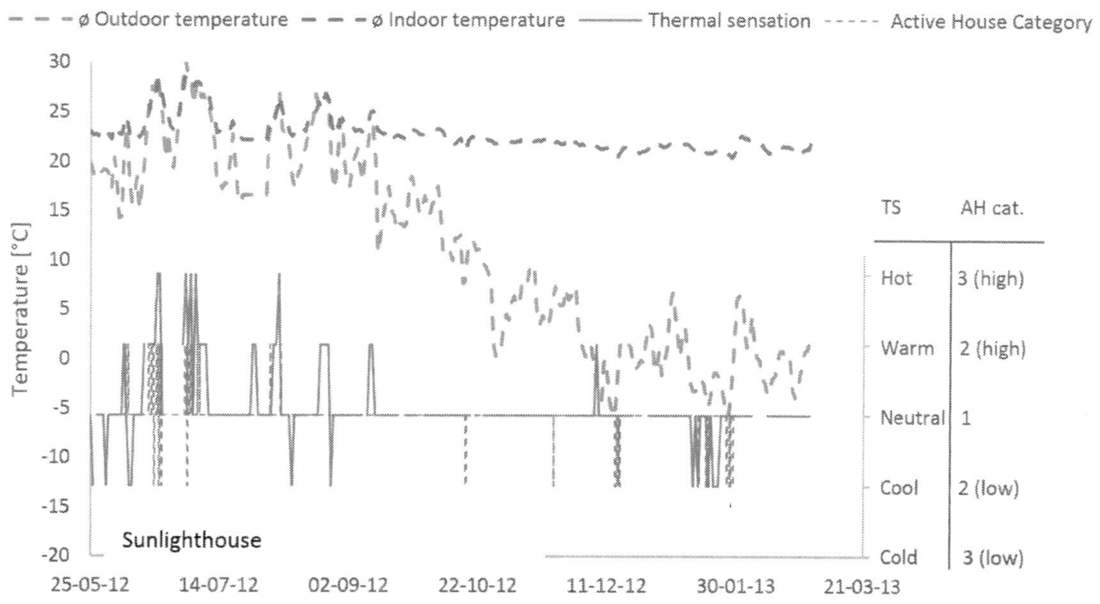

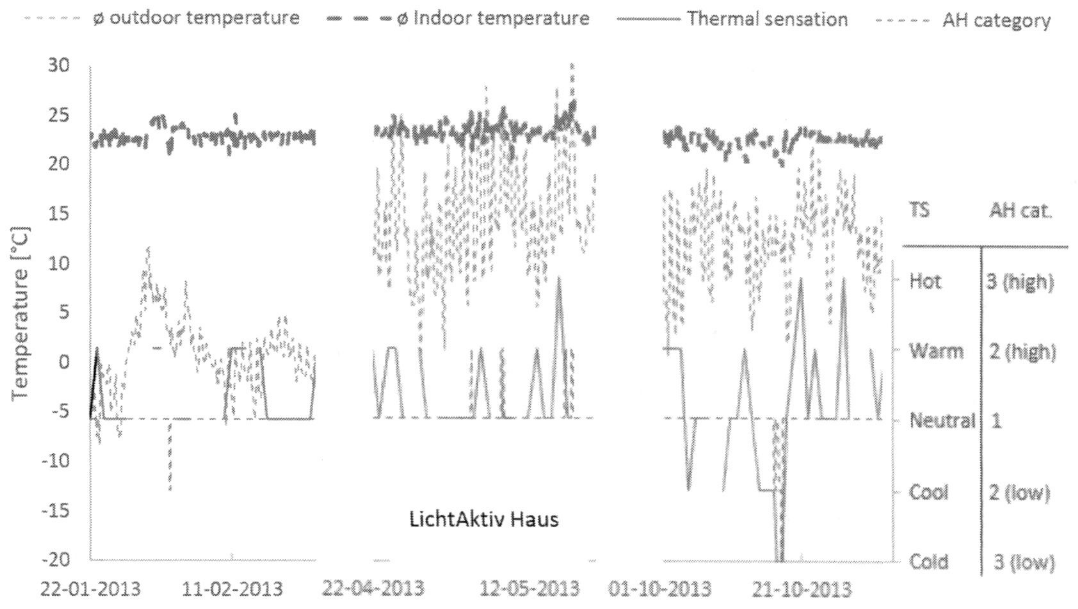

Figure 4 The dotted lines show measured indoor and outdoor temperatures in the living room of the two houses. The dotted grey line shows the thermal comfort category According to the Active House category. The solid grey line show the reported thermal sensation by the occupants. In SLH both occupants responded, for LAH, the responses of the father in the family are shown. In LAH the occupants responded during winter, spring/summer and autumn with in-between periods without responses.

In some episodes the occupants of SLH rate the thermal environment as warmer than the comfort category indicates (late July and late august). In other episodes the occupants rate the thermal environment as cooler than the comfort category (mid and late August). In general there is good correspondence between the occupant's thermal sensation vote and the thermal comfort category based on measured data.

In LAH the winter period January to February is characterised by stable indoor temperatures and more frequent votes of "warm" than "cool". The thermal comfort category is steadily 1 in this period. During the warm spring period in April to May, the outdoor temperature peaks between 25°C and 30°C. On the days with peak outdoor temperatures, the indoor temperature is generally maintained 3-5°C lower. The occupant often respond with “warm”. The thermal comfort category changes from 1 to 2 on three days in this period; there is a delay between category and thermal sensation, typically only of one day.

In the autumn period from October to November the thermal sensation of the occupant of LAH varies between “cool” and “hot”. The votes occur on the same days as the indoor temperatures increase or decrease, but the absolute variations in indoor temperatures are small, as indicated by the comfort category, which is 1 through the period with the exception of one day where it drops to 3.

3 CONCLUSIONS

The houses are evaluated according to the Active House specification, which uses the same methodology and criteria as the adaptive approach for naturally ventilated buildings in EN 15251 with regards to thermal comfort.

Despite high daylight levels, the houses experience very little overheating, and less than reported for other low energy houses. All main rooms of the houses achieve category 1 regarding overheating. Due to some hours with temperatures below 21°C during winter (by occupant preference), most main rooms achieve category 2.

Dynamic external solar shading and ventilative cooling by natural ventilation are key measures that have been used to achieve the very satisfying thermal conditions during summer.

The occupants rated the thermal environment on a 5-level thermal sensation scale. In Sunlighthouse, the ratings were made through almost a year, while in LichtAktiv Haus the ratings were made for three 30-day periods during winter, spring/summer and autumn. The thermal sensation votes are compared to the comfort category. There is good correspondence between thermal sensation votes and comfort category.

The results indicate that the adaptive approach of EN 15251 is reasonably accurate at predicting the actual thermal sensation of the occupants.

4 REFERENCES

AECOM (2012). *Investigation into overheating in homes: Literature review*. For the UK Department of Communities and Local Government

Carmichael, K , Anderson, M, Murray, V. (2011). *Overheating and health: a review into the physiological response to heat and identification of indoor heat thresholds*. UK Health Protection Agency

CEN (2007). CEN Standard EN 15251:2007. *Indoor environmental input parameters for design and assessment of energy performance of buildings addressing indoor air quality, thermal environment, lighting and acoustics*. European Committee for Standardisation.

Eriksen, K.E. et al (2011). *Active House – Specification*. Active House Alliance. 2011.

Foldbjerg, P., Rasmussen, C. and Asmussen, T. (2013). Thermal Comfort in two European Active Houses: Analysis of the Effects of Solar Shading and Ventilative Cooling. *Proceedings af Clima 2013*, Prague.

Isaksson, C. and Karlsson, F. (2006) Indoor Climate in low-energy houses – an interdisciplinary investigation. *Building and Environment*, Volume 41, Issue 12, December 2006

ISO 7730: 2005. (2005). *Moderate thermal environments - Determination of the PMV and PPD indices and specification of the conditions for thermal comfort*. ISO.

Larsen, T.S. and Jensen, R.L (2012). *The Comfort Houses - Measurements and analysis of the indoor environment and energy consumption in 8 passive houses 2008-2011*. Aalborg University.

Venticool, the International platform for ventilative cooling. Published by Venticool. Last accessed May 2014. http://venticool.eu/faqs/.

EXPERIENCES WITH VENTILATIVE COOLING IN PRACTICAL APPLICATION BASED ON EXPERIENCES WITH COMPLETED ACTIVE HOUSES

Peter Foldbjerg [*1], Kurt Emil Eriksen [2] and Karsten Duer[1]

1 VELUX A/S
Ådalsvej 99
2970 Hørsholm, Denmark
**Corresponding author: peter.foldbjerg@velux.com*

2 Active House Alliance
Groothandelsgebouw, Stationsplein 45, unit A6 0 16
3013 AK Rotterdam, Holland

Note: the contact addresses may be re-arranged

ABSTRACT

The present paper addresses experiences with ventilation and thermal comfort in the Active House concept, based on the Active House Specification and realized Active Houses. The Active House Specification is based on a holistic view on buildings including Comfort, Energy and Environment. It uses functional requirements to indoor air quality and thermal comfort. Experiences from realised Active House projects show that better airtightness than nationally required has been achieved. Indoor air quality is generally good, independently of the type of ventilation system installed (mechanical, natural and hybrid have been used). Good thermal comfort can be achieved in houses with generous daylight conditions. To succeed, natural ventilation and dynamic solar shading (ventilative cooling) must be applied and controlled to avoid overheating, which is possible under European climate conditions, where humidity is not a main issue during summer. The identified barriers have been that the current methods in standards and legislation that are used to determine the performance of ventilative cooling need to be further strengthened. And that affordable, intuitive and simple control systems for residential hybrid ventilation and dynamic solar shading are needed.

KEYWORDS

Active House, natural and hybrid ventilation, renovation, standards, controls systems

1 INTRODUCTION

Overheating is an important issue for building designers. Some demonstration houses in Scandinavia have experienced problems with overheating, often due to insufficient solar shading and use of natural ventilation (Isaksson, 2006, Larsen, 2012, Rohdin 2013). Similar results were found in a review on the situation in UK, stating that in certain cases, dwellings that were recently built or refurbished to high efficiency standards have the potential to face a significant risk of summer overheating (AECOM, 2012). Porritt et al. (Porritt, 2011) found that living room temperatures could be maintained below the CIBSE overheating thresholds, as a result of a combination of intervention measures that include external wall insulation, external surface albedo reduction (e.g. solar reflective paint), shading (e.g. external shutters) and intelligent ventilation regimes. Orme et al. found that night ventilation is a particular important measure to prevent overheating (Orme et al., 2003), and also found that the risk of overheating will increase in the future due to climate change.

The Active House Specification (Eriksen et al., 2013) has requirements in three categories, and has a main ambition that the three categories should have an equally high focus. The three categories are:

- Comfort (incl. indoor environment)
- Energy
- Environment

The Specification addresses residential ventilation as functional requirements to Indoor Air Quality (IAQ) and thermal comfort. All main rooms for occupancy must be evaluated separately. Four categories of IAQ are defined, based on the CO2-concentration, which must be achieved for minimum 95% of the occupied time:

1. 500 ppm above outdoor concentration
2. 750 ppm above outdoor concentration
3. 1000 ppm above outdoor concentration
4. 1200 ppm above outdoor concentration

It is a requirement that the air change rate can be manually influenced in all main living rooms, regardless whether mechanical, natural or hybrid ventilation is used. To reduce the risk of overheating, operable windows are recommended. Ventilation inlets, including natural ventilation openings, must be located so that draught risk is minimised.

There are no specific requirements to airtightness, or to performance of ventilation components, and no specific type of ventilation systems is required. The designer of the project can choose the most relevant system for the specific project, but to meet the ambitious requirements to energy performance, an energy efficient ventilation system and an airtight building envelope is needed (and therefore required indirectly).
Natural ventilation in combination with dynamic solar shading is a key instrument to avoid overheating with minimal use of energy. Four categories of maximum operative temperature are defined, setting requirements to air-conditioned and non air-conditioned buildings, using the definitions of EN 15251. For non-air conditioned buildings, the adaptive approach is used:

1. $T_{i,o} < 0.33 \times T_{rm} + 20.8$°C, for T_{rm} of 12°C or more
2. $T_{i,o} < 0.33 \times T_{rm} + 21.8$°C, for T_{rm} of 12°C or more
3. $T_{i,o} < 0.33 \times T_{rm} + 22.8$°C, for T_{rm} of 12°C or more
4. $T_{i,o} < 0.33 \times T_{rm} + 23.8$°C, for T_{rm} of 12°C or more

Daylight is important for humans, and the requirements are based on average daylight factors on a work plane in the main living room, which must be determined by a validated simulation tool. The criteria are:

1. DF > 5% on average
2. DF > 3% on average
3. DF > 2% on average
4. DF > 1% on average

Criteria for energy and environment are found in the Specification (Eriksen et al., 2013), which can be downloaded at no cost from the website of the Active House Alliance.

2 EXPERIENCES FROM COMPLETED ACTIVE HOUSES

2.1 Increasing number of Renovation Projects: Climate Renovation as new Paradigm

Two types of Active Houses have been completed: New buildings and Renovation projects. New buildings dominated in the first years after the launch of the Specification, but in recent years a shift towards more renovation projects have been seen. A common characteristic for the renovation projects is that they do not have improved energy performance as the only objective – improved indoor climate is often just as important. Climate Renovation has been adopted by some Active House Alliance members as the term to describe this new paradigm. Examples of recent renovation projects are (Active House, 2014):

- LichtAktiv Haus: A typical 1950's post-war, one-family house in Hamburg, Germany. The renovation has transformed it to a modern, spacious house
- De Poorters van Montfoort: Ten row-houses in a social housing corporation were renovated to offer excellent energy performance, more space, better daylight conditions and improved indoor climate.
- RenovActive: a semi-detached house in a social housing cooperation of 3.600 similar houses in the Anderlecht area of Brussels, renovated within the public financial frame and following Active House principles

2.2 Ventilation System Configurations

Many of the realized Active House have been built with demand-controlled, hybrid ventilation systems for optimal IAQ and energy performance.

An example is from the project Sunlighthouse in Austria. Natural ventilation is used during warm periods and mechanical ventilation with heat recovery is used during cold periods. The switch between mechanical and natural ventilation is controlled based on the outdoor temperature. The set point is 12,5°C with a 0,5°C hysteresis. Below the set point the ventilation is in mechanical mode, above the set point the ventilation is in natural mode. In both natural and mechanical mode, the ventilation rate is demand-controlled. CO_2 is used as indicator for IAQ, and a set point of 850 ppm CO_2 is used.

LichtAktiv Haus in Germany is an example of a house where natural ventilation is used as the only ventilation system.

2.3 Measured airtightness

It is not required in the Active House Specification to measure airtightness. But in order to meet energy efficiency targets, airtightness has been measured with blower door tests in many cases. Table 1 presents examples of measured airtightness in realised Active Houses. This is in most cases well below the requirements in the national building codes.

Table 1: Measured airtightness in five Active Houses

	Home for Life DK, 2009	Sunlighthouse AT, 2010	Maison Air et Lumiére F, 2011	Carbonlight Home GB, 2011	LichtAktiv Haus D, 2010
n50 (h^{-1})	1,5	0,52	0,60	-	1,07
l/s/m^2 @ 50 Pa	-	-	-	1,33	-

The experience from the completed Active Houses and other houses built according to other standards for low energy buildings, is that the achieved airtightness is related to the competence level of the craftsmen building the house. It is also the experience that the competence level has increased in the relatively short period from 2009 to 2011 due to the increased awareness of the importance of airtightness for low energy buildings.

2.4 Measured IAQ and Thermal Comfort

Temperatures and CO_2-concentrations have been measured on hourly level in several projects, e.g. in LichtAktiv Haus (LAH), Germany. LAH is designed with a demand controlled IAQ, with the aim to achieve category 1 (500 ppm above outdoor levels) or 2 (750 ppm above outdoor levels) (Feifer et al., 2013). The measured CO_2-concentration in the living/dining room is presented in Figure 1.

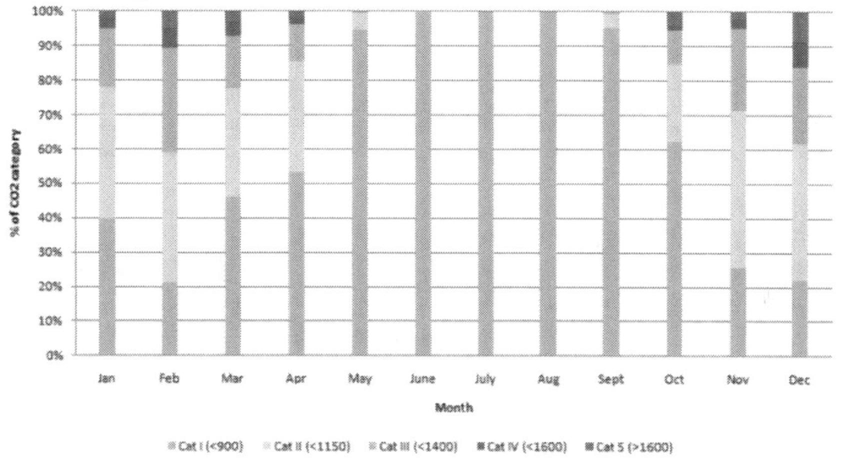

Figure 1: Measured CO_2 concentration in the kitchen/living room of LichtAktiv Haus, Germany. The data is categorized according to the Active House Specification

It is seem on Figure 1 that category 1 or 2 is achieved for 60% to 70% of the time during winter, and approx. 100% of the time during summer. The CO_2-concentration is lowest during the summer period as natural ventilation is also used to prevent overheating in this part of the year. Good summertime IAQ is thus a side-effect of applying ventilative cooling to prevent overheating. CO_2 concentration above category 2 during winter is caused by user override of automated controls. These results are similar to those seen in Active Houses with mechanical/hybrid ventilation.

It is the general experience that both natural, mechanical and hybrid ventilation systems are able to deliver the right ventilation rates and achieve the right IAQ. The key issue is that the systems must be designed, installed and maintained correctly, and most importantly, the controls must be transparent and intuitive for the occupants of the buildings.

Foldbjerg (Foldbjerg et al., 2013) reported on the thermal comfort in LAH and two other Active Houses. A typical characteristic of the realized Active Houses is that they have very generous daylight conditions. It is seen on Figure 2 that the living-dining room in LAH achieve category 1 in most months, with the exception of three summer months. Annually, the room achieves category 1. There are very few hours with temperatures below category 1. This means that there is no issues with overheating or low temperatures (undercooling).

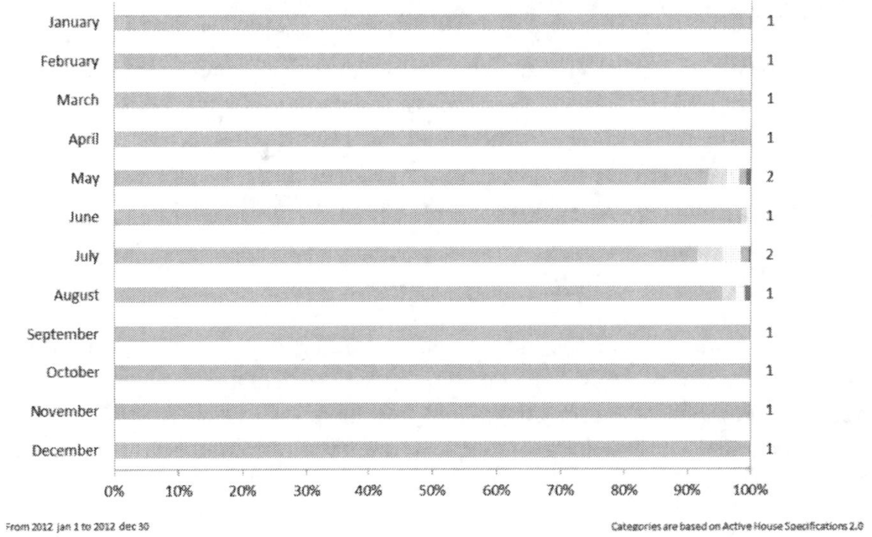

Figure 2: Measured indoor temperature in the kitchen/living room of LichtAktiv Haus, Germany. The data is categorized according to the Active House Specification. The number on the right side of the figure is the Active house category achieved for each month (max 5% of the time can exceed the category)

It is the general experience that good thermal conditions with only insignificant periods with high or low temperatures can be achieved. Prevention of overheating is a key issue, as low energy buildings can easily overheat, as reported by Larsen (Larsen, 2012) and others. The important elements to consider are natural ventilation and dynamic solar shading, as combined in ventilative cooling (venticool, 2014).

2.5 Ventilative Cooling in Standards

Peuportier (Peuportier et al, 2013) measured the air change rates achieved with natural ventilation as the means of ventilative cooling in the Active House called Maison Air et lumière near Paris, France. Air change rates in the range of 10 to 22 ACH were achieved. These results were confirmed by simulations in CONTAM. However, later calculations with the methods presented in EN 15242 show much lower results despite similar geometry and boundary conditions. This is to some extent explained by the fact that EN 15242 only includes single-sided ventilation. BS 5925:1991 presents a method that allows for a two-sided window configuration, still with very conservative results. In the on-going revision of EN 15242 it is being discussed if a more accurate and generally applicable method can be included. The work in IEA Annex 62 will further support this goal.

2.6 Ventilative Cooling in legislation

Ventilative cooling is addressed in legislation through building codes and compliance tools. Some compliance tools specify ventilative cooling with default ventilation flow rates that are difficult to change to address the performance of the actual ventilation design. To correctly account for the effect of ventilative cooling, more accurate methods are needed. There is currently work on-going in Denmark and France to improve the methodology for the calculation of ventilative cooling.

Also in France and Denmark, requirements to thermal comfort is likely to be stricter in coming revisions. This is a necessary step to prevent overheating, but requires that the underlying methodology adequately accounts for the actual performance.

2.7 Experience with Control Systems in Active Houses

Holzer (Holzer et al, 2014) is investigating the characteristics of Active Houses and particularly how the control systems should be designed to allow the houses to deliver the expected performance, and at the same time offer the occupants the experience they expect.

The preliminary conclusions are:

- Active Houses are react fast towards direct sunlight. Thus, an effective and fully automatically controlled system of dynamic shadings is obligatory for achieving good summer comfort.
- Hybrid ventilation systems stand the test, combining automated window operation, and mechanical ventilation systems as well as manual window operation. The learning is to consequently separate the operation periods of automated window and mechanical ventilation, depending from outside temperature.
- Beyond technical automation it's essential offering intuitively manually operable devices such as windows, doors, and awning blinds. Furthermore it's preferable having some devices literally manually operated than having them only manually telecommanded.
- Sun protection together with night ventilation is an effective combined strategy towards summer comfort, which turned out to be preferably automized. At least in central and northern Europe areas there turned out to be a somehow weak intuitive understanding of heat protective building operation.

There are few control systems currently available that deliver control of both mechanical and natural ventilation (as a hybrid solution), and which controls both ventilation, window openings and dynamic solar shading in a combined effort to maintain both good IAQ and good thermal comfort. Such systems should be cost-effective and are needed for the residential market.

3 CONCLUSIONS

A shift from focus on new buildings to also focus on renovation projects has been observed. The main driver is improved energy performance, but multiple benefits are actually expected, mainly related to good indoor climate. Some Active House members have used the term Climate Renovation to describe this new paradigm.

Airtightness is important to ensure the planned energy performance, and relies to a large extent on the competences of the craftsmen. There are indications that the competence level is increasing from year to year, and good results have been observed in the most recent houses. There seems to be a reduced need for focus on building airtightness now compared to the situation 5 to 15 years ago.

Good IAQ can be achieved with both natural, mechanical and hybrid ventilation systems. The important lesson is that they must be planned, installed and maintained right. This has been achieved in the investigated houses. By correct planning in the design process good IAQ can be reached with a minimum use of energy. Particular good IAQ during the summer period has been observed as a side-effect of applying ventilative cooling.

Whereas the above themes have been relatively unproblematic, some issues, mentioned below, have a greater need for increased focus regarding quality and compliance.

The realized houses are characterised by generous daylight conditions, which could potentially lead to overheating. This has not been the case. The houses show that good thermal comfort can be achieved in all seasons, regardless whether natural, hybrid or mechanical ventilation is used. But a strong relation between efficient natural ventilation in the summer (ventilative cooling) as well as dynamic solar shading has been a key element in achieving this, supported by windows being located towards more than one orientations in each room and not mainly towards the south as sometimes seen in low energy houses.

There is currently only weak support in standards and legislation to give a true and fair account of the performance of ventilative cooling and dynamic solar shading, and this needs to be improved.

There remains a need to identify and to discuss how ventilative cooling can become a standard solution in legislation and standards throughout Europe especially regarding renovation but also regarding Nearly Zero Energy Buildings.

Transparent and intuitive control systems scaled for residential buildings with regards to system architecture and price are needed. Such a control system should be able to control ventilation and dynamic solar shading to maintain both good IAQ as well as good thermal comfort.

4 REFERENCES

Active house (2014). *Case stories.*
RenovActive: http://www.activehouse.info/news/first-affordable-and-scalable-climate-renovation-brussels-based-active-house-principles
LichtAktiv Haus: http://www.activehouse.info/cases/lichtaktiv-haus
De Poorters van Montfoort: http://www.activehouse.info/cases/de-poorters-van-montfoort

Eriksen, K.E. et al. (2013). *Active House – Specification*. Active House Alliance. 2013

Feifer, L. et al. (2013) *LichtAktiv Haus – a Model for Climate Renovation*. Proceedings of PLEA Conference 2013.

Foldbjerg, P. and Asmussen, T. (2013) *Ventilative Cooling of Residential Buildings: Strategies, Measurement Results and Lessons-learned from three Active Houses in Austria, Germany and Denmark.* Proceedings of AIVC conference 2013, Athens, Greece.

Isaksson, C. and Karlsson, F. (2006) Indoor Climate in low-energy houses – an interdisciplinary investigation. *Building and Environment,* Volume 41, Issue 12, December 2006

Larsen, T.S. and Jensen, R.L (2012). *The Comfort Houses - Measurements and analysis of the indoor environment and energy consumption in 8 passive houses 2008-2011.* Aalborg University.

Orme et al. (2003). *Control of overheating in future housing, Design guidance for low energy strategies.* Hertfordshire, UK, Faber Maunsell Ltd.

Porritt, S. M. et al. (2011). Adapting dwellings for heat waves. Sustainable Cities and Society, 1(2): 81-90.

Rohdin, P, Molin A and Moshfegh, B. (2013). Experiences from nine passive houses in Sweden - Indoor thermal environment and energy use. *Building and Environment,* Volume 71.

Holzer, P. et al. (2014). *Control of indoor climate systems in Active Houses.* Submitted to World Sustainable Buildings Conference 2014, Barcelona (currently in review).

Larsen, T.S. and Jensen, R.L (2012). *The Comfort Houses - Measurements and analysis of the indoor environment and energy consumption in 8 passive houses 2008-2011.* Aalborg University.

Peuportier, B. et al. (2013) *Evaluation of Ventilative Cooling in a Single Family House.* Proceedings of AIVC conference 2013, Athens, Greece.

Venticool (2014). http://venticool.eu/

INDOOR CLIMATE IN A DANISH KINDERGARTEN BUILT ACCORDING TO ACTIVE HOUSE PRINCIPLES: MEASURED THERMAL COMFORT AND USE OF ELECTRICAL LIGHT

Peter Foldbjerg [*1], Thorbjørn Færing Asmussen[1] and Jens Christoffersen[1]

1 VELUX A/S
Ådalsvej 99
2970 Hørsholm, Denmark
**Corresponding author: peter.foldbjerg@velux.com*

ABSTRACT

The Kindergarten Solhuset is built according to the Active House vision with an emphasize of good daylight conditions and fresh air. The house was completed in 2011, and detailed measurements of the indoor environment have been performed since the completion. The daylight performance is evaluated with daylight factor simulations. The main activity rooms have daylight factors of 7%, while the innermost rooms with only roof windows achieve a high daylight factor of 4%. Electrical light is used frequently in daytime during the winter, but much less frequently during summer.

The thermal environment is evaluated according to the Active House specification (based on the adaptive method of EN 15251), and it is found that the house reaches category 1 for the summer situation. Some hours with temperatures below category 1 are observed during winter, and the building achieves category 2 for the winter situation. It is found that ventilative cooling through window openings play a particularly important role in maintaining thermal comfort and that both window openings and external solar shading is used frequently

KEYWORDS

Thermal comfort; daylight; ventilative cooling; solar shading; kindergartens

1. INTRODUCTION

Childcare centres and schools have a particular need for a good and healthy indoor climate as it strengthens wellbeing and learning capacity as well as reduces the risk of diseases.

The vision for Solhuset (The Sunhouse) was to set new standards for future sustainable childcare centres. It rests on the Active House principles (Eriksen, 2011) of buildings that give more than they take – to the children, adults, environment, and surroundings. Solhuset is showing the way; it has the framework for a healthy indoor climate where children learn to live in harmony with nature and without negative impact on the environment.

Daylight and a healthy indoor climate play a vital role in Solhuset. From the very beginning, the vision was to create a building with a positive impact on its surrounding environment and daily users. There is a particular focus on good daylight conditions and fresh air from natural ventilation

Solhuset was developed in a strategic partnership between Hørsholm Municipality, VKR Holding A/S and Lions Børnehuse, and built by Hellerup Byg A/S in co-operation with Christensen & co arkitekter a/s and Rambøll A/S.It was completed in 2011.

Solhuset is laid out as a small village with streets, lanes, small squares and niches, and is divided into three zones: an arrival zone; a small children's zone with access to group rooms, an outdoor area and an open-air shelter; and a large children's zone with access to group rooms and the outdoor area. Common exercise rooms and eating facilities are placed in the middle of the house for easy access. It has high-ceilinged rooms and strategically placed windows to ensure optimum use of daylight. The sloping roof, with roof windows that open and close automatically, creates varied ceiling heights for good air circulation in the rooms.

Intelligently controlled sun screening and window opening make the house flexible, allowing the flow of daylight and fresh air to adapt continuously to the weather conditions outside and the needs indoors. Solhuset is built of sound materials that have minimal impact on the indoor climate. Vertical windows in the southeast and south-west facades and roof windows let in more than three times as much light as in a traditional house. A weather station on the roof, together with temperature and CO_2 sensors in every room, is used to control the indoor climate – protecting against overheating, ventilating with fresh air, and switching the lights on and off according to needs and weather conditions.

Figure 1. Solhuset (exterior, interior and floor plan).

Solhuset has the following characteristics:

- All rooms get daylight from at least two sides.
- Vertical windows with iron-free glass ensure that up to 85% of the light is transmitted through the windows.
- Plenty of fresh air is ensured by a combination of automatically controlled natural and mechanical ventilation with heat recovery (hybrid ventilation).
- Strategically placed windows ensure optimal use of daylight and adequate air flow.
- The special design and volume of the rooms, together with the use of sound absorbers, ensure a good acoustic climate.
- Use of healthy materials ensures minimal impact on the indoor climate.

Measurements of IEQ include light, thermal conditions, indoor air quality and occupant presence. Measurements of energy performance include space heating, domestic hot water and electricity for appliances, lighting and technical installations.

The present paper describes the performance of the house in relation to light and thermal comfort, particularly the natural ventilation system and the solar shading. Use of natural ventilation for summer comfort is based on ventilative cooling principles (Venticool, 2014). The presented results focus on the use of electrical light and the thermal conditions.

Each room is an individual zone in the control system, and each room is controlled individually. There are sensors for humidity, temperature, CO2 and presence in each room. The building occupants can override the automatic controls, including ventilation and solar shading at any time. Override buttons are installed in each room, and no restrictions have been given to the occupants. As house owners they have reported a motivation to minimise energy use on an overall level, and to maximise IEQ on a day-to-day basis.

The data from the sensors that are used for the controls of the house is recorded. The IEQ data is recorded for each individual zone as an event log, where a new event is recorded when the value of a parameter has changed beyond a specified increment from the previously recorded value. The event log files are automatically converted to data files with fixed 15-minute time steps, which are used for the data analysis.

The recorded temperature data is evaluated according to the Active House specification (Eriksen, 2012), which is based on the adaptive approach of EN 15251 (CEN, 2007).

2. RESULTS

2.1 Light

Figure 2 presents calculated daylight factors. The building has an average daylight factor of 7% in living rooms and up to 4% in the innermost part of the rooms – even with a window area of only 28% of the floor area. This gives a score of 2 according to the Active House specification. For the Active House parameter of Direct Sunlight Availability, the score is 1.

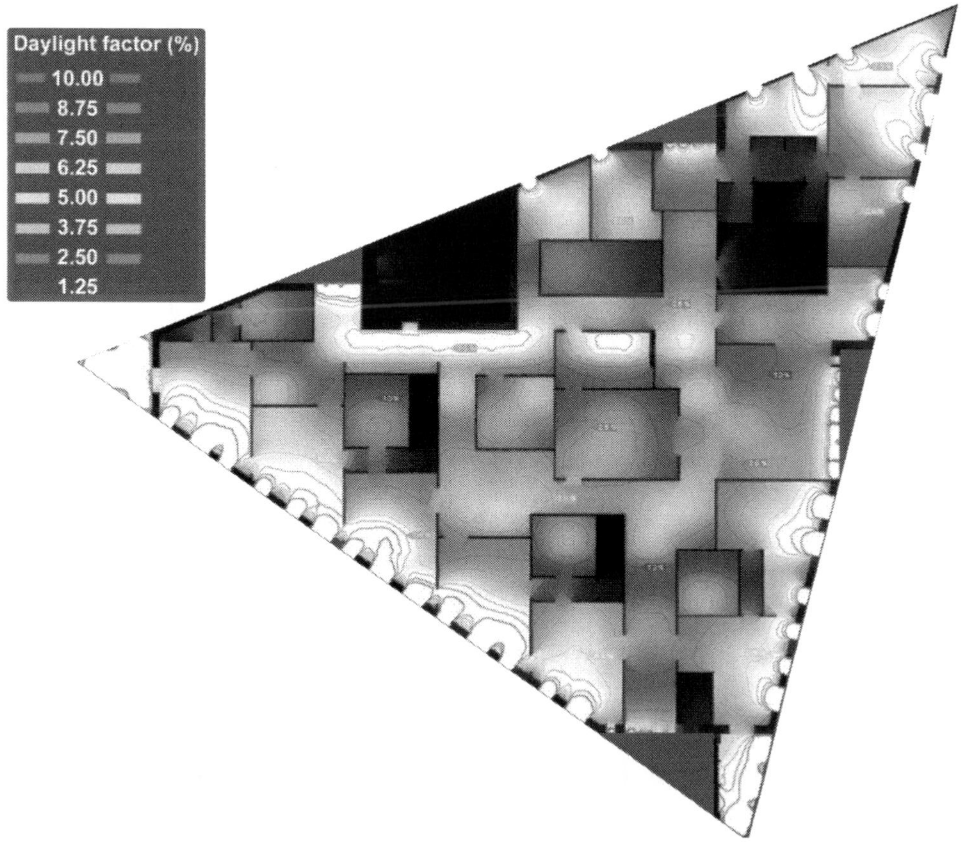

Figure 2. Calculated daylight factors in Solhuset.

The monthly lighting energy consumption varies naturally according to time of year; higher in the winter than summer. The overall lighting energy consumption in 2013 is 6100 kWh, which correspond to 4,7 kWh/m^2 (i.e. Solhuset is 1300 m^2) and only 1/3 of lighting use in commercial-sector buildings for six EU countries (average values of 15 - 18 kWh/m^2 according to Kofod. (Kofod, 2001). See Figure 3.

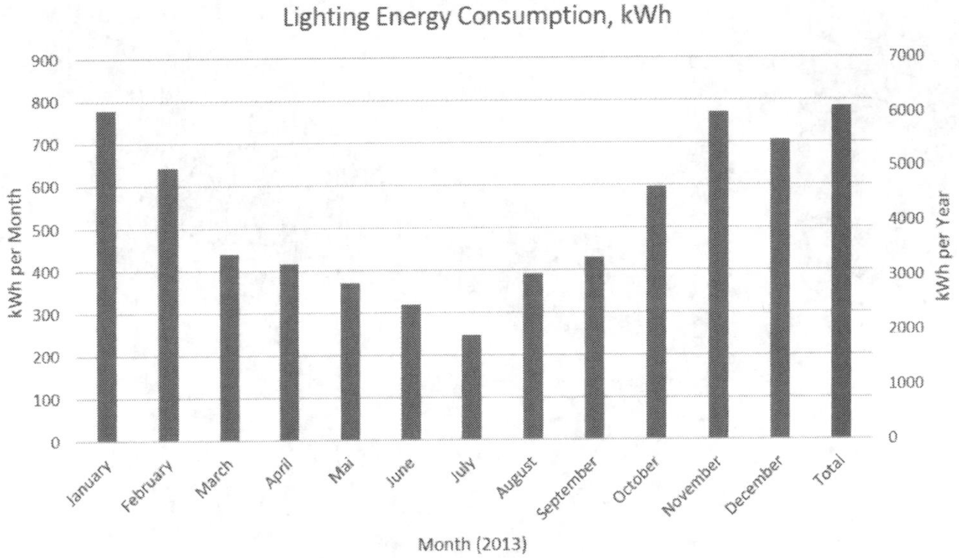

Figure 3. Monthly (blue) and total (red) lighting energy consumption (KNX_EM05_E) in kWh.

Measured lighting energy data (Figure 4) show each day of the year along the X axis and the time of day along the Y axis. In addition, the two curves mark the sunrise (blue) and sunset (red), and the curves are adjusted according to local time and Daylight Saving Time (DST). The colour map show blue colour for lighting energy consumption less than 500 Wh (mostly standby effect; around 60% of the total hours).

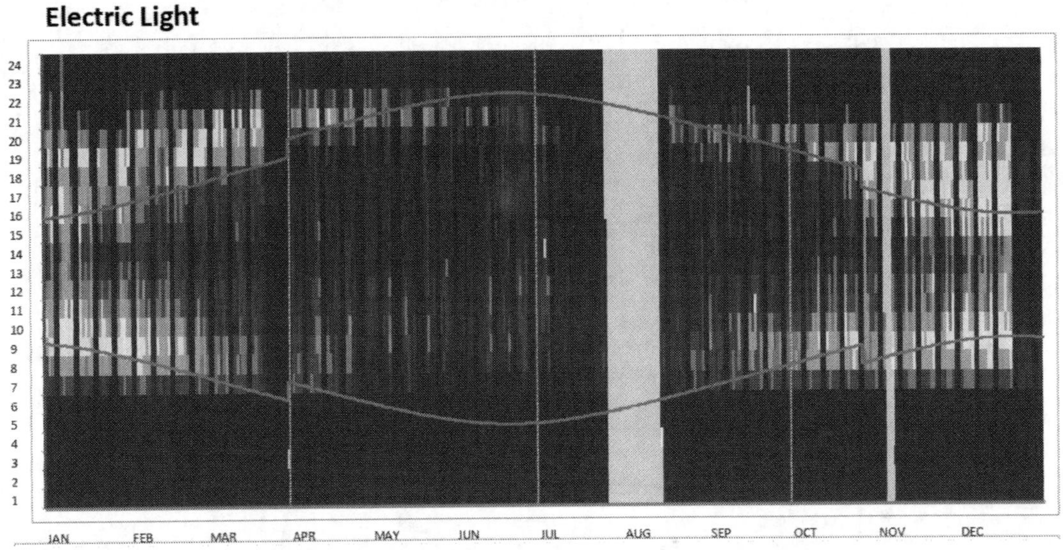

Figure 4. Temporal map of lighting use in Solhuset, 2013, including the sunrise (blue) and sunset (red). Lighting use and sunrise/sunset is according to local time, which account for Daylight Saving Time (DST). Measurements from August is missing.

The yellow and red colours reflect lighting energy use above 500 Wh, while the grey area has no measured data. There is a clear tendency that winter months have lights on most of the days (yellow/red), while the summer months show more frequent switch-off (blue). Furthermore, the switch-off in summer is more frequent in the afternoon than in the morning hours, which could be an indication that the staff of the kindergarten experience significant amount of daylight, while morning lights on is more a behavioural ritual. The results are based on the measurements from January 2013 to December 2013.

2.2 Thermal comfort

Figure 54 shows thermal comfort categories for 15 rooms in Solhuset, which represent the spread of different room types and functions in the building. Solhuset experiences practically no episodes with temperatures above category 1. This demonstrates that overheating has not occurred during the year of 2012.

Temperatures below category 1 are seen in all rooms. All rooms fall in category 2 for this reason, with the exception of two semi-outdoor rooms.

The focus of the present report is on the performance related to ventilative cooling and potential overheating. The further analyses will focus on the performance of group room 3, which is a representative room located at the perimeter of the building oriented towards south-west, which has a higher risk of overheating than other orientations.

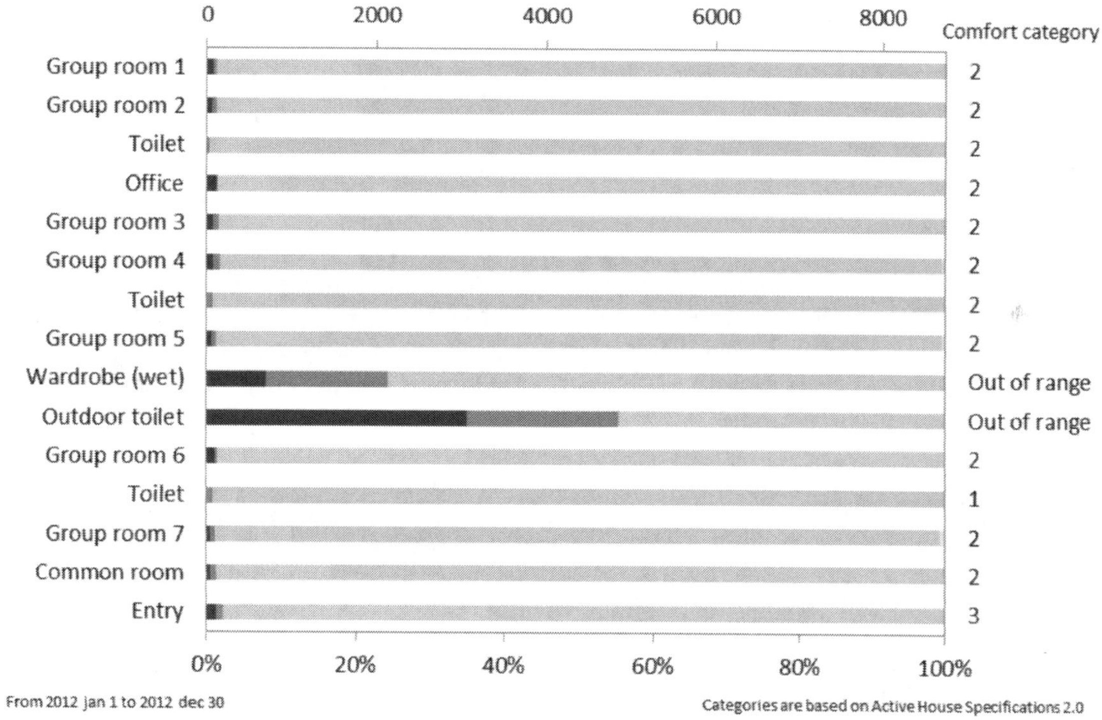

Figure 5. Thermal comfort for 15 representative rooms in Solhuset evaluated according to Active House specification (based on adaptive method of EN 15251). Criteria are differentiated between high and low temperatures.

Figure 6 shows the indoor temperature at each hour of the year plotted against the running mean outdoor temperature as defined in EN 15251. The figure shows Group room 3. The figure clearly shows that there are no hours with temperatures above category 1 (no overheating).

As seen before, the temperature in many hours during winter fall below category 1, with a few episodes with temperatures below category 2. There has been an issue with the heating system in the reported period which is part of the explanation why the temperatures are low.

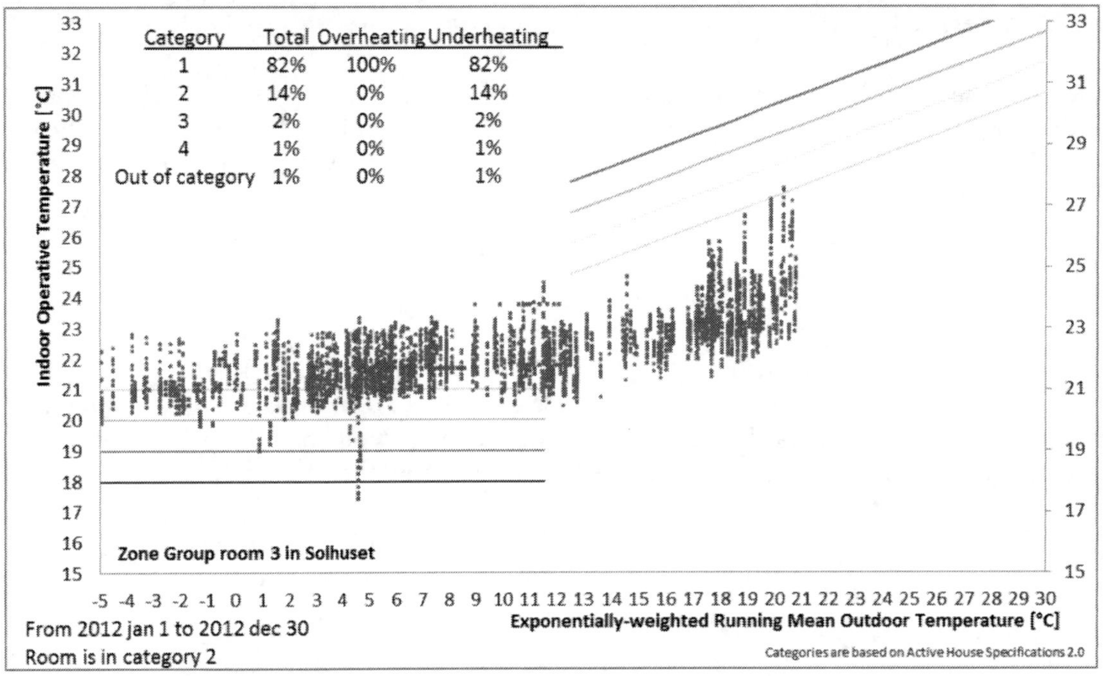

Figure 6. Measured indoor air temperature for Group room 3, plotted against the running mean outdoor temperature. Each dot represent the average temperature for an hour of the year.

The variation over time-of-day and time-of-year is further investigated in Figure 7. It is clear that the episodes with temperatures in category 4 or below are limited to two episodes, each lasting for two to four days. The hours with temperatures in category 2 (low) mainly occur during winter between 24:00 and 12:00, which indicates that the solar gains contribute to heating up the room from midday.

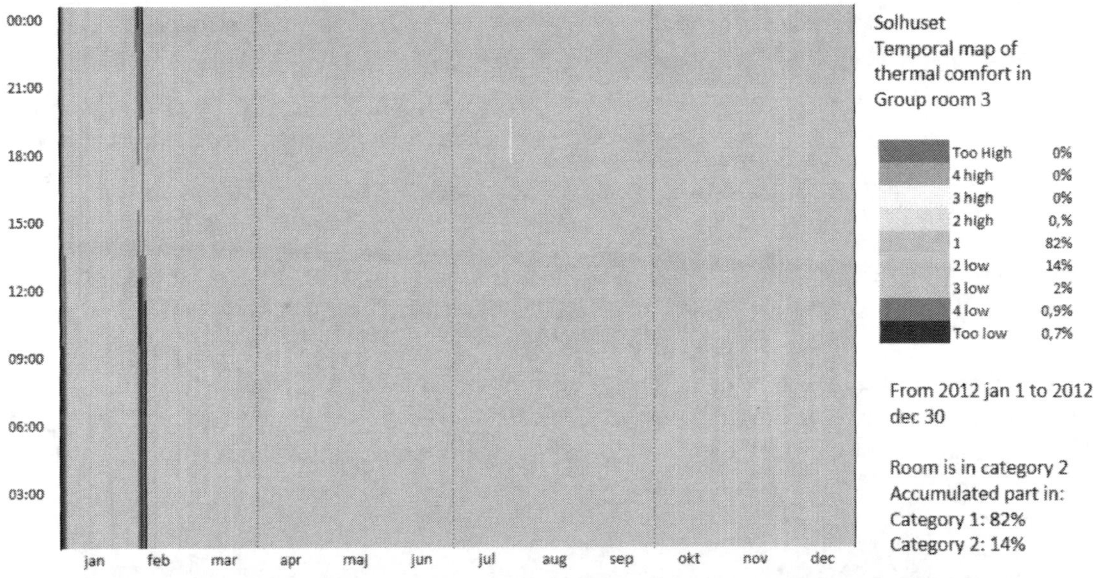

Figure 7. Group room 3. The comfort category of each hour of the year is plotted as a temporal map

Figure 8 Presents a temporal map of the same room as Figure 56, but Figure 57 also shows the use of windows in the room. Green or light green indicates that the temperature is within category 1 or 2, which is the case almost all the year. Light green indicates that windows are open in the room, dark green that they are closed. The figure shows that in the warmest period from June to the end of August, windows are open continuously from 07:00 to 19:00. Windows are also open during the night in this period, not constantly, but with airings at 22:00, 24:00 and 04:00 every day.

During almost all days of the year an airing at 07:00 seen, activated by the automatic control system. In the spring/autumn seasons, windows are used frequently during the day.
The combination of frequent use of windows and thermal comfort in the two best categories, underlines that windows have been important in maintaining good thermal comfort.

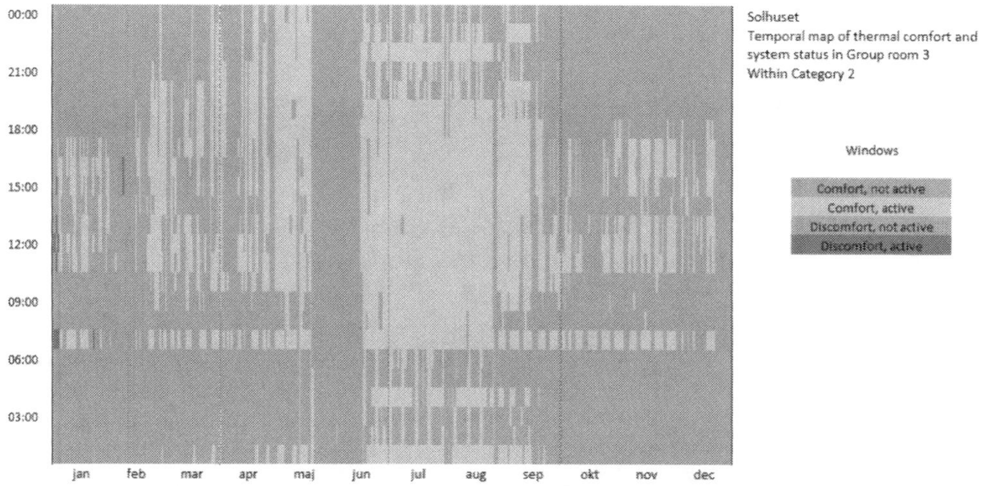

Figure 8 Temporal map for Group room 3 showing open or closed window in combination with thermal comfort category. Windows are marked as open, if one or more windows in the room are open For the sake of the illustration, category 1 and 2 are called "comfort", category 3 and 4 are "discomfort".

Figure 9 is similar to Figure 8, except that the use of awning blinds and not windows is shown. The figure shows that awning blinds are used almost constantly from January to May, which cannot be explained from the data alone. There is little or no risk of overheating during winter, so the use in this period must be caused by something else than prevention of overheating.

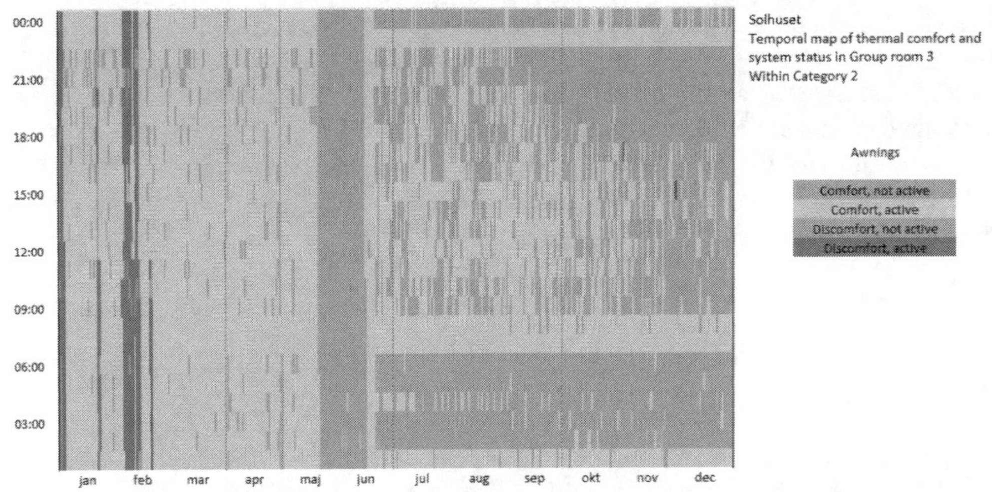

Figure 9 Temporal map for Group room 3 showing active or inactive awnings blinds in combination with thermal comfort category. Awning blinds are marked as active, if one or more awning blinds in the room are open For the sake of the illustration, category 1 and 2 are called "comfort", category 3 and 4 are "discomfort".

From June until December, the use of awning blinds is as could be expected to prevent overheating, with frequent, but not constant, use during daytime.

The frequent use of awning blinds underlines, that the awning blinds in combination with window openings were important for maintaining good thermal comfort.

3 CONCLUSIONS

Solhusets lighting energy use is low, and clear tendency of natural switch-on probability when outdoor light levels are low in the winter months. The summer months show less lighting energy use and more frequent switch-off probability due to plenty of daylight.

Daylight factors are 7% for the rooms near the facade, and 4% for the inner-most rooms, which is remarkable high for this type of building. The analysis shows Active House category 2, when the innermost part of the rooms are included, while direct sunlight availability is category 1 due to insignificant outdoor obstructions.

Solhuset is characterised with having no problems with overheating in summer. Temperatures do not exceed Active House category 1 for summer. Windows and awning blinds play an important role in maintaining good thermal comfort.

In winter, the temperatures in most main rooms are between 20°C and 21°C, which corresponds to Active House category 2. Category 2 is the standard category for normal, new buildings and considered as good performance.

4 REFERENCES

CEN. CEN Standard EN 15251:2007, *Indoor environmental input parameters for design and assessment of energy performance of buildings addressing indoor air quality, thermal environment, lighting and acoustics*. European Committee for Standardisation. 2007.

Kofod, C. (2001), *Market Research on the Use of Energy Efficient Lighting in the Commercial Sector: Final Report*, for DEFU and DG TREN of the European Commission, SAVE Contract No. 4.1031/Z/97-029, DEFU, Denmark.

K. E. Eriksen et al. *Active House – Specification*. Active House Alliance. 2011.

Venticool, the *International platform for ventilative cooling*. Published by Venticool. Last accessed July 2014. http://venticool.eu/faqs/.

AIR HEATING OF PASSIVE HOUSE OFFICE BUILDINGS IN COLD CLIMATES – HOW HIGH SUPPLY TEMPERATURE IS ACCEPTABLE?

Axel Cablé*[1], Mads Mysen[1,2], Hugo Lewi Hammer[2] and Kari Thunshelle[1]

1 SINTEF
Forskningsveien 3B
N-0314, Oslo, Norway
**Corresponding author: axel.cable@sintef.no*

2 Oslo and Akershus University College of Applied Sciences
Pilestredet 35
N-0166, Oslo, Norway

ABSTRACT

The impact of over-tempered air on the perceived indoor climate was evaluated by questionnaires filled in by the users of the first office building with passive house standard in Norway. In this building, the heating demand is covered entirely by warm air supplied into the rooms through the ventilation system.

On the coldest days of January 2014, warm ventilation air was supplied into the rooms at a constant temperature during half an hour. Each user of the building was exposed to 3 different supply temperatures (around 21.5°C, 24°C and 26°C) under the minimum ventilation rate according to the Norwegian standards (17 l/s). Questions related to both perceived thermal comfort and Sick Building Syndrome-symptoms (SBS; feeling tired, headache, etc.) were answered by all the occupants on a scale of 0 (unsatisfied) to 10 (satisfied). The data from the questionnaires were then analyzed using a random effect linear regression model.

The regression analysis did not report any significant relationship between the supply air temperature, and perceived thermal comfort and SBS. It enables to document with a 95% certainty that increasing the difference between supply air and room temperature by 1°C would cause a maximum reduction of the SBS score of 1.02 points on a scale of 190. The impact of an increase of the supply temperature on the perceived SBS seems therefore very limited.

Using air heating to completely cover the heating demand therefore appears to be a relevant solution for office buildings in cold climate with passive house standard.

KEYWORDS

Ventilation, air heating, passive house, questionnaires, indoor climate

1. INTRODUCTION

The GK environmental house is the first office building with passive house standard in Norway (see Figure 1a). It is located in Oslo, and was taken into use in August 2012. In this building, the heating demand is covered entirely by warm air supplied into the rooms through the ventilation system. Active air supply diffusers are then used to control the ventilation rate according to room temperature and occupancy (Demand-Controlled Ventilation).

While this solution sounds appealing, it is not clear whether air heating alone can ensure an acceptable indoor climate in the cold climate of Norway. In fact, short-circuiting of the warm

ventilation air may occur if the temperature difference between supply and room temperature is high, which may be the case during the coldest days.
This can be responsible for a poor perceived indoor air quality and impact the well-being and productivity of the occupants (Sundell, 2011).

In this context, the research and development project Forklima is carried out in the building (http://www.sintef.no/Projectweb/For-Klima/). The aim of this project is to assess whether it is possible to cover the heating demand with warm ventilation air exclusively in office buildings with passive house standard, while maintaining an acceptable thermal comfort and indoor air quality. In the present paper, the impact of over-tempered air on the perceived indoor climate was evaluated by questionnaires filled in by the employees located in the open plan offices of the building, see Figure 1b.

Evaluating the perceived indoor climate in buildings satisfying the passive house standard (NS3701, 2012) is crucial to validate their use in countries with cold climate. The aim of this study is to do so with occupants carrying out their daily tasks in real conditions, and therefore to provide a valuable addition to studies of the perceived indoor climate in controlled conditions.

Figure 1: a) GK environmental building, Oslo. b) View of the open plan offices.

2. METHODOLOGIES

The experiments were carried out on January 23 and January 24, 2014 which were among the coldest days of January 2014 in Oslo. In order to assess the impact of air heating on the perceived indoor climate, over-tempered ventilation air was supplied into the open plan offices through mixing ventilation diffusers located at ceiling height.

An active supply diffuser is located above each desk, covering 2 to 4 persons, and continuously recording the supply airflow rate, supply temperature, as well the air temperature in the room. The control and monitoring were carried out from a distance through the Building Management System, see Figure 2.

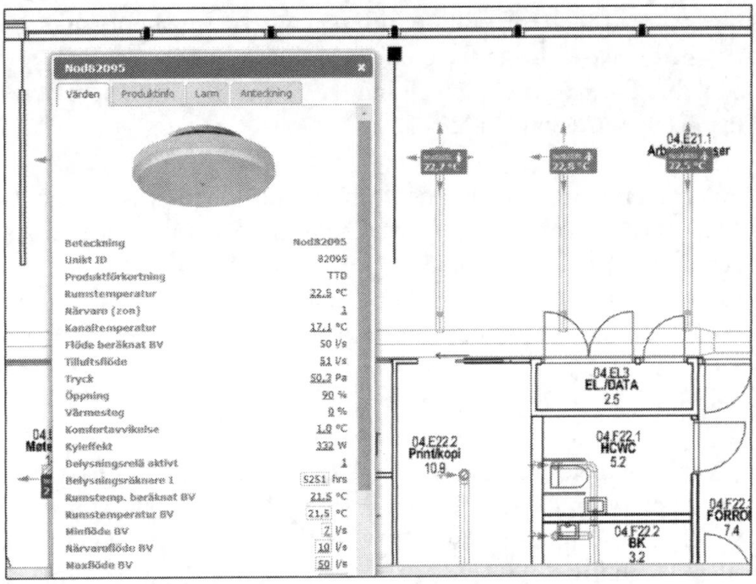

Figure 2: Remote control and monitoring of the supply airflow rate, supply temperature and room temperature for each diffuser of the building.

Interventions on the supply temperature were carried out from 10:00 to 10:30 and from 14:00 to 14:30, on both January 23 and 24. On January 23 from 10:00 to 10:30 am, the plan for the interventions were to have a supply temperature for the south and north part of the building of 24°C and 21°C, respectively (see Table 1). On January 23 2:00 – 2:30 pm, the supply temperatures were switched for the two parts of the building. On January 24, the interventions were carried out in the same way as the previous day, except that the high supply temperature was changed from 24°C to 26°C, see Table 2. In such a cross-over experiment, each user in the building was exposed to all three supply temperatures. In addition, the minimum ventilation rate according to the Norwegian standards was used (17 l/s).

A consequence of the high performance of the envelope of the building is that there is no need for a too high supply air temperature to maintain an acceptable indoor temperature. Therefore, the length of the interventions was reduced to half an hour in order to not cause a rise of the indoor temperature, and therefore study the impact of over-tempered air of the perceived indoor climate, and not the impact of the temperature level.

Table 1: Plan for the cross-over experiment on January 23

Date	23rd January 10:00-10:30		23rd January 14:00-14:30	
Part of building	South	North	South	North
Supply temperature T_s	21,5°C	24°C	24°C	21,5°C
Ventilation rate	17 l/s	17 l/s	17 l/s	17 l/s

Table 2: Plan for the cross-over experiment on January 24

Date	24th January 10:00-10:30		24th January 14:00-14:30	
Part of building	South	North	South	North
Supply temperature T_s	21,5°C	26°C	26°C	21,5°C
Ventilation rate	17 l/s	17 l/s	17 l/s	17 l/s

A questionnaire was answered by the users at the end of each of the four interventions in order to measure the users' perceived health and well-being. The questions included in the questionnaire are shown in Table 6, and the questionnaire was handled to the occupants through a web-based application. On each question, the users gave a value between one and ten ranging from very uncomfortable to very comfortable. The sum, *S,* of all the scores for a given user, represents the overall health and well-being for this user. The questionnaire also included health question like whether a used suffered from asthma or cold.

Data from the questionnaires were analyzed using a random effect linear regression model with *S* as the dependent variable. The supply air temperature was included in the model as a continuous independent variable. The effect of supply temperature on *S* was controlled for differences in gender and whether a user suffered from asthma or cold by including these variables in the model. Furthermore, we expect that repeated measures from the same user is correlated, *ie.* some users are always too cold and some users are always tired, and this was taken into account by including a random effect on the user level in the regression model. Statistical analyses were performed using the program R (R development team, 2013) and the R package lme4 (Bates, 2013).

In addition, the results obtained during both days were compared to the results obtained from a questionnaire run on November 19, 2013 under normal operating conditions for the building. That corresponds to a supply temperature and a ventilation rate varying according to the demand measured in each room, *ie.* Demand-Controlled Ventilation.

The numbers of participants answering the questionnaire for the four interventions were 34, 31, 24 and 25. On November 19, 46 persons answered.

3. RESULTS AND DISCUSSION

3.1 Actual conditions during the experiments

The measured conditions (supply and average room temperature for each part of the building) during the experiments are reported in Table 3 for January 23, and in Table 4 for January 24. The reported supply temperatures correspond to the average of the temperature at the exit of the Air Handling Unit over the 30 minutes of each intervention. The average room temperature corresponds to the average of the temperature measured by all the active diffusers in the buildings.

Table 3: Measured conditions during the test on January 23, 2014

Date	**January 23 10:00-10:30**		**January 23 14:00-14:30**	
Part of building	South	North	South	North
Supply temperature T_s [°C]	24.5	21.0	21.2	24.1
Average room temperature [°C]	22.3	22.0	22.4	22.4
Outdoor temperature [°C]	-6.2		-4.9	
Outdoor conditions	Cloudy		Cloudy	

Table 4: Measured conditions during the test on January 24, 2014

Date	**January 24 10:00-10:30**		**January 24 14:00-14:30**	
Part of building	South	North	South	North
Supply temperature T_s [°C]	26.7	21.3	21.4	26.3
Average room temperature [°C]	22.0	22.1	22.2	22.3
Outdoor temperature [°C]	-6.3		-4.5	
Outdoor conditions	Cloudy		Cloudy	

We can see that the measured supply air temperature during the tests on January 23 and 24 in both parts of the building are in good agreement with the plan for the cross-over experiments presented in Table 1 and Table 2. In addition, the outdoor conditions were similar on both days. Futhermore, the average room temperature during the experiments is similar during all experiments, ranging from 22.0°C to 22.4°C. This enables to say that the evaluated parameter is indeed the use of over-tempered air, and not the temperature level. In fact, the temperature rose only slightly during the interventions by about 0.5 °C.

The measured conditions inside of the building during the intervention on November 19 under normal operating conditions are presented in Table 5. We can see that the conditions on that day corresponds to the use of a supply air temperature below room temperature in order to compensate for the internal heat gains, and maintain an acceptable indoor temperature. There again, the average room temperature was similar to the average room temperature during the interventions on January 23 and 24.

Table 5: Measured conditions during the test on November 19, 2013

Date	November 19 06:00-14:00	
Part of building	South	North
Supply temperature T_s [°C]	15.9	15.6
Average room temperature [°C]	22.3	22.0
Outdoor temperature [°C]	4.1	
Outdoor conditions	Slightly cloudy	

3.2 Perceived indoor climate on January 23 and 24

The questionnaire handled to each occupant of the building is presented in Table 6, as well as the average score for each question of the questionnaire, describing the well-being on a scale of 1 (uncomfortable) to 10 (comfortable), for the different supply temperatures.

Table 6: Questionnaire with average score corresponding to the different ventilation strategies.

Questions	November 19	T_s=21°C	T_s=24°C	T_s=26°C
Are you tired?	6.43	7.67	7.16	8.08
Does your head feel heavy?	6.83	8.29	7.46	8.48
Do you have a headache?	8.07	8.73	8.16	8.92
Do you feel faint or dizzy?	8.26	8.79	8.65	8.60
Do you have problems concentrating?	6.87	7.52	7.46	8.20
Do your eyes feel itchy or burning?	8.39	8.65	8.46	8.64
Do you feel hoarse or does your throat feel dry?	8.17	8.90	8.43	9.24
Does your face or your hands fell itchy or burning?	9.20	8.77	8.57	9.16
Do you feel nauseous or otherwise unwell?	9.37	9.62	9.49	9.16
Is it too warm?	7.76	8.37	7.86	8.48
Is there bothersome warmth because of sunshine?	8.26	9.15	9.05	9.60
Is it too cold?	8.00	8.27	8.54	9.12
Do you feel a draught around your feet or your neck?	9.37	8.71	8.86	9.32
Are there bothersome variations of temperature?	8.85	8.19	8.59	8.56
Does the air feel heavy?	7.80	8.48	7.62	8.04
Does the air feel dry?	8.02	8.29	8.00	8.84
Is there any unpleasant smell?	9.48	9.54	9.49	9.80
Do you have a stuffy or runny nose?	9.30	9.33	8.95	9.80
Do you cough?	8.96	9.04	8.84	9.12

The users feel quite well overall, with almost all average scores above eight. Furthermore, the average scores appears to be relatively close to each other.

Table 7 shows the results from the regression analysis. All answered questionnaires (34+31+24+25=114) were included in the analysis.

Table 7: Results from the linear regression

Parameter	Estimate	St. err.	df	t value	p value
Supply temperature	-0.16	0.52	59.55	-0.31	0.755
Gender male (reference: female)	19.42	7.82	51.01	2.48	0.016 *
Asthma (reference: No asthma)	17.29	16.83	50.42	1.03	0.309
Has a cold (reference: No cold)	4.98	5.37	72.07	0.93	0.357

*Signif. codes: p-value < 0.01:**, p-value < 0.05:**

This regression analysis did not report any significant relationship between the supply air temperature, and perceived thermal comfort and SBS (*S*). On the other hand, the regression analysis documents with a 95% certainty that increasing the difference between supply air and room temperature by 1°C will cause a maximum reduction of the SBS score (*S*) of 1.02 point (-0.16 - quantile_studT(0.95, 59.55)*0.52 = -0.16 -1.67*0.52 = -1.02). This reduction is minimal, since a typical score for *S* is between 100 and 160 for the users on the questionnaire. We also observe that the perceived indoor climate is better for male users, but that there is no significant difference in the perceived indoor climate for users with asthma or cold.

3.3 Comparison with the perceived indoor climate on November 19

In a second time, the perceived indoor climate on November 19 during normal operating conditions in the building is compared to the perceived indoor climate with a supply temperature of 24°C and 26°C; see Table 8 and Table 9 for the results from the regression analysis. All answered questionnaires (34+31+ 24+25+46=160) were included in the analysis.

Table 8: Comparison of the perceived indoor climate with Ts=24°C with control Ts=21°C and control November 19 (normal operating conditions).

Parameter	Estimate	St. err.	df	t value	p value
T_s=24°C (reference: T_s=21°C)	2.836	2.361	60.640	1.201	0.2343
November 19 (reference: T_s=21°C)	-2.151	2.686	65.150	-0.801	0.4262
Gender male (reference: female)	17.996	6.818	71.520	2.640	0.0102 *
Asthma (reference: No asthma)	19.442	11.422	73.540	1.702	0.0929
Has a cold (reference: No cold)	10.005	4.427	72.080	2.260	0.0268 *

*Signif. codes: p-value < 0.01:**, p-value < 0.05:**

Table 9: Comparison of the perceived indoor climate with T_s=26°C with control T_s=21°C and control November 19 (normal operating conditions).

Parameter	Estimate	St. err.	df	t value	p value
T_s=26°C (reference: T_s=21°C)	-0.3305	2.9678	52.3600	-0.111	0.91174
November 19 (reference: T_s=21°C)	-6.9514	3.3516	57.9300	-2.074	0.04253 *
Gender male (reference: female)	20.8096	6.6642	66.8900	3.123	0.00265 **
Asthma (reference: No asthma)	20.0683	10.9914	68.3500	1.826	0.07225
Has a cold (reference: No cold)	9.6786	4.9827	70.1200	1.942	0.05610

*Signif. codes: p-value < 0.01:**, p-value < 0.05:**

The regression analysis documents with a 95% certainty that:

- the mean score *S* with supply temperature 24°C is with 95% certainty less than –2.836 + 1.67*2.361 = 1.11 worse than the mean score *S* with supply temperature 21°C, and a supply temperature of 26°C is with 95% certainty less than 5.3 worse than 21°C.

- the mean score S with supply temperature 24°C is less than 1.67 worse than the mean score S on November 19 under normal operating conditions.

- The score S with supply temperature 26°C is significantly better (higher score) then compared to November 19 with p-value 0.043.

The results indicate that the perceived indoor climate was better for short periods with supply temperature of 24°C and 26°C compared to November 19. This result may be due to the fact that the supply temperature on November 19 was of on average of 15.6°C and 15.9°C, see Table 5, which is quite low. Indeed, the scores in Table 6 indicate that the users find the environment slightly cold (with a score of 8.00 for the question "Is it too cold?"). There is however no indication that the users suffered from discomfort by draught, with a score of 9.37 to the related question. Furthermore, it can be noticed that the users were feeling more tired and had more trouble to concentrate on November 19 than on January 23 and 24. It could be possible that we observe the fact that November is a month were many people are typically feeling tired in Norway (transition to winter, poor weather, many have worked all autumn without vacations etc.).

4 CONCLUSIONS

This study enabled to document that the supply of over-tempered air as high as 4°C above room temperature over short time periods provided a good perceived indoor climate in the open plan offices of the building. Using air heating to completely cover the heating demand therefore appears to be a relevant solution for buildings in cold climate with very low heating demand.

This confirms the results presented in a previous paper (Cablé, 2014) concerning field measurements of the thermal comfort and ventilation efficiency in a cubicle office of the same building. The latter reported a good ventilation efficiency and thermal comfort even under unfavourable conditions (*ie.* high supply air temperature and low ventilation rate), provided that heat sources were present in the room.

However, winter 2013/2014 was particularly mild in the Norwegian context. The study will therefore be repeated during winter 2014/2015 in order to confirm the obtained results and conclusions.

5 ACKNOWLEDGEMENTS

This paper was written in the context of the research and development project Forklima, funded by the Norwegian Research Council. The latter, as well as the partners of the project are gratefully acknowledged.

6 REFERENCES

Standard Norge NS 3701 (2012) Criteria for passive houses and low energy buildings.

Sundell, J., Levin, H., Nazaroff, W. W., Cain, W. S., Fisk, W. J., Grimsrud, D. T., ... & Weschler, C. J. (2011). Ventilation rates and health: multidisciplinary review of the scientific literature. *Indoor air*, *21*(3), 191-204.

Cablé, A., Mysen M., and Thunshelle K. (2014). Can Demand-Controlled Ventilation replace space heating in buildings with low-energy demand? *Proceedings of Indoor Air 2014*, July 7-12 2014, Hong-Kong.

Bates, D., Maechler M., Bolker B. and Walker S. (2013) *lme4: Linear mixed-effects models using Eigen and S4*, R package version 1.0-5. Available at http://CRAN.R-project.org/package=lme4

R Development Core Team (2013) *R: A Language and Environment for Statistical Computing*, R Foundation for Statistical Computing, Vienna, Austria. Available at http://www.R-project.org/.

ENERGY-OPTIMAL VENTILATION STRATEGY OUTSIDE OF THE OPERATING TIME FOR PASSIVE HOUSE OFFICE BUILDINGS IN COLD CLIMATES

Per M Holth[1], Mads Mysen[2], Axel Cablè[3*], Kari Thunshelle[4]

1 Oslo and Akershus University College
St. Olavs plass
Box 4, Norway

2 Sintef Building and Infrastructure
Box 124 Blindern
0314 Oslo, Norway

3* Sintef Building and Infrastructure
Box 124 Blindern
0314 Oslo, Norway
Axel.Cable@sintef.no

4 Sintef Building and Infrastructure
Box 124 Blindern
0314 Oslo, Norway

Abstract

The GK environmental house is the first office building in Norway built according to the passive house concept. In such buildings, it is crucial to develop a ventilation strategy to reduce the energy use outside of the operating time. An optimal operating strategy has been developed for cold days, when the outdoor temperature falls well below 0 °C, which is presented in this paper. Indeed, these conditions correspond to the largest heat loss.

The building is mainly heated with warm air supplied through the ventilation systems, with power poles down in each area for peak load. The air handling units(AHU) can be turned into a recirculation of the indoor air mode by night, when there is no need for fresh air since there are no person presents. Data from the active supply air diffusers and AHU's are logged in a central control and monitoring system.

The approach followed in this paper, was to develop different models for running the AHU by night in the building. Energy consumption for the AHU are then downloaded from the central control and monitoring system after each test. Tests at Environmental house GK is then used as foundation for verifying calculations, by comparing the logged data and calculations.

Room temperature data from the supply air diffusers, were used to calculate the building time constant. The latter describes the thermal inertia of the building, and represents the relationship between the heat storage and heat loss capacity. The NS 3031 standard ("Calculation of energy performance of buildings – Method and data") was used as a foundation to create a dynamic model for the building. This model enabled to see the temperature development in the building and find the time constant corresponding to the measurements.

The tests are showing that turning the AHUs in recycling mode by night, enables to easily maintain the temperature throughout the night and also save energy. The results show that the studied building with very low heating demand has a long time constant compared with conventional buildings. Therefore, even for low outside temperatures, the drop in inside temperatures is slow. This makes it possible for intermittent operation without compensating for the heat loss in the off mode.

The developed strategy provides a good example of how to reduce the energy use in in office buildings with very low heating demand in a cold climate.

Keywords: Thermal inertia, cold climate, heating by ventilation, office building, low heat demand

1. Introduction

This paper is part of a bigger scientific research project by name "For Klima"(Forklima, 2012). That addresses "Simplified demand controlled air conditioning of office buildings with very low heating demand". The project is implemented by GK Norway and GK environmental house is made available as case. SINTEF building and infrastructure is project manager and will conduct field / lab studies, which will help develop and disseminate knowledge.

The overall idea of this research project is that "future well-insulated building shells will have a very low heating demand. With sufficiently low heating requirements, it is possible to simplify the current air conditioning solutions and achieve good indoor air quality with lower investment and operating costs"

As of 2015, the public policy in Norway is that all future buildings will be constructed according to the Passive House Standard NS 3701. These are buildings that use passive measures in building construction, which lowers heating demand and thus gives energy efficiency buildings(Government Stoltenberg II, 2013).

By carrying out calculations and experiments in Environmental house GK outside of the operating time, by making use of the supply air temperature of the coldest days of the year. Would it be possible to develop optimal solutions for the operation of the office buildings in the future, which maintain a good thermal comfort, good indoor air quality and lowers the energy demand.

This will provide documentation that the buildings built by the passive standard, has no need for traditional heating system such as radiators. This is also shown in other papers(Feist & Schnieders, 2009) (Feist, Schnieders, Dorer, & Haas, 2005). This will simplify air conditioning solutions in the future, lead to more sustainable buildings, simpler design and leading to lower investment and operating costs. In particular, low investment and operating costs for the building, should make this an affordable option for building owners and provide a competitive edge on other types of buildings. By utilizing the building construction and internal loads in the premises, would the energy demand for heating be lowered further. This is also documented in other works(Lehmann et al., 2010) (Karlsson, Wadso, & Oberg, 2013).

The purpose of this paper is to create "Energy-optimal ventilation strategy outside of the operating time for passive house office buildings in cold climates". This means that the heating demand in GK environmental house, must be covered by an over temperature supply air flow from ventilation units outside the operating time without use of traditional heating.

The hypothesis is therefore whether the ventilation air with the over temperature can contribute to good indoor air quality if the heating demand is sufficiently low.

The approach is therefor:

1. Can supply air from the AHU warm up the building outside of the operating time and satisfy the heating demand at low outdoor temperatures and how is this effecting the energy consumption?
2. Can recirculation of air outside of the operating time be used to keep the temperature steady in the building?
3. Under what conditions is this okay?

This with earlier works, will help to contribute to the documentation of the function and the conditions that must be the basis for future dimensions projections of passive houses. It will also prove due. stricter energy requirements, will open for ventilation solutions that provide good thermal comfort without using other heating systems.

2. Theory

In the following sections is the current theory that explains the features of the building and systems closer.

2.1 Environmental house GK

Figure 1: Environmental hlouse GK

Environmental house GK is an office building located at Ryen in Oslo. The building was opened on 23. August and was then the first office building in Norway built after the passive house standard NS 3701.
Environmental house GK is divided into 3 parts, where it's 2 balanced ventilation systems in each section that distributes air to the premises. The systems are built on the principle DCV (Demand Controlled Ventilation) and needs are controlled by temperature and the presence. In the premises the air are distributed through active supply diffusers, which may vary the slot height in the blowout sone so that the speed of the valve is kept constant regardless of the airflow.
The heating of the building occurs through heated air through ventilation systems using heat batteries installed the AHU's. Heating coils in the heating batteries is designed for a temperature of 35,0/27,9 °C, with a maximum capacity of 103.56 kW at 20,000 m3/h. The energy distributed through the heating coils is supplied by 2 pcs. reversible air / water heat pumps.
Since not all of Environmental House GK is implemented yet, will building C be the focus for further work. Building C is the southeastern part. Building C has a total floor area of 3687.4 m2. Ventilation systems are divided into a southern and northern part. Data from the active supply air diffusers and AHU's are logged in a central control and monitoring system.

2.2 Heat storage and heat loss

Since modern buildings gets better and better insulated, the heat loss will be greatly reduced. In passive houses heat losses are so low that much of the heating demand could be covered by internal loads within the operating time even at low outdoor temperatures.
Different materials have different properties, when it comes to heat storage C (J/m^2K). This tells us how much energy is required to raise the temperature by 1 K. If a material is covered by something, usually the heat storage capacity for the material is greatly reduced.
With the heat storing properties as base in combination with heat loss to the building, the temperature development of a room can be determined. This can be calculated from the building time constant, which is a term for the building's internal thermal inertia.

2.3 Demand controlled ventilation

Demand controlled ventilation is ventilation systems where airflow needs are controlled automatically based on measured demand at room level. The system is a variation of VAV (Variable Air Volume) where there is a variable air volume system over the operating time.
DCV include ventilation systems where supplied ventilation air flow is controlled automatically and in real time by measured needs at room level. That is DCV must have a room sensor, which provides a target / signal on the room air quality, and this signal is used to

control the flow in direct relation to the required quality standards(Mysen & Shcild, 2013).

3. Method

Experiments are conducted on nights when the outdoor temperature has dropped well below 0. Operating personnel at Environmental House GK is contacted to implement the desired strategy.
The paper looks at 3 different air volumes that have organized after the amount of air in each flow, V'min, V'HygMin and V'Max. V'min has an airflow of 2,5 m^3/hm^2, representing the minimum airflow by NS 15251 without present(Standard Norge, 2007a). V'HygMin are with presence, with an airflow of 4,2 m^3/hm^2. V'Max are the maximum airflow for the AHU's, with an total airflow of 10,8 m^3/hm^2. This corresponds to an airflow at 40.000 m^3/h.
Tests at Environmental house GK is then used as the basis for verifying calculations. Basic equations for heat loss by NS 3031 are as follows(Standard Norge, 2007b):
Transmissions losses (eq.1) are the heat loss caused by heat transfer through the building envelope like floor, walls, windows and roof. U represents the u-value of the relevant building part, while A is the area of heat transfer area. Ψ is the thermal bridge constant and *l* is the length of the thermal bridge part.

$$H_{Tra} = \Sigma U_i \cdot A_i + \Sigma \Psi_k \cdot l_k \;(W/K) \tag{1}$$

Ventilation losses (eq.2) are heat loss due to AHU's that extracts air from the building. V is the ventilation air rate and η is the degree of recuperation in the rotating recuperator. By recirculating the air in the AHU's, the recuperation rate will be 100%, this means that the ventilation heat loss will be 0 W/K. This would also be accomplished by turning off the AHU's. Factor 0.33 is the air heat capacity (Wh/m^3K)

$$H_V = 0,33 \cdot \bar{V} \cdot (1 - \eta_{recuperator}) (W/K) \tag{2}$$

Infiltrasion losses (eq.3) are heat loss caused by leaks in the building envelope. n_{50} is leakage number at 50 Pa, and V is the heated air volume.

$$H_{Inf} = 0,33 \cdot n_{50} \cdot V \;(W/K) \tag{3}$$

The overall heat loss is then presented in equation 4. Where heat loss figures are summed and divided by the floor area.

$$H'' = \frac{H_{Tra} + H_V + H_{Inf}}{A_{Floor}} \;(W/m^2K) \tag{4}$$

When calculating the optimal supply air temperature and heating demand for the building, is the heat balance by NS 3031 for heat gains and heat losses used (eq.5).

$$q''_{heat} + q''_{int} + q''_{sun} - q''_{loss} - q''_{vent} = C''\frac{\delta\theta_i}{\delta t} \tag{5}$$

The cumulative capacity C", refers to the accumulating layers in the building. A is the area of the material that accumulate heat, and C is the accumulating ability for the material.

$$C'' = \frac{\Sigma(A_{material} \cdot C_{material})}{A_{bra}} \;(Wh/m^2K) \tag{6}$$

According to NS 3031, the time constant is defined as

$$\tau = \frac{C''}{H''} \;(h) \tag{7}$$

4. Results

Logged indoor temperature in Environmental house GK is illustrated in Figure 2. This is a test from the night of January 24, where internal loads and the AHU's was shut down in 5 hours and the outside temperature was -8,14 °C. Results show that the temperature drop during 5 hours is minimal. The average temperature in the building falls with a total of 0,273 °C in the southern part and 0,245 °C in the northern part. Calculated average temperature drop in

Environmental house is presented by the red graph. The calculated temperature drop is almost identical to the logged temperature drop in part north and south of building C. Over a period of 5 hours, the difference between the calculated and logged temperature was 0.10 °C. The calculated temperature drop is based on heat loss and heat storage capacity of the building.

Internal loads are valued at 1 W/m^2, due. an assessment of the operational scenario. After Project Report 42 from SINTEF, the internal loads for lighting, equipment and people have a load of respectively 5, 6 and 4 W/m2 in low energy buildings - non-residential in the operating time(Dokka, Klinski, Haase, & Mysen, 2009). Outside life will lighting and personal burden be eliminated, the loads will be approximately 0. It is therefore estimated that the internal loads is 1 W/m^2, since there will always be some equipment that can affect the temperature in the building. Since the temperature drop is almost identical to the real case, input in calculations are approved.

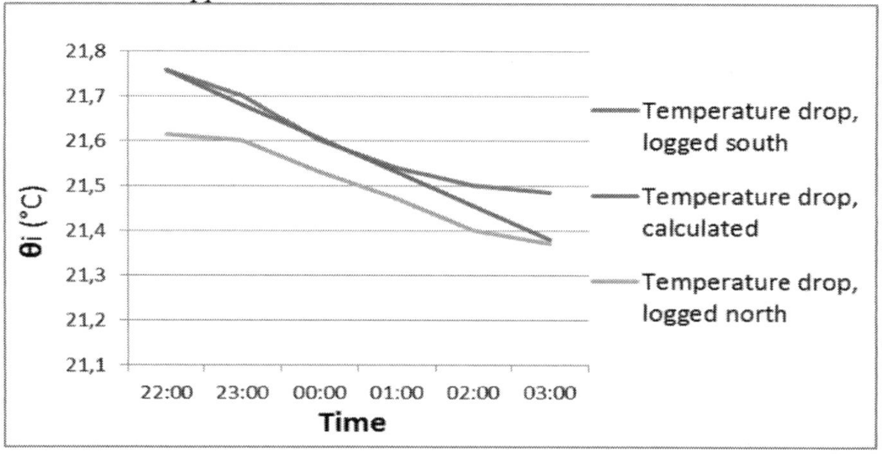

Figure 2: Logged and calculated temperature drop in Environmental house GK, building C, at an outdoor temperature of -8.14 °C.

Heat loss figure for Environmental house GK is stated to be 0.36 on their website(GK Norge AS). Table 4 shows heat loss figures for operation without recirculation mode for different airflows in night mode during winter. It shows that the heat loss increases with increasing air flow, which will have a negative impact on the heating demand.

Table 4: Heat loss figures without recirculation of air.

V''	V'' (m^3/hm^2)	η_{rec}	H'' (W/m^2K)
V'Min	2,52	0,88	0,33
V'HygMin	4,2	0,88	0,40
V'Max	10,8	0,88	0,66

Table 5: Heat loss figures with recirculation of air.

H_{tra} (W/K)	H_{inf} (W/K)	H_{vent} (W/K)	H (W/K)	H'' (W/m^2K)
793,19	51,23	0	844,42	0,23

Table 5 shows AHU's in recycling mode, it will ensure that the heat loss from ventilation will be zero. Heating will therefore only consist of transmission and infiltration. This would also be accomplished by turning off the AHU's. The overall heat loss for the building falls to 0,23 W/m^2K.

The specific power requirements at different air flows based on the heat loss figures, are shown in Figure 3. V'Min, V'HygMin and V'Max are air flows without recirculation with an efficiency for the recuperator at 88 %. Since the heat loss is unaffected by the air flow in resirculation, is this constant and only presented with a purple graph. The figure shows that the power requirement for heating increases with increasing air flow when it's not recirculation.

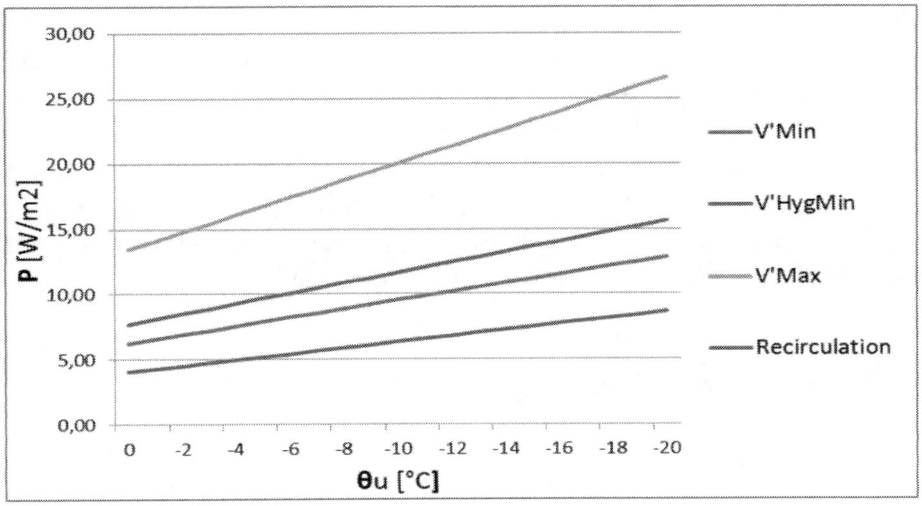

Figure 3: Power requirement by various air flow without recirculation of air and power requirement in recirculation mode.

Based on the power needed to meet the heating requirements, that is presented in Figure 62. Will this give a total power requirement at an outdoor temperature of -20 °C for building C on 32-, 47-, 57- and 98 kW respectively for recirculation, V'Min, V'HygMin and V'Max.
Optimal supply air temperature to keep an indoor temperature at 22 °C, is presented in Figure 4. Set-point temperature of the supply air will have a maximum value at an outdoor temperature of -20 °C, on 24,4-, 28,2- and 32,4 °C respectively for V'Max, V'HygMin and V'Min. With lower air flows the higher temperature of the supply air is required to meet the heating demand.

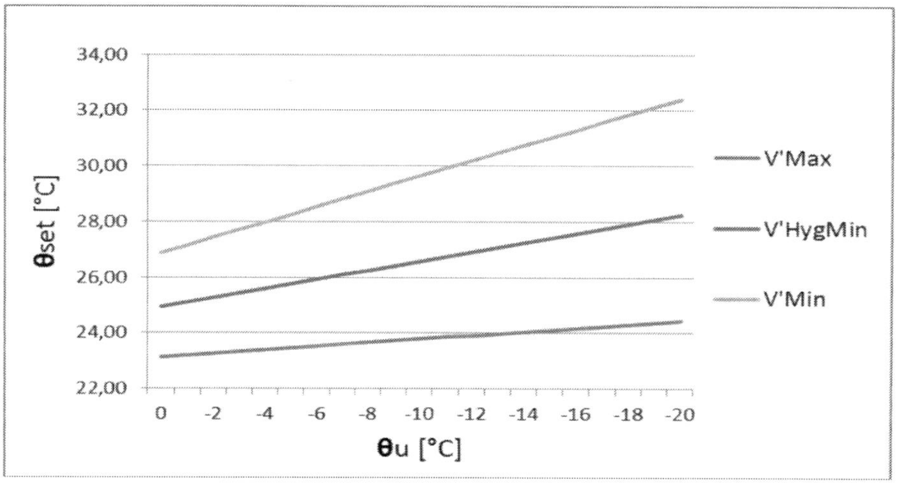

Figure 4: Optimal supply air temperature at different airflows.

Heating coils in the heating batteries is designed for a temperature of 35,0/27,9 ˚C, with a maximum capacity of 103.56 kW at 20,000 m3/h per piece. Maximum heating demand is 98 kW for the airflow V'Max, the largest optimal supply air temperature is 32,4 ˚C for V'Min. This proves that Environmental house GK covers the heating demand, by only using over tempered air from the AHU's. Since the heating demand and optimal supply air temperature is under the batteries capacity. By making use of recirculation in the AHU's, the heating demand to Environmental house GK will be reduced, compared with fresh air strategies. Recirculation would therefore be a beneficial strategy to maintain a constant temperature in the building and to reduce energy requirements for heating even further outside of the operation time. It was not chosen to look at lower airflows, since optimal supply air temperature will pass the trip temperature in the heating coil

The time constant that was calculated for Environmental house GK night of January 24, is shown in **Error! Reference source not found.** is an average of the southern and northern part of building C. The mean time constant tau that appears in the table is 285.43 hours. It is important to emphasize that this is with turned off AHU's and heating elements in the electric poles

Table 3: Time constant for logged attempts for an average of part south and north.

Date	t_{off} (h)	$t_{calculation}$ (h)	South, τ (h)	North, τ (h)	$τ_{average}$ (h)
Night of Jan 24	5	4	291,0	279,9	285,4

Table 4: Time constant for dynamic calculations.

V"	V" (m^3/hm^2)	C" (Wh/m^2K)	H" dynamic (W/m^2K)	τ (h)
V'Off	0	76,9	0,2	335,6
V'Min	2,5	76,9	1,1	72,5
V'HygMin	4,2	76,9	1,6	47,6
V'Max	10,8	76,9	3,8	20,3

By turning off the AHU's shows the dynamic calculations for the building that the building time constant is approximately around 335.63 hours (Table 4.). The difference is approximately 50 hours between dynamic calculations and calculated from the logged data. Reading error could occur for the recorded temperature profile for the building, since this is presented in a graph. When the time constant achieves this size, decimals will affect the outcome of the final result in a big scale. The difference is therefore accepted, since a small change for the read temperature profile, would make logged time constant passing the dynamic calculated time constant.

Table 4 also shows that the result around the time constant of the premises is heavily dependent on how much air flow being pulled out of the premises. V'Min V'min will have a time constant of 3 days, V'HygMin 2 days and V'Max about 1 day. This means that it requires very large volumes of air to achieve a temperature change in the building, but it will also mean that the temperature remains stable at low flows and turned off AHU's. This is due to good heat storage capacity and low heat loss figures. With such a large time constant in the off mode, will it confirm the potential for intermittent operations outside the operating time.
Figure 5 shows that the temperature drop for an outdoor temperature of -20 °C, will be approximately 0.33 °C for a time interval of 3 hours a night where the AHU's and the heating elements in the power poles is turned off. The temperature drop will only be 0.16 °C for a temperature of 0 °C. To exploit Environmental house GK's thermal inertia and low temperature drops, will be ideal to make use of intermittent operations. If one allows that the temperature in the premises to fall outside the operating time and let internal loads heat up construction during the work day, would this be an ideal solution for saving energy.

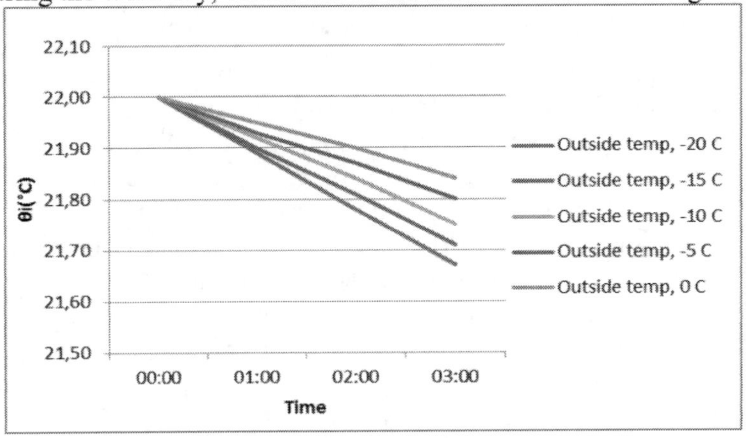

Figure 5: Average temperature drop in Environmental house GK, building C, based on calculations.

Table 5 shows the energy savings for heating, during a normal winter for Oslo. A normal winter is considered the months of December, January and February with normal outdoor temperatures. Recirculation will have a energy consumption on 13,095.34 kWh/winter, and have the lowest energy consumption compared with strategies with outdoor air. Without recirculation would the energy consumption be at 19,929.27 -, 24472.99 - 42403.95 and kWh/month at airflows V'Min, V'HygMin and V'Max.

Table 5: Energy demand for a normal winter and the difference between air flows and recirculation mode.

V''	V'' (m^3/hm^2)	$E_{Heating}$ (kWh/night)	$E_{Heating}$ (kWh/winter)	E_{diff} (kWh/winter)
V'Min	2,5	221,5	19929,4	6834,0
V'HygMin	4,2	272,0	24473,0	11377,6
V'Max	10,8	471,3	42403,9	29308,6
Reciculation	-	145,5	13095,3	0,0

Table 6 presents the potential saved energy for a deactivated AHU's, for a time interval of 3 hours a night for a winter. GK has stated on their website, they have a total heating demand on 9 kWh/m^{2}year in Environmental house GK. With the potential of intermittent operation, this could be lowered to 7.7 kWh/m^{2}year.

Table 6: Total heating demand due to intermittent operation.

Heating demand (kWh/m^2year)	Heating demand (kWh/year)	Intermittent savings (kWh/vinter)	New heating demand (kWh)	New heating demand (kWh/m^2year)
9,0	33183,9	4910,8	28273,1	7,7

5. Conclusion

Results from the paper will thus provide evidence that the office building built by the passive house standard NS 3701, are able to meet their heating demand just by making use of the supply air temperature even at outdoor temperatures down to -20 °C. This relates to office buildings with the same or better heat loss figures, or in warmer climates. It is necessary in this type of heating, that warm air will have the opportunity to flow into the room without heating or with greater heat loss.

Average heat loss figures for Environmental house GK is given to be 0.36 W/m^2K on their webpage. By using recirculation in the AHU's ensures however that the heat loss figure drops to 0.23 W/m2K, which is almost a halving of the declared value. This would also be achieved by turning off the AHU's. This is due to the heat loss of ventilation will be eliminated and will only consist of transmission and infiltration at this strategy. By running with recirculation outside operation time, will therefore ensure energy saving compared with heating with fresh air and be ideal strategy. A good recirculating solution would also even out temperature differences in the building.

The paper reveals that the future office buildings with low heating demand, could use intermittent operation as a strategy in combination with recycling mode. Due to the buildings' low heat loss figures in combination with high heat storage capacity, will cause a large time constant that makes the buildings thermal slow and causes low temperature drops. By allowing the construction dissipate heat outside the operating time by declined ventilation, will provide minimal temperature drop in the premises during the night. It is important that the AHU's do not compensate for heat loss that has occurred in off mode at a later stage, but let internal loads during the work day compensate for the heat loss during the night in the off mode.

Questionnaires related to indoor climate in the occupied open plan offices during the working day are ongoing, and will be presented in completed study. If interested, see http://www.sintef.no/Projectweb/For-Klima/ for more information.

6. Acknowledge

This paper was written in the context of the research and development project Forklima, founded by the Norwegian Research Council. The latter, as well as the partners of the project are gratefully acknowledged.

References

Dokka, T. H., Klinski, M., Haase, M., & Mysen, M. (2009). *Prosjektrapport 42. Kriterier for passivhus- og lavenergi bygg – Yrkesbygg*: Sintef Byggforsk. ISBN 978-82-536-1107-5

Feist, W., & Schnieders, J. (2009). Energy efficiency - a key to sustainable housing. *European Physical Journal-Special Topics, 176*, 141-153. doi: 10.1140/epjst/e2009-01154-y

Feist, W., Schnieders, J., Dorer, V., & Haas, A. (2005). Re-inventing air heating: Convenient and comfortable within the frame of the Passive House concept. *Energy and Buildings, 37*(11), 1186-1203. doi: 10.1016/j.enbuild.2005.06.020

Forklima. (2012). Forklima - Forenklet behovsstyrt klimatisering av kontorbygg med svært lavt oppvarmingsbehov. from http://www.sintef.no/Projectweb/For-Klima/

GK Norge AS. Demands for the building. Retrieved 01. Oct, 2013, from http://miljohuset-gk.no/passivhus/krav-til-bygningen/

Government Stoltenberg II. (2013). good experiences with passive houses. Retrieved 23 Des, 2013, from http://www.regjeringen.no/nn/dokumentarkiv/Regjeringa-Stoltenberg-II/Kommunal--og-regionaldepartementet/Nyheter-og-pressemeldinger/nyheter/2013/gode-erfaringar-med-passivhus.html?id=714278

Karlsson, J., Wadso, L., & Oberg, M. (2013). A conceptual model that simulates the influence of thermal inertia in building structures. *Energy and Buildings, 60*, 146-151. doi: 10.1016/j.enbuild.2013.01.017

Lehmann, B., Guttinger, H., Dorer, V., van Velsen, S., Thiemann, A., & Frank, T. (2010). Eawag Forum Chriesbach-Simulation and measurement of energy performance and comfort in a sustainable office building. *Energy and Buildings, 42*(10), 1958-1967. doi: 10.1016/j.enbuild.2010.06.002

Mysen, M., & Shcild, P. G. (2013). *DCV - Requirements and hand-over documentation.* (pp 11-17): ISBN 978-82-536-1369-7.

Standard Norge. (2007a). NS-EN 15251:2007. Indoor environmental input parameters for design and assessment of energy performance of buildings addressing indoor air quality, thermal environment, lighting and acoustics. (pp. 34): ICS 91.140.01

Standard Norge. (2007b). NS 3031:2007. Calculation of energy performance of building - Method and data(pp. 17-41): ICS 01.040.91;91.120.10

CAN AIR HEATING ALONE BE USED IN PASSIVE HOUSE OFFICE BUILDING IN COLD CLIMATES? REVIEW OF THE OBTAINED RESULTS

Kari Thunshelle*[1], Axel Cable[1], Mads Mysen[1,2], and Hugo Lewi Hammer[2]

1 SINTEF
Pb.124Blindern
0314 Oslo, Norway

2 Oslo and Akershus University Collage of applied science
Pb 4, St Olavs plass
0130 Oslo, Norway
**Corresponding author: kari.thunshelle@sintef.no*

ABSTRACT

The future is well-isolated buildings with low heating demand. The first office building in Norway satisfying the passive house standard, the GK environmental house in Oslo, was taken into use in August 2012.

Low energy building is the standard for new buildings according to the building codes today, and since passive house standard seems to be included in the building codes in Norway from 2015 there is a great change in the building industry. To meet the low energy concept, the ventilation industry must cope with a massive change from use of installations with constant air volume (CAV) to demand controlled ventilation.

At the same time new technology with active air handling units makes new ventilation systems more flexible to demand controlled ventilation and temperature control, air velocity and draft. The low heating demand in future buildings address the question – is it possible to use simplified heating systems, or even only air heating without any backup heating system, also for the coldest days in Norway and still have satisfied users?

However, given the cold climate in Norway, totally eliminating a backup heating system for the coldest days and only use overheated supply air would be a tough decision without proper documentation.
The R & D project "Simplified demand controlled air conditioning in office buildings with very low heating demand" will develop concepts for ventilation based space heating fit to buildings with very low heating demand, as well as develop documentation on the function and the conditions that must be fulfilled. (www.sintef.no/projectweb/For-Klima). The project period is three years, starting 2013. This paper reviews the results so far.

So far a broad range of studies have been done: Field measurements and intervention studies with user survey on perceived indoor climate at GK environmental house, laboratory measurements, calculations and theoretical evaluations. This paper discusses the results so far after the first winter period. These are preliminary results that will have to be confirmed by further studies next winter.

KEYWORDS

Ventilation, air heating, passive house, indoor climate survey, measurements

1. INTRODUCTION

The future is well-isolated buildings with low heating demand. Low energy building is the standard for new buildings according to the building codes today, and since passive house standard seems to be included in the building codes in Norway from 2015 there is a great change in the building industry. To meet the low energy concept, the ventilation industry must cope with a massive change from use of installations with constant air volume (CAV) to demand controlled ventilation.

At the same time new technology with active air handling units makes new ventilation systems more flexible to demand controlled ventilation and temperature control, air velocity and draft. The low heating demand in future buildings address the question – is it possible to use simplified heating systems, or even only air heating without any backup heating system, also for the coldest days in Norway and still have satisfied users?

Such simplification with air heating was attempted in the '70s. At that time maximum heating demand was 5-10 times greater than is the case with the upcoming building regulations on passive house level. The result was unsatisfactory indoor climate. There is reason to believe that current premises and technical solutions provide significantly better results.

Many building owners are interested in simplified systems, but without good documentations few are willing to go all in. The R & D project "Simplified demand controlled air conditioning in office buildings with very low heating demand" will develop concepts for ventilation based space heating fit to buildings with very low heating demand, as well as develop documentation on the function and the conditions that must be fulfilled. The project period is three years, starting 2013. (www.sintef.no/projectweb/For-Klima)

The first results from the project are presented in this paper.

2. OBJECTS

The object of this paper is to review the project results after the first winter and discuss what these results can indicate.

The main object of the project is to

- **Develop concepts for ventilation based space heating for buildings with very low heating demand**– knowledge about limitations and premises for ventilation with over temperature.
- **Develop knowledge about thermal comfort without traditional draft protection. –** Documentation of actual conditions the coldest days, and document what operation strategy will give the best conditions.
- **Develop documentation of function and premises that have to be met**

Then as a result recommend a documented simplified and cost efficient solution to the future office buildings.

Analyses are mainly based on the following approaches: field measurements, intervention studies with user survey on perceived indoor climate at GK environmental house, laboratory measurements, calculations and theoretical evaluations. Reports from the Central control and monitoring system will be available upon requests.

3. THE GK ENVIRONMENTAL HOUSE

Figure 1: The GK Environmental house and the active air supply diffusor used

Norway's first passive office building - the so called GK environmental house -was taken into use in August 2012. This building located in Oslo is the "home" for GK, one of the leading engineering contractor and service partner in Scandinavia for new and existing commercial buildings. Given the owner, they wanted this building to be a show case. The compact shaped building of 5 stories and 14300m^2 is well analyzed and calculated up front, and Breeam Nor certified as 'Very good'.

The net energy demand is calculated according to NS 3701 (NS 3701:2010) to 64 kWh/m^2/year, and measured delivered energy to 49 kWh/m^2/year according to NS 3031. The energy source for heating is air/water heat pump, heating of hot water is from the refrigerating machine, and there is no cooling of the space areas. The air leakage number (50Pa) is 0,23 and the u values are for external wall 0,14 W/(m2K), roof 0,10 W/(m2K), floor 0.07 W/(m2K) and doors and windows 0,78W/(m2K).

The heating demand is entirely covered by the ventilation system, using active air supply diffusors to adapt to changes in demanded ventilation rate and temperature. The diffusors can regulate the supply area according to the ventilation rate, and then keep a more stable supply air velocity than more traditional supply units. Furthermore, the system can be switched to recirculating air during night time. However, to be sure the owners have installed small electric poles as backup if needed on the coldest days of the year. The office building consists of open plan offices, meeting rooms and some separate offices. The ventilation is now controlled by presence, but can be controlled separate unit by unit. The building management system is designed with detailed logged data.

4. REVIEW OF THE PROJECT RESULTS SO FAR

4.1 Field measurements

The objective is to evaluate the indoor climate in a room heated by high supply temperature, especially on the coldest days of the year. By using a cubic office as a simple test scenario, it is possible to get principal knowledge about ventilation effectiveness and thermal comfort with this kind of solutions that can be transferred to areas with less controlled conditions. The goal is to do the evaluation on the coldest days of the year. Unfortunately the winter 2013/14 was not a very cold winter.

Thermal comfort in a standard cubic room of 9,6m^2 was assessed according to EN 15251 (2007) standard in terms of operative temperature, thermal stratification and draught rate.
In addition tracer gas tests (SF_6) according to NS-ISO 12569 were carried out in order to evaluate the ventilation effectiveness in the room.

The test were conducted as 4 different ventilation strategies (test cases)
- Low airflow rate and high supply temperature 1) without and 2) with occupants.
- High airflow rate and low supply temperature 3) without and 4) with occupants.

High airflow rate is based on maximum available airflow, and low airflow rate is the minimum needed for an office with low emitting materials and one person present.
Data for the different cases are given in the table below. Further details of method and results are given in (Cablé 2014).

Table 1: Test conditions for the four different cases studied

			Low airflow rate High supply temperature		High airflow rate Low supply temperature	
Measured parameter	**Designation**	**Unit**	**Case 1**	**Case 2**	**Case 3**	**Case 4**
Ventilation rate	Q_s	l/s	17.4	16.8	48.1	49.4
Supply temperature	T_s	°C	31.1	32.0	24.2	24.1
Temperature difference	ΔT	°C	7.5	5.0	1.2	0.8
Outside temperature	T_{out}	°C	-4.8	-7.0	-2.9	-1.8
Heating power	P_s	W/m^2	16.4	10.7	7.6	4.8
Internal heat gains	P	W/m^2	3.1	29.9	3.1	29.9

The results fulfil the best category in EN 15251 on operative temperature, above 22,9°C, but gets too high with internal gains as persons and equipment. Preheating of the supplied air above room temperature seems necessary only during colder days or when the offices are empty.

According to the standard (ISO 7730, 2005) a gradient of 4,2°C between ankle and head leads to 10% dissatisfied. For the two cases with low airflow rate, the gradient is 0,8°C and 1,4°C. Higher airflow rate gives an even more homogenous temperature in the room. The stratification seems to be marginally influenced by the presence of the internal heat gains.

Draught risk depends on local air velocity magnitude, temperature and turbulence. Using 10% dissatisfied according to the standard (ISO 7730, 2005), the velocity magnitude should be lower than 0,15 m/s. In case 4 this value is measured at ankle level. The draught risk is limited in this case. All other measurements show much lower air velocity and there is no draught risk. As expected, the air velocity is higher in the caser with higher airflow rate.

The tracer gas study indicates that short-circuiting might occur in an empty room with low airflow rate and high supply temperature. When the rooms are occupied, the mixing increase to perfect mixing and leads to good ventilation effectiveness.

4.2 Intervention studies with user survey

Two different intervention studies on user perceived health and well-beeing have been done in January 2014. The object with these studies is to evaluate which ventilation strategy that gives the best perceived indoor climate.

Each of the intervention studies compared two different ventilation strategies using cross over design. The building was divided in two parts. The first day, one part was exposed to strategy 1 as the other part was exposed to strategy 2. The next day the strategies were switched.

A questionnaire was used to measure the user's perceived health and well-being under the different cases. This was filled out at the end of each day. A reference study was conducted during the autumn. The results were analysed using a random effect linear regression model.

In the first study the initial room temperature was 21°C. Then the supply temperature was regulated slightly above room temperature for strategy 1, and slightly below room temperature for strategy 2. The outside temperature for the two test-days in January was -10.9°C and -5.5°C. More details are given by (Hammer et. al, 2014).

In the second study, a supply temperature of 24°C and 26°C was compared with supply temperature of 22°C, in both cases the room temperature was around 22 °C. The outside temperature these days were -6 °C.

Overall the users seem to feel quite well. According to (Hammer 2014 better perceived indoor climate was obtained for a supply temperature higher than room temperature than for a supply temperature slightly below room temperature. No significant discomfort regarding perceived indoor air quality was obtained compared to the control case in the autumn. This is an indication that the active supply diffuser employed display good mixing properties for a broad range of supplying conditions and that the short-circuiting of fresh and warm ventilation air is reduced to a great extent. This is corresponds to the results obtained by the measurements in the cubic office presented above. According to the results from the second study (Cablé 2014) the impact of the different supply temperature were limited.

4.3 Tracer gas measurements for different strategies on overheating, air change rates and placing of exhaust valve

In a Master thesis connected to the ForKlima project, (Aslaksen, 2014) has studied the results of ventilation with heated supply air in a laboratory test room according to NS-EN 442-2. Tracer gas studies were conducted according to NS-EN ISO 12569, in order to analyse the impact of the following parameters on the ventilation efficiency:

- Placing of exhaust outlet
- Air flow rate
- Temperature difference between supply and room air

The tests are done with a supply air temperature of 2, 4, 6 and 10°C above room temperature with ventilation rates 50l/s, 34l/s, 18l/s and 9l/s. The two first ventilation rates are, as for the field measurements, corresponding to maxima available air flow and ventilation of an office with one person present and low emitting materials. The two other values are chosen to be able to describe a trend in the results.

All tests are done both for exhaust located at floor level and at ceiling level. There were no heat gains, but a dummy and a lamp were placed in the room. Later two tests were conducted with heat gains according to Project report 42 (Mysen et al 2009), and the need for heating was calculated according to NS 3701 for Passive house office buildings.

Table 2: Heat loss and internal heat gains used for calculation of neded heating

Persons:	70W
Lights	$5W/m^2$
Technical equipment	$6W/m^2$
Heat loss	$0,50W/(m^2*K)$

The results show that exhaust at floor level is more favorable than exhaust at ceiling level. The measured ventilation effectiveness with exhaust placed at floor level is 85-90% and higher for all ventilation rates as the heating demand is low. When supply temperature exceeds room temperature with 5-6 °C the trend is reduced ventilation rate especially for the high ventilation rates.

For exhaust at ceiling level the ventilation effectiveness is around 90% and higher with supply temperature 2-3 °C higher than room temperature. With supply temperature 4-5 °C higher than room temperature the curves for the highest ventilation rates goes noticeably downwards and under 80%. With supply temperature10 °C above room temperature and 34l/s airflow it is as low as 40%. The two lowest ventilation rates keep above 80% until around 6 °C. All these results are with no internal heating from persons, light and equipment. With internal heating at ventilation rate 34l/s, the ventilation effectiveness increases considerable.

From the results of the calculating of heating demand (Aslaksen 2014) the following rough overview on need for over temperature can be made:

Table 3: Outdoor temperature where different over temperature is needed for different ventilation rates

Over temperature °C	**1**	**2**	**3**	**4**	**5**	**6**
Ventilation rate						
50l/s	-17	-23	-	-		
34l/s	-14	-19	-	-		
18l/s	-11	-14	-16	-22	-	
9l/s	-9	-11	-12	-14	-15	-16

There is a marginal heating demand before an outside temperature of -10 °C. At this temperature, a supply temperature of 1 °C above room temperature is necessary only for the lowest ventilation rates. A supply temperature of 2°C above room temperature is needed at outside temperature below -23 °C and 50l/s and at -18 °C outside temperature with 34l/s. A supply temperature of 4°C higher than room temperature is likely to be needed first at -22 °C outside temperature and 18l/s

Average number of the last 10 years of days with outside temperature at the metrological institute at Blindern , Oslo, is by (Aslaksen 2014) as given in Table 4.

Table 4: Average number of days a year with given outdoor temperature

Outside temperature	Number of days
Days with -8 °C	29,0
Days with -14 °C	5,7
Days with -20 °C	0,2

5 DISCUSSIONS

Both the field measurements in the GK environmental house and the laboratory tests show by tracer gas measurements that good ventilation effectiveness is obtained for a broad range of supply temperature and ventilation rate.

When the exhaust is located at floor level, an almost perfect mixing is achieved, while the ventilation effectiveness is a bit lower when the exhaust is located at ceiling level. Furthermore, the ventilation effectiveness is reduced when the supply air temperature exceeds the room temperature with 4-6 °C. However, according to calculations (Akselsen 2014) with an outside temperature of -20 °C, a supply air temperature a little higher than 4 °C above room temperature give good thermal comfort with a ventilation rate 18 l/s, 2 °C above room temperature with ventilation rate 34 l/s, while a supply temperature of only 1 °C above room temperature is necessary with a ventilation rate of 50 l/s. According to the statistics the last ten years, an outside temperature of -20°C has only occurred 2 days in January in Oslo. Results so far indicate that the ventilation effectiveness for these conditions should be around 90% or higher without internal loads, and increasing with internal loads. The risk for short-circuiting seems to be limited.

Furthermore the field studied measurements show that within the criteria of 10% dissatisfied users, the risk for discomfort by draught, and thermal stratification is very low. The only case that might give draught by ankle level is with high ventilation rate, 50 l/s and 2 °C and internal heat gains. This is an unusual situation, and a condition with high air velocities in the room.

The results from both of the intervention study on the users' perceived health and well-being show marginal difference in score from the reference study. Supplying air at 24 °C and 26°C with short periods of heating seem not to influence on the perceived indoor climate compared to supplying air at temperature of around 22 °C. The results correspond with the measurements showing that this degree of over temperature gives marginal reduction in ventilation effectiveness and nearly no risk for draught and thermal stratification.

6 SUMMARY

The first years of studies in the ForKlima project indicates that demand control ventilation can be used alone to cover the heating demand, thanks to the high performance of the building's envelope and the air supply diffusor. An air supply temperature of 2-4 °C is necessary for the coldest days of the year, and up to 4°C the ventilation effectiveness seems to be at a higher level. Good thermal comfort seems possible to be obtained at these conditions. Active air supply diffusors whose supply section varies according to the airflow rate seems to be an important contribution to these results.

These are preliminary results that will have to be confirmed by further studies next winter, and compared to other strategies. It is also to mention that the winter of 2013/14 was a mild winter with a measured minimum of -11 °C outside temperature.

7 FURTHER WORK

The research and development project ForKlima is now half way through a project period of 3 years. It must be emphasized that the results discussed in this paper are preliminary results. Studies at the GK environmental house will be repeated next winter 2014/15. We then hope for colder outside temperature, to test the solutions under more extreme conditions. New intervention studies will be conducted based on the results and analyzes so far.

8 ACKNOWLEDGEMENTS

This paper was written in the context of the research and development project ForKlima founded by the Norwegian Research Council. The latter, as well as the project partners of the project are gratefully acknowledged.

9 REFERENCES

Aslaksen,V (2014). *Klimatisering med overtemperatur og inneklima i kontorlokaler.* Master theses Høgskolen i Oslo og Akershus, Norway

Cablé, A. et al. (2014). *Can demand controlled ventilation replace space heating in office buildings with low heating demand?* Indoor Air conference 2014, Hong Kong

Cablé, A. et al. (2014). *Air heating of passive house office buildings in cold climates - how high supply temperature is acceptable?* AIVC conference2014 Poznan, Poland.

Hammer, HL. et al (2014). *Heating "passive house"offices in cold climate using only ventilation system – comparison of two ventilation strategies.* AIVC conference 2014 Poznan, Poland.

Mysen, M et al (2009). *Project report 42 - Criteria for passive houses and low energy buildings - non-residential buildings.* SINTEF Academic press. Oslo, Norway

NS 3701:2012.Criteria for passive houses and low energy buildings- Non-residential buildings. Standards Norway

NS 3031:2007+A1:2011. Calculation of energy performance of buildings - Method and data. Standards Norway

NS-EN 442-2: 1996. Radiators and convectors - Part 2 : Test methods and rating. Standards Norway

NS-EN 15251:2007 Indoor environmental input parameters for design and assessment of energy performance of buildings addressing indoor air quality, thermal environment, lighting and acoustics. Standards Norway

NS-EN-ISO 12569:2012 Thermal performance of buildings and materials - Determination of specific airflow rate in buildings - Tracer gas dilution method. Standards Norway.

IMPACT OF THE USE OF A FRONT DOOR ON THERMAL COMFORT IN A CLASSROOM IN A PASSIVE SCHOOL

Hilde Breesch[*1], Barbara Wauman[1,2] and Alexis Versele[1]

1 KU Leuven Campus Ghent, Faculty of Engineering Technology, Sustainable Building
Gebroeders De Smetstraat 1
B-9000 Ghent, Belgium
**Corresponding author: Hilde.Breesch@kuleuven.be*

2 KU Leuven, Department of Civil Engineering
Building Physics Section,
Kasteelpark Arenberg 40, box 2447
B-3001 Heverlee, Belgium

ABSTRACT

A new school building block in Passivehouse standard near Kortrijk (Belgium) is in use since spring 2013. The urban development regulations required that this new building did not influence the incidence of daylight in the adjacent dwellings. This results in an open corridor on the first floor and classrooms with a front door. Draught and increased energy losses are expected. This design choice is contradictory to the basic idea of a passive school that aims to be very airtight and to have very low energy use and excellent thermal comfort.

The aim of this paper is to evaluate the impact of the use of the front door on the thermal comfort in the classroom in this passive school in winter season. The requisite heating capacity at the time the door is opened is also determined and compared to the installed capacity.

This evaluation is based on a survey of pupils and teachers and on measurements in the classroom in winter 2014. The survey studies the experiences of the users. The measurements determine the use of the front door, i.e. the duration and frequency of opening, during one week. In addition, operative temperatures are measured on several positions and heights in the classroom. Finally, the airflow through the door is determined by tracer gas measurements.

The survey concluded that every respondent experiences cold draught and almost all users are annoyed by this draught. Monitoring showed that less than half the openings were expected based on the timetable of the classroom. The most occurring opening durations are 6 and 20s. The airflow was determined by tracer gas measurements for these opening durations. A significant temperature decrease was found close the door. Impact factors on the magnitude of decrease are opening duration, indoor-outdoor temperature difference, wind speed and direction, successive openings. The effect on the temperature is not evenly spread in the classroom. The temperature decrease significantly reduces as the distance to the front door increases. Temperature decrease is more pronounced close to the floor.

Survey and monitoring showed that a the use of the front door influences thermal comfort and increases infiltration losses in this classroom in this passive school. It is advised to consider this effect in the design of heating system in and the assessment of energy performance of future Passive House projects including rooms with front doors.

KEYWORDS

User impact, passive school, thermal comfort, infiltration losses

1. INTRODUCTION

Since the implementation of the European Directive 2002/91/EC on the Energy Performance of Buildings the number of passive non-residential buildings has increased significantly (Kaan, 2006). In Flanders (Belgium), the evolution towards more energy efficient schools in particular was boosted in 2009 by the approval and subsidizing of 24 passive schools,

covering almost 65000 m². The following criteria for Flemish passive schools were set forward by the government:
1° annual net energy need for heating ≤ 15 kWh/(m².a)
2° annual net energy need for cooling ≤ 15 kWh/(m².a)
3° maximum air tightness level (n_{50}) ≤ 0,6 h^{-1}
4° maximum E-level = 55 (primary energy performance level as required by EPB (EPB, 2013)

The new school building block in Passive House standard near Kortrijk (Belgium) is one of these 24 schools. The urban development regulations required that this new building did not influence the incidence of daylight in the adjacent dwellings. This results in an open corridor on the first floor and classrooms with a front door (see Figure 67). As this door is expected to be used intensively, draught and increased infiltration losses and thus heat losses are expected. This design choice is contradictory to the basic idea of a passive school that aims to be very airtight (and thus having very low infiltration losses), to have very low energy need for heating and a low primary energy performance level and an excellent thermal comfort. Infiltration losses due to the front door in a German passive school was already studied by Peper et al. (2007, 2008). Extrapolated to a full occupancy of the school, an additional heat loss of 0.5 kWh/m².a was calculated.

The objective of this paper is to evaluate the impact of the use of the front door on thermal comfort and the infiltration losses in a classroom in this passive school in winter season. This paper is based on the results of Houwen and Van Lerberghe (2014).

2. BUILDING DESCRIPTION

The new school building block in Passive House standard has a floor area of about 1500 m² and includes 3 floors with classrooms, labs, offices and a study hall. A plan of the first floor and a cross section are shown on Figure 1 and Figure 2. Dimensions of the class room and SE wall with front door are indicated on Figure 3. The classroom has a floor are of 49.8 m², dimensions of front door are 1.15m x 2.36m. Balanced mechanical ventilation is provided, the airflow in the classroom is 750 m³/h. For heating purposes, the air is preheated by an air-to-air heat recovery. Additionally, a central and local heating coil of 24 kW and 2.5 kW respectively are integrated in the supply duct. Lessons start at 8h20 and ends at the latest at 16h50. Table 1 defines in detail the use of the classroom.

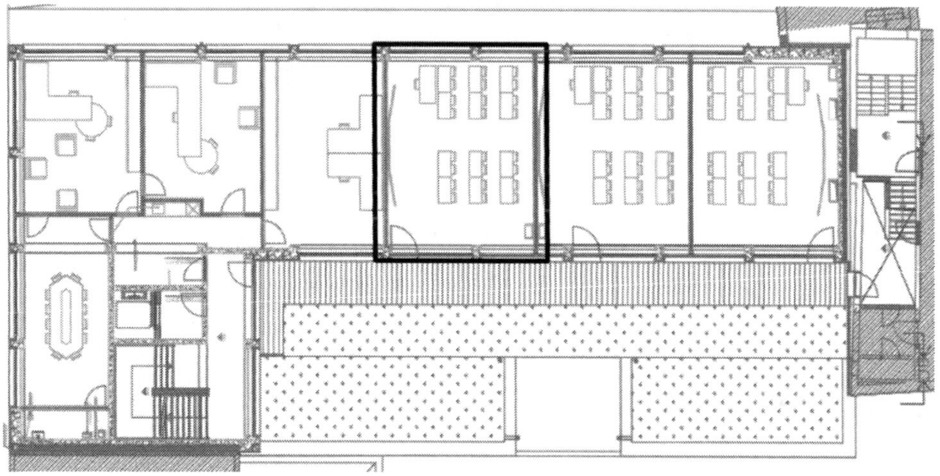

Figure 1: Floor plan of new passive school building block, classroom indicated with a black rectangular

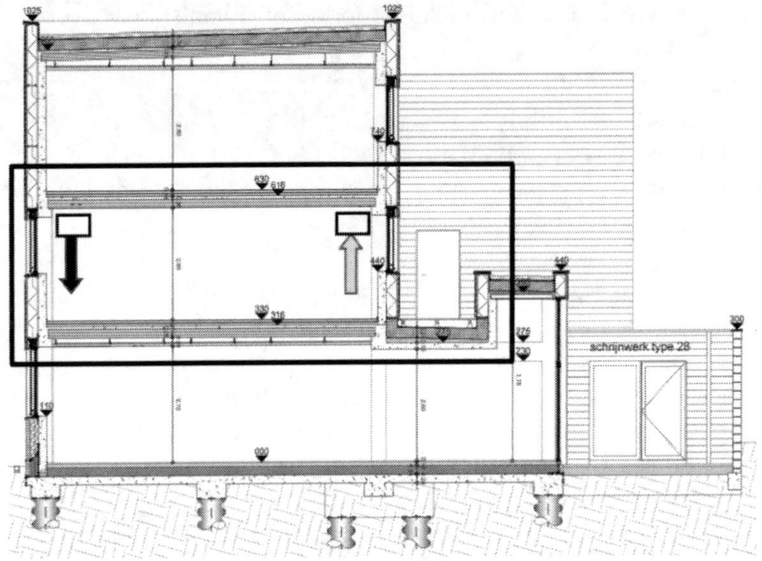

Figure 2: Cross section of new passive school building block, classroom with open corridor indicated with a black rectangular

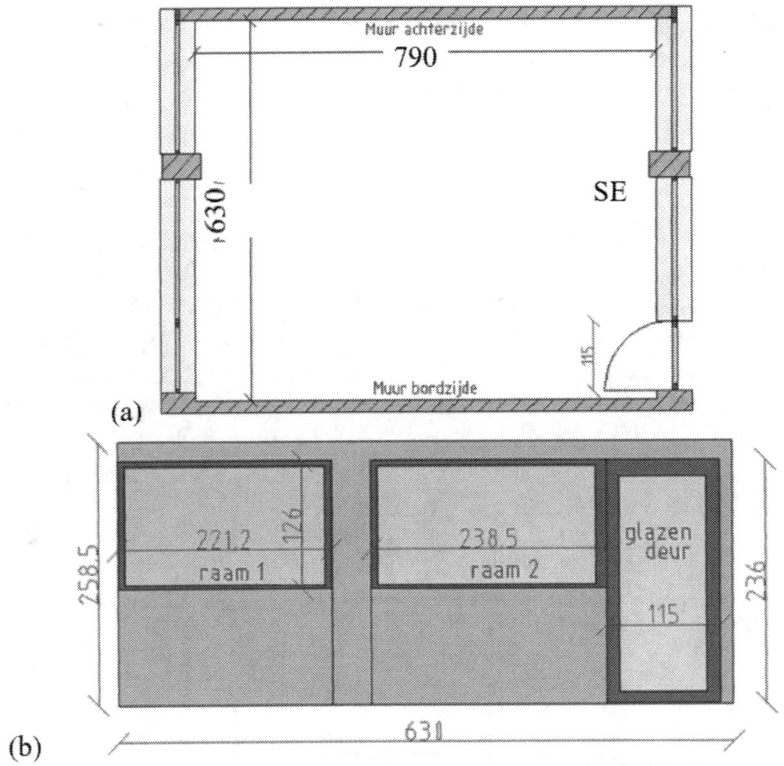

Figure 3: Floor plan (a) and SE wall (b) of the classroom

Table 1: timetable of lessons in the classroom

Day	8h20	9h10		10h15	11h05		13h15	14h05		15h10	16h
Mon	■	■		■	■		■	■			■
Tue		■		■	■					■	
Wed	■	■									
Thu							■	■		■	■
Fr				■	■		■	■		■	

3. METHODS

3.1 Survey

The evaluation of the impact of the use of the front door is based on a survey of 48 pupils and teachers. The survey was conducted during the week of January 6 to 10, 2014. The survey studies the experiences of the users and tries to answer to following questions. How do they evaluate the thermal comfort in the classroom? Do they feel a draught due to the opening the front door? Are they annoyed by this draught due to the opening of the door? Can this experience linked to their position in the classroom, the time of the day or the weather? Which actions do they take?

3.2 Measurements

Measurements determine the use of the front door, i.e. the duration and frequency of opening, during one week. An event data logger, recording the changes in state: open/closed, was put on the front door. In addition, operative temperatures are measured on 6 positions and 2 heights, i.e. on 10 cm and 60 cm as advised by EN ISO 7726, in the classroom. Figure 4 shows the measurement set up. Two aspects are studied: the local effect close to the door (sensor T1 at 10 cm above the floor) and the zone of influence in the classroom. The accuracy of temperature sensors is ± 0.35°C. The measurements were carried out every 10s from February 24 to 28 2014. Furthermore, the airflow through the door is determined by tracer gas (SF_6) measurements according to EN ISO 12569. The concentration decay test method is used. A small test room of 7.54 m³ is constructed around the front door. Tracer gas is injected at a height of 2.4 m, samples are taken at a height of 1.1 m and 1.8 m every 2 min. A fan ensures a uniform distribution of the tracer gas. The measurement is repeated 3 times for each opening duration. Outdoor temperature, wind speed and direction are obtained from the nearby meteorological station of Izegem. Measurements are conducted on March 27 2014.

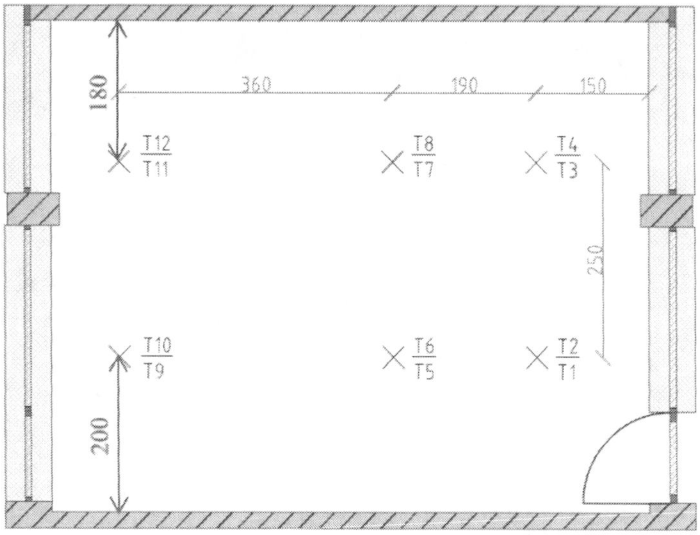

Figure 4: Position of temperature (T) sensors in the classroom

4 RESULTS AND DISCUSSION

4.1 Survey

Figure 5 evaluates the general thermal comfort in the classroom at the start of the day, classified as predicted mean vote (PMV). The majority of the respondents is satisfied with the thermal comfort in this week in January. Thermal comfort is experienced as neutral or slightly cool/warm by 79% of the respondents. Figure 6 shows the experience of and annoyance to draught due to opening of the front door. Everyone experiences cold draught of whom about half of the pupils always or often. 90% of the pupils are annoyed by this draught, 38% always or often. This means that this experience has to be examined more in detail.

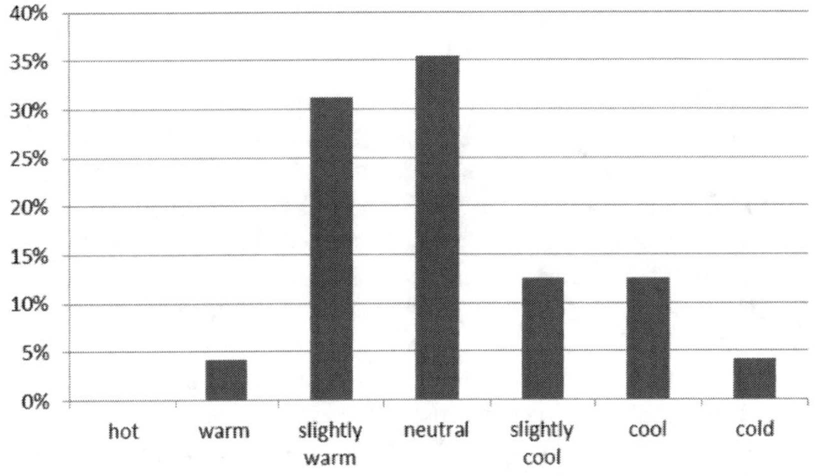

Figure 5: Experience of thermal comfort (PMV-values)

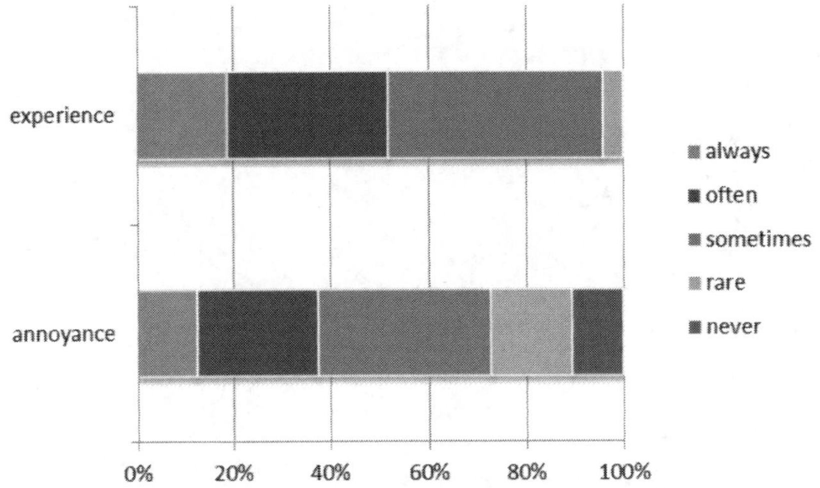

Figure 6: Experience of and annoyance to draught caused by front door opening

Pupils were also questioned about the conditions of this experience. No clear link to the position in the classroom was found. There is no significant difference in experience in pupils sitting next to the door and sitting in the rest of the classroom. In addition, respondents answer that the time of the day has no impact on their experience. By contrast, strong link is found to the following situations: 83% of the pupils experience draught when someone enters the classroom and 60% at the start of the course. Moreover, 69% of the people link their experience to the weather conditions: wind (56%) and cold weather (29%) are the most

reported. No link is reported to rain. This can be explained by the orientation of the front door (southeast) related to the dominant wind direction (south) as shown in Table 30. This table summaries the weather conditions from January 6 to 10, measured in the local weather station of Izegem, at 16 km of the school building. Weather was determined by mild maritime airflows. Temperatures were abnormally high for January. Dominant wind direction was south. Wind speed and amount of rain were normal for January (RMI, 2014).
The pupils also responded the positions on the body where they feel the draught: mostly at the head (67%), hands (54%), legs (52%) and arms (48%). This means that cold was spreaded throughout the height of the classroom. This will be checked by the temperature measurements.

Table 2: Weather conditions Jan 6-10, Feb 24-28 2014 weather station Izegem (Extreme Weather, 2014)

Date	Temperature (daily min) (°C)	Temperature (daily mean) (°C)	Wind speed (daily mean) (km/h)	Wind speed (daily max) (km/h)	Wind direction (daily mean) (°)	Rain (daily total) (mm)
Jan 6	9	12	23	65	188	2.0
Jan 7	9	10	22	61	189	0
Jan 8	7	9	10	39	180	3.3
Jan 9	5	9	13	54	174	2.5
Jan 10	3	5	10	35	190	0
Feb 24	5	9	10	43	168	0
Feb 25	5	9	16	56	182	7.9
Feb 26	3	6	7	30	189	0
Feb 27	2	5	12	69	182	4.8
Feb 28	3	4	7	30	159	7.6

4.2 Measurements

First, the use of the front door, i.e. the frequency and duration of opening, is determined. Frequency of opening is shown on Figure 7 for February 24. The door was intensively used and was opened 110 times during one week. It has to be remarked that classroom was not used on Thursday. Only 39% of the openings were expected based on the timetable in Table 1. This means that 61% of the openings were not linked to this schedule: 45% extra openings during classes, 16% before or after classes.

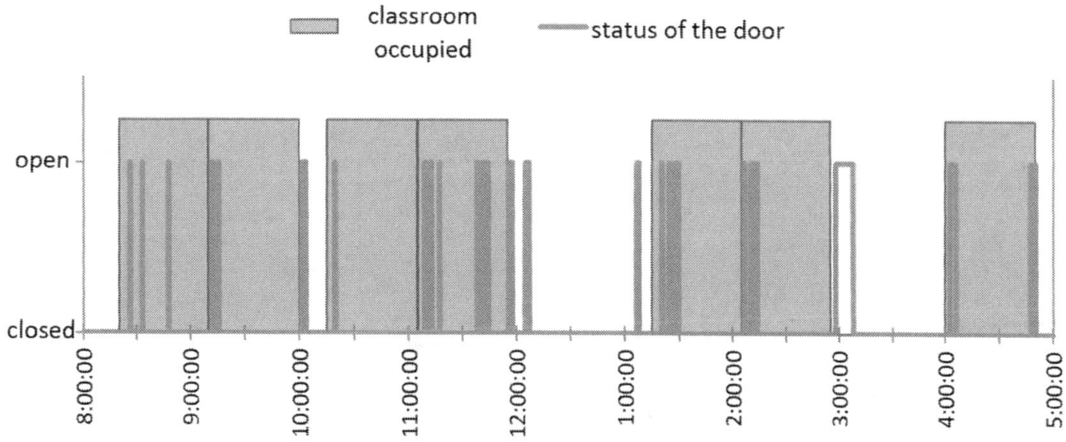

Figure 7: Use of the front door on Monday February 24 2014

In addition, Figure 8 shows the distribution of the opening durations of the front door in classes of 5s. It can be noticed that 50% of the time the door is opened less than 10s and 5% of the time even more than 200s. Distribution shows peaks at 5-10s and 20-25s. Therefore, data is studied more in detail by dividing it into classes of 2s: a peak at a duration of 6-7 s and

20s is concluded. Subsequently, the airflow is determined by tracer gas measurements for these opening durations.

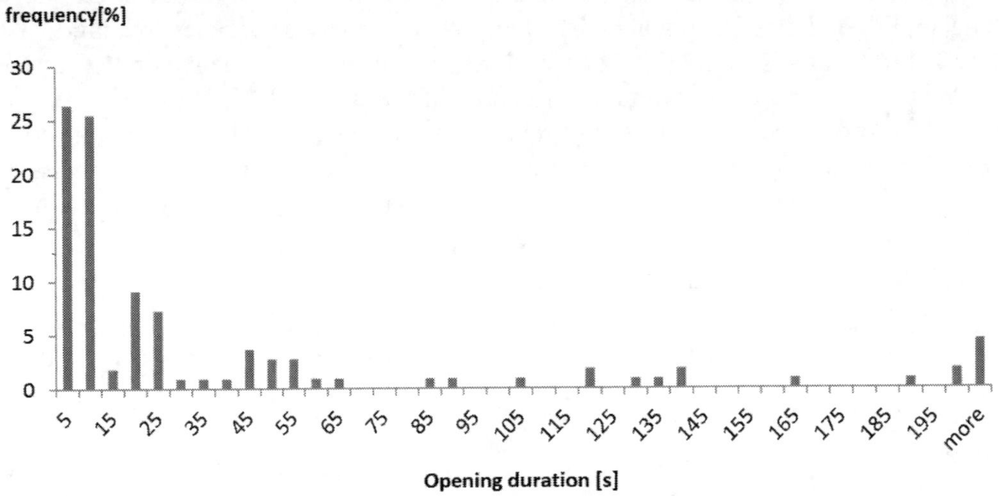

Figure 8: Duration of front door opening

Figure 9 shows the results of concentration decay method (trial 6) for an opening duration of 20s, sampled at a height of 1.1 m. The influence of door opening is clearly noticeable: the concentration decreases significantly after opening the door (indicated with red squares). The air change rate is calculated from the gradient of this curve. Figure 9 also shows that the concentration was decreasing a little bit before the door was opened (curve with blue diamonds). This means that the test room was not sealed completely airtight. This small air change rate is considered in the calculation of the airflow rate. The average airflow rates and exchanged volume for a duration of 6s respectively 20s are determined in Table 3. An average airflow rate of 72.7 and 232.8 m³/h respectively is found. The weather conditions during the tracer gas measurements are summarised in Table 4. The outdoor temperature was extremely high for March. An average indoor-outdoor temperature difference of 10.2 °C and 8.9°C respectively was noticed during measurements of opening duration 6 and 20s. The main wind direction was east, the wind speed was light to moderate.

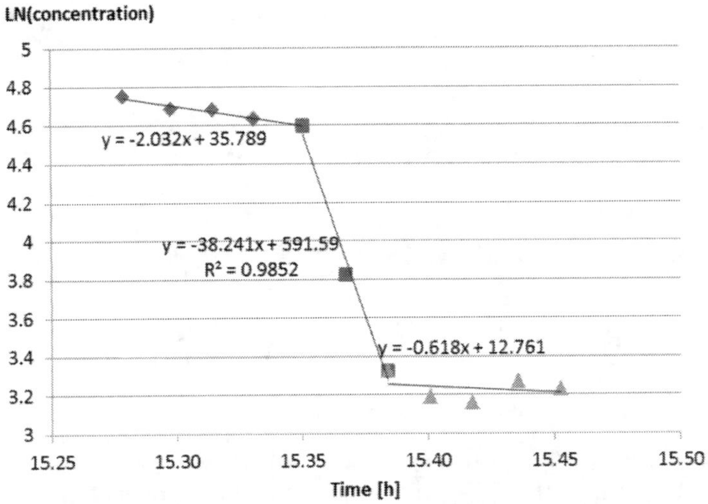

Figure 9: Result of concentration decay method of front door opening of 20s (trial 6)

Table 3: results of tracer gas measurements

Duration (s)	Air volume (m^3)	Air flow (m^3/s)	Air flow (m^3/h)
6	0.14 ± 0.06	0.020 ± 0.007	72.7 ± 26.0
20	1.29 ± 0.31	0.065 ± 0.015	232.8 ± 55.8

Table 4: Weather conditions March 27 2014 Izegem (Extreme Weather, 2014)

trial	Duration (s)	time	Temperature indoor (°C)	Temperature outdoor (°C)	Wind speed (km/h)	Wind direction (°)
1	6	11:47	21.9	10.5	11.1	88
2	7	12:29	22.4	11.5	16.7	132
3	7	14:48	21.7	13.4	7.6	88
4	20	13:33	21.8	12.1	13.0	88
5	20	14:20	21.7	13.1	10.9	88
6	20	15:19	21.5	13.0	7.4	110

Furthermore, the effect of front door opening on the temperatures in the classroom is determined. Figure 10 shows the local effect close to the door, measured in temperature sensor 1 (T1, see Figure 4) on Friday February 28 2014. A significant temperature decrease is noticed when the door is opened. Impact factors on the magnitude of decrease are opening duration, indoor-outdoor temperature difference, wind speed and direction, successive openings. Weather conditions are shown in Table 2. The average indoor-outdoor temperature was 16°C. Dominant wind direction was the same as the orientation of the front door, i.e. southeast. The wind speed was light to moderate.
Zone of influence in the classroom is also determined. Figure 76 shows the temperature decrease on every position (see Figure 4) and the duration to the minimum temperature after closing the front door for an opening duration of 47s on Friday Feb 28 2014, 3.12 pm. The effect on the temperature is not evenly spread in the classroom. The temperature decrease in T3 and T5 (closest to T1) is significantly reduced. Temperature decrease is more pronounced on a height of 10cm than of 60cm. This difference reduces as the distance from the front door increases. Duration to minimum temperature after closing the door is reached within 1 min close to the door. This time increases with increasing height and distance the door.

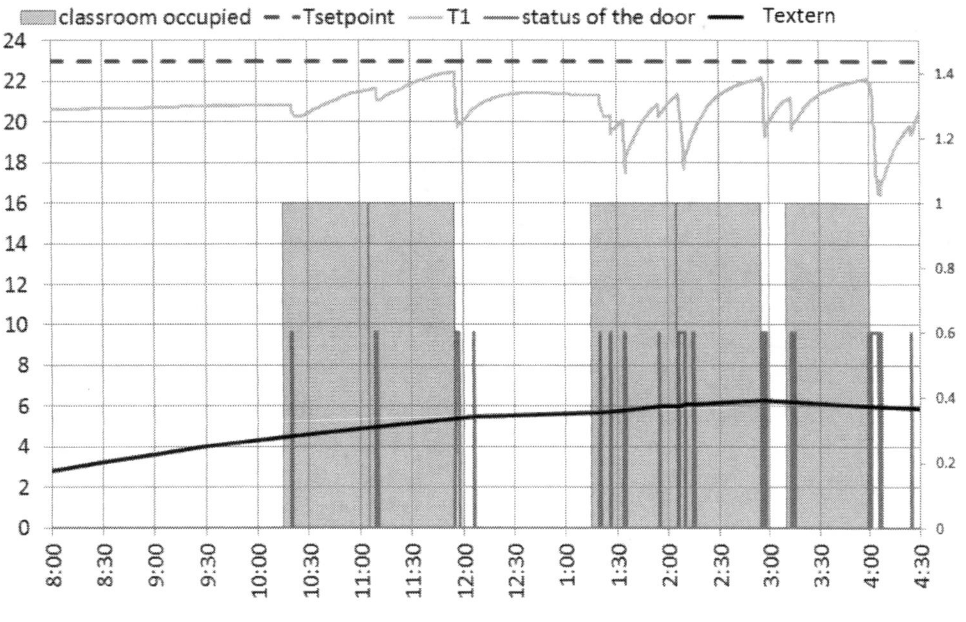

Figure 10: Temperature course measured in sensor T1 on Friday Feb 28 2014

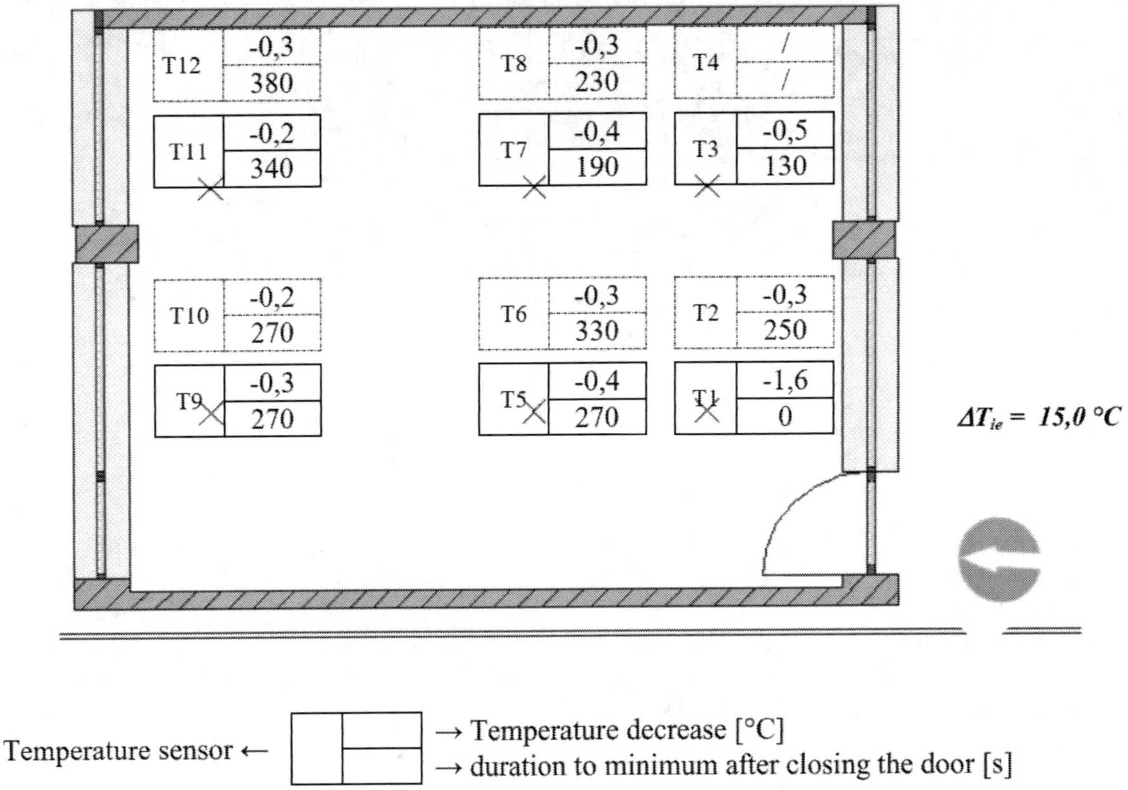

5 CONCLUSIONS

The impact of the use of the front door on thermal comfort and the infiltration losses in a classroom in a passive school in winter season was studied. The survey of pupils and teachers concluded that every respondent experiences cold draught of whom about half always or often. Almost all users are annoyed by this draught. No clear link to the position in the classroom was found. A strong link was found to situation where someone enters the classroom and to the start of the lesson. Moreover, the majority of the respondents link their experience to wind and cold weather.

The status (open/closed) of the front door was monitored during one week. It was shown that less than half the openings were expected based on the timetable of the classroom. Duration of opening was also studied: 50% of the time the door is opened less than 10s and 5% of the time even more than 200s. The most occurring opening durations are 6 and 20s. The airflow was determined by tracer gas measurements for these opening durations. An average airflow rate of 72.7 and 232.8 m^3/h was found for 6s and 20s respectively for an average indoor-outdoor temperature difference of 10.2 °C and 8.9°C and a light to moderate wind speed. The effect of front door opening on the temperatures in the classroom was determined. A significant temperature decrease was found close the door. Impact factors on the magnitude of decrease are opening duration, indoor-outdoor temperature difference, wind speed and direction, successive openings. Zone of influence in the classroom was also determined. The effect on the temperature is not evenly spread in the classroom. The temperature decrease significantly reduces as the distance to the front door increases. Temperature decrease is more pronounced close to the floor.

Survey and monitoring showed that a the use of the front door influences thermal comfort and increases infiltration losses in this classroom in this passive school. It is advised to consider this effect in the design of heating system in and the assessment of energy performance of future Passive House projects including rooms with front doors.

6 REFERENCES

EPB (2013), Calculation Method for the Characteristic Annual Primary Energy Consumption in Tertiary Buildings, Annex VI to the Flemish EPB-legislation.
Extreme Weather (2014), www.extremeweather.be
ISO (2000), Thermal insulation in buildings – Determination of air change in buildings – Tracer gas dilution method, ISO 12569:2000
ISO (1998), Ergonomics of the thermal environment -- Instruments for measuring physical quantities, ISO 7726:1998
Houwen, V., Van Lerberghe, P. (2014), Invloed van een buitedeur op het thermisch comfort in een klaslokaal van een passiefschool, Masterproof KU Leuven Campus Ghent
Kaan, H. (2006), PEP project - Passive Houses Worldwide: International Developments, 10th International Passive house conference, Hanover, Germany (May 19-20, 2006) pp.299-308
Peper, S., Kah, O., Pfluger, R., Schnieders, J. (2007). Passivhausschule Frankfurt Riedberg: Messtechnische Untersuchung und Analyse. Darmstadt: PASSIVHAUS INSTITUT
Peper, S., Kah, O., Pfluger, R., Schnieders, J. (2008) Erkenntnisse über Lüftung und Energieverbrauch sowie Bodenplattendämmung aus Monitoring-Untersuchungen an einem Passivhaus-Schulgebäude, Bauphysik 30 (2008), Heft 1
RMI (2014). Klimatologisch overzicht van Januari 2014 (in Dutch) http://www.kmi.be/meteo/view/nl/13055987-Januari+2014.html

A NOZZLE PULSE PRESSURISATION TECHNIQUE FOR MEASUREMENT OF BUILDING LEAKAGE AT LOW PRESSURE

Edward Cooper*[1], Xiaofeng Zheng[1], Mark Gillot[1], Saffa Riffat[1] and Yingqing Zu[2]

[1] Architecture, Climate and Environment Research Group, Faculty of Engineering, University of Nottingham, University Park, Nottingham NG7 2RD, UK

**Corresponding author: edward.cooper@nottingham.ac.uk*

[2] Department of Mechanics and Engineering Science, Fudan University 220 Handan Road, Shanghai, China

ABSTRACT

Air tightness is essential to building energy performance, which has been acknowledged for a long time. It plays a significant role in improving building energy efficiency by minimising the heating/cooling loss incurred during unwanted air movement through the building envelope, consequently reducing the building's energy demand and cutting down carbon emission in the building sector. A novel nozzle pulse pressurisation technique for determining the adventitious leakage of buildings at low pressure around 4 Pa, which is regarded as a more accurate indicator than conventional steady state measurement at 50 Pa, is investigated theoretically, numerically and experimentally. The investigation is based on the 'quasi-steady pulse' concept which produces a pressure pulse inside the building by introducing a certain amount of air in a very short time using an air compressor, solenoid valve, nozzle and control unit. The mass flow rate from the nozzle is obtained by measuring the transient pressure in the air receiver of the compressor during a test run. Simultaneously, the pressure difference across the building envelope is measured by differential pressure transducers. The quadratic equation, which can more closely represent the flow characteristics of adventitious openings, is used to determine the characteristic of building air leakage. Due to short time operation, the technique minimizes the effects of wind and buoyancy force and has proven to be highly repeatable. The pulse pressurisation using nozzle technique is compared with that using the piston technique. The comparison indicates that the present technique is reliable for determining building leakage at low pressure. It also gives great convenience in practical applications due to being more compact and portable. Moreover, it needs only a few seconds for a test run, barely needs to penetrate the building envelope and therefore can establish the leakage of a building very quickly and efficiently.

KEYWORDS

Air tightness, quasi steady, nozzle pulse pressurisation, low pressure;

1. INTRODUCTION

In European Directive 2010/31/EU (Directive, 2010) [1] on the energy performance of buildings, it is stated that the energy efficiency of buildings has to be calculated in the member states. Air leakage of a building envelope has a significant effect on the building's energy efficiency and making a building substantially air tight can make a considerable reduction in energy consumption and hence CO_2 production. In the UK, the adventitious leakage of buildings has received particular attention and a standard has been set in the form of a maximum value for the air permeability at a pressure of 50 Pa (Q_{50}) in Building Regulation (Building Regulation part L, 2010) [2]. The adoption of 50 Pa is a compromise, because leakage measurements at lower pressures are perceived to be subject to large errors arising from pressures generated by wind and buoyancy during the test. However, natural ventilation pressures are typically at an order of magnitude <50 Pa and therefore Q_{50} is not an ideal indicator of the infiltration potential of an envelope. In fact, a pressure difference of 4 Pa is commonly taken to be typical of natural ventilation and, ideally, the leakage at 4 Pa (Q_4) would be determined.

In order to test the leakage at 4 Pa directly, some investigations have been done on the basis of a pulse pressurisation technique by Roulet (Roulet, 1991) [3]. In early tests, a simple gravity-driven piston device was used to generate a pulse by Carey (Carey, 2001) [4]. Following this, a more practical version was devised by Cooper and Etheridge (Cooper, 2004) [5] in which the piston was driven by a supply of compressed air. Then, a quasi-steady piston pulse technique was developed by Cooper and Etheridge (Cooper, 2007) [6]. It has been demonstrated that direct measurement of Q_4 could give a much more accurate measure of the infiltration potential of an envelope than the current high-pressure technique in CIBSE (CIBSE 2000) [7]. The former could reduce the uncertainty by a factor of three or more, described by Cooper (Cooper, 2007) [8].

In this paper, a nozzle pulse pressurisation technique is explored, using a compressor nozzle to generate the pulse, so as to make the test rig more compact and portable. Unlike the piston pulse technique, which obtains the mass flow rate on the basis of the velocity of the piston, the present technique can obtain the mass flow rate from the compressor nozzle more directly and accurately.

2. PREVIOUS RESEARCH

Previous research carried out in measuring air tightness of buildings or ventilation systems can be classified into two categories, steady technique and unsteady technique, according to the way in which they approach the measurement. Steady technique does it by establishing a steady state pressure difference across the envelope and recording the induced leakage rate of airflow through the envelope.

"Blower door" is the most commonly used steady technique for measuring the building air tightness in construction industry. It typically uses a door fan to take the air in or out of the building to create a range of pressure difference (usually between 10Pa-60Pa) across the building envelope and the corresponding airflow rate of the fan is recorded. This technique was firstly used in Sweden around 1977 (Sherman 2004) [9]. The uncertainties existed in fan pressurization has been analyzed by Sherman [10], who introduced the uncertainties in measurements of airflow and pressure, and pointed out model specification errors may also contribute to the overall uncertainty in the estimation of 4Pa leakage. Cooper (Cooper, 2007) analysed and compared the uncertainties in Q_{50} measurement and Q_4 measurement and came to the conclusion that direct measurement of Q_4 can reduce the uncertainty by a factor of four.

According to the recent UK standard, the pressurisation at 50Pa has been widely recognised as the method for measuring the air permeability. It is selected at such a level because it is much higher than the level of pressure change caused by the wind and buoyancy effect so as to be able to neglect the errors caused by wind and buoyancy effects in the test. However, this method is a compromise due to the following deficiencies:

- The pressures it exerts on the building envelope are significantly higher than those experienced under natural conditions, which is typically around 4Pa. Therefore, it requires extrapolation from the measurement at high pressure level to calculate the leakage under natural condition. This adds uncertainty into the accuracy. The hydraulic characteristic of the opening is also changed by testing the pressure level which is much higher than the level given under natural conditions.

- It cannot make real-time leakage area measurements as the pressure and temperature vary with the weather condition, making the reading unstable and determination of the value unsure.
- The large volumes of air displaced by the fan can cause inconveniences such as large indoor temperature changes, which deviates from the thermal conditions in reality.
- The installation of blower door requires the removal or reposition of the window or door in the external opening where the blower door devices are installed and this changes the air leakage characteristic of the opening where the blower door is installed.
- One set of test can only obtain a single point result which is used to predict the leakage under natural condition using empirical value of C and n in the power law equation. This could cause non-negligible error to the prediction due to inaccuracy at low pressure. Multiple points test would give better accuracy than the single point test as it acquires the characteristic curve of the building air tightness (usually at an increment or decrement of 10 Pascal in the range of 10-60Pa). But it takes longer time to conduct and still needs to use the power law equation to extrapolate the leakage to that under natural condition.
- Such high level of pressure difference presents a risk of damaging the building structure or the fabric of the dwelling.
- The non-uniform pressure distribution cannot be avoided when large fan is used to pressurise buildings with large volume.

These deficiencies stand out more obviously when the measured buildings are in large scale because the required fan flow rate increases more or less in direct proportion to the volume of the building. The problem can be reduced by relaxing the requirement for 50 Pa down to 25 Pa or even lower, but another issue arises. The lower the pressure becomes, the less accurate the measurement is, due to being close to the pressure generated by buoyancy and wind. In order to control the error caused by this procedure within an acceptable range, the standard E779-03 (ASTM 2003) [11]recommends to only conduct the test when the product of the absolute value of indoor/outdoor air temperature difference multiplied by the building height, gives a result less than 200m°C. Hence, a preferable outside temperature, which is from 5°C to 35°C, is recommended in the standard to avoid any significant error caused by fan pressurisation. This also applies to the wind speed, which is under 2m/s preferably.

The unsteady technique, known as dynamic air tightness measurement technique, analyses the pressure-flow correlation when the building envelope is exposed to varying pressure. With unsteady techniques, the required information is determined indirectly by measuring the pressure response to a known disturbance (Carey 2001). It is able to accurately generate a known volume change to the building enclosure which makes the sources of error introduced by this technique less than the steady technique.

The key part of unsteady techniques is to pressurise the building enclosure or cavity to the desired pressure level by supplying air or extracting air in some occasion using pre-compressed air or outdoor air. Then the pressure is varied by devices like piston or left to decay naturally. During the pressure decay over a certain period of time, the relation between the air leakage rate and pressure difference across the building envelope is recorded. According to the pressurisation style, the unsteady technique includes three types, which are AC pressurisation, gradual pressurisation and pulse pressurisation, respectively. This paper introduces the pulse pressurisation method.

3 METHOD

3.1 Equipment and tests

The low pressure nozzle pulse technique generates an instant pressure pulse in the building enclosure by releasing compressed air into it via nozzle and the pressure pulse is left to decay naturally. Pressure decay is monitored and used to determine the building air tightness. This technique is worth attention due to the fact that relatively simple devices can be used to generate a pressure increase by adding a known volume of air to the enclosure. It only requires a volume change at the order of 0.004% to generate a pulse pressure in the order of 4 Pa. This technique can be implemented by releasing compressed air in a building enclosure to obtain the required pressure increase approximately.

A diagrammatic representation of the nozzle pulse generation unit is shown in Figure 1. It consists of a compressor with a 50 Litre tank and a maximum working pressure of 10 bar, a solenoid valve and a nozzle, shown in Figure 2.

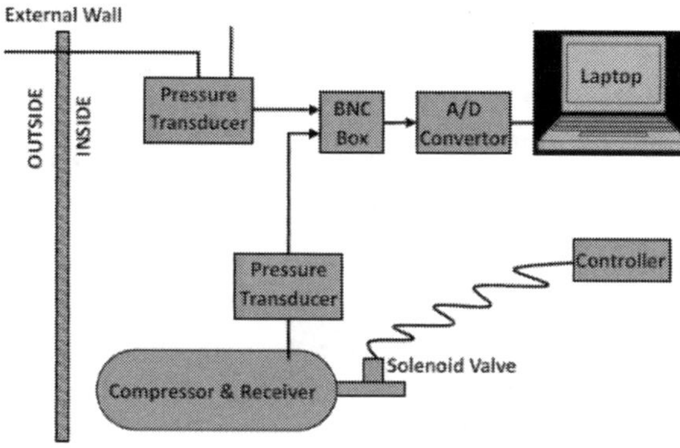

Figure 1 Schematic diagram of a single nozzle unit

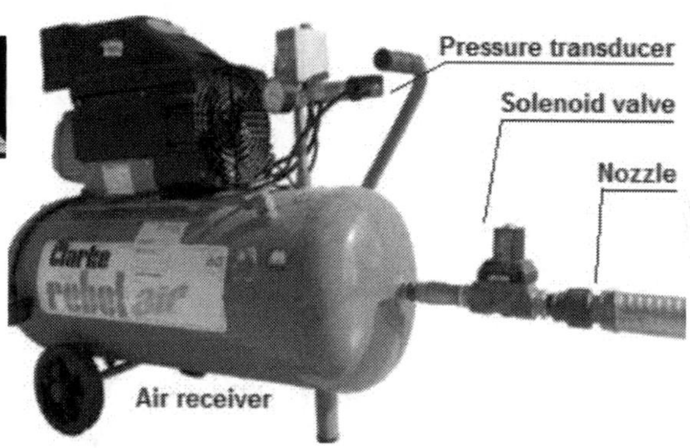

Figure 2 Set-up for one nozzle unit

The pulse generation is achieved by opening the solenoid valve for a short period. The pressure difference across building envelope and the pressure inside compressor tank are measured by differential pressure transducer and pressure transducer, respectively. At a sampling rate of 200Hz, the data is recorded by a laptop using a BNC box and A/D converter.

In the piston pulse technique, the pulse is generated by a certain volume of air which is rapidly released from a compressor tank via a solenoid valve. The solenoid valve is operated by an electronic controller which enables the air to be released in seconds. The released air is injected to the cylinder which is connected to the outlet of solenoid valve through a pipe, as shown in Figure 4. On receiving the released air from the compressor tank, the piston is moved in the cylinder due to the instant pressure increase. The piston is displaced by injecting air from the tank. A cable extension transducer (CET) is used to measure the instantaneous position of the piston. The displacement of the piston is recorded and used to calculate the volume of released air over the operating time.

The schematic diagram of piston pulse unit is presented in Figure 3. The electronic controller operates the solenoid valve by allowing it to be open for 1.5 seconds. The released air is injected to the cylinder and pushes the piston to the other side with a certain displacement. In this process, the instantaneous position of the piston, the internal pressure and external pressure are recorded by CET and differential pressure transducer respectively, and sent to the BNC terminal box via A/D converter card at a sampling rate of 200Hz.

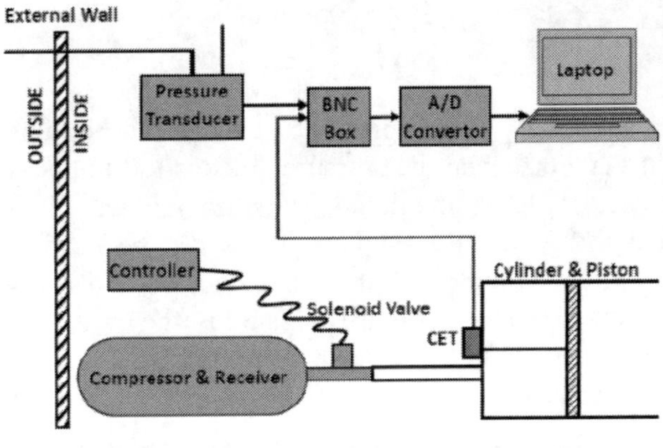

Figure 3 Schematic diagram of pulse technique using piston unit

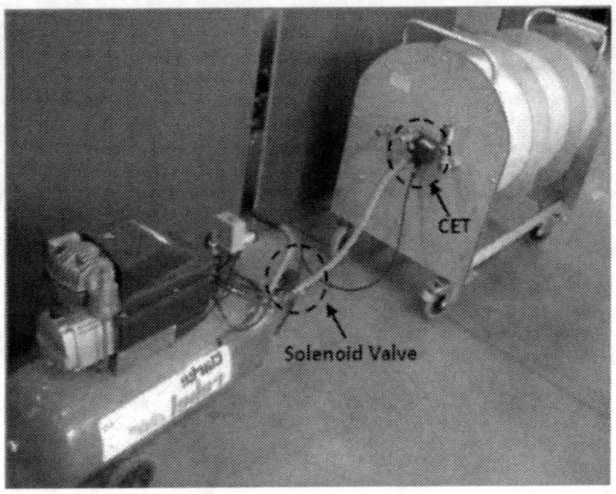

Figure 4 Installation of Pulse unit using Piston

Like the nozzle pulse unit, this technique is also designed to measure the building air leakage at low pressure. It only needs a few seconds to run the test and does not need to penetrate the building envelope. However, the airflow rate is measured indirectly through piston displacement which is driven by the injecting air from compressor tank. The recorded data of the airflow rate might be smaller than the actual value due to the unavoidable leakage at the small gap between piston and cylinder wall.

3.2 Theory

It has been shown previously that the pulse pressurisation technique creates a period of quasi-steady flow (Cooper, 2007) [6], which can be used to determine the leakage of a building. To determine it the volume flow rate of released air from compressor tank and pressure difference across the envelope must be obtained. The latter can be measured directly, but the former requires the use of a theoretical model for the nozzle technique. The air leakage rate is determined by using the gas law which correlates the time derivative of the pressure difference to a change in mass per unit time. CFD simulation (FLUENT 6.3 based on the finite volume method) has been used to numerically validate the model, specifically the following items:

1. The air pressure in the compressor tank is uniform.
2. The mass flow rate of released air via nozzle is calculated.
3. The air density in the building envelope is uniform and constant.

Due to the patent related issues, the detailed mathematical model is required to be confidential at the present stage and can't be revealed in this paper.

3.3 Results

Two separate sets of tests have been conducted to compare the nozzle pulse with the piston pulse as well as the conventional steady state technique-blower door. The building used for the comparison between the nozzle pulse technique and the piston pulse technique has a regular cube shape with a volume of 136.1 m^3 and an envelope area of 185.8 m^2. The building used for comparing the nozzle pulse technique with the conventional blower door technique has a volume of 273.1 m^3 and an envelope area of 264.3 m^2.

Figure 5 shows the pressure pulses of five repeated tests. The curves are adjusted to take account of any variation of Δp due to wind during the pulse period, for the reason given by Cooper (Cooper, 2007) [8]. This is performed by fitting a curve to the data before and after the pulse, and then subtracting the curve from the raw data. In Figure 81 the wind effect is apparent as a variation of $\Delta p(t)$ before and after the imposed pressure pulses, and the quasi-steady period occurs between 0.4 and 1.4 s. The calculated transient mass flow rate for one of the tests is shown in Figure 6.

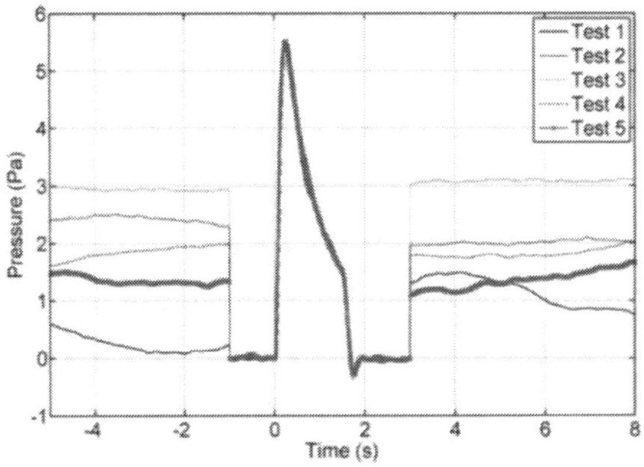

Figure 5 Five repeated pulse tests

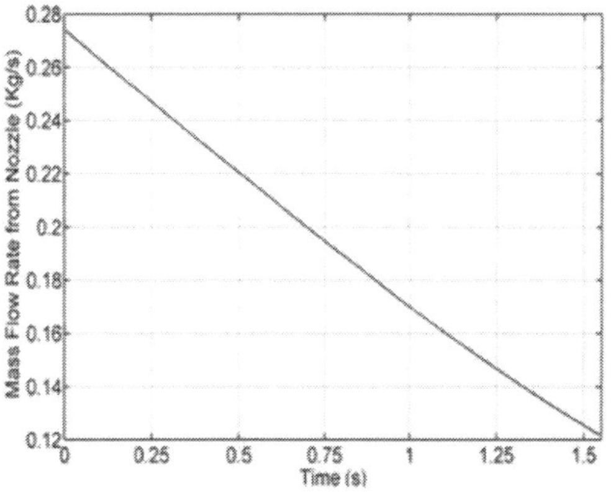

Figure 6 Transient mass flow rate of the air released from nozzle

The good repeatability of the technique can be seen in the plotted $\Delta p(t) - q(t)$ correlation curve, as shown in Figure 7, where the average $Q_4 = 0.17598$ m^3/s. Further tests were done to assess the sensitivity of the technique by sealing and unsealing the openings around the test room door. The technique measured an average difference of 0.01626 m^3/s, suggesting the technique is sufficiently sensitive to small changes in leakage.

The previously mentioned piston technique [6]was tested in the same test room under the same conditions and a comparison is shown in Figure 8. Good agreement appears and indicates that the nozzle technique is reliable for determining building leakage at low pressures. However, from Figure 8, it can be seen that the piston tests always give slightly lower values of leakage under the same pressure differences. During the piston test, there is an unavoidable leak of air from the narrow gap between the piston and cylinder wall. Therefore, the piston test may underestimate the $Q_p(t)$ slightly because it obtains $Q_p(t)$ indirectly from the velocity of the piston.

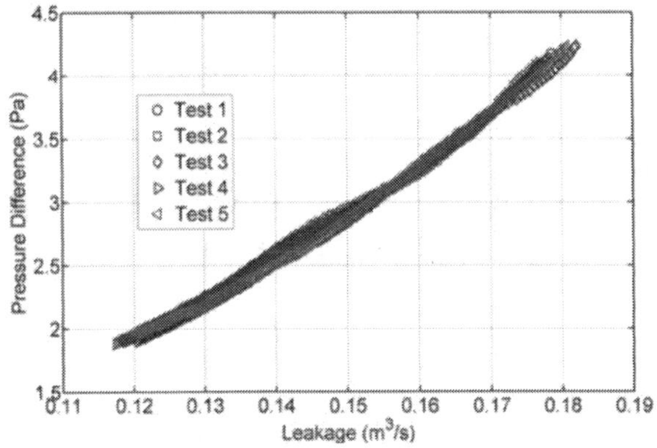

Figure 7 Nozzle test results

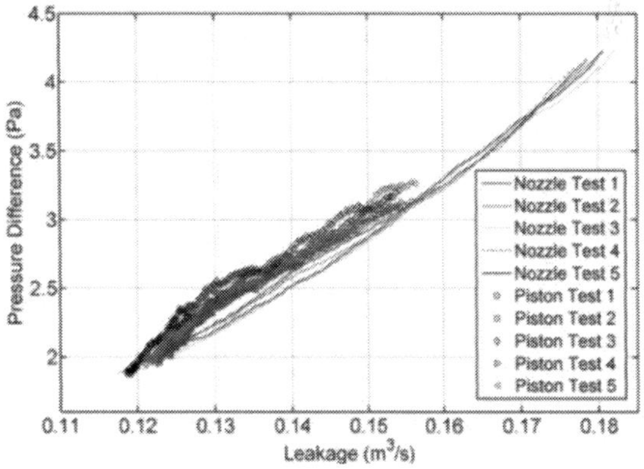

Figure 8 Comparison of nozzle and piston results

Blower door technique, which has been a globally adopted method for measuring airtightness (ISO 9972: 2006, EN 13829:2000), has been used to test the same property along with the nozzle pulse technique to see the correlation between the air leakage results given by these two techniques. There isn't a straightforward comparison between them because the air leakage rate is obtained under different pressure difference level. It is inappropriate to simply conclude the accuracy of one technique over the other. However, by plotting them together in a graph enables us to gain insight into the relation between them. As shown in Figure 9, the power law equation is used to fit the permeability-pressure obtained by both techniques. An accurate fit has been obtained because R^2 equals 0.9956, very close to 1. It gives a good indication that the low pressure pulse technique gives a good agreement with the blower door test results at high pressure (typically 50 Pa). In order to obtain the permeability at 4 Pa which is a more accurate indicator of building airtightness in reality, the airtightness characteristic obtained at high pressure by the conventional steady state technique need to be extrapolated down to 4 Pa. Extrapolating the air leakage rate at high pressure (typically 50 Pa) down to infiltration pressures (typically 4 Pa) has been shown by various authors, as summarized in the ASHRAE Fundamentals Handbook (2014) [12], to incur significant uncertainty. Murphy et al. (Murphy 1991) [13]and Cooper and Etheridge (Cooper 2007a) [8] have respectively shown experimentally and theoretically that the errors can be as high as +/- 40%.

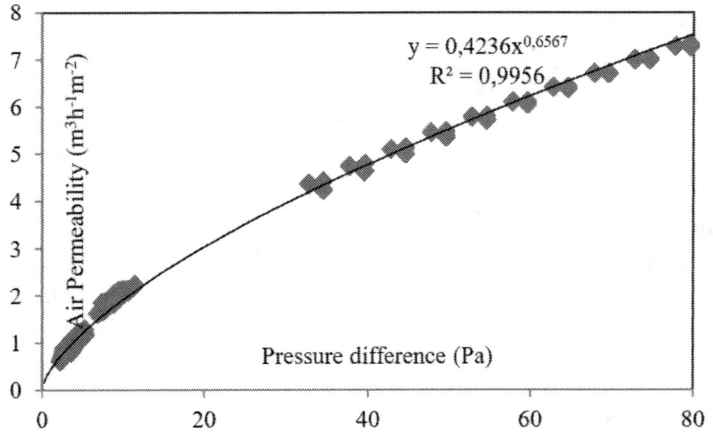

Figure 9 Power law curve fitting to the results obtained by both techniques

Moreover, at low pressure the power law equation shows a significant difference from the quadratic equation proposed by Etheridge (Etheridge 1998) [14], who believes the quadratic equation is more accurate and easier to use than the power law. The quadratic equation also gives a good representation of the flow characteristic of openings over a wide range of pressures. Etheridge pointed out that the power law can give a good fit but only over a limited range of pressures (Walker 1996) [15]. Compared to the power law, the quadratic equation should be used in preference for modelling the behaviour of adventitious openings due to the following facts:

- The quadratic equation was derived for developing flows and it does not rely on any significant length of fully developed flow within the opening. Meanwhile, it is unlikely that turbulence within the opening is a significant factor in adventitious openings (Etheridge 1996) [16].
- It is a simple matter to obtain the coefficients of the quadratic equation from existing data and the equation is easier to use compared to the power law.
- Compared to the power law, the quadratic equation relates to the unsteady behaviour of leakage openings and allows a consistent approach to be used for investigating the effects of changes to the geometry of an opening on both inertia and steady flow effects due to its reflection of the relationship between the equation and the geometry of the opening. Therefore, it can more closely represent the flow characteristics of adventitious opening.

4. CONCLUSIONS

A novel nozzle pulse pressurisation technique for determining the leakage of buildings at low pressures has been developed. The technique generates the pulse using a compressor and nozzle, so as to make the test rig compact and portable. A theoretical model for determining the pulse volume flow rate has been briefly introduced. The test results show that the present nozzle pulse technique can minimize the effects of wind and buoyancy force and has proven to be very repeatable and sensitive to small changes in leakage. In addition, a comparison between the nozzle test and piston test has been performed and indicates that the nozzle technique is more reliable, as some uncertainties in determining the volume flow rate can be avoided. In a power law fitting, a good agreement with the steady state test has been shown by the experimental results obtained in a test using the nozzle pulse technique. The mathematical representations for these two different techniques, that approach the measurement of building air tightness in different ways, are worth further experimental verifications and discussions.

5. ACKNOWLEDGEMENTS

This work is supported by the UK EPSRC (Engineering Physical Science Research Council) under grant EP/H023240/1 and led to further support from TSB to further research work towards commercialisation. The authors thank Dr. David Etheridge in the Department of Architecture and Built Environment for helpful discussion.

6. REFERENCES

[1] Directive 2010/31/EU of the European Parliament and of the Council of 19 May 2010 on the energy performance of buildings.

[2] Building Regulations: Part L. (2010) (Conservation of fuel and power). Approved Documents Part L1A (new dwellings), L2A (new buildings other than dwellings). London, UK: NBS for the Office of the Deputy Prime Minister.

[3] Roulet C, Vandaele L. (1991). Airflow patterns within buildings - measurement techniques. AIC-TN-34-1991. Coventry, UK: AIVC.

[4] Carey PS, Etheridge DW. (2001). Leakage measurements using unsteady techniques with particular reference to large buildings. Building Serv. Eng. Res. Technol; 22: 69-82.

[5] Cooper EW, Etheridge DW. (2004). Measurement of building leakage by unsteady pressurisation. AIVC Conference, Prague.

[6] Cooper EW, Etheridge DW. (2007). Determining the adventitious leakage of buildings at low pressure. Part 2: pulse technique. Building Serv. Eng. Res. Technol; 28: 81-96.

[7] CIBSE. (2000). Testing buildings for air leakage, CIBSE TM23. London, UK.

[8] Cooper EW, Etheridge DW. (2007). Determining the adventitious leakage of buildings at low pressure. Part 1: uncertainties. Building Serv. Eng. Res. Technol; 28: 71-80.

[9] M.H. Sherman, R.Chan. (2004). Building Airtightness: Research and Practice, Lawrence Berkeley National Laboratory Report No. LBNL-53356.

[10] M. Sherman. (1995). Uncertainties in fan pressurization measurements. Airflow Performance Conference 10/93. LBL-32115.

[11] ASTM. (2010). International. Standard Test Method for Determining Air Leakage Rate by Fan Pressurisation. ASTM International Designation: E779-03.

[12] ASHRAE, 2013. ASHRAE Fundamentals Handbook. ISBN 978-1-936504-46-6. pp. 420.

[13] Murphy, W.E., D.G. Colliver, and L.R. Piercy. 1991. Repeatability and reproducibility of fan pressurization devices in measuring building air leakage. ASHRAE Transactions 97(2):885-895.

[14] Etheridge, D.W., A note on crack flow equations for ventilation modelling, Building and Environment, Vol. 33, No.5, pp.325-328, 1998.

[15] Walker, I.S., Wilson, D.J. and Sherman, M.h., Does the power law rule for low pressure building envelope leakage? 1966. Proceedings of 17th AIVC Conference, Gothenburg, Sweden. 17-20 September 1996.
[16] Etheridge, D.W., Sandberg, M., Building Ventilation-Theory and Measurement, John Wiley & Sons, Chichester 1996. Pp. 78-89, 99-108 and 144-6.

EXERGY EVALUATION OF MECHANICAL VENTILATION SYSTEMS

Anna DUTKA*[1], Tomasz M. MRÓZ[2]

1 Poznań University of Technology,
ul. Piotrowo 3a,
60-965 Poznań, Poland
*Corresponding author:
anna.a.dutka@doctorate.put.poznan.pl

2 Poznań University of Technology,
ul. Piotrowo 3a,
60-965 Poznań, Poland

ABSTRACT

Energy performance of mechanical ventilation systems in modern low energy and passsive buildings is a crucial factor influencing overall energy performance of building. Energy balance is commonly used tool in evaluation of mechanical ventilation systems. In the case of low energy and passive buildings that tool might be insufficient and should be replaced by exergy analysis taking into account the first and the second Law of Thermodynamics. The paper presents principles of exergy evaluation of mechanical ventilation systems and case study calculations for an office building.

KEYWORDS

Exergy analysis, mechanical ventilation systems, demand controlled ventilation (DCV)

1. INTRODUCTION

Reduction of energy consumption is one of the biggest challenges in the modern world. Buildings consume about 40% of final energy in European Union. Heating and cooling is more than 50% of the annual energy demand of buildings in the operational phase, thus building sector requires significant energy efficiency improvement.

One of the most promising strategies in the improvement of energy performance of buildings is optimal control algorithm or control strategy of ventilation systems. Several control strategies are proposed for ventilation systems. The CO_2-based demand controlled ventilation (DCV) is one of the strategies that allow for the energy use reduction.

The performance evaluation of ventilation systems is usually based on an energy analysis (first law of thermodynamics). The exergy analysis, which takes into account both first and second law of thermodynamics, is much better evaluation tool, which can give better and more accurate indication of system inefficiencies. The results from exergy analysis can be used to assess and optimize the performance of HVAC including ventilation systems [1]. Exergy analysis is not so commonly used, the examples of exergy balance calculations for different buildings and technical systems of buildings can be found in [2,3,4,7,10,14].

The paper present the application of exergy analysis of DCV system installed in office buildings.

2 THE CONCEPT OF EXERGY ANALYSIS

All real thermodynamic processes taking place in nature are irreversible, which means that they are a source of entropy. It is described by the Gouy-Stodoli Law, whose differential form is given by equation 1 [8]:

$$\delta\dot{B} = T_0 \cdot \dot{S}_{gen} \tag{1}$$

where:

$\delta\dot{B}$ □ flux of the internal exergy loss of a thermodynamic system in an infinitely short time d□, W,

$\dot{S}_{gen}$ □ flux of total increase of entropy of a thermodynamic system in an infinitely short time d□, W K^{-1}.

The flux of increase of entropy is very often referred to as the source of entropy. There are six causes of irreversibility of real thermodynamic processes - six potential entropy sources [7, 8]:

- heat exchange caused by a finite temperature difference,
- electric current flow caused by a finite potential difference,
- chemical reactions,
- mixing of fluids,
- mechanical friction
- hydraulic friction.

The exergy balance of the thermodynamically open system can be presented in a mathematical form – equations 2, or a graphical form (Grassmann-Szargut graph) – figure 1. [8]:

$$\dot{B}_{in} = \delta\dot{B} + \dot{B}^{use}_{out} + \delta\dot{B}_{ext} \pm \Delta\dot{B}_{HS} \tag{2}$$

where:

$\dot{B}_{in}$ □ flux of exergy entering the control volume of a thermodynamically open system, W,

$\dot{B}^{use}_{out}$ □ flux of useable exergy leaving the control volume of a thermodynamically open system, W.

$\delta\dot{B}$ □ flux of the internal exergy losses of a thermodynamically open system, W,

$\delta\dot{B}_{ext}$ flux of external exergy losses of a thermodynamically open system, W,

$\Delta\dot{B}_{HS}$ – flux of change of exergy of an external heat source being in contact with a thermodynamically open system, W.

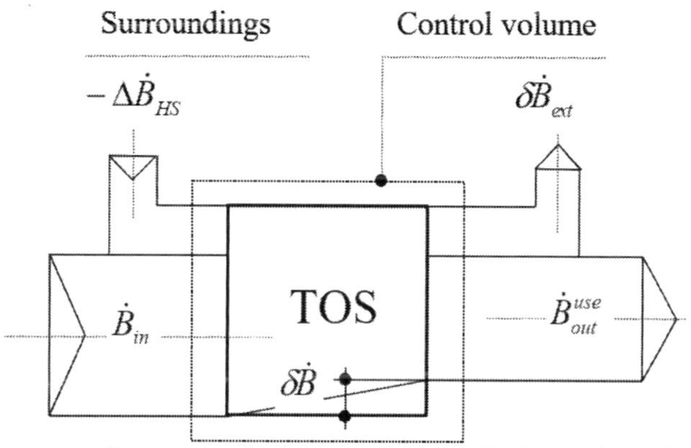

Figure 1. Grassmann-Szargut graph of exergy balance of a thermodynamically open system (TOS) [8]

Mechanical ventilation systems are dissipative systems what means, that driving exergy introduced to that systems is fully utilized for the coverage of total exergy loss.

From the exergy point of view the choice of mechanical system elements and their operating parameters has to be done in the way that allows for the minimization of driving exergy required for its exploitation [7]:

$$\dot{B}_{drv} \rightarrow \min \tag{3}$$

The following section describes case study calculations of exergy analysis of DCV system installed in the office building located in the city of Poznań, Poland.

3 CASE STUDY ANALYSIS

3.1. Description of office building

The evaluated office building is located in the city of Poznań, Poland. It consists of three stories, each of it is divided into six different zones. The basic data of evaluated building are presented in Table 1.

Table 1: The basic data of building

Parameter	Unit	
I storey	-	-
Area	[m^2]	393,3
Volume	[m^3]	1410
II storey	-	-
Area	[m^2]	441,8
Volume	[m^3]	1628
III storey	-	-
Area	[m^2]	441,4
Volume	[m^3]	1626,5
Indoor design temperature	-	-
Winter	[K]	293
Summer	[K]	295
U-value	-	-
Exterior wall	[W/m^2.K]	0,111
Roof	[W/m^2.K]]	0,095
Floor	[W/m^2.K]	0,150

3.2. Description of mechanical ventilation system

The schematic drawing of the mechanical ventilation system is shown in Figure 2. The system consists of three main items: air handling unit, air distribution system and rooms (control zones). The exergy analysis for the main part of air handling unit: rotating heat/mass regenerator, air heater, air cooler, steam humidifier, fans, air filters and air dampers has been performed. Rotating heat/mass regenerator works as long as outside air temperature is lower than ($T_{in,4}$-1K). This study calculates also air heater thermal capacity, cooling capacity, electrical power of inlet and exhaust fans in one example week in winter and summer. Typical meteorological data were taken for the city of Poznań (winter period between 1st and 7th February and summer period between 1st and 7th July) [13]. All calculations have been conducted using Matlab R2008 simulation tool.

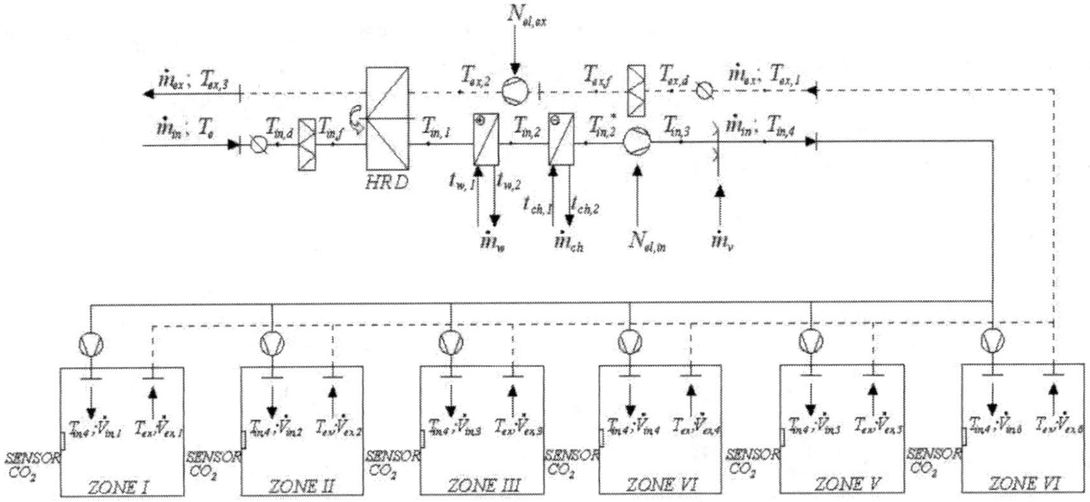

Figure 2: The schematic of mechanical ventilation system.

As the calculation experiment the simulation of exergy performance of one builing's storey divided into six separate CO_2 control zones has been performed. One zone is a conference room – the maximum occupancy is 14 peoples, the rest zones are office rooms – the maximum occupancy density is form 5 to 10 people/zone. The office works from Monday to Friday between 8 a.m. – 8 p.m. Occupancy profiles of all six zones are presented in Figure 3. The ventilation system operates continuously from 8 a.m. – 8 p.m during work day. In all the zones CO_2 sensor were installed, which measure concentration of carbon dioxide. The ASHARE 62 recommended a limit of 1000 ppm CO_2 to satisfy comfort criteria for indoor space. The outside air CO_2 concentration in the location of analyzed building has been 400 ppm.

The air flow per person was determined using equation (4) [9]:

$$\dot{V} = \frac{N}{C_S - C_O} \tag{4}$$

where:

N – CO_2 generation per person, based on activity level; N=0,02 m^3/h,

C_S – CO_2 concentration inside zone, ppm,

C_o – CO_2 concentration outside air, ppm.

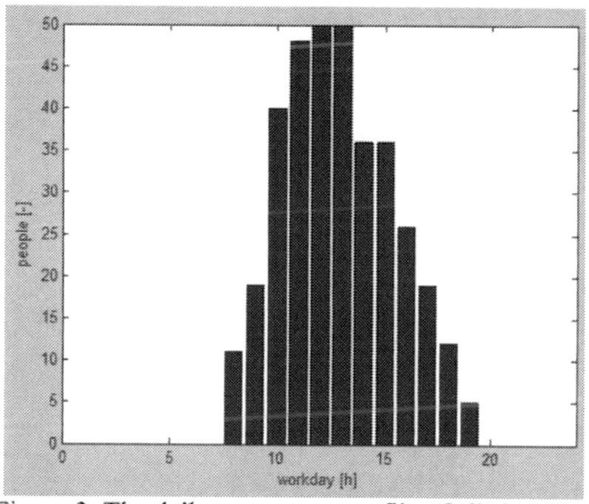

Figure 3: The daily occupancy profile of all six zones

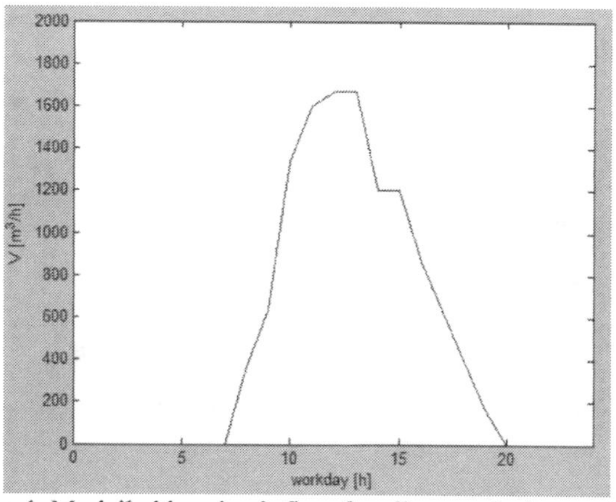

Figure 4: Modelled hourly air flow for all zones during work day

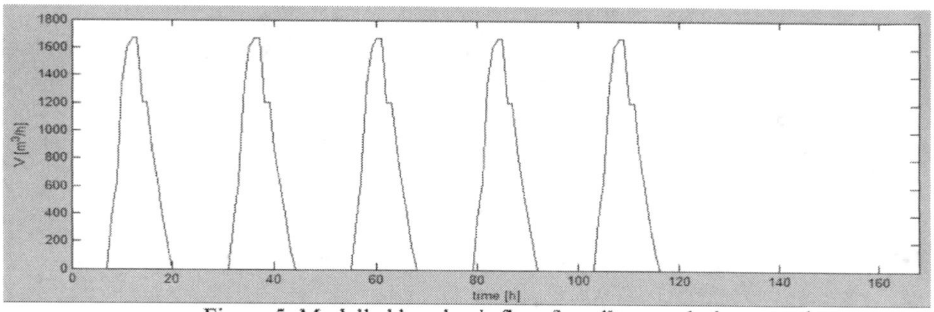

Figure 5: Modelled hourly air flow for all zones during a week

The climatic condition for winter and summer weeks taken as input data are presented in Figures 6 and 7.

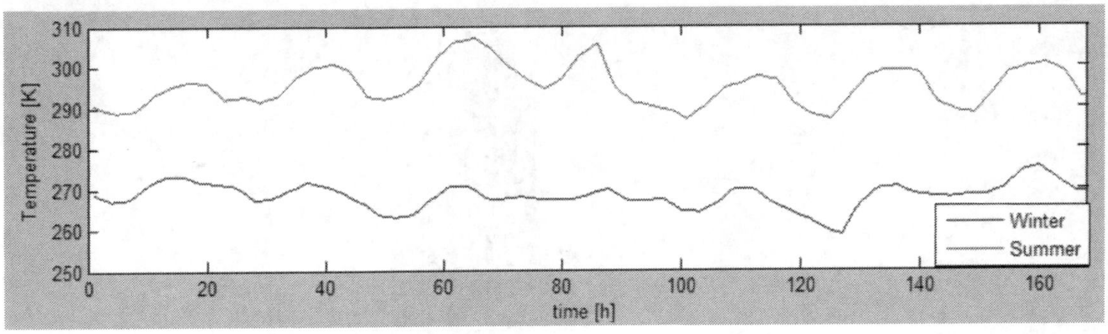

Figure 6: The ambient temperature for reference winter and summer week (data for Poznań from [13]).

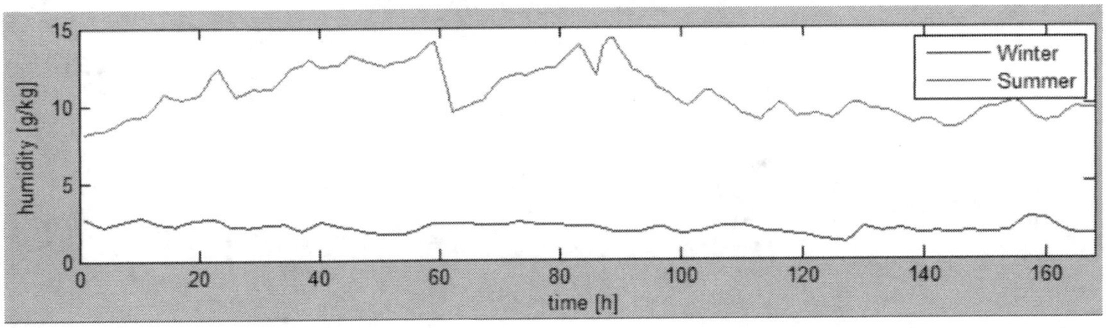

Figure 7: The ambient humidity for the reference winter and summer weeks (data for Poznań from [13]).

The other input data used in exergy simulations are given in Table 2.

Table 2: The input data

Parameter	Unit	Winter	Summer
$T_{in,4}$	[K]	291	293
$X_{in,4}$	[g/kg]	8	Resulting
Δp_{in}	[Pa]	650	650
Δp_{ex}	[Pa]	500	500
$\eta_{HR,t}$	[%]	65	65
$\eta_{HR,x}$	[%]	65	65
$\eta_{v,in,i}= \eta_{v,ex,i}$	[%]	70	70
$\eta_{v,in,em}= \eta_{v,ex,em}$	[%]	90	90
T_{w1}/T_{w2}	[K]	348,15/333,15	-
T_{ch1}/T_{ch2}	[K]	-	280,15/285,15

3.3 Theoretical model

Exergy balance equations have been constructed for each of the items of energy chain creating analyzed ventilation system. They are represented by following equations:

- The flux of the internal exergy loss in the air damper:
  - Inlet side:

$$\delta\dot{B}_{D,in} = \dot{m}_{in} \cdot \left(b_e - b_{in,d}\right) \tag{5}$$

  - Exhaust side:

$$\delta\dot{B}_{D,ex} = \dot{m}_{ex} \cdot \left(b_{ex,1} - b_{ex,d}\right) \tag{6}$$

- The flux of the internal exergy loss in the air filter:
  - Inlet side:

$$\delta\dot{B}_{F,in} = \dot{m}_{in} \cdot \left(b_{in,d} - b_{in,f}\right) \tag{7}$$

  - - Exhaust side:

$$\delta\dot{B}_{F,ex} = \dot{m}_{ex} \cdot \left(b_{ex,d} - b_{ex,f}\right) \tag{8}$$

- The flux of the internal energy loss in the heat recovery device:

$$\delta\dot{B}_{HR} = \dot{m}_{in} \cdot \left(b_{in,f} - b_{in,1}\right) + \dot{m}_{ex} \cdot \left(b_{ex,2} - b_{ex,3}\right) \tag{9}$$

- The flux of the internal exergy loss in the air heater:

$$\delta\dot{B}_{AH} = \dot{m}_{in} \cdot \left(b_{in,1} - b_{in,2}\right) + \dot{m}_{w} \cdot \left(b_{w,1} - b_{w,2}\right) \tag{10}$$

- The flux of the internal exergy loss in the cooler:

$$\delta\dot{B}_{CH} = \dot{m}_{in} \cdot \left(b_{in,2} - b^{*}_{in,2}\right) + \dot{m}_{ch} \cdot \left(b_{ch,1} - b_{ch,2}\right) \tag{11}$$

- The flux of the internal exergy loss in the fan:
  - Inlet side:

$$\delta\dot{B}_{V,in} = \dot{m}_{in} \cdot \left(b^{*}_{in,2} - b_{in,3}\right) + N_{el,in} \tag{12}$$

  - Exhaust side

$$\delta\dot{B}_{V,ex} = \dot{m}_{ex} \cdot \left(b_{ex,f} - b_{ex,2}\right) + N_{el,ex} \tag{13}$$

- Electrical power of the fan motor $N_{el,in}$ and $N_{el,ex}$ depends on the volumetric flow of humid air, its static pressure growth and total efficiency of the fan.
-

$$N_{el,in} = \dot{m}_{in} \cdot v_{in} \cdot \frac{p_{in,3} - p^{*}_{in,2}}{\eta_{V,in}} \tag{14}$$

$$N_{el,ex} = \dot{m}_{ex} \cdot v_{ex} \cdot \frac{p_{ex,f} - p_{ex,2}}{\eta_{V,in}} \tag{15}$$

- The flux of the internal exergy loss in the steam humidifier:

$$\delta\dot{B}_{SH} = \dot{m}_{in} \cdot (b_{in,3} - b_{in,4}) + \dot{m}_v \cdot b_v \quad (16)$$

Specific thermal exergy of humid air can be calculated using the equation:

$$b_{in,ex} = c_a \cdot \left[(T - T_0) - T_0 \cdot \ln \frac{T}{T_0} \right] + T_0 \cdot R_a \cdot \ln \frac{p_a}{p_{a,0}} + x \cdot \left[c_v \cdot (T - T_0) - T_0 \cdot \ln \frac{T}{T_0} \right] + T_0 \cdot R_v \cdot \ln \frac{p_v}{p_{v,0}} \quad (17)$$

Specific thermal exergy of water can be calculated using the formula:

$$b_w = c_w \cdot \left[(T - T_0) - T_0 \cdot \ln \frac{T}{T_0} \right] \quad (18)$$

Specific thermal exergy of saturated water vapor (x=1) can be calculated using the equation:

$$b_v = c_v \cdot \left[(T - T_0) - T_0 \cdot \ln \frac{T}{T_0} \right] + T_0 \cdot R_v \cdot \ln \frac{p_v}{p_{v,0}} \quad (19)$$

Heating capacity of the air heater can be calculated:

$$\dot{Q}_{AH} = \dot{m}_{in} \cdot c_a \cdot (T_{in,2} - T_{in,1}) \quad (20)$$

Cooler capacity of the air heater can be calculated:

$$\dot{Q}_{CH:} = \dot{m}_{in} \cdot c_a \cdot (T_{in,2} - T^{*}_{in,2}) \quad (21)$$

4 CALCULATION RESULTS AND DISCUSSION

The results of the exergy balance calculations, electrical power of the fan motor, air heater capacity and air cooler capacity for one week in winter and one week in summer are presented below.

The results of internal exergy loss of all energy chain in air handing unit for reference winter and summer weeks are presented in Table 3 and Figures 8 and 9.

Table 3: Case study – exergy calculation result

Parameter	Winter		Summer	
	[W/week]	[%]	[W/week]	[%]
$\delta\dot{B}_{D.in}$	451,62	0,63	351,9	1,97
$\delta\dot{B}_{F.in}$	753,0	1,05	856,4	4,8
$\delta\dot{B}_{HR}$	20165,2	28,2	687,97	3,86
$\delta\dot{B}_{AH}$	19261,8	26,94	0	0
$\delta\dot{B}_{CH}$	0	0	4177,8	23,39
$\delta\dot{B}_{V.in}$	6884,1	9,63	5928,6	33,2
$\delta\dot{B}_{SH}$	17386,8	24,31	0	0
$\delta\dot{B}_{D.ex}$	464,5	0,65	352,2	1,97
$\delta\dot{B}_{F.ex}$	753	1,05	856,4	4,8
$\delta\dot{B}_{V.ex}$	5379,0	7,52	4644,35	26,01
$\boldsymbol{\delta\dot{B}_{drv}}$	**39759,57**	**100**	**9928,96**	**100**

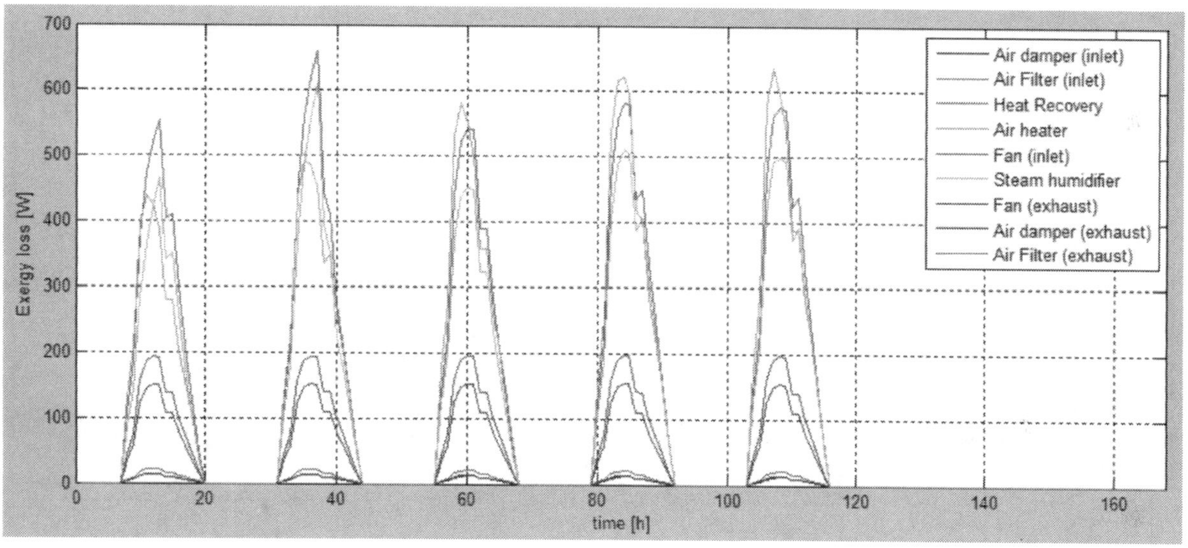

Figure 8: Calculation results – exergy loss in all analysed energy items during reference winter week.

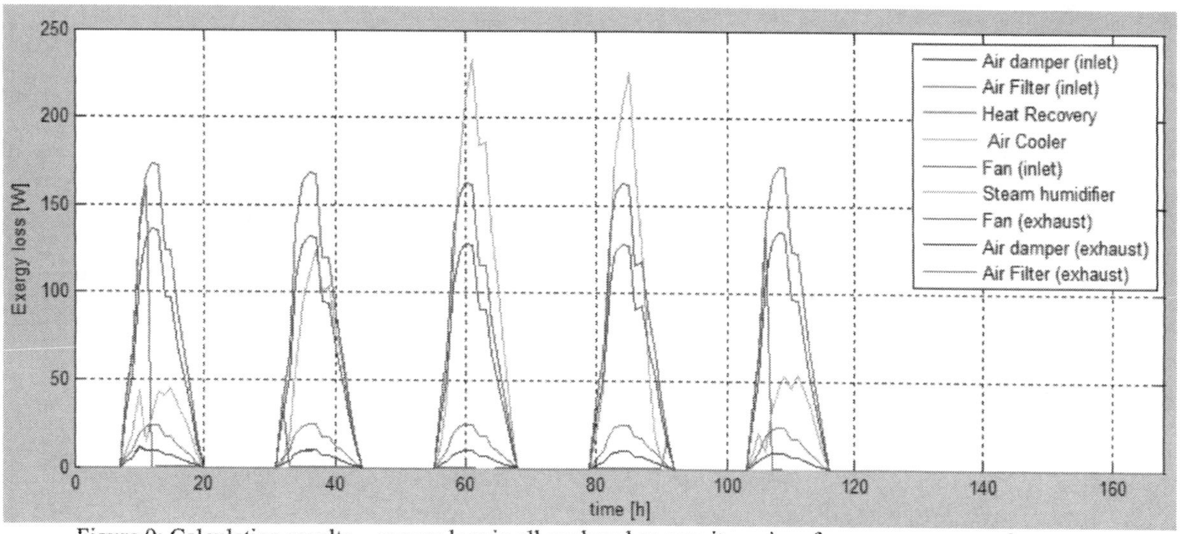

Figure 9: Calculation results – exergy loss in all analysed energy items in reference summer week.

The results of air heater thermal capacity, cooler capacity, electrical power of inlet and exhaust fans for reference winter and summer weeks are presented in Table 4 and Figures 10 and 11.

Table 4: Case study – calculation results

Parameter	Unit	Winter	Summer
Q_{AH}	[kW/week]	89,26	0
Q_{CH}	[kW/week]	0	126,37
$N_{el,in}$	[kW/week]	16,8	17,03
$N_{el,ex}$	[kW/week]	13,11	13,1

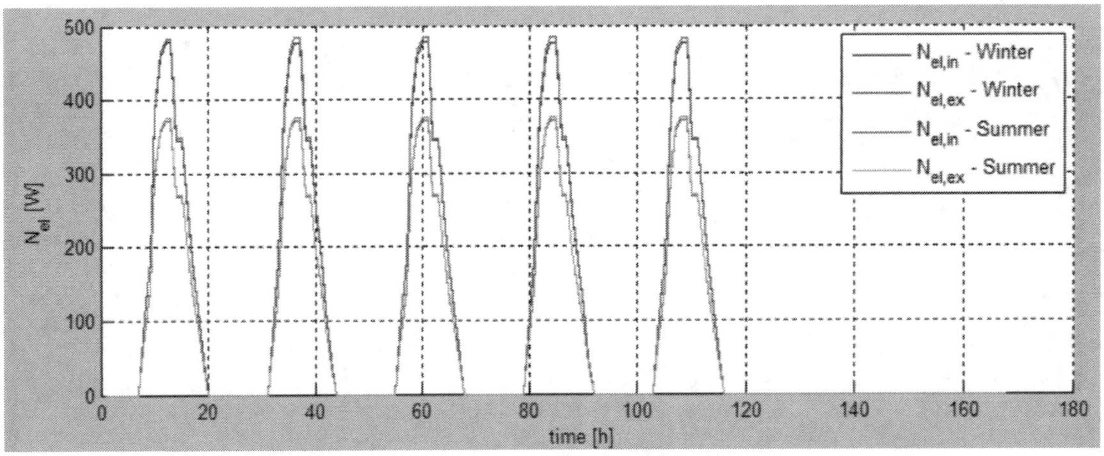

Figure 10: Electrical power of inlet and exhaust fans for reference week in winter and summer.

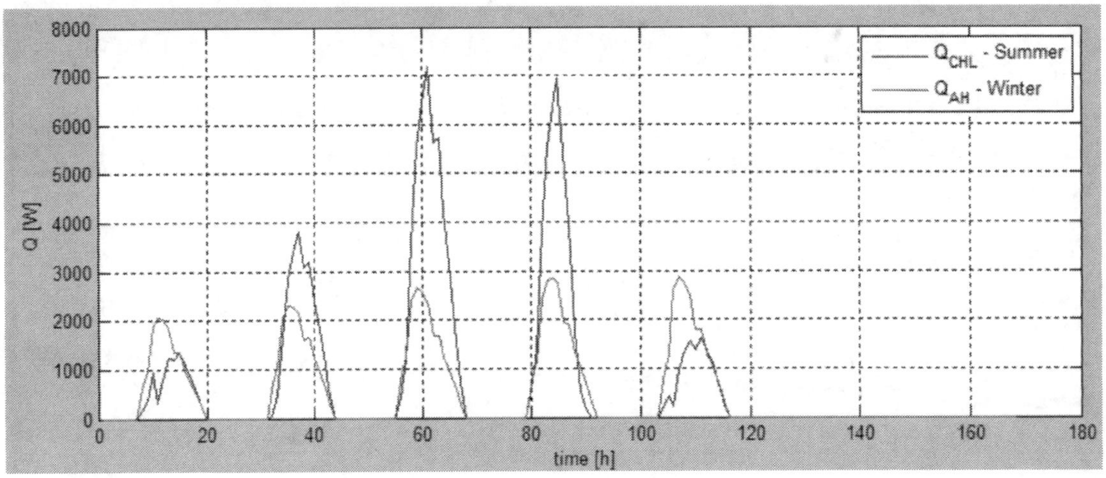

Figure 11: Air heater thermal capacity, cooling capacity for reference week in winter and summer.

Exergy analysis pinpoint the location, where the losses are the highest

In reference winter week three items (heat recovery, air heater, steam humidifier) generated more than 80% exergy losses. Exergy loss in heat recovery can not be perceived as a problem. Main part of heat is recovered from exhaust air, what can be treated in positive way, due to reduction of energy consumption of air heater and steam humidifier. The main sources of irreversibility of air heating process in cold days were: the air heater responsible for approximately 27% of the total flux of the exergy loss and air humidifier responsible for 24% of the total flux of the exergy loss.

In reference summer week the highest exergy losses were caused by fans and air cooler. Heat recovery works if temperature of outdoor air is below 292K, thus during summer season it is usually turned off, because of significant cost of work to expected profits. Supreme exergy loss during warm days is caused by air cooler, thus air handling unit needs improvements to limit exergy loss in this area. Exergy loss on fans are similar during whole year – both during summer and winter season.

During winter week exergy loss is more than four times higher than during summer season. In summer more devices are turned off, because of the high temperature and humidity and the fact that air cooler does not work as a dehumidifier – air cooler reduces the inlet air temperature only. If HVAC system controls also humidify, the exergy loss in summer would be much higher.

The above-mentioned three items of the energy chain - air heater, air humidifier in cold days and air cooler in warm days should be taken into detailed consideration in the optimization procedure.

5 CONCLUSIONS AND FUTURE WORK

Exergy analysis tool is the powerful tool for analysing, assessing, designing, improving and optimizing different systems and processes. Exergy analysis has a lot of benefits. This methods can assist for evaluation of the thermodynamic values of the energy products. Exergy losses clearly pinpoint the location, causes and sources of deviations from ideal conditions of system operation. Exergy efficienciy is the measure of the approach to ideal [11].

Subsequent steps of presented analysis should include: (i) exergy analysis for other main items in HVAC system: air distribution system and office rooms, (ii) chemical exergy analyses of the fossil fuels needed for the air handling unit operation and (iii) analysis of vetilation system for one typical year.

6. REFERENCES

1. Benz- Karlstrom P. (2010). *Calculations with SEPE – Exergy analysis of cooling systems.* Zurich Switzerland
2. Dovjak M., Shukuya M., Olesen B.W., Krainer A. (2010): *Analysis on exergy consumption patterns for space heating in Slovenian buildings*. Energy
3. Hepbasli A. (2012) *Low exergy (LowEx) heating and cooling systems for sustainable buildings and societies.* Renewable and Sustainable Energy Reviews
4. Hepbasil A., Kecebas A.(2013): *A comparative study on conventional and advanced exergetic analyses of geothermal district heating systems based on actual operational data.* Energy and Buildings
5. Iliev V., Badunski D., Seso I., Andovski S. (2014). *Direct Digital Control of HVAC System and CO2-Based Demand Controlled Ventilation.* International Journal of Innovative Technology and Exploring Engineering
6. Nassif N. (2012). *A robust CO_2–based demand – controlled ventilation control strategy for multi-zone HVAC system*. Energy and Buildings
7. Mróz T. M., Bogusławska Ż.(2012). *Exergy analysis of an air heating system*, Poland. Foundation of civil and environmental engineering. Poznań University of Technology
8. Mróz T. M.: (2013*) Energy management in built environment. Tools and evaluation procedures.* Publishing House of Poznan University of Technology

9. Shell M.B., Turner S.C., Shim O.R. *Application of CO2-Based Demand – Controlled Ventilation Using ASHRAE Standard 62: Optimizing Energy Use and Ventilation.* ASHARE Transaction Symposia
10. Sakulpipatsin P., Itard L.C.M., Kooi H.J., Boelman E.C., Luscuere P.G. (2010) *An exergy application for analysis of buildings and HVAC systems.* Energy and Buildings.
11. Sakulpipatsin P. (2008) *Exergy efficient building design.*
12. Szargut J. Petela R. (1965) *Egzergia*. Warszawa
13. http://www.mir.gov.pl
14. Zmeureanu R., Yu Wu X.(2007) *Energy and exergy performance of residential heating systems with separate mechanical ventilation*. Energy.

STRATEGIES FOR EXPLOITING CLIMATE POTENTIAL THROUGH VENTILATIVE COOLING IN A RENOVATED HISTORIC MARKET

Annamaria Belleri*[1], Federico Noris[1], and Roberto Lollini[1]

1 EURAC research
1 Druso
Bolzano, Italy
**Corresponding author: annamaria.belleri@eurac.edu*

ABSTRACT

Nearly all retail locations use ventilation and cooling systems to ensure adequate air exchange for health reasons and indoor comfort temperatures. These systems can run for over 2,000 hours per year and we expect that average operating hours will continue to rise across Europe because of the continued trend towards longer opening hours and increased number of opening days. Shopping malls often enclose large open spaces and atria with high solar and internal gains that can drive ventilative cooling. This study presents the methodology applied to a historical market located in Valladolid (Spain) for the assessment of the ventilative cooling potential and the definition of a ventilative cooling strategy. The climate suitability and potential was evaluated based on the building internal gains rate and the considered ventilation strategies options (e.g. night cooling, daytime direct ventilation). Significant performance indicators were defined for each considered ventilation strategy. Once we determined the climate suitability, we defined a ventilative cooling strategy, that exploits openings in the façade and in the skylight to promote stack effect ventilation. We sized openings area and location on the façade, taking into account design constraints, and we assessed their performances by dynamic simulations in Trnsys coupled with Trnflow airflow network. Results show the potential cooling load reduction, with the achievement of acceptable thermal comfort due to the ventilative cooling in the shopping mall. The proposed methodology can support the design decision process towards cost effective low energy shopping malls.

KEYWORDS

Climate potential, ventilative cooling, retail buildings, refurbishment, dynamic simulations

1. INTRODUCTION

Retail buildings are a significant part of the EU building stock and, among them, the wholesale and retail sector presents an average specific energy use of 200 kWh/m^2a and account for 7% of the total building stock in Europe (Atanasiu B., 2011). Nearly all retail locations use ventilation and air conditioning systems to ensure adequate air exchange and indoor comfort temperatures. These systems run for over 2,000 hours per year and average operating hours will continue to rise across Europe because of the continued trend towards longer opening hours and increased number of opening days.

According to the British Council for Shopping Centre (BCSC, 2012), natural ventilation solutions can reduce capital and maintenance cost over traditional mechanical heating, ventilation and air conditioning and they typically need less plant and equipment place.

Shopping malls often enclose large open spaces and atria with high solar and internal gains that can be drivers for ventilative cooling. However, nearly all retail buildings use mechanical ventilation systems as they require less design efforts assuring minimum requirements for indoor air quality.

The CommONEnergy EU FP7 project (www.commonenergyproject.eu) will contribute to the development of ventilative cooling strategies for the retrofit of existing shopping malls to exploit the architectural and climate drivers potential. The retrofitting solution sets developed within the project will be applied to three demo cases located in Spain, Italy and Norway, selected because of their high replication potential. This paper presents the methods and tools applied to assess the ventilative cooling potential and to define an efficient ventilative cooling strategy for one of the democases: the historic market of the city of Valladolid.

First, we evaluated the climate suitability and potential according to several building internal gains rates and ventilation strategies (e.g. night cooling, daytime direct ventilation). Once we determined the climate suitability, we defined a ventilative cooling strategy that exploits openings in the façade and in the skylight to activate stack effect ventilation. Coupled thermal and airflow building energy simulations allowed to assess the potential cooling load reduction, the air change rates and the indoor comfort conditions for several opening factors.

2. CASE STUDY

2.1 Building description

The “Mercado del Val” is an iron market located within the old town of Valladolid. Its construction was completed in 1882. Its floor plan is a rectangle of 112 meters long and 20 meters wide, with chamfered corners.

The Valladolid municipality planned a refurbishment intervention to transform the market into an innovative building that meets the contemporary commercial needs being respectful of its historic representativeness. As required by the Heritage Council, the refurbishment project aims at emphasizing the old iron structure by using glazed facade over the entire building perimeter. The new indoor layout configuration and the glazed façade will contribute to a better understanding of the global iron structure, to increase daylighting and to make the commercial activities visible from outside. The glazed façade is made by modular façade elements that aims at integrating thermal, daylighting and ventilation functions, being responsive when internal and external loads change.

Figure 1: Mercado del Val (Valladolid) building view. Source: www.commonenergyproject.eu

Figure 2: Mercado del Val (Valladolid), interior building view. Source: www.commonenergyproject.eu

2.2 Design constraints

The building shape has high potential for exploiting stack effect ventilation by integrating openings in the facade and exploiting the existing skylight openings located at 10 m height from the ground level where air can exhaust.

Therefore, the definition of a ventilative cooling strategy involves the whole building envelope as openings have to be located at opposite sides and at different height on the building envelope.
Main constraints regard inlet openings location at ground level. A large temperature difference between indoor and outdoor might cause cold draughts inside the building if the openings are placed near people's height. Therefore we decided to avoid openings within 3 m height of the building façade to prevent local discomfort situations. Inlet air at 5 m height, which is approximately the height of the upper part of the modular façade, would allow air mixing and prevent cold draughts in the occupied zones of the building. Additionally, inlet openings in the lower part of the façade might cause safety issues.
Lower inlet openings in urban environments might affect negatively indoor air quality due to transfer of outdoor pollution. Filters on the opening would cause a high pressure drop of the air reducing significantly natural ventilation effectiveness. In order to avoid water infiltration in case of precipitation, top hung windows could be used. Top hung windows have usually an opening angle of 20°, which reduces the opening effective area by 80%. Therefore, for the definition of the opening area we considered opening angles capabilities and actuators power. As enforced by the heritage council, skylights have to maintain the original design. This could reduce the effective area of the openings.

3 METHODS

The following methods were used within the first phases of the design project to analyse the climate potential and assess possible energy savings of a defined ventilative cooling strategy.

3.1 Climate suitability analysis

To estimate ventilative cooling potential we applied the method proposed by NIST (Emmerich S. J., 2011) to shopping mall building typology and fit it according to their specific needs.
This method assumes that the heating balance point temperature (T_b) establishes the outdoor air temperature below which heating must be provided to maintain indoor air temperatures at a defined internal heating set point temperature (T_{hsp}).
Therefore, when outdoor temperature (T_o) exceeds the heating balance point temperature, direct ventilation is considered useful to maintain indoor conditions within the cooling set point temperature (T_{csp}). At or below the heating balance point temperature, ventilative cooling is no longer useful but heat recovery ventilation should be used to meet minimum air change rates for indoor air quality control and reduce heat losses.
We considered the building occupied for 11 hours per day and set the cooling set point temperature to 25°C during the day and to 28°C during the night, as well as the heating set point temperature to 16°C during the day and 14°C during the night. Those set point temperatures refer to the operative temperature recommended by the standard EN 15251: 2008 for building Category II.
The heating balance point temperature (T_b) can be calculated using Equation 1. Table 33 reports the balance point temperature values calculated at different internal gains level.

$$T_b = T_{hsp} - \frac{q_i}{\dot{m}_{min} c_p + \sum UA} \tag{11}$$

where:

q_i = total internal gains [W/m²]
c_p = air capacity [J/kg-K]
$\dot{m}_{min}$ = minimum required mass flow rate [kg/s]

$\sum UA$ = envelope heat exchange [W/K]

The average U-value of the envelope was estimated at 0.49 W/m²K according to the local minimum building code requirements and the actual building shape.
The minimum ventilation rate which has been estimated as 1.7 l/s-m² (0.00204 kg/s-m²) according to the EN 15251 values for department stores (Cat. II – low polluting building).

Table 1. Heating balance point temperature calculated for different internal gain values.

	$q_i = 10$ W/m²	$q_i = 20$ W/m²	$q_i = 40$ W/m²	$q_i = 80$ W/m²
T_b	12.7 °C	9.4 °C	2.85 °C	-10.3 °C

For each hour of an annual climatic record for the Valladolid city we assessed the number of hours when:

1) Ventilative cooling is not required: when the outdoor temperature is below the heating balance point temperature no ventilative cooling can be used since heating is needed;
   If $T_o < T_b$ then $\dot{m} = 0$
2) Direct ventilative cooling with ventilation rate maintained at the minimum: when the outdoor temperature exceeds the balance point temperature, yet falls below the lower limit of the comfort zone – given the width of the comfort zone (T_{csp}-T_{hsp}) - and outdoor dew point temperature is below 17°C (or 65% relative humidity), the cooling ventilation rate may be maintained at the minimum ventilation rate required by the EN 15251;
   If $T_b \leq T_o < T_b + (T_{csp}-T_{hsp})$ and $T_{o\text{-}dp} \leq 17°C$ then $\dot{m} = 0.00204$ kg/s-m²
3) Direct ventilative cooling useful: when the outdoor temperature is within the comfort zone – given the upper limit as the cooling set point temperature - and outdoor dew point temperature is below 17°C (or 65% relative humidity), the minimum cooling ventilation rate needed to maintain indoor air conditions within the cooling set point temperature are computed as Equation 2;
   If $T_b + (T_{csp}-T_{hsp}) \leq T_o \leq T_{csp} - 1$ and $T_{o\text{-}dp} \leq 17°C$ then

$$\dot{m} = \frac{q_i}{c_P(T_{csp}-T_o)} \text{ [kg/s-m}^2\text{]} \tag{12}$$

4) Direct ventilative cooling not useful: when the outdoor temperature exceeds the cooling setpoint temperature and the dew point temperature is above 17°C (or 65% relative humidity) for at least one hour during the day, the ventilative cooling is no longer useful and nighttime cooling potential (NCP) over the following night is evaluated as the internal gains that may be offset for a nominal unit night-time air change rate have been computed as Equation 3:
   If $T_o > T_{csp} - 1$ or $T_{o\text{-}dp} > 17°C$ then

$$NCP = \frac{H\rho c_P(T_{csp-night}-T_o)}{3600} \text{ [W/m}^2\text{-ach]} \tag{13}$$

where:
H = floor height [m]
ρ = air density [kg/m³]
$T_{csp-night}$ = temperature cooling set point at night [°C]

The weather file used for the analysis derives from historical data series (2000-2009) of a weather station located in the city of Valladolid, which is part of the Meteonorm database (Weather station ID 81410).

3.2 Coupled thermal and airflow simulations

In order to assess energy savings in terms of cooling need and to evaluate thermal comfort, we performed dynamic simulations using a specific Trnsys module for airflow and thermal models coupling, Trnflow.

The proposed strategy would allow to obtain stack effect ventilation by integrating openings in the upper part of the modular facade and exploiting the existing skylight openings (Figure 3). Openings area and location fits to the modular façade design to reduce frame dividers number.

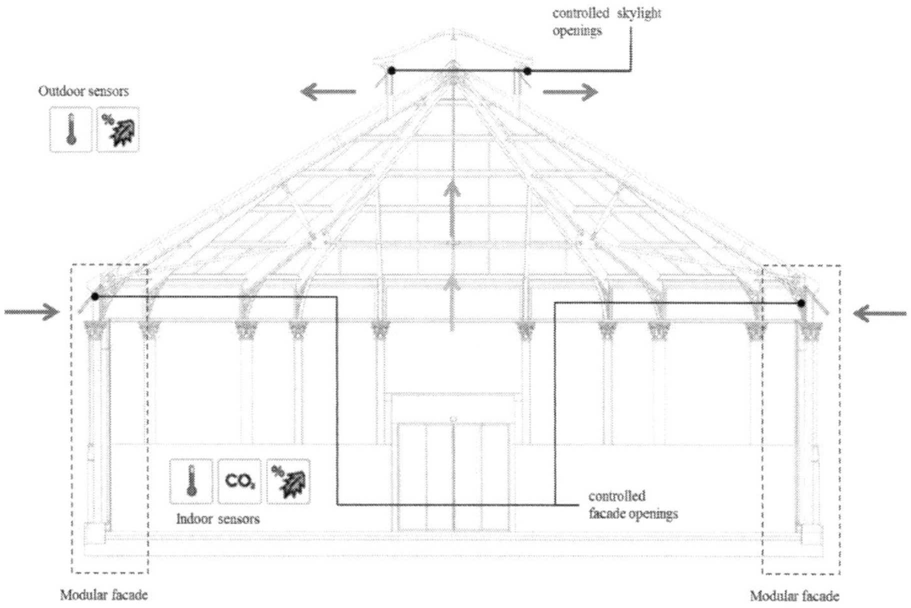

Figure 3. Building cross section with proposed strategy schema.

Table 2 reports the geometrical data about the operable area used in the simulations. Same windows are located both on south-west side and on north-east side covering 18 façade modules, all the modules directly connected to the atrium space. The window height matches the height of the upper part of the façade module. The skylight height matches the existing skylight opening height.

We considered the opening factors as varying between the standard top hung window opening angle and the maximum feasible opening factor:

- Opening factor 0.2 : top hung window with 18°C open angle
- Opening factor 0.4 : top hung window with 36°C open angle
- Opening factor 0.8 : top hung window with 72°C open angle

Figure 4: South-west front view of the building where openings locations are highlighted in red.

Table 2. Simulation input data on operable area.

	Opening type	Width [m]	Height [m]	Number of modules	Tot opening area [m²]	Height from the reference ground [m]
Window	Top hung	3.4	0.78	18	48	5.10
Skilight	Top hung	4	0.56	18	40	10.45

In each simulation model, both windows and skylights are operated in the same way and using the same opening factors. Windows are opened if outdoor temperature is above 12°C and controlled depending on the indoor - outdoor temperature difference and the control value at the previous time step.

If the controller was previously on and:

- $T_{in} \geq T_{out} + 2$ then windows stay open;
- $T_{in} < T_{out} + 2$ then windows close.

If the controller was previously off and:

- $T_{in} \geq T_{out} + 5$ then windows open;
- $T_{in} < T_{out} + 5$ then windows stay close.

However the control function is set to 0 if the outdoor temperature is lower than 12°C. Minimum run time period for window operation is set to 20 min.

When outdoor temperature conditions do not allow window opening, mechanical ventilation supply to the building the minimum air change rates required by the EN 15251 standard to maintain an acceptable level of indoor air quality:

- 7.35 kg/hr-m^2 (around 2 ach) during the occupied hours
- 3.02 kg/hr-m^2 (around 0.8 ach) during the non-occupied hours

Simulations were run from April to October both in free-floating mode to analyse comfort according to EN 15251 adaptive comfort model. To assess cooling needs, simulations were also run by setting an unlimited power cooling system with temperature set point of 25°C during the occupied hours according to the recommended indoor temperatures by EN 15251:2008 for department stores at comfort category II.

The results are compared with a baseline where mechanical ventilation always provides the minimum air change rates required by the EN 15251 standard with supplied air temperature equal to the outdoor temperature.

4 RESULTS AND DISCUSSION

In the following paragraphs we present and discuss the obtained results of the climate suitability analysis and the coupled thermal and airflow simulations.

4.1 Climate suitability analysis results

In the Valladolid weather file we identified 95 CD[1], 8567 CDH[2] and an average monthly diurnal temperature swing between 4.5 K (April) and 5.9 K (July).

The graph in Figure 5 shows the percentage of hours within a whole year when direct ventilative cooling is useful according to the classification method described in par. 3.1 and different levels of internal gains.

Depending on the internal gains level, direct ventilative cooling can be useful up to 90% of the hours within a year, whereas for around 10% of the hours the outdoor temperature is higher than the cooling set point temperature and therefore night-time ventilation potential has

[1]Cooling Days defined as the number of days with an average ambient temperature higher than 18°C.

[2]Cooling Degree Hours defined as the integral of the positive temperature differences between the hourly average ambient temperature and a base temperature of 18°C.

to be investigated.
The higher are the internal gains, the higher is the number of hours when direct ventilative cooling is required.

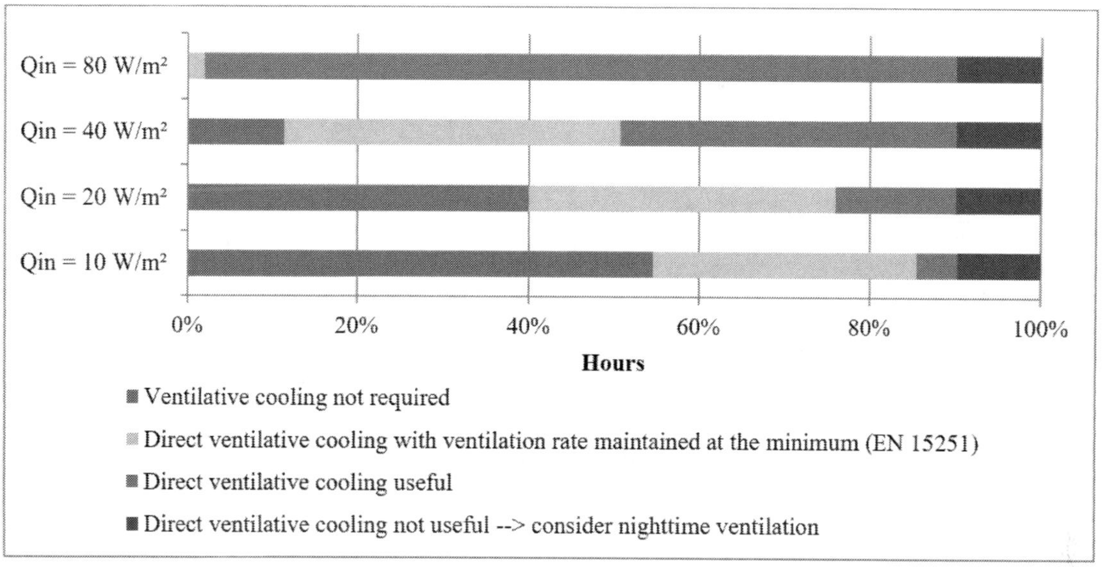

Figure 5: Percentage of hours within a year when direct ventilative cooling is required, useful or not useful in the Valladolid climate considering different values of internal gains.

Mean ventilation rates needed for direct ventilative cooling are reported in Figure 6 depending on internal gains level. For instance, considering the outdoor temperature series for Valladolid and an internal gain rate of 80 W/m^2, up to 4 ach are required on average to effectively cool the building during the occupied hours. As a consequence, the number of effective hours for direct ventilative cooling decreases with the decrease of internal gains.

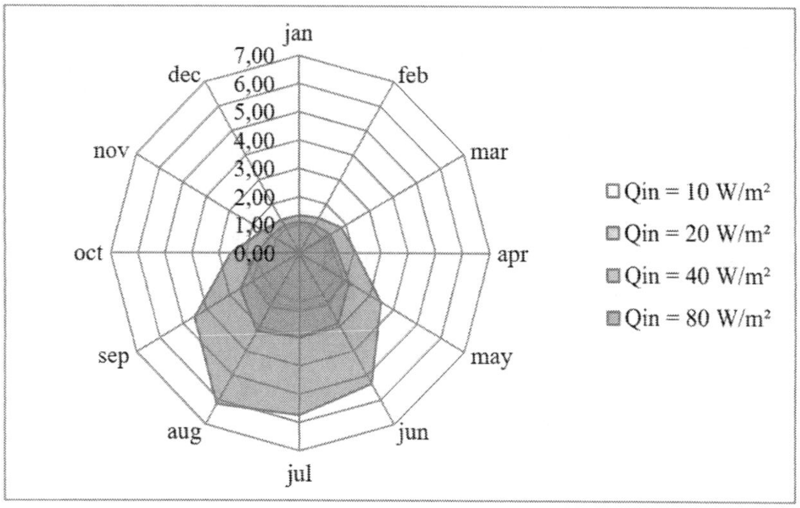

Figure 6. Average direct ventilative cooling rate (ach) for each internal gain level considered and percentage of effective hours within the year in Valladolid.

The graph in Figure 7 shows an estimation of the internal gains that may be offset for a nominal unit night-time air change rate. On average among the summer period an air change rate during the night would be able to offset around 23.5 W/m^2 of internal gains. The number of activation hours matches the percentages identified in the graph in Figure 5.

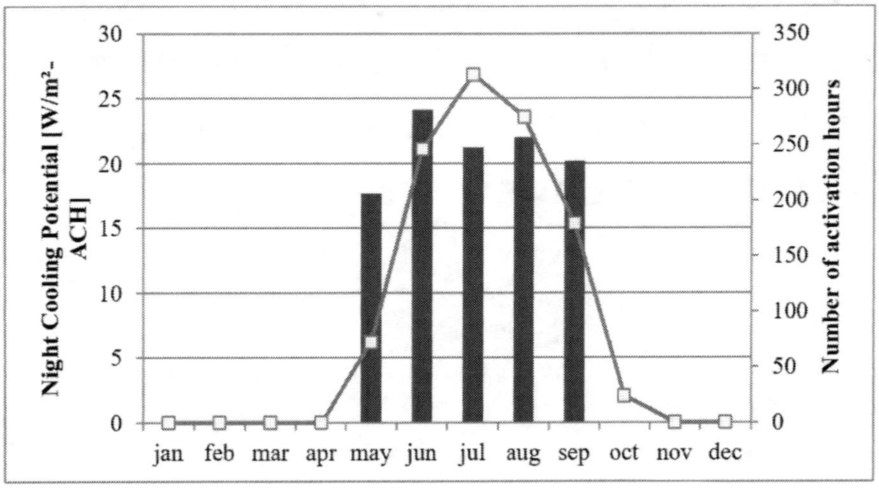

Figure 7. Night cooling potential and number of hours when night cooling can be activated in Valladolid.

4.2 Simulations results

As reported in Table 3, 41% of cooling need reduction can be obtained using an opening factor of 0.4 and 55% using an opening factor of 0.8. Windows with an opening factor of 0.2 cannot assure the minimum air change rates requirements and therefore simulation results show an increase in cooling need by 21% compared to the baseline model. In fact, in the baseline model the minimum air change rates requirements are provided at a constant rate by the mechanical ventilation system. The graph in Figure 8 represents the air change rates frequencies.

Figure 9 shows the simulations results in free floating mode. From April until October the baseline model predict temperatures within the acceptable comfort range, identified according to the adaptive comfort model (EN 15251:2008 standard), for over 85% of the occupied period. This means that, providing fresh air at the minimum air change rates required by the standard, adaptive thermal comfort requirements are met for 85% of the occupied hours. The discomfort is in this case mainly due to too hot temperatures.

Using a window opening factor of 0.8, the ventilative cooling strategy would allow us to reduce or almost zero the discomfort hours due to too hot temperature, even though the number of discomfort hours increase because of too cold temperatures.

Using a window opening factor of 0.2, the number of hours within comfort ranges is 86% compared to the 85% of the baseline case, even though the number discomfort hours is mainly due to too hot temperatures. While an opening factor of 0.2 would cause less 63% of operation hours of the mechanical ventilation system, during most of the time the air change rates are below the minimum required by the standard.

Table 3: Cooling need and number of operation hours of the mechanical ventilation system predicted by the baseline model and the models with ventilative cooling strategy.

	Cooling need [kWh/m²]	Cooling need reduction [%]	Number of operation hours of the mechanical ventilation system
open factor = 0.2	20.68	+21%	1897
open factor = 0.4	10.10	-41%	2858
open factor = 0.8	7.69	-55%	3617
baseline	17.04	-	5136

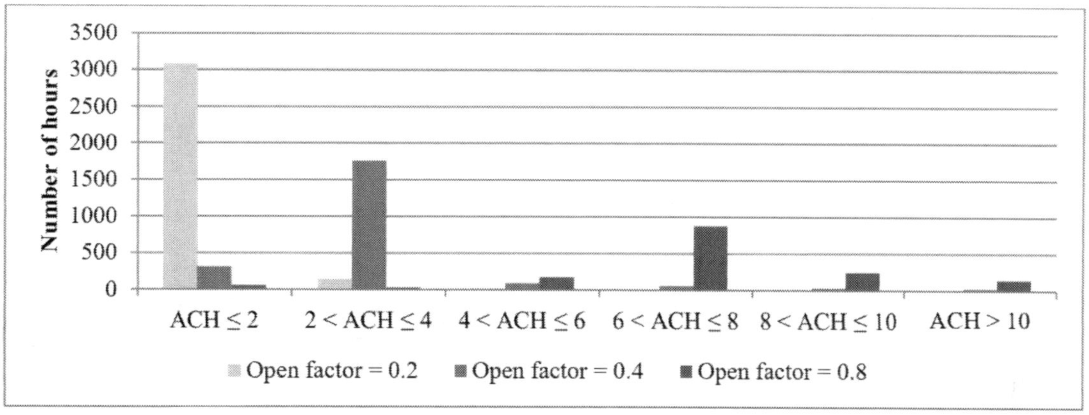

Figure 8: Frequency analysis of the predicted air changes per hour with different opening factors.

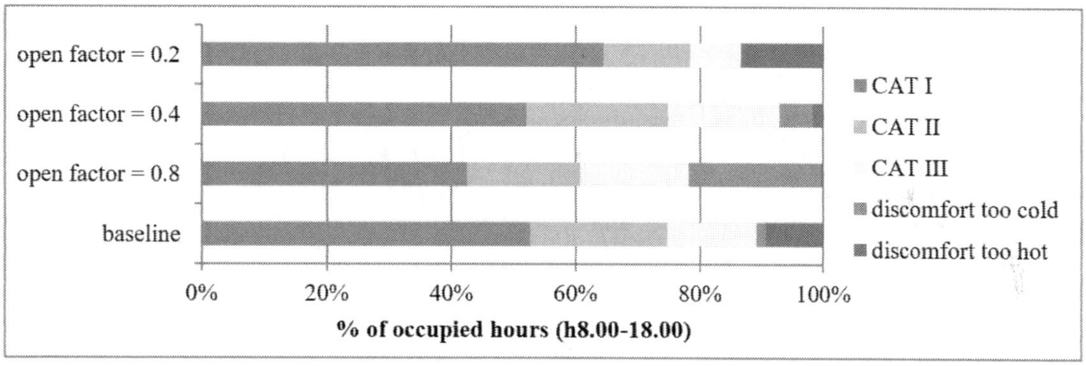

Figure 9: Percentage of occupied hours during which EN 15251:2008 requirements on thermal comfort are met by the baseline model and the models with ventilative cooling strategy.

In our simulation models, the window opening angle is not modulated according to indoor-outdoor temperature difference and therefore discomfort hours due to too cold temperatures increase by increasing the opening factors. This problem can be solved developing a more complex control strategy and using window actuators that allow different opening angles.
Therefore, our design proposal was to define a control strategy which considers opening factor modulation and optimizes it according to indoor-outdoor temperature differences.

5 ADDITIONAL CONSIDERATIONS AND NEXT DESIGN STEPS

Further integration constraints arise from the simulations results. Considering the openings geometry (W >>H) and the high opening angle needed (up to 70°), proper actuators must be selected.
In case of top hung windows, it is recommended the installation of actuators protected from water and solid bodies (dust etc.) and enough powerful to support with sufficient rigidity the maximum load that occurs when the window is completely opened.

Linear actuators are generally more powerful than chain actuators but actuator encumbrance inside the room has to be considered as it affects the aesthetic appearance. Very large windows might also require two thrust points.
Next design step will be focused on:

- the definition of a control strategy once constraints on window opening factor and operable area are defined;
- the optimization of the control strategy parameters;
- the integration of the ventilative cooling strategy within the whole building solution sets concepts (shadings, glazing system, daylighting concept) taking into account for the respective internal and solar gains;
- the integration of openings within the modular façade concept and its replicability on similar buildings;
- a CFD analysis of the building for results validation and evaluation of possible local discomfort situations.

6 CONCLUSIONS

The presented study focuses on the methodology applied to a historical market located in Valladolid (Spain) for the assessment of the ventilative cooling potential and the definition of a ventilative cooling strategy within its refurbishment project.
The climate suitability analysis showed that, depending on the internal gains level, direct ventilative cooling can be useful up to 90% of the hours within a year, whereas for 10% of the hours the outdoor temperature is higher than the cooling set point temperature and therefore night-time ventilation potential was investigated. Considering the outdoor temperature series for Valladolid and an internal gain rate of 80 W/m^2, up to 4 ach are required on average to effectively cool the building during the occupied hours.
According to the resulting climate performance indicators, we defined a ventilative cooling strategy that exploits openings in the façade and in the skylight to activate stack effect ventilation. Openings area and location on the facade have been sized taking into account design constraints and climate potential and tested by dynamic simulations in Trnsys coupled with Trnflow airflow network.
Results show up to 55% of potential cooling load reduction due to the ventilative cooling strategy in the shopping mall maintaining indoor temperatures lower than 25°C. Higher energy savings could be obtained by considering the adaptive behaviour of building occupants. Discomfort situations were mainly due to too cold indoor temperatures in case of high opening factors. Therefore, we suggested window openings to be automated, modulated and controlled depending on indoor and outdoor temperature and humidity.
The proposed methodology was applied to support the design decisions providing quantitative building performance evaluations and allowing a more robust cost estimation. The design team reacted positively at the proposed ventilative cooling solution.

7 ACKNOWLEDGEMENTS

The research leading to these results has received funding from the European Community's Seventh Framework Programme (FP7/2007-2013) under grant agreement n° 608678. The authors would like to thank Javier Antolín Gutierrez, Samaniego Muñoz J. Jesús and Luis A. Bujedo from Cartif for the simulation model production, the design team for the cooperative and positive approach to the collaboration and Magdalena Rozanska from Acciona for activities' coordination regarding the modular façade development.

8 REFERENCES

Atanasiu B., D. C. (2011). *Europe's building under the microscope. A country-by-country review of the energy performance of buildings.* Building Performance Institute Europe (BPIE).

BCSC. (20 de August de 2012). *Shopping centre natural ventilation design.* Obtido de Guidance 68: http://www.bcsc.org.uk/publication.asp?pub_id=459

Emmerich S. J., P. B. (2011). Impact of adaptive thermal comfort on climatic suitability of natural ventilation in office buildings. *Energy and Buildings, 43*(2101-2107).

Weather station ID 81410. (s.d.). Obtido de Meteonorm: http://meteonorm.com/

DURABILITY OF AIR TIGHTNESS SOLUTIONS FOR BUILDINGS

Peter Ylmén, Magnus Hansén and Jörgen Romild

SP Technical Research Institute of Sweden
Box 857
SE-501 15 Borås, Sweden

ABSTRACT

The aim of the project was to evaluate how the air tightness of buildings changes over time and how the sealing materials are affected during the expected life length of 50 years. The project was divided into two parts were one was laboratory tests of different products with accelerated ageing, and the other part were evaluation of older existing buildings. The laboratory test was conducted in a temporary room with lightweight construction in wood and different sealing products. The room was then heated to 80 °C and had changing relative moisture content in the air. The results showed that most products still maintained their function after accelerated ageing but some products considerably lost their ability to seal air through the building envelope.
In the testing of existing buildings, six single family houses that are between ten to twenty years old have been tested for air leakage. Test reports regarding air tightness from when the buildings where newly constructed were compared to new measurements. Three buildings had made changes to the building envelope while the other three had original structures. The results from the measurements showed that two of the tested buildings had considerably more air leakages than when they were new but that the rest had not changed. The change in air tightness for the two buildings are very likely due to changes made in the building envelope a few years after they were build.

KEYWORDS

Air tightness in buildings, durability, testing, infiltration, jointing, taping

1. INTRODUCTION

The method of constructing houses and buildings in Sweden changed successively during the 60's and 70's. Fiberglass and mineral wool insulation were starting to be used on a large scale instead of for example, massive wooden construction.
This change in building method led also to an increase in the use of tightening and barrier layer solutions to reduce the flow of moisture and air through the different construction sections. Different flexible barriers and seals of plastic were introduced and tested on the market, an early example of which is polythene film (PE-film), which had begun to be frequently used as air and moisture barrier in the wall building envelope.
The oil crisis in the 70's led to a requirement to reduce energy consumption. One method to help with this is to decrease thermal losses in houses by means of, for example, the use of more airtight windows and doors.
By the beginning of the 70's it was apparent that these new methods of construction resulted in some accompanying problems. Moisture problems could arise resulting in different types of moisture damage in house and buildings. Many of these problems could occur and be found very early after the completion of construction. In certain cases it was possible to see that the durability of a building had already been compromised by the use of sub-standard polythene, which had broken down, become cracked, opened up and caused damage.

In the interest of everybody, an extensive investigation with further research project was started. Statliga planverket (now Boverket), Sweden's plastic federation SPF (plastic manufacturers) and the National testing establishment (now SP Technical research institute of Sweden) all worked in cooperation on this. The aim of the project was to obtain a more reliable and durable quality of specific chosen building products (in the first case material for the air and vapour barriers) and tightening solutions. One result of this was the example that the Works Norm (VN) was designed as a basis for voluntary quality and type approval marking for a more reliable higher quality of plastic building products. Included in this was also the requirement that the built in material, not least because of the high cost involved with maintenance or even replacement, should have a minimum life length of 50 years. This means that in the Swedish building trade today it has become the "norm" to use a durable building polythene film with a minimum life length of 50 years.

One example of the works norm for air and vapour barrier using LD-polythene film (which was also the first) is:

[1]. Material requirements plus other demands are written down here, together with, amongst others, how the film should be mounted, how clamped joints and overlaps should be formed as well as requirements for jointing aids. "Joints, including joint material shall meet the minimum requirements set out in the Works Norm for the barrier layer, especially when considering age durability. Joints and joint material shall not affect the function and properties of the building film in a negative manner." [2].

A common problem today regarding the choice of building material is that the question "How long is the product expected to work without the need for maintenance or replacement?" cannot be answered by just looking at the material specification.

There can be many reasons as to why this information isn't available. It can be too expensive or impossible to find out how long the material can work. There may also not be any methods or knowledge available regarding the evaluation of the life length of a building material.

There are a number of investigations that show the importance of airtight construction together with the possibilities of energy savings if the building has good air tightness [Emmerich 2005]. It has been shown that infiltration losses in some cases are higher than the intentional losses from ventilation and much higher than transmission losses.

Air tightness in a building is created by having an airtight barrier with airtight joints and voids/services/penetrations. In many instances the airtight layer is created using a flexible material, such as polythene sheeting. The jointing for this sheeting usually involves the use of stapling, clamping or the use of joint adhesives, bands or tape. Both the polythene sheet and joint material age/degenerate with time, which can lead to a reduction in the air tightness of the building. The same principle applies to joints and voids/services/penetrations in massive constructions. The ageing of the material can be caused by many factors, some of which are heat, cold, moisture, sun (UV radiation) oxygen, ozone, chemicals and mechanical impacts. Furthermore, materials in and around the joints can affect each other by, for instance, the migration of plasticizers.

2. AIM

The aim of the project is to evaluate how the air tightness of a building changes over time, and show which air tightening solutions that are good and durable, and those which are not so good and should be avoided, and transfer this knowledge to the building industry.

2.1 Method

No literature describing durability testing for how building materials affect each other could be found for the project; therefore it was necessary to create a new test method. Since the air tightening materials are to be found inside the building construction and to a certain extent protected from air borne pollution and sunlight when construction is complete, then these parameters were deemed to have a negligible effect regarding durability. However, moisture content in the construction does vary since the relative humidity inside and outside varies throughout the year. The parameters chosen for accelerated ageing of the material were increased temperature together with varying air moisture content. A test rig with humidity and temperature controls was constructed in order to produce a representative scale for actual buildings as well as measuring as many different material combinations as possible.

Construction of test rig

It is most usual with durability testing that smaller samples of material are tested then their physical properties evaluated, for example the measurement of tensile strength. For this testing a 2.2 m square by 2.4 m high room was built, see attached figure 1, together with the different air tightening solutions built in as with a normal building. The material was therefore used in realistic amounts and lengths and subjected to realistic movements that would be expected in a real building.

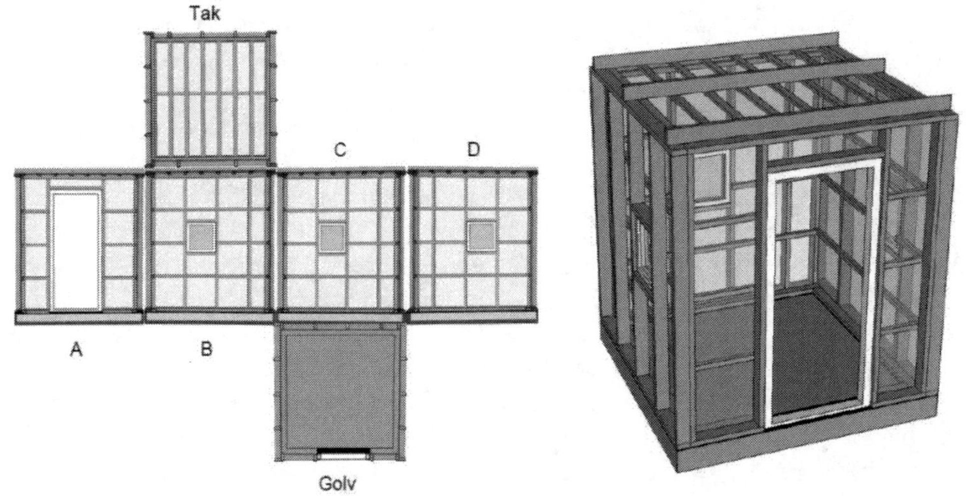

Figure 1. Diagram of the test room construction.

Different methods of jointing polythene sheeting, in particular where the sheeting meets concrete flooring/other concrete sections plus forming around doors, windows, services etc, in other words different variations of taping, jointing, stapling and sill sealing were all prioritized in the construction. This was because of the frequency of problems that arise regarding these particular points with respect to air tightness of a building. The choice of sealing products was governed by those that are readily available in Sweden. Interviews were also conducted with test and construction representatives from Skanska, NCC and Wäst bygg to ascertain which were the most commonly used air tightness solutions used in the industry. This was complemented with information collected from building site visits. The sealing products used were also ones that were described in the manufactures product information as suitable for this type of construction, and that the material could withhold it's durability up to to the 80 °C temperature used in testing.

The building framework was made from wood with fiberglass insulation. The walls were sealed with polythene plastic sheeting, and the joints taped with polythene tape. The floor was in situ cast concrete. At the sill level the polythene sheet was clamped between the inner and outer sill and folded under the inner sill. Sill insulation or joint sealant was applied between the concrete floor and the polythene sheet.

At the eaves the roof plastic sheet was drawn down behind the wall plastic sheet with an overlap of 0.5m. The sheet joints were taped at the top plate and clamped using battens and

inner wall studs. Alternative air tightening solutions were used on walls A to D, see attached figure 1.
The air tightness solutions used are explained in figure 2. Service pipes were mounted through walls B, C and D to test alternative sealing methods around services. The pipes used were:

- Zinc plated ventilation canal, marked in grey in Figure 2.
- PVC-free conduit for electricity, marked in red in Figure 2.
- PEX-pipe, used usually in heating systems, marked in green in Figure 2.
- PP-pipe, used usually in heating systems and drainage, marked in blue in Figure 2.
- Copper pipe, used in the test for chemical interest, and is most often clad in insulation, marked in yellow in Figure 5.

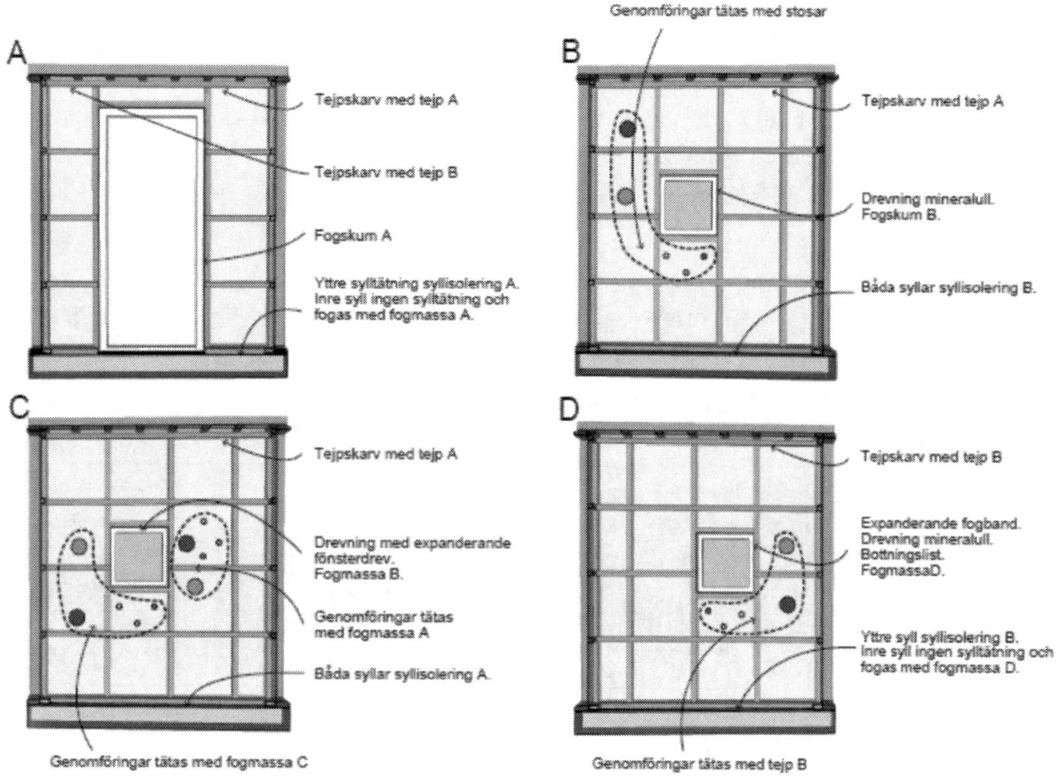

Figure 2. Air tightness solutions for walls A, B, C and D.

Three windows and one door were mounted in the test room, each with a unique individual solution for air tightening.
There are several ways of fixing the plastic sheeting around the windows. For the purpose of this testing it was chosen to fix the sheet in place on the timbers around the windows with double side butyl band. Therefore the air tightening between the window frame and the wall was not protected by the polythene sheet.

Testing and measurement

As a time horizon the durability was chosen as a minimum of 50 years, which roughly speaking was assumed to be equivalent to one year at a temperature of 80 °C and relative humidity of around 50 % in intended use. With the assumption that an increase in temperature of 10°C doubles the acceleration rate of ageing, then one year at 80°C roughly equates to a thermoxidative degradation of the material over a period of 50 years at 20°C. The relative humidity was reduced for one week to 30% to simulate the inside moisture content under the winter season. Heat in the test room was provided by two sauna heaters and a small hole was made in the wall for low ventilation so that emissions from the materials could be aired away. Temperature and relative humidity were continuously logged with a sensor placed in the middle of the room. A fan was mounted internally in order to reduce temperature variations in the room. The temperature was increased gradually after the start of the test to release any tension in the materials.

Visual inspections were carried out on a regular basis to understand how the materials were affected by the climate under the test period.

Pressure testing was carried out on the test room before and after ageing to see how much air tightness had been affected. An air speed sensor was used at the same time as the pressure testing to find any specific air leakage locations.

When the test was complete the room was dismantled and the individual tightening solutions removed and examined visually for any changes that have occurred under the 50 year period, for example, is the material still in position, has it broken down or has it had any effect on the other materials surrounding it. Specific areas we were looking at were if:

- The materials were still in position and were not loose.
- The elasticity of the materials has degraded so much that some cracks have appeared and/or have released from surrounding materials.
- The materials have shrunk resulting in a decrease in airtightness.

Free standing samples in the test room

Test samples were placed in the room of different sealing products which could be tested for tensile strength and shear after completion of the ageing process. 30 x 40 cm test specimens of the same type and material were also constructed with joint dimensions 12x12x50 mm.

Table 1. Products tested as free standing samples in the room.

Nr.	Product	Description	Test type	Evaluation
1	Tape A	Single sided tape	Film joint	Shear test
2	Tape B	Single sided tape	Film joint	Shear test
3	Spigot	Rubber spigot with single sided tape	Film joint	Shear test
4	Butyl tape	Double sided band of butyl rubber	Film joint	Shear test
5	Joint sealant A	Butyl sealant in a tube	Film joint	Shear test
6	Joint sealant B	Silane polyurethane polymer	Concrete/concrete joint	Tensile strength
7	Joint sealant D	Moisture curing MS polymer	Wood/wood joint	Tensile strength
8	Joint foam A	Moisture curing polyurethane foam	Wood/wood joint	Tensile strength

The constructed joint samples were placed and subjected to the ageing process in the test room in order to evaluate and compare them using the tensile tester with reference samples.

After exposure the joints/sealants were tested for tensile strength or shear. The evaluation of taped joints, tape against film was carried out according to SP-method 1380, third edition for materials 1-5.

Evaluation of joint sealants/joint foam was carried out according to SP-method 4372, edition 2.3 for materials 6-8.

On site measurements in existing houses

Testing was executed in older properties in order to see how air tightness is affected by time. One of the requirements for choosing which houses should be tested was that the houses were already tested for air tightness when new plus the method and documentation were still available, as well as any relevant leakages. It was therefore necessary to go through all the documentation from measurements carried out by SP on primarily small houses. For buildings made up of apartments it is very difficult in practice to measure air tightness for the whole building since all the doors of the apartments need to stand open when measurements

are taken. It has not been possible either to find sufficiently reliable documentation from earlier measurements for any apartment blocks. This means that it is only small houses that have been investigated.
A request was sent to several house owners that had a suitable test house, and six owners decided to allow testing.
Testing of the buildings climate envelope was carried out according to Europe standard EN 13829:2000.
A Minneapolis fan equipment Blower Door was used for the measurement of the buildings air tightness. A survey of air leakage at around 50 Pa inside pressure in comparison to outside pressure was undertaken. Negative pressure was formed by use of the fan to measure the buildings airtightness, plus air speed sensors and thermal cameras were used to find air leakages.
Building extensions have been included in the testing which has meant that in three cases the envelope area has been corrected. This also means that changes can have been made in the original building envelope which can give another source of error. Optimally it would have been best if no changes had been made to the houses measured. However, we had only these six houses available so measurements were carried out even if the house had an extension.

3 RESULTS

3.1 Results from the test room

Air leakage through the building envelope at 50 Pa pressure difference calculated according to EN 13829 was 0,11 l/(sm²) before ageing and 0,22 l/(sm²) after ageing.
Most of the solutions have maintained their air tightness after a representative ageing period of 50 years. Solutions which have resulted mainly in a deterioration in air tightness are tape A, tape B, spigot tape, sill insulation B plus sealants C and D.
Solutions using tape on the plastic sheeting have become less airtight because of channels formed in the tape, most likely caused by differences in shrinkage rate of the two materials, see figures 3 and 4.
A similar phenomenon occurred with the only sill insulation of rubber seals which were glued to a plastic strip, see figure 4. Tape joints in the eaves had also similar channels as the tape in other sections, but the leakage wasn't as much here because the joint was clamped by the studs and battens.

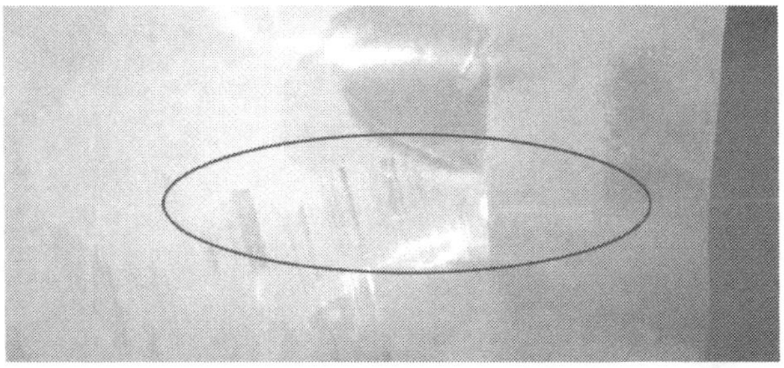

Figure 3. Air channels forming in the tape joint

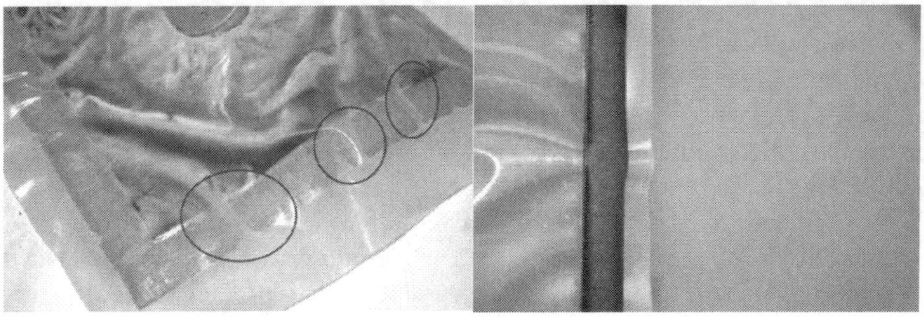

Figure 4. Air channels in the spigot tape (on the left) and sill insulation (on the right)

It is worth mentioning that the butyl band used in this test also creased together with the plastic sheet, but it was very well fixed to the plastic sheet and no air channels were formed, see figure 5

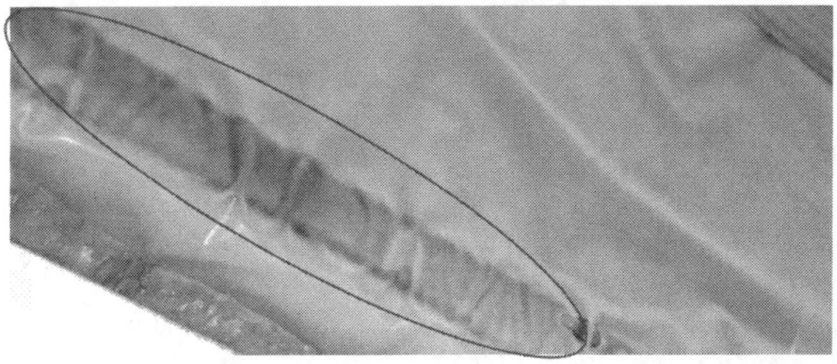

Figure 5. Butyl band had creased together but maintained an airtight joint.

The joint sealants used in this test which became dense and hardened homogenously remained so, whilst other less dense sealants had set in an uneven manner, with hard sections and soft sections..

This can be caused by the less dense joint sealants do not harden or cure sufficiently quickly and thus are not able to cope when the building sections move , with cracks occurring as a result, see figure 6.

Besides the results from the free standing samples, the visual inspection of the joints doesn't appear to show any chemical reaction with the different building materials, such as wood, concrete, frame paint and the service voids. It appears to be that it is the chemical and mechanical factors in the products themselves which lead to a decrease in air tightness, such as movements in the material and shrinkage (dimensional stability). The results from all the samples are documented in appendix 2.

Figure 6. Joint sealant around a window.

During testing the materials were checked about once a month for any noticeable changes. Channels and cracks were first noted from between one to three months, however, not from the sill insulation as this was a hidden construction. A deterioration of air tightness occurred amongst the tested products before completion of the ageing of 50 years, or not at all. One to three months of accelerated testing represents an ageing time frame of four to 13 years. Furthermore, the size of the channels and cracks increased gradually as the test progressed.

3.2 Summary of results from free standing samples

The results from the free standing samples are summarized below in table 2. The requirements that the products have been compared with in this case is described in the report. The results must be compared with the visual results obtained from the in-built products of the test room as well as the air tightness measurements of the room to obtain an overall picture. The results and requirements must also be compared with the installation method and form of the tested products. The results refer only to the tested properties of the tested materials. The raw data from the tensile strength and shear testing is reported in appendix 3.

Table 2. Summary of the results from the standardized tests of the smaller test samples

Product	Fulfils requirements of SP-method 1380/SP-method 4372
Tape A	Yes
Tape B	Yes
Rubber spigot with tape	No
Butyl tape	Yes
Joint sealant A	Yes
Joint sealant B	No
Joint sealant D	No
Joint foam A	No

Materials 1-2 and 4-5 fulfil the requirement for shear strength and appear to be "visually vapour tight" with no noticeable channels, cracks or damage. Material 3 did not fulfil the requirement, with damage and air channels clearly visible.
Materials 6-8 did not meet the requirements. Joint sealant D was already "broken" after the ageing process and before tensile strength evaluation. Joint foam A did not meet the requirement for air tightness with 3 mm elongation, and also the maximum tension requirement was not fulfilled >0.3MPa after exposure.
Joint sealant B had a reduction in elongation of 38 % at maximum peak force compared with the result from an unexposed sample of joint sealant.

3.3 Summary of results from on-site measurements

The results obtained for air tightness of the investigated houses is shown in table 3 below.

Table 3. Summary of investigated houses.

House	Year built	Air leakage when built [l/(sm²)]	New measurement of air leakage 2011-2012[l/(sm²)]	Changes made to the building/climate envelope
1	1990	0,14	0,95	Yes
2	1990	0,17	0,21	No
3	1993	0,92	1,54	Yes
4	1990	1,11	1,05	Yes
5	1990	0,64	0,57	No
6	2001	0,25	0,23	No

Houses 1, 3 and 4 have all had extensions which mean a change to the original climate envelope. The air tightness of houses 1 and 3 has decreased substantially since the construction of their extensions; however, the same result was obtained for house 4 as for 22 years ago. The reduction in air tightness in two out of the three houses that have had extensions can point towards variations in construction techniques and/or the level of precision used I carrying out these changes to the building envelope.

No changes to the climate envelope have been made over the years to houses 2, 5 and 6, and the results are roughly the same as for 10-22 years ago.

Fewer areas of poor air tightness were noted from house 6, which is 10 years younger than houses 1-5. We cannot explain exactly why this is so, but as house 6 was built as a passive house and therefore with the awareness of building airtight, then this can be a contributable factor for the better result.

To summarize, changes have been made to the climate envelop in three of the six houses, two of which (house 1 and 3) showed an increase in air leakage. In the remaining three houses no changes have been made to the climate envelope and they have maintained their original air tightness.

4 DISCUSSION

After the measurement of air leakage in six existing houses, it has been shown that it is possible to build a house without risking the durability of its air tightness. Regarding building extensions or renovations then there is a large risk that the new section doesn't come up to the same standard as the older section. This is not a surprising result, but it is worth pointing out because at some stage or other many people will extend or renovate their houses.

It would be of great value from this project if there were a few pointers as to how build a comparably airtight extension on an existing property. Above all the jointing between the old and new construction can be difficult to make airtight. The tested houses were all less than 50 years old, mainly because there were no older houses that had sufficient data to enable an evaluation of how the air tightness had changed over the years. We cannot be sure that 50 year old buildings could have contributed anything to the results or summaries as the construction techniques were quite different from the airtight solutions that are used today. It is also not possible to know exactly which solutions were used in these houses as it would involve a destructive testing of the construction.

The results from the accelerated ageing does show however that the construction material used and application methods can have a great influence on the durability of the air tightness of that building.

At the start of the 90's it was most usual to clamp the plastic sheeting between the wooden sections that make up the wall, whereas today more specialized sealing products are used.

We have not investigated in the laboratory the potential migration of volatile agents from the materials used in the test room (compatibility test). However, it is possible to evaluate each material individually to rule out "infection", and not least the joints in the vicinity of all the other materials in the test room. These migrations can both increase and decrease a material's durability properties.

We have not looked at other properties such as settlement, relaxation etc. Therefore this project must be seen as a screening of the materials involved and not the absolute truth. All the products tested in the test room could not, because of practical implications, be tested as per the standardized test methods. It is worth noting though that there was a difference in the results of the standardized tests of smaller test objects and the results from test room with regard to joint foam A, joint sealant B, tape A and tape B. There are two possible explanations as to why the joint foam and joint sealant failed the standardized tests but passed when in the test room. Firstly, only visual inspections were carried out in the test room whereas the

standardized tests requires mechanical loading. Secondly, more of the material was used in the test room, which makes it harder to form cracks through the material. In the standardized tests there is an in built safety margin to ensure that the products make the grade, so it is of no surprise that certain materials fail the standardized test but make the grade in the test room.
However, the properties of the tested tapes clearly deteriorated after the ageing process, but passed the standardized tests. The most likely explanation for this is that the smaller sample lengths do not give rise to air channels and are more dimensionally stable in comparison to longer joint length samples.
Furthermore, the plastic sheeting was clamped fast on the internal wooden framework of the walls of the test room, and as a result cannot "move with" when the tape shrinks, whereas the movement of the test joints was not restricted.
The laboratory samples tests that have been carried out in the scope for this project is just a screening and is not suitable for every product type. The laboratory tests of the different tape products in this testing show that they are not a durable solution when used in jointing, plus we cannot comment on if other tape products give a better durability than the ones tested here. With this project as a background, it would be interesting to execute further testing and study on not just tape but other joint solutions not included here. These further investigations could also be used to see if there is a requirement for the standardized methods to be developed. It can be seen to be wise though to plan the layout of the plastic sheeting so that the joints are clamped in the internal timber framework of the walls.
Since many of the products tested withstood the ageing process and kept their air tightness, then this leads us to believe that today's methods can give a life length of around 50 years. This was also apparent with the houses tested with unchanged building envelopes which had maintained their air tightness up to 22 years after construction. One should however when choosing materials be sure that they have been tested for durability.

5 CONCLUSIONS

Based on the results obtained from this project, the following conclusions can be drawn:

- It is possible to construct buildings using plastic sheeting as a vapour barrier and maintain its air tightness. Testing of actual buildings have shown that this solution is durable for at least 20 years, and the laboratory tests show that if the right materials are used then a durability of 50 years is possible.
- There is a large risk that the air tightness of a building will decrease if the building is redeveloped or extended.

There has been only a small amount of each product type used in the laboratory testing, so no general conclusions can be drawn from this, only that the results are applicable only to the tested products. The tested products do show however:

- The chosen tape products pass the durability tests for smaller test samples, but were not airtight in the test room with larger joints. This can mean that the methods used for the testing of tape should be reviewed, as well as the methods for how the tape is used.
- The chosen joint foam products failed the standardized durability testing but maintained their air tightness in the test room.
- Some of the joint sealants became more airtight through accelerated ageing/hardening whereas others had poor durability. There was a difference here also between the results obtained from the standardized tests and the results from the test room. These differences can be attributed to the thicker joints used in the test room plus the in-built safety margins of the standardized tests.

- The durability of one sill insulation became worse after the ageing process, with a similar property change as that which occurred with the tape – air channels form because of different dimensional stabilities in the different materials in the air tightening solution.
- There are no indications that the materials used in the test room lead to inferior durability of air tightening products under normal building conditions. However , the effect of large mechanical loading for example wind load, or the impact of free water on all the materials has not been investigated

6 REFERENCES

[1] SPF Sveriges Plastförbund, *SPF Verksnorm 2000 utgåva 1*, 1979

[2] SPF Sveriges Plastförbund, *SPF Verksnorm 2000/2001 utgåva 2*, 1992

[Emmerich 2005] Emmerich S., McDowell T. and Anis W., 2005. *Investigation of the impact of commercial building envelope airtightness in HVAC energy use*. NIST National Institute of Standards and Technology, NISTIR 7238, USA

MULTI-PIPE EARTH-TO-AIR HEAT EXCHANGER (EAHE) GEOMETRY INFLUENCE ON THE SPECIFIC FAN POWER (SFP) AND FAN ENERGY DEMAND IN MECHANICAL VENTILATION SYSTEMS

Łukasz Amanowicz[1], Janusz Wojtkowiak[2]

1 Poznan University of Technology
Pl. M. Skłodowskiej-Curie 5
60-965 Poznań
**Corresponding author:*
lukasz.amanowicz@put.poznan.pl

2 Poznan University of Technology
Pl. M. Skłodowskiej-Curie 5
60-965 Poznań

ABSTRACT

The energy efficiency and energy consumption of mechanical ventilation systems depend mainly on the heat and cool recovery efficiency and the operational costs of electric energy for air handling unit fans.
For free pre-heating of fresh air in winter and pre-cooling in summer and to protect the heat exchanger in the air handling unit against freezing earth-to-air heat exchangers (EAHEs) are used. For large demand of fresh air multi-pipe systems are used to diminish total pressure losses and provide required amount of thermal energy.
Total pressure losses in EAHEs and consequently the specific fan power (SFP) depends on their geometrical parameters such as: the length, diameter and number of parallel pipes, angle of connection parallel pipes to the main pipes and main pipes diameter. In this paper an influence of the angle of connection between main and parallel pipes (45 and 90 degrees) on the SFP factor and energy demand for ventilation system operation was investigated using experimentally obtained total pressure losses of EAHEs models. Results are shown in graphical form. Quite significant influence of the investigated angle of connection on the fan power, and on the SFP value is presented.

KEYWORDS

earth-to-air heat exchangers, pressure losses, specific fan power SFP, energy demand

1. INTRODUCTION

Increasing energy costs contributed to the great development of energy efficient HVAC systems (heating, ventilation and air conditioning) for buildings. For well insulated and tight buildings the mechanical ventilation system with heat recovery are often used to diminish the energy demand for the building. In cold and moderate climate in order to avoid freezing of plate-type heat exchangers in air handling units earth-to-air pipe-type heat exchangers (EAHEs) are used (Jacovides C.P., Mihalakakou, 1995 and Lee, Strand, 2008 and Bansal, Misra, 2009 and 2010).

EAHEs enable to obtain additional heat or cool gains because of the relatively stable soil temperature at a depth of about 2 m during the whole year. Unfortunately the operation of EAHEs is always connected with additional pressure losses and additional electric energy consumed by air handling unit supply fans. Total pressure losses depends on the heat exchanger structure and airflow. For higher airflow the multi-pipe EAHE structures are used (De Paepe, Janssens, 2003). The experimental investigations (Amanowicz, Wojtkowiak, 2010) show that one of the structure parameter influencing the total pressure losses in multi-pipe EAHEs is the angle of connection between main pipes and parallel pipes. In this paper results of experimental investigations into the influence of the angle of main and parallel pipes connection (45 or 90 degrees, see Fig 1) on the total pressure losses of earth-to-air multi-pipe heat exchangers and specific fan power (SFP) are presented. Fig. 1 shows the investigated multi-pipe Z-type 45^0 EAHE structure.

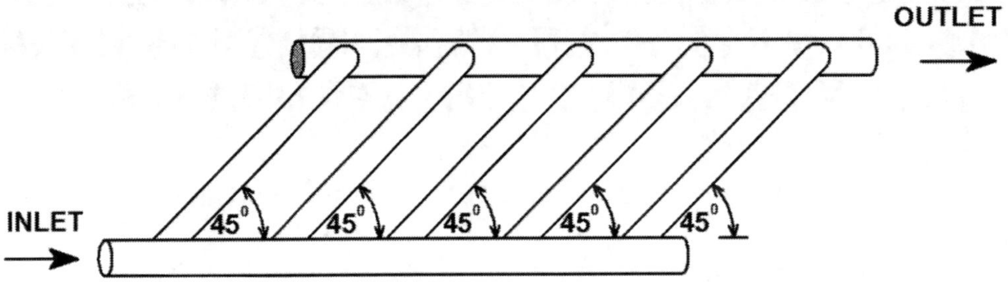

Figure 1: Multi-pipe earth-to-air heat exchanger, Z-type structure, $\alpha = 45^0$ structure

2. EXPERIMENTAL INVESTIGATIONS

Total pressure losses and total airflow were measured using multi-pipe earth-to-air heat exchanger models made in a scale 1:4 from PCV pipes of a diameter DN50 (d = 0,0461m) and length L/d = 76 for exchangers with 3, 5 and 7 parallel pipes connected to the main pipes with the angle of 45 or 90 degrees. The internal diameters of the main and parallel pipes were the same. Fig. 2 shows the schema of experimental set-up, Fig. 3 shows the view of the experimental set-up.

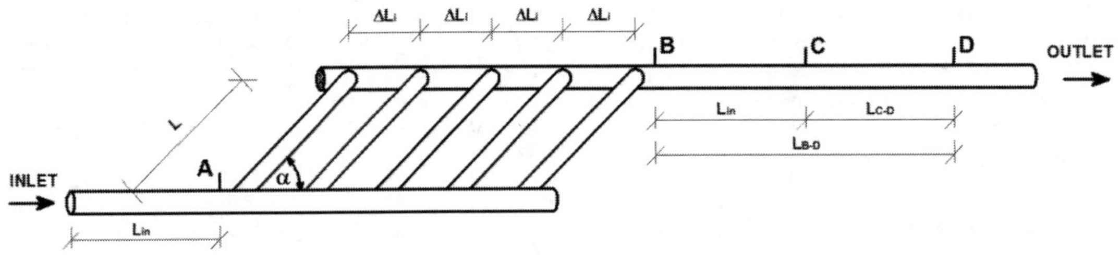

Figure 2: The schema of experimental set-up

Figure 3: View of the experimental set-up: Z-type, 7 pipes 90^0 EAHE model in a scale 1:4

Measured values:

$\Delta p_{A\text{-}D}$ – pressure losses at the measuring sector between points A and D [Pa] (Fig. 2)
$\Delta p_{C\text{-}D}$ – pressure losses at the measuring sector between points C and D [Pa] (Fig.2)
$L_{C\text{-}D}$ – length of measuring sector between C-D points [m]
$L_{B\text{-}D}$ – length of the sector between B-D points [m]
L_{in} – the developing flow region of about 20-30d, [m]
ΔL_i – distance between parallel pipes, [m]

Calculated values:

V_{tot} – total airflow as a function of friction pressure losses at the C-D sector (fully developed flow), calculated from equation (1) based on Blasius formula and Darcy-Weisbach equation,
$\Delta p_{A\text{-}B} = \Delta p_{EAHE}$ – total pressure losses on EAHE, calculated from equation (2),
$k_{EAHE,Vi}$ – total pressure loss coefficient for the EAHE for V_i = maximum or minimum value, calculated from the equation (3),
k_{EAHE} – mean value of EAHE total pressure loss coefficient calculated as an average of $k_{EAHE,Vi}$ for minimum and maximum airflows (equation (4)).

$$\dot{V}_{tot} = 3600 \cdot \left(\frac{2\Delta p_{C-D} \cdot d^{1,25}}{0,3164 \cdot \rho \cdot L_{C-D} \cdot \upsilon^{0,25}}\right)^{\frac{1}{1,75}} \cdot \frac{\pi \cdot d^2}{4} \; [(m^3/h)] \quad (1)$$

$$\Delta p_{EAHE} = \Delta p_{A-D} - \frac{L_{B-D}}{L_{C-D}} \cdot \Delta p_{C-D} \; [Pa] \quad (2)$$

$$k_{EAHE,Vi} = \frac{\Delta p_{EAHE,Vi}}{\frac{\rho \cdot w_{tot,Vi}^2}{2}} \quad (3)$$

$$k_{EAHE} = \frac{k_{EAHE,Vmin} + k_{EAHE,Vmax}}{2} \quad (4)$$

w_{tot} – average air velocity in main pipes, [m/s]
d – internal diameter of main pipes and parallel pipes, [m]

Minimum and maximum airflows were specified for the range of Reynolds number from 20 000 – 80 000 (typically values of Re for such systems). The mean value of the total pressure loss coefficient valid for the above mentioned range of Re number results in only 5-10% uncertainty, because of quite stable value of $k_{EAHE,Vi}$ (turbulent flow) what is shown in Fig. 4. Re_{tot} was calculated from equation (5).

$$Re_{tot} = \frac{w_{tot} \cdot d}{\nu}, \quad w_{tot} = \frac{\dot{V}_{tot}}{\frac{\pi \cdot d^2}{4}} \text{ [m/s]} \tag{5}$$

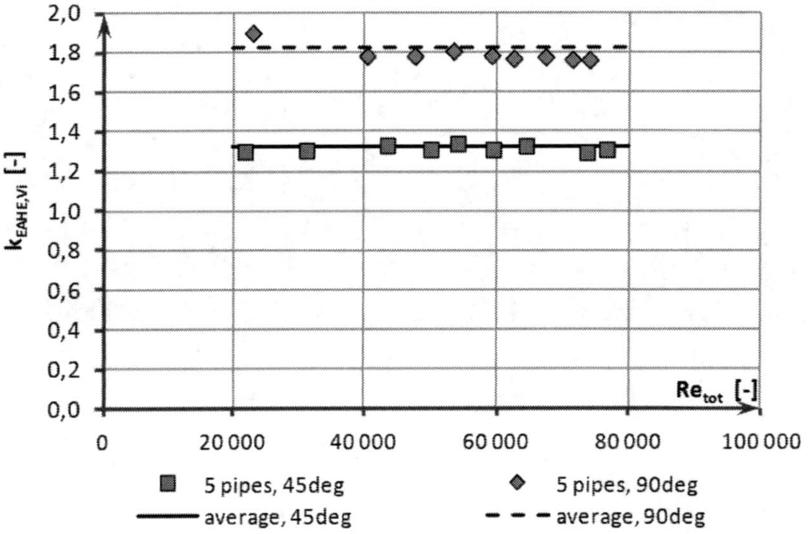

Figure 4: Total pressure losses coefficient $k_{EAHE,Vi}$ as a function of Reynolds number for exchanger with 5 parallel pipes

The mean values of total pressure loss coefficients for considered exchanger structures are listed in table 1.

Table 1: Experimental average k_{EAHE} coefficients for heat exchangers of $L/d = 76$, $\Delta L_i = 6d$

Number of pipes	k_{EAHE}, 45° [-]	k_{EAHE}, 90° [-]
3	1,26	1,77
5	1,32	1,83
7	1,42	1,87

3. RESULTS

The specific fan power (SFP) was calculated from the equation (6) both for supply and exhaust parts of the system with the assumption of total pressure losses of whole ventilation system $\Delta p_{system,Vtot,max} = 200$ Pa for the maximum (nominal) flow rate $V_{tot,max} = 600$ m^3/h (ducts, air handling unit, supply and exhaust parts). For the system with EAHE the total pressure losses were calculated using equation (7). The results are shown in Fig. 5

$$SFP = \frac{N}{\dot{V}_{tot,max}} = \frac{\Delta p_{tot,Vmax} \cdot \dot{V}_{tot,max}}{\eta \cdot \dot{V}_{tot,max}} = \frac{\Delta p_{tot,Vmax}}{\eta} \text{ [W/(m}^3\text{/s)]} \tag{6}$$

$$\Delta p_{tot,Vi} = k_{EAHE} \cdot \frac{\rho \cdot w_{tot,Vi}^2}{2} + \Delta p_{system,Vi} \tag{7}$$

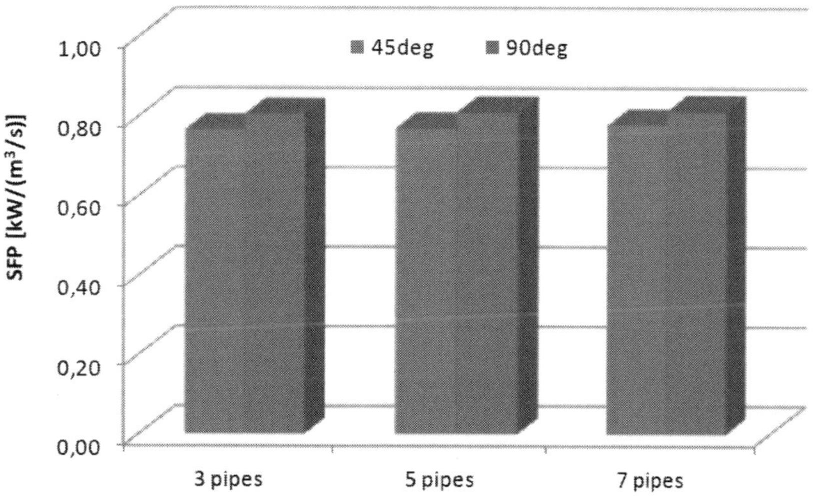

Figure 5: SFP factors for supply fan in mechanical ventilation system
with heat recovery and multi-pipe earth-to-air heat exchanger, $V_{tot,max}$ = 600 m^3/h, d = 0,184 m, L/d = 76

Calculation of a whole year energy demand for 5-pipes EAHE were done for 3 cases:

- whole system (supply and exhaust fan),
- supply only (only supply fan),
- EAHE only (only EAHE fan).

Main assumptions:

- EAHE operates during a day with various flow rates, in hours 1-6: 50%, 7-16: 100%, 17-24: 30% of maximum flow rate $V_{tot,max}$ = 600 m^3/h,
- EAHE operates 365 days per year,
- d = 0,184 m, L/d = 76,
- the total efficiency coefficient of the engine-fan-system was assumed to be constant: 0,30.

Total pressure losses of the ventilation systems for different airflows were calculated using a simplified method based on an assumption of quadratic relationship between pressure losses and air velocity (equation (8)).

$$\Delta p_{system,Vi} \approx \Delta p_{system,Vtot,max} \left(\frac{V_{tot,Vi}}{V_{tot,max}}\right)^2 \quad (8)$$

Calculation results are shown in Fig. 6. The percentage differences between SFP factors and whole year energy demand for ventilation system operation are given in table 2. The percentage differences were calculated using formulas (10) and (11).

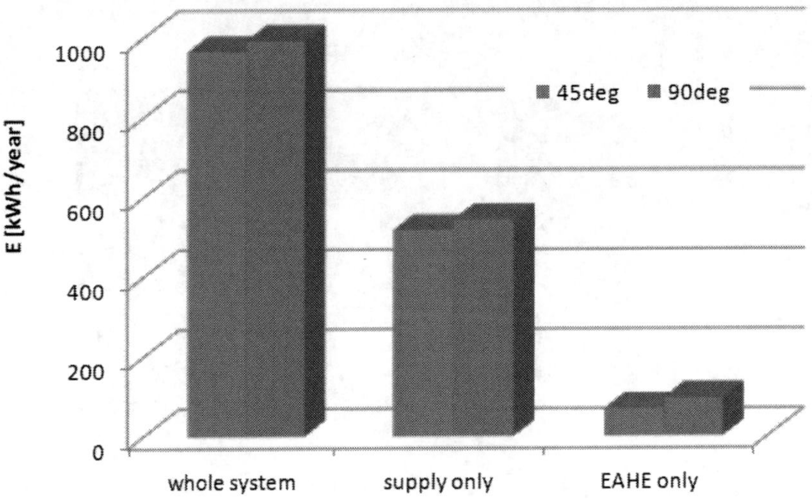

Figure 6: The whole year energy demand for mechanical ventilation system with heat recovery and 5-pipe earth-to-air heat exchanger

Table 2: Percentrage differences between SFP factors and whole year energy demand for operation of ventilaton system

Number of pipes	ΔSFP [%]		ΔE [%]
3	4,9	-	-
5	4,9	whole system	2,7
		supply only	4,9
		EAHE only	28,0
7	4,3	-	-

$$\Delta SFP = \frac{SFP_{90} - SFP_{45}}{SFP_{90}} \cdot 100\% \tag{10}$$

$$\Delta E = \frac{E_{90} - E_{45}}{E_{90}} \cdot 100\% \tag{11}$$

For mechanical ventilation system with EAHE additional stand-alone fan only for EAHE operation is used to overcome additional EAHE pressure losses. The required power of fans for considered systems is presented in the Fig. 7. Required fan power for 45 degrees structures for exchangers with 3, 5 and 7 parallel pipes is respectively: 29%, 28% and 24% lower than for 90 degrees structures. It means that there is an opportunity to apply 25-30% smaller (and cheaper) fan for 45 degrees structures of EAHE in comparison with 90 degrees structure.

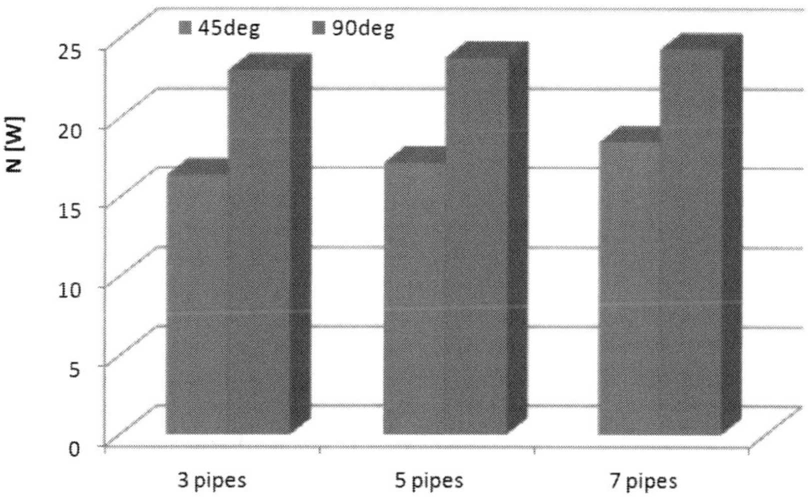

Figure 7: Required fan power for EAHEs with different angles of parallel and main pipes connections

4. CONCLUSIONS

The main conclusions are listed below:

- for the same total airflow the relative differences between SFP factor for 45 and 90 degrees EAHE are almost the same for 3, 5 and 7 pipes exchangers and equal to 4-5%,
- EAHE branches connection of 45 degrees can reduce the energy demand for the whole ventilation system operation of about 3% in comparison to 90 degrees,
- the difference in energy demand for EAHE fan for 45 and 90 degrees EAHE is 28%,
- in a case of using additional stand-alone fan for EAHE operation there is possible to use smaller (25%-30) and cheaper one for 45 degrees structures,
- EAHE structure and consequently the EAHE's total pressure losses has quite significant influence both on the SFP and the whole energy consumption of ventilation system.

5. REFERENCES

Jacovides C.P., Mihalakakou G., An underground pipe system as an energy source for cooling/heating purposes, *Renewable Energy Vol. 6, No. 9, 1995*, 893-900

Lee K. H., Strand R.K., The cooling and heating potential of an earth tube system in buildings, *Energy and Buildings 40 (2008)*, 486-494

Bansal V., Misra R., Ghansyan A.D., Mathur J., Performance analysis of earth-pipe-air heat exchanger for winter heating, *Energy and Buildings 41 (2009)*, 1151-1154

Bansal V., Misra R., Ghansyan A.D., Mathur J., Performance analysis of earth-pipe-air heat exchanger for summer cooling, *Energy and Buildings 42 (2010)*, 645-648

De Paepe M., Janssens A., Thermo-hydraulic design of earth-air heat exchangers, *Energy and Buildings 35 (2003)*, 389-397

Amanowicz, Ł., Wojtkowiak, J., Experimental investigations concerning influence of supply system of ground earth-to-air pipe-type heat exchanger on its flow characteristics. Part 1. Flow characteristics, *Ciepłownictwo Ogrzewnictwo Wentylacja 6/2010*, 263-266, 282, (in Polish)

Amanowicz, Ł., Wojtkowiak, J., Pressure losses in ground earth-to-air multi-pipe heat exchangers with connection angle of 45 degrees, *Ciepłownictwo Ogrzewnictwo Wentylacja 12/2010*, 451-454, (in Polish)

AIRFLOW MODELLING SOFTWARE DEVELOPMENT FOR NATURAL VENTILATION DESIGN

Hee Joo POH*, Meow Win TAY, Petrina Shu Hui TAY, Hoang Huy NGUYEN and Bud FOX

Institute of High Performance Computing

1 Fusionopolis Way, #16-16, Connexis North, Singapore 138632.

**Corresponding author. Tel: (65) 6419 1536; fax: (65) 6419 1580*
Email address: pohhj@ihpc.a-star.edu.sg

ABSTRACT

As the benefit of natural ventilation in reducing operational cost is well recognised, the concept of natural ventilation is becoming more received by residents and designers alike. For decades, Computational Fluid Dynamics (CFD) has been employed by the architectural and heating, ventilation and air-conditioning (HVAC) profession, as the modeling tool is able to provide detailed airflow analysis in aiding the incorporation of innovative natural ventilation concept into the building design phase. However, major bottlenecks for the widespread implementation of this tool has been the time-consuming effort required to prepare clean and compliant building geometry data for the CFD mesh generation process, as well as the high cost associated with the CFD specialists. With Singapore Building and Construction Authority's (BCA) initiative to launch the national usage of Building Information Modeling (BIM) solutions for the building industry, IHPC, together with Building Systems & Diagnostic & RightViz Solutions Pte. Ltd, has developed the Green Building Environment Simulation Technology (*GrBEST*) software to provide an integration of BIM and CFD simulation in one simple, time efficient and cost effective building airflow modeling tool.

KEYWORDS

Natural ventilation, Building Information Modeling, Computational Fluid Dynamics

1. INTRODUCTION

Singapore's current pursuit towards energy and environmentally efficient building designs requires great emphasis to be placed upon reducing the cooling load of a building. According to Singapore Energy Statistic Data, the total household electricity consumption has increased by 2% from 6,514 GWh in 2009 to 6,641 GWh in 2012 (Singapore Energy Statistics, 2012, 2013), mainly due to higher air-conditioning consumption. However, Singapore's monsoon conditions throughout the year with north or south prevailing wind direction provide an opportunity to optimize naturally ventilated facade designs and reduce the building cooling load (Wang et al., 2007). Building thermal heat gain can be minimized through a positive passive design that considers building location and orientation such as taking into consideration prevailing wind directions and the optimal planning of naturally ventilated spaces. The aim is to achieve maximum cross ventilation within built spaces and hence reduce the reliance on mechanical cooling methods. In the urban environment of

Singapore, good wind flow through the cities aid in reducing heat build-up and therefore helps to increase the thermal comfort of building occupant.

The Building Construction Authority (BCA) Green Mark Scheme (BCA, 2013) was launched in 2005 as an initiative to drive Singapore's construction industry towards more environmentally-friendly buildings and to promote sustainability in the built environment by raising awareness among developers, designers and builders at the project conceptualisation and design phase. The BCA Green Mark assessment identifies the specific energy efficient and environmental-friendly features and practices that are incorporated in the projects and points are awarded for incorporating environmental-friendly features that exceeds normal practice. Depending on the overall assessment and point scoring, the building will be certified to have met BCA Green Mark Platinum, Gold^{Plus}, Gold or Certified rating (BCA, 2013).

However, constraints associated with CFD studies have often prevented practitioners from bringing apparent value to building projects for the following reasons: 1) complexity, 2) turnaround time, 3) software cost and 4) hardware cost.

1) Complexity

Modeling and simulation tools are complex to use and usually require the domain knowledge and expertise of a CFD specialist. Building design features generally have to be simplified for CFD simulations and the validity and accuracy of these simplifications must be verified so that the results remain accurate for use. The requirement to obtain significant knowledge in a short time frame sometimes prevents architects who are involved in the design of a development from undertaking the study themselves. Moreover, engineers performing detailed CFD analyses are usually uncertain about the level of simplification required for modeling purposes without undermining the original design intent of the building.

2) Turnaround Time

The architectural design process evolves quickly and hence requires the airflow modeling and simulation analysis to be conducted in a timely manner for it to be relevant.

3) Software Cost

Commercially available software is generally costly to acquire and hence typically only well capitalized specialist companies, rather than smaller enterprises, can afford investment in applications and the necessary licences.

4) Hardware Cost

The computational time required to perform airflow simulation is often dependent on computational hardware, including, processing, memory, and data transfer resources. The larger and faster a CFD simulation, the greater the hardware cost required.

The Green Building Environment Simulation Technology (*GrBEST*) project aims to address the above constraint by developing an intuitive and cost-effective airflow modeling software for usage by the green building industry. The software enables master planners, architects, sustainability consultants, BCA Green Mark officers and general green building practitioners to perform timely CFD analyses for detailed green building conceptual design and assessment.

2. MOTIVATION AND OBJECTIVE

The *GrBEST* software is catered to comply with the BCA green mark submission criteria purpose in demonstrating a development's design of good natural ventilation. It allows users to run massively parallel computations on supercomputers for large scale computational domains. The *GrBEST* modeling and simulation software enables a seamless workflow from the early design stage, utilizing BIM data from the Autodesk *Revit* Architecture software (Autodesk Inc., 2013) to the final airflow simulation analysis.

The *GrBEST* software concept is to provide cost-effective and time-efficient CFD simulation software, through the incorporation of freely available and user-friendly third-party applications. It is targeted for use by town planners and building designers during the urban planning and early building design stages.

3. METHODOLOGY

The *GrBEST* software consists of Windows and UNIX-based components, where modeling and project management is performed on a Windows-based PC, and the more compute intensive tasks conducted on a multiprocessor UNIX-based workstation. Fig. 1 shows the flowchart diagram representing the modeling and simulation process between the Windows and UNIX machines and consists of the following key stages: 1) geometry preparation, requiring geometry conversion and checking, 2) meshing, consisting of surface triangulation and volume mesh generation, 3) flow solution, involving pre-processing, computation, and automatic post-processing, and 4) post-processing and report generation. The results of the meshing, solver and post-processing stages may be viewed on a Windows machine by using *ParaView*, an open-source multiple-platform data analysis and visualization application (Paraview, 2013).

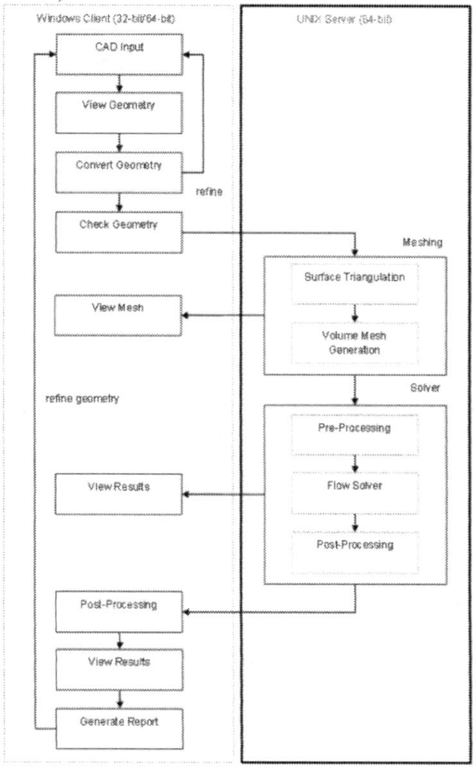

Figure 1: Windows-based client and UNIX-based server module execution workflow

3.1 Geometry Preparation

In order to conduct a CFD simulation, building geometry adhering to CFD discipline is first created using *Revit*. Buildings can only consist of walls, floors, flat roofs or wall openings. Wall thickness should be at least 0.5m. Afterwards the geometry is exported to an Industry Foundation Classes (IFC) file.

3.2 Geometry Conversion

This file is converted into the appropriate input files required of the UNIX *muSICS* system by a Geometry Converter application developed by RightViz Solutions Pte. Ltd (RightViz Solutions Private (Pte.) Limited (Ltd.), 2013). The workflow of the viewing and subsequent conversion of an IFC file to the geometrical definition, background mesh and boundary condition files required by the UNIX *muSICS* modules is shown in Fig. 2.

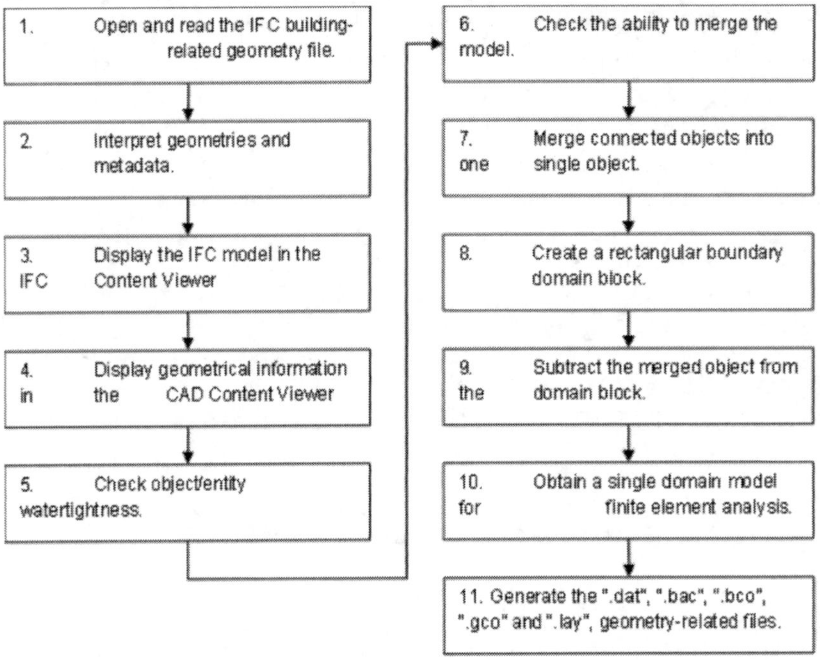

Figure 2: Overview of the IFC file viewing and conversion process

3.3 *GrBEST* Software

The computational stage involves consecutive execution of the ST, VT, Preproc, Solver, and Postproc modules, where the Solver module can perform computer number crunching using multiple processors. The Solver dialog, shown in Fig. 3, allows the user to enter the characteristic length, L (m), magnitude of the free-stream velocity, v (m s^{-1}), wind direction, number of processors and the simulation result type.

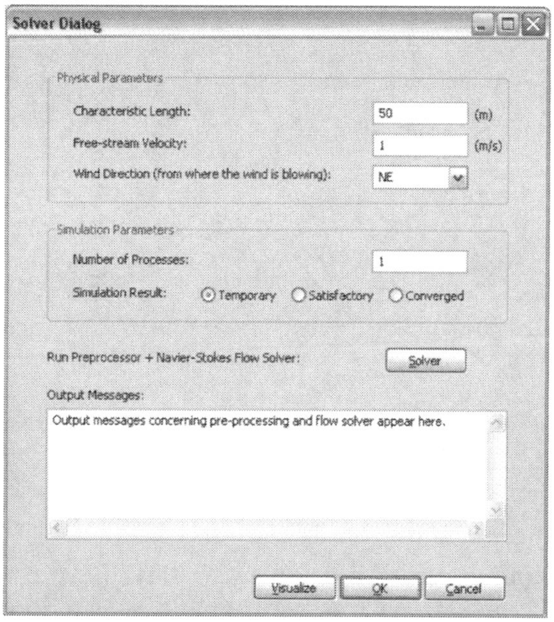

Figure 3: Solver dialog window showing default values

4. VALIDATION

The case study described in E. Simiu and R. Scalan (Simiu et al., 2011) is used as benchmark test case. The geometry is a typical building complex as shown in Figure 4. Wind tunnel studies of surface wind around this building complex have been conducted. There are three main types of surface winds: vortex flows, corner streams and through flows.

Surface winds expressed in terms of speed ratio R_H=V/V_H where V is the wind speed at pedestrian height (region A), V_H is wind speed at building height H. Dimensional analysis gives: R_H = f(L/H,W/H,D/H,H/h,Re) where Re is Reynold number. If D/H is small as in many tall buildings, then the effect of Reynolds number is insignificant.

CFD simulation using *muSICS* is depicted in Figure 4. Plots of V_A/V_H against W/H (aspect ratio) for a given L/H = 0.5, H/h = 5 on both experiment data and CFD simulations is shown in Figure 5. As can be seen from Figures 4-5, the simulation results from muSICS can capture all three types of building aerodynamic winds and are in good agreement with wind tunnel experiment data. The speed ratio R_H increases to a max with increasing W/H but changes less once W/H aspect ratio reached unity.

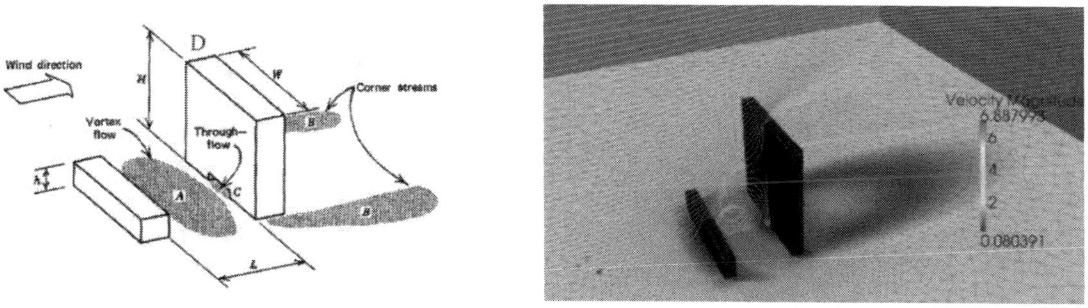

Figure 4: Geometry of building complex used for solver validation (left), CFD simulation using muSICS showing surface winds for a typical test case (right)

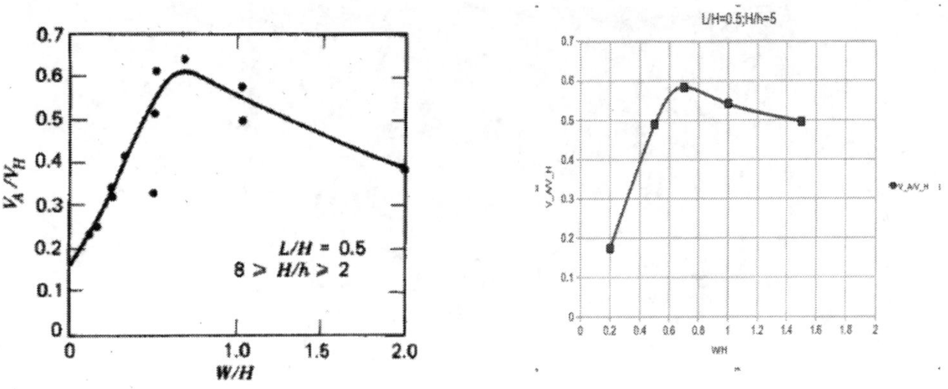

Figure 5: Speed ratio of surface wind: wind tunnel data (left), CFD simulations (right)

5. RESULTS

5.1 Natural Ventilation for External and Internal Airflow

Example 1 is presented here concerning the external airflow around a building with walls, a curved surface and a flat roof, as shown in Fig. 6. The model geometry was prepared using *Revit* to create an IFC file. The IFC file is then converted into the necessary input files for CFD simulation (Fig. 6(a),(b)) thereafter, *GrBEST* is used to perform surface triangulation followed by CFD flow solution (Fig. 6(c),(d)) using the embedded CFD solver, *muSICS*. The input parameters for the flow solver are shown in the Solver dialog (Fig. 3) and the airflow is from the northeast (NE) to the southwest, as shown by the streamlines of Fig. 6(d).

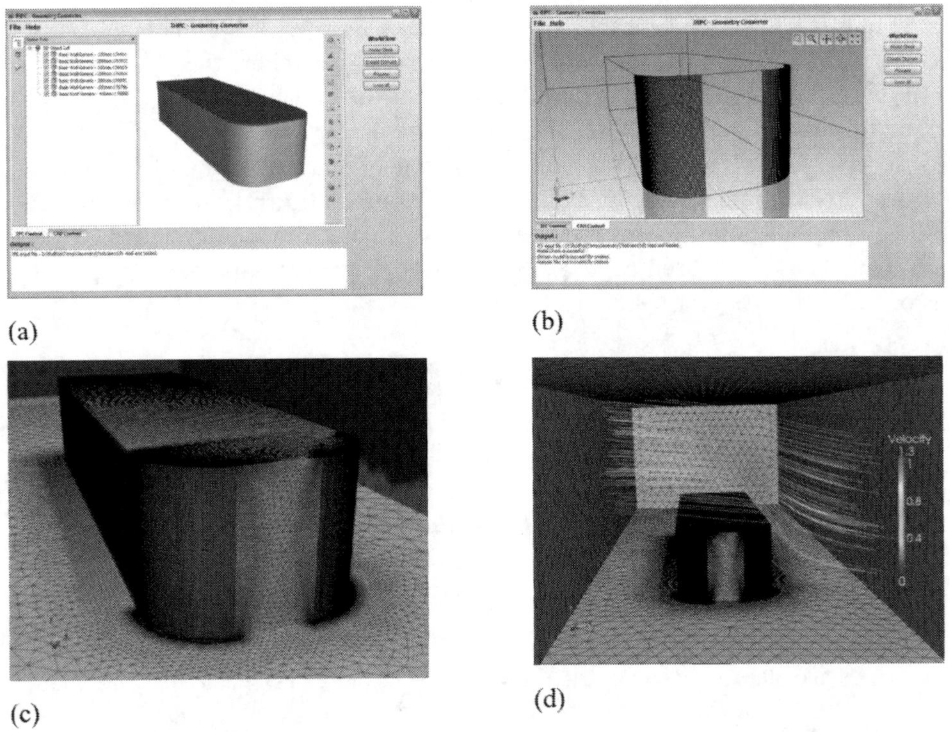

Figure 6: Simple building geometry used for external airflow simulation: (a) IFC content, (b) curve structure, (c) surface triangulation, and (d) converged flow solution

Example 2, demonstrating internal airflow is presented in Fig. 7, where the building geometry consists of two floors, external walls, one lower-level internal wall, and four openings on the front and rear walls. The workflow procedure for this example is similar to that employed in Example 1, with the only difference being that the airflow direction is from the west to the east.

A Stream Tracer filter is created in *Paraview* to visualize the airflow as a set of streamlines coloured by the magnitude of the velocity. The airflow is from the external to the internal region and the fluid flow is from the west to the east as specified (Fig. 7 (d)).

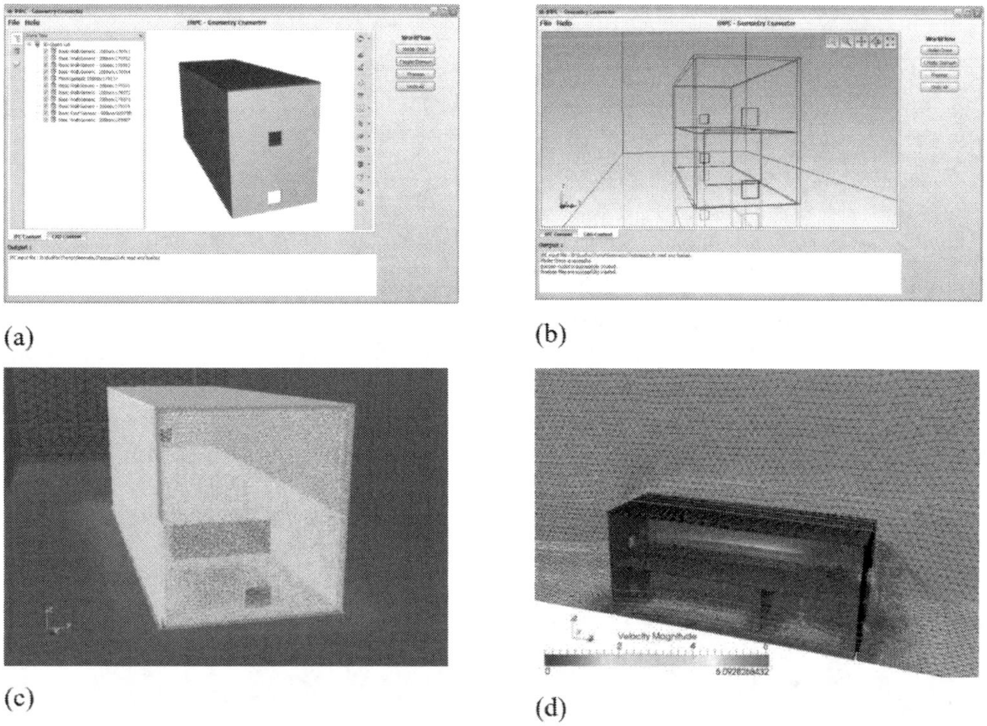

Figure 7: Two-storey building geometry with openings and an internal wall used for internal airflow simulation: (a) IFC content, (b) curve structure, (c) surface triangulation, and (d) converged flow solution

5.2 Case Study – Community Facility

An industrial project collaboration involving IHPC, BSD & RightViz with RSP Architects and Planners Pte Ltd (RSP) to generate a case study on estate airflow simulation work and evaluate the natural ventilation scenario is carried out from 1 Mar to 31 May 2014. The results are used in determining the optimal building mass, geometry, orientation and layout to achieve good natural ventilation conditions. Upon receiving the .rvt format of the model from RSP, simplification of the model to suit CFD discipline commenced with the aim of studying the natural ventilation on levels 1 and 2 in mind.

5.2.1 Geometry Simplification Steps

1. Vegetation, topology, doors, windows and basement levels are deleted.
2. Room separation lines, elevated floor, ramp, structural columns near to walls are removed. Enclosed spaces are replaced by solid blocks. Thin walls are replaced with 0.5m walls.
3. Second, third, fourth storey and roof are replaced by a solid block. Geometry of third level and roof are altered to enable successful meshing.

5.2.2 Results and Recommendations

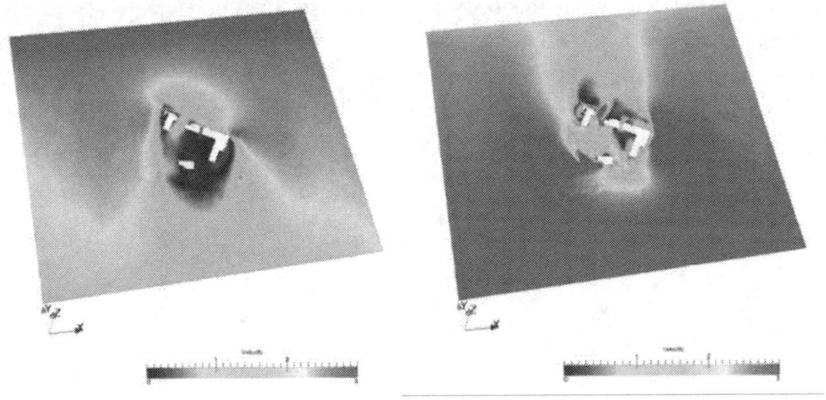

Figure 8: Simulation Result with North Wind, 3 m/s (left), South Wind, 3m/s (right)

Approximately 4.5 million cells are generated on meshing. The time taken for geometry simplification, meshing and solution generation is approximately two days. Under the present architectural design, it is observed that under North wind condition, the airflow movement on level 1 is slow or stagnant (illustrated by large regions of dark blue colour). To further improve the natural ventilation performance on level 1, it is suggested that the gap move from the present location 1, further east, to location 2 (Fig. 9). This would allow the North wind to penetrate to the central region of level 1.

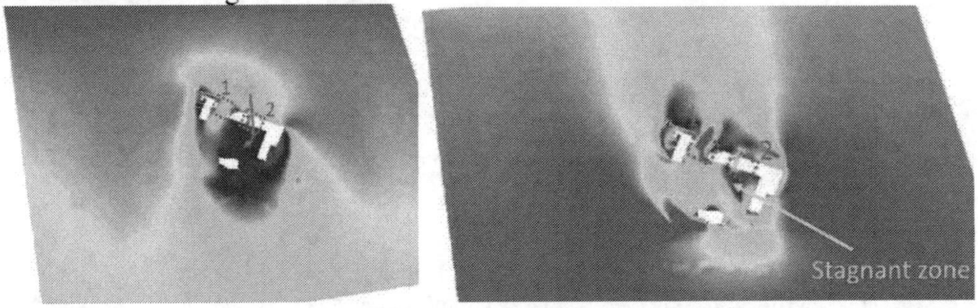

Figure 9: Recommended changes during North Wind scenario (left) South Wind scenario (right)

The overall natural ventilation performance on level 1 under South wind is better than that of North wind. If the gap is moved from the present location 1, further to the east, to location 2 (Fig. 9), then under South wind condition, it is anticipated that the highlighted stagnant zone would be eliminated.

5.2.3 Modifications

Three simulation stages are proposed by RSP. The first stage consists of the following three scenarios. In Scenario 1, the corridor is widened to 3.5m. Pantry and food preparation area remain enclosed. In Scenario 2 – Option 1, the corridor is widened to 3.5m and the walls at two ends of food preparation area are changed to 900mm high wall with opening above. In Scenario 2 – Option 2, the corridor remains at 2m and the walls at two ends of food preparation area are changed to 900mm high wall with opening above.

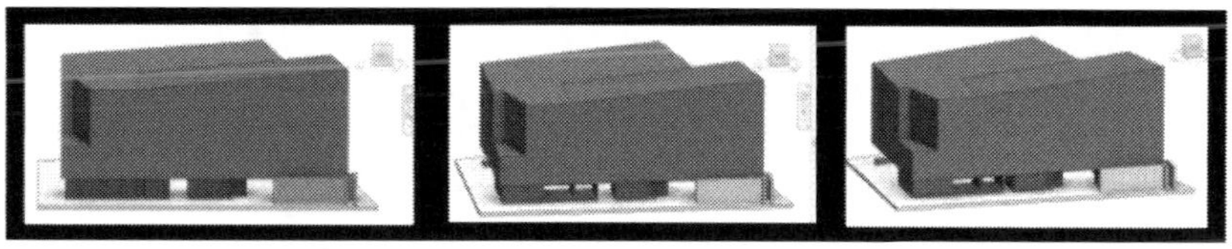

Figure 10: 3D view of Scenario 1 (left), Scenario 2 – Option 1 (centre), Scenario 2 – Option 2 (right)

In the second stage, using the geometry of the scenario which gives the best natural ventilation, that is scenario 2 – Option 1, level 2 foyer is added. In the third stage, surrounding buildings are added.

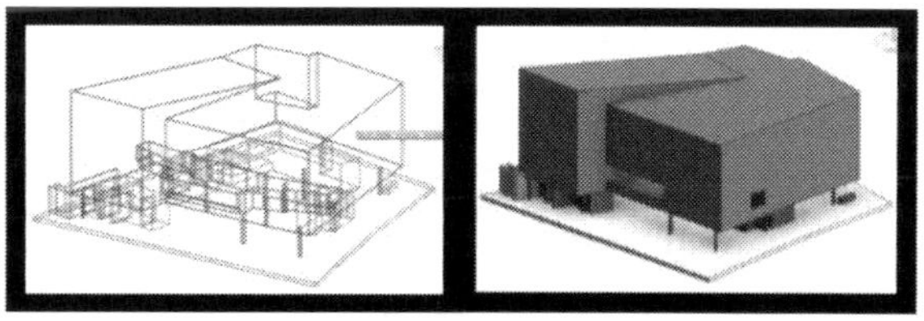

Figure 11: 3D view showing level 2 foyer

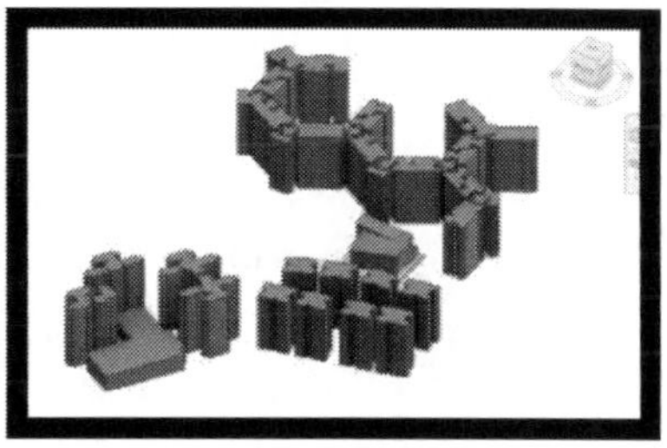

Figure 12: 3D view with surrounding buildings

5.2.4 Results

Table 1: Number of cells and mesh, solve time

	Cells	**Mesh Time**	**Solve Time**
Stage 1	4,409,997	9 min	3 hr 36 min
Stage 2	5,019,539	6 min	2 hr 14 min
Stage 3	32,475,715	1 hr 59 min	19 hr 19 min

Scenario 2 Option 1 gives the best natural ventilation scenario, as the corridor widening to 3.5m together with end walls removal are proven to be effective measure to promote cross ventilation. In this study, *GrBEST* capability to perform simulation with surrounding buildings around the site has also been demonstrated.

As compared to the current CFD software (e.g. ANSYS FLUENT, CFX, STAR-CCM+ and etc), it can be concluded the *GrBEST* can produce the estate level natural ventilation result that is of equivalent standard to the commercial and validated version. This is because our *GrBEST* estate airflow simulation is able to capture all essential features of wind aerodynamic phenomena across the buildings such as through flow, corner wind, stagnation region and vortex flow; as well as give comparable wind velocities within the natural ventilation premises. In addition, the CFD turnaround time for *GrBEST* is about 1 – 2 days, and can be rated as "fast wind modeling tool".

6 CONCLUSION

At present, the *GrBEST* software application allows the user to perform a CFD modeling and simulation to determine internal and external airflow in and around a building development. As shown in the examples and case study, *GrBEST* is able to provide airflow simulation results of simple estate building geometries with a short turnaround time, all in a single graphical user interface, simple enough to be carried out with a few clicks of a button. The case study also demonstrated the feasibility of using *GrBEST* tool to determine the optimal building mass, geometry, orientation and layout to achieve good natural ventilation conditions for a typical building project.

Moving ahead, further development work will involve improving the software to further automate the reporting and result visualization processes, and to provide an Internet-based mechanism for CFD project submission and execution on supercomputer resources. This, together with more developmental work on geometry manipulation, software development and *muSICS* computational engine extension is required to bring this software towards commercialization stage.

7 ACKNOWLEDGEMENTS

The authors would like to acknowledge Building and Construction Authority (BCA) (BCA, 2013), Singapore for the award of Ministry of National Development (MND) Core Innovation Fund (CIF) for this development work. In addition, the authors would like to thank Dr. Chiet Sing Chong from RightViz Solutions Pte. Ltd. for his work in developing Geometry Converter application and Mr. Phay Ping Tan from Building System & Diagnostics Pte. Ltd. for his CFD simulation work using commercial software. On top of that, the authors are grateful to Mr. Ngian Chung Wong and colleagues from BCA and Professor Nyuk Hien Wong from National University of Singapore (NUS) for their active involvement and technical advices during the course of the development.

8 REFERENCES

Singapore Energy Statistics 2012 & 2013, www.ema.gov.sg/media/files/publications, Energy Market Authority, (accessed in June 2014)

Wang, L and Wong, N.H., *Applying Natural Ventilation for Thermal Comfort in Residential Buildings in* Singapore, Architectural Science Review, Vol. 50 (3), pp 224-233, 2007.

Building and Construction Authority Green Mark Scheme, http://www.bca.gov.sg/greenmark/green_mark_buildings.html (accessed in Sep 2013).

Autodesk Revit Architecture (Revit), Building Information Modeling software application, http://usa.autodesk.com/revit/, Autodesk Inc., 2013.

ParaView – Parallel Visualization Application, www.paraview.org/, (accessed in 2013).

Simiu, E. and Scanlan, R., *Wind Effects on Structures: Fundamentals and Applications to Design*, Dover Publication, 2011.

REDUCING COOLING ENERGY NEEDS THROUGH AN INNOVATIVE DAILY STORAGE BASED FACADE SOLUTION

Servando Álvarez Domínguez*[1], Rafael Salmerón Lissén[1], Álvaro Ruiz-Pardo[2], José Sanchez Ramos[2], and Javier García Ramos[1]

1 Universidad de Sevilla (Seville University)
Camino de los descubrimientos S/N
Escuela Tecnica Superior de Ingenieros
Seville 41092, Spain
**Corresponding author: salvarez@us.es*

2 Universidad de Cádiz (Cadiz University)
C/ Chile,1. 11002
Cádiz, Spain

ABSTRACT

The framework of the research presented in the paper is a project oriented to promote the use of concrete solutions in buildings based on maximizing the benefits of its thermal inertia for cooling periods.

The constructive solution developed has one configuration for summer (cooling mode). This configuration is similar to a ventilated facade that is formed by a thermally insulated outer layer of concrete, an intermediate air layer and an inner layer of concrete. The inner layer is cooled at night by forced ventilation using an outdoor - outdoor scheme.

The aim of this paper is to show the preliminary results about the potential of special concrete walls as solutions to reduce energy demand in residential buildings by thermal offset in Spanish Mediterranean climates.

In summer, the concrete building facades are used as heat sinks. The aim is to cool the inner layer of concrete moving outdoor air through the air layer during night taking advantage of the low nigh-time air temperatures. The cool stored is released to the interior spaces when the maximum peak load of the space takes place.

With the aim of select a proper design of the innovative element, the influence of the thickness of the inner layer, the air velocity in the chamber and control lays to activate the ventilation are analysed. Simulations show that the use of this element is very promising for reducing energy demand in residential buildings in the Spanish climates.

Experiments are currently performing (summer 2014).

KEYWORDS

Cooling savings, Ventilated active facade, Cooling storage

1. INTRODUCTION

The framework in which this study takes place is the " Service Contract R + D + i Relating to Competence Scope of the Ministry of Public Works and Housing " with the research project

entitled "Analysis of the energy performance of closures concrete based on maximizing the benefits derived from the thermal inertia".
Meeting the "20-20-20" targets (European Commission, 2012) for reduction of CO_2 emissions necessarily involves a drastic reduction of energy consumption in buildings.
For the climatic conditions of the south of Europe, as Andalusia region in Spain, cooling is required during the summer season, since there are high daylight temperatures an high levels of solar radiation.

To reduce energy consumption in cooling and therefore CO_2 emissions, it is previously necessary to reduce energy demand.

There are several techniques that can be applied to reduce the cooling demand (Santamouris, 2007), but some of the most obvious alternatives is to use the relatively low night outdoor temperature as a heat sink. In order to use this heat sink, energy (cold) storage is required since there is a gap between the cooling needs in the building and the time when the lowest temperatures are presented in the heat sink.

One option to storage the cold is to use the thermal inertia of the structure of the building.
A simple definition of thermal inertia would say it's the ability to keep the mass of thermal energy received and gradually releasing it (Ruiz-Pardo 2008). Because of this ability, taking into account the thermal inertia of the walls of a building, can be decreased the need for air conditioning, with the consequent reduction of energy consumption and pollutant emissions.

The thermal inertia improves energy performance of buildings (Aste 2009) because it allows the damping of the temperature variation and the phase of the internal temperature relative to the outside.

For a situation with a high external temperature and solar radiation, the temperature outside the enclosure rises producing a heat transfer to the interior of the building. The evolution of the temperature of the outer face has a maximum (maximum amplitude) in a particular time of day depending on the location and orientation of the enclosure. This wave is damped outside temperature, in amplitude, crossing the enclosure, also emerging a time lag between the instants at which a peak temperature is produced. The effect of phase shift and damping allows the building to stay longer in the comfort zone without additional energy expenditure allowing free savings because they are inherent in the material (Ruiz-Pardo 2008).

The physical characteristics of the concrete gives it a high thermal inertia , which allows to predict optimum energy performance of the building in the event that this material forms the inner core (structure) and external (walls and roofs) thereof. The use of concrete as façade cladding and covered in the building:

- Reduce the energy consumption of heating.
- Softens variations in internal temperature.
- Delay maximum temperatures in offices and commercial buildings to the exit of the occupants.
- Reduces peak temperatures (maximum and minimum) and can make the air conditioning unnecessary.
- Maybe employed with nocturnal ventilation to eliminate the need for cooling during the day.
- Makes better use of sources of low temperature heating such as heat pumps for underfloor heating.

Overall the effect of thermal inertia in enclosures is a variable not usually considered in the design of the building.
In addition to its modelling difficult for designers and specifies, calculation tools were not sensitive to this parameter and the knowledge of its potential benefits have not been adequately considered by the technical and scientific community.
The main objective of the project is to parameterize the fundamental variables that characterize the thermal inertia of buildings overlooking substantially improve treatment procedures for calculating the thermal performance of buildings. This would be instrumental to value the role of concrete solutions as part of improving energy efficiency.
This setting will also allow designers can, easily, estimate the energy savings from the thermal inertia of buildings with contour and concrete structure.

2. MATERIALS AND METHODS

An experimental cell (Figure 1) was built in order to test a "thermally active wall" available to storage cold during the night and release it during the day.
The experimental cell is a reinforced concrete cubicle (3.0 x 3.0 x 3.0 m.) to which he included a special front face south to investigate several special construction solutions

Figure 1. Experimental cell

The thermally active wall (Figure 2) is a facade composed of two solid layers speared by an air ventilated layer. The inner layer is a concrete wall with high thermal inertia. The outer layer is an insulated element having low thermal transmittance. During the night the air layer is ventilated using fans located in the upper zone of the facade. During the day, the ventilation is stopped.

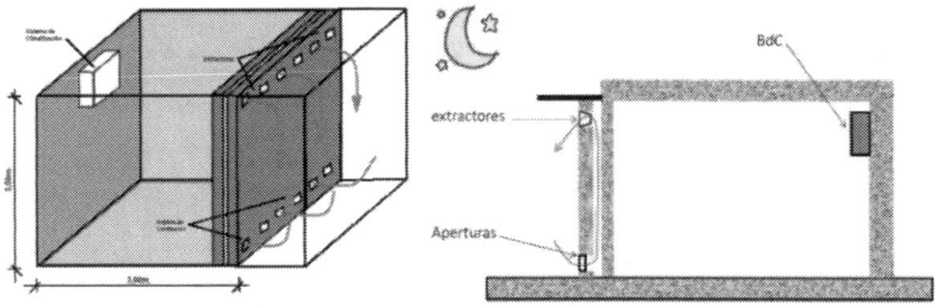

Figure 2. module test for the analysis of the influence of nocturnal cooling of the concrete wall through the chamber ventilated by forced ventilation.

A simulation model was performed with the following objectives: Dimensioning od the experimental setup, prediction of the thermal behaviour of the element under conditions different than those obtained experimentally, and to perform studies an analysis of the potential of the active facade. Thai model is based on the preiviously defined by Ruiz-Pardo (Ruiz-Pardo, 2010)

In simulation, previous to the construction of the experimental setup, the main variables affecting the thermal performance of the thermally active wall were studied: the inner layer of concrete (10-15-20 cm.), the thickness of the internal air chamber wall (2-5-10 cm.) and the air speed walking the chamber (0.5-1-2 m/s).

In the coming months the validations of the developed models will be performed using the results that currently we are obtaining.

3. RESULTS AND DISCUSSION

Previous to the campaign of experiments, some results were obtained with the aim to select a proper design and estimate the savings that can be obtained from the thermally active facade. They are shown below.

The experiments are currently in developing, so some partial results had been obtained and they are shown at the end of this section.

3.1 Simulation results

In order to select a proper design of the innovative element, several simulations were performed with different designs in two selected localities: Sevilla and Cádiz

The process performed was the following:

- Nine options were initially tested; this was the result of three air velocities in the air layer and three thicknesses of the inner wall. The results of these cases are shown in Table 1.
- From the results, the option with highest energy savings for summer was selected.
- For the selected option, some variations in the operational conditions were proved in order to maximize its behavior for winter season even though the design was optimized for summer.

For the two considered cities, it was found that the optimal design was the same. This result was not expected, since they are different climates. The only difference in the design is that for the case of Cadiz, the optimum temperature for allowing ventilation is 25°C and not 23 as in the case of Seville. The main descriptive parameters of the optimal design are shown in Table 1.

Table 1: Studied solutions to find a proper design of the innovative element in Seville

	Inner wall thickness	*Air layer velocity*	*air layer convective heat transfer coefficient (air in circulation)*
solution 5	*15cm*	*1 m/s*	*5 W/m^2K*

The overall behavior expected for the innovative element in these two cities is shown in Table 2. It can be seen that the innovative element performs cooling in summer and heating in winter.

Table 2. Gross heating gains for the innovative element in the three studied cities. Negative values means heating losses (desirable behaviour in cooling season) and positive values means heating gains (desirable behaviour in heating season). All values are in kWh/m^2

	Summer		
	month	*Monthly heat losses kWh/ m^2*	*Total heat losses kWh/ m^2*
Seville	*Jun*	*-7.57*	*-13.12*
	Jul	*-3.04*	
	Aug	*-2.51*	
Cadiz	*Jun*	*-19.81*	*-36.07*
	Jul	*-9.71*	
	Aug	*-6.54*	

The results of Table 2 show that in the two studied localities, the innovative element removes heat and in addition when it is compared with a conventional wall, its benefit is increased since the conventional wall has the opposite behaviour of the innovative element, namely, it introduces heating during the summer season.

The expected savings for the innovative element are obtained from the comparison with a conventional wall. The results of this comparison are shown in Table 3.

Table 3. Heat gains comparison between the innovative element and a conventional Wall having the same U-value in the three tested localities

	Summer		
	Innovative element Heat gains	*conventional wall heat gains*	*difference (savings)*
	kWh/ m^2	*kWh/ m^2*	*kWh/ m^2*
Seville	*-13.1*	*12.6*	*-25.8*
Cadiz	*-36.1*	*4.2*	*-40.3*

To illustrate the behaviour of the innovative element in comparison with a conventional wall, in the Figure 3, the temperature profiles are shown for the interior layer of the innovative element and for a conventional wall. These profiles were calculated for one typical day of the cooling season and for a typical day of the heating season.

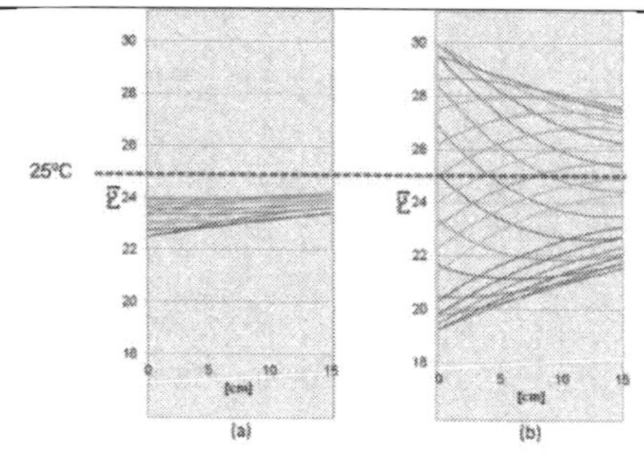

Cooling season

(a) Interior layer of the innovative element (b) Conventional wall

Figure 3. Temperature profile in: (a) the innovative element and (b) in a conventional wall

Figure 3 shows that the amplitude of the temperature oscillation obtained in the innovative element, is much lower than that obtained in the conventional wall. Additionally, it is seen how in the cooling season, the temperatures of the innovative solution are always lower than the indoor temperature (assumed as 25°C), while the conventional wall is oscillating around that temperature.

3.2 First experimental results

The first period of experiments were performed without night ventilation of the air layer. The objective was to observate the behavior of the experimental cell and compares it with the results obtained when the air layer is ventilated during the night.

In Figure 4 are shown the interior surface temperatures of the cell. The green lines are the average temperature of the conventional walls while the orange lines correspond to the surface temperatures of the active facade.

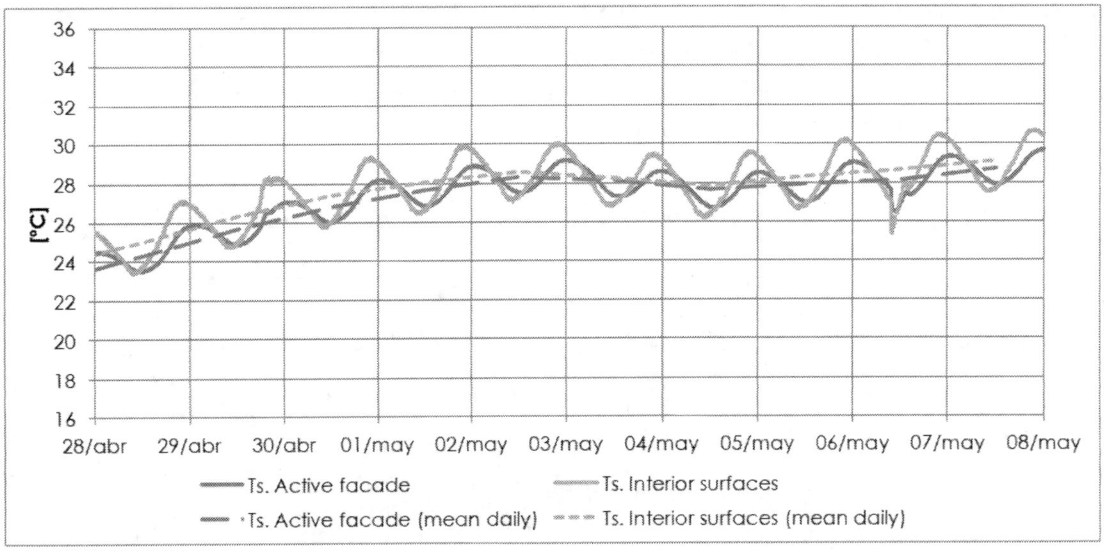

Figure 4. Temperatures of the interior surfaces of the experimental cell without night ventilation of the air layer

As it can be seen, the active façade shows les amplitude in the daily variation of temperature than the other interior surfaces. This behavior is obtained because the active facade has a greater thermal inertia. However, the mean daily temperature of the active facade is less than 0.5°C inferior than de the mean of the other surfaces when the air layer is not ventilated during the night.

In Figure 5 are shown the same interior surface temperatures of Figure 111, but with night ventilation of the air layer. In this case the daily variation of temperature is closing similar for the active façade and for the other interior surfaces. This behavior is caused by the additional excitation suffered by the active facade when the ventilation is activated. But the most important result is that the daily mean temperature of the active façade is approximately 2°C inferior than the corresponding temperature of the other walls. It is 1.5°C colder than the values obtained when the air night ventilation is not operating.

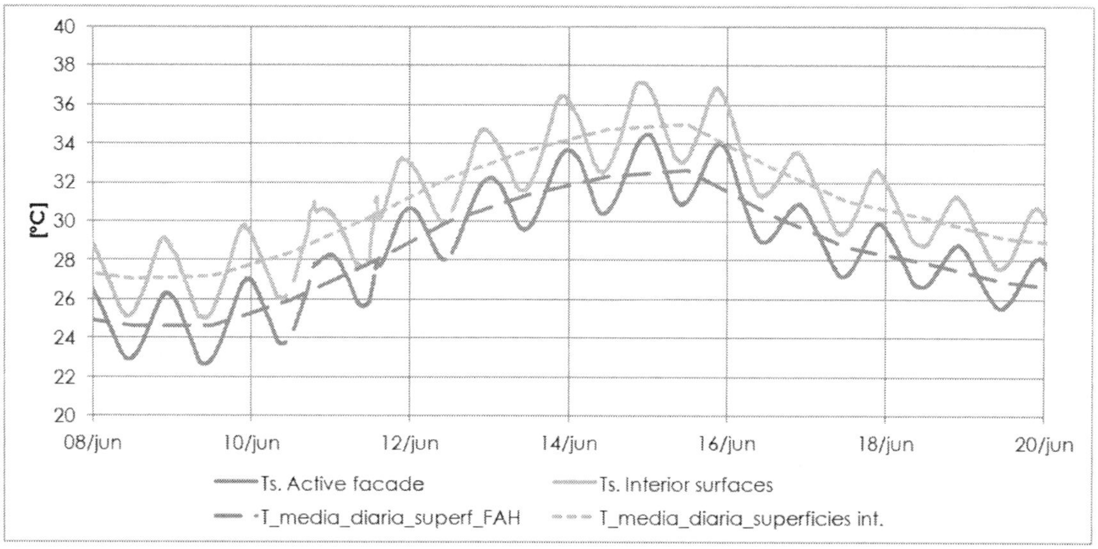

Figure 5 Temperatures of the interior surfaces of the experimental cell with night ventilation of the air layer

4 CONCLUSIONS

An innovative facade element has been designed with the objective of reduce energy cooling demand in buildings. This element is being evaluated by simulations and experiments.

Using the developed simulation model, a proper design was obtained for the cities of Seville and Cadiz. The obtained design was the same in both cases with only little differences in the adequate temperature to allow the air night ventilation.

The simulation results show that the active facade, using the selected design for the cities of Seville and Cadiz, has the potential to provide cooling in the summer season while a conventional wall provides heating. It means that the energy potential of this element shows a very interesting behaviour that has to be verified experimentally.

The very first experimental results shows a promising potential for the reduction of energy cooling consumption and seems verify the potential predicted in simulation. However, the experimental campaign is still carrying on, so definitive results and conclusions can not be taken yet.

5 ACKNOWLEDGEMENTS

The authors would like to thank the FEDER of European Union for financial support via project "Análisis del comportamiento energético de los cerramientos de hormigón en base a la maximización de las ventajas derivadas de su inercia térmica"of the "Programa operativo FEDER de Andalucia 2012-2014". We also thank all Agency of Public Works of Andalusia Regional Goberment staff and reserchers for their dedication and professionalism.
Special thanks to "Grupo Cementos Valderrivas, Alcalá de Guadaira plant" for his selfless assistance in developing the experiments.

6 REFERENCES

European Commission. (2010) Communication From the Commission to the European Parliament, the Council, the European Economic and Social Committee and the

Committee of the Regions. *Analysis of options to move beyond 20% greenhouse gas emission reductions and assessing the risk of carbon leakage*. 2010.

Santamouris,M.; Pavlou,K.; Synnefa,A.; Niachou,K.; Kolokotsa,D. (2007) *Recent progress on passive cooling techniques: Advanced technological developments to improve survivability levels in low-income households.* Energy Build., 2007, 39, 7, 859-866

Ruiz-Pardo Á.(2008) *Ahorro energético mediante el uso de elementos de doble envolvente transparente-opaco [dissertation]*. Tesis Doctoral, Sevilla: Universidad de Sevilla, ETSI; 2008.

Aste,Niccolò; Angelotti,Adriana; Buzzetti,Michela (2009) *The influence of the external walls thermal inertia on the energy performance of well insulated buildings.* Energy Build., 2009, 41, 11, 1181-1187

Ruiz-Pardo Á, Domínguez SÁ, de la Flor, Francisco José Sánchez, Lissen JMS.(2010) *Simulation model for some types of double envelope elements.* International Journal of Ventilation. 2010;9(3):227-39.

SIMULATION OF STATIC PRESSURE RESET CONTROL IN COMFORT VENTILATION

Chrysanthi Sofia Koulani [1], Remus Mihail Prunescu [2], Christian Anker Hviid [1], and Søren Terkildsen [*1]

1 Technical University of Denmark (DTU),
Department of Civil Engineering
Brovej
DK-2800 Kgs. Lyngby, Denmark
**Corresponding author: sterk@byg.dtu.dk*

2 Technical University of Denmark (DTU)
Automation and Control
Department of Electrical Engineering
Elektrovej
DK-2800 Kgs. Lyngby

ABSTRACT

Variable air volume (VAV) ventilation systems reduce fan power consumption compared to constant air volume (CAV) systems because they supply air according to the airflow demand. However VAV ventilation systems do not take fully into account the potential energy savings as the control strategy operates the terminal boxes and the air handling unit (AHU) independently without pressure integration. The pressure in the main duct is maintained at a constant static pressure (CSP) which corresponds to the pressure required under the design full load condition. Under part load conditions, the fan provides excessive static pressure which is dissipated via throttling at the terminal boxes. As a result significant fan power is wasted in mechanical energy losses. The development of sophisticated direct digital controls (DDC) creates possibilities to integrate feedback from the dampers into the building management system. In this way the operation of central plant equipment is adjusted in real time according to the actual pressure demand; this control scheme can be implemented by the static pressure reset (SPR) method. The SPR control method ensures that at least one damper remains fully opened; thus the fan generates only enough pressure to satisfy the airflow demand in the most critical zone. Consequently the airflow resistance of the ductwork is maintained at a minimum and the fan operation is optimized. There are various approaches to implement the control scheme of the SPR method; the state of the art is represented by the method of trim and respond based on pressure alarms.

This study investigates the operation of the SPR control method of trim and respond based on pressure alarms in a CO_2 demand application where large air volumes are provided to three classrooms. The investigation was based on simulations performed with a fully dynamic model of a VAV ventilation system that was developed in the Simulink programming tool which is add-on software to MATLAB mathematical programming language. The Simulink model was developed in previous research work and was built based on the International Building Physics Toolbox (IBPT), which is a library of blocks constructed for the thermal analysis in building physics. For the purpose of the current investigation the IBPT toolbox was remodelled to integrate the calculation of the airflow demand based on the CO_2 concentration occurring in the zone. The performance of the Simulink model was in previous work evaluated based on the experimental setup of a ventilation system. The investigation of the SPR control algorithm of trim and respond based on pressure alarms disclosed some issues that need to be addressed and optimized before the algorithm can effectively establish the pressure conditions that satisfy the pressure demand under high airflows. In short the algorithm must be tuned to the application beforehand or, preferably, actively learn to perform from continuous feedback before it presents a real plug-and-play solution.

KEYWORDS

CO_2 demand control ventilation, static pressure reset, modelling, Simulink, energy savings.

1. INTRODUCTION

The potential energy savings in a VAV ventilation system can be boosted by integrating the control of terminal boxes into the building management system (Hartman, 1995). In this case it is possible to implement the SPR control method, instead of operating the fan with a CSP set point which corresponds to the pressure required under the design full load condition (Wei et al, 2004). The SPR control method operates the fan with a variable pressure set point which is established based on the dynamics of the actual pressure demand occurring in the system. Since 1999 the SPR control method has been a requirement for ventilation systems equipped with terminal boxes with DDC (Taylor, 2007). In accordance with ASHRAE Standard 90.1 (ASHRAE 90.1, 2004), the fan pressure set point shall be reset based on the critical zone request. A similar requirement is included in California's Title 24 Energy Standards (Title 24, 2005). There are various approaches able to implement the control scheme of the SPR method; the state of the art is represented by the method of trim and respond based on pressure alarms as it is stable, flexible and it minimizes the impact of rogue zones (Taylor, 2007).

The objective of this paper is to investigate the operation of the SPR control method of trim and respond based on pressure alarms when the airflow demand changes in different zones in the ventilation system. The investigation is carried out by simulating the pressure and airflow conditions occurring in a system that provides air in three classrooms. The simulations are conducted with a fully dynamic model of a VAV ventilation system that was developed in Simulink (Simulink, 2013) during previous research work. The Simulink model was in previous work documented based on the experimental setup of a ventilation system that it is able to perform accurate calculations.

2 THE SIMULINK MODEL

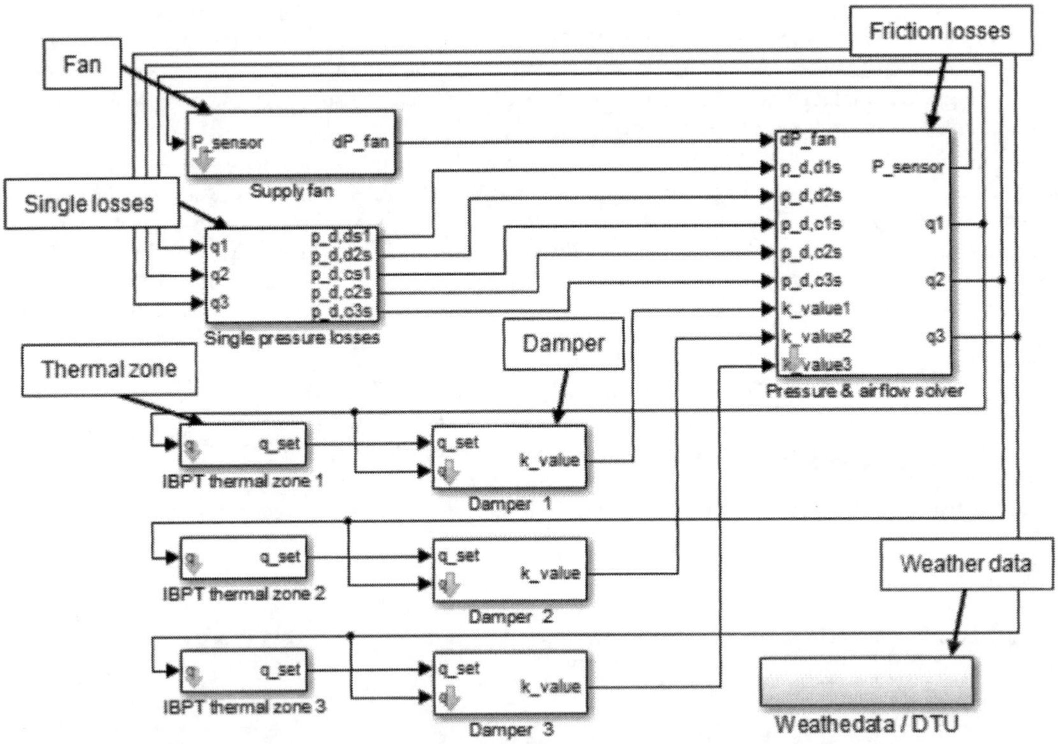

Figure 1: The Simulink model of the VAV ventilation system.

The model of the VAV ventilation system was created in the graphical environment of Simulink in Matlab (Matlab, 2013). Figure 1 illustrates the Simulink model which was built based on the blocks of the international building physics toolbox (IBPT). IBPT is library of blocks constructed for the thermal analysis in building physics (IBPT, 2012). The default IBPT blocks for the internal gains and the ventilation system were rebuilt to comply with the modelling of the VAV ventilation system (Koulani et al, 2014). A detailed validation of the model can be found in Koulani et al (Koulani et al, 2014).

2.1 The ventilation system

The IBPT ventilation system block was configured to calculate dynamically the airflow demand based on the CO_2 concentration occurring in the zone. The CO_2 concentration is approximated by equation (1) (Bekö et al, 2010) and the corresponding airflow demand is calculated dynamically by the ramp functions shown in Figure 2. The user defined data are the minimum ($q_{set,min}$) and maximum ($q_{set,max}$) airflow set point required in order to maintain a comfortable range of CO_2 concentration ($CO_{2\ set,min}$, $CO_{2\ set,max}$) in the zone.

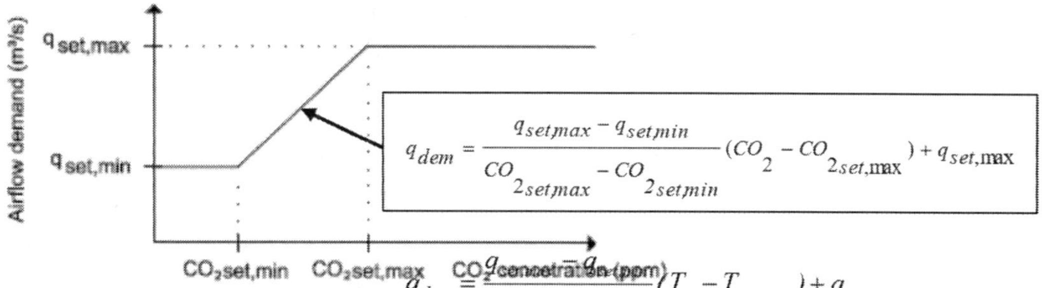

Figure 2: The dynamic calculation of the airflow demand based on the zone CO2 concentration.

$$c = (c_0 - c_i)\cdot(1 - e^{-nt}) + \frac{G}{n \cdot V}\cdot(1 - e^{-nt}) + c_i \qquad (1)$$

Where c_0 carbon dioxide concentration in the room at start, t = 0 (m^3/m^3)

c_i carbon dioxide concentration in the inlet ventilation air (m^3/m^3)

G carbon dioxide supplied to the room (m^3/h)

n air change rate (1/h)

V the volume of the zone (m^3)

t time (h)

2.2 The fan

The mathematical model of the fan represents the operation characteristics of a typical fan box ventilator. The details regarding the control of the fan are given in Koulani et al (Koulani et al, 2014). Field measurements were performed on a typical box fan with large capacity (Exhausto BESB250-4-1) in order to model a similar response. The measurements were conducted in open loop, i.e. with no feedback and aimed at identifying the built up of the fan pressure when different steps were given to the input signal (see Figure 3).

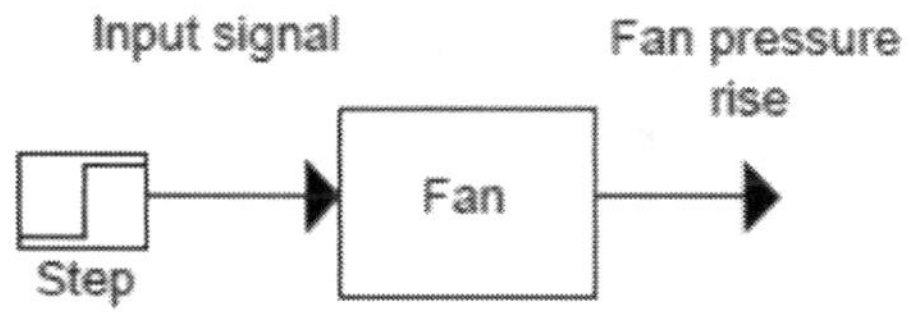

Figure 3: Setup for performing the field measurements.

The field measurements indicated that the second-order linear time-invariant (LTI) system is appropriate for modelling a fan able to provide the pressure rise required at the design full load condition. The second-order LTI system is implemented by using the state space method; the state space equations (see equation 3 & 4) calculate the derivative of the system response with respect to time (Rowell, 2002).

$$\dot{x}_1 = x_2 \tag{3}$$

$$\dot{x}_2 = \ddot{y} = \omega_n^2 \cdot u - \omega_n^2 \cdot x_1 - 2 \cdot \zeta \cdot \omega_n \cdot x_2 \tag{4}$$

Where x_1, x_2 the state vectors represent values from inside the system that can change over time

u the input vector to the system, voltage signal (V)

y the output vector from the system, fan speed (rpm)

ω_n the natural frequency is relevant to the speed response of the system (rad/s)

ζ the damping ratio is relevant to the oscillation mode of the system (-)

The modelling is performed by using the state space representation because in this way it is possible to express the operation of the fan with a variable run up time (ts). The run up time is defined as the time required for the fan to adjust the angular speed to the current input signal, thus the run up time changes according to the input signal to the fan. This behaviour can be modelled by applying equation (5) (Herring, 2005) where the damping ratio is calculated as a function of the run up time.

$$\zeta = \frac{t_s \cdot \omega_n + 1.6}{6.6} \tag{5}$$

The data for the run up time were obtained by performing open loop measurements on a typical box fan (Exhausto BESB250-4-1) when the input signal to the fan changed per 1 V within a range of 0 V to 10 V. Based on the measurement points a trend line was drawn to correlate the damping ratio with the input signal (see equation 6). The given equation was incorporated in equation (4) and thus the damping ratio became variable of the input signal. In this way the fan model is built with a variable run up time.

$$\zeta = 0.0222 \cdot V^2 - 0.4442 \cdot V + 3.5965 \tag{6}$$

Where V the input signal to the fan (Volts)

Equation (5) establishes a stable response for values of damping ratio above 0.7, for lower values oscillations occur (Herring, 2005). In this case the natural frequency can be adjusted accordingly.

The pressure rise of the fan is calculated according to the fan speed which is output of the state space system (see equation 4). The correlation between the two fan parameters is determined by collecting measurement points in open loop from a typical box fan (Exhausto BESB250-4-1). The measurements were conducted when the input signal changed per 1 V within a range of 0 V to 10 V; equation (7) was obtained based on the relevant measurement data.

$$\Delta P_{fan} = 0.0002 \cdot speed^2 - 0.0082 \cdot speed + 5.7401 \tag{7}$$

Where ΔP_{fan} the pressure rise that the fan provides (Pa)

$speed$ the angular speed that the fan rotates (rpm)

2.3 The damper

The mathematical model implementing the operation of a typical damper (D) is built according to equation (8) which corresponds to the second-order LTI system expressed in the Laplace domain (Franklin et al 1993). The details regarding the control of the damper can be found in Koulani et al (Koulani et al, 2014).

$$D(s) = \frac{k_d \cdot \omega_n}{s^2 + 2 \cdot \zeta \cdot \omega_n \cdot s + \omega_n} \tag{8}$$

Where k_d the process gain correlates the system output with the system input ($m^3/s/Pa\%$)

In the damper model the system input refers to the opening position of the damper while the system output to the corresponding resistance coefficient. The correlation between the two parameters is obtained from table values for a typical damper. The two fundamental factors that describe the response of the second-order LTI system, the damping ratio and natural frequency, are determined by trial and error. The damping ratio is set to 1 as the damper returns to equilibrium as quickly as possible without oscillating while the natural frequency is established at 50 rad/s.

2.4 The pressure and airflow solver

The friction and single pressure losses blocks illustrated in Figure 1 implement the hydraulic calculation of the VAV ventilation system according to the duct design shown in Figure 4. The unknown pressure and airflow conditions are determined by setting up a system of equations expressing the pressure losses occurring in every component of the system. The equations used for calculating the frictional and single pressure losses are described in detail by Koulani et al (Koulani et al, 2014). The hydraulic calculation determines the pressure demand (P) at the beginning and end of every component as well as the airflows (q) delivered to the different zones (see Figure 4). The system of equations derived cannot be solved analytically; therefore the Newton-Raphson numerical method is used instead.

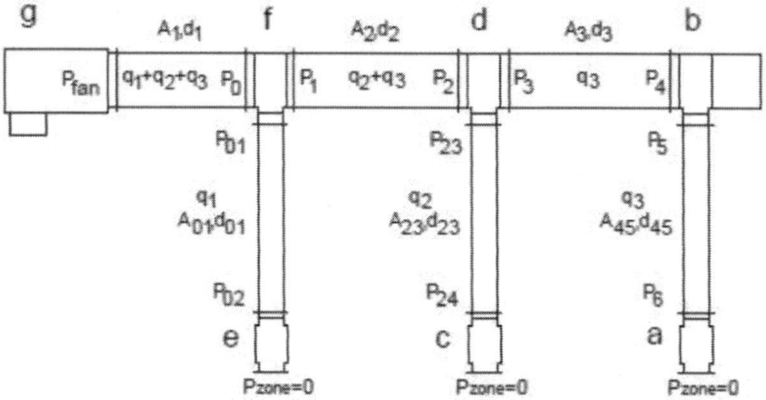

Figure 4: The duct design in the pressure and airflow solver block.

2.5 The static pressure reset algorithm

The operation principle of the SPR control method of trim and respond based on pressure alarms is presented in Figure 5. Every damper of the VAV system transmits an alarm signal when its position exceeds 85 % opening that indicates that the available pressure is critically low but still sufficient for providing the airflow demand. The zone keeps generating a pressure alarm until the damper closes to a position of 80 % opening. The alarms from all zones are summed and when at least two zones request higher pressure the operation set point of the fan is reset 10 % upwards; in the opposite case it is reset 5 % downwards. The SPR algorithm is performed within a specific pressure range; the upper limit is set equal to the CSP set point which corresponds to the pressure required under the design full load condition. The lower SPR limit is determined according to the minimum pressure demand that ensures the precise operation of the dampers. The SPR loop is iterated every 90 sec and the fan adjusts to the new pressure set point.

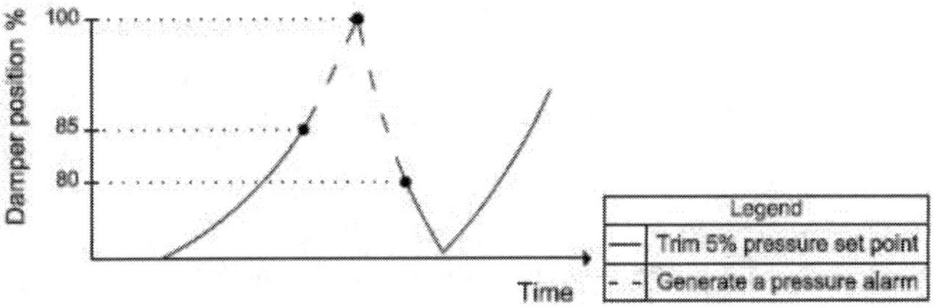

Figure 5: The control logic of the trim and respond static pressure reset method.

3 THE SIMULATION SETUP

The simulations investigate the operation of a VAV ventilation system which provides air in three classrooms, where each classroom can accommodate 20 students and has an area of 50 m^2. When no students are present in the classroom the CO_2 concentration is equal to 400 ppm ($CO_{2\ set,min}$) that corresponds to the outdoor CO_2 level. The minimum airflow demand is set to 10 l/s ($q_{set,min}$) to comply with the ventilation rate of 0.2 l/s, m^2 which is recommended during unoccupied periods in non-residential buildings (see annex B.4, EN15251, 2007). At the design full load condition a ventilation rate of 185 l/s ($q_{set,max}$) is provided to maintain the CO_2 concentration to 1000 ppm ($CO_{2\ set,max}$) that corresponds to a typical CO_2 level used in comfort ventilation. The occupancy in the three classrooms was configured as shown in Table 1.

Table 1: The occupancy profile in the three ventilated zones.

Time (min)	No of students zone 1	No of students zone 2	No of students zone 3
0 - 12	20	20	20
12 - 24	20	5	3
24 - 36	20	10	20

A simulation was performed based on the occupancy profiles given in Table 1 when the VAV ventilation system was controlled with the SPR control method. The SPR pressure set points were adjusted by trial and error to fit to the case study simulated. Table 2 summarizes the selected pressure range, within which the operation set point of the fan is reset.

Table 2: The pressure range of the SPR control method

Lower SPR pressure set point	15 Pa
Upper SPR pressure set point	145 Pa

4 RESULTS

4.1 System operation

Figure 6 and Figure 7 present the results of the simulation according to the occupancy profiles given in Table 1 when the VAV ventilation system operates with the SPR control method. The two graphs given in Figure 6 illustrate the CO_2 concentration in the three zones and the adjustment of the operation set point of the fan, respectively. The distribution of the airflows and the operation of the dampers in the three ventilated zones are shown in the graphs given in Figure 7.

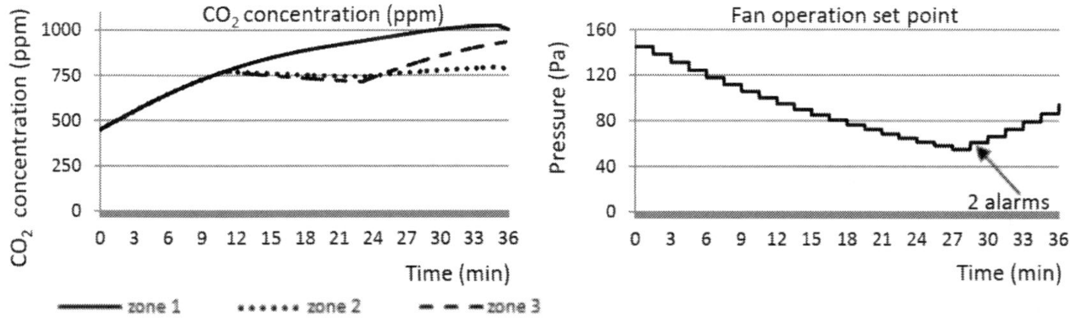

Figure 6: CO_2 concentration in the three ventilated zones and adjustment of the fan operation set point, respectively.

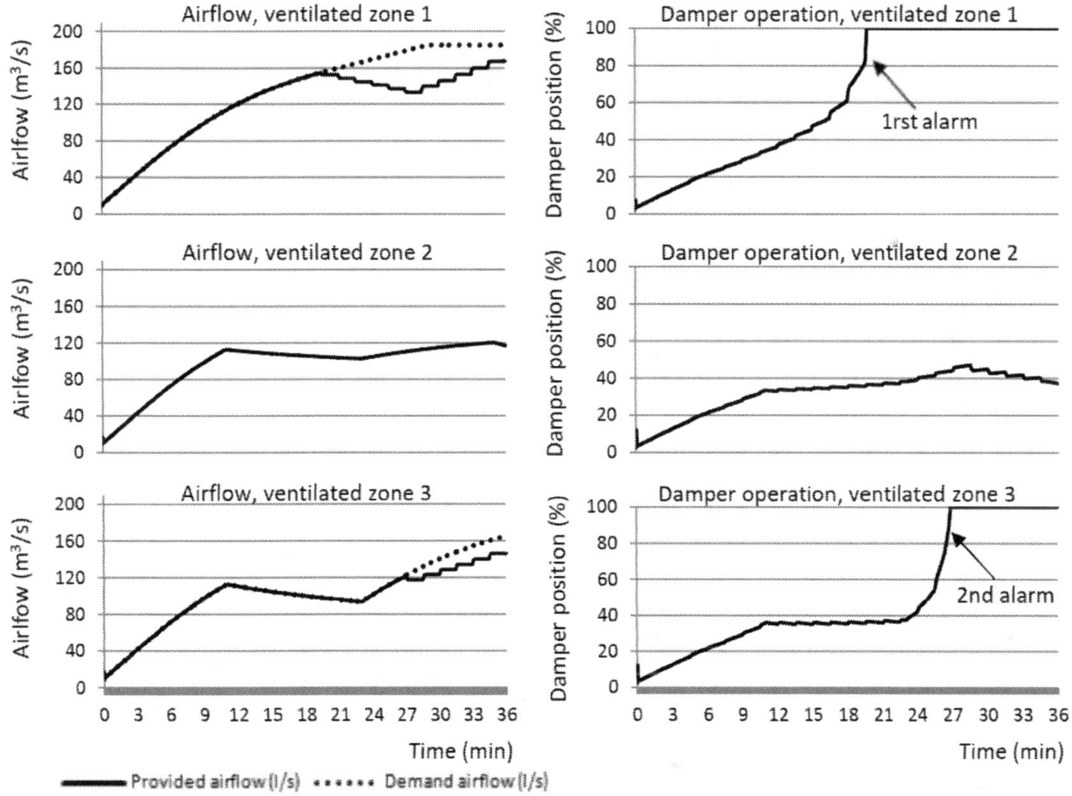

Figure 7: Airflow distribution and damper operation in the three ventilated zones.

The results are analyzed according to the different occupancy periods (see Table 1).

0 min to 12 min: When the system is set to operation the design full load is occurring in the three zones; thus the zone airflows increase and the dampers follow an upward trend. The position of the dampers is still below 85%; therefore no alarm is generated. Consequently the operation set point of the fan is decreased by 5% when the SPR control loop is iterated.

12 min to 24 min: The maximum occupancy load is maintained in zone 1 whereas in zone 2 and zone 3 the load is decreased approximately to 1/4 of the maximum. As the position of the dampers is not open enough to generate an alarm, the SPR control algorithm keeps trimming the operation set point of the fan. The pressure that is available in the system is constantly decreased and the dampers continue to open to wider positions. The rate that the damper position opens depends on the amount of airflow demanded in the zone. In zone 1, where the design full load is occurring, the position opens fast and as soon as it exceeds 85% opening an alarm is generated. Despite that the alarm indicates that the provided pressure is insufficient to satisfy the airflow demand in the zone, the fan operation set point is decreased even more. This is because the SPR control algorithm is designed to reset upwards when at least two zones are in alarm. As a result the damper saturates at 100% opening.

24 min to 36 min: Both in zone 1 and zone 3 the design full load is occurring while in zone 2 half of the maximum load is present. At the moment that the occupancy profile changes, the fan operation set point is lower than the pressure demand that satisfies the design airflow in zone 1. Therefore, once the occupancy in zone 3 becomes critical and the design airflow is requested also in this zone, the damper position rises and almost immediately saturates at 100% opening. In this case there are two zones in alarm; thus at the next iteration of the SPR control loop the operation set point of the fan is reset upwards. The pressure that is available in the system becomes higher and the airflow provided in both critical zones increases. However the increment is not enough to satisfy the pressure demand occurring in the system. This is because the SPR control algorithm is designed with a pressure increment of 10%, which is not sufficient to establish in good time the pressure conditions able to provide the design airflow in both critical zones.

5 CONCLUSIONS

The selected values for the parameters of the SPR control method of trim and respond based on pressure alarms were proven inappropriate to establish the pressure conditions that satisfy the pressure demand occurring under high airflows. The SPR control algorithm underperformed because the following parameters did not fit to the case study examined:

- The threshold number of alarms above which the fan operation set point is reset upwards
- The trim percentage used for resetting downwards the fan operation set point
- The respond percentage used for resetting upwards the fan operation set point

The parameters need to be adjusted and optimized in order to build up an algorithm able to control efficiently a VAV ventilation system where large fluctuations in the airflow demand occur. The parameters were configured based on preselected values established in previous research work where the SPR pressure range was narrow and complied with a difference of 9 Pa between the lower and the upper limit. Therefore a respond percentage of 10 % was able to reset the fan operation set point to satisfy in good time the pressure demand of the occurring airflow. In the case study examined the system is designed for providing high airflows; thus the reset is performed within a wide range. The parameters of the SPR algorithm should be adjusted to achieve an optimized solution where the system can handle efficiently large fluctuations in the airflow demand while avoiding intense dynamics and at the same time maximizing energy savings.

It is not recommended to apply a significantly high respond percentage for resetting upwards the fan operation set point because the dampers would be imposed to adjust to pressure steps that can establish unstable operation conditions in the system. The intention is to maintain a high trim percentage for resetting downwards the fan operation set point as energy savings are increased. A compromise should be found between the respond and trim percentage used for resetting the fan operation set point upwards and downwards, respectively. The higher the trimming rate, the higher the potential for energy savings as the available pressure in the system is reduced fast. However a higher value for the respond rate should also be implemented as the SPR algorithm should be able to increase in good time the fan operation set point in case that high airflow demand occurs. Thus by building up a SPR algorithm with main focus on higher energy savings may compromise stability in the system. The adjustment of the threshold number of alarms above which the fan operation set point is reset upwards is a parameter that can improve the response of the SPR algorithm to handle large fluctuations in the airflow demand without causing unstable operation conditions in the system. In the future, the algorithm, preferably, can be developed to actively learn from continuous feedback, the necessary trim and respond increments and in the end present a real plug-and-play solution.

6 ACKNOWLEDGEMENTS

The authors acknowledge the financial support from the Energy Technology Development and Demonstration Programme (EUDP), Danish Energy Agency. Moreover the authors wish to express their gratitude to Lars Kokholm Andersen, Vasilis Bellos and Christos Papoutsellis.

7 REFERENCES

ASHRAE 90.1 (2004). *Energy Standard for Buildings Except Low-Rise Residential Buildings.* American Society of Heating, Refrigerating and Air Conditioning Engineers.

Bekö, G.; Lund, T.; Nors, F.; Toftum, J. & Clausen, G. (2010). *Ventilation rates in the bedrooms of 500 Danish children.* International Journal of Building and Environment, 45(10), 2289-2295. International Centre for Indoor Environment and Energy ICIEE. Department of Civil Engineering, Technical University of Denmark.

EN15251 (2007). *Indoor environmental input parameters for design and assessment of energy performance of buildings addressing indoor air quality, thermal environment, lighting and acoustics.* 1st ed. Dansk Standard.

Franklin, G.F.; Powell, J.D. & Naeini, A.E. (1993). *Feedback control of Dynamic Systems.* 2nd ed.Addison-Wesley Longman Publishing Co., Inc. Boston, MA, USA.

Hartman, T. (1995). *Global Optimization Strategies for High-Performance Controls.* ASHRAE Transactions Volume 101.

Herring, D. (2005). *Estimating the overshoot, rise time, and settling time.* Department of Electrical and Computer Engineering. University of Illinois.

IBPT (2012). *International Building Physics in Simulink.* Available from: http://www.ibpt.org/ [accessed 15-06-2014]

Koulani, C. S.; Hviid, C. A. & Terkildsen, S.(2014). *Optimized damper control of pressure and airflow in ventilation systems.* 10th Nordic Symposium on Building Physics. Lund, Sweden.

Matlab (2013). *Mathworks, version 2012b.* Available from: http://www.mathworks.com/products/matlab/ [accessed 09-07-2014]

Rowell, D. (2002). *State-Space Representation of LTI Systems.* Analysis and Design of Feedback Control Systems. Available from: http://web.mit.edu/2.14/www/Handouts/StateSpace.pdf [accessed 09-07-2014]

Simulink (2013). *Mathworks, version 2012b.* Available from: http://www.mathworks.com/products/simulink/ [accessed 09-07-2014]

Taylor, S.T. (2007). *Increasing Efficiency with VAV System Static Pressure Set point Reset.* ASHRAE Journal 49, (6), pp24-32.

Title 24 (2005). *Title 24 Building Energy Efficiency Standards.* California Energy Commission. Code of Regulations.

Wei, G.; Liu,M. & Claridge, D.E. (2004). *Integrated damper and pressure reset for VAV supply air fan control.* ASHRAE Transactions, 110, (2).

A PROTOCOL FOR ASSESSING INDOOR AIR QUALITY IN RETROFITTED ENERGY EFFICIENT HOMES IN IRELAND

Áine Broderick[1], Marie Coggins[1], Miriam Byrne[1]

[1] *Centre for Climate and Air Pollution Studies, School of Physics, National University of Ireland, Galway, Ireland.*

**Corresponding author: a.broderick13@nuigalway.ie*

ABSTRACT

In recent years there has been much emphasis on improving the energy performance of Irish buildings. Much of this impetus stems from our requirements to implement provisions in the Energy Performance Building Directive (EPBD, 2002/91/EC, 2010/31/EU), and international targets to reduce CO_2 emissions by 2020. In Ireland, residential buildings account for 27% of Ireland's CO_2 emissions after transport. As a large proportion of the Irish building stock has already been built home owners are encouraged to retrofit existing buildings to improve the energy rating. The impact of retrofitting residential buildings to improve energy rating on indoor air quality and comfort has largely been unexplored.

This study proposes to conduct an assessment of indoor air quality and occupant comfort in 60 retrofitted energy efficient homes. Thermal comfort and occupant behaviour in retrofitted energy efficient home will also be assessed using an occupant survey. This is the first time that a full study of indoor air quality in retrofitted energy efficient homes will be completed in Ireland, and represents one of very few such studies conducted internationally.

KEYWORDS

Indoor air quality (IAQ), Indoor air pollution (IAP), Energy efficient retrofits, Retrofitted energy efficient homes, Thermal comfort

1. INTRODUCTION

In recent years there has been much emphasis on improving the energy performance of European buildings and encouraging the design, construction and market uptake of energy efficient domestic and public buildings. Much of this impetus stems from the EU commitment to decarbonise the European energy system by 2025 (European Commission, 2014a). The buildings sector accounts for 40% of the total EU energy consumption, and reducing energy in this area is an EU priority and one of the main objectives of the Energy performance Buildings Directive (European Commission, 2014b). In response to this challenge the Irish Government has published a National Energy Efficiency Action Plan (NEAAP, 2012) which defines actions to improve energy efficiency across six areas, including the residential sector, which accounts for over quarter of the primary energy used in Ireland. The Better Energy Home scheme is just one of many initiatives listed in the NEAAP, aimed at improving energy efficiency in the residential sector. To date over €300 million, or 255,000 homes have availed of house hold energy efficiency improvement under initiatives such as the Better Energy Home, or The Warmer Home Scheme (Better Energy, 2014).

It is not clear how increased building air tightness will impact on levels of indoor air pollutants (IAP), such as particulate matter or gases released from occupant activities in the home (Crump et al., 2009 and Milner et al., 2014). In most developed countries the population is known to spend upward of 80% of their time within indoor environments (Kleipeis et al., 2001), and a substantial percentage of this time at home, as a result exposure at home is likely to play a significant role in human health. Pollutants such as particulate matter ($PM_{2.5}$) play a significant role in the development and exacerbation of respiratory and cardiovascular diseases (Brunekreef and Forsberg (2005), Castillejos et al., (2000), Host et al., (2008), and de Hartog et al., (2003)). Energy efficient measures have some obvious direct health benefits such as increasing indoor temperatures and occupant comfort (Howden-Chapman et al., 2007). There are few studies which examine the potential health burden due to exposure to indoor pollutants generated in the home, and only one related to standard Irish homes (Galea et al., 2013). There have been no studies which have evaluated the health benefits of energy efficient retrofits in Europe, partly because of the lack of data on indoor air quality (IAQ) in energy efficient homes (Crump et al., 2009, Derbez et al., 2014, Arvela et al., 2013) or in energy efficiency retrofits (Milner et al., 2014).

A recent UK study (Milner et al., 2014) used mathematical models to investigate the effect that reduced home ventilation as part of a household energy efficiency retrofit would have on deaths from radon related lung cancer deaths. Results from this study suggest that increasing building air tightness could increase radon concentrations by 57%, resulting in an additional annual burden of 4700 life years and (at peak) 278 deaths. Milner and co authors conclude that reducing uncontrolled building ventilation may have a negative impact on IAQ and health and caution that energy efficient improvements in the home needs careful management requiring a ventilation strategy. A study on IAQ in newly built French homes incorporating mechanical ventilation with heat recovery (MVHR) before and after occupancy, showed that concentrations of pollutants such as benzene, $PM_{2.5}$ and radon were low and other pollutants such as CO_2, and formaldehyde were comparable to concentrations in standard French homes. The MVHR units allowed for air exchange rates of 0.5 h-1 or higher, the authors caution that IAQ could become an issue if the MVHR units are not used correctly (Derbez et al., 2014).

It is essential that consideration is given to ensure that adequate ventilation is included in the retrofit, to maintain thermal comfort parameters and to remove air pollutants generated indoors. The National Standards Authority of Ireland has provided a code of practice for the energy efficient retrofit of dwellings (National Standards authority of Ireland, 2014). Natural ventilation and wall or window trickle vents are the most common form of ventilation in Irish homes and may not be adequate for higher levels of building air tightness. Impacts of increasing the energy efficiency of households are largely positive; along with reducing energy use, helping to meet National and EU Energy targets, the building retrofit should make the home more comfortable for the occupant as it will improve indoor temperature, and reduce moisture. However there are some concerns that increasing building air tightness may have a negative effect on IAQ, which in turn could affect health.

1.1 Objectives

The objectives for this study are to characterise indoor concentrations of a number of priority pollutants, and to access occupant comfort before and after homes undergo an energy efficient retrofit. The impact of energy efficient retrofitting on indoor air pollutant concentrations, occupant comfort and energy consumption will be examined.

A questionnaire will be design and administered to the owners of the energy efficient retrofitted homes to investigate the perceived benefits of retrofitting and to explore the possibility of co-benefits from retrofitting.

2 METHODOLOGY

60 homes that are planning a household energy efficient retrofit will be recruited in collaboration with Sustainable Energy Authority of Ireland (SEAI), Energy Agencies and Local County and City Councils for the main study. A pilot study will be completed in winter 2014 and will include 12 homes. As part of the retrofit homes will have the levels of insulation will be increased, windows and doors will be replaced, and the boiler will be upgraded to at least 90% efficiency as a minimum. Homes will be recruited to participate in the study using the following selection criteria:

a) Families, 1 or 2 adults with children, occupants will be non- smoking.
b) Three bed semi-detached houses (100 – 126 m2) with Building Energy Ratings (BER) of less than D1, constructed of either hollow block or cavity wall insulation,
c) Homes which are planning to undergo energy efficient retrofit of up to B1 or B2 BER over the period September 2015 – September 2016,
d) Hollow block homes planning a retrofit upgrade to include new windows, and either external or internal insulation, and
e) Cavity wall block homes planning a retrofit upgrade to include new windows, and either internal or cavity wall insulation.

2.1 Pilot Study

The pilot study will include 12 homes that will be monitored before and after an energy efficient retrofit during the winter period of 2014 to assess worse case scenario for the generation of indoor pollutants and maximum energy use in the home. Homes will be selected for inclusion in the study by construction type as illustrated below (Figure 1).

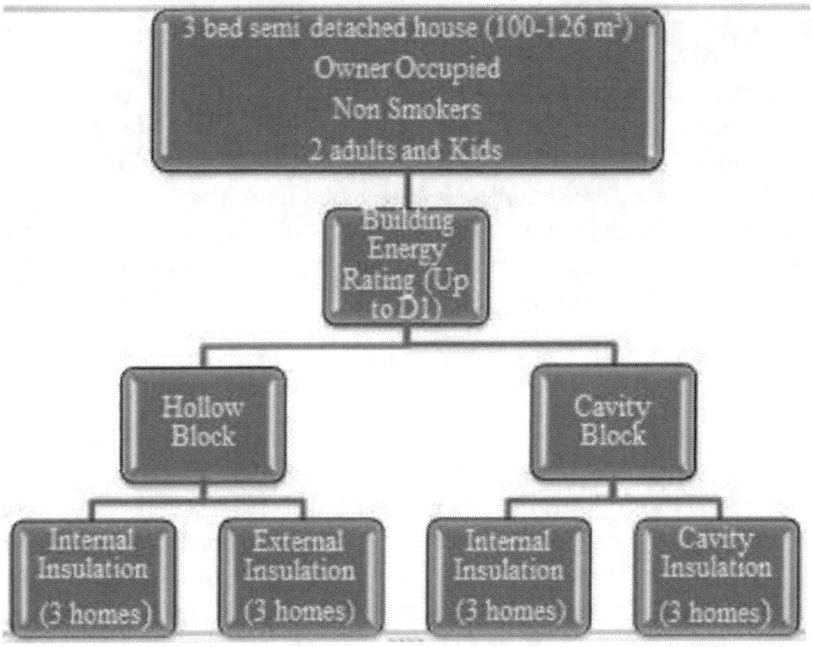

Figure 1: Breakdown of construction types and number of homes to be monitored in each group.

2.2 Types and locations of sensors

Within each participating home, samples will be collected in the area of the home where the occupants mainly use i.e. the main living room and the main bedroom.

a) A TSI SidePak AM510 Personal Aerosol Monitor fitted with a PM2.5 impactor will be used to collect and log real-time data on airborne PM2.5. The monitor will be set to log at 1 minute intervals and will allow assessment of peaks and daily variability in exposure to be determined.
b) Average indoor nitrogen dioxide levels will be measured using NO_2 Passive Diffusion Tubes (DIF100RTU-R/A) supplied by Gradko Environmental. NO_2 will only be monitored in the main living room.
c) Carbon monoxide levels will be measured using EL-USB-CO Carbon monoxide loggers
d) Radon gas levels will be measured using passive radon detectors, supplied by the Radiological Protection Institute of Ireland.
e) Total Volatile Organic Compounds and BTEX concentrations will be collected using a GrayWolf IQ-610 photoionisation detector.
f) Formaldehyde measurements will be made using a GrayWolf FM-801 formaldehyde meter. Formaldehyde will be monitored only in the main living room.
g) Temperature and Relative Humidity levels will be measured during both winter and summer, using Telaire 7001 Monitors and HOBO Data loggers and Thermo hydrographs.
h) Surface samples of dust mites will be collected from representative flooring and bed surfaces, using an SKC Carpet Tester suction sampling kit, comprising a polycarbonate filter cassette, connection to a Flite 2 High-volume sampling pump. The mites will be enumerated by microscopic analysis of the polycarbonate filters, using a protocol developed at the London School of Tropical Medicine.
i) Air exchange rates will be determined in the kitchen, living room and bedroom of each house using tracer gas analysis.

2.3 Questionnaire

A questionnaire will be designed and administered that will investigate the co-benefits of energy efficient retrofitting using previous studies in this area (Howden-Chapmen et al., 2014, Cox, 2005, Walker et al., 2014). Co-benefits of energy efficient retrofitting can include improved health and well being, energy cost savings, mould and damp reduction in the homes, and improved indoor air quality (Chapman et al., 2009). Jacob and Nutter, (2003) noted that co-benefits in energy efficient buildings such as improved comfort, better indoor air quality and enhanced noise protection may have the same monitory order of magnitude as energy-related benefits.

2.4 Data Analysis

The data obtained from before and after the retrofit will be compared to investigate the differences in levels of air quality between the dwellings. Independent samples *t* test and one-way ANOVA tests will be used to determine the differences before and after the retrofit and between the different construction types.

Thermal comfort will be assessed using the predicted mean vote (PMV) model with zero representing thermal neutrality on a scale from -3 to +3. However, although zero is the ideal scenario, anywhere from -0.5 to +0.5 is acceptable. The PMV model uses a combination of metabolic rate, clothing insulation, air temperature (dry bulb), mean radiant temperature, operator temperature, air speed, and relative humidity to determine thermal comfort (ISO EN 7730, 2005). The data for the PMV model will come from an occupant activity diary and from measurements taken from the house.

The relationship between IAQ and occupant behaviour will be assessed. Sharmen et al., (2014) suggested that occupant lifestyle and comfort can have a significant impact on the heating energy consumption in an apartment. Geurra-Santin (2013) found that although the overall energy consumption of energy efficient homes was less, this can lead to a rebound effect where the occupants tend to favour higher temperatures and less ventilation.

The rebound effect is when energy savings as a result of energy efficient improvements are negated by increased levels of energy consumption (Herring and Sorrell, 2009, Hens et al., 2010). The rebound effect will need to be taken into account when determining occupant behaviour relating to energy consumption. Factor analysis will be used to identify any underlying occupant behaviour factors that can potentially predict energy consumption and IAQ. Factor analysis will be used to investigate any possible correlation between occupant behaviour variables.

3 RESULTS AND CONCLUSIONS

The impact of increasing the energy efficiency of households are largely positive, along with reducing energy use, meeting EU Energy targets, the building retrofit should make the home more comfortable for the occupant as it will improve indoor temperature, and reduce moisture. However there are some concerns that increasing building air tightness may have a negative effect on IAQ, which in turn could affect health. Preliminary project results and proposed project methodologies will be presented here. This project aims to evaluate the impact of energy efficient retrofits on indoor air quality and occupant comfort.

4 ACKNOWLEDGEMENTS

This project is sponsored by the Irish Environmental Protection Agency under the Strive 2007- 2013 Research Programme.

5 REFERENCES

Arvela, H., Holmgren, O., Resibacka, H., Vinha, J., (2013) Review of Low-Energy Construction, Air Tightness, Ventilation Strategies and Indoor Radon: Results From Finnish Houses and Apartments, *Radiation Protection Dosimetry*, pp 1 - 13

Better Energy: National Upgrade programme, (2014) http://www.dcenr.gov.ie/Energy/Energy+Efficiency+and+Affordability+Division/Better+Energy.htm accessed July 2014

Brunekreef, B., and Forsberg, B., (2005) Epidemiological evidence of effects of coarse airborne particles on health, *European Respiratory Journa*l, 26, pp 167 - 173

Castillejos, M., Borja-Aburto, V.H., Dockery, D.W., Gold, D.R., Loomis, D., (2000) Airborne coarse particles and mortality, *Inhalation Toxicology*, 12, pp 61 – 72

Chapman, R., Davie, G. (2007) Effect of insulating existing houses on health inequality: cluster randomised study in the community, *British Medical Journa,l* bmj.39070.573032.80v1

Chapman, R., Howden-Chapman, P., O'Dea, D., Viggers, H., Kennedy, M., (2009) Retrofitting house with insulation: a cost-benefit analysis of randomised community trial, *Journal of Epidemiology Community Health*, 63, pp 271 – 277

Cox, C., (2005) Health optimisation protocol for energy-efficient buildings, Delft, The Netherlands: TNO, Final report

Crump, D., Dengel, A., and Swainson, M., (2009) Indoor air quality in highly energy efficient homes – a review, NHBC Foundation report NF18, IHS BRE Press, July 2009.

De Hartog, J.J., Hoek, G., Peters, A., Timonen, K.L., Ibald-Mulli, A., Brunekreef, B., Heinrich, J., Tiittanen, P., van Wijnen, J.H., Kreyling, W., Kulmala, M., Pekkanen ,J., (2003) Effects of fine and coarse particles on cardiorespiratory symptoms in elderly subjects with coronary heart disease – the ULTRA study, *American Journal of Epidemiology*, 157, pp 613 – 623

Derbez, M., Berthineau, B., Cochet, V., Lethrosne, M., Pignon, C., Riberon, J. and Kirchner, S. (2014) Indoor air quality and comfort in seven newly built, energy-efficient houses in France. *Building and Environment*, 72, pp 173 – 187

European Commission, (2014a), European Climate Change Programme http://ec.europa.eu/clima/policies/package/index_en.htm accessed July 2014

European Commission, (2014b), http://ec.europa.eu/energy/efficiency/eed/eed_en.htm accessed 2014

Galea, K.S., Hurley, J.F., Cowie, H., Shafrir, A.L., Sánchez Jiménez, A., Semple, S., Ayres, J.G., and Coggins, M. (2013) Using PM2.5 concentrations to estimate the health burden from solid fuel combustion, with application to Irish and Scottish homes, *Environmental Health*, 12:50

Geurra- Santin, O. (2013) Occupant behaviour in energy efficient dwellings: evidence of a rebound effect, *Journal of House and the Built Environment*, 28, pp 311 - 327

Hens, H., Parijs, W., Deurinck, M. (2010) Energy consumption for heating and rebound effects, *Energy and Buildings*, 42, pp 105 – 110

Herring, H. and Sorrell, S. (2009) *Energy efficiency and sustainable consumption. The rebound effect.*
Energy, climate and the environment series, Palgrave Macmillan, UK

Host, S., Larrieu, S., Pascal, L., Blanchard, M., Declercq, C., Fabre, P., Jusot, J.F., Chardon, B., Le Tertre, A., Wagner, V., Prouvost, H., Lefranc, A., (2008) Short-term associations between fine and coarse particles and hospital admissions for cardiorespiratory diseases in six French cities *Occup. Environ. Med.* 65, pp 544 – 551

Howden-Chapman, P., Matheson, A., Crane, J., Viggers, H., Cunningham, M., Blakely, T.,

Cunningham, C., Woodward, A., Saville-Smith, K., O'Dea, D., Kennedy M., Baker, M., Waipara, N., Chapman, R., Davie, G. (2007) Effect of insulating existing houses on health inequality: cluster randomised study in the community, *British Medical Journal*, bmj.39070.573032.80v1

ISO EN 7730, (2005) Ergonomics of the thermal environment -Analytical determination and interpretation of thermal comfort using calculation of the PMV and PPD indices and local thermal comfort criteria, International Standards Organisation, Geneva

Jacob, M., and Nutter, S., (2003) Marginal costs, cost dynamics and co-benefits of energy efficiency investments in the residential buildings sector, *ECEEE 2003 summer study – time to turn down energy demand*, pp 829 – 841

Kleipeis, N.E., Nelson, W.C., Ott, W.R., Robinson, J.P, Tsang, A.M., Switzer, P., Behar, J.V., Hern, S.C., Engelmann, W.H., (2001) The National Human Activity Pattern Survey (NHAPS): a resource for assessing exposure to environmental pollutants. *Journal of Exposure Analysis and Environmental Epidemiology*, 11(3), 231-252.

Milner, J., Shrubsole, C., Das, P., Jones, B., Ridley, I., Chalabai, Z., Hamiliton, I., Armstrong, B., Davies, M., Wilkinson, P., (2014) Home energy efficiency and radon related risk of lung cancer: modelling study, *British Medical Journal*, 10;348:f7493

National Standards Authority of Ireland, (2014) Code of Practice for energy efficient retrofit of dwellings, SR 54:2014

NEAAP, (2012) Ireland's second National Energy Efficiency Action Plan (NEAAP) to 2020, Department of communication Energy and National Resources, December 2012 Dublin 2

Sharmen, T., Gül, M., Li, X., Ganev, V., Nikalaidis, I., Al-Hussein, M., (2014) Monitoring building energy consumption, thermal performance, and indoor air quality in a cold climate region, *Sustainable Cities and Society*, 13, pp 57 - 68

Sustainable Energy Authority of Ireland, 2013 Energy in the Residential Sector, 2013 Report, SEAI

World Health Organisation, (2014) Indoor air pollution, research and evaluation http://www.euro.who.int/__data/assets/pdf_file/0009/128169/e94535.pdf [Accessed July 2014]

Walker, S.L., Lowery, D., Theobald, K., (2014) Low-carbon retrofits in social housing: Interaction with occupant behaviour, Energy Research & Social Science, 2, pp 102 - 114

TESTING FOR BUILDING COMPONENTS CONTRIBUTION TO AIRTIGHTNESS ASSESSMENT

Pedro F. Pereira[1], Ricardo M. S. F. Almeida[1,2], Nuno M. M. Ramos*[1] and Rui Sousa[3]

1 University of Porto, Faculty of Engineering,
Laboratory of Building Physics,
Porto, Portugal
**Corresponding author: nmmr@fe.up.pt*

2 Polytechnic Institute of Viseu, School of Technology & Management
Civil Engineering Department
Viseu, Portugal

3 University of Porto, Faculty of Engineering,
GEQUALTEC,
Porto, Portugal

ABSTRACT

When one intends to evaluate buildings energy efficiency their airtightness is a fundamental parameter. Airtightness is linked to undesirable and uncontrolled ventilation and, therefore, should be minimized. Quantitative characterization of expected leaks of common building elements would be useful for practitioners that intend to improve building enclosures for airtightness optimization. The most well accepted experimental procedure to evaluate in-situ buildings' airtightness is the fan pressurization method, typically making use of a "blower door" device. Individual components can be tested for air permeability in laboratory conditions according to well established standards. A systematic approach for components contribution to overall airtightness is lacking, especially due to insufficient measured data availability.

The paper presents a case study where in-situ blower door measurements were used to define the contribution of different building components to the airtightness of a small building. Several set-ups were established allowing for the individual effect of the components to be measured, namely: enclosure, windows, connection of the steel columns with the floor and ventilation ridge. An enclosure sample, including wall panel and one window, was built in the laboratory and air permeability tests were performed. The information from both in-situ and laboratory tests was combined in a methodology for airtightness prevision of a small building. The laboratory results, completed with air leakage values obtained from a published database, provide the base value for airtightness computation. That first value is then corrected and/or validated by the in-situ measurements of individual components.

Results indicate that the application of the methodology for Mediterranean countries, with the construction elements air leakage database available, could be questionable. A very different construction reality from the one in the countries that contribute for the database was found. The importance of a large number of studies in the Mediterranean countries, where constructions are less airtight, arises and should be the basis for a new database.

KEYWORDS

Airtightness, component leakage, laboratory tests, in-situ tests

1. INTRODUCTION

1.1 Motivation

Buildings airtightness represents a characteristic of the building which is fundamental for the quality of its indoor environment. Airtightness influences the heating load, the strategies of the ventilation system, the indoor air quality, the indoor acoustic comfort and, of course, the

energy efficiency of the building. Therefore, predicting airtightness is very important for both the design and the in service stages of a building (Iordache at al., 2011). The most well accepted experimental procedure to evaluate in-situ buildings' airtightness is the fan pressurization method, typically making use of a "blower door" device, which consists on an adjustable door that contains a reversible fan and that replaces one of the building's doors.
In cold climate countries, undesired ventilation, like the infiltration of air through the fabric of the building, might lead to a situation of energy waste and sometimes discomfort. Thus, an equilibrium between indoor air quality and energy efficiency must be achieved (Dimitroulopoulou, 2012).
In different climates, the desired airtightness may be different however. The values found in Italian dwellings (Alfano et al, 2012) are quite above the reference values found in EN 15242-2007 (CEN, 2007). If even warmer climates are considered, and depending on the adopted ventilation system, airtightness may even be undesirable.
Either airtightness is desirable or not, its estimation in design phase is important, and therefore a contribution to that is intended in this paper.

1.2 Airtightness estimation

The estimation of airtightness has been addressed by different authors, using different strategies to find an expected n_{50} value for a specific building or a type of building. The review by Relander et al (2012) grouped the different approaches into three categories: estimation based on multiple regression, estimation based on rough characteristics of the building and estimation based on component leakage and geometry of the building.
Component leakage can be assessed and quantified by different methods. The most common methodology is based on, step by step, artificially establishing a pressure difference (making use of the "blower door") between the two sides of the component and measure the resultant air flow. When applied for a component of a building, this methodology requires two tests, one with all the joints sealed and the other without any sealing. The flow rate difference between the tests corresponds to the component air permeability. Pinto (Pinto et al., 2011) applied this methodology to test air inlets, windows and doors of five dwellings. The studies by Relander et al (2010 and 2011) and Van den Bosshe et al. (2012) present examples of how laboratory tests were used in the assessment of component leakage and consequent influence on building airtightness.
In the present paper, the authors apply the estimation based on component leakage and geometry of the building to a very simple building used as a case study. An enclosure sample, including wall panel and one window, was built in the laboratory and air permeability tests were performed. In-situ "blower door" measurements were used to define the contribution of different building components to the overall airtightness. Several set-ups were established allowing for the individual effect of the components to be measured, namely: enclosure, window, window frame, enclosure defects and ventilation ridge. The information from both in-situ and laboratory tests is finally combined in a methodology that can estimate building airtightness.

2 IN SITU EXPERIMENTAL SET-UP

2.1 Case Study

The case study corresponds to a small building with light construction. The geometric properties are summarized in Table 1.

Table 1: Building geometric properties

Floor area [m^2]	Volume [m^3]	Envelope area [m^2]
48	150	178

The structural scheme of the building corresponds to a steel structure composed of tubular elements. Both the façade and the roof are clad with sandwich panels with 4 cm insulation in the interior, with mechanical fixation to the steel structure. The building includes a door and 4 aluminium frame windows, double glazed, distributed in pairs in two opposite façades. The building depends on natural ventilation and therefore a ventilation ridge was included in the roof.

2.2 Blower-door tests

The fan pressurization method for determination of air permeability of buildings consists of applying a known pressure difference between the two sides of a construction element or building and measure the resultant volume of air flow. This information his used to compute the "permeability law":

$$\dot{V}_{env} = C_{env} \cdot (\Delta p)^n \tag{14}$$

Where $\dot{V}_{env}$ is the is the air flow rate [m^3/h], C_{env} is the leakage coefficient [$m^3/(h.Pa^n)$], □p is the pressure difference [Pa] and n is the air low exponent [-] which characterizes the flow regime (0.5 for turbulent flow and 1.0 for laminar flow).

Typically, the experimental determination of the "permeability law" is based on several measurements on the range 10 Pa to 50 Pa with increments of 10 Pa. The procedure is repeated twice, one for pressurization and the other for depressurization. The test is performed according to standard EN:13829-2000.

The air permeability measurements were carried out using the Retrotec 1000 blower door model (Figure 1). For each scenario, both pressurization and depressurization tests were performed.

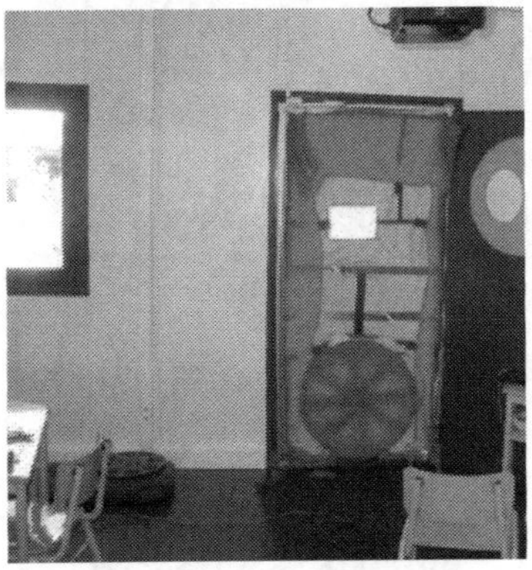

Figure 1: Experimental set-up

Exterior air temperature during tests ranged from 17.2 °C to 19.0 °C and the air velocity varied between 0.6 m/s and 0.8 m/s, which according to Beaufort scale corresponds to "1-light air" (EN:13829-2000).

2.3 Methodology

The methodology was based on the successive sealing of the following three building components: (a) ventilation ridge; (b) window frame; (c) connection of the steel columns with the floor. Therefore four testing scenarios were created as described in Figure 2 and Table 2.

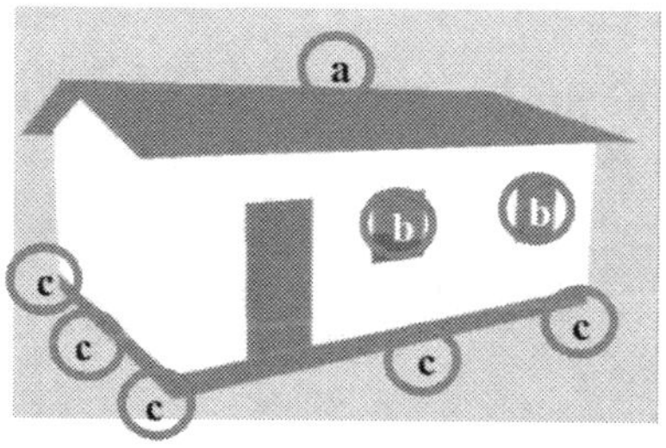

Figure 2: Building enclosure and components which were sealed

Table 2: Testing scenarios

Scenario	(a)	(b)	(c)
1	NS	NS	NS
2	S	NS	NS
3	S	S	NS
4	S	S	S

NS: not sealed; S: sealed

Consecutive differences between test scenarios allowed for the evaluation of the individual contribution of the building components for the building airtightness. In scenario 4, since the three building components were sealed, the airtightness of the enclosure is assessed.

3 LABORATORY TESTS

Two enclosure samples, one including wall panel and one window (Figure 3a)), and the other including only wall panel (Figure 3b)), were built in the laboratory and air permeability tests were performed.

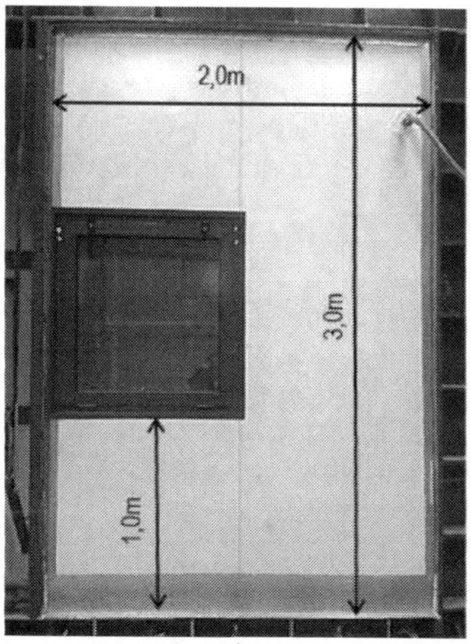

a)

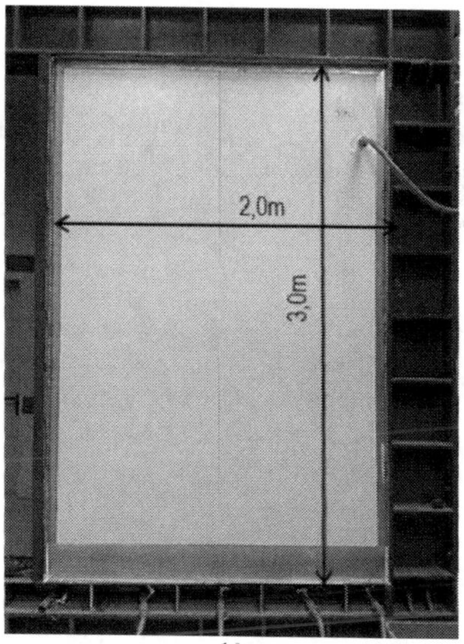

b)

Figure 3: Laboratory enclosure samples. a) wall panel and one window; b) wall panel

These samples were tested according to EN 12114-2000. Samples were built in an uncontrolled environment in the laboratory and were tested with a temperature of 16 °C and a relative humidity of 66%. Testes were carried both for positive and negative pressure differences, applying successive pressure levels starting at ±50 Pa and reaching a maximum level of ±500 Pa. In each pressure level the correspondent air flow throughout the samples was recorded.

4 RESULTS AND DISCUSSION

4.1 In-situ tests

Figure 4a) shows the result of the fan pressurization method, both for pressurization and depressurization, for scenario 1 and Figure 4b) for scenario 4. These are the two most extreme situations which were evaluated and correspond, respectively, to method A and method B described in EN:13829-2000.

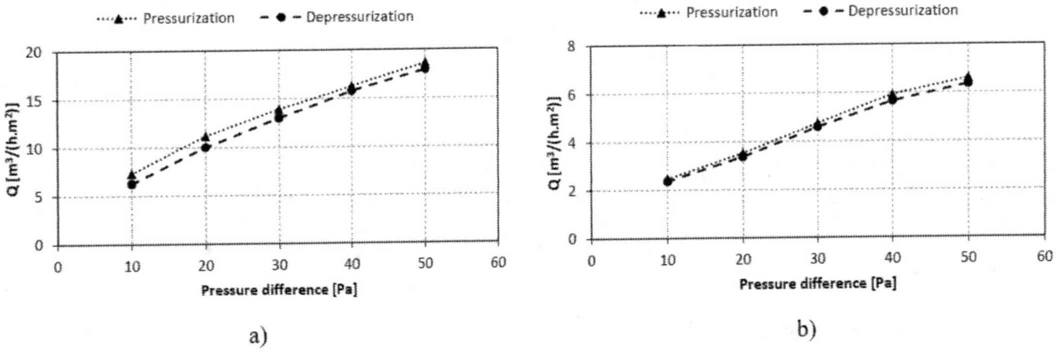

Figure 4: Air permeability tests. a) scenario 1; b) scenario 4

These results show that for a pressure difference of 10 Pa air flow rate ranges from 2.37 $m^3/(h.m^2)$, to 6.35 $m^3/(h.m^2)$, in depressurization, in the extreme scenarios. This difference is explained by the effect of the ventilation ridge, the window frame and connection of the steel columns with the floor. To evaluate their individual contribution the results of scenario 2 and 3 should also be taken into account and the consecutive differences between the test results must be computed. Figure 5 shows the individual contribution of each component for the building air permeability at 50 Pa (q_{50}).

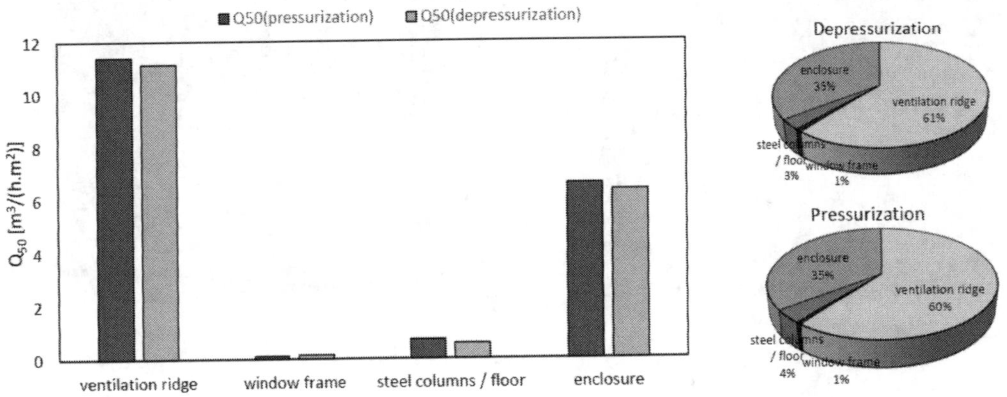

Figure 5: Contribution of the building components for the airtightness

Results revealed that the ventilation ridge is the most important component for the building permeability with a contribution of up to 61%. Building enclosure is responsible for 35% and the other two components have a residual impact (1% for the window frame and 3% for the connection of the steel columns with the floor). No significant differences between depressurization and pressurization test results were found.

Traditionally building airtightness is evaluated by the air change rate at a pressure difference of 50 Pa, n_{50}, which can be computed from the permeability law. Table 3 presents the results of n_{50} for the four scenarios.

Table 3: Results of the air change rate at a pressure difference of 50 Pa, n_{50} [h^{-1}]

	Scenario 1	Scenario 2	Scenario 3	Scenario 4
Depressurization	21.30	8.49	8.41	7.75
Pressurization	22.01	8.70	8.65	7.86

According to the current standards, regarding the energy efficiency of buildings, these are very high values of n_{50}, however it should be notice that this a small building with very specific characteristics, whose purpose is to be constructed in tropical countries and, therefore, the comfort conditions and the indoor air quality are the main preoccupation, rather than the energy efficiency.

4.2 Laboratory tests

Figure 6 shows the air permeability results of the two enclosure samples, with and without window.

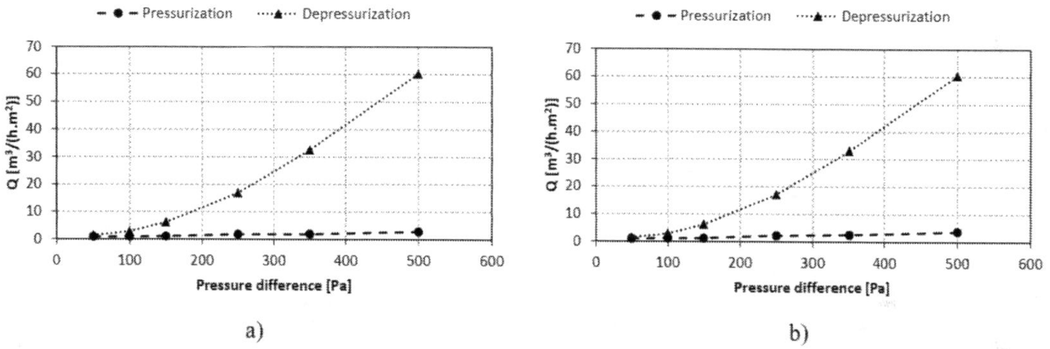

Figure 6: Wall panel air permeability tests. a) panel without window; b) panel with window

No significant differences between panels were found. This situation is in agreement with the fan pressurization tests, which indicated that windows are responsible for a very small contribution for the building airtightness. Another interesting conclusion is that results for pressurization and depressurization are very different. Additionally, this difference increases throughout the test and for the maximum pressure difference (500 Pa) reaches 59 $m^3/(h.m^2)$. However, for the lowest pressure difference (50 Pa) difference between tests was null. The explanation for this problem is related to a deformation in the wall panel connection with the steel structure (Figure 7).

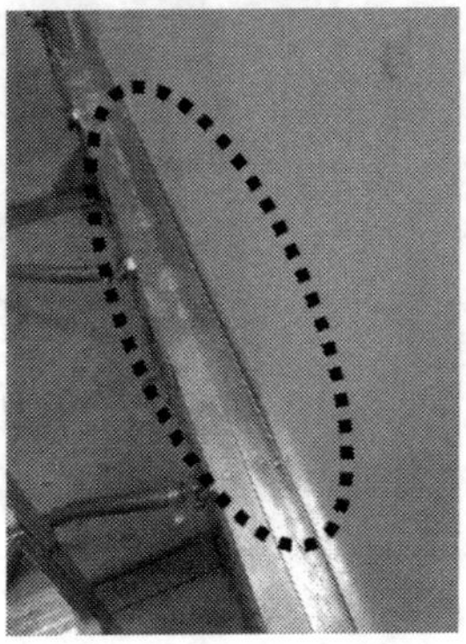

Figure 7: Wall panel deformation due to the pressure difference in depressurization test

4.3 Discussion

The information from both in-situ and laboratory tests can be compared since both include an air permeability measurement for a pressure difference level of 50 Pa. Using both is possible to obtain an estimation of the building permeability by calculating the n_{50}. Equation 2 describes the method.

$$n_{50} = \frac{(Ql_{50,wnd} \cdot L_{wnd} + Ql_{50,ridge} \cdot L_{ridge} + Q_{50,enc} \cdot A_{enc} + \dot{V}_{50,fl\_con})}{V} \quad (15)$$

Where:

n_{50} – air change rate at 50 Pa [h^{-1}]

Ql_{50},wnd – air permeability of the metal panels with and without windows [$m^3.h^{-1}.m^{-1}$]

L_{wnd} – envelope area [m]

Ql_{50},ridge – air permeability of the ventilation ridge [$m^3.h^{-1}.m^{-1}$]

L_{ridge} – Ventilation ridge length [m]

$Q_{50,enc}$ – air permeability of the enclosure, which includes construction joints such as metal panels joints, wall to ceiling/floor and wall to wall joints [$m^3.h^{-1}.m^{-2}$]

A_{enc} – Construction joints length [m^2]

$\dot{V}_{50,fl\_con}$ – Air flow rate thought connections between columns and the floor [m^3/h]

V – Volume [m^3]

The laboratory sample doesn't include all the parts and components of the building, such as joints and connections between the elements (eg. wall/roof, wall/wall, wall/floor). Therefore, to properly apply this methodology additional data is required. Several studies regarding the individual effect of the components have been published and can be used to obtain the lacking data (Orme and Leksmono, 2002). Building enclosure was sub-divided in three parcels (metal panels joints, wall to ceiling/floor joints and wall to wall corner joints). Table 3 condensates the information.

Table 3: Q values from different building elements.

	In situ	Laboratory	Literature
Ql_{50},wnd*	0.61	0.94	-
Ql_{50},ridge	244.88	-	-
Ql_{50},joint1	17.38	2.28	-
Ql_{50},joint2*		-	0.87
Ql_{50},joint3		-	-
Q_{50},col	108.75	-	-

Ql_{50},joint1 – metal panels joints

Ql_{50},joint2 – wall to ceiling/floor joints

Ql_{50},joint3 – wall to wall joints (corners)

* - the literature value was obtained for timber constructions.

The air permeability of the ventilation ridge, of the wall to wall corner joints and of the columns connection with the floor couldn't be found in the literature. Thus, only a rough approximation of the n_{50} can be established for this particular case. Table 4 shows the results.

Table 4: n_{50} values according to the two methodologies.

	In situ	Laboratory + literature
n_{50} [h^{-1}]	7.87	2.2

A large difference between the two methodologies was obtained. The approximations and limitations described previously help to explain it. A very different construction reality from the one in the countries that contribute for the database was found and the applicability of this methodology can be questioned. The importance of a large number of studies in the Mediterranean countries, where constructions are less airtight, arise and should be the basis for a new database.

5 CONCLUSIONS

The airtightness of a building can be tested using a blower door although an estimation procedure applicable in design phase could be very useful. The analysis of a small building whose components were simultaneously tested in laboratory conditions provided an opportunity to implement a strategy of airtightness estimation.

In the in-situ tests, it was found that the ventilation ridge was responsible for the highest percentage of airflow with a contribution of 61%. The permeability of the ventilation ridge was 11,0 $m^3/(h.m^2)$. The window frame and the connection of the steel columns with the floor have no significant impact on the airtightness as the permeability of the window frame and the connection of the steel columns with the floor was 0,66 $m^3/(h.m^2)$.

The laboratory tests proved that the window frame hasn't got a significant influence as the result of the test sample with or without a windows lead to the same value. The laboratory tests showed that the panel joints permeability was 1,15 $m^3/(h.m^2)$. As laboratory tests couldn't provide information for all the leakage components, information in the literature was added to the estimation process. The estimated value found was quite different from the measured one, proving that it could only be used as a lower limit n_{50} value.

To achieve the same n_{50} in the in situ tests than those achieved in laboratory, more set-ups should be made. However, some differences could occur because of problems related to construction defects and workmanship.

A very different construction reality from the one in the countries that contribute for the database was found and the applicability of this methodology can be questioned. The importance of a large number of studies in the Mediterranean countries, where constructions are less airtight, arises and should be the added to the database.

6 REFERENCES

Alfano, F., Dell'Isola, M., Ficco, G. and Tassini F. (2012). Experimental analysis of air tightness in Mediterranean buildings using the fan pressurization method. Building and Environment, 53, 16 a 25.

CEN, Thermal performance of buildings - Air permeability of building components and building elements. Laboratory test method (EN 12114-2000).

CEN, Thermal performance of buildings - Determination of air permeability of buildings - Fan pressurization method (EN 13829-2000).

CEN. Ventilation for buildings - calculation methods for the determination of air flow rates in buildings including infiltration. (EN 15242-2007).

Dimitroulopoulou, C. (2012). Ventilation in European dwellings: A review. *Building and Environment, 47*, 109-125.

Iordache, V., Nastase, I, Damian, A., Colda, I. (2011). Average permeability measurements for an individual dwelling in Romania. *Building and Environment, 46*, 1115-1124.

Orme, M., Leksmono, N., (2002) AIVC Guide 5 - Ventilation Modelling Data Guide

Pinto, M., Viegas, J., Freitas, V.P. (2011). Air permeability measurements of dwellings and building components in Portugal. Building and Environment, 46, 2480-2489.

Relander, T., Kvande, T., Thue, J. (2010). The influence of lightweight aggregate concrete element chimneys on the airtightness of wood-frame houses. Energy and Buildings, 42, 984-694.

Relander, T., Bauwens, G., Roels, S., Thue, J. and Uvsløkk, S. (2011). The influence of structural floors on the airtightness of wood-frame houses. Energy and Buildings, 43, 639-652.

Relander, T., Holos, S., Thue, J. (2012). Airtightness estimation - A state of the art review and an en route upper limit evaluation principle to increase the chances that wood-frame houses with a vapour- and wind-barrier comply with the airtightness requirements. Energy and Buildings, 54, 444-452.

Van Den Bossche, N., Huyghe, W., Moens, J., Janssens, A., Depaepe, M. (2012). Airtightness of the window–wall interface in cavity brick walls. Energy and Buildings, 45, 32-42

IMPLEMENTATION AND PERFORMANCE OF VENTILATION SYSTEMS: FIRST REVIEW OF VOLUNTARY CERTIFICATION CONTROLS IN FRANCE

Sarah Juricic*[1] and Florent Boithias[1]

1 Cerema (ex CETE de Lyon)
Direction Territoriale Centre-Est
**Corresponding author: commission.permeabilite@cerema.fr*

ABSTRACT

A voluntary certification for very low energy buildings has been implemented in 2013 in France, which requires among other the ventilation systems to be controlled by an independent technician. To ensure the expertise of these technicians, a certification scheme has been implemented for the airtightness measurement of ductwork. This certification will soon be required for the ventilation system controls.
From both these frames, inspection data has been collected. The purpose of this paper is to go through the ductwork airtightness tests and the visual inspections to establish a picture of the quality of the implementation of ventilation systems in low energy buildings in France.
This paper describes the diverse requirements in both regulation and voluntary certifications and describes the ventilation systems typically implemented in French low energy buildings, viewed from the data collected. The paper further discusses the ductwork airtightness performance and the visual controls collected from the database. The paper finally deals with some difficulties encountered today in implementing high quality ventilation systems.

KEYWORDS

Ventilation systems implementation quality, onsite performance, low energy buildings

1 INTRODUCTION

As a new voluntary building certification program has been implemented in France, ventilation ducts airtightness tests in this frame have become mandatory. After a few previous studies about ventilation performance in France, the database created from the ductwork airtightness tests results is a first opportunity to picture ventilation systems in low-energy buildings in France as well as to establish their performance and determine if and how achieving high performing systems is possible in the present state in France.

2 FRENCH REGULATION AND VOLUNTARY CERTIFICATIONS: TAKING DUCTWORK AIRTIGHTNESS PERFORMANCE INTO ACCOUNT IN LOW ENERGY BUILDING PERFORMANCE

2.1 Evolution of the airtightness requirements in the French Regulation and other certifications in France

For over 10 years, airtightness requirements have progressively been reinforced in the successive French energy performance (EP) regulations. If it were at first only true for building envelope airtightness, the possibilities to value performance in buildings in the EP calculation is now extended to ductwork airtightness.Today, as can be seen in Figure 1, the 2012 EP regulation takes ductwork airtightness measures into account.

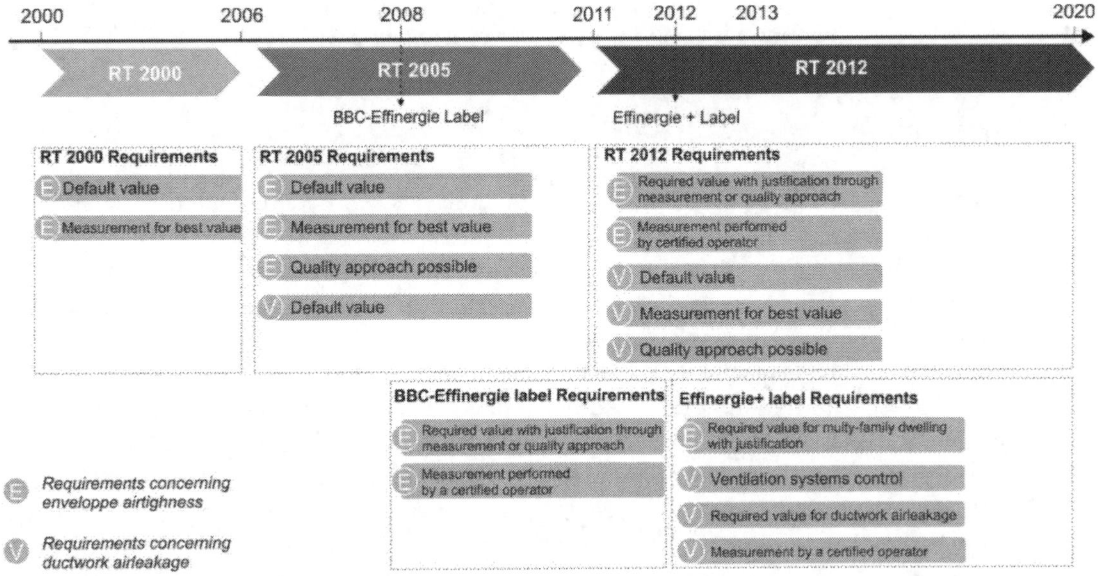

Figure 1: Evolution of French Thermal Regulation (credit : Cerema Direction Territoriale Centre-Est)

Furthermore, a voluntary building certification for very low energy buildings, Effinergie+, brought a requirement on controlling ventilation system at commissioning. A thorough inspection is required: visual control of the system as well as ductwork airtightness measurement. Without a complete inspection, a building cannot be certified Effinergie+.

2.2 Taking ductwork airtightness into account in low energy building performance

Ventilation ducts airtightness is taken into account in the energy consumption calculation of the French regulation, which is described in the Th-BCE 2012 calculation method (CSTB, 2013). Indeed, leaks have an impact on the heating and cooling calculation as well as on the energy consumption of the air handling units. Focusing on the heating and cooling energy consumption, the air leakage flow rate is proportionate to the airtightness class of the ventilation ductwork.

$$q_v = 3600 \cdot Kres \cdot A \cdot dP^{0.667} \quad (1)$$

Where qv is the air leakage flow rate, Kres a coefficient depending on the class of the ductwork, A a conventional surface area of the ductwork and dP a conventional pressure difference.

If no specific airtightness class is input in the calculation process, a default value is used. This value equals 0,0675 $m^3/s/m^2$, which is almost 3 times as leaky as a class A at 1Pa. It is however possible to voluntarily justify a better airtightness class in the energy consumption calculation, up to class C. It is possible to justify one airtightness value per ventilation zone, which can result into several measurements in one building.When it is the case, the French regulation requires a measurement by an independent technician.

2.3 Effinergie+ certification: guarantees of the quality of a ductwork airtightness measurement

The Effinergie+ certification program sets several requirements to guarantee the quality and the homogeneity of measurements. The program requires aspecifically qualified technician to fulfill the airtightness measurements of the ventilation ductwork when commissioning a newly built building.

First, technicians must have undergone a specific training, which program and contents is supervised by the Effinergie association. Secondly, to standardize the measurements procedure and adapt European norms to the French landscape, the association Effinergiehas published a measurement protocol that sets further requirements on the measurements than norms NF EN 12237, NF EN 1507, NF EN 13403, NF EN 12599 and French FD E51-767. The use and application of this protocol is mandatory to have a building certified.

Finally, technicians will soon have to fulfill a second requirement: a Qualibat qualification on measuring ductwork airtightness. To undergo the qualification, technicians have to provide proof of their experience by giving the results of at least ten measurements. Their skills are then analyzed and the quality of their reports is controlled.

3 A PICTURE OF VENTILATION SYSTEMS IN FRENCH LOW ENERGY BUILDINGS

3.1 Data collection

In the frame of the Qualibat qualification, all candidates provide proof of a sufficient and pertinent experience of at least ten measurements in different buildings by filling in a predetermined document called tests register, summarizing their results.

So far, several technicians have undergone or are still undergoing the qualification. All their measurement results have been collected from their tests register and allowed us to build a small database about which the paper further deals. Seeing the requirements set by Qualibat for the qualification, it is expected that all technicians have a homogeneous quality expertise in ductwork airtightness measurements. Therefore, it is expected that even if the data is collected from 5 different technicians, the results can be combined and compared to statistic aim.

3.2 Characteristics of the collected data

As stated in Table 1 page 323, this database contains ductwork airtightness data of 33 buildings, which is the work of 5 technicians. There are 113 test results representing 93 unique ducts. From all these buildings, 2 are certified Passivhaus and 13 have another certification.

Table 1: Characteristics of the database

Number of unique buildings in the database	33
Number of technicians who contributed	5
Number of unique tests performed	113
Number of unique ducts tested	93
Number of buildings under certification	15
Total duct surface area tested	4757,87 m^2
Mean duct surface area tested	42,11 m^2

The buildings in the database are quite recent: 15 have been built in 2013or after, 9 in 2012, 3 in 2011 and only 4 in 2010 or before. Most of these buildings could therefore be considered as recent low-energy buildings.

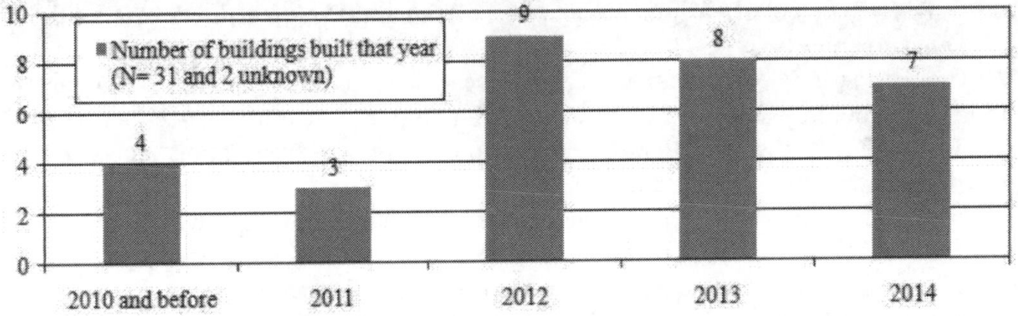

Figure 2: Year of construction (N=31 and 2 unknown year of construction)

As can be seen in Figure 3, the buildings of the database are mostly single-family dwellings, multi-family dwellings, office buildings and the remaining buildings are either education buildings, sports buildings or other/unknown.

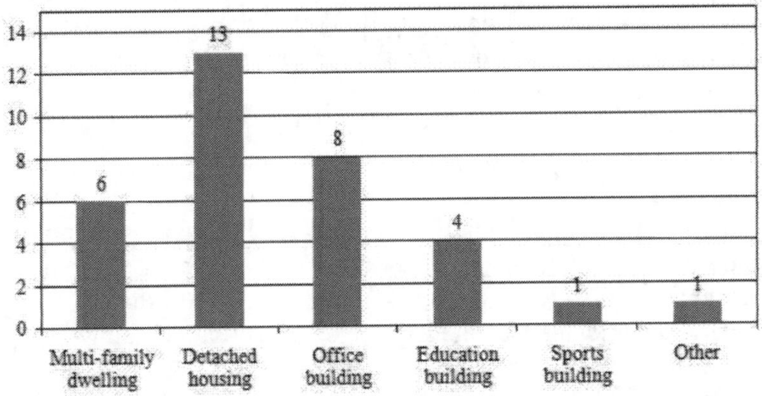

Figure 3: Type of building in the database (N=33)

31% of these buildings are located in South-East France and 61% in the East of France. This is in a certain measure a bias in the database. Indeed, newly built buildings in southern France are less dependent on energy gains through efficient ventilation than average. And on the contrary, buildings in eastern France are much more dependent on energy gains from ventilation than average in France. From this point of view, the buildings of the database collected so far can only be held representative for a part of France buildings.

3.3 Strategies in ventilation to achieve low energy buildings in France

As stated before, the data collected gives in a certain measure a picture of ventilation systems in low energy buildings in France. This section will only deal with detached housing, multi-family dwelling and office buildings for the rest of the buildings cannot be considered as representative.

Overall in the buildings of this sample, mechanical exhaust ventilation systems is used: 73% with mechanical ventilation and 18% with mechanical exhaust ventilation and natural air inlet. The other ducts tested were either non mechanical ventilation ducts or earth cooling tubes.

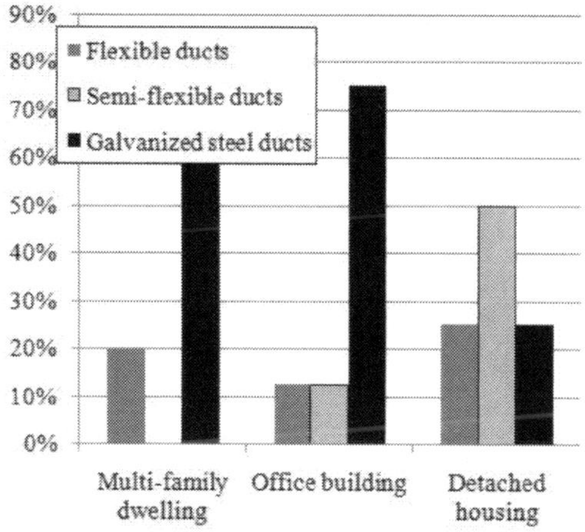

Figure 4: Ducts per building type (N=25 and 8 unknown)

As shown in Figure 4, flexible ducts, semi-flexible ducts and galvanized steel ducts are the three types of ducts used in the buildings of the database. Flexible ducts are mainly found in single family dwellings. As for galvanized steel ducts they are mainly found in office buildings but are also found in detached housing and multi-family dwellings.

It is quite unusual to find that galvanized steel ducts are found in 25% of detached housing whereas flexible ducts are actually widely used in this type of buildings. Plus, semi-flexible ducts are also found in 50% of detached housing. Since the database is constituted of tests in buildings where owners were willing to have their system controlled, we can safely assume that their goal was higher performance than basic regulation requirements. Hence we can understand that in order to achieve higher performance, craftsmen tend to use more reliable systems like galvanized and semi-flexible ducts, which are less likely to be damaged by simple manipulation and during implementation.

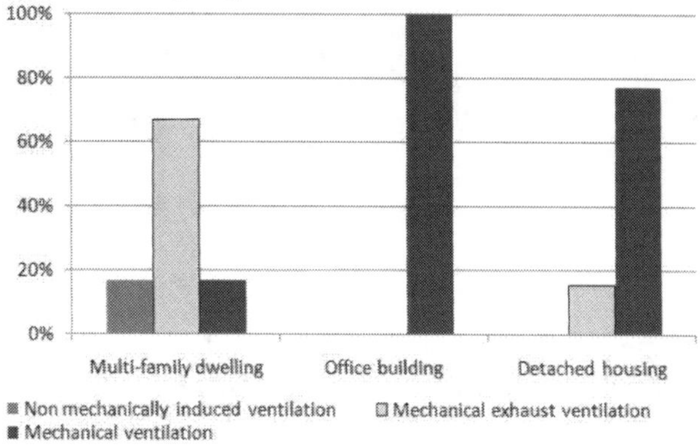

Figure 5: Type of ventilation system per type of building (1 unknown among detached housing)

Figure 5 is mostly interesting for the distribution of mechanical exhaust ventilation and total mechanical ventilation among the different types of buildings. Where mechanical exhaust ventilation is very widely used in French residential buildings (between 85 and 95% see (Promotelec,2012)), buildings in the database collected show more systems combining mechanical air inlet and mechanic outlet. Here again, this could be considered as a trend in very-low energy housing, especially when specific attention is brought to achieve performing ventilation.

3.4 Ductwork airtightness performance

As mentioned earlier, the database collects 113 tests in 36 unique buildings. When only taking into account tests that have been made at 80Pa or more in absolute, there are **103 valid tests**. The rest of the data is not taken into account and concerns tests at very low pressure because the measurement device used was not able to induce higher airflow rates. Two are measured at 20 Pa regarding a non mechanical ventilation duct which is designed to work at 20 Pa.

From Table 2 can be inferred that ducts and ducts samples achieve rather performing classes. 75 % achieve a class A or better and more than a quarter achieve a class C or D.

Table 2: Airtightness classes achieved all tests included (N=103)

Class D	Class C	Class B	Class A	3xClass A	Higher than 3xClass A
4 %	25 %	29 %	17 %	19 %	6 %
4 %	29 %	58 %	75 %	94 %	100 %

However, when put in perspective, it appears that some buildings have undergone many tests: because airtightness of all in- and outlet ducts in the buildings were measured or because the technician made many samples. Plus, some tests have been performed during construction work and is not representative of the ventilation ductwork at commissioning. This results in multiplying good results whereas most of the buildings perform rather poorly as can be seen in Figure 6.

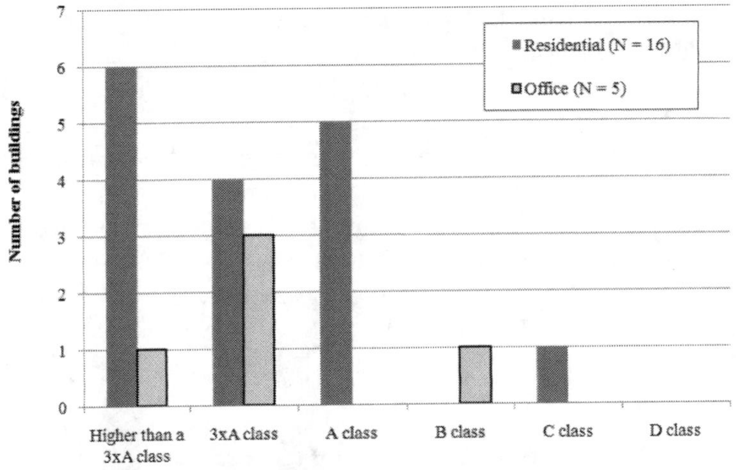

Figure 6: Final performance of residential and office buildings in the database at commissioning (N=21 buildings tested exactly at commissioning in the database)

As a summary, Figure 7 shows the number of tests for all 33 buildings (3 naturally ventilated buildings excluded)for both ductwork sample and final airtightness performance of the ventilation ducts. Some buildings have several separated ducts, which explains the 41 ducts classed.

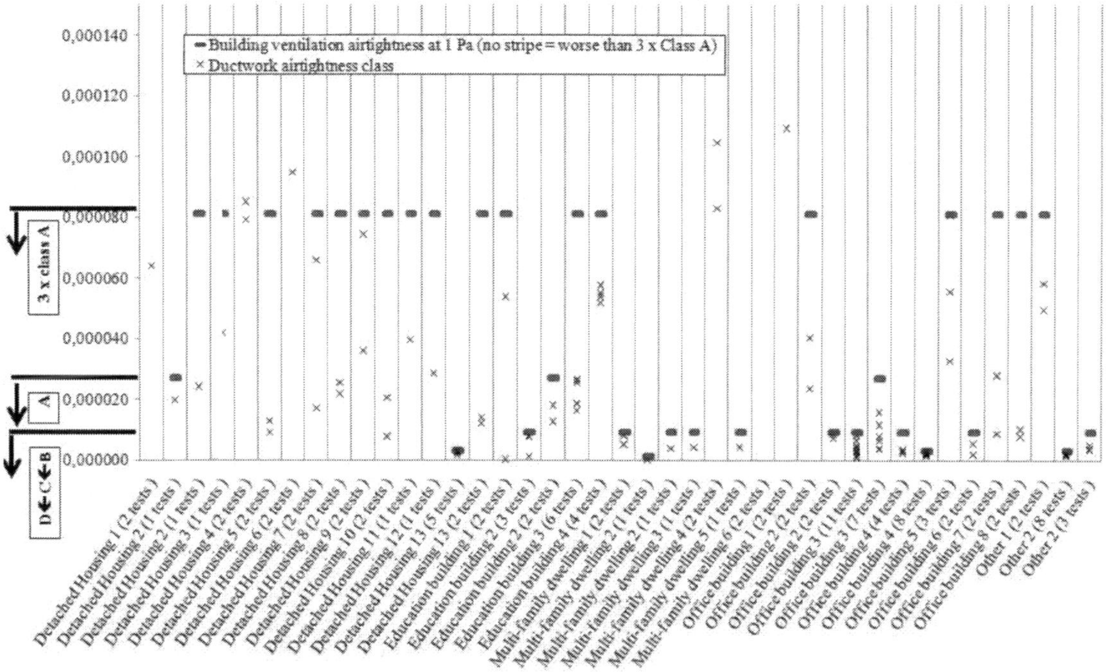

Figure 7: Ductwork airtightness at 1Pa and final building ventilation airtightness at 1Pa ($N_{buildings}$=33 for 41 ductwork performance calculated)

As stated in 2.2, the default value in the EP calculation is almost equal to 3 times a class A, a value that can be considered as a reference. Figure 6 and Figure 7 show therefore that more than half of the ducts perform worse than the default value in the EP calculation.

Furthermore, only 17 out of 41 ducts perform at a class A or better, which should be at least the reference now in very low energy construction, if not the classes B or better.

4 DISCUSSION AROUND THE IMPLEMENTATION QUALITY OF DUCTWORK

4.1 Comparison to past and present French studies results

The results acquired from the collected data show a picture of ductwork airtightness performance. These results, as representative as they can be considered, show in a certain measure the performance of ventilation ducts in low-energy newly built construction in France.

Earlier studies had shown that ventilation ducts airtightness performance in France was quite poor in comparison to other countries. For example the 1999 Save-ductstudy from AIVC (1999) showed that among 21 buildings in France, 60 % performed 3 x class A, 10 % a class A, 5% a class B and the 25% left worse than 3 x class A.

The 2009 French PREBAT Performance study showed the benefits of an integrated quality approach before and during construction with multi-family dwellings performing a class B while a second building which had not fully integrated their quality management system performed worse than 3 times the class A.

In comparison, it is obvious that the ventilation ducts in newly built buildings currently in France do not perform much better than 15 years ago according to the Save-duct project. With only a third of buildings at commissioning performing at class A or better, ductwork performance is not easily achieved.

4.2 Visual inspection for air leakage

If the Effinergie + certification program requires a detailed visual inspection including basic checks on the air handling unit, on the air inlet and outlet grids or on the state of the ducts themselves, the buildings of the database have not undergone any visual inspection. However, an air leakage inspection has been fulfilled on some of the buildings of the database and the results are described hereunder.

Ductwork air leakage in the buildings of the database can be dissociated into two types: leakage inherent to the ductwork system itself and leakage due to the implementation. In the first category, leaks inherent to the ductwork elements themselves, air leakage was found all around flexible ducts because of very small perforations of the ducts.

In the second category, leaks due to implementation and therefore manipulation are mainly found at the connection between two ducts or between a duct and an element. In these cases, leakage was found where the airtightness treatment was poorly managed. In particular, leaks were found at the junction between two different types of ducts or elements, like between flexible and galvanized steel ducts, or between flexible ducts and the plastic sleeve joining the flexible duct and the air in- or outlet grid. Leaks were also found at the connection between rectangular galvanized steel ducts.

4.3 Achieving high performance in ventilation airtightness today

As stated in 3.3, some configurations found in the buildings of the database were quite unusual. We suspected a tendency among craftsmen for the use of particular systems or type of ducts to achieve high performance. The results of the airtightness measurements of the database corroborate the idea in a certain measure.

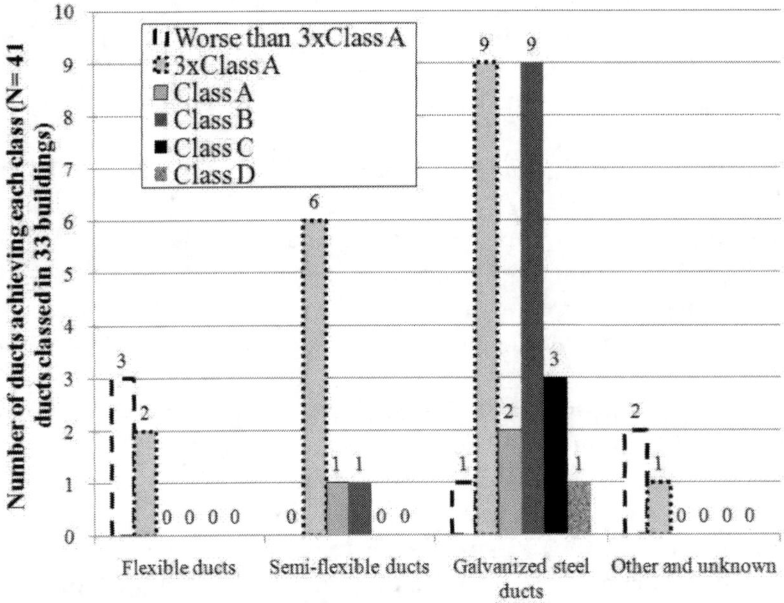

Figure 8: Class achieved by the 41 ducts tested (33 buildings)

Figure 8 shows the classes achieved by the ducts tested. Admittedly, the amount of ducts tested is not statistically representative but it shows a certain tendency. The flexible ducts did not achieve better than three times a class A. Semi-flexible ducts achieved higher performances with at least three times a class A. Galvanized steel ducts achieved the highest performances with more than a half of the ducts achieving a class A or better.

The results from the airtightness tests in Figure 8 and from the leakage analysis in caption 4.2 might therefore indicate that the use of flexible ducts is a risk for airtightness of the system. On the contrary, we can assume that it is less risky to use galvanized steel ducts, combined of course with the adaptedjunction and connection systems, in order to achieve good airtightness of ductwork.

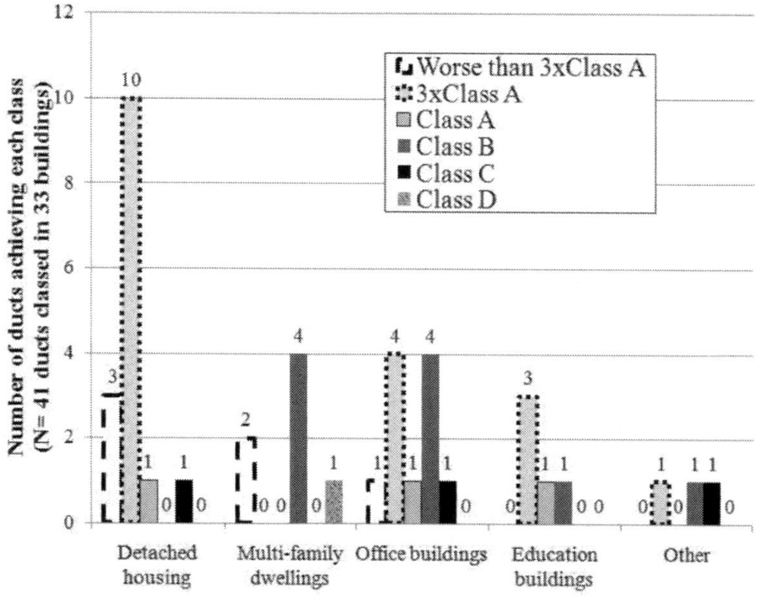

Figure 9: Class achieved for each type of buildings

Figure 9 pictures the classes achieved for each type of buildings: detached housing, multi-family dwelling, office building or education building. It is clear that detached housing dos not perform as well as multi-family dwellings or office buildings. This confirms the lack in France of real craftsmanship for ventilation, particularly in detached housing. Most of the time, the ventilation system is implemented by the plumber, the electrician or the plasterer. In this frame, the lack of knowledge, training and practice is highly probable and results into various problems on site regarding the quality of the implementation of ventilation systems. These results underline once again the need for the rise of an expert craftsmanship.

4.4 Discussion on the sample representativeness for French low-energy buildings

The database collected cannot quite be considered as a perfectly representative sample of new constructions in France. First because the amount of data is not large enough to draw reliable conclusions. Also because it is issued from ductwork airtightness data, which unsurprisingly requires ventilation ducts in a building. So the database does not account for naturally ventilated building which might be used in very-low energy buildings today. Plus, building owners probably order tests more oftenfor fully mechanical ventilation than for mechanical exhaust ventilation, which is today largely used in France. If we can surely define trends, it is not possible to draw definite conclusions yet.

5 CONCLUSIONS

The database issues from the collection of data from technicians undergoing a Qualibat qualification, which only a few underwent today. However this number is expected to drastically rise which means that a larger amount of data will enrich the database and allow exploiting it with a better statistical significance and accuracy.

Exploiting these tests still showed valuable results on both the picture of low-energy buildings today and airtightness performance of ventilation ducts in these buildings. If the amount of data does not allow definite conclusions, it shows quite concerning trends in the performance achieved today in France, which is not much better than 5 or 15 years ago, according to previous studies. There is therefore reasonable doubt about the air renewal achieved in buildings where ventilation systems are indeed not properly implemented.

This is a serious concern because in the meantime, the 2012 French Energy Performance Regulation has set a high requirement on the airtightness of the envelope of residential buildings. All of them undergo an airtightness test at commissioning and have to perform better than 0,6 $m^3/h/m^2$ at 4 Pa. At this level of airtightness, it is required to guarantee a good air quality by a performing ventilation system and therefore avoid sick building syndromes in newly-built buildings. There is today an urgent need in creating an expertise area around ventilation craftsmanship and around control technicians to check the performance at commissioning.

6 ACKNOWLEDGEMENTS

We would like to thank ALLIE'AIR for their kind contribution without which this paper could not have been published.

7 REFERENCES

CSTB (2013), Méthode de calcul Th-BCE 2012, Annexe à l'arrêté du 30 avril 2013 portant approbation de la méthode de calcul Th-BCE 2012

AIVC (1999)and SAVE-DUCT project partners, Improving ductwork – A time for tighter air distribution systems – A status report on ductwork airtightness in various countries with recommendations for future designs and regulations

Promotelec (2012), La construction de bâtiments basse consommation – Retour d'expérience et analyse de Promotelec

Air-h (2009), Performance de la ventilation et du bâti, Rapport final

AIR LEAKAGES IN A RETROFITTED BUILDING FROM 1930: MEASUREMENTS AND NUMERICAL SIMULATIONS

Pär Johansson*[1], Angela Sasic Kalagasidis[1]

[1]*Chalmers University of Technology*
Sven Hultins gata 8, 412 96 Gothenburg, Sweden
**Corresponding author: par.johansson@chalmers.se*

ABSTRACT

Many buildings in Sweden are in need of renovation in order to meet the current standards of energy use in buildings. Particularly challenging are old listed buildings, the majority built before 1950, because the renovation is restricted to the parts that do not change the appearance of the building. This paper presents experiences gained during the renovation of a listed building where aging of materials, movements and settlements during the building operation have left trails in form of air leakage paths that are difficult to cover by renovation. The aim of the work is to bring up certain issues related to air movements in old houses that are not necessarily present in new ones. In the renovation project in question, efforts were made to seal the exterior walls with polyethylene foil before additional insulation in form of vacuum insulating panels was installed. The blower door tests taken in the apartments showed that the air permeability of the building remained basically unchanged after the retrofit. Complementary diagnostics with thermal imaging camera revealed that air leakage paths in intermediate floors and interior walls substantially contributed to the overall air leakage of the building. It was then concluded that the blower door tests alone were not sufficient for the estimation of the air permeability of the building. In addition, hygrothermal measurements in exterior walls showed that new air leakage paths were created between the old wall and the additional insulation. By using numerical simulations, it has been proven that it is the outdoor air that flows through these new air paths, which is advantageous from the moisture safety point of view. However, the flow of outdoor air through the exterior wall increases the overall thermal transmittance of the wall and thus decreases the effects of additional insulation.

KEYWORDS

Retrofit, listed buildings, hidden air leakage paths, increased thermal transmittance, moisture safety

1. INTRODUCTION

There are approximately 2.1 million buildings in Sweden distributed on 1.9 million single family houses, 165 000 multi-family buildings and 46 000 commercial buildings (Boverket, 2009). These buildings are from different time periods which mean they were built with different building techniques and technical solutions. A majority (47%) of the buildings were built before 1960, while 32% were built during 1961-1975. Buildings from the following 10 years comprise 12% of the building stock and thereafter 31% from 1986-1995 and 12% from 1995-2005 (Boverket, 2010).

Building codes were first implemented in 1946 in Sweden (IEA, 2013) and the first energy use requirements were introduced in 1975 after the oil crisis in 1973-1974. The requirements were specified with maximum U-values and demands on the airtightness for different building parts. The codes have been developed during the following years, tightening the demands on the energy use. The latest performance based energy codes are aiming at reducing the energy use further by introducing the same demands on retrofitted buildings as for new developments (Boverket, 2011). The requirements on airtightness have also varied over the years. Currently there are no requirements on maximum air permeability of the buildings in Sweden, except for specifications of passive houses (less than 0.3 l/(m^{2}s)). Figure 1 shows the air permeability in a selection of newly constructed buildings in Sweden. As can be seen, the majority of the buildings have an air permeability of around 0.8 l/(m^{2}s), which was the air permeability requirement in the Swedish building code until 2008. However, there are buildings with much larger air leakages (air permeability above 1.5 l/(m^{2}s)).

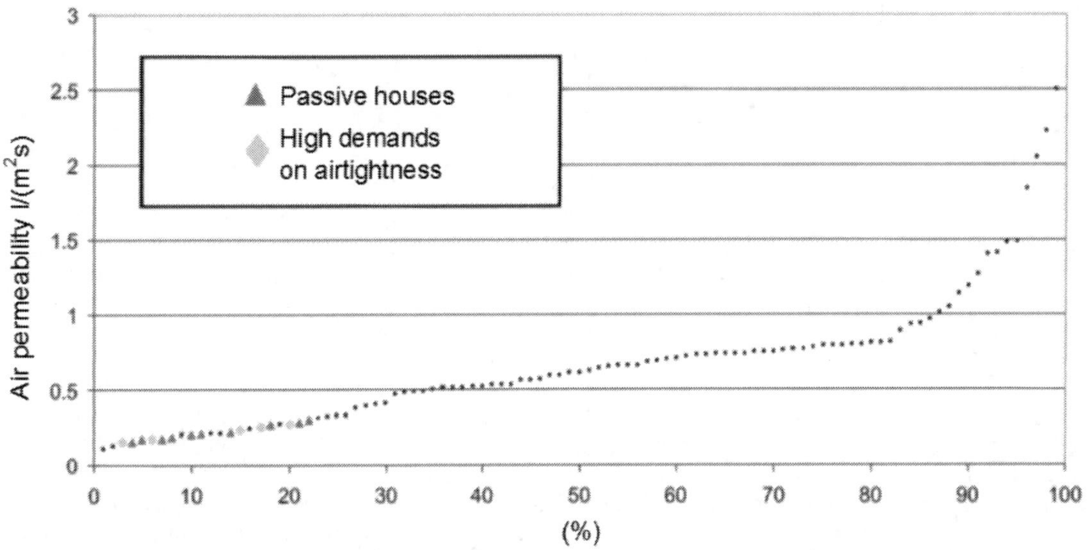

Figure 1: Airtighness in 100 newly constructed Swedish buildings. The sample includes single and multi-family buildings, (individual apartments are measured) and schools; both wooden and concrete structures (Wahlgren, 2010).

Low air permeability of a building envelope is essential to avoid unwanted consequences such as high energy use and insufficient thermal comfort, caused by draft and cold surface temperatures. Other consequences by a leaky building is that spreading of odour, particles and radioactive gases, such as radon from the ground, can take place between apartments. Also the acoustic insulation between apartments is affected by air leakages. Sandberg et al. (2007) discussed the economic consequences by insufficient air tightness. They concluded that a building with a good air tightness is, with a large probability, profitable for the builder and leads to increased well-being, less spreading of odours and better acoustic insulation.

There is much knowledge and techniques on how to achieve good airtightness in buildings, but one gets the impression that the focus is mainly on new buildings. As presented above, a majority of the building stock in Sweden is older buildings. This is also the case for the rest of Europe. One of the keys to successful retrofit is to achieve control over the air movements in these buildings. Natural or climate induced aging of materials and movements and settlements during the building operation have left trails on these buildings. These are typically visible as angle and surface distortions, cracks, and displacements. A large number of air flow paths through the building can be created by these geometrical changes. Since many of them are hidden inside the building envelope, they need to be treated in renovation projects to prevent undesired heat losses, comfort disturbances, or moisture damages.

The aim of this paper is to bring up certain issues related to air movements in old houses that are not necessarily present in new ones. Therefore, we present here the findings from a field study on airtightness and air movements in an old house, before and after the retrofitting. Unexpected air paths in the building envelope are revealed by hygrothermal measurements. These are further analysed and explained by numerical simulations.

2. CASE STUDY BUILDING

The building chosen for the study is a landshövdingehus "County governor's house" built in 1930 in Gothenburg, Sweden, see Figure 2. Most of these buildings were constructed during 1876-1936 when fire regulations limited the height of wood buildings to two floors. Today there are over 1 400 similar buildings in the Gothenburg region. The exterior aesthetics of the building are protected by Swedish legislation as a cultural environment of national interest. The building contain rented apartments and there are many complaints on draught and insufficient thermal comfort from the occupants. Therefore the building is in great need of retrofitting measures (Johansson et al., 2014).

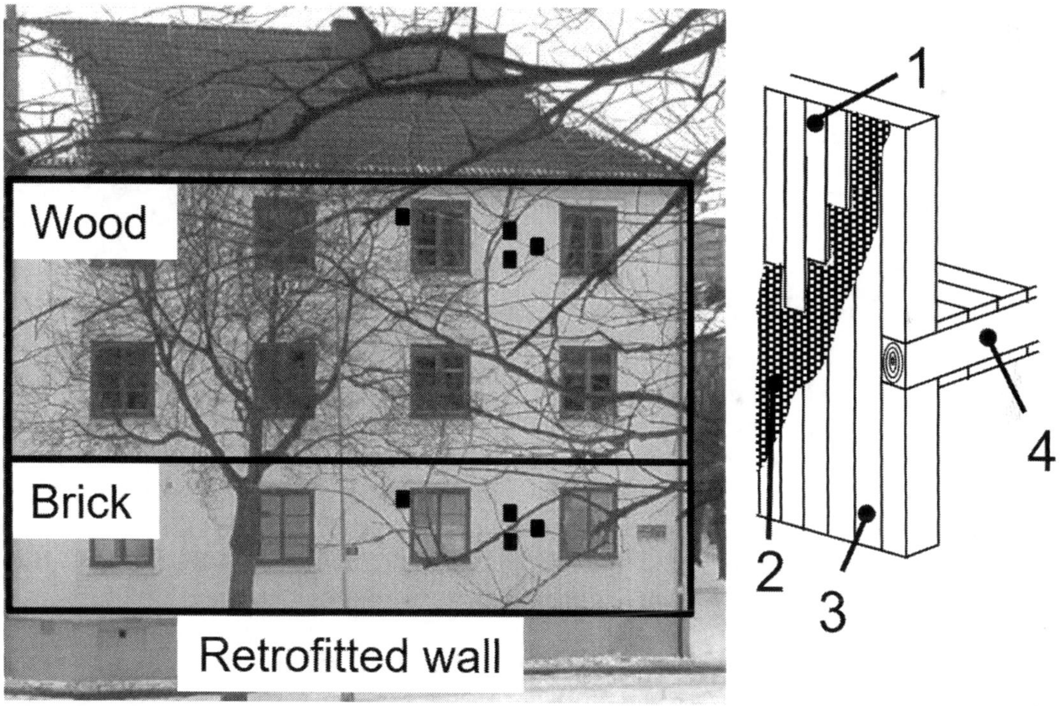

Figure 2: Left: The studied Landshövdingehus "County governor's house" with approximate location of the hygrothermal sensors in the walls. Right: construction of the original wood wall. 1) Wooden cover boarding, 2) Tar paper, 3) Structural wood or brick, 4) Intermediate wooden floor

In the time when the building was constructed, thermal insulation was normally not used in the walls. This building was no exception with brick walls of 1.5 stone thickness, approximately 340 mm, in the ground floor. In the two upper floors, the walls were made of 80 mm wooden planks in three layers. Flax fibres were used between the boards to decrease the air permeability of the wall. On the exterior, a 22 mm thick vertical wooden cover board with rib flanges was installed on top of a wind and waterproof tar paper as shown in Figure 2.

The interior side of the walls was originally covered with a thin finish of plaster on reed (Larsson & Lönnroth, 1972), which was later replaced by modern plaster boards.

Due to the very limited space for additional insulation in the wall, the retrofitting, presented in Figure 3, was done with 20 mm thick vacuum insulation panels (VIPs) placed on the exterior side of the wall and protected by 30 mm of glass wool. Between the old wall and the VIPs, a polyethylene foil was applied as an air barrier to prevent indoor air entering the wall. An air space, 28 mm thick, was added to the façade which makes the total additional thickness of the wall to be 80 mm (Johansson, 2014).

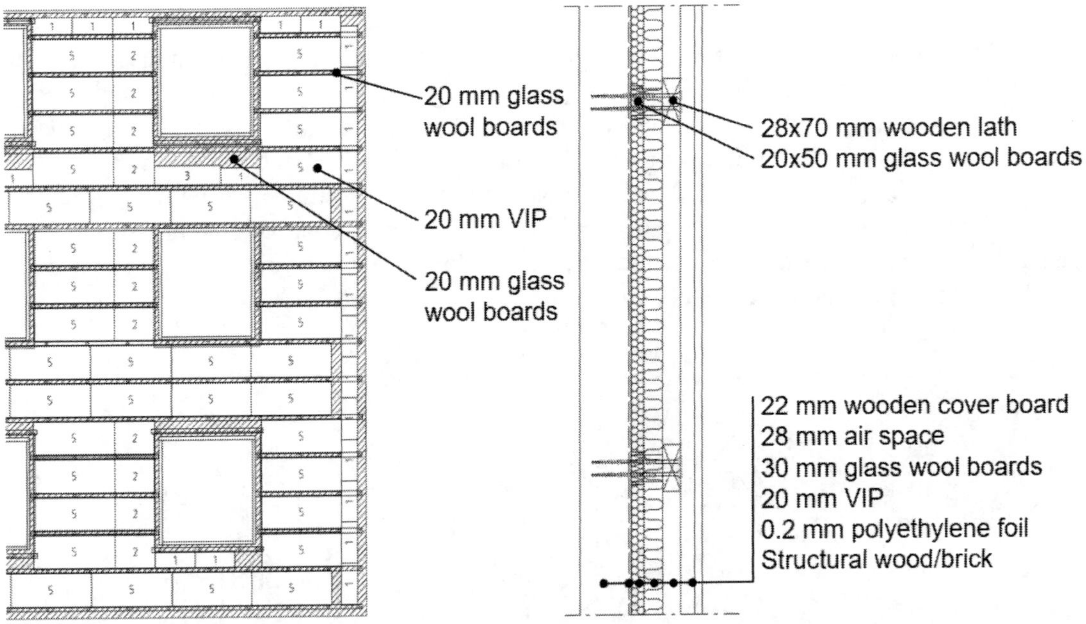

Figure 3: Wall layout after retrofitting with 20 mm VIPs and 30 mm glass wool boards.

The energy use for heating and domestic hot water before retrofitting was estimated to 160 kWh/m^2 (exact figures are not possible to obtain because the energy use in this building is measured together with many other buildings in the area). It is estimated that the additional insulation of the wall has reduced the energy use by 20% (with all VIPs functioning and all thermal bridges included). As a comparison, changing the windows to windows with a U-value of 1 W/m^{2}K gave an energy use reduction of 15%. The combination of changing windows and installing VIPs gave an energy use reduction of 34% (Johansson, 2014).

To evaluate the hygrothermal performance of the wall after the retrofitting, four temperature and relative humidity (RH) sensors were installed in the brick and wood wall, respectively. The sensors were located on the exterior of the existing wall, before the polyethylene foil. To study the influence of the indoor climate in the building, one sensor was located in one of the kitchens closest to the monitored part of the wall in the brick and wood wall, respectively. The outdoor temperature and RH at the building site was monitored by a sensor located in a perforated plastic box placed underneath the roof eave facing southwest (Johansson, et al., 2014).

3. RESULTS FROM AIR PERMEABILITY MEASUREMENTS

There were complaints from the occupants on an insufficient thermal comfort in the building due to draft and low surface temperatures. Thus, the building was investigated before and after the retrofitting using blower door and infrared thermography (Svensson, 2010, 2011). It was found that the interior wall surface temperature was around 1.5-2.4°C lower than the indoor air temperature before the retrofitting. After the retrofitting the surface temperature was only 0.3-0.8°C lower than the indoor air.

The blower door measurements were performed in single rooms since only one of the façades was equipped with VIPs. The results before and after the retrofitting are presented in Table 1. The air leakage paths were dominated by leakages around the windows, at the connections between the interior and exterior wall and along the floor (see Figure 4). These air leakage paths remained after the retrofitting but from the air temperature we can conclude that it is the indoor air that infiltrates through these leakages. Any further general conclusions of changed air tightness after the retrofitting could not be made based on the blower door measurements performed in single rooms. A better picture would be obtained if the blower door measurements were supplemented with, for example, tracer gas measurements that could separate infiltration of outdoor air from the air leakage paths through interior walls and floors.

Table 1: Results from blower door measurements at ±50 Pa (Svensson, 2010, 2011). According to Swedish regulations, the leakage area is the area that separates the indoor from the outdoor environment. The ground floor apartment is above the unheated basement, therefore, the leakage area includes both the wall and floor.

Floor	Size (length x width x height)	Leakage area (m^2)	Room volume (m^3)	Before retrofitting ($l/(s \cdot m^2)$)	Before retrofitting (1/h)	After retrofitting ($l/(s \cdot m^2)$)	After retrofitting (1/h)
Ground floor	3.9 x 4.4 x 2.7	27.7	46.3	2.6	5.6	2.7	5.8
1st floor	3.8 x 4.8 x 2.7	10.3	49.2	4.3	3.2	3.8	2.9

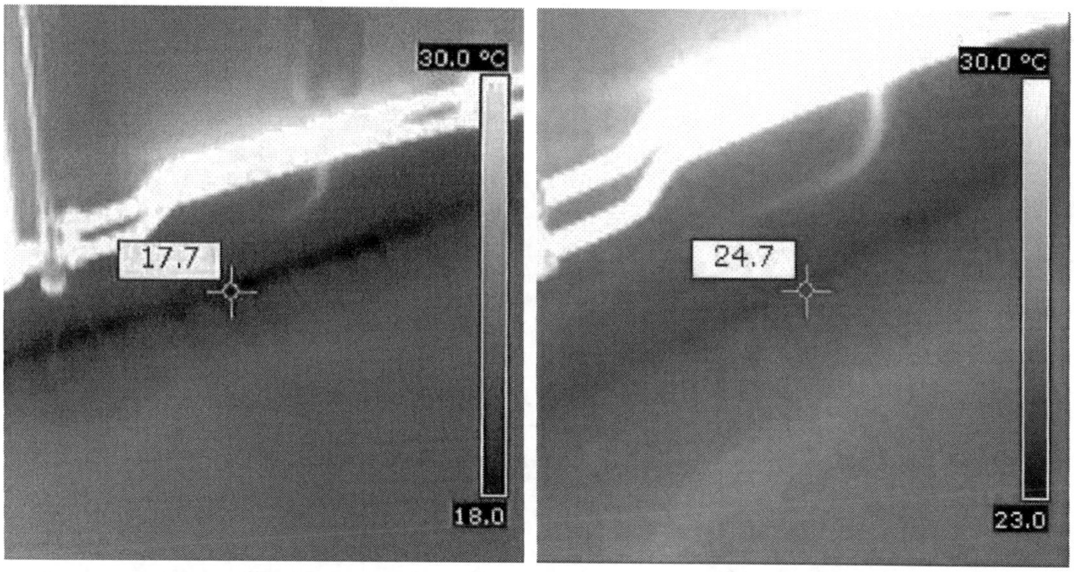

Figure 4: Thermograms showing air leakages in an apartment on the 2nd floor along the floor to wall connection before (left) and after (right) the retrofitting. The air leakage is less after the retrofitting (Svensson, 2010, 2011). At the site visits before and after the retrofitting, the indoor temperature was 21°C and 23°C and the outdoor temperature was 4°C and 6°C, respectively. The temperature of the air passing through the air leakage is around 24°C, which indicates that it is indoor air, possibly from the apartment below.

4. COMPARISON BETWEEN MEASUREMENTS AND HYGROTHERMAL SIMULATIONS

The temperature and RH was measured in the wall by the sensors described in Section 2. The measurements were compared to numerical simulations in the hygrothermal calculation tool WUFI 2D (Fraunhofer IBP, 2010). This software solves coupled heat and moisture transport equations by finite volumes where the temperature and RH are the driving potential for the heat and moisture transport through the construction. The numerical simulations were performed using the measured indoor and outdoor climate together with material data for the materials in the construction, see Table 2. Air flow in the wall is simulated by adding a coupling for the heat and moisture transfer between the wall and the indoor or outdoor air, respectively. The resulting temperature and RH in the wall without air leakages is presented in Figure 5.

Table 2: Material data used in the numerical simulations based on data from WUFI 2D material database (Fraunhofer IBP, 2010).

Material	*d (mm)*	λ (mW/(m·K))	ρ (kg/m^3)	c_p (J/(kg·K))	μ (-)
Gypsum board	20	200	625	850	8.33
Spruce, tangential	80	200	430	1600	83.3
Spruce, radial	80	200	455	1500	130
PE membrane	0.2	1 650	130	2200	$8.7 \cdot 10^4$
Glass wool	-	33	60	850	1.3
Glass wool board	20/30	33	115	850	3.4
Air layer	30	25	1.3	1000	0.46
Evacuated VIP core	18	5	200	850	1.3
Air filled VIP core	18	20	200	850	1.3
VIP laminate, tangential	1	540	189	134	Inf.
VIP laminate, radial	1	200 000	189	134	Inf.

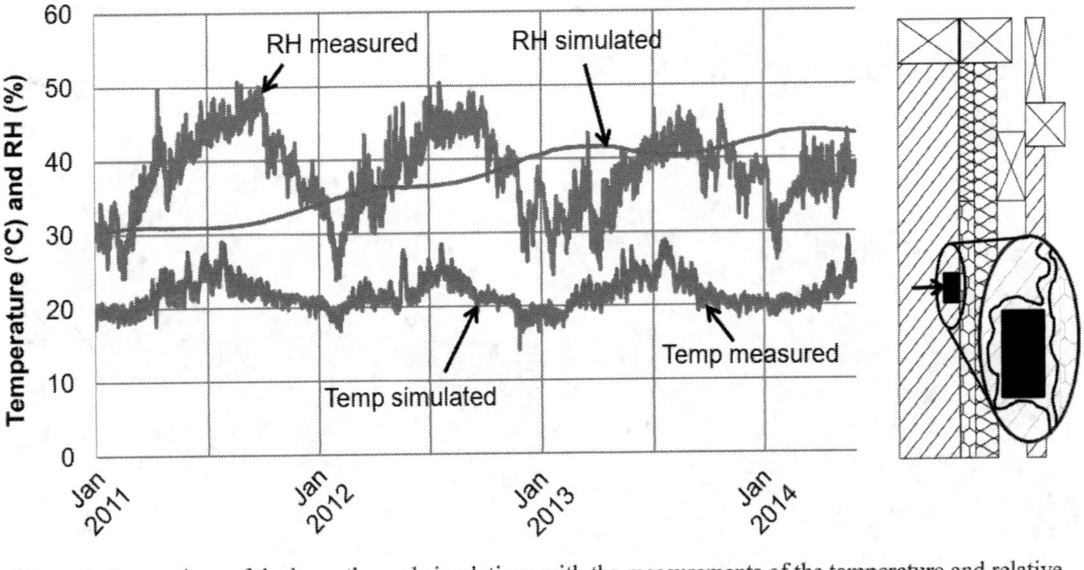

Figure 5: Comparison of the hygrothermal simulations with the measurements of the temperature and relative humidity behind the VIP in the wood wall. The simulation was based on the measured indoor and outdoor climate from January, 2011, to June, 2014. The location of the sensor is marked by the black box with a black arrow pointing from the left.

The average simulated temperature and RH at the sensor position in the wall was 21.9°C and 37.4% while the measurements gave 21.5°C and 37.8% on average during January, 2011, to June, 2014. It was noted that the vapor content was on average 1.3 g/m^3 higher in the numerical simulations than in the measurements. This suggested that there was an additional drying process in the wall. The sensor is located in an air filled void in the wall, measuring the temperature and RH of the air closest to the sensor. The high fluctuation in the measured RH indicated a leaky construction with air leakage from the interior or exterior side of the vapor barrier into the construction. It is anticipated that the air could enter e.g. through details around windows and flow along the interior of the vapor barrier to the sensor position.

Although measures were taken to make the exterior wall air tight with the polyethylene foil, due to uneven surfaces certain connections have remained loose. Therefore, an air exchange with the outdoor air was added to the model between the vapor barrier and wood. The resulting temperature and RH is presented in Figure 6.

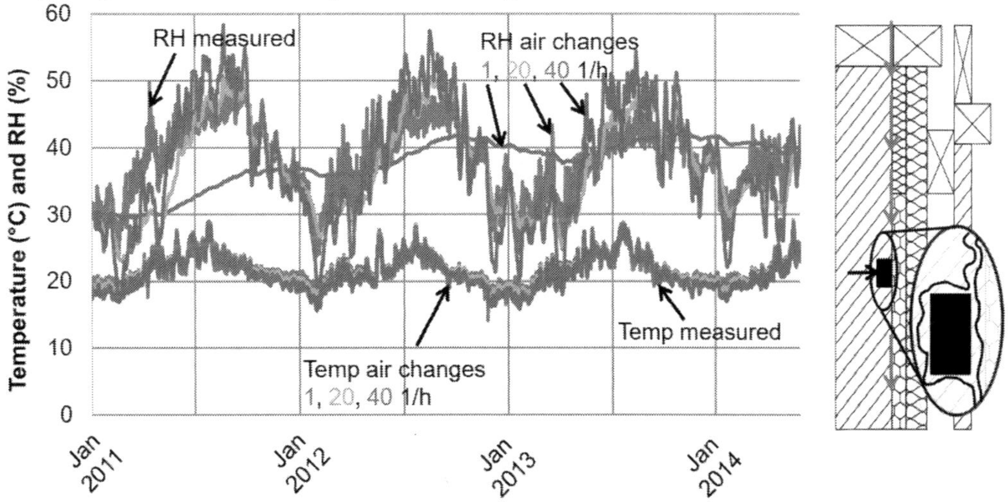

Figure 6: Comparison of the hygrothermal simulations with the measurements of the temperature and relative humidity behind the VIP in the wood wall. The air changes per hour with the outdoor air were varied from 1 1/h to 40 1/h. The location of the sensor is marked by the black box with a black arrow pointing from the left. The material surfaces in the wall are rather uneven which makes air flow through the small voids between them possible.

When an outdoor air exchange rate of 40 1/h was used, the simulated average temperature was 21.2°C. This is close to the measured average temperature 21.5°C. The simulated average RH was 37.6% which is also close to the measured average RH 37.8%. In addition, the simulated variations during the year corresponded well with the measurements. However, one cannot be entirely sure that the air is really coming from the exterior without testing this. Figure 7 presents the temperature and RH in the wall given that the air is coming from the interior.

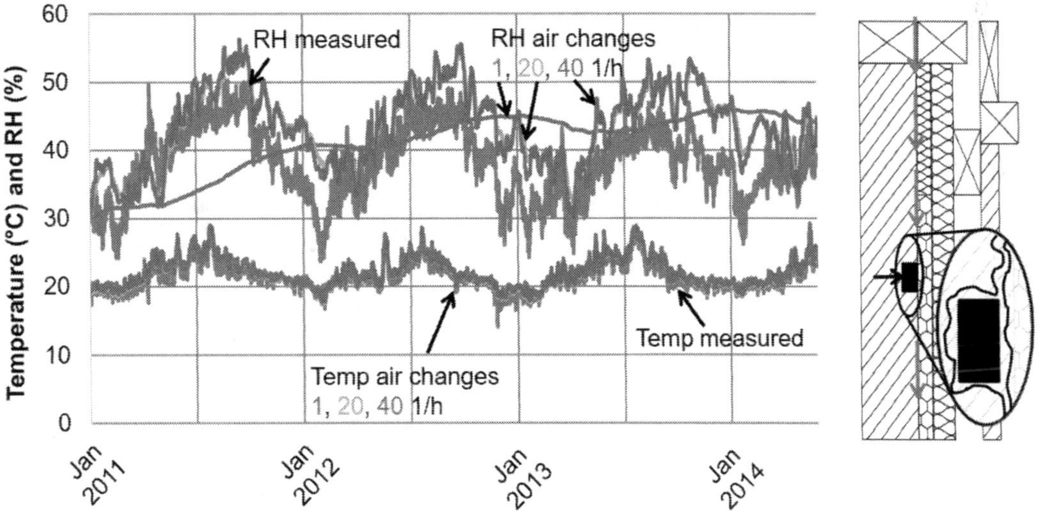

Figure 7: Comparison of the hygrothermal simulations with the measurements of the temperature and relative humidity behind the VIP in the wood wall. The air changes per hour with the indoor air were varied from 1 1/h to 40 1/h. The location of the sensor is marked by the black box with a black arrow pointing from the left. The material surfaces in the wall are rather uneven which makes air flow through the small voids between them possible.

The simulated temperature and RH deviated more when the air was coming from the interior than in the previous case. The simulation gave an average temperature 22.1°C and 43.4% RH for the case with 40 l/h, compared to the measurements that were 21.5°C and 37.8% on average. This shows that the air leaking into the wall is most certainly coming from the exterior.

5. CONCLUSIONS

This work presents the findings from a field study on airtightness and air movements in an old listed building. The building was retrofitted on the exterior with vacuum insulating panels. Although efforts were made to seal the wall with polyethylene foil on the interior of the additional insulation, blower door tests showed that the air permeability of the building remained basically the same before and after the retrofitting. Complementary diagnostics with thermal imaging camera revealed that air leakage paths at the intermediate floors and interior walls substantially contributed to the overall air leakage of the building. This implies that the blower door tests alone is not sufficient for estimation of the air permeability of buildings with large internal air leakage paths, and when the blower test cannot cover the whole building at once.

The hygrothermal measurements in the exterior wall revealed that new air leakage paths were created between the existing wall and the additional insulation. By using numerical simulations, it has been proven that it is the outdoor air that flows through these new air paths. This is advantageous from the moisture safety point of view. However, the flow of outdoor air through the exterior wall increases the overall thermal transmittance of the wall and thus decreases the effects of the additional insulation.

Based on these findings, we conclude that one of the keys to successful retrofit of old buildings is to achieve control over the air movements in these buildings. Knowledge of how potential air flow paths could affect the wall is crucial for making predictions of the long-term performance of a construction based on hygrothermal simulations.

6. ACKNOWLEDGEMENTS

The work is supported by the Swedish Research Council for Environment, Agricultural Sciences and Spatial Planning (FORMAS) and the public housing corporation Familjebostäder i Göteborg AB.

7. REFERENCES

Boverket. (2009). *Så mår våra hus - redovisning av regeringsuppdrag beträffande byggnaders tekniska utformning m.m. (The state of our buildings - report of governmental mission on the technical design of buildings etc.). [In Swedish]*. Karlskrona, Sweden: Boverket.

Boverket. (2010). *Energi i bebyggelsen - tekniska egenskaper och beräkningar - resultat från projektet BETSI (Energy in the built environment - technical properties and calculations - results from the BETSI study). [In Swedish]*. Karlskrona, Sweden: Boverket.

Boverket. (2011). *Regelsamling för byggande, BBR 2012 (Regulations for construction, BBR 2012). [In Swedish]*. Karlskrona, Sweden: Boverket.

Fraunhofer IBP. (2010). WUFI 2D Transient Heat and Moisture Transport (Version 3.3.2) [Computer Program]. Holzkirchen, Germany: Fraunhofer IBP.

IEA. (2013). *Policy Pathway: Modernising Building Energy Codes*. Paris, France: OECD/IEA and New York, NY, USA: United Nations Development Programme (UNDP).

Johansson, P. (2014). *Building Retrofit using Vacuum Insulation Panels: Hygrothermal Performance and Durability*. PhD Dissertation. Gothenburg, Sweden: Chalmers University of Technology, Department of Civil and Environmental Engineering.

Johansson, P., Sasic Kalagasidis, A., Hagentoft, C.-E. (2014). Retrofitting of a listed brick and wood building using vacuum insulation panels on the exterior of the facade: Measurements and simulations. *Energy and Buildings, 73*(April 2014), 92-104.

Larsson, U., Lönnroth, G. (1972). Landshövdingehus och trähus i Göteborg (County governor's houses and wooden houses in Gothenburg). [In Swedish]. *Den nordiske træstad*(28).

Sandberg, P. I., Sikander, E., Wahlgren, P., Larsson, B. (2007). *Lufttäthetsfrågorna i byggprocessen - Etapp B. Tekniska konsekvenser och lönsamhetskalkyler. SP Rapport 2007:23. (Air tightness in the buliding process - Stage B. Technical consequences and cost benefit calculations)*. Borås, Sweden: SP Technical Research Institute of Sweden.

Svensson, O. (2010). *Termografering och lufttäthetsmätning, Wärnsköldsgatan 9B-9C, Göteborg. (Thermography and air permeability measurement, Wärnsköldsgatan 9B-9C, Gothenburg). Unpublished report*. Borås, Sweden: SP Technical Research Institute of Sweden.

Svensson, O. (2011). *Uppföljande Termografering och lufttäthetsmätning, Wärnsköldsgatan 9B-9C, Göteborg. (Follow up thermography and air permeability measurement, Wärnsköldsgatan 9B-9C, Gothenburg). Unpublished report*. Borås, Sweden: SP Technical Research Institute of Sweden.

Wahlgren, P. (2010). *Goda exempel på lufttäta konstruktionslösningar. SP Rapport 2010:09. (Good examples of airtight details)*. Borås, Sweden: SP Technical Research Institute of Sweden.

INDOOR ENVIRONMENTAL QUALITY-GLOBAL ALLIANCE: THE VALUE CHAIN OF IEQ

Bjarne W. Olesen

ABSTRACT

Recently a global alliance on Indoor Environmental Quality (IEQ) has been formed. This alliance consists of several international societies dealing with the indoor environment. The intend is that the alliance should cover all partners involved in the value chain of IEQ from research, design, manufacturing, construction (building properties), installation (building service systems), operation, inspection to the behaviour of the occupants. The presentation will present the intend and scope of this alliance and discuss the value chain of indoor environmental quality.

DO EXISTING INTERNATIONAL STANDARDS SUPPORT VENTILATIVE COOLING?

Bjarne W. Olesen

ABSTRACT

The use of ventilative cooling must meet general criteria for indoor environmental quality like thermal comfort, indoor air quality and noise. These criteria are often expressed in standards, building codes and guidelines. This presentation will discuss existing standards like EN15251, EN13779, ASHRAE-55, ASHRAE-62.1, ASHRAE 62.2 and others with regard to the use of ventilative cooling. The limitation for the use of ventilative cooling can be the specified acceptable comfort range for room temperatures, draft sensation, noise, safety and others. For acceptable room temperatures the standards includes different concepts for buildings with and without mechanical cooling (adapted approach). The presentation will discuss the difference in these concepts and the relation to ventilative cooling.

THE INFLUENCE OF TRAFFIC EMISSION ON IAQ, ESPECIALLY IN STREET CANYONS

Zbigniew Bagieński
Institute of Environmental Engineering, Poznan University of Technology, ul. Piotrowo 3a, 60-965 Poznan, Poland,
e-mail: zbigniew.bagienski@put.poznan.pl

ABSTRACT

Traffic emissions have a significant impact on urban air quality, which particularly concerns street canyons, i.e. spaces with limited air exchange. Traffic emissions in street canyons create high concentrations of air pollutants. Based on measurements carried out for selected routes and model experiments conducted in a wind tunnel, it is shown that a roadway's urban structure has a significant impact on concentrations of pollutants from traffic emission which enter buildings by means of mechanical and natural ventilation systems. In the case of buildings in a narrow street canyon pollution concentrations close to ventilation air intakes may be several times higher than in buildings in wide roadways. The paper also describes a possibility of identifying building surfaces that are particularly exposed to being infiltrated by traffic pollutants from adjacent streets.

KEYWORDS

Vehicle emission, street canyon, urban air quality

1 INTRODUCTION

Among factors that determine indoor air quality (IAQ) chemical composition of air and concentration of fine particles $PM_{2,5}$ have a special significance as quantities which are difficult to control, which is the case for temperature and humidity. If it is assumed that the subject of analysis does not concern industrial sites indoor air pollution concentration values are influenced by:

- Internal factors, such as furnace emissions (e.g. gas cookers) and other technological processes, emissions from dividing structures and fittings, human emissions, flows from other closed spaces,
- External conditions such as local air pollution concentration values and their changeability in time and space, meteorological conditions including wind direction and velocity,
- Ventilation systems,
- Urban structures of buildings and their surroundings.

Indoor air pollution levels are of particular concern because most people (especially in towns and cities) spend 80 to 90% of their time indoors. According to the Environmental Protection Agency (EPA, 2011) indoor pollution levels may be 2-5 times, and occasionally, much higher than outdoors levels. A rise in concentration of certain air pollutants is caused by increased use of more synthetic materials for building and furnishings, use of more chemically formulated personal products in the more air tight buildings with reduced ventilation rates to save energy. Pollutants generated indoors are accompanied by those coming from the outdoor environment, including sulphur dioxide, nitrous oxides, fine and ultra fine PM, volatile organic compounds VOCs, and tropospheric ozone. Ventilation systems used do not alter the chemical composition of inlet (external) air. For ventilation of houses by opening the windows Schembari et al. (2013) found a high statistically significant correlation between personal exposure to NO_x, NO_2 and $PM_{2,5}$ and indoor and outdoor levels of these pollutants in Barcelona. Jensen et al. (2009) presents the results of NO_2, NO_x and $PM_{2.5}$ concentration measurements in various New York City's boroughs and demonstrates slight differences in concentration values of outdoor and indoor air with considerable differences among particular boroughs. Vincent et al. (1997) compares indoor air quality in office buildings with different ventilation systems and pays attention to approximate values of average concentrations of CO_2, CO and particles for systems with natural ventilation, fan coil units and HVAC, and the approximate value of the indoor CO_2/outdoor CO_2 ratio equalling 1.9-2.1 for particular systems.

Urban air quality is diversified and depends not only on the kind and value of pollution emissions, but also on the technical conditions of emissions and the urban structure of a given area. Buildings are responsible for the disturbances in the air flow. The horseshoe vortices can be generated over the upper and side edges of a building as well as the lower part of the windward side. At the lee side the wake region and the cavity recirculation region are generated with a reduced air change. According to Peterka (Peterka et al., 1985), the cavity region can approach z=(2-2.5)H_B in the range x=6H_B, while the boundaries of the wake region can approach z=3H_B in the range x=16H_B, where H_B is the height of a building. Point-source and line-source pollutions (short chimneys and vehicle routes, respectively) emitted in this space cannot disperse freely, which may lead to high pollution concentrations (Bagieński, 2008, Ahmad et al., 2005, Meroney et al., 1966). Figure 1 shows a visualisation of a fume flow in a wake effect carried out in a wind tunnel (Bagieński, 2010).

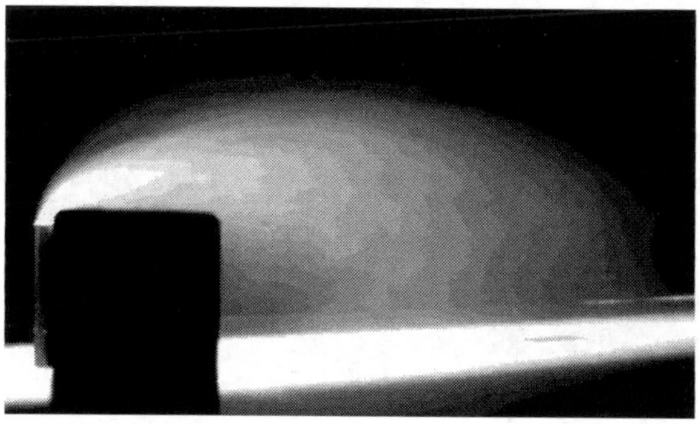

Fig. 1 Disturbed fume flow, short chimney on the windward side of a building

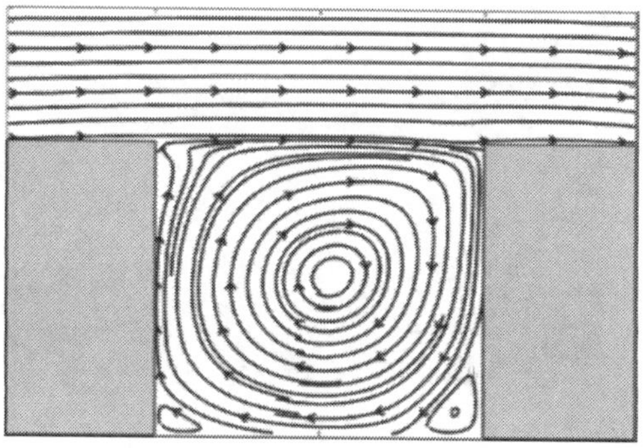

Fig.2 Streamline field in the street canyon for F_C>0.8 (Kwak, 2012)

Traffic emissions are the major outdoor source of directly emitted pollutants in urban regions. Many authors (e.g. Sini et al.,1996, Baldauf et al.,2013, Carpentieri et al., 2012) pay attention to elevated air pollutant exposures and adverse human health effects for population living close to streets with heavy transport traffic. It is especially crucial in densely-built and high-traffic areas in street canyons. Street canyons are not infinitely long or isolated entities in cities, but they are often segmented and connected to other canyons at urban intersections. High pollution levels are often associated with street canyons.

The street canyon factor Fc is the ratio of the average height of buildings delimiting the canyon at the length L>2.5H_B to the total width W of the street.

$$F_C=H_{Bav}/W \tag{1}$$

Depending on the value of the ratio Fc, three types of airflow perpendicular to the canyon symmetry axis can be classified (Ahmad et al., 2005; Oke, 1988):

- skimming flow for F_C>0.8, which is characterized by a significant reduction of air exchange inside the canyon, Fig. 2 (Kwak et al., 2012)
- wake interference flow for F_C=0.20-0.8, where the stream is circulated inside the canyon and part of pollution is transported to adjacent canyons (mainly within the wake region)

- isolated roughness flow for $F_C<0.20$, when the influence of reciprocal interaction of the canyon walls on the interference can be ignored.

The value of F_C has a decisive influence on diffusion conditions of pollution from traffic sources, but also from adjacent low stationary sources (Bagieński, 2006; Chang Cheng-Hsin, Meroney, 2003). When the street canyon factor $Fc>0.8$, high concentration of traffic pollutants can also occur on level of the upper storeys of a building delimiting the canyon (Park et al., 2004, Kwak et al., 2012).

The aim of the paper is to determine the influence of traffic emissions on indoor air quality especially in urban areas of compact settlement and intensive vehicle traffic in street canyons. The assessment is based on two streets with different urban structures in Poznań, a Polish city with over 500 000 inhabitants. For these roadways, concentrations of basic traffic pollution were specified, measurements were taken to determine concentrations of selected substances close to the streets and to ventilation air supplied to adjacent buildings and areas with increased pollution concentrations were identified. Fig. 3 and Fig. 4 show pictures of the analysed streets.

Fig. 3 Street G – street canyon

Fig. 4 Street P – four-line street

2 VEHICLE EMISSIONS

Total emission from road transport is the sum of exhaust emissions, evaporative emissions and *road vehicle tire and brake wear emissions*. Exhaust emissions include *hot* and *cold-start emissions.*

$$E_{Total} = E_{Hot} + E_{Cold} + E_{Ev} \tag{2}$$

Hot emissions depend on the kind and amount of fuel burned and the kind and age of an engine, i.e. motion velocity and a vehicle's technical data such as the category of a vehicle (passenger cars, light- and heavy-duty vehicles, buses), engine capacity, the kind of fuel, time of certification (permissible emission according to EURO standards). Cold-start emissions concern emissions of monoxide and hydrocarbons during the initial stage of the (cold) motion of a vehicle and depend on air temperature. In calculations they are treated as additional emissions considered solely for urban traffic. Evaporation *emissions stem from fuel evaporating from a hot engine, carburettor and fuel tank. They are considered for fuel engine cars and depend on air temperature, humidity and fuel tank pressure related to these values. Vehicle emissions include pollutants such as* CO, NO, NO_2, $PM_{2.5}$, PM_{10}, CO_2, VOCs including mono- and polycyclic aromatic hydrocarbons (MAH, PAH) and benzene C_6H_6, and to CH_4, N_2O, metals like Cd, Cr, Cu, Ni. Values of emission indexes for particular categories of vehicles in road transport are specified in the EMEP/EEA air pollutant emission inventory guidebook (European Environment Agency, 2009).

In recent years particular attention has been paid to PM emission. Fine particles $PM_{2.5}$ (diameter ≤2.5μm) and ultra fine particles (diameter ≤0.1μm) emissions in urban regions are mainly emitted by diesel vehicles (Kim, Y. et al., 2011). Directly emitted elemental carbon soot (black carbon) from diesel vehicles, re-suspended dust from roads, and secondary acidic particles are major contributors of PM in urban roadside areas. The problem is particularly serious in cities with a large proportion of diesel-powered vehicles, e.g. Barcelona – about 50% (Schembari et al., 2013). According to the Air quality in Europe –report_(2013), between 2009 and 2011, up to 96% of city dwellers were exposed to fine particulate matter ($PM_{2.5}$) concentrations above WHO guidelines.

Emission values may be referred to the linear dimension E_{ti} (mg km^{-1} s^{-1}) or the surface as traffic emission strength of a particular area with the i-th substance E_{tia} (mg m^{-2} s^{-1}). Traffic emission strength values were determined for selected pollutants based on real conditions of two streets in Poznan.

Street structure and car traffic variations:

- Street G – a one-way traffic (two-line) street with the traffic volume of 1600 vehicles per hour (v/h), passenger traffic with a 10% share of light-duty trucks, value of factor: F_c>0.8 – street canyon
- Street P – a four-line street with the traffic volume of 3800 v/h, passenger traffic with a 15% share of light-duty trucks, value of factor: F_c <0.2
- Vehicle traffic volume was averaged for the time interval from 7 a.m. to 7 p.m.
- Time intervals for which calculations were made:
  - Summer period: averaging Poznan climatic conditions for June, July, August; t=18.0°C, φ=45%.
  - Winter period: averaging climatic conditions for December, January, February; t=0.5°C, φ=90%

Tables 1 and 2 show emission values of particular pollutants for summer and winter periods for the analyzed streets.

Table 1 Values of substance emission and heat emission for summer.

street	vehicles	CO	NMVOC	NO_x	$PM_{2.5}$	benzene	CO_2	Q_{dL}
	v/h	mg $km^{-1}s^{-1}$					g $km^{-1}s^{-1}$	kW km^{-1}
G	1600	2876.1	382.8	296.7	16.7	18.0	108.0	1594
P	3800	4921.2	643.4	760.6	35.2	30.8	209.2	3090

Table 2 Values of substance emission and heat emission for winter.

street	vehicles	CO	NMVOC	NO_x	$PM_{2.5}$	benzene	CO_2	Q_{dL}
	v/h	mg $km^{-1}s^{-1}$					g $km^{-1}s^{-1}$	kW km^{-1}
G	1600	5212.2	560.2	320.0	25.7	29.1	119.4	1762
P	3800	8965.8	1158.6	782.6	53.0	52.0	231.4	3416

A higher value of the average traffic speed and a lower share of cold emission in street P in relation to street G contribute to lower values of CO, NMVOC and C_6H_6 emissions with higher values of NOx emissions. However, in winter emission of incomplete combustion products (CO, NMVOC, benzene) is much higher compared to summer, which results from an increase in fuel consumption, but first of all from an increase in cold emission indexes at a low air temperature.

The quantity Q_{dL} determines heat flux emitted by vehicles as one of the components that generate the urban heat island (UHI). The values are relatively high. In summer as solar radiation flux density in the UV-VIS range is higher, additional significant anthropogenic heat emission has a considerable impact on chemical reactions that create secondary air pollution.

For example, Taha (Taha, 1996) states that an increase in the Los Angeles local air temperature from 22°C to 32°C caused a rise in ozone concentration from 0.120 ppm to 0.240 ppm. Such a situation takes place especially in narrow street canyons, or areas with limited ventilation conditions, a low *SVF* (sky view factor) value and high radiation absorption values.

3 INFLUENCE OF A STREET'S STRUCTURE ON VENTILATION AIR QUALITY

3.1 Influence of a street's structure on pollution concentrations close to the street

A street's urban structure, often to a greater degree than traffic volumes, contributes to the level of its nuisance defined as concentrations of polluted air – both external (close to the ground) and internal air of the adjacent buildings. An experimental indicator of how a street's structure contributes to its nuisance is the ratio of pollution concentrations measured directly close to the street to the calculated value of the emission of the pollution C/E .

CO concentrations were measured for two streets: street G and street P in the meteorological conditions of summer and winter. The values were measured using the non-dispersive infrared method with a detection limit of 50 ppb. CO analyzers were located approximately 2 m from the street edge at the level of about 2 m, i.e. where the emission source directly affects people. Measurements were taken on working days between 10 a.m. and 12 noon (moderate traffic volume) and 4 p.m. and 6 p.m. (large traffic volume). Wind velocity did not exceed 1 m/s. Measured CO concentration values C_{CO} were referred to CO emissions E_{CO} calculated according to COPERT IV procedure. Tables 3 and 4 present the meteorological conditions of the measurements, traffic volume N, share of light- and heavy-duty trucks (LDT, HDT), the results of measuring the concentrations of C_{COb} and computational analysis of the results comprising the concentration value of CO after deducting the background of C_{CO}, the calculated emission value of CO, the value of the ratio C_{CO}/E_{CO} for particular measurements.

Table 3 CO concentration measurements and analysis of the results – street G

Temp.	**Humid.**	**Time**	N	**LDT**	C_{COb}	C_{CO}	E_{CO}	C_{CO}/E_{CO}
C	%	hours	v/h	%	µg m^{-3}	µg m^{-3}	µg m^{-1} s^{-1}	s m^{-2}
21.5	42	10-12	1310	8	1400	1250	2355	0.53
26.5	34	4-6 p.m	1870	10	2100	1950	3361	0.58
20.0	42	10-12	1230	7	1100	950	2005	0.47
23.5	38	4-6 p.m	1690	12	2300	2150	3190	0.67
1.2	85	10-12	1405	10	2550	2300	4577	0.50
-0.5	90	4-6 p.m	1780	9	3500	3250	5799	0.56
2.5	78	10-12	1520	6	2270	2020	4120	0.49
0.5	85	4-6 p.m	1730	8	3450	3200	5520	0.58
							average	0.55

Table 4. CO concentration measurements and analysis of the results – street P

Temp. C	Humid. %	Time hours	N v/h	L/H-DT %	C_{COb} μg m^{-3}	C_{CO} μg m^{-3}	E_{CO} μg m^{-1} s^{-1}	C_{CO}/E_{CO} s m^{-2}
20.5	40	10-12	2850	18	600	500	4132	0.12
22.0	41	4-6 p.m	3920	15	740	640	4938	0.13
21.5	38	10-12	2950	13	450	350	3835	0.09
23.5	35	4-6 p.m	4020	14	600	740	4928	0.15
2.4	85	10-12	3050	15	800	700	6405	0.11
-0.5	90	4-6 p.m	3840	12	1400	1250	7680	0.16
1.5	88	10-12	2710	14	880	730	6070	0.12
-2.0	92	4-6 p.m	3350	20	1200	1050	8207	0.13
							average	0.13

The canyon structure of street G causes emitted traffic pollution to accumulate within the canyon (Fig. 2), which is indicated by the high value of $(C_{CO}/E_{CO})av$=0.55 being over four times higher than that for street P. Such a structure of the street contributes to the fact that meteorological factors such as wind velocity and direction, temperature and solar radiation have a greater influence on average concentrations of particular pollutants and on vertical profiles of concentrations (Kwak et al., 2012; Park et al., 2004; Ahmad et al., 2005; Liu et al. 2003; Gartmann et al., 2012).

3.2 Influence of a street structure on pollution concentration values of ventilation air supplied to buildings

Two buildings were examined – one in street canyon G (Fig. 3) and the other in street P (Fig. 4). The one in street G has natural ventilation by opening windows and fan coil units. A measurement sensor was positioned outside a window at the level of 4.5 m above the street on its western side. The building in street P has a HVAC system. The air intakes were located on the roof of the building at the height of 17 m above the street and at the horizontal distance of 22 m from the edge of the street on its western side. A measurement sensor was placed in an air intake duct. Based on Gartman's research programme concerning the contribution of traffic emissions to air quality (Gartman et al., 2012), CO_2 concentrations were accepted as a traffic pollution indicator. Measurements were taken continuously for a fortnight from in June, 2014. On Sunday and Monday the wind blew with the speed of 2.0-3.5 m/s from the west, i.e. perpendicular to the axes of both streets. On the other days wind velocity did not exceed 1.2 m/s, which means that it did not have a substantial impact on traffic emission dispersion, especially within the street canyon.

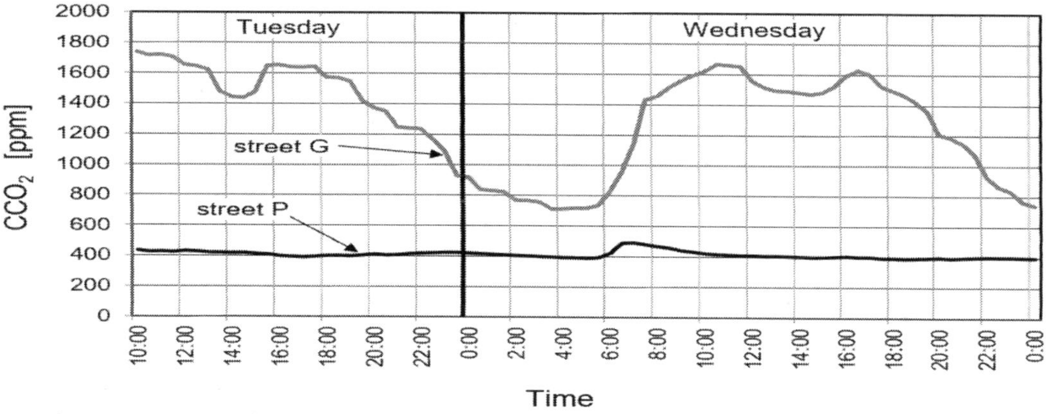

Fig. 5 CO_2 concentration values at the positions in streets G and P on Tuesday and Wednesday in June, wind velocity u<1.2 m/s

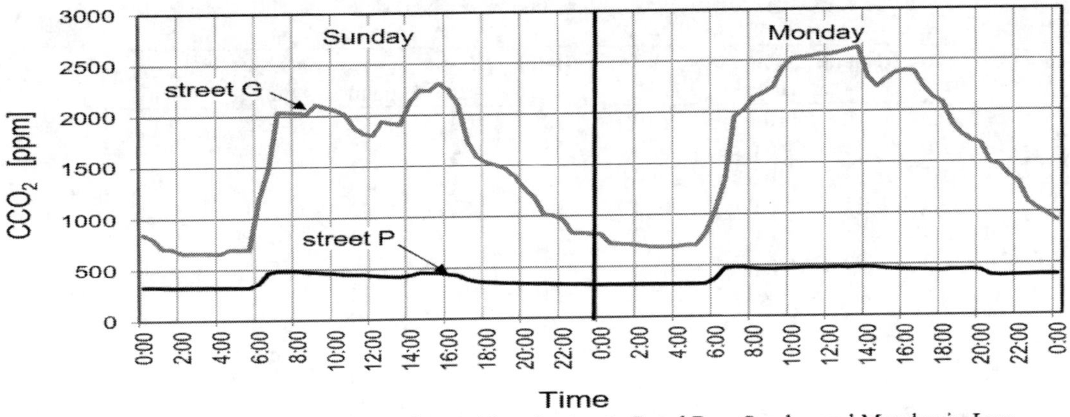

Fig. 6 CO_2 concentration values at the positions in streets G and P on Sunday and Monday in June, wind velocity u=2-3 m/s

Fig. 5 and Fig. 6 show CO_2 concentration values for selected days with low and higher wind velocities measured at positions close to streets G and P. Because of its placement concentrations at the position close to street P may be considered as those approaching the background concentration C_{bac} for Poznań. Concentrations in street G were 3-5 times higher (up to 2600 ppm), especially on days with increased wind velocity. With the wind blowing from the west the whole canyon was in the cavity recirculation region. The increased velocity of the wind blowing from the west contributed to a rise in CO_2 concentrations in the air intake on the roof of the building in street P, which is located in the wake region. Concentration values varied within 24 hours and for weekdays and weekends with the variations being higher for the street canyon.

In order to identify building surfaces that are particularly exposed to being infiltrated by traffic pollutants from roadways a series of model experiments were conducted in a wind tunnel. The wind tunnel used for testing the diffusion of gaseous pollutants emitted from the short-point and line sources was constructed at the Institute of Environmental Engineering, Poznan University of Technology, Poland. A sketch of the experiment set-up is shown in Fig. 7. Each time particular elements of the set-up and the measurement system are adapted to an accepted program of the investigation. The explanation of such set-ups can be found in papers by Bagienski (2008).

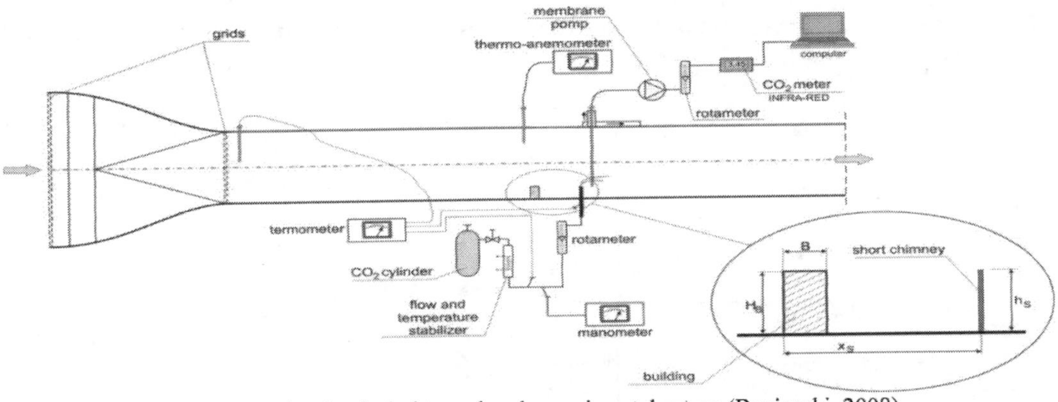

Fig.7 A sketch of wind tunnel and experimental set-up (Bagienski, 2008)

A set of nozzles with horizontal vents for fume from the collector was used as a line source simulating traffic emissions. Figures 8 and 9 show a visualization of dispersion of fume emitted from line sources in the street canyon. Air flow with the velocity u_H=2 m/s was perpendicular to the axis of the street. The highest pollution concentrations were recorded at the leeward side of the building limiting the canyon from the windward side. Fumes were also identified close to the roof of the building, which means that pollution may enter an air intake even when they are on roofs. In the case of the canyon with the index F_C>0.8 high pollution concentrations may fill the whole of the canyon, which suggests that a considerable share of traffic emissions may enter buildings by means of natural or mechanical ventilation systems.

Fig. 8 Line source of emission in the street canyon when F_C=1.0

Fig. 9 Line source of emission in the street canyon when F_C=0.3

4 CONCLUSIONS

Traffic emission has a considerable impact on urban air quality. It may cause high concentrations of air pollution to appear especially in street canyons – much higher than in wide and open roadways. Building solids cause disturbances in wind flow by generating the cavity recirculation region and the wake region on the leeward side that limit the possibility of air pollution emitted in the areas being dispersed freely. This leads to higher concentrations of traffic emissions at different heights of the canyon, also on roofs. The pollutants may enter buildings by means of natural and mechanical ventilation systems. In a narrow street canyon pollution concentrations in an air ventilation intake of a building may be 3-5 times higher than in that located on the roof of a building in wide roadways. However, it needs to be taken into consideration that also in these cases traffic emissions contribute considerably to ventilation air quality. Before the location and the kind of ventilation air intake is designed for buildings, traffic pollution dispersion from adjacent roadways need to be analysed. This is particularly the case for buildings in street canyons.

5 REFERENCES

Ahmad K., Khare M., Chaudhry K.(2005). Wind tunnel simulation studies on dispersion at urban street canyons and intersections – a review. *J. of Wind Eng. and Ind. Aerodyn., 93*, 697-717

Bagieński Z. (2010). *The influence of structure of energy consumption on air quality in urban area in temperate climate*. Dissertation 440, Poznan University of Technology, Poland (in Polish with English abstract)

Bagieński Z. (2006). The analysis of dispersion of pollutants from short point sources – wind tunnel experimental investigation. *Environment Protection Engineering, 32*(4), 37-45

Bagieński Z. (2008). Analysis of diffusion within cavity region of pollutants from short-point sources – wind tunnel experimental investigation. *Environment Protection Engineering, 34* (4), 43-50

Baldauf, R.W., Heist, D., Isakov, V., Perry, S., Hagler, G., Kimbrough, S., Shores, R., Black, K., Brixey, L. (2013). Air quality variability near a highway in a complex urban environment. *Atmospheric Environment. 64*, 169-178.

Carpentieri, M., Hayden, P., Robin, A.G. (2012). Wind tunnel measurements of pollutant turbulent fluxes in urban intersections. *Atmospheric Environment. 46*, 669-674

Chang, C.H., Meroney R.N. (2003). Concentration and flow distributions in urban street canyons: wind tunnel and computational data. *J. of Wind Eng. and Ind. Aerodyn., 91*, 1141-1154

European Environment Agency (2009). Air pollutant emission inventory guidebook

European Environment Agency (2013). Air quality in Europe –report

Gartmann, A., Müller M.D., Parlow, E., Vogt, R. (2012). Evaluation of numerical simulations of CO_2 transport in a city block with field measurements. *Environ. Fluid Mech.,12*, 185-200

Jensen, S. S., Larson, T., Deepti, K.C., Kaufman, J.D. (2009). Modeling traffic air pollution in street canyons in New York City for intra-urban exposure assessment in the US Multi-Ethnic, Study of atherosclerosis and air pollution. *Atmospheric Environment, 43*, 4544–4556

Kim, Y., Gudmann, J.M. (2011). Impact of traffic flows and wind directions on the pollution concentrations in Seoul, Korea, *Atmospheric Environment, 45*, 2803-2810

Kwak, K.H., Baik, J.J. (2012). A CFD modeling study of the impacts of NO_x and VOC emissions on reactive pollutant dispersion in and above a street canyon. *Atmospheric Environment, 46,* 71-80

Liu, H., Liang, B., Zhu, F., Zhang, B., Sang, J. (2003). A Laboratory Model for the Flow in Urban Street Canyons Induced by Bottom Heating. *Advances in Atmospheric Sciences, 20(4)*, 554-564

Meroney, R.N., Pavageau, M., Radalidis, S., Schatzmann, M. (1996). Study of line source characteristics for 2-D physical modeling of pollutant dispersion in street canyons. *J. of Wind Eng. and Ind. Aerodyn., 62*, 37-56

Oke, T.R. (1988). Street design and urban canopy layer climate. *Energy Buildings, 11*, 103-113

Park S.K., Kim S.D., Lee H. (2004). Dispersion characteristics of vehicle emission in an urban street canyon. *Science of the Total Environment, 323*, 263-271

Peterka, J.A., Meroney R.N., Kothari K.M. (1985). Wind flow patterns about buildings. *J. of Wind Eng. Ind. Aeroden., 21*, 21-38

Schembari, A., Triguero-Mas, M., de Nazelle, A., Dadvand, P., Vrijheid, M., Cirach, M., Martinez, D., Figueras, F., Querol, X., Basagana, X., Eeftens, M., Meliefste, K., Nieuwenhuijsen, M. (2013). Personal, indoor and outdoor air pollution levels among pregnant women. *Atmospheric Environment, 64*, 287-295

Sini, J.F., Anquetin, S., Mestayer, P. (1996). Pollutant dispersion and thermal effects in urban street canyons. *Atmospheric Environment, 30*, 2659-2677

Taha, H. (1996). Modeling the impacts of increased Urban vegetation on the ozone air quality in the south coast air basin. *Atmospheric Environment, 30*, 3423-3433

Vincent, D., Annesi, I., Festy, B.,Lambrozo, J. (1997). Ventilation System, Indoor Air Quality, and Health Outcomes in Parisian Modern Office Workers. *Environmental Research, 75*,100-112

SIMULATION OF NIGHT VENTILATION PERFORMANCE AS A SUPPORT FOR AN INTEGRATED DESIGN OF BUILDINGS

Jerzy Sowa*[1], Maciej Mijakowski[1], Jerzy Kwiatkowski[1,2]

1 Warsaw University of Technology
Faculty of Environmental Engineering
Nowowiejska St. 20, 00-653 Warsaw, Poland
**Corresponding author: jerzy.sowa@is.pw.edu.pl*

2 National Energy Conservation Agency
Świętokrzyska St. 20,
00-002 Warsaw, Poland

ABSTRACT

Passive cooling by night ventilation is one of the most promising approaches to reduce cooling energy demand of office buildings in moderate climates. However, the effectiveness of this system depends on many parameters. Some of them are related to characteristics of building (percentage of glass in building envelope, shading devices, accessible thermal mass of partitions, height of building), some to installed ventilation (type of ventilation, cross-section of ducts, location of air intake and exhaust) and finally some are related to use of building (profile of occupation, heat gains form lighting and equipment, required indoor parameters etc.).
The existing case-study buildings with night ventilation indicate that close to optimal performance of night ventilation requires detailed analysis performed at very early design stage when change of some input parameters is still possible. That is typical for concept of integrated design of buildings.
The paper describes simulations of night ventilation for the office building (floor area 3640 m^2 for approximately 400 persons) located in Warsaw, Poland. The calculations were performed with utilisation of quite simple but satisfactory accurate method 6R1C that is a modification of simply hourly method described in EN ISO 13790 (Energy performance of buildings - Calculation of energy use for space heating and cooling).
Achieved results confirmed big impact of night ventilation intensity and thermal capacitance of building on energy use in office buildings. Taking into account integrated design of buildings the paper provides also basics of methodology for selection of the optimal level of thermal capacitance and optimal intensity of night ventilation.

KEYWORDS

Night ventilation, Integrated design, Simulation, Energy demand

1. INTRODUCTION

The implementation of EU policies related to energy efficiency forces scientists and technicians to pay more attention to integrated design of buildings. Collaboration in multi-disciplinary teams, discussion and evaluation of multiple design concepts at early design stage are of great importance. In case of buildings with high energy and environmental ambitions integrated design seems to be necessity. Experiences from projects that applied integrated design show that the investment costs may be slightly higher about (5 %), while at the same time the annual running costs are reduced by as much as 40-90 %.

Meeting the EU requirement of nearly zero emission energy buildings (nZEB) in 2020 calls for intensive studies on practical application of low energy cooling of buildings. In moderate climates one of the most promising approaches to reduce cooling energy demand of office buildings is passive cooling by night ventilation. As the effectiveness of this system depends on many parameters, night ventilation is a very good example for integrated design approach.

The potential of night ventilation have been proved in number of studies and in many countries including USA (e.g. Keeney and Braun 1997 or Braun et al. 2001), UK (e.g. Kolokotroni and Aronis, 1999), France (e.g. Blondeau et al., 1997), Greece (e.g. Geros et al., 2005) or China (e.g. Wang et al. 2009).

After analysis of climatic data from 259 weather stations all over Europe Artman et al, (Artman et al, 2007) proved a high potential for night-time ventilative cooling over the whole of Northern Europe and still significant potential in Central, Eastern and even some regions of Southern Europe. However, in later publication Artman et al, (Artmann, et al. 2008) presented high sensitivity of simulation results to climatic conditions that should draw attention of the scientists to quality of climatic data for building energy simulations. They additionally state that simulations based on commonly used semi-synthetic 1 year data sets such as DRY or Meteonorm data tend to underestimate the extent of overheating compared to measured weather data.

2. PERFORMANCE OF THE NIGHT VENTIALTION IN POLISH CLIMATE

2.1 Concept of the Study

Unluckily, despite huge interest of scientists all over the world in night ventilation there is very limited information on similar studies for Polish conditions (e.g. Górzeński and Odyjas, 2003). Study of Dębowczyk and Sowa (Dębowczyk, 2012) presented huge potential for energy saving due to night-time ventilation of office buildings. It should be pointed out that, according to EU estimations (Pardo et al., 2012) the useful energy demands for office buildings in Poland for 2009 are quite high: 35.2 PJ for space heating, 9.4 PJ for water heating and 24.1 PJ for space cooling. Moreover, trends on construction market leads to the situation that the share of demand for space cooling in recently constructed office buildings is increasing. The aim of the study is to provide information on potential savings due to night ventilation and to promote the concept of integrated design that can help to reverse observed trend of rapidly rising cooling demand of office buildings.

2.2 Short description of applied simulation tool 6R1C +AHU

There are number of simple simulation tools capable to perform energy calculation for buildings with night ventilation. Balaras (Balaras, 1996) presented 16 simplified models for estimating the cooling load of a building, taking into account the building's thermal mass. The progress is quite fast and some of the models have been incorporated to Standard ISO 13790:2007 (ISO 2007). One of described methods, simplified hourly method utilizes the analogy between heat flow in buildings and current flow within electric circuits and is often called 5R1C (5 thermal resistances and 1 thermal capacity). The authors developed further this concept into 6R1C model (thermal resistances of controlled ventilation and uncontrolled infiltration have been split) and extend it by equations modelling behaviour of air handling units 6R1C +AHU (Narowski et al., 2009).

Depending on the type of air conditioning system the following processes can be taken into account:

- heat recovery (sensible and latent) during winter and summer,
- heating,
- humidifying,
- cooling,
- dehumidifying,
- preheating and precooling of air in ground-air heat exchanger.

Although the equations describing air conditioning processes are simple and well know (e.g. from EN 15241 (CEN, 2007)) the annual behaviour of AHU may be quite complex. The advanced logical analysis (the substitution of control system modelling) is often necessary. The model has an open structure and may be extended by other air treatment processes (e.g. evaporative cooling). The model introduced to Microsoft Excel spreadsheet may be utilized in personal computers.

2.3 Description of analysed building

The annual energy consumption simulations were performed for a virtual office building located in Warsaw, Poland. The building of total area of 3640 m^2 and volume of 10920 m^3 is occupied by 400 persons. The building is equipped with water based 4 pipe air conditioning system and constant air volume mechanical ventilation. Total ventilation rate for whole building is 20000 m^3/h (50 m^3/h per person). Additionally it has been assumed that the air tightness test (blower door test) gave the result $n_{50} = 2\ h^{-1}$. Heating and cooling loads were calculated assuming profiles of operation (from 7.00 a.m. to 8 p.m.). Peak internal heat gains from people lighting and office equipment reach 25.6 W/m^2. Simulations took into account weekends and free days with the assumption that in these cases indoor temperature is running freely with lower limit of 16°C.

The analysed building is quite well insulated and is characterised be the following parameters (symbols according to ISO 13790:2008, ISO, 2008):

- total thermal transmission coefficient of opaque building elements H_{tr_op}=1651 W/K
- total thermal transmission coefficient of doors, windows, curtain walls and glazed walls H_{tr_w}=861 W/K

It was assumed that the basic model of the building with active cooling can be additionally equipped with night ventilation, which operation can start an hour after leaving office by employees and can be switched off an hour before the regular operation of the building. In addition, the following restrictions for temperatures were considered:

- night ventilation is activated when indoor air temperature at the end of the work day exceeds 25°C,
- in the cooling period, minimum indoor air temperature during the operating hours is 22°C,
- in the cooling period, maximal indoor air temperature during the operating hours is 28°C.

Simulation were performed for different intensity of the night ventilation 5 ACH, 10 ACH and 15 ACH and different thermal capacitances of the building corresponding to medium, heavy and very heavy classes of dynamic parameters of building, according to ISO 13790:2008. In medium class the area of effective thermal mass is 2.5 times higher than the conditioned floor area, while in heavy and very heavy classes this coefficient is equal 3 and

3.5 respectively. At the same time internal heat capacity divided by the conditioned floor area equals 165 kJ/(K m^2) for medium class, 260 kJ/(K m^2) for heavy class and 370 kJ/(K m^2) for very heavy class.

3 DISCUSSION OF SIMULATION RESULTS

3.1 Analysis of the distribution of power demand over the year

Reference building for comparisons in presented study is a building with moderate thermal capacitance and without night ventilation. Simulations performed for this building show very high cooling power in comparison with heating power and long cooling period in comparison with heating period. The results can be astonishing but the experiences from other projects confirm that this phenomenon is observed in number of office building in Poland.

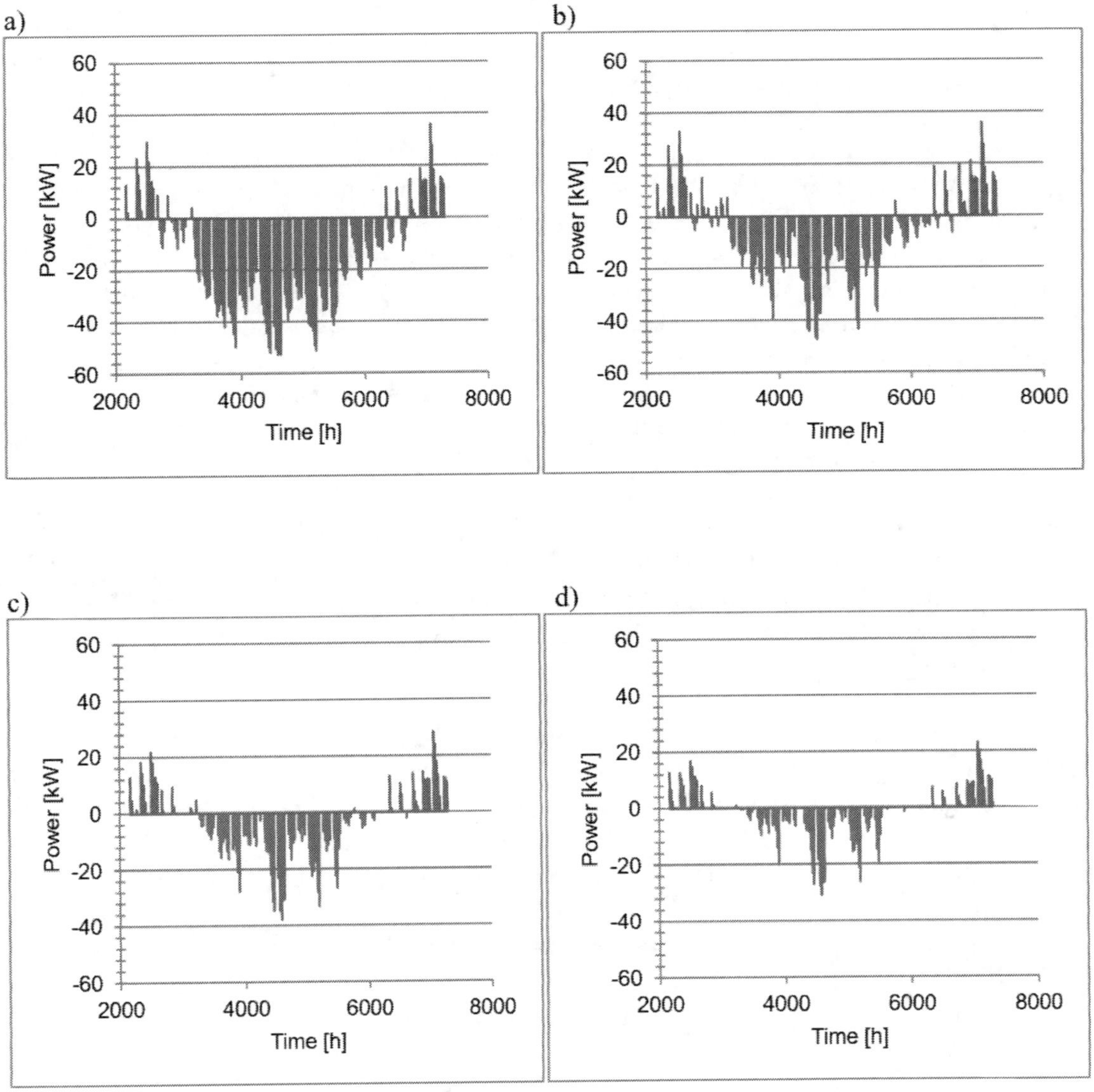

Fig. 1. Distribution of cooling power and heating power necessary to maintain comfort conditions in analysed office building, a) dynamic class – medium, no night ventilation, b) dynamic class – medium, night ventilation 5 ACH c) dynamic class – heavy, night ventilation 5 ACH, d) dynamic class – very heavy, night ventilation 5 ACH, (red line – heating, blue line – cooling).

The set of graphs presented at Fig. 1 allows the analysis of the influence of night ventilation with intensity 5 ACH on variable cooling and heating power for buildings with different types of construction. The mere fact of the introduction of night ventilation to building with medium thermal capacitance can noticeably reduce energy consumption for cooling. Maximum demand for cooling power decreases just by 10% after the introduction of night ventilation, however the cooling system operating time is reduced from 1,254 to 795 hours for the entire year. Increasing the thermal capacitance of the building allows further reduction of both cooling power and working time of the cooling system. In building corresponding to heavy thermal class cooling system operates only 602 hours per year while in case of building with very high thermal capacitance (very heavy class) operating hours are reduced just to 446 hours per year. In building with medium and high thermal capacitance the use of night ventilation slightly increases the operation period of heating system.

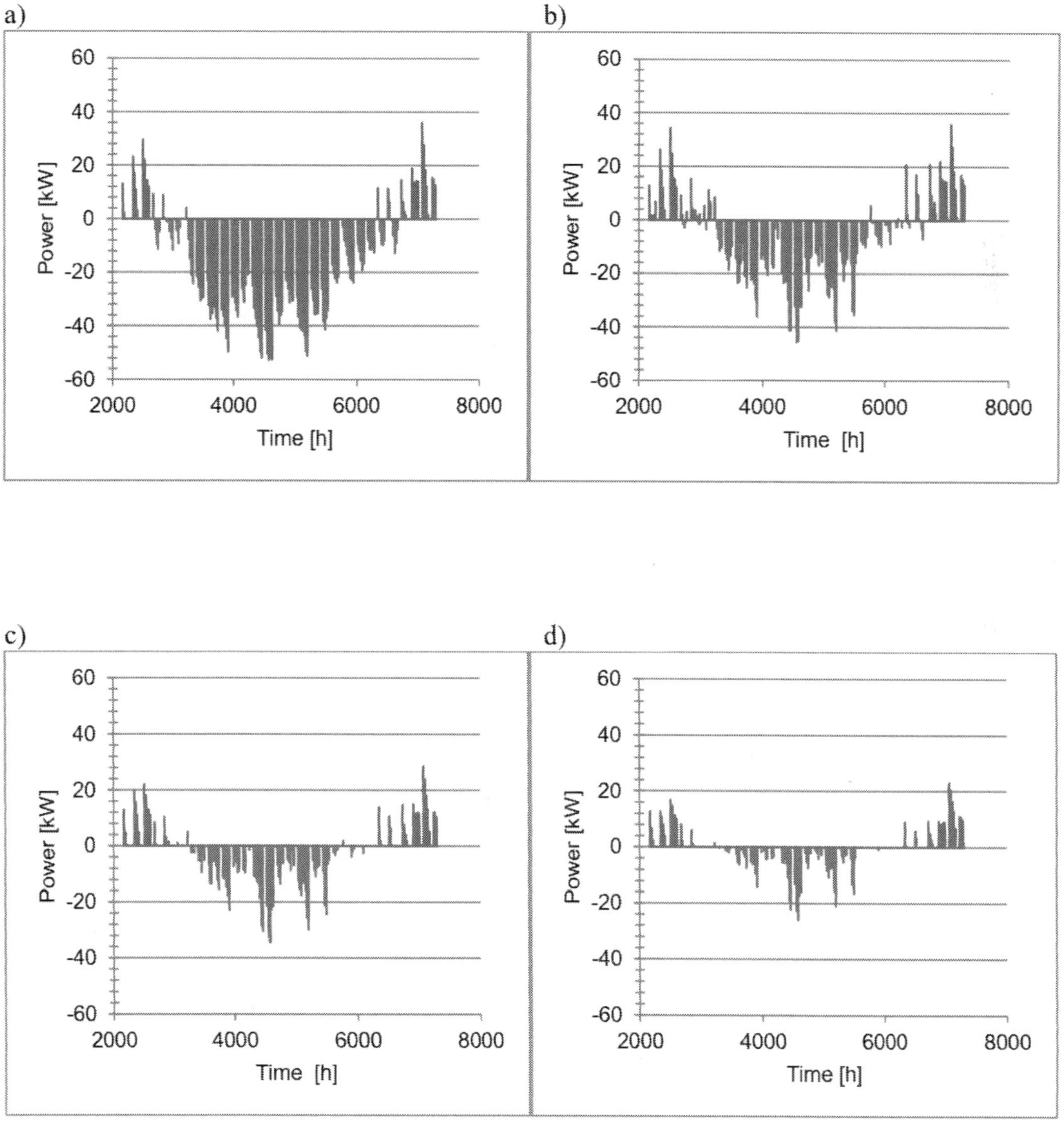

Fig. 2. Distribution of cooling power and heating power necessary to maintain comfort conditions in analysed office building, a) dynamic class – medium, no night ventilation, b) dynamic class – medium, night ventilation 10 ACH c) dynamic class – heavy, night ventilation 10 ACH, d) dynamic class – very heavy, night ventilation 10 ACH, (red line – heating, blue line – cooling).

The graphs presented at Fig. 2 illustrate the impact of night ventilation with intensity of 10 ACH on power demand in analysed building assuming its different dynamic properties (different thermal capacitances). A significant reduction in the power required for cooling is observed as well as the reduction of the working time of the cooling system from 1254 to 707 hours can be notices. For heavy building with night ventilation 10 ACH peak demand for cooling power is 34.53 kW that is an important reduction in comparison with reference variant (52.76 kW). Increasing the weight of the building structure to a level corresponding to a very heavy class (370 kJ/(Km^2)), allows to reduce the peak energy consumption by 51% in comparison with the reference variant. Cooling system operating time is reduced to 312 hours per year.

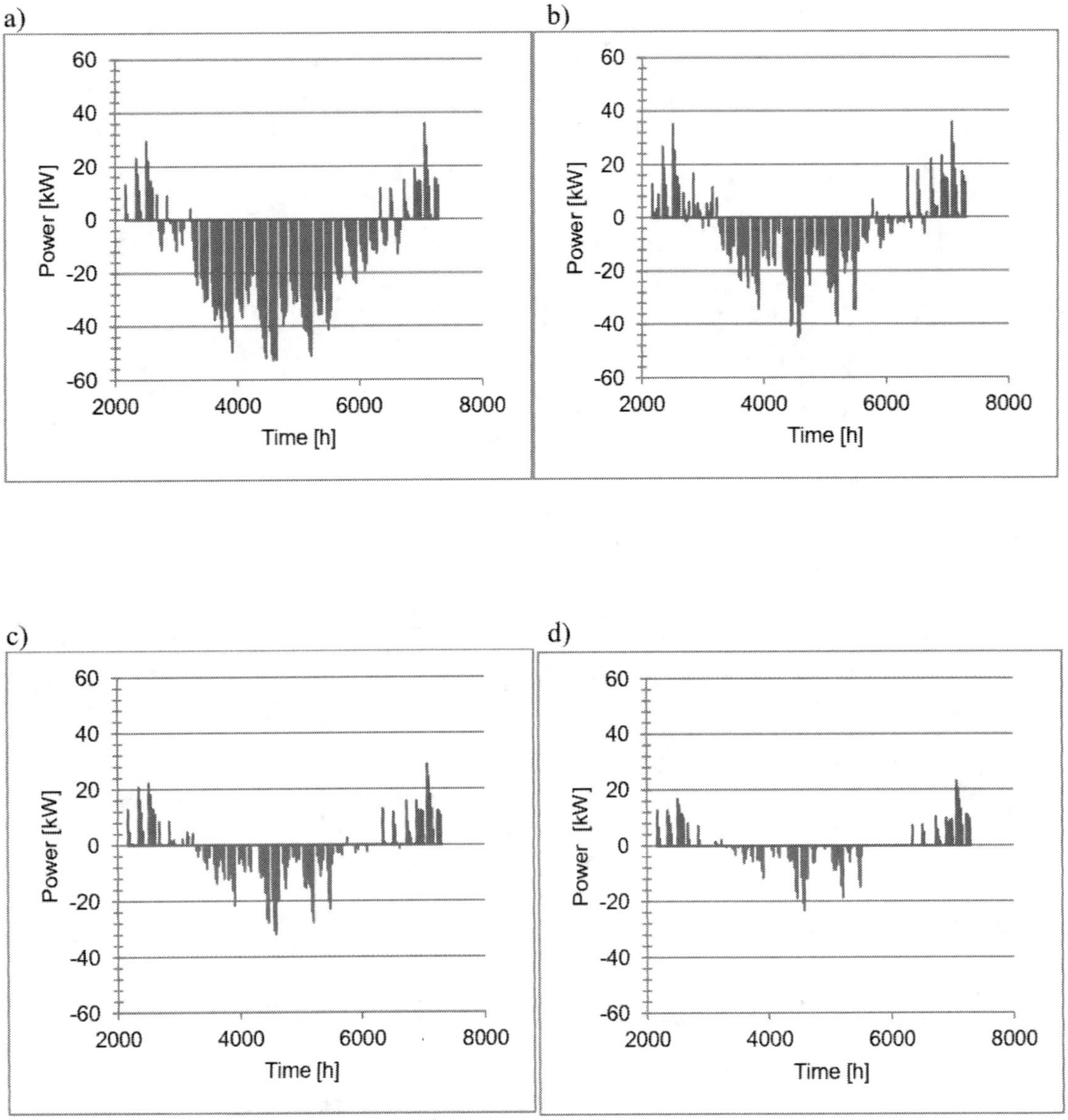

Fig. 3. Distribution of cooling power and heating power necessary to maintain comfort conditions in analysed office building, a) dynamic class – medium, no night ventilation, b) dynamic class – medium, night ventilation 15 ACH c) dynamic class – heavy, night ventilation 15 ACH, d) dynamic class – very heavy, night ventilation 15 ACH, (red line – heating, blue line – cooling).

Charts presented at fig 3 allow reader to analyse how different dynamic properties of the building influence cooling and heating power in case when night ventilation with intensity 15

ACH is used. For the moderate thermal capacitance the average working time of the cooling system is reduced by almost 50%. However, for the heavy class building cooling time is limited to 457 hours per year and peak demand for cooling power is just 32.00 kW. In case of very heavy class operating time is equal 264 hours per year, which is a very good result compared to the base building without night ventilation (1 254 hours).Moreover, peak cooling power demand decreased by 56%, and the demand for thermal power has decreased by more than 11% in relation to the reference variant.

3.2 Comparison of annual energy demand

Performed simulations can be summarised by presentation of energy use (fig 4). For the reference variant (building with medium class of thermal capacitance, no night ventilation), the total energy required for cooling in the reference year is 27 631 kWh and the demand for energy for heating at 2 024 kWh. Application of night ventilation with intensity 5 ACH leads to 60% reduction of total energy use for cooling, while the energy use for heating is higher by approximately 20% (fig. 4). Increasing the intensity of night ventilation to 10 ACH allows to reduce the energy use for cooling by an additional 5% while the heating energy requirement of supplementary 4%. Night ventilation with intensity 15 ACH can drop cooling energy use to below 30% of reference variant.

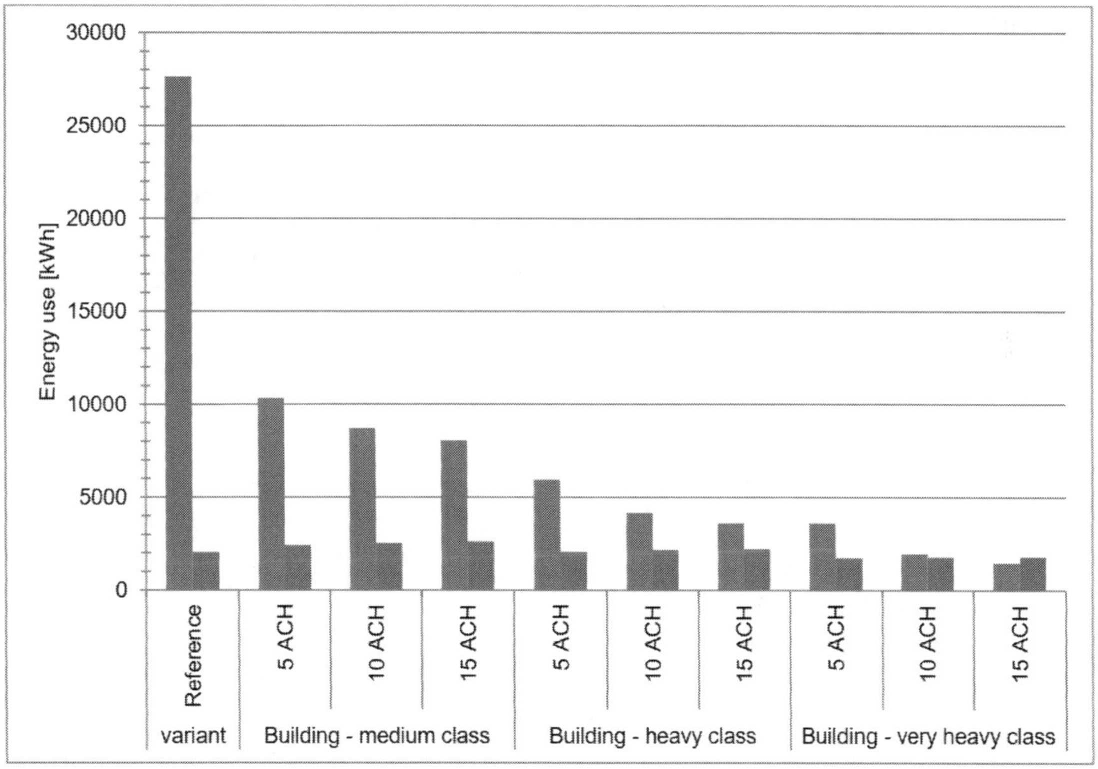

Fig. 4. The energy demand for cooling and heating for the presented variants in analysed building (red - heating, blue - cooling).

An increase of thermal mass of the building to 260 kJ/(K m^2) can give even higher reductions. Additional reductions can reach 5% to 10% compared to the reference building, depending on the intensity of the night ventilation. It is worth to point out that this scenario also, decreases energy use for heating by approximately 5-15% when compared to the moderate building and corresponding intensity of night ventilation.

The best results in terms of energy have been achieved for building with very heavy construction (very high thermal capacitance) with additional ventilation night. Cooling energy use for such a building with night ventilation at 5 ACH is reduced by more than 80% in relation to the reference building (medium construction and no night ventilation). The use of night ventilation of 10 ACH is characterised by more than 90% reduction in cooling energy use. High intensity of night ventilation 15 ACH leads to impressive 95% reduction of cooling energy use to 1 464 kWh per year. Furthermore the energy demand for heating of the building is reduced by more than 10% compared to the reference building and by approximately 20-30% compared to the building with moderate thermal capacity and active night ventilation.

4 FURTHER ECONOMIC ANALYSIS IN INTEGRATED DESIGN

The analysis described above indicated the big impact of the night ventilation intensity and thermal capacitance of building on energy use in office buildings also in Polish climate. The experiences of the authors indicate that similar trends can be observed in other office buildings excluding those with low thermal capacitance.
Parameters used in simulations (intensity of night ventilation and thermal capacitance of building) can be obtained due to variety of technical solutions at different investment costs. Also achieved savings in energy use can be converted into different costs depending on sources and prices of energy, efficiencies of devices etc. Further analyses are therefore presented just as a general scheme.

As recommended by EU (EU, 2012) total costs for buildings and/or building elements shall be calculated as a sum of the different types of costs after applying to them the discount rate (it means expressing them in terms of value in the starting year), and taking into account discounted residual values (equation 1).

$$C_g(\tau) = C_I + \sum_j \left[\sum_{i=1}^{\tau} \left(C_{a,i}(j) \times R_d(i) \right) - V_{f,\tau}(j) \right] \qquad (1)$$

where:

- τ the calculation period
- $C_g(\tau)$ global cost (referred to starting year τ 0) over the calculation period
- C_I initial investment costs for measure or set of measures j
- $C_{a,i}(j)$ means annual cost during year i for measure or set of measures j
- $V_{f,\tau}(j)$ means residual value of measure or set of measures j at the end of the calculation period (discounted to the starting year τ 0).
- $R_d(i)$ means discount factor for year i based on discount rate r to be calculated by equation 2

$$R_d(p) = \left(\frac{1}{1 + r/100} \right)^p \qquad (2)$$

where p means the number of years from the starting period and r means the real discount rate.

Than final selection of different combinations technical solutions (e.g. technical solutions for night ventilation, materials and technology for inner surfaces influencing thermal capacitance of the building) can be made taking into account:

- the lowest energy use
- the lowest primary energy consumption
- the lowest payback time
- the highest net present value of the investment
- the lowest life-cycle cost

or any other criteria agreed upon the designing team.

5 CONCLUSIONS

Integrated design can be a valuable approach to reduce the complexity of the design process and to facilitate the interactions between the members of the design team. This procedure allows the design team to provide the best solution for the whole building. The design team should have an access to simple but accurate simulation tools. Performed analysis using calculations with 1 hour time step gave much more information (e.g. peak power demand or operating hours) than monthly method.

Night ventilation can be a very attractive technology for low cost cooling of office buildings. Active systems (mechanical ventilation operating during favourable conditions during nights) require additional energy for fans but can be more efficiently controlled in comparison with ventilation systems run by natural forces. Buildings with high thermal capacitance are more appropriate for application of night ventilation. Achieved results confirm potential for huge energy savings during building operation. Presented approach is quite universal, however one should remember that recommendations for optimal variants can be different in similar cases as economic analyses are strongly dependent on local contexts (construction technologies, local materials, prices of energy etc.).

6 ACKNOWLEDGEMENTS

The paper was written due to the support of the project MaTrID "Market Transformation Towards Nearly Zero Energy Buildings Through Widespread Use of Integrated Energy Design" co-funded by the Intelligent Energy Europe Programme for European Union.

7 REFERENCES

Artmann, N., Manz, H., Heiselberg, P. (2008). Parameter study on performance of building cooling by night-time ventilation. *Renewable Energy*33.12, 2589-2598.

Artmann, N., Manz, H., Heiselberg, P. (2007). Climatic potential for passive cooling of buildings by night-time ventilation in Europe. *Applied Energy* 84, 187-201.

Balaras, C. A. (1996). The role of thermal mass on the cooling load of buildings. An overview of computational methods. *Energy and Buildings*, *24*(1), 1-10.

Blondeau, P., Spérandio, M., Allard, F. (1997). Night ventilation for building cooling in summer. *Solar energy* 61.5, 327-335.

Braun, J. E., Montgomery, K. W., Chaturvedi, N. (2001). Evaluating the performance of building thermal mass control strategies. *HVAC&R Research*, *7*(4), 403-428.

CEN (2007) EN 15241:2007 Ventilation for buildings — Calculation methods for energy losses due to ventilation and infiltration in commercial buildings.

Debowczyk, M. (2012), Night-time ventilation as a passive cooling method of office buildings. M.Sc. Thesis, Warsaw University of Technology, faculty of Environmental Engineering, supervisor Jerzy Sowa. (in Polish).

EU (2012) DELEGATED REGULATION (EU) No 244/2012 of 16 January 2012 supplementing Directive 2010/31/EU of the European Parliament and of the Council on the energy performance of buildings by establishing a comparative methodology framework for calculating cost-optimal levels of minimum energy performance requirements for buildings and building elements.

Geros, V., et al. (2005). On the cooling potential of night ventilation techniques in the urban environment. *Energy and Buildings* 37.3, 243-257.

Górzeński, R., Odyjas, A. (2003). Wykorzystanie "nocnego chłodzenia" w klimatyzacji pomieszczeń. *Chłodnictwo i Klimatyzacja*, no 8, 39-43 (in Polish).

ISO (2008), ISO 13790: 2008 Energy performance of buildings - Calculation of energy use for space heating and cooling.

Keeney, K. R., Braun, J. E. (1997). Application of building precooling to reduce peak cooling requirements. *ASHRAE transactions*, *103*(1), 463-469.

Kolokotroni, M., Aronis A. (1999). Cooling-energy reduction in air-conditioned offices by using night ventilation. *Applied Energy* 63.4, 241-253.

Narowski, P. Mijakowski, M. Sowa, J. (2009) Integrated calculations of thermal behavior of both buildings and ventilation and air conditioning systems, *Proceedings of the 9th International Healthy Buildings Conference and Exhibition.* Eds.: Santanam, S., Bogucz, E.A., Peters, C, and Benson, T., Healthy Buildings 2009, Syracuse, NY, USA, Paper No: 628.

Pardo, N., et al. (2012). Heat and Cooling Demand and Market Perspective. European Commission, Joint Research Centre, Institute for Energy and Transport, Publications Office.

Wang, Z., Yi, L., & Gao, F. (2009). Night ventilation control strategies in office buildings. Solar Energy, 83(10), 1902-1913..

ESTIMATING THE IMPACT OF INCOMPLETE TRACER GAS MIXING ON INFILTRATION RATE MEASUREMENTS

Peter D McDowall* and Manfred Plagmann

BRANZ
1222 Moonshine Rd,
Porirua,
New Zealand
**Corresponding author: peter.mcdowall@branz.co.nz*

ABSTRACT

The mixing of a tracer gas with zonal air was compared between two zones in an unoccupied test building in both the horizontal and vertical direction. A constant injection of sulphur hexafluoride (SF_6) tracer gas was released into each zone separately and its concentration was measured at different positions within the zone. Variations in concentration were observed for different horizontal positions in the southern zone indicating incomplete mixing. The impact of incomplete mixing on the accuracy of subsequent infiltration measurements was determined and compared to the case where good mixing was achieved. Absence of direct solar radiation was identified as the predominant cause of incomplete tracer gas mixing observed in our experiments.

KEYWORDS

Tracer Gas, Mixing, Ventilation, Infiltration.

1. INTRODUCTION

The removal of unwanted moisture and pollutants from indoor air is an important issue. Natural ventilation, termed infiltration, plays a vital role in the removal of these contaminants, which has led to numerous techniques being developed that measure the air infiltration rate of a building. Common amongst most infiltration rate measurement techniques is the use of a tracer gas such as carbon dioxide, nitrous oxide or sulphur hexafluoride (SF_6) which is used in the experiments presented in this paper, to name a few.

Tracer gases can be used in various ways to determine the infiltration rate of a room or building but all methods rely on the same conservation laws to infer the air infiltration rate. Specifically, it is assumed the air flow into and out of a zone obeys the following mass flow balance equation (Sandberg et al, 1989; Sherman, 1990; Sherman et al, 2014):

$$V\dot{C} + QC = Q_T \tag{16}$$

Where V is the volume of the zone, C is the instantaneous tracer gas concentration, Q is the ventilation rate of the zone, and Q_T is the injection rate of the tracer gas into the zone. Using modern mass flow controllers, it is straight forward to accurately know Q_T, whereas C can be measured down to low levels from different sampling points within the zone using a calibrated gas analyser.

Tracer gas concentration can only be sampled at discreet locations within a zone leading to the assumption that the tracer gas is homogeneously mixed with both the air already present in the zone and the air that is entering the zone (Sherman et al., 2014). However, it has been shown that tracer gases do not always perfectly mix within a zone such that infiltration rate measurements can be influenced by tracer gas sampling positions (Barber et al, 1984; Lunden et a., 2012; Maldonado et al, 1983; Van Buggenhout et al, 2009).

A common solution to incomplete tracer gas mixing is to incorporate a fan into the setup to artificially mix the air and ensure concentration homogeneity of the tracer gas (Chao, 1994; Lunden et al, 2012). However, mixing the air by means of a device such as a fan can introduce errors in the infiltration measurement if the device is driving ventilation as reported by Shao et al, 1994.

We examined the horizontal and vertical mixing of a tracer gas released inside a test building without the use of an artificial mixing device. The influence of solar radiation on the mixing process was investigated and an estimated error associated with incomplete mixing for this scenario was obtained.

2. EXPERIMENTAL PROCEDURE

Figure 1(a) shows the floor plan of the unoccupied test house used to conduct the tracer gas mixing experiments. Only the northern (sun-facing) zone, labelled Zone 1, and the south-facing zone, labelled Zone 2, were used. Separate experiments were conducted in the two zones to determine the tracer gas mixing for different vertical heights and for different horizontal positions across each zone. The airtightness of each zone was controlled through removable ports in the walls, ceiling and floor of each zone while the zone door was kept closed during experimentation. No artificial mixing devices were used and the test building was unoccupied while the experiment was running. Wind, sun and outside temperature data was collected via an external weather station located next to the test building.

To prevent cross contamination of tracer gas between zones, experiments conducted in the two zones illustrated in Figure 1(b) were carried out at different times. Similarly, horizontal and vertical mixing experiments were also conducted separately.

The zone was initially dosed with enough SF_6 to ensure the overall concentration within it was several times above background levels. The flow of SF_6 into the zone was then reduced to a constant rate to try to maintain a constant concentration. A mass flow controller allowed us to accurately measure the injection rate. Tracer gas concentrations were measured at each sampling point by removing a small sample of air from a given location and passing it through an Innova 1412 Photoacoustic Field Gas Analyser. Once the concentration of the tracer gas at that sampling point was ascertained, the air sample was released back into the zone. Samples were taken at each sampling point every 10-15 minutes over a couple of days.

For vertical mixing tests, sampling tubes measured SF_6 concentration at the centre of the zone at three different heights; 55 mm, 1065 mm and 2065 mm from the floor. Figure 1(b) illustrates the locations of the sampling points for the horizontal mixing tests. Each sampling point in the horizontal tests was 1600 mm above the floor of the zone.

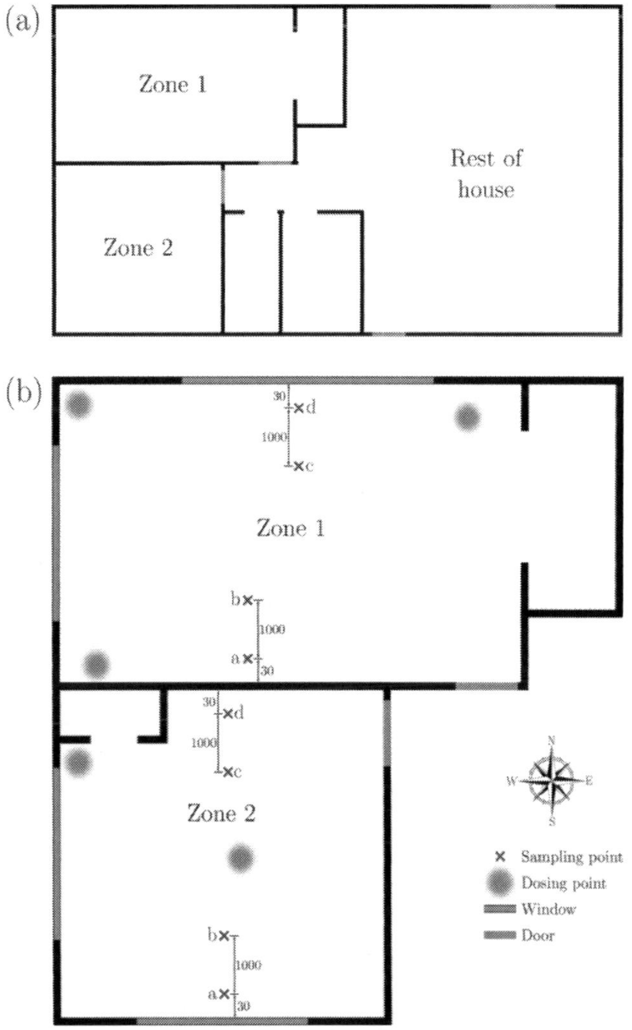

Figure 1: (a) Floor plan of the entire test house; (b) Locations of the sampling and dosing points in Zones 1 and 2 for horizontal mixing experiment. All sampling points were at a height of 1600 mm from the floor. Distances indicated in the figure are in units of mm.

3. RESULTS

3.1 Vertical Mixing

Figure 2 shows the difference in tracer gas concentration between three different heights above the zone floor for the two zones of interest together with the relevant environmental data. We observed negligible variation between tracer gas concentrations at different heights for the two zones. The airtightness of each zone was approximately 1 ACH at 50 Pa with major changes in infiltration rate being wind-driven. For the majority of the two experiments, wind speed was less than 4 ms^{-1} with no discernible impact on the vertical mixing of the tracer gas.

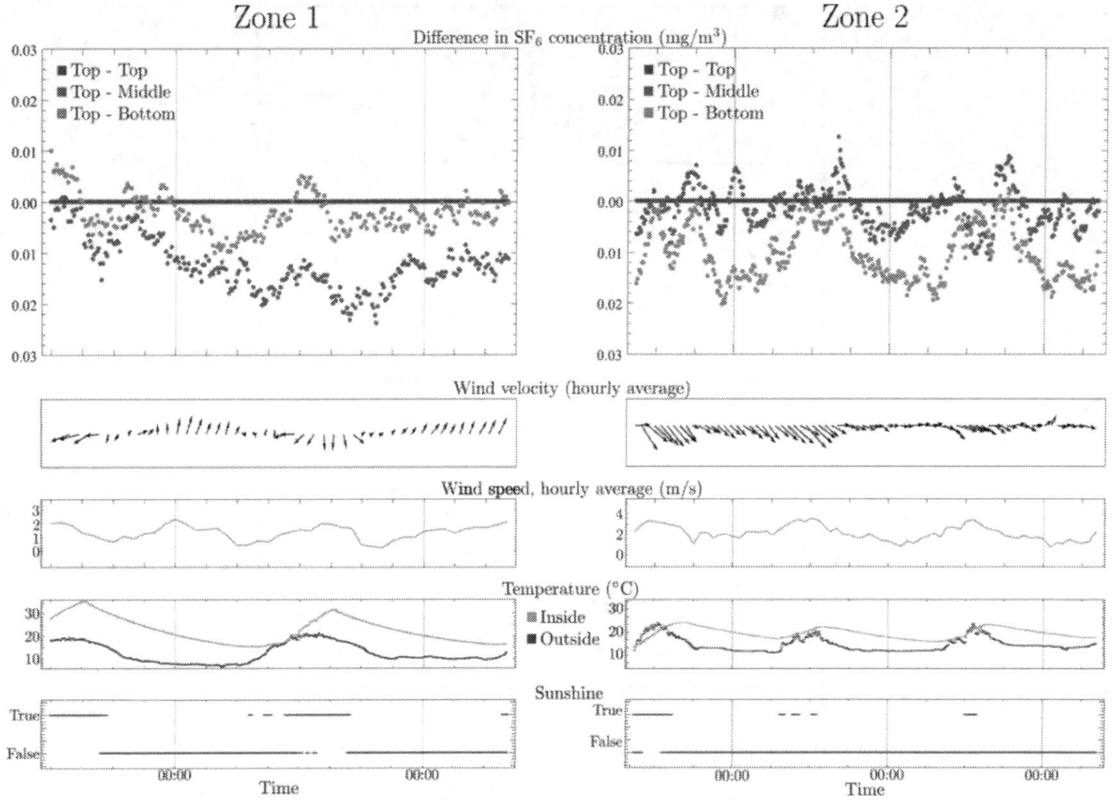

Figure 2: Results of vertical mixing experiment for Zones 1 and 2. Three sample points were used; 2065 mm (top), 1065 mm (middle) and 55 mm (bottom) above the zone floor. The two uppermost plots show the difference in SF_6 concentration between the top sample point and the middle and bottom sample points.

3.2 Horizontal Mixing

Figure 3 shows the difference in tracer gas concentration between four different locations of equal height in each of the two zones together with the relevant environmental data. Zone 1 shows no discernible variation between tracer gas concentrations at the different sampling locations suggesting good mixing in that zone.

A discrepancy between gas concentrations at different sampling points in Zone 2 indicated incomplete mixing within that zone. This feature appears more prominently at certain times of the day when tracer gas concentrations are greater at the southernmost sampling points than at the northern most sampling points in the room.

The airtightness of Zones 1 and 2 during the horizontal mixing experiments was approximately 8 ACH at 50 Pa. This was achieved by opening ports in the walls and ceiling of the zone as described in Section 2.

Both zones were measured over periods that had similar wind velocities and sunshine hours. However, we do note that the ambient air temperature inside the zone was on average warmer in Zone 1 than in Zone 2. This discrepancy was expected since Zone 1 is north (sun)-facing whereas Zone 2 faces away from the sun.

The infiltration rate of a given zone can be determined from Equation 1. Using this expression and the concentration data for horizontal mixing in Figure 3 we were able to estimate the error associated with an infiltration measurement under the current conditions. We did this by first determining the infiltration rate based on the concentration data from each sampling point individually averaged over 24 hours. We considered the true infiltration rate to be the mean of the four values (one for each sampling point) with an error given by the standard deviation.

Incomplete mixing in Zone 2 in the horizontal plane was found to give an uncertainty in the infiltration measurement of approximately 4%. This is in comparison to the same experiment in Zone 1 that displayed good tracer gas mixing, where we found the infiltration rate uncertainty to be approximately 1%.

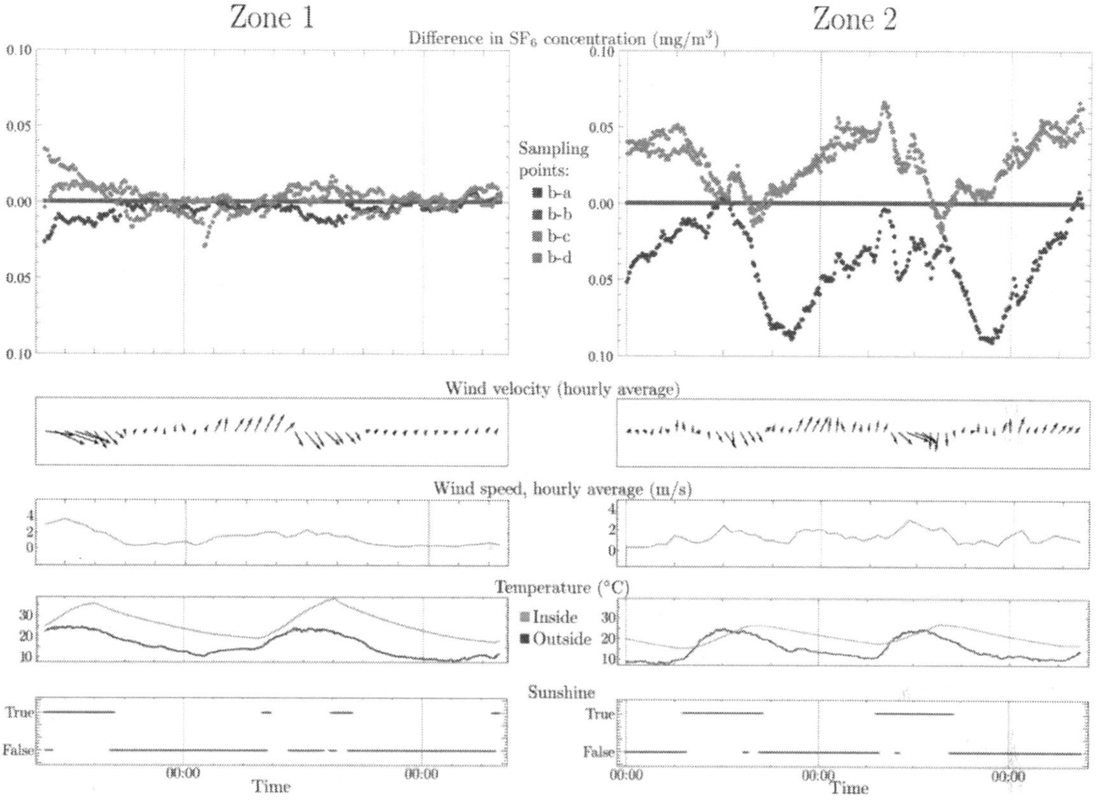

Figure 3: Results of horizontal mixing experiment for Zones 1 and 2. Four sampling points were used in each zone as illustrated in Figure 1(b). A consistent SF_6 concentration across all sampling points in Zone 1 indicates good mixing in the horizontal plane within that zone. Variation in tracer gas concentration at different sampling points in Zone 2 suggests incomplete mixing within that zone.

4 DISCUSSION

The tracer gas appeared to mix reasonably well in the vertical plane for both the north (Zone 1) and south (Zone 2) facing zones regardless of external environmental conditions. The same conclusions can also be drawn for horizontal mixing in the north-facing zone. However, horizontal mixing in the south-facing zone appears to suffer from stratification at certain times of the day. This stratification leads to a greater uncertainty when determining the infiltration rate and as such is a problem that deserves consideration.

The collective mixing effects of diffusion, convection, direct solar radiation convection, and air movement through infiltration all contribute to mixing of a tracer gas within a zone. The two zones we considered were of roughly equal air tightness and tested under similar environmental conditions. From this we conclude that direct solar radiation was the limiting mechanism for mixing. In the south-facing zone, where we observed incomplete tracer gas mixing, there was no direct solar radiation into the zone which is in contrast to the north-facing zone.

Convections within a zone can be driven by temperature gradients caused by solar radiation. Air within a zone can absorb heat when in contact with direct solar radiation through say a window. Buoyancy forces cause the warm air to rise due to the stack effect (Straube et al, 1995) creating a convective process as cold air moves to fill the void left by the warm air. This convective process increases mixing of the tracer gas with the air present in the zone. The absence of direct solar radiation in Zone 2 meant the occurrence of this process was minimal, leading to incomplete mixing.

We have assumed here that the stack effect is the dominant source of natural mixing within the zone. However, it should be noted that a less airtight house or one that is more susceptible to wind-driven ventilation (e.g. in an area of greater wind speeds) may have enough natural ventilation to adequately mix all the air within the zone and hence not exhibit the same behaviour we observed here.

The magnitude of infiltration rate errors resulting from incomplete mixing are due partly to the geography of the zone being measured and the ventilation within it. Van Buggenhout et al., 2009 observed an error of 86% between the measured and the actual ventilation rate for a mechanically-ventilated test zone with openings at each end. Tracer gas concentration differences of up to 44% have been observed within a single zone of a three-storey test house (Maldonado et al, 1983). Both of these experiments were performed under different conditions to the experiments presented here and therefore do not allow for direct comparison of errors. However, they do show that incomplete tracer gas mixing is not a feature specific to just our test building.

5 CONCLUSIONS

Lack of direct solar radiation can lead to incomplete mixing of a tracer gas within a zone when no artificial mixing apparatus is used. We showed that for the test house used in our experiments this incomplete mixing caused an increase in the uncertainty of our infiltration measurements from approximately 1% in a sun-facing zone to 4% in a non-sun-facing zone in the horizontal plane. Tracer gas concentrations in the vertical plane were observed to display good mixing in both zones.

6 ACKNOWLEDGEMENTS

This work was funded by the New Zealand Building Research Levy and the New Zealand Ministry of Business, Innovation and Employment.

7 REFERENCES

Barber, E. M., and Ogilvie, J. R. (1984). Incomplete Mixing in Ventilated Airspaces. Part II: Scale Model Study. *Canadian Agriculture Engineering, 26*(2), 189-196.

Chao, C. Y., Wan, M. P., and Law, A. K. (2004). Ventilation Performance Measurement Using Constant Concentation Dosing Strategy. *Building and Environment, 39*(3), 1277-1288.

Lunden, M., Faulkner, D., Heredia, E., Cohn, S., Dickeroff, D. J., Noris, F., Louge, J., Toshifumi, H., Singer, B., and Sherman, M. (2012). Experiments to Evaluate and Implement Passive Tracer Gas Methods to Measure Ventilation Rates in Homes. *Lawrence Berkeley Laboratory Report LBNL-5984E.*

Maldonado, E. A. B., Woods, J. E. (1983). A Method to Select Locations for Indoor Air Quality Sampling. *Building and Environment, 18*(3), 171-180.

Sandberg, M., and Stymne, H. (1989). The Constant Tracer Flow Technique. *Building and Environment, 24*(3), 209-219.

Shao, L., and Riffat, S. B. (1994). Tracer-Gas Mixing with Air: Effect of Tracer Species. *Applied Energy, 49*, 197-211.

Sherman, M. H. (1990). Tracer-gas Techniques for Measuring Ventilation in a Single Zone. *Building and Environment, 24*(4), 365-374.

Sherman, M. H., Walker, I. S., and Lunden, M. M. (2014). Uncertainties in Air Exchange Using Continuous-Injection Long-Term Sampling Tracer-Gas Methods. *International Journal of Ventilation, 13*(1), 13-27.

Van Buggenhout, S., Van Brecht, A., Eren Özcan, S., Vranken, E., Van Malcot, W., and Berckmans, D. (2009). Influence of Sampling Positions on Accuracy of Tracer Gas Measurements in Ventilated Spaces. *Biosystems Engineering, 104*, 216-223.

VENTILATIVE COOLING IN NATIONAL ENERGY PERFORMANCE REGULATIONS:

REQUIREMENTS AND SENSITIVITY ANALYSIS

I. Pollet[*1,2], S. Germonpré[1] and A. Vens[1]

1 Renson Ventilation
Maalbeekstraat 10
Waregem, Belgium

2 Ghent University
Department of Biosystems Engineering
Coupure Links 653
Gent, Belgium
**Corresponding author: ivan.pollet@ugent.be*

ABSTRACT

Higher insulation and air tightness levels of buildings, increase the risk on overheating. Ventilative cooling as passive technique can limit overheating and decrease cooling energy consumption. The national energy performance regulations (EPBD) determine whether, how and under which requirements ventilative cooling can assist to reduce cooling demand and overheating. Therefore, those regulations are a key factor in the market uptake of ventilative cooling. Without a realistic and achievable approach, ventilative cooling will marginally be applied in buildings.

In this study, actual and possible requirements imposed on devices or systems for ventilative cooling are described. Besides, a sensitivity analysis is performed to assess the impact of parameter variation on the ventilative cooling effect. One reference dwelling is selected and introduced into the EP software of Belgium, The Netherlands and France. In that way, differences in buildings characteristics between countries on the output are ignored. Of course, the presented results are only valid for the selected reference dwelling.

In the three countries ventilative cooling by means of openable windows can be taken into account, in Belgium and France as a percentage of openable windows, in The Netherlands as present or not. Burglary resistance and water tightness of ventilative cooling devices are requirements (for The Netherlands) or have an impact on the performance of the ventilative cooling device (for Belgium). In France, openable windows are not allowed in case of active cooling. When there are openable windows in France, they are supposed to be opened during the heating season as well, resulting in an increased heating demand due to ventilative cooling.

As can be expected, ventilative cooling always decreases the cooling demand (26 to 96% in Belgium, 10 to 35% in The Netherlands), especially in combination with mobile solar shading.

Similar to the cooling demand, the risk on overheating in Belgium decreases by applying ventilative cooling. In France, summer indoor temperature can be strongly reduced by using openable windows, although the fraction openable windows has no effect. Next to ventilative cooling, thermal capacity as well as solar shading can have a considerable impact on summer comfort and should be considered as complementary means.

The overall primary energy consumption of the reference dwelling is lower when ventilative cooling is applied (5 to 12% in Belgium, up to 4% in The Netherlands) except for France where openable windows remarkably increase the heating demand (up to 38%). The lowest primary energy consumption is achieved by applying ventilative cooling in combination with sun shading.

KEYWORDS

Energy performance regulation, Overheating, Sensitivity analysis, Ventilative cooling

1. INTRODUCTION

Higher insulation and air tightness levels of buildings reduce the energy demand of dwellings, but increase the risk of overheating due to lower heat losses even in intermediate season. This has been highlighted in a number of Northern, Central and Southern European reports as listed by McLeod [1]. Experience shows that active cooling is too often considered to resolve these overheating problems, while other options should be prioritized in building design when relevant [2]. Possible design strategies affecting the heat balance (heat gains = heat losses + cooling demand) are:

heat gains: - solar gains: window surface, solar protection
- internal heat gains (if variable)
heat losses: - transmission losses: U-value
- controlled ventilation losses: air flow rate
- uncontrolled ventilation losses: air tightness level n_{50}
- ventilative cooling: air flow rate
thermal mass

The impact of some of the afore mentioned design strategies is studied before (f.i. [1], [3], [4]) and are also integrated in the national Energy Performance (EP) calculation procedures. Although ventilative cooling is a known technique since ancient times, this passive cooling technique is only included in a few national EP calculations, such as Belgium, France and The Netherlands. Ventilative cooling refers to the use of natural or mechanical ventilation strategies to cool indoor spaces with air flow rates higher than hygienic ones. This effective use of outside air reduces or cancels the energy consumption of cooling systems, while increasing thermal comfort. Ventilative cooling is relevant in a wide range of buildings and may even be necessary to realize renovated or new NZEB [5].

The national energy performance regulations (EPBD) determine whether, how and under which requirements ventilative cooling can assist to reduce cooling demand and overheating. Therefore, those regulations are a key factor in the market uptake of ventilative cooling. Without a realistic and achievable approach, ventilative cooling will marginally be applied in buildings.

In this study, actual and possible requirements imposed on devices or systems for ventilative cooling were described in three countries: Belgium, France and The Netherlands. Besides, a sensitivity analysis was performed to assess the impact of ventilative cooling (simulated by a openable windows), while varying other building characteristics, on net yearly heating demand, net yearly cooling demand, summer comfort and yearly primary energy consumption. To this end, one reference house was selected and modelled with the Energy Performance (EP) software of Belgium, France and The Netherlands. In this way, the effect of different building characteristics between countries on the output was ignored.

2. EP CALCULATION METHOD

In the three countries studied in this paper, Belgium, France and The Netherlands, heating demand, cooling demand as well as summer comfort is expressed in a different way (Table 1):

- Heating as well as cooling set points are different in each country (Table 2), and even function of occupation. In that way, differences in absolute energy demands between countries are enhanced.
- A (fictive) cooling demand cannot be determined within the French calculation procedure.
- In The Netherlands, no summer comfort parameter is used.
- The overall energy performance of the dwelling (the primary energy consumption) is in Belgium and The Netherlands expressed as a relative value, in France an absolute value is used.

The way cooling demand and summer comfort is treated in the national methodologies is described hereafter.

Table 1: Output of different national EP calculations

Output	Belgium	France	The Netherlands
Heating demand	Net yearly heating demand [MJ]	Net yearly heating demand [MJ]	Net yearly heating demand [MJ]
Cooling demand	Fictive net yearly cooling demand [MJ]	---	Fictive net yearly cooling demand [MJ]
Summer comfort	Overheating indicator [Kh]	Indoor temperature exceed T_{ic} [K]	---
Primary energy consumption	E-level [-] $PEF_{el}=2.5$ $PEF_{fuel}=1.0$	C_{ep} [kWh/m²] $PEF_{el}=2.58$ $PEF_{fuel}=1.0$	EPC [-] $PEF_{el}=2.56$ $PEF_{fuel}=1.0$

Table 2: National boundary conditions for heating and cooling demand calculations

Country	Heating set point	Cooling set point
Belgium	18°C	23°C
France	19°C / 16°C (non occ.)	28°C / 30°C (non occ.)
The Netherlands	20°C	24°C

2.1 Belgium

The Belgian procedure to calculate the so-called "overheating indicator" for dwellings is described in annex V of the EPB regulation [6]. The risk on overheating depends on the overheating indicator ($I_{overheat}$) which is determined by the normalized excessive heat gains (Eq. 1). When the overheating indicator does not exceed the maximum allowed value, chances of overheating are limited, but not impossible.

$$I_{overheat} = Q_{excessnorm,a} = \sum_{m=1}^{12} Q_{excessnorm,m} \tag{1}$$

$$Q_{excessnorm,m} \frac{\left(1-\eta_{util,overh,m}\right) \cdot Q_{g,overh,m}}{H_{T,overh} + H_{V,overh,m}} \cdot \frac{1000}{3.6} \tag{2}$$

With:

$I_{overheat}$	Overheating indicator	[Kh]
$Q_{excessnorm,m}$	Monthly normalized unwanted heat gains	[Kh]
$\eta_{util,overh,m}$	Utilization factor of the monthly heat gains	[-]
$Q_{g,overh,m}$	Monthly heat gains by insolation and internal heat production	[MJ]
$H_{T,overh}$	Heat transfer coefficient by transmission	[W/K]
$H_{V,overh,m}$	Monthly heat transfer coefficient by ventilation	[W/K]

In accordance with the overheating indicator, a conventional probability of placing an active cooling installation (p_{cool}) in a later stage, is defined (Eq. 3). This probability augments linearly with the overheating indicator (1 in case of active cooling).

$$p_{cool} = max\left\{0, min\left(\frac{I_{overheat} - I_{overheat,thresh}}{I_{overheat,max} - I_{overheat,thresh}}, 1\right)\right\} \quad (3)$$

With:

$I_{overheat,thresh}$	Threshold value overheating indicator (above this value the risk on placing active cooling afterwards grows)	[1000 Kh]
$I_{overheat,max}$	Maximum allowed value overheating indicator (when the overheating indicator exceeds this value, a fine is imposed)	[6500 Kh]

When a risk on overheating occurs, without installing an active cooling system, a fictive cooling demand is calculated to take into account a possible installation of an active cooling installation afterwards. The fictive cooling demand equals the conventional probability multiplied with the net energy demand for cooling ($Q_{cool,net,princ}$) (Eq. 4):

$$Q_{cool,net,m} = p_{cool} \cdot Q_{cool,net,princ,m} \quad (4)$$

With:

$Q_{cool,net,m}$	Monthly fictive net energy cooling demand	[MJ]
$Q_{cool,net,princ,m}$	Monthly net energy need for cooling per month	[MJ]

In case of no active cooling, a fixed system efficiency of 90% and a fixed EER of 2.5 is taken into account to determine the secondary and primary energy consumption.

2.2 France

The French procedure to calculate the so-called "T_{ic}" parameter is described in [7]. T_{ic} has to be lower than a reference value $T_{ic,ref}$. T_{ic} is the conventional indoor room temperature reached during a reference hot day in summer. The value of T_{ic} depends on the climate zone, the building type and the characteristics of the building envelope.

$T_{ic,ref}$ is determined by substituting the actual building characteristics by reference characteristics, such as:

- close obstacles are ignored;
- all sun screens and windows are supposed to be opened manually;
- the solar factor of the windows are fixed according to the region and altitude;
- ...

In contrast with Belgium and The Netherlands, it seems that in France no fictive cooling is taken into account when no active cooling is installed.

2.3 The Netherlands

The Dutch procedure to calculate and evaluate the overheating risk in buildings is described in [8].

For buildings without active cooling, the fictive net cooling demand ($Q_{C;nd}$) is calculated:

$$Q_{C,nd} = a_{C;red} \cdot (Q_{C;gn} - \eta_{C;ls} \cdot Q_{C;ht}) \tag{5}$$

With:

$Q_{C;nd}$	Cooling demand	[MJ]
$Q_{C;gn}$	Total heat gains	[MJ]
$Q_{C;ht}$	Total heat losses	[MJ]
$\eta_{C;ls}$	Utilization factor for heat losses	[-]
$a_{C;red}$	Reduction factor for non-continuous cooling (= 1 in case of residential function)	[-]

In case of no active cooling, a fixed EER of 3.0 is taken into account.

3 DEVICES AND REQUIREMENTS

A lot of ventilative cooling means can be thought of, but besides openable windows (and vents in The Netherlands) none of them is integrated in the EP calculation procedure for residential buildings (Table 3). Additional properties of ventilative cooling devices are, in case of natural devices: burglary resistance, insect proof, water tightness and acoustic attenuation. In case of mechanical devices power consumption and noise production are important properties to take into account. These properties improve the potential of ventilative cooling and guarantee that the devices will be used. As shown in Table 4 almost none of these parameters are integrated in the current calculation procedure. Only in Belgium burglary proof is integrated in the EP calculation procedure. For burglary proof devices the ventilative cooling impact is 100% considered. When a device is not burglary resistance, it is considered as not having a cooling impact. When it is moderate burglary proof, it is considered as having 1/3 of its cooling potential. The definition of (moderate) burglary proof is not defined yet but will be determining the potential of ventilative cooling.

Table 3: Ventilative cooling means for dwellings in Belgian, French and Dutch EPBD regulation

Ventilative cooling means	Belgium	France	The Netherlands
Openable windows			
Turn windows	Net area (m²)	Max. opening ratio (%)	Yes / no
Tilt windows	Net area (m²)	Max. opening ratio (%)	Yes / no
Sliding windows	---	Max. opening ratio (%)	Yes / no
Roof windows	Net area (m²)	Max. opening ratio (%)	Yes / no
Vents integrated in/around windows (~ window grills)	---	---	Yes / no
Wall louvres	---	---	---
Natural extract chimney	---	---	---
Mechanical extract and/or supply fans	---	---	---

Table 4: Parameters regarding natural ventilative cooling integrated in current residential EP calculation procedure (yes / no) and current requirements (Req) and recommendations (Rec) in national EPBD regulation

	Belgium	France	The Netherlands
Burglary proof	Yes / ---	---	No / Rec
Insect proof	---	---	---
Water tightness	---	---	No / Rec
Acoustic attenuation	---	---	---

4 SENSITIVITY ANALYSIS

In this study, the effect of ventilative cooling is investigated by means of openable windows. The objective of the sensitivity analysis is to check how openable windows in combination with other building characteristics are assessed in Belgium, France and The Netherlands since these countries have a similar climate (in France, simulations are performed for the most northern climate zone, namely H1a). Therefore, net yearly heating demand, net yearly cooling demand, summer comfort and yearly primary energy consumption are determined and analysed.

4.1 Methodology

Different parameters are varied in the EP software of the three considered countries (Table 5). Each time the simulations are run on the same reference dwelling (Figure 1). The characteristics of the reference dwelling are listed in Table 6, which is also the reference to check the influence of the parameters listed in Table 7 on the ventilative cooling performance of the dwelling in the different countries. Actually there are two reference situations depending on the fraction openable windows. The first reference situation (REF 1) has no openable windows, referring to a dwelling without ventilative cooling techniques. The second reference situation (REF 2) has 50% openable windows, referring to a dwelling with ventilative cooling devices (namely openable windows).

As stated in Table 1 there are different output parameters to evaluate the building performance. Each of the output parameters of Table 1 is determined within the national EP software to be able to perform an overall comparison of the effect of the parameters listed in Table 7.

Table 5: Used EP software

Country	EP software	Version
Belgium	3G-Software	5.0.5
France	Clima-Win	2.0
The Netherlands	Enorm	1.5

Figure 1: Reference dwelling used in the simulations: a semi-detached two story building

Table 6: Characteristics reference dwelling

Geometric characteristics		
External volume	472	m^3
External heat loss surface	313	m^2
Compactness (external volume/external heat loss surface)	1,51	m
Total floor surface	160	m^2
Window surface (south window living room = 13,2 m^2)	30,7	m^2
Net floor surface	137	m^2
Insulation of building envelope		
Average U-value of building envelope	0,39	W/m^2K
Average U-value of the windows	1,60	W/m^2K
Air tightness of the building envelope		
Belgium:	2,00	m^3/hm^2
France:	0,37	m^3/hm^2
The Netherlands:	0,19	dm^3/sm^2
Space heating and hot water		
Heat generation system	Condensing boiler (107%)	
Heat delivery system	Floor heating	
Mobile external sun protection		
Belgium:	No sun protection	
France:	No sun protection	
The Netherlands:	No sun protection	
Thermal mass		
Belgium:	Moderate heavy	
France:	Moderate heavy	
The Netherlands:	Average inertia	
Openable window fraction		
Belgium:	0% - 50%	
France:	0% - 50%	
The Netherlands:	Fixed windows – openable windows	
Active cooling		
Belgium:	No active cooling	
France:	No active cooling	
The Netherlands:	No active cooling	

Table 7: Parameters and variations (default (reference dwelling) values bold) and how it is simulated in the national EP software

n°	Parameter	Variations	
1	Fraction openable windows	**0** / 25 / **50** / 75 / 100%	
		Belgium:	0 / 25 / 50 / 75 / 100%
		France:	0 / 25 / 50 / 75 / 100%
		The Netherlands:	Openable windows
2	Thermal mass	**moderate heavy** / light	
		Belgium:	Light construction
		France:	Very light
		The Netherlands:	Timber framed
3	Solar shading and solar control	**no sun shading** / manual operated on South / automatically operated on South and East	
		Belgium:	$\square_{\square}$ = 3,9% ; $\square_{\square}$ = 8,1%
		France:	$\square_{\square}$ = 3,9% ; $\square_{\square}$ = 8,1%
		The Netherlands:	lump value of g_{tot}
4	Window-to-floor area ratio	15 / **22%**	(= 20,7 / 30,7 m^2 window surface)
5	Building air tightness	**2** / 10 $m^3/(h.m^2)$	
		Belgium:	2 / 10 $m^3/(h.m^2)$
		France:	0,57 / 2,83 $m^3/(h.m^2)$
		The Netherlands:	0,25 / 1,24 $dm^3/(s.m^2)$
6	Active cooling	Yes (EER = 2.5) / **no**	

4.2 Results

The output mentioned in Table 1 is shown for each parameter variation for Belgium (Be, red columns), France (Fr, blew columns) and The Netherlands (Nl, orange columns) in Figure 2 to Figure 13.. In Figure 14 to Figure 17, the results are expressed relatively to REF 1. For visibility reasons the Dutch output of primary energy consumption (EPC) is multiplied by 100 in the plotted results.

4.2.1 Fraction openable windows

For the three countries, the effect of the fraction openable windows on heating and cooling demand, summer comfort and primary energy consumption is illustrated in Figure 2 to Figure 5. In Belgium and The Netherlands, openable windows are supposed to be closed during the heating season in contrast to France. As can be deduced from Figure 2, in France, the heating demand can differ considerably depending on the fraction openable windows. In this case, the heating demand increased with 30% between no and all openable windows (100%). This increasing heating demand is a serious disadvantage for ventilative cooling in France.

It is remarkable that the heating demand in The Netherlands is 40% lower compared to Belgium, although the setpoint for heating is 2°C higher in The Netherlands and the outdoor climate is slightly colder. Heating demands in France are situated in between (even though the windows are opened during heating season).

As illustrated in Figure 3, openable windows have a great effect on the fictive cooling demand in Belgium and The Netherlands. Unfortunately, the effect on the cooling demand is not known in France. In for dwelling considered, a fraction of 100% openable windows can nearly eliminate the cooling demand in Belgium. In The Netherlands, a reduction in cooling demand of about 30% is observed independent of the fraction openable windows since this cannot be varied in the software.

The risk on overheating in Belgium (Figure 4) follows a similar trend as the cooling demand (Figure 3). When no openable windows are present, the risk on overheating is one. In France, the summer indoor temperature exceed is strongly reduced by placing openable windows. The fraction openable windows, however, has no effect.

Figure 5 illustrates the yearly primary energy consumption in the three countries. In Belgium, ventilative cooling by means of openable windows can have a big impact on the total energy consumption. In this case, a reduction of up to 15% is found when the fraction openable windows is increased to 100%. In The Netherlands, the maximum reduction is 3%, which is much smaller than in Belgium due to a smaller effect on the cooling demand (Figure 3). However, with 25% openable windows, the reduction in primary energy consumption is similar as in Belgium.

In France, it is remarkable that an increased fraction openable windows leads to an increased primary energy consumption. This means that the increased heating demand (Figure 2) is not compensated by the decreased cooling demand (not shown, but can be deduced from the reduced indoor temperature exceed, Figure 4). Is seems that in the French EPBD, the air flow rate through openable windows is similar in winter as in summer time. In practice, however, windows are less opened in cold periods.
Furthermore, it is not allowed to compare the primary energy consumption for the reference dwelling between the three countries due to difference in its determination.

4.2.2 Thermal mass

For the three considered countries, the effect of the building thermal mass is illustrated by comparing the reference situations with the situations marked with “Light” on Figure 10 to Figure 13.

In Belgium and France, less thermal mass leads to an increased heating demand, whereas in The Netherlands, the heating demand decreases with decreasing thermal mass (Figure 10).. According to building physics, however, heating demand increases with decreasing thermal mass. In France, the light building with openable windows (Light 50% OW, Figure 10) shows a significant increase in heating demand in comparison with the other presented cases. This can be explained by the fact that windows are also opened during winter period, causing high energy losses due to lack of thermal capacity.

The cooling demand clearly increases when there is less thermal capacity available (Figure 11). Again, the effect is logically higher with an increased fraction openable windows. With openable windows, cooling demands can be more than doubled in a light construction compared with moderate one. Especially in The Netherlands, a major impact is found.

As can be observed in Figure 12, the risk on overheating in Belgium clearly increases with decreasing thermal capacity, similarly to the cooling demand. A similar trend is observed for the French summer indoor temperature exceed. Logically, the temperature increase due to less thermal mass is much smaller in the dwelling with 50% openable windows.

For the three countries, primary energy consumption increases with reduced thermal mass (Figure 13). In The Netherlands, the relative impact of thermal mass on the total energy consumption seems to be independent of the fraction openable windows. In Belgium and France, however, the effect of thermal mass is more or strongly pronounced with an increased fraction openable windows. In The Netherlands, for lower thermal capacities, the decreased heating demand is compensated by the strongly increased cooling demand.

4.2.3 Solar shading and control strategy

Effects of solar shading and solar control is illustrated by comparing the reference situations with the situations marked with "SS" (solar shading). In Belgium and The Netherlands, mobile solar shading (Autom. SS, Figure 10) is supposed to be permanently open during the heating season, causing no increase in heating demand. In France, however, especially automated solar shading increases the net heating demand due to the supposed use of solar shading during the heating season. Without openable windows (0% OW), the heating demand is 50% higher when placing automated solar shading.

In Belgium, the effect of manual south oriented shading (Manual SS-S, Figure 11) on the cooling demand is similar to the effect of 25% openable windows (Figure 3). With automated solar shading on the south and east façades (Autom. SS on S & E), the reduction in cooling demand is similar to nearly 50% openable windows. In case of ventilative cooling (50% OW), the relative effect of solar shading on the cooling need is similar as when there were no openable windows, as long as a cooling demand exists.

In The Netherlands, the effect of solar shading on the cooling demand is similar. Since the effect of openable windows is limited, absolute effects of mobile solar shading can be significantly higher than ventilative cooling.

Figure 12 illustrates how the overheating risk decreases with more advanced solar shading control strategies, in Belgium and France. In contrast to the other parameters studied, adjustable solar protection devices give rise to maximum indoor temperatures lower than the reference value for France (negative indoor temperature exceed, Figure 12).

In Belgium and The Netherlands, the primary energy consumption (Figure 13) follows the same trend as the cooling demand (Figure 11). This means less energy consumption with increasing control of solar shading. Oppositely, in France, the primary energy consumption follows the course of the heating demand (Figure 10). This means, higher energy consumptions with more advanced solar shading. This illogical effect is similar to the impact of an increasing number of openable windows and can only be explained by a cooling demand which doesn't decrease by using solar shading (although the indoor temperature exceed becomes negative, Figure 12).

4.2.4 Window-to-floor area ratio

Windows have an impact on the heating and cooling demand via the solar gains and the transmission losses. Traditionally a major part of the windows are south-oriented to have maximal solar gains during heating season. In this study, the window area of the south oriented window is lowered, marked with "$A_w/A_{fl} = 15\%$" on Figure 10 to Figure 13.

For the reference dwelling, the heating demand increases and the cooling demand decreases with reduced window area, due to a reduction in solar gains which is higher than the reduction in heat losses (Figure 10 and Figure 11). The effect is similar for different fractions of openable windows.

The risk on overheating or indoor temperature exceed decreases logically with lower window surface (Figure 12).

The primary energy consumption in Belgium and The Netherlands decreases with lower window surface (Figure 13), due to a stronger effect on the cooling demand than on the heating demand. In France, depending on the fraction openable windows, the primary energy consumption increases or stays constant. Similar to what is observed for other parameters, the impact on the cooling demand is much smaller.

4.2.5 Building air tightness

On Figure 10 to Figure 13 the simulations marked with "h_{50} = 10" show the influence of a worse building airtightness on the reference dwelling. As can be expected, the tighter the building envelope, the smaller the heating demand and a the higher the cooling demand (Figure 10 and Figure 13.). Regarding ventilative cooling performance, the dwelling doesn't need to be very air tight. Although, for this case, in Belgium, the building air tightness, has no influence on the overheating risk (Figure 12).

In Belgium and The Netherlands, the heating demand augments by 20 to 35% when the air tightness (n_{50}) changes from 2 to 10 $m^3/(h.m^2)$ (Figure 10.). Openable windows don't influence the heating demand. Whereas in France the openable windows have a large influence (as stated before). When there are no openable windows an augmentation of the heating demand of 100% occurs, with 50% openable windows an augmentation of only 40% occurs.

Regarding primary energy consumption (Figure 13.), the effect of the higher heating demand is larger than the effect of the decreased cooling demand. This effect is greater in Belgium than in The Netherlands. In France, when the air tightness is bad (n_{50} = 10 $m^3/(h.m^2)$), the influence of openable windows becomes negligible.

4.2.6 Active cooling

The main purpose of applying ventilative cooling techniques in a dwelling is to avoid the installation of an active cooling system. On Figure 6 to Figure 9 the influence of applying an active cooling installation is shown. In France the combination of an active cooling installation and openable windows is impossible.

Logically, active cooling has no influence on the heating demand (Figure 6). The cooling demand in Belgium can only be lowered by applying openable windows since the risk on overheating is one when there are no openable windows (Figure 7 and Figure 8). When applying active cooling, the risk on overheating is always one (= chance active cooling is used), resulting in a higher cooling demand in case of openable windows.

In The Netherlands, applying an active cooling system gives rise to an increased cooling demand (Figure 7), meaning that active cooling is also penalized. Remarking on Figure 8 is the fact that when there are no openable windows in France, the indoor temperature exceed stays the same, despite the active cooling system.

Of course applying an active cooling systems leads to a higher primary energy consumption (Figure 9). Because REF 1 has an overheating risk of one in Belgium and a same EER for fictive and active cooling of 2.5 is used in the calculations, this augmentation of primary energy consumption does not appear. In the Netherlands, the increase in primary energy consumption due to active cooling is enlarged by the smaller efficiency of 2.5 for active cooling instead of 3.0 for fictive cooling.

4.3 Conclusion

As stated before the goal of the sensitivity analysis is to check for which building characteristics ventilative cooling can be applied so the cooling demand and risk on overheating can be sufficiently reduced without influencing the heating demand. To this end, the effect of the parameters studied, is shown relatively to REF 1 on Figure 14 to Figure 17.

In Belgium and The Netherlands, applying ventilative cooling does not result in a modified heating demand (Figure14.). In France this is not the case, since the windows are also opened during the heating season. A less air tight building, leads in all considered countries to the highest heating demand. With the exception of a lightweight dwelling with 50% openable windows in France.

Regarding cooling demand (Figure 15.), each calculated variation with ventilative cooling results in a significant reduction of the cooling demand (26 to 96% in Belgium, 10 to 35% in The Netherlands). When ventilative cooling is applied, the cooling demand increases when there is less thermal mass available. For all other varied parameters, the cooling demand is decreased compared to the reference situation with ventilative cooling (REF 2). The lowest cooling demand is achieved by combining ventilative cooling with sun shading.

The risk on overheating can clearly be significantly lowered by applying ventilative cooling (Figure 16.). Yet again, a lightweight dwelling performs worse than the reference situation. Applying sun shading in combination with ventilative cooling results in the lowest risk on overheating and temperature exceed. For this case, in Belgium, without applying ventilative cooling, the calculated risk on overheating is only lower than one when automated sun shading is applied.

An overall assessment of the performance of the dwelling is obtained by considering the primary energy consumption (Figure 17.). Except for France where openable windows increase the heating demand, the primary energy consumption, is lower when ventilative cooling is applied. The lowest primary energy consumption is achieved by applying ventilative cooling in combination with sun shading.

5 CONCLUSIONS

In the three countries ventilative cooling by means of openable windows can be taken into account, in Belgium and France as a percentage of openable windows, in The Netherlands as present or not. Burglary resistance and water tightness of ventilative cooling devices are requirements (for The Netherlands) or have an impact on the performance of the ventilative cooling device (for Belgium). In France, no cooling demand could be determined, whereas in The Netherlands no summer comfort criterion is defined. In Belgium and The Netherlands, a kind of fictive cooling is calculated when no active cooling is present. In France, openable windows are not allowed in case of active cooling.

With respect to the heating demand, openable windows have a negative effect in France due to supposed openable windows during heating season. It is also remarkable that in contrast to Belgium and France, less thermal mass decreases the heating demand in The Netherlands.

As can be expected, ventilative cooling always decreases the cooling demand (26 to 96% in Belgium, 10 to 35% in The Netherlands), especially in combination with mobile solar shading. The effect is more pronounced with increasing thermal mass.

Similar to the cooling demand, the risk on overheating in Belgium is decreased by applying ventilative cooling. Also in France, summer indoor temperature can be strongly reduced by using openable windows, although the fraction of openable windows has no effect. Next to ventilative cooling, thermal capacity as well as solar shading can have a considerable impact on summer comfort and should be considered as complementary means.

Except for France, where openable windows remarkably increase the heating demand (up to 38%), the primary energy consumption, is lower when ventilative cooling is applied (5 to 12% in Belgium, up to 4% in The Netherlands). The lowest primary energy consumption is achieved by applying ventilative cooling in combination with sun shading.

Although the building simulated was the same for the three countries and the outdoor climate quite similar, huge differences in output values can be found for the same set of parameter values. It is important to realize that these specific conclusions are based on simulations run on one building type. Further research on different buildings types should be carried out, to have more general conclusions.

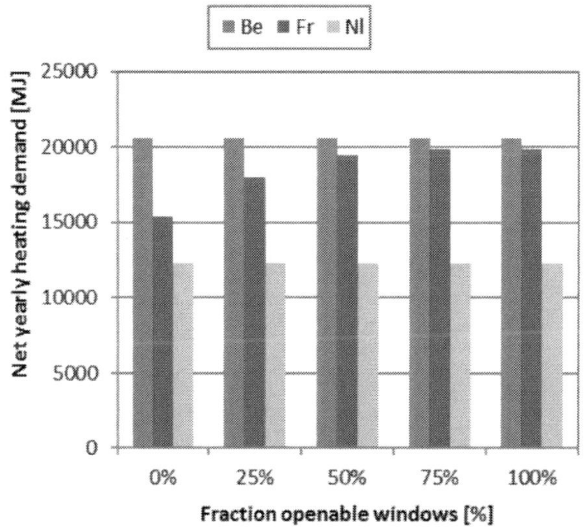

Figure 2: Net yearly heating demand [MJ] as a function of the fraction openable windows for Be, Fr and Nl

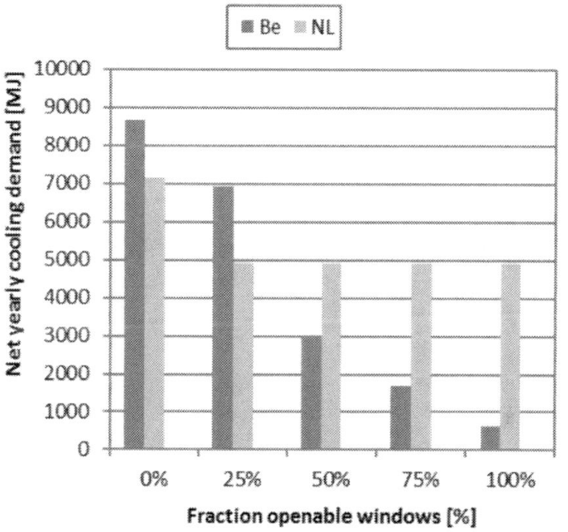

Figure 3: Fictive net yearly cooling demand [MJ] as a function of the fraction openable windows for Be and Nl

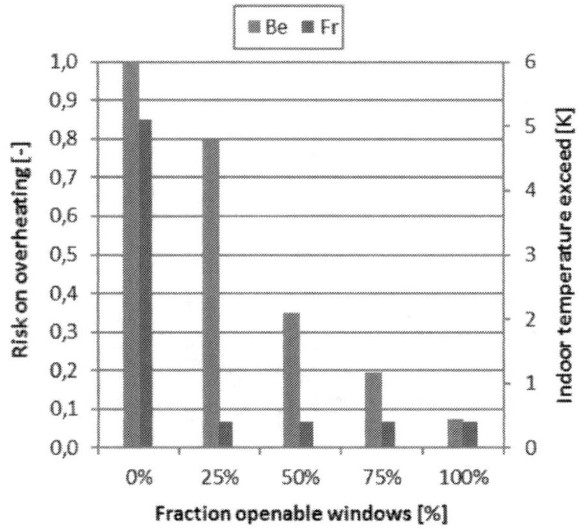

Figure 4: Risk on overheating (Be, left axis) and indoor temperature exceed (Fr, right axis) as a function of the fraction openable windows

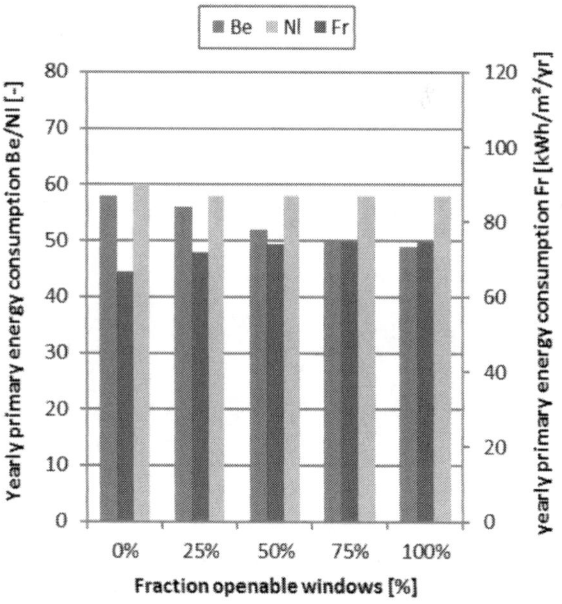

Figure 5: Yearly primary energy consumption [-] or [kWh/m²/yr] as a function of the fraction openable windows for Be, Fr and Nl

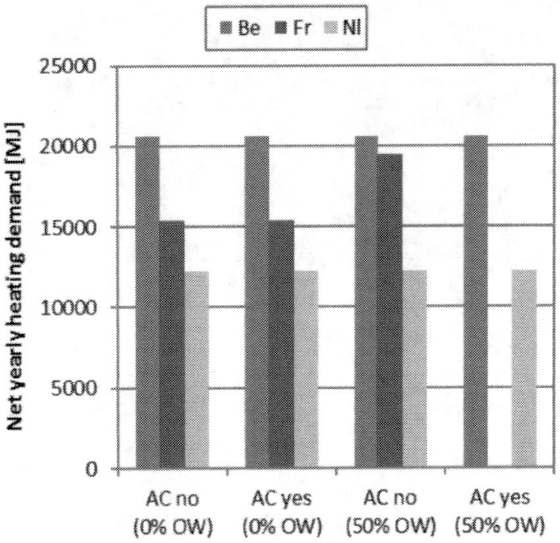

Figure 6: Net yearly heating demand [MJ] as a function of presence of active cooling (AC) for Be, Fr and Nl

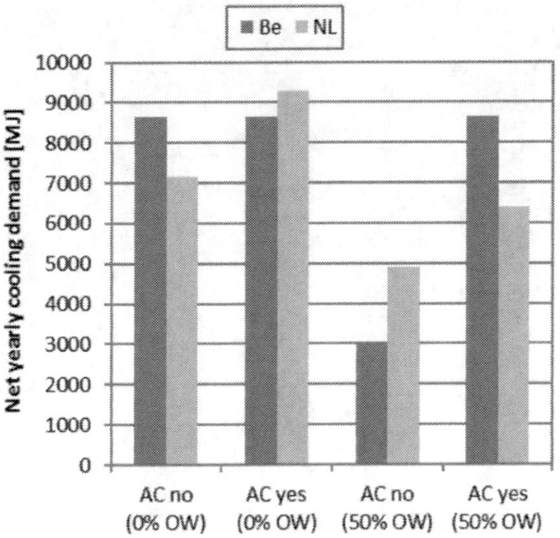

Figure 7: Fictive net yearly cooling demand [MJ] as a function of presence of active cooling (AC) for Be and Nl

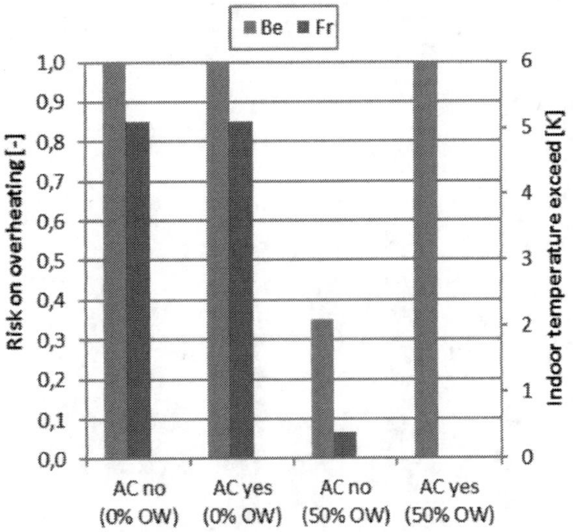

Figure 8: Risk on overheating (Be, left axis) and indoor temperature exceed (Fr, right axis) as a function of presence of active cooling (AC)

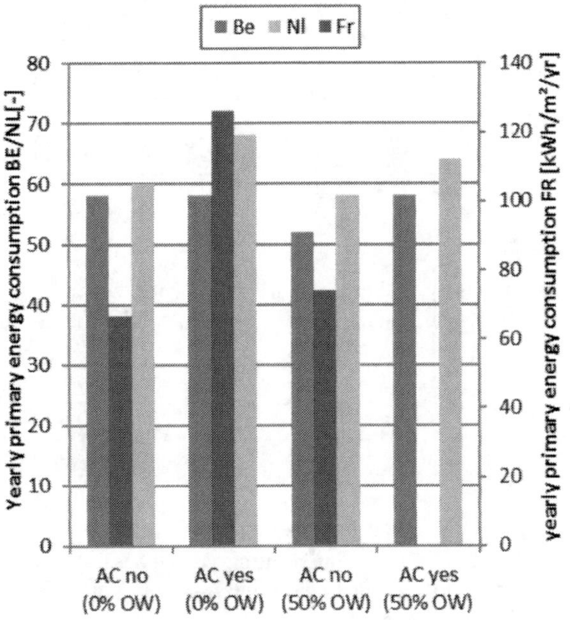

Figure 9: Yearly primary energy consumption [-] or [kWh/m²/yr] as a function of presence of active cooling (AC)

6 REFERENCES

[1] R. S. McLeod, C. J. Hopfe and A. Kwan, "An investigation into future performance and overheating risks in Passivhaus dwellings," *Building and Environment,* no. 70, pp. 189-209, 2013.

[2] "Ventilative cooling: need, potential, challenges, strategies," Venticool, Selection of papers from the Proceedings of the 33rd AIVC- 2nd TightVent Conference, 2013.

[3] I. Pollet, D. Dolmans and A. Van Eycken, "Solar shading solutions to secure sustainable summer comfort and to reduce energy consumption in buildings," in *PassiveHouse Symposium 2011*, Brussels, 2011.

[4] F. Encinas and A. De Herde, "Sensitivity analysis in building performance simulation for summer comfort assessment of apartments from the real estate market," *Energy and Buildings,* no. 65, pp. 55-65, 2013.

[5] "What is ventilative cooling?," Venticool, 2014. [Online]. Available: http://venticool.eu/faqs/what-is-ventilative-cooling/.

[6] "Bijlage V Bepalingsmethode van het peil van primair energieverbruik van woongebouwen," 2014.

[7] "Réglementation Thermique 2012," Centre Scientifique et Technique du Bâtiments (CSTB), 2012.

[8] NEN 7120+C2:2012 Energy performance of buildings - Determination method, Delft: The Netherlands Standardization Institute (NEN), 2012.

[9] NEN 8088-1+C1:2012 Ventilatie en luchtdoorlatendheid van gebouwen - Bepalingsmethode voor de toevoerluchttemperatuur gecorrigeerde ventilatie- en infiltratieluchtvolumestromen voor energieprestatieberekeningen - Deel 1: Rekenmethode, Delft: The Netherlands Standardization Institute (NEN), 2012.

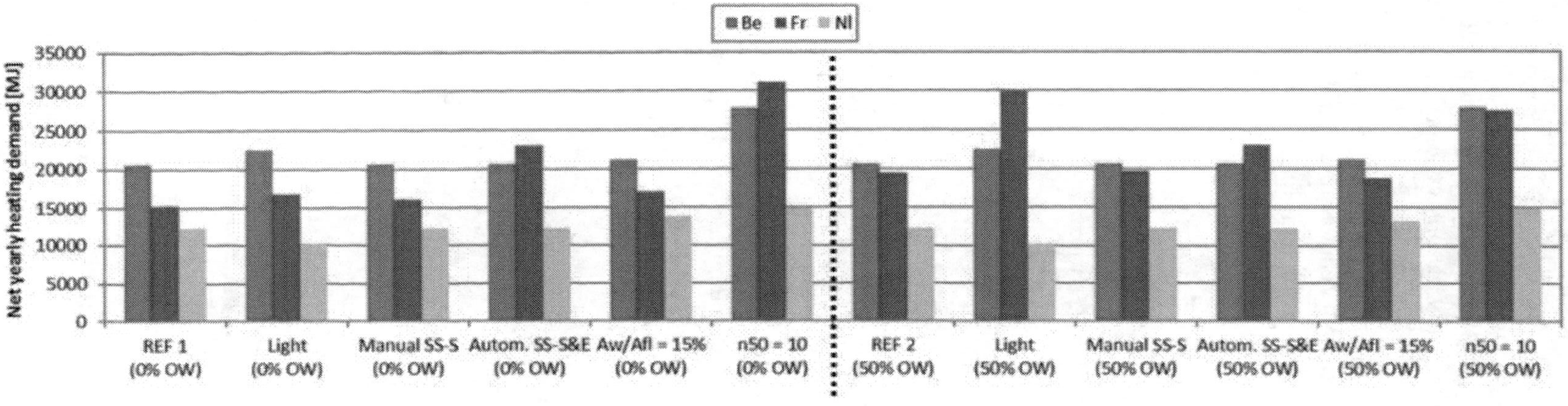

Figure 10: Net yearly heating demand [MJ] for Be, Fr and Nl. At the left side of the dotted line the reference dwelling without ventilative cooling (0% OW) and its variations, at the right side of the dotted line the reference dwelling with ventilative cooling (50% OW) and its variations.

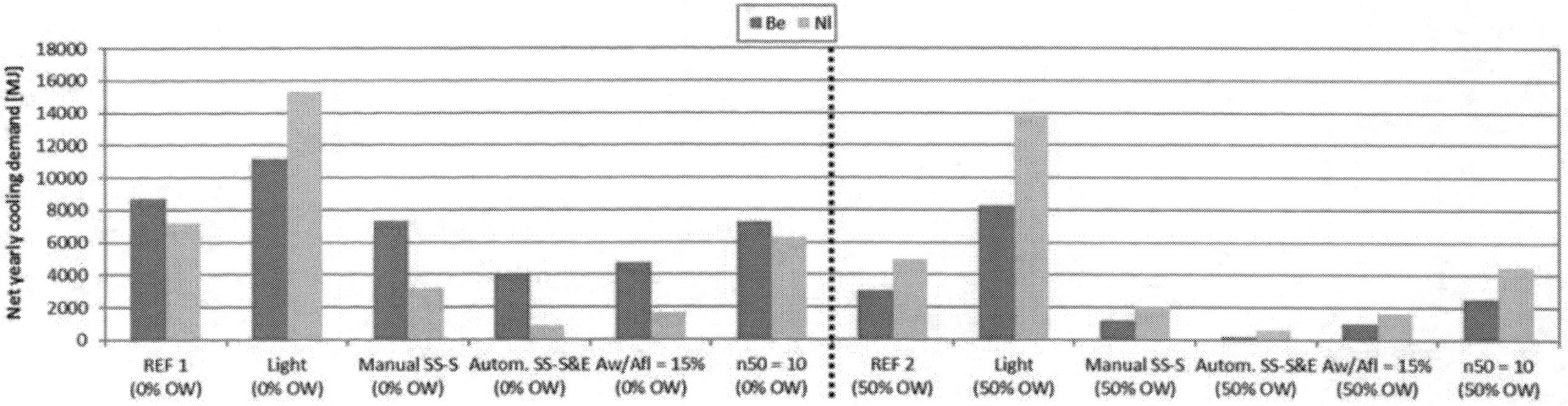

Figure 11: Fictive net yearly cooling demand [MJ] for Be and Nl. At the left side of the dotted line the reference dwelling without ventilative cooling (0% OW) and its variations, at the right side of the dotted line the reference dwelling with ventilative cooling (50% OW) and its variations.

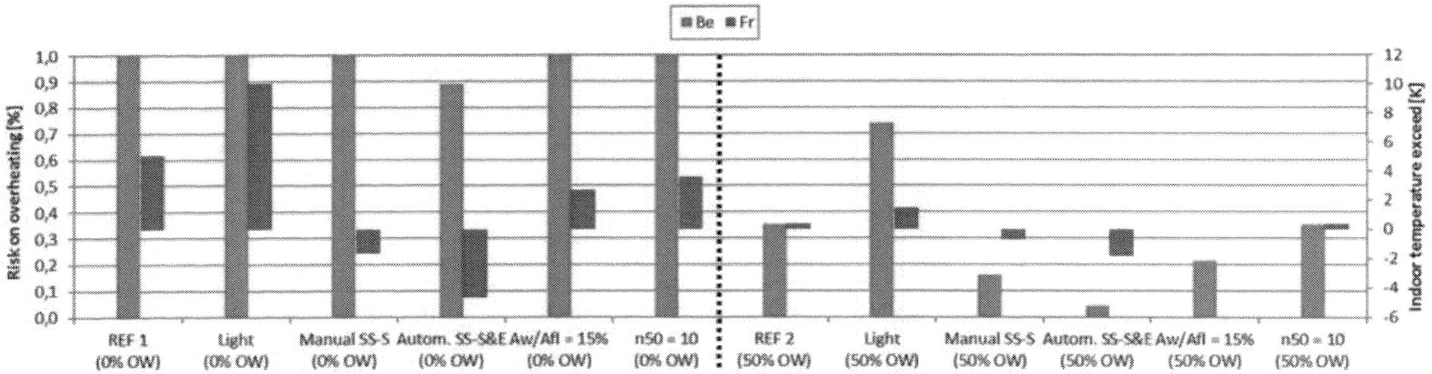

Figure 12: Risk on overheating (Be, left axis) and indoor temperature exceed (Fr, right axis). At the left side of the dotted line the reference dwelling without ventilative cooling (0% OW) and its variations, at the right side of the dotted line the reference dwelling with ventilative cooling (50% OW) and its variations.

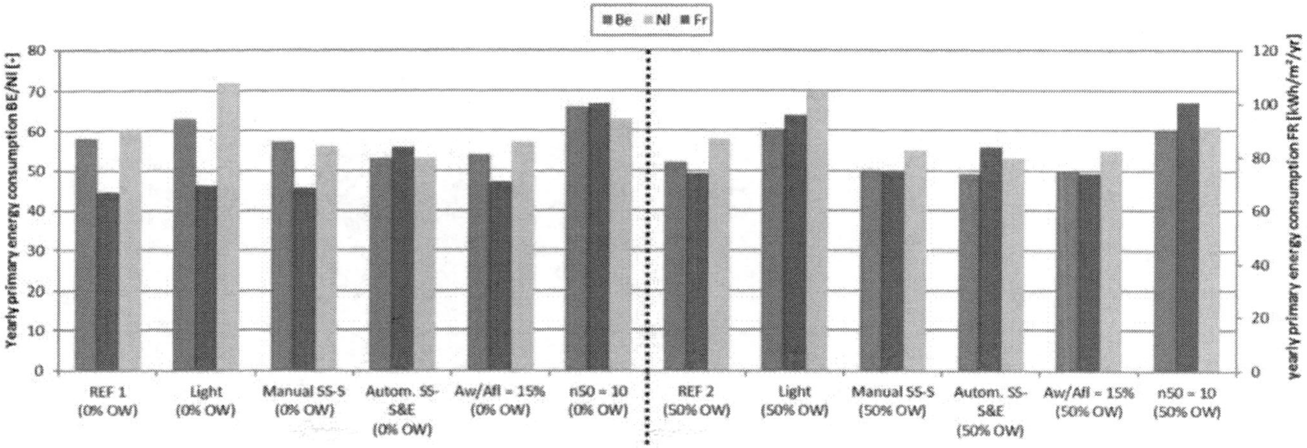

Figure 13: Yearly primary energy consumption [-] or [kWh/m²/yr] for Be, Fr and Nl. At the left side of the dotted line the reference dwelling without ventilative cooling (0% OW) and its variations, at the right side of the dotted line the reference dwelling with ventilative cooling (50% OW) and its variations.

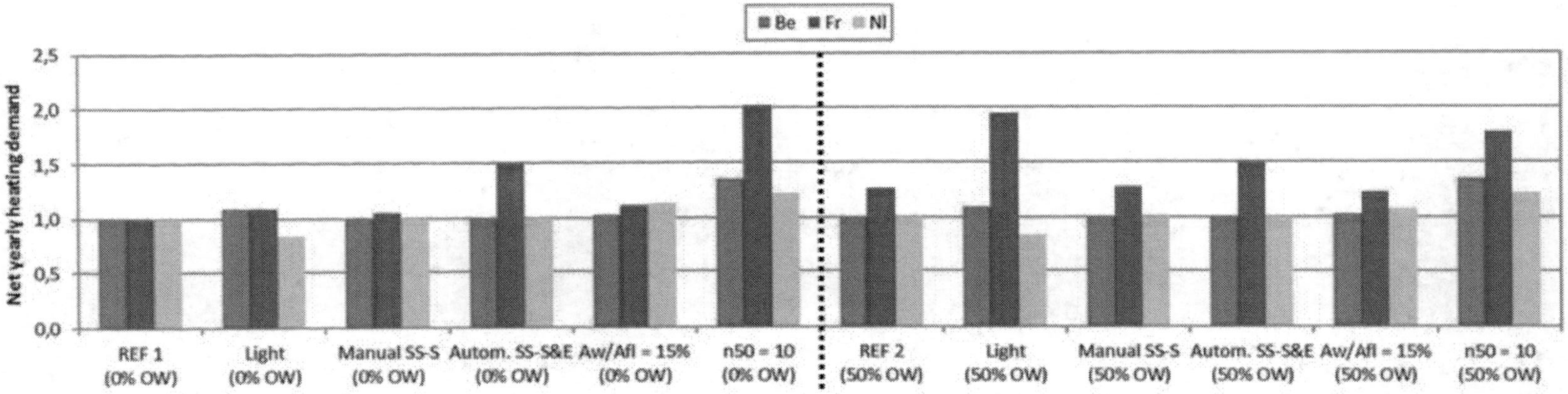

Figure 14: Relative net yearly heating demand with respect to REF 1 for Be, Fr and Nl. At the left side of the dotted line the reference dwelling without ventilative cooling (0% OW) and its variations, at the right side of the dotted line the reference dwelling with ventilative cooling (50% OW) and its variations.

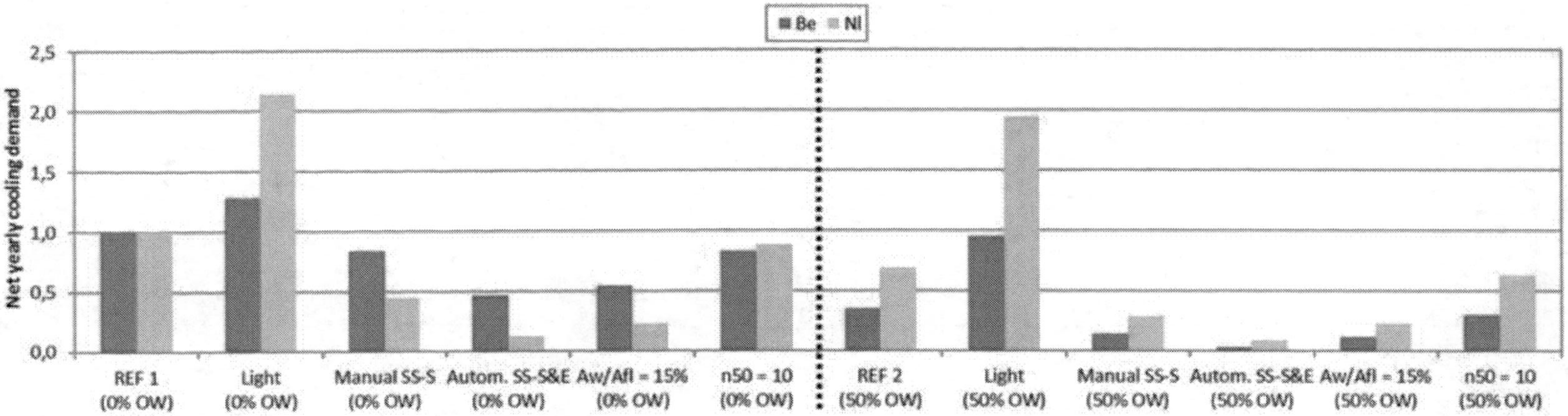

Figure 15: Relative fictive net yearly cooling demand with respect to REF 1 for Be and Nl. At the left side of the dotted line the reference dwelling without ventilative cooling (0% OW) and its variations, at the right side of the dotted line the reference dwelling with ventilative cooling (50% OW) and its variations.

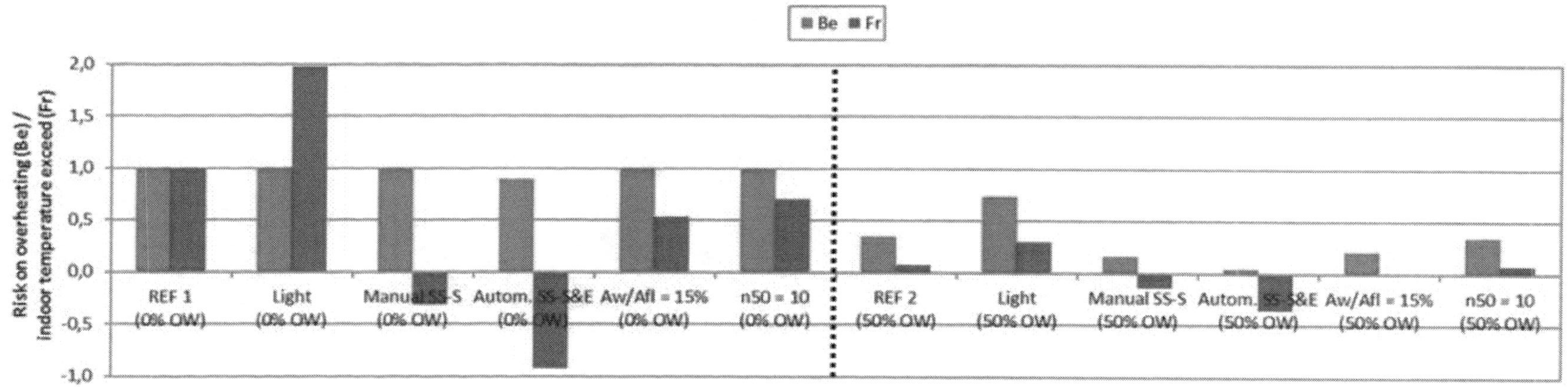

Figure 16: Relative risk on overheating (Be) and relative indoor temperature exceed (Fr) with respect to REF 1. At the left side of the dotted line the reference dwelling without ventilative cooling (0% OW) and its variations, at the right side of the dotted line the reference dwelling with ventilative cooling (50% OW) and its variations.

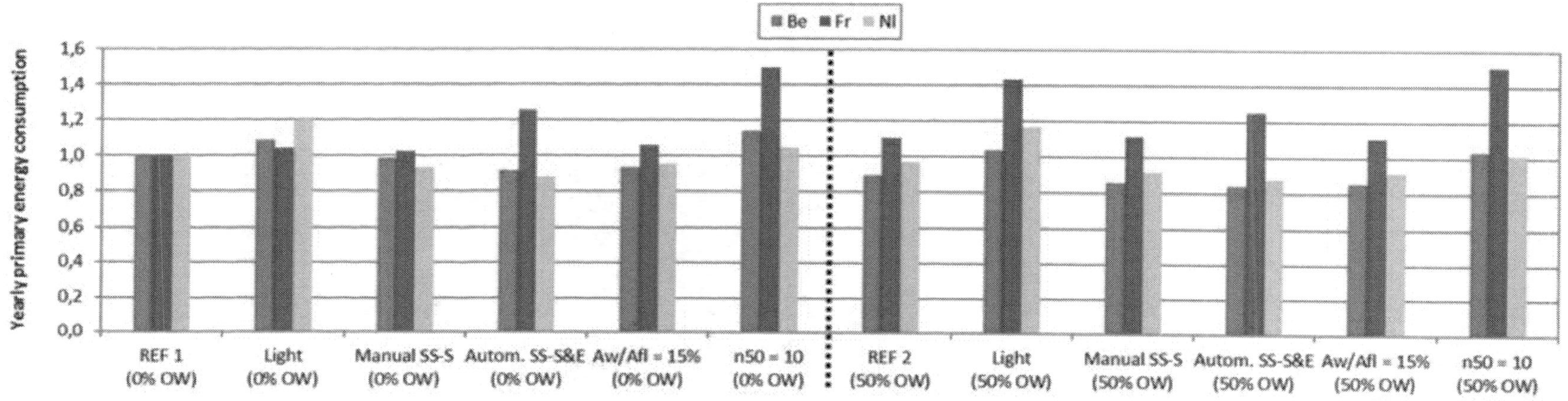

Figure 17: Relative yearly primary energy consumption with respect to REF 1 for Be, Fr and Nl. At the left side of the dotted line the reference dwelling without ventilative cooling (0% OW) and its variations, at the right side of the dotted line the reference dwelling with ventilative cooling (50% OW) and its variations.

THE 10 STEPS TO CONCEIVE AND BUILD AIRTIGHT BUILDINGS

Clarisse Mees[1*], Christophe Delmotte[1], Xavier Loncour[1] and Yves Martin[1]

1 Belgian Building Research Institute
(BBRI)
Av. Pierre Holoffe, 21
B-1342 Limelette
Belgium
www.bbri.be

*Corresponding author: *cme@bbri.be*

ABSTRACT

Airtightness becomes a more and more important parameter in the rationalization of the energy consumption. The quality of the works during the construction process is essential. However, this particular step is on itself absolutely not sufficient to build airtight buildings. Airtightness has to be taken into account from the pre-project. For that, architects have to deal with a large bunch of items. Steps as the definition of the ambition level, the precise positioning of the airtightness barrier into the building are essential. Possible contradictions with permanent ventilation requirements have to be taken into account. For instance, in Belgium, there are natural ventilation requirements for fire safety reasons or for open heating devices (as stoves) and it could be difficult to conciliate these requirements with the performance of airtightness. In addition, some equipment such as distribution board must be adequately placed to avoid some air leakages. At last, the choice of equipment types and their position could be crucial to achieve the airtightness level considered. The construction process on itself, intermediate blower door tests during the construction process, the final blower door test as well as the attention points during the occupancy have also to be taken into account.
Thereby, a path of 10 steps starting from the very beginning of the conception up to the utilization of the building have been identified to facilitate the realisation of affordable airtight buildings.

KEYWORDS

Airtightness of buildings, fan pressurization test, conception, Belgium, air permeability class

1. INTRODUCTION

Airtightness becomes a more and more important parameter in the rationalization of the energy consumption. The quality of the works during the construction process is essential. However, this particular step is on itself absolutely not sufficient to build airtight buildings. Airtightness has to be taken into account from the pre-project. For that, architects have to deal with a large bunch of items. The present paper proposes a path decomposed on 10 practical steps to support the architect in the design and the build of an airtight building. The 8 first steps refer to the conception itself and must be carried before the start of the construction site. The exact contain of each step depends on national norms or regulations but the general path could be applied in every country.

2. THE 10 STEPS

2.1 Step 1: Define the ambition level

Currently in Belgium, there is no requirement regarding the airtightness in the EP regulations. But this criterion is taken into account into the EP calculation for new buildings. Without pressurization test, a default value is applied. This v50 default value is equal to 12 $m^3/(h.m^2)$ for the calculation of the heating consumption. The ambition level is generally chosen by the customer.

Belgium consists of three regions which are independently responsible for the rational use of energy in buildings on their territory. The three regions are: the Flemish Region, the Walloon Region and the Brussels-Capital Region. Recent information indicates that the average air permeability (v_{50}) is 2.9 $m^3/(h.m^2)$ in the Walloon Region and 3.7 $m^3/(h.m^2)$ in the Flemish Region.

In any case, the targeted value of airtightness should be fixed as soon as the pre-project. This value should have an impact on the first choices of the architect. The different steps of the process are all important but may be crucial if the airtightness level is severe.

2.2 Step 2: Define the airtight zone within the building – The protected volume

From the first stage of the draft, the architect should define and keep in mind the protected volume. The protected volume is defined in the Energy Performance of Buildings regulations as the volume including all heated spaces (and/or cooled) directly or indirectly, either continuously or intermittently, within the considered building. The architect must know the way that the pressurisation test will be realized in that building and which part of the building is involved by the ambition level of airtightness. He should also check the consistency between the thermal insulated volume and the airtight volume.

2.3 Step 3: Choose equipment types and their position regarding the airtight zone

Many technical appliances are integrated into our buildings, whether for heating, ventilation, gas distribution, electrical and telecommunication, active cooling system, ...

Most of the rooms including these technical installations require direct outside air, e.g. for combustion purposes.

A representative example of this step's issue concerns the heating appliance. Belgian standards require natural ventilation for the technical room including open heating devices. Non-sealable openings through the outside must be provided for air supply and to evacuate the combustion gases. These openings have a significant impact on the airtightness of the building. Indeed, for a single-family house of 500 m^3 (interior volume) with an open heating device type B (25kW) in the protected volume, the required openings lead to an increasing of 2,4 h^{-1} of the ACH50. Thus, the use of this type of device is not recommended within the protected volume. If such kind of heating system is chosen, architects should forecast an installation space outside the protected volume.

However, room sealed appliance doesn't require openings through the envelope. The architect should privilege room sealed appliance and should place the installation space in the protected volume.

Thus, for heating device and for other appliances, it is possible to conciliate fire safety, ventilation requirements and airtightness by choosing appropriate equipment and by positioning them relative to the protected volume. These decisions have to be taken early enough in the process of conception.
These choices are dependant to the norms and regulations of each country. In Belgium, summary tables are available as tools for designer. One of this tables is given below as example.

Table 1: Examples of advices for equipment positioning relative to the airtight zone

Equipment / rooms	**Recommended position**
Garages	Foresee specific ventilation system or place them outside the protected volume
Technical shafts	Depending on the fire regulations. If not applied in the considered building: inside the protected volume If applied in the considered building: outside the protected volume or provide a partitioning of the shafts
Elevator shafts	Inside the protected volume and provide a ventilation system with motorized valves Or Outisde the protected volume
Counters gas	Outside the protected volume

2.4 Step 4: Place piping and ducts

The architect should place piping and ducts in order to minimize the openings through the airtight envelope. He also should try to keep ducts in the protected volume.

2.5 Step 5: Choose the good material to achieve an airtight envelope

We could consider a material as airtight if its air permeability is below 0,1 $m^3/(h.m^2)$ for a pressure difference of 50 Pascal. This criterion could be useful to choose wood panel, for example. In Belgium, plaster (generally at least 12mm) is traditionally used on the inner surface. Laboratory tests have shown that the air permeability of this kind of product whatever the masonry is below the limit value. Recent research[1] has studied the initial air permeability performance of different types and thicknesses of plaster and their durability.

2.6 Step 6: Correctly choose doors and windows

Doors and windows are one of the only building components whose the airtightness performance is tested in laboratory. These tests are realized for CE marking reasons and are realized according to European standards. The architect should privilege a class 4 air permeability. However, in some cases of high level of ambition, it could be not enough efficient.
An analysis of 300 windows tested in the laboratory of the BBRI (2008-2009) shows that 87% of tested products corresponding to the class 4, whatever the material of the frame. Therefore, a subdivision of the class 4 seems needed to help architects and builders choose the most appropriate product.

[1] The research project DREAM funded by the Walloon Region (Belgium) investigates the durability of air barrier by testing them in laboratory. A paper on this project is available: "Assessment of the durability of the airtightness of building elements via laboratory tests"

A subdivision of the class 4 into two supplementary classes (5 and 6) illustrated by the figure 1 will be proposed to the Committee for Standardization (CEN).

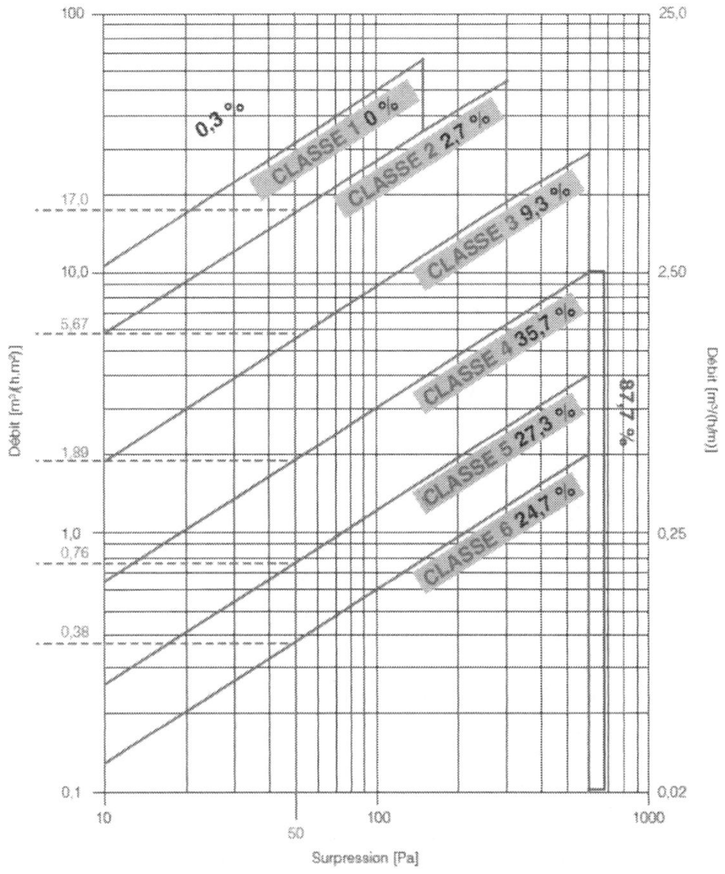

Figure 1: Distribution of tested windows between different classes of air permeability

Currently, to achieve high airtightness level and without classes 5 and 6, it is recommended to ask the test report and estimate if the air permeability is compatible with the ambition level of the considered project.

2.7 Step 7: Prioritize the constructive nodes

To guide the designer, the path includes tables giving an indication of the potential leakage if constructive nodes are not properly treated. These tables have been developed during the research project Etanch'air. One of its goals was the quantization of common air leakages.
This kind of table is useful for the designer in order to identify the priorities he has to give to different constructive nodes.
Note that these priorities should be considered depending upon the construction system of the project, its geometry and its level of ambition. The cavity wall example, the mainly used constructive system in Belgium, as well as the timber frame construction system are presented in the next tables.

Table 2: Illustration of the priorities

Prioritization order	
****	Priority 1
***	Priority 2
**	Priority 3
*	Priority 4

Table 3: Prioritization of the constructive nodes to be treated – Cavity walls

Constructive nodes	Priority
Ground bearing floor	* → **
Junction between separating wall and façade	*
Junction between flat roof and façade	*
Junction between intermediate floor and façade	**
Pitched roof: Purlins	*** → ****
Pitched roof : Gable	****
Pitched roof : Eaves	*** → ****
Pitched roof : Separating wall	*** → ****
Service penetration through roof	***
Junction between window and façade	** → ***

Table 4: Prioritization of the constructive nodes to be treated – Timber frame

Constructive nodes	Priority
Ground bearing floor	****
Junction between separating wall and façade	*** → ****
Junction between intermediate floor and façade	*** → ****
Pitched roof: Purlins	*** → ****
Pitched roof : Gable	*** → ****
Pitched roof : Eaves	*** → ****
Pitched roof : Separating wall	*** → ****
Service penetration through roof	***
Junction between window and façade	** → ***

2.8 Step 8: Choose technical solutions for each constructive nodes

A library of construction details in linked to the proposed path. An example is given with the figure 1 illustrating the treatment of a junction between façade and floor with an airtight membrane.

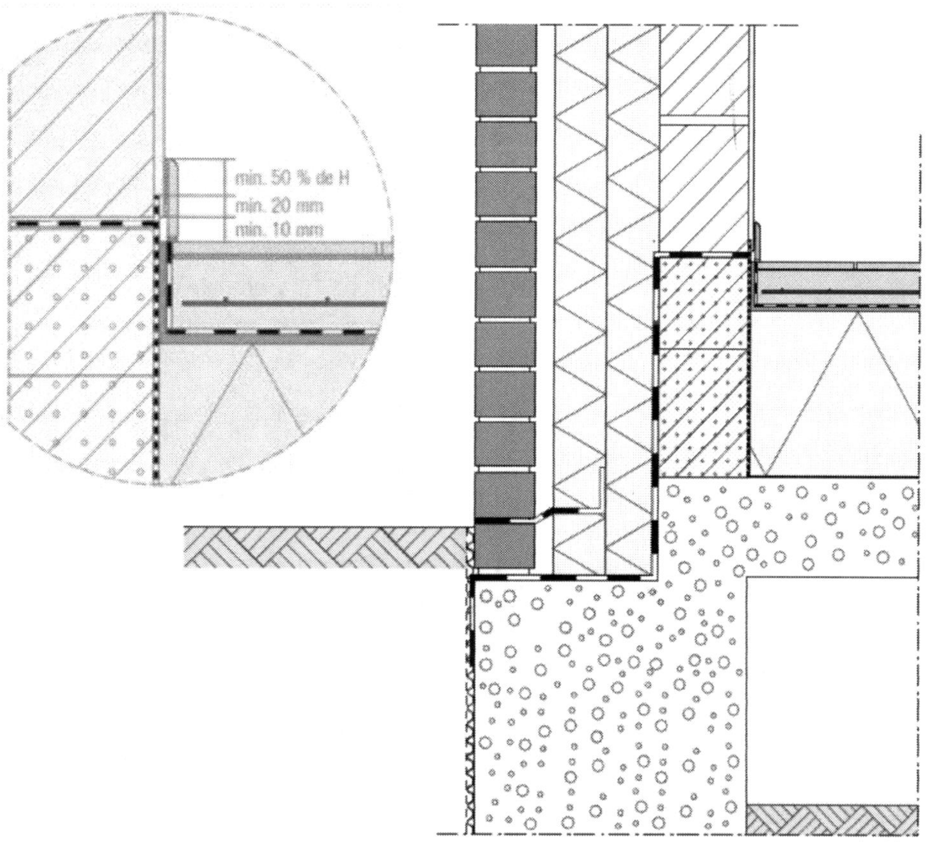

Figure 2: Illustration of a technical detail: junction between façade and floor using airtight membrane

Most of the time, several solutions are possible to resolve the nodes. Thus, a comparative table is proposed to support architect or builder in their choice. The following table compares three technical solutions available to treat the junction between façade and floor. This comparison is qualitative and uses three criteria: initial performance, implementation issue and durability. In the proposed path, each solution is illustrated by a construction detail or a picture.

Table 3: Qualitative comparison between different technical solutions for the junction between façade and floor.

	Initial performance	Implementation issue	Durability
No specific treatment	Good if the vertical joints of the masonry are properly filled	Importance of filling vertical joints	Possible cracks due to the building movements
Junction made with airtight membrane (figure 1)	Very good	Risk of significant damage to the membrane during construction	Junction products could be better in terms of durability for this node
Junction made cementing	Very good	Easy implementation	Possible cracks due to the building movements

2.9 Step 9: Check the coordination and communication between all the builders

Each involved builder has to be informed about the targeted value of airtightness, the protected volume and the nature of air barriers. The constructive details elaborated by the architects should be given to the concerned people as soon as possible.

2.10 Step 10: Provide an intermediate pressurization test

This intermediate test realized during the construction is highly recommended especially if the ambition level is important. The exact moment of the test will be chosen in order to be able to correct leakages that would be detected (e.g. when airtight layers are still accessible in roofs). The architect (of the builder) should plan it. A final test has of course also to be planned.

3 CONCLUSIONS

To achieve an airtight building, architects have to deal with a large bunch of items. Steps as the definition of the ambition level, the precise positioning of the airtightness barrier into the building are essential. Possible contradictions with permanent ventilation requirements have to be taken into account. In addition, some equipment such as distribution board must be adequately placed to avoid some air leakages. At last, the construction process on itself, intermediate blower door tests during the construction process, the final blower door test as well as the attention points during the occupancy have also to be taken into account.

Thereby, a path of 10 steps starting from the very beginning of the conception up to the utilization of the building have been identified to facilitate the realisation of affordable airtight buildings. The goal is to use these steps as a daily tool by the architects.

4 ACKNOWLEDGEMENTS

The work described in this paper was partly developed within the research project ETANCH'AIR supported by the Walloon Region in Belgium and the project LUCHTDICHT BOUWEN VAN A TOT Z supported by the Flemish Region.

5 REFERENCES

NBN EN 13829 (2001) – Thermal performance of buildings — Determination of air permeability of buildings — Fan pressurization method (ISO 9972:1996, modified)

NBN B61-001 – Générateurs de chaleur d'une puissance totale installée supérieure ou égale à 70 kW - Exigences et prescriptions en matière d'amenée d'air, d'évacuation d'air et d'évacuation des fumées dans les chaufferies

NBN B61-002 - Chaudières de chauffage central dont la puissance nominale est inférieure à 70 kW - Prescriptions concernant leur espace d'installation, leur amenée d'air et leur évacuation de fumée (+ AC:2008)

NBN D51-001 - Chauffage central, ventilation et conditionnement d'air. Locaux pour poste de détente de gaz naturel

NBN D51-003 - Installations intérieures alimentées en gaz naturel et placement des appareils d'utilisation - Dispositions générales

NBN D51-006-3 - Installations intérieures alimentées en butane ou propane commercial en phase gazeuse à une pression maximale de service de 5 bar et placement des appareils d'utilisation - Dispositions générales - Partie 3: Placement des appareils d'utilisation.

NBN EN 14351-1+A1 : Fenêtres et portes - Norme produit, caractéristiques de performance - Partie 1: Fenêtres et blocs portes extérieurs pour piétons sans caractéristiques de résistance au feu et/ou dégagement de fumée.

NBN EN 1026 : Fenêtres et portes - Perméabilité à l'air - Méthode d'essai.

BELGIAN FRAMEWORK FOR RELIABLE FAN PRESSURIZATION TESTS FOR BUILDINGS

Xavier Loncour[1], Christophe Delmotte[1*], Clarisse Mees[1] and Maarten De Strycker[2]

1 Belgian Building Research Institute
(BBRI)
Av. Pierre Holoffe, 21
B-1342 Limelette
Belgium
www.bbri.be

2 Belgian Construction Certification Association
(BCCA)
Rue d'Arlon, 53
B-1040 Bruxelles
Belgium
www.bcca.be

*Corresponding author: *cde@bbri.be*

ABSTRACT

This paper presents the new framework for the realization of reliable pressurization tests in Belgium and the provisions taken to widen the number of buildings where a valid pressurization test can be realized.
The pressurization test is described in standards. In Belgium, the measured airtightness performance can be used in the regional Energy Performance (EP) regulations in order to improve the calculated performance. A few years ago, supplementary specifications to the test standard (NBN EN 13829) describing detailed specific elements of the test, were added to the regulations. Testing is not required by the regulations but a relatively unfavourable default value for the airtightness has to be used if no test is carried out. With the progressive strengthening of the energy standards included in the regulations, the airtightness performance has become more and more important. Measuring the airtightness is a common practice nowadays. In some Regions, over 50% of the buildings is now tested. Up to 2014, in Belgium, requirements have not been imposed on the testers. Everyone in possession of the test equipment could realize a pressurization test and could use the test result in the EP regulation. Moreover, several years of experience with the regulations show specific situations where the strict respect of the test standard is not always possible (e.g. in high buildings).

KEYWORDS

Airtightness of buildings, Fan pressurization test, Quality, Large building, Competent tester

1. INTRODUCTION

Since 2006 in Belgium, the airtightness of buildings determined by fan pressurization tests can be used on a voluntary basis as input data in the Energy Performance (EP) regulations for new residential buildings, offices and schools. Even though the rules for these tests are well defined, some questions are still open. Considering the increasing number of tests and evidences that some of them are not performed as good as they should be, updated rules including also the possibility of a quality framework imposing a competent tester scheme have been developed.

2 SITUATION REGARDING THE PRESSURIZATION TESTS IN BELGIUM

2.1 Energy performance regulations and use of the results of pressurization tests

The regional EP regulations[1] require the introduction of a final declaration at the end of the works taking into account the actual characteristics of the buildings in order to prove the compliance with the requirements. The result of a pressurization test can be used as input data for new buildings (residential buildings, offices and schools). It has to be mentioned that in these buildings, the installation of a ventilation system is mandatory.

At present, there is no requirement regarding the airtightness performance of new buildings. Making a pressurization test is also not mandatory and no requirements are set on testers. The Brussels Region however has announced the introduction of an explicit requirement on the airtightness performance of new buildings from 2018 ($n_{50} \leq 0.6\ h^{-1}$). A similar evolution is not planned in Flanders and in the Walloon Region.

If a test is not realized, the energy consumption for heating and cooling is calculated with quite unfavourable $\dot{v}_{50}$ air permeability default values[2]. These values are equal to 12 $m^3/(h\ m^2)$ for heating calculations and 0 $m^3/(h\ m^2)$ for the risk of overheating and cooling calculations.

The airtightness has an important impact on the calculated energy performance of buildings. When the first EP requirement was set in 2006, this impact was about 5% to 15% for residential buildings. However, according to the definition of Nearly Zero Energy Buildings (NZEB), it could increase up to about 30% and make it nearly impossible to comply with regulation using the air permeability default value. Even if a pressurization test is not required by the regulations, it is becoming indispensable in practice.

2.2 Increasing number of pressurization tests in new buildings

EP regulation taking into account the airtightness of buildings has been introduced in 2006 in the Flemish Region and in 2010 in the Walloon Region. Both regions have developed databases with data included in the final EP declarations and they are able to produce statistics regarding airtightness of buildings (VEA, 2013).

In the Walloon Region, 22% of the final declarations include measured air permeability (information from early 2014). The average air permeability ($\dot{v}_{50}$) is 2.9 $m^3/h\ m^2$.

In the Flemish Region, more than 14 700 final declarations include measured air permeability (information from June 2014).

The average air permeability ($\dot{v}_{50}$) is 3.7 $m^3/h\ m^2$. The percentage of final declarations including measured air permeability is increasing every year (see Figure 1)[3].

1 Belgium consists of three regions which are independently responsible for the rational use of energy in buildings on their territory. The three regions are: the Flemish Region, the Walloon Region and the Brussels-Capital Region.

2 The air permeability (v'_50) is equal to the mean air leakage at 50 Pa (V'_50) divided by the envelope area of the building based on external dimensions. The energy performance regulations define the exact way of calculating it.

3 These figures are changing with the number of final declarations introduced. The statistics of the last year presented (2012) are based on a limited number of final declarations (3000).

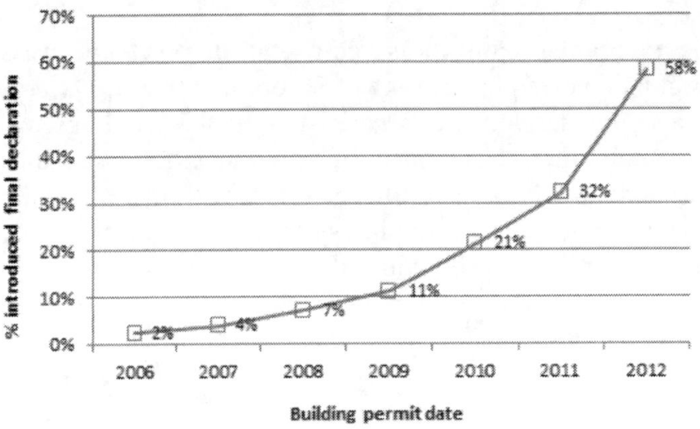

Figure 1: In the Flemish Region, the percentage of final EP declaration including measured air permeability of buildings is increasing since 2006

2.3 Rules for the realization of the pressurization test

The pressurization test has to be performed according to NBN EN 13829 (2001). Clarifications needed in the context of the EP regulations were published by the regions in 2008. These clarifications include, for example, the preparation of the buildings in coherence with the EP calculation procedure, the type of test (type A), specific requirements regarding the testing material or the content of the test report. Updates published in 2010 and 2013 include clarifications about closing or opening of internal doors and prohibition of simultaneous pressurization of adjacent spaces not included in the test volume.

When analysing reports, it appears that some tests are not performed with sufficient quality. Though a detailed study has not been carried out, different explanations can be presented:

- Modern measuring equipment's have automatic testing options which make it very easy to obtain results without specific knowledge of the test procedure;
- Many small companies propose pressurization tests just as a secondary service;
- There are no requirements on the tester.

2.4 Example of problematic situations with pressurization tests

The tester measures leakage airflow. The client is generally more interested in specific airtightness indicators, as the n_{50} or the $\dot{v}_{50}$-values, than in the air leakage flow on itself. The calculations of these indicators require knowledge of the parameters such as the volumes or areas related to the tested building. The conventions to calculate these parameters are well defined. However, the responsibility to determine these values was up to now not really well defined. The common practice was that this information, communicated by the client, was integrated into the test report and used to calculate the airtightness indicators. In the past, some testers have been asked to update their test reports and modifying the surfaces or volumes earlier communicated by the client.

In case of measurements in high buildings, it is not always evident to strictly follow the requirements of the test standard mainly regarding the zero-flow pressure difference criteria. The stack effect caused by temperature differences limit in a very substantial way the acceptable conditions the realization of a pressurization test in accordance with the rules of the measuring standard.

In large buildings, the highest pressure difference to reach may also require many fans, potentially installed in all available external doors. The test standard is ambiguous on the acceptable conditions related to this highest pressure difference.

In Belgium, the EP-regulations require the tests to be made in the two measuring modes (overpressure and underpressure). This is more restricting than NBN EN 13829 and is sometimes not evident or even impossible. For example, it can be very difficult in presence of non-accessible valves that progressively open with increasing overpressure. This kind of valve can be present in exhaust ducts connected to cooker hoods. Measurements in the two modes can also be difficult with measurement devices like large fans fixed on a trailer.

2.5 Belgian Technical Specifications

The supplementary specifications for the pressurization tests were updated in early 2014 to take into account the development of the new STS-P 71-3.

The STS documents[4] are published by the Federal Public Service Economy in Belgium[5]. The main objective of a STS is to help clients or architects to establish the technical specifications for a specific construction project. The STS documents describe how a specific product or construction technique has to be specified, installed and controlled. They are developed in collaboration with all interested parties of the construction sector and are based on a consensus. They are applicable if reference is made to them for a specific construction project and are mandatory if this reference is made by a regulation. In case of optional elements within the STS as recommendations or informative annexes, specific reference to these particular elements has to be made to make them applicable.

3 UPDATED RULES FOR PRESSURIZATION TESTS IN BELGIUM – STS-P 71-3

The STS-P 71-3 developed in 2014 considers specifically pressurization tests. The schematic content of this document is shown in Figure 2. An informative annex describes general requirements regarding the organization of a quality framework. The complete scheme including this informative annex will be applicable in the Flemish Region for every airtightness measurement realized from January 2015.

The STS clearly indicates which part of the process is the responsibility of the tester. Some elements within this process are the responsibility of the client.

4 STS stands for « Spécifications techniques unifiées – Eengemaakte technische specificaties » which means « unified technical specifications »

5 Website : http://economie.fgov.be/en

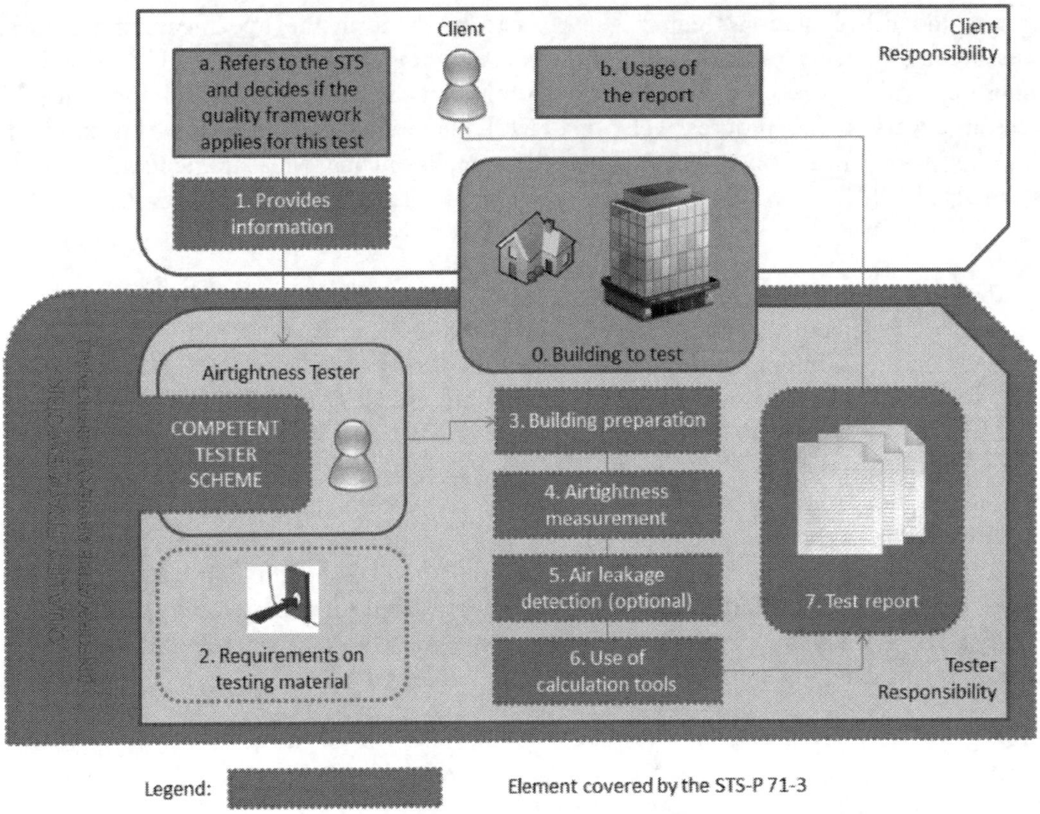

Figure 2: Steps of a pressurization test and elements covered by the STS-P 71-3

3.1 Responsibilities of the client

If not mandatory by the regulation, the client has first to decide to make the STS applicable by making a reference to it (step a. - Figure 2).

According to the objective of the test, different requirements are imposed on the state of the building (step 0 - Figure 2) for the realization of the pressurization test (Figure 3). In the context of the EP regulation, a “standard” test has to be realized. This test can only be carried out if the building is totally closed (all windows and doors installed). No temporary closing system is allowed. It is however recommended to wait until the end of the construction works to perform the test.

When ordering a pressurization test, the client has to provide the tester with information relative to the building (step 1 - Figure 2). The client has to define the tested zone in accordance with all applicable specifications. The test report has to contain one of the following parameters relative to the tested zone: total area of the walls around the tested zone (information used for the determination of the $\dot{v}_{50}$-value) or internal volume (information used for the determination of n_{50}-value). The client is responsible for providing the information and for its correctness.

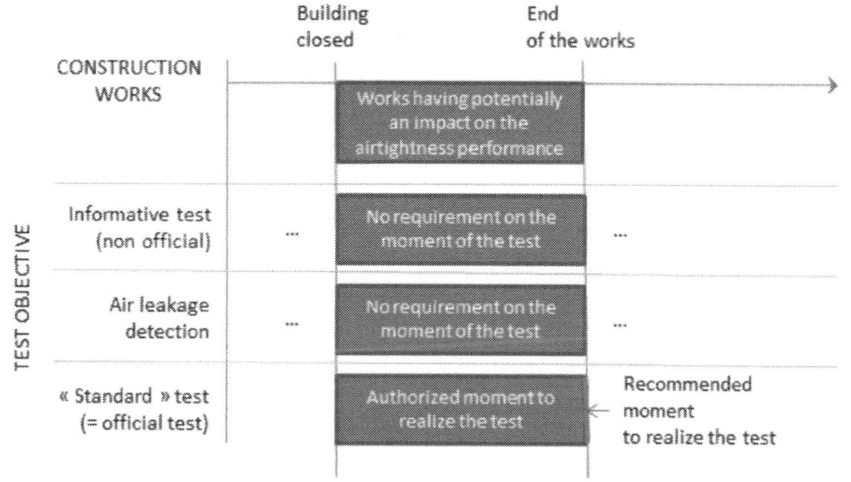

Figure 3: Specifications relative to the state of the building to realize the test according the test objective

In some cases, the person in charge of the building preparation has to be defined (step 3 - Figure 2). In small buildings, it is by default the responsibility of the tester. In large building, it may require a significant effort and it is not part of the usual occupation of the tester. The client has also to decide whether air leakage detection has to be realized (step 5 - Figure 2). Finally, the client is responsible for the use of the test report (e.g. in the context of the EP regulation) (step b. - Figure 2).

3.2 Responsibilities of the tester

Based on the information communicated by the client, the tester is responsible for the realization of the steps necessary to deliver the test report. Specific steps e.g. the building preparation in case of large buildings, can be delegated to third parties.

3.2.1 Requirements on the testing material

Requirements on the testing material going beyond the requirements of the test standard have been introduced (step 2 - Figure 2). These requirements are mainly related to calibration aspects. The verification of the calibration of measuring devices[6] has to be realized with a frequency of two years. This frequency is imposed on manometers and thermometers. If the verification of the calibration shows unacceptable results, the measuring devices have to be effectively calibrated.

In the absence of evidence that regular calibration of hardware e.g. a blower door is necessary, a specific requirement is not imposed on the frequency of calibration of such equipment.

3.2.2 Building preparation for the test

The building preparations for tests A and B are described in the STS (step 3 - Figure 2). In the case of a "standard test", the building preparation requirements are closely related to the conventions presented in the EP regulation, in order to consistently take into account the energy losses related to infiltration and ventilation. A summary of the requirements for the method A is given in Table 1.

In large buildings, the building preparation can be delegated to third parties. The tester has anyway the responsibility to check that the building preparation complies with the applicable specifications.

[6] Definition given in [JCGM 200]

Table 1: Overview of the treatment of intentional openings (method A)

Component	Status	Example
Opening **inside** the measured zone		
Doors, windows, stairs and other intentional openings (except special dispensation)	Opened	Door to a technical room, a boiler room, etc. Stair, greater than 1 m², to a room, accessible for purposes of technical maintenance of installations. Etc.
Opening **in the envelope** of the measured zone		
Mechanical ventilation openings	Sealed	Inside air terminal devices or ducts or outside air terminal devices
Natural ventilation system opening **with** closing device	Closing required Sealing allowed (1)	Adjustable supply or extract air terminal device (for natural ventilation only)
Other opening **with** closing device	Closed (2)	Exterior doors and windows Doors to a space outside of the measured zone: to a basement, garage, an attic, a crawl space Letter slot, cat-flap Evacuation of used water (3) Air outlet with closing device for tumble-dryer, kitchen hood (4) Chimney with closing device (open fire place, boiler, stove, etc.) (4) (5)
Other opening **without** closing device	Open	Unsealable air inlet for an open combustion appliance, etc. Aeration of waste water discharges Lock, openings for the belts of the shutters Other air outlets and chimneys without closing device (4) (5) Etc.

(1) Sealing is allowed (but not mandatory) in order to be consistent with the calculation of the infiltration and ventilation losses in the EP regulation.
(2) By using the closing device(s) present in the opening, but no sealing.
(3) Filling of the siphon = closing.
(4) If there is no closing device in the opening itself but an apparatus is connected to the opening, the apparatus must be closed (example: valve of a kitchen hood, door of a dryer, door of a stove, etc.).
(5) All the combustion appliances concerned must imperatively be stopped before any intervention. Note that it is not necessary to take sealing measures for appliances.

3.2.3 Requirements regarding the pressurization test

The STS-P 71-3 introduces adapted requirements regarding the pressurization test (step 4 - Figure 2).

3.2.3.1 Criteria to realize a measurement in accordance with the test standard in large buildings

Among all the criteria to realize a measurement in accordance with the test standard, specific requirements on the following three criteria are imposed:

- the pressure differences at zero-flow (□P0),
- the pressure difference for the first measurement point
- the highest pressure difference to realize.

Figure 4 presents the measurements conditions in accordance with the test standard. These conditions have been unchanged for measurement realized in small buildings.

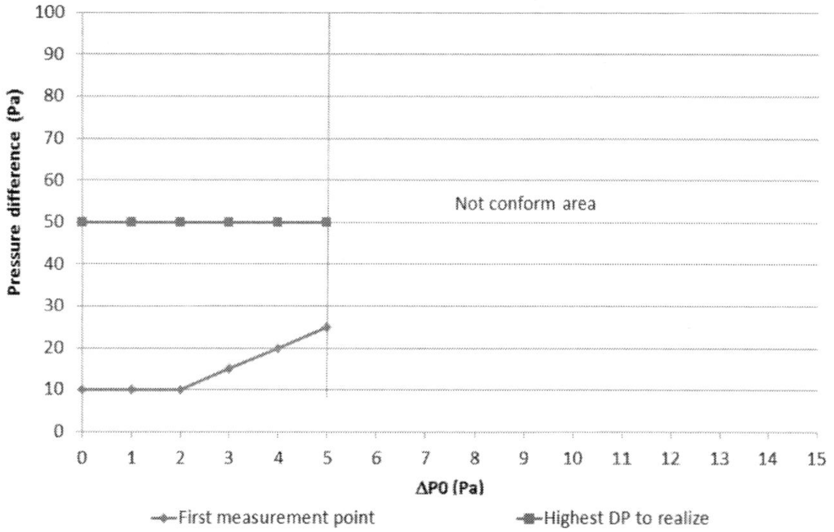

Figure 4: Conditions for a measurement in accordance with the NBN EN 13829 – criteria of the STS-P71-3 for small buildings

It is not always possible to realize a pressurization test fully in accordance with the test standard, especially in high buildings. In particular, the criteria on maximal pressure difference at zero-flow (|□P0|≤ 5Pa) can be problematic. This pressure difference is influenced by the wind conditions and the stack-effect. In low-rise buildings, respecting the criteria is generally not an issue. In high-rise buildings, even small temperature differences between inside and outside can make the measurement not conform to the test standard. Figure 5 presents the maximal allowed temperature difference between inside and outside according to the height of the building. In building higher than 45m, only 5K temperature difference can cause problems for the test.

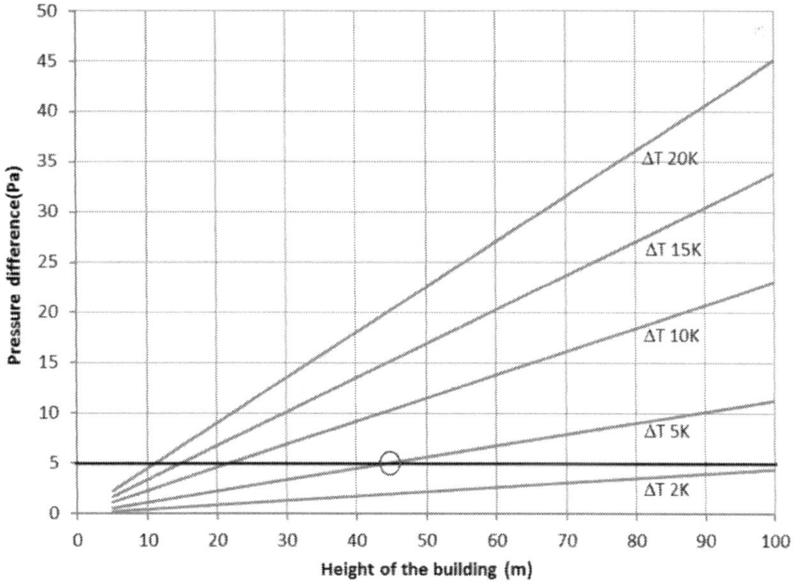

Figure 5: Maximal allowed temperature difference according to the building height to fulfil the maximal pressure difference at zero flow criteria (□P0 < 5Pa)

In high-rise buildings, the existing criterion on □P0 limits the possibilities to realize conformable measurements during periods of time where the temperature difference inside-outside is very limited.
In large buildings, in some cases, the availability of a sufficient number of openings where blower doors can be installed can also be an issue.
In order to enlarge the number of conformable pressurization tests, adaptations to the allowed combinations of the above mentioned three criteria have been introduced for large buildings. Small and large buildings are defined according to the building volume (threshold of 4000m³). Adaptations to these criteria have been proposed while keeping a sufficient accuracy of the measurements. The following elements have been considered:

- The pressure difference at zero flow (□P0) should not be higher than 20% of the lowest test pressure,
- A minimal range of 25Pa should be obtained between the lowest and the highest test pressure (= minimal criteria accepted by the test standard).
- The highest pressure difference should not exceed 100Pa,
- Extrapolate of the values at 50Pa is not recommended, especially if high pressure differences at zero flow are measured,
- Based on the above mentioned conditions, the measuring conditions could be relaxed when exceeding 85.000m³/h (corresponding to about 10 to 15 blower doors).

These considerations resulted in the definition of a new set of criteria presented in Figure 6.

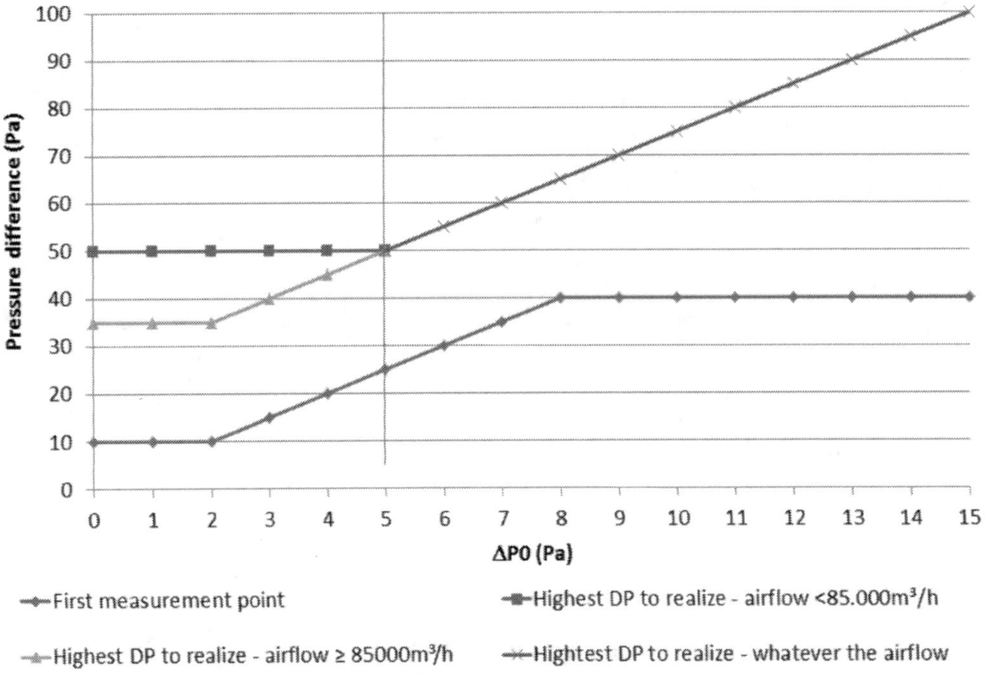

Figure 6: Conditions for a conform measurement in large buildings – adaptation of the criteria of the test standard

3.2.3.2 Measurements in the two modes

In the context of the EP regulations, it is required to realize the pressurization test in the two modes (pressurization and depressurization). The average performance has to be used for the calculation. However, in specific cases, it may happen that the measurement in one of the two modes is problematic or even impossible. The new STS-P 71-3 still requires realizing the measurement in the two modes. Though, if only a measurement in one mode should give conformable results, the average performance to consider is the result of this conformable measurement increased by 20%. Given the fact that differences of 40% between the two modes hardly occur; this approach is considered to be at the safe side to effectively stimulate the realization of the measurement in the two modes.

3.2.4 Air leakage detection

The air leakage detection (step 5 - Figure 2) is not mandatory. Discussions about the opportunity to systematically realize this detection took place with sets of arguments in favour of the two options. The main arguments against the mandatory air leakage detection were:

- This detection has an impact on the cost of the test. In many cases, the clients are not interested in this detection. They should in this case not have to pay for it.
- It is difficult to describe what precisely an air leakage detection is and which level of detail would be required. This part of the test cannot be controlled.
- The air leakage detection raises questions related to responsibilities of the actors – who is responsible for the leakage and is it necessary to fix (or reduce) it.
- Even in airtight buildings build by experienced building contractors, very limited air leakages always exist. What is the sense of this detection is such cases?

The tester can realize this air leakage detection. He is not responsible for the identification of the cause of the leakages and is also not responsible to fix them.

A standardised list of air leakage sites has been added as an informative annex to this STS. This list was developed and tested in the scope of a national research project.

3.2.5 Use of calculation tool

Specific requirements have been introduced regarding the correctness of calculation tools (step 6 - Figure 2). A specific annex to the STS contains a set of input data: indoor and outdoor temperature, pressure difference at zero flow as well as the airflows corresponding to different pressure differences. This annex also contains the main calculation and indicators calculated with these input data. The calculation tools must have been tested according to this annex and produce the correct calculation results.

3.2.6 Content of the report

In addition to the requirements on the report of the test standard, supplementary requirements have been introduced into the STS (step 7 - Figure 2). The following elements have been introduced:

- Attestation of conformity of the test with the requirements of the STS-P 71-3
- Presence of pictures of the installed blower door: one picture taken from inside the building as well as one picture taken from outside has to be included in the report. The tested building has to be recognizable. This element guarantees that a test has effectively been realized. It can also be helpful to easily identify the building (or a specific apartment inside a building).
- To respond to an identified existing problem (see §12.4), the report has to contain the surface of the tested zone or the internal volume. The origin of this information has to be mentioned. It is generally communicated by the client. The tester can on a voluntary basis add to the report the derived quantities ($\dot{v}_{50}$, n_{50}) based on these values.

3.3 A new quality framework

The STS-P 71-3 contains an informative annex describing in general terms the requirements regarding a quality framework for the realization of pressurization tests. A certification

organization may organize a system allowing to show the compliance with the STS and this particular annex.
Quality systems demonstrating the compliance with the STS should contain a competent tester scheme as well as a control system of all the steps of the pressurization tests described in Figure 2. Central databases of the test reports will be developed.
From January 2015, in the context of the EP regulation, the Flemish region imposes the respect of the STS including the annex related to the quality framework for every new pressurization test. Systems for demonstrating the compliance to the STS are operational by this date.

4 CONCLUSIONS

The result of a pressurization test is an input value in the context of the Energy Performance regulations for new buildings in Belgium. The impact of this parameter is growing with the strengthening of the EP requirements. Testing is not mandatory. Since the introduction of the first EP regulation in 2006 in Belgium, measuring the airtightness has become a common practice and more than half of the new buildings are nowadays tested in the Flemish Region.
Although the rules to realize these tests are already well defined; specific problems regarding the tests exist and quality problems of the test have also been identified. To solve these problems, updated rules for the realization of the test have been defined in the STS-P 71-3. Specific rules have i.a. been defined for large buildings in order to widen the number of situations where a test in accordance with the standard can be realized. The steps of the test the tester is responsible for have been defined. The STS introduces also the possibility of a quality framework imposing a competent tester scheme.
In the Flemish region, the rules of the STS including the quality framework will become applicable in the scope of the EP regulation for every new test in 2015. Operational systems enabling the conformity demonstration to this STS are developed.

5 ACKNOWLEDGEMENTS

The STS-P 71-3 described in this paper benefits from the results of the research project ETANCH'AIR supported by the Walloon Region in Belgium.
This paper has been written in the scope of the IEE project QUALICHECK (IEE/13/610/SI 2.675574) funded by the European Union.
The sole responsibility of the content of this paper lies with the author. The Executive Agency for Small and Medium-sized Enterprises (EASME) is not responsible for any use that may be made of the information contained therein

6 REFERENCES

VEA. (2013). Cijferrapport energieprestatieregelgeving, Procedures, resultaten en energetische karakteristieken van het Vlaamse gebouwenbestand - periode 2006 – 2012, – available on www.energiesparen.be

FPS economy. (2014). STS-P 71-3 – Airtightness of Buildings – pressurization tests – (Not published yet)

NBN EN 13829 (2001) – Thermal performance of buildings — Determination of air permeability of buildings — Fan pressurization method (ISO 9972:1996, modified)

JCGM 200:2012 - International vocabulary of metrology – Basic and general concepts and associated terms (VIM), Joint Committee for Guides in Metrology (JCGM)

CO-HEATING TEST AND COMFORT ASSESSMENT OF A COUPLED SYSTEM MADE BY A VENTILATED WINDOW AND A HEAT RECOVERY UNIT

Ludovico Danza[1], Benedetta Barozzi[1], Lorenzo Belussi[1], Francesco Salamone[1]

1 Construction Technologies Institute of the Italian National Research Council
Via Lombardia 49
San Giuliano Milanese, Milano, Italy

ABSTRACT

The article describes the results of an experimental campaign carried out at ITC-CNR in outdoor test cells to evaluate the energy performance and the related comfort level achieved through a coupled system made up of a dynamic window and a heat recovery unit.
The test was carried out on two calibrated test cells with the same geometric, constructive and thermo-physical characteristics. They were appropriately monitored, throughout the year 2013. In the reference cell, called C2, a traditional window with double glazing, aluminium frame and indoor blind was installed, while the air circulation was provided by a centrifugal extractor. In the second one, called C3, a dynamic window with integrated blind was installed and the air circulation was provided by a heat exchanger. The air conditioning systems consist of electric heaters in winter and heat pumps in summer.
The different operating configurations allowed the trends of the dynamic system to be assessed in two different phases: analysis of seasonal energy behaviour and analysis of the thermal comfort conditions.
The first phase consisted in the assessment of the energy consumption of the two test cells using the co-heating methodology. The results showed an overall lower consumption of C3 compared to C2, both in winter and in summer, with 20% and 15% peak energy savings, respectively. The results were also confirmed by the analysis of Energy Signature.
During the second phase, the psychrometric analysis was introduced to better understand the complex heat fluxes management actions carried out by the dynamic window-heat recovery integrated system: action of heat recovery unit; dual action to reduce heat transfer in the dynamic window; air pre-heating action by the dynamic window, before entering in the exchanger.
The transformed patterns confirmed the positive synergy during the winter season (maximum yield equal to 1.9), while, for the summer season, they provided a clear interpretation of the better operation of the system only with the heat recovery unit turned on (average yield minimum of 0.7).
The analysis of the PMV and PPD indices showed that, when the maximum solar radiation is less than 600 W/m2, the C3 recorded a slightly higher PPD. With average external temperatures equal to about 20 °C and maximum solar radiation of approximately 900 W/m2, the values of PPD and PMV of the two cells were equivalent. When the dynamic glass was turned off and only the recovery unit worked, the C3 provided the best comfort conditions.
Finally, the Fusion Tables, a web service provided by Google, were used to extend the results to the Italian provinces taking into account the standard monthly climate data. It was showed that with decreasing latitude, the energy savings of the combined system increased and reached the maximum value calculated for cities located in the south.

KEYWORDS

Dynamic system, ventilation, heat recovery, co-heating, thermal comfort

1 INTRODUCTION

This article describes the results of an experimental campaign carried out on a dynamic window, called VetroVentilato®, consisting of a ventilated window connected to a heat recovery unit. The system is an evolution of a previous configuration consisting of just the ventilated glass (Lollini et a., 2012) that showed a good energy behaviour compared to a traditional configuration.
The experiment, carried out on two test cells, involves two phases: the former aimed to analyse the energy savings due to the installation of the dynamic system, the latter aimed to assess the indoor thermo-hygrometric comfort. In particular, in the first phase, the energy savings have been assessed in terms of cumulated consumption and the Energy Signature method (Belussi and Danza, 2012) has been applied, aimed to assess the energy behaviour of the dynamic system under variable external climatic conditions.

2 EXPERIMENTAL CAMPAIGN AND CO-HEATING TEST

The experiment includes two external test cells with the same dimensional and thermo-physical characteristics, properly calibrated and equipped with the appropriate monitoring devices (Stamp, 2012). The dynamic system has been installed in one cell (hereinafter C3), while a window with a traditional double-glazing has been installed in the other cell (hereinafter C2). The experiment was carried out with different configurations, as shown in Table 1.

Table 1: Configurations

Code	Heating	Cooling	Dynamic window	Heat exchanger	Flow Rate
Heat 1	**On**	Off	**On**	**On**	45 m^3/h
Heat 2	**On**	Off	**On**	**On**	60 m^3/h
Heat 3	**On**	Off	**On**	Off	-
Cool	Off	**On**	**On**	**On**	45 m^3/h
Comf 1	Off	Off	Off	**On**	45 m^3/h
Comf 2	Off	Off	**On**	**On**	45 m^3/h

2.1 Brief description of the dynamic system

The dynamic system consists of a ventilated window connected to a traditional heat recovery unit placed above the window. The window consists of a double-glazing (3+3-16-3) and an internal micro-drilled blind placed 2 cm from the glass so as to create an internal air gap. The exchanger is connected to the dynamic system so the indoor inlet flow allows the air to be drawn from the air gap.
The operation of the combined system allows the air of the internal environment to be got in the air gap, the air pre-heats thanks to the solar radiation and the heat is recovered by the exchanger before being ejected. The heat recovered is used to heat/cool the air from the outdoor (Figure 1a). In C2, a centrifugal extractor is placed on the north wall (Figure 1b).

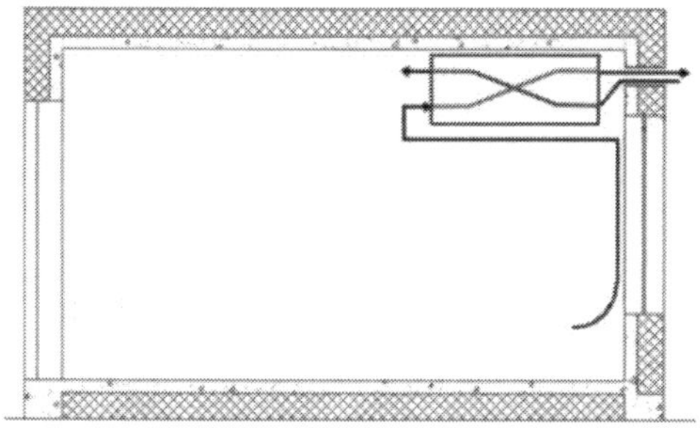

Figure 1a: Test cell section - ventilation with heat recovery

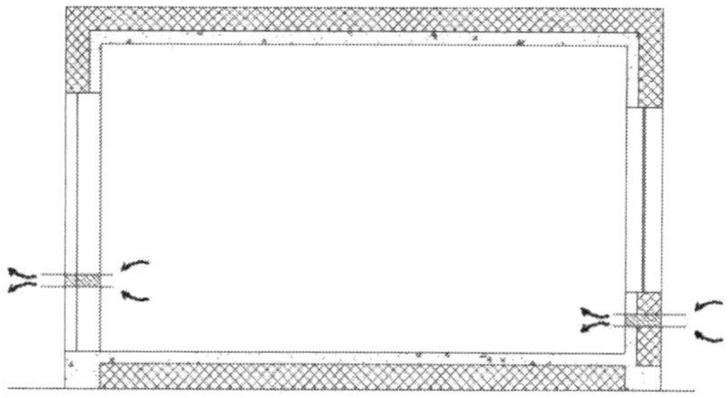

Figure 1b: Test cell section - direct extraction

3 FIRST PHASE: THE ASSESSMENT OF THE ENERGY CONSUMPTION

The first phase of the experimental campaign consists of the analysis of the energy consumption of the two test cells aimed to assess the energy performance of the dynamic system compared to a traditional double glazing. In particular, the energy consumption has been analysed following two approaches:

- comparison of the cumulated consumption;
- comparison of the Energy Signature of the test cells.

In this article, the Energy Signature method has been used in an alternative way compared to that described in Annex B of the international standard EN 15603:2008. Normally, in fact, the Energy Signature is used to assess the energy consumption and consequently the energy behaviour of a building. In this article, instead, the method is used as an indirect empirical tool to assess the energy performance of the dynamic system. Cumulated consumption and Energy Signature have been evaluated in the period shown in Table 2.

Table 2. Heating and cooling period

Code	Period
Heat 1a	07/01/14 - 11/02/14
Heat 1b	04/04/14 - 30/04
Cool	01/07/13 - 21/07/13

3.1 Consumption and energy savings

The analysis of the cumulated consumption allows the energy savings to be identified due to the dynamic system in a given period, compared to the consumption of C2.

Configuration "Heat" (a+b)

Figure 2a and 2b show the consumption trend (red and blue lines) of the two test cells and the mean daily percentage difference detected in the configuration "Heat 1" (a+b). The overall energy consumption of C3 is constantly lower than that of C2, in both the periods. In particular, in the configuration "Heat 1a" the consumption of C3 is on the average 18% lower than C2 (dotted line in Figure 2a), with a final difference equal to 109 kWh. In the configuration "Heat 1b", characterized by higher temperatures, the energy behaviour of the test cells diverges to a greater extent and the mean percentage deviation increases up to about 30% with peaks exceeding 40% (Figure 2b).
This trend is explained by observing in more detail the absolute consumption values of the test cells: the greatest energy savings, in absolute value, and the lowest percentage deviation are detected in the first, cooler, period; vice versa, the lowest energy savings, in absolute value, and the greatest percentage deviation are detected in the second, warmer, period.

The dynamic system allows a lower utilization of the heating plant, in the warmer period. The increase of the ventilation rate, as in configuration "Heat 2", determines the increase of the percentage deviation of the two test cells.

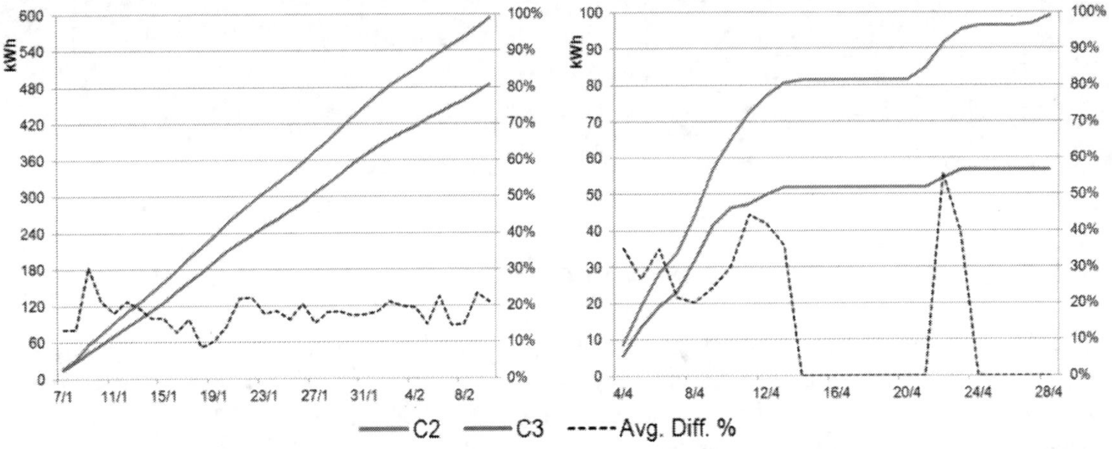

Figure 2a: Cumulated consumption configuration "Heat 1a"

Figure 2b: Cumulated consumption configuration "Heat 1b"

Configuration "Cool"

Figure 3 shows the trend of the cumulated consumption of the test cells in the "Cool" configuration. C3 shows the best energy behaviour and the lowest consumption, in percentage terms (about 13%) and in absolute terms (about 25 kWh) than C2. However, in the cooling phase, the combination of the dynamic system and the heat recovery unit does not allow the flux to be optimally managed, because exhaust air that comes into the exchanger from the indoor environment transfers heat to the inlet air, heating it.
In the cooling season, the reduction of the consumption is due only to the dynamic glazing.

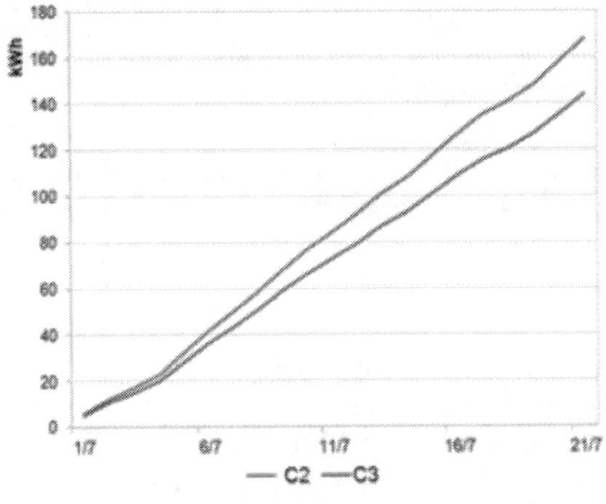

Figure 3: Cumulated consumption configuration "Cool"

3.2 Energy Signature

The Energy Signature has been applied in order to determine the energy behaviour of the dynamic system when the external climatic conditions vary. The Energy Signatures of the two test cells have been compared in the different configurations. The contribution of the dynamic system to the overall performance of C3 could be assessed thanks to this method.

Configuration "Heat"

The comparison of the Energy Signatures in the configuration "Heat 1a" (Figure 4a) shows a better energy behaviour of C3 than C2, confirming the performance of the dynamic system both in terms of heat losses (related to the slope of the straight lines) and in terms of switch-off temperature of the heating system (intersection of the straight line with the x-axis). The slope and the switch-off temperature of C3 are about 3.5 W/K and 2°C lower than those of C2, respectively.

The same reasoning can be made for the configuration "Heat 2". Figure 4b shows how the energy behaviour of C3 is further improved than that of C2. This is due to the performance of the two single components, dynamic glazing and heat recovery unit, which are more efficient with the increasing of the airflow rate. In C3, the increasing of the airflow rate, that involves greater ventilation losses, is balanced by the combined action of the dynamic system. In C2, the ventilation losses are not recovered, with a consequent increasing of the overall consumption.

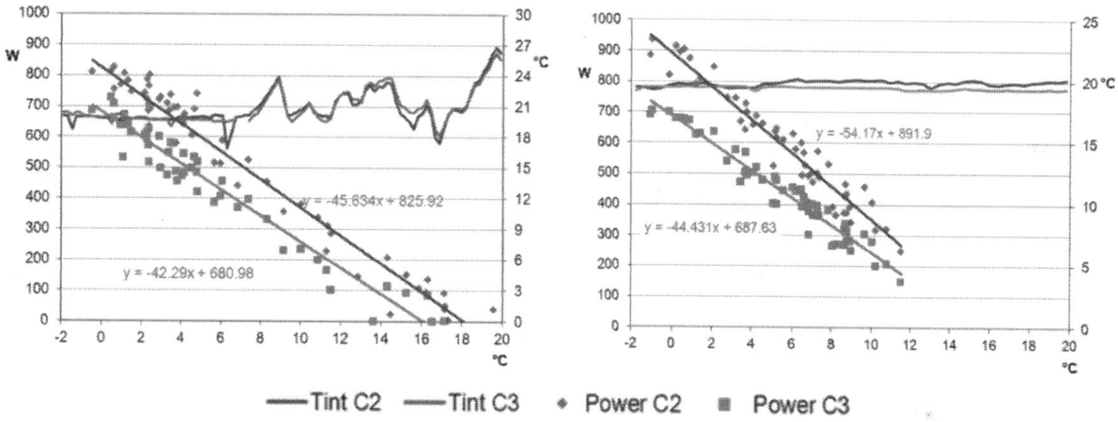

Figure 4a: Energy Signature configuration "Heat 1 (a+b)" - comparison

Figure 4b: Energy Signature configuration "Heat 2" - comparison

Configuration "Cool"

Similarly to the winter situation, it is possible to analyse the dynamic system behaviour in summer through the Energy Signature.

The outcome of the cumulative consumption analysis is confirmed by the assessment of the Energy Signature. The comparison, in fact, shows both a lower slope and a downward translation of the Energy Signature of C3 with respect to C2, due to a better management of the thermal loads of the combined system compared to the reference one.

Thus, also in summer, the dynamic system has the best energy behaviour.

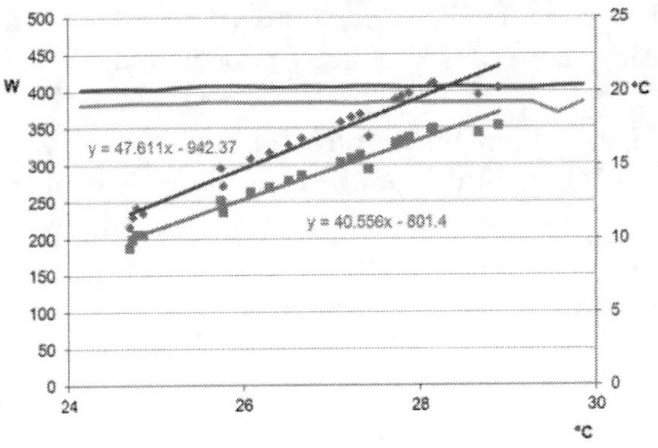

Figure 5: Energy Signature configuration “cool” - comparison

3.3 Extention of results

The Energy Signature method allowed the slope and the constant term of the straight lines of the test cells to be identified. Taking into account the average monthly temperature, it is possible to extend the obtained results to the whole Italian territory, using an experimental approach, aimed to theoretically assess the energy behaviour in different climatic conditions.
In this article, and by way of example, two configurations are considered: “Heat 1” and “Heat 3”. The mapping has been applied to all the Italian provinces.
Starting from the external temperature data and the Energy Signature equations (see Figure 4a), the theoretical consumption of the cells and the percentage difference of consumption between the two cells, C2 and C3, are calculated.
Five classes, each characterized by a specific colour, have been identified as a function of the energy savings guaranteed by the dynamic system, compared to the reference system.
The two resulting thematic maps (Figure 6 and Figure 7) allow the two cases to be compared and the average energy savings to be evaluated. From configuration “Heat 3” to “Heat 1”, the following savings are possible:

- Northern Provinces: from 6.2% to 16.5%;
- Central Provinces: from 9.8% to 19.7%;
- Southern Provinces: form 13.7% to 26.5%.

The warmer the conditions, the greater the saving.

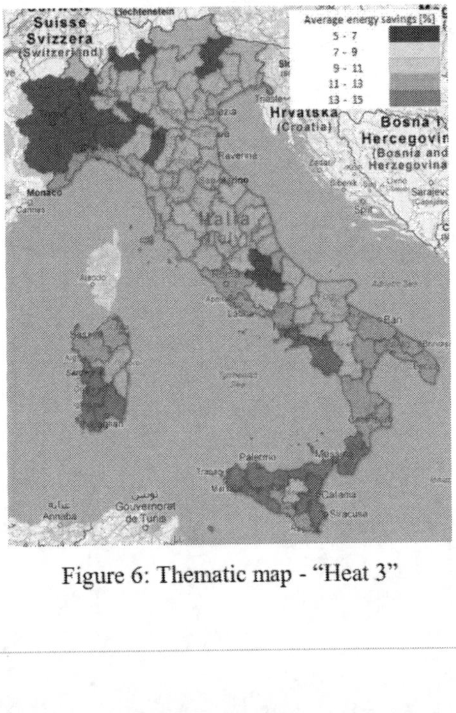

Figure 6: Thematic map - “Heat 3”

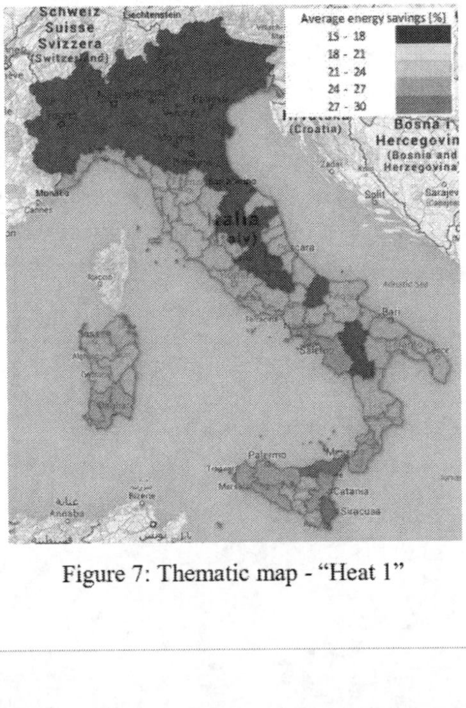

Figure 7: Thematic map - “Heat 1”

4 SECOND PHASE: THE EVALUATION OF INDOOR IGRO-THERMAL CONDITIONS AND COMFORT

4.1 The psychrometric analysis

The current operating logics of the dynamic system play a key role in the management of heat flows, carrying out three different actions. The first takes place at the heat recovery level: the cross-flow exchange of thermal energy between the exhaust air outgoing from the conditioned test cell and the incoming air from the external environment involves a pre-heating action during the winter season.

Then, the transparent component involves a reduction in heat exchanges with the external environment: the air, drawn in by the indoor environment and passing through the ventilated cavity, allows the surface temperature of the inner glass as close as possible to the temperature of the indoor environment to be maintained. This action involves a reduction of the thermal transmittance of the component as a function of the increasing flux in the air gap. At the same time, the integrated preheating action due to the solar radiation hitting the transparent component pre-heats the air that passes through the gap before reaching the recovery unit (McEnvoy et al., 2003).

In this synergic plant context, the psychrometric analyses (ASHRAE 55), particularly focused on the environment line and the straight patterns of the transforms, allow the results of the energy evaluations to be represented and to be interpreted.

Table 3 shows the overall average data efficiency of the two test cells, related to representative detailed analyses.

During the winter testing period, the indoor environmental monitoring data show the positive synergy between the dynamic system and the heat recovery unit, which confirms the energy consumption data. In fact, at constant indoor conditions (Temperature and RH%), C3 shows an average efficiency improvement compared to C2 of approximately 7%.

On the contrary, during the summer, the combined system does not allow to effectively heat flows to be managed.

Table 3: Heating & Cooling configurations

Season & Plant Configurations	Reference period	Reference day	DT (C3-C2)	D%RH (C3-C2)	Avg. Efficiency (C3-C2)
Heat1	22-29 April 2013	22 April	<1°C	<5%	7% ca
Cool	01-07 July 2013	03 July	≈1°C	≈1%	--

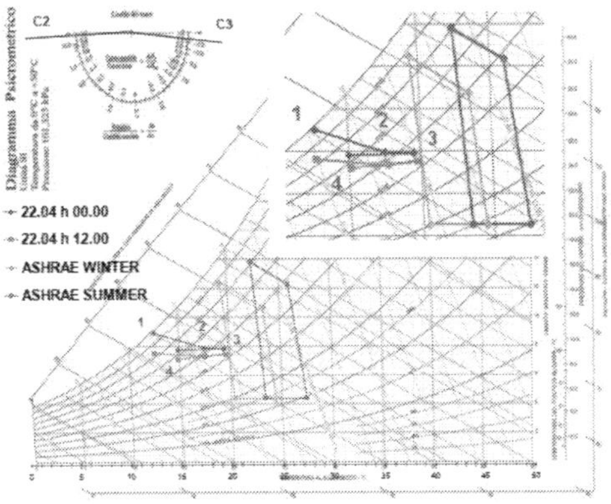

Figure 7: Psychrometric chart related to the operating condition factor and the plant-related transforms for "Heat1" configuration

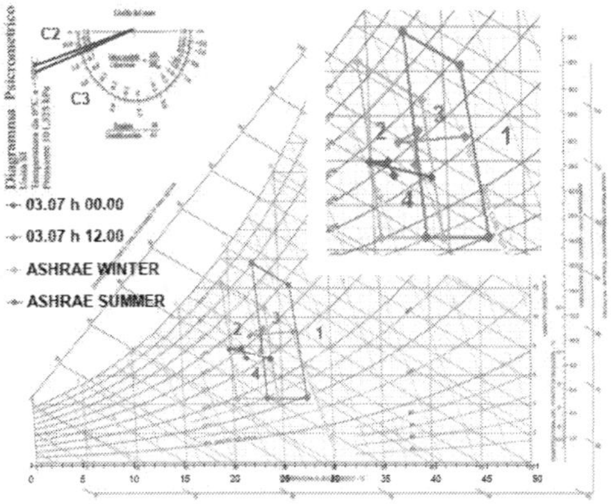

Figure 8: Psychrometric chart related to the operating condition factor and the plant-related transforms for "Cool1" configuration

Figure 7 and Figure 8 show the points of the plant-related transforms:

- point 1 is the inlet airflow from the outside;
- point 2 is the heat recovery unit airflow to the inside;
- point 3 is the intake airflow of the heat recovery unit from the inside, after passing through the ventilated cavity of the glass;
- point 4 represents the heat recovery unit airflow to the outside.

During the winter tests, as shown in Figure 7, the combined operation involves that the slope of the operating condition factor corresponds of the slope of the line connecting points 1-3.
So, the combined system operates in synergy, generating a virtuous cycle in energy savings through the combined action.
Instead, concerning the summer season, the possible considerations are very different.
Even if the indoor conditions in both the test cells are the same, as shown in Table 66, the psychrometric chart in Figure 8 shows how in this configuration the slope of the operating condition factor radically differs from the slope of the line connecting points 1-3. This difference is more evident during direct solar radiation phases.
During winter testing the combined system contributed to the increase of the overall efficiency, allowing to achieve the set point conditions provided for by the indoor environment line with a reduced energy expenditure by the heating system.
The operation of the dynamic window, represented by the segment 2-3 of the orange polyline (Figure 8), involves a daily overheating of the air pre-treated in C3, reducing the heat recovery unit efficiency (1-2 segment). In fact the outgoing air is cooled by the exchanger (3-4 segment).
In the summer season the combined system involves a worse phase shift than expected, which is made evident by the psychrometric chart.

4.2 Comfort analysis

The thermo-hygrometric comfort level of the two test cells has been evaluated using the Fanger method (EN ISO 7730:2006). The analysis has been carried out in the configurations "Comf 1" and "Comf 2", in different weeks (Table 4).

Table 4: Cases comfort analysis

Code	Period
Comf 1	03/06/13 – 09/06/13
Comf 2	14/05/13 – 24/05/13

Configuration "Comf 1"

In Figure 9, the Predicted Percentage of Dissatisfied values (PPD), calculated assuming a clothing value equal to 0.75 clo, are better in C3 than in C2, with a maximum deviation slightly more than 30%. In Figure 10 the Predicted Mean Vote index (PMV) indicates, in both cells, a thermo-hygrometric sensation average comprised between the "neutral" class and the "slightly warm" class. In C2, higher temperature and humidity values are detected.

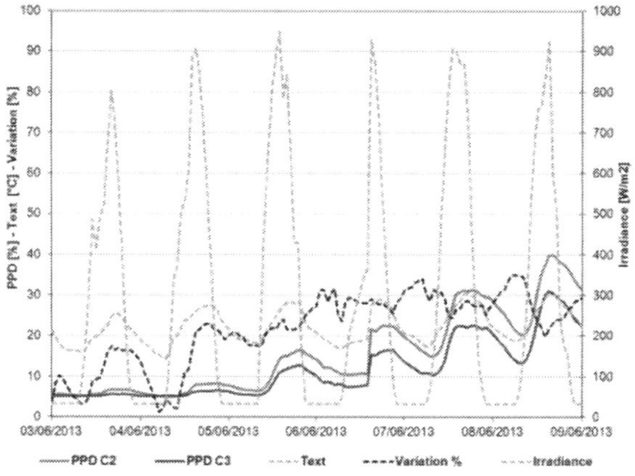

Figure 169: PPD trend - "Comf 1"

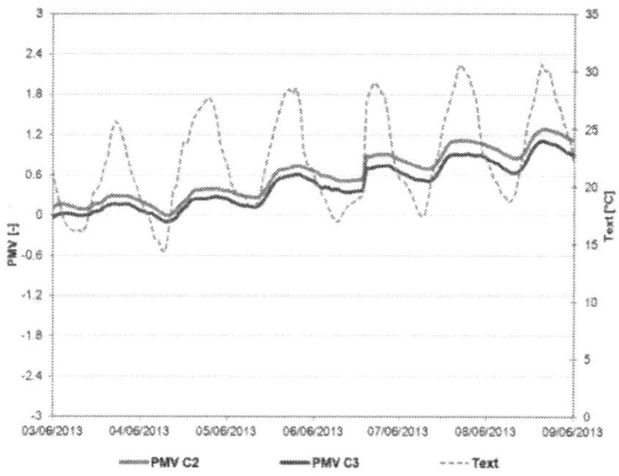

Figure 170: PMV trend - "Comf 1"

Configuration "Comf 2"

In the configuration "Comf C2" it is possible to divide the considered period in two parts. From May 15 to 17, the average external temperature is equal to about 17 °C and the average solar radiation is about 400 W/m^2. In these conditions, no significant changes in thermo-hygrometric comfort between the two cells are recorded. From May 18 to 24, instead, the external temperature and the solar radiation are lower (about 15°C) and higher (about 600 W/m^2), respectively, compared to the previous period. In these conditions, C3 provides the best thermo-hygrometric comfort (Figure 11). The PMV values highlight the differences in the two periods: in the former, the thermo-hygrometric sensation is on the average "neutral", for both cells, while in the latter the thermo-hygrometric sensation is on the average "slightly cool", with C3 values higher than C2 and close to "neutral".

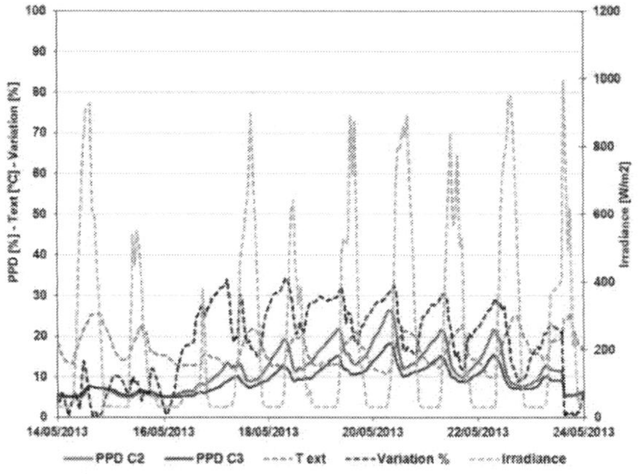

Figure 11: PPD trend - "Comf 2"

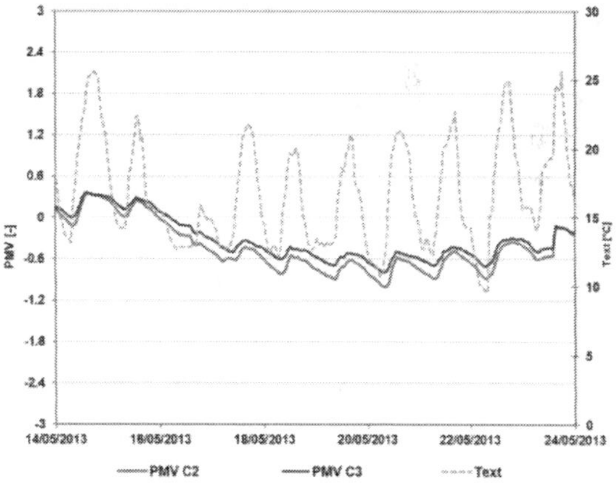

Figure 12: PMV trend - "Comf 2"

6 CONCLUSIONS

This article has investigated the energy behaviour and the comfort level of a dynamic system, consisting of the VetroVentilato® system and a commercial heat recovery unit. The results of the experimentation show that, in winter, the combined system allows for interesting energy savings. When the external temperature is equal to 10 °C, the energy savings that may be achieved with the dynamic system amount to about 38% compared to 8% obtained by the single system alone (without the heat recovery unit) and compared to 20% obtained by the heat recovery unit alone. The analysis shows an improvement of the performances of the

combined system. The combined operation of the dynamic system is not the sum of the single contributions; in fact, the system detects higher values due to the pre-heating function of the window.
In summer, the energy savings of the combined system are about 13%, which is lower than the operation of VetroVentilato® alone.

ACKNOWLEDGEMENTS

This work was carried out according to an industrial research project with the VetroVentilato society.

REFERENCES

Lollini, R., Danza, L., Meroni, I. (2010). Energy efficiency of a dynamic glazing system. Solar Energy, 84(4), 526-537.

Belussi, L., Danza, L. (2012). Method for the prediction of malfunctions of buildings through real energy consumption analysis: Holistic and multidisciplinary approach of Energy Signature. Energy and Building, 55, 715-720..

Stamp, S. (2012) A Summary of the Co-heating test procedure and analysis method, Technical Report.

ANSI/ASHRAE 55:2013. Thermal environmental conditions for human occupancy.

EN ISO 7730:2006. Ergonomics of the thermal environment - Analytical determination and interpretation of thermal comfort using calculation of the PMV and PPD indices and local thermal comfort criteria.

EN 15603:2008, For space heating - Overall energy use and definition of energy ratings.

McEvoy M.E., Southall R.G., Baker P.H., Test cell evaluation of supply air windows to characterise their optimum performance and its verification by the use of modelling techniques, Energy and buildings, 35 (2003), pp. 1009-1020.

J.U. Sjogren , S. Andersson, T. Olofsson, Sensitivity of the total heat loss coefficient determined by the energy signature approach to different time periods and gained previous energy, Energy and Buildings, 41 (2009), pp. 801–808.

DERIVATION OF EQUATION FOR PERSONAL CARBON DIOXIDE IN EXHALED BREATH INTENDED TO ESTIMATION OF BUILDING VENTILATION

Masaki TAJIMA*[1], Takayuki INOUE[1], and Yuji OHNISHI[1]

[1] *Kochi University of Technology*
Tosayamada
Kami-city, Kochi, 782-8502 Japan
* *tajima.masaki@kochi-tech.ac.jp*

ABSTRACT

Carbon dioxide included in exhaled breath is often used as a tracer gas when estimation of ventilation aspect in buildings with occupants is performed. Carbon dioxide produced by occupants is the key for the estimation. JIS A 1406 and ASTM D6245-12 refer personal carbon dioxide production rate. However JIS does not take into account personal attribute like as body height and weight. On the other hand, ASTM does not take into account gender difference and based on average westerner adult data. Hence, by using Douglas bag method with approximately total 70 voluntary Japanese subjects, a prediction equation for occupants' carbon dioxide production rate included in exhaled breath is developed. Moreover, the equation is tested in single zone with occupants aiming at accuracy testing. The results can allow predicting more correct carbon dioxide concentration produced by occupants', exhaled breath in single zone compared with the standards.

KEYWORDS

Ventilation, Exhaled breath, Carbon dioxide, Single zone, Douglas bag method

1. INTRODUCTION

Carbon dioxide included in exhaled breath is often used as a tracer gas when estimation of ventilation aspect in buildings with occupants is performed. Amount of carbon dioxide produced by the occupants is a key for this kind of estimation. Carbon dioxide production rate is shown in e.g. a Japanese Industrial Standard (JIS A 1406[1974], 2010) and an American Society for Testing and Materials International Standard (ASTM D6245-12, 2012).The JIS standard indicates a relationship between personal carbon dioxide production rate and RMR (relative metabolic rate) including gender difference, which shows female's production rate is 0.9 times of male's. The ASTM standard gives a relationship between personal carbon dioxide production rate and parameters, as body surface area and metabolic rate. These personal carbon dioxide production rates were referenced in other standard (e.g. SHASE-S102, 2011) and have been used for computing ventilation rate or air change rate in some previous works (e.g. Saito et al., 2002, Stavova et al., 2006). These carbon dioxide production rates in the standards are described simply, therefore they are useful for the estimation. However, the JIS's rate is not considered body height and weight. On the other hands, the ASTM's rate is not considered gender difference and is based on average westerner adult data with a fixed respiratory quotient value ($RQ = 0.83$).

Because of these backgrounds, this study aims at derivation of a prediction equation using variables such as *Met* and personal attribute for personal carbon dioxide production rate included in exhaled breath with Japanese subjects, aims and at accuracy testing of this equation at rooms with occupants in real use conditions.

2. DERIVATION OF EQUATION FOR PERSONAL CARBON DIOXIDE PRODUCTION RATE

2.1 Outline of the derivation

The derivation of an equation for carbon dioxide production rate was performed with obtained carbon dioxide production rate, Met, and subjects' personal attribute. Then, with using these obtained data, an equation was derived trough regression analysis.

2.2 Measurement

Using Douglas bag method, the measurement of carbon dioxide production rate was performed with total of approximately 70 voluntary people who were mainly university students. The testing activities and number of subjects are shown in Table 1 and testing instruments are shown in Table 2.

Metabolic value M and carbon dioxide production value V_{CO_2} are determined by the following equations (ISO8996, 2004) with using obtained data which are V_{O_2} and V_{CO_2}. Moreover, personal attribute, age, gender, body height and weight, were also obtained.

Table 1: testing activities and number of subjects

Activity	Seated, quiet	Walking (2km/h)	Walking(4km/h)
Work period	10 minutes	5 minutes	3 minutes
Number of Subjects*	23(9)	23(9)	23(9)

*total 69 people, ( ) means internal numbers of female

Table 2: testing instruments

Douglas bag	Volume: 100L	TK-11288
Gas analyser	CO_2 NDIR 0-5/10/20 vol% O_2 zirconia process 0-5/10/25 vol%	PG-240

$$RQ = \frac{V_{CO_2}}{V_{O_2}} \quad (1)$$

$$\mathrm{EE} = (0.23RQ + 0.77)5.88 \quad (2)$$

$$M = EE \times V_{O_2} \quad (3)$$

Where

RQ is the respiratory quotient
V_{O_2} is the oxygen consumption rate [L/h]
V_{CO_2} is the carbon dioxide production rate [L/h]
EE is the energetic equivalent [Wh/LO$_2$]
M is the metabolic rate [W]

2.3 Regression analysis

The measured data was selected into for 51 subjects' data for the regression analysis, by considering to the aspect of the RQ value referred in former works like as the followings. RQ varies from 0.7 to 1.0(e.g. Herman, 2007). RQ equals 0.83 for an average adult engaged in light or sedentary activities (ASTM D6245-12 2012). $RQ = 0.83$ applies to a normal diet mix of fat, carbohydrate, and protein (ASHRAE Standard 62.1, 2010)

The whole measured values range approximately from 0.6 to 1.2. Only the data ranges from 0.7 to 1.0 were used as the selected data. Therefore, the selected RQ values finally range from 0.78 to 0.99 and the average value is 0.93 for seated quiet, 0.92 for walking at 2km/h and 0.91 for walking at 4km/h. Hence these measured RQ values are larger than the ASTM's and ASHRAE's theoretical fixed values.

(1) Relationship between carbon dioxide production rate and metabolic rate

The relationship between carbon dioxide production rate P_{CO_2} in m^3/h and metabolic value M, simple correlation, is shown in Figure 1. The equations (1), (2) and (3) show that the metabolic rate M is determined by using oxygen consumption rate V_{O_2} and carbon dioxide production rate V_{CO_2}. Even without oxygen consumption rate V_{O_2} this relationship has strong correlation and the relationship can be expressed by equation (4). Thus this equation within metabolic rate as a variable can be considered as basic equation for the prediction of personal carbon dioxide production rate.

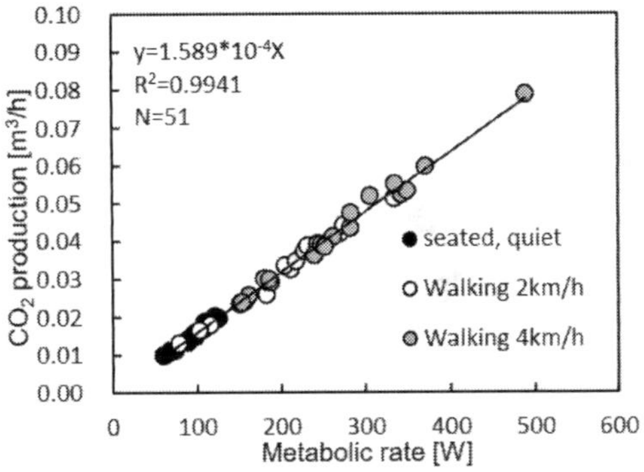

Figure 1: Relationship between carbon dioxide production rate and metabolic value

$$P_{CO_2} = 1.589 \times 10^{-4} M \qquad (4)$$

Where

P_{CO_2} is the carbon dioxide production rate [m^3/h]

(2) Relationship between metabolic rate and body height and weight

Relationship between metabolic rate M and body surface area A_D is shown in Figure 2. The body surface area is given by equation (5) (SHASE Handbook, 2010). This equation is based on Japanese adults' data.

From regression analysis results with explanatory variables as the BMI (body-mass index), height and weight themselves, and body surface area, the body surface area A_D has the strongest correlation with metabolic rate as same as previous works (e.g. ASTM D6245-12 2012, Persily, 1997) .

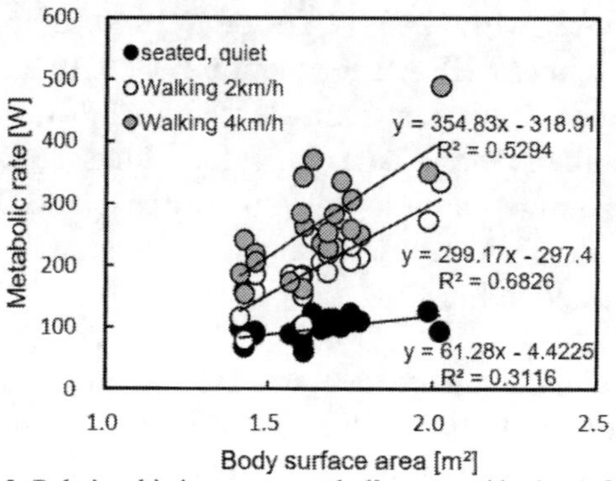

Figure 2: Relationship between metabolic rate and body surface area

$$A_D = 0.007246 \times W_b^{0.425} \times H_b^{0.725} \quad (5)$$

Where

A_D is the body surface area [m^2] (for Japanese adult)
W_b is the body weight [kg]
H_b is the body height [cm]

(3) Relationship between metabolic rate and Met

Figure 3 indicates relationship between metabolic rate and *Met* given by equation (6). *Met*, which is generally used as an index of occupants' activity, is calculated by substituting each subject's metabolic rate obtained under the condition of "seated, quiet" as M_S into this equation. The coefficient of determination R^2 shown in Figure 3 is approximately 0.82.

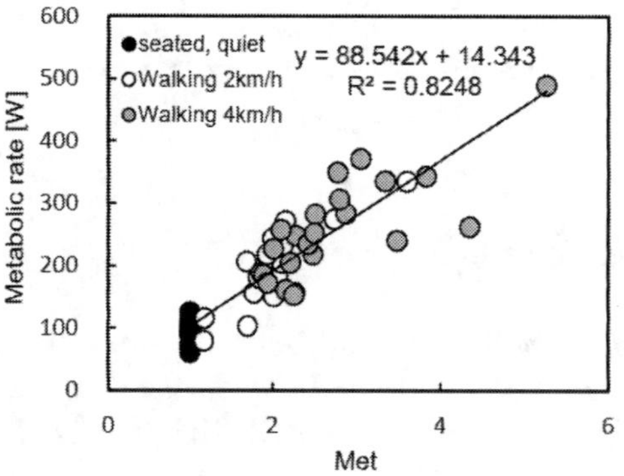

Figure 3: Relationship between metabolic rate and Met

$$Met = \frac{M}{M_S} \quad (6)$$

Where

M_S is the metabolic rate of relax seated [W]

2.4 Determination of equation for personal carbon dioxide production rate

Metabolic rate is given by equation (7) which also has coefficient of gender as a variable. Subject's age was not employed in the equation, because most of subjects were similar ages. The coefficient of determination did not increase even when subject's age was taken as variables.

Relationship between measured metabolic rate and calculated metabolic rate using equation (7) is shown in Figure 4, in which R^2 is approximately 0.88 and the RMSE is 33.1.

$$M = 94.4A_D + 83.9Met + 21.0C_g - 149.7 \quad (7)$$

Where

C_g is the coefficient of gender (0: female, 1: male)

Finally, the prediction equation for personal carbon dioxide production rate is derived as equation (8) using equation (4) and (7). Figure 5 represents relationship between measured carbon dioxide production rate and calculated carbon dioxide production rate by using equation (8). The trends agree with measured result as shown in the figure in which the R^2 is approximately 0.88 and the RMSE is 0.006.

$$P_{CO_2} = 1.589 \times 10^{-4}(94.4A_D + 83.9Met + 21.0C_g - 149.7) \quad (8)$$

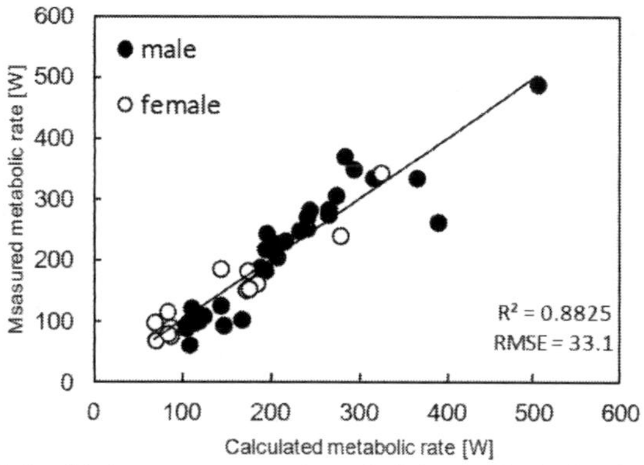

Figure 4: Relationship between measured metabolic rate and calculated metabolic rate

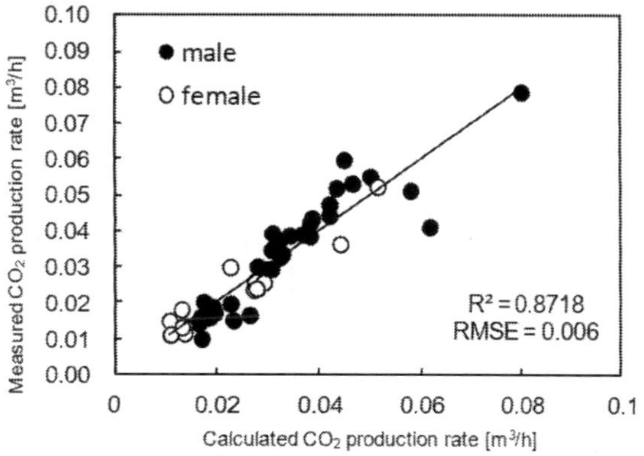

Figure 5: Relationship between measured carbon dioxide production rate and calculated carbon dioxide production rate

2.5 Comparison of carbon dioxide production rate of the equation, ASTM and JIS

Figure 6 shows carbon dioxide production rate of present work calculated by using the equation (8), of ASTM and of JIS. The ASTM's rate is calculated by equation (9) and (10) with RQ is 0.83 as the fixed value. Japanese average adult's body height and weight are put into the equation (8) and (10) to calculate the rate. Originally, the JIS shows relationship between carbon dioxide production rate and *RMR* (relative metabolic rate), which can be transformed into *Met* by using equation (11) (Herman, 2007).

Figure 6 indicates that present work's carbon dioxide production rate is larger than JIS's rate and the difference is $0.008m^3/h$ (8L/h) at 1.0Met. The present work's rate is also larger than ASTM's rate under the range of $Met = 3$. The difference at 1.0 Met is $0.004m^3/h$ (4L/h).

$$F = RQ\frac{0.00276 A_{DB} Met}{(0.23RQ + 0.77)} \quad (9)$$

$$A_{DB} = 0.203 \times W^{0.425} \times H^{0.725} \quad (10)$$

Where
- A_{DB} is the DuBois body surface area [m^2]
- F the carbon dioxide production rate [L/h]
- W is the body weight [kg]
- H is the body height [m]

$$RMR = 1.2 \times (Met - 1) \quad (11)$$

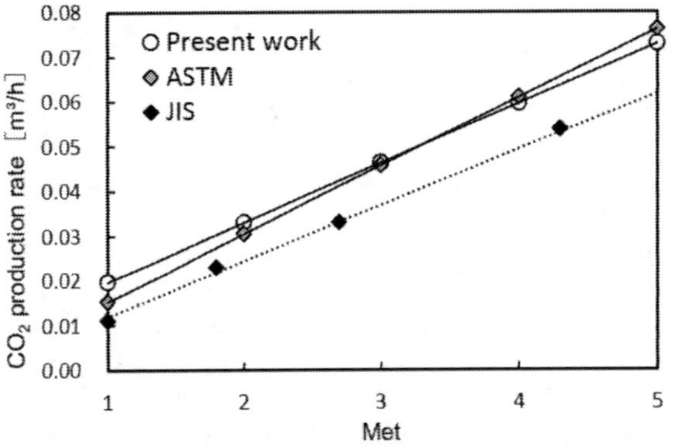

Figure 6: carbon dioxide production rate of present work, ASTM 6245-12 and JIS A1406 (using average Japanese adult's body height and weight)

3 ACCURACY TESTING IN SINGLE ZONE

3.1 Measurement for accuracy testing

As shown in the above, there are certain differences of carbon dioxide production rate between the equation and shown in the ASTM and the JIS. Therefore, by measuring carbon dioxides concentration under condition of known ventilation rate in rooms considered as single zone, the accuracy of the equation was tested.

Condition of the measurement is shown in Table 3. The room-A was occupied by a lecturer and university students for lecture. The room-B was occupied by university students for self-schooling. The room-A employs mechanical ventilation fans included in an air handling unit and its ventilation rate was measured by decay method using carbon dioxide concentration after leaving all occupants. The room-B employs mechanical fans, however, only an experimental exhaust fan, whose air flow rate can be given even in situ situation, was operated during the measurement.

Personal attribute, which are age, gender, body height and weight, of occupants were obtained. Additionally, Occupants' activities, which were $Met = 1.0$ and 1.2, were determined by visual judgment.

Table 3: Condition of measurement

Case	A-1	A-2	B-1
Name of room	room-A		room-B
air volume of room [m³]	468		141
Ventilation rate of room [m³/h]	1,300*		140**
Number of occupants	58	61	6
Outdoor CO_2 conc. [ppm]	413	403	400

*rate of mechanical ventilation employed in AHU measured by decay method
**set by using additional experimental exhaust fan

3.2 Measurement results and discussion

Measured carbon dioxide concentration and computed results given by using equation (12) are shown in Figure 7, 8 and 9. Carbon dioxide concentration measurement instruments were placed at the centre of the room-A and next to the suction part of the experimental exhaust fan in the room-B. The measurement interval was 30 seconds.

$$\frac{dC_i}{dt}V_i = \sum P_{CO_2} + Q_{io}(C_o - C_i) \qquad (12)$$

Where

C_i concentration of carbon dioxide of target room [m³/m³]
C_o concentration of carbon dioxide of outdoor [m³/m³]
Q_{io} ventilation rate [m³/h]

As shown in these diagrams, the trends of present work's concentration are almost similar to the measurement values except occasional risings especially in the room A compared with other computed values. The occasional risings are considered to be caused by non-uniform mixing state of carbon dioxide and incompleteness of visual judgement for occupants' activities. However, the present work's concentration of the room B, considered as being nearly uniform concentration condition, is almost accordance with measured concentration. In these diagrams, RMSE of the each computed concentration is shown, and they indicate that present work using equation (8) gives the most correct carbon dioxide concentration compared with the other ways.

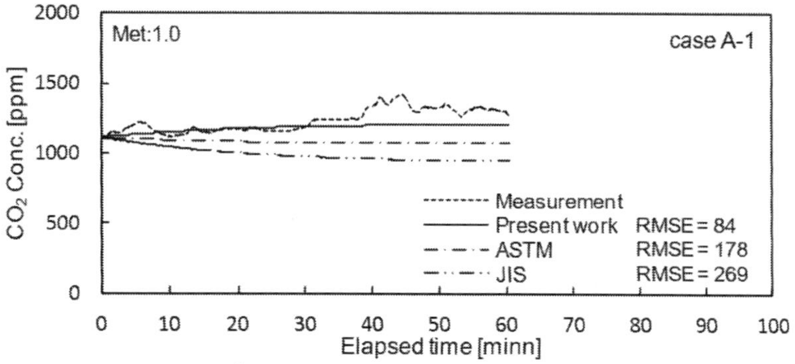

Figure 7: CO_2 concentration in case A-1

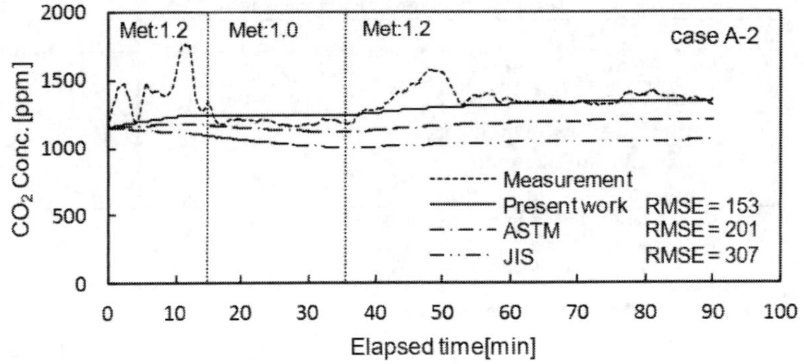

Figure 8: CO_2 concentration in case A-2

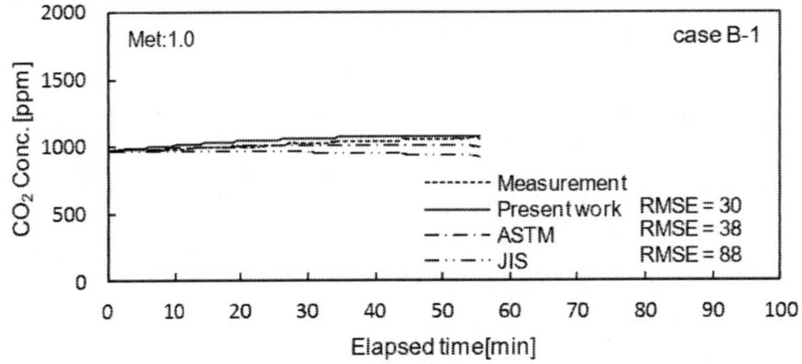

Figure 9: CO_2 concentration in case B-1

4 CONCLUSION

Based on data obtained by Douglas bag method with approximately total 70 voluntary Japanese subjects, a prediction equation for occupants' carbon dioxide production rate included in exhaled breath is developed. This equation has variables, such as *Met* and personal attribute (body height, body weight and gender). Therefore, the equation has certain advantages, like as taking gender difference into the formula and based on multiple Japanese subjects' data, compared with the personal carbon dioxide production rates shown in JIS A 1406[1974] and ASTM D6245-12.

From measurement results operated at single zones, computed carbon dioxide concentration of present work shows the closest to the measurement value compared with the other standards' value under the condition of occupants' *Met* is 1.0 or 1.2. Therefore these results can allow predicting more correct carbon dioxide concentration produced by occupants', especially Japanese occupants', exhaled breath in single zone. Higher *Met* testing, and more measurement in buildings including within multi zone will be warranted.

5 REFERENCES

JIS (2010), JIS A 1406[1974] Method for Measuring Amount of Room Ventilation (Carbon Dioxide Method), Japanese Industrial Standards Committee

ASTM (2012). ASTM D6245-12 Standard Guide for Using Indoor Carbon Dioxide Concentrations to Evaluate Indoor Air Quality and Ventilation. American Society for Testing and Materials International

SHASE (2011). SHASE-S102-2011 Ventilation Requirements for Acceptable Indoor Air Quality. The Society of Heating, Air-Conditioning and Sanitary Engineers of Japan

Saito, M. Ishii, A. et al., (2002). Comfort and Amenity of Classroom Environment. Part 7 Carbon-dioxide Concentration and Ventilation. summaries of Technical Papers of Annual Meeting, Architectural Institute of Japan, D-1, 1071-1072

Stavova, P. Melikov, A. Sundell, J. Naydenov, K. (2006). A New Approach for Ventilation Measurement in Homes Based on CO2 Produced by People. 17th Air-conditioning and Ventilation Conference 2006, 291-296

ISO. (2004). ISO 8996 Determination of metabolic rate. International Organization for Standardization

Herman,I.(2007). Physics of the Human Body. Biological and Medical Physics, Biomedical Engineering, Springer-Verlag GmbH & CO.KG

ASHRAE. (2013). ASHRAE Handbook Fundamentals 2013. American Society of Heating, Refrigerating and Air-Conditioning Engineers, 9.6

SHASE. (2010). Handbook of air-conditioning and sanitary engineering. Vol.1, The Society of Heating, Air-Conditioning and Sanitary Engineers of Japan

Persily, A. (1997). Evaluating building IAQ and ventilation with indoor carbon dioxide. ASHRAE Transactions, vol.103, part2, American Society of Heating, Refrigerating and Air-Conditioning Engineers

PASSIVE COOLING THROUGH VENTILATION SHAFTS IN HIGH-DENSITY ZERO ENERGY BUILDINGS: A DESIGN STRATEGY TO INTEGRATE NATURAL AND MECHANICAL VENTILATION IN TEMPERATE CLIMATES.

Luca Guardigli*[1], Paolo Cappellacci[1], Fausto Barbolini[1]

1 School of Engineering and Architecture, University of Bologna,
Via del Risorgimento 2,
Bologna, Italy

ABSTRACT

Zero Energy Buildings require airtightness and mechanical ventilation systems to provide air changes and energy saving. These requirements contrast with the principles of natural ventilation. Through a case study located in Modena, Italy, a design strategy is proposed as a solution to integrate natural and mechanical ventilation systems at different times of the year to reduce the energy consumption in a newly designed high-density ZEB. The internal comfort evaluation for the warm season is then verified with a multizone dynamic simulation and a CFD analysis.

The proposal consists of two different approaches, the cold season and the warm one. For the cold season, a mechanical ventilation system with earth tubes and heat recovery has been designed, together with airtightness, solar greenhouses and high thermal mass and insulation. For the warm season the design allows a free-running use: open trickle ventilators applied to windows which provide background ventilation, mass and insulation mitigate the heat loads, vertical ventilation shafts support natural ventilation and free night cooling. The ventilation shafts have been designed with aerodynamic principles to provide each apartment with additional (and maximised) differences of pressure due to the stack effect. The indoor comfort conditions in the warm season are then evaluated according to the ASHRAE 55 adaptive model for free-running buildings.

The results of the study confirm that in the warm season acceptable indoor comfort conditions can be achieved in a free running building. The ventilation shaft has an important role for the free cooling of a ZEB and can also be adopted in the renovation of existing buildings.

Keywords

Integrated design, hybrid ventilation system, Zero Energy Building, adaptive thermal comfort, temperate climate

1. INTRODUCTION

A Zero Energy Building (*ZEB*) is a residential or commercial building with greatly reduced energy needs through efficiency gains such that the balance of energy needs can be supplied with renewable technologies.

The efficiency required to achieve the zero energy goal leads to many design strategies, including airtightness to avoid infiltration and mechanical ventilation systems with heat recovery to provide air conditioning and indoor air quality (*IAQ*). Natural ventilation systems generally contrast with the principle of mechanical control of indoor environment.

The designed energy balance of a *ZEB* can however be invalidated with an improper use of technologies by occupants, such as opening windows, changing the operative temperature or not providing the right maintenance of systems.

Thermal comfort is defined as that condition of mind which expresses satisfaction with the thermal environment and depends on physiological and psychological aspects. The former have been widely investigated by Fanger and other scholars, the latter seem to be neglected, at least in the current design. Psychological aspects of comfort involve the interaction of occupants with the environment and vary with latitude, cultural and social factors. Adaptive thermal models, essentially valid for free running buildings, are based on the assumption by Humphreys and Nicol: '*if a change occurs such as to produce discomfort, people react in ways which tend to restore their comfort*'. These models are entirely focused on psychological aspects and allow wider tolerances of indoor thermal comfort conditions than the physiological-only ones.
The mechanical control of thermal comfort, much emphasized in *ZEBs*, aims to reduce the interaction of users with the outdoor environment and this contrasts with the principles of psychological comfort. Monitoring of ultra-low energy buildings in Italy have revealed that occupants rarely use mechanical systems properly and the energy consumption often exceeds the expected results.

The aim of this paper is to propose an integrated design strategy to integrate natural and mechanical ventilation systems in a high density *ZEB* at different times of the year, overcoming airtightness related problems and thermal comfort ones. Energy and comfort evaluations are then estimated according to current standards.

2. PRESENTATION OF THE CASE STUDY

The Case Study is located in Modena, Italy, in a typical temperate climate. Winter is typically not very cold, with temperatures rarely under 0°C. Summer can instead be slightly hot, with the average of daily temperatures in the warmest months around 30°C. Wind speed is generally low at all times of the year, with values around 1,5 m/s, except for some gust at 5-8 m/s. The site of the building has 2258 Degree Days.

A high-density building is designed according to the best practice principles for a ZEB:
- Energy Saving through Building Design;
- Energy Efficiency of Mechanics and HVAC;
- Energy Production from Renewable Sources.

The integrated design of a building concerns all these principles at the same time, with choices that have influences on the whole system. The aim of the case study is to investigate the possibilities to adopt a natural ventilation strategy of use of high-density residential buildings, reducing or cancelling the energy consumption in the warm season of temperate climates, in a Zero Energy Building viewpoint. The strategy adopted for the case study are listed below.

2.1 Building Design

The Energy Saving goal is achieved through several passive Building Design strategies, as shown in Figure 1, which include both formal and technical features:
- Orientation, shape of building, tilt angle of roof and shadow analysis;
- The exposed surfaces of facades are relatively limited compared to their inner volume, with a ratio of 0,3;

- Optimisation of solar gains through greenhouses;
- Strategies for passive cooling: cross ventilation for all apartments and shafts for additional ventilation just in the warm season;
- Increase of green surfaces for climate mitigation;
- Air tightness windows with additional trickle ventilators(to be opened in the warm season);
- Choice of materials as reported in the table 1.

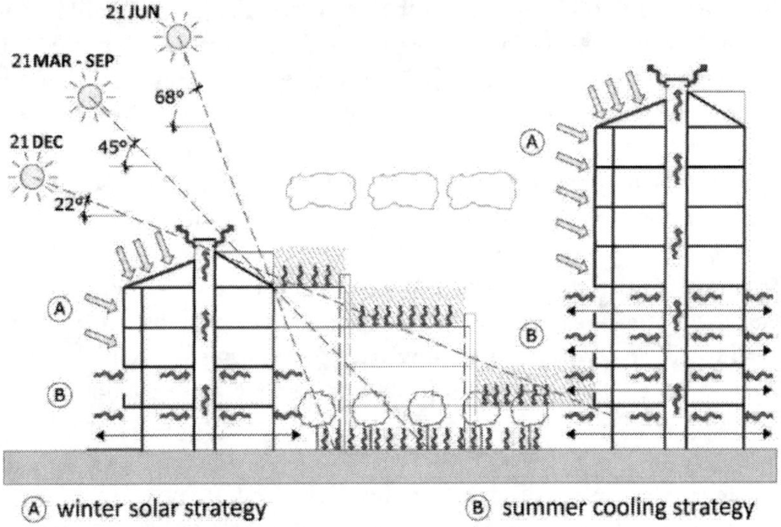

Figure 1: Building Design Strategies

Table 1: Materials

Element	Materials	U-Value (W/m²K)
External wall	10 mm internal plaster 250 mm perforated bricks 320 mm rockwool panels 20 mm reinforced external plaster	0,107
Outer floor	30 mm finishing flooring 50 mm sand-cement screed 240 mm concrete floor 320 mm rockwool panels 20 mm reinforced external plaster	0,109
Roof	70 mm terracotta shingles 320 mm rigid rockwool panels 240 mm concrete floor 10 mm internal plaster	0,110
Windows (general)	Triple glazed 3x4mm, 14 mm air gap	1,00
Windows (on greenhouses)	Double glazed 3x4mm, 14 mm air gap	1,30

The ventilation shaft has been designed to maximise the indoor-outdoor difference of pressure in all the apartments for a better ventilation in the warm season, applying the stack.effect equation:

$$\Delta p_s = (\rho - \rho_i) \cdot g \cdot (H_{NPL} - H) = \rho \cdot \left(\frac{T_i - T_e}{T_i}\right) \cdot g \cdot (H_{NPL} - H) \qquad (1)$$

The height of H_{NPL} has been estimated according to Ashrae guidelines, with values typically of 0,7 of total height of the shaft. To optimise its effect, the shaft has been divided in two separate portions, the former for the lower stories, the latter for the two upper ones, as shown in Figure 2. The difference of temperature from the inlet to the outlet of the shaft has been initially settled at 25 °C. The pressure of wind has been also calculated at the various stories:

$$\Delta p_w = C_p \cdot \rho \cdot \frac{U^2}{2} \qquad (2)$$

C_p coefficient has been estimated through both empirical tables and CpCalc+ software. As shown in Figure 2, in the climate of the building site, the effects of wind pressure have little relevance compared to the stack effect ones.

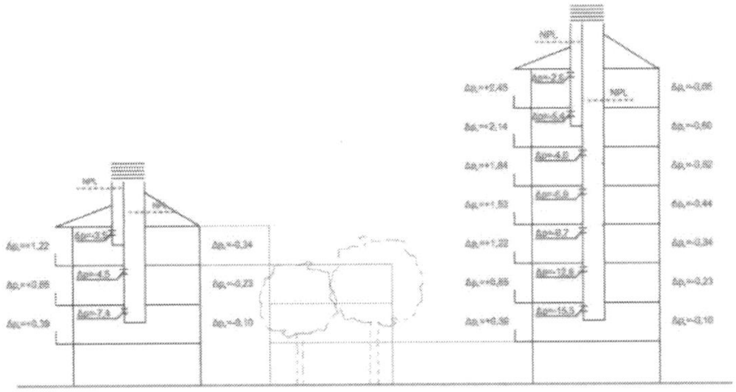

Figure 2: Differences of pressure due to wind and stack effect. Shaft optimisation.

Once Δp have been known, the airflows are calculated by the equation:

$$Q = C_d \cdot A \cdot \left(\frac{2 \cdot \Delta p}{\rho}\right)^{0,5} \qquad (3)$$

C_d is a discharge coefficient for the sharp-edged orifice, taken as 0,61. The areas of the inlet openings in the shaft are set as 0,06 m^2 (30cm x 20 cm). By applying the equation (3), a first empirical verification shows that the airflow through the shaft at each story allows a air change rate of at least 1 volume per hour. The same results are given for a reduced difference of temperature between inlets and the outlet of the shaft of 10°C.
The results suggest that a free running use of the building guarantees at a good IAQ.

2.2 Mechanics and HVAC

Energy Efficiency is achieved by the choice of high performance systems together with the entire design of the building. Designing a ZEB means to choose technologies apt to reduce, save and optimise the energy consumption, and the choice of HVAC systems is strictly related to this scope. A first overview of the project highlights questions to be solved:

- The surfaces for PV and solar thermal collectors of roofs are relatively small related to the numbers of apartments, and the energy production does not supply all the needs of the building;
- An additional and renewable system of energy production has thus to be provided to achieve the Zero Energy goal;
- The heating system of the cold season should be able to operate in reverse even in the warm season, so that a single mechanical system serves the building;

- HVAC should be shut down in the warm season in behalf of natural ventilation;
- IAQ has to be guaranteed at all times of the year, despites air tightness of windows.

The choice adopted for the building consist of a mechanical ventilation system equipped with earth tubes, heat recovery and a heating unit. During the cold season the system provides heating and ventilation in each apartment (the ventilation shaft is closed). During the warm season HVAC system should be off in behalf of natural ventilation: the background ventilation is then provided by the trickle ventilators in windows and the open shaft. In case of extreme climate conditions the HVAC system can be turned on and the ground-coupled heat exchanger provides cool and de-humidified fresh air. Hot water for domestic use and heating and electricity are then provided by a Combined Heating Power System (CHP), where PV and solar thermal collectors are not sufficient.
The Figures 2 and 3 show the operational scheme of the system in the cold and warm seasons. Figure 4 shows a relevant section of the building, Figure 5 the ventilation system scheme.

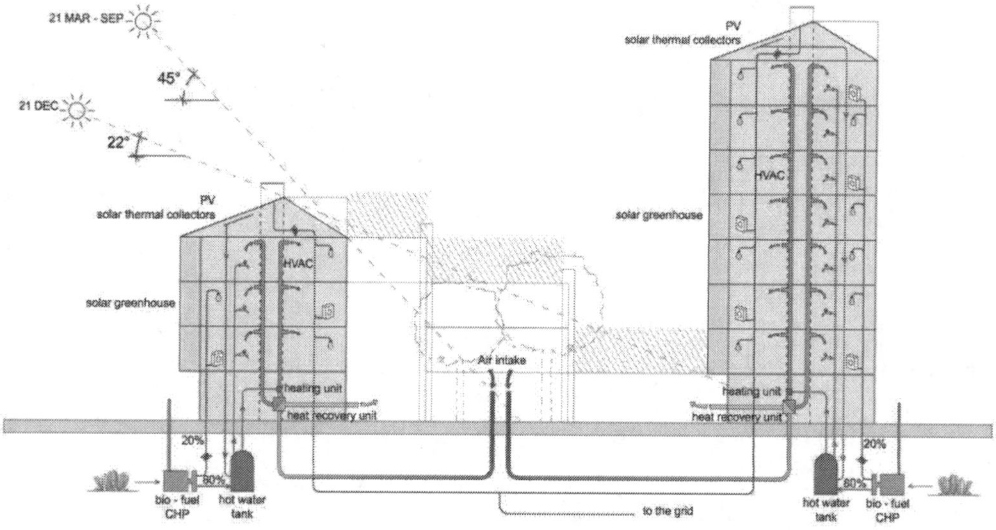

Figure 2. Operational scheme in the cold season.

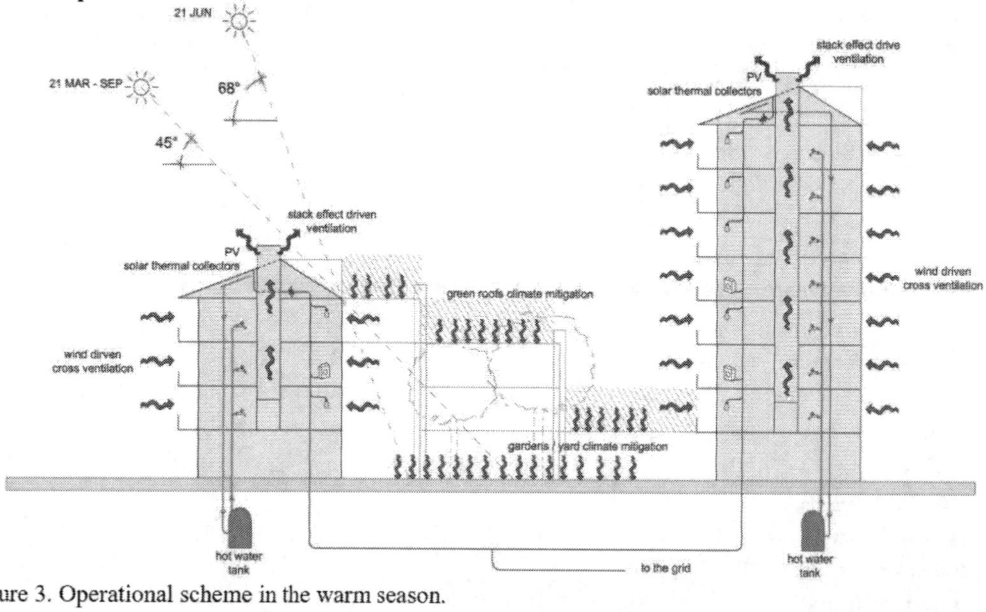

Figure 3. Operational scheme in the warm season.

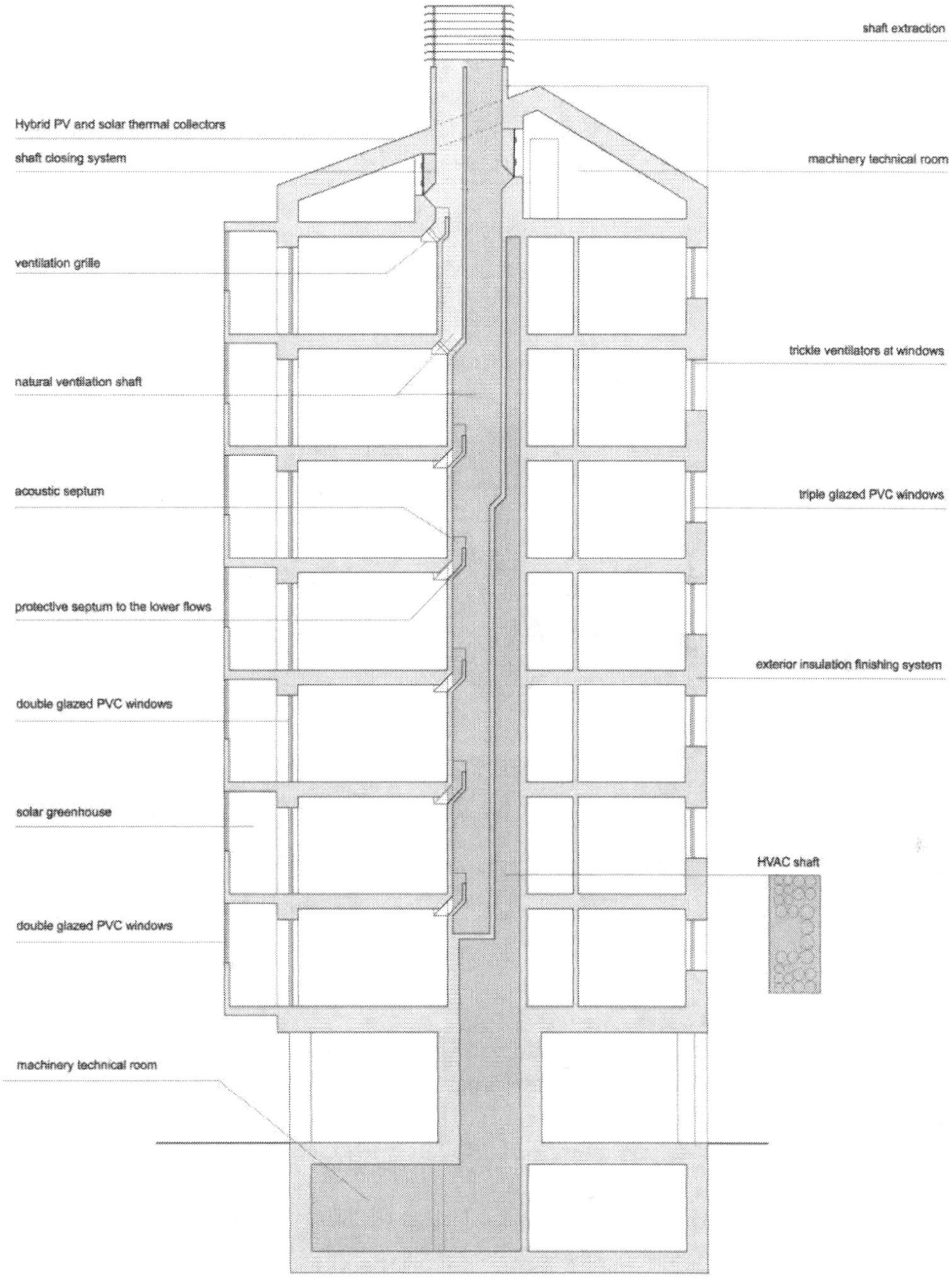

Figure 4. Relevant section of the ventilation shaft and the HVAC system shaft.

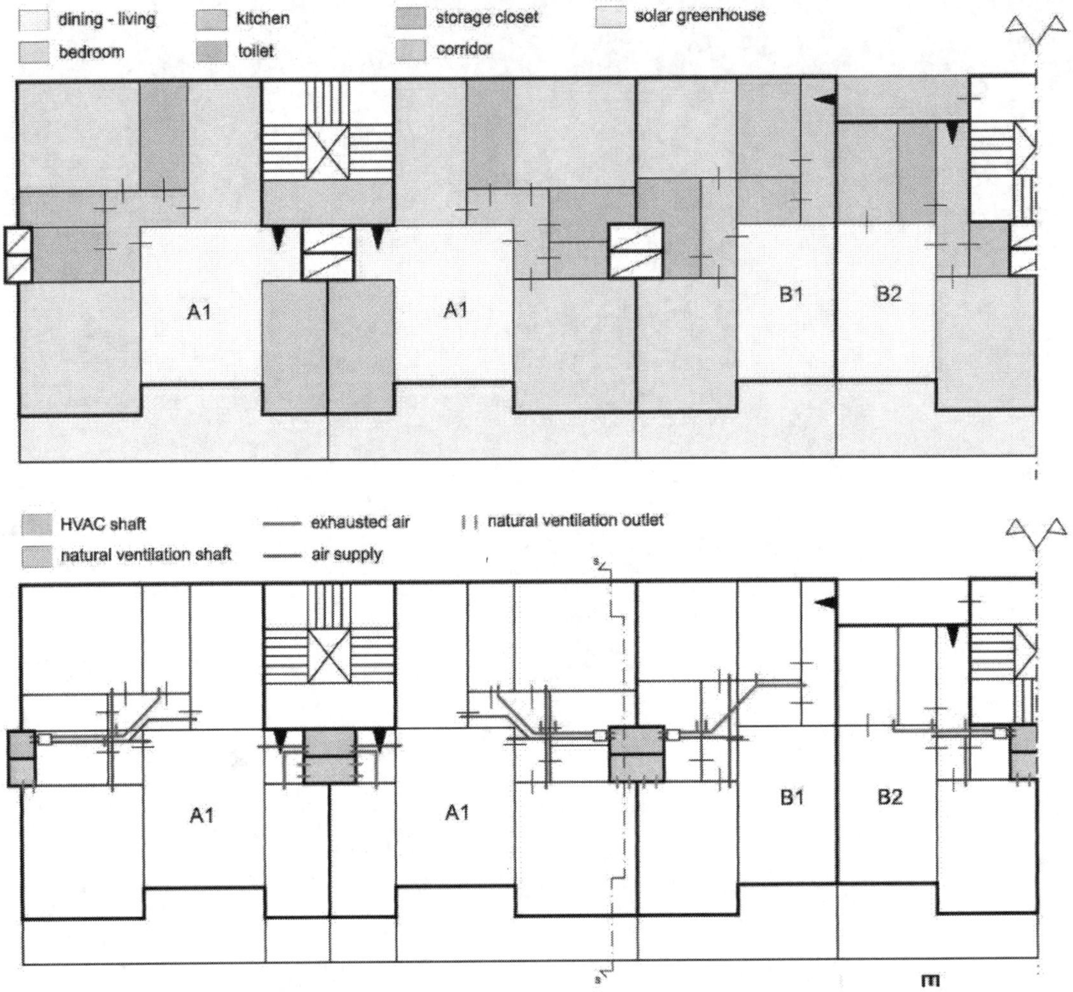

Figure 5. Ventilation system scheme.

2.3 Renewable Sources of Energy

The Energy Production in ZEBs should balance the consumption. To optimise the sun-exposed surfaces of the roofs, PV is designed to be settled for 604 m^2 while photovoltaic thermal hybrid solar collectors are designed for the remaining 250 m^2. The total energy production is about 155100 kWh per year, and the hot water production supplies all the needs of the building for the most part of the year.

The electric needs of the building are estimated in 171000 kWh per year through the SVI index *(StromVerbrauchsIndex)*, related to the number of occupants of every single apartment according to the equation 4:

N. of occupants x 500 kWh + 500 kWh = goal value per apartment in kWh (4)

A CHP system supplies the electric deficit of PV and also provides hot water for heating and domestic use in the cold season. A power of 80 kW is sufficient to furnish hot water and the residual electric energy needed.

Building energy rating is evaluated according to EN ISO 13790:2008 in 7,7 kWh/m^2 per year and all the energy consumption is balanced by the production from renewable sources also with a slight surplus.

3 ANALYSIS AND RESULTS

Thermodynamic and a CFD analyses have been carried out to evaluate the effectiveness of the design strategies adopted. The main purpose of the analyses has been to estimate the indoor thermal comfort conditions, especially in the warm season when a free running use of the building is desired. The results have been so checked through the Ashrae 55 standard for buildings with no HVAC systems working.

3.1 Multizone Thermodynamic Analysis

A relevant portion of the building has been modelled and analysed in the typical design summer week. The internal environments have been modelled in different ways to evaluate the influence of the shaft on performances. The settings of the simulation included air infiltration for 0,5 volumes per hour in each apartment, a night cooling program of 5% windows openings between 8pm and 7am and an advanced mathematical pattern for ventilation calculation. The output results of the simulation are copious and detailed. Hence the main and interesting outputs are presented. A global overview of the internal temperatures, presented in Figure 6, shows that the average values range between 25°C and 28°C, despites the daily temperature fluctuations.

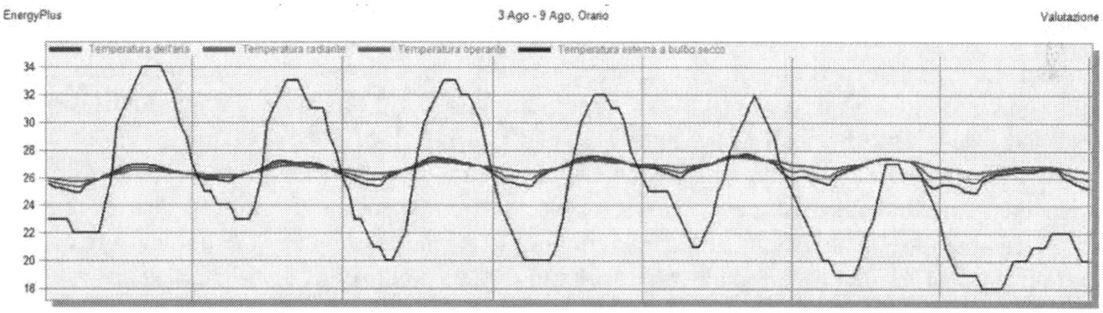

Figure 6. Internal and external temperatures in the typical design summer week.

A detailed analysis of the various zones shows that local temperatures vary at each story. If the lower stories are the cooler ones, with temperatures of 1°C below the building average, the upper floors reach in the hottest hours of the day temperatures of 29°C.

The entire results for each story and zone of the model have been evaluated according to Ashrae 55 standard and are reported in Figure 7. As it is clearly visible, thermal comfort conditions are satisfied within a 80% of acceptability limit in the hottest days of the year.

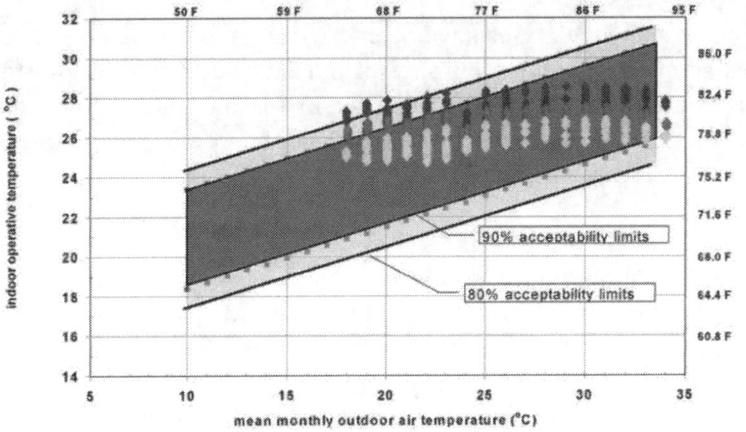

Figure 7. Internal comfort conditions in Ashrae 55 adaptive model chart. In blue the upper floor, in yellow the lower floor, and in red the average of the whole building.

3.2 CFD Analysis

A more accurate analysis has been led for the rooms served by the shaft to check the internal conditions of comfort. A simplified model of seven rooms, one for each story, and the annexed ventilation shaft has been analysed. The boundary conditions have been set adopting the output of the thermodynamic simulation in a typical situation of night cooling, at 4am of 7th August, when the outdoor air temperature is about 20°C and the average of the whole building is 26,5°C. The grid used for the model has steps of 0,1m, with 0,05m subdivisions close to the openings, the shaft and the surfaces of walls and floors.

The results of the simulation show the temperature gradient across the height of the building, as expected, and the air speed, never exceeding 0,25m/s in the rooms (Figure 8). The air speed and the temperature in the shaft are instead much higher and this causes the cooling effect in the rooms. A further detailed look at the results in every single room of the model shows the convective movements of air just adjacent the floor and the walls, with the air speed in the middle of the rooms quite little.

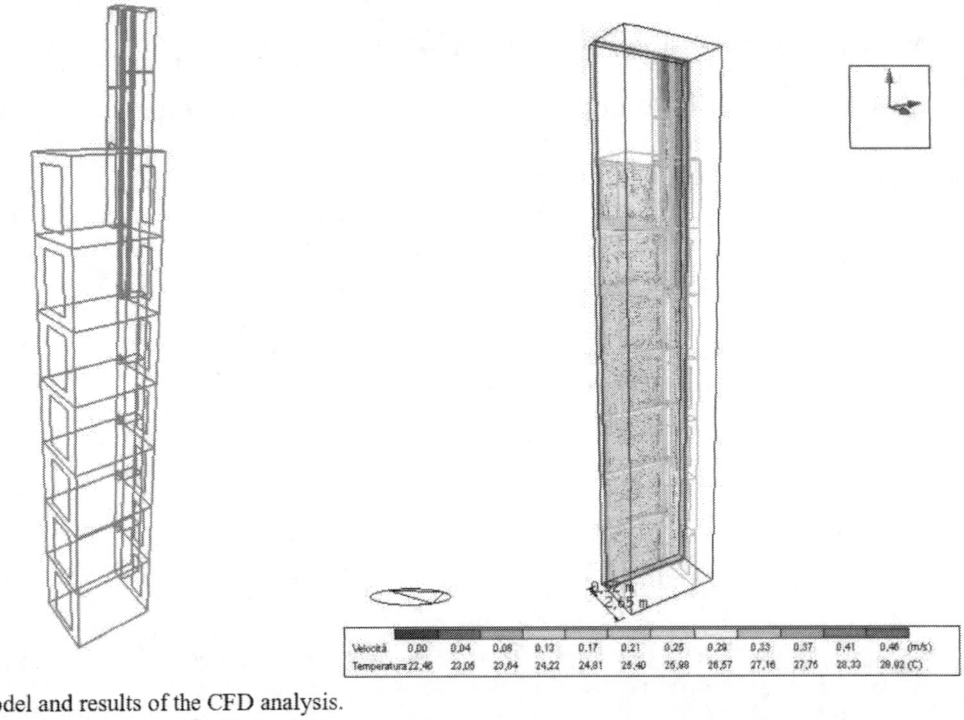

Figure 8. Model and results of the CFD analysis.

4 CONCLUSIONS

The case study presented shows the possibility to achieve good levels of comfort in Zero Energy Buildings in temperate climates through passive strategies of design, in the perspective of minimum impact on the environment and resources. The results are the sum of various systems that work together, from building design to mechanics and HVAC systems. The ventilation shaft, which is an element that characterise the project, is a part of this integrated process of design. Indeed adopting passive strategies of cooling is not a requirement to design a ZEB, as the zero energy goal can be achieved in many other ways.

If in winter heating is and ventilation are required, as the analyses have shown acceptable indoor thermal comfort conditions are still achievable even without the use of HVAC systems in the warm season. Adaptive comfort models indeed guarantee high psychological and physical satisfaction by occupants, moreover admitting wider ranges of temperature. The outputs of thermal simulations show acceptable conditions and are still preventive: indeed, a simple change of a parameter can adulterate the results. The settings have so been precautionary, e.g. the percentage of opening of the windows have just been set at 5% and could be more, or the timing of windows opening has not intentionally set in the best desirable conditions. A consideration should be also done on the possibility of using the HVAC system in the hottest days: indeed the position adopted for the project is not absolute, and in extreme climate conditions mechanics can be turned on at any time.

Hence the case study presents a strategy of integrated design of a high-density ZEB and suggests an use of hybrid ventilation in buildings: mechanical in the cold season, just natural in the warm one. In extremely hot conditions in summer natural ventilation can be assisted by HVAC systems. The ventilation shaft represents an option among the various possible. In the site of the case study its contribution to the final result is significant, as it increase the airspeed in the room and accelerates the cooling, but it is part of a complex design. For higher heights and windier locations its contribution for free cooling could be decisive.

Finally, the use of the ventilation shaft can be also adopted in the renovation of old buildings.

5 REFERENCE

ASHRAE, *2009 ASHRAE Handbook: Heating, Ventilating and Air Conditioning Systems and Equipments*, Atlanta, 2009.

Allard F., *Natural ventilation in buildings – A Design Handbook*, James & James, London, 1998.

Axley J. W., *Residential Passive Ventilation Systems: Evaluation and Design (AIVC Technical Report n°54)*, INIVE EEIG, Sint-Stevens-Woluwe (Bruxelles), 2001.

Liddament M. V., *Occupant Impact on Ventilation (AIVC Technical Report n°53)*, INIVE EEIG, Sint-Stevens-Woluwe (Bruxelles), 2001.

Santamouris M., Asimankopoulos D., *Passive cooling of buildings*, James & James Ltd, London, 1996.

AIVC Centre – *AIVC, Ventilation Information Paper n° 12 – Adaptive Thermal Comfort and Ventilation*, INIVE EEIG, Sint-Stevens-Woluwe (Bruxelles), 2006.

Brager G. S., De Dear R. J., *Thermal comfort in naturally ventilate buildings: revisions to ASHRAE Standard 55*, 2002.

Grosso M., Marino D., Parisi E., *A wind pressure distribution calculation program for multi zone air flow models*, Turin, 1994.

Prajongsan P., Sharples S., *The ventilation shaft: An alternative passive cooling strategy for high-rise residential buildings in hot-humid climates*, Lima, 2012.

Torcellini P., Pless S., Deru M., *Zero Energy Buildings: a critical look at the definition*, Pacific Grove, California, 2006.

American Society of Heating, Refrigerating and Air Conditioning Engineers, *ANSI/ASHRAE 55-2004, Thermal Enviromental Conditions for Human Occupancy*, Atlanta, 2004.

EN ISO 13790:2008 – Energy performance of buildings – calculation of Energy use for space, heating and cooling, 2008.

ENERGY SAVING AND THERMAL COMFORT IN RESIDENTIAL BUILDINGS WITH DYNAMIC INSULATION WINDOWS

Daisuke Kawahara[1], Shinsuke Kato[*2]

1 School of Engineering, The University of Tokyo,
IIS Cw403,
4-6-1, Komaba, Meguro, Tokyo, Japan
kawa-d@iis.u-tokyo.ac.jp

2 IIS, The University of Tokyo
IIS Cw403,
4-6-1, Komaba, Meguro, Tokyo, Japan

ABSTRACT

To realize the concept of low-energy buildings, an increase in the thermal insulation performance of building parts, especially the openings that show poor insulation performance, is necessary. In addition, an adequate level of thermal comfort is needed within residential buildings. We have developed window-applied dynamic insulation (DI), and verified thermal insulation performance in chamber and field tests. The purpose of this study is to verify the thermal comfort level within residential buildings using DI windows, via measurements obtained from an equivalent temperature-calculated thermal manikin, and to verify energy consumption values in consideration of disturbances by dynamic simulation. The thermal comfort evaluation results showed that, with DI window installation, the equivalent room temperature decreased to approximately 5% of the temperature difference between the exterior and interior. This implies that the effect of the DI window on thermal comfort was small. The dynamic simulation results also showed that the energy consumption reduction reached a value of approximately 1.5% during a 5-month period regardless of location. We confirmed that the proposed window system did not cause great damage for thermal comfort in a real environment and realized the energy saving.

KEYWORDS

Dynamic insulation, thermal comfort, ventilation, energy savings, windows

1 INTRODUCTION

There is an urgent demand worldwide for the construction of low-energy buildings. Designs for low-energy buildings generally require thicker walls or the use of thermal insulation. However, these methods have drawbacks, such as decreased floor space and high costs. Dynamic insulation (DI) systems are gaining prominence because they can mitigate such problems and provide effective ventilation. A cross section of a DI window is shown in Figure 1. A DI system has a heat recovery mechanism that uses the airflow produced by the pressure difference between the interior and exterior of a room. This airflow transports into the room heat that would otherwise be lost, because the airflow direction is opposite to that of heat loss. In previous studies of DI systems, Baily (1987), Qiu (2007), Gan (2000), and Imbabi (2013) focused on walls. However, a few studies applied DI systems to building parts other than walls. Because building openings generally have low thermal insulation performance, we produced a DI window prototype and used it in an existing building, and we verified its thermal insulation performance (0.73 W/m^2K) using a chamber test.

The temperature of a DI windows' inner surface decreases during the winter, due to the passage of air through the window. For this report, we evaluated the thermal comfort in a residential building with DI windows installed, because the air inlet of DI windows are set at a relatively lower position than vent holes. The experimental model house is located in Sapporo, Hokkaido, which is a cold region in Japan. This residence is an existing house for which the thermal insulation performance and airtightness were improved during the winter of 2012. The thermal comfort evaluation criterion was equivalent temperature, which is based on local drift and air temperature. We measured the local equivalent temperature using a thermal manikin in the winter of 2014. The indoor air temperature was set to approximately 20°C–22°C, which is the daily average interior temperature range in this area. We also verified the energy consumption by dynamic simulation using a chamber test result, which is a regression equation between the airflow amount and the thermal insulation performance. The dynamic simulation period is one-year, and we compared two different climate locations.

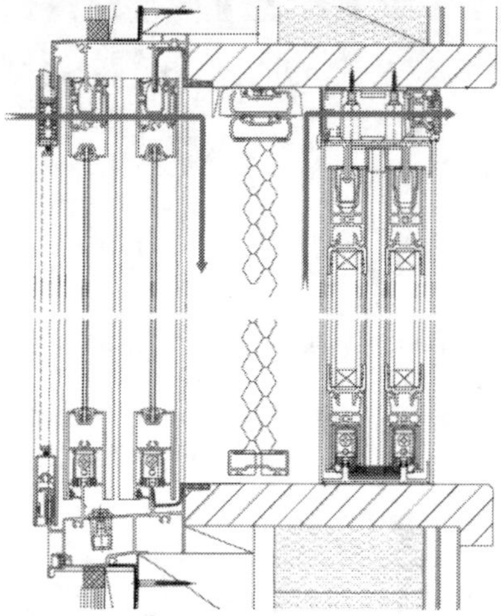

Figure 1: Cross section of a DI Window

2 THERMAL COMFORT ANALYSIS

The ventilation inlet of the DI window is in a relatively low position, although the general ventilation inlet is arranged in a higher position. In this study, we measured thermal comfort by using a thermal manikin and compared the general vent cap to the DI window.

2.1 Measurement

To verify the effects of DI windows on thermal comfort, we utilized a thermal manikin. The manikin has a total of 16 body parts, as shown in Figure 2, which control the surface temperature and heat flux. The heat regulation mode of the manikin that controlled surface temperature satisfied the below thermal neutral equation:

$$\Theta_i = 36.4 - 0.054 Q_i \quad (1)$$

where Θ_i (°C) is the area-weighted average surface temperature of a certain part i, and Q_i (°C) is the area-weighted average heat flux on a certain part i. An index of this measurement, equivalent temperature $\Theta_{eq,i}$ (°C) as defined by ISO 14505-2 2006, is stated as the "temperature of a homogenous space, with mean radiant temperature equal to air temperature and zero air velocity, in which a person exchanges the same heat loss by convection and radiation as in the actual conditions under assessment" (ISO 13731). The associated equation is shown below:

$$\Theta_{eq,i} = \Theta_i - Q_i / \alpha_i \quad (2)$$

where α_i (W/m^{2}K) is the area-weighted average total heat transfer coefficient on a certain part i.

A measurement in which the thermal manikin is nude effectively evaluates the draft from the DI window, and another measurement in which the thermal manikin is clothed evaluates thermal comfort in a normal mode of life.

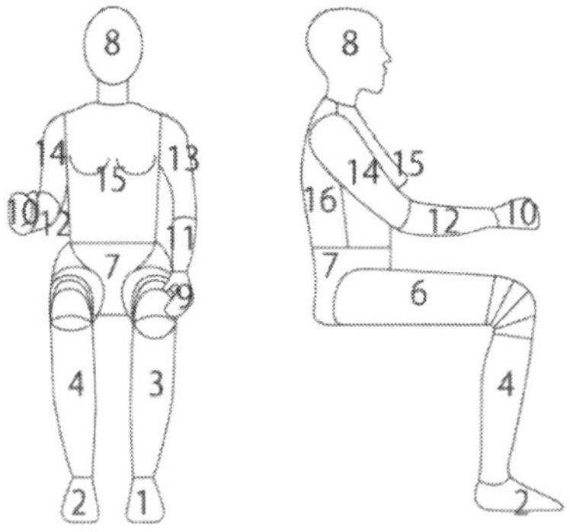

Figure 2: Thermal manikin parts

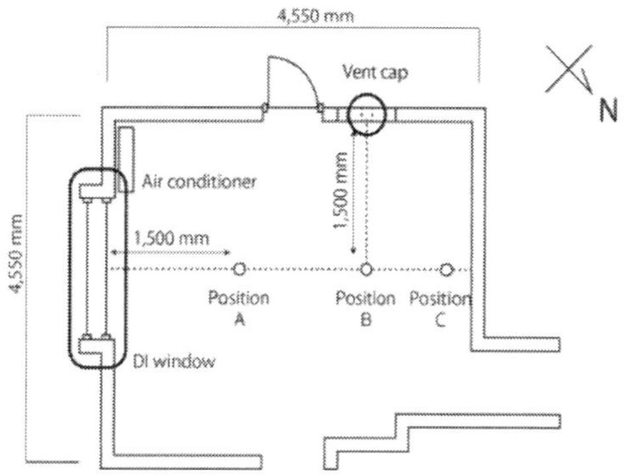

Figure 3: Position of the thermal manikin within the room

Note: 1:Left foot, 2:Right foot, 3:Left leg, 4:Right leg, 5:Left thigh, 6:Right thigh, 7:Pelvis, 8:Head, 9:Left hand, 10:Right hand, 11:Left forearm, 12:Right forearm、13:Left shoulder, 14:Right shoulder, 15:Chest, 16:Back

Figure 3 shows the position of the thermal manikin within the room; position A is 1.5 m, position B is 3 m, and position C is 4 m from the DI window. Table 1 shows measurement cases, which study the equivalent temperature changes based on the following effective factors: position, ventilation type, and airflow amount. The measurements were carried out in January of 2004 in Sapporo, Hokkaido. The exterior temperature ranged from -5.1°C to -2.1°C during the measurement period, and the interior room temperature was approximately 20°C with utilization of an air conditioner. However, the indoor room temperature fluctuated because of changes in the exterior temperature and solar radiation. The temperature of the interior environment, prior to measurements, ranged from 19°C to 20°C.

Table 1: Measurement cases

	15th Jan		16th Jan				17th Jan			
Position		B	A	B	B	C	A	B	B	C
Ventilation type	DI	DI※	DI	D	General	DI	DI※	DI※	General※	DI※
Air change rate	0 m^3/h		0.5 ACH				0.5 ACH			

※denotes the addition of clothes (=approximately 0.6 clo)

Table 2. Measurement equipment

Sensor	Type	Range	Accuracy
Surface temperature Inlet air temperature	Thermocouple type T	-200°C to 400°C	±1.0
Pressure difference between interior and exterior	SDP1000-L025	-62 Pa to 62 Pa	0.5% FS /1.5% m.v.
Data logger	GRAPHTEC mini logger GL 820		

The measurement results are shown in Figures 4 to 7 and the measurement conditions are shown in Tables 2 and 3. The equivalent temperature with DI window installation decreased to approximately 5% of the temperature difference between the room interior and exterior, regardless of whether the manikin was clothed or nude, showing that the effect of the DI window on thermal comfort was minor.

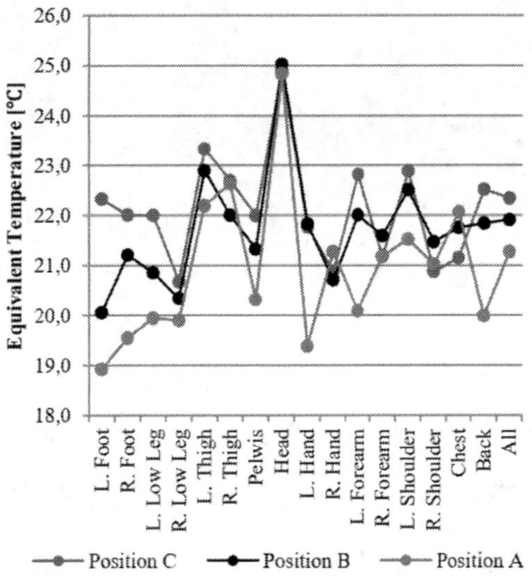

Figure 4: Equivalent Temperature "nude-position"

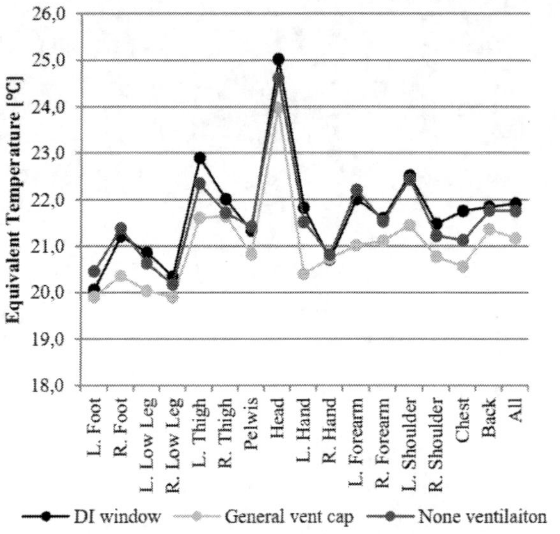

Figure 5: Equivalent Temperature "nude-ventilation type"

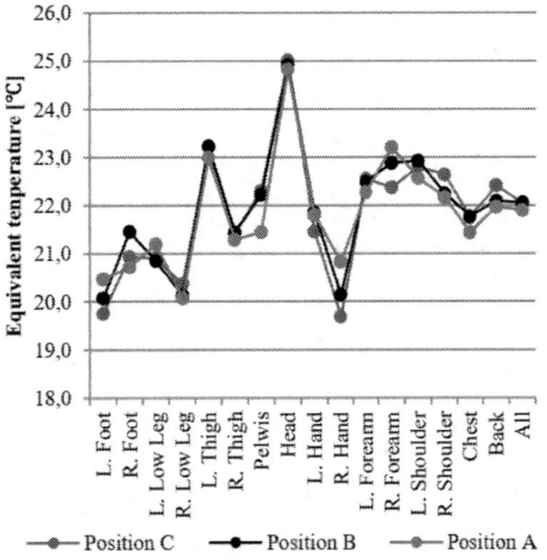

Figure 6: Equivalent Temperature "cloth-position"

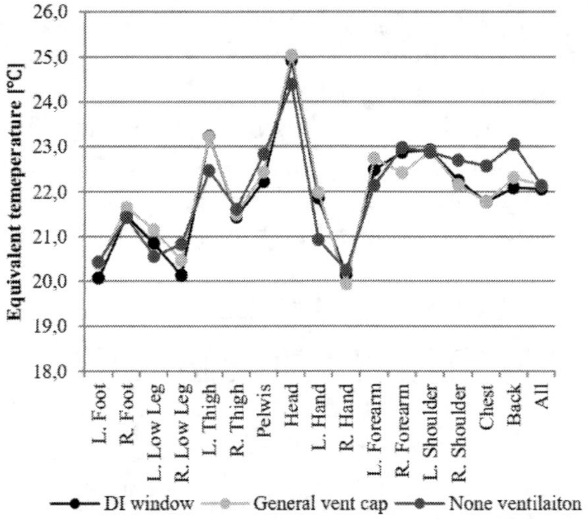

Figure 7: Equivalent Temperature "cloth-ventilation type"

Table 2: Measurement Condition "nude"

	Position A	Position B	Position C	Position B
Ventilation type	DI window	DI window	DI window	General vent cap
Airflow amount (m^3/h) (Standard deviation)	20.8 (1.1)	19.6 (3.0)	21.2 (0.3)	21
Inlet air temperature (°C) (Standard deviation)	22.3 (0.8)	23.0 (0.4)	7.7 (0.1)	-4.9 (0.1)
Exterior temperature (°C)	-3.3 (0.2)	-2.1 (0.1)	-2.9 (0.1)	

Table 3. Measurement Condition "cloth"

	Position A	Position B	Position C	Position B
Ventilation type	DI window	DI window	DI window	General vent cap
Airflow amount (m^3/h) (Standard deviation)	21.6 (0.4)	21.2 (0.4)	21.0 (0.6)	21※
Inlet air temperature (°C) (Standard deviation)	10.0 (1.5)	12.8 (0.5)	7.2 (0.2)	-4.8 (0.1)
Exterior temperature (°C)	-5.6 (0.1)	-3.7 (0.1)	-3.7 (0.1)	

※denotes the additi

3 DYNAMIC SIMULATION FOR ENERGY SAVINGS

3.1 Dynamic simulation conditions

We also simulated differences in energy consumption and comfort as compared to static insulation, and then evaluated these results in relation to the additional cost of DI windows. The thermal insulation performance of DI windows depends on airflow amount, which is determined by the temperature difference between the interior and exterior of the room, and the outdoor wind velocity. In consideration of disturbances by dynamic simulation, we used TRAFlow as based on COMIS 3.1, which is an add-on program of TRNSYS. The model we used was a standard Japanese residential building, as shown in Figure 8.

The simulation conditions were determined by a number of reports and measurement data. In this study, the clearances of the DI window and residential building were both 60 cm^2, and the clearance was partitioned into major sections. The partition ratios, as reported by Seiichi et al. (2001), were 39% for the building frame, 28% for the opening, and 33% for the ventilation equipment. In this study, the clearance of the residential building limits building frame and ventilation, because the clearance of the DI window, including that around the window frame, was made so purposely. The air link in the residential building is shown in Figure 10. In the case with DI window installation, one ventilation fan in an attic space exhausted 120 m^3/h to the exterior, and two ventilation fans in the restrooms exhausted a total of 40 m^3/h, which were ineluctable ventilation amounts in the residential building. On the other hand, in the case with general window installation, the air change rate was 0.5ACRand this ventilation was present in each room. To simulate the ventilation under practical conditions, wind pressure coefficients were determined through reports by Shin-ichi et al. (1994). There were no internal loads and ventilation control, and the regression equation used in this simulation, which determines thermal insulation performance, assumed a DI window with blind installation. The thermal performance of the residential building is shown in Table 4. The positions of the DI windows were arranged at the four points of the compass, as shown in Figure 9. The diagram of airflow direction shows in Figure 10.

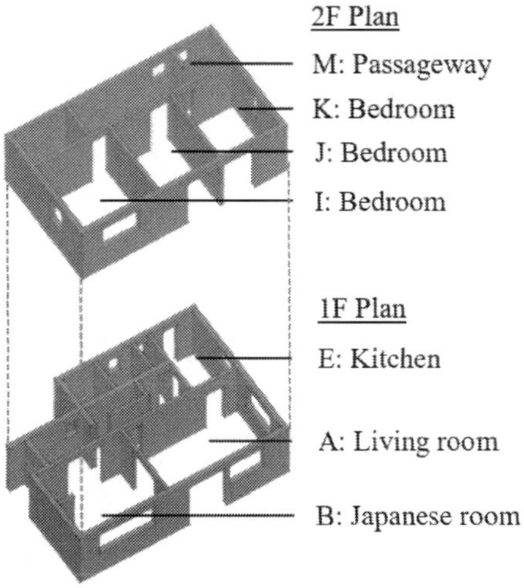

Figure 8: Diagram of a standard residential building.

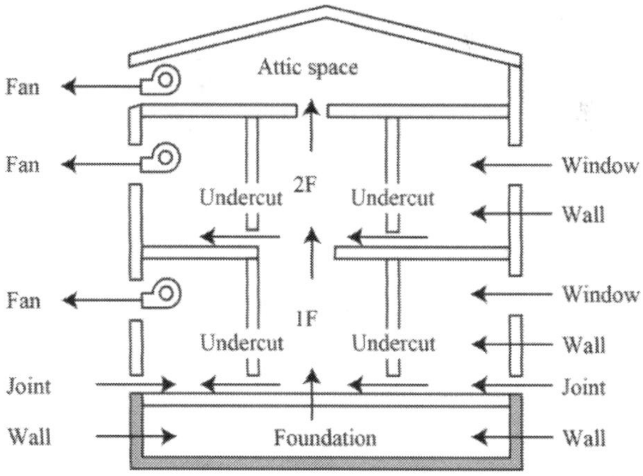

Figure 9: Diagram of airflow direction.

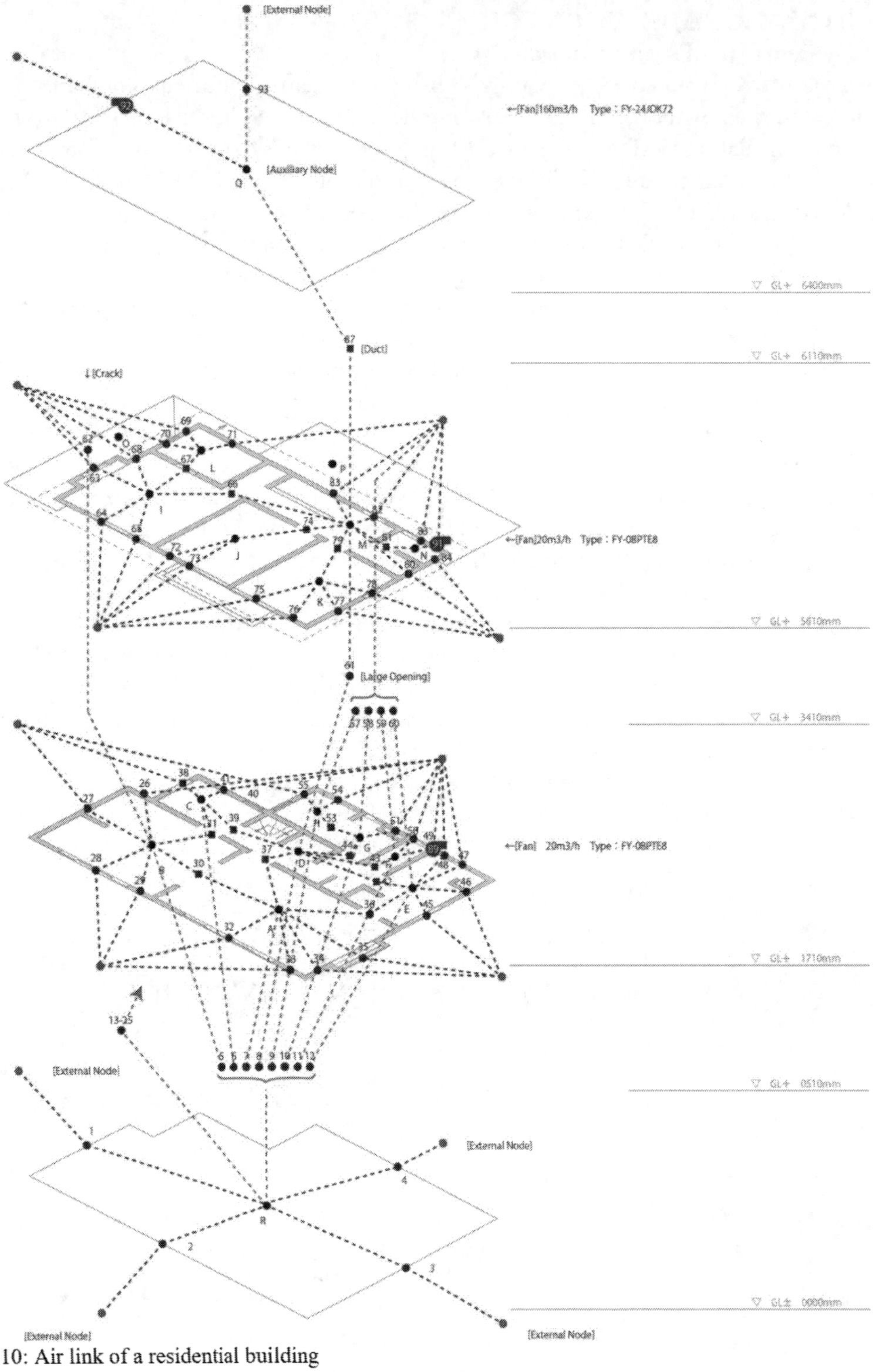

Figure 10: Air link of a residential building

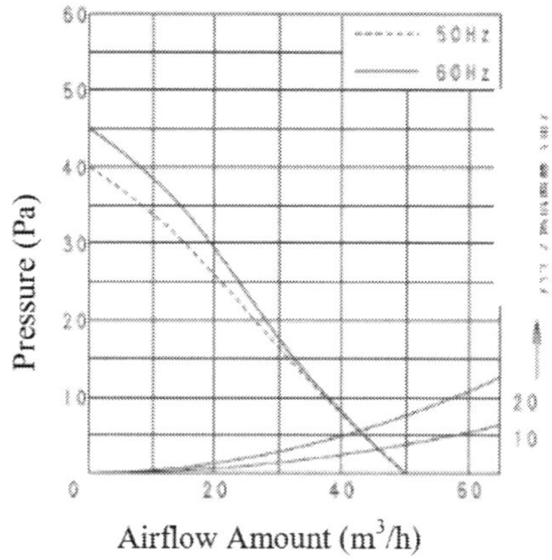

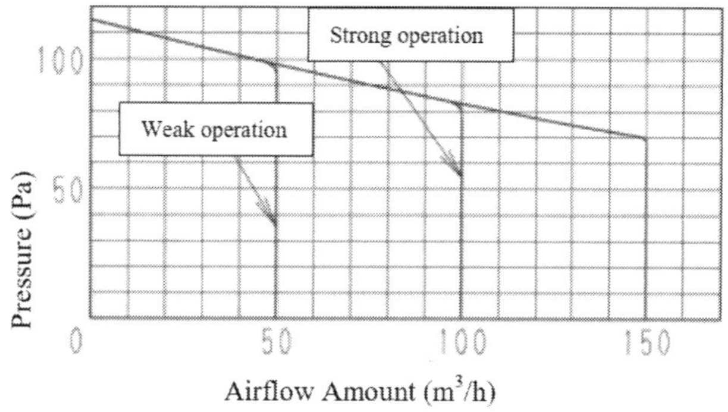

Figure 11: Data from the fan located in the rest room

Figure 12: Data from the fan located in the attic space

Table 4: The simulated thermal performance of the residential building

Building parts		Thermal performance [W/m^{2}K]
Wall	External	0.254
	Internal	3.583
Ceiling		0.167
Floor		0.208
Window		1.240

Table 5: Conditions of residential building clearance

	Building Parts	Percentage of total clearance [%]		[%]		[%]
Clearance of the residential building	Wall	2.10	North 1F	0.33	B	0.03
					C	0.08
					H	0.05
					G	0.05
					F	0.03
					E	0.08
			North 2F	0.3	N	0.06
					M	0.16
					L	0.09
			South 1F	0.33	B	0.14
					A	0.19
			South 2F	0.3	I	0.12
					J	0.09
					K	0.09
			East 1F	0.24	E	0.12
					A	0.12
			East 2F	0.22	N	0.04
					M	0.04
					K	0.15
			West 1F	0.23	H	0.06
					C	0.06
					B	0.11
			West 2F	0.17	L	0.06
					I	0.12
	Ceiling	1.29	Q			0.99
			P			0.18
			O			0.12
	Floor	0.39	A			0.12
			B			0.09
			C			0.03
			D			0.05

				E		0.05
				F		0.01
				G		0.02
				H		0.02
	Joint between foundation and floor	50.37	North	14.85	B	1.24
					C	3.71
					H	2.47
					G	2.47
					F	1.24
					E	3.73
			South	14.85	B	6.46
					A	8.39
			East	10.33	E	5.17
					A	5.17
			West	10.33	H	2.58
					C	2.58
					B	5.17
Ventilation Equipment	Kitchen	45.84				45.8

3.2 Dynamic simulation results

Figure 13 shows outdoor temperature at Sapporo (cold region) and Tokyo. The temperature of Sapporo is smaller than that of Tokyo by approximately 10 °C. Figure 14 shows differences of airflow amount that come into interior through DI window. Buoyancy effect affected the difference between 1F and 2F, especially during the period of large demands for heating. Figures 15 and 16 show the energy consumption reduction for heating in a 5-month period, from November to April. The reduction rate in Sapporo (a cold region) and Tokyo are 1.6% and 1.5%, respectively. The energy consumption of Sapporo was higher than that of Tokyo, because the heat loss was proportional to the temperature difference between the room exterior and interior.

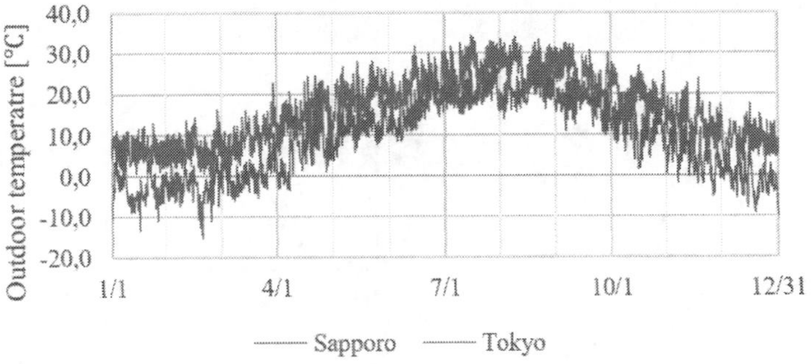

Figure 13: Outdoor air temperature at Sapporo (cold region) and Tokyo

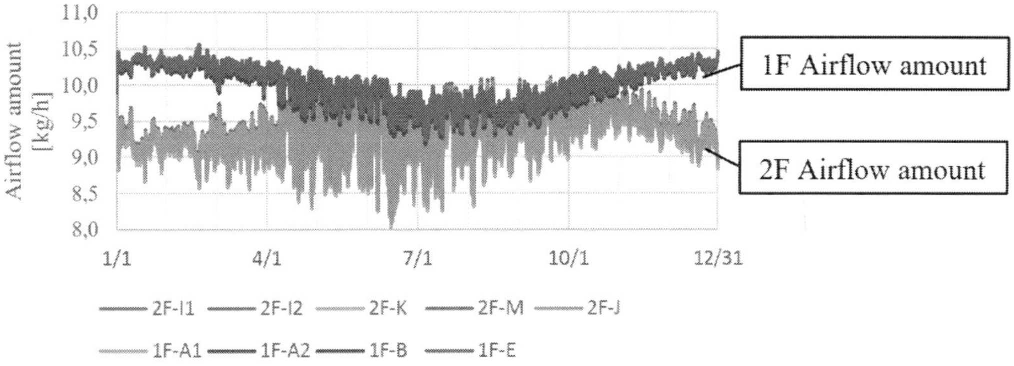

Figure 14: Differences of airflow amount among each DI window

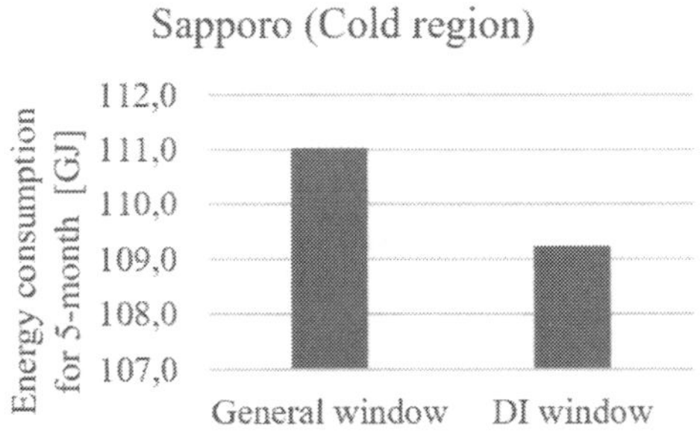

Figure 15: Energy consumption in Sapporo

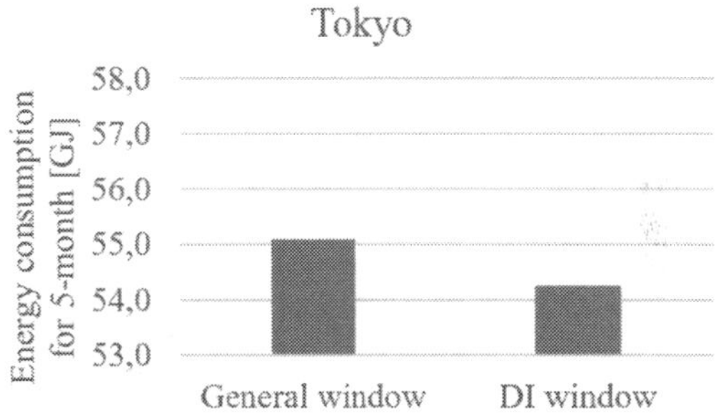

Figure 16: Energy consumption in Tokyo

4. CONCLUSIONS

To verify the thermal comfort of residential buildings using DI windows and energy consumption reduction, we performed various measurements and a dynamic simulation. The result of the thermal comfort measurements showed that the equivalent temperature decreased to approximately 5% of the temperature difference between the room interior and exterior. This suggests that the effect of the DI window on thermal comfort was minor. As well, the dynamic simulation results show that the energy consumption reduction was approximately 1.5 % during 5-month in winter regardless of location.

5. REFERENCES

Annex44 (2009). *Expert guide RBE, APPENDIX 8A,* 3.

American Society of Heating Refrigeration and Air Conditioning Engineers (2001). ASHRAE Handbook. *FUNDAMENTALS*, 8.

Baily, R. (1987). Dynamic insulation systems and energy conservation in buildings. *ASHRAE Transactions*, 93(1), 447-466.

Gan, G. (2000). Numerical evaluation of thermal comfort in rooms with dynamic insulation. *Building and Environment*, 35(5), 445-453.

Imbabi, M. (2013). A passive-active dynamic insulation system for all climates. *International Journal of Sustainable Built Environment*, 1(2), 247-258.

Seiichi, H. (2001). Study on airtight performance of various building material joints: Part 2, Characteristics of air gap on envelope parts and comparison between measurement and estimation. *Summaries of technical papers of annual meeting Architectural Institute of Japan*, 11-12.

Shin-ichi, A. (1994). Wind tunnel test on surface wind pressure and prediction of air change rate of detached house: Experimental study on natural ventilation of detached house Part1. *Journal of Architecture, Planning and Environment Engineering, AIJ*, No.456, 9-16.

Qiu, K. (2007). Modeling the combined conduction-air infiltration through diffusive building envelope. *Energy and Buildings*, 39, 1140-1150.

EXPERIENCES IN THE AIRTIGHTNESS OF RENOVATED TERTIARY EXEMPLARY BUILDINGS IN THE BRUSSELS CAPITAL REGION

Bram De Meester*[1], Thibaut Hermans[2] and Hendrik-Jan Steeman[1]

1 ARCADIS Belgium
pp Brussels Environment
Rue Royale 80
Brussels, Belgium
**Corresponding author:*
b.demeester@arcadisbelgium.be

2 Brussels Environment
Gulledelle 100
Brussels, Belgium

ABSTRACT

In the "Exemplary Buildings" program of the Brussels Capital Region, building owners and designers are challenged to realise building projects of both high architectural quality and superior environmental performance. After a project competition phase in which the Exemplary Buildings are selected, winning projects are supported by grants and expert guidance throughout further design development and construction. Building envelope airtightness is an important aspect during the follow-up, given its influence on the net energy demand.

This article describes through cases and general trends, the airtightness of renovated non-residential buildings. Several cases studies demonstrate that the resulting airtightness for renovated non-residential buildings, can be as low as the current standards for new-built buildings. The case studies are supported by photographs and technical information, showing the design approach and construction methods which were applied to obtain good airtightness. Airtightness testing will be discussed for the studied cases. This is particularly interesting as intermediate testing has been undertaken in order to enable airtightness improvement works well before final building acceptance. These tests also allow to determine the importance of some of the imperfections before they were mitigated.
General trends can be made by analysing the dataset of renovated tertiary buildings, realised within the Exemplary Buildings program. Keeping in mind that these projects aimed – and were supported – to reach superior environmental performance, the results show that it is possible to obtain excellent airtightness levels when renovating tertiary buildings. Although the dataset for this particular type of buildings is small, overall results of the Exemplary Buildings program indicate a learning effect throughout the years.

The cases and general trends demonstrate that excellent building airtightness can be realised when renovating non-residential buildings, when a coherent design approach, building methods and follow-up during construction are maintained. These results are important, given the necessity of renovating existing building stock in Brussels and the energy performance legislation evolutions, proving feasibility of gradually moving towards airtightness performance requirements similar to passivehouse criteria.

KEYWORDS

Renovation, airtightness, tertiary buildings, practical experiences

1. INTRODUCTION

The Brussels Capital Region started the Exemplary Buildings program in 2007 to illustrate leverage the building and renovation of buildings with high environmental performance. The aim was to demonstrate that both high architectural quality and superior environmental performance could be reached, even if financial means are limited. After a project competition phase in which the Exemplary Buildings are selected, winning projects are supported by grants and expert guidance throughout further design development and construction. The competitions of 2007, 2008, 2009, 2011, 2012 and 2013 yielded 243 projects of all sizes (Deprez et al, 2012), of a total built surface of over 621.000 m² which were supported by 33 million euros in total, according to the Brussels Environment administration (Brussels Environment, 2014). Over the years, the required levels of performance set forward as an ambition in competition phase were elevated, in order to challenge the market further.
To illustrate the leverage effect of this program, it suffices to mention that in 2007 no building in the Brussels Capital Region answered to the passivehouse criteria, while the Region will have 350.000m² of passivehouse compliant buildings by 2017 thanks to the program.

Part of the program criteria and the passivehouse requirements is the respect of a very high airtightness standard. Next to being key to come to a low energy demand, airtightness also has considerable influence on the thermal and acoustical comfort of the future building owners, by avoiding draught and limiting airborne noise nuisance, which is particularly interesting in an urban environment. However, during technical design and construction works, the aspect of airtightness is an element raising a considerable number of issues. Certainly in the case of renovation, where existing building components with unknown airtightness performance are reused and possibilities in terms of joints and connections between components are limited.

This article describes, by explaining two case studies and an analysis of the exemplary buildings data, the airtightness of renovated tertiary buildings in particular.

2. CASE STUDY 1: EUROPEAN FOUNDATION HOUSE

2.1 Project description

The European Foundation House project (hereafter referred to as EFH), was an existing office building built in 1988, which is renovated and currently occupied by the European Foundation Center (EFC), an international membership association of foundations and corporate funders promoting maecenatism. The 5-story building houses a conference room, meeting rooms, exposition area, office floors (floors 2 to 4) and a cafeteria.

It was decided to maintain the existing façade of natural white stone (see Figure 1) and the existing concrete structure, including the roof structure. Windows and doors were replaced. In order to obtain the targeted energy performance, the existing 6 cm of mineral wool wall cavity insulation were insufficient and internal insulation had to be added. 20cm of insufflated cellulose were added and an interior finishing of a wooden panel and plasterboard was applied. Extensive research was done on thermal bridging and hygrothermal effects, which were reported by Roger France (2013). More details can be found in Table 69 below.

Figure 1: Façade of the EFH building during an airtightness measurement

2.2 Airtightness approach

Airtightness measures taken were taping of joints of panels (Figure 2), use of airtight membranes around the new windows (Figure 3), the implementation of a motorized ventilation grill in the elevator shaft (Figure 4). An intermediate airtightness test has been realised in an upper corner of the building using a temporary wall, which gave the opportunity of testing both exterior wall and roof (Figure 5). These elements are illustrated in the photographs below. As a conclusion of this intermediate and partial testing, it was found that the airtightness membranes around the window frames were able to minimize infiltration, only minor infiltrations in the corner joints were experienced. It was also shown that the bored holes to screw the metal studs to the concrete did not endanger the airtightness. It was difficult to quantitatively predict the airtightness of the building, given the presence of the (not entirely airtight) temporary wall and its temporary joints with walls and roof. In Belgium, airtightness is usually quantitatively characterized as a n_{50} value, which is the measured infiltration rate (in m^3/h) at a pressure difference of 50 Pa between interior and exterior divided by the interior heated volume (in m^3) of the building. As this worst case intermediate testing yielded a n_{50} value of about 1 (h^{-1}) for a building part with several types of walls, corners and joints, it was deemed possible to obtain the same value for the rest of the envelope.

A second and third intermediate testing were done on the entire heated volume of the building (internal volume of about 5750 m^3), which were not yet conclusive as the entrance door was missing, some of the automatic drop seals of the doors appeared to not be adjusted yet, the motorized grills in the elevator shafts was not responding correctly to building controls, etc. For some of the issues, a test was done before and after closing a certain opening by tape or membranes, in order to evaluate the impact on the final airtightness.

Figure 2: Airtightness tapes on the panel joints - Figure 3: Membranes around window frame edges

Figure 4: Motorized grill for ventilation shaft - Figure 5: intermediate testing in a corner of the top floor

2.3 Results and lessons learned

The final figure of the finished building yielded a n_{50} value of 0,91 h^{-1}, which is considerably better than the 1,5 h^{-1} originally put forward as a target by the design team during the Exemplary Building competition stage. Given the availability of intermediate results, the influence of certain elements can be estimated. Firstly, the effect of the motorized grills in the elevator shaft can be estimated. These grills are opened whenever there is a need for ventilation in the elevator shaft (elevator movement, rising temperatures, smoke evacuation), but in general they are closed. This is a considerable reduction of infiltration compared to the standard grills which are opened all the time. During the test, the motorized grills can be closed according to the technical guidelines for airtightness testing within the framework of energy performance legislation (Three Belgian Regions, 2013). During one of the intermediate tests, the grill was tested in opened state. At that moment, the blowerdoor was unable to build up the necessary 50 Pa for official testing, and only about 30 Pa was reached at an maximal airflow through the blowerdoor of about 8000 m³/h. After closing the grill with a membrane, the measured airflow at 50 Pa was about 6600 m³/h. Assuming a square root dependency of airflow to pressure difference, the influence of the motorized grills on the n50

value is 0,3 h^{-1}, corresponding to an airflow of 1800 m^3/h at 50 Pa. This proves the importance of this kind of device to achieve airtight buildings.
A second influence was the effect of the air-water heat pump. The air intake of this device was combined with the air intake of the ventilation air handling unit. While the latter can be sealed off during testing according to the method A of the European norm (NBN, 2001) which is to be used for official testing in Belgium (Three Belgians Regions, 2013), this is not the case for the former. As a result, this combined air intake had to be left open during measurement, giving rise to infiltration via the metal casing of the heat pump device. By measuring both with opened and closed air intake, this effect could be quantified to an airflow of about 550 m^3/h at 50 Pa, corresponding to a difference of about 0,1 h^{-1}on the n_{50} value.
Thirdly, the infiltration airflow through the chinks around doors has been evaluated. By taping all around the doors (in this case, the double entrance door and 3 doors in the basement), it could be established that an airflow of about 700 m^3/h at 50 Pa passed through these openings around the – at that time unfinished – doors. This shows the importance of taking the necessary measures (automatic drop seals, flexible draught strips, brushes, filling up voids around door locks) for doors.

Table 1: Key data on the case studies

	European Foundation House (EFH)	**Police office Tritomas**
Building owners	European Foundation House	Zone de police 5342 : Uccle, Watermael-Boitsfort et Auderghem
Architects	GREENARCH architecture + environment sprl	bg&k associati
Engineers	MATRIciel, Delvaux	Crea-tec sprl
Contractors	Valens-Eiffage	CFE
Location	Brussels	Watermael-Boitsfort
Gross building area	2.229 m²	2.327 m²
Start of works	1/12/2012	1/1/2013
Completion of works	9/10/2013	1/1/2014
Energy demand for heating	15 kWh/m²/yr	24 kWh/m²/yr
Energy saving strategy	Interior insulation, window replacement, mechanical ventilation with heat recovery, air-water heat pump, use of radiant ceilings and thermal inertia of the structure	Interior insulation, window replacement, mechanical ventilation with heat recovery, PV panels, geothermal heating and cooling, chilled beams

3 CASE STUDY 2: POLICE OFFICE TRITOMAS

3.1 Project description

Converting a former telephone exchange building into a modern and sustainable police station, was the challenge of the Tritomas project. Given the modernist architecture, which can be considered as part of Brussels architectural partrimony, the choice was made to maximally preserve the building's exterior, shown in Figure 6. More details of the building are listed in Table 1.

Figure 6: Entrance of the Police Office Tritomas

3.2 Airtightness approach

The technical measures to come to an airtight building were very similar to the first case study, with the use of internal insulation, new windows provided with airtight membranes,… Particular to this project, was the fact that part of the existing building was not touched during the renovation project, moreover, this was in use throughout the construction works by another user (grey zones in Figures 7 and 8). As a consequence, in order to establish good airtightness for the police station, not only external building envelope, but also internal walls and doors had to be airtight. In this project, an intermediate airtightness test was performed on one level (on the first floor, red zone in Figure 8) in order to check the façade concept (internal plasterboard finishing) and its joints. At the end of the project, a measurement was performed on the entire project.

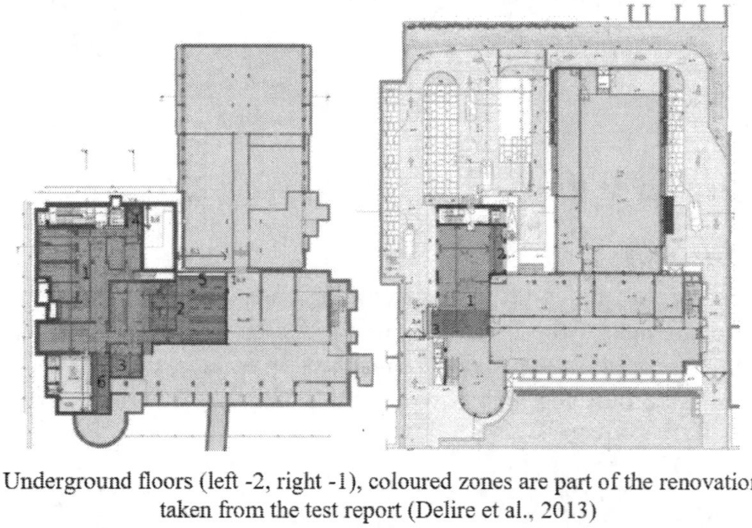

Figure 7: Underground floors (left -2, right -1), coloured zones are part of the renovation project, taken from the test report (Delire et al., 2013)

Figure 8: Building floors (left ground floor, right +1), coloured zones are part of the renovation project, taken from the test report (Delire et al., 2013)

3.3 Results and lessons learned

The result of the intermediate test showed clearly that the applied concepts of façade insulation and window-wall joints for the tested floor were sound, yielding a n_{50} of 0,73 for the tested volume. However, during the testing of the entire building, interaction with the untouched building volume influenced the results. Indeed, during the testing, all doors within the tested volume were opened to enable internal air circulation during the test, and doors to the untouched volume were closed, as shown in Figure 9. As the previous case study showed, doors can influence the airtightness to an important extent. Even when care is taken during selection and installation of the doors – most of the doors were escape exits and had to comply to fire regulations as well – some infiltration losses are inevitable. Given the complex geometrical interface between the tested volume and the untouched volume and the number of doors, the final result of 1,08 h^{-1} is a very good result, compared to the initial target of 1.5 h^{-1}.

Figure 9: Opened and closed internal doors on the first floor, taken from the test report (Delire et al., 2013)

4 GENERAL TRENDS

The dataset of the Exemplary Buildings currently contains information on the measured airtightness of 26 completed tertiary Exemplary Buildings, of which 11 buildings are renovation projects. The entire Exemplary Buildings dataset considers residential buildings as well, however they are not considered in the scope of this article. Also, the design and construction time of tertiary buildings, generally several thousands of square meters, overspans several years, explaining the limited number of completed buildings.

Figure 10 shows the range of building airtightness values measured in the completed buildings, for both new built and renovation projects. It is expected and clearly visible that new built projects have better airtightness, however, our case studies (boxed in the graph) and other renovation projects show that also for renovation projects low n_{50} values are feasible.

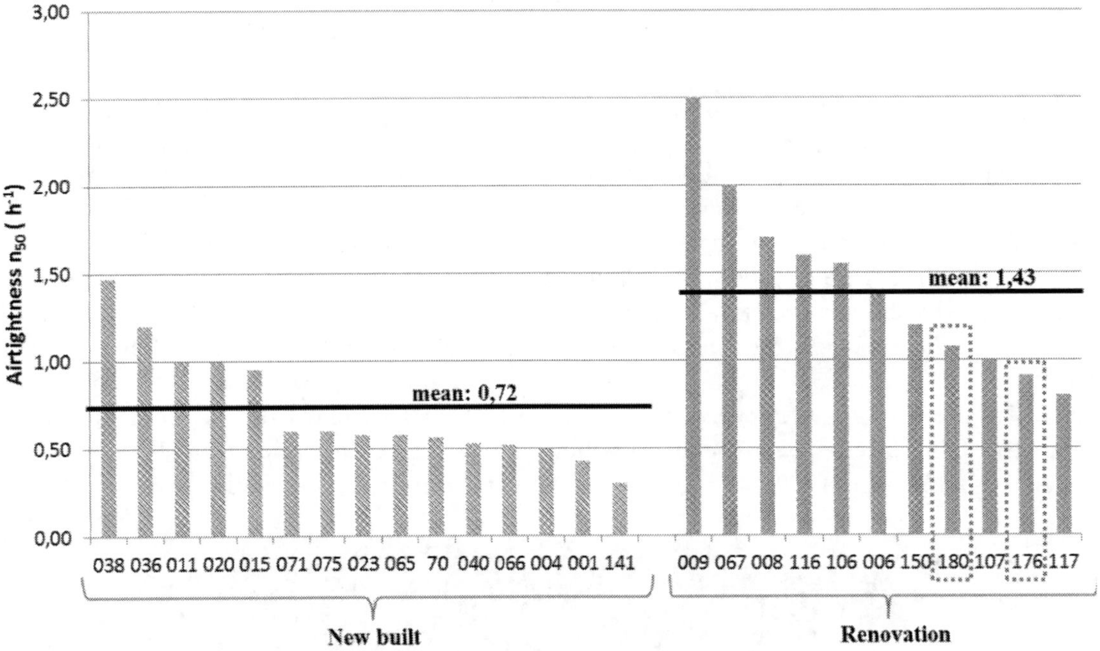

Figure 10: Range of measured building airtightness for new built and renovation in the Exemplary Buildings program. Case studies are indicated in the dotted boxes (case study EFH = number 176, Tritomas = number 180)

Although the dataset for this particular type of buildings is relatively small – too limited to perform full statistical analysis – the overall results of the Exemplary Buildings program indicate a learning effect throughout the years, which is shown in Figure 11. As a complement of information, this effect can also be noticed in the dataset for new built Exemplary projects in Figure 12.

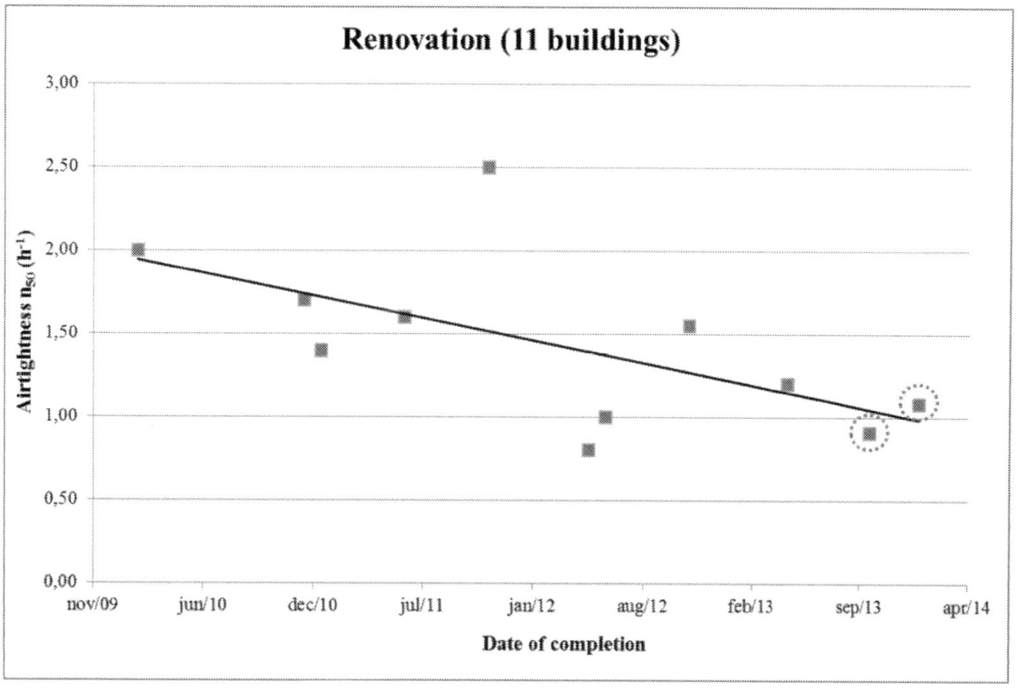

Figure 11: Evolution of airtightness of renovated tertiary buildings in time, dataset of the Exemplary Buildings program

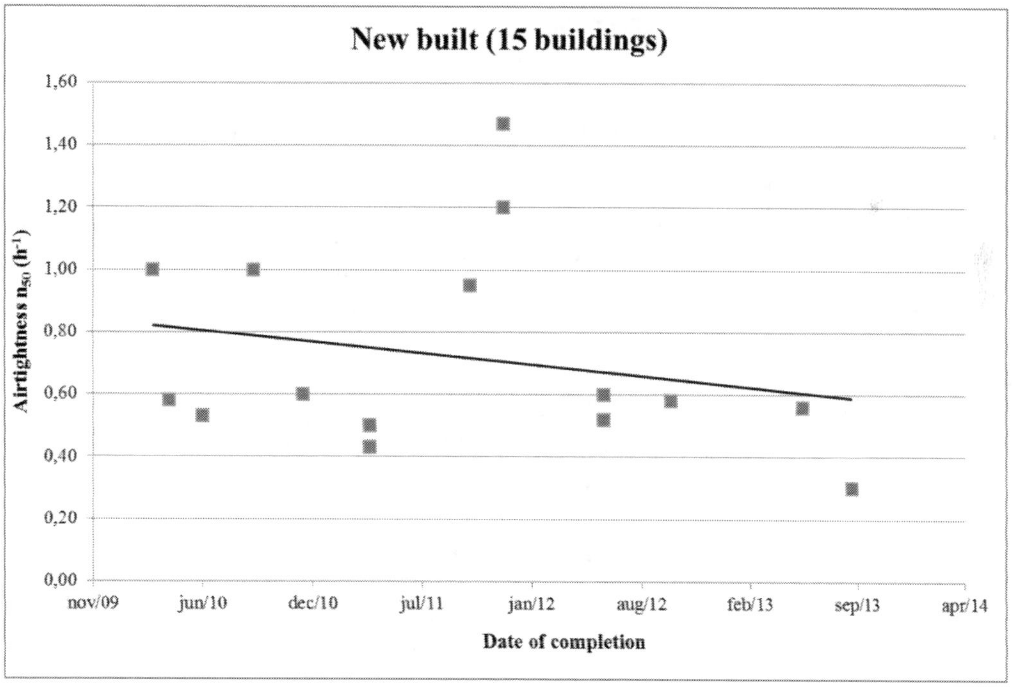

Figure 12: Evolution of airtightness of new built tertiary buildings in time, dataset of the Exemplary Buildings program

5 CONCLUSIONS

The airtightness is an important factor in the energy demand of buildings, but it is a rather challenging aspect of building performance, certainly in the case of renovation projects. The Exemplary Buildings program of the Brussels Capital Region provides – through case studies and the compiled data of realised buildings – a unique insight in the feasibility of high airtightness performance in renovated tertiary buildings. This article proves that – even if a difference remains with new built projects, good to excellent airtightness can be realised in renovated tertiary buildings.

6 ACKNOWLEDGEMENTS

This paper was supported by Brussels Environment, Energy department.
The authors wish to acknowledge the building owners and designers of the cited Exemplary Buildings: for the Rue des Tritomas project: Zone de police 5342 : Uccle, Watermael-Boitsfort et Auderghem (building owner) and bg&k associati (architects); For the Rue Royale 94 project: European Foundation House (building owner) and GREENARCH architecture + environement sprl (architects).

7 REFERENCES

Brussels Environment (2014). http://www.bruxellesenvironnement.be/Templates/Professionnels/niveau-dossiers.aspx?maintaxid=11660&taxid=11706&langtype=2060, (consulted June 23th, 2014).

Delire, J., Bouche, S. (2013) Rapport de contrôle – étanchéité à l'air, chantier Tritomas, DEL Thermographie, 25 September 2013, internal document.

Deprez, B., Cech, J. (2012). Het verhaal achter de voorbeeldgebouwen (in Brussel). Racine, EAN 9782873867997.

NBN – Bureau for Standardisation (2001). NBN EN 13829: Thermal performance of buildings - Determination of air permeability of buildings - Fan pressurization method (ISO 9972:1996, modified).

Roger France, J. (2013). The European Foundation House: The first passive office building renovation in Belgium, Proceedings of PLEA 2103, Munich, Germany, 10-12 September 2013. Available on http://plea-arch.org/plea-proceedings/, (consulted June 23th, 2014).

Three Belgian Regions - the Flemish region, the Walloon region and the Brussels-Capital region (2013). Bijkomende specificities voor de meting van de luchtdichtheid van gebouwen in het kader van de EPB-regelgeving, versie 3, 28 mei 2013. Available on http://www.epbd.be/index.cfm?n01=air, (consulted June 23th, 2014).

MONITORING THE ENERGY- & IAQ PERFORMANCE OF VENTILATION SYSTEMS IN DUTCH RESIDENTIAL DWELLINGS

Rob C.A. van Holsteijn, William L. Li,

Van Holsteijn en Kemna BV (VHK)
Elektronicaweg 14
2628 XG Delft, Netherlands
Consortium member and project manager MONICAIR
e-mail address: r.van.holsteijn@vhk.nl

ABSTRACT

MONICAIR --MONItoring & Control of Air quality in Individual Rooms-- is a pre-competitive field research project of a broad consortium of Dutch ventilation unit manufacturers and research institutes, supported by the Dutch government. The aim is to investigate the indoor air quality (IAQ) performance and energy characteristics of 9 different mechanical ventilation solutions in dwellings that meet strict air-tightness standards and comply with current building regulations. Over a full year 62 residential dwellings were monitored, with in each habitable room sensors for occupancy, CO_2, relative humidity and air temperature. Energy consumption of the mechanical ventilation units and space heating boilers were also continuously monitored. The ventilation solutions included mechanical exhaust ventilation (MEV) systems (mechanical exhaust in the wet rooms, and with natural air supply in habitable rooms), systems with mechanical exhaust in all rooms, as well as local and central balanced (MVHR) systems with various types of control systems. Sampling and data evaluation were handled by specialists. The monitoring part of the MONICAIR project is nearing completion and the first results are now available.

The data show that (from the ventilation systems under investigation) systems with *mechanical* air supply and/or exhaust provisions for each individual room are more robust in maintaining certain IAQ-levels, whereas systems with only *natural* air exchange provisions in the habitable rooms show larger variations in IAQ and depend to a large degree on the number and behaviour of inhabitants. In dwellings with MEV-systems the CO_2 concentration in the average habitable room is 380 ppm above the limit value of 1200 ppm during 10% of the heating season (515 of 5088 hours). If inhabitants spend only half of the time at home, then 20% of this time the air exchange is insufficient. Looking at individual houses these percentages vary from 2% to over 50%. In dwellings with MVHR systems the CO_2 concentration in the average habitable room is 320 ppm above the limit value of 1200 ppm during 4.6% of the heating season (237 of 5088 hours). Again, if inhabitants spend half of the heating season indoors and at home, 9.2% of these hours have insufficient air exchange. Per dwelling this figure varies from 0.1% to more than 20%.

Looking at the energy performance of the ventilation systems under investigation, monitoring data show that also in this respect significant differences occur. The average primary energy consumption for mechanical ventilation without heat recovery is 125 MJ/m^2. For the ventilation systems with heat recovery the average primary energy consumption is 22 MJ/m^2.

KEYWORDS

Ventilation systems, Monitoring, IAQ-performance, energy-performance

1. INTRODUCTION

The built environment accounts for approximately 40% of the EU total energy consumption with space heating as the largest contributor (>50%). Not surprisingly, EU-legislation focuses on increasing the insulation levels and improving the air tightness of buildings. As a result, the contribution of air infiltration to Indoor Air Quality (IAQ) is reduced and ventilation systems become more and more the sole responsible for the IAQ. Relevant question in the light of this development is: "How well do ventilation systems perform in terms of IAQ and energy consumption in well insulated and air tight dwellings *in real life,* and how can their performance be further improved?" Past research has focussed on compliance with standards at nominal conditions -- and improvements are certainly necessary there [1] – but research on IAQ and energy performance at real life conditions is definitely scarce [2,3]. While housing stock is being improved in terms of insulation and air tightness as we speak, we do not know whether ventilations systems sufficiently perform and are capable to compensate for the reduced infiltration.

To answer these key questions, a consortium of manufacturers and specialized consultancy firms from the Dutch ventilation industry was established that initiated the long term monitoring project "MONICAIR" (MONItoring & Control of Air quality in Individual Rooms). MONICAIR is a 1,6 million euro research project, partly financed by the consortium members and partly by the Dutch Ministry of Economic Affairs within the framework of TKI (Top consortia for Knowledge & Innovation). One of the aims of the monitoring project is to assess the real life IAQ performance in all individual rooms during presence of the occupants and the related primary energy consumed by the ventilation system. The assumption is that with a full analysis of the acquired monitoring data, system parameters can be identified that can help the participating manufacturers further improve their ventilation systems on both the IAQ- and energy performance. This paper reports the first results from a full year monitoring, covering nine commonly used ventilation systems in Dutch residential dwellings. Before the start of the monitoring project, all ventilation systems were checked and – when necessary - adjusted according to building code practice, thus securing correct system specifications and ensuring that possible design- or installation errors do not influence performance.

2. SELECTION OF VENTILATION SYSTEMS AND DWELLINGS

Ventilation systems

The manufacturing partners in the MONICAIR consortium produce both type C (MEV) and type D (MVHR) ventilation systems. Most commonly used in the Netherlands is ventilation system type C (mechanical extraction in wet rooms combined with natural air supply vents in habitable rooms). Ventilation systems type D (mechanical air extraction and mechanical supply) are applied in the new built sector when the budget allows for more expensive ventilation systems. Both types of ventilation systems are selected for the monitoring study, both with their specific variants and combinations as described in Table 1. The table describes per type of ventilation system 1) the air exchange provisions in both the wet rooms (bathroom, kitchen, toilets) as well as the habitable rooms (living room, bedrooms), 2) whether heat recovery is applicable and 3) what type of controls are used for the exhaust and supply provisions. The system type numbers refer to classification used in the Netherlands Standard NEN 8088-1, 2011 [4].

Table 1: Type of ventilation systems selected for MONICAIR

System type	Section of house that is served	Air exchange provisions			Controls	
		Exhaust	Supply	Heat Recovery	Exhaust	Supply
Type C ventilation systems						
C.1	Whole house	Mech. extraction in wet-rooms	Stnrd nat.supply vents in hab.rooms	No	Manual 3-pos. switch	Manual
C.2c	Whole house	Mech. extraction in wet-rooms	wind contrl. nat. supply in hab.rooms	No	Manual 3-pos. switch	Manual
C.4a	Whole house	Mech. extraction in wet-rooms	wind contrl. nat. supply in hab.rooms	No	CO_2-sensor in living room	Manual
C.4c	Whole house	Mech. extraction in all rooms	wind contrl. nat. supply in hab.rooms	No	CO_2-control on all hab.rooms	Manual
Type D ventilation systems						
D.2	Whole house	Mech. extraction in wet-rooms	Mech. supply in hab.rooms	Yes	Manual 3-pos. switch	
D.5a	Whole house	Mech. extraction in wet-rooms	Mech. supply in hab.rooms	Yes	Manual 3-pos. switch with CO_2-contrl in 2 zones	
D.5b	Whole house	Mech. extraction in all rooms	Mech. supply in hab.rooms	Yes	CO_2 and RH -controlled ventilation rate in all rooms	
Combination of systems						
X1/C	Living section: D	Mech. extraction in hab.rooms	Mech. supply in hab.rooms	Yes	CO_2 and RH -controlled ventilation rate in living room	
	Sleeping section:C.2c	Mech. extraction in wet-rooms	wind contrl. nat. supply in bedrooms	No	Manual 2-pos. switch	Manual
X1/A	Living section: D	Mech. extraction in hab.rooms	Mech. supply in hab.rooms	Yes	CO_2 and RH -controlled ventilation rate in living room	
	Sleeping section: A	Nat. extraction in wet-rooms	wind contrl. nat. supply in bedrooms	No	No control	Manual

Dwellings

Selecting and finding suitable dwellings for the monitoring project proved to be challenging and could not have been done without the help of the housing corporations. The first requirement was to find groups of 5 to 6 identical clustered terraced houses with an air tightness of $q_{v;10} \leq 1.0$ dm^3/s.m^2, that all have the same system, which also had to be one of the ventilation systems selected for the project. Initially the air tightness requirements were set at a higher level, but blower door tests proved that – although dwelling specifications stated otherwise – these higher levels of air tightness were not always achieved. The second requirement was to find inhabitants that were prepared to allow sensors being installed in all rooms that monitor the IAQ and their behaviour for more than a year. The last requirement was that all dwellings needed to have the same condensing combi boiler that enable the monitoring of gas consumption for space heating and domestic hot water (dhw). In dwelling clusters that complied with the first two requirements but had a different combi boiler, the combi boilers were replaced. Eventually 62 families with as many dwellings were found prepared to participate in the MONICAIR project. Before the start of the actual monitoring, all ventilation systems were checked and adjusted to the latest building code requirements. The monitoring results can thus also be used to re-evaluate the existing building code requirements for ventilation.

3. DATA MONITORING SYSTEM

In every dwelling participating in the MONICAIR project the following sensors were installed:

- Sensors that monitor the CO_2 concentration, relative humidity and the air temperature in all habitable rooms (living room, kitchen, bedrooms, study and utility room (if applicable) every five minutes. As far as possible these sensors were mounted in the middle of the room approximately 1.5 m above the floor. The CO_2-sensor used is a non-dispersive infrared (NDIR) type sensor with an accuracy of ± 50 ppm.
- Presence sensors (PIR-type) in all habitable rooms.
- A sensor monitoring the relative humidity in the bathroom every five minutes.
- Sensors that monitor the power consumption of the mechanical ventilations units and kitchen hoods (and with it the corresponding airflow) every 10 minutes as well as every change within this period.

Apart from these room-specific sensors, the following appliances are monitored:

- The condensing combi boilers, monitoring the gas consumption for space heating and dhw through IM protocol, at least once in every six minutes.
- In dwelling clusters that use ventilation units equipped with rf-communication, their data on power consumption and airflow was also logged every five minutes.
- Finally, meteorological data on outdoor temperature, relative humidity, wind speed, wind direction and air pressure of the most nearby weather station was gathered.

Per cluster of dwellings all data of each house is gathered through rf-communication and stored on a local PC. Through an FTP connection the data stored on local PCs is regularly copied onto the centralized MONICAIR SQL database.

4. DATA ANALYSES

For a period of more than a year an enormous amount of data was gathered about the ventilation systems and their real life performance on indoor air quality and energy consumption, as well as data on consumer behaviour regarding the preferred temperature per habitable room, operation of ventilation units, use of kitchen hoods, and hot water consumption pattern. With these data various angles of data analysis are possible. Given the focus and resources related to the first part (WP1a) of the MONICAIR project, the analysis here is restricted to the IAQ- performance and the energy performance during heating season. At the moment four of the nine ventilation systems were analysed and will be discussed here.

Indicator for the IAQ performance

To assess the IAQ performance of ventilation systems the measured CO_2-concentration in the individual habitable rooms will be the leading parameter. The CO_2-concentration is generally accepted as the key indicator for the existing ventilation rate during presence [5] and consequently for the occurring IAQ levels. The following procedure is therefore used to assess the IAQ in the various habitable rooms:

a) Determine for each habitable room in each dwelling the number of hours per day that the CO2 concentration is above 1200 ppm in unit [hours/day]. The limit value of 1200 ppm corresponds to the IAQ category IV (= the lowest IAQ level) as described in EN 15251 [6].
b) Determine for each habitable room in each dwelling the average concentration with which the CO_2 limit of 1200 ppm is exceeded in unit [ppm/h].

c) Calculate the CO_2 excess dose per heating season in kppmh by multiplying the outcome of a) with the outcome of b) and then multiplying the result with the number of days in a heating season (212) and dividing it with 1000 to convert the ppm-figure to kppm.
d) Add all the CO_2 excess doses per habitable room per dwelling and divide it with the number of rooms of that specific dwelling to determine the average achieved IAQ level per dwelling.

(Note 1: The calculations for a) and b) are made with a resolution of five minutes and summarised to obtain either the number of hours per day or the excess value per hour. Note 2: This procedure is comparable to the methodology used in the declaration of equivalence of the VLA [7])

Indicator for the Energy performance

The indicator for the Energy performance of the ventilation systems is obtained according to the following procedure:

I) Determine the total average hourly mechanical ventilation rate of all ventilation units in the dwelling (kitchen hood excluded).
II) Determine the hourly average of the indoor temperatures and indoor humidity of the sleeping section and the living section separately and use these two values to calculate the average indoor climate conditions.
III) Determine the hourly average of the outdoor climate conditions, in casu air temperature, relative humidity and air pressure (data from the Royal Netherlands Meteorological Institute)
IV) Calculate the thermal energy exchange per hour on the basis of I), II) and III) in relation to the hourly temperature difference between indoors and outdoors ($\Delta T_{in\text{-}out}$).
For systems with heat recovery this energy figure is corrected with the heat recovery efficiency as determined according to EN 13141-7 and -8.
V) Calculate the total daily thermal energy exchange in relation to the daily average $\Delta T_{in\text{-}out}$.
VI) Calculate the total primary energy consumption for mechanical ventilation for an average heating season by multiplying the daily average thermal energy exchange (the value corresponding to the annual average $\Delta T_{in\text{-}out}$ of 13°C) by 212 days (duration of the heating season) and divide this figure by the average system efficiency of 85% (for condensing heating system with limited distribution losses); add to this figure the total power consumption of all ventilation units during heating season, after converting it to primary energy.
VII) Divide the outcome of VI) with the total heated surface area of the dwelling to obtain the average energy consumption of the mechanical ventilation system per m^2 during heating season.

Note : Energy consumption related to cross ventilation (air that enters the dwelling through infiltration or airing on one façade and leaves through another façade) is not included; also internal and solar gains are not included in this assessment of the energy performance.

5. RESULTS

IAQ-performance

So far, data of four ventilation systems have been analysed on IAQ performance and energy performance. The following graphs give an overview of the results.

Ventilation system C.2c

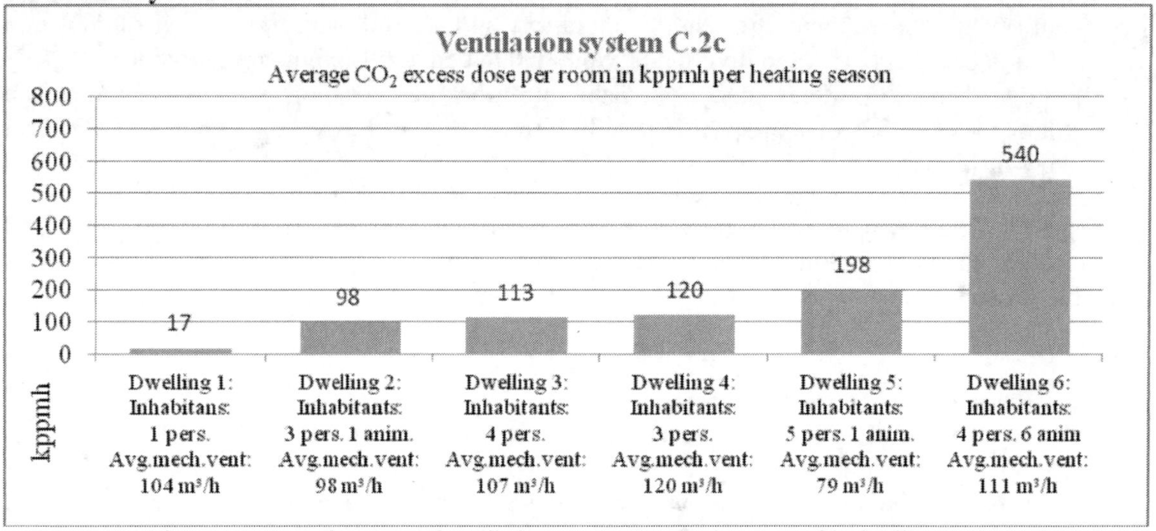

Figure 1. IAQ performance of ventilation system C.2c in six comparable dwellings

The arithmetic mean of the IAQ performance of these six houses with ventilation system C.2c is 181 kppmh. This average is very much influenced by dwelling number 6. The qualitative data that was gathered through face-to-face interviews before the start of the monitoring project however show, that there are valid reasons for the high kppmh figure. The intake interview reveals that, compared to the other dwellings, people in house nr. 6 stay at home for much longer periods per day, they keep the vents in the bedrooms closed during presence and airing of the room is less frequent than in the other houses.

Ventilation system C.4a

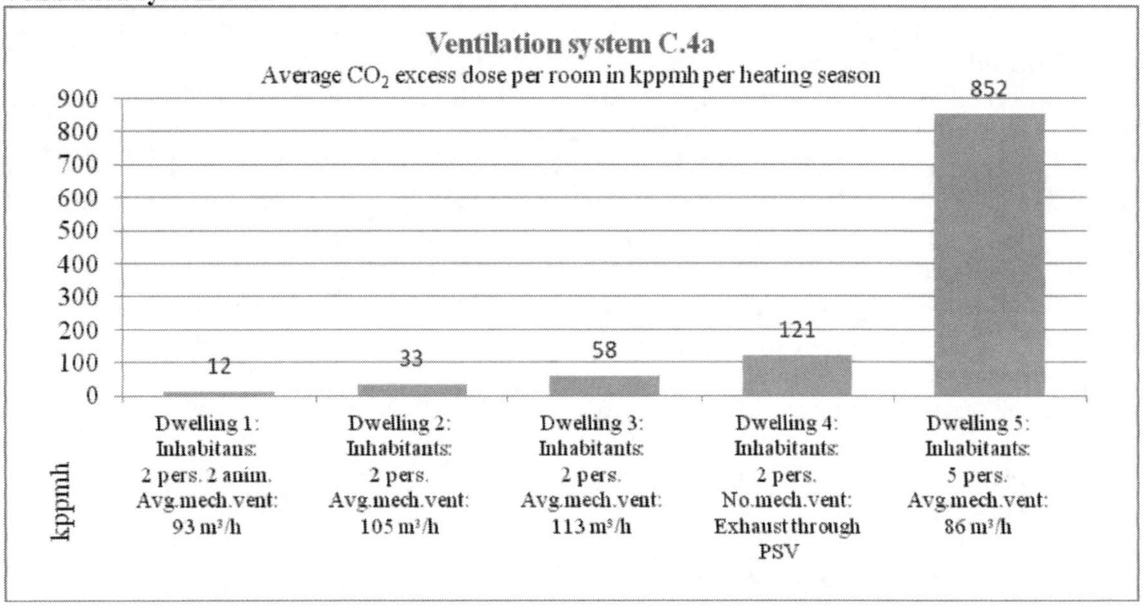

Figure 2. IAQ performance of ventilation system C.4a in five comparable dwellings

The arithmetic mean of the IAQ performance of these five houses with ventilation system C.4a is 215 kppmh. Without dwelling nr.4 (Passive Stack Ventilation (PSV) system) the mean value is 238 kppmh. Also in this cluster the average is largely influenced by dwelling nr. 5 where people also stay at home for much longer periods per day. The inhabitants state in the intake interview that the air vents in the bedrooms are open, but the CO_2 concentrations of the bedrooms suggest that they are closed when people are present.

Ventilation system D.2

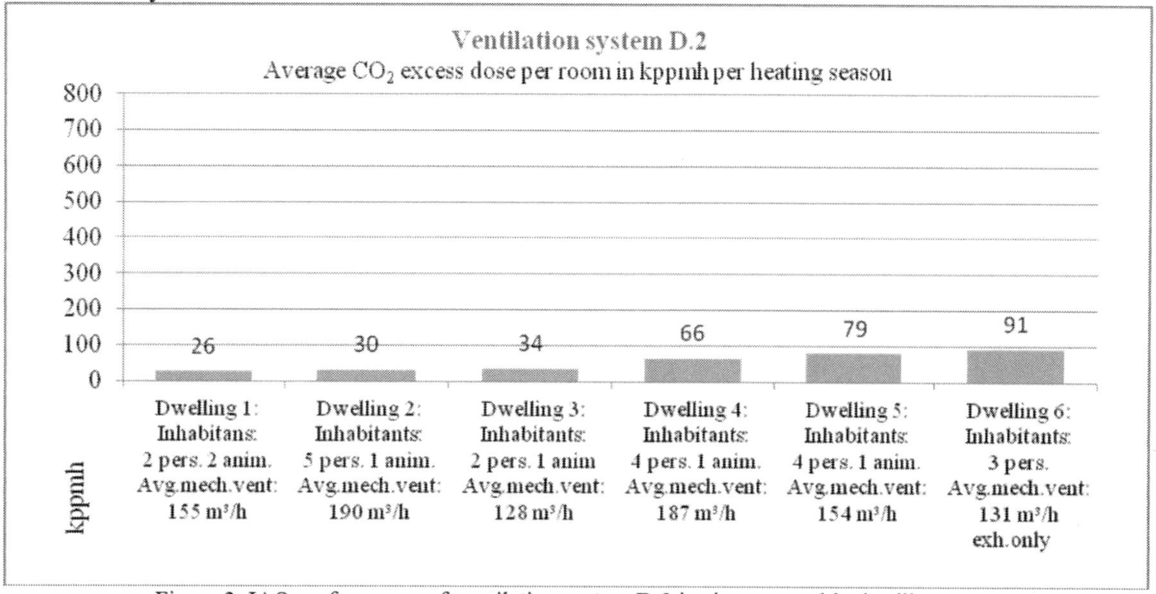

Figure 3. IAQ performance of ventilation system D.2 in six comparable dwellings

The arithmetic mean of the IAQ performance of these six houses with ventilation system D.2 is 54 kppmh.

Ventilation system D.5a

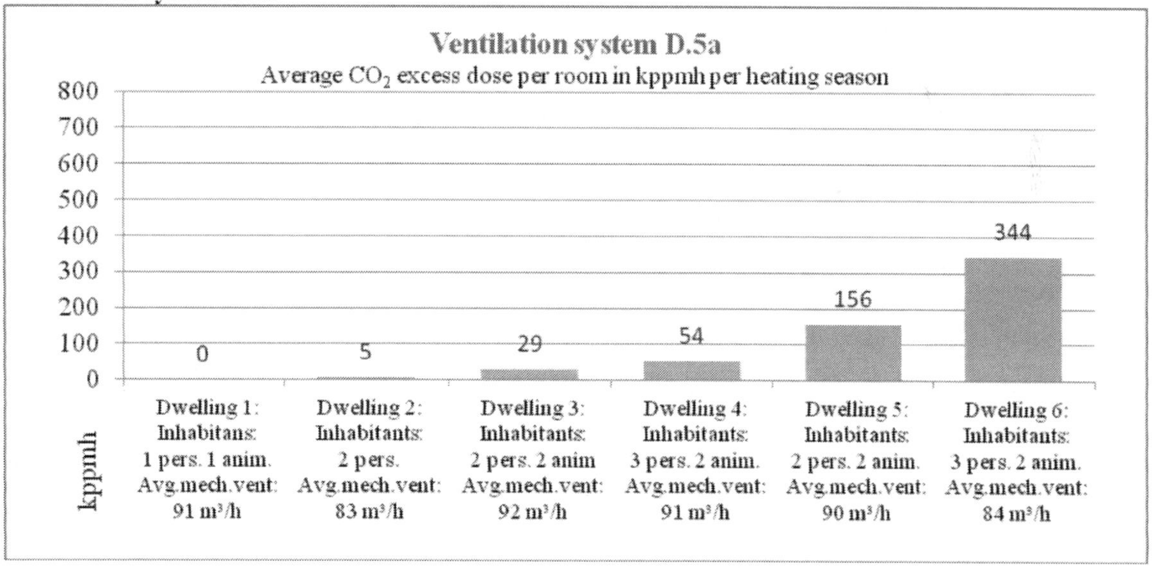

Figure 4. IAQ performance of ventilation system D.5a in six comparable dwellings

The arithmetic mean of the IAQ performance of these six houses with ventilation system D.5a is 98 kppmh. In dwelling 6 the size of the air supply opening in bedroom 3 was reduced after the start of the project, resulting in a higher CO_2-excess dose for that particular room and consequently for the whole dwelling.

Energy performance

The energy performance of each ventilation system is determined according to the method described under the paragraph 'Data analyses', and presented in the graph below. The electricity consumption of the mechanical ventilation units during a heating season is displayed separately from energy related to the air exchange.

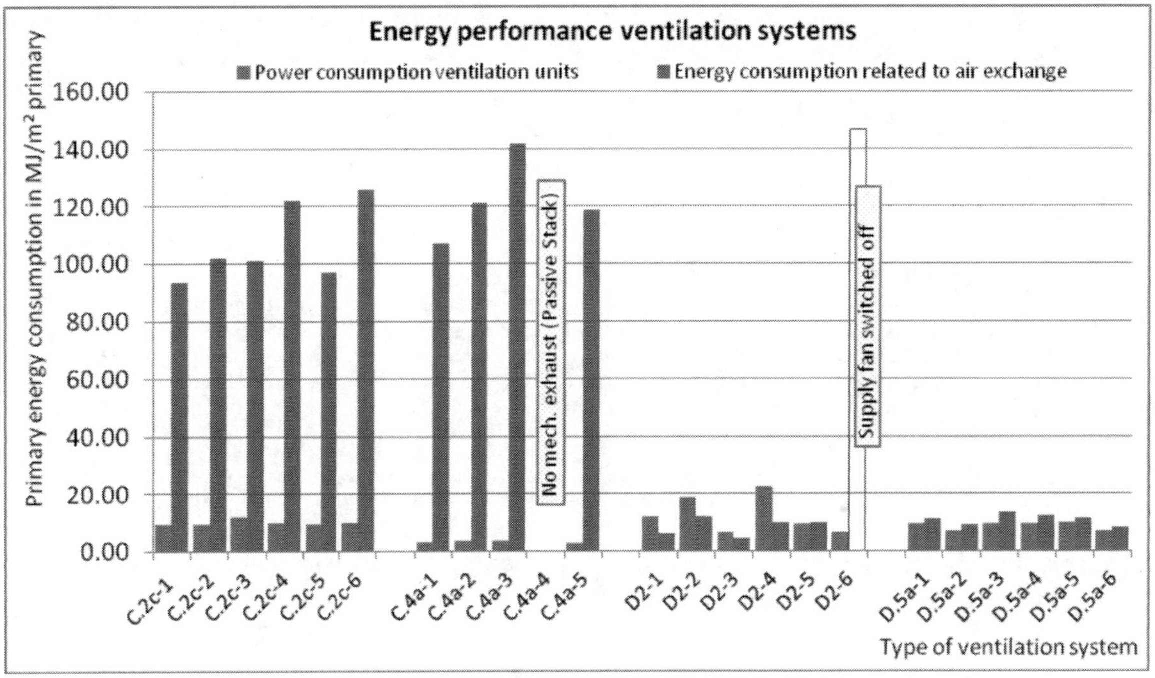

Figure 5. Energy performance of the ventilation systems

6. CONCLUSIONS

Data analyses will continue for the other five ventilation systems (thirty-nine dwellings), but based on the results of the four systems analysed so far, it can already be concluded that there are significant differences in both the real-life IAQ- and energy performance of ventilation systems.

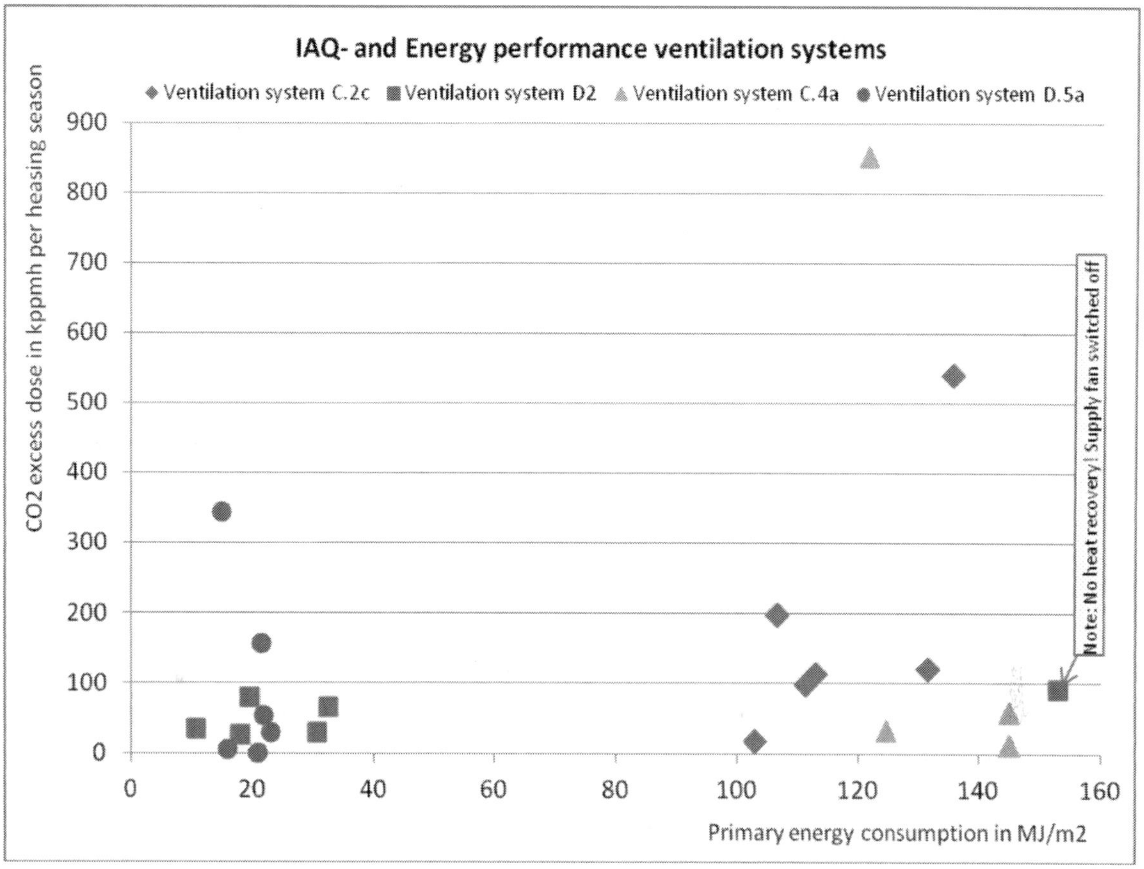

Figure 6. IAQ- and energy performance of the ventilation systems

The differences in energy performance are evident. For the average EU-heating season ($\Delta T_{in\text{-}out}$ = 13°C) the average primary energy consumption for ventilation is 22 MJ/m^2 for systems with heat recovery and 125 MJ/m^2 for systems without heat recovery. Further analysis will be necessary to determine if the use of IAQ-sensors has an effect on the energy performance.

The preliminary analyses furthermore suggest that systems with mechanical air supply and/or exhaust provisions in each room are more robust in maintaining a certain IAQ- performance compared to systems that use only natural air exchange provision in habitable rooms. Systems with natural air exchange provisions in habitable rooms achieve an average CO_2 excess dose of 196 kppmh in each habitable room per heating season, whereas systems with a mechanical component serving either the air supply and/or exhaust in each room achieve 76 kppmh.

In terms of time spent in rooms with insufficient air exchange, this means that:

- In dwellings with MEV-systems combined with only natural air exchange provisions in habitable rooms, the CO_2 concentration in the average habitable room is 380 ppm above the limit value of 1200 ppm during 10% of the heating season (515 of 5088 hours). Since inhabitants are not continuously at home, this percentage will be higher when related to the actual time spent at home. If the occupants are only 12 hours per day at home (2544 hours per heating season), then 20% (515/2544) of this time the air exchange is insufficient. Looking at individual houses these percentages vary from 2% over 50%.
- In dwellings with MVHR-systems the CO_2 concentration in the average habitable room is 320 ppm above the limit value of 1200 ppm during 4.6% of the heating season (237 of 5088 hours). Again, if inhabitants spend half of the heating season indoors and at home, 9.2% of these hours have insufficient air exchange. Per dwelling this figure varies from 0.1% to more than 20%.

7. DISCUSSION

Current practice is that ventilation systems are primarily selected on the basis of their purchase- and installation costs. Building codes determine the minimal level of air exchange rates that must be installed as well as its minimal energy performance level (EPBD-legislation). But whether ventilation systems actually deliver the required air exchange rates in the right rooms at the right times is not a topic or a criterion in selecting ventilation systems. More monitoring studies will certainly be necessary but the preliminary results argue in favour of an approach that looks more specifically at the *real-life* IAQ-performance of the various systems, especially with a building stock that is becoming more insulated and more air tight. Furthermore, like with all other energy using products, the energy performance of ventilation systems can only be compared between systems that have equal IAQ-performance. Without an IAQ assessment we are comparing apples with oranges. Further discussion on the method for determining the IAQ-performance will be necessary as well as discussions on the limit values and IAQ- classes than can be set as guidelines.

8. ACKNOWLEDGEMENTS

This paper was prepared within the MONICAIR project part WP1a. The project is partly financed by the consortium members and partly by the Dutch Ministry of Economic Affairs within the framework of TKI (Top consortia for Knowledge & Innovation). Particular thanks are due to the consortium members and their representatives involved in this part of the MONICAIR study:

Jelmer de Jong and Rudolf Tolj (Brink Climate Systems BV), Peter Schabos, Rutger Vasters and Rolf-Jan Hans (ClimaRad BV), Hans van Klooster and Roelof Ziengs (Honeywell Customized Consumer Products), Leon van Bohemen, Jan Rijnbeek and Douglas Budde (Itho Daalderop BV), Harm Valk (Nieman Raadgevende Ingenieurs), Wim Kornaat (TNO), Bart Cremers and Rick Bruins (Zehnder Group Nederland BV).

The MONICAIR consortium further owes thanks to Peter Cool (Intergas Verwarming BV) for his support on the data monitoring of the boilers, to Robèrt de Haas (Alparo) for his patience and support in developing the MONICAIR database and queries, to Wilbert van den Eijnde (Consumersvoice) for the face-to-face interviews with all the inhabitants and finally to the housing corporations for their support in the selection of dwelling for this part WP1a of the MONICAIR project: Barry Top (De Woonplaats), Gerben Heering (WBO) and Winfried van Emmerik (WSDH).

9. REFERENCES

[1] Janssens, A., Carrié F.R., Durier, F. ed. (2013). Securing the quality of ventilation systems in residential buildings: existing approaches in various countries, Edited proceedings of the AIVC-Tightvent international workshop, Brussels.

[2] Savin J.L., Bernard A., (2009). Improvement of the ventilation and building airtightness performance in occupied dwellings in France, 4th International Symposium on Building and Ductwork Air Tightness – BUILDAIR, Berlin.

[3] De Gids, W.F., Op 't Veld, P.J.M., (2004). Onderzoek naar ventilatie in relatie tot gezondheidsaspecten en energieverbruik voor een representatieve steekproef van het Nederlandse woningbestand.

[4] NEN 8088, (2011). Ventilation and infiltration for buildings – Calculation method for the supply air temperature corrected ventilation and infiltration are volume rates for calculating energy performance – Part 1: Calculation method.

[5] De Gids, W.F., Wouters, P., (2010). CO_2 as indicator for the indoor air quality, General principles, AIVC Paper N° 33.

[6] EN15251 (2007). Indoor environmental input parameters for design and assessment of energy performance of buildings addressing indoor air quality, thermal environment, lighting and acoustics.

[7] Vereniging Leveranciers Luchttechnische Apparaten, VLA, (2012). VLA methodiek gelijkwaardigheid voor energiebesparende ventilatieoplossingen in woningen.

DUCTWORK AIRTIGHTNESS: RELIABILITY OF MEASUREMENTS AND IMPACT ON VENTILATION FLOWRATE AND FAN ENERGY CONSUMPTION

Sylvain BERTHAULT*[1], Florent BOITHIAS[1], Valérie LEPRINCE[2]

1 CEREMA
Direction Territoriale Centre-Est
1 Boulevard Giberstein
71404 Autun, FRANCE
*Corresponding author: sylvain.berthault@cerema.fr

2 PLEIAQ
84 c Avenue de la Libération
69330 Meyzieu, FRANCE

ABSTRACT

Reduction of energy consumption and green house gas emissions of buildings is a great challenge in Europe. In this context French energy performance regulation, RT2012, requires an improvement of the buildings' airtightness. In airtight buildings, ventilation must be perfectly controlled to ensure good indoor air quality. However, many failures are observed when ventilation systems are inspected (Jobert, 2012). They are mainly due to bad conception, poor implementation and lack of maintenance. Most of the time, dysfunction leads to a reduction of the ventilation flowrates and to poor indoor air quality.
To improve the ventilation systems' reliability, French energy label "Effinergie +" imposes the ductwork airtightness to be at least Class A. This leads to various issues regarding ductwork airtightness:

- What is the impact of each component of a ductwork on the leakage rate? Is good airtightness reachable with existing products?
- What is the impact of airtightness on ventilation flowrate and fan energy consumption?
- Are ductwork pressurization tests reliable? Do the position of the measurement devices, leakage distribution and ductwork pressure drop have an impact on the result of the test?

To answer those questions a full-scale exhaust ductwork of multi-family building was set in our laboratory. The impact of each components of the ductwork (rigid, circular GALVA) was tested. Holes were then drilled in the ductwork to vary airtightness.
Our study showed that:

- Even with good implementation it is difficult to reach class C for ductwork airtightness because of some leaky components in the ductwork;
- A decrease of the airtightness (from class C to 1,6*class A) may either reduce flowrates by 10% at the air terminal devices or increase the energy consumption of the fan by 10 to 40% ;
- Neither the position of the fan, nor the leakage distribution, nor the pressure drop has an impact on the result of the airtightness test (at class C and 1,6*class A).

KEYWORDS

Ventilation – Airtighness – Ductwork – Implementation – Measurement

1 INTRODUCTION

The improvement of thermal regulation in France leads to more and more efficient and airtight buildings. Therefore, an issue which was forgotten till recently is becoming crucial: how to ventilate buildings well enough to ensure healthy indoor air (Kirchner et al., 2011) and durable building structure while keeping good energy performance?

To achieve such opposite targets, new mechanical ventilation systems were designed with growing complexity. These systems make it possible to meet both sanitary and energy targets provided that they are very well implemented and maintained.
However, literature shows that design, implementation and maintenance are often neglected (Jobert, 2012; Lucas et al., 2009; Save-Duct, 1999; Observatoire CSTB-ORTEC, 2009). Ventilation systems' quality therefore becomes a growing concern because of observed dysfunctions, which highlight a need for information and technical support to professionals together with control of the proper operation of these systems. The ductwork airtightness is one of the various implementation issues.
In such a context, this study aims to better understand:

1. The airtightness of ductworks and its impact on flowrates and fan consumption
2. The reliability of ductwork airtightness tests

by implementing a simple flux ventilation ductwork in our laboratory. This ductwork represents a small collective housing's system which is described in the following section.

2 DESIGN OF THE DUCTWORK STUDIED

The ductwork is made of two main columns which are supposed to connect to four dwellings of different sizes (figure 1 and 2). Each dwelling included a kitchen and a bathroom both connected to ventilation.
The ductwork was made of:

- Circular GALVA ducts
- Specific elements with joint such as curves, branch connections, dampers, T-connections, inspection hatches, conical reduction, ... (figure 3)
- A "C.VEC micro-watt +" extraction fan and automatic constant-flow regulated terminal devices

The extraction fan is a constant pressure one. This means that it automatically adapts its speed to maintain a constant pressure when the flowrate varies (when kitchen terminal devices open and close).

Figure 1: The ductwork studied

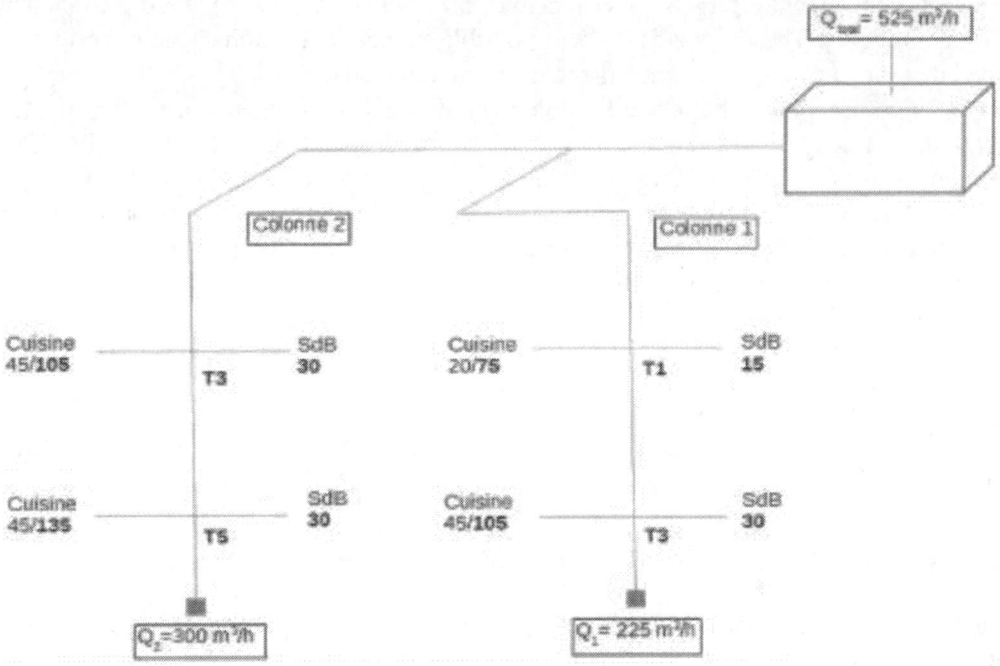

Figure 2: Outline of the ductwork

3 METHODOLOGY

3.1 Impact of airtightness

3.1.1 Impact of each specific element of the ductwork on airtightness

The aim was to quantify the leakages of each specific element of the ductwork (figure 3). The methodology was as follows:

1. Connect each specific element one by one
2. Measure airtightness after each element is connected without being screwed (20 measurements were carried out)
3. Measure airtightness after each element is connected and screwed (20 measurements were carried out)

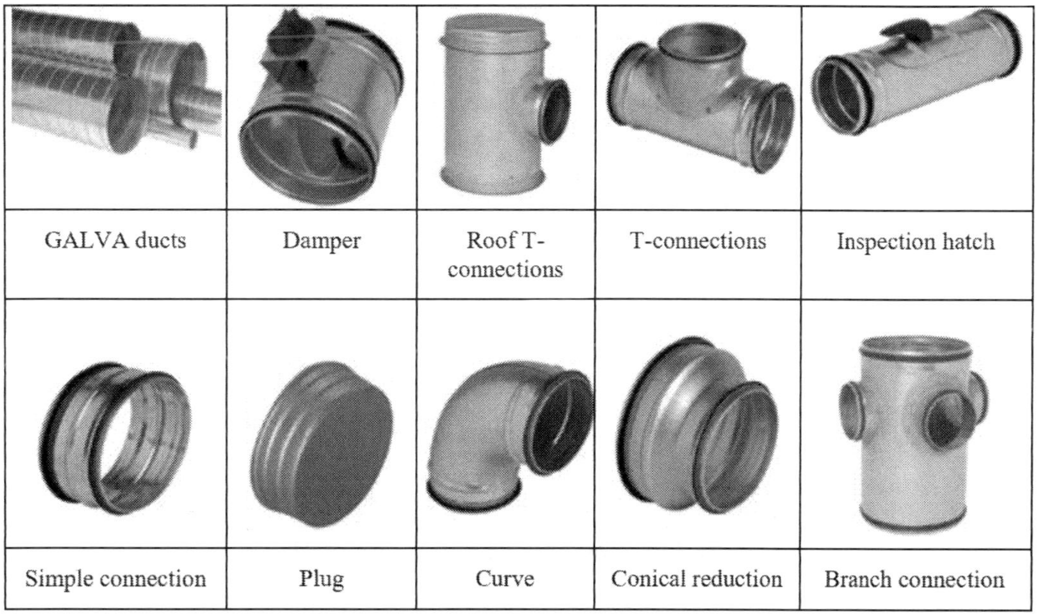

Figure 3: Specific elements tested in this study

3.1.2 Impact of the airtightness level on ventilation flowrates and fan energy consumption

The aim was here to quantify the fan's energy overconsumption due to a decrease of the ductwork's airtightness, and the associated impact on ventilation flowrates observed at air terminal devices.
The methodology consisted in measuring flowrates and the fan's energy consumption in the following conditions:

1. Ductwork airtightness: class C, B, A and 1.5 * class A (from class C, the ductwork was drilled to reach other classes)
2. Pressure setpoint of the fan: 90 Pa and 120 Pa

3.2 Reliability of ductwork airtightness test

3.2.1 Impact of pressure drop in the ductwork

The aim was to quantify the impact of pressure drop in the ductwork on the result of the airtightness test. To induce pressure drop, two dampers were placed in the ductwork (see figure 4). Measurements were carried out in the following configurations:

1. Ductwork airtightness: class C and 1.5 * class A
2. Dampers' position: 7 different positions from closed to open position

The pressure drop due to the damper was estimated by measuring the pressure variation at an air terminal device. Leakages were equally distributed along the ductwork.

3.2.2 Impact of leakage distribution and location of the measurement device

The aim was to quantify the impact of leakage distribution and location of the measurement device on the result of the airtightness test.
Four different leakage distributions inducing approximately the same airtightness level were tested:

1. Concentrated leaks close to the fan
2. Concentrated leaks at the bottom of columns 1 and 2
3. Concentrated leaks at the bottom of column 2
4. Equally distributed leaks all along the ductwork

Leakages were created by drilling.
Tests were carried out by placing the measuring device at three different locations:

1. By the fan
2. On column 1
3. On column 2

4 RESULTS

4.1 Impact of airtightness

4.1.1 Impact of each specific element of the ductwork on airtightness

Main leaks were observed at (figure 4):

- Roof T-connections
- Dampers

Even with careful implementation, the ductwork was only class B if the plug of the roof T-connection was not tightened with tape. With tape on the plug, the ductwork reached class C.
It was also observed that elements were slightly leakier when screwed than when just connected together.

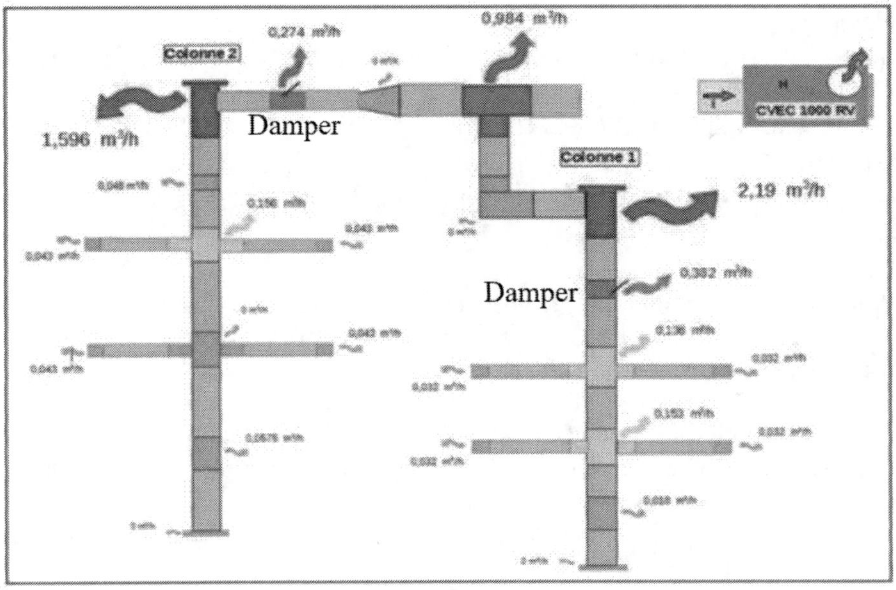

Figure 4: Leakage rates of specific elements

4.1.2 Impact of the airtightness level on ventilation flowrates and fan energy consumption

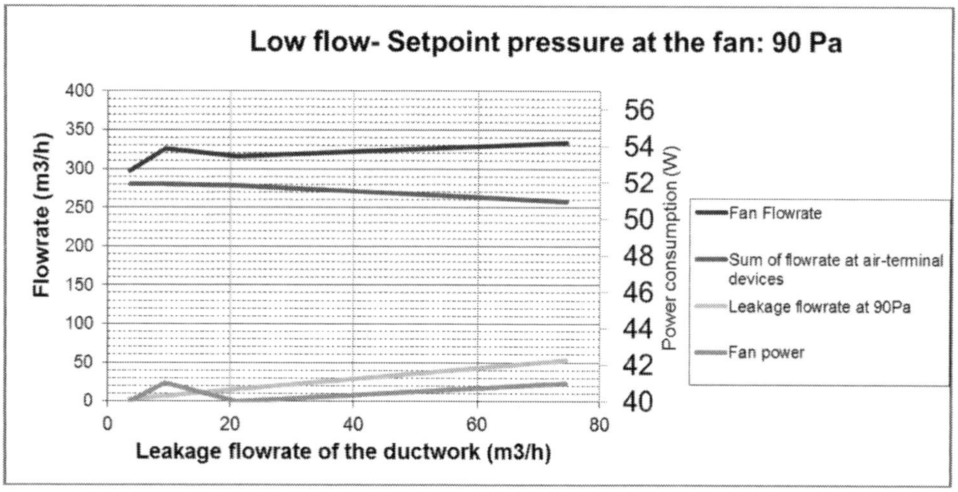

Figure 5: Operating conditions of the ductwork with airtightness varying from class C to 1.5 * class A (pressure setpoint at the fan: 90 Pa)

The first result, shown on Figure 7, is that flowrate targets are no longer met for 4 of the terminal devices for a 90 Pa pressure setpoint at the fan if the ductwork is class 1,5 * A, while it is met when it is class C. For this configuration, the leaky ductwork would lead to a lack of ventilation and maybe poor indoor air quality.

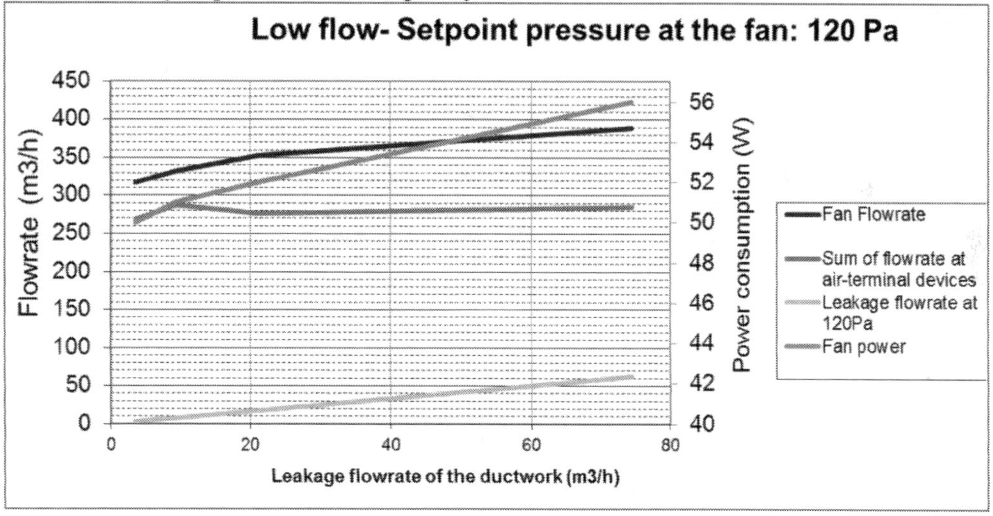

Figure 6: Operating conditions of the ductwork with airtightness varying from class C to 1.5 * class A (pressure setpoint at the fan: 120 Pa)

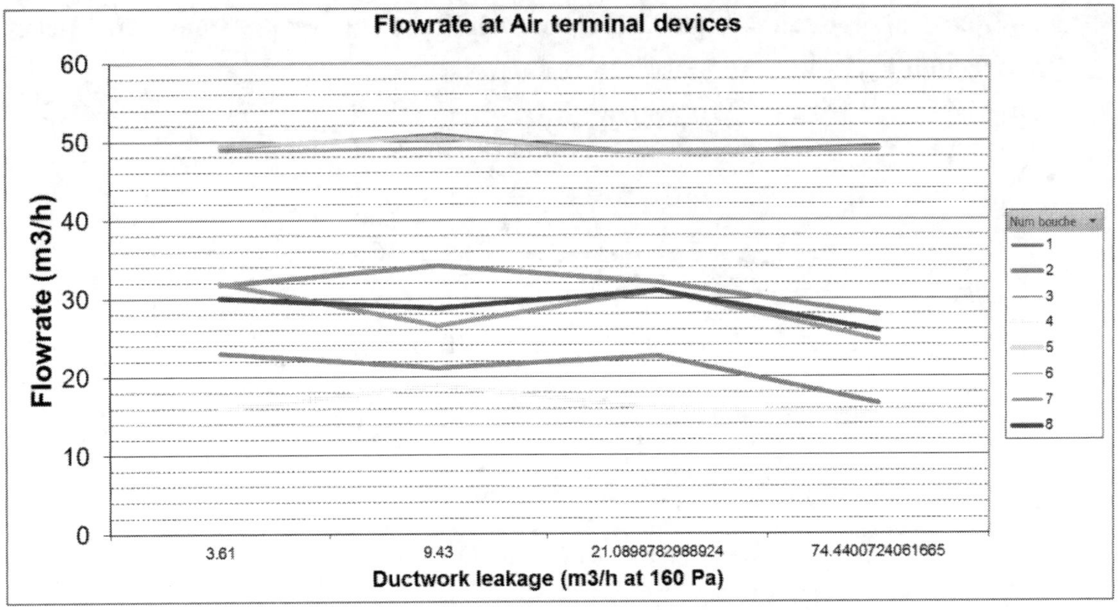

Figure 7: Flowrates at terminal devices with airtightness varying from class C to 1.5 * class A

Flowrate targets:

- Terminal devices 1, 5 and 7: 45 m^3/h
- Terminal devices 2, 6 and 8: 30 m^3/h
- Terminal device 3: 20 m^3/h
- Terminal device 4: 15 m^3/h

Thus, to maintain the flowrate target at air terminal devices, when the airtightness of the ductwork decreases, it is necessary to change the setpoint of the fan from 90 Pa to 120 Pa. We can see from figures 5 and 6 that it leads to an increase of the fan's power from 40 W to 50 W (comparing with class C consumption).
Moreover, when airtightness varies from class C to 1.5 * class A and the fan is set at 120 Pa, the fan's energy consumption increases by 12%.
Therefore, for this configuration, poor airtightness can either lead to an overconsumption of fan of 40% (16W) or to flowrates not complying with the regulation.

4.2 Reliability of ductwork airtightness test

4.2.1 Impact of pressure drop in the ductwork

Even if the damper's position induces a high pressure drop in the ductwork, as can be seen in figure 8, it does not significantly affect the result of the ductwork airtightness test, as shown in figures 9 and 10, at airtightness levels of class C and 1.5 * class A.
For this ductwork, at class C, the damper's position has no impact at all (see figure 10). At 1.5 * class A, the results of the test is 15% smaller when the two dampers are closed compared to when they are open.
An interpretation could be that it depends to the tightness of the damper with respect to the tightness of the ductwork. A mathematical model could be set to explain this phenomenon.

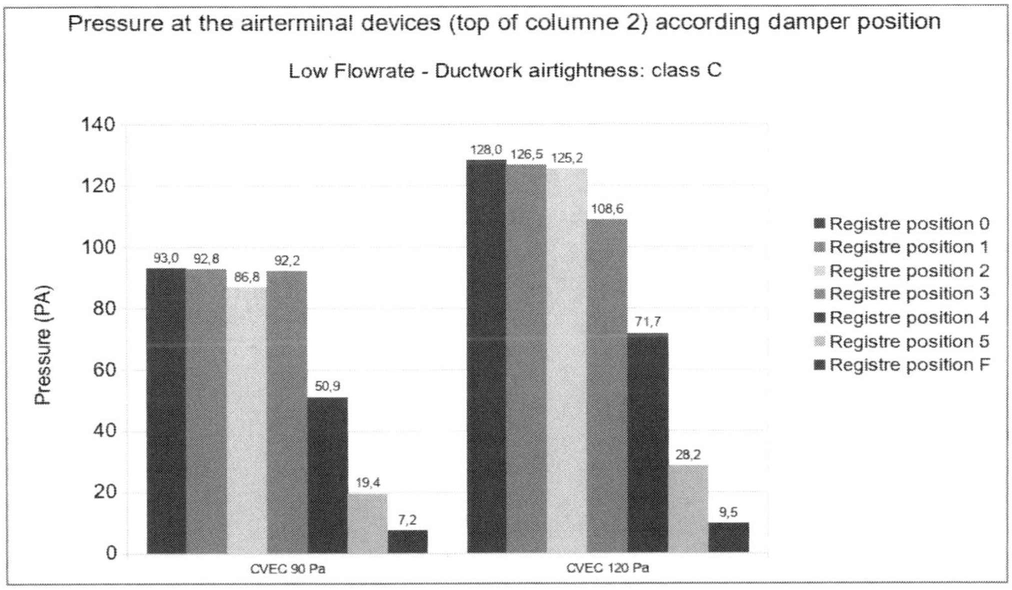

Figure 8: Pressure at the kitchen's terminal device (top of column 2) as a function of the damper's position

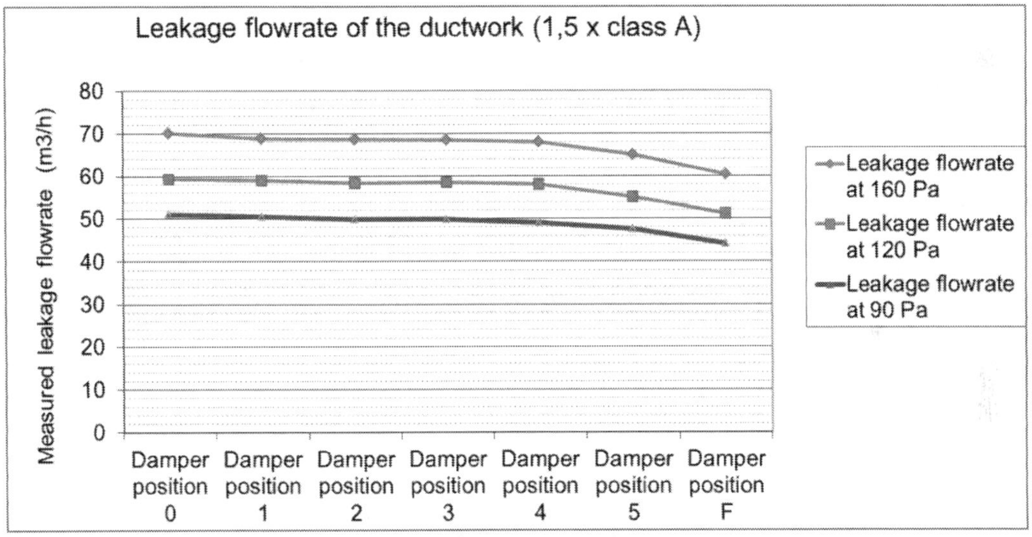

Figure 9: Leakage flowrate of the ductwork as a function of the damper's position at 1.5 * class A (position 0: open; position F: closed)

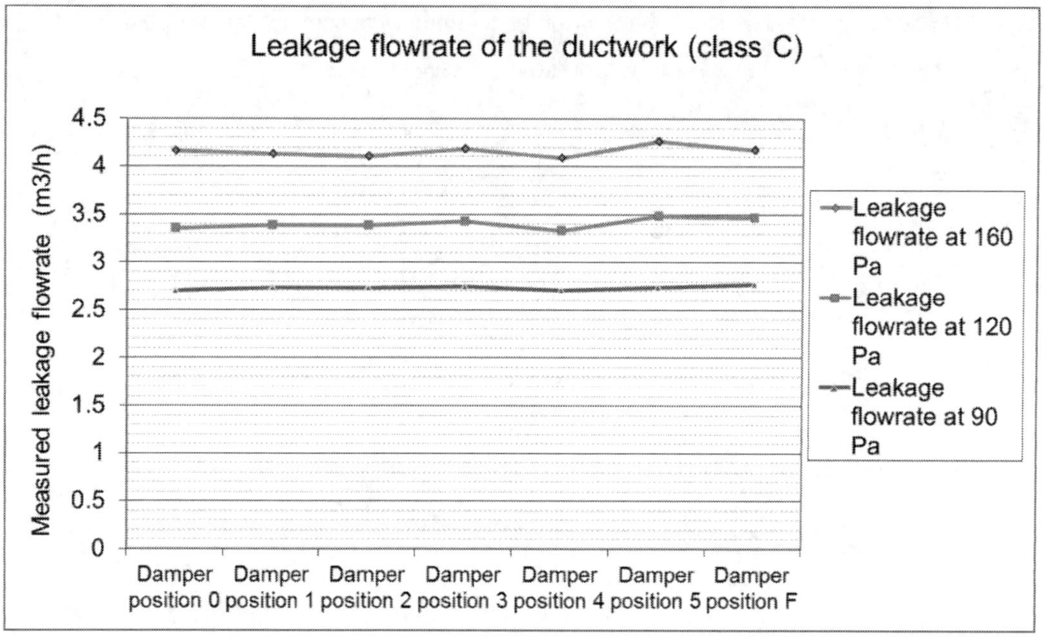

Figure 10: Leakage flowrate of the ductwork as a function of the damper's position at class C (position 0: open; position F: closed)

4.2.2 Impact of leakage distribution and location of the measurement device

Figure 11 shows that the location of the measurement device does not modify the result of the ductwork airtightness test whatever the leakage distribution is.

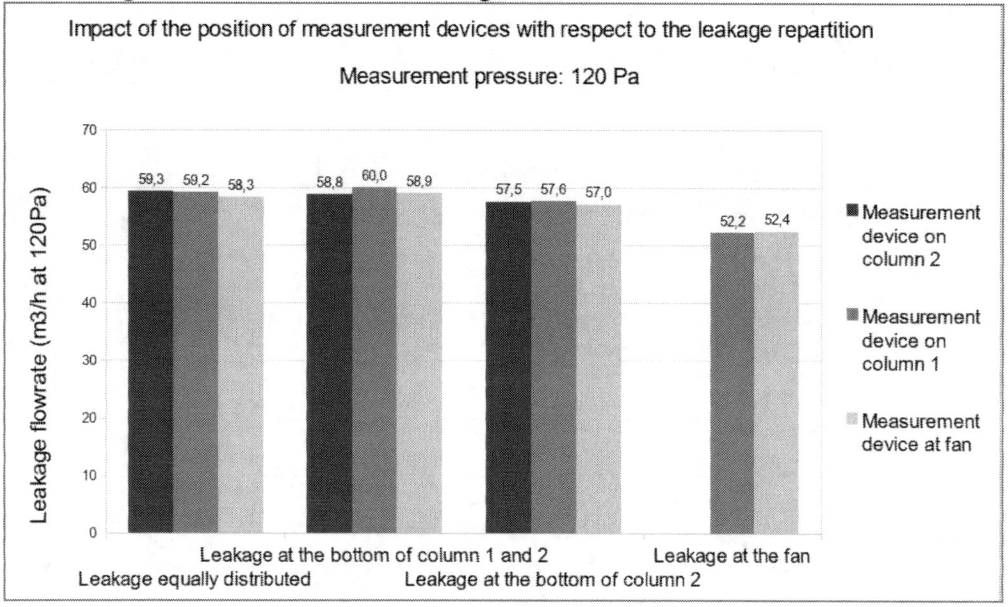

Figure 11: Leakage flowrate as a function of leakage distribution and location of the measurement device at 1.5 * class A

5 CONCLUSIONS

In its first part, this study showed that a careful implementation of the ductwork only makes it possible to reach class B. To achieve class C it is necessary to treat roof T-connections with tape.

Moreover, the airtightness level of the ductwork has proved to have a major impact on the fan consumption (till 40% in this case). Furthermore, if the ductwork is too leaky the flowrate target at air terminal devices may not be reached which may have a major impact on indoor air quality. In its second part, regarding the ductwork airtightness test, this study concluded that for this ductwork, the result of the test is neither influenced by the pressure drop in the ductwork nor by the location of the measurement device whatever the leakage distribution is. This study could now be completed by a mathematical model to explain those results.

6 ACKNOWLEDGEMENTS

This study was carried out as an internship for a master degree. A defense took place in June 2014 at Joseph Fourrier University in Grenoble (France). The authors would like to thank Simon Salager, from Joseph Fourrier University, for his help and the board of examiners.
The contribution of Cerema is funded by the French ministry for ecology, sustainable development, and energy (MEDDE). The sole responsibility for the content of this publication lies with the authors. It does not necessarily reflect the opinion of the ministry.

7 REFERENCES

Jobert, R. (2012). *La ventilation mécanique des bâtiments résidentiels neufs.* Rapport ENTPE –GBBV, 2012, 91p.

Kirchner et al. (2011). *Qualité de l'air intérieur - Dix ans de recherche pour mieux respirer.* Observatoire de la qualité de l'air intérieur (OQAI). Éditions CSTB, 208p.

Lucas et al. (2009). *État de la ventilation dans le parc de logements français.* Rapport final DESE/SB, Éditions OQAI, 84 p.

SAVE-DUCT European Project (1999). *Improving ductwork – A time for tighter air distribution systems – Final Report CE DG XVII.*

Observatoire CSTB-ORTEC (2005-2009). *Rapport de synthèse de la rubrique aération.*

AIRTIGHTNESS OF BUILDING PENETRATIONS: AIR SEALING SOLUTIONS, DURABILITY EFFECTS AND MEASUREMENT UNCERTAINTY

Wolf Bracke[*1], Nathan Van Den Bossche[1], Arnold Janssens[1]

1 Ghent University
Jozef Plateaustraat 22
9000 Gent, Belgium

**Corresponding author: wolf.bracke@ugent.be*

ABSTRACT

During field measurements on the airtightness of passive houses, ventilations system's roof penetrations showed to be one of the major leakage paths, as they were not sealed using the appropriate, durable techniques. Therefore, a series of laboratory measurements was conducted on wood-frame walls to study different air sealing solutions. The use of special airtight gaskets is compared to less advanced sealing methods such as sprayed polyurethane foam and the use of pieces of tape.

The airtightness of different solutions is tested on two different test setups at the Test Centre for Façade Elements, Ghent University. One large-scale setup, designed to measure the air leakage rate of windows, was used and an additional smaller-scale test setup was built. For both setups, measurement principle, test setup and results are reported.

As workmanship quality is an essential aspect for achieving airtight building connections, the repeatability of the sealing methods was addressed by installing multiple identical setups. The quality was reviewed with the use of a smoke generator and the samples were subjected to multiple water tightness tests under static and pulsating pressure. To evaluate the durability of the different solutions, the impact of the water tightness tests is discussed.

A classification for the airtightness of building penetrations is reviewed together with the impact of the air leakage through building penetrations on the overall air leakage of buildings. In the laboratory, all tested solutions could be classified in the best airtightness classification, though large differences between executions methods were evident. The use of standard, rigid airtightness tape is not recommended for sealing 3-dimensional connections such as building penetrations, and is not regarded as a durable sealing method. However, specific flexible tapes are on the market which perform very well and sprayed PUR also showed excellent results when executed correctly. EPDM gaskets showed a higher leakage than flexible tape and PUR, but are also a very good choice. The adaptability of the penetration is a great advantage regarding the durability of the connection.

KEYWORDS

Airtightness, penetrations, laboratory test results, durability

1. INTRODUCTION

A blowerdoor measurement campaign in Belgium on the durability of airtightness of passive houses ($ACH_{50} < 0.6$ upon completion), showed that the average air leakage of the houses had raised by 30% only two years after completion. Though this raise can be attributed to durability effects of various building components (windows, doors, …) and interventions of the inhabitants (installation of outdoor lighting, hanging of paintings, …), the roof penetrations were responsible for a significant part of the additional leakage. Large air

leakages were observed around the air supply and exhaust pipes of the ventilation system (Bracke et al., 2013).
In one house, which was tested on a regular basis during 18 months to investigate seasonal and durability effects, the leaky roof penetration of the air exhaust was replaced with an airtight, flexible EPDM gasket. This small intervention caused the total air leakage of the house to drop from 254 to 239 m^3/h @ 50 Pa, meaning this one penetration was responsible for 6% of the overall air leakage.
Even though an air leakage rate of 15 m^3/h @ 50 Pa might seem relatively small, as multiple other roof and wall penetrations can exist (e.g. ventilation ductwork, solar boiler connections, drain-waist-vents) these can be responsible for a significant share of the total air leakage, causing energy losses, malfunctioning of the ventilation system, draft and sound insulation problems.
Though it is clear that problematic connection such as wall-wall, wall-roof and window-wall interfaces are typically responsible for a larger share of the total air leakage, the leakage around roof and wall penetrations should not be disregarded. From field measurements it became clear that with very little effort and investment, a significant improvement could be achieved.
This paper starts with a review of the research that was published in literature on this topic. Laboratory measurements were performed on two different scales for which the test method, test setup and results are reported. Subsequently, test results are compared to an airtightness classification scale found in literature, and finally workmanship reproducibility and durability are discussed.

2. LITERATURE REVIEW

A good overall airtightness is considered as an essential requirement to obtain energy efficient and comfortable buildings. A lot of research has been done on the air leakage of buildings but little to none focusses on the air leakage of individual building components or building envelope interfaces. As the total air leakage is the sum of various smaller leakages, the isolation of a particular building element can help to acquire certain insights so the overall air leakage can be improved. This approach was also used by Van Den Bossche (Van Den Bossche et al., 2012) to propose an airtightness classification for different solutions to seal the window-wall interface, which is often a weak spot in the airtightness of buildings.
Very little is written on the airtightness of building penetrations. In one study performed by Kalamees (Kalamees et al., 2008), the air leakage distribution of 32 houses in Finland was analysed qualitatively using thermography and smoke detection. Air barrier penetrations for ductwork were held responsible for 3 to 10% of the total air leakage, depending on the building type (detached/apartment) and the presence of other critical connections such as the junction between intermediate floors and external walls.
To the knowledge of the authors, no test standards, classification systems or performance requirements exist for air barrier penetrations in the European framework for standards and codes. The American standard ASTM E2357 describes a test method for the assembly of air barrier systems, wherein two penetrations are part of a standard test wall, but are not tested separately.
In the Netherlands, three airtightness performance levels are specified according to NEN 2687: Class 1 (basic), Class 2 (good) and Class 3 (very good, passive houses). SBR, a Dutch research foundation, has published practical information on how to achieve the different air tightness classes. For several building components and interfaces, maximum air leakage rates are available to ensure the feasibility of the aspired maximum overall leakage of the building. Table 1 shows the maximum leakages for roof, wall and floor penetrations according to different penetration diameters. The values were derived from an area-based leakage rate for a gap of a few millimetres around the penetration.

Table 1: Maximum penetration leakages according to SBR

Diameter (mm)	Air leakage @ 50 Pa (m^3/h)		
	Class 1	*Class 2*	*Class 3*
15	9.3	3.4	0.7
25	15.6	5.4	1.1
50	27.8	10.7	2.1
80	91.9	17.1	3.4
100	112.9	21.5	4.3
125	139.7	27.4	5.5
150	167.1	32.2	6.4
200	220.4	43.0	8.6

The air flow of 15 m^3/h @ 50 Pa through the 200 mm roof penetration of the air exhaust, as mentioned in the introduction, can thus be regarded as too high for a Class 3 passive building. The air sealing probably complied with airtightness Class 3 after installation, but was not executed in a durable way. The question rises whether durability aspects should be taken into account when classifying airtightness solutions. Currently, very little is known about the durability of sprayed PUR foam, tapes and glues and no normative documents exist which prescribes a test protocol. It is still unclear how artificial ageing correlates with the expected aging in practice and how mechanical impacts (wind pressures, contractors working on the penetrations, …) should be translated in standard test protocols. In Germany, the FLiB (Fachverbandes Luftdichtheit im Bauwesen) is working on a test method to evaluate the durability of tapes and glues, which should result in a future normative document DIN 4108 part 11. At Ghent University, about 5000 tests on the peel and tear strength of the bonding of different tapes to different substrates were executed (Van Den Bossche et al., 2009). The samples were subjected to different temperatures and relative humidity levels in a climate chamber to evaluate the durability. Large differences in bonding strength were observed, not only between different products and different substrates, but also between different manufacturers of very similar products..

3. EXPERIMENTAL RESEARCH

Appropriate sealing materials are available on the market, but in practice, the poorly sealing of air barrier penetrations can cause significant amounts of air leakage due to the use of wrong materials and/or bad workmanship. The sealing of certain penetrations is often done by a variety of contractors, which may have a lack of awareness about the importance of airtightness.

In the context of a Belgian research project 'DO-IT' on wood-frame constructions, laboratory measurements on the airtightness of wood-frame building components were performed in the Test Centre for Façade Elements at Ghent University. A part of these measurements focus on the air leakage of wall penetrations, for which the results are reported in this paper.

Two test setups were used: one large-scale setup where a wood-frame wall was built and a variety of penetrations were installed together, and one small-scale setup where different specimens could be installed with one penetration in each specimen.

3.1 Test method and error analysis

A. Large scale test setup

The airtightness of the wall penetrations was measured in a standard calibrated test rig according to EN 12114. Because the setup is unable to obtain a certain pre-set pressure, the airflow was measured at 10 random pressure differences in the interval 0 – 600 Pa. The airflow was derived by measuring the pressure difference over a calibrated orifice opening.

These measurement points are controlled for outliers using Chauvenet's criterion (Taylor, 1982), a power law function is fitted through the results after which the leakage is calculated for a 50 – 100 – … – 500 Pa pressure difference. A standard deviation is defined, taking into account the error on airflow measurements and the fitting to a power law.

The air leakage through the tested wall penetrations was not measured separately for every penetration. Instead, a wall was built in a steel test rig, in which different penetration diameters and sealing solutions were installed. After each newly installed penetration, an airtightness test was conducted according to EN 12114. The air leakage of a certain penetration can be derived by subtracting the air leakage of the previous measurement from the new measurement. The error on the leakage through the tested penetration will thus be a combination of the error on the previous and on the new measurement. As the errors are primarily the result of the fitting of the power law, which is uncorrelated for different measurements, the total error is calculated by adding in quadrature the errors of both measurements.

The penetrations tested in Figure 1 are almost perfectly airtight. The power laws fitted through the measurement points of the previous and the new measurement cross each other, meaning the test setup is unable to measure the difference between both measurements accurately.

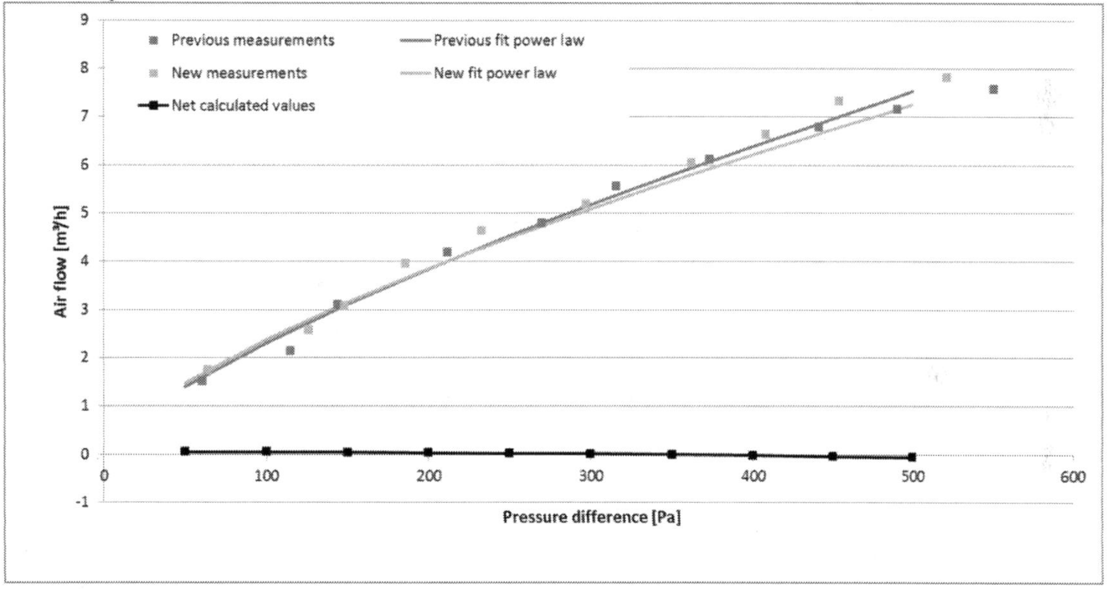

Figure 1: Measurement principle and calculation of net leakage (2 x EPDM gasket Ø 75 mm)

B. Small scale test setup

The measurements are conducted using a Lindab LT600, a test instrument designed to measure the airtightness of ductwork and chimneys. The instrument calculates the airflow by measuring a temperature difference over a hotplate, which is heated and cools down as the air passes over it. The hot plate principle is a more accurate solution to measure very small airflows than measuring a pressure difference over an orifice opening.

As the Lindab LT600 is able to measure the airflow at a given pressure difference, no complete pressure sequence was performed and no power law was derived. Measurements were taken at 3 pressure differences: 50 Pa, 250 Pa and 500 Pa.

Because the device is designed to test the air leakage of ductwork, it follows EN 12237, which states that the pressure difference should be applied for 5 minutes to obtain a stable airflow. The measurements were logged to Excel, and the average airflow is calculated over the last 2 minutes of the measurement. Every measurement was executed 5 times and the average value and standard deviation are calculated.

Before the measurement of the wall penetrations, the extraneous air leakage through the test setup is determined. The net leakage through the different specimens is derived by subtracting the extraneous air leakage from the measured leakage and the standard deviation is calculated by adding in quadrature the standard deviation of both measurements.

3.2 Test setup

A. Large scale test setup

The tests were executed on a full-scale setup built in a steel test rig. A wood-frame wall of 228 cm by 196 cm was built consisting of a framework and exterior sheathing of 18 mm bituminous impregnated fibreboard. As the fibreboard contains more bitumen than most similar products, it does not only act as a rain screen, but also as an additional air barrier next to the interior barrier and provides enough structural stiffness to avoid the need of OSB sheathing on the interior side. To make an abstraction of the real-world situation, the use of OSB panels and interior insulation was omitted in the test setup.

Four different penetration diameters were tested: 4 mm (single electrical cable), 20 mm (ribbed conduit pipe for electrical cables), 75 mm (PVC pipe) and 130 mm (PVC pipe). The openings for the different penetrations were respectively 6 mm, 22 mm, 85 mm and 140 mm. Every case was sealed using three different methods

- gasket: A piece of flexible EPDM rubber with an opening smaller than the penetration's diameter is stretched over the penetration. The EPDM slab is taped to the fibreboard using a standard airtightness tape.
- PUR: Flexible polyurethane foam is sprayed with a PU-pistol on the joint between the penetration and the fibreboard. As the board has a limited thickness, the PUR is not truly applied in the cavity between the penetration and the wall, but is rather sprayed on the surface of the fibreboard. Water was sprayed to increase the humidity, which enhances the expansion of the foam and the bonding to surrounding surfaces.
- tape: Multiple pieces of tape were radially applied on the fibreboard and penetration. The tape is designed to seal the 2-dimensional joints of vapour retarder membrane and consists of a reinforced PE membrane and a acrylate adhesive.

The last method, which is not typically recommended, was executed to compare test results with better solutions, and investigate durability effects and workmanship quality. To test the worst-case scenario, the air leakage without applying any air sealing was also evaluated. Especially in this case, the diameter of the openings in the fibreboard is of great influence. In most of the test cases, the penetration was executed multiple times to extract a reliable average value for the air leakage rate.

Figure 2 shows the tested wall with different penetration diameters and sealing solutions. The wall is connected airtight to the test rig using butyl-aluminium tape and the joints between the different fibreboards are also sealed to minimize extraneous air loss and resulting measurement uncertainty.

Figure 2: Wood-frame wall with different air sealing solutions in steel test rig

B. Small scale test setup

A small test box measuring 300 x 300 x 300 mm was built using plywood with an epoxy coating. To minimize extraneous air leakage, silicone sealant and butyl tape are used to seal every joint. The different test specimens were clamped against the test box and the connection was made airtight by compressing a closed-cell foam band attached to the specimens.

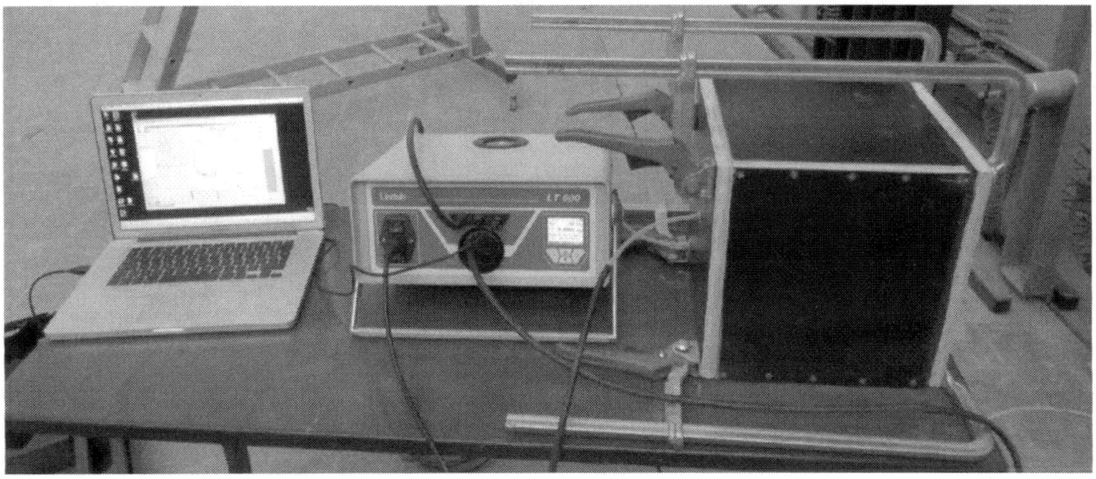

Figure 3: Small scale test setup

Four specimens were built to test different methods to seal a 75 mm PVC pipe wall penetration in a 90 mm opening.

- PUR: PUR foam was sprayed in the gap (7,5 mm wide) between the PVC pipe and the wall element. Plywood with an epoxy coating was attached to a PUR foam board to provide enough thickness for the expansion of the sprayed PUR.
- gasket: An EPDM gasket with an opening of 55 mm was pulled over the PVC pipe and taped to the plywood.
- tape (1): The PVC pipe is sealed to the plywood using pieces of tape. The tape is designed to seal the 2-dimensional joints of vapour retarder membrane and consists of a reinforced PE membrane and an acrylate adhesive.

tape (2): The PVC pipe is sealed to the plywood using pieces of tape. The elastic tape is designed to seal difficult corners and 3-dimensional connections and consists of an elastic PE membrane and a butyl rubber adhesive.

Figure 4: Test specimens: tape (1), tape (2), PUR, gasket

3.3 Measurement results

A. Large scale test setup

As explained in paragraph 3.1, the calculated net leakages are accompanied by large error intervals. As different penetrations are tested, and thus additional leakages are created during the testing sequence, the total air leakage and accompanying errors are increasing. Sometimes, due to measurement uncertainty, small negative net leakages are calculated for a certain pressure difference as Figure 1 illustrates. For a better interpretation and comparison, the calculated leakages at 50 Pa, 250 Pa and 500 Pa and associated standard deviations are presented in the following tables.

For a first test, the air leakage through the bituminous fibreboard was measured. Joints around the wall and between the boards were sealed so only the air leakage through the material itself was measured.

Table 2: Test results bituminous impregnated fibreboard

	50 Pa		250 Pa		500 Pa	
	Air leakage ($m^3/h.m^2$)	**σ** ($m^3/h.m^2$)	**Air leakage** ($m^3/h.m^2$)	**σ** ($m^3/h.m^2$)	**Air leakage** ($m^3/h.m^2$)	**σ** ($m^3/h.m^2$)
fibreboard	*0.266*	0.028	*1.249*	0.091	*2.341*	0.160

When a hole is cut out of the fibreboard, airtight tape or EPDM is placed over the board and the leakage through the test wall will be lower due to a smaller fibreboard surface. In the following tests, the covered surfaces are measured and multiplied with the measured characteristic air leakage through the fibreboard. The results are subtracted from the newly measured air leakage to correct for this smaller fibreboard surface.

Table 3 shows the test results for the 4 mm electrical cables. For every case, 16 cables were installed, the total air leakage is measured and the average leakage rate for one penetration is calculated. At 50 Pa, both the tape and gasket seem to perform very well, but for higher pressure differences the EPDM gasket is clearly the best solution. Applying the sprayed foam around the electrical cables was very inconvenient and resulted in a large air leakage rate.

Table 3: Test results 4 mm electrical cable

	#	50 Pa		250 Pa		500 Pa	
		Air leakage (m^3/h)	**σ** (m^3/h)	**Air leakage** (m^3/h)	**σ** (m^3/h)	**Air leakage** (m^3/h)	**σ** (m^3/h)
no seal	16	*0.030*	0.006	*0.115*	0.026	*0.206*	0.049
tape	16	*0.001*	0.005	*0.017*	0.018	*0.039*	0.033
PUR	16	*0.026*	0.006	*0.077*	0.022	*0.123*	0.040
gasket	16	*0.001*	0.004	*0.003*	0.017	*0.005*	0.032

The leakage results for the 20 mm conduit pipes are reported in Table 4. Two types of EPDM gaskets were tested: 8 single gaskets (gasket_1) for 1 penetration, and 1 gasket for 9 penetrations (gasket_9). The gasket with multiple penetrations was harder to install correctly, because the holes in de EPDM might not be completely aligned with the conduit pipes coming through the fibreboard. Probably this caused a higher leakage than the single gaskets, but both types are a better solution than using pieces of tape or sprayed PUR foam.

Table 4: Test results 20 mm conduit pipe

	#	**50 Pa**		**250 Pa**		**500 Pa**	
		Air leakage (m^3/h)	**σ** (m^3/h)	**Air leakage** (m^3/h)	**σ** (m^3/h)	**Air leakage** (m^3/h)	**σ** (m^3/h)
no seal	9	*0.600*	0.012	*1.788*	0.043	*2.850*	0.076
tape	16	*0.140*	0.042	*0.343*	0.154	*0.488*	0.275
PUR	16	*0.118*	0.014	*0.327*	0.050	*0.501*	0.088
gasket_1	8	*0.001*	0.007	*0.003*	0.027	*0.005*	0.049
gasket_9	9	*0.005*	0.007	*0.017*	0.029	*0.030*	0.053

The results of the 75 mm PVC pipe penetrations are shown in Table 5. Again, tape and PUR show similar results and perform worse than the gasket. The net air leakage rate for the gasket calculated at 50 Pa is higher than the leakage rate at 250 Pa, and for 500 Pa, a negative result is calculated. Because the gasket is almost perfectly airtight, the difference between both measurements could not be quantified by the test setup and subtracting both fitted power laws rendered these physically impossible results. The power laws and calculated net leakage are visualised in Figure 1.

Table 5: Test results 75 mm PVC pipe

	#	**50 Pa**		**250 Pa**		**500 Pa**	
		Air leakage (m^3/h)	**σ** (m^3/h)	**Air leakage** (m^3/h)	**σ** (m^3/h)	**Air leakage** (m^3/h)	**σ** (m^3/h)
no seal	2	*6.275*	0.186	*16.130*	0.566	*24.148*	0.930
tape	2	*0.128*	0.043	*0.325*	0.106	*0.477*	0.176
PUR	2	*0.077*	0.056	*0.298*	0.119	*0.526*	0.194
gasket	2	*0.049*	0.068	*0.022*	0.131	*-0.057*	0.208

For the 130 mm PVC pipe, two pipes were again installed and sealed with tape, PUR and a gasket. The pipes sealed with pieces of tape and sprayed PUR foam were installed and tested one by one, as visible in Table 6. Although the intention was to execute the sealing in exactly the same way, a large spread is evident between the results. This is an indication that workmanship quality is a very important factor but also that the solutions are very prone to errors. The results from the EPDM gasket are again unreliable, as air leakage at 500 Pa is lower than that at 50 and 250 Pa. It is clear however that the sealing with a gasket is by far superior over the sealing with PUR or pieces of tape.

Table 6: Test results 130 mm PVC pipe

	#	**50 Pa**		**250 Pa**		**500 Pa**	
		Air leakage (m^3/h)	**σ** (m^3/h)	**Air leakage** (m^3/h)	**σ** (m^3/h)	**Air leakage** (m^3/h)	**σ** (m^3/h)

tape (1)	1	*1.847*	0.115	*4.698*	0.333	*6.988*	0.552
tape (2)	1	*0.837*	0.219	*1.581*	0.636	*1.982*	1.026
PUR (1)	1	*1.475*	0.046	*4.252*	0.109	*6.695*	0.181
PUR (2)	1	*0.048*	0.090	*0.302*	0.214	*0.603*	0.360
gasket	2	*0.092*	0.092	*0.113*	0.174	*0.045*	0.286

B. Small scale test setup

The net air leakage of the different specimens is shown in Table 7, together with the extraneous leakage through the test setup. Contrary to the results from setup 1, the specimen with sprayed PUR foam showed the lowest air leakage, followed by the specimen with the elastic butyl tape. The leakage through the EPDM gasket is relatively higher but in absolute value it is still negligible. The penetration with reinforced, rigid tape is clearly the worst solution. Only 1 test was conducted, as additional leakages were created during the 5 min tests at 50, 250 and 500 Pa.

Table 7: Test results 75 mm PVC pipe (setup 2)

	# tests	50 Pa		250 Pa		500 Pa	
		Air leakage (m^3/h)	**σ** (m^3/h)	**Air leakage** (m^3/h)	**σ** (m^3/h)	**Air leakage** (m^3/h)	**σ** (m^3/h)
extraneous	5	*0.0006*	0.0001	*0.0020*	0.0003	*0.0038*	0.0008
PUR	5	*0.0003*	0.0002	*0.0005*	0.0006	*0.0021*	0.0013
gasket	5	*0.0019*	0.0001	*0.0108*	0.0010	*0.0211*	0.0018
tape (1)	1	*0.0211*		*0.0712*		*0.1308*	
tape (2)	5	*0.0016*	0.0002	*0.0042*	0.0005	*0.0057*	0.0011

4. AIRTIGHTNESS CLASSIFICATION

According to the classification as proposed by SBR, all tested solutions belong to Class 3 and are thus appropriate for application in a passive house. Even a 20 mm conduit pipe in a 22 mm opening without any air sealing can be classified in Class 3. The 75 mm PVC pipe in a 85 mm opening without air sealing resulted in an air leakage rate of 6.3 m^3/h @ 50 Pa, whereas 3.4 m^3/h is allowed for a 80 mm Class 3 penetration. Class 3 seems achievable without special airtightness solutions, if the opening in the airtightness layer is not too large. Consequently, it cannot be regarded as a very ambitious airtightness level.
To evaluate the maximum leakages according to SBR, a fictional passive house with a volume of 500 m^3 is considered. With a maximum air change rate of 0.6 h^{-1}, the allowed air leakage @ 50 Pa is 300 m^3/h. Such a house typically has the following building penetrations: 3 x 200 mm (ventilation inlet and exhaust, chimney), 2x 125 mm (sewage pipes), 1 x 80 mm (waist-drain-vent), 4 x 50 mm (solar boiler inlet and outlet, water supply, gas supply), 10 x 25 mm (electricity for solar panels, outdoor lighting, ...). When summarizing the maximum penetration leakages, the total air leakage for the three different classes is given in Table 8.

Table 8: Maximum total penetration leakages according to SBR

	Air leakage @ 50 Pa (m^3/h)		
	Class 1	*Class 2*	*Class 3*
Total air leakage	744.6	178.3	35.7
share of allowed leakage	321.1%	99.2%	19.8%

For this example, the penetrations executed with an airtightness according to Class 3, are responsible for 19.8% of the total allowed leakage. This does not seem dramatic, but could be improved with little effort and investment, as only 15 local and easily detectable spots have to be sealed using the appropriate techniques.

5 WORKMANSHIP REPRODUCIBILITY

In the wood-frame research project, of which a part of the results are reported in this paper, the water tightness of building components was evaluated next to the air leakage rate. The results of these tests are also useful to evaluate the difference in workmanship quality between identical sealing solutions, as leakages can be visualised. Water tightness tests were executed according to

- EN 1027: water is sprayed at a rate of 120 $l/(h.m^2)$, first 15 min without pressure, then a static pressure is applied in consecutive 5 min time steps at 50 Pa, 100 Pa, 150 Pa, 200 Pa, 250, 300 Pa, 450 Pa, 600 Pa, …
- EN 12865: water is sprayed at a rate of 120 $l/(h.m^2)$, first 20 min without pressure, then a pulsating pressure (10 seconds with pressure, 5 seconds without pressure) is applied for 10 minutes at 150 Pa, 300 Pa, 450 Pa, 600 Pa, …

As the wall has an area of 4 m^2, a total spray rate of 480 l/h was applied. The following water infiltrations were observed for the water tightness tests. Due to practical reasons, the complete pressure sequence was not executed during some tests, which is indicated with a '/' in Table 9.

Table 9: Infiltration pressures

			static pressure							pulsating pressure					
	Pressure (Pa)	#	0	50	100	150	200	300	900	0	150	300	450	600	750
4 mm	tape	16	2			1	1	/	/	1		10		1	
	PUR	16					2	/	/			7	3	3	
	gasket	32						/	/			2			
20 mm	tape	4	2				2	/	/	1	2		/	/	/
	PUR	16	6			3	1	/	/	4	12		/	/	/
	gasket_1	4		1			1	/	/		2	2	/	/	/
	gasket_9	9	1					/	/		1	9	/	/	/
75 mm	tape	2	1					1	/	1		1			
	PUR	2	2						/	2					
	gasket	4			2		1	1	/		1	3			
130 mm	tape	2	2							2					
	PUR	1	1							1					
	gasket	2			1							1			1

A large spread in the pressure at which infiltration occurred was visible during most of the tests, suggesting workmanship quality plays an important role in the leakage around penetrations. Applying a positive pressure in combination with a smoke generator confirmed the differences in air leakages which accompany the differences in water infiltration, in particular when PUR or tape were used to seal the penetrations. As the PUR was sprayed on the fibreboard, the foam did not expand into the cavity and in some cases, large openings were present around the penetrations, which could easily be detected using the smoke generator. The installation with rigid tape was also prone to errors, as the overlapping parts of the pieces of tape showed significant leakages.
In general, the water infiltration around the penetrations sealed with EPDM gaskets was much smaller than the penetrations sealed with tape or PUR.

6 DURABILITY

The air leakage rate through the complete wall was measured after the execution of the water tightness tests reported above. In this way, the effect of the sprayed water on the bonding of the tapes and the mechanical impacts of the pressure pulsations were evaluated.

Table 10: Additional air leakage after water tightness tests

	50 Pa		**250 Pa**		**500 Pa**	
	Air leakage (m^3/h)	**σ** (m^3/h)	**Air leakage** (m^3/h)	**σ** (m^3/h)	**Air leakage** (m^3/h)	**σ** (m^3/h)
combined	*0.581*	0.168	*1.188*	0.284	*1.658*	0.462

As the air leakage rate of the installed penetrations could not be measured separately, the individual increase in air leakage was not determined. As shown in Table 10, the overall increase in air leakage was limited, and visual inspection indicated which air sealing solutions were problematic.
No tapes were detached from the surface of the fibreboard as a result of the large amounts of sprayed water, but the impact of the pressure pulsations on the taped PVC pipes was clearly visible. As Figure 5 illustrates, the pipes are pressed through the opening in the fibreboard, and additional leakages are created.

Figure 5: Durability effects taped PVC pipe: before and after water tightness testing

The flexible PUR foam should be able to tolerate mechanical impacts up to a certain level and no degradation was detected after the water tightness tests. However, one can reasonably assume that contractors working on a penetration could damage the foam or the adhesion between the foam and the wall. The execution with an EPDM gasket is less sensitive to these problems. As the penetration is not attached to the wall, it can be moved back and forth through the gasket or even be replaced without damaging the air sealing.

7 CONCLUSIONS

Three different methods to ensure an airtight connection around building penetrations were evaluated in this paper: airtightness tape, sprayed polyurethane foam and EPDM gaskets. Air leakage tests were executed on two setups of a different scale. In one setup, a wall was built and a variety of sealing methods and penetration diameters were installed. This setup was able to compare the quality of the different tested solutions and to investigate differences in workmanship quality between identically installed elements. Due to the measurement principle of the test setup and the calculation method of the net leakages through the installed penetrations, a high measurement uncertainty is involved. A second setup was built on a much smaller scale to test individual components, which was able to measure the air leakage rate through different sealing methods very accurately.

In the large-scale setup, PUR foam was sprayed around the penetration on the surface of the wall. As the foam could not expand into the cavity and fill the connection, significant local leakages were observed. In the small-scale setup, a thicker wall element was used, and the foam could expand in the cavity between the wall and the penetration, resulting in an almost perfect airtight connection.

The use of standard airtightness tape, which is designed to seal joints between panels, is not recommended to seal 3-dimensional connections such as building penetrations. As the tape is very rigid, it is unable to follow the shape of the pipes and large leakages can arise. Air and water tightness tests showed these connections can be very tight, but most of the times they result in large air leakage rates. Furthermore, the use of rigid tape in this situation is very sensitive to mechanical impacts and cannot be regarded as durable.

In the smaller setup, a flexible butyl tape was tested which is designed for 3-dimensonal connections, which proved to be very airtight.

EPDM gaskets showed to be less airtight then sprayed PUR, or flexible butyl tape, but in absolute values, the air leakage rate is still negligible. Contrary to PUR or tape, the connection is not permanent and can be adapted after installation, which is a very important aspect in respect to long-term durability.

8 REFERENCES

ASTM E2357-11. Standard Test Method for Determining Air Leakage of Air Barrier Assemblies.

Bracke W., Laverge J., Van Den Bossche N., Janssens A. (2013). Durability and measurement uncertainty of airtightness in extremely airtight dwellings. *34th AIVC conference and 3th TightVent Conference: Towards Optimal Airtightness Performance,*

Athens.

Kalamees T., Korpi M., Eskola L., Kurnitski J., Vinha J. (2008). The distribution of the air leakage places and thermal bridges in Finnish detached houses and apartment buildings. *Proceedings of the 8th Symposium on Building Physics in the Nordic Countries NSB2008,* Copenhagen, 1095-1102.

NBN EN 12114 (2000). Thermal Performance of Buildings – Air Permeability of Building Components and Building Elements – Laboratory Test Method, CEN, Brussels, Belgium.

NBN EN 12237 (2003). Ventilation for buildings - Ductwork - Strength and leakage of circular sheet metal ducts, CEN, Brussels, Belgium.

NBN EN 1027 (2000). Windows and doors - Watertightness - Test method, CEN, Brussels, Belgium.

NBN EN 12865 (2001). Hygrothermal performance of building components and building elements - Determination of the resistance of external wall systems to driving rain under pulsating air pressure, CEN, Brussels, Belgium.

NEN 2687. Luchtdoorlatendheid van woningen - Eisen.

SBRCURnet (2007). Luchtdicht bouwen: Theorie – ontwerp – praktijk.

Taylor J.R. (1982). An introduction to error analysis – The study of uncertainties in physical measurements, 2nd ed., University Science Books, Sausalito, Canada.

Van Den Bossche N. (2012). Airtightness of the window-wall interface in cavity brick, Energy and buildings 45, 32-42.

Van Den Bossche, N., Janssens, A., Moens, J., & Ost, T. (2009). Performance assessment of building envelope interfaces: self-adhering flashings. *The future is in the balance* (pp. 113–126). Presented at 'The future is in the balance: symposium on building envelope sustainability', Washington, DC, USA: Roof Consultants Institute Foundation.

COMPARISON OF BUILDING PREPARATION RULES FOR AIRTIGHTNESS TESTING IN 11 EUROPEAN COUNTRIES

Valérie LEPRINCE[1*], François-Rémi Carrié[2]

1 PLEIAQ
84 C Avenue de la Libération
69330 Meyzieu, FRANCE
**Corresponding author: Valerie.leprince@pleiaq.net*

2 ICEE
93 rue Molière
69003 Lyon, FRANCE

ABSTRACT

Mandatory building airtightness testing came gradually into force in the UK, France, Ireland and Denmark. It is considered in many other European countries because of the increasing weight of the building leakage energy impact on the overall energy performance of low-energy buildings. Therefore, because of related legal and financial issues, the building airtightness testing protocol and reporting have become crucial issues to have confidence in the test results as well as the consistency between the measurement results and values used in the energy performance calculation method.
The reference testing protocol in Europe is described in EN 13829, but many countries have developed specific guidelines to detail or adapt EN 13829 requirements.
This study compares building preparation rules for airtightness testing in 11 European countries.
Information has been collected through a questionnaire sent to TAAC (TightVent Airtightness Associations Committee) members. We found that building preparation differs significantly from one country to another and that the two methods described in the standard are either too detailed or insufficiently described to fit with the specificities of each country regarding building preparation. It concludes on possible improvement for EN 13829 including:

- In one hand describing more precisely the basic principles of the preparation to avoid ambiguities and,
- In another hand, allowing some flexibility to the countries to specify rules consistent with their energy performance calculation method.

KEYWORDS

Airtightness testing, building preparation

1. INTRODUCTION

Building airtightness is a key issue to reach low- and very low-energy targets. Therefore an increasing number of tests are performed in European countries for various reasons: compliance to the energy performance regulation; compliance to a specific energy programme; or will of the building owner. For instance, to our knowledge, measuring the airtightness of all new buildings or at least part of them is required by the energy performance regulation in UK, France, Ireland and Denmark. Besides, specific energy programs (such as Passivhaus or Minergie) that require or encourage building airtightness testing are increasingly popular in many other countries. Likely, within a few years, over a million tests will be performed every year in Europe.

Airtightness measurements are useful to check that the building envelope complies with a given requirement. The measurement protocol is described in the European standard EN 13829 adapted from ISO 9972 which is now under revision.

Measurement uncertainty has been discussed in various publications (Bailly, 2012) (Delmotte C. , 2013) (Delmotte C. L., 2011) (Rolfsmeier, 2010) (Sherman, 1995) and it is widely accepted that the building preparation (i.e., the way openings and vents are sealed, closed, or left as during the airtightness test) represents one of the main sources of uncertainty. Rolfsmeier et al (Rolfsmeier, 2010) have had tests performed by 17 testers on the same building, firstly with a preparation made by the tester himself, and secondly with homogenous preparation rules. 70% of testers obtained a V_{50} within 7% of the right result, but for the 30% the error could reach 63% because of an inappropriate preparation and a misinterpretation of the volume to be tested. When the preparation was done correctly, results stayed within 6% of the right results for all testers. The consequences of such discrepancies can be crucial for building professionals or owners seeking e.g., if the subsidies are given on condition to meet a given airtightness level, or if the remedial actions are costly and impractical to implement.

Moreover, if the result is used in energy performance calculation method, the building preparation has to be consistent with the calculation method. Logically, one should seal or close the openings through which the airflow rate is taken into account elsewhere in the calculation method and leave as what is not. In principle, this is clear for instance for the air terminal devices of a mechanical ventilation system, but it may be ambiguous for devices such as unvented combustion appliances.

Therefore, to address the question raised by the increasing number of airtightness tests, we have collected and analysed information on building preparation.

2. OBJECTIVES

This study aims at answering the following questions:

- How is EN 13829 used and interpreted in European countries?
- Does EN 13829 fit with countries' needs in terms of protocol description?
- How can the standard be improved to better march those needs?

3. APPROACH

This work has been done in the context of the TightVent Airtightness Associations Committee (TAAC). TAAC is a European working group, set up and hosted within TightVent. The scope of this working group includes various aspects such as:

- Airtightness requirements in the countries involved
- Competent tester schemes in the countries involved
- Applicable standards and guidelines for testing
- Collection of relevant guidance and training documents.

At present, the participants are from Belgium, Czech Republic, Denmark, Estonia, France, Germany, Latvia, Ireland, Poland, Sweden, UK and contacts have been established with other European countries.

A questionnaire has been developed within the group to compare building preparation with the objective of being useful for someone who:

- plans to perform a test in a foreign country
- would compare the test results from different countries
- prepares guidelines for airtightness measurements in its country

A representative from each country has kindly accepted to answer the questionnaire.
This document summarizes their answer.
Note that the scope of the answers differs between countries:

- For the UK, Denmark, Germany and Ireland, the answers concern only the energy performance regulations
- For France and Belgium, the answers concern both energy performance regulation and specific energy savings programs.
- For Czech Republic, Poland and Sweden, they concern specific energy saving programs only.

3.1 Analysis of questionnaire results

Airtightness requirements and guidelines

Out of the 11 countries who have answered the questionnaire only four of them have requirements in the building regulation to measure airtightness (Denmark, France, Ireland and UK)[1]. Those four countries have all developed specific guidelines to detail airtightness testing conditions.

Three other countries (Belgium, Czech Republic and Germany) have also developed specific guidelines for airtightness testing in the context of specific energy saving program or for voluntary testing in the context of energy performance regulation.

This means that 7 out of the 11 countries have developed specific guidelines beyond the European standard EN 13829.

[1] The regulation does not impose to measure every buildings but at least part of them have to be

Time of measurement

This question concerns the requested completion of the building. EN 13829 specifies that even if preliminary measurement may be done during the constructions, the effective measurement can take place only when the envelope of the building is completed.

We noticed large differences between countries:

- In Sweden, for example, as building regulation and even energy performance programs give very few guidelines to perform airtightness tests, most of the time it is the client himself who chooses time and condition of measurements. As a consequence, most of the time, he prefers to make the test when it is still possible to reach the airtight layer in case leakages have to be repaired. Thus the building envelope may not even be completed (some doors or traps could be missing and sealed instead). Most often, a final test is not performed.
- In Belgium, test has to be performed when the envelope is completed. It is recommended to have it done when all works are finished.
- In Denmark and UK the building envelope has to be in its final complete state.
- Germany requires that the building envelope – all airtightness concerning parts – shall be completed.
- In France and Czech Republic, the tests have to be done after all work that may affect airtightness is completed.
- In Ireland, the building must be in "practical completion", i.e., almost ready for occupation (possibly just need to have painting done and furniture placed)

General rules for the building preparation

Testing according to EN 13829 requires that all intentional exterior openings of the building or part of the building to be tested (windows, doors, fireguard) be closed. Besides all interconnecting doors in the part of the building to be tested shall be opened to have a homogeneous pressure in the building (except for cupboards and closets, which should be closed).

Beyond these basic rules, "method A" of the standard does not allow further improvement of the air tightness of building components. On the other hand, with "method B", all adjustable openings shall be closed and remaining intentional openings shall be sealed.

The underlying ideas behind the methods are the following:

- Method A: the building as prepared represents its condition during the season in which heating or cooling systems are used.
- Method B: the leakage measured represents the leakage through the envelope excluding those due to the HVAC equipment and its penetrations.

Method A is used in the energy performance regulation only in two countries (Ireland and Belgium). Three countries (Czech Republic, France and Germany) have developed a specific method. However, in the Czech Republic and Germany, this method is used only for specific energy saving program and method B is required by energy performance regulation. Therefore, in 8 out of the 11 countries, method B is used to estimate the building airtightness in the energy performance regulation context.

To further describe the building preparation, we have defined six categories to classify how each opening shall be processed:

- "Sealed": make hermetic by any appropriate means (adhesive, inflatable balloon, stopper, etc.)
- "Depends on EP-calculation": sealed if the associated flow rate is already taken into account in the energy performance calculation method, left as if not
- "Closed": Closed if there is a closing device, left as if not
- "Held with tape": the device is closed with a tape to prevent it from moving during the test, but the tape should not sealed the opening
- "Left as": left in normal position of us (this can be closed if the normal position is close like some fire-door or open of if normal position during heating or cooling period is "open")
- "Open"

Those categories were re-constructed after the survey to fit with respondents' answers but there is still some specificity. In UK the rule can be "closed if closing device exists, sealed if not": in those cases both categories are selected in the analysis. In Sweden, the rule can be "closed if the closing device leads to a perfect airtightness of the opening, sealed if not", thus in those cases, for Sweden, the category "sealed" has been selected.

The category "depends on EP-calculation" is only used in Czech Republic and France.

Openings for ventilation

EN 13829 requires to seal air terminal devices of mechanical ventilation or air conditioning system and to close other ventilation openings for purposes of method A and seal them for method B.

The general requirements of EN 13829 (§5.2.3) does not detail what is included in air terminal devices of mechanical ventilation (local, general, permanent, intermittent…) and do not detail what shall be done with openings for natural ventilation that do not have closing devices.
To overcome this problem, we have agreed on the following additional definitions:

- Openings for whole building mechanical ventilation or air conditioning: openings that have one side plugged on a duct linked to a fan that provides general and permanent ventilation for the building. "Permanent" means that it cannot be turned off manually; this includes permanently regulated system which may sometimes be automatically turned off (for example when offices are unoccupied).
- opening for local mechanical ventilation of air conditioning with intermittent use: openings that have one side linked to a local fan that can be turned off manually (not including kitchen hood)
- ventilation openings for natural ventilation: straight openings from outside to inside (or vice-versa) including opening that contribute to the general ventilation system in association with openings for mechanical ventilation

All respondents agreed that openings for whole building mechanical ventilation or air conditioning have to be sealed.
Regarding openings for local mechanical ventilation of air conditioning with intermittent use, the answer is "sealed" for 7 out of the 9 respondents (or closed in UK) and "Depends on the EP-calculation" for 2 out of the 9.
Last, regarding ventilation openings for natural ventilation, the answer is sealed for 7 countries out of 9 (or closed in UK), "depends on EP-Calculation" for one and closed for Belgium (which is consistent with Method A).

Opening for the heating system and smoke exhaust

EN 13829 only requests to avoid exhaust hazards from heating systems.
Inlet and outlet openings for the main heating system (this may include air intake of a combustion appliance, chimney flue, air handling unit…) shall be sealed in Estonia, Denmark, France, Germany and UK and closed in Ireland and Belgium.
Inlet and outlet openings for additional heating system shall be sealed in Estonia, Denmark, Germany and UK, prepared consistently with energy performance calculation method in France and closed in Belgium and Ireland.
If we take a closer look at the preparation of various specific components we can see that the interpretation on general rules differs (Figure 1 and Figure 2).
For example, chimney flue either have to be sealed (5 out of 9) or it "depends on EP-calculation" (1 out of 9), or it shall be closed (1 out of 9) or it shall be left as (2 out of 9). And it is about the same with chimney rear ventilation opening.

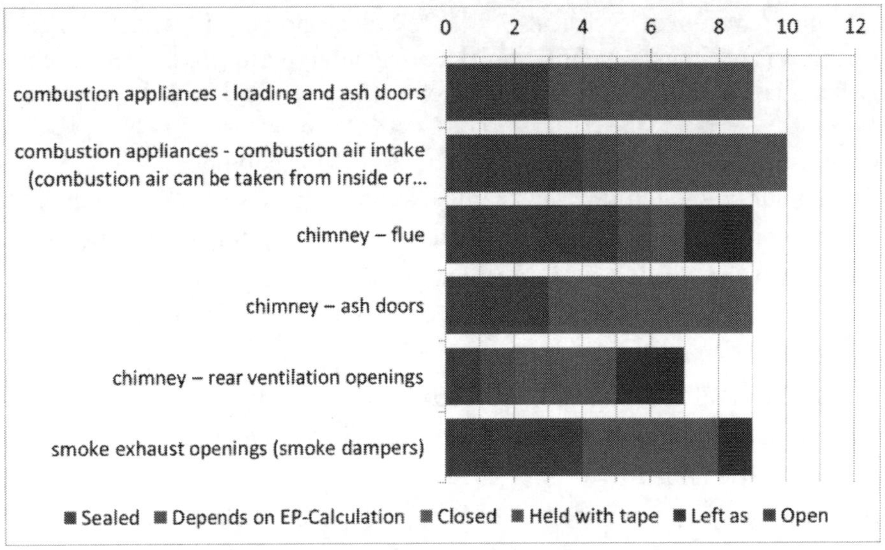

Figure 1: Preparation of openings for heating system and smoke exhaust

Specific openings

We have listed 6 "other envelope openings" but this list is not exhaustive.

Belgium remains consistent with method A: those openings are "left open" or only "held with a tape". France mostly recommends to leave them in normal position of use or to be consistent with energy performance calculation. Estonia is consistent with method B and seals all of them. Other countries (Czech Republic, Denmark, Germany, Ireland, Sweden and UK) have a preparation depending on the kind of opening Figure 2 .

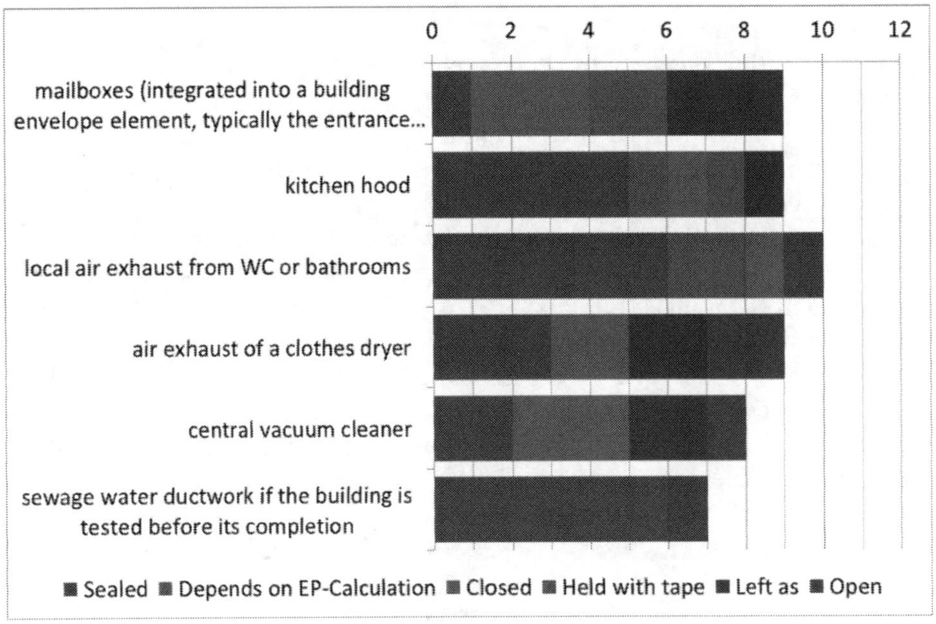

Figure 2: Preparation for "other openings"

4 DISCUSSION

The analysis of the questionnaires brings to light the importance of the purpose of the measurement. Regarding the time of measurement, if the purpose of the test is to estimate the energy losses of the building by leakage it seems obvious that the test shall be done after all work that may affect airtightness is completed.

Nevertheless, requiring test after the completion of the building can lead to organization problem in some cases --e.g., in multi-family buildings some dwelling may be occupied while others are not completed yet. Furthermore when the work is completed it is hard to improve efficiently the airtightness. This probably explains why some countries (BE, SE) have chosen not to require the completion of the building before the test is performed.

Of course, some designs allow testing before completion without compromising the validity of the test for regulatory compliance purposes. This could apply for instance when interior membranes are used as air barriers and the design is such that there are little penetrations of this barrier after it is installed. On the other hand, other systems may be severely affected by works until completion—e.g., interior gypsum boards used as air barriers. It seems difficult to go into that level of detail in a regulatory framework.

Regarding the building preparation, the survey shows that most countries agree on basic principles such as closing windows, doors and trapdoors and leaving internal doors open. On the other hand, the interpretation of the purpose and details of methods A and B varies significantly.

To check the compliance to the energy regulation, the general rule should be to seal all openings for which the air flow is taken into account in the calculation, and leave in normal position of use all other openings. The underlying idea is closer to method A but it can deviate significantly depending on aspects and components taken into account in the calculation method.

Method B is appropriate if the purpose is to impose a minimum requirement for the envelope only, excluding all penetrations (due to HVAC equipment's, mailbox, etc.). However, it is unclear to us how useful this requirement would be. It would be similar to a minimum requirement for the level of insulation of some of the walls but not all of them. Therefore, motivations behind the choice of method B versus method A are rather unclear to us. It seems that the issue of consistency with the energy performance calculation method is not well understood. Method B seems mostly chosen by default because method A is too restrictive.

Those variations in interpretations mean that it is impossible to compare measurements and building airtightness performance from one country to another, unless prior check of the consistency of the building preparation on the samples analyzed. Moreover, it appears challenging for a tester to perform tests in a foreign country unless he well acquainted with the specificities of building preparation in that context.

Nevertheless, there are two obstacles to harmonize pressurization test methods between countries:

- In each country the test method shall be consistent with the energy performance calculation method. But, there are significant differences in calculation methods between countries in the way airflow rates are taken into account.
- There are specific devices and construction traditions in each country that may require tailored rules.

As an illustration of the last obstacle, closing devices are forbidden on air inlets in France. However, the test can only be done with those inlets sealed as most of them self-adjust their position according to the pressure difference. This is nevertheless consistent with the calculation method which takes into account the contribution of these devices to the airflow rates.

Maybe, the standard could leave the possibility to define a measurement protocol at national level consistent with the purpose of airtightness testing.

Nevertheless, to help decision makers setting it, a protocol schemes could be defined in the standard. This scheme may give a detailed list of openings and how to determine the way they should be prepared to be consistent with the test purpose.

5 CONCLUSION

This study shows that building preparation in the 11 countries investigated differs significantly from one country to another even when the same method is used and without clear and technically-sound motivations behind the choices. Therefore, it is clear that building airtightness should not be compared between countries unless precaution is taken in the sample analyzed and in the interpretation of the results.

To address this issue, the revision of the standard shall, in one hand, describes more precisely the basic principles and avoid ambiguities and, in another hand, allows some flexibility to the countries to specify rules consistent with the energy performance calculation method. At the same time, it should allow the identification of tests performed under comparable conditions to allow meaningful comparative analysis between countries.

6 ANNEX

The following table gives references of specific guidelines for each country

Country	**France**	**UK**	**Czech republic**
Name of the document where the guidelines are specified	GA –P 50-784 (application guide of NF EN 13829)	ATTMA Technical Standard L1 – Measuring Air Permeability of Building Envelopes (Dwellings)	Metodický pokyn - Pravidla pro měření průvzdušnosti obálky budovy (Methodical instruction – Rules for measurement of air permeability of building envelopes, in Czech)
Author or editor of the document	Afnor	The Air Tightness Testing and Measurement Association (ATTMA) Technical Committee	Association Blower Door CZ, www.asociaceblowerdoor.cz
Institution which issued the document	Working group set by the ministry of ecology	The Air Tightness Testing and Measurement Association (ATTMA)	Státní fond životního prostředí České republiky (State environmental fund of the Czech Republic)
Date of issue	February 2010	October 2010	not yet issued, prepared for publication (state from September 2013)
The document is available (for downloading or purchase) at:	http://www.boutique.afnor.org	www.attma.org	http://www.nzu2013.cz/zadatele-o-dotaci/rodinne-domy/dokumenty/metodicke-pokyny/
Comments:	This document is under revision		Document issued as a part of instructions for subsidy applicants from the New green savings programme

Country	Belgium	Denmark	Germany	Ireland
Name of the document where the guidelines are specified	Spécifications supplémentaires	"Metoder for certificerede virksomheders trykprøvning efter EN 13829 og gældende bygningsreglement" version 1,	EnEG_Staffel 11 – Auslegungsfragen zur Energieeinsparverordnung – Teil 11	ATTMA publication "Measuring air permeability of Building Envelopes" & EN 13829
Author or editor of the document	Governments of the three regions	Klimaskærm	Dr. Justus Achelis, DIBt	ATTMA/CEN
Institution which issued the document	Governments of the three regions	Klimaskærm	Fachkommission Bautechnik der Bauministerkonferenz	ATTMA/CEN
Date of issue		5. September 2013	9.3.2004; 9.12.2009	October 2010 Issue - 18 October 2000
The document is available (for downloading or purchase) at:	www.epbd.be	www.klimaskaerm.dk	https://www.dibt.de/de/Service/Dokumente-Listen-EnEV.html	http://www.attma.org/wp-content/uploads/2013/10/ATTMA-TSL1-Issue-1.pdf

7 ACKNOWLEDGEMENT

Authors address a special thanks to the TAAC members who have kindly accepted to answer the questionnaire:

- Clarisse Mees, Belgium
- Jiri Novak, Czech Republic
- Walter Sebastian, Denmark
- Valerie Leprince, France
- Oliver Solcher, Germany
- Paul Carling , United Kingdom
- Targo Kalamees , Estonia
- Andrejs Nitijevskis, Latvia
- Andrzej Gorka, Poland
- Owe Svensson, Sweeden
- Mark A. Shirley, Ireland

This work was supported by INIVE and BCCA.

8 REFERENCES

Bailly, A. L. (2012). *Numerical evaluation of airtightness measurement protocols.* Copenhagen: AIVC conference 2012.

Delmotte, C. (2013). *Airtightness of buildings - Calculation of combined standard uncertainty.* Athens: AIVC conference 2013.

Delmotte, C. L. (2011). *Interlaboratory Test for the Determination of repeatability and Reproducibility of Buildings Airtightness Measurements.* Brussels, Belgium: AIVC Conferences, 32nd, Proceedings.

Rolfsmeier, S. V. (2010). *Ringversuche zu Luftdurchlässigkeitsmessungen vom Fachverband Luftdichtheit im Bauwesen e.V.,.* Buildair.

Sherman, M. a. (1995). *Uncertainty in Fan Pressurization Measurements.* Lawrence Berkeley National Laboratory: ReportLBNL-32115.

THE IMPACT OF AIR-TIGHTNESS IN THE RETROFITTING PRACTICE OF LOW TEMPERATURE HEATING

Qian Wang*[1], Sture Holmberg[1],

[1] *Division of Fluid and Climate Technology, Department of Civil and Architectural Engineering, Royal Institute of Technology (KTH)*

Brinellvägen 23, 100 44 Stockholm, Sweden
Corresponding author:* ***qianwang@kth.se*

ABSTRACT

In Sweden, the energy usage in existing residential buildings amounted to 147 TWh in 2012, equivalent to almost 40 % of the final overall national energy usage. Among all the end users in building service sectors, 60 % of the final energy in Sweden is used for space heating and domestic hot water (DHW) production in 2013. In order to reduce the supply temperature for space heating in existing buildings, combined approaches are favorably adopted: to reduce the net energy demand by air-tightness and insulation retrofits; and renovate the conventional high temperature heating to low temperature heating (LTH) systems. As an energy-efficiency alternative, LTH technology has shown promising advantages and shortcuts to improve the coefficient of performance (COP) of heat pump system, which further saves primary energy. However, existing modeling achievements and field testing reveal that the attained application of LTH has a relatively high requirement to the air-tightness in new constructed single-family houses. Moreover, in some leaky multi-family building stock with low envelope surface temperature, LTH may have limited energy saving potentials. How to evaluate the impact of air-tightness for the LTH implementation and energy saving potentials in existing houses are not sufficiently attained so far. This paper presents a modeling approach combining LTH simulation with air-tightness evaluation, aimed to estimate whether the selected existing building types can cope with LTH with upgraded primary energy savings. In addition, the impact of air-tightness retrofits for LTH implementation in selected Swedish residential buildings is of interests.

In the simulation Consoli Retro are employed to simulate the energy performance. It is revealed that the combined effect of floor heating/ ventilation radiators and air-tightness retrofits to 1/1.5 ACH can contribute 19 % to 36 % primary energy savings in total. However, different LTH systems and archetypes have varies sensitivities to air-tightness retrofits. Benchmark the impact of air-tightness to different LTH systems needs further investigations among other archetypes and on-site measures for future application of LTH on a larger scope.

KEYWORDS

Air-tightness, retrofitting, energy savings, low temperature heating, Swedish residential buildings

Nomenclatures	
Acronyms	
ACH	Air changes rate, h^{-1}
BBR	Swedish building regulations
CHP	Combined heat and power
COP	Coefficient of performance
DH	District heating
DHW	Domestic hot water
FH	Floor heating (hydraulic)
HP	Heat pump
LTH	Low temperature heating
PE	Primary energy
PEF	Primary energy factor
VR	Ventilation radiator (low-temperature)

T_1	Building type 1, Swedish slab houses (low raise), before 1950
T_2	Building type 2, Swedish slab house (three- to four storeys), 1960–1975
T_3	Building type 3, Swedish slab house (high raise), 1970–1975
Symbols	
$U-value$	Heat transfer coefficient of building elements, W/m²K
t_{op}	Operative temperature, °C
t_a	Air temperature, °C
t_r	Mean radiant temperature, °C
$E_{F,HOB}(i)$	Energy for fuel type *i* during heat production provided in heat boilers
$E_{F,CHP}(i)$	Energy for fuel type *i* during heat production provided in CHP
$PEF_{HOB(i)}$	Primary energy factor for fuel type *i* during heat production provided in heat boilers
$PEF_{CHP(i)}$	Primary energy factor for fuel type *i* during heat production provided in CHP
$\alpha_{h,i}$	Allocation factor for on-site or off-site production for fuel type *I*
Φ_E	Total energy demand, kWh/m²
Φ_H	Monthly heating demand, kWh/m²
Φ_{EL}	Electricity demand, kWh/m²
Φ_{DHW}	Domestic hot water energy demand, kWh/m²
$\Phi_{H,T}$	Transmission heat loss, kWh/m²

1. INTRODUCTION

In Sweden, existing residential building stock comprises approximately 2.5 million dwellings, including apartment units and multi-family houses, and approximately 2 million detached or semi-detached single-family houses/villas [1]. The energy usage in this part amounted to 147 TWh in 2012, equivalent to almost 40 % of the final overall national energy usage [2]. As a baseline and essential technique, energy retrofitting is considered as an effective way to accelerate the sustainable transformation of existing Swedish building stock [3]. However, the industry approach and pilot project typically oriented with operational energy costs savings or tap-water savings/treatments, therefore, the retrofitting solutions tend to be highly case-specific and conventional, the primary energy saving potentials are limited [4][5][6].

As an energy-efficiency alternative, low temperature heating (LTH) technology has shown promising advantages and shortcuts to improve the efficiency of heat supply in terms of improved coefficient of performance (COP) with HP (heat pump), thermal comfort contributions and easily installed solution [7]. The advantages further provides more renewable based heating solutions with upgraded primary energy (PE) savings [8]. Nevertheless, theological studies have revealed that the air-tightness and energy demand of the existing buildings play major roles for the energy performance of LTH [9][10]. More importantly, little is known about the impact of air-tightness, particularly to those buildings which are planning to be renovated by low-temperature ventilation radiator (VR) or hydraulic floor heating (FH). Available models and studies from other countries are mainly based on local building energy codes and national heating directives. For example, Hasan et al. [8] and Cellura et al. [11] investigated both the delivered and primary energy saving potential of FH and low temperature radiators with respect to the Finnish climate condition. The findings revealed that with a reduction of supply/return temperature to 40/35°C, the LTH can save both delivered energy and PE without compensations on thermal comforts (1.0 m-1.3 m elevations). Furthermore, the energy usage of FH in the bathroom of the studied buildings can be amounted up to 33% to 43% of the total energy use. Specific to Swedish residential buildings, Energy Europe TABULA project [12] performed a general energy retrofitting guideline based on 44 typology categories of existing Swedish residential buildings for simplified heating system alternatives with respect to energy demand retrofits. Zou [13] developed a bottom-up approach to classify and assess existing Swedish buildings by improving the air infiltration database and construction techniques. Hesaraki and Holmberg [14] and Myhren and Homberg [7] evaluated long-term energy savings by low-temperature

VR in Swedish multi-family houses. It is found that with the air-tightness level of 0.68 l/(s m^2), annual on-site measurements shows 48 kWh/(yr m ^{2}) to 55 (kWh/yr m ^{2}) energy usages for both space heating and DHW can be achieved when the buildings are equipped with LTH and HP. Gustavsson, Dodoo, Truong and Danielski [15][16] modeled the combined effects of heat supply and demand retrofits considering four major types of heat production systems in Sweden. It is found that the PE savings are largely dominated by the heat producing systems and the capacities to reduce the existing energy demand, in which air-infiltrations are commonly one of the most sensitive parameters for the studied archetypes. Other possible software and modelling techniques, including IDA ICE, Design Builder/EnergyPlus, Trnsys, eQuest[@], have been employed in some LTH practices to evaluate different heating parameters that impact the energy performance and thermal comfort before and after retrofitting [8][17][18][19]. The models are capable of providing relatively accurate one- or multi-zone air temperature and radiant temperature simulations for the reference buildings. However, these tools have had limited usage in retrofitting Swedish residential buildings and are not easily adapted to larger contingents of similar archetypes under Swedish climate conditions. Based on the target, this study simulated the PE saving potentials led by LTH retrofits and further defines the impact from air-tightness variances based on the current air penetration levels in the selected archetypes.

2 METHODOLOGY AND SIMULATION MODEL

2.1 Energy performance model

The main advantages of installing LTH in retrofits are the potential of reducing primary energy and providing more sustainable heating energy alternatives along with thermal comfort contributions. Designed with Excel tools, Consolis Retro is employed in the study. The model is based on the simplified calculation and parametric analysis of energy usage, applying EN ISO 13790 calculation methodologies [20]. The model is capable of handling 1 or 2-zones at the same time for the reference building. The building block was set heat balanced with variable major parametric factors that impact the heat loss and heat distributions in the calculation zone. Parameters are set previously to indicate the building archetypes. The total net energy usage Φ_E is calculated from Equation (1):

$$\Phi_E = \Phi_H + \Phi_{DHW} + \Phi_{EL} \tag{1}$$

To simplify the calculation process, the transmission heat loss $\Phi_{H.T}$ is calculated by building envelope parameters and linear thermal bridges. Old Swedish slab houses might be constructed with no insulations with cold surface temperatures, this will lead to rather high differences between operative and air temperature [11]. In another word, occupants may feel colder than the air temperature is set 20 °C. As a result, 22°C is set for air temperature of heated space in the modeled archetypes. The operative temperature of the buildings are gained by Equation (2)

$$t_{op} = (t_a + t_r)/2 \tag{2}$$

In the model, delivered energy are calculated as net building demand of the selected archetypes, primary energy saving potentials is calculated by both the delivered energy and primary energy factor (PEF) variances before and after LTH retrofitting, which is obtained as Equation (3) [21]:

$$\text{PEF} = \frac{\sum_{i=1}^{n}(E_{F,HOB\,(i)} * PEF_{HOB(i)}) + \sum_{i=1}^{n}(\alpha_{h,i} * E_{F,CHP\,(i)} * PEF_{CHP(i)})}{\sum_{j=1}^{n} Q_{del,j}} \tag{3}$$

In Sweden, district heating (DH) accounted for around 60 TWh in 2013, which is considered as the most common space heating system for existing multi-family houses and apartment blocks. In single family houses, district heating and electricity are used in 7 % and 22 % of all detached and semi-detached houses, respectively[22]. Swedish district heating are mainly produced by combined heat and power (CHP) for residential buildings, it has a PEF of 0.5-1.3 depending on the energy sources (waste heat, biomass, coal and natural gas, etc) [23].Within the last decades, heat pump (HP) shows increasing competences with DH because of its PE saving potentials when designed with LTH. Up to 2012, Sweden has the largest application of HP systems for both new and retrofitting buildings among EU [24]. The PEF of LTH combined HP systems are calculated based on the supply temperature and the COP of the heat pump [25]. Validation and testing of the calculation model was conducted [26]. The tool was compared with IDA ICE and EnergyPlus for accuracy analysis, with an acceptable agreement of 0 to 8 % error [27]

2.2 Air-tightness retrofits in existing Swedish residential buildings

Air-tightness is one of the most significant parameters to not only provide hygienic protection for the occupants but also reduce the operational energy usage[28]. Sweden has a relatively long heating season (6-7 months), it is found that the impact of air-tightness can be higher than transmission heat loss through building envelopes in some Swedish detached/semi-detached houses. In addition, the large application of exhaust ventilation systems in Sweden led to a relatively higher air-infiltration compared with balanced ventilation systems [29]. The air-tightness retrofits have been commonly recommended in some conventional renovation projects, nevertheless, despite measurements and blow-door tests have been conducted in pilot houses, the existing information on ait-tightness and its impact to energy usage are still scarce, particularly for those buildings heated with reduced supply temperature lower than 50°C [30]. In 2012, the revised BBR (Swedish building regulations) provides no specific limit values in respect of tightness, but the significance of good ventilation is stressed in an advisory in order to decrease the moisture damage and hygienic issues. To obtain a well performance of LTH, a good guideline minimum value for 0.80 l/(s m^2) and 0.35 l/(s m^2) surface area at a pressure difference of +/- 50 Pa are recommended for existing and new Swedish residential buildings, respectively [31]. However, 1400 existing Swedish building stock statistics from derived field studies shows that the actual air-tightness level ranges from less than 0.3 l/(s m^2) to approximately 1.5 l/(s m^2), at pressure difference of 50 Pa [13]. And it varies largely among different archetypes and exhaust/balanced ventilation systems. Figure 1 shows the air-tightness level of existing residential buildings in Sweden compared with other countries [13][29].

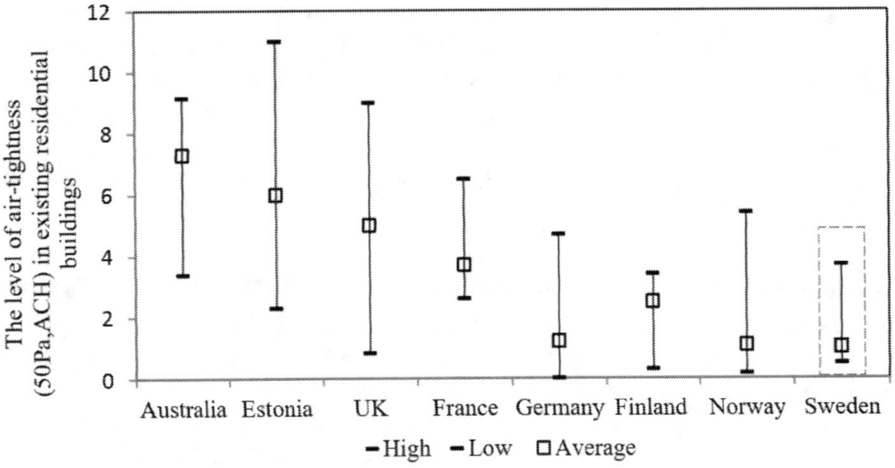

Figure 1 The air-tightness level (50 Pa) of Swedish residential buildings compared with other countries.

The existing retrofitting techniques in Swedish residential buildings are based on the following three perspectives in the models [32]:

- Insulate the air gaps existed in joints between ceiling/floor/balcony to the walls, particularly for two or three storey slab houses.
- Install more efficient mechanical (balanced preferably) ventilation systems.
- Insulate the ventilation studs and piping systems.

The improvements and variances of air-tightness level are based on the existing performance of in the selected building types.

2.3 LTH retrofits and selected archetypes

The heating system in Swedish slab houses is usually district heating (DH), occasionally heated partially by electricity, gas, oil and renewable sources in some renovated cases [23]. To standardize the archetypes, low-rise slab houses are classified in this study by age according to three periods: pre-1950 (T_1), 1951-1960 (T_2). Additionally, special booming time 1965-1975 for high slab apartments (T_3) is chosen in the category. Two types of LTH are selected as the retrofitting alternatives: FH and VR, the structure and components are shown in Figure 2. For FH, in order to fit the existing old slab floor in renovation practice, overlay floor panels are installed with embedded PEX tubing circus, shown in Figure 2, left. FH is set as 100 W/m^2 heat outputs with design temperature 35 °C/ 29 °C. The coverage area is 12 m^2/circuit. The new floor layers are set as tiles in bathroom and laminate in other rooms. For VR (shown in Figure 2, right), the cold air is preheated by the radiator through the slab wall vents and filtered as clean warm air. Because most of the selected multi-family houses have installed exhaust ventilation, 10 P pressure drop between indoor and outdoor are set as the driven force for the cold air , no extra energy is needed for the convection [33]. The supply/return temperature is designed with 45 °C/ 35 °C in the study for VR.

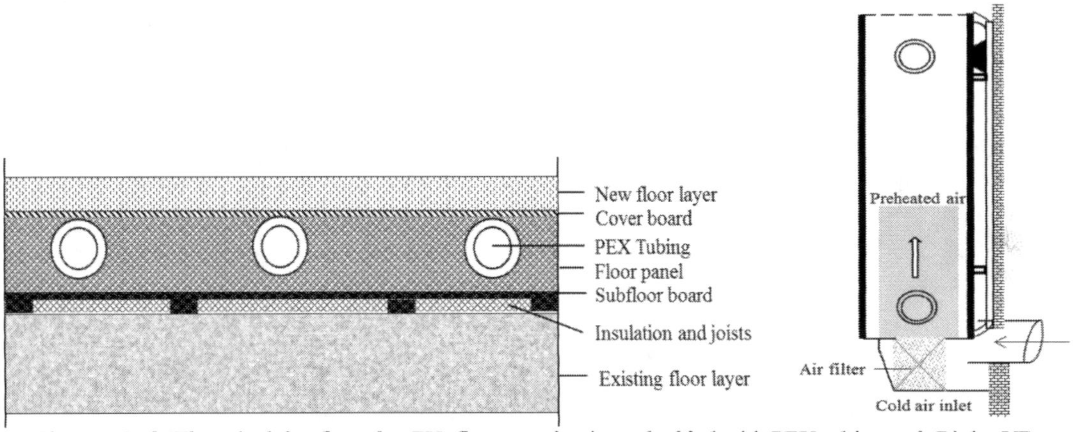

Figure 2 Left, The principle of overlay FH, floor panel units embedded with PEX tubing and, Right, VR, designed as preheated air and low-temperature radiator

In each archetype, four retrofits are designed and compared for implementing LTH, these are indicated in Table 1. For system 1, living rooms are renovated with VR under the windows, two-pipe hydraulic existing radiators are kept in the rest rooms. For system 3, hydraulic FH are implemented only in bathrooms, the rest rooms are kept with existing high temperature radiators. For system 2 and 4, the whole buildings are renovated by VR and FH, respectively. The parameters of the LTH retrofits are indicated in Table 1.

The building material for floors and ceilings in the studied archetypes was primarily 10- 50-centimeter-thick, reinforced concrete, and relatively thinner slabs were applied as exterior paving and coating [6][1]. Although Swedish slab houses may have different facades and terraces, the buildings' main elements and service systems are similar. Three types of Swedish slab houses (T_1, T_2, T_3) constructed during different years were selected for this retrofitting investigation. The archetypes, building features and parameters were generalised and collected through surveying the statistics; these are presented in Table 2. The corresponding energy system retrofits are designed as water-to-water HP. The PEF are calculated by the supply temperature and the COP of HP [34], shown in Table 3.

Table 1: LTH retrofitting designs

System	Retrofit room	Supply/return temperature (°C)	Non-retrofit room	Supply/return temperature (°C)
System 1	VR in living rooms	45/35	Conventional radiator	55/45
System 2	Whole building VR	45/35	-	
System 3	FH only in bathroom	35/30	Conventional radiator	55/45
System 4	Whole room FH	35/29	-	-

Table 2: Selected archetypes and energy systems for retrofitting analysis

	T1	T2	T3
Archetypes			
Dwelling types	Single family house	Multi-family house	Apartment block
Age	Before 1950	1960-1975	1970-1975
Foundation	Lightweight concrete	Concrete slab	Polished concrete
External wall	10 cm mineral wool insulation	13 cm mineral wool insulation	15 cm mineral wool insulation
Window	Double glazing, aluminium frame	Double glazing, timber frame with ventilation fan	Double glazing, aluminium frame with one-side ventilation fan
Roof /ceiling	Brick and cutter coke ash insulation	Flat roof covered with cardboard and mineral wool	Concrete foundation with galvanized sheet metals, mineral wool insulation
Ground floor	Linoleum and coke ash	Slab covered with linoleum mats or plastic board	Slab covered with mineral wool or linoleum
Heating	Furnaces /electricity	District heating	District heating
Radiator	Furnaces/electricity	Single-pipe hydraulic radiator	Two-pipe hydraulic radiator
Ventilation	Natural ventilation	Exhaust ventilation	Exhaust ventilation
Energy mix	Gas/oil/partly el.	CHP	CHP
Air-tightness	2 ACH	2 ACH	5 ACH

Table 3: PEF modeling of LTH retrofits

Supply/return Temperature (°C)	Heating system	COP	Energy mix	PEF
70/60	Conventional supply temperature District heating	-	CHP	0,90
50/40	Medium supply temperature output District heating	3.1	CHP	0.80
45/35	Low supply temperature output	3.5	HP	0.68
40/30	Low supply temperature output	3.6	HP, Nordic mix	0.68
35/29	Low supply temperature output	3.8	HP, Nordic mix	0.60

3 REUSLTS AND DISCUSSION

All archetypes were selected within the same Swedish climate zone III, Stockholm, for comparison. Figure 3 shows the monthly energy flow before and after implementing retrofitting for the selected archetypes. Among the four archetypes, system 4 (whole building with FH) shows the largest energy savings from 26 % to 33 % after retrofitting compared with other systems. Among all the archetypes, relatively new archetype (T_3) shows the highest PE saving potentials. Followed by system 2 and system 1 (whole building VR and living room VR), PE savings range from 20 % to 25 % and 15 % to 18 %, respectively. Among the three archetypes, older houses (T_1) show lower saving potentials compared with apartment block.

The reason could be that the old multi-family houses are more sensitive to the air-infiltrations due to their existing leaky conditions. The exhaust ventilation system installed in T_2 and natural ventilation in T_1 make the envelope leakage larger when installed with VR, compared with other archetypes. System 3 shows the lowest energy saving potentials compared with other types in both FH and VR. It didn't show promising energy savings in selected single family houses and low-raise multi-family houses. Furthermore, T_1 shows the lowest energy savings when renovated only in bathroom FH (system 3). The reason can be its limited heated bathroom areas. Among the archetypes, the bathroom FH retrofits shows the greatest savings for apartment block T_3 (5 % savings). Attention should be paid that the operative temperature in bathrooms sometimes can be 3°C to 5 °C higher than the rest rooms, practically. With respect to the occupations, this will lead to an increased uncertainties when focusing on modeling the energy savings only in bathrooms and its conjunct effects with other heated zones [8].

Figure 4 shows the primary energy saving potential after implementing both LTH and air-tightness retrofits. The impact of air-tightness shows linear reduction of PE. Due to the high variety of ventilation vent designs in the windows, the high slab house (T_3) is set with greater variances. The rests of the archetypes are set as approximately from 2.0 ACH to 1.5/1.0 ACH before and after retrofitting. The combined effect of LTH and air-tightness shows that the energy saving potentials from 28 % to 36 % in most of the archetypes can be achieved. Among all the archetypes, T_2 shows the highest sensitivity to the air-tightness retrofits, particularly for VR. When the air-tightness level is reduced to 1.5 ACH, 38 % of PE savings can be achieved by VR retrofits. In addition, T_3 shows the lowest impact by air-tightness to the LTH in general (approximately 6 % to 19 %). The reason can be the existing ventilation systems have a relatively higher performance in the archetypes. In addition, the modern constructions made the building envelope much less sensitive by the air-infiltration through joists and ventilation ducts.

4 CONCLUSIONS

In the study, a simplified calculation model is developed and integrated with parametric investigations to three major Swedish archetypes that are planned to be renovated with FH and VR. PE (Primary energy) saving potentials led by LTH retrofits is in focus. The variation in terms of air-tightness levels among the archetypes are of interests. It is revealed that the PE savings can be up to 33 % depending on the LTH systems. FH in all rooms shows the highest savings in most archetypes while VR shows high savings in relatively modern archetypes. The air-tightness retrofits shows 4.5 % to 6 % energy savings in most archetypes except the highest savings 18 % in T_2. The combined effect of air-tightness and LTH retrofits can contribute 19 % to 36 % PE savings in total; however, the VR retrofits shows high limitation and sensitivities in T_2. Furthermore, high slab houses T_3 shows the relatively stable PE saving levels and the impact by air-tightness is relatively low among all the studied archetypes. Given the limited data sources and the basic target for performing the air-tightness impact analysis, retrofits from building demand sides in terms of wall insulation, windows and on-site measurements are not included in the current analysis, which will be further performed and verified.

5 ACKNOWLEDGEMENTS

The authors are grateful to Nordic Innovation (project NB 13339), Formas for providing financial support, as well as the building owners and radiator industries for contributing valuable information and empirical documents.

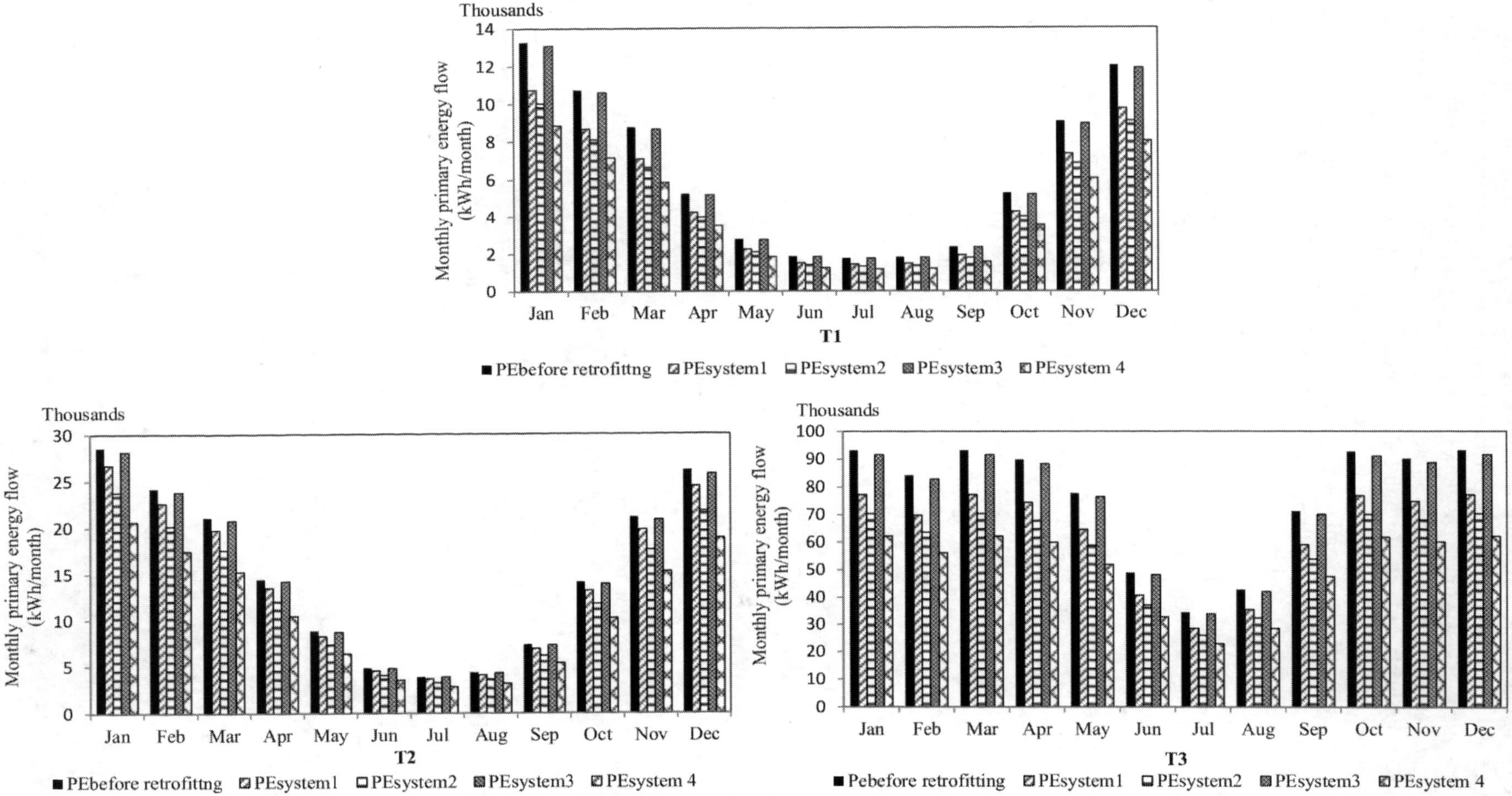

Figure 3 Energy monthly flow of the archetypes installed with LTH systems before and after retrofitting

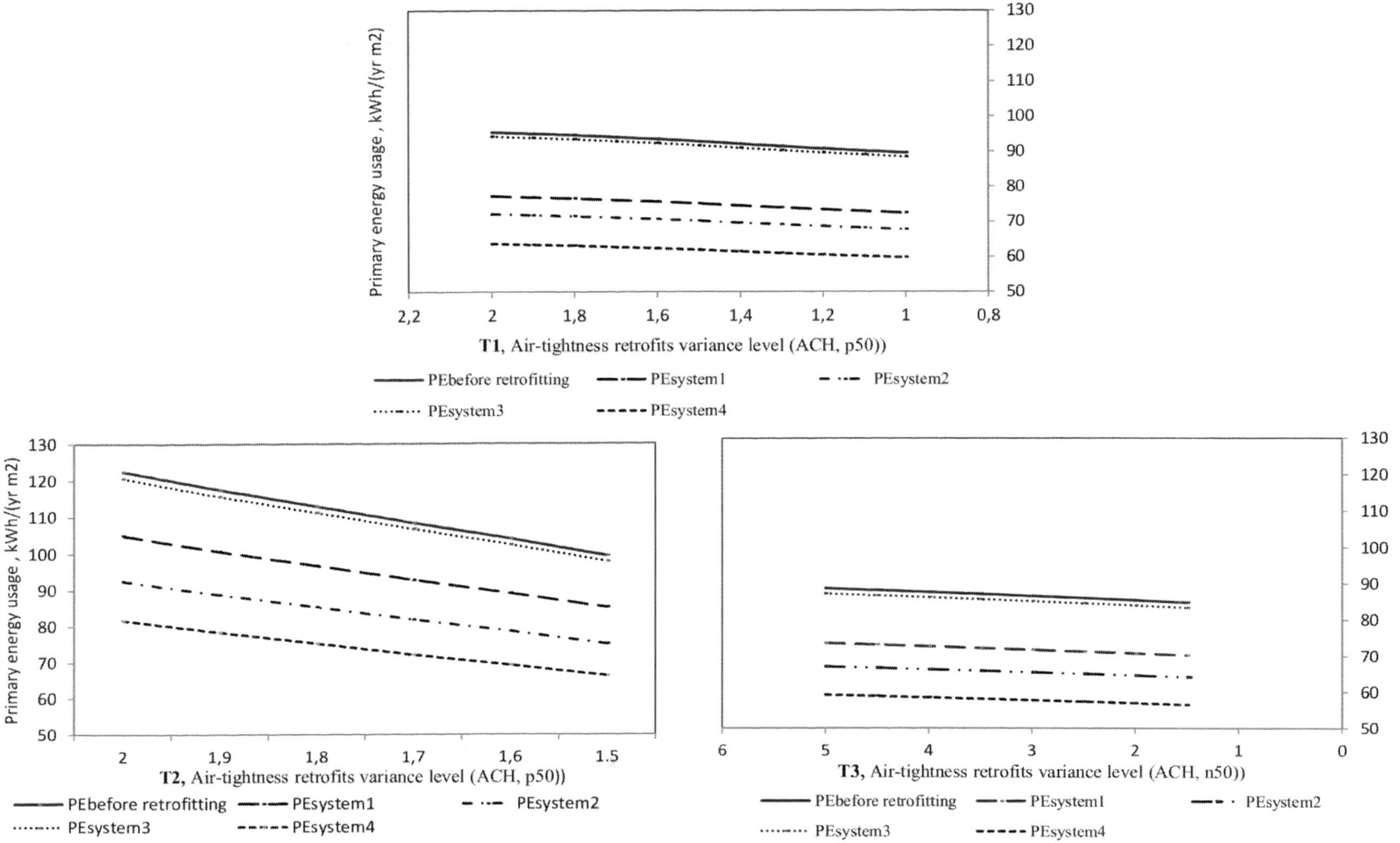

Figure 4 The impact of air-tightness variance levels to the primary energy usage of studied archetypes (kWh/yr m^2)

6 REFERENCES

[1] C. Björk, P. Kallstenius, and L. Reppen, *Så byggdes husen 1880-2000 (As built houses 1880-2000)*. Forskningsrådet Formas, 2002.

[2] Energimyndigheten (Energy Agency), "Energy in Sweden," Swedish Energy Agency, Eskilstuna, Sweden, 2012.

[3] V V S Företagen, *Renoveringshandboken - för hus byggda 1950-75 (Renovation handbook for houses built in 1950-75)*. Stockholm: Wallén Grafiska AB, 2009.

[4] A. Göransson and B. Pettersson, "Energieffektiviseringspotential i bostäder och lokaler - Med fokus på effektiviseringsåtgärder 2005 – 2016," Chalmers University of Technology, Report, 2008.

[5] M. Economidou, B. Atanasiu, C. Despret, J. Maio, I. Nolte, and O. Rapf, "Europe's Buildings under the Microscope. A Country-by-Country Review of the Energy Performance of Buildings," Buildings Performance Institute Europe (BPIE), 2011.

[6] Q. Wang, "Toward Industrialized Retrofitting : Accelerating the Transformation of the Residential Building Stock in Sweden," 2013.

[7] J. A. Myhren and S. Holmberg, "Improving the thermal performance of ventilation radiators – The role of internal convection fins," *Int. J. Therm. Sci.*, vol. 50, no. 2, pp. 115–123, Feb. 2011.

[8] A. Hasan, J. Kurnitski, and K. Jokiranta, "A combined low temperature water heating system consisting of radiators and floor heating," *Energy Build.*, vol. 41, no. 5, pp. 470–479, May 2009.

[9] J. A. Myhren and S. Holmberg, "Flow patterns and thermal comfort in a room with panel, floor and wall heating," *Energy Build.*, vol. 40, no. 4, pp. 524–536, 2008.

[10] K. Högdal and WSP Environmental, "Halvera Mera,Slutrapport (Half more, final report)," BeBo (Energy Agency purchaser group for energy efficient apartment buildings), Stockholm, 2013.

[11] M. Brand and S. Svendsen, "Renewable-based low-temperature district heating for existing buildings in various stages of refurbishment," *Energy*, vol. 62, pp. 311–319, Dec. 2013.

[12] TABULA, "Byggnadstypologier Sverige (Building typology in Sweden)." Mälardalen University Sweden, 2009.

[13] Y. Zou, "Classification of buildings with regard to airtightness," 2010.

[14] A. Hesaraki and S. Holmberg, "Energy performance of low temperature heating systems in five new-built Swedish dwellings: A case study using simulations and on-site measurements," *Build. Environ.*, vol. 64, pp. 85–93, Jun. 2013.

[15] L. Gustavsson, A. Dodoo, N. L. Truong, and I. Danielski, "Primary energy implications of end-use energy efficiency measures in district heated buildings," *Energy Build.*, vol. 43, no. 1, pp. 38–48, Jan. 2011.

[16] N. L. Truong, A. Dodoo, and L. Gustavsson, "Effects of heat and electricity saving measures in district-heated multistory residential buildings," *Appl. Energy*, vol. 118, pp. 57–67, Apr. 2014.

[17] F. Ardente, M. Beccali, M. Cellura, and M. Mistretta, "Energy and environmental benefits in public buildings as a result of retrofit actions," *Renew. Sustain. Energy Rev.*, vol. 15, no. 1, pp. 460–470, Jan. 2011.

[18] S. V. Russell-Smith, M. D. Lepech, R. Fruchter, and Y. B. Meyer, "Sustainable target value design: integrating life cycle assessment and target value design to improve building energy and environmental performance," *J. Clean. Prod.*

[19] T. Y. Chen, J. Burnett, and C. K. Chau, "Analysis of embodied energy use in the residential building of Hong Kong," *Energy*, vol. 26, no. 4, pp. 323–340, Apr. 2001.

[20] ISO, EN, "Energy performance of buildings–Calculation of energy use for space heating and cooling (EN ISO 13790: 2008)." European Committee for Standardization (CEN), Brussels, 2008.
[21] Sweden Green Building Council, "Treatment of Scandinavian District Energy Systems in LEED v1 2012, Energy models for LEED EA credit 1," Stockholm, 2012.
[22] Boverket (Swedish National Board of Housing, Building and Planning), *Teknisk status i den svenska bebyggelsen: resultat från projektet BETSI (Technical status of the Swedish settlements: results from the project BETSI)*. Boverket, 2010.
[23] K. Ericsson, "Introduction and development of the Swedish district heating systems - Critical factors and lessons learned," *RES-HC Policy Proj. Rep. D23*, 2009.
[24] K. Klein, K. Huchtemann, and D. Müller, "Numerical study on hybrid heat pump systems in existing buildings," *Energy Build.*, vol. 69, pp. 193–201, Feb. 2014.
[25] A. Hepbasli and Y. Kalinci, "A review of heat pump water heating systems," *Renew. Sustain. Energy Rev.*, vol. 13, no. 6–7, pp. 1211–1229, Aug. 2009.
[26] G. Jóhannesson, "Building energy - a design tool meeting the requirements for energy performance standards and early design - validation," presented at the Research in Building Physics and Building Engineering, 2006, pp. 627–634.
[27] T. Kalema, G. Jóhannesson, P. Pylsy, and P. Hagengran, "Accuracy of Energy Analysis of Buildings: A Comparison of a Monthly Energy Balance Method and Simulation Methods in Calculating the Energy Consumption and the Effect of Thermal Mass," *J. Build. Phys.*, vol. 32, no. 2, pp. 101–130, Oct. 2008.
[28] T.-O. Relander, S. Holøs, and J. V. Thue, "Airtightness estimation—A state of the art review and an en route upper limit evaluation principle to increase the chances that wood-frame houses with a vapour- and wind-barrier comply with the airtightness requirements," *Energy Build.*, vol. 54, pp. 444–452, Nov. 2012.
[29] P. I. Sandberg, P. Wahlgren, C. Bankvall, B. Larsson, and E. Sikander, "The effect and cost impact of poor airtightness-Information for developers and clients," in *Proceedings of the 10th Conference on the Thermal Performance of the Exterior Envelopes of Buildings, Clearwater Beach, Florida*, 2007.
[30] P. Levin, A. Jidinger, and A. Larsson, "Rekorderlig renovering, Demonstrationsprojekt för energieffektivisering i befintliga flerbostadshus från miljonprogramstiden, Objektrapport för Norrbacka – Sigtunahem Etapp 1 & 2 (Record our renovation, Demonstration projects for energy efficiency in existing apartment buildings from the million program, Object Report for Norrbacka - Sigtunahem Stage 1 & 2)," Jul. 2010.
[31] Boverket (Swedish National Board of Housing, Building and Planning), "Swedish Building Regulations-Building codes BBR." 2012.
[32] M. Dahl, M. Ekman, T. Kuldkepp, and N. Sommerfeldt, *Energy Saving Measures in Existing Swedish Buildings : Material for a book to be published by VVS Företagen*. 2011.
[33] S. Holmberg, J. A. Myhren, and A. Ploskic, "Low-temperature heat emission with integrated ventilation air supply," presented at the Proceedings of International Conference Clima 2010, 2010.
[34] A. Ploskic, "Low - Temperature Basedboard Heaters in Built Environments," 2010.

THE USE OF A ZONAL MODEL TO CALCULATE THE STRATIFICATION IN A LARGE BUILDING

Lien De Backer*[1], Jelle Laverge[2], Arnold Janssens[3], and Michel De Paepe[4]

1 Ghent University- Architecture and Urban planning
Sint-Pietersnieuwstraat 41B4
9000 Ghent, Belgium
**Corresponding author: lien.debacker@ugent.be*

2 Ghent University- Architecture and Urban planning
Sint-Pietersnieuwstraat 41B4
9000 Ghent, Belgium

3 Ghent University- Architecture and Urban planning
Sint-Pietersnieuwstraat 41B4
9000 Ghent, Belgium

4 Ghent University-Flow, Heat and Combustion Mechanics
Sint-Pietersnieuwstraat 41B4
9000 Ghent, Belgium

ABSTRACT

In the past, many churches were raised and in a church building no heating no heating system was installed, except a simple individual coal or peat stove, which could be rented by the churchgoers. The thick high stone walls of the church alleviated the fluctuations of the ambient air temperature and relative humidity. Accordingly, the indoor climate in the church building was quite stable. After the Second World War the living standard of the people increased and the increased prosperity also led to higher comfort demands in churches. As a consequence, many "local" heating systems were often replaced by rugged air inlets, designed to quickly heat the space to increase the thermal comfort. These were designed and operated without taking the effect of the fluctuating temperature and relative humidity on the artwork in the church into account. Consequently, the many artworks in the church building like the organ, the pulpit and the panel paintings are also exposed to this changing climate, leading to faster deterioration or even to damage. Evaluating the conservation conditions for the artworks in a large space requires knowledge of the stratification in temperature and humidity inside a large space that occurs during heating. The use of Computational Fluid Dynamics (CFD) is probably the most suitable method to predict the airflow pattern, but it is quite time consuming and requires a powerful computer. As an alternative a zonal model is a suitable method to predict the airflow in a large space in a simplified way. These models can be linked to a BES- software in which each zone is assumed to be perfectly mixed. By this coupling, the influence of the airflow on the temperature distribution and vice versa can be calculated, in order to judge the thermal comfort and preservation conditions in one zone. This paper presents the coupling of the existing thermal zonal model of Togari with the BES-software TRNSYS. In addition a moisture preservation equation was added to the thermal zonal model to predict the vertical relative humidity gradient in a large space. Further the model also was extended with a EMPD-model to include the moisture buffering of the walls in a simplified way.

KEYWORDS

Thermal stratification, zonal airflow model, dynamic simulation

1. INTRODUCTION

In the past , church buildings were usually unheated, except for a simple individual coal or peat stove, which could be rented by the churchgoers. The thick high stone walls of the church alleviated the fluctuations of the ambient air temperature and relative humidity. As a result, the indoor climate in the church building was quite stable (Schellen, 2002). After the Second World War the living standard of the people increased and the increased prosperity lead to higher comfort expectations (Camuffo et al., 2010), not only in residential dwellings, but also in churches. As a consequence many "local" heating systems were replaced by heating systems that heated the whole indoor air volume of the church. Often these systems were rugged air inlets, designed to quickly heat the space and achieve thermal comfort during services. These systems were designed and operated without taking the effect of the fluctuating temperature and relative humidity they caused on the artworks in the building into account. Consequently, the many wood based artworks usually present in the church, like the pulpit, the organ and the panel paintings are also exposed to this changing climate. According to the preservation needs determined by the ASHRAE conservation classes (Anon, 2011), changes in temperature and relative humidity or space gradients could lead to faster deterioration or even to damage at these artworks.

Table 1: ASHRAE conservation classes (Anon, 2011) related to the risk on damage at wooden artworks

Class		Temperature ΔT_{short} Space gradients	$\Delta T_{seasonal}$	Relative Humidity ΔRH_{short} Space gradients	$\Delta RH_{seasonal}$	Risks
AA		±2°C	±5°C	±5%	No changes	No risk of mech. damage
A	As	±2°C	+5°C -10°C	±5%	±10%	Small risk of mech. damage
	A	±2°C	+5°C -10°C	±10%	No changes	
B		±5°C	+10°C max. 30°C	±10%	±10%	Moderate risk to high-vulnerability artefacts but tiny risk to most paintings
C			<25°C (<30°C)		25-75%	High risk of mech. damage
D		-		<75%		Prevent dampness

To reconcile the heating demand due to comfort expectations of the visitors with the preservation needs for the artworks in designing a heating system for such churches including artwork, computer simulations can be helpful. During heating, thermal stratification occurs and in order to make a correct assessment of the stratification, in addition to the calculation of the energy exchange, the calculation of the airflow in the space is necessary. To predict the temperature and humidity distribution, different modelling approaches can be used: namely Computational Fluid Dynamics (CFD) models and the linking of a Building Energy Simulation (BES) software with an airflow model (the so-called zonal models).
The CFD-method is a widespread approach to model the airflow with a computational simulation. Models, based on this method, predict the temperature, velocity and other flow

parameters by numerically solving the Navier-Stokes equations with a high degree of accuracy. However, for this research, the fluctuations in temperature and humidity over a longer time period and the effects of heating on the temperature and humidity distribution in large historical buildings, related to damage occurring at wooden artifacts are of interest. To simulate longer time periods, the mentioned CFD models are less suitable because of their need for powerful computers with a large amount of memory. In reality, these resources are often not available. To be able to predict the airflow in a building in a fast way and for a longer time period, the use of a macroscopic airflow model offers a solution. These so-called zonal models are an intermediate approach between the CFD and the multizone models used in BES, which consider the air as perfectly mixed. The difference with a CFD model is that the grid is much coarser and the differential equations between the cells are simplified. Based on the simplifications made, the zonal models can be further subdivided into pressure-zonal, temperature-zonal and momentum-zonal models.

As pointed out above, a BES-software considers a zone perfectly mixed. However, a BES software is commonly used for the evaluation or comparison of different heating techniques to meet sizing and energy operating cost requirements (Bouia & Dalicieux, 1991). Though, a BES software, because of this assumption of a well-mixed zone is unable to calculate the temperature stratification in the air. So to determine the thermal comfort and the preservation conditions the BES model has to be extended with an airflow model. Because the emphasis lies on predicting the airflow in a building in a fast way, we chose to implement a temperature-zonal model in the BES-software. For this purpose the thermal-zonal model of Togari(Togari, Arai, & Milura, 1993) was recoded in C++ and coupled to the commercial BES simulation environment TRNSYS (v17). In this paper the possibilities of the model and its shortcomings are discussed by applying it on the case of a typical church building.

2 THE COUPLED THERMAL ZONAL MODEL

The model proposed by Togari et al.(Togari et al., 1993) is a simplified model for calculating the vertical temperature distribution in a single large space. They assumed that the main components of the air movement in the large space are airflows along the vertical wall surfaces (so-called wall currents) and supply airstreams. Further, they also assumed that the horizontal temperature is uniform, except for the regions affected by supply air jet ventilation. Based on these assumptions a method was proposed to calculate the stratification.

First the space is divided into a finite number of horizontal layers or blocks. Each layer consist of a core cell and walls cells. The core cell represents a layer and when the layer is bounded by a wall, a wall cell is defined which accounts for the mass flows in the boundary layer. The method considers three major paths for the heat and mass transfer: (1) heat and mass transfer between the core cell and the wall cells, (2) heat transfer and air movement between the different core cells and (3) heat transfer through the wall. To calculate these transfers some decision were made:

- To predict the air and heat transport to the wall elements, the aid of the boundary layer theory is applied.
- Since the heat transfer through the walls, which is affected by both outdoor and indoor conditions, is rather complicated, a heat balance equation for coupled thermal conduction, convection and radiation is necessary to predict the wall surface temperature. For this the implementation of the model in a BES-software was used.

2.1 The wall currents

When the air in a space is warmed up or cooled down , the air is cooled at the colder walls or warmed up at the hotter walls. By consequence an airflow along the vertical walls is induced. The heat and air mass transfer along the wall is modelled using a so-called wall current model. This model assumes that the heat convection drives mass flow $m_{out(I,K)}$ with an average temperature $T_{D(i,K)}$ from layer i to its related boundary layer.

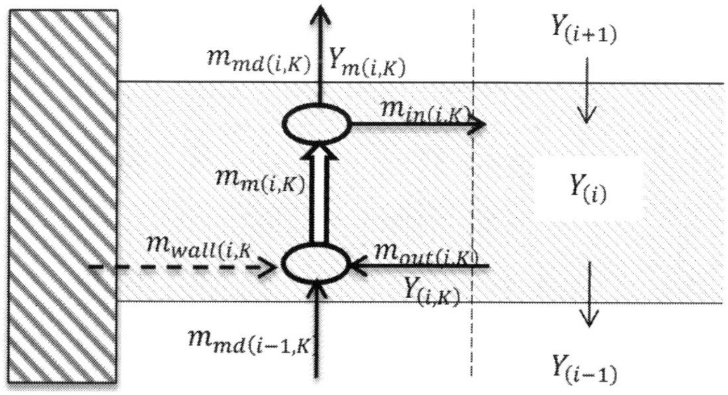

Figure 1: *Schematic representation of the composition of the wall currents and the mass flow in and between the wall cell and the core cells*

To calculate $m_{out(I,K)}$with temperature $T_{D(i,K)}$ following equations were applied:

$$T_{D(i,K)} = 0.75T_{(i)} + 0.25T_{w(i,K)} \quad (17)$$

$$m_{out(I,K)} = 4\,\frac{\alpha_{C(i,K)} \cdot A_{w(I,K)}}{C} \quad (18)$$

The current flow from layer i will combine with the current flow $m_{md(i+1,K)}$ from layer i+1 to form a total flow with mass $m_{m(i,K)}$ with average temperature $T_{m(i,K)}$, yielding:

$$m_{m(i,K)} = m_{out(i,K)} + m_{md(i+1,K)} \quad (19)$$

$$T_{m(i,K)} = \frac{m_{out(i,K)}T_{D(i,K)} + m_{md(i+1,K)}T_{m(i+1,K)}}{m_{m(i,K)}} \quad (20)$$

Some of the air of this current $m_{m(i,K)}$ returns to the air layer i ($m_{in(i,K)}$) and some continues to the cell down or up ($m_{md(i,K)}$), depending on the direction of the current flow. The splitting of the mass $m_{m(i,K)}$ into $m_{in(i,K)}$ and $m_{md(i,K)}$ is calculated by the ratio P(i,K):

$$m_{in(i,K)} = (1 - P_{i,K})m_{m(i,K)} \quad (21)$$

$$m_{md(i,K)} = P_{i,K}m_{m(i,K)} \quad (22)$$

The ratio P(i,K): is dependent on the relationship between the air temperature in zone i and in zone i+1 or zone i-1, associated with the current temperature $T_{m(i,K)}$. Table 1 presents the judgment criteria for the airflow pattern.

Table 2: judgment criteria for the airflow pattern

Flow direction	Temperature conditions	$P_{i,K}$
Descending	$T_{m(i,K)} \geq T_{(i)}$	0
	$T_{(i)} > T_{m(i,K)} > T_{(i-1)}$	$\frac{T_{(i)} - T_{m(i,K)}}{T_{(i)} - T_{(i-1)}}$
	$T_{m(i,K)} \leq T_{(i-1)}$	1
Ascending	$T_{m(i,K)} \leq T_{(i)}$	0
	$T_{(i)} < T_{m(i,K)} < T_{(i+1)}$	$\frac{T_{m(i,K)} - T_{(i)}}{T_{(i+1)} - T_{(i)}}$
	$T_{m(i,K)} \geq T_{(i+1)}$	1

2.2 Conservation equations

The calculation of the mass balance in every zone, was started from the lowermost zone. Flows are defined as positive in the upward direction and from the wall cells to the layers. The mass balance for a layer i is then calculated by:

$$0 = m_{source} + \sum_{k=1}^{m} (m_{in,i,K} - m_{out,i,K}) + m_{i-1} - m_i \qquad (23)$$

Where m_i is the mass flow from layer i to i+1 [kg/s] and m_{source} is the mass flow coming from a source [kg/s].

In the air the heat transfer equation can be written as:

$$V_i \frac{(\rho C)^{t+\Delta t,m} T_i^{t+\Delta t,m} - (\rho C)^t T_i^{\ t}}{\Delta t} = Q_{source} + Q_{layer} + Q_{currents} + Q_b \qquad (24)$$

Where Q_{source} is the heat coming from a source, Q_{layer} means the heat transfer between the layers and $Q_{currents}$represents the heat flow from or to the wall currents. $Q_b$expresses the heat transfer between the layers due to thermal stability yielding:

$$q_{b(i)} = C_{b,i} A_b (T_{i-1} - T_i) \qquad (25)$$

Where $A_b$is the boundary area between zone i-1 and zone i. $C_{b(i)}$ is the heat transfer factor and is defined as “the value obtained when a stable (Cb = 2,3W/m²K) or unstable (Cb = 112W/m²K) temperature stratification is formed.

The model of Togari was extended by an equation for moisture conservation. Hereby the air is assumed as a mixture of dry air and water vapour, each component (vapour and dry air) obeying the ideal gas equation. Taking into account the time dependency of the moisture content of the air, the moisture balance equation for a layer i can be expressed as:

$$\rho_a V_i \frac{(Y_i^{\ t+\Delta t,m} - Y_i^{\ t})}{\Delta t} = G_{layer} + G_{currents} + G_{source} \qquad (26)$$

G_{source} represents the vapour flow produced by people, systems activities such as washinh,...and G_{layer} is the vapour mass flow between the layers. The vapour mass flow $G_{currents}$ from the wall current is calculated by:

$$G_{currents} = \sum_{k=1}^{m} m_{in(i,K)} Y_{m(i,K)} - \sum_{k=1}^{m} m_{out(i,K)} Y_i^{\ t+\Delta t,m} \qquad (27)$$

Where $Y_{m(i,K)}$ is the humidity ratio of wall current and is determined by the moisture flux $m_{wall(I,K)}$coming or going to the wall, the humidity ratio of the mass $m_{out,i,K}$coming from the adjacent layer and the humidity ratio from the wall current under or above (see Figure 197). The equation to determine the humidity ratio of the wall current $Y_{m(i,K)}$ is:

$$Y_{m(i,K)} = \frac{{Y_{i(i,K)}}^{p-1} m_{out(i,K)} + {Y_{m(i-1,K)}}^{p-1} m_{md(i-1,K)} + m_{wall(i,K)}}{m_{m(i,K)}} \tag{28}$$

The amount of water vapour exchange between room air and the wall (i,K) was modelled with a simplified algorithm. The chosen model was an effective moisture penetration depth (EMPD) model (Janssens, Rode, De Paepe, Woloszyn, & Sasic-Kalagasidis, 2008). This approach assumes that the moisture transfer takes place between the zone air and a thin fictitious layer of a uniform moisture content with a thickness $d_{buf,}$ which is related to the variation of water vapour pressure at the material surface. The effective penetration depth d_{buf} is calculated by equation 13, in which t_p is the period of cyclic variation (s).

$$d_{buf} = \sqrt{\frac{\delta(\varphi) P_{v,sat}(T) t_p}{\rho\xi(\varphi)\pi}} \tag{29}$$

Where $\delta(\varphi)$ is vapour permeability [s], $\rho\xi(\varphi)$ is the moisture capacity in terms of humidity derived from the material sorption isotherm [kg/m³] and $P_{v,sat}(T)$ represents the saturation water vapour pressure at temperature T [Pa]. The following equation is then solved together with the moisture balance equation for indoor air within a space under the non-steady-state.

$$m_{wall} = \frac{A(P_{v,i} - P_{v,buf})}{\frac{1}{\beta_i} + Z_{buf}} = A\rho\xi(\varphi) d_{buf} \frac{\left(\frac{P_{v,buf}}{P_{v,sat}(T_{buf})}\right)^{t+\Delta t} - \left(\frac{P_{v,buf}}{P_{v,sat}(T_{buf})}\right)^{t}}{\Delta t} \tag{30}$$

Where $P_{v,i}$ represents the water vapour pressure indoor and $P_{v,buf}$ is the average water vapour pressure in the layer[Pa]. β_i is the surface convection coefficient of the wall [m/s] and Z_{buf} is the vapour diffusion resistance between the surface and the moisture storage centre of the layer.

2.3 Adding infiltration

Infiltration is due to wind-driven or buoyancy-driven ventilation.With buoyancy-driven ventilation the pressure differences are due to air density differences, which in turn are due to temperature differences. Following equation was used to calculate the infiltration rate:

$$u_{inf} = \frac{u_{50}}{2}\left(\frac{\Delta\bar{p}}{50}\right)^n \tag{31}$$

Where u_{50} is the infiltration defined in air changes per hour; i.e. zone volume per hour, when the pressure difference between inside and outside is 50 Pa. 'n' is equal to 2/3 and $\Delta\bar{p}$ is the pressure difference[Pa] related to the height H [m] of the zone and the temperature difference [°C] which is calculated by:

$$\Delta\bar{p} = 0{,}01H(T_{indoor} - T_{outdoor}) \tag{32}$$

2.4 Adding a jet flow model

To model a free jet stream, the following assumptions were made:

- **First the trajectory of the path was determined.**

For horizontally compact free jets, the trajectory of the centreline is described by Koestel (Awbi, 2003), yielding:

$$\frac{y}{\sqrt{A_0}} = 0{,}0522\left(\frac{x}{\sqrt{A_0}}\right)^3 Ar \tag{33}$$

The centreline velocity decay of compact jets can be described by the following equation.

$$\frac{U_{centerline}}{U_0} = \frac{K_1\sqrt{A_0}}{x} \tag{34}$$

Theoretical values of the characteristic K1 depend upon the type of velocity profile equation and supply conditions assumed (Goodfellow & Tahti, 2001). For a 3D-jet the value of K_1 equals 7,2 (Awbi, 2003). The temperature decay in the centreline of the jet was calculated with the following equation.

$$\frac{T_x - T_i}{T_0 - T_i} = \frac{K_2\sqrt{A_0}}{x} \tag{35}$$

In this equation the value for K2 was calculated by the formula of Shepelev (sec (Goodfellow & Tahti, 2001)):

$$K_2 = \frac{1}{K_1}\frac{(1 + Pr)}{2\pi 0{,}082^2} \tag{36}$$

- **To evaluate the total airflow rate transported by the jet to some distance from the diffuser, the entrainment rate was determined.**

For a given opening area A_0, the entrainment ratio is proportional to the distance x. For a compact jet, the following equation by Grimithlyn was used (sec (Goodfellow & Tahti, 2001)):

$$\frac{\dot{Q}_x}{\dot{Q}_0} = 0{,}29\frac{x}{\sqrt{A_0}} \tag{37}$$

3 VALIDATION OF THE MODEL

The validation case studied in this paper is the case that can be found in the report of Togari et al. (Togari et al., 1993) and that of Arai et al.(Arai, Togari, & Miura, 1994). The geometrically simple test room has a ground plane of 3m x 3m and measures 2,5m in height. The room consist of insulated boards (three vertical walls, ceiling and floor) and 1 glass wall. In the wall opposite to the glass wall, two openings were made in the symmetry plane: a supply inlet at 0,625m above the floor and a return outlet at 0,250m above the floor. The simulated cases are visualized in Table 3.

Table 3: Overview of the simulated case studies

Name Case	Supply air condition					Inlet location	Outside temperature
	outlet size [mm x mm]	air volume [m³/h]	velocity [m/s]	temperature [°C]	momentum [kg m/s²]		
N-11	500 x 74						42 (14)
N-10	500 x 74						12 (42)
H-100	500 x 74	150	1,13	40,7	0,053	zone 1	11

The room described above was fully modelled in TRNSYS. The space was vertically divided into 5 layers. In the model described by Togari the calculations of the stratification started based on measured wall surface temperatures, while in the model in TRNYS only outdoor conditions can be used. As a consequence the behaviour of the walls and the inside surface temperature is calculated by TRNSYS. For the case of natural convection (case N-11 and case N-10) good agreement was found (De Backer, Laverge, Janssens, & De Paepe, 2014). On Figure 198 results are shown for the case H-100 where the measured results by Togari were compared with the calculated results by the thermal-zonal model. The measured results were predicted well enough by the jet-model.

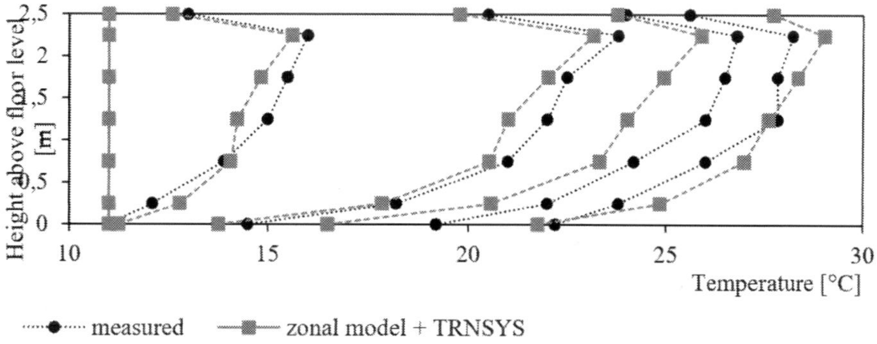

Figure 2: *The progress of the temperature in case of hot air supply condition (jet)*

4 A CASE STUDY

The goal of this study was to look how a church building could be modelled with the thermal zonal model. As already pointed out, the thermal zonal model had some limitations. One of the limitations is that it can only deal with simple geometries and the geometry can only be subdivided into horizontal layers. For that, the geometry of the church building had to be simplified.

For this study, a church was modelled with a floor plan and dimensions typical for a church in Belgium. The geometry of the church includes a rectangular nave of 32m in length, 17m in width and 15m in height (internal dimensions). The floor plan also containes side aisles. Each side aisle measures 5x5m. The church has 0,9m thick masonry walls . The windows are single-glazed. Based on literature (Fawcett, 2001), the following structure was assumed for the build-up of the church floor: a plain square paving stone, laid on a bed of lime mortar. The floor was modelled using the standard ISO 13370:2007. Figure 199 shows the model used in the simulation study, where the church building was divided into 5 layers. Initially, the attics of the church were not included in the model.

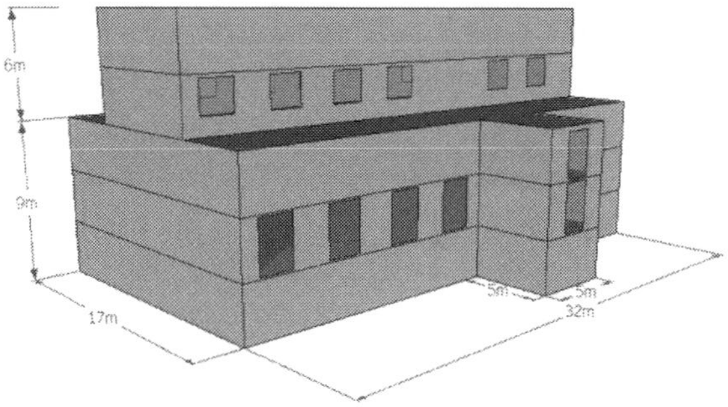

Figure 3: *Simplified geometrical model of a church building*

In the simulation study, 24 hours were simulated during which heating only occurred once, during service. The service started at 10h00 and ended at 11h00, so the heating device was operated from 9h30 till 11h30 with a set point temperature of 16°C. The temperature at the supply inlet was 45°C and the flow rate was 30000kg/h. The initial conditions in the church building were 8°C and 80%RH. Figure 200 and Figure 201 show results for the temperature and relative humidity profile in the church building. The temperature is the lowest zone rises to 16°C and achieves the set comfort temperature. The relative humidity decreased to 50%. In contrast, the temperature in the highest zone rose to 21°C, while relative humidity dropped down to 35%. For an artwork that encompasses several subsequent layers, like an organ, large relative humidity gradients occur during heating. Furthermore, there also can be concluded that the ASHREA class B , the highest class that can be reached in a church building (Ankersmit, december 2009), cannot be achieved if the church is only heated during a church service and one wants to achieve comfort temperature.

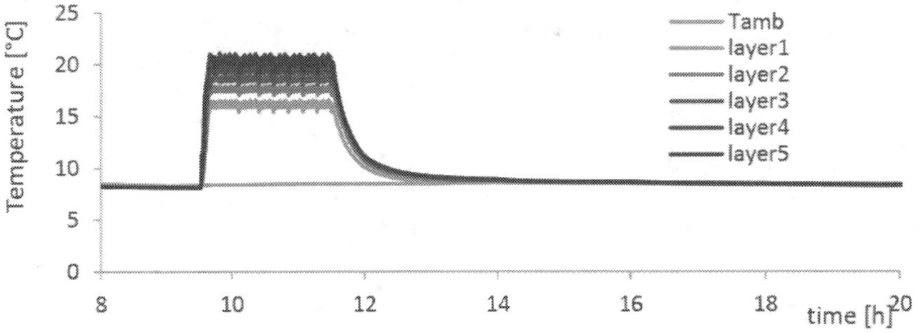

Figure 4: *Temperature distribution in the church building*

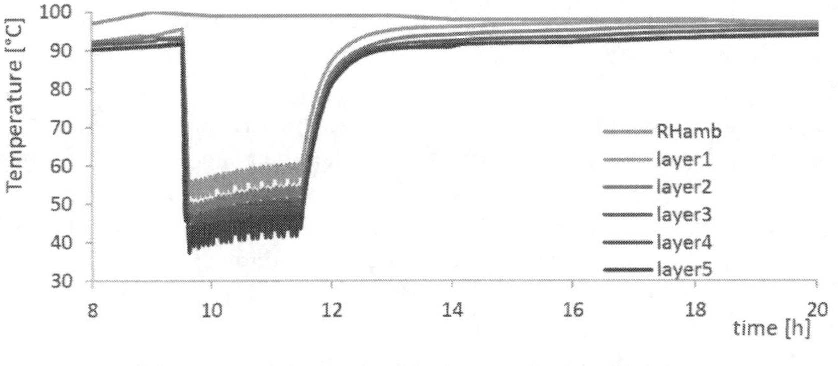

Figure 5: *Relative humidity distribution in the church building*

5 CONCLUSIONS

As an alternative to the complex CFD models, zonal models are a suitable method to predict the airflow in a large space in a simplified way. These zonal models can be linked to a BES-software, in which each zone is assumed to be perfectly mixed. By this coupling, the influence of the airflow on the temperature distribution and vice versa can be calculated, in order to assess the thermal comfort and preservation conditions in the zone. This paper presents the coupling of the existing thermal zonal model of Togari with the BES-software TRNSYS. In addition a moisture preservation equation was added to the thermal zonal model to predict the vertical relative humidity gradient in a large space. The model was additionally extended with a EMPD-model to include the moisture buffering of the walls in a simplified way. To validate the model, the cases described in the paper of Togari were simulated. Good agreement was found between the measurements and the thermal-zonal model. A case study of a typical church building was modelled. The results showed how stratification in the church building occurs during heating with a typical rugged air inlet, demonstrating that the ASHREA conservation class B, cannot be achieved when heating solely during church services.

6 REFERENCES

Ankersmit, B. (december 2009). *Klimaatwerk: richtlijnen voor het museale binnenklimaat*: Amsterdam University Press.

Anon. (2011). ASHRAE handbook: Heating, ventilating, and air-conditioning applications, SI edition (pp. 23.21-23.22): American Society of Heating, Refrigerating and Air-Conditioning Engineers.

Arai, Y., Togari, S., & Miura, K. (1994). Unsteady-state thermal analysis of a large space with vertical temperature distribution. *ASHRAE Transactions, 100*(part 2), 396-411.

Awbi, H. B. (2003). *Ventilation of Buildings*. London: Spon Press

Taylor & Francis.

Bouia, H., & Dalicieux, P. (1991). *Simplified modeling of air movements inside dwelling room.* Paper presented at the IBPSA (The International Building Performance Simulation Association), Nice, France.

Camuffo, D., Pagan, E., Rissanen, S., Bratasz, L., Kozlowski, R., Camuffo, M., & della Valle, A. (2010). An advanced church heating system favourable to artworks: A contribution to European standardisation. *Journal of Cultural Heritage, 11*(2), 205-219. doi: DOI: 10.1016/j.culher.2009.02.008

De Backer, L., Laverge, J., Janssens, A., & De Paepe, M. (2014). *On the coupling of a zonal model with a multizone building energy simulation model.* Paper presented at the NSB 2014, Lund, Sweden.

Fawcett, J. (2001). *Historic floors: their history and conservation*: Butterworth-Heinemann.

Goodfellow, H. D., & Tahti, E. (2001). *Industrial Ventilation Design Guidebook* California

San Diego: Academic Press.

Janssens, A., Rode, C., De Paepe, M., Woloszyn, M., & Sasic-Kalagasidis, A. (2008). From EMPD to CFD – overview of different approaches for Heat Air and Moisture modeling in IEA Annex 41 Copenhagen: IEA ECBCS Annex 41, Closing Seminar.

Schellen, H. L. (2002). *Heating monumental churches : Indoor Climate and Preservation of Cultural Heritage.* (PhD thesis), Technische Universiteit Eindhoven.

Togari, S., Arai, Y., & Milura, K. (1993). A Simplified Model for Predicting Vertical Temperature Distribution in a Large Space. *ASHRAE Transaction, 99*(part 1), 84-90.

RECENT ADVANCES ON FACTORS INFLUENCING HUMAN RESPONSES AND PERFORMANCE IN BUILDINGS AND POTENTIAL IMPACTS ON VENTILATION REQUIREMENTS

Pawel Wargocki

International Centre for Indoor Environment and Energy,
Department of Civil Engineering,
Technical University of Denmark (paw@byg.dtu.dk)

Implementation of sustainability and energy efficiency measures creates opportunity to revisit some requirements regarding indoor environmental quality and also necessitates examining, whether these measure do not incidentally increase the risk for human health, and do not compromise human comfort and work performance. This presentation attempts to address these issues.

Brief review is provided of the effort to define principles for building ventilation in commercial and residential buildings based on health requirements and on the role of carbon dioxide in setting ventilation requirements with focus on potential negative effects on work performance. Building and environmental factors, which influence human comfort and performance are reviewed, as well as the newest research summarized on the role of bedroom ventilation for sleep quality and the next day performance. An account is given on whether humans can tolerate slowly increasing temperatures and what potential consequences are brought for health and human performance when this is taking place in buildings.

It is postulated that future ventilation standards should be based on health requirements and should to a much larger extent map and be related to the actual exposures to pollutants. The source control should be considered as the main mitigation measure of poor air quality (both indoors and outdoors) and ventilation used only when source control methods are ineffective; ventilation can be supplemented (not replaced) by air cleaning. There is no evidence that carbon dioxide is detrimental to comfort and health at levels typically occurring indoors, though there is potential for negative effects on some aspects of performance; carbon dioxide is a good proxy for pollution from humans which do reduce perceived air quality and increase acute health symptoms. To maximize overall satisfaction with one's personal workspace, investments should first be made that lead to increasing satisfaction with the amount of space, noise level and visual privacy. These investment will not guarantee the highest output as regards work performance. This can be achieved by improving satisfaction with temperature, noise level and air quality. Results suggest that insufficient bedroom ventilation does negatively affect sleep quality and the next day performance though the results are preliminary and need further verification. There are some seasonal differences in human responses to increasing temperatures, subjects being slightly more tolerant to slowly increasing temperature in warm season rather than in cold season. Too high temperatures may compromise work performance.

The above results provide an important and rational supplement to already existing large body of evidence on the factors influencing human responses in indoor environments and advance our understanding of the underlying mechanisms. They allow enhancing the process of design and operation of the buildings by providing further guidance to building stakeholders. However, there is still very little systematic evidence on the impact of certification schemes, which promote sustainability and energy, on the indoor environmental quality and human responses; these schemes become widespread standard in case of new construction and major renovations. Such studies are undergoing and will provide input to revise and improve the schemes, so that the objectives of human centric design of buildings, one of the major reason why we have buildings, can be accomplished.

STRATEGIES FOR EFFICIENT KITCHEN VENTILATION

Bas Knoll[1] & Wouter Borsboom [1]

1 TNO
Netherlands

1 HEALTH RISKS

Cooking devices are a major source of contaminants in dwellings. They cause exposure to combustion products and vapors. The type and production rate of contaminants depend on the heating type (gas vs. electric cooking) and cooking process (frying vs. boiling).
Major health risks appear to be fine dust particles (which may contain carcinogenic PAH), NOx, CO, CO2, VOC like acrolein and micro-biological contaminants due to excess moisture.
Apart from this, odors from the cooking appliances may be annoying as well as excessive heat, which (especially in combination with a high humidity) even may be a physiological burden.

2 MINIMIZE DISPERSION FLOW AND CHARACTERIZE IT

To control the cooking contaminants, it is important first to restrict their dispersion and to know their characteristics. This depends on:

- Type of heat source, e.g. gas burning, electric hot plate, electric induction plate;
- Pan type (casserole, saucepan, frying pan, griddle, pressure cooker), pan height, lid type;
- Status of the cooking or frying process (warming, cooking or frying stage) and typical temperature (depending on the major substance, e.g. water, oil, sugar);
- Free flow space (height to cooker hood, closed back or corner or island arrangement), flow conduction (Coanda effect) and local flow disturbance (nearby windows, doors and pathways);
- Cook behavior (stirring, temperature control, lid use, moving along).

The first step to control contamination is to try and minimize the production rate and dispersion forces by altering said factors (processes and geometry).
For the design of an effective cooker hood, the flow characteristics (flow velocities, flow directions and flow field geometry) needs to be measured in critical situations.

3 LOCAL CAPTURE IS PREFERRED OVER DILUTION

Local capture is by far the most effective way to control the dispersion of cooking contaminants. Once dispersed into the room a larger air volume becomes contaminated and needs to be replaced by a much larger ventilation air volume during a longer period. In the meantime people present in the kitchen are still exposed to the resulting concentrations. Hence, local capture potentially is much more efficient, concerning both exposure and energy use.

In airtight, energy efficient buildings with minimized ventilation due to demand control, the low background ventilation level leads to long lasting, high concentrations of (cooking) contaminants. Therefore, in general the need for a high capture efficiency rises with increasing energy efficiency of dwellings.

4 MAJOR SPECIFICATIONS FOR USER-FRIENDLY COOKER HOODS

For users two major drawbacks prevent an unrestricted use of the cooker hood:

1. It does not sufficiently reduce odor spread;
2. There is a high noise nuisance.

Odor sensation has a logarithmic relation with pollutant concentration (e.g. 90% reduction of odor concentration will reduce odor complaints by maximal half). So, only a nearly perfect capture efficiency will meet the requirements for odor control. Since the design of an 80 or 90% effective hood only slightly alters from a 100% efficient one, the latter should be the design goal.

For the cooker hoods noise production a threshold value of 40 dB is proposed.

Apart from this, important demands are an attractive design and no adverse effect on the usability of the cooking appliances.

5 IMPROVE LOCAL CAPTURE EFFICIENCY

For common cooker hoods the capture efficiency is poor. Recommended steps to improve this are:

- Redesign the cooker hood, i.e. minimize the capture distance (hood height), maximize the overhang, move the suction slots to the critical contaminant escape spots at the perimeter, avoid large internal pressure drops;
- Enclose the contaminant flow by conducting plates and/or direct it to the hood by support flows. Automatically close down openings when no activities (like stirring) are ongoing;
- Derive the correct capture (and support) flow rates from the flow characteristics and the new hood geometry, using known design rules. Evaluate and tune this with tests;
- Control the flows and the distribution over the hood depending on production spots and intensity, favorably automated control using sensors.

There are interesting examples of improving capture efficiency in professional kitchens.

6 RECIRCULATION OR EXHAUST?

If captured contaminated air is exhausted from the kitchen area the flow equilibrium means that provisions for comparable supply flows are necessary, especially in airtight dwellings. Some imbalance (overflow from surrounding rooms) may be useful to prevent dispersion of poor quality kitchen air to surrounding rooms. In general this (temporary applied) supply flow is quiet large compared to normal ventilation system flows. Therefore, a special kitchen supply provision is recommended.

This kitchen supply needs special attention concerning draught risk. Furthermore a temporary high energy loss may occur due to the temperature deviation of the incoming outside air. Heat recovery, favorably including latent heat exchange to maximize the re-gain, enables the re-use of the heat of the exhaust air.

Another option is to recirculate contaminated air after filtration. However, filter demands will be high due to the complex mixture of (fine) particles, aerosols, gaseous components and odors. To filter the water vapor, condensation on an additional cold heat pump element may be required. Also this element may extract excess heat to control the kitchen temperature.

Finally, necessary periodic filter replacement will have its drawback on user-friendliness and filter inefficiency will still require (possibly extra) background ventilation to reduce rest exposure.

7 BACKGROUND VENTILATION

Some rest contamination into the kitchen may be unavoidable, e.g. because of incidentally escaping flares of cooking air or due to exhalation of food that is further processed after it is taken of the stove. Therefore some background ventilation is considered to be necessary. Also after preparing food some low level elongation of exhaust fan operation time is recommended.

COUPLING HYGROTHERMAL WHOLE BUILDING SIMULATION AND AIR-FLOW MODELLING TO DETERMINE STRATEGIES FOR OPTIMIZED NATURAL VENTILATION

Matthias Pazold[*1], Florian Antretter[1], Marcus Hermes[2]

1 Fraunhofer Institute for Building Physics
Fraunhoferstrasse 10
83626 Valley, Germany
**Corresponding author:*
matthias.pazold@ibp.fraunhofer.de

2 Fraunhofer Institute for Building Physics
Nobelstrasse 12
70569 Stuttgart, Germany

ABSTRACT

In both, newly built and renovated buildings the building air-tightness has to be ensured. With a tight building envelope comes a low infiltration air-exchange. A minimum outdoor air exchange to ensure acceptable moisture and indoor air quality levels must be maintained.
A model is introduced, that couples hygrothermal whole building simulation with a multi-zone air-flow simulation. This coupling allows assessing the combined effects of air-flow on building energy use, comfort conditions, air quality and possible hygric issues.
An application example shows the use of the software for a low rise residential building. In a first step the parameterization of the air-flow network is checked by computing the resulting infiltration air-flow with closed windows. The maximum achievable natural air exchange for different window types and openings, like pivot-hung windows, windows opened with a small parallel gap around and side-hung windows is assessed. Based on these numbers it is assessed if any of the above mentioned opening options alone with a certain opening duration is able to achieve good indoor air quality and acceptable indoor moisture levels to avoid any hygric problems or if a combination of some of the measures is required.
The presented dynamic hygrothermal whole building simulation coupled with a multi-zone air-flow model is capable of simulating the effects of different window opening options and strategies on building energy demand, indoor comfort conditions and air quality and on the hygrothermal building component performance. It is shown, that a sufficient air exchange can be realized even with a parallel-action window with small opening gaps which can be run with motor drives. Especially demand controlled opening strategies can be developed that ensure high indoor air quality and comfort conditions while also controlling humidity indoors to avoid moisture damage. Also an improvement for summer conditions can be achieved by ventilative cooling with this type of window opening.

KEYWORDS

hygrothermal building simulation, air-flow-model, WUFI® Plus, natural ventilation

1 INTRODUCTION

The model development bases on a building component tool to simulate the coupled heat and moisture transport across building components, like walls, roofing and floors (Künzel, 1994). It simulates the unsteady temporal development of the heat and moisture profile within a component and, of course the heat and moisture exchange on the component surfaces. Combining simulations for all components of the building envelope and inner walls with a zone model leads to the hygrothermal whole building simulation tool, called WUFI® Plus (Holm, Künzel, & Sedlbauer, 2003). One or different zones describe one or more rooms within the building with equal air properties for each simulation time step. The boundaries of each zone are the building components. Regarding this, the inner and the outer surface of a component are assigned to zones. Besides the heat and moisture transport across the components, solar radiation through windows, inner sources or sinks, HVAC systems and of course the air exchange influence the indoor climate and are taken into account. The indoor climate within zones is calculated with a heat and moisture balance for the whole building model.

An assumed or measured air exchange can be considered as constant, or time scheduled before running the hygrothermal building simulation. But, in fact, the air flow depends on the simulated indoor climate. It also depends on the wind velocity and direction mostly delivered with the outdoor climate data. Therefore a pressure driven multi-zone airflow model (Pazold & Antretter, 2013), quite similar to the model used within CONTAMN (Walton & Dols, 2008), is implemented, to simulate this air flow for each time step. The multi-zone definition of the airflow model is equal to the definition of the zones within the whole building software. If a building component is not fully airtight, is has to be defined as air flow path. Furthermore air openings and fans have to be defined as air flow paths. Therefore different types of sub-models, to calculate the mass flow rate depending on the pressure difference, are used for the different flow patterns. The wind pressure on the outer side of the building envelope and the hydrostatic air pressure are calculated for each path in a respective relative height, to simulate stack effects.

The introduced hygrothermal building simulation tool coupled with the described multi-zone air flow model is used for the investigation of ventilation strategies to determine the optimized natural ventilation for an example low rise building. This paper describes the application of the tool. Different window opening options and strategies are assumed to assess their effect on the indoor comfort conditions, the air quality and the hygrothermal building component performance of the building.

2 LOW RISE RESIDENTIAL BUIDLING MODEL

The visualisation of the investigated two-storey residential building is shown in Figure 1. Each room is regarded as one zone; hence the building exists of 10 heated zones. The attic and the basement excluded the basement stairwell, is regarded unheated. The zone separation enables the assessment of the differing indoor climates within the building because of the inter-zone air flow and among other things, like different inner sources. The treated floor area of the building is 148.51 m^2 and the net volume (air volume) is 349.3 m^3. The characteristic length of the building is 10m and the width 9 m.

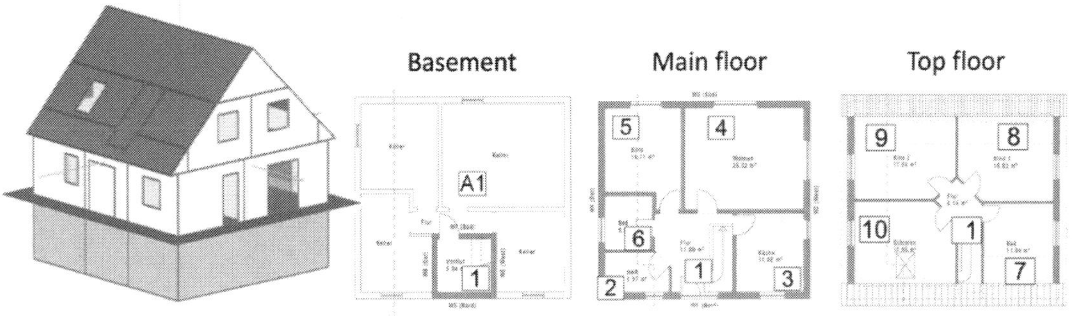

Figure 1: 3-D view and floor plans with zoning of the modelled building

2.1 General simulation assumptions

The location of the building is Holzkirchen in Germany. The used reference outdoor climate data is representative for the temperate climate region. The average outdoor temperature is 6.6°C with a range from -20.1°C up to 32.1°C in summer. The average outdoor relative humidity is 81 %. The average wind velocity is 2.2 m/s mostly from west. An open terrain is assumed to calculate the wind pressure at building height and solar shading.
The investigated simulation period is one year, beginning in January. The result time step is one hour. Built-in moisture effects are taken into account by initializing the envelope components to realistic built-in moisture contents. But 2 months are pre-calculated and not assessed to allow a short dry-out period because of differing boundary conditions after the completion of the building and the following usage. During the investigated simulation period, the building is continuously occupied by four persons, two adults and two kids. Different inner heat-, moisture, and CO_2- daily profiles for each room reflect the day-presence of the residents (Figure 2 and Figure 3). In sum the overall internal heat gain is 11.14 kWh per day; the moisture production is 8.96 kg per day and the CO_2-gain is 3 kg per day for the whole building.

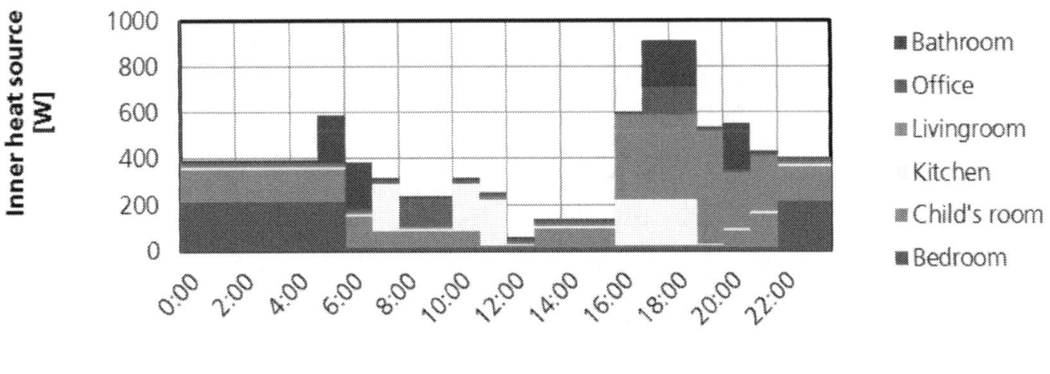

Figure 2: Daily-profile of inner heat gains (summarized for all zones)

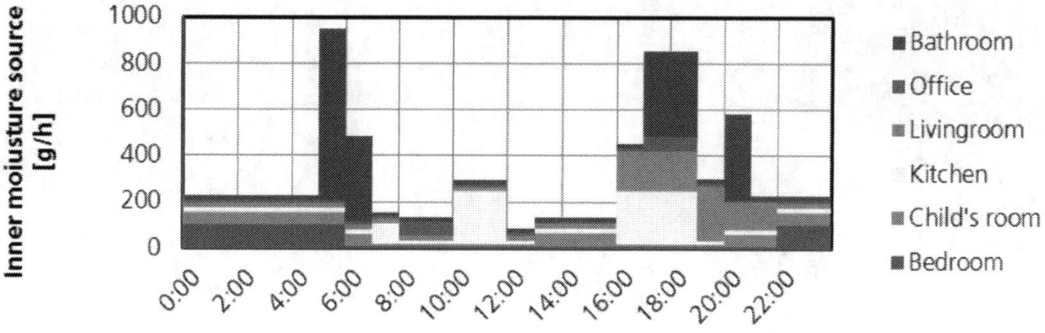

Figure 3: Daily profile of inner moisture sources (summarized for all zones)

An ideal heating system is used to keep the indoor climate at a minimal design temperature of 20°C. If the inner temperature would decrease below this design condition, because of more heat loss than gains within a zone, the actual heating demand is calculated to keep the minimal temperature. The maximal design temperature of 24°C is used for activating the temporary sunscreen device. If the inner temperature would increase above the design condition, the incident solar radiation through the windows is decreased to 10 % due to an activated sunscreen device.

2.2 Building envelope

The simulation tool calculates the coupled heat and moisture transfer through multi-layer building components. Within this paper only a few parameters of the components are presented. The exterior wall is made of aerated concrete with a mineral plaster at the outer side and a gypsum board at the inner side. The resulting thermal resistance is 3.6 m²K/W. The thermal resistance of the roofing is 5.0 m²K/W and for the basement slab 2.85 m²K/W. The U-Value of the windows is 1.3 W/(m²K) and the solar heat gain coefficient for the transparent area 0.6. Those values meet the current German energy saving ordinance requirements.

2.3 Air flow parameters

The whole building envelope is assumed to be not air-tight. As described in the introduction the air flow across the different components is calculated with the pressure driven multi-zone air flow model for each time step, taking into account the actual inner climate and wind pressure. The mass flow rate through the components is mostly calculated with a power law, fitted for different kinds of flow patterns. The opening state of windows and inner doors can vary during the simulation. The opening times and durations are predefined for different natural ventilation strategies. The model calculating the air flow depends on the opening condition and can change for each time step.

Inner doors can be open or closed. Three states of the windows are investigated. They can be closed or opened in turn-position. As a third position, windows with parallel action fittings are investigated. They can be opened by parallel offsetting the sash by approx. 6 mm to the frame. This creates an air gap on all sides of the casement. With a motor fitting drive the parallel offsetting of the window can be automated, without any user interaction.

The parameters used for each component are given in Table 1. Some are based on (Orme, 1999) Appendix E but adjusted to get a requested air change rate at 50 Pa pressure difference n_{50} = 1.5 h^{-1}. This is checked by a simulated pressure difference test and a whole year simulation with a closed envelope (closed windows and open inner doors). For the difference pressure test, a constant volume flow rate fan is added in the entry door. With an iterative method, the volume flow rate of the fan is adjusted to reach different pressure differences between the inside and outside air. For the test the wind pressure is neglected and an equal temperature of 20°C, also in the outdoor zone is assumed. With those different pressure differences and their associated volume flow rates, a n_{50}-value of 1.48 h^{-1} is calculated for the closed building air flow network with the defined air flow parameters. This is in good agreement to the requested air permeability, so the chosen air flow parameters seem to be reasonable with this first check.

Table 1: Used air flow parameters

Building Component	Air flow parameters	Related to
Exterior walls	Flow coefficient 0.021 $dm^3/(s\ m^2\ Pa^n)$; Flow exponent 0.84	Wall area
Inner walls	Flow coefficient 0.021 $dm^3/(s\ m^2\ Pa^n)$; Flow exponent 0.84	Wall area
Roofing	Effective leakage area 20cm² on each roof side	Roof sides
Entry door, closed	Flow coefficient 0.12 $dm^3/(s\ m\ Pa^n)$; Flow exponent 0.6	Joint length
Inner doors, closed	Flow coefficient 1 $dm^3/(s\ m\ Pa^n)$; Flow exponent 0.6	Joint length
Inner doors, open	Two-way-flow pattern; Discharge coefficient 0.6; Exponent 0.5	Opening area
Windows, closed	Flow coefficient 0.12 $dm^3/(s\ m\ Pa^n)$; Flow exponent 0.6	Joint length
Windows, open	Two-way-flow pattern; Discharge coefficient 0.6; Exponent 0.5	Opening area
Windows, parallel open	Two-openings; Opening Areas 0.058m² (at the top and bottom of the window); Discharge coefficient 0.162; Flow exponent 0.5	Opening area

3 NATURAL VENTILATION STRATEGIES AND RESULTS

Different natural ventilation strategies are analysed by designing cases with different opening times and states for the windows. Beside the inner air temperature, relative humidity and CO_2-concentration, the air change rate with the outside climate and the inter-zonal air change rates are calculated for each time step and every defined zone. The huge amount of results is evaluated for each case and some average annual values are presented in this paper. Some additional assessments available with the coupled hygrothermal building simulation with the air flow model are shown.

3.1 Closed envelope

In a first step, the natural infiltration due to air leakage was investigated. The windows are kept close for the whole simulation time. In fact, this is not a realistic case, but should also prove the designed air flow network with the air flow parameters. As a result, the average annual natural air change rate for this case is 0.11 h^{-1}. This value can be expected for the requested air permeably of n_{50} = 1.5 h^{-1} and the used outer climate. In 25 % of the year, the air change rate is below 0.04 h^{-1}. The room with the lowest outside air change rate is the upstairs bedroom facing north-east with an average annual value of 0.01 h^{-1}. The inter-zonal air change rate for this room is on average 0.2 h^{-1}. The most outside air ventilated room is the kitchen, (0.26 h^{-1}) facing north-west. The heating demand due to air leakage and without any other ventilation is 51 kWh/(m² year).

3.2 Normal window turn-opening

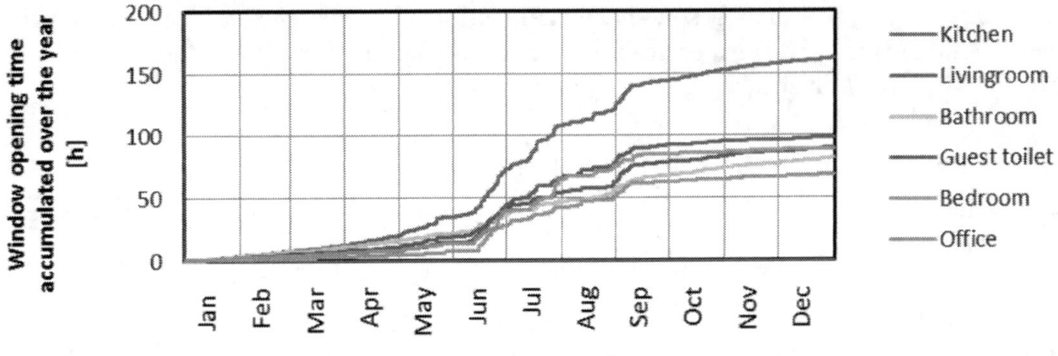

Figure 4: Accumulated window opening time over the simulation year for normal window turn-opening

This case should represent normal user ventilation behaviour by turn-open the windows. The opening times are pre-defined by calculating possible opening durations for each simulated time step depending on climate conditions and user presence. If the possible opening duration is more than zero, the actual opening is randomly calculated with a probability of 20 %.

Only when there is a very low rain load (below 0.1 liter/(m^2 h)), or when a person is presence in a room the possible opening duration for each window is calculated. Further, if the outdoor temperature in is above 20°C (for the window in the bedroom above 15°C) the window can be opened for the whole time step, one hour. Is the outdoor temperature below this boundary, a short opening for 10 minutes (0 – 20°C) or 5 minutes (below 0°C) can only happen, when a person enters or before leaving a room. With this algorithm, the window in the kitchen is averagely opened twice a day and the windows in the other rooms averagely once a day. In the heating period, the opening duration is 5 to 10 minutes. In the summer a window can stand open for one or more hours. The accumulated window opening times for each window in the different rooms are shown in Figure 4.

The resulting outer air and inter-zonal air change rate for each zone is shown on Figure 5. The resulting average annual air change rate for the whole building is 0.31 h^{-1}, the moving average is shown in Figure 6.

As indicator for the air quality the CO_2-concentration within the rooms is used. For this case the average concentration in the whole building is 44 % of the year above 1500 ppm. Within the worst room, a child's room, it is 73 % of the year above 1500 ppm. Following this the air quality is rather poor. As indicator for the comfort condition the inner air temperature is used. It is 4.3 % of the year above 26°C. The heating demand in this case is 59 kWh/(m^2 year).

As a first indicator for the hygric issues the relative humidity within the rooms is used. In the winter months the average relative humidity is about 50 %. However, especially in the bathroom, with short high moisture loads, the relative humidity is more than half of the year above 60 % and even one third of the year above 70 % up to 99 %. So the first assumption is that there will be hygric issues at least within the bathroom. Further investigations in order to assess the risk of mould growth under transient conditions can be made with a bio-hygrothermal method. For this investigation, the tool WUFI Bio (the basics are discussed in (Sedlbauer, 2001)), is used to predict mould growth on the interior surfaces. It calculates the moisture content available for mould and compares it with a critical water content which allows a spore to germinate. On the inner surface on the exterior wall of the bathroom, for the substrate category II (materials with porous structure such as the gypsum board) the predicted mould growth rate is 4.6 mm/year and the mould growth index is about 0.008, which seems to be no issue. However, the simulation is made with an interior thermal resistance of 0.13 (m²K)/W which might have to be increased due to less convection in corners or behind cupboards. Then the surface temperature and the risk of mould growth would increase.

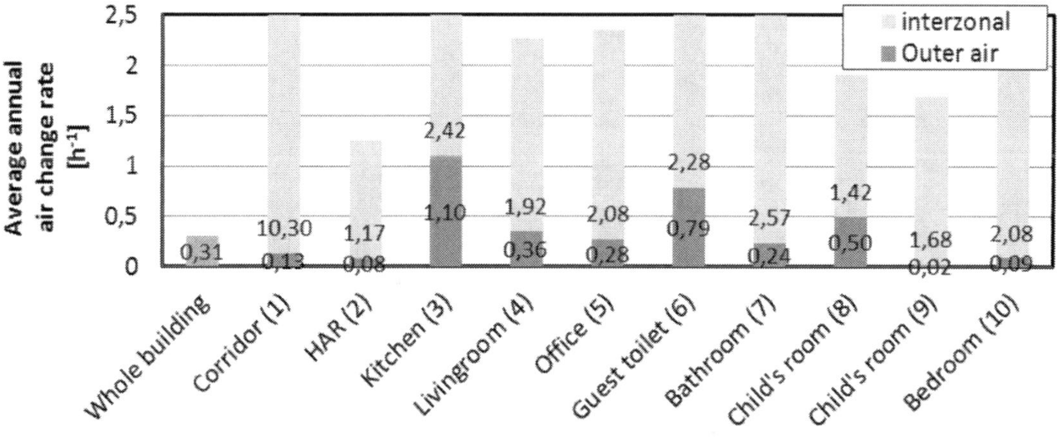

Figure 5: Resulting average annual outer air and inter-zonal air change rates for the different rooms. Results for normal window turn-opening

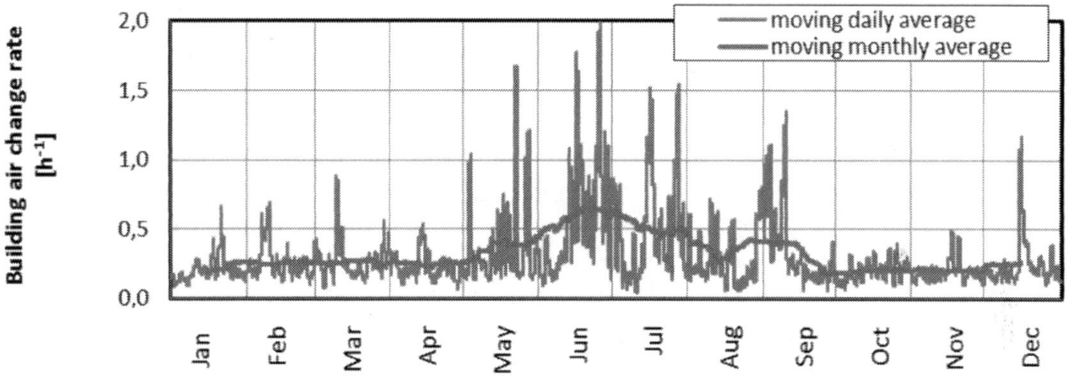

Figure 6: Moving daily and monthly average air change rate within the whole building. Results for normal window turn-opening

3.3 Frequent window turn-opening

This case should represent a frequently user ventilation behaviour by turn-open the windows up to four times a day for 5 till 60 minutes in summer, shown in Figure 7. The algorithm to calculate the opening durations, depending on climate conditions and person presence is equal to the case before with the assumed normal user behaviour. However, the probability calculating the actual opening randomly is set to 50 %.

The average annual air change rate within the whole building is 0.69 h^{-1} for this case. The moving monthly average is mostly above 0.5 h^{-1} and increase in summer up to1.0 h^{-1}. In the kitchen, where the window is west-faced and most frequent turn-opened, the average annual outside air change is 2.5 h^{-1}. The higher infiltration rate increase the calculated heat demand to 84 kWh/(m² year) but the CO_2-concentration is only for 12 % of the year above 1500 ppm and mostly below 2000 ppm. Only in the sleeping rooms, with the high continuous CO_2-source during night time, the CO_2-concentration is about 30 % of the year above 1500 ppm. The average inner air temperature used for assessing the comfort condition is only for a short time during summer, when the outer temperature is quite high, above 26°C, for 2.4 % of the year. The average relative humidity is decreased to 36 % during the winter months and in the bathroom for quite a few hours above 70 %. The bio-hygrothermal model predicts no mould growth on the inner surfaces, the moisture content available for spores is significantly below the critical water content for spore germination.

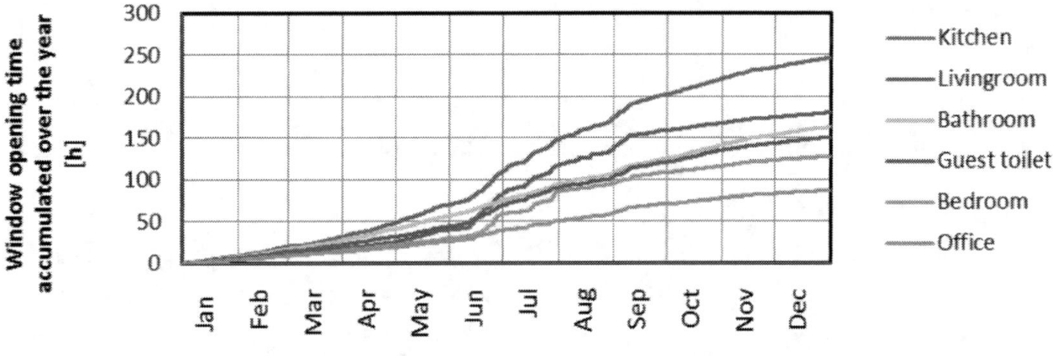

Figure 7: Accumulated window opening time over the simulation year for frequent window turn-opening

3.4 Automatic parallel action window

This presented natural ventilation strategy uses the described automated parallel offsetting of the windows with a parallel action fitting. This ventilation method should require no further occupant interaction to get a good air quality, good comfort conditions, no risk for mould growth and a low heating demand. This shall be checked with this simulation case. The parallel offset of the windows is set to last four hours a day nearby high moisture and CO_2 contribution in the rooms. It is assumed that they automatically open from 5:30 to 7:30, 12:30 to 13:00 and from 18:30 to 20:00. During summer time, when there is no heating demand and the outer air is useful for night cooling, the parallel offsetting of the windows should automatically last longer, even the whole night. For this case the user behaviour is not neglected. It is additionally assumed that the occupants turn-open the windows at least every 4th day for 5 to 10 minutes.

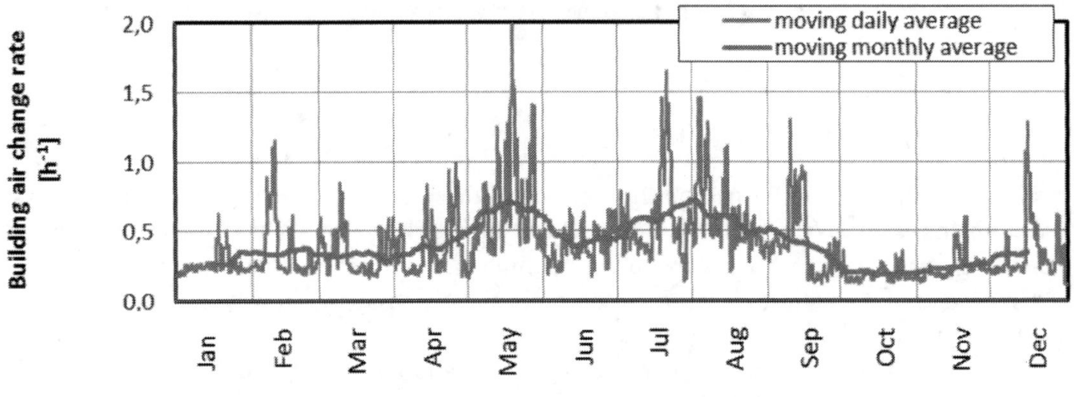

Figure 8: Moving daily and monthly average air change rate within the whole building. Results for Automated window opening by parallel offsetting the casement

The resulting average annual air change rate is 0.4 h^{-1}, with the moving daily and monthly averages shown in Figure 8. The calculated heating demand, to keep the inner air temperature above 20°C is 67.5 kWh/(m² year). Even with this lower air change rate, and lower heating demand, compared to the case before, the CO_2-concentration within the rooms is on average only 7 % of the year above 1500 ppm and in the sleeping rooms only for 15 % of the year. The inner air temperature is 3.8 % of the year above 26°C. The relative humidity during winter is averaged for the whole building 44 % and in the bathroom lower than in the cases before. The bio-hygrothermal model predicts no mould growth on the inner surfaces.

4 CONCLUSIONS

At first, the coupled hygrothermal whole building simulation with a multi-zone air flow model is introduced. In the following parts of this paper the application of this coupled simulation method is shown exemplary for a low rise building. It is used to determine natural ventilation strategies to optimize the air quality and the comfort conditions by minimizing the heating demand and the risk of mould growth. Only the opening times and durations of the windows are investigated in three presented cases. The first case assumes a normal window opening occupant behaviour. For the second case the turn-opening durations of the windows are more frequent. In fact, for longer turn-opening times, the air quality getting better and the risk of mould growth is decreased, but also the heating demand is increased about 40 %, for the well-insulated building. Both named cases require user interaction. This is not necessary in the last case where the windows should be opened automatically by parallel offsetting the sash to get a 6 mm air gap around the casement. The parallel offsetting of the windows for at least four hours a day show better indoor conditions by less heating demand (14 % more compared to the first case) due to the lower but more continuous air change rates. Of course, there is a lot of input data and results not presented in this paper, but at least some of the results that can be expected by the application of such a coupled whole building simulation tool are presented. Hygrothermal whole building simulation coupled with an airflow model enables a holistic air-flow, air-quality, comfort and energy assessment while also being able to determine possible hygric issues on and in the building components.

5 ACKNOWLEDGEMENTS

This study, especially the coupling of the hygrothermal building software with the air flow model was funded by the German Federal Ministry of Economics and Technology (BMWi 0329663L). The application cases, especially the last with the parallel action fitting windows originated in a project called "Natürliche Lüftung" realised with D+H Mechatronic AG, Hella Automation GmbH, H.O. Schlüter GmbH und Aug. Winkhaus GmbH & Co. KG.

6 REFERENCES

Holm, A., Künzel, H. M., & Sedlbauer, K. (2003). The hygrothermal behaviour of rooms : combining thermal building simulation and hygrothermal envelope calculation. *Proceedings of the 8th International IBPSA Conference.* Eindhoven, Netherlands.

Künzel, H. M. (1994). Simultaneous Heat and Moisture Transport in Building Components. *Dissertation.* University of Stuttgart.

Orme, M. (1999). TN 51: Applicable Models for Air Infiltration and Ventilation Calculations. AIVC Technical Note 51.

Pazold, M., & Antretter, F. (2013). Hygrothermische Gebäudesimulation mit Multizonen-Gebäudedurchströmungsmodell. *Bauphysik, 35.*(2), pp. 86-92.

Sedlbauer, K. (2001). Prediction of mould fungus formation on the surface of and inside building components. *Dissertation*. University of Stuttgart.

Walton, G. N., & Dols, W. S. (2008). CONTAM 2.4c User Guide and Program Documentation. NISTIR 7251, National Institute of Standards and Technologiy.

EXPERIMENTAL CHARACTERISATION OF DOMINANT DRIVING FORCES AND FLUCTUATING VENTILATION RATES FOR A SINGLE SIDED SLOT LOUVER VENTILATION SYSTEM

Paul D O'Sullivan[1, 2], Maria Kolokotroni [2]

1. *School of Mechanical, Electrical & Process Engineering, Cork Institute of Technology, Cork, Rossa Avenue, Bishopstown, Cork, Ireland*
2. *Howell Building, Mechanical Engineering, School of Engineering and Design, Brunel University, Uxbridge UB8 3PH, United Kingdom*

ABSTRACT

Adopting natural ventilation as a retrofit strategy for cooling, due to the low impact nature of the installation, is attractive due to the cooling potential of untreated outdoor air for large periods of the extended cooling season, particularly in northern climates. In line with this it is important to characterise the performance of natural ventilation components in low energy buildings in successfully transferring the cooling potential of outdoor air to the occupied zone. This paper presents an analysis of the results from 25 individual ventilation rate tests of a single sided slot louver ventilation system installed in a low energy retrofit application and 13 tests from a pre-retrofit window opening, taken as a control space for comparative purposes. Results for 3 opening configurations for the slot louver ventilation system are compared. Parameters characterising momentum and buoyancy driving forces during each test were also recorded. A number of different permutations for combined wind and buoyancy effects were recorded allowing an investigation of the existence of any underlying patterns as well as the relative effect of the different opening configurations. Owing to the nature of single sided ventilation and the primary airflow exchange mechanisms normally present, the transient evolution of the normalised tracer gas concentration during tests is also discussed and compared. Analysis shows that different patterns emerge for the dominant driving forces depending on opening configuration in the slot louver system. The slot louver ventilation system has led to steadier ventilation rates. Opening height and geometry is shown to have a significant effect on the net contribution from momentum driving forces and the fluctuation amplitude of the ventilation rate and this effect is wind direction dependant. Ventilation rates are shown to correlate well with fluctuation amplitude. The nature of the ventilation rate during tests for different wind directions is shown to vary depending on wind patterns at the building envelope.

KEYWORDS

Single sided ventilation, dominant forces, warren plot, buoyancy asymptote, fluctuating ventilation rate

1 INTRODUCTION

While experimental data exists for single sided ventilation rates, (Dascalaki et al 1996) (Dascalaki et al. 1995) (de Gids and Pfaff. 1982) (Caciolo et al 2011), information is not exhaustive for opening types other than common window types. Single sided ventilation techniques are generally reserved for single cell (Irving et al 2005), isolated spaces and when considering older office buildings that need retrofitting, the floor plan can often be cubicle and not intended as open plan spaces. Developing ventilation components that can be externally applied, provide sufficient weather protection and are effective at ensuring good ventilation rates by responding to contributing airflow mechanisms is central to ensuring successful implementation of climate change adaptation strategies. This paper presents an analysis of the mechanisms contributing to time average single sided ventilation rates from test results for a slot louvre ventilation component operated as part of a single sided ventilation strategy. It considers two key aspects of the ventilation rate; the combined effect of momentum and buoyancy forces on mean ventilation rates and analysis of the nature of the ventilation rate during tracer decay tests using a fluctuation parameter, σ_c. The objective is to

Nomenclature

Symbols		*Subscripts*	
Ar	Archimedes Number	i	inside
F	Flow Number	o	outside
K	constant related to C_d for opening	ie	internal to external
C_d	discharge coefficient	ACH	Air change rate
Re	Reynolds Number	t	Tracer, total
T	temperature (K)	th	Thermal, stack effect
H	height, (m)	ope	opening
g	gravitational constant, $(\mathrm{m/s^2})$	eff	effective
v	velocity, $(\mathrm{m/s})$	w	wind , test space envelope wall
q_{ACH}	time averaged air change rate, $(\mathrm{h^{-1}})$	N	Normalised concentration
A	opening area $(\mathrm{m^2})$	int	Zone interior
β	Power law exponent	h	hydraulic
L	Characteristic length (m)	c	concentration (relating to fluctuations)
ρ	Density $(\mathrm{kg/m^3})$		
C	tracer gas concentration (ppm)	*Abbreviations*	
t	time, (h)	CS	Control space
P	Total pressure ($\mathrm{kg/ms^2}$)	RS	Retrofit space
σ	standard error of estimate of predictions	ACH	Air change rate
		P	Parallel incidence direction
		L	Leeward incidence direction
		W	Windward incidence direction
		WD	Wind direction

investigate the conditions contributing to mean ventilation rates for a slot louvre system used in single sided ventilation. Data presented was recorded in a full scale test room for different opening configurations. Literature review has revealed little reported work of full scale experiments characterising how slot louvre systems with low hydraulic diameters perform within ventilation strategies where mechanisms such as turbulent eddy diffusion play an important function.

2 EXPERIMENTAL SETUP

A total of 25 full scale ventilation rate tests from the retrofit space for different component opening configurations were compared to 13 tests in a control space in the existing building with dynamically similar characteristics. Ventilation components tested are described in section 3. Experimental measurements were recorded using a tracer gas decay technique with the regression method applied to spatially averaged and normalised concentration values. All tests were completed under normal operating mode for the ventilation system resulting in the inclusion of effects from some complex geometry at the openings as well as a number of openings serving the test space being included in the recordings.. A detailed summary of the experimental setup and test conditions for both retrofit space and control space has recently been published by the authors and is not repeated here (O'Sullivan and Kolokotroni, 2014).

3 VENTILATION COMPONENT DETAILS

3.1 Slot louvre ventilation system (RS.02, RS.03, RS.04)

The installed slot louvre system has a net 50% free open area for airflow and overall structural opening dimensions are 0.30m (w) x 1.60m (h) with a net opening area of 0.102 m^2 (2 openings at low level and 2 openings at high level in the test space). On the internal side of the slot louvres there are automated higher level insulated doors and manual lower level insulated doors providing different control mechanisms. The ventilation system forms part of an externally applied retrofit fenestration module supplied with two glazed sections and two ventilation sections. The louvres are manufactured in anodized aluminium alloy 6063-T6 with a resulting smooth surface finish, see Figure 1. Each of the ventilation openings has 17 airflow slots across the louvre bank. Taken individually the louvre slots have an extremely low porosity at 0.057%. Table 1 summarises key information regarding the slot louvre system.

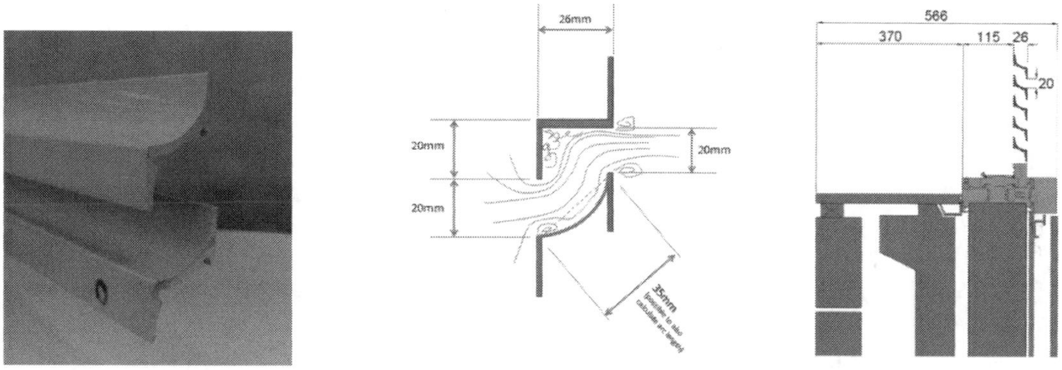

Figure 1: Slot louvre component details (a) sample louvre profile, (b) opening dimensions (c) installation

3.2 Top hung outward opening window (CS.01)

The control space ventilation component consists of an outward opening top hung window unit. This was used as a basis for comparison of time average ventilation rates and ventilation unsteadiness during tests with the slot louvre system. Details are summarised in Table 1 and in (O'Sullivan and Kolokotroni. 2014). There appears to be limited data available on full scale performance of this type of window in the literature. Recently Grabe has done some work characterising flow resistance (Grabe 2013).

Table 1: comparison of purpose provided ventilation opening types in CS & RS

Parameter	CS	RS	Units
Plain structural opening dimensions (W x H)	0.92 x 1.14	0.30 x 1.60	m
Total opening area A_t	0.32	0.42	m^2
Porosity	0.18	0.057	(%)
Total opening "wetted" perimeter	5.96	0.99	m
Hydraulic diameter (d_h) (A_t & based on "wetted" perimeter)	0.214	0.02	m
Aspect Ratio (L/d_h)	1.070*	1.075	(-)
Opening type (categories according to Etheridge 2011)	short	short	(-)

*Note: length dimension for measured along perimeter of opening window section

3.3 Definition of F_{th} asymptote for ventilation components

Dimensional Analysis using Warren plots is predicated on the correct selection of the asymptote through the origin defining flow number due to buoyancy alone, F_{th}. The gradient of this asymptote is sensitive to correct selection of both the still air discharge coefficient, C_d and the exponent used in the power law relationship for pressure and flow (i.e. $\beta = 0.5$ for orifice flow). C_d is an important parameter for a ventilation opening as it depends on the geometry of the opening and the Reynolds number, R_e of the flow, and is normally taken as

0.61 for a flush faced sharp edged orifice. Various research studies have considered the effect on C_d of different opening types and their geometry, shape and porosity under wind driven flow (Karava et al 2004) (Heiselberg et al 2001), while other studies have concentrated on the effects of wind in terms of direction and the effects of dynamic pressure (Chiu and Etheridge 2007). Caciolo et al (Caciolo et al 2011) used a C_d value of 0.6 to describe the flow characteristic when calculating F_{th} for various open window geometries while a C_d of 1.0 is used in (Dascalaki et al 1996). Grabe has recently presented findings specific to flow conditions under buoyancy alone that relates to the definition of C_d for top hung outward opening windows (Grabe 2013; von Grabe et al 2014). The value selected for C_d for the control space window opening to facilitate analysis is 0.55 and is largely based on Grabe et al. Both Pinnock (Pinnock 2000) and Sharples et al (Sharples and Chilengwe 2006) have carried out experimental work considering the use of alternative exponent values in the power law equation when dealing with buoyancy alone case and slot louvres respectively. Pinnock proposed $\beta = 0.6348$ and Sharples suggested $\beta = 0.9301$. The orifice flow equation is retained for analysis here although this is something that warrants further investigation given the significant impact on $Ar^{\beta=0.5}$ this would have. The slot louvre system used here is a flush faced sharp edged orifice at the inlet and flow is likely unidirectional through the individual slot openings. There is a length component of the louvre in a circular shape (see Figure 1) that might promote some flow reattachment allowing viscous forces and a boundary layer to develop, reducing the flow separation that normally results in C_d being independent of Re. However, for the purposes of establishing a C_d value under buoyancy alone conditions 0.61 may be a little low but acceptable for RS.04 and RS.02 for the purposes of initial analysis. The RS.03 configuration has the combined effect of slot louvre and inward opening ventilation door due to its restricted opening angle and a C_d probably lower than 0.61 as a result. A C_d value of 0.55 is assumed for RS.03. Quantifying a correct C_d value and power law exponent for the slot louvre system is still under investigation and is not presented here.

4 CHARACTERISING DRIVING FORCES FOR SINGLE SIDED VENTILATION

The mechanisms that produce airflow in single sided ventilation are generated through varying combinations of wind and buoyancy forces acting at the opening. Depending on wind incidence angle the dominant mechanisms are either from a pulsating flow due to pressure difference at the opening, turbulent diffusion through a mixing layer at the opening plane or a combination of both. When due to buoyancy forces alone, the flow will be bidirectional with a neutral pressure at the opening mid height point. The temperature difference at the opening results in a buoyancy effect that produces a stable airflow exchange. However, at low wind speeds and a leeward incidence direction the effective enveloped temperature has been shown to be reduced due to a recirculation zone counteracting buoyancy effects resulting in air change rates lower than in the absence of wind (Caciolo et al 2013). When wind is normal to the opening plane a pulsation airflow effect will dominate increasing compression of the air mass but not necessarily adding to ventilation rate. Cockroft and Robertson suggested that 37% of the volume flowrate across the opening due to pulsation will contribute to an air change rate (Cockroft and Robertson. 1976). The local wind speed at the opening is very much dependant on wind direction due to changes in flow patterns along the envelope given different wind incidence directions. Larsen shows how air change rate depends on wind incidence direction with effect more pronounced at low wind speeds and the dominating force differs between wind speed and ΔT_{ie} depending on the ratio between these forces and the wind direction (Larsen and Heiselberg 2008). A number of semi empirical models exist that use Bernoulli flow theory and can also account for contributions from wind effects. Warren and Perkins (Warren and Parkins 1985) proposed 2 separate correlations for buoyancy and wind effect, taking the larger of the two to quantify ventilation rate. Dascalaki (Dascalaki et al

1996) proposed an alternative correlation to take account of wind effects. See also for example (Crommelin and Vrins. 1988), (De Gids and Pfaff. 1976), (Wang and Chen 2012). When studying permutations of contributing forces for ventilation rate tests Warren used the relationship between a dimensionless ventilation parameter, Flow Number F, and an adjusted Archimedes Number, $Ar^{0.5}$. The purpose of the Warren plot is to separate out the data dominated by buoyancy effect. Warren plots have been used by researchers to analyse air change rate data, for example see (Warren and Parkins 1985) (Van Der Mass. 1992), (Caciolo et al 2011). Archimedes Number, Ar, is used as a measure of the relative magnitudes of the buoyancy (gravity) forces and the momentum (inertial) forces acting on elements of fluid. This ratio can be expressed in (1) where L is a characteristic height.

$$\frac{gL\Delta\rho}{P_w} \tag{1}$$

For dynamically similar flows substituting ρv^2 for total wind pressure, P_w, and taking the square root one obtains a dimensionless parameter which is basically the same as that known as Ar (Etheridge, 2011):

$$Ar \equiv \sqrt{\frac{\Delta\rho g L}{\rho v^2}} \tag{2}$$

The ratio $\Delta\rho/\rho$ can be replaced with $\Delta T/T$ for cases of interest. It should be noted that the square root in the definition of Ar is only used in connection with envelope flows, when it can be interpreted as a velocity ratio. For large Ar values buoyancy forces will dominate. $Ar^{0.5}$ can be defined in (3) as where H is the opening height in question.

$$Ar = \frac{\Delta T_{ie}\, g\, H}{\bar{T}_i\, v_w{}^2} \tag{3}$$

The ventilation parameter, Flow Number F, is a practical dimensionless number to describe wind induced ventilation. Generally where flow is buoyancy dominated F should approach the asymptote defined by F_{th}. When wind dominates $Ar^{0.5}$ tends to zero and F becomes independent of $Ar^{0.5}$. For parallel flows F should be approximately constant at 0.03. F can be defined as:

$$F = \frac{q_{ACH}}{A_{eff} v_w} \tag{4}$$

Plotting these vales for each test allows an interpretation of the influence wind forces have on the buoyancy effect, either assisting or opposing its contributing force.

5 TEST CONDITIONS & RESULTS

5.1 Test conditions

Figure 2 presents polar frequency plots of wind speeds distributed according to wind direction for each test configuration summarised in Table 3 of (O'Sullivan and Kolokotroni. 2014). An analysis of wind data for each test using directional statistics (Mardia. 1972) shows that the mean resultant length of direction vectors, a measure of "concentration" for circular data such as wind direction, were in many instances close to 1.0 with low dispersion of wind orientation during individual tests. This suggests that wind direction was consistent during a given test and the resulting analysis can assume to represent the effects from flow phenomena present for this type of orientation relative to the ventilation opening. We have taken Parallel flow to occur between wind directions of 347.5° – 22.5° & 157.5° – 202.5°. 270° wind direction is normal to the ventilation opening.

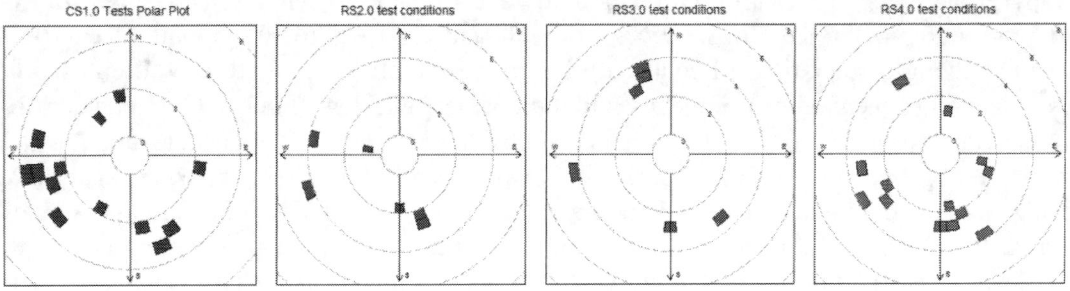

Figure 2: Polar plots for each set of test configuration for different wind speed and wind direction combinations

5.2 Dominant driving forces

Figure 3 presents measured air changes rates as a function of wind direction categorised according to envelope temperature difference and grouped according to test configuration. Figure 4 presents Warren plots for CS.01 & RS.04 and figure 5 presents Warren plots for RS depending on configuration and wind orientation. The purpose of the different plots is to highlight any trends that relate to opening geometry, test environment or driving forces present.

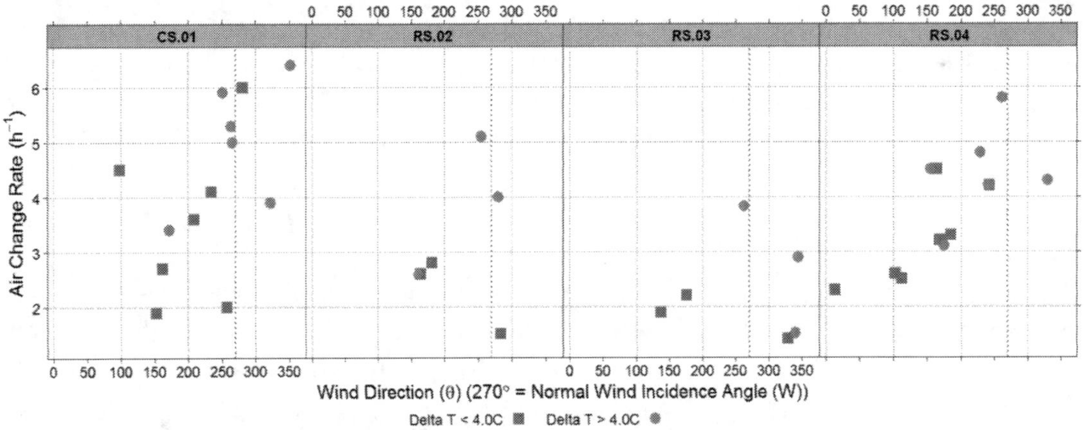

Figure 3: Test Config ventilation rates (h^{-1}) as a function of wind direction grouped according to ΔT_{ie}

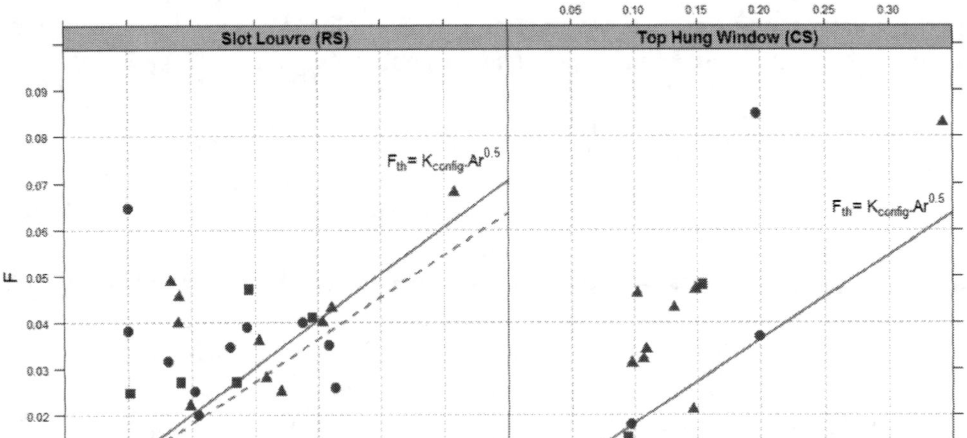

Figure 4 Warren plots for (a) slot louvre system in RS and (b) top hung window in CS. (F_{th} shown in red)

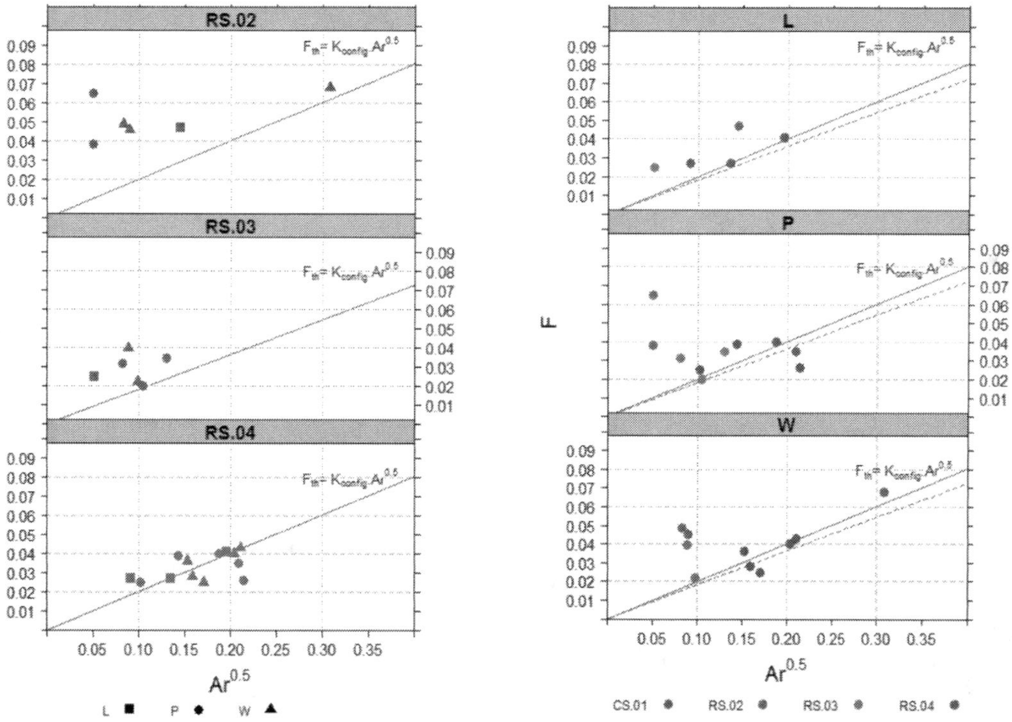

Figure 5 Warren plot categorised according to (a) configuration and (b) wind direction. (F_{th} shown in grey)

5.3 Transient ventilation rates

Single sided ventilation strategies rely on a number of low and high frequency unsteady flow phenomena relating to wind pressure, gustiness and turbulence. Tracer decay rates were measured at a frequency of 1Hz during each test and figure 6 presents ACH values as a function of wind direction, (270° = Normal to surface), grouped according to, σ_c, that is based on the estimate of error in prediction from the regression model fitted to the normalised concentration decay, C_N, to determine the ACH values (O'Sullivan and Kolokotroni. 2014). This is taken as an indicator of the level of unsteadiness present during each test.

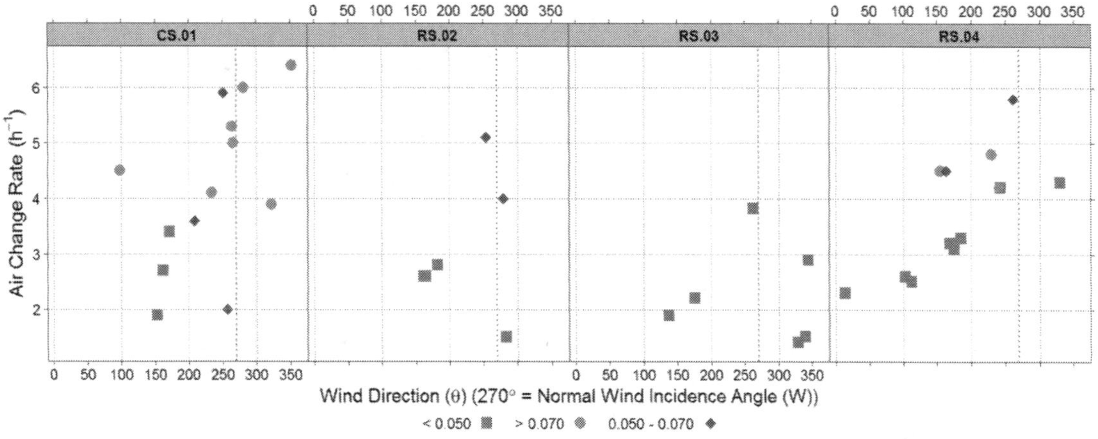

Figure 6: Air Change rate as a function of wind direction grouped according to magnitude of σ_c

6 ANALYSIS & DISCUSSION

6.1 Dominant driving forces

Considering figure 3 in all instances the highest recorded air change rates occurred with ΔT_{ie} greater than 4°C and a windward incidence direction. This occurs in both control and retrofit spaces. Leeward conditions lead to the lowest ventilation rates (although the test population is small for this wind incidence) with parallel incidence directions showing less defined patterns. Low ΔT_{ie} generally resulted in lower ventilation rates for low opening height RS.02 & RS.03 and some CS.01 & RS.04 tests have ACH values higher than 4.0 h^{-1} even at ΔT_{ie} below 4.0°C. It appears that where the wind incidence angle is not approximately normal to the surface then it is more likely to have lower ventilation rates even at relatively high ΔT_{ie} values. This suggests that leeward and parallel flows at the opening are more likely to oppose buoyancy forces. Considering Figure 4(a) & 4(b) we see a comparable spread of F values for both spaces with the retrofit $Ar^{0.5}$ range extending further towards zero ordinate. F seems to display slightly higher dependency on $Ar^{0.5}$ in CS.01. No clear patterns are visible from review of the combined datasets. However, when looking at figure 5(a), when the data in figure 4(a) is split according to RS opening configuration three different patterns emerge. RS.04, ($H_{ope} = 1.60m$), shows a clear pattern of buoyancy dominant ventilation irrespective of the wind incidence angle, which seems to agree with data in figure 3 where the highest RS.04 ventilation rates had high ΔT_{ie} values. Alternatively RS.03 has $H_{ope} = 0.76m$ (H_{int}= 2.43m (above floor level)) and also has a potentially lower C_d value than RS.02. It exhibits increased contributions from wind forces with similar F values but consistently lower $Ar^{0.5}$ values compared to RS.04. It appears that with lower opening height buoyancy forces have less ability to establish resulting in lower $Ar^{0.5}$ values and as a result F values appear more independent of $Ar^{0.5}$ in lower ranges. RS.02 ($H_{ope} = 0.76m$) (($H_{int} = 1.59m$) has a greater internal door opening angle than RS.03. It exhibits the least dependency on buoyancy forces with nearly all tests showing wind dominant F values. The clear variation in response of each opening location presents a situation where for the slot louvre system the contribution from momentum forces is a function of the overall structural opening height and vertical location on the envelope. Considering figure 4(b) leeward tests seem to display a good range of $Ar^{0.5}$ values and a general trend of F values near the asymptote F_{th}. According to (Caciolo et al. 2013) leeward conditions should be counteractive to buoyancy forces. This is not the case here with the slot louvre opening. Only one of the five leeward tests had $\Delta T_{ie} < 4.0$°C. Both Parallel and Windward incidence angles show a pattern of reducing dependency of F on $Ar^{0.5}$ as its value tends to zero. At low $Ar^{0.5}$ F is a function of wind incidence direction. At higher values these incidence angles show a tendency towards the asymptote and the wind direction effects are diminished. These results are in line with (Larsen and Heiselberg, 2008) in showing the interdependence of F on both $Ar^{0.5}$ and wind direction and suggest that at low $Ar^{0.5}$ wind incidence angle becomes important in determining whether or not wind becomes dominant.

6.2 Ventilation rate characteristics

While it is difficult to draw any clear quantitative conclusions regarding high frequency fluctuating phenomena when examining time averaged test data, the concentration fluctuation parameter, σ_c (based on 10 second averages of readings taken at 1Hz), can still give a measureable indication of the overall unsteadiness present in a ventilated space during a tracer decay test. Figure 6 groups ACH data according to σ_c. The causes of this unsteadiness can be due to several factors: pulsation flow, penetration of eddies, and static or molecular diffusion. It is difficult to isolate the specific causes without detailed measurements of these various phenomena and an assessment of overall trends is the most that can be achieved at this point. 10 of the 13 CS.01 window opening tests exhibited unsteadiness at the upper end of all recorded σ_c values and these appeared to happen primarily under windward incidence directions. These tests had averaged wind speeds > 3.0 ms^{-1}, ΔT_{ie} > 5.0°C and appear wind dominant according to the Warren plot analysis. This would suggest that unsteady ventilation rates are more pronounced when wind is normal to the outward opening window. This may be due to the nature of flow impingement with the window obstructing entry resulting in increased turbulence. The highest fluctuation value for parallel flow in CS.01 happened with one of the lowest average test wind speeds (1.8 ms^{-1}) while the two other parallel flow tests showed lower unsteadiness profiles with lower F values suggesting that with parallel flow and an outward opening window fluctuating components of ventilation rate are less pronounced and buoyancy driven flow is able to better establish. This may be due to the fact the opening section of the window does not obstruct parallel flow at the boundary layer of the structural opening. More parallel flow tests are needed before this is conclusive. In the slot louvre system only RS.04 exhibits tests with significant unsteadiness and all these tests occurred at windward incidence angles. Higher fluctuation parameters are associated with higher ventilation rates in RS.04 suggesting the importance of turbulence and wind induced phenomena to single sided ventilation rates. These also occurred at windward incidence directions. For RS.03 & RS.02, irrespective of wind incidence angle fluctuation parameters were always low as were ventilation rates. Warren plot analyses suggest these are generally wind dominant with low $Ar^{0.5}$ but they consistently exhibited low σ_c values appearing counter intuitive. When comparing fluctuation rates for parallel flow in the RS configuration they are generally lower than windward incidence angles. What is apparent from the test data is the opening geometry has a direct influence on the nature of ventilation rate and wind incidence direction also plays an important role with its effect dependant on the opening geometry.

7 CONCLUSIONS

Previous work presented in (O'Sullivan and Kolokotroni, 2014) have highlighted the potential interdependence of F on both $Ar^{0.5}$ and wind direction, particularly at low $Ar^{0.5}$ values. This paper investigated this further and has shown that the magnitude of $Ar^{0.5}$ to be dependent on opening geometry. In physical terms this interdependence highlights the challenges in predicting the correct airflow phenomena driving ventilation for different types of natural ventilation components. An allowance for wind incidence direction, opening height and aspect ratio and opening location in relation to the internal space will affect ventilation rates. If σ_c is taken as an indicator of internal airflow environment for single sided ventilation this suggests the interdependence of opening geometry and wind incidence direction has a large influence on the penetration of wind-generated air exchange mechanisms. Modification of thermophysical properties due to the retrofit have reduced the magnitude of $Ar^{0.5}$ for similar F values suggesting wind driving forces have an increased contribution even though the slot louvre system is perceived as restricting flow. Further tests including measurements directly at the external and internal faces of the opening will increase understanding of the providing a more definitive allocation of the effect of the parameters investigated; such measurements are under progress. Further work will include CFD model of the slot louvre systems and this would allow further analysis of the effect of opening geometry on unsteadiness magnitudes.

8 ACKNOWLEDGEMENTS

The authors wish to thank Dr Peter Warren for his very helpful discussions and advice regarding the use of the Warren plots. The original pilot project works was supported through a grant from the Department of Education and Skills, Ireland. The authors wish to acknowledge the co-work of the pilot project design and research team in the design and development phases of the pilot project

9 REFERENCES

Caciolo, M., Cui, S., Stabat, P., Marchio, D. (2013) "Development of a new correlation for single-sided natural ventilation adapted to leeward conditions," Energy & Buildings, 60, 372–382,

Caciolo, M., Stabat, P., Marchio, D. (2011) "Full scale experimental study of single-sided ventilation : Analysis of stack and wind effects," *Energy & Buildings*, 43(7), 1765–1773,

Chiu, Y.-H., Etheridge, D.W. (2007) "External flow effects on the discharge coefficients of two types of ventilation opening," *Journal of Wind Engineering and Industrial Aerodynamics*, 95(4), 225–252,

Cockroft, J. P., & Robertson, P. (1976), Ventilation of an enclosure through a single opening, *Building and Environment,* 11(1), 29-35

Crommelin, R. D., & Vrins, E. M. H. (1988), Ventilation through a single opening in a scale model, *Air Infiltration Review*, 9(3), 11-15

Dascalaki, E., Santamouris, M., Argiriou, A., Helmis, C., Asimakopoulos, D.N., Papadopoulos, K., Soilemes, A. (1996) "On the combination of air velocity and flow measurements in single sided natural ventilation configurations," *Energy and Buildings*, 24(2), 155–165

Dascalaki, E., Santamouris, M., Argiriou, A., Helmis, C., Asimakopoulos, D. N., Papadopoulos, K., & Soilemes, A. (1995). Predicting single-sided natural ventilation rates in buildings, *Solar Energy*, 55(5), 327-341

de Gids, W., & Phaff, H. (1982). Ventilation rates and energy consumption due to open windows: a brief overview of research in the Netherlands. *Air infiltration review*, 4(1), 4-5

Etheridge, D., 2011. *Natural Ventilation of Buildings: Theory, Measurement and Design*, Wiley-Blackwell.

Grabe, J. Von (2013) "Flow resistance for different types of windows in the case of buoyancy ventilation," *Energy and Buildings*, 65, 516–522,

Von Grabe, J., Svoboda, P., Bäumler, A. (2014) "Window ventilation efficiency in the case of buoyancy ventilation," *Energy and Buildings*, 72, 203–211

Heiselberg, P., Svidt, K., Nielsen, P. (2001) "Characteristics of airflow from open windows," *Building and Environment*, 36, 859–869

Irving, S., Etheridge, D., Ford, B. (2005) *Natural Ventilation in Non-Domestic Buildings CIBSE AM10*, The Chartered Institution of Building Services Engineers, CIBSE.

Karava, P., Stathopoulos, T., Athienitis, A. (2004) "Wind driven flow through openings–a review of discharge coefficients," *International Journal of …*, 3(3)

Larsen, T.S., Heiselberg, P. (2008) "Single-sided natural ventilation driven by wind pressure and temperature difference," *Energy and Buildings*, 40(6), 1031–1040

Mardia, K. V. (1972). Statistics of directional data, Academic Press

O'Sullivan, P., Kolokotroni, M. (2014) "Time-averaged single sided ventilation rates and thermal environment in cooling mode for a low energy retrofit envelope system." *International Journal of Ventilation* (In Press)

Pinnock, D. (2000) *An Investigation into the Influence of Wind in Single-Sided Natural Ventilation*, available: https://dspace.lboro.ac.uk/xmlui/handle/2134/7465

Sharples, S., Chilengwe, N. (2006) "Performance of ventilator components for natural ventilation applications," *Building and Environment*, 41(12), 1821–1830,

Van der Maas, J. (1992). *Air Flow Through Large Openings in Buildings (Subtask 2): IEA Annex 20–Air Flow Patterns Within Buildings*, LESO-PB, EPEL, CH-1015. International Energy Agency, Lausanne, Switzerland

Wang, H., Chen, Q. (2012) "A new empirical model for predicting single-sided , wind-driven natural ventilation in buildings," *Energy & Buildings*, 54, 386–394,

Warren, P., Parkins, L. (1985) "Single-sided ventilation through open windows," *ASHRAE SP49*, (1),

A STUDY OF CARBON DIOXIDE CONCENTRATIONS IN ELEMENTARY SCHOOLS

John Sifnaios*, Paraskevi Vivian Dorizas, Margarita-Niki Assimakopoulos

Faculty of Physics, Departments of Environmental Physics and Meteorology, University of Athens, University Campus, Athens, 157 84, Greece
**Corresponding author: john_sif@hotmail.com*

ABSTRACT

The present study aims at investigating carbon dioxide (CO_2) concentrations inside elementary schools' classrooms and how students' productivity is affected. Measurements were conducted in 9 naturally ventilated schools of Attica from April to May 2013. Monitoring lasted for 7 hours per day, for a period of one to five days per school. CO_2 concentrations were monitored simultaneously in the inside and the outside environment of the classrooms. Indoor concentrations of CO_2 in almost all schools were higher than the ASHRAE threshold limit values. In order to examine the effect of CO_2 concentrations on students' productivity, they were asked to take two productivity tests during their stay at school. Results showed that, higher concentrations of CO_2 have a negative impact on students' productivity. Also, the effect of window opening on CO_2 concentration was further examined and there has been an effort to correlate indoor and outdoor CO_2 concentrations. Finally, the correlation between the occupancy density and the increase of CO_2 concentrations was also investigated.

KEYWORDS

Schools, student performance, ventilation, occupant density, CO_2

1 INTRODUCTION

Air quality in school buildings is a matter of great concern with students spending almost 30% of their time inside classrooms, more time than in any other building environment except their home (Mendell and Heath 2005; Grimsrud et al. 2006). Furthermore, schools present a much higher occupancy than any other building, having four times as many occupants per unit of area compared to office buildings (EPA 1995). In addition, students belong to a quite vulnerable group of people who can easily be affected by chemical factors emitted in schools' premises (EFA 2001).

For the past decade there has been a scientific interest on the effects of the degraded indoor environment to students and office workers performance and productivity (Wargocki et al. 1999; Wargocki et al. 2000; Witterseh et al. 2004; Mendell and Heath 2005) however knowledge on that area is insufficient.

The main objectives of this study are to examine: 1. the effect of CO_2 to students' productivity, 2. the effect of occupancy density, window opening and ventilation rate to the CO_2 concentration, 3. the correlation between indoor and outdoor CO_2 concentration and 4. to evaluate the schools' ventilation rate.

2 METHODOLOGY

In this paper secondary data analysis is conducted, based on existing data, in order to investigate research questions other than the ones raised during the original study. Details on the original methodology can be found in Dorizas' papers (Dorizas et al. 2013a; Dorizas et al. 2013b; Dorizas et al. 2013c). In brief the above mentioned studies were carried out in 9 naturally ventilated primary schools of Attica, Greece (Figure 1), for a 32 days period, between April and May 2013. Details on the schools and the classrooms where the measurements were conducted are shown in Table 1. In total 193 students participated in the survey and undertook 1310 performance tests.
The current study comprises the effect of CO_2 concentration on students' performance and the correlation of occupancy density to CO_2 concentrations and diurnal CO_2 concentrations. Also it involves the comparison between indoor to outdoor CO_2 concentrations, the correlation of CO_2 concentration to window opening, ventilation rate and lastly the evaluation of the schools' ventilation rate.

Figure 1 : Map of Attica (left) and locations of schools under study (right)

Table 1: Schools' characteristics

School Code	School Name	Classroom's Volume (m^3)	Number of Students in the Classroom	Measurement Period
School 1	Thrakomakedones 1	198	25	8-12/4/13 (5 days)
School 2	Thrakomakedones 2	162	25	20-24/5/13 (5 days)
School 4	Acharnae 4	155	24	14-18/4/13& 24/4/13 (5 days)
School 8	Acharnae 8	159	19	27-29/5/13 (3 days)
School 11	Acharnae 11	172	15	31/5/13 (1day)
School 12	Acharnae 12	157	25	13-17/5/13 (5 days)
School 14	Acharnae 14	165	17	1-5/4/13 (5 days)
School 18	Acharnae 18	138	18	23/4/13 (1 day)
School Pal	Pallini	137	25	19&22/4/13 (3 days)

2.1 Measurements of CO_2 and Ventilation

The concentrations of CO_2 were measured in one classroom of each school from 7.30 a.m. to 2.30 p.m. Inside the classrooms the measurements were conducted using the MultiRAE IR (RAE Systems) while outside the classrooms, the concentrations were measured using the Innova 1312 & 1303. Both instruments calculated CO_2 concentrations in units of parts per million (ppm). In order to be able to compare the indoor to the outdoor concentration, lab tests were conducted, under controlled conditions, so as to calculate a correction coefficient between the two instruments. Classroom ventilation was also measured with Innova 1312 & 1303. The "Rate of Decay" method (Hitchin & Wilson 1967) was used in which, a quantity of sulfur hexafluoride (SF6) was released in the classrooms near a low speed fan, after school hours. The purpose of the fan was to produce a uniform concentration inside the classrooms. The concentration decay was exponential and by sampling the air at several times the apparent ventilation rate was determined in air changes per hour (ach).

2.2 Performance tests

The performance tests and operative protocol were taken from the SINPHONIE project, the Schools Indoor Pollution and Health: Observation Network in Europe aiming to improve the air quality in schools. Students were given a 'code' of symbols, in which each symbol was associated to a digit number. They had 120 sec. to complete the relevant symbols at a given series of numbers. They were asked to fulfil two tests each day, one in 10.00 a.m. and one in 13.30 p.m. Full detail of the methodology followed for the productivity tests is described in Dorizas' paper (Dorizas et al. 2013c).

3 RESULTS AND DISCUSSION

3.1 Performance Scores vs CO_2 Concentrations

In order to be able to compare the concentration of CO_2 to the results of the productivity test, a 15 minutes average was calculated (approximately 15 minutes was the duration of the test). Figure 2 shows the average test scores of each day of each school for the code test (1st and 2nd). Also the figure illustrates the average concentration of CO_2 at the time students took the test. From the graph we can see that, in 26 out of the 28 days the score of the second test was higher compared to the first. Also, the concentration of CO_2 during the second test was lower than the CO_2 concentration of the first test. As a result, in the great majority of the cases (24 out of 28), lower CO_2 concentration led to a higher test score. Taking into consideration the above finding, we can conclude that there is a negative correlation between the performance scores and CO_2 concentrations. This result comes in agreement with Dorizas' result (Dorizas et al. 2013c) where, there had been a similar analysis calculating the average productivity test score for each school (not for each day of the measurements).

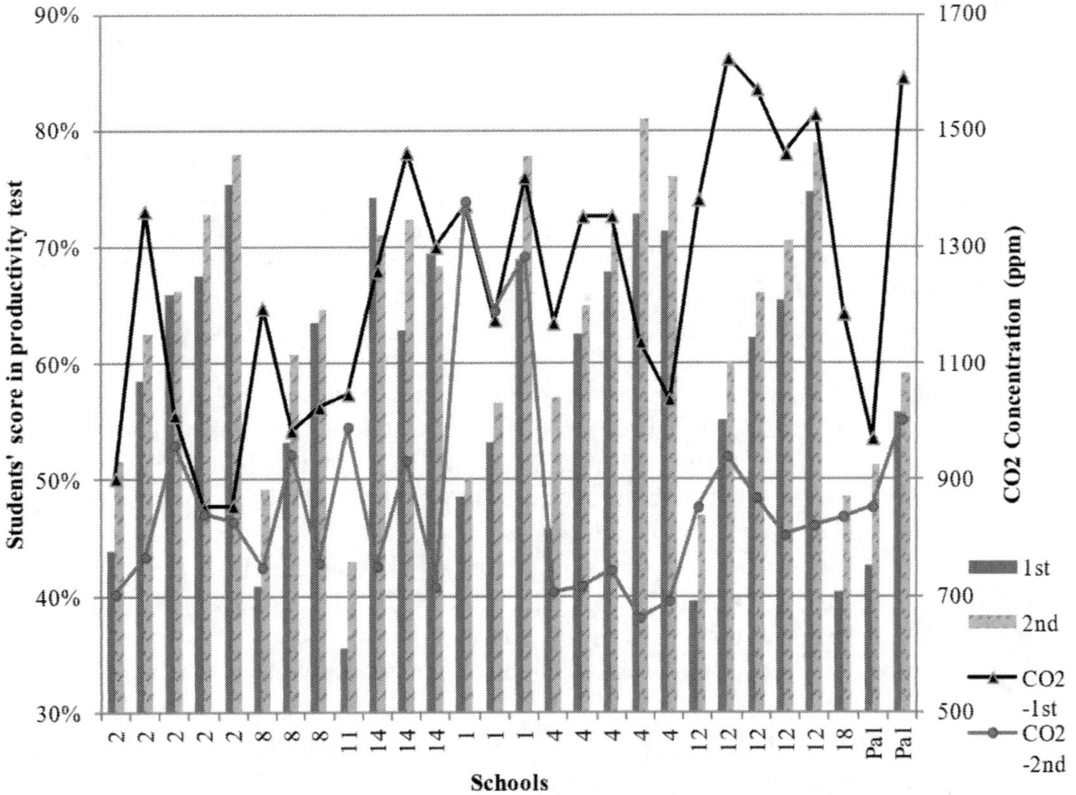

Figure 2 : The trend of test scores per school per day and corresponding CO2 concentrations

3.2 Increase of CO_2 Concentration due to Occupancy Density

Occupancy density was correlated to CO_2 concentration so as to investigate the impact of the class' area available to each student to the increase of CO_2. In order to calculate the effect of students' presence on CO_2 concentrations, we subtracted from the CO_2 peak the background concentration (the one that was measured before the arrival of the students). The results were divided into three groups depending on the window opening surface in square meters (m^2) during the lessons. In order to do that division, the surface area of the windows of each class was measured and then the average value for each group was found. The first group comprises the cases that the classroom's windows were closed during the lesson and opened in the break (Figure 3). In the second group, one window was opened halfway the lesson hour (Figure 4). This scenario corresponds to a mean opened window surface of 1,56 m^2. The third group includes the cases during which two windows were open throughout the lesson time (Figure 5). This scenario corresponds to a mean opened window surface of 3,28 m^2.

It is obvious from all the diagrams that, there is a clear correlation between occupancy density and CO_2 concentrations. All figures indicate that the higher the occupancy density, the lower the CO_2 concentration, meaning CO_2 concentrations reach lower rates when there is more space available per person. It is also found that the value of the correlation coefficient decreases with the window opening. That indicates that, window opening plays a key role in CO_2 concentrations. From Figure 3 it can be seen that, when the windows are closed there is a strong correlation between CO_2 and occupancy density (R^2 = 0,68) with occupancy density affecting primarily CO_2 concentrations when outdoor air is not entering the classroom. For the cases where air from the outside of the classroom comes to the inside the figures indicate that the occupancy density does not significantly affect the CO_2 concentrations. In particular, a moderate correlation is observed for a window opening of 1,56 m^2 (one window open) in Figure 4 (R^2 = 0,57) while a weak correlation is observed for a window opening of 3,28 m^2 (two windows open) in Figure 5 (R^2 = 0,32).

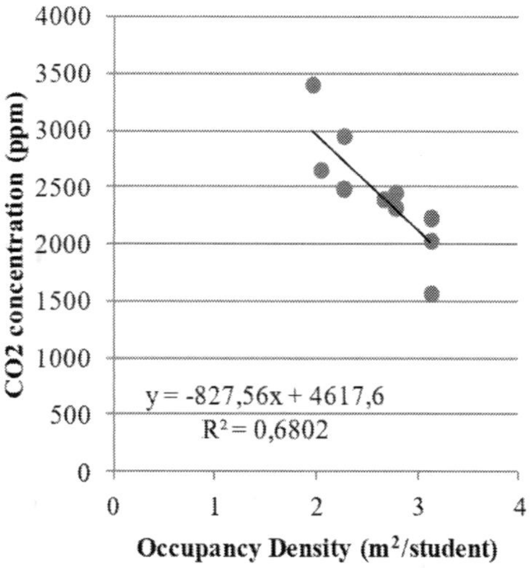

Figure 3 : Occupancy density correlated to CO_2 concentration for closed windows during the lesson time

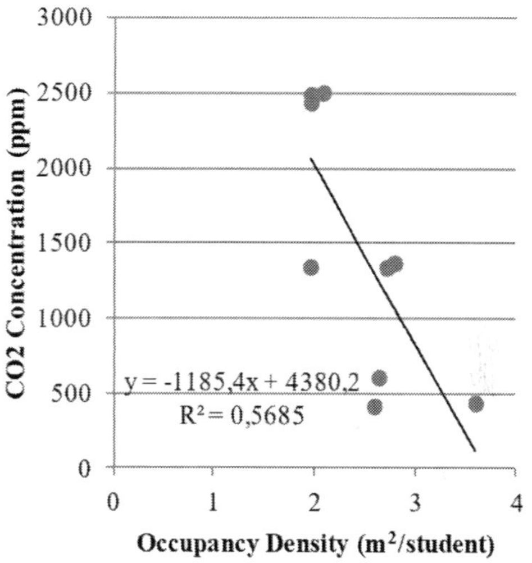

Figure 4 : Occupancy density correlated to CO_2 concentration for opened window surface of 1,56 m^2

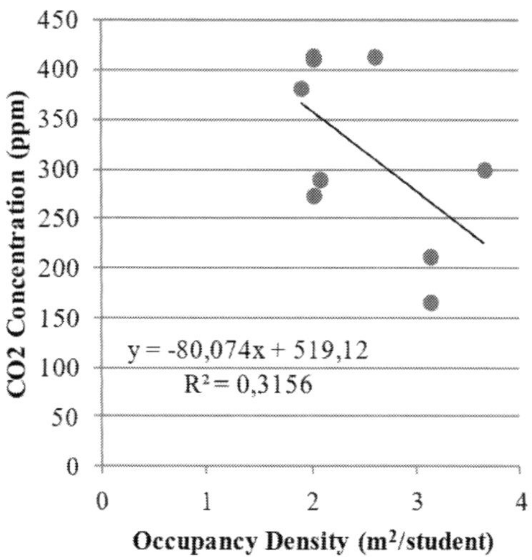

Figure 5 : Occupancy density correlated to CO_2 concentration for opened window surface of 3,28 m^2

The average contribution of each student to the concentration of CO_2 was also estimated for a 1 hour period. Results are shown in table 3. To calculate the average student contribution to the CO_2 concentrations, we subtracted from the CO_2 peak the background concentration (the one that was measured before the arrival of the students) and we divided the result with the number of the people that were present during the lesson. This process was done for all three categories mentioned above (closed windows, one open window and two open windows). The results of the table come to an agreement with the results of the above figures, as they indicate that when the open window area gets bigger, the average student contribution to the CO_2 concentration drops significantly.

Table 3 : The average student contribution to the CO_2 concentration for a period of one hour

Window Opening	Average Student Contribution to CO_2 concentration (ppm)
All windows closed	116,8
One window opens halfway the lesson hour (open window surface = 1,56 m^2)	77,5
Windows open the whole lesson (open window surface = 3,28 m^2)	13,8

3.3 Diurnal CO_2 Concentration and Ventilation Rate

In order to examine the effect of ventilation rate to the concentration of CO_2, the diurnal variation of CO_2 was plotted in the same graph with the ventilation rate. Figure 6 presents data from 5/4/2013 for School 14. That day was chosen because the variation of CO2 is representative compared to the ones that were measured in the other schools. The coloured areas indicate the times of the day that 17 students were inside the classroom. The changes in ventilation occurred because the teacher of the class changed the window opening. For a ventilation rate of 0,53 ach all windows were closed, for 5,9 ach one of the two windows of the classroom was open and for a rate of 15,6 ach both windows were open.

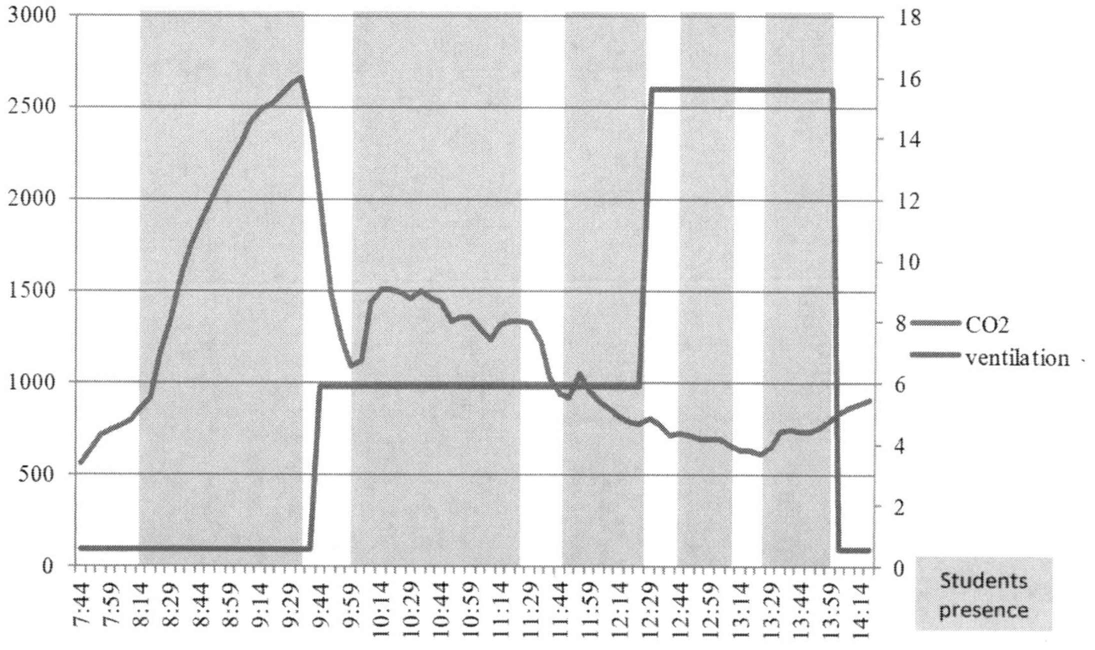

Figure 6 : Diurnal variation of CO_2 concentration correlated to ventilation rate

From Figure 6 it is obvious the highest CO_2 concentrations were achieved for closed windows and students presence. Around 9:30 a.m. (when the students had their first break and the teacher opened one of the classroom's windows) there is a sharp drop in concentration and then a rise at 10:00 a.m. when the students returned to the classroom. The lowest concentration of CO_2 was reached for both classroom's windows opened and an empty classroom. The conclusions from the graph are that the students' presence increases the CO_2 concentration regardless the window opening. Also, there is an inverse proportion between CO_2 concentration and ventilation rate. The above conclusions were observed for all the measured school days.

3.4 Indoor to Outdoor Concentration of CO_2

The indoor and outdoor concentrations of CO_2 were measured in order to identify if the indoor concentration is affected primarily by the outside concentration or by indoor sources. From Figure 7 it is easy to see that, excluding School 2, all daily mean indoor concentrations are above ASHRAE's threshold limit value (1000 ppm). Also, the indoor concentration of CO_2 is larger than the outdoor for all schools (Figure 8). The I/O ratio has its lowest rate in School 2 because in that school the windows were open during the whole lesson. As a result, there is neither a big difference in the CO_2 concentrations nor a very high indoor concentration. On the other hand, the I/O ratio has its highest rate in School 1 which was the only school where the windows remained closed during the whole lesson and were opened only during breaks. Due to the fact there had been no chemical analysis of the pollutants in order to identify their sources, we can't be certain if the high concentration in School 1 occurs primarily from the number of students inside the classroom (25) or not. What we can assume from the graphs is that, because the outdoor concentration is much lower than the indoor, it is highly likely the increased CO_2 concentration to originate from the lack of ventilation in the lesson hours.

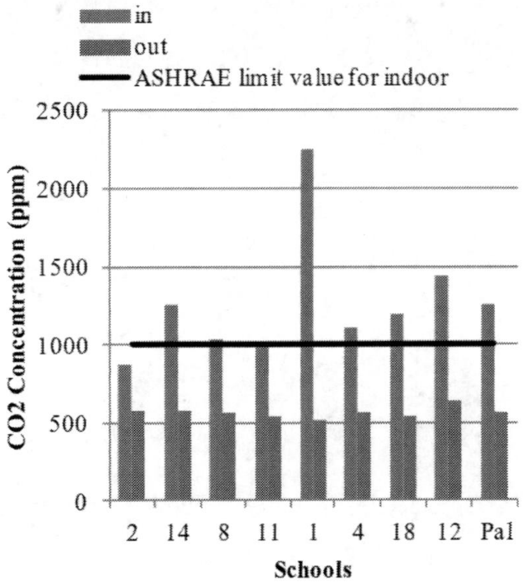

Figure 7 : Daily mean CO_2 concentration for the inside and outside environment of the schools

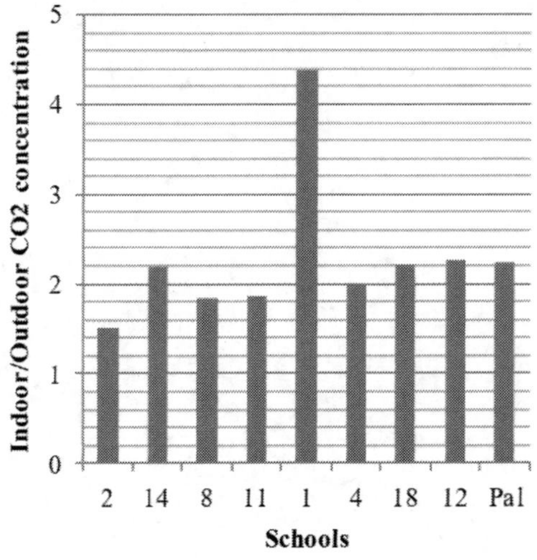

Figure 8 : Indoor/Outdoor concentration for each school

For the rest seven schools, the windows may had all three positions throughout the day (all windows closed, one window open or two windows open), so we assumed in general the window positions were the same for these schools. The correlation between indoor and outdoor concentration for these schools is shown in Figure 9, where the correlation coefficient is $R^2 \approx 0{,}63$ which indicates a significant correlation between indoor and outdoor CO_2 concentration.

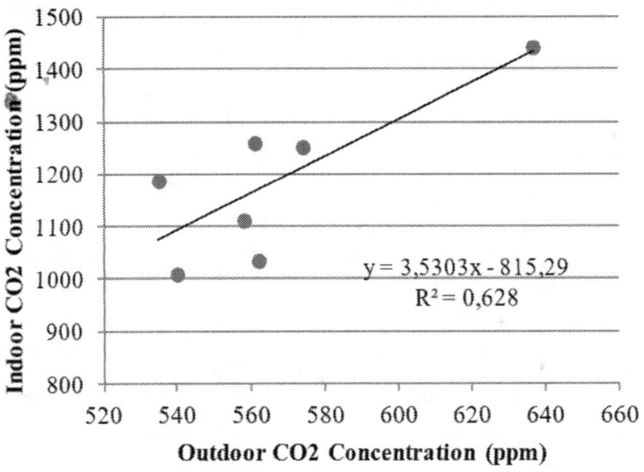

Figure 9 : Correlation of indoor CO2 concentration to the outdoor CO2 concentration

3.5 Drop of CO_2 vs Window Opening and Ventilation Rate

The drop of CO_2 was calculated during brakes where no students were inside the classroom and all schools had at least one open window. To calculate the drop (ppm/min), CO_2 concentrations were plotted over time (for a period of 15 minutes) in order to find the gradient of that curve for each school. That process was done in order to identify if the drop of CO_2 concentration over time is primarily affected by the window opening (m^2) or the ventilation rate (ach). Figures 10 and 11 illustrate the two correlations for all nine schools.

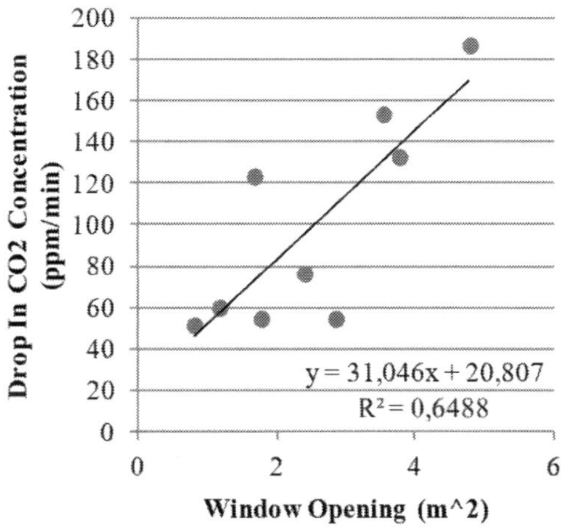

Figure 10 : Correlation of the drop of the concentration of CO_2 to the window opening

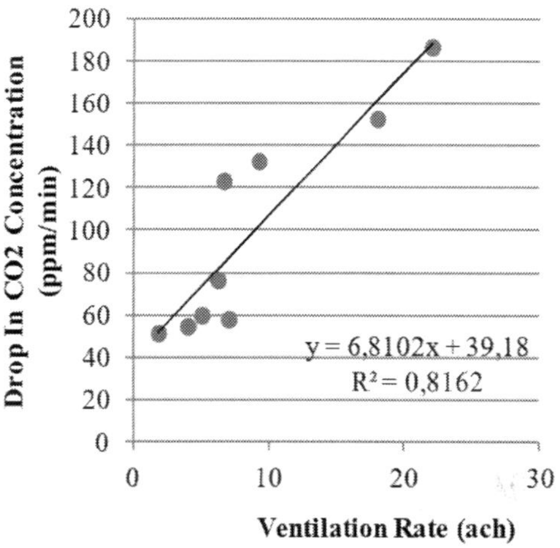

Figure 11 : Correlation of the drop of the concentration of CO_2 to the ventilation rate

It can be seen that, there is a strong correlation between the ventilation rate and the drop of CO_2 concentration (R^2 = 0,82) and a moderate correlation between the window opening and the drop of CO_2 concentration (R^2 = 0,65) indicating ventilation rate has a greater impact on the decrease of CO_2 concentration than just the window open surface. In general, the lowest drop rate for the CO_2 concentration is achieved for the lowest ventilation rate and the smallest window opening as well as the highest drop rate for the CO_2 concentration is achieved for the highest ventilation rate and the largest window opening. It must be noted for the case where the CO_2 concentration drops with a rate of 120 ppm/min, there is a rapid decrease in CO_2 concentration despite the fact that there is a low ventilation rate and a relatively small window opening. This paradox could be justified probably by a strong wind current that was present at the day of the measurement.

3.6 Evaluation of the School's Ventilation

Finally, there has been an evaluation of the ventilation rates for the different cases that occurred during teaching hours. The ventilation rates, although they were primarily measured in ach, they were converted to l/p/s in order to be compared to the ventilation's rate threshold as it is set by ASHRAE (ASHRAE 2007), which is 8 l/p/s. The results are shown in Table 4. All cases with closed widows during the lesson had inadequate ventilation rate (below 8 l/p/s). The cases of the second category had all, except one, sufficient ventilation rates. Also sufficient ventilation rate was observed when two windows were open throughout the lesson.

Table 4 : Ventilation rates (l/p/s) for different window opening for each school

Window Opening	School	Ventilation Rate (l/p/s)
Closed windows the whole lesson time	14	1,43
	1	0,99
	4	0,57
	12	1,18
One window opens halfway the lesson hour (open window surface = 1,56 m^2)	8	11,72
	14	12,98
	4	11,09
	12	3,19
	18	16,55
Two windows open the whole lesson (open window surface = 3,28 m^2)	2	15,85
	4	39,31
	Pal	14,52
	11	9,39

4 CONCLUSIONS

The conclusions from this paper are:

- There is a negative correlation between the students' scores in productivity tests and the concentration of CO_2.
- Occupancy density affects primarily CO_2 concentrations if outdoor air is not entering the classrooms. For the cases that air from the outside of the classrooms comes to the inside, occupancy density does not significantly affect CO_2 concentrations.
- As the open window area gets bigger, the average student contribution to the CO_2 concentration drops significantly.
- Students' presence increases CO_2 concentration regardless the window opening. Also, there is an inverse proportion between CO_2 concentration and ventilation rate.
- Excluding one school, all daily mean indoor concentrations are above ASHRAE's threshold limit value (1000 ppm). Also, the indoor concentration of CO_2 is larger than the outdoor for all schools. In addition there is a significant correlation between indoor and outdoor CO_2 concentration.
- Ventilation rate has a greater impact on the decrease of CO_2 concentration than the window open surface.
- All cases with closed widows during the lesson time had inadequate ventilation rate (below 8 l/p/s), while sufficient ventilation rates were observed when there were open windows throughout the lesson.

5 ACKNOWLEDGEMENTS

We are greatly indebted to the school directors, pupils and parents. Without their consent this study would have not been possible.

6 REFERENCES

ASHARE (2007), American Society of Heating, Refrigerating and Air-Conditioning Engineers Inc., *Standard 62.1.2007: Ventilation for acceptable Indoor Air Quality*, Atlanta

Dorizas P.V., Assimakopoulos M.N., Helmis C., Santamouris M. (2013a), *Analysis of the indoor air quality in Greek primary schools,* 34th AIVC-3rd TightVent-2nd Cool Roofs'-1st Venticool Conference, Athens, 25-26 September 2013, p. 110-113.
Dorizas P.V., Assimakopoulos M.N., Helmis C., Santamouris M. (2013b), *A study on the thermal environment in Greek primary schools based on questionnaires and concurrent measurements,* 34th AIVC-3rd TightVent-2nd Cool Roofs'-1st Venticool Conference, Athens, 25-26 September 2013, p. 13-15.
Dorizas P.V., Assimakopoulos M.N., Helmis C., Santamouris M. Sifnaios J., Stathi K. (2013c), *Does indoor environmental quality affect students performance?*,34th AIVC-3rd TightVent-2nd Cool Roofs'-1st Venticool Conference, Athens, 25-26 September 2013, p. 113-115
EFA (2001), European Federation of Asthma and Allergy Associations, *Indoor Air Pollution in Schools*, Sweden
EPA (1995), U.S. Environmental Protection Agency, *Indoor Air Quality Tools for Schools. IAQ Coordinator's Guide*, EPA 402-K-95-001, Washington DC.
Grimsrud D., Bridges B., Schulte R. (2006), *Continuous measurements of air quality parameters in schools*, Building Research and Information 34 (5), 447– 458.
Hitchin E.R. and Wilson C.B. (1967), *A Review of Experimental Techniques for the Investigation of Natural Ventilation in Buildings*, Building Science vol. 2, pp. 59 – 82
Mendell M., Heath G.A. (2005), *Do indoor pollutants and thermal conditions in schools influence student performance? A critical review of the literature*, Indoor Air – International Journal of Indoor Air Quality and Climate 15 (1), 27–52.
SINPHONIE project, Schools Indoor Pollution and Health: Observatory Network in Europe Available at: http://www.sinphonie.eu/about, accessed on June 14th, 2014.
Wargocki, P. Wyon, D.P., Baik, Y.K., Clausen G. and Fanger, O. (1999*). Perceived Air Quality, Sick Building Syndrome (SBS) and Productivity in an Office with Two Different Pollution Loads*, Indoor Air, 9:165-179.
Wargocki, P., Wyon, D.P., Sundell, J., Clausen, G. and Fanger, O. (2000*). The effects of Outdoor Air Supply Rate in an Office on Perceived Air Quality, Sick Building Syndrome (SBS) Symptoms and Productivity*, Indoor Air, 10:222-236
Witterseh, T., Wyon, T. and Clausen, G. (2004). *The effects of moderate heat stress and open-plan office noise distraction on SBS symptoms and on the performance of the office work*, Indoor Air, 14: 30-40

MEASUREMENT OF INFILTRATION RATES FROM DAILY CYCLE OF AMBIENT CO_2

João Dias Carrilho[a,*], Mário Mateus[a], Stuart Batterman[b] and Manuel Gameiro da Silva[a]

[a]ADAI-LAETA
Dep. Mechanical Engineering, University of Coimbra
3030-788 Coimbra, Portugal

[b]Department of Environmental Health Sciences
University of Michigan
Ann Arbor, MI 48105 USA

**Corresponding author: joao.carrilho@dem.uc.pt*

ABSTRACT

We propose a new approach for measuring air infiltration rates in buildings. The method belongs to the class of tracer gas techniques but, unlike conventional CO_2 based methods that assume the outdoor ambient CO_2 concentration is constant, the proposed method recognizes that photosynthesis and respiration cycle of plants and processes associated with fuel combustion produce daily, quasi-periodic, variations in the ambient CO_2 concentrations. These daily variations, which are within the detection range of existing monitoring equipment, are utilized for estimating ventilation rates without the need of a source of CO_2 in the building. The new method has the advantages that no tracer gas injection is needed, and time resolved results are easily obtained.

KEYWORDS

Tracer gas techniques; Air infiltration; Atmospheric CO_2; Discrete Fourier transform

1 INTRODUCTION

There are two main methodologies available to assess infiltration rates in buildings: one based on fan pressurization techniques; and another based on tracer gas techniques. The methods based on fan pressurization techniques have the potential advantage for providing quick results, but can be costly to implement an may not accurately represent infiltration rates at the lower pressure differentials found during normal operation of the building. The methods based on tracer gas techniques can be comparatively cheap and easy to implement but rely heavily on the assumptions of uniform mixing of the tracer gas and the indoor air. Carbon dioxide (CO_2), which is naturally present in the atmosphere and is also produced by the building occupants, has been shown to satisfy the uniform mixing assumption and can be used effectively in determining ventilation rates in buildings (Barankova, Naydenov, Melikov & Sundell, 2004). Metabolic CO_2 produced by the building occupants is commonly used as the source of the tracer gas, and an estimate of infiltration rates can be obtained from the step response of a first order system, either in concentration rise or in concentration decay situations. A fundamental assumption in the conventional CO_2 tracer gas techniques, however, is that the outdoor ambient concentration is constant (Sherman, 1990; Persily, 1997).

In the present paper, the authors propose a new approach for determining infiltration rates, which does not require the presence of building occupants as the source of CO_2. The method belongs to the class of tracer gas techniques, but instead of relying on a step forcing function, it uses a naturally occurring periodic forcing function. Despite being a common assumption in the existing methods, the outdoor ambient CO_2 concentration is not constant (e.g., see Refs. (Massen, Kies & Harpes, 2003) and (Miyaoka, Yoshikawa Inoue, Sawa, Matsueda & Taguchi, 2007)). The photosynthesis and respiration cycle of plants, and other processes associated with burning of fuels, are the cause of variations in the ambient CO_2 concentration which can be of the order of 100 ppm in amplitude, on a daily basis. This quasi-periodic variation is taken into advantage for estimating infiltration rates without the need of a source of CO_2 in the building. This method is particularly suited for measurement during extended periods of no occupancy, e.g., during the pre-commissioning test phase of a new or a renovated building, when the building is ready for operation but has not yet been occupied. The application of the new method is demonstrated with real data obtained in a residential building.

2 THEORETICAL FORMULATION

Consider a single zone with volume V, such that air is exchanged with the outdoor environment, through one or more of its boundaries, at a volume flow rate q cubic meters per hour. Assuming complete mixing, and in the absence of filtering mechanisms, deposition and absorption processes, the time evolution of the CO_2 concentration within the enclosure, $C_{\text{int}}(t)$, is well described by the mass balance equation (Etheridge & Sandberg, 1996, p. 277)

$$VC'_{\text{int}}(t) + q[C_{\text{int}}(t) - C_{\text{ext}}] = g(t), \tag{1}$$

where the prime denotes differentiation with respect to time, C_{ext} is the outdoor CO_2 concentration, $g(t)$ is the rate of CO_2 generation within the enclosure, and q is the volume flow rate of new air entering the enclosure or, equivalently, the volume flow rate of old air leaving the enclosure.

In writing (1), it was implicitly assumed that the outdoor CO_2 concentration, C_{ext}, does not vary in time. Allowing for $C_{\text{ext}}(t)$ to be now an explicit function of time, setting the rate of generation of CO_2 to zero and writing $\lambda = q/V$, (1) becomes

$$C_{\text{int}}'(t) + \lambda[C_{\text{int}}(t) - C_{\text{ext}}(t)] = 0. \tag{2}$$

Dividing both sides of the equation by λ and moving $C_{\text{ext}}(t)$ to the right-hand side gives

$$\frac{1}{\lambda}C_{\text{int}}'(t) + C_{\text{int}}(t) = C_{\text{ext}}(t), \tag{3}$$

which is a mathematical representation of a first order Linear Time Invariant (LTI) system with the input $C_{\text{ext}}(t)$, output $C_{\text{int}}(t)$, and time constant $\tau = 1/\lambda$.

Assuming that the input to the system is the outdoor concentration time series, represented by a periodic sinusoidal function, with an amplitude A_{ext} and an angular velocity $\omega = 2\pi f$

$$C_{\text{ext}}(t) = A_{\text{ext}} \sin(\omega t)\,, \tag{4}$$

the output, after a long enough time such that transients have died away, will be another sinusoidal function attenuated in amplitude and phase lagged and will have the form

$$C_{\text{int}}(t) = A_{\text{int}} \sin(\omega t + \phi_{\text{int}})\,, \tag{5}$$

where $\phi_{\text{int}} = \text{atan}(\omega\tau)$ is the phase lag angle. The ratio of amplitudes $A_{\text{int}}/A_{\text{ext}}$, the time constant τ, and the frequency of excitation ω are related by the well-known expression for first order, linear time invariant (LTI) systems (Holman, 1994, pp. 23)

$$\frac{A_{\text{int}}}{A_{\text{ext}}} = \frac{1}{\sqrt{1+(\omega\tau)^2}}. \tag{6}$$

So, in the case of sinusoidal excitation, both the time delay between input / output and the amplitude attenuation depend on the system time constant, that for our situation has the physical interpretation of the mean age of air inside the compartment. By rearranging (6), the air infiltration rate, which is the inverse of the mean age of air ($\lambda = 1/\tau$) can be calculated from

$$\lambda = \frac{\omega}{\sqrt{\left(\frac{1}{A_F}\right)^2 - 1}}, \tag{7}$$

where $A_F = A_{\text{int}}/A_{\text{ext}}$. To estimate A_{ext} and A_{int}, and associated uncertainty bounds, we make use of the Discrete Fourier Transform (DFT):

$$X_k = \sum_{n=0}^{N-1} x_n e^{-i2\pi kn/N}, \quad k = 0, \ldots, N-1 \tag{8}$$

where the complex coefficient X_k carries information on the amplitude and relative phase of the kth sinusoidal component of the sample time series x_n. The frequency of the sinusoidal component is $f_k = kf_s/N$, where f_s is the sampling frequency, and its amplitude is

$$A_k = \frac{2|X_k|}{N}. \tag{9}$$

Assuming that most of the signal's energy in x_n is concentrated in a single frequency component, in this case the frequency corresponding to a 24 h period, a model for x_n is

$$\hat{x}_n = \frac{1}{N}\left[X_1 e^{i2\pi n/N} + X_{N-1} e^{i2\pi n(N-1)/N}\right] + \varepsilon_n, \tag{10}$$

where N is the number of sample in a 24 h period and ε_n accounts for noise, i.e., it is the time series of the residuals obtained by subtracting the deterministic part of the model in (10) from the measured data x_n. Letting $p = \text{var}(\varepsilon_n)$ be the power in the residuals in a complete period, an estimate of the standard uncertainty associated with the determination of the Fourier coefficient X_k is (Thibos, 2003)

$$u(X_k) \approx \frac{F_{2,2R,0.318}}{R} p, \tag{11}$$

where $R = (N-3)/2$ and $F_{a,b,\alpha}$ is the F-distribution with a numerator degrees of freedom, b denominator degrees of freedom, and α is the critical value such that $1-\alpha$ corresponds to the level of confidence. For a sampling rate of one sample every 30 min, the number of samples in 24 h is $N = 48$, so that $R = 22.5$ and, for a level of confidence of 68.2%, $F_{2,2R,0.318} = 1.1754$. In practice, (11) has the meaning of a radius defining a circle in the complex plane, centred in X_k (Figure 1). With the specified level of confidence of 68.2%, we can say that the true value of the Fourier coefficient is inside this circle.

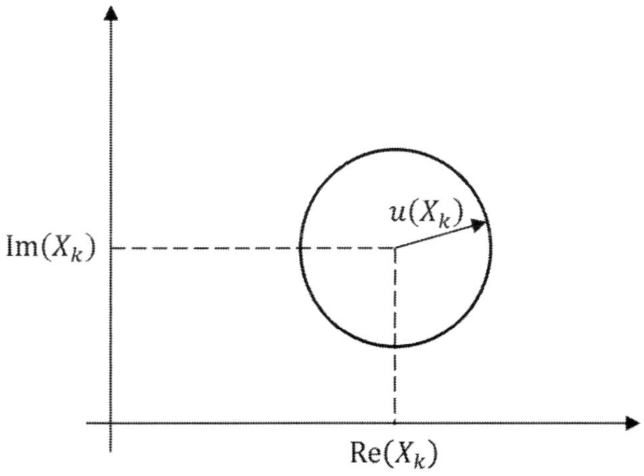

Figure 1 - Graphical representation of the standard uncertainty associated with the determination of the Fourier coefficient X_k, estimated from the confidence bound (11).

The standard uncertainty associated with the determination of λ can then be estimated from

$$u(\lambda) \approx \sqrt{\left(\frac{\partial \lambda}{\partial A_F}\right)^2 u(A_F)^2}, \tag{12}$$

where

$$\frac{\partial \lambda}{\partial A_F} = \frac{\omega A_F}{\left(1 - {A_F}^2\right)^{\frac{3}{2}}}, \tag{13}$$

and $u(A_F)^2 \approx \mathrm{var}(A_{\mathrm{int}}/A_{\mathrm{ext}})$ is estimated from the first order Taylor expansion approximation to the variance of the ratio by

$$\mathrm{var}\left(\frac{A_{\mathrm{int}}}{A_{\mathrm{ext}}}\right) \approx \frac{\mathrm{E}(A_{\mathrm{int}})^2}{\mathrm{E}(A_{\mathrm{ext}})^2}\left[\frac{\mathrm{var}(A_{\mathrm{int}})}{\mathrm{E}(A_{\mathrm{int}})^2} - 2\frac{\mathrm{cov}(A_{\mathrm{int}}, A_{\mathrm{ext}})}{\mathrm{E}(A_{\mathrm{int}})\mathrm{E}(A_{\mathrm{ext}})} + \frac{\mathrm{var}(A_{\mathrm{ext}})}{\mathrm{E}(A_{\mathrm{ext}})^2}\right], \tag{14}$$

where $\mathrm{E}(\chi)$ denotes the expected value of χ and $\mathrm{cov}(\chi, \gamma)$ denotes the covariance between χ and γ. Since the covariance between the output and the input amplitudes of a first order LTI system is always positive, we can safely neglect the covariance term in (14) to obtain an upper bound for the variance of the ratio:

$$\mathrm{var}\left(\frac{A_{\mathrm{int}}}{A_{\mathrm{ext}}}\right) \leq \frac{\mathrm{E}(A_{\mathrm{int}})^2}{\mathrm{E}(A_{\mathrm{ext}})^2}\left[\frac{\mathrm{var}(A_{\mathrm{int}})}{\mathrm{E}(A_{\mathrm{int}})^2} + \frac{\mathrm{var}(A_{\mathrm{ext}})}{\mathrm{E}(A_{\mathrm{ext}})^2}\right], \tag{15}$$

Finally, we identify the variance of the amplitude with the square of the standard uncertainty obtained from (11).

3 MATERIALS AND METHODS

Two Extech SD800 devices were used to record CO_2 concentration, at a rate of one sample every 30 min, in a two bedroom flat, built in the 1990s, located in a rural village near Oliveira do Bairro, Portugal. The flat has an interior floor area of approximately 88 m^2 and height 2.5 m, and is at the first floor above ground of a three level building. The exterior device was placed in the east-facing balcony, shielded from direct solar radiation, whereas the interior device was placed in the living-room, leading to the same balcony. Figure 2 shows a floor plan of the flat, with the approximate locations of the measuring devices.

All windows and exterior doors were left fully closed, while all interior doors were left fully open. There were no occupants or other sources of CO2 inside the flat, and there was no heating, cooling or mechanical ventilation in operation during the entire measurement period.

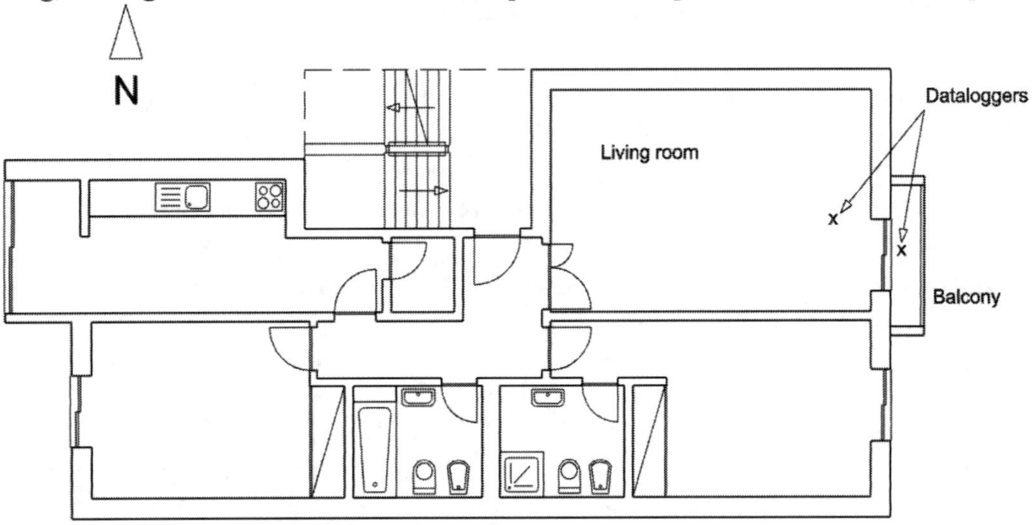

Figure 2 - Sketch of the residential flat were the measurement campaign took place. The locations of the interior and exterior measuring devices are shown with crosses, in the living room and in the balcony. All windows and exterior doors were left closed during the measurement period

4 RESULTS AND DISCUSSION

Continuous CO_2 concentration time series were obtained between 15:49 on August 26 and 15:19 on August 28, 2013. Figure 3 shows the raw data obtained from both the exterior and the interior devices. The time series are shown with an artificial vertical offset for better visualization. The shaded areas indicate the time periods from 20:00 to 07:00, which correspond approximately to the night periods.

The daily quasi-periodic nature of the atmospheric CO_2 is well visible in Figure 3. During the night periods, plants enter the respiration phase and produce CO_2, which reaches a maximum concentration around the early hours of the morning, before the sunrise. In the morning, as the sun rises above the horizon, plants enter the photosynthesis phase and the exterior CO_2 concentration starts decreasing, until it reaches a minimum around midday. As the sun sets in the evening, the exterior concentration starts to increase again. The interior time series clearly follows the trend of the exterior time series, with a lower amplitude and a phase lag. This behaviour is characteristic of the input-output relation of a first order system.

Both exterior and interior time series where segmented in overlapping periods of 24 h, with 48 samples each, i.e. the first segment starts at sample 1 and ends at sample 48; the second segment starts at sample 2 and ends at sample 49, and so on. The DFT was then applied to each segment, from which the amplitude of the second Fourier component (corresponding to the 24 h period) was calculated. The infiltration rate corresponding to each 24 h segment was then calculated from (7). The infiltration rate, as a function of time, is shown as a solid line in Figure 4, where each point on the line represents an average infiltration rate for the last 24 h. The dashed lines represent upper and lower expanded uncertainty estimates (coverage factor $k = 2$). The infiltration rate is seen to vary randomly about a mean value of 0.19, which is in agreement with typical values for infiltration rates in Portuguese residential buildings built in the last 30 years (Gomes, Gameiro da Silva & Simões, 2013). The expanded uncertainty varies from 10% to 20% of the estimated infiltration rate.

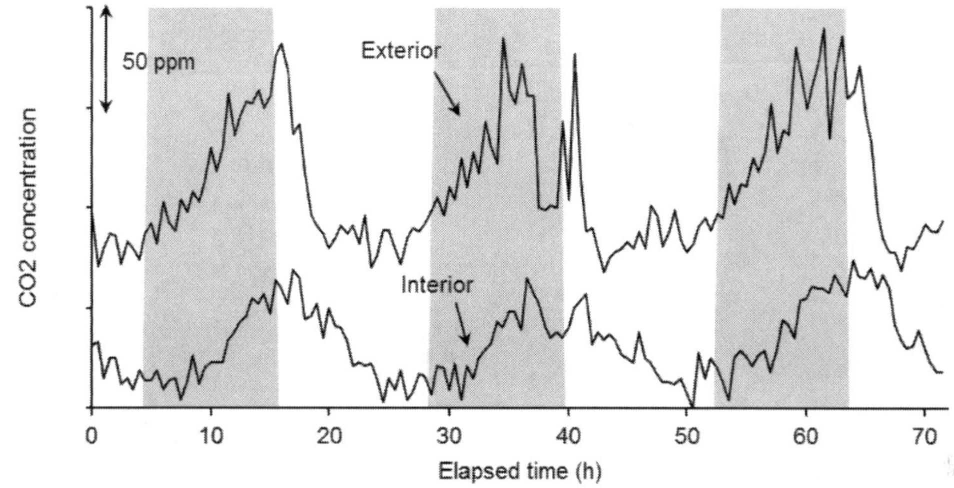

Figure 3 - Time series, recorded over 3 days, of exterior and interior CO_2 concentrations: raw data shown with an artificial vertical offset for better visualization. Ticks on the vertical axis are 50 ppm apart and the shaded areas identify night periods, between 20:00 and 07:00.

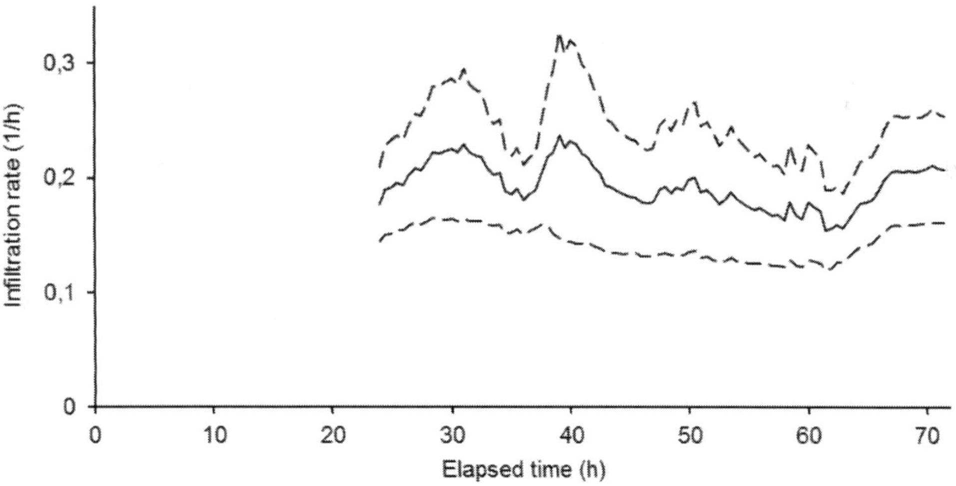

Figure 4 – Air infiltration rate results obtained from the raw data in Figure 3. Each point on the solid line represents an average infiltration rate corresponding to the previous 24 h. The dashed lines are estimates of the expanded uncertainty associated with the determination of the infiltration rate (coverage factor $k = 2$).

5 CONCLUSION

A new method to determine air infiltration rates in buildings has been proposed. The method belongs to the class of tracer gas techniques but, contrary to established methods that are based on the step response to the sudden introduction or extinction of a source of tracer gas, it uses the naturally occurring variation in atmospheric ambient CO_2 concentrations to derive 24 h moving average estimates of the infiltration rate. This novel approach has several advantages: it does not rely on the injection of a tracer gas or the use of metabolic CO_2 generated by the building occupants; it produces 24 h moving average time series of infiltration rates with the time resolution dictated only by the sampling frequency; and it may be less sensitive to mixing assumptions compared to methods which require the injection or generation of a tracer gas inside the building. The new method may provide a very useful tool for studying the dynamic behaviour of ventilation in buildings, which remains a main source of uncertainty in modelling building systems and, consequently, in the assessment of the energy balance in buildings as well as the quality of the indoor environment.

6 ACKNOWLEDGEMENTS

The presented work was framed under the Energy for Sustainability Initiative of the University of Coimbra and LAETA (Associated Laboratory for Energy, Transports and Aeronautics) Project Pest E/EME/LA0022/2011. The first author wishes to acknowledge the Portuguese funding institution FCT – Fundação para a Ciência e Tecnologia, for supporting his research through the Ph.D. grant SFRH/BD/77911/2011.

7 REFERENCES

Barankova, P. Naydenov, K. G. Melikov, A. K. and Sundell, J. (2004). Distribution of carbon dioxide produced by people in a room: Part 1 – Laboratory study. in *CD of Proceedings of the 9th International Conference on Air Distribution in Rooms (ROOMVENT 2004)*, 5-8 September 2004, Coimbra, Portugal.

Sherman, M. H. (1990). Tracer-gas techniques for measuring ventilation in a single zone. *Building and Environment*, 25, 365-374.

Persily, A. K. (1997). Evaluating building IAQ and ventilation with indoor carbon dioxide. *ASHRAE Transactions*, 103, 193-204.

Massen, F. Kies, A. Harpes, N. (2003). Seasonal and Diurnal CO2 Patterns at Diekirch, LU 2003 – 2005. Retrieved June 10, 2014, from http://meteo.lcd.lu/papers/co2_patterns/co2_patterns.html.

Miyaoka, Y. Yoshikawa Inoue, H. Sawa, Y. Matsueda, H. Taguchi, S. (2007). Diurnal and seasonal variations in atmospheric CO2 in Sapporo, Japan: Anthropogenic sources and biogenic sinks. *Geochemical Journal* 41, 429-436.

Etheridge, D. W. Sandberg, M. (1996) *Building ventilation: theory and measurement*. Chichester: Wiley.

Holman, J. P. (1994). *Experimental methods for engineers*. New York: McGraw-Hill.

Thibos, L. N. (2003). *Fourier analysis for beginners*. Retrieved July 14, 2014, from http://research.opt.indiana.edu/Library/FourierBook/ch09.html#E14

Gomes, M. R. Gameiro da Silva, M. Simões, N. (2013). Assessment of air infiltration rates in residential buildings in Portugal. In *Proceedings of CLIMA 2013, 11th REHVA world congress – Energy efficient, smart and healthy buildings*, Prague, Czech Republic, 16-19 June 2013.

PREDICTING THE OPTIMUM AIR PERMABILIY OF A STOCK OF DETACHED ENGLISH DWELLINGS

Benjamin Jones[*1], Robert Lowe[2]

1 Department of Architecture and Built Environment
Lenton Firs House, University of Nottingham
NG7 2RD, United Kingdom
**Corresponding author:*
benjamin.jones@nottingham.ac.uk

2 Energy Institute
University College London
Central House, 14 Upper Woburn Place
London, WC1H 0NN, United Kingdom

ABSTRACT

Mechanical positive input and extract ventilation are common strategies employed in English houses, generally because they provide adequate indoor air quality and specifically because they are effective at minimizing mould growth and its associated negative health consequences. Air is either exclusively supplied or extracted (never both) by a mechanical system at a prescribed airflow rate designed to ensure adequate indoor air quality. The remaining airflow, required to balance mass through the building, occurs through the fabric, through a mix of purpose provided and adventitious airflow paths. However, there is an important need to reduce adventitious airflow during the heating season in order to save energy and this is highlighted by the building standards of many countries who advocate increasing the airtightness of their housing stocks. Increasing a fabric's airtightness also increases its resistance to airflow and so a positive displacement or extract ventilation system must balance the corresponding increase in the fan power required to overcome this resistance with the need to provide an adequate supply of fresh air. Accordingly, the total energy consumption attributable to these ventilation systems during the heating season is a function of the dwelling's infiltration rate and heating system efficiency, and the fan's hydraulic power and efficiency.

This paper explores the possibility of an optimum air permeability for detached (single family) English houses ventilated by a mechanical positive input or an extract ventilation strategy. Firstly a theoretical model of air infiltration and exfiltration is used to investigate the underlying relationship between the fabric airflow rate, the mechanical ventilation rate (supplied or extracted), and the air permeability in a single detached dwelling. A meta-model of the total ventilation rate and heating system energy demand is developed. Secondly, the infiltration model then utilises a sample of over 16,000 English houses located in 10 different regions to investigate the relationship between the fan power and air permeability in a randomly chosen sub-sample of detached English houses. Thirdly, it is shown that there is indeed an optimum permeability for detached English houses with such systems and the effects of heating system and fan efficiencies on this value are identified. Finally, these predictions are contextualised by exploring their relevance to the current English housing stock.

KEYWORDS

Ventilation, infiltration, mechanical, positive-input, DOMVENT

1 INTRODUCTION

Mechanical ventilation systems are found in many European dwellings, particularly in the kitchen and bathroom where they are used maintain adequate indoor air quality (IAQ). A mechanical extract ventilation (MEV) fans reduce the internal air pressure below ambient and causes fresh air to be drawn through openings (commonly adventitious air leakage paths (ALPs) (Stephen, 2000)) in the thermal envelope. Positive input ventilation (PIV) reverses this process by supplying a steady flow of fresh air through a single opening located at the top of a stairwell.

The need to save energy is highlighted by the building codes of many countries that advocate increasing the airtightness of their housing stocks to save energy by reducing adventitious infiltration and exfiltration during the heating season. However, increasing the airtightness of a dwelling's fabric increases its resistance to airflow and when using a continuous CMEV (CMEV) or PIV system one must balance the corresponding increase in the fan power required to overcome this fabric resistance with the consequent reduction in spatial and temporal variation in air flow (Lowe, 2000). Accordingly, the total energy consumption attributable to these ventilation systems during the heating season is a function of the dwelling's permeability, the heating system efficiency, and the fan's specific power.
An understanding of the interaction between CMEV and PIV, and infiltration and exfiltration is critical to the optimization of their performance and the minimization of energy consumption. Accordingly, this paper uses a theoretical approach to explore these relationships and asks the question: is there an optimum permeability for English houses ventilated using a CMEV or PIV system?

2 A MODEL OF VENTILATION AND ASSOCIATED HEAT LOSS

This paper applies DOMVENT3D, a model of infiltration and exfiltration through any number of façades that incorporates a mechanical ventilation system. The model was conceived by Lyberg (2000), initially developed and applied by Lowe (2000) and extended significantly by Jones *et al.* (2013a, 2014). Its assumptions, merits, limitations, and the corroboration of its predictions are discussed widely by Jones *et al.* (2013a, 2014), and an analysis of the sensitivity of its predictions to its inputs is undertaken by Jones *et al.* (2013b). The model, which is implemented using bespoke MATLAB (MathWorks, 2013) code, is discussed in Section 2.1. It requires inputs that may be unique to each dwelling or are general to a geographic region. Unique inputs comprise dwelling geometry, the flow exponent, internal air density, scaled wind speed, and façade wind pressure coefficients. General inputs are the ambient air temperature, regional wind speed, and wind orientation. Sources of data are discussed in Section 2.2.

2.1 Model of Ventilation and Associated Heat Losses

DOMVENT3D makes two assumptions about a façade: (i) it is uniformly porous; (ii) the pressure distribution over it is linear. It integrates the airflow rate in the vertical plane to predict the total airflow rate Q_i (m^3/s) through the ith of j façades (Jones *et al.*, 2014) where

$$Q_i = \alpha(|\Delta p|/\Delta p) \int_{z_{min}}^{z_{max}} (|\Delta p|)^{\beta}\, dz \tag{1}$$

Here, z_{max} and z_{min} are the upper and lower height of a façade, Δp (Pa) is the pressure difference across an infinitesimal section dz (m) of a façade in the vertical plane, and α (Pa$^{-\beta}$) is a flow coefficient. Jones *et al.* (2014) show that $\alpha = E(2\bar{\rho})^{0.5}\, W$, where E is a dimensionless relative leakage area calculated from a known value of air permeability Q_{50} (m^3/h/m^2) or an equivalent metric, $\bar{\rho}$ (kg/m^3) is the mean of the internal and external air densities, and W (m) is the façade width. The net airflow through all j façades is zero

$$\textstyle\sum_{i=1}^{j} Q_i + Q_M = 0 \tag{2}$$

where Q_M (m^3/s) is the airflow through a mechanical system and the positive sign indicates airflow into a building. An important feature of the model that makes it suitable for this study is its dynamic neutral height z_0 (m) which is described by

$$z_0 = \frac{\frac{1}{2}\rho_E u^2 c_p - p}{(\rho_{ext} - \rho_{int})g}. \tag{3}$$

Here, ρ is the air density (kg/m^3), u is the wind speed at building height (m/s), c_p is a mean façade wind pressure coefficient, g is the gravitational acceleration (m/s^2), p (Pa) is the internal air pressure (gauge), and the subscripts int and ext represent internal and external conditions, respectively. DOMVENT3D makes two further assumptions about the dwelling: (i) all rooms are interconnected and internal flow resistances are small so that a dwelling can be treated as a single-zone (Jones *et al.*, 2013a); (ii) each horizontal and vertical surface of the external envelope requires only a single flow equation linked by a continuity equation; Once the total ventilation rate at an instant in time $Q_I(t)$ (m^3/s) is predicted by DOMVENT 3D, the ventilation heat loss (W) is calculated by

$$H(t) = Q_I(t)\bar{\rho}c\Delta T(t)|_{T_{ext}\leq(T_{int}-3)} \quad (4)$$

where c is the specific heat capacity of air ($Jkg^{-1}K^{-1}$), $\bar{\rho}$ is the mean of ρ_{ext} and ρ_{int}, and ΔT is the difference between the internal and ambient air temperatures. The internal air temperature, T_{int} (°C), of an average unheated English house is, on average, 3°C higher than the ambient air temperature, T_{ext} (°C), and so the heating system is assumed to function only when the ambient air temperature is T_{ext} ≤3°C below the internal air temperature (Hamilton *et al.*, 2011). Heat loss is only calculated when the heating system is "on" and so Equation (4) is integrated over the entire heating season to estimate the total energy required by a heating system, H_I (MWh). Further energy may be required by a fan whose hydraulic power $\mathcal{P}_M = |p\dot{Q}_M|$ (W) is also integrated over the heating season to estimate the total fan energy demand, H_M (MWh). The total energy demand H_T (MWh) is given by $H_T = H_I/\mu_H + H_M/\mu_M$, where μ_H and μ_M are the heating system and fan efficiencies, respectively.

3 SOURCES OF DATA

The model is now used to predict the total heat lost from a typical detached (single family) dwelling located in an urban area of London and to investigate the underlying relationships between a CMEV or PIV system and adventitious airflow. The dwelling's geometry, physical, and environmental properties are largely obtained from the 2009 English Housing Survey (EHS) (DCLG, 2011), a statistically representative sample of 16,150 dwellings that represents the 22.3 million dwellings of the English Housing stock. The sources of data are documented by Jones *et al.* (2013a) and so are only briefly discussed here.

Data inputs to DOMVENT3D may be divided into three distinct types: geometric, physical, and environmental. We now discuss each data type in turn beginning with geometric data. The EHS assumes that two connecting cuboids can reasonably represent the geometry of ~98% of English dwellings. The cuboid model applied here directly follows the Cambridge Housing Model (CHM) (Hughes *et al.*, 2012) used to estimate energy use and CO_2 emissions in the English stock for the UK government. For each façade the model requires the parameters z_{max}, z_{min}, and W. Dwelling orientation is not given by the EHS and so it is assumed to be a uniformly distributed random variable between 0 and 359 degrees (°).

DOMVENT3D requires three key physical parameters: Q_{50}, dwelling β, and façade c_p. The former is varied between limits of 0.1 and $20m^3/h/m^2$ in order to evaluate its effect on Q_T and H_T; see Sections 4 and 5. The flow exponent variable characterises the airflow regime through an ALP and is a function of its geometry and surface roughness. Its value affects both the pressure difference across an ALP and the airflow rate through it. Sherman & Dickerhoff (1998) show that a mean value of μ=0.65 with a standard deviation of σ=0.08 best represents more than 1900 measurements made in U.S. dwellings. This distribution is similar to the smaller international AIVC data set (Orme *et al.*, 1998) and so is applied with confidence as a Gaussian random variable. Wind pressure coefficients (c_p) are defined for the horizontal and vertical surfaces. For the latter, the algorithm of Swami and Chandra (1987) gives a normalized average wind pressure coefficient for long-walled low-rise dwellings and is a function of the angle of incidence of the wind, and local sheltering. The coefficient is then scaled to account for local shielding. Horizontal surfaces are assumed to be completely shielded from the effects of the wind following Sherman and Grimsrud (1980).

DOMVENT3D requires seven key environmental parameters: geographic location, local wind speed and direction, T_{int} and T_{ext}, terrain type, and shielding. The EHS indicates the region in which each sample is located and allows suitable weather data to be chosen from the CIBSE Test Reference Year (TRY) weather data set (CISBE, 2002). This source provides synthesised typical weather years for 10 English regions and is suitable for analysing the environmental performance of buildings. Accordingly, each EHS region is mapped to an appropriate CIBSE TRY region and where more than one CIBSE region is located in an EHS region the CIBSE region is chosen randomly (with equal probability) from the set of possible regions. The TRY weather data provides local wind speed, wind direction, and T_{ext} at hourly intervals. The wind speed is be scaled according to the terrain and dwelling height using a standard power law formula (BSI, 1991). Dwelling height is obtained from the cuboid model and the terrain type is indicated by the EHS. The four BSI terrain types and the local wind pressure shielding coefficients of Deru and Burns (2002) are mapped to the six EHS terrain types with format EHS (BSI){Deru and Burns}: city (city){very heavy}, urban (urban){heavy}, suburban town (urban){heavy}, rural residential (urban){moderate}, village centre (urban){moderate}, rural (country with scattered wind breaks){light}. Finally, DOMVENT3D is not a thermal model and so T_{int} must be prescribed. Here, a gaussian distribution of thermostat temperatures is chosen with μ=21.1°C and σ=2.5°C following Shipworth et al. (2009) who calculate these values from measurements made in a representative sample of 196 English dwellings.

4 PREDICTING VENTILATION ENERGY DEMAN OF A SINGLE DWELLING

The model is now used to investigate the underlying relationships between the CMEV and PIV systems and adventitious airflow using a detached (single family) dwelling located in an urban area of London. The dwelling is selected at random from the EHS database and its geometry, physical, and environmental properties and are given in Table 1. Their sources are described in Section 3, but for random variables, such as orientation, β, and T_{int}, typical values are chosen, where possible, for illustrative purposes. The fan provides an normalized (by dwelling volume) airflow rate of $N_M = \pm 0.5\text{h}^{-1}$, which is recognised by many European countries as a threshold ventilation rate below which some negative health effects increase (Jones *et al.*, 2013a). The sign indicates the flow direction, which is dependent on the mechanical system. Both CMEV and PIV systems are investigated and predictions of heating season ventilation air change rate and ventilation heat loss (excluded fan energy use) with varying Q_{50} are given in Figure 1 and compared against those for naturally occurring infiltration and exfiltration (with no mechanical system). Here we note two things. Firstly, the predictions of heating season ventilation rate made with and without a mechanical system are both described by the terms N_I or Q_I, where the former term is a normalized version of the latter using dwelling volume, but for clarity and brevity we use the subscript *NV* to describe the airflow solely attributable to natural airflow forces when $Q_M = 0$, and the subscript *T* to describe airflow where a mechanical system is present when $Q_M > 0$. Secondly, the terms Q_{50} and N_{50} (the air change rate at a pressure differential of 50Pa) can be considered almost identical because the sample dwelling's surface area to volume ratio is approximately unity; see Table 1.

Figure 1 shows that N_{NV} and its corresponding heat loss increase linearly with Q_{50}. Ventilation rates for CMEV and PIV are similar, differing by ~2% at 20 m^3/h/m^2. At the limits of Q_{50}. As $Q_{50} \rightarrow 0$m^3/h/m^2 the ventilation rate is $N_T \approx N_M$, and the ventilation heat loss $H_I \propto N_M$. These relationships are explained by considering Figure 2, which gives the variation of z_0', the mean dimensionless relative neutral height (relative to the dwelling height) for all façades during the heating season, with Q_{50}. It shows that $|z_0'| < 1$ when $Q_{50} \lesssim 6$ m^3/h/m^2 indicating that, depending on the sign of Q_m, z_0 is either above the building ($z_0 > z_{max}$) or below the base of the building ($z_0 < z_{min}$) for most of the heating season. In these circumstances, the mechanical airflow rate is balanced by infiltration or exfiltration over the whole envelope. The location of z_0 is a function of p and so by rearranging Equation (3) it is possible to identify the environmental conditions when $z_0 > z_{max}$ or $z_0 < z_{min}$. Here, Equation (3) shows that a CMEV system raises $z_0 > z_{max}$ on any façade when $p < \left[\frac{1}{2}\rho_E u^2 c_p - (\rho_{ext} - \rho_{int})g z_{max}\right]$. For the rare situation where $T_{ext} = T_{int}$ and $(\rho_{ext} - \rho_{int}) = 0$, the pressure difference across a façade does not vary with height, but the inequality still applies. Instead it describes the p required to overcome the prevailing environmental conditions and reverse the airflow direction through an entire leeward façade (when $c_p < 0$) so that air flows into a building. For the inequality to be true for a specific façade, either the $u^2 c_p$ term is small in magnitude or $p \ll 0$. The latter occurs in airtight dwellings; for example, in the sample detached dwelling, Figure 1 suggests this is when $Q_{50} < 6$ m^3/h/m^2. Conversely, a PIV system lowers $z_0 < z_{min}$ when $p > \left[\frac{1}{2}\rho_E u^2 c_p - (\rho_{ext} - \rho_{int})g z_{min}\right]$. When $(\rho_{ext} - \rho_{int}) = 0$, the inequality still applies and describes the p required to overcome the prevailing environmental conditions and reverse the airflow direction through a windward façade (when $c_p > 0$) so that air flows out of a building. For the inequality to be true, either the $u^2 c_p$ term is small in magnitude or $p \gg 0$. The latter also occurs in airtight dwellings, which for the sample detached dwelling Figure 1 shows is when $Q_{50} \lesssim 6$ m^3/h/m^2. This value of Q_{50} is identified by Lowe (2000) as a *critical* air permeability because it is the point where the gradients of N_T and H_I for PIV and MEV systems increase. It is interesting to note that Lowe also predicts the critical N_{50} to be around 6 h^{-1}, which is equivalent to $Q_{50} \approx 6$ m^3/h/m^2 when the surface area to volume ratio is unity, although it is likely to vary across a housing stock.

When $Q_{50} \to 20$ m^3/h/m^2 the dwelling can be considered to be *leaky* because Figure 1 shows that the total ventilation rate with mechanical assistance through the dwelling is predicted to be significantly more than the mechanical ventilation rate, N_M. It is also clear from Figure 1 that the total ventilation rate with mechanical assistance is $N_T \approx N_{NV} + C$, where C is the change in the ventilation rate attributable to the mechanical system. In these circumstances a CMEV system with a small airflow rate is responsible for a slight increase in the z_0 of all façades (see Figure 2) and a concurrent increase in the infiltration rate increases by ΔQ_{inf} (m^3/s) and decrease in the exfiltration rate by ΔQ_{exf} (m^3/s). The ΔQ_{inf} and ΔQ_{exf} terms are equal in magnitude but opposite in sign and so the total change attributable to the mechanical system is $|Q_M| \approx 2\Delta Q_{inf}$, or $\Delta Q_{inf} \approx |Q_M|/2 \approx C$. The same proof applies to a leaky dwelling ventilated by a PIV system, where z_0 decreases in all façades and when $T_{ext} = T_{int}$, where the CMEV or PIV system amends the magnitude of p and so that ΔQ_{exf} and ΔQ_{inf} are similarly equal in magnitude but opposite in sign. Here we note that the change in the total heating season space heating ΔH_I (MWh) required to compensate for the increased airflow rate caused by the addition of a CMEV or PIV system can be described in the same way, and so in the limit of small mechanically induced airflows $\Delta H_I \propto Q_M/2$. A simple function that fits the simulated data over the whole range of air leakage is given by

$$N_T = \left\{|N_M^n| + \left(N_{NV} + \frac{|N_M|}{2}\right)^n\right\}^{\frac{1}{n}} \tag{5}$$

where $n = 8$ gives a maximum absolute error (MAE) of 1% over the entire range of Q_{50} for a CMEV system, and $n = 7$ gives a MAE of 3% over the entire range of $0 < Q_{50} \leq 20$ for a PIV system, where the MAE is a measure of the largest deviation of the model's outputs. Figure 4 shows that the relationship between n and the MAE is for the sample detached dwelling is non-symmetric about a minimum MAE. However, when the value of n is greater than the location of the minimum MAE, the gradient is small and tends to an asymptote. Accordingly, if the most appropriate value of n is unknown, a high value should be selected

so that $n \gg 20$. For the sample detached dwelling, $n = 20$ gives a MAE of <0.03h^{-1} or <5% across the range of $0 < Q_{50} \leq 20$.
The model can be augmented by applying another simplified models that predicts N_{NV} from Q_{50} or N_{50}; see Jones *et al.* (2013). Further work is required to investigate the suitability of Equation (5) for predicting N_T in English houses that accounts for the variance of the inputs to the DOMVENT3D model (see Section 3). This is undertaken in Section 5.
Figure 5 shows the total predicted fan energy demand during the heating season, H_M. There is little difference between the predictions for the CMEV and PIV systems, but Figure 5 shows that there is a minimum value that occurs in the region of the *critical* permeability, where

Table 1: Properties of sample detached dwelling.

EHS sample	G0322417
Dwelling type	Detached
Location	London, England
Terrain type	Urban
Sheltering	Heavy
Orientation	North/south
Height (m)	5.1
Width (m)	6.4
Depth (m)	7.2
Volume, V (m^3)	232.7
Surface area, A_{surf} (m^2)	229.5
A_{surf}: V (m^2/m^3)	1.01
β	0.6
T_{int} (°C)	21.1

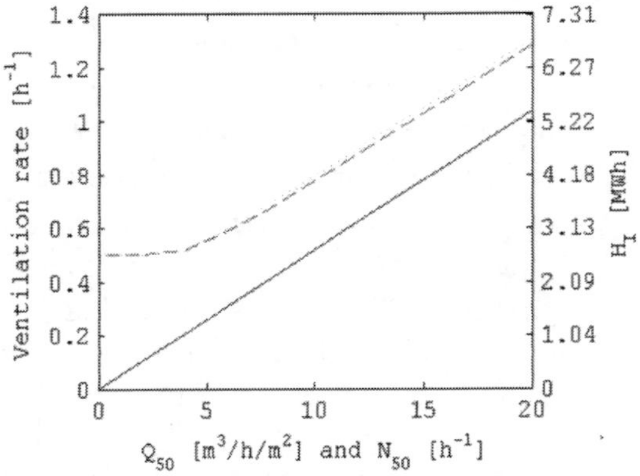

Figure 1: Predicted ventilation rate and heat loss, H_l (MWh), during the heating season.
—, natural infiltration; ----, CMEV; ·····, PIV

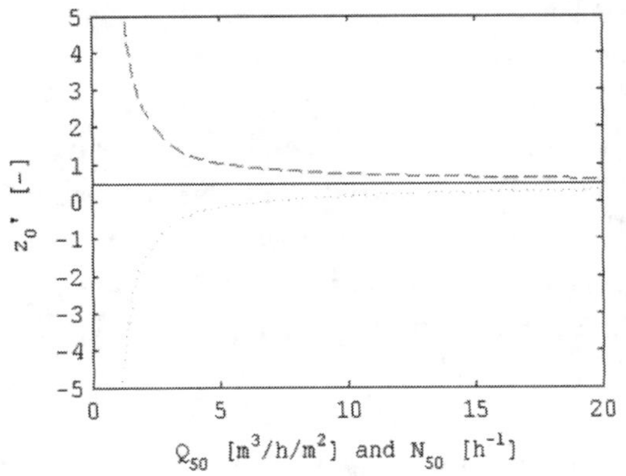

Figure 2: Predicted mean relative neutral height, z_0', for all façades during the heating season.
—, natural infiltration; ----, CMEV; ·····, PIV.

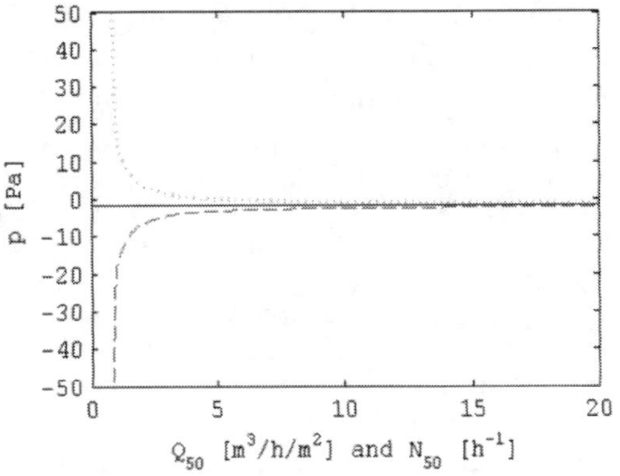

Figure 3: Predicted mean internal air pressure, p (Pa), for all façades during the heating season.
—, natural infiltration; ----, CMEV; ·····, PIV.

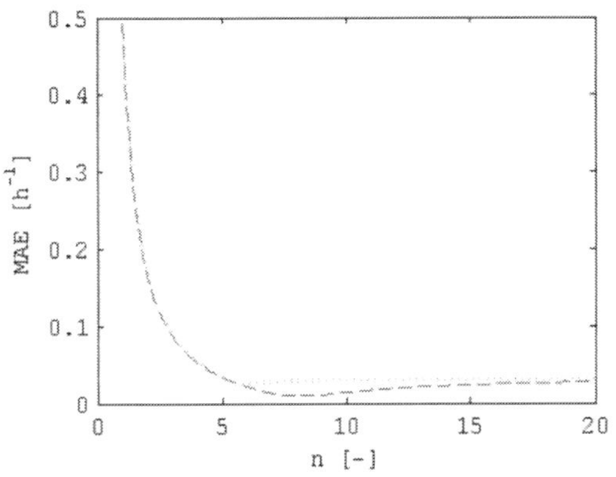

Figure 4: Change in maximum absolute error with n in English detached dwellings during the heating season. ----, CMEV; ·····, PIV.

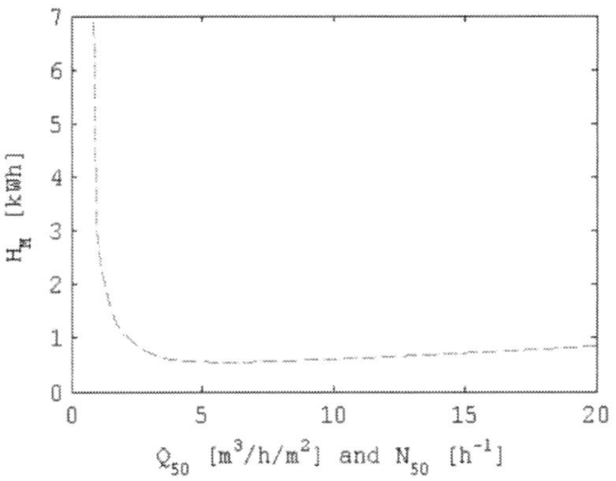

Figure 5: Predicted fan energy demand, H_M (kWh), during the heating season.
----, CMEV; ·····, PIV.

$Q_{50} \approx 6$ m^3/h/m^2. Section 2.1 shows that $H_M = f(N_T, p)$, whose respective relationships with Q_{50} are given in Figures 1 and 3. When $Q_{50} < 6$ m^3/h/m^2 the internal pressure $|p| \to \infty$ as $Q_{50} \to 0$, and the airflow rate tends to an asymptote where $N_T \to N_M$. Accordingly, the air pressure p, strongly influences the hydraulic power because its gradient is steep. Conversely, when $Q_{50} > 6$ m^3/h/m^2 Figure 3 shows that the internal pressure p, approaches a small asymptote and N_T increases linearly with Q_{50}. The airflow rate N_T, only has a moderate influence over the hydraulic power because its gradient is modest. Pressure losses within the system are unaccounted for in this study because they are manufacturer-specific but it is possible they would increase the hydraulic power, particularly when Q_{50} is greater than its critical value. The application of multiple fans is also unaccounted for and, because their efficiencies may vary, they can reach a stalling point leading to an unstable airflow rate.

In Figure 7, H_M and H_I are summed to calculate H_T where the boiler and fan efficiencies are unity, $\mu_H = \mu_M = 1$. A minimum H_T is observed between $1 < Q_{50} < 2$ m^3/h/m^2. A root finding technique that utilizes the Newton Raphson method (see Verbeke & Cools, 1995) finds the minimum or *optimum* permeability at $Q_{50} = 1.4$ m^3/h/m^2 for both CMEV and PIV systems. Figure 6 shows the effect of systematically varying μ_M between 0.1 and 1, and μ_H between 0.1 and 4 (where $\mu_H > 1$ simulates the coefficient of performance of a heat pump) on the optimum permeability for a CMEV system. Figure 7 shows that the variation in the optimum permeability is non-linear and limited, increasing with μ_H and decreasing with μ_M. Although the performance of the heating and ventilation equipment is shown to influence the optimum permeability of a dwelling, it is likely that general and specific dwelling parameters (described in Section 3) affect the optimum permeability. Accordingly, this is investigated in Section 5.

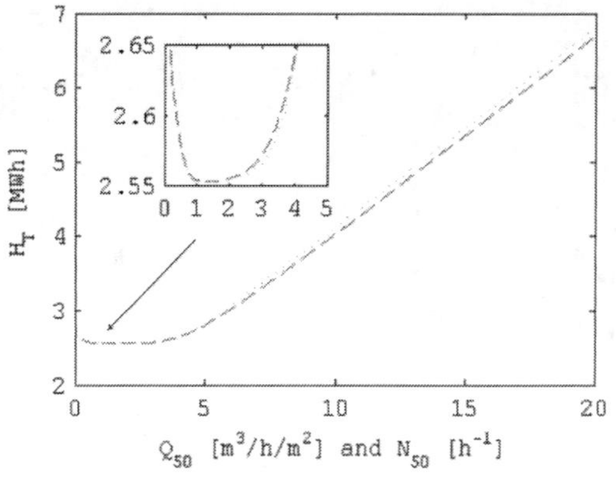

Figure 6: Predicted total energy demand, H_T (MWh), during the heating season. ----, CMEV; ·····, PIV.

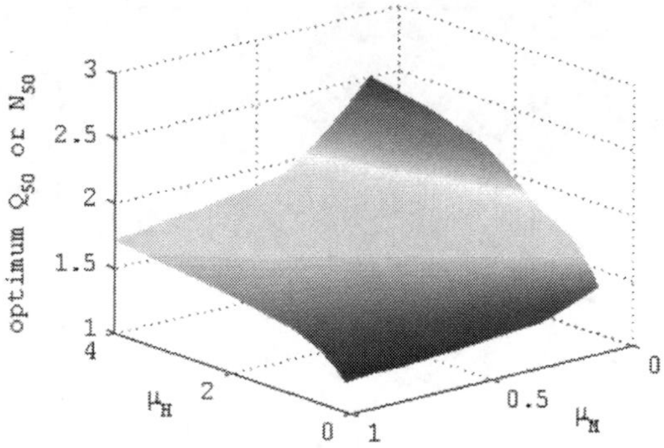

Figure 7: Optimum heating season Q_{50} and $N_{50}$with varying μ_H and μ_M for a CMEV system.

5 CONSIDERING STOCK VARIANCE IN ENERGY DEMAND

Section 4 considered the use of a CMEV or PIV in a single detached dwelling and identified a simple equation to predict the total ventilation rate from known mechanical and natural airflow rates; see Equation (5). However, it is desirable to know what the variance of the exponent n, is across the stock of English detached dwellings so that a general value can be identified. Section 4 shows that n differs according to the system employed and so both CMEV and PIV types are investigated using DOMVENT3D and a Monte Carlo (MC) sampling approach. There are four stochastic inputs to DOMVENT3D: the EHS sample (using dwelling weight), dwelling orientation, indoor air temperature, a flow exponent, and air permeability. A set of twenty samples of the 1st four inputs (hence excluding Q_{50}) are chosen at a time using a Latin Hypercube. However, the Q_{50} is varied systematically for each sample at values of 0.1, 2, 4, 6, 8, 10, 15, 20, and 50 m^3/h/m^2 (a total of 9 values). Accordingly, with a fan airflow rate of $N_M = \pm 0.5$h^{-1} (see Section 4) nine predictions of N_T are made by DOMVENT3D for each sample to identify its relationship with Q_{50}; see Figure 1. In addition, Figure 1 shows that the natural ventilation rate N_{NV}, during the heating season varies linearly with Q_{50} and so only a single prediction of N_{NV} is needed (apply the sample inputs used to estimate N_T) and is scaled by Q_{50}. The predictions of DOMVENT3D are then compared to those of Equation (5) for a range of exponents where $1 \leq n \leq 30$. The MAE for each value of n is used to identify the best n for each dwelling. The total sample size increases incrementally according to the set size. After each set of predictions is made, the mean (μ) and standard deviation (σ) of n for the whole sample are calculated and used to decide if a stopping criterion has been met. The number of samples is deemed adequate if the change in μ and σ from one set of 20 samples to the next is less than 0.2% after a minimum of 5 sets (100 samples) have been obtained. The model is run twice because a distribution is required for each mechanical system.

Figure 8 gives distributions of n (see Equation (5)) for CMEV and PIV systems estimated using MATLAB's ksdensity kernel smoothing function (with a moderated bandwidth of 1.6 for the PIV system). The CMEV sample required 27 sets (540 samples) to converge, whereas the PIV sample required 11 sets (220 samples). Here, the mean and standard deviation of n for a PIV system are μ=4.5 and σ=1.5, and the mode value is approximately equal to the mean value. A two-sample Kolmogorov-Smirnov (KS) test is used to test the null hypothesis that the data given in Figure 8 for a PIV system is Gaussian (with μ=4.5 and σ=1.5) is rejected at a 5% significance level (p=1.09×10^{-4}). Accordingly, the mean value of n =5.1 is more appropriate for this stock of detached dwellings. The need for simplicity and brevity often necessitates the use of whole numbers, rather than fractions. Accordingly, Figure 8 suggests that an exponent of n =5 is broadly appropriate for detached English dwellings. Applying n =5 to the sample dwelling discussed in Section 4 gives a MAE of 0.03h^{-1} or 6% within the limits of $0.1 \leq Q_{50} \leq 20$.

Figure 8 shows that the distribution of n for a CMEV system is tri-model, although it is clear that the majority of the data is located where $n \approx 9$. Here, μ=15.6, σ=9.6, and the median value is n =11. A two-sample Kolmogorov-Smirnov (KS) test rejects the null hypothesis (at 5% significance) that the distribution is Gaussian (with μ=15.6 and σ=9.6) (p=2.86×10^{-8}). Applying a rounded value of $n = 11$ to the sample dwelling discussed in Section 4 gives a MAE of 0.02h^{-1} or 3% within the limits of $0.1 \leq Q_{50} \leq 20$. Here note that Figure 3 shows the relationship between n and the MAE to be asymmetric and so if the most appropriate value of n is unknown, a high value should be selected. Figure 8 suggests that the value of n propose in Section 4 should be increased to $n \gg 30$ in order to minimize the MAE for the whole stock of English detached dwellings.

The sampling method described earlier in this Section is now applied to estimate the distribution of the optimum permeability across the English stock of detached dwellings, although some small amendments are required. Firstly, Figure 5 shows that the minimum is likely to occur at low Q_{50} and so it is varied using a root finding technique (see Section 4) that searches for the minimum value of H_T. In addition, the boiler and fan efficiencies are given typical values. Here, Lowe (2000) suggests that a boiler efficiency of 90% ($\mu_H = 0.9$) is appropriate for most UK houses because many now contain condensing gas-fired systems,

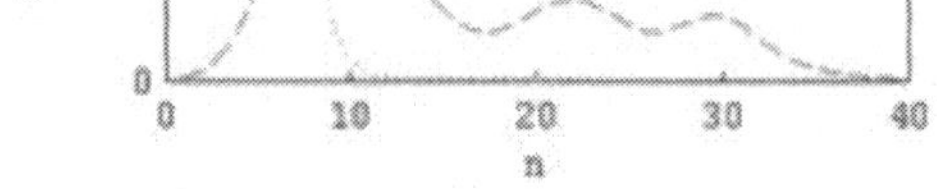

Figure 8: Distribution of exponent, n, in English detached dwellings during the heating season.
—, CMEV; ---, PIV.

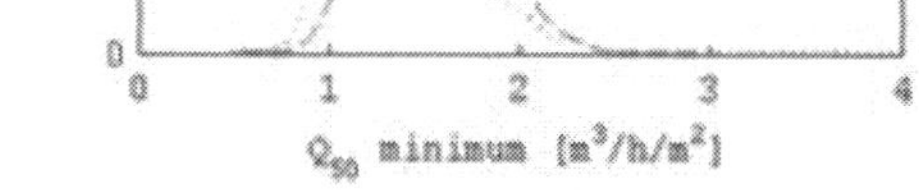

Figure 9: Distribution of optimum permeability in the English housing stock for the heating season.
—, CMEV; ---, PIV.

although extending the approach to other heating and electrical supply systems in the future would be trivial. CIBSE (2005) suggest that the total efficiency of an axial-flow fan (with or without guiding vanes) is approximately 65% ($\mu_M = 0.65$). The model is run twice because a distribution is required for each mechanical system.
Figure 9 shows predicted distributions of optimum Q_{50} in English detached dwellings during the heating season, when $\mu_H = 0.9$ and $\mu_M = 0.65$. The CMEV sample required 26 sets (520 samples) to converge, whereas the PIV sample required 24 set (480 samples). The optimum permeability for CMEV and PIV are respectively found to occur at 1.5 and 1.4m^3/h/m^2, respectively, with most of the date lying between 1 and 2m^3/h/m^2. These airtightness values are considered to be very tight because they and comprise <1% of UK dwellings built before 2000 (Stephen, 1998) and <2% of UK dwellings built after 2006 (Pan, 2010). Furthermore, Figure 3 shows that the required change in internal air pressure is up to 18Pa, and so non-standard fan may be required. Lowe (2000) notes that the default system of an airtight dwelling is a balanced mechanical ventilation with heat recovery (MVHR), but shows that a badly designed, low-efficiency, system will always be outperformed by a CMEV system. This study confirms that there is merit in increasing the airtightness of dwellings that apply CMEV or PIV strategies to reduce ventilation heat loss. A demand control strategy could further reduce energy demand. Increasing fabric tightness is also important to ensure adequate IAQ because it is shown (Stephen, 2000) that a PIV system can force stale, warm, and moist air into the loft space where the moisture condenses and where it is recirculated back into the house reducing IAQ. A mechanical system can help to provide adequate indoor air quality in a dwelling and reduce the negative health consequences associated with under-ventilation. The increase in total energy demand that occurs at very low permeabilities is unlikely to be a problem in practice.

6 CONCLUSIONS

This paper utilises an existing model of infiltration and exfiltration and databases of dwelling geometry, dwellings' physical parameters, and environmental information to investigate the performance of mechanical extract ventilation and positive input ventilation systems in English detached houses during the heating season.
A sample detached house is used to investigate the underlying physics of the problem and finds that at low detached houses values of dwelling air permeability the total infiltration rate is approximately equal to the fan airflow rate. As the infiltration rate increases significantly, the total ventilation rate of the dwelling is approximately equal to the sum of the airflow rate found the same detached dwelling without a fan (naturally occurring infiltration and exfiltration) and half the fan's airflow rate. The same relationship applies to the demand placed on a heating system to replace lost warm air during the heating season. The relationship is described using a simple equation that includes an exponent. For the sample detached house the exponent is found to be different for the CMEV and PIV systems. Then, a stochastic method is used to predict distributions of the exponent for a stock of detached English dwellings. They distributions show that geometric, physical, and environmental dwelling information does influence the most appropriate choice of exponent and a method that minimises the maximum absolute error is used to enables an appropriate value to be chosen for each of the CMEV and PIV systems. The model is of interest to policy makers of any country who use simplified energy models, such as the UK's standard assessment procedure, to make informed decisions.
When the energy demand of the mechanical demand ventilation system is considered in a sampled detached dwelling, an optimum air-permeability is found where the total energy demand is minimized. Then, a similar stochastic approach is used to investigate the location of the optimum air-permeability in the stock of English detached dwellings. It is found to

occur at between 1 and 2 $m^3/h/m^2$, which is only found in <1% of all UK dwellings constructed before 2000. Accordingly, although theoretically possible, the reality of achieving an optimum permeability in a large number of houses ventilated by CMEV and PIV systems is unlikely. However, this highlights that the increase in total energy demand that occurs at very low air permeability is unlikely to be a problem in practice and that it is difficult to make a house too airtight.

7 REFERENCES

CIBSE. (2002). Guide J - Weather, Solar and Illuminance data. London, CIBSE Publications.

CIBSE. (2005). Guide B - Heating, Ventilating, and Air Conditioning. London, CIBSE.

DCLG (2011). English Housing Survey: Headline report 2009-10. London, Department for Communities and Local Government.

Deru, M.P. and Burns, P.J. (2003). Infiltration and natural ventilation model for whole building energy simulation of residential buildings. *ASHRAE Transactions* 109(2), 801-811.

Hamilton, I., *et al.* (2011). The impact of housing energy efficiency improvements on reduced exposure to cold — the 'temperature take back factor'. *BSER&T* 32(1), 85-98.

Hughes, M., *et al.* . (2012). Converting English Housing Survey Data for Use in Energy Models, Cambridge Architectural Research Ltd. and University College London.

Jones, B.M., *et al.* (2013a). The Effect of Party Wall Permeability on Estimations of Infiltration from Air Leakage. *International Journal of Ventilation* 12(1): 17-29.

Jones, B.M., *et al.* (2013b). A stochastic approach to predicting the relationship between dwelling permeability and infiltration in English apartments. 34th Air Infiltration and Ventilation Centre, 3rd TightVent, 2nd Cool Roofs', and 1st Venticool Conference. Athens, Greece: 199-209.

Jones, B.M., *et al.* (2014). Modelling uniformly porous façades to predict dwelling infiltration rates. *Building Services Engineering Research and Technology*. 35(4).

Lowe, R.J. (2000). Ventilation strategy, energy use and CO_2 emissions in dwellings – a theoretical approach. *Building Services Engineering Research and Technology* **21**(3), 179-185.

Lyberg, M. (1997). Basic air infiltration. *Building and Environment* **32**(2), 95-100.

MathWorks (2013). MATLAB Version 8.3.0.532 (R2014a). The MathWorks Inc.

Orme, M., Liddament, M.W. and Wilson, A. (1998). TN 44: Numerical Data for Air Infiltration and Natural Ventilation Calculations, Air Infiltration and Ventilation Centre.

Pan, W. (2010). Relationships between air-tightness and its influencing factors of post-2006 new-build dwellings in the UK. *Building and Environment* 45(11), 2387-2399.

Sherman, M.H. and Dickerhoff, D.J. (1998). Airtightness of U.S. dwellings. *ASHRAE Transactions* 104(4), 1359-1367.

Sherman, M.H. and Grimsrud, D.T. (1980). Infiltration–pressurization correlation: simplified physical modelling. Lawrence Berkeley Laboratory Report LBL-10163.

Shipworth, M., *et al.* (2009). Central heating thermostat settings and timing: building demographics. Building Research & Information 38(1): 50-69.

Stephen, R.K. (2000). IP 12/00: Positive Input Ventilation in Dwellings. BRE, Watford, UK.

Stephen, R.K. (1998). Airtightness in UK dwellings: BRE's test results and their significance, Building Research Establishment.

Swami, M. and Chandra, S. (1987). Procedures for Calculating Natural Ventilation Airflow Rates in Buildings, ASHRAE.

Verbeke, J. and Cools, R. (1995). The Newton-Raphson method. International Journal of Mathematical Education in Science and Technology 26(2): 177-193.

DEVELOPMENT OF AN EVALUATION METHODOLOGY TO QUANTIFY THE ENERGY POTENTIAL OF DEMAND CONTROLLED VENTILATION STRATEGIES

Samuel Caillou*[1], Nicolas Heijmans[1], Jelle Laverge[2], and Arnold Janssens[2]

1 Belgian Building Research Institute (BBRI)
Avenue P. Holoffe 21, B-1342 Limelette, Belgium

2 Ghent University (UGent)
Department of Architecture and Urban Planning
Faculty of Engineering
Jozef Plateaustraat 22, B-9000 Gent, Belgium

**Corresponding author: samuel.caillou@bbri.be*

ABSTRACT

Demand controlled ventilation (DCV) is seen more and more as a promising way to limit the energy consumption due to ventilation in buildings. However DCV is always a compromise between decreasing the ventilation flow rates and assuring the indoor air quality (IAQ). Ventilation requirements are usually expressed as required air flow rates in the ventilation standards and regulations. Up to now, no consensus for an absolute criterion of IAQ exists in the international scientific community. Quantifying the energy potential of DCV strategies taking into account the indoor air quality is still a big challenge. This paper describes a preliminary study for the development of such an assessment method for DCV strategies in the context of the regulation for the energy performance of buildings in Belgium.

This method was based on numerical simulations using the CONTAM software, with a Monte Carlo approach and representative occupation profiles.

An important question in this study was the choice of the reference system and/or flow rate used to quantify the energy potential of DCV strategies. Different possible references have been compared and discussed. In the absence of an absolute IAQ criterion and because manual regulation strategies cannot be considered as DCV, we propose to quantify DCV strategies by comparing the DCV strategy to be tested with a system, working at constant flow rate, achieving the same IAQ-level as the tested DCV system.

Different DCV strategies have been evaluated using this methodology. The following types of detection have been studied: CO2 or presence detection in the living spaces, and/or relative humidity (RH) detection in the service spaces. The regulation strategies ranged from very complex, with local detection and local regulation in each room independently, to very crude, with central detection and central regulation for the whole dwelling, and a series of intermediary strategies.

The main results can be summarized as follows. Applying this methodology to a large range of different DCV strategies, the simulation results lead to reduction factors of 0.61 for a complete DCV system combining CO2 detection in the living spaces and RH detection in the service spaces, 0.64 for the DCV system with only CO2 detection, and around 1 for DCV system with only RH detection in the service spaces. DCV strategies with only RH detection in the service rooms showed a poor energy potential because this kind of detection is not sufficient to control the IAQ in the living spaces. In contrast, DCV strategies with only CO2 detection in the living spaces gave a higher energy potential, depending also on the number of living spaces equipped with a detection sensor.

KEYWORDS

Manually regulated ventilation systems, CO_2 detection, relative humidity detection, presence detection, Contam simulations

1 INTRODUCTION

1.1 Context and objective

Demand Controlled Ventilation (DCV) is one of the promising solutions to decrease the energy impact of the building ventilation. The aim of DCV is to automatically adapt the ventilation flow rates to the ventilation needs in the building. By decreasing the ventilation flow rates, DCV also possibly influences the indoor air quality (IAQ). DCV is then always a compromise between energy savings (lower flow rates) and IAQ (higher flow rates). In the context of the energy performance of buildings (EPB), taking this trade off into account is crucial to carefully quantify the potential of energy savings for DCV systems.

The aim of this study was (1) to develop a methodology to quantify the energy savings of DCV and (2) to apply this methodology to a range of different DCV strategies as complete as possible, in order to quantify de energy savings of these DCV strategies in the form of reduction factors (value lower or equal to 1). To achieve this aim, numerical simulations using the software Contam have been carried out.

1.2 What is DCV?

First of all, a DCV system is a ventilation system, aiming to assure a sufficient air exchange rate of the building in order to maintain a sufficient level of IAQ at all times.
Specifically, a DCV system is an automatic control system aiming to modulate the ventilation flow rates by adapting the ventilation flow rates to the real, varying ventilation needs in the building. To achieve this, a DCV system is composed of:

- One or several sensors (CO_2, humidity, etc.) to measure the ventilation needs in the building,
- One or several regulation devices (adjustable fans, control valves, etc.) to regulate the ventilation flow rates in the building.

1.3 Origin of the energy potential of DCV

The energy potential of DCV is related to the real ventilation needs which are sometimes lower than the ventilation requirements for which the building has been designed:

- Lower real occupancy compared to the design capacity of the building (for example 2 occupants are living in a dwelling designed for 4 occupants);
- Variable occupancy in time (for example the occupants are not present during the day because they are at work or at school).

The key goal of DCV is to achieve flow rate reduction due to lower ventilation needs and DCV should be evaluated as such, independently of other ventilation parameters, such as:

- Effect of natural ventilation versus mechanical ventilation;
- Effect of the airtightness of the building and possible interactions between ventilation and in/exfiltration or adventitious ventilation.

Although some of these aspects of ventilation systems are sometimes discussed and evaluated together with DCV, they should be considered separately. In order to evaluate DCV independently from these others aspects, the evaluation of DCV was carried out in the following conditions:

- Mechanical supply and exhaust system (the ventilation system which provides the most controlled flow rates),
- Airtight building envelope (avoiding interaction between ventilation and airtightness),
- Model dwelling situated on one floor (avoid effect of buoyancy related adventitious ventilation).

2 METHODOLOGY

2.1 Modelling

The performances of the reference and DCV ventilation systems were evaluated using numerical simulations in the multi-zone airflow simulation package Contam[1]. The modelling used in this study was based on those used up to now in the context of EPB procedure in Belgium (equivalence procedure for innovative products), which is described in details in the following references (Heijmans et al, 2007[2]; Laverge et al, 2011[3]) and summarized as follows.

The model dwelling has been adapted and Table 1 lists the dimensions (m²) and design ventilation flow rates of the spaces in the model dwelling. For the results presented in this paper, the building envelope was airtight.

Table 1: Composition of the model dwelling

Spaces	Dimension (m²)	Supply flow rate (m³/h)	Exhaust flow rate (m³/h)
Living room	30	108	
Kitchen	13.5		60
Laundry	7.7		60
Toilet	1.6		30
Hall	10.9		16
Bathroom	7.5		60
Sleeping room 1	12.2	44	
Sleeping room 2	9.8	35	
Sleeping room 3	10.8	39	
TOTAL	104	226	226

To take the sensitivity of the simulation results to variations in the input data into account, a Monte Carlo approach has been used in this study. In this approach, instead of fixing 1 value for each input data, a distribution is determined for the key parameters and multiple simulations are carried out with different values of these parameters. In this study, this approach was carried out with 100 simulations with distributions of input data, such as different occupancy profiles. The number of occupants in the dwelling varies from one to six (1: 3%, 2: 21%, 3: 31%, 4: 32%, 5: 10%, 6: 3%), with an average of 3.34 persons per building.

2.2 IAQ criteria

In dwellings, the main reasons to ventilate are:

- To evacuate polluted air due to human occupancy for comfort and health reasons,
- To evacuate moisture produced by human activities such as cooking, baths and showers, clothes drying, etc., to avoid moisture and mould problems.

For the first ventilation rationale, CO_2 concentration is usually used as tracer of the human exposure to unpleasant IAQ. A first criterion is therefore based on the CO_2 exposure of the occupants. The cumulative CO_2 concentration above 600 ppm to which the occupants are exposed during their time of residence in the dwelling is used in this study. For the second ventilation rationale, the relative humidity is a useful indicator. The second IAQ criterion is based on the risk of high humidity levels on thermal bridges. The monthly averaged relative humidity on a thermal bridge with a temperature factor of 0.7 cannot be higher than 80%, which is a typical threshold for higher risk of mould development.

Two other criteria were also used in the study but not detailed in this paper: dispersion of odours due to the use of the toilets and exposure to pollutant emissions from building materials and furniture.

The main outputs from the simulations are (1) the total average ventilation heat loss, (2) the total exposure of the occupants to CO_2, and (3) the relative humidity in the service spaces.

Thanks to this method, the performance of different DCV strategies can be compared to different reference ventilation systems.

3 RESULTS AND DISCUSSION

3.1 Reference systems to quantify DCV

The most important question of this study was the choice of the reference system and/or flow rate used to quantify the energy potential of DCV strategies. This choice is particularly complex because DCV is always a compromise between flow rate reduction and IAQ. However, up to now, no consensus for an absolute criterion of IAQ exists in the international scientific community. Different possible references have been compared (see Figure 1):

- A mechanical supply and exhaust system working permanently at the design flow rates as required in the EPB regulation in Belgium for the model dwelling (red spot 100 % in the figure), as a result of the simulation methodology used in this study;
- The same system working at the flow rate used in the EPB calculation in Belgium for the model dwelling, which is around 150 m³/h (red arrow in the figure);
- Mechanical supply and exhaust systems working permanently at a constant flow rate, ranging from 100% to 30% of the design flow rate as required in the EPB regulation in Belgium for this model dwelling (blue curve in the figure), as a result of the simulation methodology used in this study.

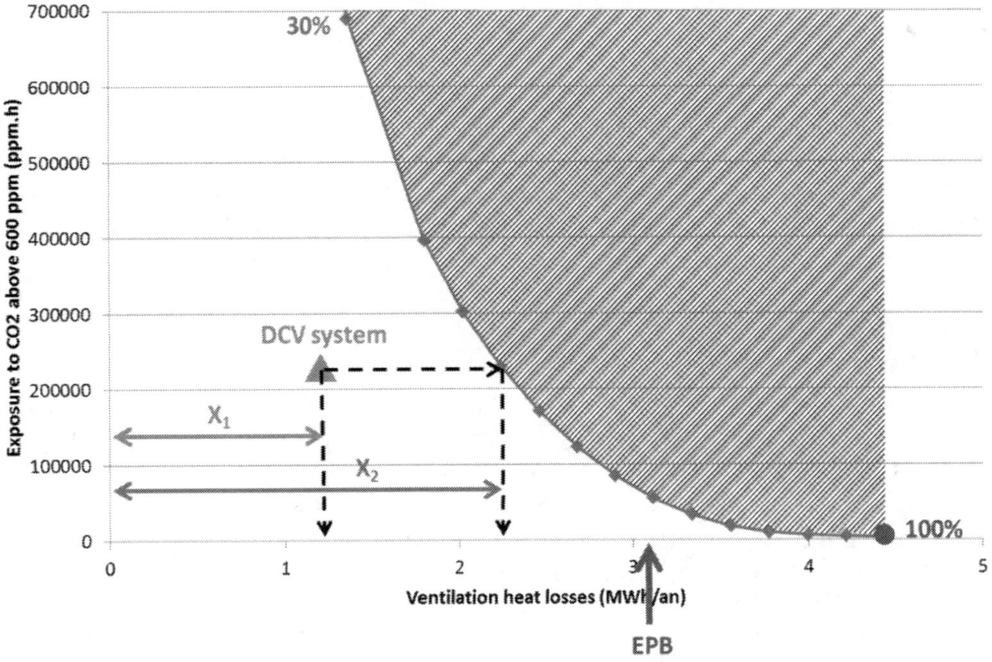

Figure 1: Comparison of different possible references: mechanical supply and exhaust system working at 100 % of required flow rate (red spot 100%), the flow rate used in the EPB calculation in Belgium for the model dwelling (red arrow), and the curve of different mechanical supply and exhaust systems working permanently at a constant flow rate, ranging from 100% to 30% of the required flow rate (blue curve).

As in most of the ventilation regulation and standards, there is no requirement on the IAQ in the ventilation regulation in Belgium, neither does it prescribe an average operational flow rate, but there are requirements on the flow rates for the design capacity of the ventilation systems.[4] The required flow rate (see red spot 100% in the figure) corresponds to a minimum capacity of the ventilation system, but at the same time, it is allowed to regulate the ventilation by adjusting, the flow rates, manually or automatically. The flow rate used in the EPB calculation in Belgium is around 2/3 of the required design flow rate (see red arrow in the figure), and takes already into account a certain regulation of the flow rate, such as manual regulation for example, as an estimate of the average operational flow rate.

The blue curve presented in Figure 1 for a mechanical supply and exhaust systems working permanently at a constant flow rate, ranging from 100% to 30% of the required flow rate, shows that a simple fixed reduction of the flow rates allows already a certain reduction of ventilation heat losses, by degrading the IAQ first very slightly (down to flow rates of around 60%) and then more significantly (flow rate between 60% and 30%). However, it is clear that such a fixed regulation of the flow rates is not a DCV system because DCV means, by definition, an **automatic** regulation of the flow rate. Nevertheless, a DCV system that does not performs better than such a fixed reduction seems of little value.

Required flow rate as reference.
One could consider that the required flow rate (red spot 100% in the figure) should be the reference to quantify DCV strategies. Such an approach is used in the current methodology to evaluate DCV in Belgium (but with additional interpolation between the results of the different ventilation systems, natural or mechanical, allowed in the ventilation regulation in Belgium).
But the main limitation of this approach is that there is no IAQ requirement in the ventilation regulation. A given DCV strategy can then provide a lower IAQ (and a lower flow rate) in this figure, while this strategy is perfectly authorized by the ventilation standard in Belgium. In the absence of an absolute IAQ criterion, this approach therefore leads to more favourable reduction factors for DCV strategies decreasing drastically both the flow rate **and** the IAQ.
A simple fixed reduction evaluated with this methodology and reference could lead to large energy savings, up to around 50%! However it would not be convincing to attribute such favourable reduction factors to a simple manual regulation system, because the calculation of ventilation heat losses in Belgium takes already into account a certain regulation of the flow rate, for example with manual regulation. This approach therefore leads to an embarrassing contradiction.

Flow rate used in EPB calculation as reference.
One could then consider the flow rate used in EPB calculation (red arrow in the figure) as reference to quantify DCV strategies, because this flow rate assumes implicitly a certain reduction of the flow rates, for example thanks to manual regulation strategies. Nevertheless, this only offsets the absolute value of the reduction factor, without addressing the problem of neglecting IAQ in the assessment.

Curve of systems with fixed flow rate reduction as reference.
Finally one could consider quantifying the DCV strategies by comparison with the curve of mechanical supply and exhaust systems working permanently at flow rates from 100% to 30% of the required flow rate (blue curve in the figure).
In this approach, the manual regulation strategies become the reference and correspond logically to a reduction factor equal to 1.0 (default value).This approach takes into account not only the flow rate reduction but also the IAQ provided by the DCV strategy to be tested. That is why this approach has been adopted in this study. The following arguments are also in favour of this approach:

- Most of the ventilation systems today on the market are equipped, as standard, with a manual regulation (for example a switch with 3 positions) and most AHU units allow to adjust the total flow rate in several steps.

- The flow rate used in the EPB calculation in Belgium takes already into account a certain reduction of the flow rates, for example using manual regulation.
- To be really effective, a given DCV strategy should be able to reduce the flow rate compared to a system working permanently at a constant flow rate and providing an equivalent IAQ.

It is particularly important to repeat at this point that the different ventilation systems used in this reference curve are systems working permanently at a constant flow rate. They are not systems manually adjusted by the occupants according to their needs.

This curve allows comparing a DCV system with a system, working at constant flow rate, giving the same IAQ than the DCV system tested. In the following of this study, the DCV systems are evaluated by comparison with this curve and the reduction factor of a given DCV strategy is calculated by X_1/X_2 as illustrated in Figure 1.

3.2 Tested DCV strategies

Different DCV strategies have been investigated and can be classified according to the following parameters:

- Type of **detection**: type of sensor, spaces where detection takes place (living spaces and/or service spaces), local detection (sensor in each space) or centralised detection (sensor in a plenum for example), etc.
- Type of **regulation**: regulation of the supply and/or the exhaust flows, local regulation (the flow rates are adjusted in each space) or centralised (the flow rates are adjusted at the fan for example), etc.

The types of sensor simulated in this study were:

- For the living spaces: CO_2 sensors or presence detection;
- For the service spaces: humidity (RH) detection in the kitchen, the bathroom and the laundry, and presence detection in the toilets.

For each DCV strategy to be tested, different variants of the regulation algorithms have been tested, to check the impact of, for example, the setpoint (for example 950 ppm CO_2), the shape (eg. linear, hysteresis, PID) and the minimum/maximum flow rates (for example 10/100% or 30/100%). The investigation of these variants allowed to identify additional conditions corresponding to each DCV strategy to assure a sufficient effectiveness (for example, conditions on the threshold, the minimum/maximum flow rates, etc.). For the different types of algorithms tested, the main influencing parameters have been found to be the threshold values in terms of flow rates and detected concentrations (for example 950 ppm for CO2 or 70% for RH), rather than the shape of the relation between the detected concentration and the flow rate.

3.3 Maximum potential of DCV

The most complete DCV system investigated combines:

- CO_2 detection locally in each living space, and local regulation of the supply flow rate in each living space;
- RH detection in each service space (presence detection in toilets), and local regulation of the extract flow rate in each service space.

According to the methodology used in this study, and compared to a system working permanently at a constant flow rate and giving the same IAQ, the average reduction factor for this most complete system was 0,61. This result is somewhat higher than but comparable to reduction factors mentioned in other references. For example, the most favourable reduction factor for a DCV system (similar to the one tested in this study) is 0,52 in the EPB regulation in The Netherlands (standard NEN 8088). [5]In the context of the preparation of the Ecodesign

Directive for ventilation units, the value of 0,50 is also mentioned as the most favourable reduction factor (for multi-variable DCV) in the draft of Directive. [6]

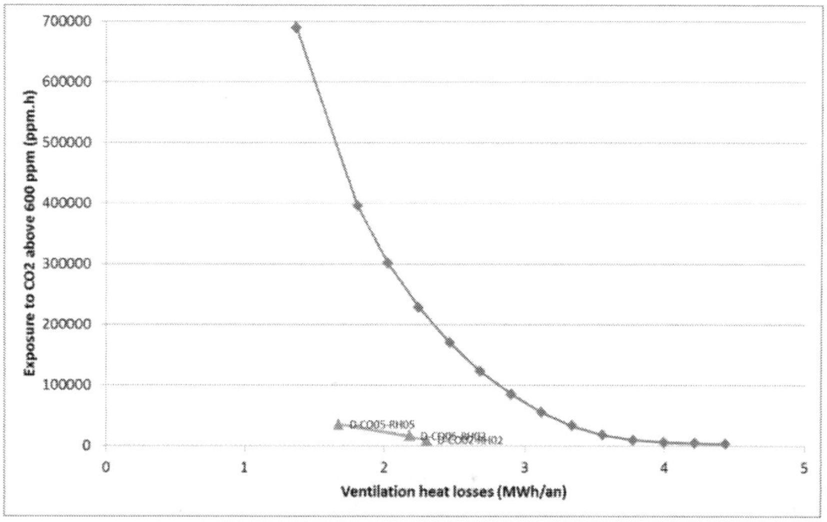

Figure 2: Result of DCV strategies combining CO_2 in the living spaces (local detection and local regulation) with RH in the service spaces (local detection and local regulation). The blue curve is the reference.

3.4 DCV with only CO_2 detection in the living spaces

The most complete DCV system with only detection and regulation in the living space combines:

- CO_2 detection locally in each living space, and local regulation of the supply flow rate in each living space;
- Total flow rate in the service spaces adjusted to the total flow rate for supply in the living spaces.

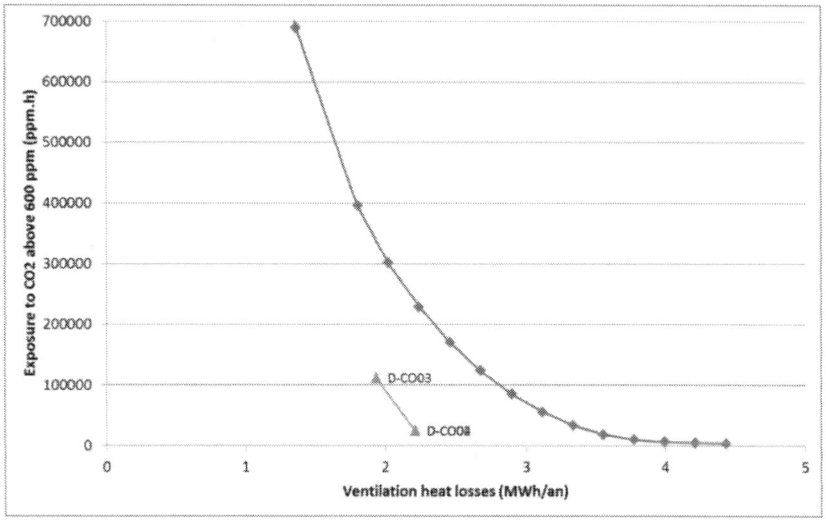

Figure 3: Result of DCV strategies with CO_2 in the living spaces (local detection and local regulation). The blue curve is the reference.

The DCV systems with only CO_2 detection in the living spaces showed an important potential of energy savings compared to a system working permanently at a constant flow rate and giving the same IAQ. But they showed also a higher risk for the humidity criteria in some cases. For the CO_2 criteria, these systems are, as expected, quite effective because the regulation is directly based on the IAQ detected in the living spaces although most of the occupancy time is concentrated in these spaces. Nevertheless, excessive reduction of the flow rates can lead to systems that cannot meet the humidity criterion, while the CO_2 criterion is easily fulfilled. Thus, to meet the humidity criterion and limit the risk of mould development, the results showed that a minimum flow rate of 30-35% for the supply flow rates (compared to the required flow rates) is necessary. The average reduction factor obtained for the DCV systems satisfying the humidity criteria was 0,64.

3.5 DCV with only presence detection in the living spaces

For the detection of the need in the living spaces, presence detection was also investigated as an alternative to CO_2 detection. The results of the simulations showed that the DCV strategies with presence detection are less effective than those with CO_2 detection, with reduction factors 33% higher compared to those for CO_2 detection. The lower performance obtained for the presence detection can be explained by the binary working of the presence detection ("on" or "off"), while the CO_2 detection takes into account variation of demand during occupancy, for example as a function of the number of persons in the space.

3.6 DCV with only RH detection in the service spaces

The most complete DCV system with only RH detection in the service spaces combines:

- RH detection in each service space (presence detection in toilets), and local regulation of the extract flow rate in each service space.
- Total flow rate in the living spaces adjusted to the total flow rate for extraction in the service spaces.

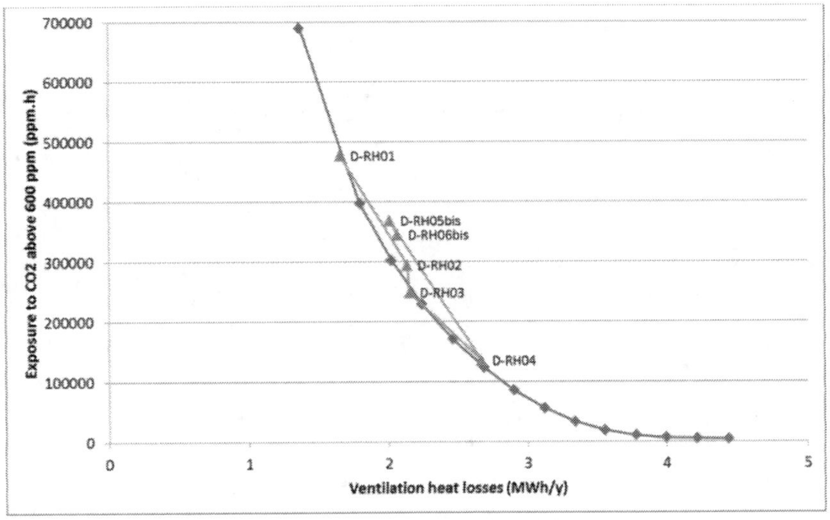

Figure 4: Result of DCV strategies for RH in the service spaces (local detection and local regulation). The blue curve is the reference.

According to the methodology used in this study, and compared to a system working permanently at a constant flow rate and giving the same IAQ, the average reduction factor for the DCV strategies with only RH detection in the service spaces was around 1. This means no advantage compared to a system working permanently at a constant flow rate and giving the same IAQ. This result is maybe not as surprising as it seems and can be explained among other by the too weak link between the RH detected in the service spaces and the human occupancy in the living spaces. This type of DCV strategies is mainly working when humidity is released in the services spaces (shower, cooking, etc.). But it is working on the lowest regime when there are little or no humidity sources, during periods of possibly higher occupancy in the living spaces, for example during the night. In contrast to this DCV strategy with RH only, the systems working permanently at a constant flow rate (the reference curve) assure permanently minimum ventilation:

- By supplying outdoor air to the living spaces, assuring sufficient IAQ for the occupants;
- By evacuating continuously humidity from the service spaces.

3.7 Intermediate DCV strategies

Besides the most complete DCV strategies (local detection and local regulation) presented in § 3.3 to § 3.6, different intermediate DCV strategies have been investigated with a lower number of sensors and moving from local regulation to zonal and centralised regulation.
Based on the simulation results, the most promising intermediate DCV strategies were identified as follows (results presented for CO_2 detection in the living spaces).

Table 2: Results for the most promising intermediate DCV strategies with CO_2 detection in the living spaces. Hatched zones indicate irrelevant combination of detection and regulation.

Type of detection in the living spaces	Type of regulation in the living spaces	Reduction factor
Locally : One sensor or more in each space	Locally	0.64
	2 zones (day/night) or more	0.70
	Centralised	0.80
Partially locally : One sensor or more in each sleeping room	Locally	
	2 zones (day/night) or more	
	Centralised	0.88
Partially locally : One sensor or more in the living room and one sensor or more in the sleeping room	Locally	
	2 zones (day/night) or more	0.73
	Centralised	1.03
Other or no detection in the living spaces	aucune, locale, par zone, ou centrale	1.00 = default value

3.8 Influence of the type of ventilation system (natural versus mechanical)

The results presented above are for the mechanical supply and exhaust system (type D in Belgium). To evaluate the influence of the type of ventilation system (for example the natural ventilation system, type A; or the mechanical exhaust system, type C), DCV systems of type A have been compared to a reference curve of type A, and DCV systems of type C have been compared to a reference curve of type C. The results of the simulations showed that the reduction factor, determined by this way, were close to each other whatever the system type (A, C or D) for a given DCV strategy. The difference between the system types was lower than 10%, sometimes positive, sometimes negative.

4 CONCLUSIONS

This study aimed to develop a methodology to quantify the energy savings of DCV strategies. In the absence of an absolute IAQ criterion and because manual regulation strategies cannot be considered as DCV, it was proposed to quantify DCV strategies by comparing the DCV strategy to be tested with a system, working at constant flow rate, achieving the same IAQ-level as the tested DCV system.
Applying this methodology to a large range of different DCV strategies, the simulation results lead to reduction factors of 0.61 for a complete DCV system combining CO_2 detection in the living spaces and RH detection in the service spaces, 0.64 for the DCV system with only CO_2 detection, and around 1 for DCV system with only RH detection in the service spaces.

5 ACKNOWLEDGEMENTS

This study was funded by the Brussels Capital Region, which is one of the 3 Regions in Belgium, responsible for the EPB regulation.

6 REFERENCES

[1] "Multizone Airflow and Contaminant Transport Analysis Software" developed by the "National Institute of Standards and Technology" (NIST) in the years 2002-2003 (Dols en Walton, 2002; Walton en Dols, 2003)

[2] N. Heijmans, N. Van Den Bossche, A. Janssens. (2007). *Berekeningsmethode gelijkwaardigheid voor innovatieve ventilatiesystemen in het kader van de EPB-regelgeving Toegepast op SYSTEEM C+ RENSON Ventilation NV.* Eindrapport – maart 2007,UGent-CSTC.

[3] J. Laverge, N. Van Den Bossche, N. Heijmans, A. Janssens. (2011). *Energy saving potential and repercussions on indoor air quality of demand controlled residential ventilation strategies.* Building and Environment, 46 (7) 1497-1503.

[4] NBN D 50-001. (1991). *Dispositifs de ventilation dans les bâtiments d'habitation.*

[5] NEN 8088-1. (2011). *Ventilatie en luchtdoorlatendheid van gebouwen - Bepalingsmethode voor de toevoerluchttemperatuur gecorrigeerde ventilatie- en infiltratieluchtvolumestromen voor energieprestatieberekeningen - Deel 1: Rekenmethode (Ventilation and infiltration for buildings - Calculation method for the supply air temperature corrected ventilation and infiltration air volume rates for calculating energy performance - Part 1: Calculation method).*

[6] Draft Working Document - with draft Ecodesign and draft Energy Labelling Requirements for Ventilation Units. Downloaded on 10/04/14 from: http://ec.europa.eu/enterprise/policies/sustainable-business/ecodesign/product-groups/airco-vent/index_en.htm

SYSTEM FOR CONTROLLING VARIABLE AMOUNT OF AIR ENSURING APPROPRIATE INDOOR AIR QUALITY IN LOW-ENERGY AND PASSIVE BUILDINGS

Kamil Szkarłat PhD, Eng.[1], Prof. Tomasz Mróz PhD, Eng.[2]

1 Applied Information Technology Institute
Department of Physics
Adam Mickiewicz University in Poznań

2 Environmental Engineering Institute
Department of Building and Environmental Engineering, Technical University in Poznań

ABSTRACT

In low energy buildings and passive houses due to very low heating demands integrated heating and ventilation (VAV or DCV) systems are used to provide proper indoor climate conditions – thermal comfort and indoor air quality. Dynamic changes of indoor conditions result in permanent changes in air flow. Control systems have to follow those changes and on the other hand have to minimize energy consumption for the system as a whole.
The article presents advanced simulation methods and the results of experimental investigation of control of variable volume system installed in passive house located in Poznań, Poland.

KEYWORDS

Low energy and passive buildings, VAV and DCV systems, Simulation

1 INTRODUCTION

In low-energy buildings, ventilation and heating systems are the main elements that form the microclimate due to a very low heat load. This requires the application of integrated systems that form indoor climate quality, i.e. thermal comfort (TC) and indoor air quality (IAQ). The systems of Variable Air Volume (VAV) and Demand Control Ventilation (DCV) are the examples of such solutions. In these systems dynamically changing internal conditions enforce continual change of air volume. Therefore, on one hand control systems must maintain a given CO_2 concentration at a predefined level and on the other hand they must minimize energy consumption. At the same time variable air volume along the whole operated ventilation line must result in stable and optimal performance of the source itself (air handling unit). Undoubtedly, proper and predefined usage schedules for given zones and algorithms supporting e.g. footfall systems are helpful in case of control systems for CO_2 concentration.
The article presents the application of advanced calculation and simulation methods together with the results of real measurements of variable air volume control using different configurations of connections between controllers.

2 CLIMATE COMFORT IN LOW-ENERGY AND PASSIVE BUILDINGS

Nowadays over 80 percent of the time is spent by a human being in confined spaces (houses, offices, shops, schools, etc.) [31], the objective function of buildings for people should be to ensure and maintain, in variable conditions, climate comfort and user's comfort. It is the main task of heating, ventilation and air condition devices and systems to ensure wellbeing in a building. There are several various physical and chemical parameters, very often time-changing, which determine the microclimate and have direct impact on living organisms. Microclimate parameters are usually grouped in two categories:

- *heating and humidity* conditions, creating the feeling of thermal sensation related to heat exchange between a human being and close environment,
- *health and sanitary* conditions covering indoor air quality in a broad sense [19].

One of the main components determining climate comfort is thermal comfort. It is an ideal temperature condition of a room that guarantees people in there that a sustainable thermal balance is kept [11,27,28,29,32]. Creating certain microclimate in rooms facilitates such thermal conditions in which every present person would feel thermal comfort, i.e. satisfaction from thermal conditions of the environment. [1,2,3,4].

The next important element of climate comfort is good quality of indoor air (IAQ - Indoor Air Quality) that defines such an air condition which is free from contaminants and does not cause any irritation, discomfort or illnesses of the users [3,16].

The role of contamination of a given kind which occurs individually or in coexistence, is decisive for Sick Building Syndrome or Building Related Illness to occur or not [17].

2.1 Correlation of main climate comfort parameters

The results of the microclimate research proves that as a consequence of correlation of individual parameters being the components of a thermal balance equation, it is necessary to steer and control several factors concurrently. Only in this way it is possible to improve the perceived climate comfort. In recent years there have been many unconventional approaches to microclimate issues. As an example, in air quality research the influence of temperature and humidity is considered jointly, as the effect of enthalpy of humid air on the perception of the environment quality [8]. A similar approach to analyses can be noticed when taking into consideration the effect of carbon dioxide concentration which defines air freshness of a given space. The effects and mutual influences of the main regulation parameters may lead to establishing a "golden mean" that gives the required comfort. At the same time it may result in elimination of at least one of the regulation parameters and therefore significantly reduce costs already at the stage of investment.

One of the dependencies in correlations with other basic parameters for climate comfort regulation is CO_2 concentration which, when changing along the changes of temperature and relative air humidity in the space (therefore identifying the presence of a human being), has a great influence on indoor microclimate and comfort perception. As an example the papers [7,12] present these dependencies in detail. Figure 1 presents percentage share of increasing satisfaction from climate comfort in a ventilated space depending on CO_2 concentration.

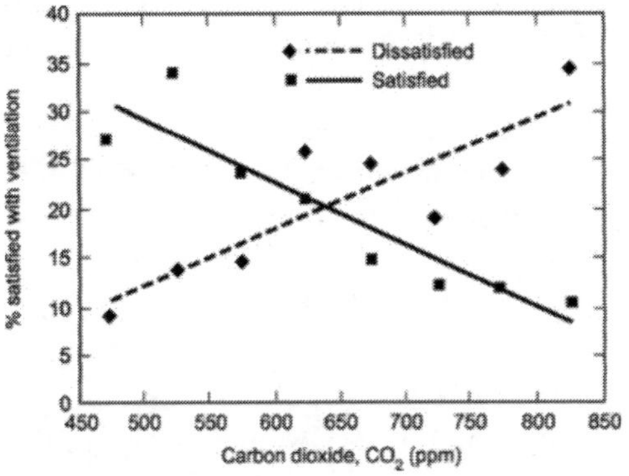

Figure 1: *Percentage of occupants that were satisfied or dissatisfied with ventilation at each level of carbon dioxide concentration* [39].

3 SYSTEMS FOR MAINTAINING CLIMATE COMFORT IN LOW-ENERGY AND PASSIVE BUILDINGS

Thanks to significant reduction of heat demand, in low-energy buildings, and first and foremost in passive ones, it is possible to eliminate a traditional static heating system – a convection-radiant or radiant system – to use an air heating system. Therefore a ventilation system is the main structure that provides climate comfort in passive building spaces. Its task is to keep thermal parameters at an appropriate level and at the same time to provide fresh air. Thanks to good insulating power and tightness of a passive building, combined supply and exhaust ventilation has a defined and targeted cascading flow. Fresh air is supplied to so-called occupant's "clean" spheres – bedroom, living room, guests room, office, laboratory, and it is removed from these spaces where the level of contamination is higher – kitchens, bathrooms, toilets. In passive buildings a ventilation system delivers all functions of climate comfort at the same time. This is a system that provides fresh air through ventilation itself but concurrently ensures thermal comfort by air heating in winter or providing cool air in summer (e.g. using a ground-coupled heat exchanger).

When analysing ventilation systems in passive and low-energy buildings in terms of regulation and control, two solutions can be distinguished:

- ***Centralized heat-ventilation system***: where the supplied air is prepared to suit appropriate technological parameters, most frequently in a compact device, based on one measurement from so called "equivalent sphere", or a point of reference representative for the whole building and supplied in a cascade manner to all regulation spheres; the operation of such a system ensures rather a permanent air flow fulfilling maybe several saving functions (such as e.g. reduction planning, etc.),
- ***decentralized heat-ventilation system***: where every regulation sphere e.g. a space or a group of spaces that have the same or similar functions is measured separately; additionally in each of the regulation spheres there is a separate system that prepares the supplied air (local heater) and a controller, e.g. VAV (Variable Air Volume) that makes it possible to provide proper doses of air stream to a given sphere; the system is fully integrated with the central unit where the supplied air is initially prepared (central heater) based on adaptive regulation connected with air temperature sucked from the outside and current parameters of individual regulation spheres (spaces in a passive building); additionally, the central unit by maintaining a pressurization level in a central ventilation duct operates in a master (air supply)-slave (air exhaust) configuration, changing the supplied air stream in the central duct depending on current needs of air streams in all spheres (forced by the location of VAV controllers).

The first solution of the centralized heat-ventilation system gives quite small possibilities of control optimization with a concurrent climate comfort maintained in every of control spheres. Maintaining the supplied air stream at the permanent level that ensures appropriate temperature and fresh air may on the other hand lead to dry air. However, limiting the overall air stream will automatically cause the increase of CO_2 concentration in occupied spaces or worsen thermal comfort.

In this respect it seems reasonable to use a decentralized heat-ventilation system. This solution gives much bigger possibilities of steering and control optimization, making it possible to dynamically influence climate comfort in a given regulated sphere (spaces of a passive building) depending on a current load.

From the control point of view a given system of ventilation, heating and air conditioning should be perceived as a dynamic multidimensional system which at a given point of time is influenced by any internal ($X_{interni} \div X_{extern}$) and external ($X_{interni} \div X_{extern}$) input signals as well as by disturbances ($Z_I \div Z_O$), but also as a system which at the same time should generate output signals ($Y_1 \div Y_k$) causing first and foremost stability of performance and the control optimum in terms of any control factors. In Figure no.2 such a dynamic system has been symbolically depicted. [based on 36].

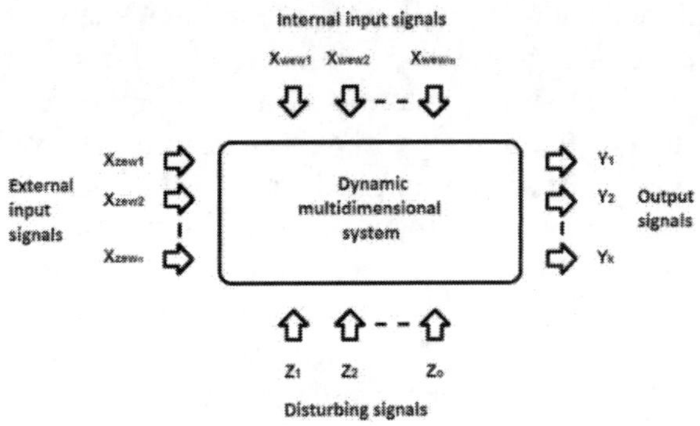

Figure 2: Symbol of HVAC system as a dynamic multidimensional system.

4 SYSTEM FOR CONTROLLING CLIMATE COMFORT PARAMETERS IN PASSIVE AND LOW-ENERGY BUILDINGS

When talking about control and all the more its optimization, at the very beginning a question must be asked: *„What is in fact to be controlled and how the regulation itself can be conducted?"*. When it comes to passive and low-energy buildings, it is necessary to start the analysis from a definition of the required climate comfort parameters, meeting of which is the main task when using technical building equipment [30].

When designing the technology of the air preparation process only two courses of the process are analyzed for two calculation parameters of internal and external air: one for winter and one for summer. In both cases the condition of external air and nominal values of heat gains, humidity, CO_2 emission and other contaminants are applied as input parameters and output parameters encompass a target status of supplied air and air status in a space [37,38].

In order to provide climate comfort in passive and low-energy building spacer, first and foremost appropriate regulation must be performed:

- of the temperature in a room,
- of relative air humidity in a room,
- of CO_2 concentration in the air,

by delivering the prepared air for the mechanical ventilation system of a building. This process needs technically advanced systems together with an automatic steering and control system already in case of several or more spaces. Variable Air Volume Systems (VAV) or Demand Control Ventilation Systems (DCV) are the examples of such structures.

Air parameters in spaces are a derivative of internal and external disturbances [6] which must be taken into consideration by the control system.

The whole control system of a given parameter should be resistant to any internal and external disturbances which may include, among others:

- changes of external air parameters,
- changes of thermal and humidity load of chambers,
- changes of energy parameters of factors which feed the air supply system itself.

An example of a block diagram showing an automatic control system for climate comfort in a passive building has been outlined in Figure no. 3.

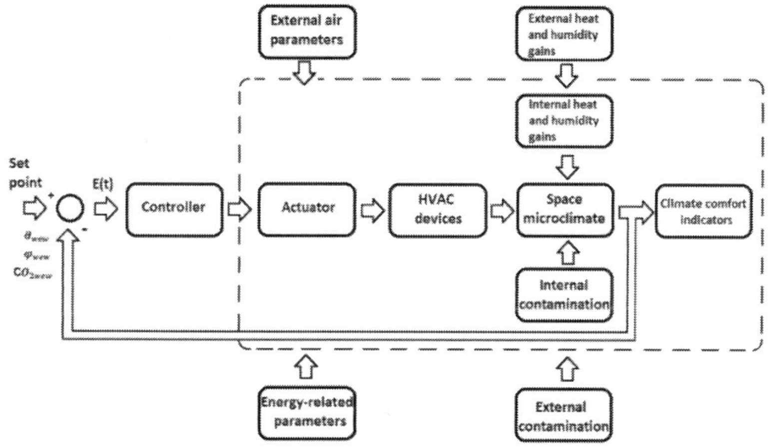

Figure 3: An example of an automatic climate comfort control system in passive and low-energy buildings.

5 MODELLING OF THE STRUCTURE AND OPTIMIZATION OF CONTROL SYSTEMS

If a fragment, for which it is possible to define the input and output signal, is separated from the automatic control system, then this fragment can be called an element of the system. It can be a single set or a subset, it can be a measuring instrument or a larger element of the system. It is important to ensure that the operation of the element is triggered by a certain reason and results in the required effect. For every element defined this way, one can define its intended use because it has a strictly defined distinctive function. In order to be able to assign a given element in the appropriate manner, its static or dynamic characteristics must be known as well as its function defined as a quotient of the function which defines the output signal and the function which defines the input signal.

The mathematical description of the control system encompasses descriptions of it components. The mathematical description of continuous element or automation system covers two parts:

- **an equation** or **a diagram of static characteristics** defining the dependency of the output signal on the input signal in defined situations.
- **a differential** or **an operator equation** that describes static and dynamic properties in the environment of performance point selected on a static characteristics.

Dynamic properties are evaluated based on y(t) curve as a response to the defined input signal x(t). Theoretically, plotting these curves, known as responses to a typical force, requires a solution of an equation of a dynamic system, which can be carried out by two methods:

- **classical method** consists in calculation of the root of an equation and setting constants based on initial conditions;
- **operator method** – commonly used in automation – consists in application of a transformation which would make it possible to replace an integral-differential equation with an ordinary algebraic equation. This is a Laplace transform. It assigns a transform (a transformation image) to a given function and the other way round $f(t) \leftrightarrow F(s)$.

5.1 Modelling of an automatic control system using a simplified transfer function

Each complex system may be converted to a simplified transfer function. It can be performed by dividing the system into its individual parts. Then for each of the created subsystem individual elements are described and combined using appropriate controllers, which results in modelling of the automatic regulation system. For the climate comfort systems a model consisting of a pure time delay and a series of n intertial elements of 1st order is very often used [34,36]. Taking advantage of the fact that a Laplace transform of the function adjusted by τ time units equals:

$$L(f(t-\tau))=e^{-\tau s} \tag{1}$$

which, when using Padé approximation, can be converted to adopt the following form:

$$e^{-\tau s} \approx \frac{1-\frac{\tau}{2}\cdot s}{1+\frac{\tau}{2}\cdot s} \tag{2}$$

the equivalent transfer function of such an element adopts the form of:

$$G_{sys}(s)=\frac{k_0}{(T\cdot s+1)^n}\cdot e^{-T_0 s} \tag{3}$$

or

$$G_{sys}(s)\approx\frac{k_0}{(T\cdot s+1)^n}\cdot\frac{1-\frac{T_0}{2}\cdot s}{1+\frac{T_0}{2}\cdot s} \tag{4}$$

where:
k_0 – equivalent gain of the plant
T – equivalent time constant of the plant,
T_0 – equivalent time delay of the plant,
n – interial order.

By adopting basic data in a simplified way, time constants and time delay of individual elements of the whole system have been set and then the parameters of the simplified equivalent transfer function of the whole system have been defined. For the controlled system defined in this way, there have been simulations carried out of the automatic control systems with classical controllers (P, PI, PID) as well as fuzzy controllers (Mamdani type P, PI and Tagaki-Sugeno type P, PI).

5.1.1 Classically controlled systems

Looking for optimal settings for classical systems of automatic control, advanced optimization methods have been used – namely gradient and hessian algorithms with the application of NCD-Outport Block in the Matlab/Simulink software. In the simulation tests for all modelled automatic control systems a unit step has been given as a standard signal which activate automation systems.
When analysing the received quality of control, apart from linear indexes read from the control run, in the simulation process additionally integral indexes have been used in a dynamic calculation of an integral from the product of a control error signal and an integral from the module of a control error signal.
An example of a model of a continuous automatic control with a PID controller has been depicted in Figure 4a and Figure 4b shows the response of the control system.

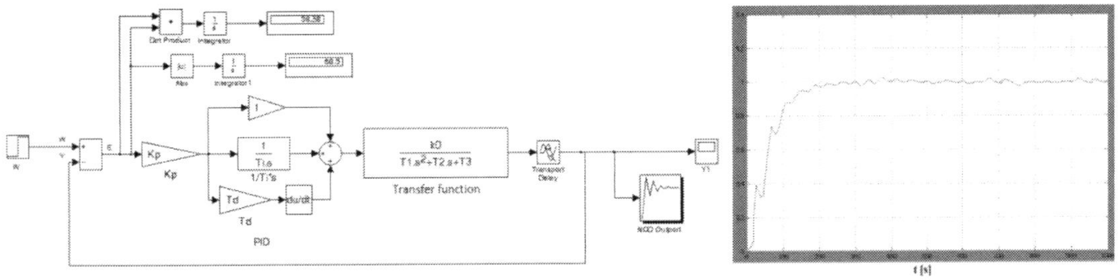

Figure 4. Model of the tested automatic control system with a PID-type controller (a) and the response of the control system as a simulation result (b).

5.1.2 Control system with unconventional control (fuzzy logic control)

Publications [8,10,23] show the modelling of fuzzy logic control dedicated to the process of forming the microclimate of the room.
In the first stage, when creating an automatic control system with a fuzzy logic controller, the construction of the FLC model (Fuzzy Logic Control) is important. Depending on a type of a given controller, input and output data must undergo fuzzification.
Figure 5a depicts an example of an automatic control system with FCL of PI type (Mamdami) modelled in Simulink (Matlab). Figure 5b however shows a response signal of given control during ongoing simulation.

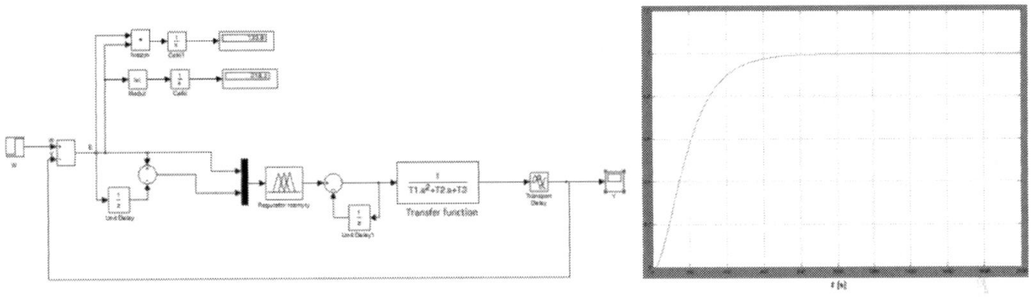

Figure 5. Model of the tested automatic control system with a fuzzy logic controller PI (a) and the response of the control system as a simulation result (b).

Simulation tests that have been conducted explicitly indicate that it is possible for continuous PID controllers with a precise selection of the parameters of the controller itself and fuzzy logic controllers with fuzzy values of the control process to be optimally controlled for the systems under test.

5.2 Modelling automatic control systems using system equations

Every automatic control system can be described in terms of equation constituents, equations describing process that take place in the system as well as equations describing the whole analysed system. In case of the climate comfort analysis in a passive building, energy balance equations based on individual spheres' load can become a starting point to mathematical modelling and simulation.
The whole modelling and simulation process has been performed using Matlab v 7.10 and its subprogram Simulink v.7.5 of Mathworks for dynamic system modelling.
Based on the VDI2078 guidelines [33], the load of the whole building can be determined using two methods:

- ***simplified*** – statically calculating individual internal and external loads, and
- ***advanced*** – allowing for dynamics and based on combination of recursive filters.

5.2.1 Building loads – simplified method

According to VDI2078 [33] the whole building load will be in line with the equation no. 5.

$$\dot{Q}_{KR} = \sum_{j=1}^{n} \dot{Q}_{KRj}(t) \tag{5}$$

where: $\dot{Q}_{KRj}$ – loads of a given j-th room,

n – number of rooms in a building.

The load of a given space consists of internal loads ($\dot{Q}_I$) and external loads ($\dot{Q}_A$). This relation is depicted in the equation no. 6:

$$\dot{Q}_{KR} = \dot{Q}_I + \dot{Q}_A \tag{6}$$

where:

$\dot{Q}_I$ – cooling loads coming from internal sources,

$\dot{Q}_A$ – cooling loads coming from external sources.

Load calculations for all spaces in a given passive building have been programmed and modelled based on the algorithm as defined in VDI2078 guidelines, using all climate data and also in line with the character of a passive building under test. An example of individual load constituents for one room and for a selected period of time has been depicted as daily and hourly diagrams in Figure 6.

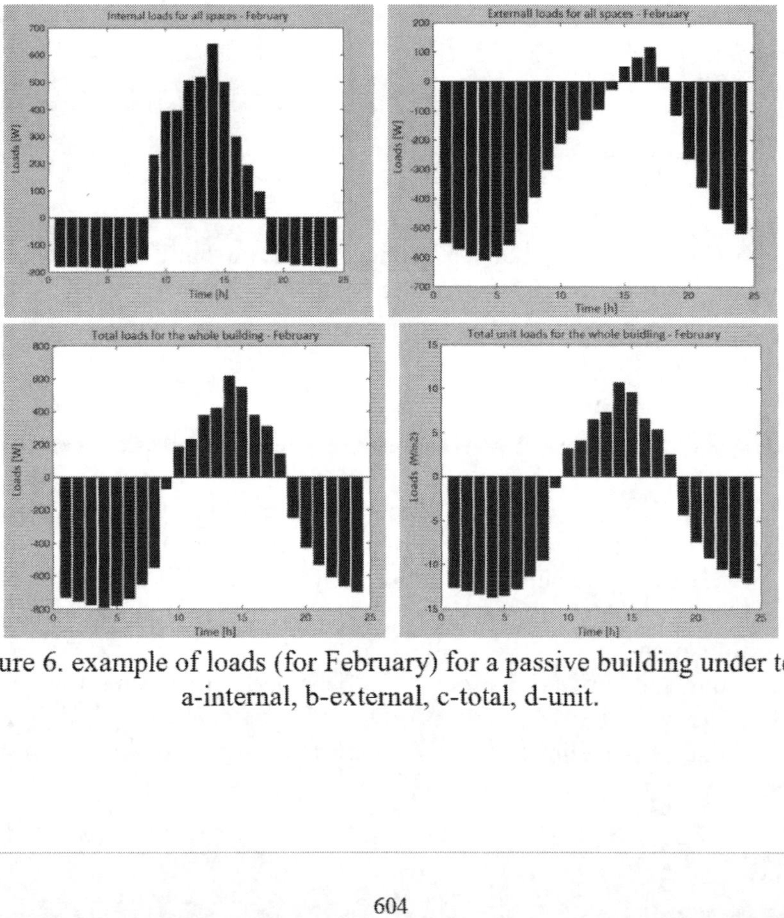

Figure 6. example of loads (for February) for a passive building under test: a-internal, b-external, c-total, d-unit.

Calculations and simulations that have been carried out explicitly imply a huge domination of internal loads. The results of the simulations for the whole calculation year together with the use of standard data for calculations, the application of use profiles in line with the requirements in the norm and the adoption of given accumulation matrices imply that basically the examined passive building should be cooled all year round as a result of higher values of internal load. This however does not correspond to the actual situation. Therefore it should be explicitly stated that it is a mistake to adopt standard calculation data connected with cooling and heating loads for general building industry also in case of low-energy buildings and especially for passive ones.
Very good thermal insulation of all bulkheads and high tightness are the reason why the biggest influence on formation of the microclimate in passive buildings refers to internal loads. Therefore, all constituents which have this impact on internal load must be very precisely defined for all analyses, research and calculations.
Similar conclusions can be drawn when applying the methods for calculating design heat load for passive buildings as set forth in PN-EN 12831 [21,22,24,25,26]. In the publication [20] calculations made based on this method show a design heat load for a passive building of about 2500W and in case of a unit heat load – of about 35 $\frac{W}{m^2}$.
Therefore, as far as a passive house is concerned, it would be necessary to properly modify accumulation matrices providing a detailed description of internal loads which give the possibility to make calculations in a continuous utilization profile. Regardless of accumulation matrices, this utilization profile of a given room in a passive building must be precisely defined: its occupation, using of lighting as well as any machines and devices.

5.2.2 Building loads - advanced method

Another possibility of comprehensive load setting is the simulation method suggested in [33] and its modification included in [18].
This calculation method is based on the combination of so called recursive filters. The definition of the combination defines the values of the output function (in a given point) as an integral of the weighted function multiplied by the input function calculated one stage earlier:

$$y(t) = \int_0^{\infty} g(\tau) \cdot u(t-\tau) d\tau \qquad (7)$$

where:
- u – input function,
- y – output function,
- g – weighted function,
- t – calculation time (in a given moment).

Transforming the equation (7) to the discrete sum we will receive:

$$y_k = \sum_{i=0}^{\infty} g_i \cdot u_{k-1} \qquad (8)$$

where:
- g – weighted coefficient,
- y – calculation time (in a given moment).

By using recursive filters the number of necessary weighted coefficients is significantly reduced. If dynamic conversion is approximated by the second-grade model, then a discrete filter equation is obtained (9).

$$y_k = \sum_{m=0}^{3} a_m \cdot u_{k-m} + \sum_{n=1}^{2} b_n \cdot y_{k-n} \tag{9}$$

where:

a – load coefficient for u,

b – load coefficient for y.

The results of the simulations based on the method described above have been depicted in Figure 7 as daily diagrams of internal and external loads as well as total and unit loads for the whole year (all months separately) for the whole building in question. The share of convection and radiation has been adopted according to [35].

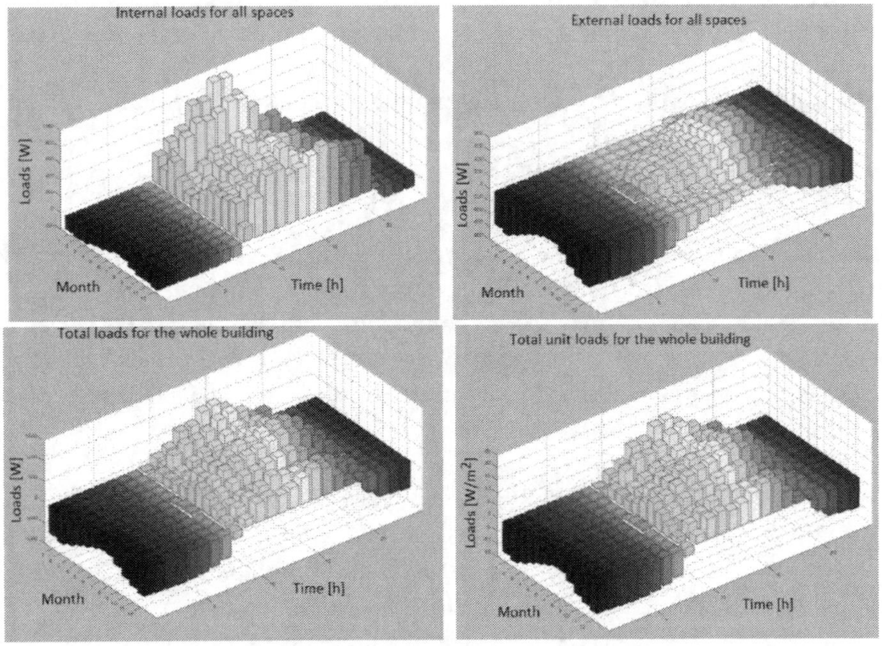

Figure 7. Daily loads of a passive building in question set for the whole year: a-internal, b-external, c-total, d-unit.

5.3 Setting an air stream that provides the climate comfort

Having all loads set for each and every rooms in a given passive building using a load criterion and when it is wished to control a given value, it is possible to set an optimal air stream variable in terms of demand:

- Based on sensible heat balance of temperature control,
- Based on latent heat balance of relative humidity control,
- Based on contamination balance (CO_2 emission) for control of carbon dioxide (CO_2) content in the air.

For example, in control for which carbon dioxide concentration is the main parameter, first and foremost one should take into consideration CO_2 concentration emitted by a human being and contamination of the air in the inlet duct of a ventilation system.

Taking these factors into account, the process of setting a ventilating air stream will adopt the following form:

$$\dot{V} = \frac{C}{k_w - k_n} \left[\frac{m^3}{h}\right] \tag{10}$$

where:

$$C = n \cdot s_{co} \cdot S_i \tag{11}$$

k_n – CO_2 concentration in the supply air,

k_w – CO_2 concentration in the exhaust air (in a controlled room),

C – amount of CO_2 concentration emitted by a human being,

S_i – human co-presence coefficient,

n – number of people in a room [person],

s_{co} – carbon dioxide concentration emitted by a human being $[\frac{dm^3}{h\ os.}]$.

It has been assumed in calculations that a carbon dioxide concentration emitted by a human being is proportional to the emitted heat stream [14] and equals $4 \cdot 10^{-5}\ \frac{dm^3}{sW}$ [9,13] 0. The calculation is made based on the assumption of a unit CO_2 concentration coming from persons at the level of 18 $\frac{dm^3}{h\ os.}$. In the simulation examination the inlet air contamination has been assumed at the level of 350ppm and as an expected value of CO_2 concentration inside a room the value of 750ppm has been adopted.

Carbon dioxide concentration in a room after τ time can be set from the relation (12) [15]

$$k_w(\tau) = \left(\frac{C}{\dot{V}} + k_n\right) \cdot \left(1 - e^{-n\tau}\right) + k_n \cdot e^{-n\tau} \tag{12}$$

where n – is the multiplication of air exchange after τ time.

Figure 8 shows a model for CO_2 concentration control with classical PID control and with PI fuzzy control. Figure 9a shows a defined optimal air stream and subsequently in Figure 9b – the result of stream control simulation with classical PID control and in Figure 9c – the result of stream simulation with a FLC fuzzy control (Mamdani type).

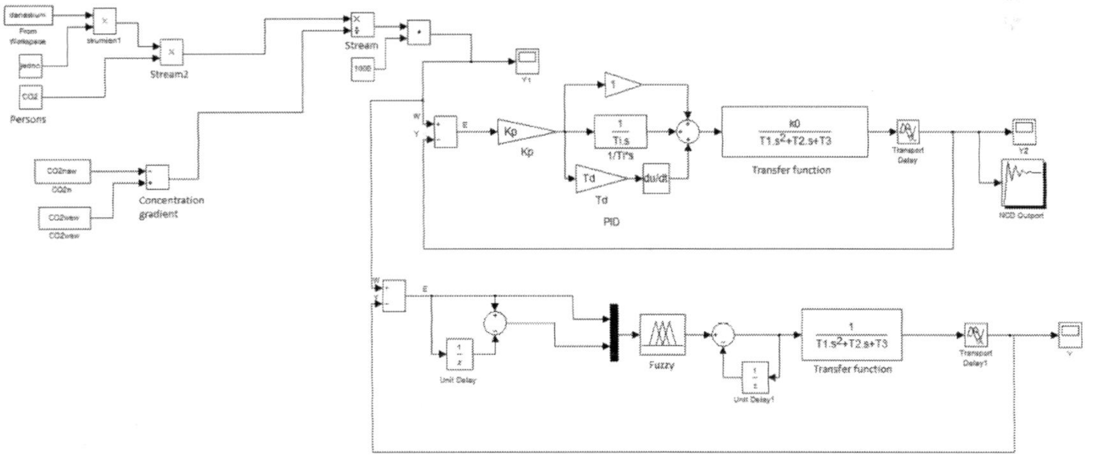

Figure 8. CO_2 concentration control model with classical PID control and PI fuzzy control.

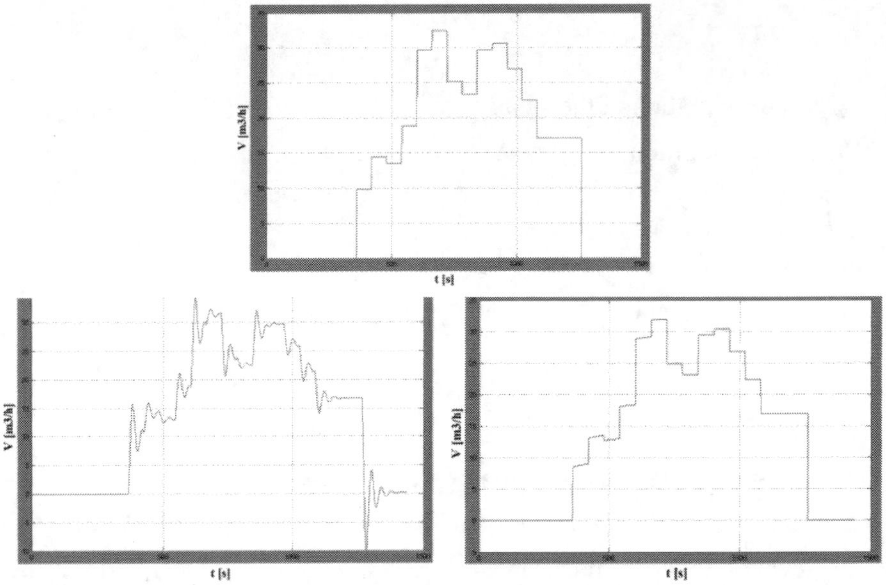

Figure 9. Setting the optimal air stream based on the CO_2 concentration balance (a) and the result of the simulation of the requested air stream control with a PID controller (b) and FLC Mamdami fuzzy controller (c).

6 EXPERIMENTAL TEST

As part of the experimental test a number of measurements have been carried out on a real controlled system – in a passive building located at the premises of the Technical University in Poznan. The research has been divided into two parts: in the previously existing centralized structure and in a decentralized structure specially created for the purpose of this test. When configuring the centralized system the attention has been drawn to the measurements only, but in case of decentralized system, apart from continuous daily measurements (within one year), the control system has been tested in different configurations.

6.1 Description of the test site

The real passive building is located at the campus of the Technical University in Poznan. It is a one-storey detached building. It has a basement where all devices have been installed for the preparation of the appropriate air to maintain the climate comfort in the rooms of the ground floor which function as offices for the employees of the Environmental Engineering Institute of the Technical University in Poznan.

The building has a wooden construction. DZ-3 technology has been used for the ceiling over the cellar. The roof has a wooden construction with a metal roofing thatch. The foundation grillage made of wooden beams with the size of 100 x 60 mm forms the structural elements of the walls. The spaces between the beams have been filled with mineral wool. The external walls have been constructed using ecological and recyclable products [5].

The ground floor level after the modernization of the building in question has been constructed in the passive building technology. It has 5 office spaces (rooms), a bathroom, a toilet and a hall.

Each of the office spaces has become a separate VAV control sphere and the rooms 4 and 5 create so called "an open space" with an "artificial" separation of the rooms from one another and from the hall.

The projection at the level of the ground floor is depicted in Figure 10a. The photograph of the building itself in presented in Figure 10b.

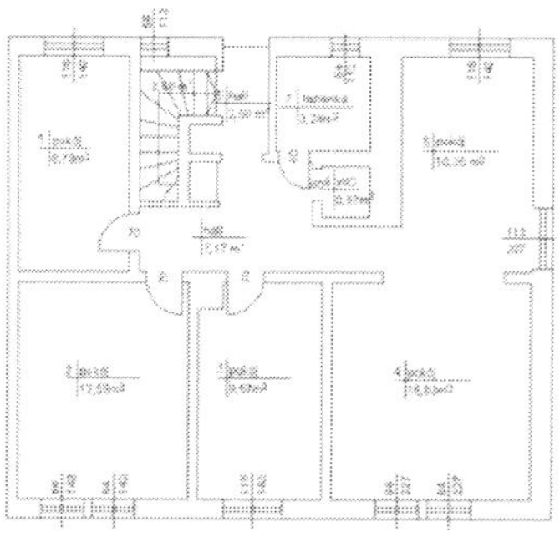 

Figure 10. Projection of the ground floor (a) a photograph (b) of a real passive building.

In the passive building in question at the level of the ground floor there are five rooms which function as office spaces. Each office has been treated as a separate control area.

6.2 Real measurements

Research has been carried out based on two measuring and ventilation systems:

- Viessmann's Vitotres 343 heat and ventilation unit as an example of a centralized system,
- Swegon GOLD air handling unit with a control, measuring and operating system by Siemens and Trox – as an example of a decentralized system.

6.2.1 Centralized system

The first variant – so called a measuring and control variant – referred to the analysis of the centralized control based on a representative measurement in the selected representative sphere of the whole passive building. The heat and ventilation central unit is located in the cellar of the building in question. It provides power supply to the air heating system and ventilation unit.
The control algorithm in the first variant consisted in measurement of the representative temperature, theoretically corresponding with the one present in the whole building. Then, based on this temperature, the central unit adjusted the temperature of the air supply in the central ventilation duct according to internal control algorithms with a continuous air stream of 150m3/h for the whole passive building.
An example of the changes of measured CO_2 concentration for each of the sphere has been presented in Figure 11.

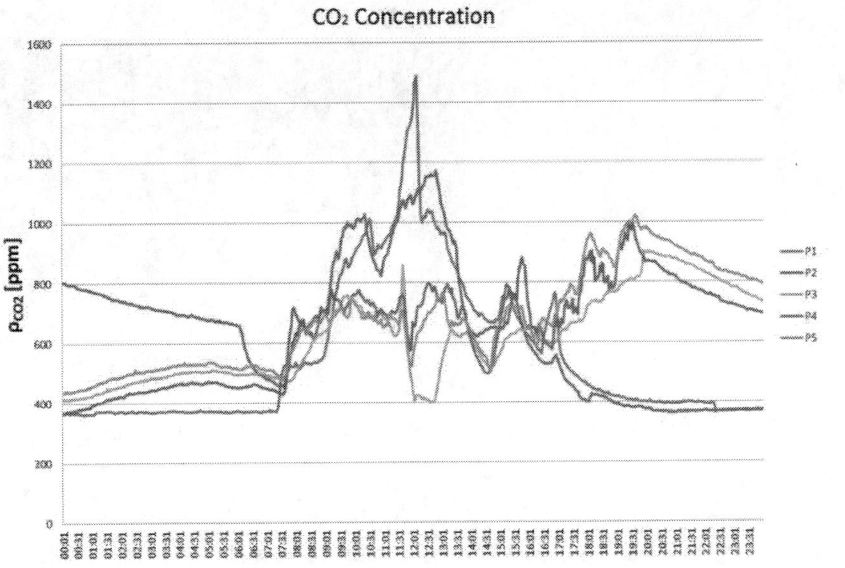

Figure 11. Example of daily changes of CO_2 concentration for each sphere (room)

6.2.2 Decentralized system

In the second measurement and control variant the analysis has been focused on the decentralized system. Each sphere (office room) has been treated as a separate controlled system. Therefore the whole system has been significantly extended. The comprehensively modernized system has been symbolically depicted in Figure 12.

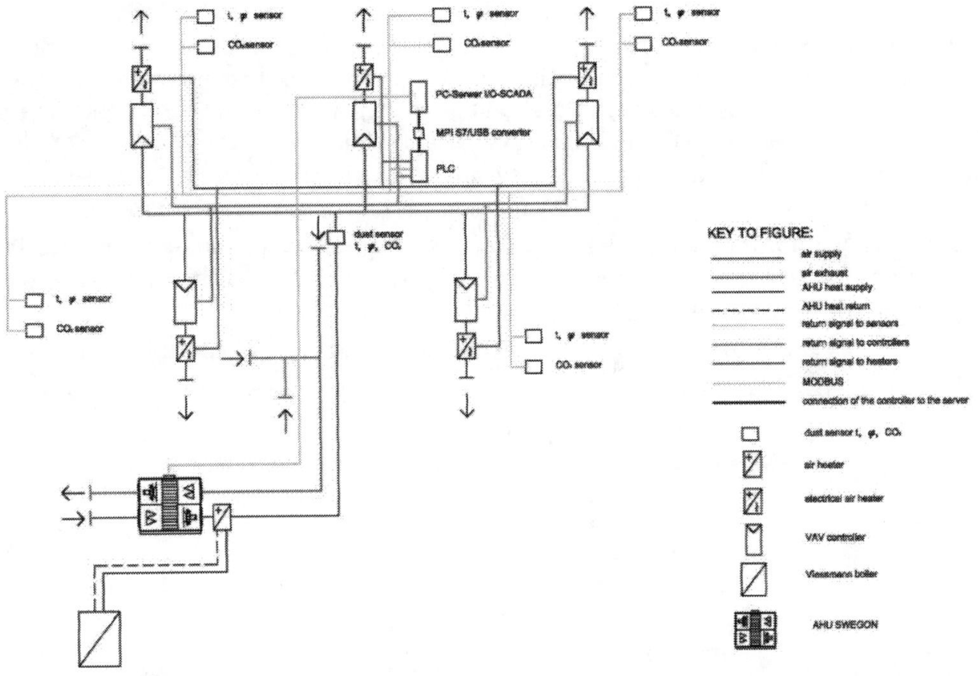

Figure 12. Symbolic diagram of a modernized system with all measurement devices, actuators and control

For control with CO_2 concentration priority as a controlled value, the algorithm checks the occupation of the building in question. If there are no people inside, then VAV controllers are closed to leave minimum air flows, so called representative air flows, that only keep the airing of given spheres. In the opposite situation, what is compared is the current CO_2 concentration with the set one and then VAV controllers are used to ensure the air flow control. An example of the CO_2 concentration regulation curve for the examined spheres has been depicted in Figure 13a (April) and 13b (May).

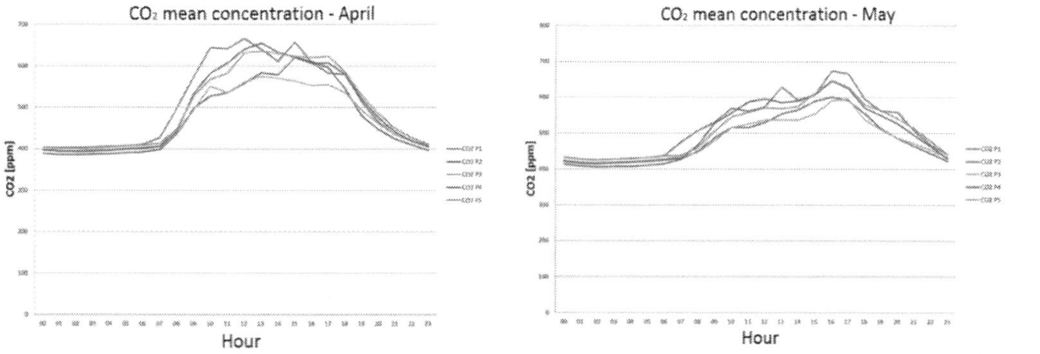

Figure 13. Daily changes of CO_2 concentration in each of the controlled spheres of a passive building: April (a) and may (b).

So far conducted measurements and research unequivocally show that the decentralized system used gives the possibility of more stable performance, fixed set point control of the parameters, thus it allows for optimal formation of the climate comfort in a given space. In case of the decentralized system with so-called measurement in an equivalent sphere, it often happened in the heating period that in case of temporary load caused by a larger group of people, continuous air stream delivered to a given room was insufficient. Therefore, the CO_2 concentration value of 1000ppm was very often exceeded and sometimes even reached the value of 1500ppm or more.

7 SUMMARY AND CONCLUSIONS

The article has presented the analysis of the process of variable air volume control which was to ensure the required freshness of air as one of the main elements of climate comfort in low-energy and passive buildings. Concurrently it must be stated that heat and ventilation systems can shape the microclimate in rooms of passive buildings because of very low heat demand.
The starting point to optimization of the control process of the main components of the climate comfort in the presented research was modelling of heat load of the passive building in question. The standard data adopted from different calculation methods have shown that internal loads are dominant as a result of big insulation and tightness of a passive building.
Generating internal heat gains have such a huge influence that standard calculation methods proves insufficient. Therefore, what should be very precisely defined is the profile for using a particular sphere in a passive building with a concurrent indication of internal gains emission.
Simulation research that has been carried out show that for a proper and optimal control of the parameters that define the climate comfort, continuous PID control must be used together with control parameters precisely set or control with the use of e.g. fuzzy logic controllers (FLC). Application of fuzzy logic controllers with the appropriate fuzziness of signals used in the control process gives much more stable results (without overshoot), however, sometimes with longer control time.

Undoubtedly, more stable and more precise reflection of signals increases the efficiency of control itself and at the same time results in bigger savings in energy consumption.
So far conducted measurements and research unequivocally show that the decentralized system used offers the possibility of more stable performance, fixed set point control of the parameters, thus it allows for optimal formation of the climate comfort in a given space.
The application of the decentralized system facilitated precise control separately for each of the spheres. In every controlled sphere the user was able to define different requirements. The possibility of dynamically variable air volume coming to the sphere on one hand ensures given air freshness kept at a high level (below the assumed level of 750ppm) and on the other hand reduces energy consumption of useful energy of a given system and in case of lack of load the flow decreased to the set level, so called equivalent level, fulfilling only the function of airing of a given sphere.

8 REFERENCES:

[1] Alexei A., Geoegescu S.: *Komfort cieplny w budynkach*, Arkady Warszawa 1971,

[2] ASHRAE: *Thermal environmental Conditions for Human Occupancy*, Atlanta GA (1992),

[3] ASHRAE: *Fundamentals Handbook, Thermal comfort*, Atlanta 2005,

[4] ASHRAE: *HVAC Fundamentals Handbook*, 2001,

[5] Basińska M., Górka A., Górzeński R., Kaczorek D., Szymański M.: *Materiały instalacyjno remontowe doświadczalnego budynku pasywnego DoPas Politechniki Poznańskiej*; wewnętrzne materiały niepublikowane,

[6] Basińska M., Krzyżaniak G.: *Zintegrowany układ grzewczo-chłodzący z wymiennikiem gruntowym. Koncepcja stanowiska doświadczalnego*, Chłodnictwo 11/2008,

[7] Charles K., Reardon T., Magee R.J.: *Indoor Air Quality and Thermal Comfort in Open-Plan Offices*; NRC-CNRC, Construction Technology Up. No. 64, Ottawa,

[8] Chen K., Jiao Y., Lee S.: *Fuzzy adaptive networks in thermal comfort*, Applied Mathematics Letters 19/2006,

[9] Clark J. et al: *ESP-r User Guide*, Glasgow, 2002,

[10] Dounis A., Santamouris M., Lefas C., Argiriou A.: *Design of a fuzzy set environment comfort system*, Energy and Buildings 22 1995,

[11] Fanger P.O.: *Komfort cieplny*, 1974,

[12] Gładyszewska-Fiedoruk K.: *Badania stężenia dwutlenku węgla w sali dydaktycznej*, Ciepłownictwo, Ogrzewnictwo, Wentylacja nr 5/2009,

[13] Górzeński R.: *Sterowanie pracą układu wentylacji mechanicznej w budynku mieszkalnym w funkcji zapotrzebowania*, PP Poznań 2011,

[14] Jokl .: *Indoor Air Quality Assessment Based on Human Physiology*, Acta Polytechnica 43/2003,

[15] Jones W.: *Klimatyzacja* Arkady Warszawa 2001,

[16] Kabza Z. Kostryko K. i zespół: *Regulacja mikroklimatu pomieszczenia*, Agenda Wydawnicza PAK, Warszawa 2005,

[17] Melikov A.K.: Indoor Environmental Requirements and Assessment; Proceedings of Workshop „*Measurement and control Techniques for HVAC Systems and Indoor Climate*" – April 2005,

[18] Nadler N.: *Korrekturvorschläge zum EDV-Verfahren der VDI2078*, Oranienburg 2004

[19] Niedzielko J.: *Wybrane problemy związane z mikroklimatem w budynkach edukacyjnych*; Problemy jakości powietrza wewnętrznego w Polsce, 2001,

[20] Pańczak M.; Basińska M.: *Budynek pasywny DoPas – ocena w czasie jego eksploatacji*;

Politechnika Poznańska 2010,

[21] PN-EN 12831:2004 *Instalacje ogrzewcze w budynkach. Metoda obliczania projektowego obciążenia cieplnego,*

[22] Ruszel F. *Obciążenie cieplne pomieszczeń wg norm PN-EN 12831 i PN-B-03406*, Ciepłownictwo, Ogrzewnictwo, Wentylacja nr 3/2008,

[23] Saade J., Ramadan A. *Control of thermal-visual comfort and air quality in indoor environments through a fuzzy inference-based approach*, International Journal of Mathematical Models and Method in Applied Sciences vol.2 2008,

[24] Strzeszewski M. *Obliczanie projektowej straty ciepła przez przenikanie*, Ciepłownictwo, Ogrzewnictwo, Wentylacja nr 1/2007,

[25] Strzeszewski M. *Obliczanie projektowej wentylacyjnej straty ciepła w przypadku instalacji wentylacyjnej wg PN-EN 12831,*

[26] Strzeszewski M., Wereszczyński P.: *Norma PN-EN 12831 Nowa metoda obliczania projektowego obciążenia cieplnego*. Poradnik. Purmo Warszawa 2007,

[27] Sudoł-Szopińska I., Chojnacka A.: *Komfort termiczny w pomieszczeniach biurowych w aspekcie norm*, Bezpieczeństwo pracy 6/2007,

[28] Sudoł-Szopińska I., Chojnacka A.: *Określenie warunków komfortu termicznego w pomieszczeniach za pomocą wskaźników PMV i PPD*, Bezpieczeństwo pracy 5/2007,

[29] Sudoł-Szopińska I., Chojnacka A.: *Wybrane aspekty dotyczące komfortu termicznego w pomieszczeniach*, Ciepłownictwo. Ogrzewnictwo, Wentylacja 4/2007,

[30] Szkarłat K., Mróz T.: *Strategia optymalnego sterowania układami utrzymania komfortu klimatycznego w budynku pasywnym*, INSTAL 6/2009, Warszawa 2009,

[31] Szymański T. Wasiluk W.: *Ilość i jakość powietrza wewnątrz pomieszczeń w świetle obowiązujących norm*, Ciepłownictwo, Ogrzewnictwo, Wentylacja nr 12/2008,

[32] Śliwowski L.: *Mikroklimat wnętrz i komfort cieplny ludzi w pomieszczeniach*, Wrocław 1999,

[33] VDI2078 Berechnung der Kühllast klimatisierter Räume, Juli 1996,

[34] Würstlin D.: *Regulacja urządzeń ogrzewczych, wentylacyjnych i klimatyzacyjnych*, Arkady 1978,

[35] Xu Y., Hosni M., Jones B., Sipes J.: *Total heat gain and the split between radiant and convective heat gain from office and laboratory equipment in buildings*, ASHRAE Transactions: Research SF-98-29-3,

[36] Zawada B.: *Układy sterowania w systemach wentylacji i klimatyzacji*, Oficyna Wydawnicza Politechniki Warszawskiej, Warszawa 2006,

[37] Zawada B.: *Układy w systemach wentylacji i klimatyzacji. Od klimatyzatorów indywidualnych po systemy VAV* (cz.I), Chłodnictwo & Klimatyzacja nr 4/2004,

[38] Zawada B.: *Układy w systemach wentylacji i klimatyzacji. Od klimatyzatorów indywidualnych po systemy VAV* (cz.II), Chłodnictwo & Klimatyzacja nr 5/2004,

[39] ZCharles K., Reardon T., Magee R.J.: *Indoor Air Quality and Thermal Comfort in Open-Plan Offices*; NRC-CNRC, Construction Technology Up. No. 64, Ottawa

MULTI-ZONE DEMAND-CONTROLLED VENTILATION IN RESIDENTIAL BUILDINGS: AN EXPERIMENTAL CASE STUDY

Arefeh Hesaraki[1*], Jonn Are Myhren[2] and Sture Holmberg[1]

1 Division of Fluid and Climate Technology, School of Architecture and the Built Environment, KTH Royal Institute of Technology, Stockholm, Sweden
**Corresponding author: arefeh@kth.se*

2 Construction Technology, School of Technology and Business Studies, Högskolan Dalarna, Falun, Sweden

ABSTRACT

Numerous studies have investigated the application of multi-zone demand-controlled ventilation for office buildings. However, although Swedish regulations allow ventilation rates in residential buildings to be decreased by 70 % during non-occupancy, this system is not very common in the sector. The main focus of the present study was to experimentally investigate the indoor air quality and energy consumption when using multi-zone demand-controlled ventilation in a residential building. The building studied was located in Borlänge, Sweden. This building was recently renovated with better windows with low U values, together with internally-added insulation materials. The building had natural ventilation, which decreased significantly after retrofitting and resulted in poor indoor air quality. Therefore, a controllable mechanical ventilation system was installed. The ventilation rate was controlled according to the demand in each zone of the building by CO_2 concentration as an indicator of indoor air quality in habitable spaces and relative humidity and VOC level in the toilet and bathroom. The study showed that multi-zone demand-controlled ventilation significantly reduced the CO_2 concentration leading to improvement in indoor air quality. However, building with demand-controlled ventilation consumed more energy than natural ventilation as it increases the ventilation loss by forcing more air into the building. Nevertheless, in the demand-controlled ventilation system, the energy consumption for the ventilation fan and ventilation loss was almost half of the constant high rate ventilation flow.

KEYWORDS

Demand-controlled ventilation, Indoor air quality, Energy consumption

1 INTRODUCTION

Approximately 73 % of the total energy demand in north European residential buildings is used for space and ventilation heating (Anisimova 2011). Poor insulation in old houses means that the share of space heating is higher than it is for ventilation heating. On the other hand, the addition of more insulation layers with low heat transfer coefficient (U value) in new and retrofitted buildings means that space heating demand is lower than it is for ventilation heating. Nevertheless, indoor air quality should not be sacrificed for energy saving through retrofitting; in other words, in buildings with natural ventilation, air circulation should not be lost by renovation. In Sweden 90 % of single family houses built before 1975 have natural ventilation (boverket 2009).

People spend around 90 % of their time indoors (Höppe 2002), and 60 % of it is spent in residential buildings. Previous studies (Joint Research Center, JRC) have shown that indoor contaminants are much more dangerous than outdoor pollutant. Therefore, it is essential to have enough ventilation rates to achieve acceptable indoor air quality in residential buildings. In order to decrease ventilation heat loss, energy-efficient ventilation systems, including heat recovery, or demand-controlled ventilation systems are advised. In heat recovery system, the heat from exhaust air is recycled and transferred to the incoming air. This system is more profitable in cold climates with high number of degree hours. In a demand-controlled ventilation system, the amount of ventilation air is adjusted to the occupants' needs, in addition to the constant rate for removing pollutants from furniture or building materials. This system decreases ventilation heat loss and energy for driving ventilation fans during un-occupancy compared to constant air volume. According to the Swedish building regulations (BBR), ventilation rates can be decreased from 0.35 to 0.10 $l \cdot s^{-1} \cdot m^{-2}$ when no one is home. A constant low rate of 0.1 $l \cdot s^{-1} \cdot m^{-2}$ is used to remove pollutants from furniture and building materials.

Demand-controlled ventilation systems are more common in office buildings due to more unpredictable occupancy levels. However, the need to reduce energy demand has led to this system also becoming interesting in residential buildings. Usually, if occupants are the main pollutants in the room, the CO_2 level is used as an indicator of indoor air quality. The general threshold for CO_2 is 1000 ppm. If the CO_2 concentration is greater than this threshold, the ventilation rate is increased. In residential buildings, odours from occupants, relative humidity (RH), and emissions from building materials, i.e. volatile organic compounds (VOC), all have a role in polluting the indoor air. Therefore, the efficient controlled set point for ventilation rates can be sensitive to VOC, RH and CO_2 variables. A system in which the ventilation rate is adjusted based on the demand in each zone is called a multi-zone demand-controlled ventilation system.

A previous study (Hesaraki and Holmberg, 2013) has shown that in new buildings with mechanical exhaust ventilation systems, reducing the ventilation rate during un-occupancy for eight hours caused a problem with indoor air quality (IAQ) regarding VOC concentration; specifically it was more than 0.1 ppm when the occupants returned home. This was due to high emission rates from building materials in initial years of construction. Therefore, a suggestion was made to increase the ventilation rate two hours prior to the occupants arriving home. In addition, in investigated time-controlled ventilation system the heating requirements for ventilation air and electricity consumption for the ventilation fan were decreased by 20 % and 30 %, respectively.

In old buildings usually there is no problem with VOC concentration. Therefore, as previous study (Hesaraki and Holmberg, 2013) suggested time-controlled ventilation is not needed in old houses to increase the ventilation rate before occupants arrive home. The aim of this study was to investigate the performance of multi-zone demand-controlled ventilation system in a retrofitted old house with respect to energy consumption and IAQ.

2 DESCRIPTION OF THE STUDIED BUILDING

A 92 m^2 two-story single-family house built in the 1950s, located in Borlänge, Sweden, was chosen for this study (see Fig. 1). To make the building more environmentally-friendly, the heating system of this building was changed from furnace to ground-source heat pump. In addition, to save energy this building was renovated with better windows with low U values, together with internally-added insulation materials. The building had natural ventilation, which decreased significantly after retrofitting. This was due to reducing the pressure difference caused by chimney (wind-effect) and decreasing the temperature difference (stack-effect) between indoors and outdoors by adding more insulation layers in the building envelope. Hence, because of insufficient air circulation in this building, mechanical ventilation was mounted to force the air to provide acceptable indoor air quality. To decrease heat loss and provide the exact amount of ventilation needed in each zone in this building, multi-zone demand-controlled ventilation system was installed. The controlled variables were VOC and relative humidity (RH) in the toilet and bathroom, and CO_2 for habitable spaces (see Fig. 2).

Figure 2 The two-story single family house selected for study in Borlänge, Sweden

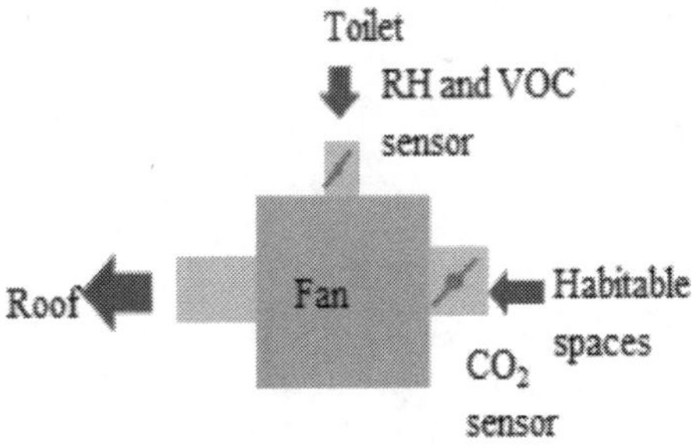

Figure 2 Multi-zone demand-controlled ventilation system for the studied building with RH, VOC and CO_2 sensor

3 METHOD

Experimental investigation was the main method used in this study. For measurements, concentration of CO_2 and RH was monitored as indicators of IAQ before and after installing demand-controlled ventilation. The indoor CO_2, RH and temperature is not uniform in the whole room (Seppanen et al. 1999). Therefore, to detect a mean value of CO_2, RH or temperature, the sensor should be placed in the breathing zone or in the exhuast valve. The latter was chosen for this study. In addition, with regard to energy consumption, ventilation fan energy consumption and ventilation heat loss were measured and calculated. In the ventilation heat loss calculation shown in Eq. (1), the ventilation flow rate was estimated from ventilation fan electrical energy consumption by having flow chart of installed fan in this system (see Fig. 3). The specific power for this fan was 0.14 W / ($m^3 \cdot h^{-1}$) at pressure 100 Pa and airflow 250 $m^3 \cdot h^{-1}$.

$$P_{ventloss} = q \cdot \rho \cdot c_p \cdot (T_{in} - T_{out}) \qquad (1)$$

where q is ventilation rate ($m^3 \cdot s^{-1}$), ρ is air density ($kg \cdot m^{-3}$), c_p is specific heat capacity ($J \cdot kg^{-1} \cdot °C^{-1}$) and T_{in} and T_{out} are indoor and outdoor temperature (°C), respectively.

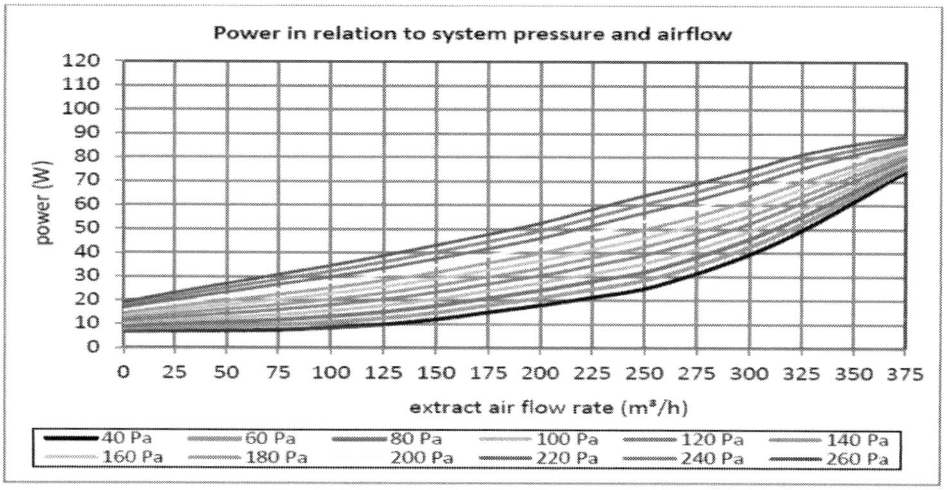

Figure 3 Power consumption of ventilation fan in relation to system pressure and airflow from device manual

4 RESULTS AND DISCUSSION

The results of the experiment are provided in two parts: indoor air quality and energy consumption.

4.1 Indoor air quality

For indoor air quality, two parameters were considered: RH and CO_2. Measurements were performed before and after installing the demand-controlled ventilation system. According to the field tests, the CO_2 level was higher than 1000 ppm most of the time when people were home before improving the ventilation system; that is, natural ventilation. Figs. 4 and 5 show the results during a weekday and a weekend, respectively.

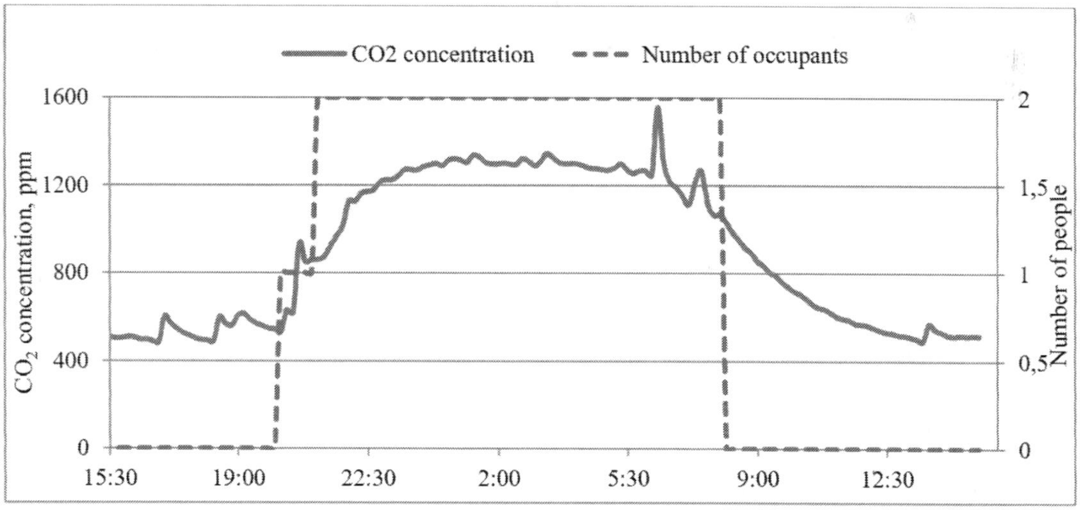

Figure 4 CO_2 level during a working day without demand-controlled ventilation system and with natural ventilation

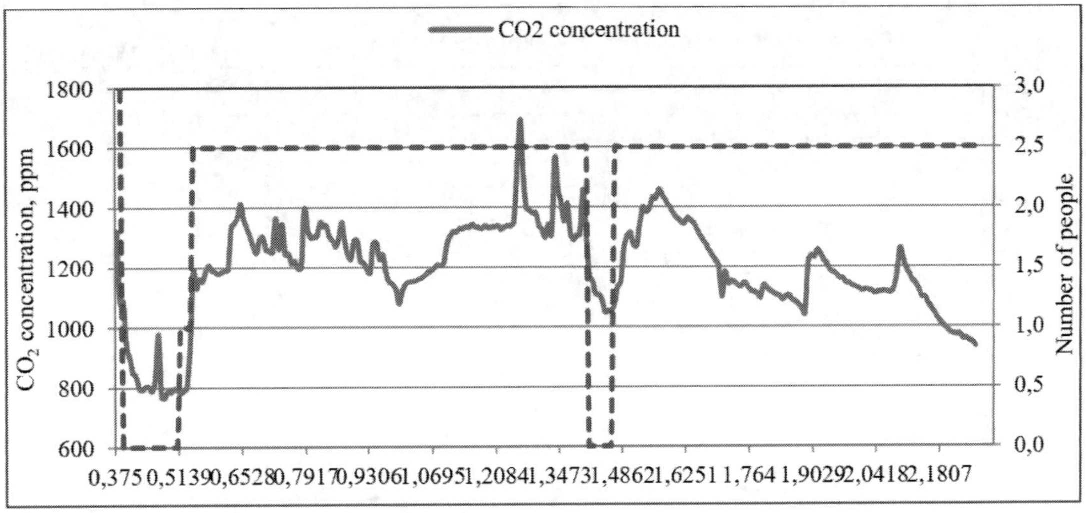

Figure 5 CO_2 concentrations during weekend with natural ventilation and without demand-controlled ventilation system

In the demand-controlled ventilation system, the CO_2 level was the main factor determining the ventilation flow rate. The threshold level was set to 800±50 ppm. As Fig. 6 shows, the CO_2 level never exceeded 800 ppm when using the demand-controlled ventilation system.

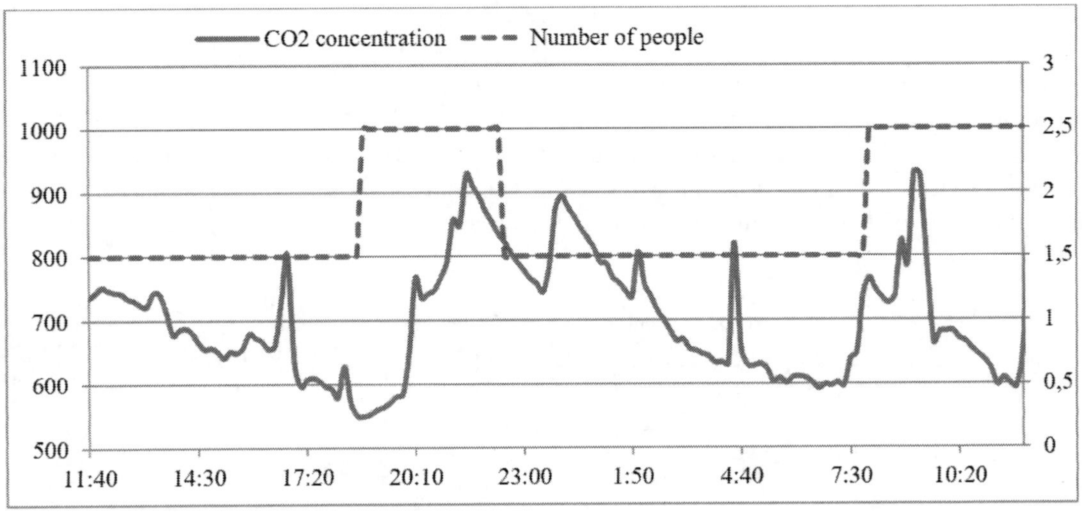

Figure 6 CO_2 concentrations during a working day with demand-controlled ventilation system

Measurements showed that the RH only affected flow rates during showering when using the demand-controlled ventilation system. For the rest of the time, the relative humidity was 40–60 % within healthy range.

4.2 Energy consumption

The degree hours method was used to estimate the ventilation heat loss. This is a simplified method of calculating the building energy demand for active heating. The degree hours depends on the building location, the chosen indoor temperature and the indirect/passive heat supply. The heating contribution from the limit temperature to comfort temperature is given by indirect/passive heating. Degree hours is 115080 °C·h· year^{-1} for an assumed active temperature of 17 °C and a mean outdoor temperature of 4.2 °C in Borlänge (Warfvinge and Dahlblom 2010).

ventilation heat loss and ventilation fan energy consumption for high constant ventilation rates, which is dominant in most multi-family houses in Sweden, was also calculated. Having degree hours, ventilation heat loss for natural, demand-controlled ventilation and high rate constant ventilation was calculated for this building (see Fig. 7). As can be seen, the lowest consumption is for natural ventilation due to the low flow rate and not having any mechanical ventilation. In the demand-controlled ventilation system, the energy consumption for the ventilation fan and ventilation loss was almost half of the constant high rate ventilation flow; in other words, 45 % and 22 % of ventilation heat loss and ventilation fan, respectively, are saved by demand-controlled ventilation. Due to the very low specific fan power of ventilation fan used in this system, there was not a great deal of saving in terms of ventilation fan power by using the demand-controlled ventilation system. However, savings for ventilation heat loss was significant compared to the constant high-rate ventilation.

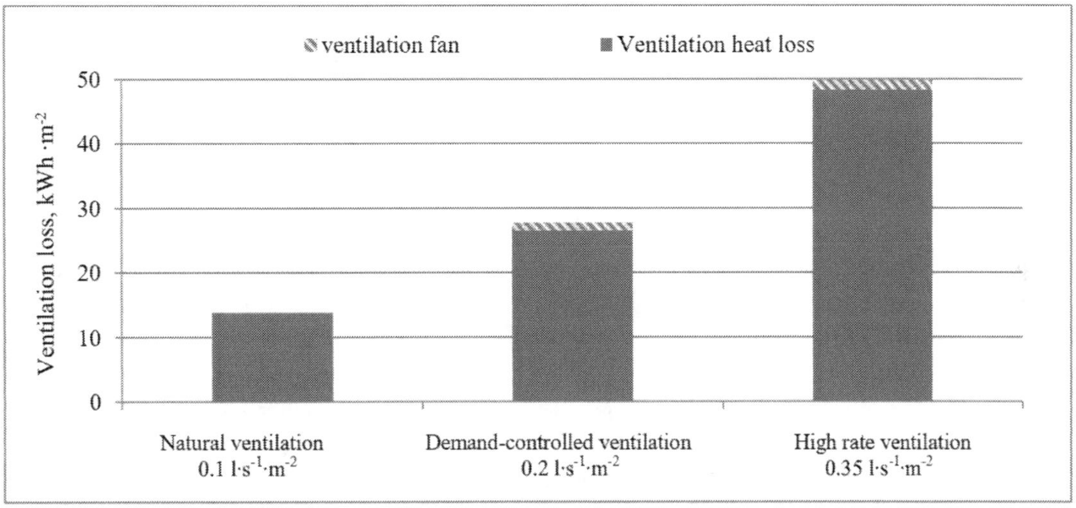

Figure 7 Energy consumption for ventilation heat loss and ventilation fan electricity consumption for natural, high, and multi-zone demand-controlled ventilation system

5 CONCLUSION

This study investigated the performance of multi-zone demand-controlled ventilation in a single-family house in Sweden. The studied building had been renovated with improved windows with lower leakage and insulated internally with more insulation layers. In addition, the heat source was changed from furnace to ground source heat pump. All these changes resulted in lower ventilation flow rate in this building, leading to poor indoor air quality. This was proved by measuring the CO_2 level, which was approximately 1400 ppm when the occupants were at home. This value is much higher than the common threshold level of CO_2 (that is 1000 ppm) as an indicator for acceptable IAQ. This indicated the need to improve the ventilation system in this building. For this purpose, a multi-zone demand-controlled ventilation system was installed in this building. The control set point for ventilation flow rate was based on the CO_2 level in habitable spaces and the relative humidity and VOC for the toilet and bathroom. Measurements showed significant reduction in CO_2 concentration leading to improvement in IAQ by using demand-controlled ventilation system. However, in demand-controlled ventilation energy consumption for ventilation heating was almost twice compared to natural ventilation due to higher ventilation rate. Nevertheless, using demand-controlled ventilation resulted in 44 % saving in ventilation heating and ventilation fan consumption compared to constant ventilation system. More experiments to measure actual heating consumption in the building studied are planned for autumn. The need to decrease energy demand in residential buildings without sacrificing indoor air quality highlights the application of demand-controlled ventilation system.

6 ACKNOWLEDGEMENTS

We are grateful to the Swedish Energy Agency (Energimyndigheten) and SBUF, The Development Fund of the Swedish Construction Industry for financial support.

7 REFERENCES

Anisimova N (2011). The capability to reduce primary energy demand in EU housing. CzechTechnical University in Prague. Energy and Buildings 43. Issue 10. Pages 2747–2751

Boverkets byggregler (BBR 2009), Swedish building regulations

Höppe P (2002). Different aspects of assessing indoor and outdoor thermal comfort, Energy and Buildings (34) 661–665

Joint Research Center (JRC). [Online]. http://ec.europa.eu/dgs/jrc/index.cfm

Hesaraki A, Holmberg S (2013) Demand-controlled ventilation in new residential buildings: consequences on indoor airquality and energy savings," *Indoor and Built Environment,* DOI: 10.1177/1420326X13508565.

Warfvinge C and Dahlblom M 2010, Projektering av VVS-installationer. Lund, Sweden

Seppanen OA, Fisk WJ and Mendell MJ (1999) Association of Ventilation Rates and CO2 Concentrations with Health and Other Responses in Commercial and Institutional Buildings," *Indoor Air*, vol. 9, pp. 226-252, 1999

SEASONAL VARIATION IN AIRTIGHTNESS

Paula Wahlgren

Chalmers University of Technology
Building Technology
SE-41296 Göteborg, Sweden
paula.wahlgren@chalmers.se

ABSTRACT

Airtightness of buildings is necessary to obtain healthy, sustainable and energy efficient buildings. Measuring the airtightness of a building has become more common lately, much due to the higher energy use in leaky buildings. The airtightness of a building can for example be measured in order to attain a certification, or on demand from a developer.

In some studies, there have been large seasonal variations in airtightness. In most cases, the buildings are more leaky in wintertime, but there are also some investigations that show the opposite. In the current project, the aim is to investigate how, and if, the airtightness varies over the year. The airtightness is measured, using a blower door, approximately eight times a year, in three different buildings. The air leakages in the buildings are also detected and the air velocities at a number of leakages are measured. Two of the buildings are one-family, two story, wooden frame houses and one is a multi-family, concrete building (where one apartment is measured). In Swedish wooden houses, the air barrier is often a polyethylene foil, which is also the case in these two buildings.

The measurements have been analyzed with respect to indoor/outdoor temperature and indoor/outdoor relative humidity. The trend in the measurements is that the airtightness is lower (more leaky envelope) when the indoor air is drier (low relative humidity). Consequently, the air leakage is largest during the winter measurements. The decrease in airtightness from summer to winter is in the order of 8-10%.

KEYWORDS

Airtightness, seasonal variation, air leakage, fan pressurization method

1 INTRODUCTION

The airtightness of a building has an impact on the energy use and on the moisture safety of a building. It also affects the thermal comfort, the air quality in a building, sound insulation and the spread of fire gases (Sandberg et al. 2007). Measuring the airtightness of a building has become more common lately, much due to the increased energy use in leakier buildings. The airtightness of a building can be measured in order to attain a certification or on demand from a developer. The consequences if failing the target can sometimes be severe. Therefore, it is of great importance to obtain a correct and representative measure of the airtightness.

The airtightness in a building is created by a continuous and airtight thermal envelope. The airtight layer in a thermal envelope can be either a thin layer, such as a polyethylene foil, a board, such as plywood, a homogeneous construction (e.g. a concrete component) or an outer coating, such as rendering. In all examples it is of great importance that the joints are properly sealed.

Airtightness measurements are usually performed in accordance with EN 13829:2000 (Fan pressurization method). In this standard there are limitations with respect to the climatic conditions during measurements. There is for example a limit on the maximum allowed wind speed and the maximum allowed temperature difference over the thermal envelope. The purpose of the limitations is to assure a correct measured airtightness. Nevertheless, measurements have shown that there is a variation in the measured airtightness with respect to the time of year for the measurement. Yoshino (2012) described variations of ±20% over the year. Boorsboom et al. (2012) analyzed airtightness measurements, from the 80ies, made on 21 window frames mounted in masonry or concrete walls. The average difference in air tightness between summer and winter was about 30% (higher leakage during winter) and the maximum seasonal difference was 120%. To be noted, some window frames showed a lower leakage rate during winter. Boorsboom et al. suggest measurements during three subsequent seasons in order to obtain correct values. Also Kim and Shaw (1986) showed increased leakages during winter time. The highest leakages occurred in winter and early spring, and the lowest in late summer and fall. Two wood frame constructions were studied and the effect was more pronounced in the leakier building. The measurements indicate that there is a correlation between indoor humidity and envelope leakage.

There are also measurements showing a higher leakage during the summer. An example is Dickinson et al. (1986) who, in one out of three residential, wooden frame houses, measured a lower air leakage in winter time. They speculate in the influence of snow and ice on the airtightness. Alev et al. (2014) also show a lower leakage in wintertime. The investigated houses are log houses in Estonia. They give as possible explanation that the weight of the snow in wintertime tightens the log house.

Bracke et al. (2013) measured the airtightness in two new buildings during almost four months (December to April). In the masonry building, there was an increase in air leakage over time. A possible explanation is the different thermal expansion of the masonry/concrete structure and the plaster that assures most of the airtightness in these buildings. This difference in thermal expansion could create cracks in the plaster. Another possible explanation is a gradual deterioration of the ventilation ductwork due to repeated dismantling for the preparation of the pressurization tests.

In this project, full scale measurements and numerical simulations have been performed in order to investigate the possible variations in airtightness at different seasons, and the relation to climate.

2 METHOD

The air tightness' variation over season and climate is studied by measuring the air tightness of three buildings, during one year, and by performing numerical simulations on the climate and the effect on airtightness. The first measurements, on two one-family wooden buildings, started in June 2013. Measurements on a multi-family concrete building stared in March 2014. The measurements have been performed by SP, Technical Research Institute of Sweden. Initial numerical calculations have been made on the influence of wind, and on air properties.

The first airtightness measurements were performed on two residential one family houses, both located in the south west part of Sweden, one house in Landvetter and one in Sevred, located outside Borås. The houses are light weight wooden houses in two floors (plain wood beams/joists in Landvetter and light weight wooden beams/joists in Sevred) and they both have slab on ground and cold, ventilated attics. Both buildings have mineral wool insulation and polyethylene foil on the inside (between insulation and interior board) as air barrier and moisture barrier. The house in Landvetter is built in 2004 and the house in Sevred is built in 1993. Both houses have mechanical exhaust ventilation systems. The multi-family concrete building was finished early 2014 and one apartment has been measured. The apartment has 3.5 external walls.

Figure 1. The tested wooden buildings, Landvetter (2004) to the left and Sevred (1993) to the right.

The airtightness quantity used is air permeability, q50 (l/sm²). It is the amount of air that passes through the thermal envelope at a pressure difference of 50 Pa, per area of thermal envelope. The air flow is measured both when the building is pressurized and depressurized and the mean value is used. The airtightness measurements are made according to EN 13829:2000, using a Minneapolis BlowerDoor. Temperature and relative humidity, indoor and outdoor, is measured at each airtightness measurement occasion, as well as the outdoor wind speed. In addition, the temperatures and relative humidities are continuously logged, and these measurements will be analyzed at the end of the study. The airtightness measurements are performed 6-8 times a year to study the different seasons and climate conditions.

3 NUMERICAL SIMULATIONS

Simulations have been performed in order to study how the measured air flow is affected by different densities of the air. The density of the air is different due to variations in temperature and relative humidity over the year. Both high temperatures and high relative humidities result in low air densities. Numerical investigations have also been performed on the effect of wind.

During a pressurization test, the air is drawn into the building through leakages in the air barrier of the thermal envelope. These leakages can have different geometry and surface roughness and the air flow can also pass through an air permeable material. The magnitude of the air flow depends on the pressure difference over the leakage path, of the characteristics of the leakage path, but also on the characteristics of the air. The density of the air changes due to temperature and relative humidity, and is thus not the same for all measurement conditions. When measuring according to standard EN 13829:2000, the density change that affects the measurement equipment is corrected for. However, there is also the air flow through the leakages, which is slightly different at different air densities. This is investigated with numerical simulations.

The geometry of the building in the calculations is a box with an, initially, equal amount of leakages on all sides (including the roof), see Figure 2. The leakages are assumed to be long gaps, having a width of 2 mm, in walls with a thickness of 120 mm. The air leakages are calculated according to Hagentoft (2001). The effect of air temperature and relative humidity (RH) is determined by investigating two temperature conditions, -20°C and 30°C (RH=40%), and two relative humidity conditions, 0 and 100% (T_{mean}=10°C). A pressure of 50 Pa is used for the calculations and the resulting airtightness of the building is approximately 0.8 l/sm² (varying slightly in the different cases). By investigating extreme temperatures and relative humidities, the factors that possibly influence the airtightness measurements are determined.

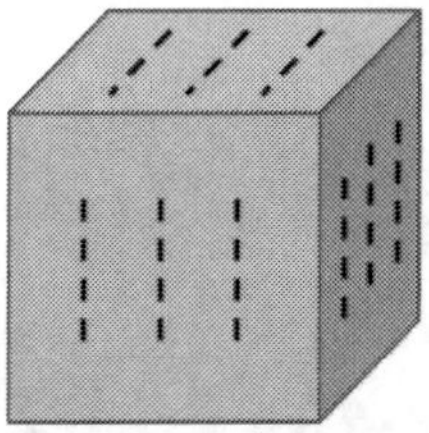

Figure 2: Simulated box with evenly distributed gaps.

The investigation shows that the temperature of the air can affect the measurements by affecting the air flow through the air gaps, see Table 1. Using -20°C, the measured airtightness is 0.76 l/sm² and at a temperature of 30°C the measured airtightness is 0.81 l/sm². Consequently, the difference between the two extreme measurement situations is 5.7%.

Different relative humidities, however, have minor importance. At 0% relative humidity the airtightness is 0.789 l/sm² and at 100% it is 0.791 l/sm². The difference is 0.14%, thus negligible. The effect of the air density on the measurement equipment (not leakages as above) is compensated for when measuring according to EN 13823:2000.

TABLE 1. Airtightness at different temperatures and relative humidities.

Air tightness at minimum temperature (l/sm²)	Air tightness at maximum temperature (l/sm²)	Air tightness at minimum relative humidity (l/sm²)	Air tightness at maximum relative humidity (l/sm²)
0.763	0.807	0.789	0.791
5.7%		0.14%	

The simulations on wind are made to investigate if different wind speeds give different airtightness results. Since the average wind speed can be different during different seasons of the year, this could be a part in explaining why different seasons have different airtightness.

In the standard, it is noted that if the meteorological wind speed exceeds 6 m/s or reaches 3 on the Beaufort scale, it is unlikely that a satisfactory zero flow pressure difference will be obtained. Three kinds of zero flow pressure is measured before and after the pressurization test and if either of these zero flow pressures is over 5 Pa, the test does not meet test conditions according to EN 13823:2000.

The shape of the simulated building is quadratic, with a flat roof. The shape factor C_p (-), that determines the pressure difference over a wall at a certain wind influence, is for the windward wall, 0.4, of the leeward wall, -0.2, and for the other walls, -0.3. The roof is -0.6 and the shape factor of the inside of the building, C_{pi}, is determined by a mass air flow balance. The pressure difference, ΔP (Pa), over a wall subjected to wind, v (m/s), is

$$\Delta P = (C_p - C_{pi}) \cdot \frac{v^2}{2} \tag{1}$$

The building is first simulated with equally distributed leakages, and then with a windward side that is twice as leaky as the other sides.

The results from the simulations show that there in many cases is a small difference in the measured airtightness values for pressurization and depressurization when wind is present. However, the average value is not affected until the wind speed increases. For example, at a wind speed of 9 m/s, the building is estimated 2% more airtight with wind than without wind. At a wind speed of 9 m/s, the zero flow pressure difference is most likely exceeding the value accepted in EN13823:2000. In the simulations, a higher wind speed resulted in lower calculated air permeability (more airtight building).

For the case of a non-uniform air leakage distribution, simulations were made for a wind speeds up to 9 m/s. The results are similar to those of the equal leakage distribution, i.e. unless the wind speeds are high there is little error due to wind.

4 AIRTIGHTNESS MEASUREMENT RESULTS

Full scale airtightness measurements have been performed in three buildings, two light-weight wooden buildings and one apartment in a residential concrete building. Only three measurements have, so far, been made on the concrete buildings, March, May and July 2014. Even though this is a newly erected building, there is no change in the air permeability. All measurement occasions showed an air permeability of 0.35 l/sm^2. This building will be further studied when more measurements have been made, in particular winter measurements.

The light-weight wooden buildings have, so far, been measured 5 or 6 times each. The lowest measured relative humidity outdoor is 54% (July, Landvetter) and it reaches 100% in November in Landvetter. Indoor, the relative humidity ranges from 23% (January, Landvetter) to 60% (September, Landvetter).

In the house in Sevred, there is a constant decrease in relative humidity indoor from July to November, while the house in Landvetter has the highest indoor relative humidity at the measurement in September. Both buildings have the lowest relative humidity indoor at the winter measurements. The wind speed is low during the measurements.

Both buildings have the lowest airtightness (highest air permeability) at the winter measurement (January-February). This coincides with the lowest indoor air relative humidity. The measured airtightness (expressed as air permeability) as a function of time is shown in Figure 3, and as a function of indoor air relative humidity in Figure 4.

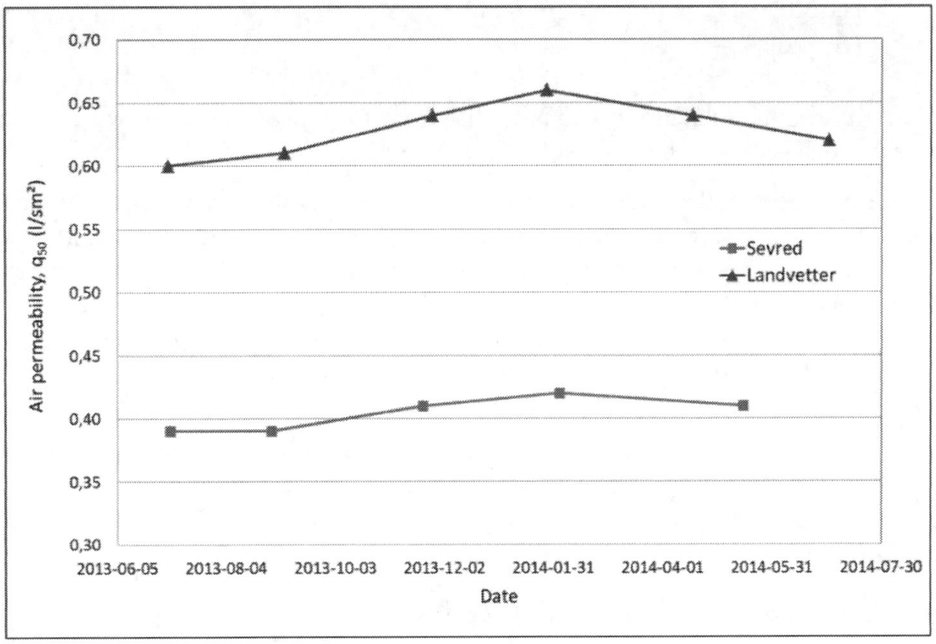

Figure 3: Measured air permeability as a function of time for the two wooden buildings.

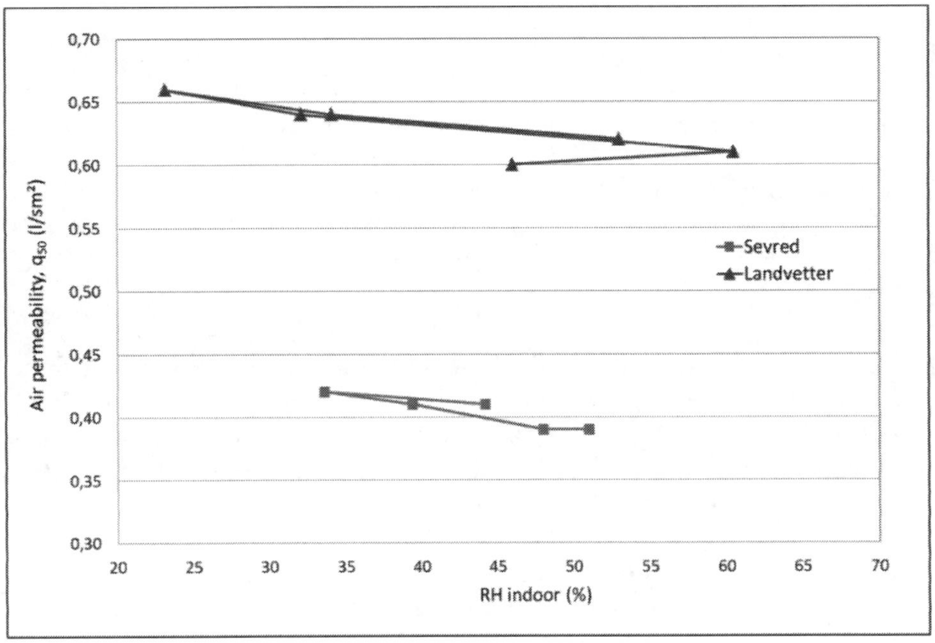

Figure 4: Measured air permeability as a function of relative humidity indoor.

There is an increase in air leakage from summer to winter. The increase from summer to winter is 10% for the building in Sevred and 7.7% for the building in Landvetter. For both buildings, the airtightness is less (higher air permeability) when the indoor relative humidity is the lowest. There is no clear correlation with indoor or outdoor temperature, or outdoor relative humidity. The correlation between air tightness and indoor air relative humidity will be more thoroughly investigated when the loggers that continuously measure indoor air relative humidity are collected. The variations in airtightness from summer to winter can be noticeable for stakeholders aiming for a certification.

In the two measured wooden buildings, the reason for the increase in permeability during winter could be that the wooden construction dries when the relative humidity indoor decreases. This, in combination with a poor connection to the polyethylene foil, causes the leakages to increase when the wood, exposed to indoor air, is dried.

The main leakages in the leakiest building (Landvetter) are found around the attic hatch, at the connection between the upper and lower floor (see Figure 5) and over a window in the kitchen at the bottom floor, Figure 6. The reasons are probably a poor connection between window and polyethylene foil for the windows, both a poor connection between hatch and polyethylene foil plus a leaky hatch (poor seal) for the attic hatch. The leaky connection between the upper and lower floor, that is evident in the stairs, is probably caused by a discontinuous air barrier that does not pass the floor/wall connection.

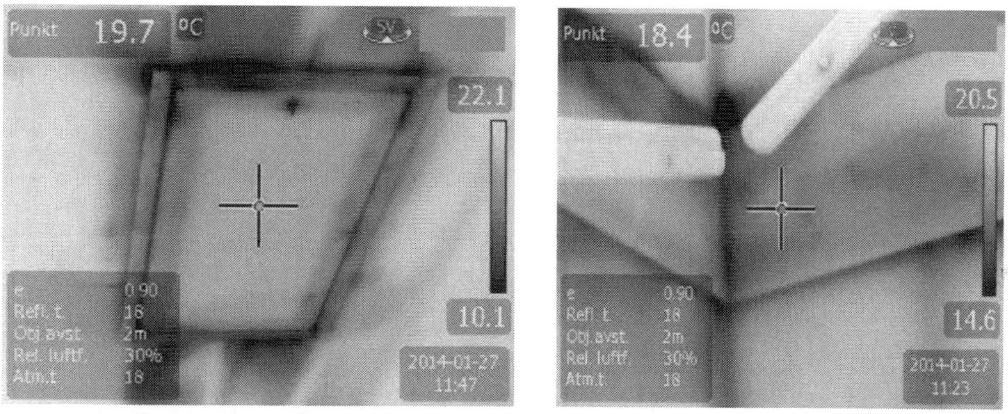

Figure 5: Thermographic image of attic hatch (left) and hand rail (right) in corner of stairs in January 2014. The hand rail is at the position of the connection between the top and bottom floor.

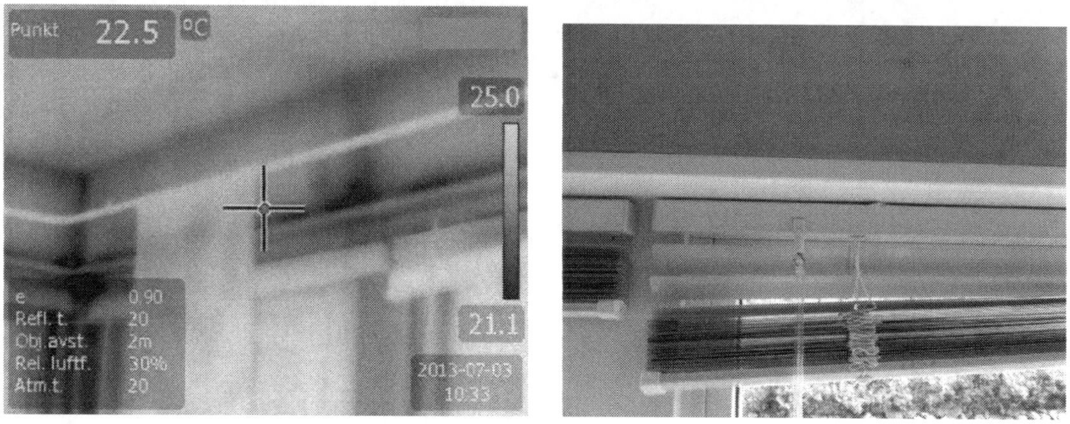

Figure 6: Thermographic image of kitchen window in July, and photo of the same.

5 CONCLUSIONS

Airtightness has been measured for ten months in two residential wooden buildings. The trend in the airtightness measurements is that the airtightness is lower when indoor air is drier. The winter measurements have the lowest airtightness of all measurements. The decrease in airtightness from summer to winter is in the order of 8-10% (from July to February). The measurements will continue so that a whole year will be covered for all buildings.

The numerical simulations show a small change in air flow through the leakages due to high or low air temperatures (affecting the air density of the air that flow in the leakages), but no change in air flow due to different relative humidities. The effect of the air density on the measurement equipment (not leakages) is compensated for when measuring according to EN 13823:2000.

6 ACKNOWLEDGEMENTS

The research and the measurements have been financed by SBUF, the Development Fund of the Swedish Construction Industry, and supported by FoU-väst (regional comittee), which is greatly appreciated.

7 REFERENCES

Alev, U. Uus, A., Teder, M., Miljan, M-J., Kalamees, T. (2014). Air leakage and hygrothermal performance of an internally insulated log house, *Proceedings from the 10th Nordic Symposium on Building Physics*, Lund, Sweden, June 2014.

Borsboom, W., de Gids W. (2012). Seasonal variation of facade airtightness: field observations and potential impact in NZEB. *Proceedings of the AIVC-TightVent International Workshop*, Brussels, Belgium, March 2012.

Bracke, W., Laverge, J., Van Den Bossche, N., Janssens, A. (2013). Durability and measurement uncertainty of airtightness in extremely airtight dwellings. *Proceedings of the AIVC-TightVent International Conference*, Athens, Greece, September 2013.

Dickinson, J.B., Feustel, H.E. (1986). Seasonal variations in effective leakage area, *Thermal performance of the exterior envelops of buildings III*, Atlanta,,USA, ASHRAE, 144-160.

Hagentoft, C-E., (2001). *Introduction to Building Physics*, Studentlitteratur, Lund, 2001.

Kim, A.K., Shaw, C.Y. (1986). Seasonal Variation in Airtightness of Two Detached Houses, *Measured Air Leakage of Buildings, ASTM STP 904*, 1986.

Sandberg, P-I., Bankvall, C. Sikander, E., Wahlgren, P., Larsson, B. (2007). The effects and cost impact of poor airtightness- Information for developers and clients. *Proceedings of the Thermal Performance of the Exterior Envelopes of Whole Buildings X*, Florida, USA, December 2007.

Yoshino, H. (2012). System for ensuring reliable airtightness level in Japan. *Proceedings of the AIVC-TightVent International Workshop*, Bryssel, Belgium, March 2012.

EN 13829:2000 (Thermal performance of buildings - Determination of air permeability of buildings - Fan pressurization method (ISO 9972:1996, modified))

OPTIMIZATION OF DATA CENTER CHILLED WATER COOLING SYSTEM ACCORDING TO ANNUAL POWER CONSUMPTION CRITERION

Piotr Kowalski[1], Mieczysław Porowski[2]

[1]Poznan University of Technology
Piotrowo 5
60-695 Poznan, Poland
piotr.kowalski@emerson.com

[2]Poznan University of Technology
Piotrowo 5
60-695 Poznan, Poland
mieczyslaw.porowski@put.poznan.pl

ABSTRACT

The paper presents optimization model of the chilled water based data center cooling system. The optimization procedure includes system technological and mathematical model, limiting conditions and optimization criterion, which in this case is annual power consumption minimum. The cooling system model is defined by constant parameters and decision variables and consists of aircooled chiller, independent external freecooling heat exchanger (drycooler), computer room air handling unit (CRAH) and constant flow chilled water system with circulation pump. Influence of the server racks architecture (open or closed rack aisle), server inlet temperature and chilled water regime on the annual power consumption of the cooling system has been shown. Case study calculation based on the described model has been presented, including optimum variant designation and power usage effectiveness ratio of mechanical system calculation.

KEYWORDS

data center, efficiency, optimization, precision cooling, chilled water

NOMENCLATURE

c_p – specific heat, [kJ/kgK]

CRAH – computer room air handler

E_{CR} – CRAH unit fan power consumption, [kWh]

E_{CWP} – chilled water pump power consumption, [kWh]

E_{DX} – compressor power consumption, [kWh]

E_{FC} – freecooling coil fan power consumption, [kWh]

E_{IT} – server power consumption, [kWh]

E_K – condenser fan power consumption, [kWh]

E_{TOT} – cooling system power consumption, [kWh]

h_{CRS}, h_{CRR} – CRAH unit supply specific enthalpy, [kJ/kg]

h_{KA}, h_{Ke} – condensing coil supply/return specific enthalpy, [kJ/kg]

h_{FA}, h_{Fe} – freecooling coil supply/return specific enthalpy, [kJ/kg]

l – hydraulic system length, [m]

p – absolute air pressure, [Pa]

P_{CR} – CRAH unit fan power input, [kW]

P_{CWP} – chilled water pump power input, [kW]

P_{DX} – compressor power input, [kW]

P_{FC} – freecooling coil fan power input, [kW]

P_{IT} – IT systems power input, [kW]

P_K – condenser fan power input, [kW]

P_{TOT} – mechanical systems total power input, [kW]

PUE(m) – power usage effectiveness of mechanical systems, [-]

Q_{CW} – chilled water cooling capacity, [kW]

Q_E – evaporator cooling capacity, [kW]

Q_{FC} – freecooling coil cooling capacity, [kW]

Q_{IT} – server room cooling load, [kW]

Q_K – condenser capacity, [kW]

R – pressure loss coefficient, [Pa/m]

R_p – gas constant of air, 287,05 [J/kgK]

t_A – ambient temperature, [°C]

t_{Amax} – max amb. temperature for a given profile, [°C]

t_{Amin} – min amb. temperature for a given profile, [°C]
t_{CHR} – condensing coil return air temperature, [°C]
t_{CHS} – condensing coil supply air temperature, [°C]
t_{CRR} – CRAH unit return air temperature, [°C]
t_{CRS} – CRAH unit supply air temperature, [°C]
t_{CWE} – chilled water temperature before evaporator, [°C]
t_{CWR} – chilled water return temperature, [°C]
t_{CWS} – chilled water supply temperature, [°C]
t_E – evaporating temperature, [°C]
t_{FA}, t_{Fe} – freecooling coil supply/ return air temperature, [°C]
$t_{FC,0\%}$ – freecooling start ambient temperature, [°C]
$t_{FC,100\%}$ – full freecooling ambient temperature, [°C]
t_{FCR} – freecooling coil return air temperature, [°C]
t_{FCS} – freecooling coil supply air temperature, [°C]
t_{KA}, t_{Ke} – condensing coil supply/return air temperature, [°C]
T_E – evaporating temperature, [K]
T_K – condensing temperature, [K]
V_{CR} – CRAH unit airflow volume rate, [m^3/s]
V_{CW} – chilled water flow volume rate, [m^3/s]
V_{FC} – freecooling coil airflow volume rate, [m^3/s]
V_K – condensing coil airflow volume rate, [m^3/s]
Δp_{CR} – CRAH unit pressure drop, [kPa]
Δp_{CWP} – chilled water pump available pressure, [kPa]
Δp_E – evaporator pressure drop, [kPa]
Δp_{FC} – freecooling coil pressure drop, [kPa]
Δp_p – hydraulic system pressure drop, [kPa]
ΔQ_{FC} – freecooling coil capacity difference, [kW/°C]
Δt_{CR} – CRAH unit air temperature difference, [°C]
Δt_E – evaporating and chilled water supply temp. difference, [°C]
Δt_{EXV} – min. evaporating and condensing temp. difference, [°C]
$\Delta t_{FC,100\%}$ – 100% freecooling capacity temp. difference, [°C]
Δt_{IT} – server air temperature difference, [°C]
Δt_K – condensing and ambient temp. difference, [°C]
$\Delta\tau$ – duration of a given ambient temperature, [h]
η_{CR} – CRAH unit fan efficiency, [%]
η_{CWP} – chilled water pump efficiency, [%]
η_{DX} – compressor efficiency, [%]
η_{FC} – freecooling coil fan efficiency, [%]
η_K – condenser fan efficiency, [%]
ρ_{CRS} – mass density, CRAH unit supply air temperature [kg/m^3]

1 INTRODUCTION

Data Center cooling is one of the most energy consuming mechanical systems available. There are several factors that contribute to high power demand including continuous operation independently of ambient conditions (cooling load depends mostly on IT equipment and external heat gains have minor impact on it), high expenditures on cooling and movement of load carriers (especially air and water) and constant growth of heat gains and density of servers. Power consumption of mechanical systems has large impact on total cost of ownership and carbon footprint of data center class buildings, so many different researches to minimize this impact have been carried and the problem is still valid.

Application of optimization methods to design chilled water based HVAC systems have been presented by Lu (Lu, Cai, Chai, 2005; Lu, Cai, Xie, 2005), while Porowski (Porowski, 2011) presented composite strategy to select optimum ventilation and air-conditioning system based on energy consumption criterion. Research on data center cooling system optimization have been carried by i.e. Shah et al., (Breen, Walsh, Shah, 2010; Breen, Walsh, Punch, 2010; Breen, Walsh, Punch, Shah, Kumari, 2010) Iyengar et al. (Iyengar, 2009) and Demetriou et al. (Demetriou, 2011), who presented holistic, analytic models taking into account the heat flow from servers to ambient air with a watercooled chiller and cooling tower as a cooling source.

The subject of presented paper is optimization model of the data center chilled water based precision cooling system with external, aircooled chiller and additional freecooling heat exchanger (water economizer) as a mechanical cooling source.

2 PROBLEM FORMULATION

Data center cooling system optimization problem includes formulation of the system model (including constant parameters and decision variables), definition of the limiting conditions and optimization criterion. Simplified technological scheme of the analyzed system is shown on Fig. 1.

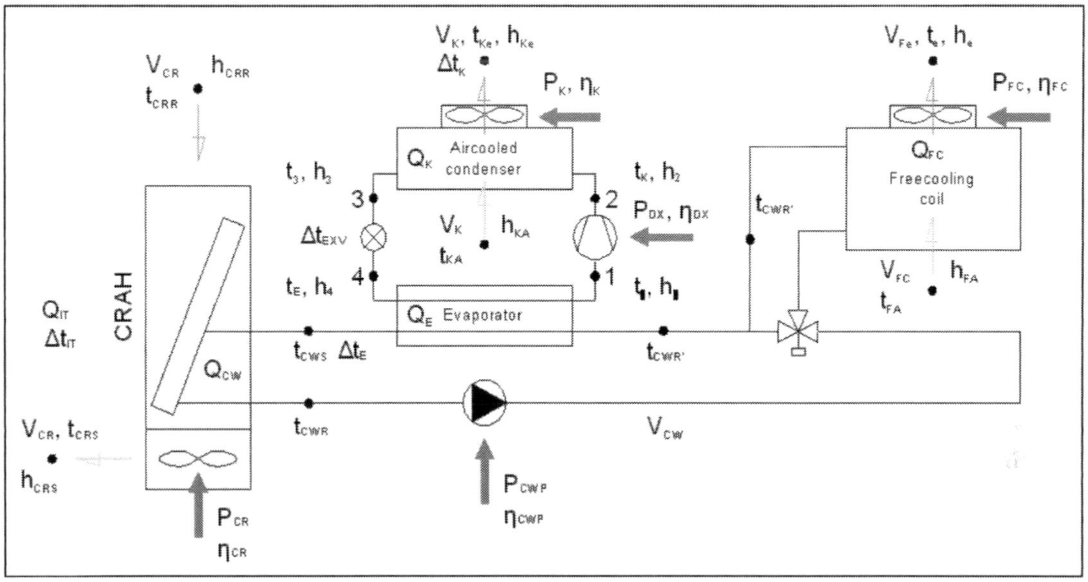

Figure 1. Technological scheme of the chilled water based data center cooling system

The data center cooling system model is described by the constant parameters and decision variables. Constant parameters, by definition, remain constant throughout optimization procedure (in general they can be a function of time). Decision variables, by definition, are changing throughout optimization procedure. Cooling system model described in this paper has following constant parameters: system structure, thermo-dynamical parameters of load carriers (air, water, refrigerant), ambient temperature distribution profile, server room heat load, mechanical efficiency of various components (pump, fans, compressor), difference between chilled water supply and evaporating temperature, minimum difference between refrigerant evaporating and condensing temperature, difference between ambient and refrigerant condensing temperature, length of the chilled water piping.

There are two types of decision variables used in the described model:

- incommensurable variables - server rack architecture. There are two variables of this kind included in the model: closed aisle architecture (separation of the hot and cold aisle, cooling system control according to supply air temperature, mass flow of the CRAH air as a function of server temperature rise) or open aisle architecture (no cold/hot aisle separation, cooling system control according to the return air temperature and CRAH unit airflow 30% higher compared to closed aisle architecture),
- measurable variables: server air temperature raise, supply and return chilled water temperatures (thus chilled water mass flow), freecooling coil operation starting ambient temperature, temperature difference to reach full required capacity of the freecooling coil.

There are following limiting conditions used in the optimization procedure: CRAH unit return air temperature (for open aisle option) or server inlet air temperature, thus supply air temperature range according to ASHRAE 2011 A1 class recommended envelope – for closed aisle architecture.

The optimization criterion used in the described model was minimum annual electrical power consumption of the cooling system according to the below account:

$$E_{TOT} = \min(E_{TOT1}, \ldots E_{TOTj}, \ldots E_{TOTj\,max}) \tag{1}$$

3 MODEL ASSUMPTIONS AND SOLVING PROCEDURE

For each of the components of the cooling system model (CRAH, freecooling coil, water chiller) balance equations have been formulated including power input of the components, i.e. cooling capacity delivered by the chilled water system includes the additional heat dissipated by the CRAH unit fans according to the equation:

$$Q_{CW} = Q_{IT} + P_{CR}(1-\eta_{CR}) \tag{2}$$

CRAH unit fan power input depends on the volume flow of the server room air, which further depends on the mass flow and supply air temperature. It has been calculated as a function of the reference values of the fan power input and volume airflow according to the following equations:

$$P_{CR} = P_{CR\,ref}\left(\frac{\dot{V}_{CR}}{\dot{V}_{CR\,ref}}\right)^3 \tag{3}$$

$$\dot{m}_{CR} = \frac{Q_{IT}}{C_P(t_{CRR}-t_{CRS})} \tag{4}$$

$$\dot{V}_{CR} = \frac{\dot{m}_{CR}}{\rho_{CRS}} \tag{5}$$

$$\rho_{CRS} = \frac{p}{R_p T_{CRS}} \tag{6}$$

Cooling capacity of the external freecooling coil has been calculated in a simplified way, including decision variables $\Delta t_{100\%FC}$ and $t_{A,0\%FC}$ according to the equations:

$$t_{FC,100\%} = \frac{t_{CWS}+t_{CWR}}{2} - \Delta t_{FC,100\%} \tag{7}$$

$$Q_{FC} = \begin{cases} 0, & t_A \geq t_{A,0\%FC} \\ Q_{CW} - (t_A - t_{FC,100\%})\Delta Q_{CF}, & t_{FC,0\%} > t_A > t_{FC,100\%} \\ Q_{FC} = Q_{CW}, & t_A \leq t_{FC,100\%} \end{cases} \tag{8}$$

$$\Delta Q_{FC} = \frac{Q_{CW}}{t_{FC,0\%} - t_{FC,\,100\%}} \tag{9}$$

Power input of the freecooling coil fan has been calculated using equations (3,4,5,6) described for the CRAH unit, with the assumption of constant ambient temperature raise across the coil.

Cooling load of the evaporator has been calculated according to the equation:

$$Q_E=Q_{CW}-Q_{FC} \tag{10}$$

Heatload of the aircooled condenser includes power input of the compressor and has been calculated according to the equation:

$$Q_K=Q_E+P_{DX} \tag{11}$$

Power input of the compressor has been calculated according to following equations:

$$P_{DX}=\frac{Q_E}{COP_{DX}} \tag{12}$$

$$COP_{DX}=\left(\frac{T_E}{T_K-T_E}\right)\eta_{DX} \tag{13}$$

$$t_E=t_{CWS}-\Delta t_E \tag{14}$$

$$t_K=\begin{cases} t_A+\Delta t_K, (t_K-t_E)\geq\Delta t_{EXV} \\ t_E+\Delta t_{EXV}, \ (t_K-t_E)<\Delta t_{EXV} \end{cases} \tag{15}$$

where η_{DX}, Δt_{EXV}, Δt_K are decision variables based upon chiller manufacturer data.

Power input of the aircooled condenser fan has been calculated using equations (3, 4, 5, 6) with the assumption of constant temperature rise of the ambient air. Power consumption of the circulating pump has been calculated using following equations:

$$P_{CWP}=\frac{\dot{V}_{CW}\Delta p_{CWP}}{\eta_{CWP}} \tag{16}$$

$$\Delta p_{CWP}=\Delta p_P+\Delta p_{CR}+\Delta p_{FC}+\Delta p_E \tag{17}$$

$$\Delta p_P=1{,}3Rl \tag{18}$$

with the simplifying assumption, that local pressure drop is 30% of the linear. Pressure drop of the CRAH unit, freecooling coil and evaporator has been calculated using below equations:

$$\Delta p_{CR}=\Delta p_{CR\ ref}\left(\frac{\dot{V}_{CW}}{\dot{V}_{CW\ ref}}\right)^2 \tag{19}$$

$$\Delta p_{FC}=\Delta p_{FC\ ref}\left(\frac{\dot{V}_{CW}}{\dot{V}_{CW\ ref}}\right)^2 \tag{20}$$

$$\Delta p_E=\Delta p_{E\ ref}\left(\frac{\dot{V}_{CW}}{\dot{V}_{CW\ ref}}\right)^2 \tag{21}$$

where values of reference pressure drops have been taken from manufacturers datasheets.

To find optimum mix of the design parameters according to the given criterion a complete review optimization method was used. Starting point of the method was creation of the mathematical models of the particular components of the cooling system, which influence the total power consumption (CRAH, chiller, freecooling coil, hydraulic system and pump). For each of the components system of the equations for the load carriers and energy balance were created. After applying the limiting conditions, a set of permissible variants was formulated, including the mix of design parameters (aisle architecture, chilled water supply and return temperatures, supply and return CRAH unit air temperatures).

For each of the permissible variants, according to the ambient air temperature profile in the $<t_{A,min}, t_{A,max}>$ range with a step of 1°C and assigned duration (time) of each temperature, power consumption of the particular components has been calculated according to the following accounts:

$$E_{TOT}(t_A)=E_{CR}(t_A)+E_{FC}(t_A)+ +E_{DX}(t_A)+E_K(t_A)+E_{CWP}(t_A) \quad (22)$$

$$E_{CR}(t_A)=P_{CR}\Delta\tau \quad (23)$$

$$E_{FC}(t_A)=P_{FC}(t_A)\Delta\tau \quad (24)$$

$$E_{DX}(t_A)=P_{DX}(t_A)\Delta\tau \quad (25)$$

$$E_K(t_A)=P_K(t_A)\Delta\tau \quad (26)$$

$$E_{CWP}(t_A)=P_{CWP}\Delta\tau \quad (27)$$

$$E_{TOT}=\sum_{t_{A\,min}}^{t_{A\,max}} E_{TOT}(t_A) \quad (28)$$

For each of the permissible variants power usage effectiveness ratio of the mechanical systems has been calculated according to following equations:

$$PUE(m)=\frac{E_{TOT}+E_{IT}}{E_{IT}} \quad (29)$$

$$E_{IT}=8760P_{IT} \quad (30)$$

Optimum variant according to minimum annual power consumption criterion (1) has been determined. General algorithm of the optimization procedure is shown on figure 2.

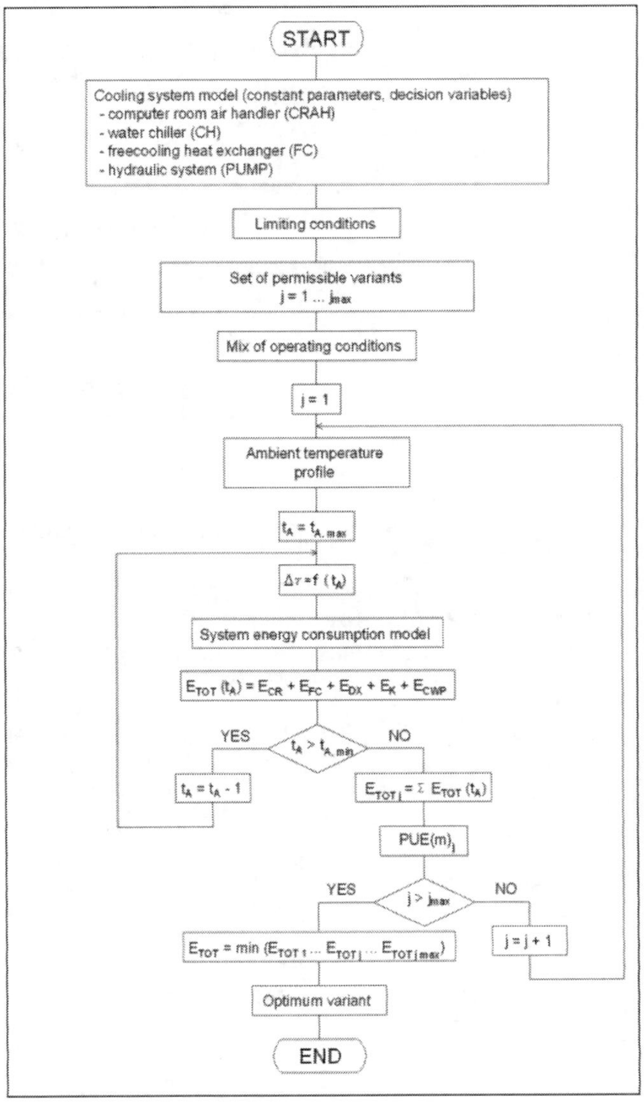

Figure 1. General algorithm of the optimization procedure

4 CASE STUDY INPUT DATA AND RESULTS

The permissible variants (mix of decision variables and limiting conditions) calculated for the case study purposes include:

- open aisle room architecture: chilled water regime range t_{CWS}/t_{CWR}= (5-10/13-15)°C, CRAH unit return temperature t_{CRR}=24°C, CRAH unit temperature difference Δt_{CR}= 11,5°C,

- closed aisle room architecture: chilled water regime range t_{CWS}/t_{CWR}= (12-21/20-26)°C, temperature of the cold aisle (server inler) according to ASHRAE 2011 A1 Data Center class recommended range (A. T. C. TC 9.9., 2011), CRAH unit temperature difference Δt_{CR}= 15°C.

There have been 24 permissible variants calculated. List of decision variables of the selected five calculations for the case study example is shown in Table 1.

Table 1. Case study calculation selected input data

Variant No.	Architecture	Q_{IT}	t_{CWS}	t_{CWR}	Δt_{CR}	$\Delta t_{FC0\%}$	$\Delta t_{FC100\%}$
[-]	[-]	[kW]	[°C]	[°C]	[°C]	[°C]	[°C]
1	open	200	8	13	11,5	1	12
2	open	200	8	15	11,5	1	12
3	closed	200	15	22	15	1	12
4	closed	200	16	24	15	1	12
5	closed	200	18	26	15	1	12

The case study calculation results of the selected permissible variants are shown in Table 2., which includes annual power consumption of the respective components of the modelled system, total annual power consumption of the system and PUE(m) ratio. Energy consumption of the optimum variant (No. 5) based on the minimum annual energy consumption criterion (variant No. 5) is 60% lower compared to most energy consuming system (variant No. 5). Figure 3. shows energy consumption split of the optimum calculated system, while Figure 4. shows energy consumption split of the maximum calculated variant (No. 1).

Table 2. Case study calculation selected results

t_{CWS}/t_{CWR}	E_{DX}	E_K	E_{FC}	E_{CWP}	E_{CR}	E_{TOT}	PUE(m)
[°C]	[kWh/year]	[kWh/year]	[kWh/year]	[kWh/year]	[kWh/year]	[kWh/year]	[-]
8/13	176020	26541	43332	50189	83134	379215	1,22
8/15	165007	23416	47376	22547	83861	342208	1,20
15/22	77413	10382	46793	22547	40514	197649	1,11
16/24	63051	7687	44699	16887	41108	173432	1,10
18/26	46337	5265	40973	16887	41958	151420	1,09

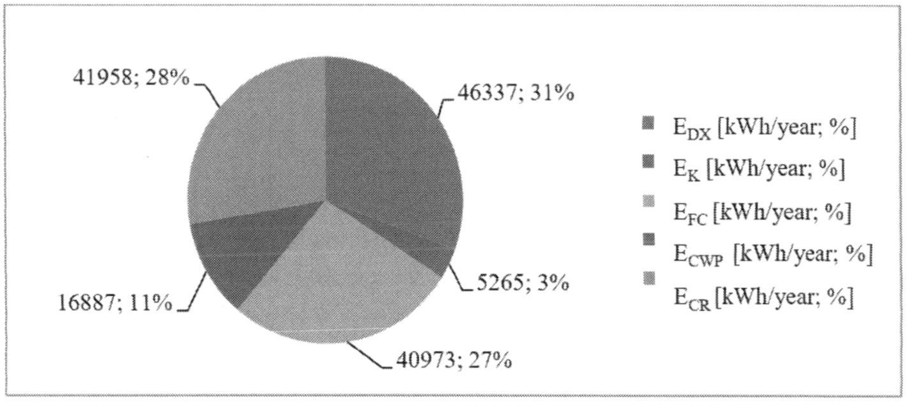

Figure 3. Cooling system energy consumption split (optimum variant – No. 5)

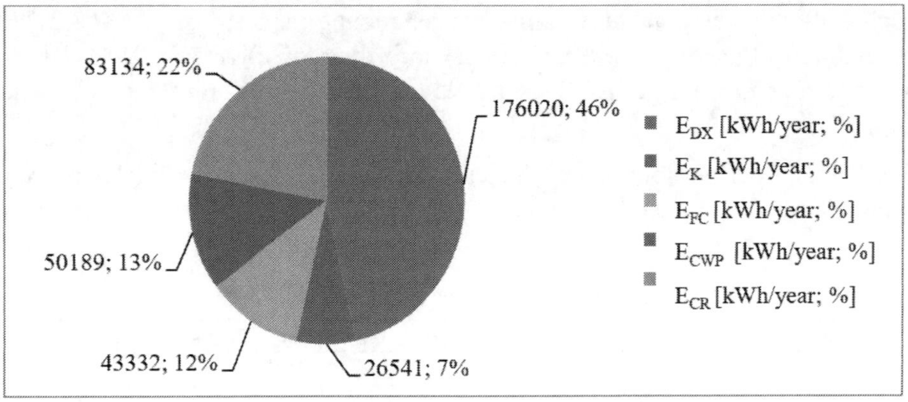

Figure 4. Cooling system energy consumption split (maximum variant – No. 1)

Figure 5. shows power input of the respective components of the cooling system, total power input and annual energy consumption of the optimum calculated system versus ambient temperature range.

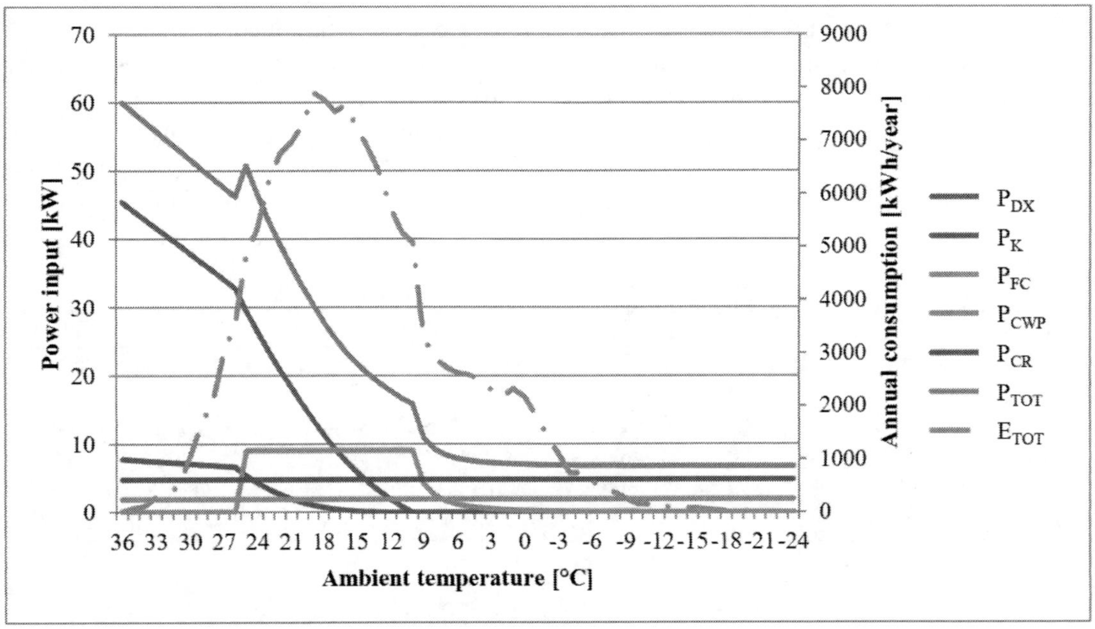

Figure 5. System components power input and total energy consumption versus ambient temperature – – optimum variant – No. 5

5 CONCLUSIONS

The presented paper shows universal model of the optimization procedure and partial results of the calculations of an example case study. Complete review of the permissible variants is used as an optimization method. Based on a case study calculations is has been shown, that presented algorithm can provide an energy-optimized mix of design parameters of the data center cooling system, including server rack architecture (open/closed), chilled water temperature regime, chilled water mass flow and server inlet air temperature. Example calculation results presented in this paper are showing big potential of possible energy conservation depending on design parameters mix. For the case study calculation the difference between maximum and minimum calculated annual power consumption of the cooling system is 60%. Authors are conducting further research on the discussed topic, purpose of the research is further refinement of the cooling system particular components and introduction of more complex optimization procedures.

6 REFERENCES

A. T. C. TC 9.9. (2011). Thermal Guidelines for Data Processing Environments – Expanded Data Center Classes and Usage Guidance. *American Society of Heating, Refrigerating and Air-Conditioning Engineers, Inc.*

Breen, T. J.; Walsh, E. J.; Punch, J.; Shah, A. J.; Bash, C. E. (2010). From chip to cooling tower data center modeling: Part II Influence of chip temperature control philosophy. *IEEE.*

Breen, T. J.; Walsh, E. J.; Punch, J.; Shah, A. J.; Kumari, N.; Bash, C. E.; Lyon, G. (2010). From Chip to Cooling Tower Data Center Modeling: Validation of a Multi-Scale Energy Management Model. *13th IEEE ITHERM Conference.*

Breen, T. J.; Walsh, E. J. P. J.; Shah, A. J.; Bash. C. E. (2010). From chip to cooling tower data center modeling: Part I Influence of server inlet temperature and temperature rise across cabinet. *IEEE.*

Demetriou, D. W.; Khalifa, H. E.; Iyengar, M.; Schmidt, R. R. (2011). Development and experimental validation of a thermo-hydraulic model for data centers. *HVAC&R Research,* vol. 17:4, pp. 540-555.

Iyengar, M.; Schmidt, R. R. (2009). Analytical Modeling for Thermodynamic Characterization of Data Center Cooling Systems. *Journal of Electronic Packaging,* vol. 131, no. June, pp. 021009-1 - 021009-9.

Lu, L.; Cai, W.; Chai, Y. S.; Xie, L. (2005). Global optimization for overall HVAC systems - Part I problem formulation and analysis. *Energy Conversion & Management,* vol. 46, pp. 999-1014.

Lu, L.; Cai, W.; Xie, L.; Li, S.; Soh. C. (2005). HVAC system optimization - in-building section. *Energy and Buildings,* vol. 37, pp. 11-22.

Porowski, M. (2011). Strategia wyboru energooptymalnego systemu wentylacyjnego lub klimatyzacyjnego. Wydawnictwo Politechniki Poznańskiej, Poznań.

NUMERICAL SIMULATION OF INDOOR AIR QUALITY - MECHANICAL VENTILATION SYSTEM SUPPLIED PERIODICALLY VS. NATURAL VENTILATION.

Ewelina Kubacka[1], Dariusz Heim[2]

[1] *Lodz University of Technology*
Department of Heat and Mass Transfer
Wolczanska 213, 90-924, Poland
**Corresponding author:*
ewelina.kubacka@wipos.p.lodz.pl

[2] *Lodz University of Technology*
Department of Heat and Mass Transfer
Wolczanska 213, 90-924, Poland
90-924, Poland

ABSTRACT

Building integrated renewable energy sources e.g. photovoltaic system is one of the promised solution for improving energy efficiency in building. However such kind of the system is restrained by irregular power supplied and necessity to convert current from direct to altering form. Therefore, very often the electrical energy generated by photovoltaic system cannot be effectively utilised to supply building devices, e.g. components of HVAC or lighting system. The other problem is energy storage during the day to be successfully used during the night.

The paper is devoted to analysed numerically overall effectiveness of two types of ventilation system: low-power, direct current mechanical ventilation and natural ventilation. Both systems were applied for office building, generally occupied during a day. A comparison of both system were conducted for the whole calendar year.

Mechanical system considered in the paper is fully supplied from the photovoltaic panels integrated with façade cladding system. Panels covers a total opaque area of external wall, transferring solar energy into electricity with constant performance 15%. The proposed, low-power, mechanical ventilation system can be treated as an alternative for natural ventilation, because both are free from exploitation costs. On the other hand the effectiveness in providing proper indoor air quality of both systems is expected to be different. Efficacy of natural ventilation strongly depends on environmental conditions, while efficiency of proposed mechanical system can productively work during the day only.

Taking into account the changeably character of both ventilation systems work a dynamic simulation methods were proposed to analysed their effectiveness. The model of one office zone, occupied during a day were modelled in ESP-r system. The geometry of the room was discretized by control volume elements. The air flow inside the room was defined by a nodal network method. In the case of natural ventilation the window cracks and openings characterized the inlet, while outlet was defined as a ventilation chimney. On the other hand, while mechanical system is considered the inlet and outlet were defined as a mechanical ventilation devices.

Presented results shows the effect of both system on indoor air quality and thermal comfort. The results are limited to the period of occupancy only, during the whole calendar year.

KEYWORDS

Mechanical, natural, thermal comfort, photovoltaic, efficiency, simulation.

1 INTRODUCTION

Building integrated PV, e.g. façade system can be one of the renewable sources of electricity in zero or plus energy buildings. In the first place PV power should be used by any devices providing proper indoor environment conditions according to the main idea of sustainability building. Therefore, any technical building systems that use electrical energy for heating, cooling, ventilation and/or domestic water to satisfy energy needs can be supply directly from renewable electricity sources.

Photovoltaic is one of the most promised technology, because of easy installation, building integration and cost effectiveness. On the other hand, the periodical, changeable character of its operation generates the following two problem. Renewable energy is providing during the day only and in some characteristic periods of time the supply exceeds the demand. Therefore, the proper size and arrangement of electrical network components including battery, can provide an appropriate and effective operation of the whole system. One of the most precisions method of system design is dynamic numerical simulation using relatively short, e.g. one hour time step. For the purpose of this work the numerical model were developed using ESP-r (Energy System Performance) software.

The main goal of presented work was to investigate the effectiveness of proposed mechanical ventilation system supplied from photovoltaic modules under local, moderate weather conditions. It was assumed that analysed office building is occupied from 8:00 to 16:00, while façade ventilation unit is furthermore active one hour before and after that time. Outside working hours, the ventilation system is switch off. The only source of electrical energy for mechanical ventilation is energy from PV panels and reserve energy stored in battery. For the purpose of presented analysis different amount of panels was considered taking into account solar irradiation according to Typical Meteorological Year, TMY weather file. The effectiveness of solar energy conversion (15%) and stored (85%) were considered in analysis.

The second objective of the study was devoted to compare effectiveness of natural and proposed mechanical system supply from PV. Both systems works at no cost but theirs effectiveness depends on local climatic conditions: wind and temperature for natural or solar radiation for mechanical ventilation. Comparison of both systems were done based on the following parameters: air flow rates in analysed room (pollutant removal efficiency) and comfort indices (user thermal sensation).

2 METHODOLOGY

To conduct the analysis of indoor air quality, the computational model had to be subjected to a dynamic simulation, requiring the use of a suitable model of air flow. In order to determine the appropriate coupling phenomena of the heat and mass transfer, an Air Flow Network method was applied. In the presented approach, all of the components are defined by the nodes representing the volume of solids or the volume of air and the associated heat capacities [6]. The thermal model is based on the finite-volume discretization heat balance method, in which elements of the building structure, zones and associated systems are represented via nodes. Nodes with different physical parameters are connected by the flow path and they are remaining in the thermodynamic equilibrium state. A system of nonlinear differential equations characterise the properties and the form of the flow and they are describing the network connection. The flow through each connection is performed in such a manner that the amount of air flowing into and flowing out of each zone remains at equilibrium (as the principle of conservation of mass).

3 SIMULATION MODEL

3.1 Geometry and materials

The simulation was performed for the model of a single office room. Climate database was built on the basis of meteorological data type WYEC (Weather Year for Energy Calculation, Version 2) [5], developed for the city of Lodz. The room has been defined based on the actual area dimensions (depth x width x height) 5.8 x 3.0 x 3.6 m. All partitions were designed with the assumption of good thermal insulation of the area. The mineral wool with a thickness of 200 mm was used as a basic insulation material and it was located on the inside of both vertical and horizontal partitions. Window with dimensions of 1.2 x 1.2 m and a thickness of 0.01 m oriented to the west was located at the central part of the outer wall. In order to obtain accurate simulation results, the area of the room was divided into 3 zones with finite volumes equal 19.44 m^3 and 21.6 m^3.(Figure 1). The article authors assumed that the work spot is placed in the central part of Zone A.

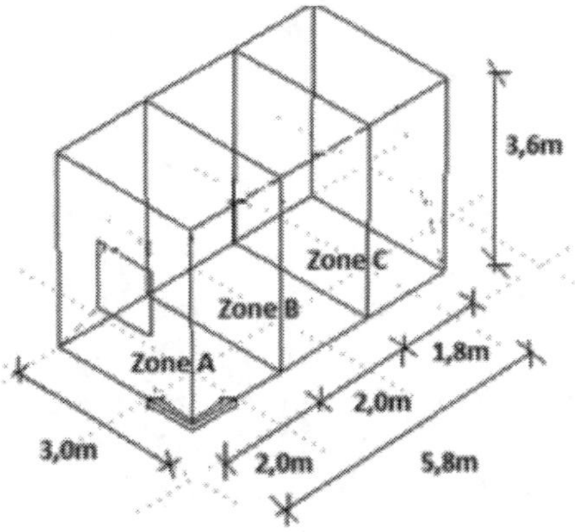

Figure 1: Model geometry

3.2 Computational assumptions

For the purposes of the article indoor air quality evaluation was made by using two thermal comfort indices: PMV and PPD. The parameters necessary for their designation were adopted in accordance with the standards for the required parameters of the proposed internal environment [4]. It was assumed that during cooling season (from October to March) the thermal resistance of clothing is I_{cl} = 0.5 clo (clothing in the form of thin trousers and a shirt with short sleeves), the activity level of 1.2 met = 70 W/m^2 (the characteristic value for office work). During the heating season (from April to September) thermal resistance of clothing assumed equal to 1.0 clo (shirt, pants, jacket). The level of activity remains unchanged and the air velocity was taken as the average air velocity value for each month.
Analyses were performed with the assumption the zone is occupied by one person from Monday to Friday from 8:00 ÷ 16:00. The total heat load being defined as 215 W/person [3], where 100 W is derived from the equipment room (electronic devices). It was assumed that solar irradiation is sufficient light source from 8:00 ÷ 15:00, and therefore heat gain from artificial lighting assumed equal to 10 W/m^2 between 15:00 ÷ 16:00.

3.3 Air Flow Network

The Air Flow Network is designed by five nodes (three internal - A, B and C and two external - 1 and 2, where the flow is caused by the effects of wind pressure on the surface elevation) (Figure 2). Wind pressure distribution coefficient is specified as for the partially exposed walls in the inlet node (1), while in the outlet node (2) as for the roof with inclination angle of less 10 ° [1]. It was assumed that the distance from the roof to the floor of the considered room is 15.0 m The flows in the nodes are a function of pressure and the properties of the components to which the nodes are attached. The area of the hole and the coefficient of discharge are two features of the components that were crucial for the analysis.

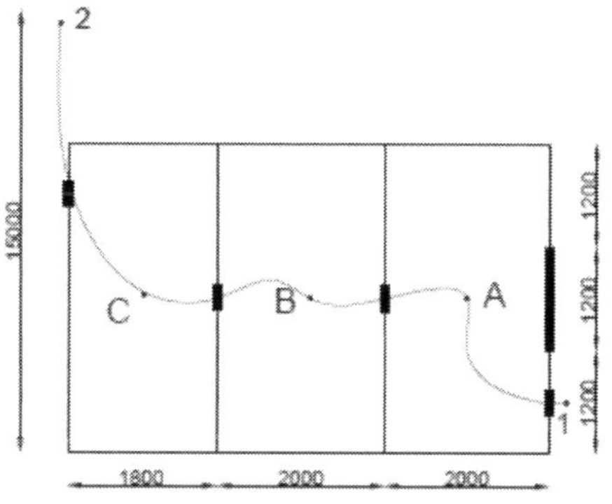

Figure 2: The Air Flow Network in the model

3.4 Adopted ventilation systems

For the purposes of the article two different types of ventilation systems was modelled to compare the indoor air quality - low-power, direct current mechanical ventilation and natural ventilation. Both systems were applied for an office room described above.

Mechanical system consists of the ventilation unit with constant volume of supply air at the level of 90 m^3/h which works from 7 am to 5 pm to fully ventilate the room before and after it is being occupied. It is fully supplied from the photovoltaic panels integrated with façade cladding system. Building integrated photovoltaic components were implemented into ESP-r by Clarke *et al.* [2] as a multi-layer construction model. Photovoltaic facade consisted of outer glazing, PV element, resin binder and back sheet glazing. Photovoltaic material was formed as a layer of number of solar cells. The PV properties were determined in form of specific components using 'special material' database. The 13 PV panels, 0,72 m^2 surface area each, covers a total opaque area of external wall that equals 9.36 m^2. It was assumed that panels are transferring solar energy into electricity with constant efficiency 15%. Whole system is supplied with a battery which efficiency is equal to 85%. It was assumed that the battery can accumulate up to 1320 Watts of the power.

In the case of natural ventilation the window cracks and openings characterized the inlet, therefore its efficiency strongly depends on environmental conditions. In both cases the outlet was defined as a ventilation chimney.

4 RESULTS

Simulations were performed for the whole calendar year with the time step of 60 minutes. Results were presented in the form of monthly averaged and daily distribution for selected weeks of summer and winter.

4.1 PV energy supply

The energy comes from PV panels needs to cover the power demanded by a ventilation unit which stands at the level of 37 Watts. Efficiency of proposed mechanical system supplied from the photovoltaic panels strongly depends on intensity of solar radiation throughout the year. Periods when the solar radiation intensity is the highest and the lowest can be distinguished. According to climate database two occupancy weeks were highlighted: week with the highest and the lowest values of solar radiation intensity – respectively from 23rd to 27th of June and from 8th to 12th of December.

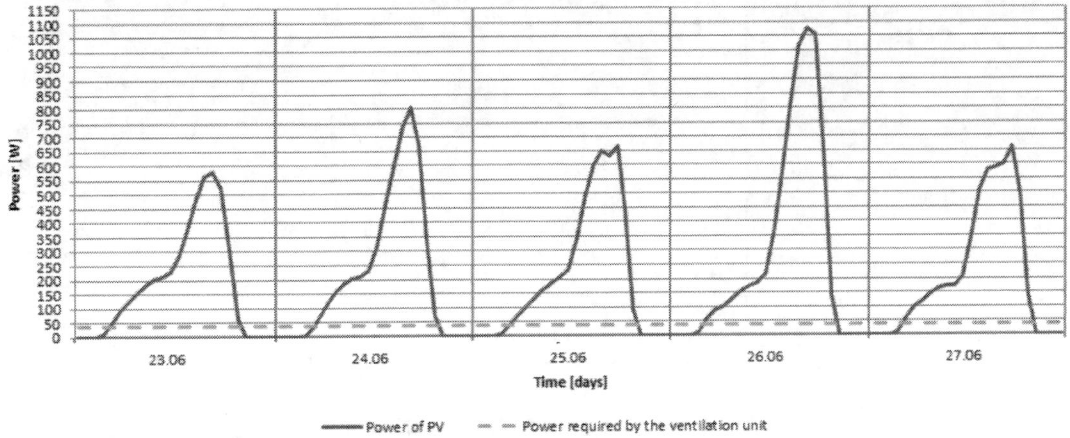

Figure 3: Power generated from PV panels during week the highest values of solar radiation intensity

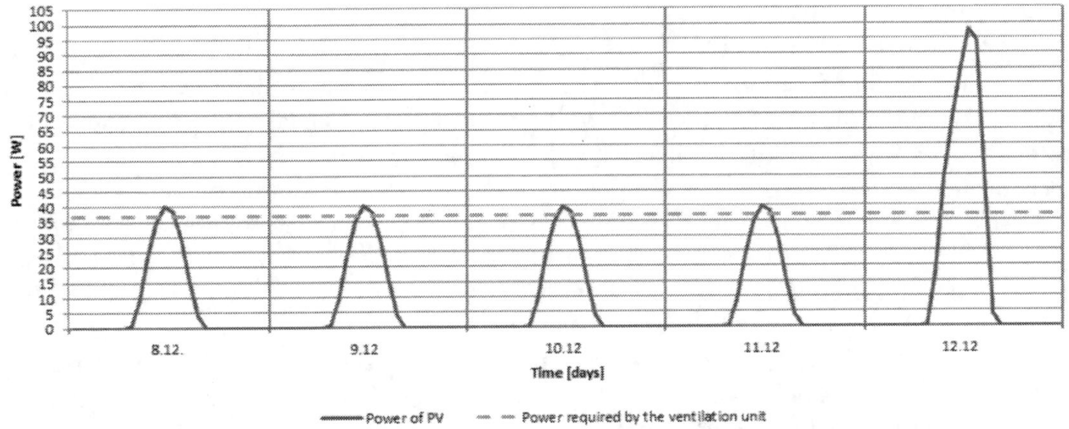

Figure 4: Power generated from PV panels during week the lowest values of solar radiation intensity

Figures 3 and 4 presents power generated from PV (including the battery efficiency 85%) during these weeks with an indication of power required by the ventilation unit.

The graphs shows that during the whole calendar year seasons in which the PV panels cover the whole energy demanded for efficient work of ventilation unit as well as those over which accumulated power there is required appear. Maximum value of the excess of the energy occurs on 26th of June and it equals 1040 W. The amount of the energy supplied by PV panels is not sufficient to provide work of the ventilation unit from 8th to 11th of December.

Amount of the energy that can be stored depends on the number (surface area) of PV panels. Figure 5 presents possible amount of the solar energy to be storage for different number of panels installed in the building, including energy consumption of ventilation unit.

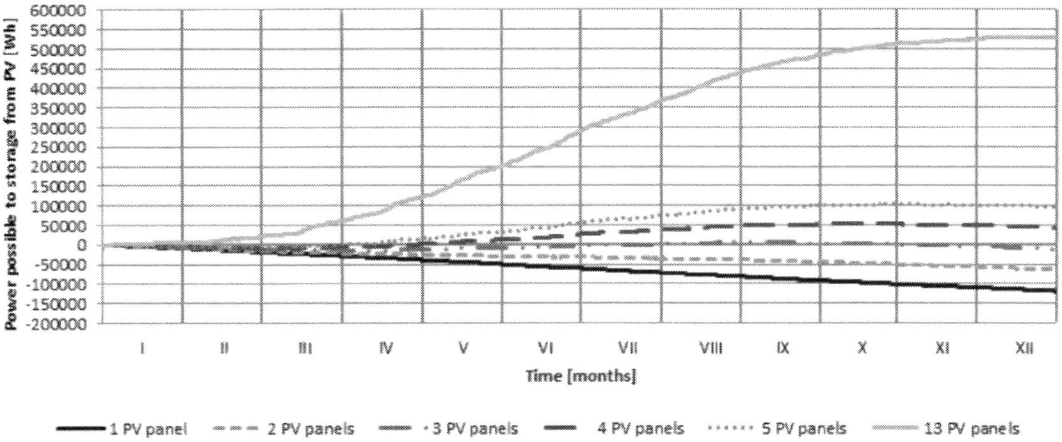

Figure 5: Power generated from PV panels during week the lowest values of solar radiation intensity

Figure 5 shows that installing from one to three PV panels cannot provide a sufficient powering of the ventilation unit from solar energy. Application of four PV panels on the building wall will cause the possibility energy storage in the second half of the year at the level of 50 kWh. However, the great amount of the solar energy cannot be stored and exploited due to the low capacity of the storage cell.

4.2 Indoor air quality

Presented results for indoor air quality are restricted to the hours and days of the room occupancy – from 8 am to 4 pm and from Monday to Friday each week.

Mean values of interior temperature in Zone A, air velocity in Zone A, PMV and PPD rate for each month of the year for both models – room with natural and mechanical ventilation system are presented in Table 1.

Table 1: Monthly mean values indoor air quality indices in Zone A in both models – room with natural and mechanical ventilation system

Month		I	II	III	IV	V	VI	VII	VIII	IX	X	XI	XII
Temperature [oC]	Room with natural ventilation	20.0	20.0	20.0	20.6	22.2	23.0	22.7	23.6	21.8	20.1	20.1	20.0
	Room with mechanical ventilation	21.4	21.5	21.7	22.1	23.5	23.8	24.1	24.4	22.8	21.9	21.6	21.4
Air velocity [m/s]	Room with natural ventilation	0.005	0.003	0.004	0.003	0.003	0.003	0.003	0.002	0.003	0.005	0.003	0.005
	Room with mechanical ventilation	0.011	0.010	0.010	0.010	0.010	0.009	0.009	0.009	0.009	0.010	0.010	0.011
PMV [-]	Room with natural ventilation	-0.32	-0.29	-0.25	-1.11	-0.56	-0.27	-0.35	0.16	-0.69	-0.20	-0.24	-0.32
	Room with mechanical ventilation	-0.04	0.01	0.08	-0.66	-0.18	-0.04	0.05	-0.07	-0.43	0.16	0.06	0.00
PPD [%]	Room with natural ventilation	7.18	6.81	6.45	33.30	17.80	10.71	12.99	8.72	12.95	6.36	6.45	7.20
	Room with mechanical ventilation	5.83	5.91	6.17	16.96	10.67	10.61	10.22	7.40	12.29	6.03	6.04	5.98

The results presented in the Table 1 show that indoor air quality in room supplied in mechanical ventilation system is on higher level than in room ventilated by natural ventilation system. Differences between mean values of the interior .temperature reach to 1.8°C and differences between mean values of the air velocity are at the level of 0.006 m/s. Almost all monthly mean values of PMV in room with mechanical ventilation system are included in the interval $-0.5<PMV<+0.5$, mean value in April is the exception. The most unfavorable conditions of thermal comfort occur in period from April to September. The reason for this occurrence is insufficient cooling during the heating season.

Figures 6 and 7 presents daily mean values of PMV and PPD rates for week with the highest and the lowest values of solar radiation intensity – from 23rd to 27th of June and from 8th to 12th of December.

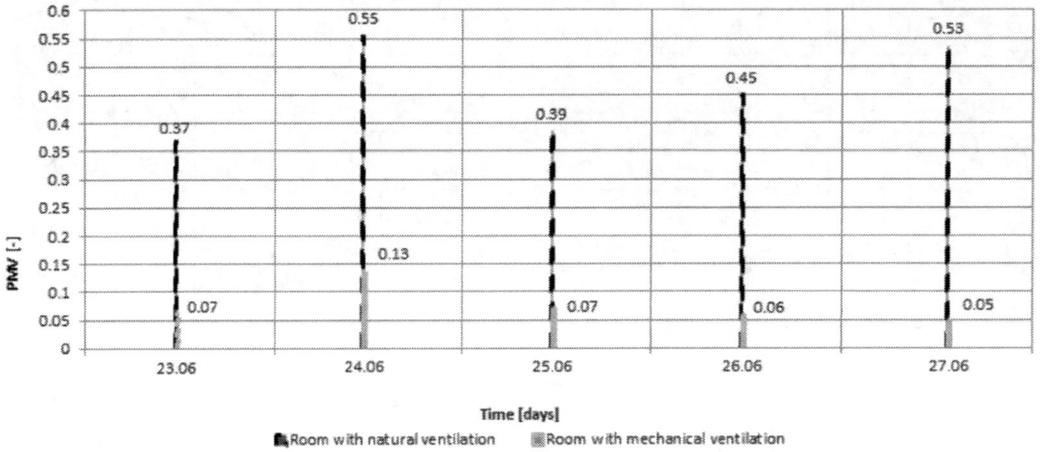

a)

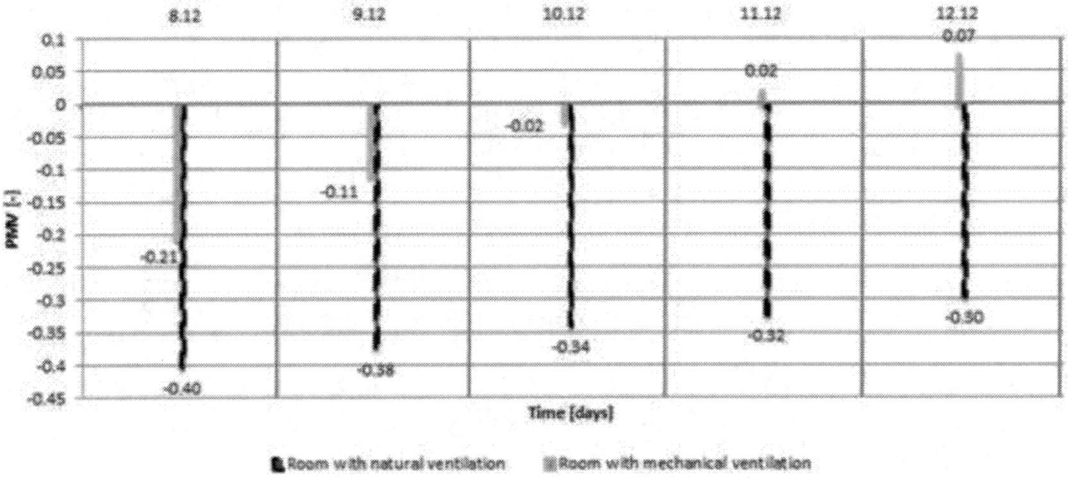

b)

Figure 6: Daily mean values of PMV rate for a) 23rd to 27th of June b) 8th to 12th of December

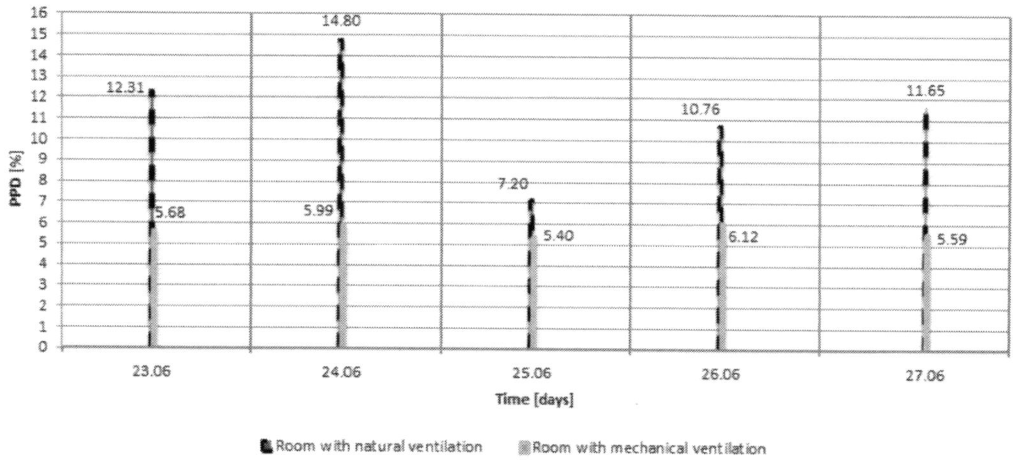

a)

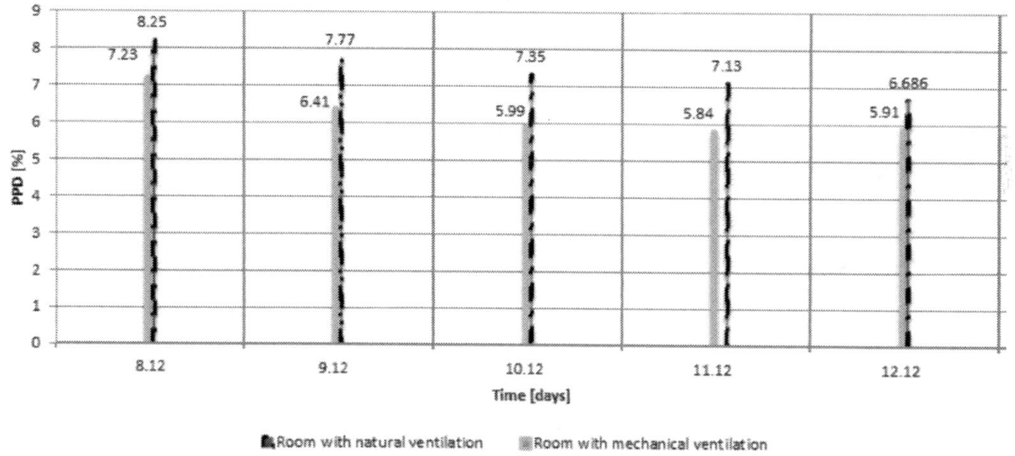

b)

Figure 7: Daily mean values of PPD rate for a) 23rd to 27th of June b) 8th to 12th of December

Results show that values of PMV and PPD rate are within the standard in the model with the room equipped with mechanical ventilation system for both distinctive periods. The number of people dissatisfied with indoor air quality in the room does not exceed 7.23%. For the naturally ventilated room the PPD rate exceeds the optimal value even by 50% in the week of June.

Figure 8 presents volume flow rate of the internal air in area of work spot for two selected weeks for both models of ventilated office room.

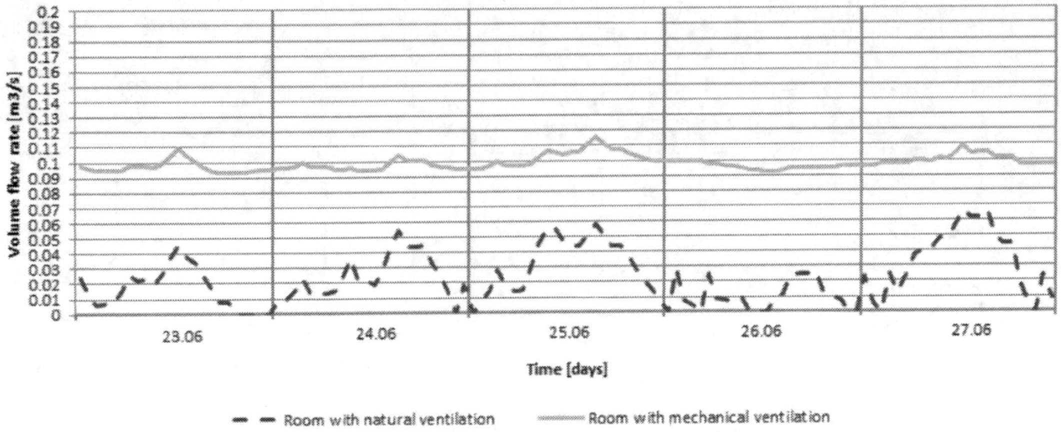

a)

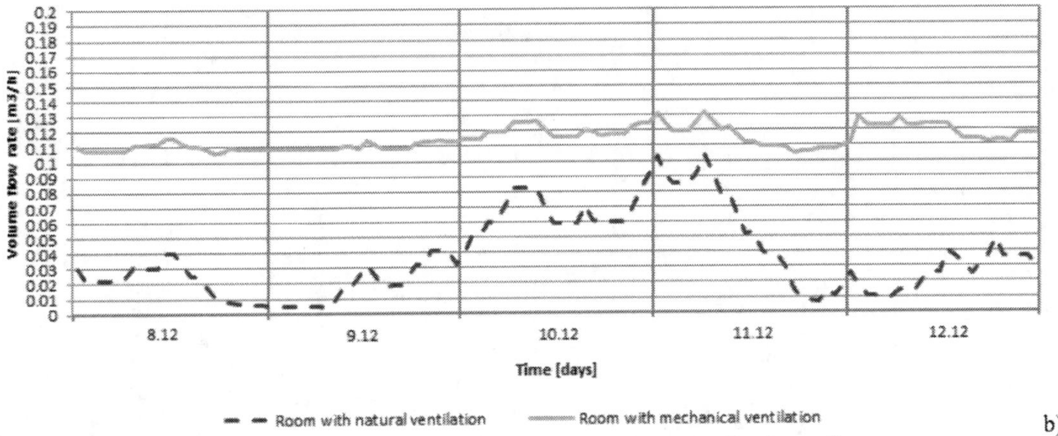

b)

Figure 8 : Volume flow rate in Zone A for a) 23rd to 27th of June b) 8th to 12th of December

Graphs show that volume flow rate is higher during the winter week. For room with natural ventilation values of the volume flow rate in occupied zone are lower than for room equipped with mechanical ventilation system. Differences vary from $0.002m^3/s$ in December to $0.009m^3/s$ in June. It is easy to observe that the volume flow rate of the air supplied form the mechanical ventilation unit is maintained constant level while volume flow rate of the air in natural ventilated room is variable and its values equal from 0.000 to $0.009m^3/s$. The unstable character of air flow for natural ventilation system effects from changeable wind speed and direction as well as indoor and outdoor temperature differences for analysed period.

5 CONCLUSIONS

Conducted simulations showed that PV panels as a source of renewable energy acquired from solar radiation characterise with a very high efficiency. Installation of PV panels in the building has the potential to supply it with enough power to power one or more electrical device. The amount of energy supplied from the panel of a surface area about $3m^2$ greatly exceeds the capacity of an average battery. Installing the appropriate number of panels gives the possibility of obtaining energy amounting to several hundred thousand Watts per hour during the calendar year. A proper project of PV system equipped with a battery of high efficiency and capacity allows to power high number of electrical devices in building that are used in everyday life. One of these appliance can be a mechanical ventilation unit supplying building in the proper amount of high quality air. This is especially useful for office buildings which are workplaces for many people. Employee satisfaction with thermal conditions and indoor air quality increases his work efficiency and reduces the likelihood of diseases. The results of the calculations confirm that the mechanical ventilation system works much better than the natural ventilation system, even it is supply from renewable energy source. It provides significantly better thermal comfort and indoor air quality thanks to higher volume of the air stream supplied during the entire calendar year. Nowadays, in order to gain a high comfort conditions currently existing and new buildings are being equipped with more and more mechanical ventilation devices. Buildings supplied with such an installation have a great opportunity to obtain the status of a zero-energy building with high indoor air quality.

6 ACKNOWLEDGEMENTS

This work was funded by The National Centre for Research and Development as part of the project entitled: "*Promoting Sustainable Approaches Towards Energy Efficiency in Buildings as Tools Towards Climate Protection in German and Polish Cities: developing facade technology for zero-emission buildings*" (acronym: GPEE).

7 REFERENCES

[1]Clarke, J.A. (2001). *Energy simulation in building design,* Oxford/2nd-edition/Butterworth-Heineman.

[2] Clarke, J.A., Kelly, N.J. (2001). Integrating power flow modelling with building simulation, *Energy and Building,* (33), 333-340.

[3] Comstock, W.S. (2001). Chapter 29, Nonresidential Cooling and Heating Load Calculations, *ASHRAE Handbook – Fundamental.*

[4] Olesen, B.W., Parsons, K.C. (2002). Energy and Buildings, vol. 34, 537-548, *Introduction to thermal comfort standards and to the proposed new version of EN ISO 7730.*

[5] Narowski, P., Janicki, M., Heim, D. (2013). *Comparison of Untypical Meteorological Years (UMY) and their influence on building energy performance simulations*, Le Bourget-du-Lac/Proc. of Conference "Building Simulation - BS2013", 1414-1421.

[6] Hensen J. L. M., Lamberts R. (editors) (2011), *Building Performance Simulation for Design and Operation*", Routledge, UK.

DEVELOPMENT OF A DECENTRALIZED AND COMPACT COMFORT VENTILATION SYSTEM WITH HIGHLY EFFICIENT HEAT RECOVERY FOR THE MINIMAL INVASIVE REFURBISHMENT OF BUILDINGS

Janez Zgaga*[1], David Lanthaler[1], Christoph Speer[2], and Rainer Pfluger[2]

1) Fraunhofer Innovation Engineering Center,
Fraunhofer Italia Research
Via Macello 57
Bolzano 39100, Italy
**Corresponding author: janez.zgaga@fraunhofer.it*

2) Institute for Structural Engineering and Material Sciences, University of Innsbruck
Technikerstraße 13
Innsbruck 6020, Austria

ABSTRACT

To ensure adequate indoor air quality, ventilation is necessary in new constructions as well as in modernized existing buildings. In order to minimize energy losses, ventilation systems with integrated heat recovery should be used. Particularly in building refurbishment, ventilation systems need to be designed as compact as possible, to allow a subsequent integration in the existing building stock. Ventilation systems in which one component is responsible for ventilation and simultaneously for heat recovery are well suited for this application area. Already existing systems like the Heat Recovery Centrifugal Fan (HRCF) have a systemic performance limitation regarding ventilation and heat recovery efficiency. These disadvantages still prevent the wide usage in building renovation.

Based on the already known HRCF principle and the unsatisfying market-ready solutions, the "Interreg IV" project VENT4RENO, in which the Fraunhofer Innovation Engineering Center (Italy) and the University of Innsbruck (Austria) collaborate, has the aim to improve the efficiency of such systems and its management, so that they can compete with other systems offered on the market. Beside the design of a compact device, heat recovery, anti-freeze protection, absorption of condensate as well as device control management should be enhanced. In this regard, the University of Innsbruck optimizes the aero- and thermodynamic aspects of the system, while the Fraunhofer Innovation Engineering Center implements the enhanced automatic control as well as the usability of the device.

The modified concept of the Counterflow Heat Recovery Fan (CHRF) allows enhancing ventilation and heat recovery performance at the same time, by using only one cross flow fan for the generation of both flows. Thereby the blades work simultaneously as fan and heat exchanger. Moreover, this modified concept has an integrated systemic anti-freeze protection and allows moisture recovery, making a condensate drain superfluous. A summer bypass is integrated in order to switch off the heat recovery to avoid overheating in the warm season.

Besides measuring and reacting to air parameters like temperature, humidity and carbon dioxide levels, the device needs airflow balancing. By using just one fan, a balance mechanism must control the airflow rate between the intakes. This is essential to avoid over or under pressure, which can cause low efficiency due to leakages or even mold formation. Different operating scenarios were identified, which the device handles automatically or semi automatically. In the latter case, the user can intervene, if wanted, via control panel.

The characteristics of this CHRF concept makes it well suited for decentralized ventilation in refurbished building stock. The simple and compact design allows wall and envelope integration as well as low running costs.

KEYWORDS

Heat Recovery Centrifugal Fan (HRCF)
Counterflow Heat Recovery Fan (CHRF)
Sustainable Building Refurbishment
Demand-controlled Ventilation Strategy
Airflow Balancing Mechanism

1 INTRODUCTION

Nowadays good climate comfort in new buildings is a matter of course, but having the same standards also in refurbished buildings, requires substantial interventions together with continuous and major energy wastage. Since both the number of buildings and their life cycle duration grow, reducing their energy usage has become a major priority and challenge. In the European Union 40% of the total energy consumption is caused by buildings (European Parliament and the Council, 2010). The directive 2010/31 published by the European Parliament and Council in May 2010 (based on the Kyoto-Protocol), provides that the entire energy consumption in the European Union has to be reduced by 20%, compared to the energy consumption in 1990, until 2020 (European Parliament and the Council, 2010). To fulfill this objective while ensuring climate comfort and efficiency, ventilation systems should use integrated heat recovery. Given that the existing building stock outnumbers by far the number of new buildings, ventilation systems should be designed in such a way, to allow a minimally invasive installation and therefore ensure low renovation costs.
The components of such a system should be multifunctional to allow a very compact system design. There are already existing HRCF systems (Heat Recovery Centrifugal Fan) based on this concept, but these market ready products have several weaknesses in terms of thermal and electric efficiency as well as usability. These disadvantages still prevent a wide market penetration of this type of system.
Based on the already known HRCF principle and unsatisfying market-ready solutions, the "Interreg IV" project VENT4RENO, in which the Fraunhofer Innovation Engineering Center (Italy) and the University of Innsbruck (Austria) collaborate, has the aim to improve the efficiency of such systems and its management, so that they can compete with other systems offered on the market. Beside the design of a compact device, heat recovery, anti-freeze protection, absorption of condensate as well as device control management should be enhanced.

2 OVERALL DESIGN CONCEPT AND FEATURE DESCRIPTION

The main components of state of the art ventilation systems with heat recovery are two fans for extract/exhaust air and outdoor/supply air as well as a heat exchanger. Combining these three components into one single component enables the chance to develop very compact ventilation systems. Based on the studies of Dr.-Ing. Sprenger and De Fries (De Fries, 1969) for the development of a rotating heat exchanger (Heat Recovery Centrifugal Fan, HRCF) the company Josef Friedl GmbH has developed the "Frivent®-Wärmerückgewinner" (Josef Friedl GmbH, 2011). As shown in Figure 1, this system aspirates the cold outdoor air as well as the warm extract air separately and axially in the center and blows out the supply and exhaust air radially using the centrifugal force of a rotating porous foam. While the airstreams pass the foam, thermal energy is transferred from one to the other airflow.

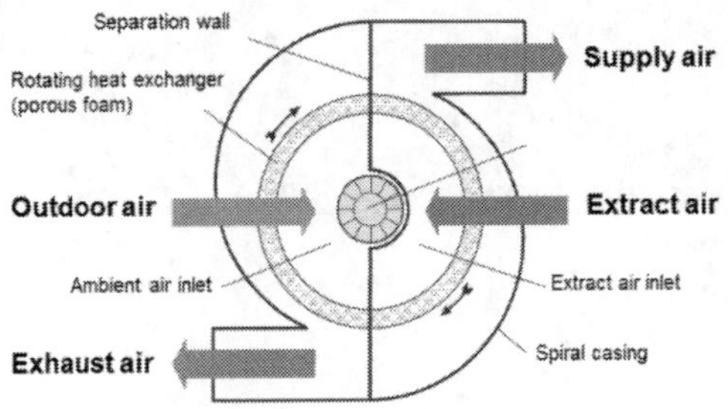

Figure 1: Functional diagramm of the cross section of the Frivent® HRCF (Josef Friedl GmbH, 2011).

The main advantages of this concept are a very simple, cost-efficient and compact ventilation unit design. The disadvantages, like limited fluid-mechanical and thermal efficiency, on the other hand prevent the widespread use in building modernization. The objective of the research project is the development of an innovative decentralized ventilation unit with heat recovery, which maintains the advantages of the HRCF and, at the same time, increases the fluid-mechanical and thermal efficiency. An initial concept idea, where the porous foam was replaced by a cross flow fan and the flow conduction was modified, has been presented in (Pfluger, et al., 2013). Through these modifications, the heat recovery rate can be increased from less than 50% (Josef Friedl GmbH, 2011) to more than 80% (Pfluger, et al., 2013) because the flow regime is switched to a counterflow principle. In addition, the fluid-mechanical efficiency can be improved because the cross flow fan is much better suited to generate airflows. Both air aspiration and air blow out is done radially through the fan blades. Walls in the center of the cross flow fan separate the two airflows and also induce them to change level, in order to use the counterflow heat recovery twice. The results of the numerical CFD simulation of the modified concept can be seen in Figure 2. A detailed description of the flow behavior can be found in (Pfluger, et al., 2013).

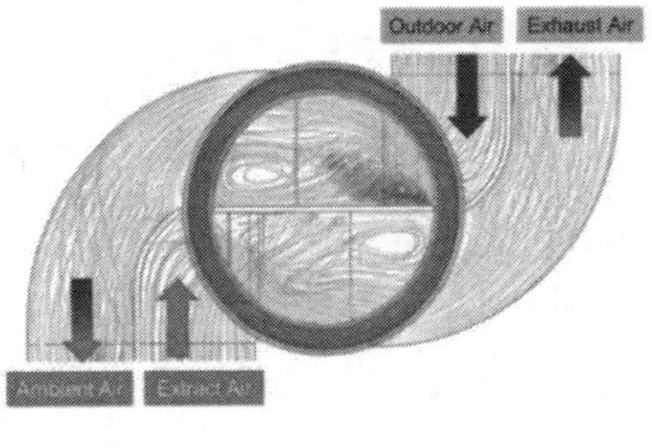

Figure 2: Numerical CFD simulation of the modified concept.

Due to different flow resistances caused by filters or ventilation ducts, the airflows might not be equally balanced. An additional adjustable bypass airflow, which passes through the unit without level change is implemented to enable a volumetric flow balance. To avoid over or under pressure inside the building, the additional bypass airflow can compensate for eventual pressure drops. Moreover, during the warm season the system can operate in pure bypass mode to disable the heat recovery. In this case, both airflows pass the unit without changing level; in this way no heat transfer from the warmer to the colder air stream takes place. An additional advantage of the system is the automated moisture recovery during cold periods. Condensed water of cooled extract air can be absorbed by heated outdoor air, which on the one hand prevents dry supply air and on the other hand prevents the icing of the cross flow fan.

3 MECHATRONIC COMPONENTS AND VENTILATION CONTROL

Before implementing the design concept into a physical prototype, a topology of the system has been created. This gives an overview of all the information (flow rate, temperature, pressure, motor voltage, etc.) and mechatronic components (sensors and actuators) that are needed to control and manage the device. Figure 3 shows the system topology and the correlations, flows, connections and interaction between airflows, mechanical and electrical components.

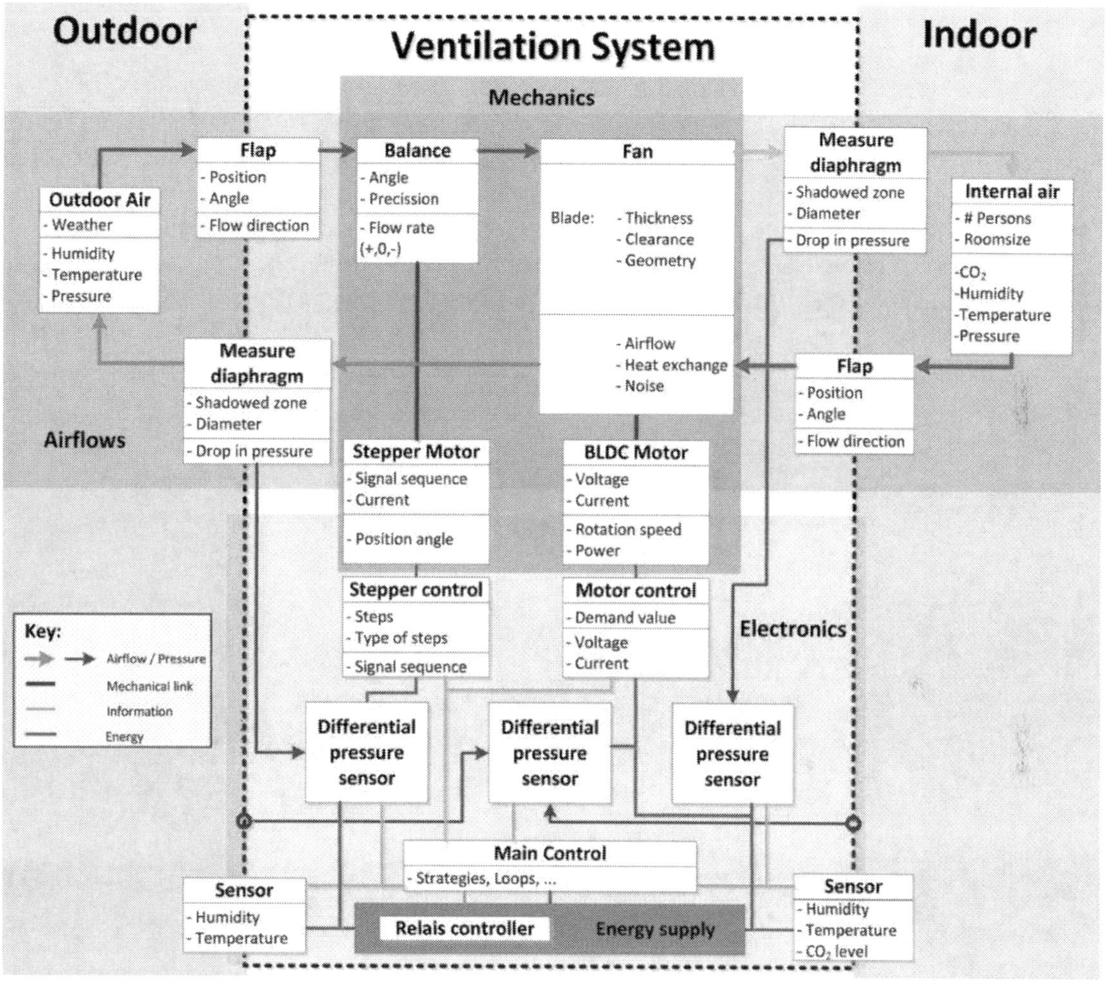

Figure 3: System topology of the design concept.

3.1 Sensors, actuators and ventilation strategy approaches

Depending on the usage scenario and the information given from the sensors (temperature, humidity, CO_2 level, etc.), a certain target airflow rate is preset. The fan motor is led by this target airflow rate and any measured actual/target deviation is dynamically compensated in an inner control loop. The balanced ratio between exhaust and outdoor airflow rate is fundamental for an efficient indoor ventilation system. To guaranty a symmetrical flow rate, the modified design concept uses measuring diaphragms at each inlet (supply and exhaust port) with differential pressure sensors. The respective airflow rates can be derived from the dynamic pressure, which the differential pressure sensors can determine. In addition, to prevent dangerous scenario like fireplaces interacting with the ventilation system, a highly sensitive differential pressure gauge monitors the barometric differential pressure between inside and outside.

The most commonly used control strategies are airflow based; alternative strategies to realize a CO_2-based demand-controlled ventilation are discusses in the work of Nabil Nassif (Nassif, 2012). This concept implies the monitoring of the indoor air quality (IAQ) by the use of low-cost sensors, in order to add the CO_2 level as process variable. Based on the geometry of the fan and due to superior energy efficiency a speed variable high torque brushless direct current (BLDC) motor with integrated hall-sensors is used.

3.2 Airflow balancing mechanism

It is important to ensure equal flow rates in both directions to avoid low efficiency due to leakages or even high moisture in walls (mold formation) caused by over or under pressure. The airflow balancing between both airflows is realized in a non-common way, by introducing a small imbalance of the intake apertures. This non-adequate difference causes a significant offset flow in one direction. In combination with a flow-regulating iris diaphragm in the other intake aperture, positive, zero or even negative balance rates can be obtained. Normally the system tries to stabilize the two synchronous flows to prevent the deficiencies. The iris diaphragm was chosen because round apertures are most suitable concerning flow rates.

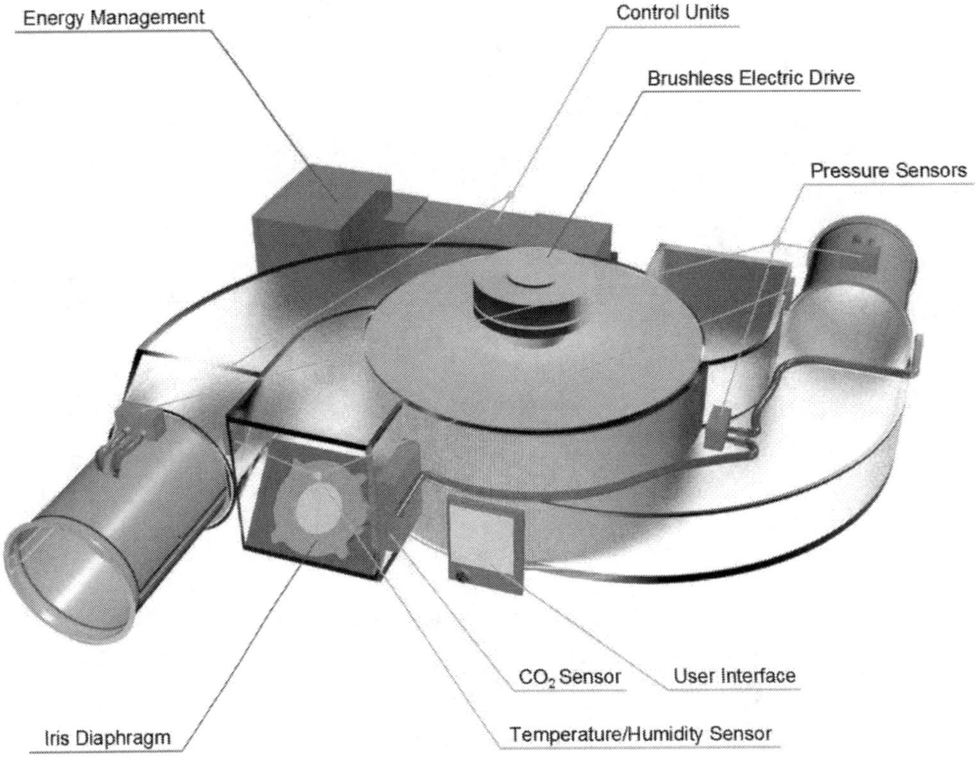

Figure 4: Concept drawing and positioning of the mechatronic components.

A concept drawing and the preliminary positioning of the mechatronic components are shown in Figure 4. The colors used in Figure 4 correspond to the cluster colors in Figure 3.

4 SCENARIO-BASED VENTILATION STRATEGY AND DEVICE USABILITY

The usage of this kind of ventilation system depends on various and some non-daily scenarios. This requires that several scenarios can occur without failure of the ventilation system. Also after installation, only a minimum of maintenance should be required. In the following three scenarios will be discussed.

4.1 Ventilation in combination with a fireplace

If the system is installed in a room near an integrated fireplace, the sensors must be able to detect a sparked fire. The fire will be detect by recognizing an underinflation caused by outdoor back pressure. The system detects this condition within a few seconds and switches itself off.

4.2 Summer bypass

In the warm season, it is important to have an air exchanger without heat recovery. In summer nights, when the outdoor temperature drops and gets lower than the indoor temperature the ventilation system can be used as indoor air cooler, exchanging fresh outdoor air with warm indoor air. The design concept includes two additional ducts and two switchable flaps, which in summer mode position only allow the airflows to pass through the fan on the upper, respectively lower side. Thus, heat exchange can only occur by the thermal longitudinal conduction of each blade, but no more by convection. This - in this case - parasitic heat exchange is reduced to a minimum by manufacturing the fan out of material with low thermal conductivity. This does not affect the normal heat exchange ability, because the thermal conductivity coefficient has no impact on thermal storage capability.

4.3 Indoor air quality (IAQ) control strategy

Besides discussing common controlled ventilation strategies, (Mossoly, et al., 2009) also present two new ventilation strategies using a genetic algorithm. The goal of the first strategy is to maintain the temperature set point while assuring indoor air quality (IAQ). The second strategy controls the supply air rate and temperature to ensure an acceptable thermal comfort and IAQ. Comparing these two strategies with conventional control strategies that only maintain a temperature set point results in considerable energy savings. Conventional control strategies use only a constant airflow rate to guarantee an acceptable IAQ.

The multi strategy modes of the modified design concept for the ventilation system are based on IAQ (more specific, the indoor CO_2 level), thermal and overall efficiency and maximum airflow (for non-common situations).

While running in the IAQ-based mode, the carbon-dioxide concentration is measured and ventilation runs to keep the CO_2 level under a certain level. Lowest possible energy consumption is the target of the efficiency-based mode. The ventilation system tries to run in the most efficient way regarding the total energy consumption, but still monitoring the IAQ and remaining below the maximum acceptable carbon-dioxide concentration. The maximum airflow mode, often called "party mode", keeps the fan motor at full speed. This is used for rooms and situations that needs intensive ventilation.

On top of these modes a control loop monitors that the air exchange always complies with the legal required minimum and the CO_2 level doesn't exceed legal limits. At the highest level, a control loop guarantees maximum safety by shutting down the ventilation system in potential failure-cases. A graphical user interface, consisting of a color display with joystick, allows the change of usage modes and displays warning messages in failure-cases.

5 CONCLUSION

An important part of this research project is of course the construction and realization of a functional prototype to verify and validate the calculated results. The components used in the modified concept are shown in Figure 4: Concept drawing and positioning of the mechatronic components. At the moment the prototype is not yet fully engineered, nevertheless Table 1 gives an overview of the expected characteristics. The major improvements of the modified design concept are an enhanced counterflow concept in combination with an efficient IAQ-based ventilation control. The theoretical efficiency improvement between the modified counterflow concept and the existing co-flow concept is expected to be more than 30%.

Table 1: Comparison between the existing system and the modified design concept

Feature	Existing system	Modified design concept
Dimensions	570 x 460 x 160 mm	560 x 350 x 160 mm
Fan outer diameter	Ø 250 mm	Ø 260 mm
Fan width	60 mm	120 mm
Air flow	100 m^3 per hour	100 m^3 per hour
Rotational speed	15,2 rotations per second	20 rotations per second
Heat recover efficiency up to	48%	80%
Summer bypass	No	Integrated

For the modified design concept a controlled balance mechanism, which regulates the flow rate with a maximum differential flow rate of 10%, is essential to avoid the risk of permanent excess or negative pressure. It is hardly possible to give a forecast about the real efficiency of the two control strategies. Upcoming tests in the climatic chamber of the University of Innsbruck will demonstrate and prove how close the measured values of the prototype will meet the theoretical calculated characteristics.

6 ACKNOWLEDGEMENTS

This work was carried out as part of the research project VENT4RENO within the framework of the funding program "Interreg IV" of the European Regional Development Fund.

7 REFERENCES

De Fries, J. R., 1969. *Heat Recovery Centrifugal Fan.* US, Patent No. 3456718.

European Parliament and the Council, 2010. *Directive 2010/31/EU of the European Parliament and of the Council of 19 may 2012 on the energy performance of buildings,* s.l.: Official journal of the European Union.

Josef Friedl GmbH, 2011. *Frivent-Wärmerückgewinner - Technische Daten Abmessungen.* [Online] Available at: http://www.frivent.com/userdata/4034/uploads/Produkte/pdf-files/WRG-Katalog-DE-2011.pdf

Mossoly, M., Ghali, K. & Ghaddar, N., 2009. Optimal control strategy for a multi-zone air conditioning system using a genetic algorithm. *Energy 34*, pp. 58-66.

Nassif, N., 2012. A robust CO2-based demand-controlled ventilation control strategy for multi-zone HVAC systems. *Energy and Buildings vol. 45*, February, pp. 72-81.

Pfluger, R. et al., 2013. Entwicklung eines hocheffizienten Ventilators mit integrierter Gegenstromwärmerückgewinnung für den Einsatz in der Gebäudemodernisierung. *Bauphysik,* August, pp. 242-249.

CFD SIMULATION OF AN OFFICE HEATED BY A CEILING MOUNTED DIFFUSER

Bård Venås*, Trond Thorgeir Harsem, and Bent A. Børresen

Norconsult AS
Vestfjordgaten 4
1338 Sandvika, Norway
**Corresponding author: bard.venas@norconsult.com*

ABSTRACT

The paper investigates the possibility for using a traditional ventilation system with ceiling mounted diffusers to provide heating under winter time conditions in relatively cold climates – in buildings with low transmition losses such as "passive houses". The analysis is done through a number of CFD simulations of a simplified office. It is shown that even small over-temperatures reduce the Air Change Efficiency substantially. On the other hand even very small internal heat sources increase the efficiency. It is concluded that the principle as it is works fairly well when the building is occupied, i.e. for providing peak load heating, but quite poorly without internal heat sources.

KEYWORDS

Ventilation, Heating, Air Change Efficiency, Passive House, CFD

1. INTRODUCTION

It is a general conception that using ventilation for room heating may lead to low ventilation efficiency and reduce the quality of the indoor climate. This, as buoyancy will tend to gather the introduced warm air below the ceiling with limited mixing into the occupied zone.

In most cases ventilation air is therefore introduced with lower temperature than the one that is desired for the room. Heating is instead provided by a separate system, traditionally being radiators mounted below windows in the exterior walls. In addition to the heating of room air, radiators are located in such positions to improve indoor climate, by 1) reducing cold drafts, and 2) counter the effect of cold windows on operative temperature.

However, new buildings for instance adhering to "the passive house standard" have highly insulating and airtight building envelopes. Standards Norway NS 3701 (2012) states that for an office building the U-values of windows should be lower than 0.8 W/m^{2}K, including the frame, and the remaining wall below 0.12 W/m^{2}K.

From simple hand calculations one may conclude that internal heat loads (from occupants, lightning, computers and other technical equipment) are likely to provide most of the heating during the day when office buildings are occupied. In addition to the specific use of the building and the heat loads, this evidently also depends on the ambient temperatures in the region.

This paper investigates the possibility for using a traditional ventilation system, with ceiling mounted diffusers, to provide heating under typical winter time conditions in relatively cold climates. The work is part of the research and development project "ForKlima" funded by the Research Council of Norway.

2 METHOD

2.1 Test room model

A Computational Fluid Dynamics (CFD) analysis has been performed for an "office", based on a set-up in a laboratory test, performed by another group in the project (data have not yet been studied in detail and comparisons with measurements are not presented in this paper).

The test room has a floor area of 4 m × 4 m and a nominal height of 3 m. The walls are highly insulated and equipped with a cooling system able to control the different wall temperatures. In the laboratory test, suspended ceiling plates were mounted 2.4 meters above the floor, and the 0.6 meter high volume above the plates used for the ducting and mounting the diffuser. Experiments were then performed by introducing air of varying temperature from the diffuser, and cooling down one of the walls in the room to represent an exterior wall.

2.2 Computational Fluid Dynamics

For simplicity, and generality, the suspended ceiling was set as the upper boundary for the CFD model, and the cavity above the ceiling plates neglected.

All walls, including the suspended ceiling, were defined as adiabatic. Adiabatic conditions imply that there is no conduction through the wall, and thus, local balance along the wall between convective heat flux and radiation heat flux:

$$\text{Convective Wall Heat Flux} + \text{Radiative Wall Heat Flux} = 0 \tag{38}$$

This is a simplification, as there in practice always will be some thermal interactions with the surrounding volumes (such as the cavity above the suspended ceiling). For non-stationary conditions thermal storage in the building itself will also be paramount.

Figure 1 shows the computational domain and the numerical mesh along the surfaces. There are in addition internal heat loads in the form of a "laptop" (a point heat source, not represented by the mesh) and a thermal dummy representing a person in the laboratory tests.

As in the experiments, to mimic an exterior wall, one of the surfaces was chilled to a specified temperature which was kept constant for each simulation. All other walls were adiabatic.

The simulations were performed using ANSYS CFX 14.5. Due to the unsteady nature of the flow the simulations were run transient to achieve proper convergence. The time steps were chosen in order to reach convergence within three inner-loop iterations for every time step. The simulations were allowed to run until sufficiently stationary conditions were judged to have been achieved. This took three hours room model time or more, depending on the ventilation rate and heat loads in the room (typically 3-5000 time steps).

The mesh is a variable density hexahedral mesh consisting of a total of 460.000 cells. The cell sizes expand from 1 mm in the normal direction close to the walls, through an inflation layer of 500 mm, up to 10 cm × 10 cm in the core of the room.

Turbulence was modelled using the two-equation SST (Shear Stress Transport) model and automatic wall functions accounting for viscous effects in low Reynolds number regions. The fine mesh resolution close to the walls was chosen to obtain y_+ values of the order of 1, to

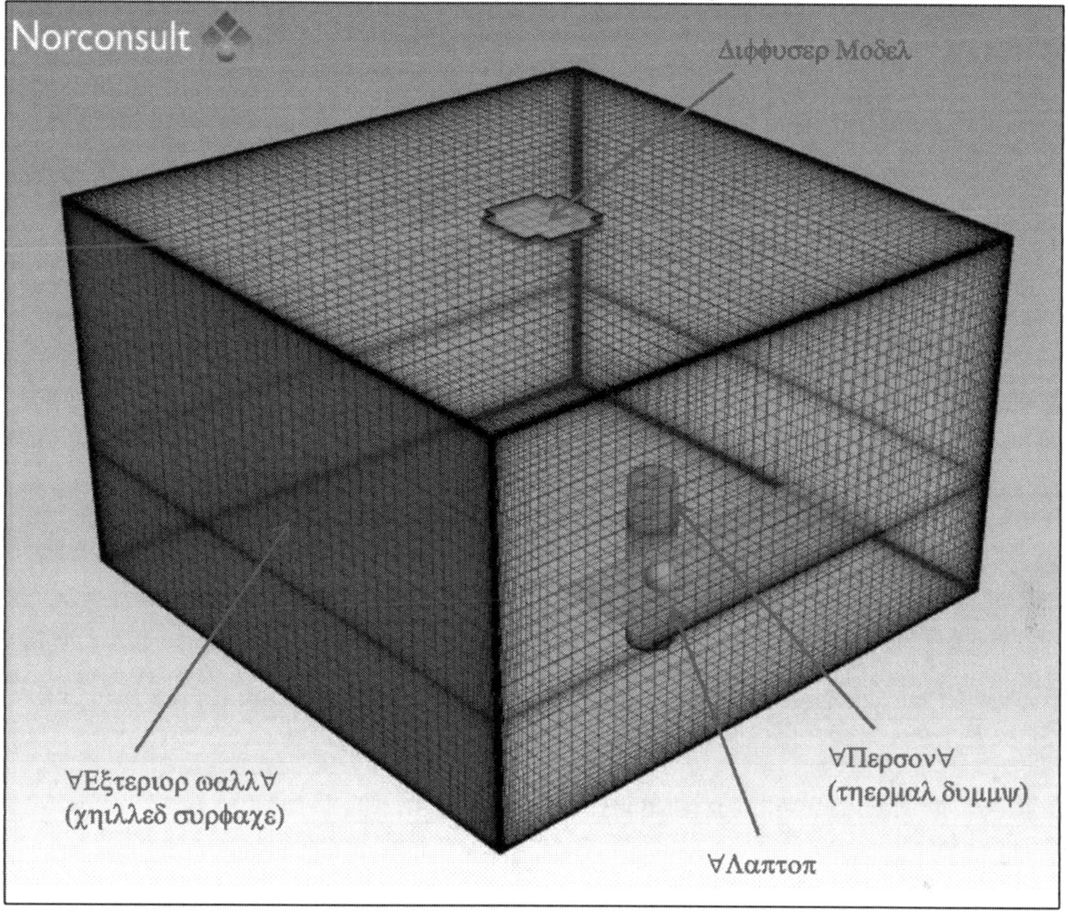

Figure 1: Geometry, boundary conditions and mesh used in the simulations. Grey walls are adiabatic.

better represent the "non-boundary layer" structure of natural convection (see e.g. Venås, B.& Børresen, B.A. (2008)) and wall-jets, flow features that are very important for the case under consideration.

Thermal radiation was modelled using the Discrete Transfer Model with 32^2 rays for ca. 13000 separate radiation elements. The turbulent Schmidt number was kept at its default value of 0.9 for all scalars.

2.3 Diffuser model

A specific type of radial VAV diffuser was mounted in the test room. This diffuser consists of a number of parallel baffles whose internal separation is controlled by the unit as a function of desired air flow rate and air pressure in the supply duct.

Due to the complexity of this geometry the CFD simulations were performed in two steps. First initial runs were performed with a highly resolved 2D diffuser-only model. These results were then loaded into the test room model to provide "effective" diffuser boundary conditions for the room model.

Figure 2 shows results from the diffuser model. One can notice how the exit jets between the individual baffles, first joins together, and then attaches as one to the ceiling due to the

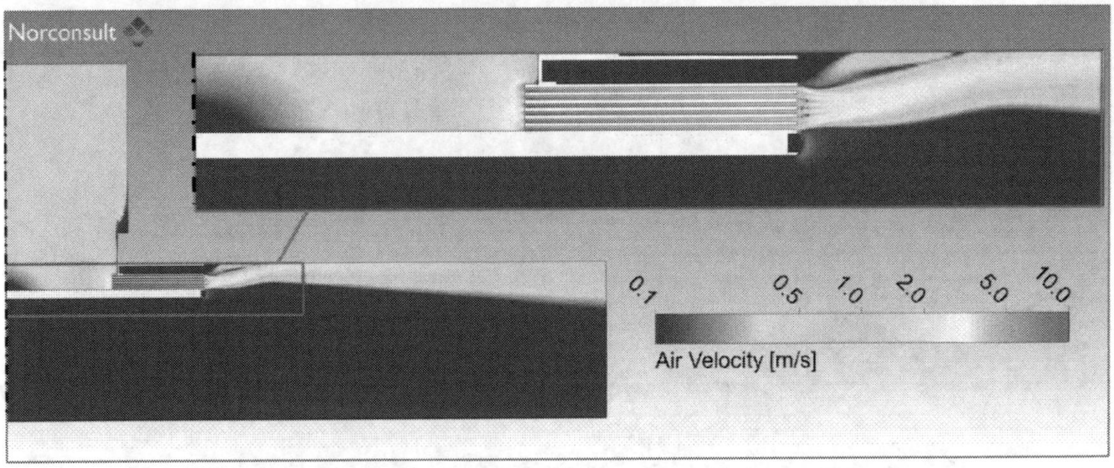

Figure 2: Air Velocity from CFD diffusor model. The centreline of the diffuser (rotational symmetry axis) follows the left edge of the figure. Note that a logarithmic velocity scale is been used.

Coanda effect. The flow then continues outwards, as a radial wall-jet, sticking to the ceiling as it expands away from the diffuser.

Two volume flows have been investigated in this work, respectively 18 litres per second (l/s) and 34 litres per second (the case shown Figure 2).

The diffuser models were run isothermally and the mixing of inlet air and entrained air was represented based on the concentration of an inlet air scalar at the diffuser model interface. This concentration marker was used to evaluate the scalar quantities in the effective diffuser inlet for the room model (temperature and age of air), i.e. determining the mixing between inlet air and entrained air near the diffuser.

In the current simulations the individual baffle separation was kept at "82.5 % opening ratio" (equal to 1.23 mm), as this was typically the opening ratio in the lab measurements.

Figure 3 shows air velocity in a cut through the room for an isothermal case. In this situation there were no heat loads or wall cooling, so that the room temperature perfectly equals the inlet air temperature. The inserted lower left frame shows how the diffuser model is interfaced with the room model.

In the figure one can notice how the air flow from the diffuser spreads out along the ceiling, as in the diffuser model, but is diverted downwards when the flow hits the vertical walls. A distance down along the wall a fluctuating separation of the flow from the wall takes place, and large vortices are formed in the room. The air velocities are generally low, typically less than 5 cm/s, except for close to the ceiling and the upper parts of the walls.

2.4 Air Change Efficiency

To investigate the efficiency of the ventilation system in replacing the air in the room – while simultaneously providing heating – the *Air Change Efficiency* has been evaluated. This measure represents how efficient the ventilation works; i.e. relative to the specific amount of ventilation.

Air Change Efficiency (or *Ventilation Efficiency* in general) can be evaluated in different ways. The method employed here is to calculate the local *Mean Age of Air* based on the CFD data for flow pattern advection and turbulent mixing in the room. In practice this is done by introducing a constant scalar flux throughout the volume inside the room, and letting the scalar accumulate until stationary conditions are achieved. The scalar then represents the time the air has been in a certain position.

The nominal time constant τ_n is used as reference. This time always corresponds to the mean age of the air exiting the room

$$\tau_n = \frac{V}{Q} \tag{39}$$

The overall room Air Change Efficiency is then found as

$$\varepsilon_{\text{room}} = \frac{\tau_n}{2 \cdot \langle\tau\rangle} \tag{40}$$

where $\langle\tau\rangle$ is the volumetric average of the mean age of air. The Air Change Efficiency in a specific location equals

$$\varepsilon_{\text{point}} = \frac{\tau_n}{2 \cdot \tau_{point}} \tag{41}$$

where τ_{point} is the local mean age of air in that location.

The factor "2" in the denominator of eq. (2) and (3) normalizes the expression so that perfect "piston flow" corresponds to $\varepsilon = 100$ %, whereas "full mix" equals $\varepsilon = 50$ %.

In theory the full mix definition stems from an idealized situation where the mean age of air is equal in the whole room (and thus also at the outlet). This is typically assumed to be the best practically achievable value for small rooms where there is not any possibility of obtaining efficient displacement ventilation.

The scale to the left in Figure 4 shows the mean age of air for the isothermal simulation in Figure 3. As the Air Change Efficiency simply is an inverse of the mean age of air it can be represented by the "inverse scale" which is shown to the right in the figure.

In the figure one can notice that new air inserted from the diffusor flows along the ceiling and "ages" with distance from the diffuser due to the elapsing time, and due to mixing with older air in the central parts of the room.

For more on Air Change efficiency see e.g. Novoselac & Srebric (2003).

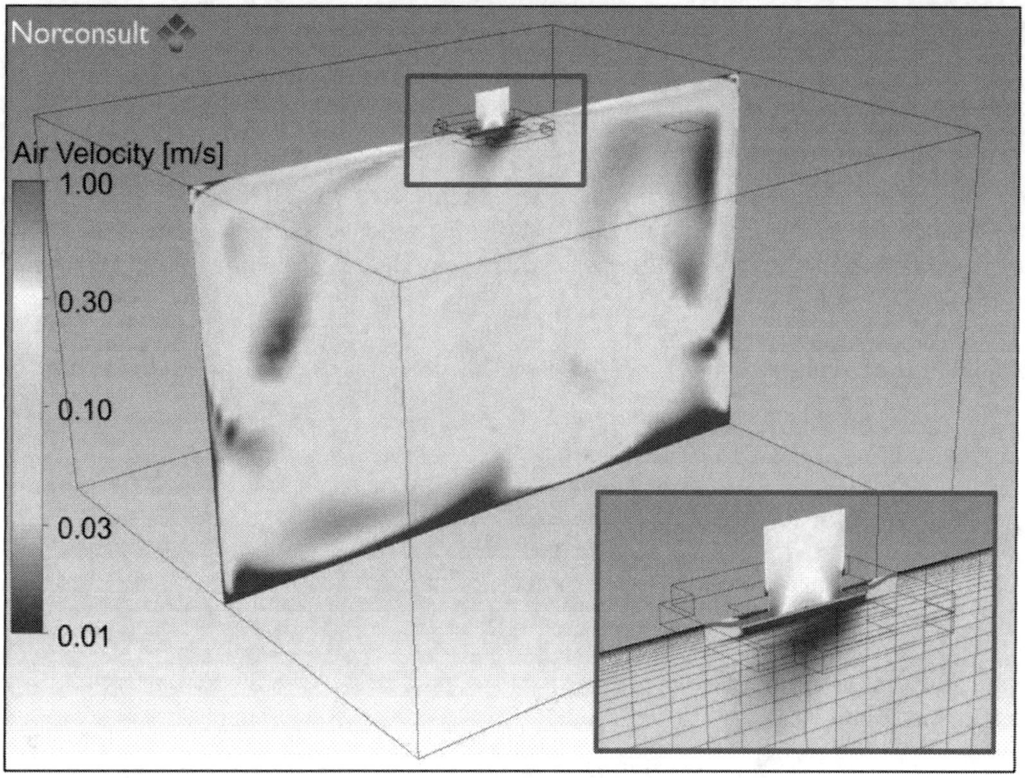

Figure 3: Air velocity in a cut through the room (note: logarithmic scale). Isothermal case with 18l/s ventilation. Close-up on diffuser model inserted lower right.

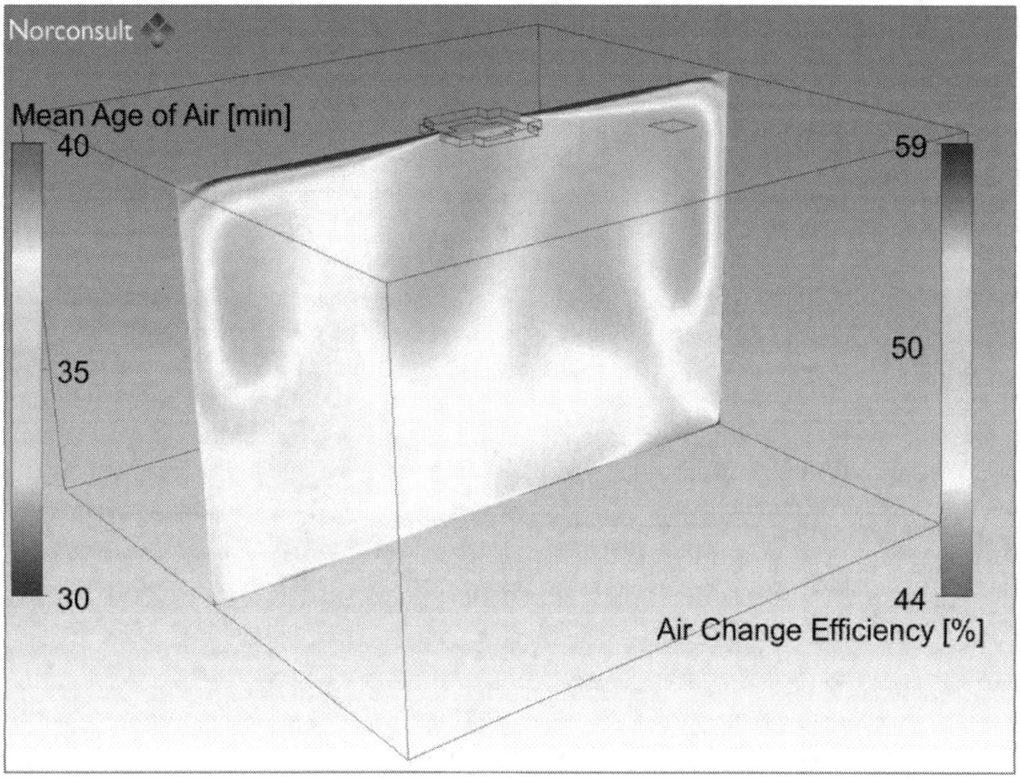

Figure 4: Local *Mean Age of Air* and *Air Change Efficiency* (shown by inverse scale to the right). Isothermal case with 18l/s ventilation.

3 RESULTS

A total number of 30 simulations have been performed, combining varying inlet temperatures, with different internal temperatures on the "exterior wall" and different internal heat loads.

Figure 5 shows Air Change Efficiency, as function of over-temperature, for the simulations performed without any internal heat loads. The isothermal cases here correspond to the leftmost measuring points in the figure. The fully drawn lines represent the efficiency in a measuring point 1.1 m above the floor next to the "desk" in model, whereas the dashed lines represent results for the room as a whole.

We can see that the efficiency is 46 to 47 % for the isothermal cases, both for the occupied zone and the whole room. Thus all being close to the full mix value.

When increasing the over-temperature (by increasing inlet temperature and decreasing the temperature of the chilled surface) the efficiency drops rapidly. For the 18 l/s ventilation case just 1 K over-temperature is enough to reduce the efficiency considerably. For 34 l/s one degree does not significantly change the efficiency, but it starts dropping for 2 K.

Another way of interpreting these results, which may be even more relevant for the problem at hand, is to calculate which ambient temperatures the total heating corresponds to. This has been done by using the upper limiting U-values for a passive house (see page 1), combined with an assumption that 2/3 of the exterior wall consists of windows.

The upper part of Figure 6 shows the same results as Figure 5 presented in this way. One can see that when heating using only ventilation at ambient temperatures of e.g. -10°C or -20°C the efficiency is low. Increasing the ventilation rate to 34 l/s improves the results, and the natural trend seems to be that using more ventilation at lower temperature is more efficient.

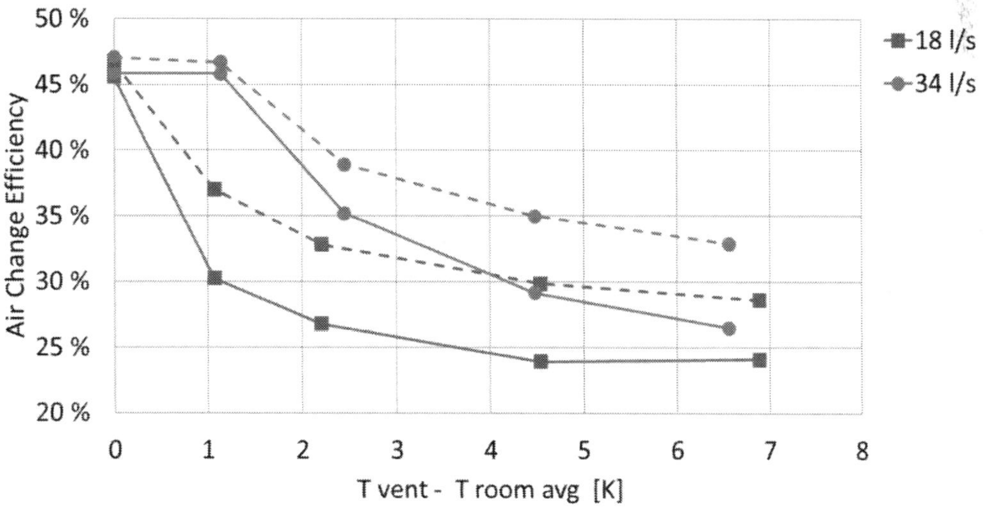

Figure 5: Air Change Efficiency without any heat load in room. Dotted lines are room averages.

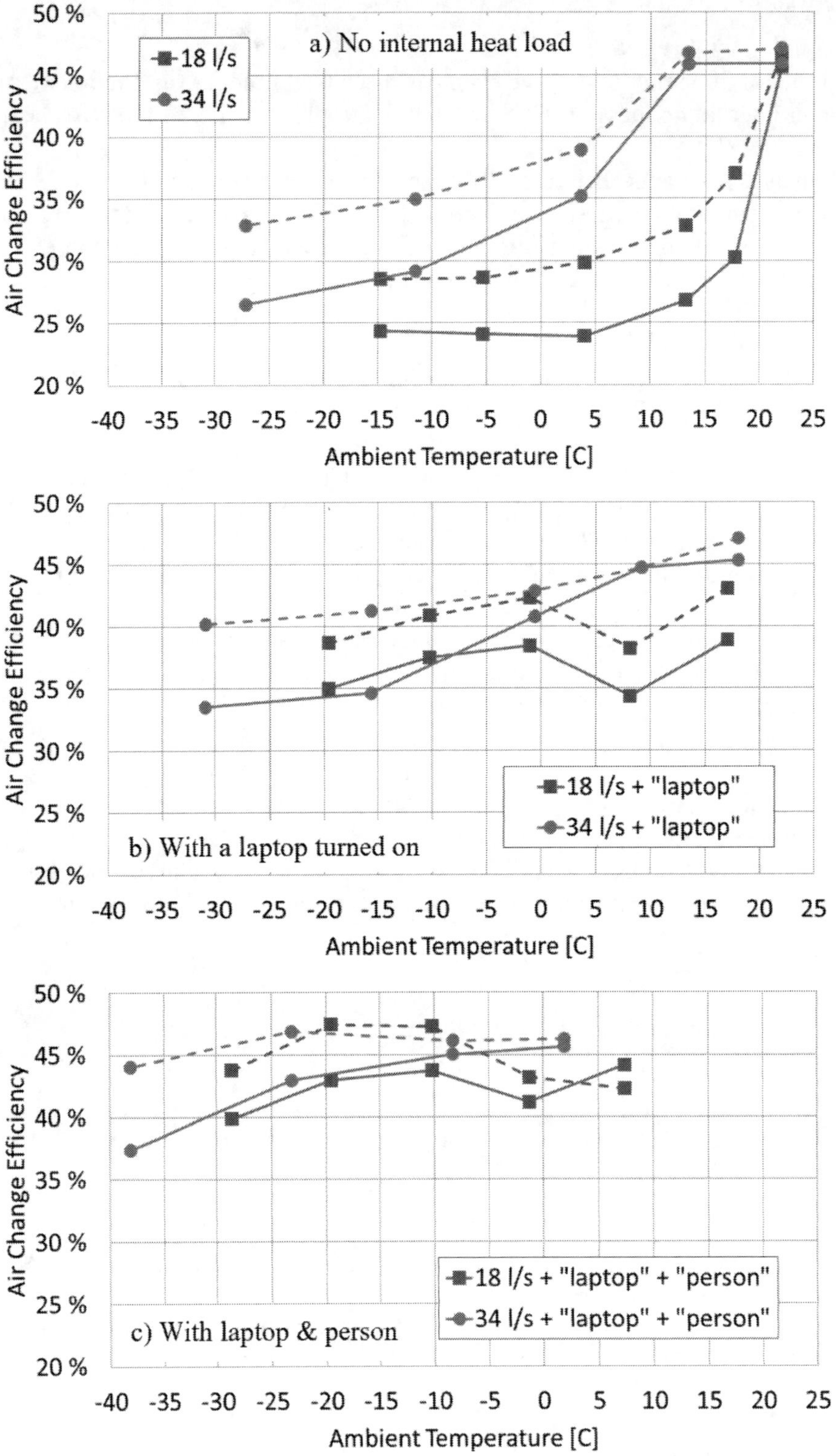

Figure 6: Air Change Efficiency at function of ambient temperatures when ventilation is used for heating – and for situations with respectively: a) no internal heat load (upper); b) with a laptop turned on (middle); and c) with laptop and person (bottom). Fully drawn lines are results from the occupied zone, and dotted lines room averages.

The middle and lower parts of Figure 6 show the corresponding curves when introducing, first a "laptop" computer releasing 40 Watt of convective heat, and then adding a person releasing 35 Watt convective and 35 Watt radiative heat.

One can see that the Air Change Efficiency increases substantially as more of the heat is added from internal sources. It is considered probably to be an important factor that the heat is introduced at low heights, as the rising plume in this case entrains more room air and thus increases the circulation in the occupied zone, than if the they were located high (e.g. lights).

Figure 7 shows plane cuts of Air Change Efficiency through the location of the "laptop" and the "thermal dummy" for different simulations with 18 l/s ventilation rate.

The upper left part (a) shows the results for the fully isothermal case: the flow is circulating and being distributed in the whole room (same as Figure 4 but shown with a different scale).

The upper right part (b) represents the situation for an over-temperature of ca. 2 K without internal heat gains. One can see that there is high efficiency near the ceiling and low efficiency in the occupied zone. The downdraft along the chilled wall brings some new air into the occupied zone.

In the lower left part (c) the plume caused by the 40 W released by the laptop increases the mixing in the occupied zone considerably, bringing old air upwards, and a consequently new fresher air into the zone. The Air Change Efficiency thus increases.

The lower right figure (d) exhibits even more improvement. It is mostly the convective part of the heat that sets up the plume updraft, and this is roughly doubled by introducing a person in addition to the laptop.

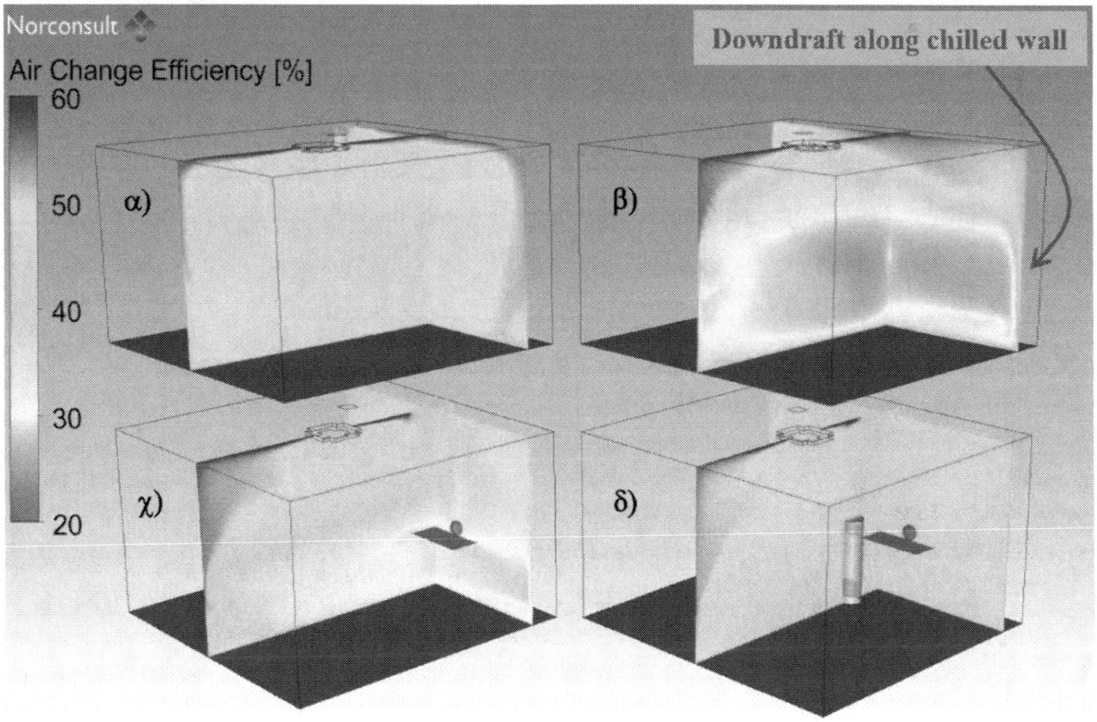

Figure 7: Comparison of Air Change Efficiency with 18 l/s ventilation, combined with respectively:
a) isothermal; b) ΔT ≈ -2K, no internal loads; c) ΔT ≈ -2K; "laptop" only; and d) ΔT ≈ -2K, "laptop"+ "person".

4 DISCUSSION AND CONCLUSION

The main findings considering the flow physics are summarized in Figure 8. The radial wall-jet flows outwards and is bent downwards by the side walls. The downwards "penetration depth" is reduced when thermal stratification in the room increases. At the same time all simulations with a cold wall exhibit a low-velocity downdraft that flows down the entire wall and into the room along the floor, and thus brings fresh air to the occupied zone.

Introducing internal heat loads, which represent low-level heat sources, increase the mixing by setting up thermal plumes which improve the ventilation efficiency considerably. A laptop computer or a person is enough.

When it comes to conclusions for using this type of ventilation for heating it seems like adding additional heat for peak loads is fully possible when the building is occupied, i.e. during work days and work hours for an office.

On the other hand the efficiency is quite low when using over-temperature in a situation without internal heat loads. Work continues on finding ways to improve this.

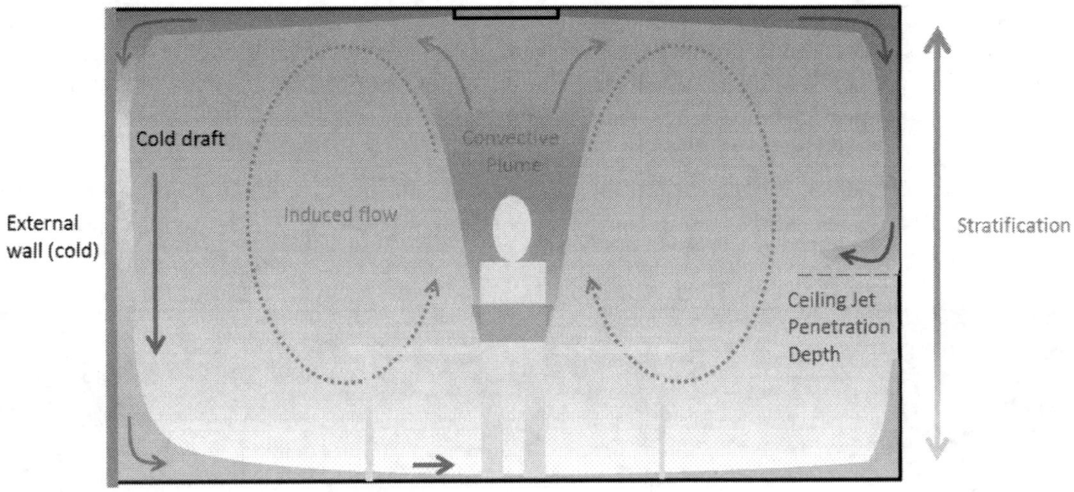

Figure 8: Sketch showing the interpretation of working principle of the studied flows.

5 ACKNOWLEDGEMENTS

This paper was written in the context of the research and development project ForKlima, funded by the Research Council of Norway. RCN as well as the partners of the project are gratefully acknowledged.

6 REFERENCES

Novoselac, A., & Srebric, J. (2003). Comparison of Air Exchange Efficiency and Contaminant Removal Effectiveness as IAQ Indices. *ASHRAE Transactions 109 (2)*, pp. 339-349.

Standards Norway NS 3701. (2012). *Standards Norway NS 3701:2012: Criteria for passive houses and low energy buildings - Non-residential buildings* (in Norwegian).

Venås, B., & Børresen, B. (2008). On the Use of Downdraft Spoilers in Glazed Atria. *29th AIVC Conference, Kyoto, Japan, 14-16 October 2008.*

OPTIMAL POSITIONING OF AIR-EXHAUST OPENINGS IN AN OPERATING ROOM BASED ON RECOVERY TEST: A NUMERICAL STUDY

Sasan Sadrizadeh*, Sture Holmberg

Division of Fluid and Climate Technology, School of Architecture and the Built Environment, KTH Royal Institute of Technology, Stockholm, Sweden

Corresponding author: sasan.sadrizadeh@byv.kth.se

ABSTRACT

This study investigates the influence of outlet location on conventional, turbulent-mixing operating-room (OR) ventilation performance. This was done by numerical simulation using computational fluid dynamics. Multiple configurations of OR outlets, both at floor and ceiling level, were examined, and the results were compared. OR ventilation-system performance in each case was examined by conducting a tracer-recovery test. Two common anesthetic gases, halothane ($C_2HBrClF_3$) and desflurane ($C_3H_2F_6O$), were used to perform the test. Particle simulation was also considered to simulate bacteria-carrying particles.

Based on achieved results, the floor-level exhaust outlets effectively removed anesthetic gases and other odors that might be released in the OR during surgery. Such gases are most likely found at floor level, since they are usually heavier than the air. On the other hand, air-exhaust openings at ceiling level very efficiently evacuated any airborne particles carrying microorganisms, lighter anesthetic gases, or other chemical substances. It is found that floor-ceiling mounted exhaust outlets at every corner of the OR is the optimal arrangement.

KEYWORDS

Hospital operating room, ventilation system, particle distribution, exhaust outlets

1 INTRODUCTION

It is very important to preserve good indoor air quality in the operating room (OR) to ensure health and safety for both surgical team members and the patient. It is well-known that the surgical staff and the patient are the main sources of airborne particles. Human skin is continually being renewed, and the outer skin is shed as squames, or scales. The size distribution of such particles are anywhere between 4 μm and 60 μm, with a mean size of 12 μm in diameter (Noble et al., 1963; Noble, 1975). Microorganisms or bacteria are carried on more than 10 percent of these skin particles, which drift around in the air until they settle down or are evacuated. A substantial amount of anesthetic gases and odors are also released into the OR air from equipment. The highest concentrations of gases during surgery are at floor level. Nevertheless, these gases can be mixed with OR air due to staff movement, and be inhaled by both staff and patient.

The ventilation system of a hospital OR preserves indoor air quality, dilutes and reduces microorganisms concentration to acceptable levels, and removes anesthetic gases and odors released during an operation. OR ventilation-system performance depends on several factors, including type of ventilation system (such as mixing, displacement, or laminar), the air exchange rate, and position of air-exhaust grills (Cao et al., 2014; Kruppa & Ruden, 1996). The air-exhaust openings within ORs vary in size, number and geometry, depending on type of ventilation system and the function of the space. Typically, there should be at least two air discharge outlets inside the OR, usually at floor level. Different studies use different numbers and locations of air-exhaust openings (Chow & Yang, 2003; Chow & Wang, 2012; Sadrizadeh et al., 2014). However, a great deal of care should be given to considering possible airflow obstructions, such as improper positioning of staff members, equipment, and other furniture.

This study assesses and explores the influence of different numbers and locations of air-exhaust openings on OR mixing-ventilation system performance.

Use of commercially available computational fluid dynamics (CFD) tools is now commonplace in the OR design process. Using these techniques, the airflow and temperature distribution in a room can be predicted before the room is built, making it possible to base design decisions on predicted flow conditions.

2 METHOD

A numerical simulation by the CFD technique was performed to investigate the influence of outlet positions on ventilation performance. Three-dimensional airflow and pollutant dispersion were modeled using the RNG k–ε turbulence model, which was numerically solved using the finite-volume method. ICEM CFD was used for grid generation, and double-precision FLUENT 14.5 was used to solve the flow field. The airflow simulation was subjected to grid-sensitivity analysis to find the grid-independent solution.

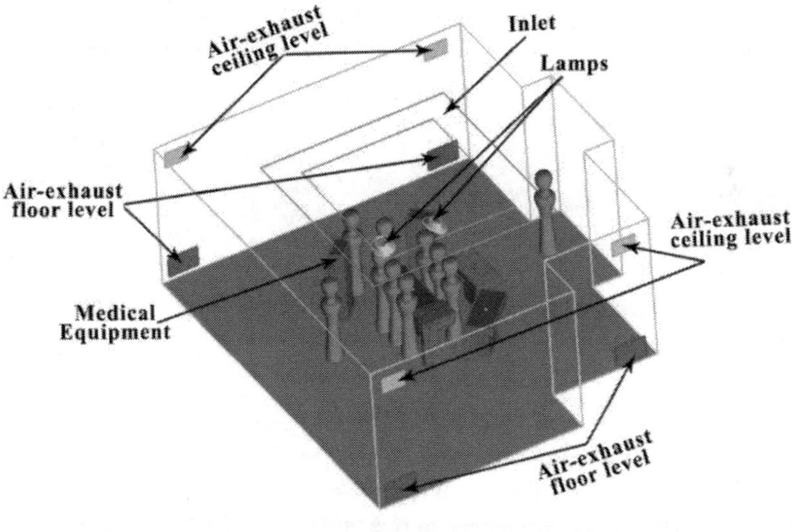

Figure 1: Geometric configuration of operating room and different configuration of air-exhaust outlets

Figure 1 shows the physical arrangement of the OR in this study, based on a Nya Karolinska hospital in Stockholm, Sweden. The overall dimensions were 8.5 m (L) × 7.7 m (W) ×3.2 m (H). Ten surgical staff members, with 100 W/m^2 heat flux, were considered in upright, stationary positions surrounding the patient. In this model, staff members, medical lamps, and equipment were located relative to the operating table, according to the DIN 1946-4 (2008) specification. Incoming air was introduced to the OR at a total airflow rate of 2 m^3/s in a turbulent-mixing flow pattern. Three different configurations of exhaust opening were considered, at both ceiling and floor levels:

- Case 1: Four air-exhaust outlets considered at floor level

- Case 2: Four air-exhaust outlets considered at ceiling level
- Case 3: Eight air-exhaust outlets considered at both ceiling and floor level

The air-exhaust outlets were mounted in every corner of the OR, for a total of four floor and four ceiling exhausts. Bacteria-carrying particles, at a source strength of four colony-forming units (CFUs)/s per person, were simulated in a Lagrangian framework to evaluate transmission risk of airborne particles (Sadrizadeh et al., 2014). Particle-carrying microorganisms, with an aerodynamic diameter of 12 μm, were emitted from staff skin to the OR air. The particle trajectory terminated when particles reached rigid surfaces (trap boundary) or were evacuated from the OR air through the exhaust opening (escape boundary). No recycling from the walls was considered.

2.1 Recovery Test

ISO 14644-3 (14644-3:2005, 2006) describes methods, called *recovery tests*, that show the ability of the OR ventilation system to remove particles and gases. It also states if the OR can change from a dirty to clean state within the specified time. *Recovery time* is the time, in minutes, to decrease particle concentrations by two orders of magnitude (100:1). *Recovery performance* can also be determined from the slope of the particle concentration decay-curve. Cleanliness recovery rate in minute^{-1} is determined by measuring the effect of ventilation on the decay rate of test particles introduced into the OR, calculated from the following:

$$n = \left(-2.3 \times t^{-1}\right) \log\left(\frac{C_1}{C_0}\right) \quad (42),$$

where n is the cleanliness recovery rate (/min), C_0 is the initial concentration, C_1 is the final concentration, and t is the time taken from initial to final concentration.

In this study, based on the German DIN 1946-4 standard (2008), the OR was exposed to 3,500 particles/m^3 of air (0.5 μm). The time taken to reduce the particle concentration by two orders of magnitude (100:1) was then calculated and listed as recovery time.

According to Zhao and Wu (2005), this size of particles can be treated as a passive contaminant. Therefore, to examine the recovery test, we sampled two commonly used gases in anesthetic practice, halothane ($C_2HBrClF_3$) and desflurane ($C_3H_2F_6O$), to perform the tracer-based methodology (Habre et al., 2001).

3 RESULTS AND DISCUSSION

To investigate the effect of numbers and locations of exhaust outlets on the airflow field and particle distribution during surgical procedures, different configurations of exhaust openings were simulated, and results were compared. The steady airflow field was first obtained by steady-state simulation. Convection and diffusion of contaminants within the OR were then modeled by solving conservation equations in transient simulation. Moreover, BCP distribution was examined using Lagrangian particle tracking method.

Figure 2 shows the recovery test result for all three cases. The cleanliness recovery rates respectively were -2.6/min, -3.2/min and -3.5/min for ceiling, floor, and ceiling-floor air-exhaust outlets.

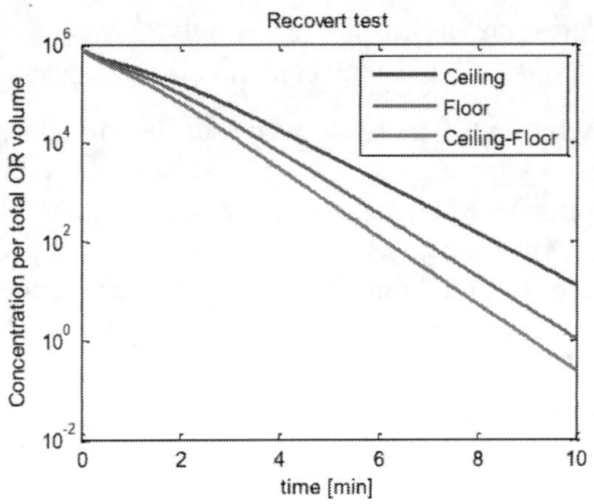

Figure 2: Recovery test results for three different air-exhaust outlet configurations

The lowest possible recovery time was achieved with Case 3 (ceiling-floor exhaust). When the exhaust openings were at both floor and ceiling levels, all the released gases were easily evacuated. Desflurane gas concentration was much higher at the floor-level outlets, since it is heavier than air. Conversely, Halothane concentration was higher at the ceiling level outlets, as it is lighter than OR air.

Figure 3 shows normalized particle concentration contour plots on a vertical plane ($x = 1$ m) for all three cases. Particle concentration was normalized with outlet concentration. Particle density ($\rho = 1000$) was much higher than OR air. However, the lowest possible particle concentration was achieved by positioning multiple outlets (ceiling-floor exhaust opening). This layout may contribute to more uniform air movement within the OR. Fewer eddies were generated by the air circulation inside the OR, so particles may not be trapped by such eddies and evacuated from the OR.

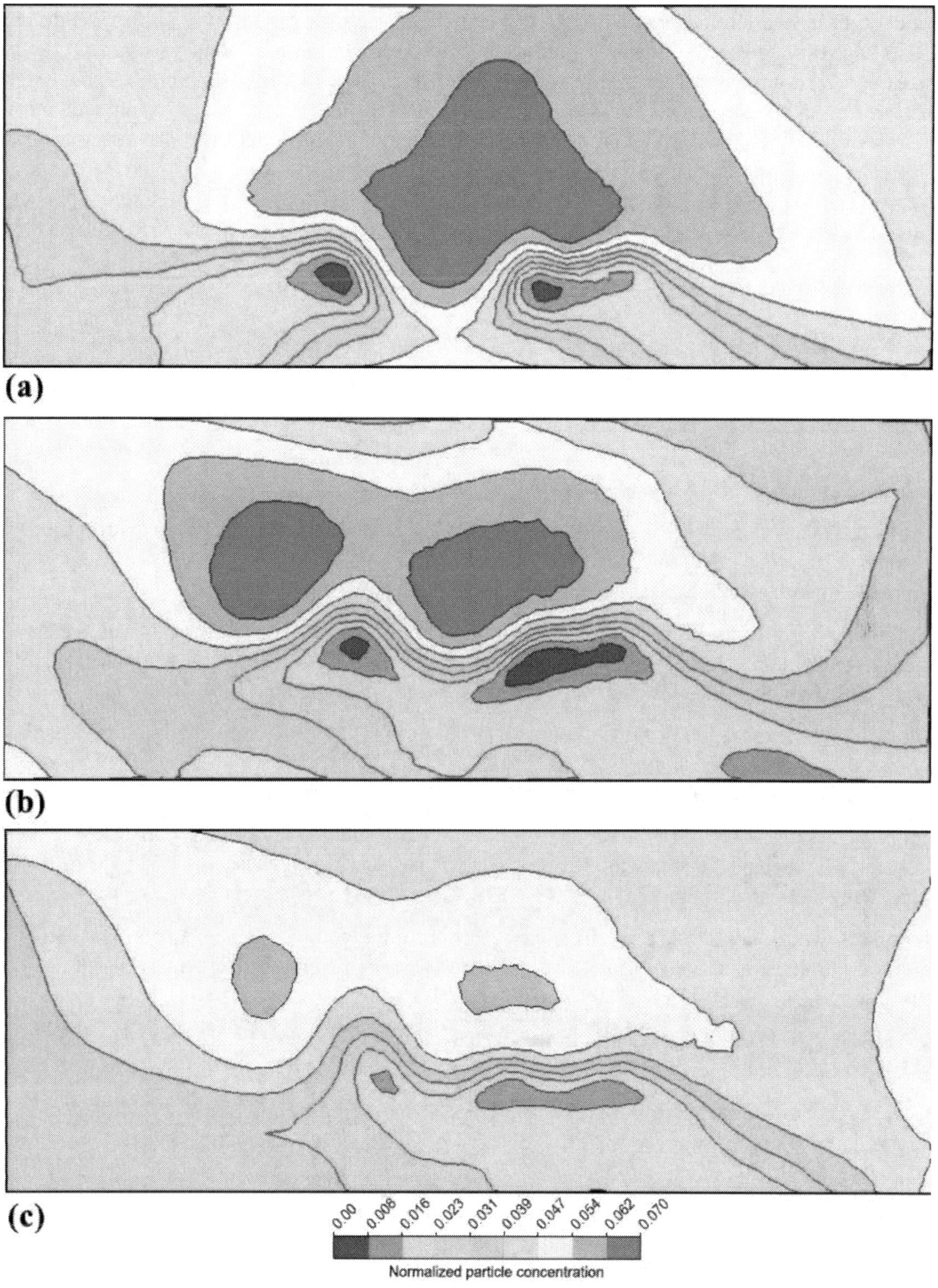

Figure 3: Normalized particle concentration over a vertical plane, x = 1 m . Air-exhaust opening at: **a)** ceiling level, **b)** floor level, **c)** ceiling-floor level

4 CONCLUSIONS

Ceiling-floor exhaust opening had the best performance in terms of recovery-test and particle simulation. This exhaust-mounting strategy achieved the lowest possible particle concentration and shortest recovery time. Positioning multiple air-exhaust outlets inside the OR is highly suggested. Floor-level outlets are more efficient at removing various anesthetic gases and infectious particles that may be heavier than the air. The ceiling exhaust openings that are mounted high on the wall are the most effective at removing any bacteria or other pathogen-containing particles and chemical substances that may become airborne during a surgical procedure. The optimum layout is both ceiling- and floor-height fixed exhausts at every corner of the OR. That makes a total of four floor- and four ceiling-mounted exhausts. This arrangement contributes to a more uniform airflow pattern within the OR. This layout avoids producing big eddies that may trap particles for a long time.

5 REFERENCES

14644-3:2005, EN ISO (2006). *Cleanrooms and associated controlled environments: Part 3: Test methods (ISO 14644-3:2005).*

Anon (2008). *DIN standard 1946-4, Ventilation and air conditioning – Part 4: VAC systems in buildings and rooms used in the health care sector.*

Cao G (2014). A review of the performance of different ventilation and airflow distribution systems in buildings. *Building and Environment*, 73:171–186. http://linkinghub.elsevier.com/retrieve/pii/S036013231300365X

Chow TT, Wang J (2012). Dynamic simulation on impact of surgeon bending movement on bacteria-carrying particles distribution in operating theatre. *Building and Environment*, 57:68–80. http://linkinghub.elsevier.com/retrieve/pii/S0360132312001266

Chow TT, Yang XY (2003). Performance of ventilation system in a non-standard operating room.*Building and Environment*, 38:1401–1411. http://linkinghub.elsevier.com/retrieve/pii/S0360132303001550

Habre W, Asztalos T, Sly PD, Petak F (2001). Viscosity and density of common anaesthetic gases: implications for flow measurements. *British Journal of Anaesthesia* 87:602–7. http://www.ncbi.nlm.nih.gov/pubmed/11878731

Kruppa B, Ruden H (1996). The influence of various air exchange rates on airborne particles and microorganisms in conventionally ventilated operating rooms. *Indoor Air* 6:93–100 http://doi.wiley.com/10.1111/j.1600-0668.1996.t01-2-00005.x

Noble WC (1975). Dispersal of skin microorganisms. *British Journal of Dermatology* 93: 477–485. http://doi.wiley.com/10.1111/j.1365-2133.1975.tb06527.x

Noble WC, Lidwell OM, Kingston D (1963). The size distribution of airborne particles carrying micro-organisms. *J Hyg (Lond)* 385–391.

Sadrizadeh S, Tammelin A, Ekolind P, Holmberg S (2014). Influence of staff number and internal constellation on surgical site infection in an operating room. *Particuology* 13:42–51. http://linkinghub.elsevier.com/retrieve/pii/S167420011300223X

Zhao B, Wu J (2005). Numerical investigation of particle diffusion in a clean room. *Indoor and Built Environment* 14:469–479. http://ibe.sagepub.com/cgi/doi/10.1177/1420326X05060190

STRATEGIES FOR THE PLANNING AND IMPLEMENTATION OF AIRTIGHTNESS ON EXISTING SLOPED ROOFS

Wilfried Walther

Energie- und Umweltzentrum am Deister (e.u.[z.])
Zum Energie- u. Umweltzentrum 1
D-31832 Springe, Germany

ABSTRACT

A project at the Energie- und Umweltzentrum (e.u.[z].) Springe looked into strategies how insulation and sealing components can be installed in existing constructions to improve the best airtightness.

KEYWORDS

Air tightness issues, building physics, vapor retarder, Quality assurance

1 INTRODUCTION

Figure 1: Melting snow in some regions shows a bad airtightness

Sloped roofs on existing buildings have a number of penetrations, connecting walls and beams with poor airtightness.
The attics are completely developed, with a layer of plaster applied to baseboards on the inner side of the rafters. Unfortunately, connecting light-construction interior walls and the visible beam construction so frequently penetrate the air tight layer that an n_{50}-value of app. 3.5 1/h was the best that could be done despite comprehensive improvements.

Now, roof renovation from the outside should considerably improve that value.

Figure 2: View Eastern side – main e.u.[z.] building, constructed in 1925

2 FOCUS AND AIRTIGHTNESS CONCEPT

When adding insulation from the outside (externally) of a roof U-values below 0.20 W/(m²K) are relatively easy to attain with a combination of completely insulated rafters and rooftop insulation. Though, it is not self-evident that the airtightness is improving. In many cases we don't know, whether there are problems or not. A measurement of the building's airtightness before renovation is a big help for the planning process. By looking for leaks based in the old airtightness layers, we can determine:

- Which connection and which layers are sufficient?
- Which ones need to be changed and improved?

Figure 3: Fog in a room and pressurizing shows the air passage from the inside to the outside

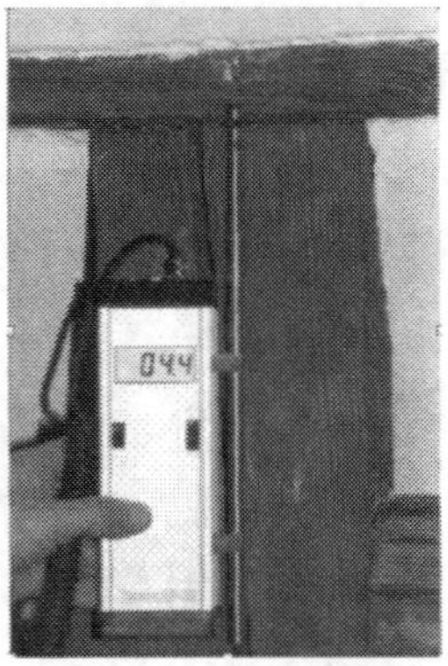

Figure 4: Depressurizing shows airflow on gaps of timber construction

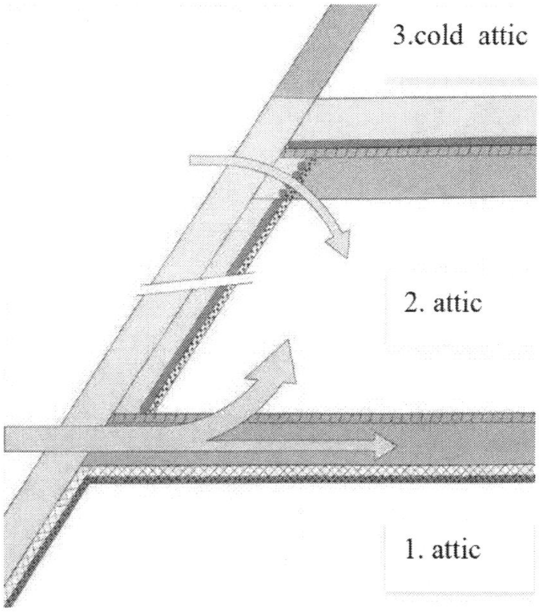

Figure 5: Airpath from the outside through the leaky construction

In our object we find following materials air tight:
- Plaster of walls in the ground floor
- Plaster of walls on eaves storage space
- Concrete ceiling of the ground floor

Not air tight:
- sealing to roof windows
- sealing to light-construction and
- Gaps from timber that penetrates the wood-wool slab
- Gaps from electric cables and pipesac
- Grove and tongue boards in the attic floor

It is quite obvious, that the airtightness must be achieved in the surface area from the exterior and not through sealing from indoor.

Figure 6: After removing the roof tiles you can see the big cavity where the floor structure between the 1. attic und 2. attic is located.

To characterize the air permeability of the building it is not so important to calculate the n_{50}-Value but rather air permeability with reference to the envelope's surface (q_{50}). Values above 3.0 $m^3/(hm^2)$ show us a surface that should be upgraded.

In our object, we measured a permeability of $q_{50} = 4,1\ m^3/(h\ m^2)$.

3 PLANNING AND INSTALLATION TIPS FOR ROOF SURFACE

The planning concept shows an additional airtightness layer over the whole roof. This layer must be bond to the ground floor ceiling, gable wall, roof windows, penetrations e.g.. The thermal insulation is situated between and over the rafters. Above the new timber there is a sheathing membrane, or wooden fiber sheeting plate with vapor permeable quality planned. A number of manufacturers of vapor barriers and airtight materials offer three variants to improve airtightness.

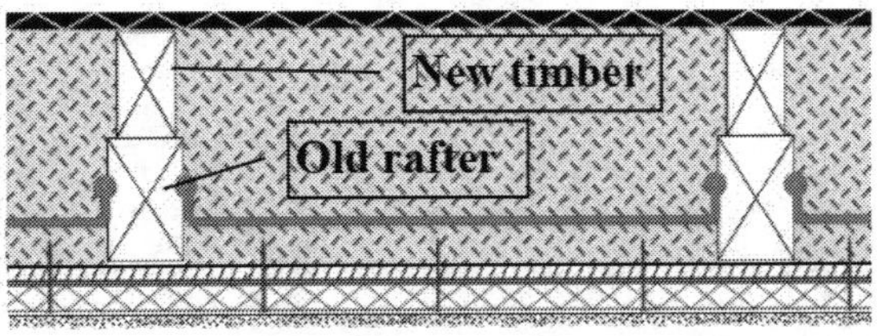

Figure 7: Var. I – airtight and vapor barrier (red) fixed with adhesive bonding compound from rafter to rafter.

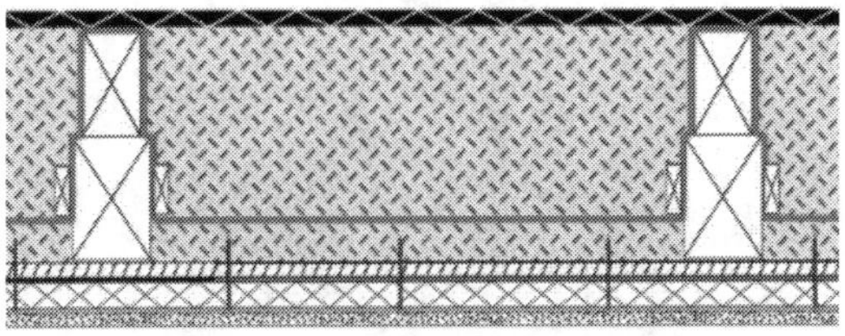

Figure 8: Var. II – airtight and variable vapor barrier (red) "sub and top" system. The layer can as well situate between the rafter and timber.

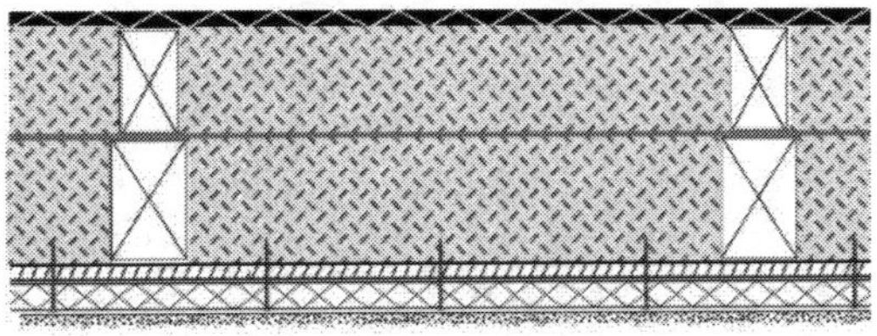

Figure 9: Var. III – airtight layer and water vapor permeable on the old timber, covered with insulation.

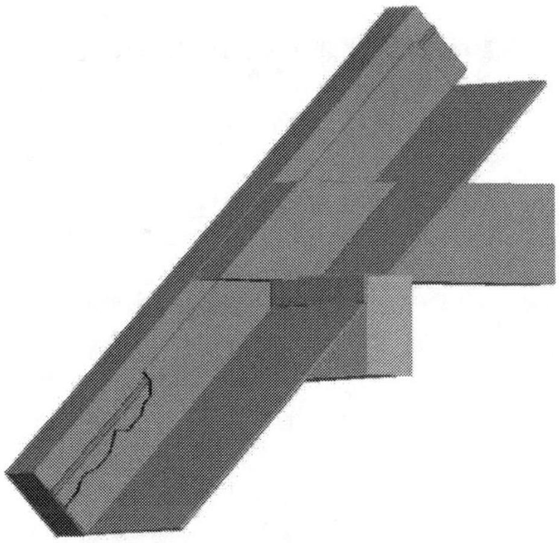

Figure 10: Var. I – Airtightness from rafter to rafter with collar beams (Source: WTA, Airtightness of buildings)

Var. I needs a lot of bonding-material for sealing gaps and jags from rafter to rafter (most expensive). It will only succeed across small surfaces without collar beams, roof windows or dormers.
For this reason this is not a useful solution.

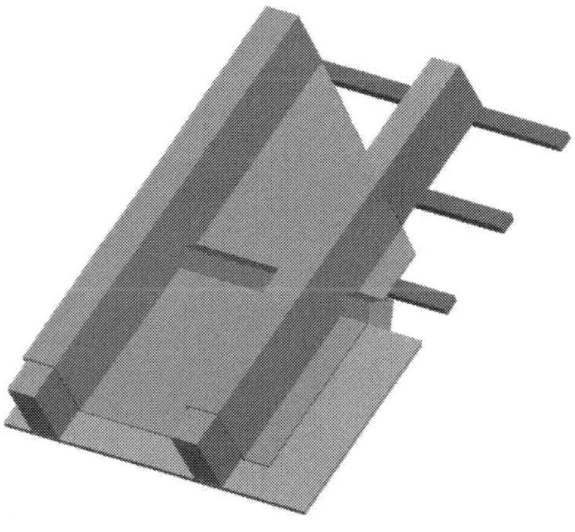

Figure 11: Var. II – Airtightness with sub-top implementation (Source: WTA, Airtightness of buildings)

Var. II solves this problem of sealing gaps and jags. The air barrier, installed on the bottom, runs by the rafter to the cold site covering the whole rafter und runs to the warm site to the bottom. In terms of building physics the layer must be a variable vapor retarder. This material becomes more permeable if there is a hight relative humidity around it. This characteristic is needed on the outside (cold side) of a construction. The same material reacts on the warm site (dry condition) like a vapor retarder.

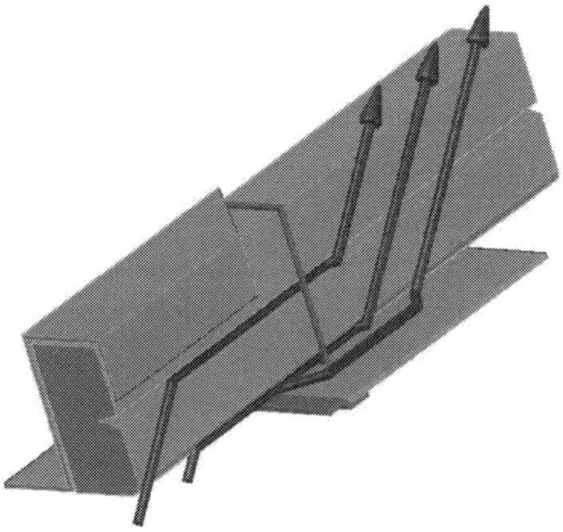

Figure 12: Gaps by alteration the airtightness with sub-top to a system of indoor airtightness. There are a lot of air paths (Source: WTA, Airtightness of buildings)

The "sub and top" system has three disadvantages:

- The bonding to roof-Windows is difficult
- The bonding and sealing to the eaves is difficult, because the rafters do penetrate the airtight layer
- The bonding and sealing by alteration to the inside airtightness (fig. 12)

Figure 13: Var. III Airtightlayer above the timber

A retaining ledge on the bottom of the rafter makes sure that the layer is fixed to the rafter and there are no gaps where warm air moves fast to the cold site of the timber. Experience on the construction site has shown that there are considerable tricky situations in applying adhesives for this variant and that the success of sealing systems should not be taken for granted.
Var. II is possible with a modest roof area with few roof windows and dormers,

Figure 14: Var. III (left) and Var.II (right) on site. In Var.III the rafter space is already filled with insulation (cellulose-fiber)

Var. III uses this opportunity: In terms of building physics it is possible to separate airtightness-layer and vapor barrier. The old construction with plaster and wood-wool slab has the character of the vapor retarder und the new layer above rafter has the character of the airtightness layer.

Rafter heads have to be cut off in variants II and III, and the eaves and roof overhang have to be constructed new.

Figure 15: Var. III airtightlayer above the original rafters

Var. III is advantageous in all respects. Connections and folds are minimized.

Figure 16: Var.III the rafter space is quiet filled with insulation (cellulose-fibre) and now the timber cavity will be filled with cellulose fibre.

The airtight layer must be covered with an insulation material, so there will be no condensation if warm air circulation from the inside moves to this layer.

4 PLANNING AND INSTALLATION TIPS FOR DORMER

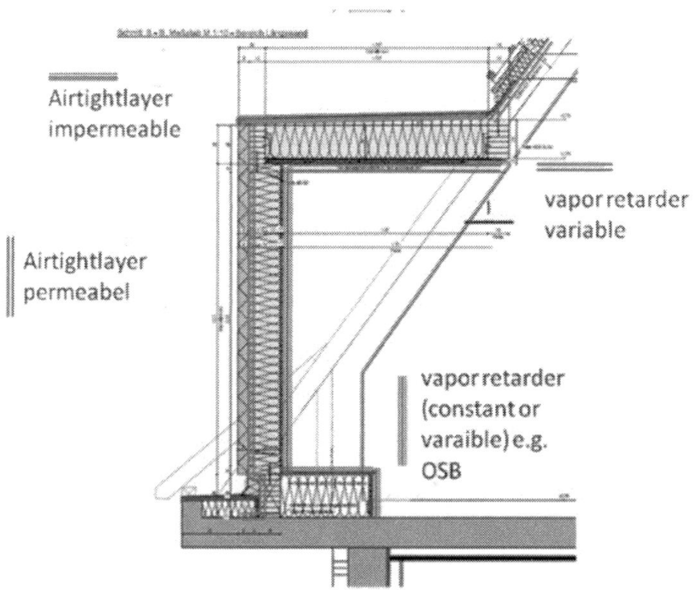

Figure 17: Horizontal section of a dormer, different kinds of vapor barriers are necessary

The new dormer elements (front, sides) also now have an airtight layer outside in accordance with the airtightness concept. It is covered with 8 cm wood fiber insulating board. Inside, the wall elements have a variable vapor retarder in the flat roof and an OSB panel, like those generally used in timber frame construction as a moderate vapor retarder in the wall.
The flat dormer roof has a impermeable airtight layer outside. In terms of building physics this construction is free of failure, if there is a variable vapor retarder inside and the roof gets solar heat in the summertime. The surface must be grey or of dark colour, a green roof can not be applied.

Figure 18: Connection of the dormer-base to the concrete Ceiling, different kinds of vapor barriers are necessary

The airtight layer must be sealed to the threshold and then to the concrete of the ceiling.
To realize that, it is necessary to sweep and vacuum-clean the ground. A primer is needed when the surface is still draggled. If the surface is to rough it is necessary to smooth the surface with plaster.

5 QUALITY ASSURANCE

The success of adhesives and sealants largely depends on whether the people to use them are informed. In advance of plans to review the targets for the maximum air permeability value q_{50} must be provided to support quality assurance. After renovation, measurements can be taken to determine success.

WTA Workgroup 6.14 (Wissenschaftlich-Technische Arbeitsgemeinschaft für Denkmal-pflege und Bauwerkserhaltung e.V.) "Luftdichtheit im Bestand" (Airtightness of buildings) aims to work up measurement procedures, target values for air permeability and planning tips. These issues were studied in this project.

The existing envelope permeability is measured with a value q_{50} = 4,2 m³/hm².

The new airtightness specifications of the layer should go below the value q_{50} = 1,5 m³/hm².

Under the assumption that the air permeability with reference to the envelope's surface (q_{50}) of the existing structure is equally spread over the surface of the building envelope, we can generate a new value q_{50} by relating the reverence q_{50} (1,5 m³/hm²) to the new/renovated surface area to the surface area that has not been changed.

The new default values q_{50} for the whole building are calculated according to the following equation (1):

$$q50\ (requir.) = \frac{\text{q50 (existence)} \times \text{Area(existance)} + \text{q50 (new)} * \text{Area (new)}}{\text{Area (existance)} + \text{Area (new)}} \quad (1)$$

Figure 19: After renovation, 2011

After the completion of the reconstruction we measure a n_{50} of 1,5 1/h just reaching the target-setting. The high demands of Passivhouse-Standard with a q_{50} of 0.6 till 0.8 m³/hm² could not be achieved.

Here we see the limits of what can be done in current practice and on the other hand the need of quality control.

6 CONCLUSIONS

In the project the discussion with planners and craftsman has show us, that the success of adhesives and sealants largely depends on whether people are well informed. The knowledge about which material is the airtight layer and which sealants are not tight helps construction workers a lot.

To inspire planners and craftsman it is of advantage that targets for the maximum air permeability value q50 are provided. After the renovation work has finished, measurements should be taken to determine successful airtightness.

7 ACKNOWLEDGEMENTS

This project has been realized in different workshops with the help of voluntary craftsmen and planners during 2006 until 2011 where we developed good practice solutions. Many thanks to all of you.

8 REFERENCES

WTA Guideline E-6-9 (2014). Airtightness of buildings, Part 1: General principles of planning; WTA Publications, email: wta@wta.de; (available in English 2015)

WTA Guideline E-6-10 (2014). Airtightness of buildings, Part 2: Planning details; WTA Publications, email: wta@wta.de; (available in English 2015)

WTA Guideline E-6-11 (2014). Airtightness of buildings, Part 3: Measuring procedure; WTA Publications, email: wta@wta.de; (available in English 2015)

ACH AND AIR TIGHTNESS TEST RESULTS IN THE CROATIAN AND HUNGARIAN BORDER REGION

Dr. László Fülöp[1], György Polics[2],

1 Faculty of Engineering and Information Technology, University of Pécs, Hungary
fulop@pmmik.pte.hu

2 Faculty of Engineering and Information Technology, University of Pécs, Hungary
gyorgy.polics@pmmik.pte.hu

ABSTRACT

The article presents the results of our research, which was realized under a cooperation project between the University of Pécs, Hungary and the University of Osijek, Croatia. The aim was to gather 50 Pa ACH, air tightness and spontaneous ACH information of residential houses by the Croatian and Hungarian border. The budget of the project allowed approximately 50 tests for each university; these summarized results are presented together with correlations found between the results.

KEYWORDS

Blower Door; tracer gas, ACH; leakage; air tightness

1. INTRODUCTION

For determining ACH and air tightness of the buildings, we have used Blower Door tests, which have two types:

Type "A" which can serve for checking the building in use, and provides ACH information in case of 50 Pa pressure difference. In this case, the purposely made vents on the building envelope (e.g.: chimneys, ventilation shafts, etc.) are not sealed. Concerning this case there have been measurements conducted on both sides of the border.

Type "B" Blower Door, which provides information on the air tightness of the building, at standard 50 Pa pressure difference. Before the test, each of the purposely installed vents on the building envelope must be sealed. The result can be applied for qualification, according to the air tightness limits. Currently, in Hungary air tightness testing is a requirement only for passive house qualification, but in the future it is expected to be introduced also as part of the standard certification procedure for all new buildings. Type "B" tests have been conducted only on the Hungarian side of the border. This test, with a little complementation, could be suitable for exploring air tightness faults as well,: using e.g. smoke test, thermography, air velocity meter, etc.

Blower Door tests presented above are valid at 50 Pa pressure difference, which shall not provide information about actual spontaneous ACH. An interesting field of research is to examine how we can derive the spontaneous ACH number of the building from 50 Pa ACH, which can be derived for Hungary as 4 Pa for annual average pressure difference.

Another possible testing method for measuring spontaneous ACH is using tracer gas, which has an advantage as compared to Blower Door is that this test method does influences the test results to the least extent. During our research we have conducted tracer gas tests as well on the Hungarian side of the border, and we have chosen SF_6 gas for this purpose.

2. TYPE "A" BLOWER DOOR TEST REUSLT

During choosing the buildings in order to gain somewhat representative results for a certain area, we have focused only on residential houses on both sides of the border, although we have conducted some tests for a single room, or office.
As far as the construction method of the buildings concerned, residential houses belonging to a period of 100 years have been tested; therefore they included brick, adobe, panel-block, and light-weight buildings.

Table 1 presents the usual ACH rates according to the recommendation of the German Passive House Institute:

Table 1. Usual ACH rates

magnitude of n50 ACH	n50 [1/h]
old building	7..
today's new buildings HU	5..10
today's new buildings DE	2..6
low energy consumption house	0,17..5
passive house	0,17..0,6

The summary of test results according to table 3, as presenting Croatian results and table 4. presenting Hungarian results in a similar form. The buildings are shown in chronological order in the table, indicating their year of reconstruction (first of all referring to the change of windows and doors), and the 50 Pa type "A-type" Blower Door test results. In order to offer better overview, according to figure 1, the summarized results of the 2 countries are shown by point charts, where we have fitted trend lines on the summarized ACH results. It can be observed that there is a significant deviation for a more than one hundred years period: at 50 Pa ACH it is changing between 1 and 20. However, evaluating the trend lines, a consequent decrease can be observed in parallel with the increase of the age of the buildings; this decrease met our previous assumptions, since, as time went on construction standards became ever stricter and together with energetic requirements and more up-to-date construction technologies together resulted in the decrease of ACH.

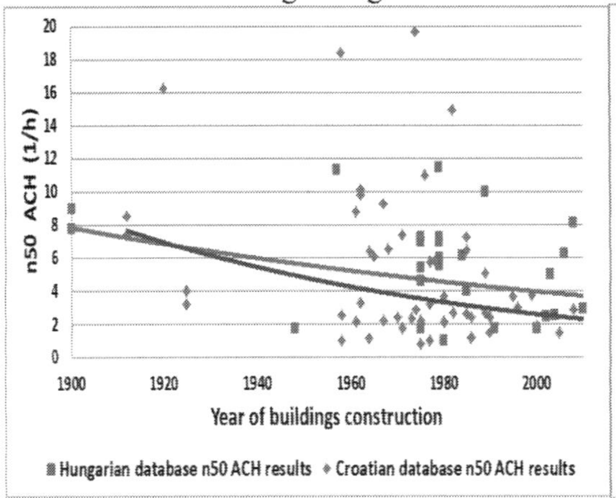

Fig 1. n50 ACH values in all tested houses

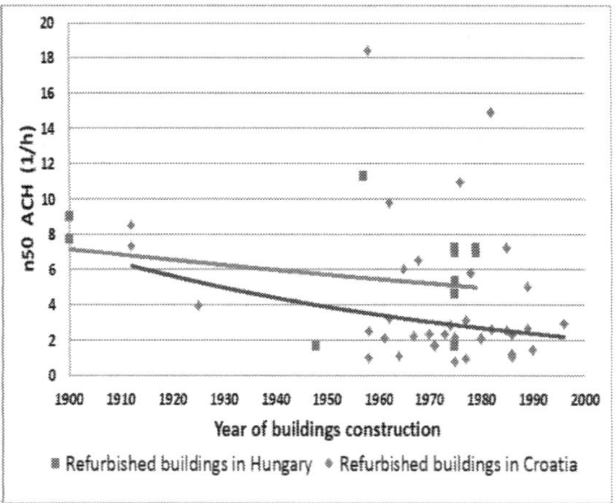

Fig 2. n50 ACH values in refurbished houses

Table 2. Croatian test results

Year of construction	NO.	Type of building	Year of construction	Year of reconstruction	ACH, n50 [1/h]	Average ACH, n50 [1/h]
Before 1945	1.1	Apartment house	1912	2007	7,35	7,84
	1.2	Apartment house	1912	2009	8,53	
	1.3	House	1920	-	16,21	
	1.4	Terraced House	1925	2006	3,94	
	1.5	House	1925	-	3,16	
1945.-1965.	2.1	Terraced House	1958	2011	2,51	6,31
	2.2	Apartment house	1958	1981	18,39	
	2.3	Apartment house	1958	2001	0,99	
	2.4	Terraced House	1961	2007	2,12	
	2.5	Terraced House	1961	-	8,77	
	2.6	House	1962	1994	3,22	
	2.7	House	1962	-	10,12	
	2.8	House	1962	1993	9,77	
	2.9	Apartment house	1964	-	6,39	
	2.10	Apartment house	1964	2004	1,13	
	2.11	Terraced House	1965	1975	6,01	
1966.-1975.	3.1	Apartment house	1967	2010	2,2	5,08
	3.2	Apartment house	1967	-	9,26	
	3.3	House	1968	2008	6,49	
	3.4	Apartment house	1970	2010	2,35	
	3.5	Apartment house	1971	2010	1,71	
	3.6	Apartment house	1971	2005	1,69	
	3.7	Apartment house	1971	-	7,35	
	3.8	Apartment house	1973	2013	2,32	
	3.9	House	1974	1998	2,86	
	3.10	Apartment house	1974	-	19,64	
	3.11	House	1975	2003	0,76	
	3.12	Apartment house	1975	-	7,23	
	3.13	House	1975	1998	2,18	
1976. - 1985.	4.1	House	1976	2010	10,95	5,24
	4.2	Apartment house	1977	2007	0,94	
	4.3	Apartment house	1977	2006	3,14	
	4.4	Apartment house	1977	-	5,73	
	4.5	Apartment house	1978	2003	5,77	
	4.6	Apartment house	1980	2002	2,13	
	4.7	Apartment house	1980	-	3,63	
	4.8	Apartment house	1980	2009	2,07	
	4.9	House	1982	2012	2,63	
	4.10	Terraced House	1982	2012	14,91	
	4.11	Terraced House	1985	1994	2,56	
	4.12	Apartment house	1985	1994	7,21	
	4.13	Apartment house	1985	-	6,41	
1986.-1995.	5.1	Apartment house	1986	2011	2,34	2,47
	5.2	Apartment house	1986	2008	1,07	
	5.3	Apartment house	1986	2004	1,2	
	5.4	Apartment house	1989	2007	2,64	
	5.5	Apartment house	1989	2007	5,02	
	5.6	House	1990	2009	1,45	
	5.7	House	1990	-	2,37	
	5.8	Apartment house	1995	-	3,63	
1996. - 2005.	6.1	Apartment house	1996	2007	2,96	2,48
	6.2	House	1999	-	3,69	
	6.3	House	2000	-	1,81	
	6.4	House	2005	-	1,45	
From 2006	7.1	House	2008	-	2,82	2,99
	7.2	House	2011	-	0,86	
	7.3	House	2011	-	3,4	
	7.4	House	2013	-	4,88	

Table 3. Hungarian test results

Year of construction	NO.	Type of building	Year of construction	Year of reconstruction	ACH, n50 [1/h]	Average ACH, n50 [1/h]
before 1945	1.1	House	1900	2012	7,75	8,38
	1.2	Room	1900	2012	9	
1945.- 1965.	2.1	House	1948	2012	1,68	6,49
	2.2	House	1957	1994	11,3	
1966.-1975.	3.1	Office room	1975	2008	6,94	5,16
	3.2	Office room	1975	2008	4,59	
	3.3	Office room	1975	2008	5,36	
	3.4	Library	1975	2008	1,68	
	3.5	Classroom	1975	2008	7,21	
1976. - 1985.	4.1	House	1979	-	11,42	6,02
	4.2	Room	1979	-	6	
	4.3	Room	1979	-	5,5	
	4.4	House	1979	2010	7,23	
	4.5	Room	1979	2010	6,96	
	4.6	Room	1980	-	1	
	4.7	House	1984	-	6,08	
	4.8	Flat	1985	2008	4	
1986.- 1995.	5.1	House	1989	-	10	5,84
	5.2	House	1991	2011	1,68	
1996. - 2005.	6.1	House	2000		1,68	2,89
	6.2	House	2002	-	2,41	
	6.3	House	2003	-	5	
	6.4	House	2004	-	2,48	
from 2006	7.1	House	2006	-	6,25	4,65
	7.2	House	2008	-	8,1	
	7.3	House	2010	-	2,89	
	7.4	House	2011	-	3,19	
	7.5	House	2011	-	8	
	7.6	House	2011	-	1,86	
	7.7	Room	2011	-	3,45	
	7.8	Room	2012	-	3,49	

In case of the flats where ACH is outstandingly high (over 10/h) this phenomenon is explained by the use of open-chamber furnaces or fire places, or the existence of uncontrolled ventilation shafts. At the same time, the too low ACH figures can be concerning, too (i.e. under 2/h) in case the flat did not have artificial ventilation system, but they can be considered as traditional window-ventilation buildings. According to the trend lines, as time went on the Croatian ACH results are lower than those of measured on the Hungarian side of the border, however, studying tables 2 and 3, it is clearly seen, that following the Bosnian war, nearly all of the Croatian flats were renovated, unlike the ones on the other side of the border. Therefore the interim renovation of flats can be a significant factor of influence. This uncertainty can be eliminated by limiting the results, and presenting only the results of the flats that had been renovated earlier, according to the results of figure 2. In the figure presented this way deviance can still be observed, and the results still show a decreasing tendency in favour of the Croatians. However, this statement is based on a sample of relatively small number of flats that had been renovated on the Hungarian side of the border.

3. DETERMINING SPONTANEOUS ACH BY TYPE "A" 50 PA BLOWER DOOR TEST

Tectitle 3.6 software provides test results together with trend line as shown in figure 3 (presenting the results of the house 7.6 in table 3.)

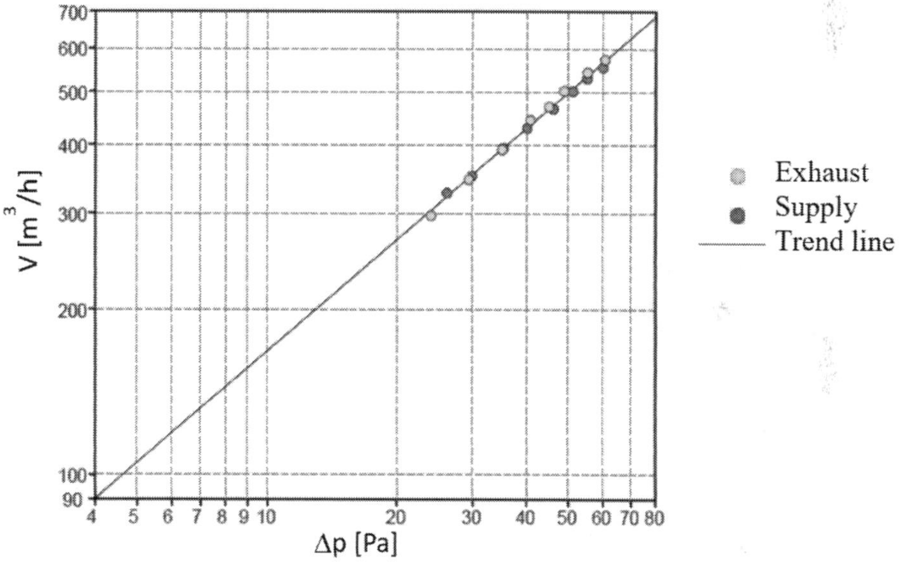

Averages:								
Δpn [Pa]	60	55	50	45	40	35	30	25
Δp [Pa]	60,6	55,3	51	44,7	39,9	35,4	30,2	25,3
V [m3/h]	760.8	722.3	685.5	634.5	587.0	542.4	484.1	430.7

Fig 3. Extrapolating ACH values by the trend line of the test results

The software provides both the multiplying factor and the exponent of the describing equation belonging to the trend lines created for both pressurization and depressurization, or else they can be even derived exporting the software results (measured pressure difference and the related air flow) by an excel program, such as seen in figure 4.

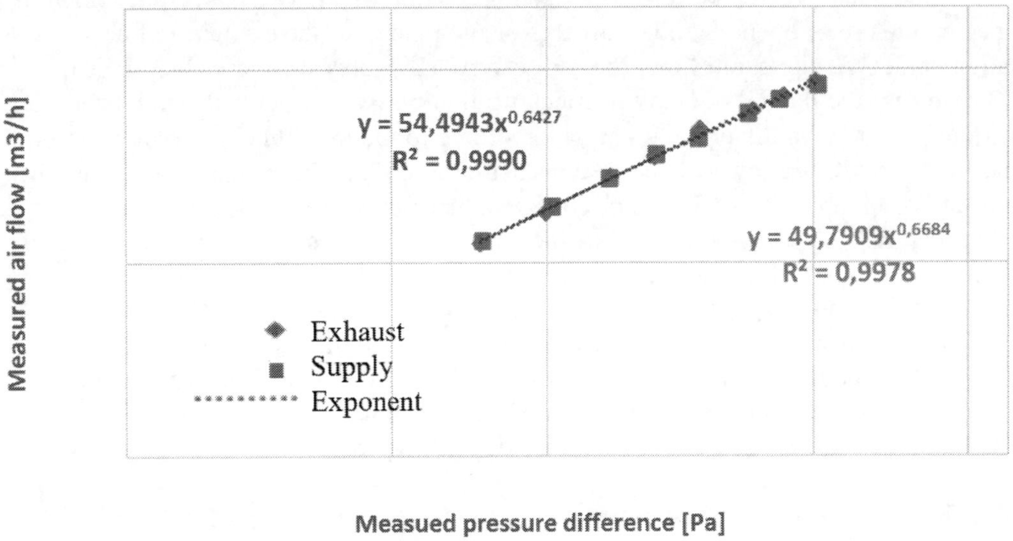

4. figure

Next, the values of air change number typical for the building can be calculated. The following results were yielded for the detached house tested:

- In case of overpressure (supply):
  The exponent of the descriptive equation: n=0,6427
  The multiplying factor of the describing equation: C=54,4943
  ACH for 50 Pa pressure difference: n_{50}=1,85 1/h
- In case of depression (exhaust):
  The exponent of the descriptive equation: n=0,6684
  The multiplying factor of the describing equation: C=49,7909
  ACH for 50 Pa pressure difference: n_{50}=1,87 1/h
- In case of the mean of depression (exhaust)and overpressure (supply):
  The exponent of the descriptive equation: n=0,6501
  The multiplying factor of the describing equation: C=53,1409
  ACH for 50 Pa pressure difference: n_{50}=1,86 1/h

The equation describing the results:

$$\dot{V} = C \cdot \Delta p^{n} \quad [m3/h] \qquad (1)$$

Where:

$\dot{V}$ - the measured air flow by the Blower Door
C – the multiplying factor of the describing equation
Δp – pressure difference at the point of testing
n - The exponent of the descriptive equation

Following the 50 Pa type “A” Blower Door test, knowing the multiplication factor and the exponent of the describing equation depending on the characteristics of the buildings, ACH can be calculated of any pressure difference. A good approximation by statistic methods achieved despite the fact that some leaks can behave quite differently in case of diverse pressure differences. The most important factors are: turbulent or laminar flow or opening pressure difference.

For testing spontaneous ACH the annual natural pressure difference should be known which is quite dependent on the weather conditions, such as actual indoor and outdoor temperature difference or the prevailing wind force. In Hungary, the annual average pressure difference is 4 Pa, and taking it into account when using the equation (1) we could get quite accurate results for spontaneous ACH. Regarding the initial example, substituting the data:

With 4 Pa spontaneous pressure difference, the filtration ACH is:

$$\dot{V}_{4Pa} = C_L \cdot \Delta p^n = 53{,}14 \cdot 4^{0{,}65} = 130{,}8 \quad [m3/h]$$

With 4 Pa spontaneous pressure difference, ACH value is:

$$n_4 = \frac{\dot{V}_4}{V} = \frac{130{,}8}{363{,}2} = 0{,}36 \quad [1/h]$$

Completing the data of table 3 by the 4 Pa ACH calculated for brickwork houses:

Table 4. ACH values of traditional brickwork houses

No.	**Year of construction**	**Type of building**	**C**	**n**	**n50**	**n4**
			-	-	**[1/h]**	**[1/h]**
1	1900	House	96,9	0,563	7,75	1,87
2	1979	House	150,4	0,569	11,42	2,71
3	1979	House	243,9	0,595	7,23	1,61
4	1980	Room	6	0,632	1,10	0,23
5	1984	House	135,9	0,59	6,08	1,37
6	1985	Flat	87,5	0,649	4,85	0,94
7	1989	House	471,7	0,596	10,00	2,22
8	2002	House	24,5	0,596	2,41	0,54
9	2003	House	192,5	0,609	5,02	1,08
10	2004	House	64,1	0,664	2,48	0,46
11	2006	House	89,1	0,566	6,25	1,49
12	2008	House	162,9	0,601	8,10	1,52
13	2010	House	75,3	0,591	2,89	0,65
14	2011	House	57,7	0,636	3,19	0,64
15	2011	House	171,7	0,689	8,28	1,45
16	2011	House	53,1	0,65	1,86	0,36
17	2012	Room	15	0,628	3,49	0,72

It could be important to know that the expected 4 Pa spontaneous ACH can be determined quite precisely from the result of type "A" 50 Pa Blower Door ACH test results without forming any functions. For this, as shown in figure 5. using the results of table 4, imaging the measured 50 Pa and calculated 4 Pa pairs of figures, whereby a trend line is fitted indicating the equation describing the function concerned. Using this equation, any the 4 Pa spontaneous ACH can be determined from the relevant 50 Pa ACH result. The high standard deviation figure, R^2=0,9819 suggests that the results can be considered reliable. In order to provide better illustration the results are also shown by a bar chart by figure 6, compared to the calculated 4 Pa results, and the 4 Pa results calculated by the describing equation of figure 5. It can be seen in figure 6 that 4 Pa ACH can be calculated by approximately 90% accuracy.

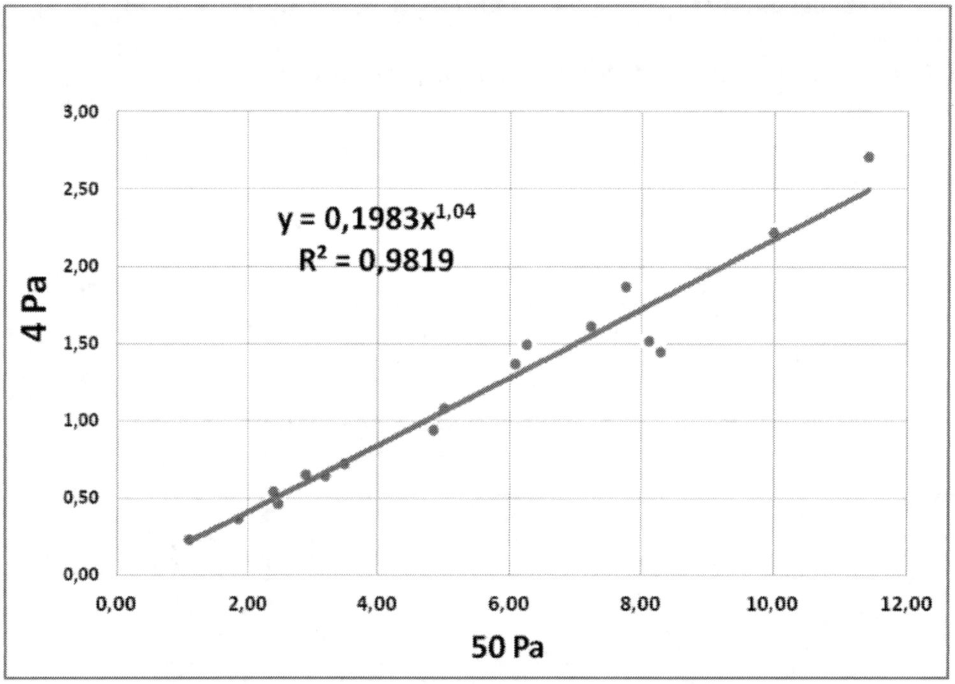

Figure 5. n4 ACH values as a function of n50 values

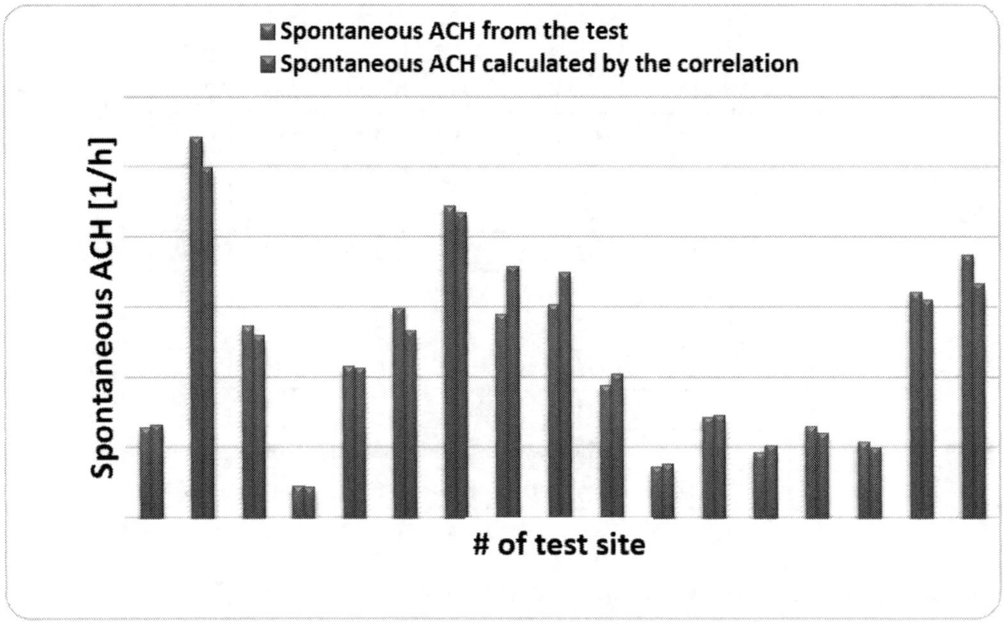

Fig 6. Comparison of test results and calculated ACH values

4. DETERMINING SPONTANEOUS ACH BY TRACER GAS:

For carrying out the tests, we used an INNOVA 1412i Photoacoustic Gas Monitor sold by LumaSense Technologies. The gas chosen for the tests is Sulphur-hexafluorid (SF6). A specific characteristic of the test is that it is hard to apply for measuring the ACH of multi-space buildings due to the lack of appropriate mixture of air, however, it is perfectly suitable for measuring the spontaneous ACH of only one space. SF_6 gas is approximately 5 times heavier than air, therefore during the test the appropriate mixture of the gas inserted should be provided by a fan. The test cycles as a function of time and their evaluation is presented in figure 7.

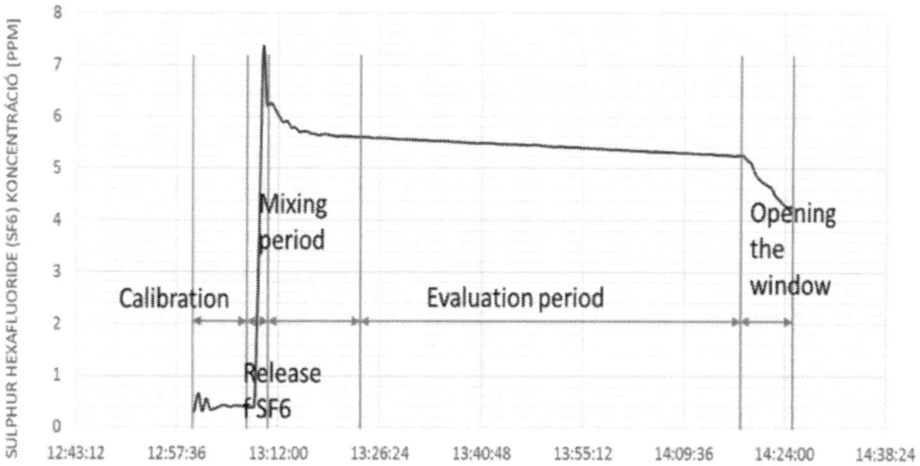

Fig 7. Procedure of tracer gas test

The evaluable period of the change of gas concentration is seen by the enlarged figure 8.

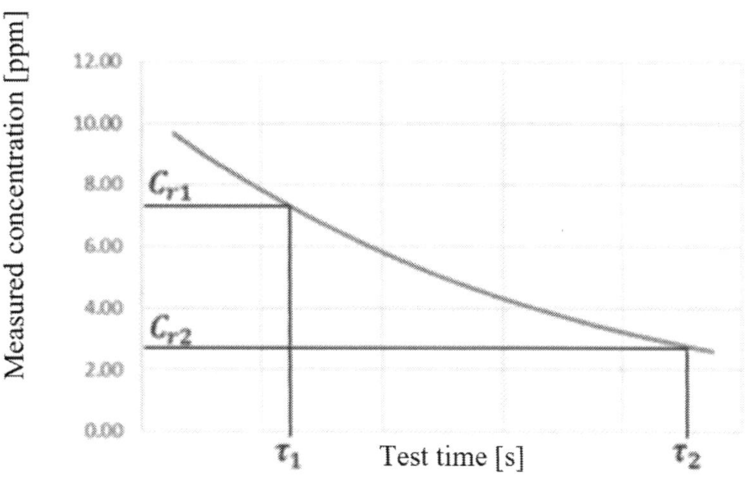

Fig 8. Concentration decay test method

The spontaneous ACH can be calculated if concentration decline is divided by the time elapsed:

$$n = \frac{\ln(C_{r1}) - \ln(C_{r2})}{\Delta\tau}\left[\frac{1}{h}\right] \qquad (2)$$

Where:

Cr1: is the concentration at the beginning of the phase suitable for evaluation
Cr2: a is the concentration at the end of the phase suitable for evaluation
Δτ: is the time elapsed in between

5. CORELATION BETWEEN BLOWER DOOR TYPE "A" 50 PA AND BLOWER DOOR TYPE "B" 50 PA

Blower Door Tests of both type "A" and "B" have been completed only in Hungary. The summarized results concerned are included in table 5 and figure 9.

Table 5. Comparison of A-type and B-type test results

no	Year of Construction	Type of building	A-type Blower Door n50 ACH [1/h]	B-type Blower Door n50 Air tightness [1/h]
1	1900	room	9,28	9,28
2	1979	house	11,42	10,11
3	1979	room	5,5	5,5
4	1979	room	6	6
5	1979	room	6,96	6,96
6	1980	room	1,1	1,1
7	1984	house	6,08	5,31
8	1984	flat	8,91	7,14
9	1989	house	10	10
10	2002	house	2,41	1,45
11	2004	house	2,48	2,4
12	2006	house	6,25	5,25
13	2008	house	8,1	6,92
14	2011	house	3,19	2,75
15	2011	house	1,86	1,38
16	2011	room	3,45	3,45
17	2012	room	3,49	3,49

Evaluating the results of the above table, it can be seen that if air tightness or ACH measures were completed only for one room, generally the results show no differences. The reason for this lies in the fact that in case of traditional window-ventilation spaces usually there are not further artificial leaks on the envelope of the building which should be sealed according to the requirements of air-tightness tests, and which would show changes in the results. This occurs in every such a case even in case of complete house test where there no artificial leaks on the building envelope that should be sealed. Limiting the data of table 5 focusing only the figures which show differences between type "A" and "B" test results, we have presented these pair of data in a graph in figure 10 where, by means of curve fitting, we have determined the equation describing the trend line. Since the deviation $R^2=0.9893$ is high, the equation can be considered reliable, thus, we have found a quasi correlation between the "A" and "B" ACH tests at 50 Pa. However, it is important to point out that there are several factors which can influence the outcome of the results e.g. the shape and size of the leak can result in significant differences as compared to their determined equation.

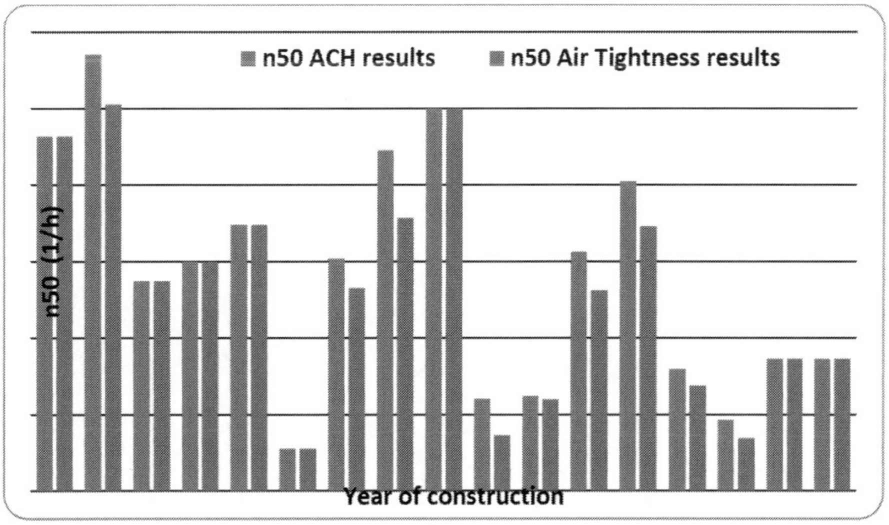

Fig 9. Comparison of A-type and B-type test results

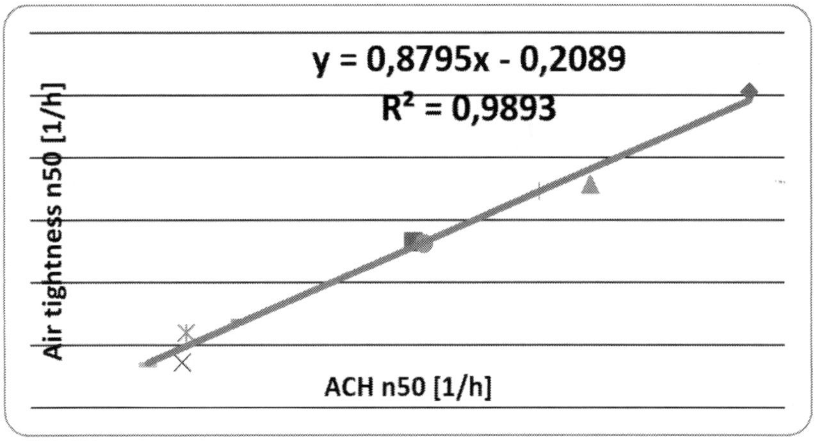

Fig 10. Comparison of A-type (ACH) and B-type (air tightness) test results

6. CONCLUSION

Results of air tightness and ACH tests in Hungary and in Croatia presented. The project was supported by the EU IPA cooperation project. Correlations found between the spontaneous ACH and Blower Door type "A" 50 Pa ACH results.

Correlations found between the results Blower Door type "A" 50 Pa and Blower Door type "B" 50 Pa tests. The research and results presented could be helpful for conducting measurements and evaluating their results.

Furthermore the tracer gas concentration decay test method is presented which is most suitable to determine spontaneous ACH. This method does not influence the tested parameters. However it cannot be applied to measure complex spaces simultaneously and it is especially sensitive to the changes of the external conditions.

7. REFERENCES

Fülöp, László, et al: Air tighness tests of buildings from the point of view of energy and comfort, IPA HUHR/1001/2.1.3/0009 project summary, Pécs, HU, 2013.

Polics, György PTE-PMMIK ÉG Tanszék, Summary and comparison of test results. IPA HUHR/1001/2.1.3/0009 project closing conference. Pécs, HU 27-28 June 2013

Luftdichte Projektierung von Passivhausen, CEPHEUS, Passivhaus Institut, 2002.

MEASURED MOISTURE BUFFERING AND LATENT HEAT CAPACITIES IN CLT TEST HOUSES

Ivana Katavic[1], Kristine Nore[*2], and Tormod Aurlien[1]

1 Department of Mathematical Sciences and Technology, Norwegian University of Life Sciences
Universitetstunet 3
1430 Ås, Norway

2 Norwegian Institute of Wood Technology
Forskningsveien 3B
0373 Oslo, Norway
**kristine.nore@treteknisk.no*

ABSTRACT

This research investigates the significance of the moisture buffering and latent heat capacities in exposed cross-laminated timber (CLT) walls with the respect to indoor climate and energy consumption. Hygroscopic materials have the ability to accumulate and release moisture due to change in the surrounding humidity. The moisture buffer capacity is regarded as this ability to moderate, or buffer, the indoor humidity variations. Latent heat refers to the heat of sorption due to the phase change from vapour to bound water in the material and the other way around. The indoor relative humidity (RH) is closely related to indoor comfort, more specifically to thermal and respiratory comfort and perceived air quality. Both persistently low RH (<20 %) and high RH (>75 %) can cause health threats for humans such as respiratory infections or the growth of mould. Wood is naturally hygroscopic, which enables it to act passive and efficient to stabilize the indoor humidity and thus, temperature variations. A better understanding and more deliberate implementation of these properties could potentially reduce the need for ventilation and heating without compromising the indoor comfort. A full-scale experimental study compares the responses of two 25 m2 test houses to an applied moisture load. The test houses are identical constructions made of CLT elements, where the exposed spruce interior in module A is kept as is, while the interior in module B is covered with PE foil emulating an impermeable surface. The moisture load is applied as continuous flow of mist under several various conditions, including different magnitudes of the load and altered air change rates of the ventilation fan. The responses are measured in terms of RH, moisture content in the wood, changes in surface and air temperature and the time needed for the system to restore itself to the initial state. The software WUFI®Plus is utilized to perform a hygrothermal building performance simulation for comparison and evaluation. Both the experimental and calculated results show that exposed wood is an efficient moisture buffer capable of reducing daily fluctuations in RH. The results also show a rise in surface temperature, which is a contribution from the latent heat of sorption.

KEYWORDS

Moisture buffer capacity, latent heat, indoor climate, energy, buildings, exposed wood surface, Cross Laminated Timber, (CLT)

1 INTRODUCTION

In the last decades a global "green awakening" has led to new standards and requirements concerning environmental awareness and sustainable development. According to the International Energy Agency (IEA), commercial and residential buildings represent 32 % of the worlds total final energy consumption (IEA 2014). In reducing this figure, the potential is mainly related to reducing the need for heating and cooling. The public and legislative demands for energy efficient building design have led to increased air tightness and insulation in modern dwellings. In these modern buildings, the ventilation accounts for a large amount of the heat loss through air renewal. One solution to keep the energy consumption down whilst providing a comfortable and healthy indoor climate is the use of passive design strategies. In its simplest form, passive methods include considerations such as shape and orientation of buildings to reduce the ventilation, cooling and/or heating demands. Thermal inertia associated with heavy walls is another type of passive design that is actively implemented. Thermal mass design takes advantage of the heat capacity, which enables heavy buildings to heat up and cool down up to three times slower compared to light buildings (Karlsson 2012). Well-controlled HVAC systems or ventilation that is RH or CO2 sensitive are other options.

Studies on the moisture buffer capacity of hygroscopic materials have shown that these materials have a great ability in terms of moderating the indoor RH levels (Rode & Grau 2008). Moisture buffering is described as the ability of porous materials to buffer changes in the RH by absorbing and desorbing water vapour from the surrounding air. Indoor humidity varies significantly through the day and seasons. Materials that can store and release moisture can reduce the extreme values of these fluctuating humidity levels. This results in improved thermal comfort while keeping the energy consumption low.

Another phenomena closely connected to the moisture buffer capacity, is the latent heat of sorption. In the process of moisture exchange between the hygroscopic material and surrounding air, the humidity undergoes a phase change. The energy associated with this phase change from vapor state in air to liquid water in hygroscopic pores, or vice versa, is the latent heat of sorption. (Osanyintola. & Simonson, 2006).

This paper aims to measure the significance of the moisture buffering capacity and latent heat of sorption of exposed spruce CLT in full-scale experimental facilities. The experimental facilities include two test houses: one with exposed spruce interior surface; and one control house with impermeable surface where moisture buffering and latent heat are limited. The responses of the test houses are compared to one another, as well as to results from a computer performed hygrothermal whole building performance simulation using WUFI®Plus (2007) in order to evaluate these models.

2 BACKGROUND

2.1 Indoor climate

The main purpose of air ventilation is to ensure adequate indoor air quality (IAQ) for users with regard to health and comfort (TEK10, 2010), and to keep the indoor humidity at correct levels to maintain the building structure and envelope. The indoor RH is affected by different factors, such as internal moisture sources (human activity and respiration, household appliances and equipment), airflow, leakages and external air moisture content. There is also a significant seasonal impact on the RH, more dependent on outdoor temperature than outdoor humidity. Heated homes are usually dry because the infiltrating cold winter air contains little water vapour. RH reaches its lowest levels during the coldest days. During hot summer days, the outdoor air contains plenty of vapour even if the RH is lower than in winter. But because of high vapour pressure outdoor, vapour generated inside hardly migrates out. (Kubler 1982)

Daily indoor moisture loads due to normal life activities leads to fluctuations and peaks in RH, which easily reaches 80-100 % in airtight spaces, like bathrooms. The moisture production in residences differs among people depending on habits and behaviour. The total moisture generation (sum of respiration and transpiration) of an adult at rest is in the range of

0.8-1.7 kg/day, while the total water vapour production (including daily chores, plants, pets etc.) for a family of five is in the range of 6.6-10.2 kg/day (TenWolde & Pilon 2007).

Keeping the RH at correct levels is vital for both the durability of the building materials and the indoor climate comfort. Surveys show that humans feel most comfortable at certain temperatures and humidities (ANSI/ASHRAE; ISO). The RH is important for skin humidity (Toftum et al. 1998a), respiratory comfort (Toftum et al. 1998b) and perceived IAQ. High RH can be associated with moisture problems in the building envelope, such as mould growth, as well as human health problems, including asthma and allergies. Arlian (Arlian et al. 1999) suggests maintaining mean daily RH below 50 % to effectively restrict population growth of house dust mites.

In a wide variety of commercial buildings the right levels of RH can be of even higher importance. Swimming halls and laundry facilities often have excess humidity, while offices and production facilities with heavy machines tend to be perceived as too dry. Museum and gallery artefacts require specific and steady humidity and temperature levels to minimize deterioration (Janssen & Christensen 2013).

2.2 Energy efficient design strategies

By 2020, the EU Energy Performance of Buildings Directive (2010/31/EU) aims for a 20 % reduction in European primary energy consumption and Nearly Zero-Energy Building norm for all new buildings. Mandatory energy performance certificates are already implemented in commercial property development, and recent studies show that better energy efficiency is rewarded in the market.

The energy consumption in dwellings can be divided into three main categories: electricity for lightning and equipment, room heating/cooling and water heating. In Norwegian households, an estimated 60 % of the consumed energy is used on room heating alone. (Edvardsen et al. 2006) In more southern climates, the cooling and air conditioning in the hot season is the main energy drain. Energy efficient development means utilizing plan strategies where the total purchased energy need for a building is kept at a minimum.

Sustainable development requires that the choice of materials for building take into account the environmental impact of the materials being used, the energy consumption of building and the indoor environment. Manufacture, use and disposal of wood is associated with low energy cost and low emissions. Furthermore, wood used as indoor surface material enables its hygrothermal inertia to act as a passive system regulating the temperature and moisture fluctuations. Hameury and Lundström (2003) describe an experimental study performed in four occupied apartments in Sweden with large areas of exposed massive wood surfaces. The results show evidence of the wood contributing to buffer the indoor temperature.

A Canadian study estimated that applying hygroscopic materials in combination with RH and user presence control of HVAC systems reduces the heating energy consumption with 2-3 % and the cooling energy consumption with 5-30 % in moderate climates. (Osanyintola & Simonson 2006).

Compared with other materials, the heat conductivity of wood is low, especially perpendicular to the fiber axis. This makes (dry) wood a good insulator, but also poses as a limitation for heat storage purposes. Nevertheless, because wood is excellent at holding water, this captured moisture increases its basic heat storage capacity. This gives wood a beneficial compromise between insulation and heat storage. Since the thermal and hygroscopic behaviors of building physics are closely related, the wood moisture content also affects the thermal fluctuations through the material. A high moisture content enhances the heat flux.

2.3 Hygroscopic potential

The total moisture buffer capacity depends on the moisture buffer capacities of each material and furniture in the room together with the moisture production, air change rate and the ratio between the material surface area and the air volume. The materials active thickness, vapor permeability and moisture storage capacity, as well as the thickness of the boundary air layer are factors that determine the moisture buffering.

Moisture content in wood is expressed as a percentage of dry wood mass. Most hygroscopic properties in wood, including the wood surface – ambient air moisture exchange, are related to the hygroscopic moisture range where minimal capillary forces occur (Wood moisture content <30 %). The hygroscopic water is bound to the wood cells via hydrogen bonds, and its amount is limited by the number of sorption sites available and how many water molecules each site can hold. More energy is needed to release bound, as opposed to free, capillary water. As the bound-water content increases, the physical properties of the wood are altered: swelling, decrease in the mechanical strength, increase in thermal and electrical conductivity and higher rate of bound water diffusion (Siau, 1984).

In a hygrothermal simulation, Korsnes (2012) attempted to identify the magnitude of the latent heat exchange in a small bathroom under normal conditions. The comparison of case 1 with wood panels on the walls and ceiling to case 2 with solely impermeable surfaces shows huge advantages to the former. Not only are the values of RH lower and more stable, but there is also an increase in indoor air temperature of 2.5 °C due to the latent heat. In a follow-up laboratory test conducted to verify the hygrothermal simulation, Nore (2014) subjected wood samples in a climate chamber to rapid increase in RH (from 20 % to 90 % RH). Thermography was used to measure the surface temperature change. In a few minutes, the exposed wood sample reached its maximum surface temperature nearly 2 °C higher than the reference sample covered in low-emitting PE-foil.

The latent heat is naturally user controlled in the sense that a space has to be occupied or employed, with a following moisture load, in order to be released. This has potential as local heating in spaces when it is actually needed. For instance, both kitchens, bathrooms and laundry rooms, are only occasionally used during the day. If the operating temperature of a room can be raised with 2 °C upon being occupied, the set temperature can be lowered accordingly when the room is not in use.

The latent heat of desorption may also be of practical use in building design. Desorption is the opposite process of sorption; here excess moisture in the wood is dried out by energy contribution from the ambient air. This may be applied to buildings which need cooling during the day in hot seasons. By airing the building during the night when the temperature is lower and humidity higher, humidity can accumulate in the indoor surfaces. During day time, this moisture will require heat from the indoor air to vaporize thus lowering the indoor temperature. A known problem to cabin owners in Norway, is the time it takes to heat the wood cabin during winter times; lot of energy is needed to firstly dry out the walls before the indoor temperature rises.

3 METHODOLOGY

This paper assesses the issues of moisture buffering and latent heat of sorption of CLT in two different ways; measurements under controlled conditions in experimental field facilities; and hygrothermal whole building performance simulation with WUFI®Plus. The test houses have been exposed to large moisture loads, 616 g/h and 1232 g/h over a 9 h 25 min. period, in an attempt to aggravate a response with as high RH as possible. The studied scenarios are presented in Table 1. Case I and II are conducted in multiple runs. Case III is only performed in module A, meaning the permeable hygroscopic module. The initial conditions differ in each case, and are here only approximately presented.

Table 1: Presentation of test cases and initial conditions

	Moisture load		Initial conditions				Experimental		Simulated	
Case	Diffusion rate [g/h]	Total load [kg]	Operative temperature [°C]	Indoor RH [%]	WME [%]	ACH [h^{-1}]	A	B	A	B
I	616	5.8	20.2	23-24	7.4	0.5	x	x	x	x
II	616	5.8	20.1-20.5	32-37	7.4	0.3	x	x	x	x
III	1232	11.6	20.2-20.5	31	7.4	0.5	x		x	x

3.1 Field experiment

Figure 1: View of test the houses from northwest. Module B is seen in the front; module A in the rear.

The field facilities consists of two identical test houses, module A and module B (see Figure 1) situated in a meteorological field, Søråsfeltet, affiliated to the Norwegian University of Life Sciences (NMBU) in Ås, Norway. The test houses are constructed of CLT made of spruce. The walls are made up 100 mm CLT and externally insulated with 100 mm mineral wool in the south and east directions, and 150 mm mineral wool in the west and north directions. There is a layer of weather resistive barrier between the CLT and insulation, and on the exterior side of the insulation, which can be seen in Figure 1. This barrier is vapour diffusion-open. The ceiling and roof consist of 140 mm CLT and 250 mm mineral wool insulation with a wind barrier in-between, and sheet metal roofing on top. The floor has 14 mm oak parquet over 22 mm chipboard and 100 mm mineral wool insulation. The modules are placed on top of 200 mm Rockwool with vapour barrier inhibiting any moisture to penetrate from the ground. The internal dimensions of the modules are 7,.0 x 3,.6 metres and 2,.2 metres height from floor to ceiling. Table 2 shows the exact materials of the wall assembly.

The test houses do not fulfil the requirements demanded by the Norwegian Building Regulations (TEK 10) in terms of insulation, and are thus not comparable within these standards. Blower door tests ran in module A have shown that the air infiltration n is 1.63 and meets the national guideline requirements for residential houses with n < 2.5 (Olaussen 2014).

In module A the spruce walls and ceiling are kept exposed and untreated, in direct contact with the indoor climate. In module B the spruce surface has been upholstered with PE-foil to make the ceiling and walls impermeable and limit the effects of moisture buffering and latent heat (Figure 2).

The instrumentation inside the modules includes heat and ventilation flow control; registration of temperature and RH of air into and out the modules; wood moisture content sensors; and energy meter. The instrumentation system is limited to 9 nine parameters in each test house. An additional 5 five freestanding moisture sensors are mounted on small timber blocks of spruce and placed in the test houses, 3 three in module A and 2 two in module B. Each of these sensors monitors the RH, ambient air temperature and wood moisture content. The 3rd sensor in module A is mounted directly on the CLT element of the south wall. Table 3 displays all the measured parameters and Figure 4 depicts the floor plan.

Table 2: Wall assembly

Wall layer	Material
Surface coating	Module B: PE, 0,2 mm, s_d =1500
Indoor surface	Spruce CLT, 100 mm
Weather barrier	Tyvek: water resistant, diffusion open
Insulation	100/150 mm GLAVA mineral wool
Weather barrier	Tyvek, UV resistant

Figure 2: Interior view of module B where the walls and ceiling are upholstered in PE foil

The indoor air temperature was held at 20°C using an electric heater with temperature control. The extract ventilation system supplies constant air change, controlled by differential pressure sensors. The ventilation is pre-set to n = 0.5 h-1 for both modules, in accordance with Norwegian building regulations (TEK 10), but later adjusted in Case 2. The moisture loads were introduced by ultrasonic evaporative humidifiers, each with a total capacity of 5.8 litres and diffusion rate 616 g/h. To prevent the humidity of being extracted straight out through the ventilation canal, a small table fan was placed on the floor to stir up the air and distribute the moist more evenly in the space. This fan was directed towards the floor to avoid disturbing the laminar boundary layer.

Table 3: Measured parameters. The placement of the sensors is shown in Figure 3.

	Module A	Specification	Module B
1	RH indoor	DT043 sensor Range 0-100%±1,5%	RH indoor
2	Extract temperature	DT043 sensor Range -50-100°C ±0,5°C	Extract temperature
3	Air flow	Sensiron SDP	Air flow
4	Operative room temperature	MCP9700/01	Operative room temperature
5	Surface temperature, ceiling	Temperature sensor Range -40-125°C, ±4°C	Surface temperature, ceiling
6	Surface temperature, north wall		Surface temperature, north wall
7	Surface temperature, south wall	DT043 sensor. Range 0-100%, ±1,5%	RH outdoor
8	Surface temperature, floor	DT043 sensor. Range - 50-100°C, ±0,5°C	Temperature intake
9	Energy consumption	Energy meter	Energy consumption
10	A1 Surface temperature, RH, WME	Hygrotrac S-900-1, ±0.4°C, ±3.5%RH	B1 Surface temperature, RH, "WME"
11	A2 Extract temperature, RH, WME		B2 Extract temperature, RH, "WME"
12	A3 Surface temperature, RH, WME		

(Figure 3) Some assumptions have been made regarding the experimental setup: the moisture load is applied with steady flow. The diffusion rate was measured in a laboratory and divided on the water container capacity to get the moisture load duration.

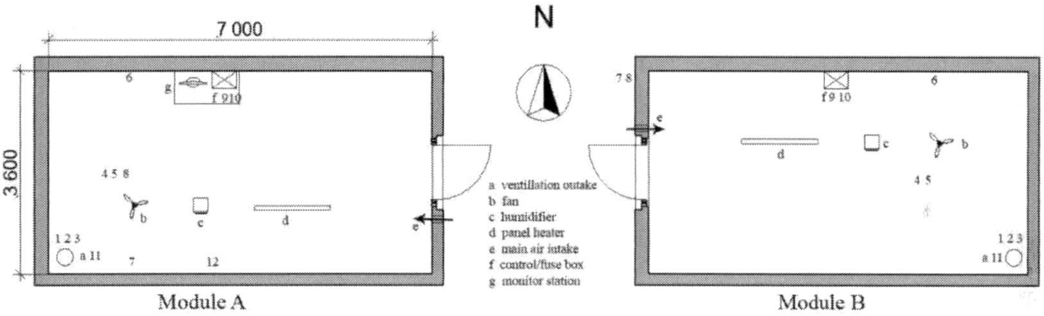

Figure 3: The ground plan of the modules and a schematic figuration of the instrumentation placement.

3.2 Simulation set-up

In the simulation part of this study, the hygrothermal building simulation software WUFI®Plus (2007) is used. This software is developed by Fraunhofer Institute IBP, and has been validated through experimental studies. The simulation cases are based on and designed after the real life test houses from the experimental study. The outdoor boundary conditions are the recorded weather data provided from the meteorological station in which the test houses are situated. The indoor initial conditions, including operating temperature, RH and wood moisture content, are set equal to the measured initial conditions from the equivalent experimental case.

4 RESULTS AND DISCUSSION

The results are presented in Figures 4-6, with the permeable cases on the left and the impermeable on the right. The experimental data is depicted with full line and the calculated results with a dotted. A summary of the complete results is presented in Table 34. The values are recorded with 1 minute intervals. A moving average data treatment of 10 steps have been applied to smoothen the all the curves. The exception is the temperature measurements from module B which have been treated with a 30 step moving average smoothing. The reason being an unsteady meter controller in this module making the recorded data flutter.

The RH curve starts rising as soon as the moisture is applied. The RH in the impermeable cases reaches 100 % in a short amount of time. The curves are steep, both increasing and decreasing. Under the permeable conditions, the RH has a lower interval, a gentler slope and delayed peak. In case 1, which is the first case conducted in the wood active module A, the initial RH is the same in both modules due to the fact that they both where dried out a long time ahead. In following experiments the initial RH differs with about 5 % percentage points less in the impermeable module, demonstrating clearly the hygroscopic inertia of wood. In Figure 4b the PE covered module B doesn't seem to reach 100 % RH, which has to do with the built-in hygrostat in the humidifier kicking in. In case 4, which has the same setup, but with the built-in hygrostat being disconnected, this curve looks like the B module RH curve in Figure 5b. The peak in RH in the permeable module is in all cases at the max (end point) of the moisture load. The wood moisture content also reaches its highest value when RH is max. The surface temperature curves follow the same path as the curves for wood moisture content, showing the effect of latent heat of sorption. After case 3 (last case) is conducted and the module is left closed, the wood moisture content reaches its initial state of 7.4 % 60 hours after the moisture load is fully applied.

There is a drop in temperature at the start of every case which is caused by the entrance door opening and shutting while launching the experiment. Temperature prior to this decrease is considered as the initial temperature, thus the increase in temperature due to the latent heat of sorption is quite conservatively estimated.

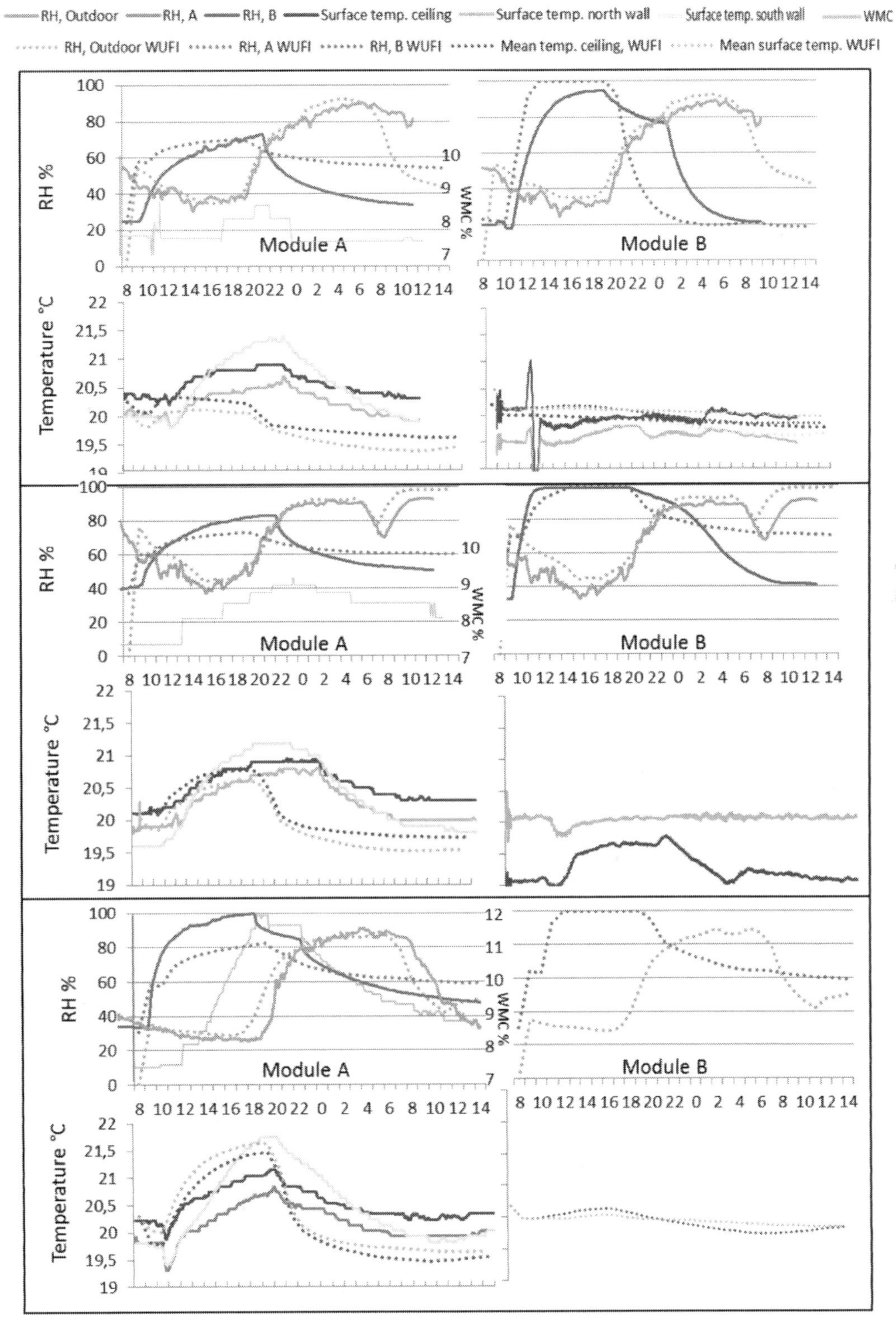

Figure 4-6, a-d: Results from case 1-3 from top to bottom. The wood active module A is depicted on the left side and the PE-covered module B on the right side.

The summarized Table 4 shows the initial/max temperatures from the ceiling surface sensor in the experimental tests. This curve was representative in all the cases, not being extreme either ways, as well as being given by WUFI. The temperatures from the different measurement points differ with at least 1 °C within each case. There are certainly some temperature differences from point to point due to air circulation and humidity not being distributed evenly. The latent heat effect from the south (See attachment)wall is higher in every case, compared to the other surfaces. It is reasonable to assume that the south walls are more dried out than the other surfaces by solar radiation, as well as being less insulated than the north wall. The increases in temperature due to latent heat lies around 0.5-1° C, not taking into account the temperature drop at the start. Nevertheless, there are clear trends regarding the permeable vs. impermeable cases when it comes to temperature rise.

Concerning the simulated cases; the accordance is quite accurate between the experimental and calculated impermeable cases. In the impermablepermeable cases however, there is more discrepancy. The max RH is too high in the simulations, and the RH decline to gentle compared to the measured cases. This problem might be linked to the values for the heat and moisture transfer coefficients used by the simulation program. Furthermore, the thickness of the CLT might cause the permeability to become too complex a process. The surfaces temperature curves returned from WUFI®Plus (mean surface and mean ceiling temperatures) do look more cohesive than the experimental results, as can be expected from an ideal computer simulation.

When RH is the primary factor for ventilation, energy savings can be made by reducing the air exchange rate. In situations where humidifying/dehumidifying is necessary, exposed wood can help stabilize the environment and decrease this need. Another option for energy profit includes applying moisture buffering to cooling strategies; moistened exposed wooden surfaces absorb heat from the ambient air during hot days.

Table 4: Summary of the test results

Case				Moisture load				Experimental test					Simulation test			
#	Setup	Date	Ventillation rate [h^{-1}]	Diffusion rate [g/h]	Duration [t]	Total load [litres]		Permeable			Impermeable		Permeable		Impermeable	
							initial	23.9	20.2	7.4	23.1	20.2	23.9	20.2	23.1	20.2
1	I	31.3.	0.54	616	09:25	5.8	max	75.4	20.9	8.6	98.1	20.2	70.4	21	100	20.2
							variance	+51.5	+0.7	+1.6	+75	0	+46.	+0.8	+76.9	0
							initial	37.3	20.1	7.4	32.4	20.5	37.3	20.1	32.4	20.5
2	II	11.4.	0.32/0.31	616	09:25	5.8	max	83.8	21	9.2	100	20.2	73.8	20.8	100	20.5
							variance	+46.5	+0.9	+1.8	+67.6	-0.3	+36.	+0.7	+67.6	0
							initial	30.8	20.3	7.4			30.8	20.3	30.8	20.3
3	III	15.4.	0.54	1232	09:25	11.6	max	97.1	21.2	12			82.3	21.6	100	20.3
							variance	+66.3	+0.9	+4.6			+51.	+1.3	+69.2	0
							initial	34	20.2	7.4	27.1	20.5	34	20.2	27.1	20.5
4	I	9.4.	0.53	616	09:25	5.8	max	74	20.4	8.8	100	20.5	72.4	20.3	100	20.5
							variance	+40	+0.2	1.4	+72.9	0	+38.	+0.9	+72.9	0
								RH	Temp.	WMC	RH	Temp.	RH	Temp	RH	Temp

5 CONCLUSIONS

The results from this paper show that the buffering effect from large areas of exposed wood surfaces helps keep the RH within a closer interval, with slower alteration upon moisture load being applied/removed compared to impermeable surfaces. The latent heat exchange gives rise to the indoor temperature. There is a significant potential regarding exposed wood and indoor climate, especially in combination with a well-controlled HVAC system. For this to become practically applicable there is however still need for research and engineering.

6 ACKNOWLEDGEMENTS

This study was conducted within the research project Wood – Energy, Emission, Experience (WEEE). All participants are gratefully acknowledged. A thank to Florian Antretter and Fraunhofer IBP for great assistance. The financial assistance from Innovation Norway is greatly appreciated.

7 REFERENCES

ANSI/ASHRAE. Standard 55 - Thermal environmental conditions for human occupancy.

Arlian, L. G., Neal, J. S. & Vyszenski-Moher, D. L. (1999). Reducing relative humidity to control the house dust mite Dermatophagoides farinae. Journal of Allergy and Clinical Immunology, 104 (4): 852-856.

Edvardsen, K. I., Haug, T. & Ramstad, T. Ø. (2006). Trehus. 9. utg. ed. Håndbok / [Norges byggforskningsinstitutt]. Oslo: Norges byggforskningsinstitutt. 333 s. pp.

Geving, S. & Thue, J. V. (2002). Fukt i bygninger. Håndbok (Norges byggforskningsinstitutt : trykt utg.). Oslo: Norges byggforskningsinstitutt.

Hameury, S. and Lundström (2003) Contribution of indoor exposed massive wood to a good indoor climate: in situ measurement campaign. Energy and Buildings, 36 (3): 281-292.

International Energy Agency. (2014). IEA. 2014 [ONLINE] Available at: http://eur-lex.europa.eu/ LexUriServ/LexUriServ.do?uri=OJ:L:2010:153:0013:0035:EN:PDF. [Accessed 2014].

ISO. 7730:2005 (en) Ergonomics of the thermal environment - Analytical determination and interpretation of thermal comfort using calculation of the PMV and PPD indices and local thermal comfort criteria.

Janssen, H. & Christensen, J. E. (2013). Hygrothermal optimisation of museum storage spaces. Energy and Buildings, 56 (0): 169-178.

Karlsson, Jonathan (2012) Possibilities of using thermal mass in buildings to save energy, cut power consumption peaks and increase the thermal comfort. Licentiate thesis, Lund Institute of Technology, Division of Building Materials, Lund, Sweden

Korsnes, S. & Nore, K. (2012). Moisture buffering and the influence of exposed wooden surfaces on the indoor environment. Master thesis, Department of Civil and Transport Engineering, Norwegian University of Science and Technology, Trondheim, Norway

Kubler, Hans (1982) Effect of wood on humidities in homes. Forest Products Journal, Vol 32, No 6: 47-50.

Nore, K. Englund F, Aurlien, T. and Nyrud, A.Q. (2014) Wood construction: Energy, Emissions and Experience. In: 10th Nordic symposium of building physics. Lund, Sweden June, 2014.

Osanyintola, O. F. & Simonson, C. J. (2006). Moisture buffering capacity of hygroscopic building materials: Experimental facilities and energy impact. Energy and Buildings, 38 (10): 1270-1282.

Rode, C. & Grau, K. (2008). Moisture buffering and its consequence in whole building hygrothermal modeling. Journal of Building Physics, 31 (4): 333-360.

TEK 10, (2010). Norwegian building regulations.

TenWolde, A. & Pilon L. C. (2007) The effect of indoor humiduty on water vapor release in homes. Proceedings of Thermal Performance of the Exterior Envelopes of Whole Buildings X, Atlanta, USA

Thue-Hansen V. and Grimenes A.A. Meteorological data for Ås, Norwegian University of Life Sciences, 2014, Ås, Norway

Olaussen, P. A. (2014) Master Thesis (In progress), Department of Mathematical Sciences and Technology, Norwegian University of Life Sciences, Ås, Norway

Toftum, J., Jorgensen, A. S. & Fanger, P. O. (1998a). Upper limits for indoor air humidity to avoid uncomfortably humid skin. Energy and Buildings, 28 (1): 1-13.

Toftum, J., Jorgensen, A. S. & Fanger, P. O. (1998b). Upper limits of air humidity for preventing warm respiratory discomfort. Energy and Buildings, 28 (1): 15-23.

WUFI®Plus (2007) Fraunhofer Institut für Bauphysik, Holzkrichen, Germany.

BREATHING FEATURES ASSESSMENT OF POROUS WALL UNUITS IN RELATION TO INDOOR AIR QUALITY

Başak Yüncü*[1], Ayşe Tavukçuoğlu[2] and Emine N. Caner-Saltık[3]

1 Middle East Technical University
Department of Architecture
Graduate Program in Building Science
Dumlupinar Bulvari, No:1
Ankara, Turkey
*Corresponding author: basakyuncu@gmail.com

2,3 Middle East Technical University
Department of Architecture
Materials Conservation Laboratory
Dumlupinar Bulvari, No:1
Ankara, Turkey

ABSTRACT

Traditional building technologies establishing highly-breathing multi-layered wall systems provide healthy indoor environment and energy efficiency in buildings due to the use of lightweight, porous, water vapour permeable and thermal resistive building materials. The breathing performance of traditional buildings and materials that contribute to the healthy indoor conditions and air quality are needed to be investigated in detail. That knowledge on breathing performance has also vital importance for the improvement of the contemporary building and materials technologies.

The study was conducted on three kinds of porous masonry units: mud brick, collected from the sound parts of the traditional houses of Hamzalı Village, Kırıkkale (Turkey) and autoclaved aerated concrete (AAC) type G2 and type G4, the commonly used lightweight concrete material which are also produced in Kırıkkale. These three groups of materials were examined in terms of their air exchange properties by means of laboratory analyses. Some supportive laboratory analyses were also done on material characterization of the samples. The experimental setup is developed for the analyses of the samples, particularly for the assessment of air flow through the material by using CO_2 as tracer gas.

The interpretation of the results were done in order to compare the porous building materials in terms of their effects on indoor air quality. The double-zone experimental setup based on concentration decay procedure was found to be useful to better-understand the air exchange features of a material in terms of rates of concentration decrease and increase in neighbouring zones by monitoring the concentration of outgoing air. The data achieved is expected to improve the contemporary building walls by benefitting from the self-ventilation capability of building materials and to provide healthier indoor conditions for the occupants. The results are also useful to discuss the airtightness aspect of passive house technology and to minimize the mechanical ventilation needs for fresh air intake by benefitting from the self-ventilation performance of air permeable skin.

KEYWORDS

Indoor air quality, breathing walls, mud brick, autoclaved aerated concrete, air exchange properties.

1. INTRODUCTION

Traditional building technology provides energy efficiency and good indoor air quality in houses by using porous, water vapour permeable building materials establishing highly-breathing multi-layered wall systems (Meriç *et al.*, 2013; in press). Each layer of a mud brick traditional wall section has particular performances, such as water impermeable exterior finishing layers and thermal resistive intermediate layers while each layer allows continuous passage of water vapour along the overall wall section. The breathing performance of traditional buildings that contribute to the healthy indoor conditions and air quality needs to be investigated in detail. That knowledge is essential for the improvement of the contemporary building materials and construction technologies. Due to the high porous and breathing characteristics of autoclaved aerated concrete (AAC), it appears to be an alternative material to establish contemporary breathing walls. The potentials and restrictions of AAC, therefore, need to be examined in terms of its breathing properties. In this regard, the experimental study was shaped to discover the air exchange characteristics of traditional mud brick and AAC unit samples. Supportive laboratory analyses were done to define the raw material characteristics of traditional mud brick.

The concept of airtightness is mostly defined as absence of air leakage and excluded from the issues of ventilation in buildings (Fennell and Haehnel, 2005). Some others focus on minimum airtightness levels required for energy saving and support that the higher airtightness levels the lower energy costs (Katunsky *et al,*.2013). The studies that support maximum airtightness propose controlled mechanical ventilation systems. The operating and maintenance expenses of mechanical ventilation systems as well as the risk of health problems associated with the mechanical ventilation systems, such as Legionnaire's disease, are some of main reasons of why natural self-ventilation properties of building materials have vital importance on the healthy indoor conditions and air quality (US EPA, 1991; Hodgson, 1992). On the other hand, most studies point out that indoor air quality would suffer in case of over-airtightness (Sherman and Chan, 2004; Kirch, 2008a; Kirch, 2008b). The inherent breathing features of building components can provide self-ventilation at a certain level while minimizing the energy cost for mechanical heating, ventilation, and air conditioning (HVAC) systems. The breathing features of porous building materials and their contribution to the indoor air quality, therefore, are the key concerns of this study.

2. MATERIAL AND METHOD

The study was conducted on mainly mud brick and AAC samples, both of which were porous masonry units and well-known with their good breathing capability (Andolsun, 2006; Andolsun *et al.*, 2006; Jacobs *et al.*, 1992; Kömürcüoğlu, 1962; Meric *et al.* 2013; Meric *et al.* in press; Naranyan *et al.*, 2000, Torraca, 2009): The mud brick units were collected from the sound parts of the traditional houses of Hamzalı Village, Kırıkkale (Turkey), together with mud mortar and mud plaster samples complementing the mud brick wall section. Those houses are about 70 year old. The AAC units are manufactured in Kırıkkale (Turkey). The commonly-used types of AAC blocks, type G2 (infill unit) and type G4 (load-bearing unit), were selected for the study. The breathing features of these materials were examined with an emphasis on their air exchange rate. Supportive analyses were also done on some basic physical and physicomechanical characteristics of mud brick and AAC samples, such as density (ρ, $g.cm^{-3}$), effective porosity (ϕ, % by volume), ultrasonic pulse velocity (UPV, $m.s^{-1}$) and modulus of elasticity (MoE, GPa) as well as on some raw materials characteristics of original mud brick samples, such as silt & clay ratio, binder-aggregate ratio and fibre ratio.

The air exchange properties of each sample were determined by measuring the air flow through the material by using CO_2 as tracer gas and an experimental setup was developed for this purpose by adapting the concentration decay method defined in the international standard ASTM E 741-11 (2011). This method was assumed to represent the case for an indoor having a very high carbon dioxide concentration level (after it has been occupied) and ventilating itself only by the porous wall section. The experimental setup developed for the determination of air change rate was composed of mud brick or AAC block samples positioned in between two chambers/zones (Figure 1). Each block sample with the sizes of 180mm x 125mm x

310mm (thickness x height x length) sealed two plexiglass chambers positioned at its both sides. The volume of each chamber was $0.016m^3$ with the dimensions of 130mm x 390mm x 310mm. The concentration of CO_2 (in ppm) in each chamber was measured by using a CO_2 analyser "Testo 480" with two indoor air quality measuring probes. One probe was placed in each chamber. During the experiment, the ambient temperature, relative humidity and absolute pressure in the chambers were also recorded for the analyses. The solution composed of 2g sodium hydrogen carbonate powder in 50ml pure acetic acid which was put in a glass beaker acted as the source of CO_2 in the Chamber 1(Ch-1) and provided a CO_2 concentration about 15000 ppm. This high concentration level was assumed to simulate over-occupied indoor environment. The CO_2 concentration in the Chamber 2 (Ch-2) was filled with fresh air which has a CO_2 concentration level around 400 ppm. The change in the CO_2 levels in the chambers in time was recorded at 5s intervals by the CO_2 analyser for a period of 24 hours. The concentration decay curves were produced showing the CO_2 concentration in the chamber, C, as a function of square root of time. The slope of the linear regression presented the rate of decrease in CO_2 concentration in Ch-1, R_D and the rate of increase CO_2 concentration in Ch-2, R_I, for each sample block.

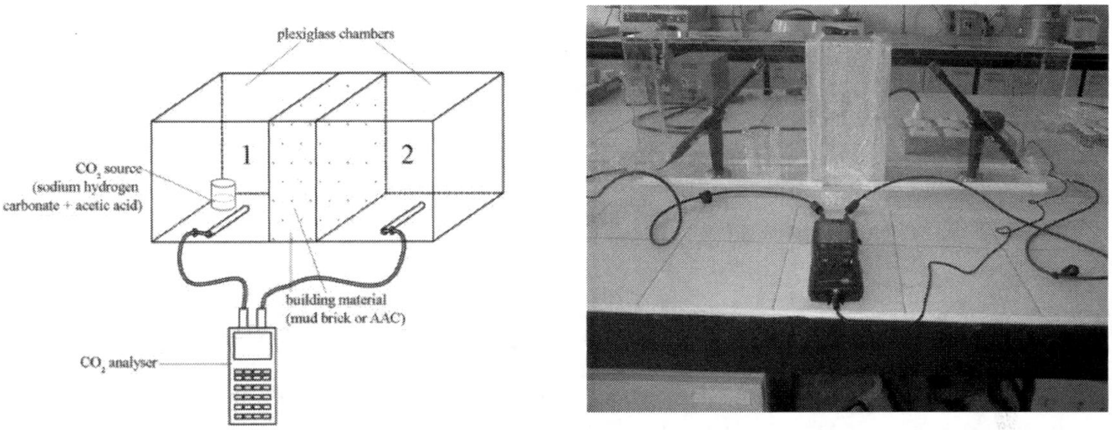

Figure 1: Experimental setup for the analyses of air exchange in double zone by using CO_2 as tracer gas.

Density and porosity of the material samples were examined by the standard analyses described in RILEM (1980). The modulus of elasticity of the samples were determined indirectly by means of equations described in ASTM D 2845-08 (2008) and RILEM (1980), using their ultrasonic pulse velocity (UPV) and density values. The UPV values were measured in the direct transmission mode (cross direction) by using a portable PUNDIT Plus CNS Farnell Instrument with 220 kHz transducers.

For the determination of binder-aggregate ratio, fibre ratio and clay-silt content of mud brick, mud plaster and mud mortar samples were examined by sieve analysis (Teutonico, 1986). The samples were kept in water. First, fibre ingredients suspended in water were separated. Following the drying out of the samples, they were sieved by using a set of sieves with specific sizes of 4mm, 2mm, 1mm, 0.500mm, 0.250mm, 0.125mm and 0.063mm.

3. RESULTS AND DISCUSSION

The results were evaluated to compare the air exchange characteristics of mud brick and AAC block samples. The relevant material characteristics of the samples were also determined for the joint interpretation of the results.

The mud brick samples were found to have considerable high ratio of silt & clay (the grain size below 0.063mm) with the value of 48.9% by weight (Table 1). The largest portion of the aggregate was determined to be within the grain sizes of 0.25mm and 0.063mm with 41.8% by weight while 10% of belonging to the particles between 4mm and 0.25mm. The ratios of the aggregates and fibres were determined to be 52.8% and 0.2%, respectively, by weight (Table 1). The mud plaster and mud mortar samples complementing the mud brick masonry wall section seemed to have similar binder-aggregate ratios which may signal the use of same local and natural raw material sources.

Table 1: The results of some raw materials characteristics for the mud brick, mud plaster and mud mortar samples.

Sample Type	Silt&Clay Ratio (%)	Aggregate Ratio (%)	Fibre Ratio (%)
Mud brick	48.86 ± 4.40	52.75 ± 5.53	0.15 ± 0.12
Mud plaster	43.03 + 6.14	56.97 ± 6.14	0.89 ± 0.06
Mud mortar	43.27 ± 0.87	56.73 ± 0.87	1.57 ± 0.21

The data on air exchange characteristics of mud brick and AAC block samples are presented in Table 1 together with their density, porosity, modulus of elasticity and water vapour diffusion resistance index properties. All samples were less dense and highly porous materials while the mud brick had higher density and lower effective porosity values G2 and G4 types of AAC. The mud brick seemed to have higher MoE than the AAC samples while having the lowest UPV value. The G2 type of AAC used as infill block presented very low MoE among all samples. All samples were highly-breathable materials due to their very low resistance to water vapour permeation while the mud brick presented noticeable higher breathing capability than the AC samples. The physical and physicomechanical properties of the samples signalled the differences in the pore structure of the mud brick and AAC samples.

The results of air exchange properties of the samples were summarized below (Figures 2-4):

- The concentration decay of CO_2 in Ch-1 was observed to be the fastest at G4 type of AAC block with the R_D of 98.4 ppm.$s^{-1/2}$ while a very slight passage of CO_2 to the other side (Ch-2) was observed with the R_I value of 2.3 ppm.$s^{-1/2}$ (Figure 4). This signalled that G4-AAC sample have tendency to absorb CO_2 while not permitting its passage to the other side. The G4-AAC material may filter CO_2 particles in its pore structure.
- The reduction in CO_2 concentration seemed to be fast in G2 type of AAC block with the R_D value of 70.9 ppm.$s^{-1/2}$, however, that rate is slower when compared with the R_D of G4-AAC sample (Figure 3 and Figure 4). G2-AAC sample presented faster increase in CO_2 concentration in Ch-2 with the R_I value of 37.6 ppm.$s^{-1/2}$. That designated the air permeability G2-AAC sample from one side to the other is more than G4-AAC sample while both AAC samples have tendency to absorb CO_2 at certain extents.
- The highest transmission of CO_2 from Ch-1 to Ch-2 was observed in mud brick sample with the R_I value of 61.6 ppm.$s^{-1/2}$. The reduction in CO_2 concentration in Ch-1, on the other hand, was lower than expected with the R_D value of 31.5 ppm.$s^{-1/2}$. The very fast increase in CO_2 concentration in Ch-2 may decrease the difference in the CO_2

concentration between two chambers. This phenomenon may be one of the reasons for the slowing down the R_D in Ch-1.

It is known that the main mineral in the composition of AAC is tobermorite-11Å (Narayanan and Ramamurthy, 2000), which may react with atmospheric carbon dioxide (CO2) gas in the presence of moisture and be converted to silica and calcium carbonate (Matsushita, *et al.*, 2000; 2004). AAC blocks was observed to be attractive for CO_2 gas by keeping it in its porous fabric and acting like an indoor air cleaning material, that could be attributed to its inherent material characteristics.

Table 2: The data on density (ρ), effective porosity (ϕ), ultrasonic pulse velocity (UPV), modulus of elasticity (MoE), water vapour diffusion resistance index (μ) and the rates of CO_2 concentration decrease (R_D) and increase (R_I) of the mud brick and AAC samples

Sample Type	ρ (g.cm^{-3})	ϕ (%)	UPV (m.s^{-1})	MoE (GPa)	μ (unitless)*	R_D in Ch-1 (ppm.s$^{-1/2}$)	R_I in Ch-2 (ppm.s$^{-1/2}$)
Mud brick	1.60±0.03	42.37±0.34	1321±65	2.569±0.242	1.4 – 1.6	-31.539	61.595
G2-AAC	0.42±0.00	74.10±1.23	1703±20	1.109±0.017	3.8 - 5.7	-70.908	37.635
G4-AAC	0.62±0.02	67.67±2.49	1955±30	2.168±0.119	3.2 - 6.4	-98.370	2.265

* The μ values of mudbrick and AAC samples was obtained from the literature, Meric et al. (2013) and Andolsun (2006), respectively.

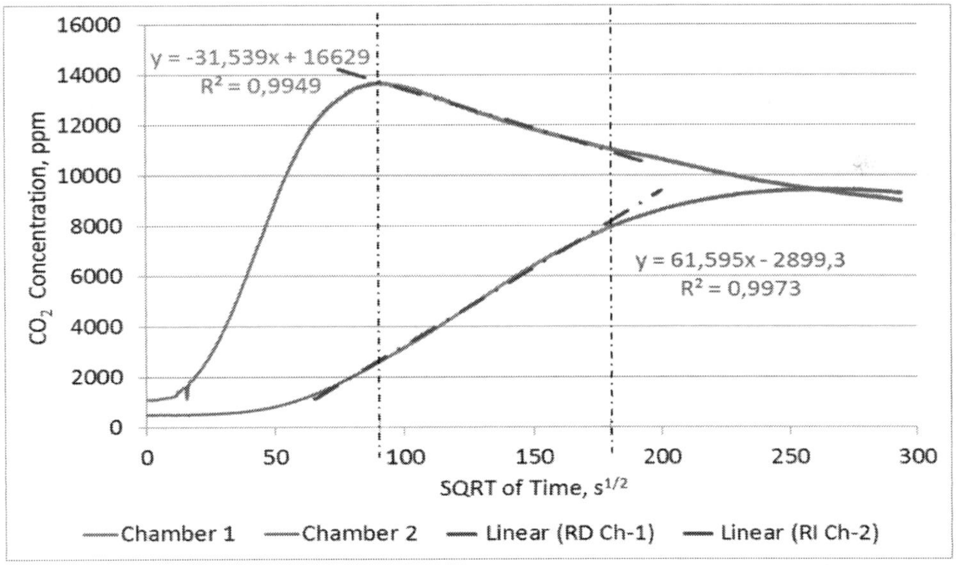

Figure 2: The CO_2 concentration curves in Ch-1 and Ch-2 during the 24 hours examination period of mud brick block sample: The linear fitting of CO_2 concentration versus square root of time showing the rates of concentration decrease in Ch-1 (RD Ch-1) and concentration increase in Ch-2 (RI Ch-2) as the slope of the regression line.

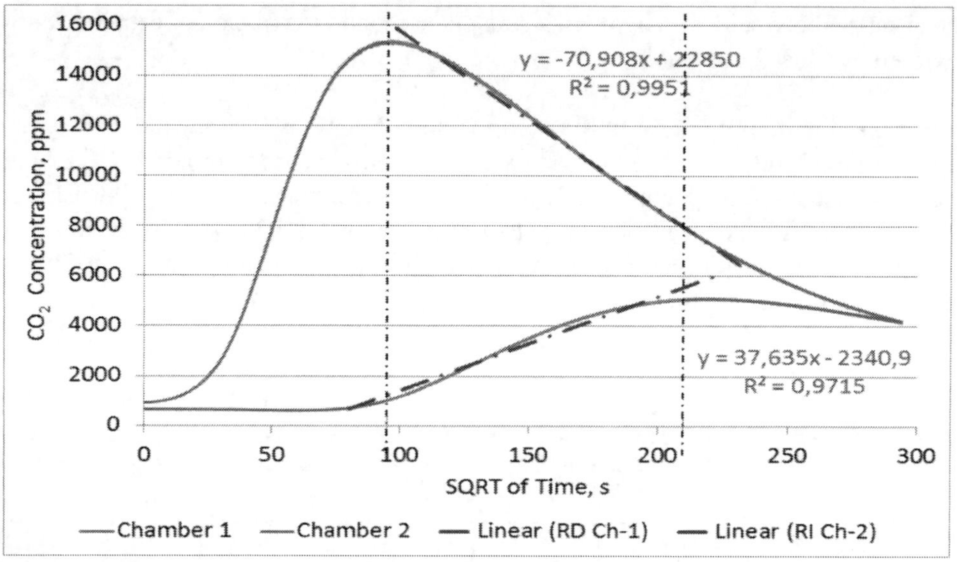

Figure 3: The CO_2 concentration curves in Ch-1 and Ch-2 during the 24 hours examination period of G2 type of AAC block sample (G2-AAC): The linear fitting of CO_2 concentration versus square root of time showing the rates of concentration decrease in Ch-1 (RD Ch-1) and concentration increase in Ch-2 (RI Ch-2) as the slope of the regression line.

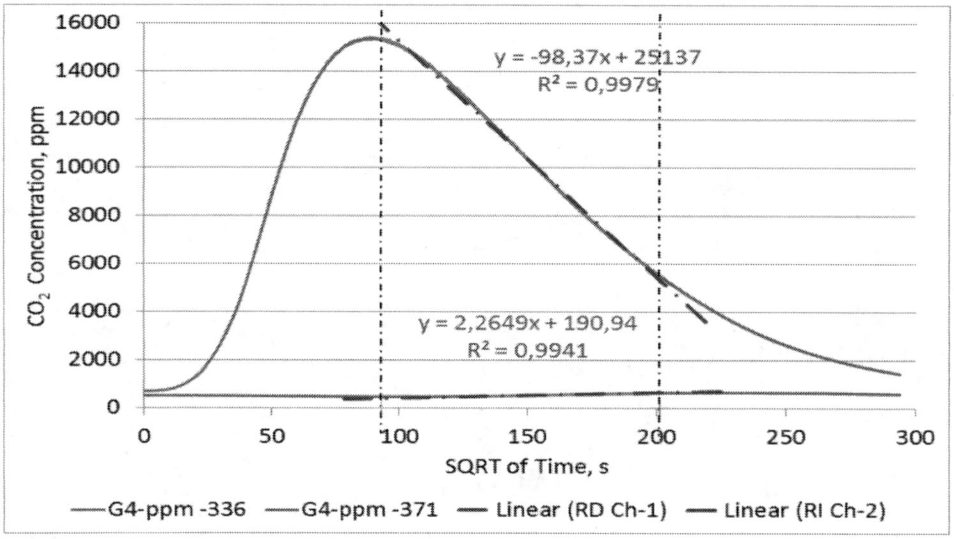

Figure 4: The CO_2 concentration curves in Ch-1 and Ch-2 during the 24 hours examination period of G4 typeof AAC block sample (G4-AAC): The linear fitting of CO_2 concentration versus square root of time showing the rates of concentration decrease in Ch-1 (RD Ch-1) and concentration increase in Ch-2 (RI Ch-2) as the slope of the regression line.

4. CONCLUSIONS

Breathing features of building envelopes should be taken into consideration for the improvement of contemporary building walls by benefitting from air exchange ventilation through building materials and to provide healthier and sustainable indoor conditions for the occupants. The preliminary results of the study are also expected to discuss the airtightness aspect of passive house technology and to minimize the mechanical ventilation needs for fresh air intake by benefitting from the self-ventilation performance of air permeable skin.

Comprehensive studies are required to benefit more from the inherent breathing features of porous building materials.

The preliminary examination on the air exchange properties of traditional mud brick and autoclaved concrete masonry blocks have shown that those highly-breathable materials have different air exchange properties depending on their physical, physicomechanical and compositional properties, all of which shape their particular porosity characteristics. The air exchange capability of highly breathing materials may contribute to the indoor air quality by considering their self-ventilating capacities during the design stage of buildings and by improving the consciousness on materials selection for constructions.

The double-zone experimental setup based on concentration decay procedure was found to be useful to better-understand the air exchange features of a material in terms of rates of concentration decrease and increase in neighbouring zones by monitoring the concentration of outgoing air. That method allowed differentiating particular performances of porous materials. For instance, the relatively rapid reduction of CO_2 level in a room may signal the higher air exchange characteristics of a material or may be due to the reacting capability of the material composition with CO_2 particles.

Further studies on measuring self-ventilation capacity of building materials are required, especially on measurable parameters that can be used to assess and estimate their self-ventilation characteristics.

5. ACKNOWLEDGEMENTS

The authors would like to acknowledge the project METU Research Grant No. BAP 02-01-2014-003: Breathing features of building walls and their contribution to the indoor air quality: testing and assessment; and to thank Dr. B. Alp Güney and Dr. Evin Caner for their invaluable technical support throughout the experiments, Mr. Ömer Taşkazan and AKG Gazbeton Company who provided the necessary AAC block samples as well as Mr. Mert Öcal who provided original mud brick samples.

6. REFERENCES

Andolsun, S. (2006). A study on material properties of autoclaved aerated concrete (AAC) and its complementary wall elements: Their compatibility in contemporary and historical wall sections. *Unpublished Master Thesis*. Graduate Program in Building Science, Department of Architecture, Middle East Technical University, Ankara, Turkey.

Andolsun, S., Tavukcuoglu, A., Caner-Saltik, E.N, Düzgüneş, A. (2006) A study on the compatibility of contemporary neighbouring plasters with autoclaved aerated concrete (AAC). In, S. Andolsun, A. Temizsoy & M. Ucar (Eds), *Proceedings of the First International CIB Endorsed METU Postgraduate Conference on Built Environment and Information Technologies (PGRC'06), Ankara, Turkey*, 17 – 18 March 2006 (pp. 587-599). Ankara: METU Faculty of Architecture.

ASTM D 2845-08, Standard Test Method for Laboratory Determination of Pulse Velocities and Ultrasonic Elastic Constants of Rock. (2008). American Society for Testing and Materials, Philadelphia.

ASTM E 741-11, Standard Test Method for Determining Air Change in a Single Zone by Means of a Tracer Gas Dilution. (2011). American Society for Testing and Materials, Philadelphia.

Fennell, H.C. and Haehnel, J. (2005). Setting Airtightness Standards, *ASHRAE Journal*, 47(9), 26-30.

Hodgson, M. (1992). Field Studies on the Sick Building Syndrome, *Annals of the New York Academy of Science,* 641(1), 21-36.

Jacobs, F. and Mayer, G. (1992). Porosity and permeability of autoclaved aerated concrete, *Proceedings of the Third RILEM International Symposium on Autoclaved Aerated Concrete* (p.71-75). 14-16 October, 1992. Advances in Autoclaved Aerated Concrete, F.H. Wittmann, ed. A.A. Balkema, Rotterdam, Brookfield.

Katunsky, D., Nemec, M.,and Kamensky, M. (2013). Airtightness of buildings, *Advanced Materials Research*, 649, 3-6.

Kirch, W. (2008a). *Sick Building Syndrome, Encyclopedia of Public Health, v.2*, pp.1303, New York, USA: Springer Science+Business Media, LLC.

Kirch, W. (2008b). *Chemical Sensitivity Syndromes, Encyclopedia of Public Health, v.1*, pp.107, New York, USA: Springer Science+Business Media, LLC.

Kömürcüoğlu, E. A. (1962). *Yapı Malzemesi olarak Kerpiç ve Kerpiç İnşaat Sistemleri.* İstanbul: İTÜ Matbaası.

Meric, I., Erdil, M., Madani, N., Alam, B. A., Tavukcuoglu, A. and Caner Saltik, E. (2013). Technological Properties of Earthen Building Materials in Traditional Timber Frame Structures. In: *Kerpic'13 3rd International Conference – New Generation Earthern Architecture: Learning from Heritage*, 11-15 September 2013 (pp.465-473),. Istanbul, Turkey: Istanbul Aydın University

Meriç, I., Erdil, M., Madani, N., Alam, B. A, Tavukçuoğlu, A.and Caner Saltık, E.N. (In Press). Material characterization of mudbrick and neighbouring plasters in traditional timber framed structures. *9th International Symposium on Conservation of Monuments in Mediterranean Basin*, Ankara, Turkey, 2-5 June 2014. Ankara, Turkey: Middle East Technical University.

Narayanan, N., and Ramamurthy, K. (2000). Structure and properties of aerated autoclaved concrete: a review, *Cement and Concrete Composites*, 22, 321-329.

Matsushita, F., Aono, Y. and Shibata, S. (2000). Carbonation Degree of Autoclaved Aerated Concrete, in: *Cement and Concrete Research* (p.1741- 1745, v.30).

Matsushita, F., Aono, Y., and Shibata, S. Calcium silicate structure and carbonation shrinkage of a tobermorite-based material. *Cement and Concrete Research, 34* (2004), 1251-1257

RILEM (1980) Tentative Recommendations, in: Commission — 25 - PEM, Recommended Test to Measure the Deterioration of Stone and to Assess the Effectiveness of Treatment Methods. *Materials and Structures* (13:73) 173-253.

Sherman, M.H. and Chan, Z. (2004). *Building Airtightness: Research and Practice.* Lawrence Berkeley National Laboratory Report No. LBNL-53356.

Teutonico, J.M. (1986). *A Laboratory Manual for Architectural Conservators*, ICCROM, Rome.

Torracca, G. (2009). *Earth as a building material, Lectures on Materials Science for Architectural Conservation*, pp.38-41, Los Angeles, USA: The Getty Conservation Institute.

US EPA (1991). *Indoor Air Facts No.4 (revised), Sick Building Syndrome, Fact Sheets.* United States Environmental Protection Agency. Retrieved from www.epa.gov/iaq/pdfs/sick_building_factsheet.pdf

LARGE BUILDINGS AIRTIGHTNESS MEASUREMENTS USING VENTILATION SYSTEMS

Szymański Michał*[1], Górka Andrzej[1], Górzeński Radosław[1]

[1] *Poznan University of Technology*
Pl. Sklodowskiej – Curie 5
60-965 Poznan, Poland
**Corresponding author: michal.szymanski@put.poznan.pl*

ABSTRACT

The airtightness test of the building is one of a few building envelope measurements used in practice, which is quantitative, not just qualitative as e.g. infrared thermography. The so-called blower-door test result may be a measure of the building design and construction quality and could also be used for the energy demand for heating and cooling analyses.

With the development of energy-efficient buildings, the awareness of the airtightness importance increases, which results in a growing number of measurements of small and mid-size structures. Due to the high cost and limitations in availability of measurement equipment, tests of large buildings are relatively uncommon. Sometimes in the existing buildings the leakage measurements could be conducted with the use of ventilation systems, which is allowed by the (EN 13829) standard. For this kind of tests one can use the measurement of the air flow rate at fans (fan nozzles), at variable air volume terminals (VAV) or in the duct system. The accuracy of the air flow measurement is crucial in these cases.

Attention should be paid to a number of aspects related to the building preparation and the test procedure. The problems of large buildings leakage tests are discussed in this paper, as well as typical design and workmanship mistakes. This paper presents the methodology and results of airtightness measurements of the selected facilities envelopes using a ventilation system in comparison to blower-door set.

In the case of multi-zone airtightness measurements, the ventilation system is suitable to generate an appropriate pressure in the adjacent zones. Another application is the coarse testing of the airtightness of the building, when the result of the measurement can be approximate. In the case of final measurement of the airtightness of the building with the use of ventilation system, the biggest problem is the reliable measurement of the air flow rate with the accuracy required by the standards. Measurement of airtightness of large buildings with use of existing ventilation system can be completed correctly in some cases, however, to obtain accurate results more expertise and effort are required than in the case of measurement with a set of typical blower-door fans.

KEYWORDS

large building airtightness, envelope air permeability, air leakage, blower door, air handling unit (AHU)

1. INTRODUCTION

Measurements of airtightness of large buildings usually require huge air streams during the test. In the case of standard blower door fans (BD) use for testing of a large building, a set of multiple devices must be built. Large fans (100.000 – 300.000 m³/h pro unit), powered by combustion engine and mounted on truck or trailer, are rarely used.

In the case of multi-zone buildings, each zone can be tested separately. Thereby the necessary number of fans is reduced, but air pressure in all adjacent zones must be equal to the tested zone, to eliminate the air flows between zones. Existing ventilation system can be effectively applied for this task, the air flow does not need to be measured, however, the possibility of

individual control of dampers and fans in the air handling units (AHU) is necessary, which often requires the cooperation with the facility technical staff.
Reduction of the test cost leads to a use of ventilation systems also directly to the measurements. Here, however, arises the question of air flow measurement accuracy. According to the European Standard (EN 13829), the accuracy of the device for measuring of air flow rate should be ± 7% of the measured value or better. In practice, the airflow can be determined, e.g. as a function of differential pressure measured at the inlet nozzle of the fan, but this applies almost exclusively to diagonal fans (axial-radial fans). Manufacturers of this type of fans declare the measurement accuracy worse than required by the standard (EN 13829) and the declared measurement accuracy is achieved only if the manufacturer of air handling unit (AHU) follows the guidelines for the clearances from other elements of AHU (Ziehl-Abegg, 2013). Another problem is the lack of continuous reading and recording of the measured air flow rate.
When using air handling unit fans to pressurize/depressurize the building, a problem regarding continuous reading and recording of the measured air flow may occur. In the case of AHU integrated in a Building Management System (BMS), recording of the variables with use of BMS is usually possible, but mainly with some constraints – e.g. low refresh rate and discretization of the measured values.

2. LARGE BUILDINGS AIRTIGHTNESS MEASUREMENTS

Large buildings are often much more complicated than single-family houses and equipped with lots of installations (except of warehouse buildings).
Compared to measurement of small buildings, testing of large buildings - especially in the case of public buildings – require:

a) more equipment and
b) higher labor input,

but also

c) better familiarity with a building and its technical equipment and
d) deeper understanding of the laws governing the air flows in the building.

To avoid the costs associated with the use of a large number of fans (BD), some measuring teams try to perform an air leakage test using the existing ventilation systems in the building, which is permitted by the standard (EN 13829). Sometimes, even with an adequate amount of equipment, one can forget about the points b) ÷ d) and this could be a source of measurement errors many times larger than the result of allowable measurement errors of air flow and differential pressure.

2.1 Preparation of the building

The influence of building preparation on the result of measurement is high. In the case of large buildings, the provisions of the standard (EN 13829) raise objections concerning the sealing of mechanical ventilation systems. Recommendation of diffusers and exhaust terminals sealing is sometimes very difficult to perform or practically impossible.
In the case of method A (testing of an operated building), most commonly the installation is cut off by closing the dampers on the side of the air intake and extract. This corresponds to the period of standstill of the building ventilation system (e.g. during the night), and do not introduces an error resulting from the ducts leakages, which in the case of ducts within the enclosure of the building, does not affect the air infiltration at normal operation.
Comprehensive preparation of the large building consumes in most cases much more time than just testing and leakage detection.

In practice, due to the lack of education in the area of building construction and installations, you can meet the contractors, who underestimate the role of this stage and concentrate only on achieving of appropriate levels of positive and negative pressure during the measurement.
Effect of this approach is getting results differing from the real airtightness of the object, an example of which is described in Section 4.

2.2 Location of air leakages

Large buildings are often characterized by a large number and variety of air leakages, also located in difficult to reach or inaccessible places. This requires leakage detection techniques different than for small buildings. Instead of smoke tubes and hot-wire anemometers, rather infrared cameras, as well as powerful smoke generators are mainly used.
Appropriate lifting equipment and permissions required by legal regulations to work at heights are also often necessary.

3. SPECIFIC NATURE OF AIRTIGHTNESS MEASUREMENT USING EXISTING VENTILATION SYSTEM

3.1 Introduction

Measurements of the building airtightness using air handling unit fans are rarely reported in the literature. There is a dedicated Canadian standard on this issue (CGSB-149.15) and some of the literature, of which the most important may be considered (Jeong, 2008) and (Kim, 2013).
The airtightness test performed with use of the ventilation system differs from standard measurement using BD fans mainly by the method of movement of the air (AHU fans instead of BD fans) and by the method of the air flow rate measurement. The measurement of the pressure inside the building envelope can be carried out in a standard way. In large or complex, low-energy commercial/educational buildings, the maximum air flow rate of the existing mechanical ventilation is more than enough for the realization of airtightness test. In the cases described below, the air change rate induced by the ventilation system were 2.8 h^{-1} and 2.1 h^{-1}, and the results of the airtightness test n_{50} were 0.6 h^{-1} and 0.3 h^{-1}. This means that the performance of the ventilation system was 4 ÷ 7 times higher than the performance needed to conduct an airtightness test of the building, what is opposite to *Kim* (Kim, 2013).
The remaining problem is the precise measurement air flow rate supplied to the building. This measurement can be performed in the following locations:

a) on the air intake or air outlet,
b) on the elements of the air handling unit,
c) on the ventilation ducts upstream or downstream of the air handling unit,
d) on the supply or exhaust grilles (terminals of the ventilation system).

In the cases b) ÷ d) one should bear in mind the impact of airtightness of ventilation components and properly to take into account/ in the results.

3.2 Measurement of the air flow inside the air handling units (AHU)

The air flow passing through the air handling unit can be determined on the basis of e.g.:

a) measurement of the dynamic pressure at the inlet to the fan,
b) measurement of the differential pressure between the static pressures in front of the inlet ring and the static pressure in the inlet ring of the narrowest point (Fig. 1) (Ziehl-Abegg, 2013),

c) measurement of the average dynamic pressure at the place of the air filter replaced temporarily by measuring element (Fig. 3) (The Energy Conservatory, 2014).

Case b) has occurred in the building analyzed in section 5. The accuracy of the air flow measurement, declared by the manufacturer of the fan is 8%, and does not meet requirements of the standard (EN 13829). Moreover the declared accuracy of 8% is achieved only when the air velocity at the inlet to the fan is more than 9 m/s and when required clearances between the fan and other elements of the air handling unit are preserved (Fig. 1).
Case c) is a solution of one of the companies operating in the field of measurement of airtightness of buildings (Fig. 3) (The Energy Conservatory, 2014). The declared accuracy is 7% of the measured value, which meets the requirements of (EN 13829). This measurement can be useful also in other locations of the ventilation system, because of very small required clearances: 15 cm upstream and 5 cm downstream.

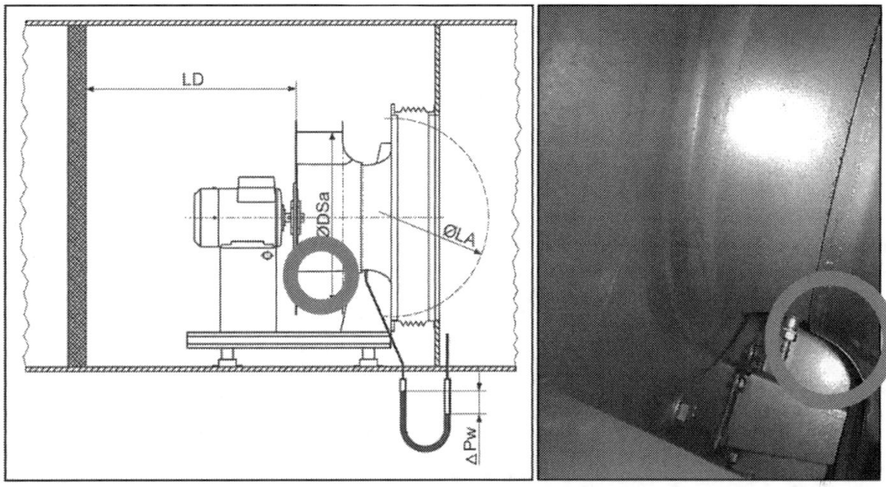

Fig. 1. Method of the air flow measurement inside AHU and location of the measuring connector (Ziehl-Abegg, 2013)

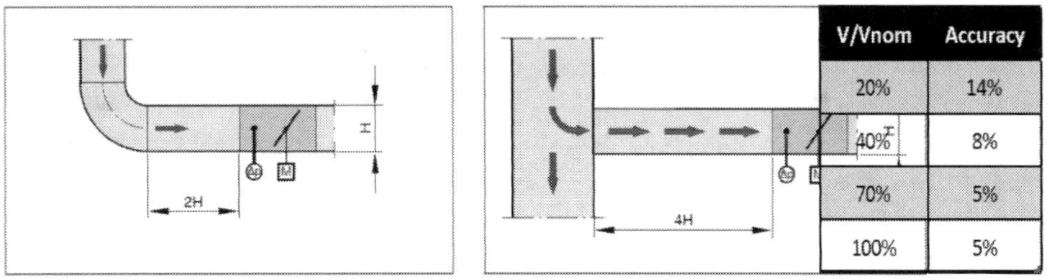

Fig. 2. Mounting conditions of VAV controller for correct air flow measurement and its accuracy (Trox)

Fig. 3. Air handler flowmeter temporarily mounted in the place of air filter (The Energy Conservatory, 2014)

3.3 Measurement of the air flow using VAV terminals

The measurement of the air flow on the VAV controllers is attractive because these devices are factory-equipped with a sensing element that generates an analog output signal proportional to air flow. However, the accuracy is variable and satisfies the requirements of (EN 13829) only when the elements are correctly mounted and when the flow is higher than 70% of the nominal flow (Fig. 2). Additionally, the measurement of the air flow using the terminal devices of ventilation system should take into account the possible leakage of air in the ventilation system – it was estimated for the case described in Section 5.

3.4 Application of the Building Management System (BMS)

In a modern public building, equipped with mechanical ventilation and appropriately designed control system and BMS, it is theoretically possible to perform the airtightness test even remotely. The only task to do on site is to close external doors and windows, and open the inner door, which does not require skilled staff. With the use of BMS, appropriate dampers have to be closed and selected fans should be operated at variable speed to reach the desired pressure in the building envelope. The results of the measurement is obtained from BMS as a record of key parameters: air flow rates, differential pressures etc. Such a measurement would be performed according to the method A of (EN 13829) and would correspond to the actual state of unused building, which takes place at night or on weekends. In practice, such a method of measurement encounters serious problems, examples of which are described in Section 5. Nevertheless BMS system in a building can be an effective support when carrying out airtightness test using AHU fans.

4. AQUAPARK - POZNAŃ

The tested building is a new, large waterpark, with an internal volume of about 200.000 m^3 and the A/V ratio of 0.20 m^{-1} – see Figure 4.

Fig. 4. Aquapark (sports and leisure complex) – Poznań, 2011

The initial air leakage test was ordered by the investor in 2012 to measurement company, which used the existing mechanical ventilation system. The contractor declared the flow measurement accuracy at AHUs of ± 2÷5%, despite no such a data from the manufacturer. During the building preparation and the airtightness test, several mistakes were made. The most important measurement mistakes were as follows:

- three-zone test with no simultaneous pressurization of adjacent zones (uncontrolled flows),
- airflow measurement was performed with unknown accuracy and several ventilation systems (not involved in the test) operating,
- building not prepared properly for the test: conducted with no control on inter-zonal opening positions (doors), some windows were opened.

The final, control measurement with use of BD set-up was carried out in 2013, as a part of the research of Institute of Environmental Engineering, Poznań University of Technology (IEE PUT). The facility was divided for this purpose into two zones: A – recreation area, B – sports pool hall and office area (Fig. 5).The pressurization in the tested zone was made by seven blower door fans set. In the adjacent zone the pressure was maintained at the same level (±3,0 Pa) with use of AHUs. Measurement of the flow rates on the fans (registered by the central BMS), gave the opportunity to compare the results for both zones, using concurrently Blower Door units and air handling units of unknown accuracy.

Table 1 presents the results of the primary and control measurements.

Table 1. The zone measurements results of the aquapark building V≈200.000 m^3

Year of measurement	Method of measurement	Accuracy of measurement	Adjacent zone pressurized?	Measuring team	n_{50} (h^{-1})	q_{50} ($m^3/(h \cdot m^2)$)
2012	AHU	unknown	No	External, 2 pers.	1,30	6,46
2013	BD	5%	Yes	IEE PUT*⁾, 10 pers.	**0,64**	3,17
2013	AHU	unknown	Yes	IEE PUT*⁾, 10 pers.	0,83	4,11

*⁾ IEE PUT = Institute of Environmental Engineering, Poznań University of Technology

Considering the measurement with use of BD devices as a reference, for the entire facility the following indicators were obtained: n_{50} = 0,64 h^{-1} and q_{50} = 3,17 $m^3/(h \cdot m^2)$.

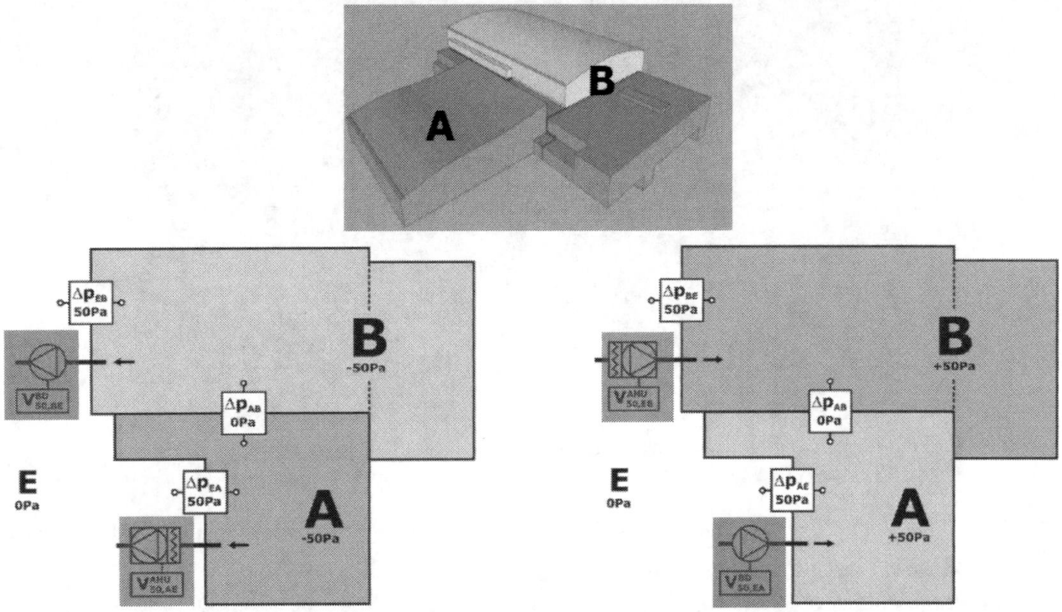

Fig. 5. Air tightness measurement with pressurization of adjacent zones (description: A, B - building zones, E - environment, AHU - Air Handling Units, BD - Blower Door); 3D model (middle); Sample measurements using a Blower Door for underpressure in zone B (left) and the overpressure in zone A (right)

The results obtained with the ventilation system use, with adjacent zones pressurized, are about 30% higher (n_{50} = 0,83 h^{-1}). It should be noted however, that the accuracy of airflow measurement in air handling units was not determined.

The primary measurement, performed in 2012 by the external test company, with the mistakes listed above, gave the result (n_{50} = 1,30 h^{-1}) of more than twice higher than reference one.

5. EDUCATIONAL AND RESEARCH BUILDING - POZNAŃ

The Mechatronics, Biomechanics and Nanoengineering Centre (MB&N) is a facility used for educational and scientific (research and development) purposes, with an internal volume of about 50 000 m³ and the A/V ratio of 0.20 m^{-1} – see Fig. 6.

Fig. 6. Educational and research building – Poznań, 2011

The building is equipped with 36 supply-exhaust ventilation lines and 40 exhaust lines, the total supply air flow is approximately 118.000 m³/h (2,3 h^{-1}). During the building commissioning process, the airtightness test with use of BD devices was properly carried out. After three years, in 2014, as part of research conducted by IEE PUT, the airtightness of the building was re-tested in two ways:

a) with use of BD set-up,
b) with use of ventilation system and BMS (air flow readings from VAV terminals).

The summary of measurements results is presented in Table 2.

Table 2. The measurements results of the educational building V≈50.000 m^3

Year of measurement	Method of measurement	n_{50} (h^{-1})	q_{50} ($m^3/(h \cdot m^2)$)
2011	BD	**0,30**	1,18
2014	BD	**0,44**	1,70
2014	VAV	0,54	2,07

Leakages in the building, during the first three years of operation increased by almost 50%. Major leaks were caused by ventilation dampers and their actuators damages as well as windows and doors damages.

The accuracy of VAV controllers installed in the building, according to the manufacturer's declaration, is 5-14% when properly assembled - Figure 2.

The measurement error resulting from the ventilation ducts leakages was estimated, based on the surface of ventilation ducts and their air tightness class, for about 2%. Possible leakages of the duct system would cause an underestimation of the test results in both cases, pressurization and depressurization.

The measurement with VAVs was more labor-intensive than using the BD, because of the need to manually adjust the ventilation system components. It was not possible to use the full capabilities of the BMS according to the approach described in section 3.4, mostly because of:

a) air intake and extract dampers are controlled by a common signal from the BMS - it is not possible e.g. remotely open intake damper and while closing the extract,
b) air handling units have safeguards to ensure equal supply and exhaust air flows - for example, it is not possible remotely turn on a supply fan with exhaust fan turned off.

Accordingly, dampers and fans control was carried out mainly in manual mode, and the BMS system was used primarily for data acquisition, on the air flow and pressure in the building – Figure 7.

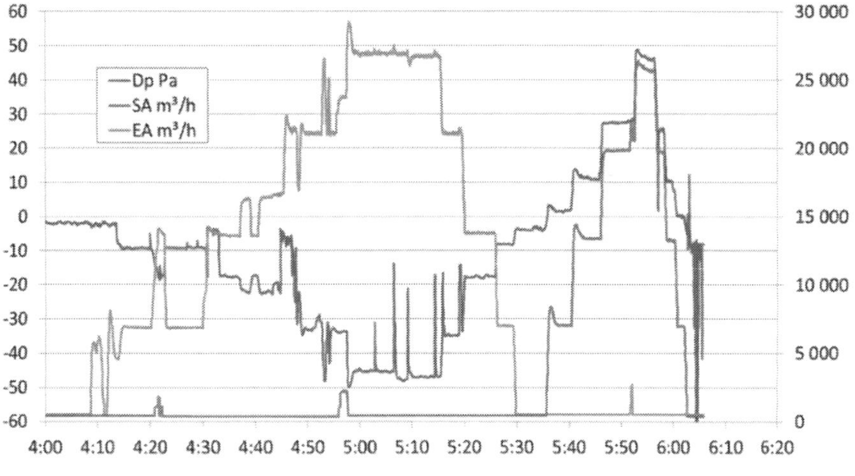

Fig. 7. Example data: air flows (SA – supply air, EA – exhaust air) and pressure difference (Dp) recorded by the BMS during the airtightness test.

6. CONCLUSIONS

Large buildings airtightness measurements using ventilation systems is a method acceptable by the (EN 13829) standard. It allows to conduct the airtightness test of a large building without application of blower-door fan set-ups, which can reduce the cost of the test. The accuracy of the air flow measurement is crucial in these cases. For this kind of tests the air flow rate can be measured at air intake, at fans (fan nozzles), at variable air volume terminals (VAV) or in the duct system. Standard equipment of the air handling units and of the ventilation system is often insufficient to achieve the desired accuracy of the measurement.

In the case of multi-zone airtightness measurements, the ventilation system is very suitable to generate a corresponding pressure in the adjacent zones. Another application is the coarse testing of the airtightness of the building, when the result of the measurement can be approximate. Measurement of airtightness of large buildings with the use of the existing ventilation system can be completed correctly in some cases, however, to obtain accurate results more expertise and effort are required than in the case of measurement with a set of typical Blower Door fans.

Lack of knowledge of airtightness testing procedures on the side of clients results in accepting the measurements which are conducted in the wrong way and whose results are incorrect. Establishment of an association of persons or companies, carrying out airtightness measurements could ensure a proper quality of services provided on the market.

7. REFERENCES

Canadian General Standards Board, CGSB-149.15-96, *Determination of the overall envelope airtightness of buildings by the fan pressurization method using the building's air handing systems.*

European Committee for Standardization (2002), EN 13829, *Determination of Air Permeability of Buildings. Fan Pressurization Method.*

European Committee for Standardization (2006), EN 1507, *Ventilation for buildings - Sheet metal air ducts with rectangular section - Requirements for strength and leakage.*

Górzeński R., Szymański M., Górka A., Mróz T., (2014), Airtightness of Buildings in Poland, The International Journal of Ventilation, 12/4, 391–400.

Jeong J.W., Firrantello J., Freihaut J., Bahnfleth W., Musser A., (2008), Case studies of building envelope leakage measurement using an air-handler fan pressurization approach, *Building Services Engineering Research and Technology 29*, 137–155.

Kim M.H., Jo J.H., Jeong J.W., (2013), Feasibility of building envelope air leakage measurement using combination of air-handler and blower door, *Energy and Buildings, 62*, 436–441.

The Energy Conservatory, *TrueFlow® Air Handler Flow Meter*, http://www.energyconservatory.com/products/trueflow%C2%AE-air-handler-flow-meter, (accessed 2014-07-15)

Trox Technik (2012), *Regelgeräte 2012*, http://www.trox.at/xpool/download/de/technische_unterlagen/regelgeraete/regelgeraete.pdf, (accessed 2014-07-15)

Ziehl-Abegg (2013), *Centrifugal Fans. Main Catalogue Part 1*, Edition 2013.

SIMULATION ANALYSIS FOR INDOOR TEMPERATURE INCREASE AND REDUCTION OF HEATING LOAD IN THE DETACHED HOUSE WITH BUOYANCY VENTILATED WALL IN WINTER

Kan LIN[1], Shinsuke KATO[2]

1 Institute of Industrial Science,
The University of Tokyo
Room Cw403, Building of IIS, 4-6-1 Komaba,
Meguro-ku, Tokyo, 153-8505 Japan

2 Institute of Industrial Science,
The University of Tokyo
Room Cw403, Building of IIS, 4-6-1 Komaba,
Meguro-ku, Tokyo, 153-8505 Japan
**Corresponding author: kato@iis.u-tokyo.ac.jp*

ABSTRACT

Detached residential wooden houses are a common type of housing in Japan. Decay of wooden components within the walls is easily caused by condensation or defective flushing. To solve this problem, a double-skin system with a room-side air gap was developed. In this system, during winter, the airflow in the ventilated wall circulates freely around the whole house. Therefore, during daytime, the airflow moves solar heat to base, and releases heat to the house at night which can increase indoor temperature. The purpose of this study is to evaluate the flow rate in the ventilated wall and evaluate indoor temperature increase and reduction of heating load in winter. Therefore, it is important to develop an expression for flow rate in the wall. An airflow-energy simulation program was used to predict flow rates in the ventilated wall, and the performance of air-flows in several exterior walls of the house was investigated. The results verify that the flow rate in the ventilated walls increase with solar radiation. By using the double-skin system with a room-side air gap, the heating load in winter was reduced by 5%.

KEYWORDS

Double-skin system, Simulation, Buoyancy ventilation, Heating load, Reduction

INTRODUCTION

Wooden detached houses, which have wood-based structural insulation materials for walls and flooring, are widely used in Japan. However, many problems exist with this type of home construction. For example, during the summer period, incoming solar radiation is not equally distributed, leading to uneven indoor temperature distribution. In the winter period, indoor humidity is closely related to the durability of the building envelope. Moisture leaking from the interior rooms increases the risk of condensation on the walls, because the surface temperature of the wall is low. Condensation on the walls may cause the wood used in the houses' construction to decay. In addition to condensation

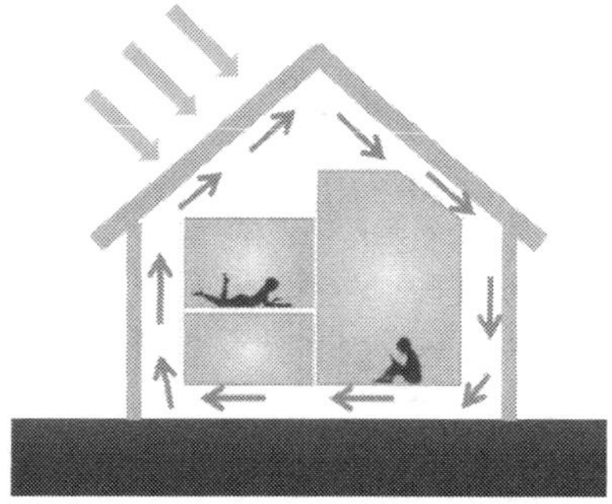

Figure 3: Double-skin system of room-side air gap with buoyancy ventilated wall applied to detached house in winter

from interior rooms, water vapour condenses on wooden window frames, and rainwater can leak through the roof and walls. The increased moisture is absorbed by the wooden structure of the house, and reduces the endurance and strength of the wood, and also accelerates its decay.

To remove the moisture between the external and internal walls, a double-skin system is commonly installed to detached houses. A double-skin system consists of a multi-layered façade, with a buffer space used for ventilation. This system can integrate the mechanical ventilation with natural ventilation so that air can move freely in the buffer space.

In summer, air vents in the basement are opened and the ventilation fans on top of the roof are turned on to facilitate airflow through the ventilated wall, and out the roof. Airflow in the ventilated wall removes heat from solar radiation and from the internal wall. The airflow discharges heat in the wall and attic space, while obtaining cold heat from the ground under the floor. This ensures that the rooms are kept in moderate temperature and comfort.

In winter, the airflow in the ventilated wall circulates freely around the whole house. Therefore, during daytime, the airflow moves solar heat to base, and releases heat to the house at night which can increase indoor temperature.

Although houses using this system have been built for about 30 years in Japan, only a few studies have been conducted on them. Using a simulation of a model house from previous research, Ozaki et al. analyzed the airflow speed in the wall, as well as the airflow in the space under the floor during the summer period. However, there is hardly any information available regarding the actual airflow and thermal effects of this system. For this reason, it is difficult to select appropriate strategies to maintain airflow rate in the ventilated wall.

The objective of this study is to understand the basic properties of the double-skin system, temperature increase, and reduction in heating load during the summer period by performing a ventilation network simulation. To perform calculations during the winter period, we used a network simulation by creating two building models with spaces in the air gap and rooms of the house that were regarded as independent nodes. The impact of each element to the airflow rate was estimated, and the heating load reduction effect of the system was examined in the simulation.

The results show airflow rate and speed for each wall and reduction in the heating load when buoyancy ventilated wall was used to detached house. The reduction in the heating load increased significantly when buoyancy wall was used. In addition, natural indoor temperature increased slightly.

METHODOLOGIES

Simulation Software

To calculate the airflow in the buffer space, we used the variable energy calculation software TRNSYS17, and its add-on program TRNFlow. TRNFlow is designed based on COMIS3.1, a ventilation network calculation software, and can perform the iterative calculations by resolving the movement of the air and heat at the same time in TRNSYS.

Building Model

In this research, the house model (4th region) proposed by IBEC, Institute for Building Environment and Energy Conservation, as a standard detached house in Japan was used in the simulation. This house model is a two-story house with 120.07 m^2 total floor area, and it is intended for a four-person family. There are several spaces divided by rooms, as shown in figure 2. Occupants' schedules as well as the lighting, equipment, ventilation, and air conditioning are based on the investigation. The dimensions and the basic configuration of the building were set based on this model. In addition, the dimensions of the air gap in the ventilated wall were set based on the real house.

The air gap in the ventilated wall is shown in figure 3. The area of the air gap in which air moves upwards is represented as the blue portion. Grooves are arranged on the surface of the insulation material in equal intervals. These create openings between the insulation and the other components, so that air flows freely. The air gap is divided into several zones by resistance parts, such as vents or these openings. In such zones, the flow path is narrower than the open air gap. For the calculations by TRNFlow, we modeled the air gap as an "airnode," and defined any resistance parts as an "airlink." For each airlink, the relationship of the mass flow rate and the pressure difference is calculated based on equation (1). In equation (1), a common formula for ventilation, the value of C_S, for each airlink, was decided by fixing α to 0.6 as a typical value. Also determined was the opening area A of the airlink, from the actual specification of this house. The value of n was determined to be 0.5, from the above substitution for the opening area, which is reflected in equation (2).

$$\dot{m} = C_s \cdot (\Delta p)^n \qquad (1)$$

$$\dot{m} = \alpha A \sqrt{2\rho \cdot \Delta p} \qquad (2)$$

$\dot{m}$: mass flow rate [kg / h], α: flow coefficient [-], A: opening area [m^2],
ρ: air density [kg/m^3] , Δp: pressure difference [Pa]

In this simulation, a simple building model that does not take into account the air infiltration, was used; neither air leaking into the rooms, nor outside air from the gas passage is considered. A simple network that connects the air gap to the outside air was created, as shown in figure 4.

With this model, it is possible to estimate the potential of the maximum amount of airflow generated by mechanical ventilation, regardless of the influence of air infiltration. On the other hand, because heat from solar radiation and internal heat cannot be discharged by infiltration, simulation results may show that the room temperature is higher than the actual environment.

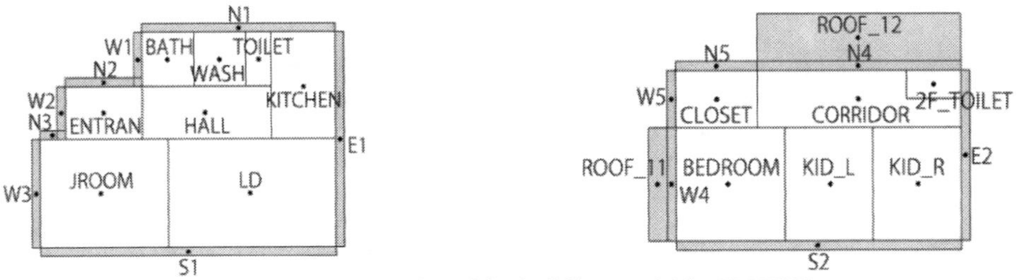

Figure 2: Zones and airnodes of the building model in TRNSYS

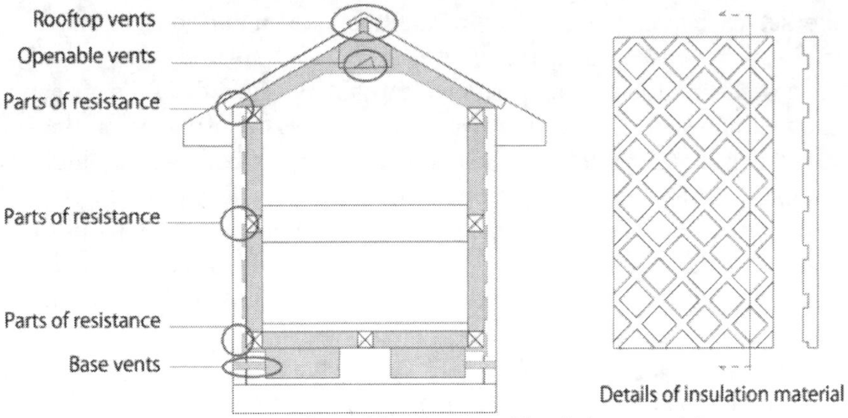

Figure 3: Overview of the air gap and insulation material

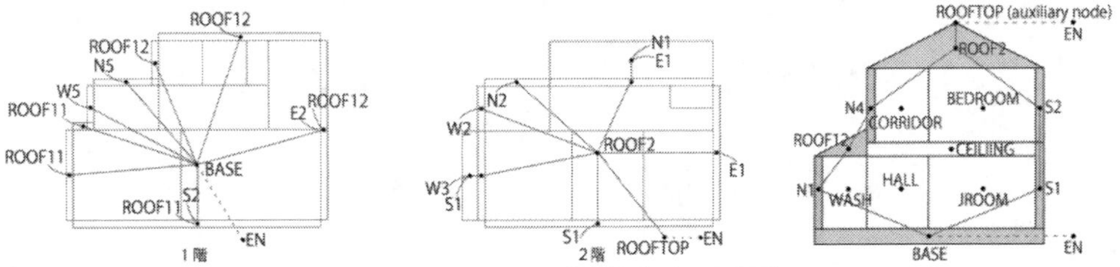

Figure 4 : Overview of airlinks in TRNFlow

Table 1: Building Properties

Wall of external side	Extruded polystyrene foam = 50 mm + Air layer = 15 mm + Tile = 10 mm
Wall of internal side	Gypsum board = 13 mm
Air gap thickness	120 mm
Roof	Extruded polystyrene foam = 50 mm + Air layer = 30 mm + Plywood = 10 mm + Slate = 5 mm
Base	EPS 50 mm + RC 150 mm + EPS 50 mm
Internal wall, Ceiling, Floor	U = 3.125 W/m^{2}K, 4.082 W/m^{2}K, 4.082 W/m^{2}K
Window, double-glazing	U = 1.8 W/m^{2}K, Shading coefficient = 0.5
Infiltration	None

Table 2: Ventilated Wall Setting

Opening area A [m^2]	Part of resistance in the air gap	0.00396 m^2/m (opening area per unit length wall)
	Ventilation unit at the rooftop to outside	0.0919
	Attic space to the ventilation unit at the rooftop	0.0755
	Outside to under-floor space	0.0720
	Constants	$\alpha = 0.6$, $\rho = 1.2$, n = 0.5

Table 3: Calculation Conditions

Internal heat gain	Exists (based on the schedule by IBEC)
Ventilation	None
Air-conditioning	None, Exists (based on the schedule by IBEC)
Ground temperature at 1m depth	9°C
Weather data	Expanded AMeDAS standard data (2000) , Tokyo
External wind	None
Calculation period	11/1 - 2/28 (plus 3 days for run-up period)

Calculation Conditions

The calculation conditions used in this study are listed in Table 1 through Table 4. Table 1 lists the properties of the walls, air gap, windows, etc. The conditions of the ventilated wall are listed in Table 2, and the calculation conditions are listed in Table 3. For evaluating the airflow rate and reduction effects, a case with buoyancy ventilated wall and a case without buoyancy ventilated wall are supposed.

To estimate the properties of airflow in the ventilated wall at natural room temperature, the simulation was first performed without air-conditioning. Then, calculations were carried out with air conditioning, and the effect of heating load reduction due to heat carried by airflow in the wall was estimated. The air-conditioning temperature was set to 20°C, and a total of five rooms—living room, kitchen, bedroom, and two children's rooms—were regarded as air-conditioned.

As shown in Table 3, the schedule of internal heat generation, ventilation, and air-conditioning was set according to the criteria of the IBEC standard. The ground temperature at 1m depth was set to 9°C, in reference to the results of past measurements. For weather data, including external wind, AMeDAS' expansion 2000 standard data in Tokyo was used. The calculation period was 11/1 - 2/28 (plus 3 days for run-up period) in the study.

RESULTS AND DISCUSSION

The temperature decrease during representative day is shown in Figure 5. Indoor temperature and the temperature of base change along with the variation of ambient temperature. The figure shows the average temperature of base is around 12°C when the ground temperature at 1m depth was set to 9°C.

Figure 6 shows the temperature decrease during representative day. The effect of the temperature decrease in 2 rooms is slight. The temperature in the living room increases by 0.3°C, when buoyancy ventilated wall is used. Temperature decrease in the bedroom is smaller than temperature decrease in the living room. This is because the heating schedule in the living room concentrates during the daytime while heating schedules of the bedroom concentrate in the nighttime.

Figure 7 shows flow rate for each wall during representative days, by direction. The figure shows both positive airflow and negative airflow in walls, and flow direction along a wall (upward direction) is considered positive. Based on the graph, the flow rate for each wall is affected by the outdoor temperature. The flow rates in the wall N1and N4 change drastically, and the biggest flow rate is as much as $2m^3/h$. The flow rates in the walls N2, N3, and N5 are low when compared to the walls N1 and N4, which range from -0.5 to $0.5m^3/h$. In east walls and south walls, all walls change drastically due to the influence of solar radiation during daytime. The biggest flow rate is as much as $-2m^3/h$, which range from -2 to $0.5m^3/h$. In west walls, all walls change slightly because these walls are not seriously affected at solar radiation during daytime.

Figure 8 shows flow speed for each wall during representative days. The figure shows both positive airflow and negative airflow in walls, and flow direction along a wall (upward direction) is considered positive. Based on the graph, the flow rate for each wall is affected by the outdoor temperature. The flow rates in the wall N3and N5 change drastically. The flow speed in the walls N1, and N4 change gently when compared to the walls N3 and N5, which range from 1 to 2.5cm/s. In east walls and south walls, all walls change drastically due to the influence of solar radiation during daytime. The biggest flow rate is as much as 2cm/s, which range from -1 to 2cm/s.

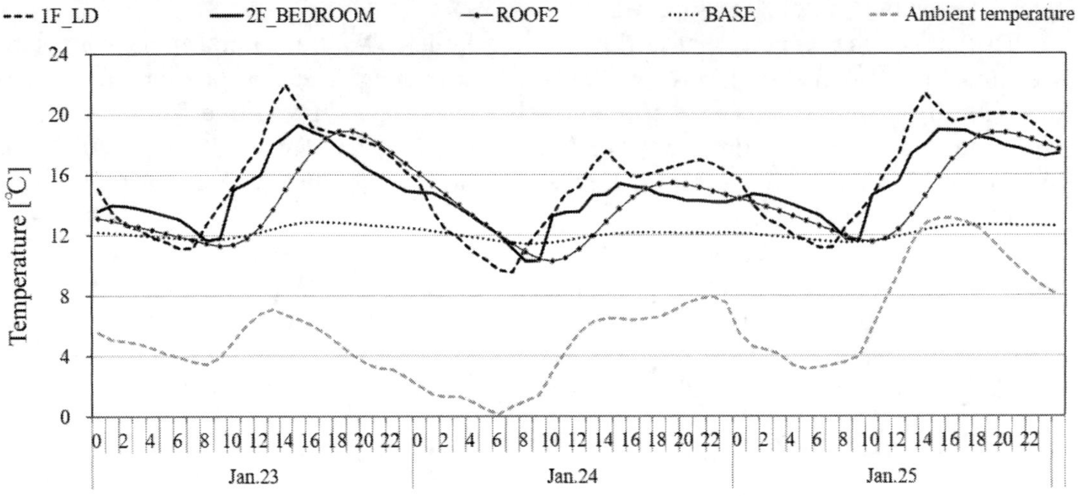

Figure 5: Temperature during representative days

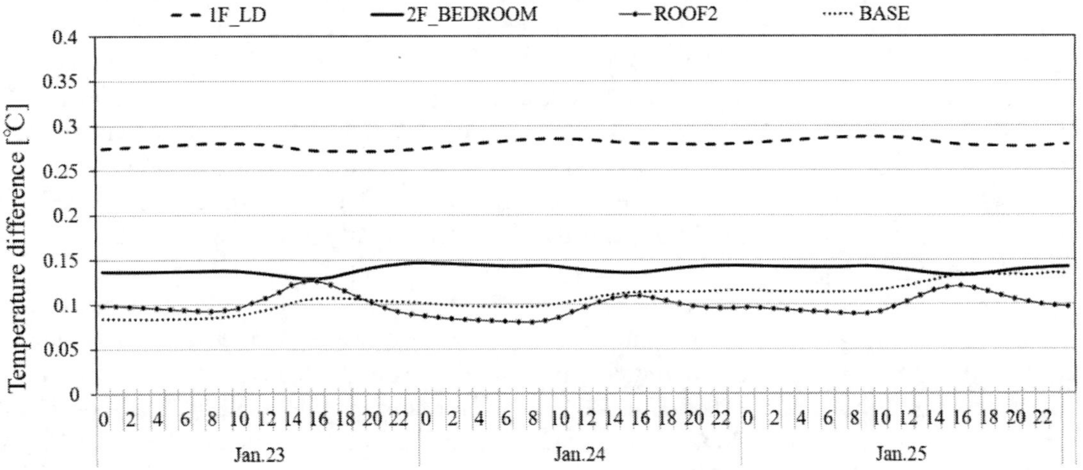

Figure 6: Temperature decrease during representative days

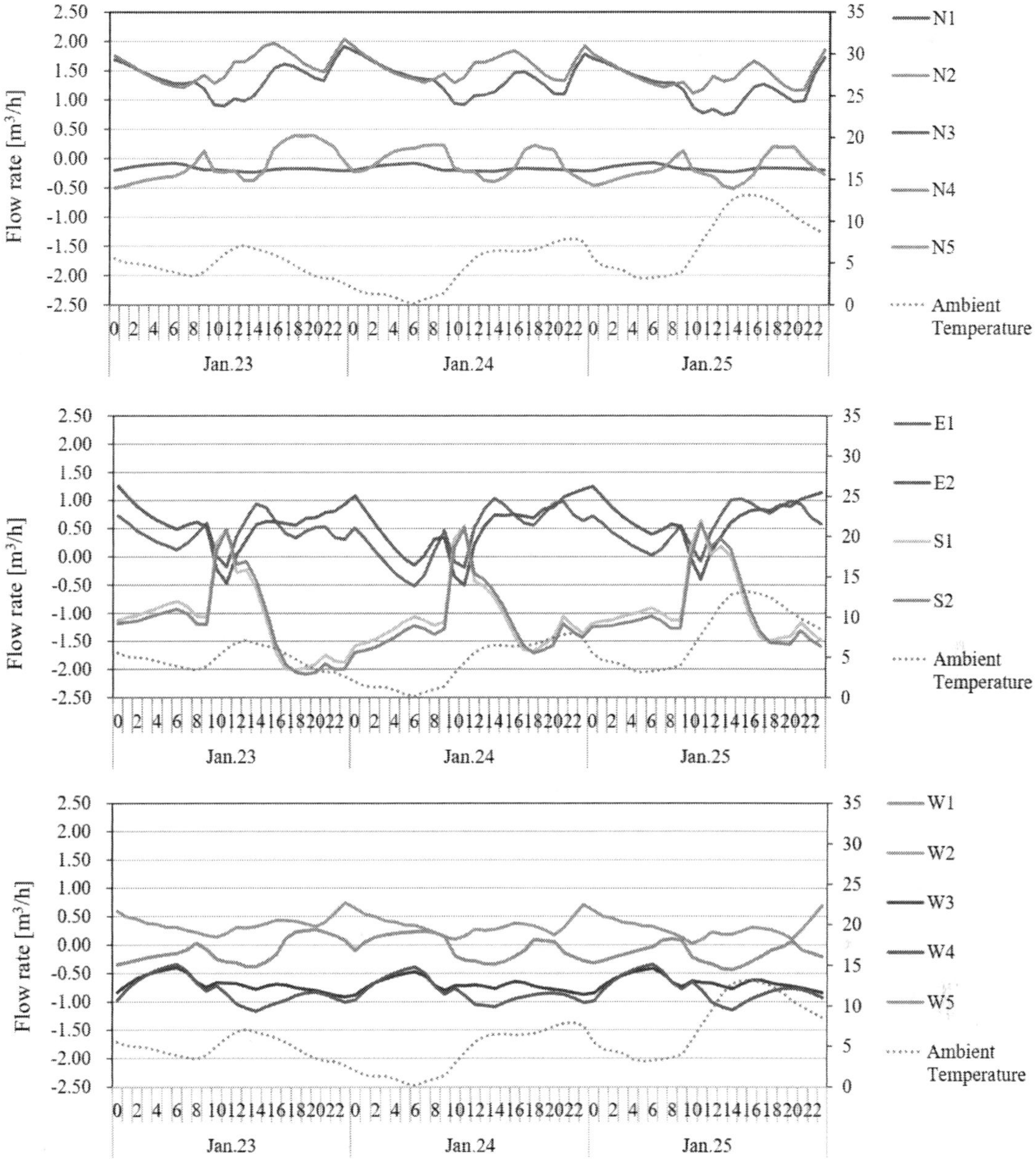

Figure 7: Flow rate for each wall during representative days

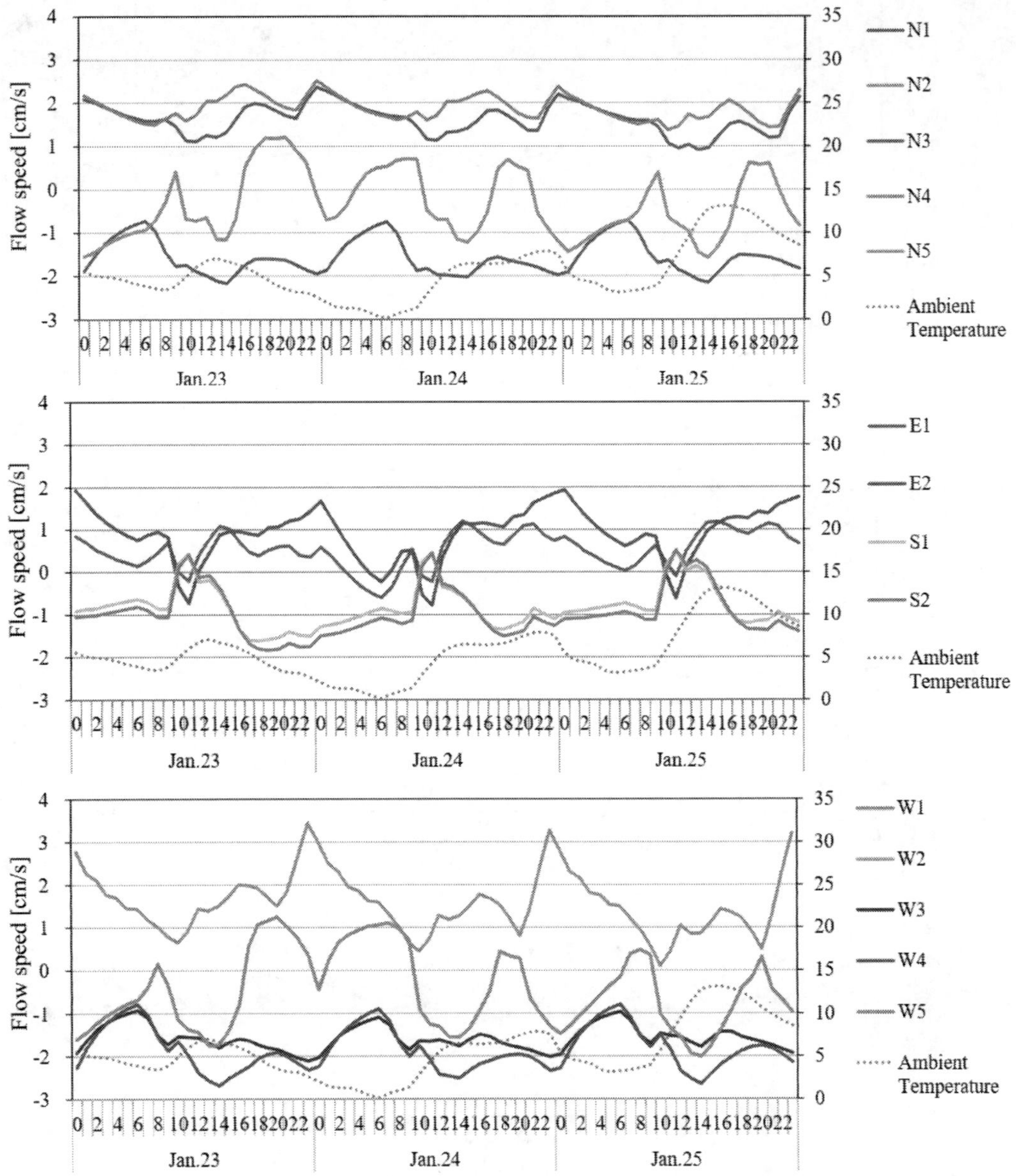

Figure 8: Flow speed for each wall during representative days

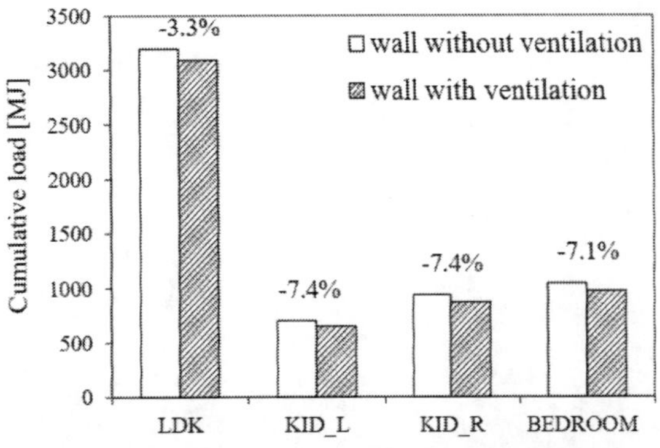

Figure 9: Total heating load in each room

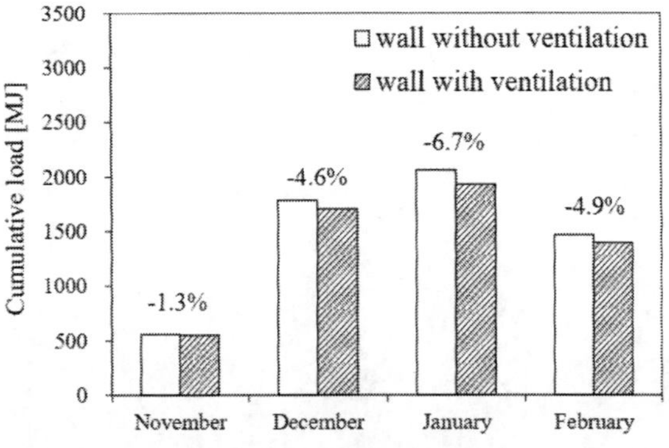

Figure 10: Total heating load in each month

Figure 9 shows the total heating load in the ventilated walls for each room, for the case with air-conditioning set at 20°C. Results for the four rooms are shown in the graph. The two children's rooms receive the largest heating load reduction because it consumes the least energy among the four rooms. The average heating load reduction for the dining room from June to September was reduced by 5.1%.

Figure 10 shows the total heating load from November to February for each month. The heating load during the period from November to February was reduced by 1.3%, 4.6%, 6.7%, and 4.9%, respectively. The reduction was largest in January in which the temperature was lowest.

CONCLUSIONS

In this study, we evaluated the airflow rate in the air gap of a double-skin system, the heating load reduction, and the temperature increase by performing a ventilation network simulation. The temperature in the living room increases by 0.3°C, when buoyancy ventilated wall is used. Temperature decrease in the bedroom is smaller than temperature decrease in the living room. This is because the heating schedule in the living room concentrates during the daytime while heating schedules of the bedroom concentrate in the nighttime.

In general, the average heating load reduction for the dining room from June to September was reduced by 5.1%. The two children's rooms receive the largest heating load reduction because it consumes the least energy among the four rooms. The heating load during the period from November to February was reduced by 1.3%, 4.6%, 6.7%, and 4.9%, respectively. The reduction was largest in January in which the temperature was lowest.

It was found that the total period heating load is reduced while the ventilated wall is working and the air-conditioning is on. It seems that this effect is mainly because the air-conditioning settling time was reduced by the room temperature increases. As a result, it is expected that the heating load supplied from subfloor ground surface Therefore, a further simulation of geothermal heat is necessary for future study. Furthermore, because heat from solar radiation and internal heat cannot be discharged by infiltration, the simulation result shows the room temperature is higher than that of the actual environment. To evaluate the influence of infiltration, a simulation that considers infiltration between the interior-exterior wall interface and the wall-room interface is needed in further research.

REFERENCES

Ozaki, A. (1992). Heat and water transfer through an ventilated structure using urethane panels [in Japanese]. *Summaries of technical papers of Annual Meeting Architectural Institute of Japan, D,* 945-946.

Institute for Building Environment and Energy Conservation. (2009). *Description of the energy consumption calculation method in the standards of judgment of the housing business building owners* [in Japanese].

Kurabuchi, T. (2009). Wind tunnel experiments on surface wind pressure and cross-ventilation flow rate in densely populated residential area: study on proper design method of locating windows aiming at utilization of cross-ventilation in densely populated residential area (part 1) [in Japanese]. *Journal of environmental engineering 74(642), 951-956.*

DEVELOPMENT OF A UNIQUE THERMAL AND INDOOR AIR QUALITY PROBABILISTIC MODELLING TOOL FOR ASSESSING THE IMPACT OF LOWERING BUILDING VENTILATION RATES

James McGrath[1], Miriam Byrne[*1]

1 School of Physics and Centre for Climate and Air Pollution Studies,
Ryan Institute,
National University of Ireland, Galway,
University Road,
Galway
Ireland

ABSTRACT

Adequate ventilation is necessary to maintain thermal comfort and remove indoor air pollutant concentrations (Crump et al., 2005). Indoor pollutant concentrations vary considerably depending on occupants' behaviour patterns, building characteristics and meteorological parameters and seasonal effects. Experimental measurements are time consuming and expensive to carry out, while computational models are regarded as a valid complement. The team at NUI, Galway have recently developed the IAPPEM model (described in (McGrath et al., 2014a)representing the state-of-the-art in probabilistic modelling of indoor air quality; the model is currently capable of highlighting locations of different pollutant concentrations within the same building by considering the infiltration of outdoor air pollutants, meteorological parameters and seasonal effects, indoor activities of occupants, emissions from indoor air pollutants, removal by deposition, the dilution of indoor air pollutants through external and internal air exchange and internal house layout.

To date no computational model exists that examines both inter-zonal and external temperatures in combination with assessing indoor air pollutants, and the enhancement of the IAPPEM source code framework to incorporate these features constitutes the main tasks of the current work. This will allow an assessment of individuals' thermal comfort and exposure to airborne pollutants, and will also allow an assessment of the heat loss to surroundings as a result of changing ventilation rates, a feature that is highly relevant in the context of energy efficient homes. While previous studies have examined the effect of heat loss due to ventilation, no model to date has combined the temperature and airborne pollutant components in a single computational structure.

In this paper, a plan to incorporate a temperature parameter into the existing model is described; this would allow the unique capability to assess, for the first time, the effects of ventilation and occupant behaviour on both an individual's thermal comfort and airborne pollutant exposure. The basis of the IAPPEM code adaptation strategy will be the addition of a source term and a loss term to account for localised heating sources, such as radiators in each room, solid fuel fires, and cooking events. A loss term will account for the removal of heat through the building walls and windows. Ventilation rates will define the transfer of heat between rooms but also between indoor and outdoor, allowing the assessment of heat loss as well as thermal comfort. This would provide a vital tool in defining optimum building air exchange rates that do not have a negative impact on human health.

.

KEYWORDS

Indoor Air Quality, Thermal Comfort, Probabilistic modelling, emission sources,

1 INTRODUCTION

Particulate Matter (PM) has become a major environmental concern because of its known impacts on human health (COMEAP, 2009, COMEAP, 2010). A WHO project REVIHAAP (WHO, 2013) reported a large body of evidence linking the effects of long term PM exposure on both cardiovascular mortality and all causes of mortality. PM related morbidity outcomes include aggravated asthma, chronic respiratory disease, increased emergency room visits and hospital admissions, acute respiratory symptoms, decrease in lung function and even premature mortality. (Brook et al., 2010) reported that short-term PM2.5 exposure over a few hours to weeks can trigger cardiovascular morbidity and mortality events. WHO (2013) reported that significant evidence, based on toxicological and clinical studies, linking peak exposures to combustion-derived particles of short duration (ranging from less than an hour to a few hours) leads to immediate physiological changes.

Most of the general population in North America or Europe spent 89% of their time indoors, with 69% spent in the residential indoor environment (Klepeis et al., 2001, Schweizer et al., 2006), and homemakers or the elderly spending up to 90% of their time in the residential indoor environment (Torfs et al., 2008). Therefore, the residential environment deserves particular attention.

The need to achieve carbon and greenhouse gas emissions has promoted energy efficiency improvements because of the potential saving in energy costs. Reduction in energy usage in buildings is a significant component of the national CO_2 reduction strategies. For example, the Technical Guidance Document L of Part L of the Irish Building Regulations (SI 259 of 2011) sets out new performance levels for air permeability to be 7 $m^3/hr/m^2$ at 50Pa. Under the European directive (2010/31/EU, 2010), all new building must adhere to minimum energy performance requirements. Irish legislation (S.I. No. 259/2008, 2008) refers to reducing heat loss in building through increasing the air tightness.

However, as buildings air-tightness increases, a reduction occurs in the air exchange rates (AERs) reducing heat loss, while increasing the indoor pollutant concentrations if significant indoor sources are present (Gens et al., 2014). Wilkinson et al. (2009) reported that changes in energy efficiencies in domestic homes may impact on indoor air quality, due to increasing air-tightness without alternative methods of reducing indoor generated pollutants. Bone et al. (2010) highlights that the driver for more energy efficiency homes will harm occupant health.

The World Health Organization's (WHO, 2010) guidance on thermal comfort ensures satisfaction with the ambient temperature, but also links to human health. The guidance for the home environment aims at protecting human health particularly those most susceptible, such as the very young and the elderly (Ormandy and Ezratty, 2012). Additionally, thermal comfort has been reported to influence individuals' behavioural patterns; Andersen et al. (2009) reported that window opening behaviour in Danish dwellings is strongly linked to the outdoor temperature. (Rijal et al., 2007, 2008) found relationships between the thermal environment and the occupants' behaviour in relation to windows/doors opening patterns and the usage of HVAC system. McGrath et al. (2014b) investigated variations in PM concentrations due to interzonal airflow in a six-room apartment layout. The study found that peak concentrations in adjoining rooms were directly linked to the time of the door opening/closing events.

Andersen et al. (2009) reported that the use of internal heating was correlated with the outdoor temperature and the presence of a wood-burning stove within the dwelling. In Ireland, the use of solid fuel burning in fireplaces is often used as means of secondary heating source (McGrath et al., 2011, Semple et al., 2012).

Indoor PM concentrations are affected by infiltration of outdoor particles, meteorological parameters and seasonal effects, indoor activities of occupants, emissions from indoor PM sources, removal of particulates by deposition, the dilution of indoor PM through external and internal airflow and internal house layout (Ferro et al., 2009, Singer et al., 2002, Ott, 1999). Numerous studies have highlighted the indoor activities that contribute to PM concentrations, such as smoking, frying, solid fuel fire and use of incense and candles (He et al., 2004, See and Balasubramanian, 2011, Semple and Latif, 2014) and also resuspension activities such as walking, dusting or sitting on furniture (Ferro et al., 2004, Jing et al., 2008).

It is often impractical or expensive to obtain large-scale indoor or personal exposure measurements to account for the range of indoor factors (indoor emission sources, indoor deposition and resuspension, and the occupant's behaviour). Computational modelling is an effective tool at separating out the contribution of indoor emission source from outdoor generated air pollution; this approach allows for effective exposure reduction strategies to be devised. For the proper management of air quality to be fully evaluated, indoor exposures need to be further separated into the contributions of indoor sources and the penetration of outdoor pollution. Modelling approaches are essential if this is be effectively achieved (Dimitroulopoulou et al., 2000).

Steinle et al. (2013) highlights that exposure models are an appropriate tool to increase sample sizes and reduce the cost of experimental exposure studies. Exposure models need to integrate a wide range of factors (e.g. economic, social and demographic), to assess relations and associations between human exposure to environmental air pollutants and potential health effects. In recent years, a number of models have predicted indoor pollutant concentrations in indoor environments (Dimitroulopoulou et al., 2006, Fabian et al., 2012, Sohn et al., 2007, Parker and Bowman, 2011). Fabian et al., (2012) reported challenges imposed on simulations due to the large variation in emission strengths. Limitations in these physical pollutant models have led to poorer estimations of indoor air pollutant exposure assessment.

Probabilistic modelling is one approach that overcomes these issue. McGrath et al. (2014a) developed IAPPEM, a state-of-the-art probabilistic model for indoor PM concentration estimation, capable of simulating up to 15 interconnecting rooms with the incorporation of up to 12 simultaneously-operating emission sources. IAPPEM proved to be capable of simulating the large variation in emission rates, predicting both peak and mean PM_{10} and $PM_{2.5}$ concentrations. IAPPEM provided a detailed analysis of overall PM contribution from multiple different emission sources, in a variety of different internal locations in a dwelling, and the effect that both emission source location and internal household configuration has on PM transfer throughout a dwelling has been quantified.

The current work plans to incorporate temperature parameters into the existing model; this would allow the unique capability to assess, for the first time, the effects of ventilation and occupant behaviour on both an individual's thermal comfort and airborne pollutant exposure. This would provide a vital computational tool that has wide relevance in epidemiology studies, building design, ventilation studies, traffic management studies, and many other fields.

2. METHODOLOGY

2.1 Air pollutant model

The current model, titled Indoor Air Pollutant Passive Exposure Model (IAPPEM), is an advanced probabilistic modelling tool that evaluates the contribution of both indoor and outdoor sources to the air pollution concentration in the indoor environment. IAPPEM calculates the change in indoor pollutant concentrations by solving the differential equation 1; considering the infiltration of outdoor air pollution, the generation of air pollution indoors and its transport between rooms, and the indoor deposition of air pollution. At each time interval, probability density functions are used to simulate a range of possible values for each parameter, assessing the range of likely outcomes when specific details are unavailable. This can encompass uncertainties in experimental obtained data, but can also encompass uncertainties in the selection of appropriate modelling parameters between studies. Further details on the parameterisation of the pollutant model are available in McGrath et al. (2014a).

$$\frac{dC_k}{dt} = \frac{(\lambda_{0k})}{V_k}(f_k * (C_0 - C_k)) - v_g(\frac{A_k}{V_k})C_k + \frac{Q_k}{V_k} + \sum_{i=1}^{n} \frac{\lambda_{ik}}{V_k}(C_i - C_o) \quad (1)$$

Equation 1 is solved for each k, where k represents each individual room. Subscripts of 0, 1 and 2 are used to represent outside, room 1 and room 2 for different parameters. C_k represents the concentrations of the pollutant in that room □g m^{-3}, where $C_□$ represents the outdoor concentration (□g m^{-3}). f_k represents the building filtration factors between the outdoor and that room. v_g is the deposition velocity of the pollutant (m hr^{-1}) $□_{ik}$ is the interzonal airflow between internal rooms, e.g. $□_{12}$ represents the transport of pollutants from room 1 into room 2, and $□_{0k}$ the external airflow between the outside and room k (m^3 hr^{-1}). A_k is the surface area of room k (m^2). V_k is the volume of room k (m^3). Q_k is the indoor emission rate of the pollutant in room k (□g hr^{-1}).

2.2 Addition of a Temperature Parameter

The current work comprises of altering the model's source code to include the temperature parameters, allocating a source file to account for time-varying outdoor temperature concentrations, as well as incorporating a temperature parameter to each room. A 'source' term will allow for internal heating sources (solid-fuel burning, central heating, electric fan heater or HVAC). A 'decay' term will allowing for heat losses through the building fabric through external and internal walls, as well as floors and ceilings. Airflow rates in Equation 1 will be used to assess the heat loss to the outdoors but also the internal transfer of heat between zones. Additionally, a heat 'recovery' term will be added to account for any heat recovery systems.

The model's parameterisation will account for variation in building fabrics; window types and wall and roof insulation. Altering the model's parameterisation will allow assessments of performance pre and post retrofitting a building.

3. RESULTS

The application of the air pollutant model is demonstrated by a number of simulations below. The simulation are based on the 11-room dwelling described in McGrath et al. (2014a). The emission scenarios comprised of two frying events in the kitchen and five smoking events in the living room while internal doors remained closed. Each frying event lasts for 15-minutes, while each smoking event lasts for nine minutes. Figure 1 and 2 shows a section of the 24hour time-series profile for $PM_{2.5}$ concentrations in both the kitchen and the living room.

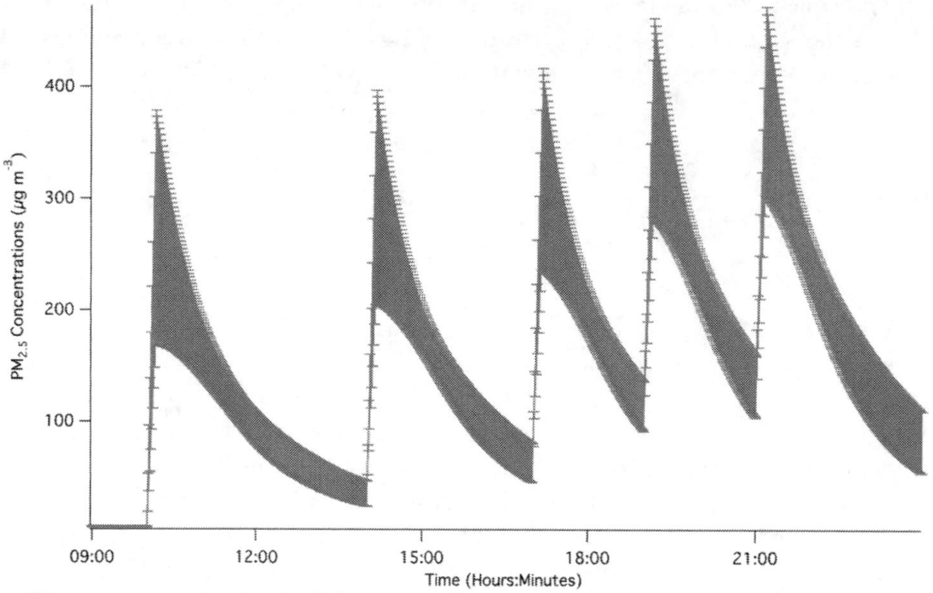

Figure 1. $PM_{2.5}$ concentrations in the living room. The time axes have been scaled to focus on the emission periods. The 'y' bars represent one standard deviation at each time step, highlighting the probabilistic nature of the model. Each of the peak concentrations refers to the end of the smoking events.

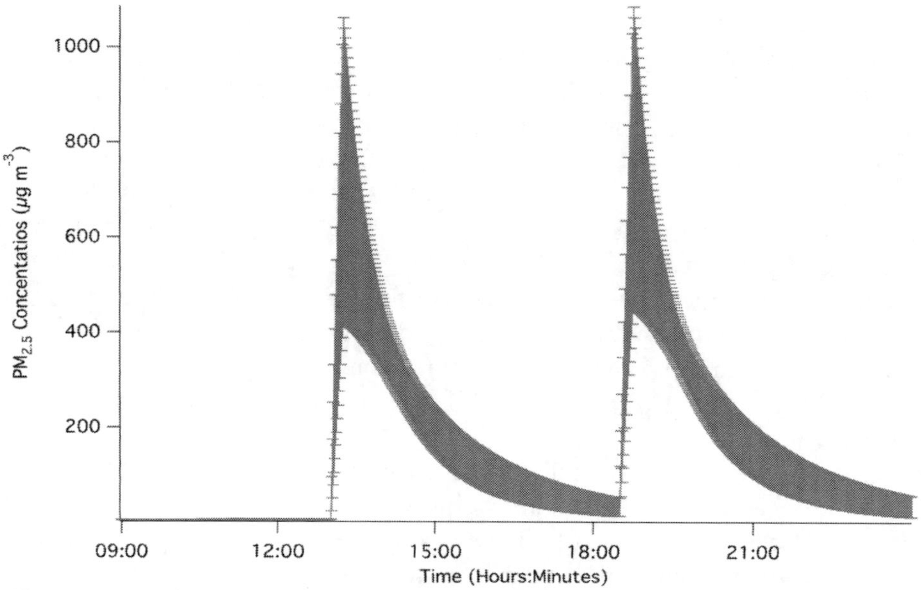

Figure 2. $PM_{2.5}$ concentrations in the kitchen. The time axes have been scaled to focus on the emission periods. The 'y' bars represent one standard deviation at each time step, highlighting the probabilistic nature of the model. Each of the peak concentrations refers to the end of the frying event.

Table 1 shows the effect that reducing AERs have on indoor $PM_{2.5}$ concentrations. The values of 0.30, 0.44, 0.70 ACH^{-1} reflect similar AERs reported in literature for European dwelling (Crump et al., 2005, Dimitroulopoulou, 2012). It can be seen that PM concentrations increase with decreasing AERs; this is a major concern considering that the World Heath Organisation (WHO, 2006, 2013) recommends a 24-hour mean $PM_{2.5}$ concentration guideline of 25 □g m^{-3}.

Table 1 shows the 24-mean PM2.5 concentrations in both the kitchen and sitting room under three different air exchange rates in units of Air Changes per Hour (ACH^{-1}).

Location	0.7 ACH^{-1}	0.44 ACH^{-1}	0.3 ACH^{-1}
Kitchen	70 ± 6□□g m^{-3}	95 ± 9 □g m^{-3}	112 ± 12 □g m^{-3}
Sitting Room	74 ± 4□□g m^{-3}	100 ± 7□□g m^{-3}	122 ± 10 □g m^{-3}

The proposed enhancement to the model will simultaneously allow time-series analysis of internal room temperatures in combination with indoor air quality. Any proposed changes in AERs can then be assessed in terms of potential energy saving strategies to analyse the effect on the reduction in heat loss but also the effect on indoor air quality.

4. CONCLUSIONS

The modified model will make simulations to assess both thermal comfort and indoor air pollutants, resulting in a time-series analysis of both parameters. This will allow simulations of variations in household characteristics, occupants behavioral patterns, mechanical and natural ventilation.

The results of this project will inform policy makers and the public on energy usage reduction in the built environment sector, and any mis-understanding regarding potentially adverse effects of reduced ventilation on indoor air quality could be extremely detrimental in this regard. The model can assess the potential benefits of retrofitting a residential dwelling while ensuring there is no reduction in indoor air quality.

This unique computational tool can be used to define the optimum ventilation rates for energy efficient homes ensuring thermal comfort and also an appropriate level of indoor air quality. One potential strategy could examine increasing the external airflow in rooms that experience higher air pollutant concentrations while reducing the external airflow in rooms that experience lower air pollutant concentrations.

5 ACKNOWLEDGEMENTS

This work was funded by Sustainable Energy Authority of Ireland (SEAI) under the SEAI Renewable RD&D Programme 2014.

6 REFERENCES

ANDERSEN, R. V., TOFTUM, J., ANDERSEN, K. K. & OLESEN, B. W. 2009. Survey of occupant behaviour and control of indoor environment in Danish dwellings. *Energy and Buildings,* 41, 11-16.

BONE, A., MURRAY, V., MYERS, I., DENGEL, A. & CRUMP, D. 2010. Will drivers for home energy efficiency harm occupant health? *Perspectives in public health,* 130, 233-238.

BROOK, R. D., RAJAGOPALAN, S., POPE, C. A., BROOK, J. R., BHATNAGAR, A., DIEZ-ROUX, A. V., HOLGUIN, F., HONG, Y., LUEPKER, R. V. & MITTLEMAN, M. A. 2010. Particulate matter air pollution and cardiovascular disease an update to the scientific statement from the American Heart Association. *Circulation,* 121, 2331-2378.

COMEAP 2009. Long term exposure to air pollution: Effect on mortality.: Committee on the Medical Effects of Air Pollutants.

COMEAP 2010. Review of Evidence on Health Aspects of Air Pollution REVIHAAP Project: Final technical report.

CRUMP, D., DIMITROULOPOULOU, S., SQUIRE, R., ROSS, D., PIERCE, B., WHITE, M., BROWN, V. & COWARD, S. 2005. Ventilation and indoor air quality in new homes. *Pollution atmosphérique,* 1, 71.

DIMITROULOPOULOU, C. 2012. Ventilation in European dwellings: A review. *Building and Environment,* 47, 109-125.

DIMITROULOPOULOU, C., ASHMORE, M. R., BYRNE, M. A., HILL, M. T. R., KINNERSLEY, R. P., MARK, D. & NI RIAIN, C. 2000. Modelling of indoor aerosol concentrations in UK buildings. *Journal of Aerosol Science,* 31.

DIMITROULOPOULOU, C., ASHMORE, M. R., HILL, M. T. R., BYRNE, M. A. & KINNERSLEY, R. 2006. INDAIR: A probabilistic model of indoor air pollution in UK homes. *Atmospheric Environment,* 40, 6362-6379.

FABIAN, P., ADAMKIEWICZ, G. & LEVY, J. I. 2012. Simulating indoor concentrations of NO 2 and PM 2.5 in multifamily housing for use in health-based intervention modeling. *Indoor Air,* 22, 12-23.

FERRO, A. R., KLEPEIS, N. E., OTT, W. R., NAZAROFF, W. W., HILDEMANN, L. M. & SWITZER, P. 2009. Effect of interior door position on room-to-room differences in residential pollutant concentrations after short-term releases. *Atmospheric Environment,* 43, 706-714.

FERRO, A. R., KOPPERUD, R. J. & HILDEMANN, L. M. 2004. Source strengths for indoor human activities that resuspend particulate matter. *Environmental science & technology,* 38, 1759-1764.

GENS, A., HURLEY, J. F., TUOMISTO, J. T. & FRIEDRICH, R. 2014. Health impacts due to personal exposure to fine particles caused by insulation of residential buildings in Europe. *Atmospheric Environment,* 84, 213-221.

HE, C., MORAWSKA, L., HITCHINS, J. & GILBERT, D. 2004. Contribution from indoor sources to particle number and mass concentrations in residential houses. *Atmospheric Environment,* 38, 3405-3415.

JING, Q., FERRO, A. R. & FOWLER, K. R. 2008. Estimating the Resuspension Rate and Residence Time of Indoor Particles. *Journal of the Air & Waste Management Association (Air & Waste Management Association),* 58, 502-516.

KLEPEIS, N. E., NELSON, W. C., OTT, W. R., ROBINSON, J. P., TSANG, A. M., SWITZER, P., BEHAR, J. V., HERN, S. C. & ENGELMANN, W. H. 2001. The National Human Activity Pattern Survey (NHAPS): a resource for assessing exposure to environmental pollutants. *Journal of Exposure Analysis and Environmental Epidemiology,* 11, 231-252.

MCGRATH, J., BYRNE, M., ASHMORE, M., TERRY, A. & DIMITROULOPOULOU, C. 2014a. Development of a probabilistic multi-zone multi-source computational model and demonstration of its applications in predicting PM concentrations indoors. *Science of The Total Environment,* 490, 798-806.

MCGRATH, J. A., BYRNE, M. A., ASHMORE, M. R., TERRY, A. C. & DIMITROULOPOULOU, C. Simulation of Solid Fuel Burning Events in Irish Fireplaces under Varying Air Exchange Rates. The 12th International Conference on Indoor Air Quality and Climate. Indoor Air 2011., 2011 Austin, Texas.

MCGRATH, J. A., BYRNE, M. A., ASHMORE, M. R., TERRY, A. C. & DIMITROULOPOULOU, C. 2014b. A simulation study of the changes in PM2.5 concentrations due to interzonal airflow variations caused by internal door opening patterns. *Atmospheric Environment,* 87, 183-188.

ORMANDY, D. & EZRATTY, V. 2012. Health and thermal comfort: From WHO guidance to housing strategies. *Energy Policy,* 49, 116-121.

OTT, W. R. 1999. Mathematical models for predicting indoor air quality from smoking activity. *Environmental Health Perspectives,* 107, 375-381.

PARKER, S. & BOWMAN, V. 2011. State-space methods for calculating concentration dynamics in multizone buildings. *Building and Environment,* 46, 1567-1577.

RIJAL, H. B., TUOHY, P., HUMPHREYS, M. A., NICOL, J. F., SAMUEL, A. & CLARKE, J. 2007. Using results from field surveys to predict the effect of open windows on thermal comfort and energy use in buildings. *Energy and Buildings,* 39, 823-836.

RIJAL, H. B., TUOHY, P., NICOL, J., HUMPHREYS, M. A., SAMUEL, A. & CLARKE, J. 2008. Development of adaptive algorithms for the operation of windows, fans and doors to predict thermal comfort and energy use in Pakistani buildings. *ASHRAE transactions,* 114, 555-573.

SCHWEIZER, C., EDWARDS, R. D., BAYER-OGLESBY, L., GAUDERMAN, W. J., ILACQUA, V., JANTUNEN, M. J., LAI, H. K., NIEUWENHUIJSEN, M. & KÜNZLI, N. 2006. Indoor time–microenvironment–activity patterns in seven regions of Europe. *Journal of Exposure Science and Environmental Epidemiology,* 17, 170-181.

SEE, S. W. & BALASUBRAMANIAN, R. 2011. Characterization of fine particle emissions from incense burning. *Building and Environment,* 46, 1074-1080.

SEMPLE, S., GARDEN, C., COGGINS, M., GALEA, K. S., WHELAN, P., COWIE, H., SANCHEZ-JIMENEZ, A., THORNE, P. S., HURLEY, J. F. & AYRES, J. G. 2012. Contribution of solid fuel, gas combustion, or tobacco smoke to indoor air pollutant concentrations in Irish and Scottish homes. *Indoor Air,* 22, 212-223.

SEMPLE, S. & LATIF, N. 2014. How Long Does Secondhand Smoke Remain in Household Air: Analysis of PM2. 5 Data From Smokers' Homes. *Nicotine & Tobacco Research,* ntu089.

SINGER, B. C., HODGSON, A. T., GUEVARRA, K. S., HAWLEY, E. L. & NAZAROFF, W. W. 2002. Gas-phase organics in environmental tobacco smoke. 1. Effects of smoking rate, ventilation, and furnishing level on emission factors. *Environmental Science & Technology,* 36, 846-853.

SOHN, M. D., APTE, M. G., SEXTRO, R. G. & LAI, A. C. 2007. Predicting size-resolved particle behavior in multizone buildings. *Atmospheric Environment,* 41, 1473-1482.

STEINLE, S., REIS, S. & SABEL, C. E. 2013. Quantifying human exposure to air pollution—moving from static monitoring to spatio-temporally resolved personal exposure assessment. *Science of the Total Environment,* 443, 184-193.

TORFS, R., BROUWERE, K. D., SPRUYT, M., GOELEN, E., NICKMILDER, M. & BERNARD, A. 2008. Exposure and risk assessment of air fresheners. VITO.

WHO 2006. WHO Air quality guidelines for particulate matter, ozone, nitrogen dioxide and sulfur dioxide. *World Health Organisation.*

WHO 2010. Housing, Energy and Thermal Comfort. A review of 10 countries within the WHO European Region. *World Health Organisation.*

WHO 2013. Review of Evidence on Health Aspects of Air Pollution REVIHAAP Project: Final technical report. World Health Organization.

WILKINSON, P., SMITH, K. R., DAVIES, M., ADAIR, H., ARMSTRONG, B. G., BARRETT, M., BRUCE, N., HAINES, A., HAMILTON, I. & ORESZCZYN, T. 2009. Public health benefits of strategies to reduce greenhouse-gas emissions: household energy. *The Lancet,* 374, 1917-1929.

ASSESSMENT OF THE DURABILITY OF THE AIRTIGHTNESS OF BUILDING ELEMENTS VIA LABORATORY TESTS

Benoît Michaux*, Clarisse Mees, Evelyne Nguyen, Xavier Loncour

Belgian Building Research Institute
Lombard Street, 42
BE 1000 Brussels, Belgium
**Corresponding author: e-mail address*

ABSTRACT

The airtightness just after the end of a building phase is assumed to be relevant criteria for high energy performance. Testing on site the initial performance of the airtightness via the blower door test has become nowadays a common practice. This test is generally realized at the end of the construction works. What about the influence of ageing on the airtightness? Many questions exist on the durability of this initial performance. Even if retesting a building a few years after the initial test can provide a general view on the evolution of this performance, this could generate adding cost and couldn't give information on the origin of potential changes. Another approach may be to validate technology and building technics as sustainable solutions. In order to quantitatively evaluate the durability of the airtightness of building elements as well as building technics, a research realized in Belgium has tested in laboratory the initial performance of more than 50 building walls and their materials.
This performance has afterwards been tested after an accelerated ageing process including a.o. exposure to wind cycles (storms) corresponding to 10, 25 and 40 years of lifetime. The paper presents the main results of this research and points out the principal recommendations to guarantee the sustainability of this performance.

KEYWORDS

Durability, Airtightness, Building Technology

1 INTRODUCTION

Even we focus more and more on the airtightness of the new building to reach high energy standards, a lot of questions are still remaining targeting the sustainability of the performances. As of now, the blower door test confirms the initial airtightness just at the end of the building phase. This test will not provide any information on the long term performances of the airtightness. To consider the sustainability of the airtightness we should use some different ways.
The monitoring of the performances may cost and is intrinsically not a quality guarantee but only a control. But it's more efficient to be proactive and to choose combinations of products which will therefore ensure the durability of these performances.
The research project conducted from 2012 to 2013 by the Belgian Building Research Institute in partnership with the University of Liège has targeted the basic criteria and useful technologies insuring the durability of the airtightness.
This article will not analyse all the results of the research but provide the important learnings from the research.

2 THE « DREAM » RESEARCH PROJECT

The research project called "DREAM" aims to evaluate and improve the durability of the airtightness of buildings quantifying the airtightness performance of different materials before and after ageing for 46 different walls (divided into 4 families).
Four different families of system ensuring the airtightness have been targeted in this project.

- Walls of blocks / bricks whose airtightness is ensured by coatings/plastering. Seventeen different walls were analysed in this family (inside plastering or outside coating);
- Walls whose airtightness is ensured by wood panels. Eleven different walls were analysed in this family
- Walls whose airtightness is ensured by a membrane. Eleven different walls were analysed in this family;
- Walls composed by industrialized systems (sandwich panels, architectural concrete panels, ...). Seven different walls have been tested for this family.

2.1 Ageing types

All the samples were tested before and after ageing phases. Three types of ageing were implemented successively on the samples.

- Ageing representing wind effects and storms.
- Ageing representing the variation of moisture
- Ageing representing the variation of temperature

The succession of ageing are adapted to the influence of the ageing on the type of walls
The ageing representing the wind effect is implemented through storms. Each storm has

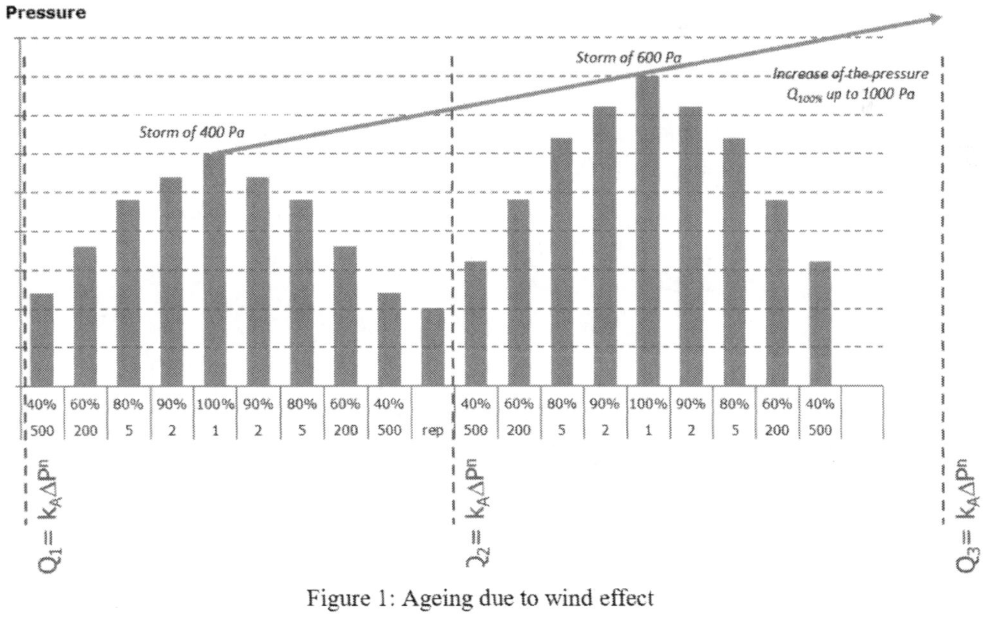

Figure 1: Ageing due to wind effect

1,415 wind cycles ranging from 40% to 100% of the maximum value of wind pressure (as described on figure 1). Four different storms were applied on the walls. After each storm, the air permeability has been measured.

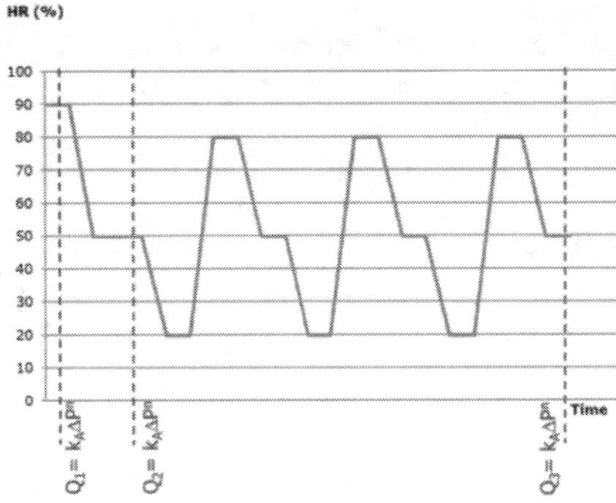

Figure 2: Ageing due to Moisture effect

The ageing representing the variation of moisture is implemented through rapid variation of the air moisture. After a first stabilisation up to 50% HR, three cycles from 20 to 80% are applied on the inside part of the walls (as described on figure 2). Each cycle lasts one day. It represents short-term variations therefrom that may occur in buildings with indoor climate control (for example in bathrooms, ...).

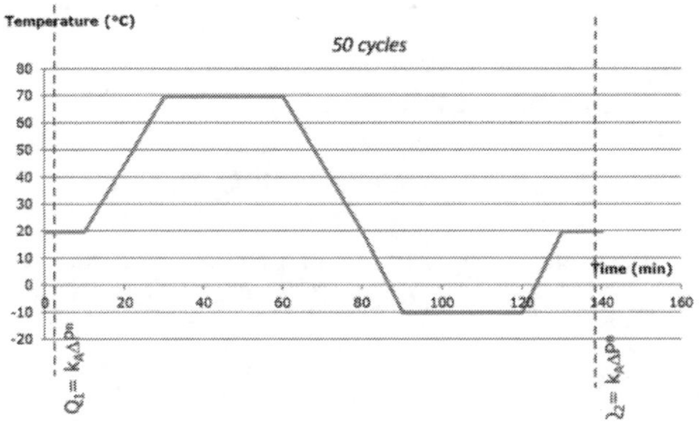

Figure 3: Ageing due to temperature effect

Ageing types due to temperature effect are applied to the outer side of the walls. 50 temperature cycles are performed (as described on figure 3). An airtightness tests is realized after at the end of the ageing procedure.

All the permeability tests are realized both in over- and underpressure. Based on these tests, it is possible to determine the following equations Q_n= k_A□P^n (as described on figure 4).

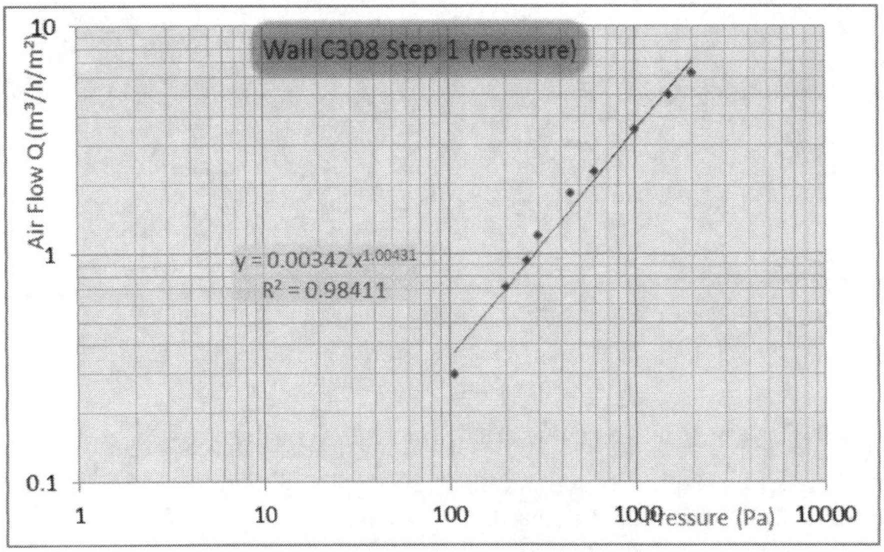

Figure 4: Graph (Q,□P) – Sampling (plastering)

46walls were tested representing more than 700 graphs (Q,□P). All the ageing tests realized on a specific wall are summarized on one graph (Q_{50}, Ageing test) to represent the evolution of the airflow at 50Pa after each step of the ageing process. The figure 5shows an example of the evolution for walls whose airtightness is ensured by a wood panel with tape.

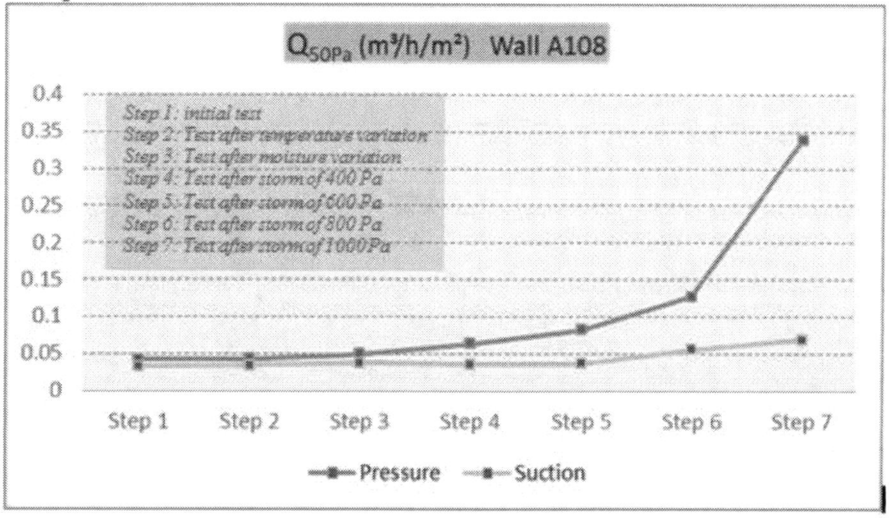

Figure5:Evolution of the Q50Pa (case of OSB12mm panel with tape (type 3)and lathing)

The tested samples are representative of the existing building method in Belgium. The results of this research are not exhaustive and could not cover all the existing technics.
The samples were built carefully in laboratory conditions in order to start in optimal situation before proceeding with ageing. The purpose of this research is to confirm the durability of the performance of the investigated systems.

2.2 « Walls of blocks / bricks whose airtightness is ensured by coatings/plastering »

2.2.1 Walls of blocks / bricks whose airtightness is ensured by internal plastering

On those samples the ageing steps are established by a first application of moisture variation, temperature variations and finally storm cycles.

The interior coating systems achieve good airtightness performance, as well the initial performance as after ageing ($q50<0.1 m^3/h/m^2$). However we can note some learnings:

- Only moisture and thermal ageing influence the airtightness;
- The thickness of the plastering and the smoothness influence the initial performance;
- The storms ageing up to 1000 Pa haven't any influence on the performances;
- Thin plastering (thickness ≤ 6mm) shows a significant loss of airtightness after hygrothermal cycles. This loss is due to micro-cracks in the plastering. The effect of hygrothermal cycles in steps 2 (moisture variation) and 3 (thermal variation) of a 6 mm plaster (evolution shown in figure 6) are higher than the influence from wind effect ageing. Performance after ageing are still remaining acceptable;

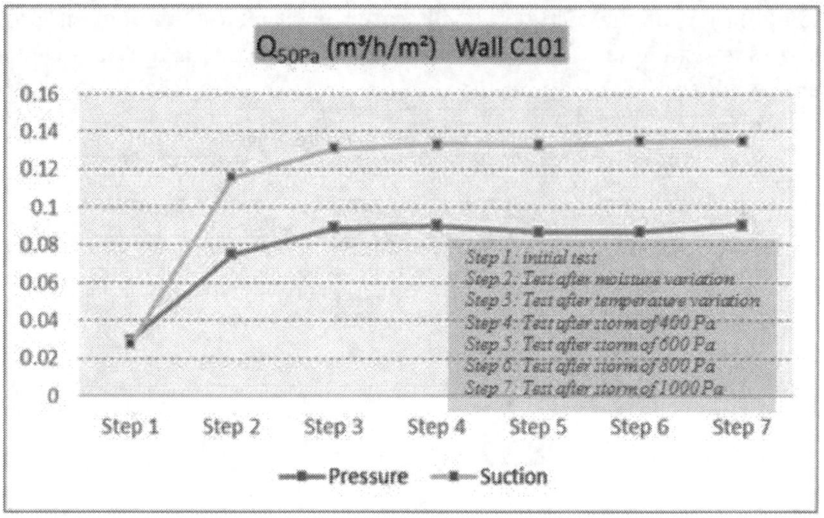

Figure 6: Evolution of the Q50Pa (case of 6mm plastering)

- It's also interesting to note the quality of the surface (finishing) influences the level of the initial performance but does not influence the risk of micro-cracking (this is even more pronounced in thin plastering). Some samples have required the application of new plastering to start with better initial performances;
- Applying a double plastering influences the initial airtightness performance. However the risk of degradation of the plastering is influenced by the thicknesses of the layers.

In the case of clay-based coatings, the airtightness is influenced by the thickness.

- The grain size fluctuations of the clay are compensated by the thickness of the applied clay (in the research : 28 and 51 mm); For the 28mm thick, the Q_{50Pa} stays around 0.1m³/h/m² all along the aging procedure. For the 51mm thick, the Q_{50Pa} stays between 0.012 and 0.02 m³/h/m² all along the aging procedure.
- For both thicknesses, with clay-based coatings, the storms cycles have no significant influence on the evolution of the performances. The action of hygrothermal ageing has less influence on the airtightness especially for the higher thickness. This could be explained because of the test procedure (the thermal and moisture variation are quite rapid – the clay did not have any time during humidity cycles to stabilize the humidity in the material). The initial airtightness performance of thicker clay-based coating was maintained. A consequent loss of airtightness could come into consideration for buildings where moisture conditions are not mastered quickly.

2.2.2 Walls of blocks / bricks whose airtightness is guaranteed by external coatings/plastering

The same ageing procedure was followed (see§2.2.1).The following outputs emerge:

- The influence of humidity cycles depends strongly on the used coating system (as well as the methods of implementation). For each product, an initial performance testing by the fabricant should be required;

- The overriding factors influencing the initial performance are:
    - The particle size of the products used
    - Pre-layers and preparation layers
    - The thickness of the layers
    - The nature of the binder
- The initial values (Q_{50Pa}) are varying between 0.004 up to 0.8 $m^3/h/m^2$. After ageing procedure, the Q_{50Pa} is going up to 0.01 $m^3/h/m^2$ for the best airtight one and up to 1.7$m^3/h/m^2$ for the most degraded one.
- The action of pressure cycles is only marked from high pressures storms (over 800 Pa) but it should be noted that the influence of ageing is relatively limited on the coatings. These coatings have been developed to meet the external application and watertight requirements. So the number of cracks is rather limited with these products.

2.3 Walls whose airtightness is ensured by panels

The fixing of panels and techniques of airtightness (continuity between panels) can significantly influence the performance of the wall.

The ageing procedure is defined by a first application of temperature variations and moisture variations and finally ageing due to pressure.

We can note the following findings:

Figure 7: Sample with panel

When walls with joints between panels are glued, they delivers a performance for which ageing has limited influence. Note that the (PU based) glue in the grooves was carried out with particular care along the grooves to ensure the continuity of all the collages. The analysis does not include the variability of implementations. The effect of ageing is limited in time and humidity fluctuations were fast enough (cycles less than 24 hours). This glue must meet the criteria for structural bonding.

When sealing between panels is ensured by tape, it is essential to ensure the compatibility between the panels and the tape. Strengthen the tape by a lathing ensures the continuity of the airtightness. Indeed, tests have shown that the storms 800 and 1000 Pa generate significant impairments in tape (although this is limited for optimal combinations tape / panel). The effect of hygrothermal cycles (short period) remains low on these systems (in case of rapid management of internal conditions). However, tests realized by realizing longer hygrothermal cycles (no active control) generate higher damage. The following figure shows the evolution of airtightness for complete walls whose airtightness is ensured by panels (OSB3-15mm) and tapes between panels without lathing.

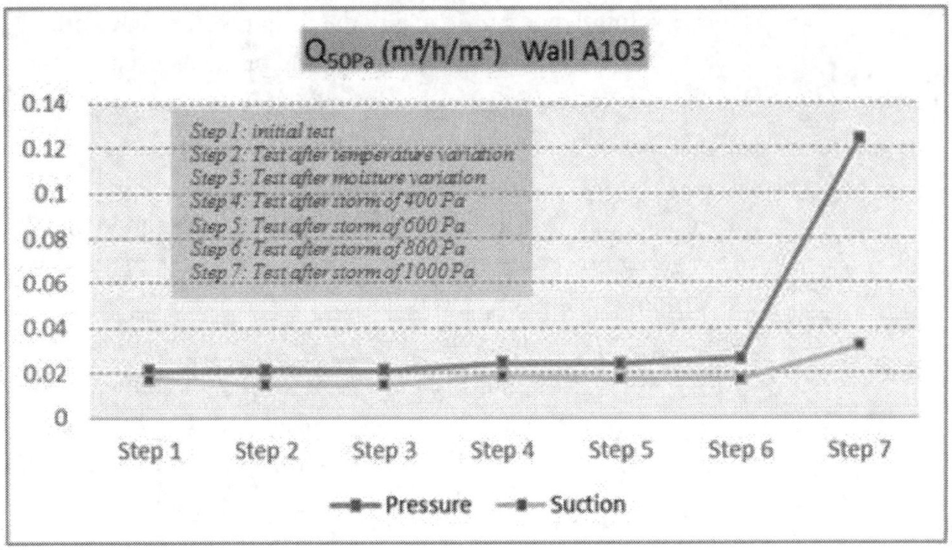

Figure 8: Evolution of the Q50Pa (case of panels with tape)

When a strapping protects these tapes, the degradation is significantly reduced. This construction method ensures the sustainable performances. It is also clear that in the case of tapes placed on non-reinforced seams (upper parts of frames), the tapes are strongly solicited and damages appear with high pressure storms (from 800Pa).

When tightness between panels is provided by sealants under lathing, the solution offers a sustainable solution. The following figure gives the evolution of the Q50 for panels whose continuity is ensured by a compressed sealant.

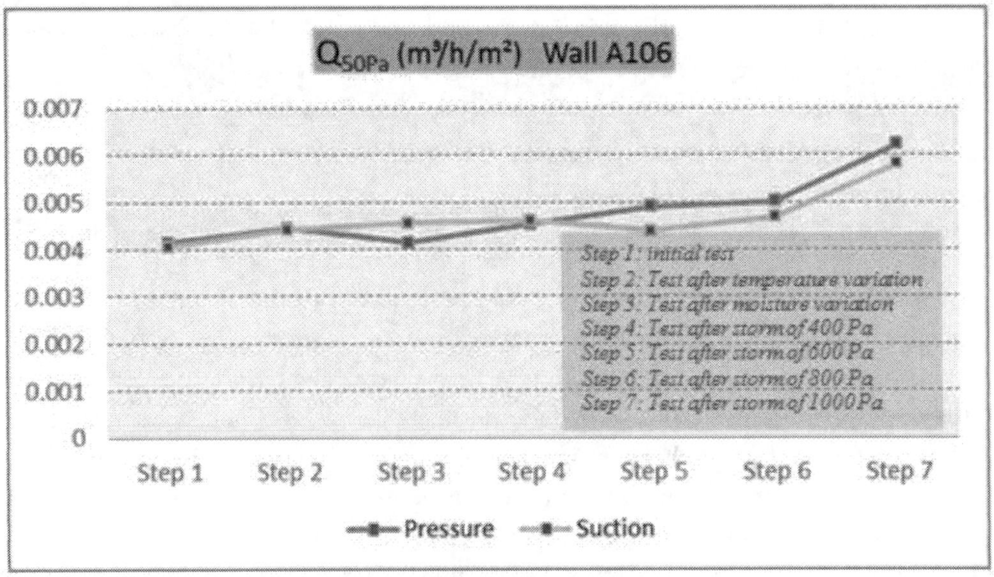

Figure 9: Evolution of the Q50Pa (case panel + compressed sealant)

2.4 Walls whose airtightness is guaranteed by a membrane

The ageing procedure is defined by a first application of temperature variations and moisture variations and finally ageing due to pressure.

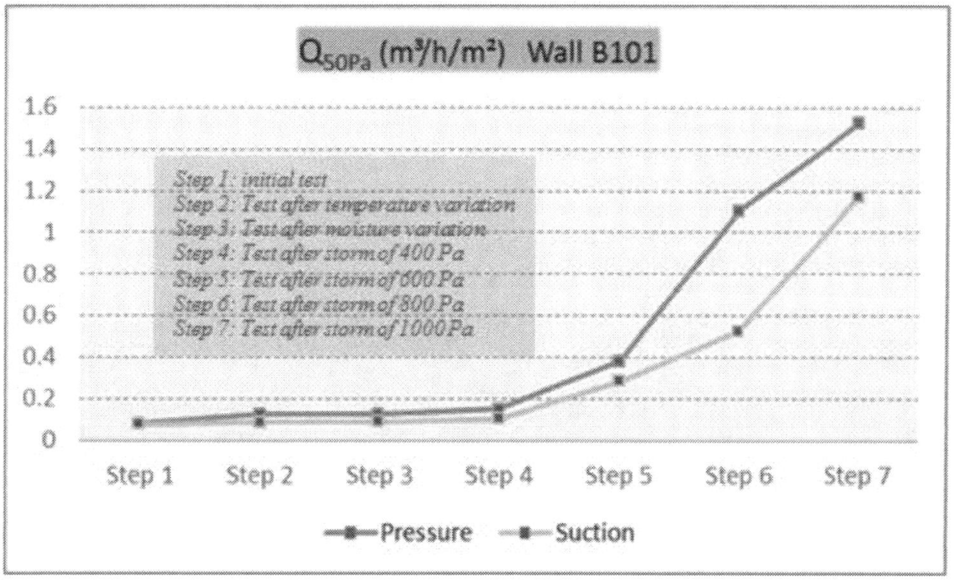

Figure 10: Evolution of the Q50Pa (case of stapling membrane)

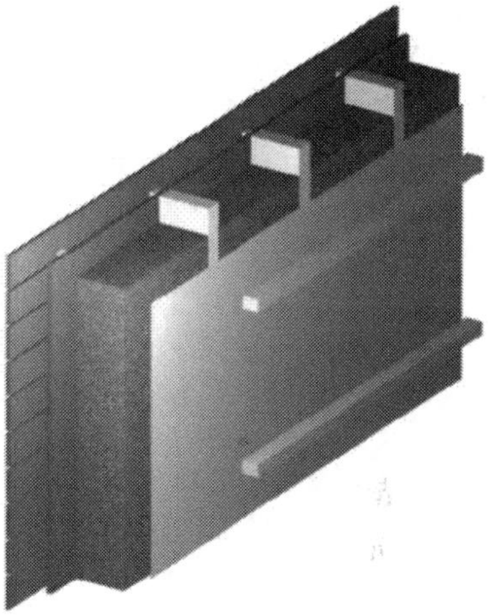

Figure 11: Sample with membrane 1

- Stapling the membrane is a risky method, the ageing effect has a consequent impact if the stitching is not supported by lathing or a compatible tape;
- Connections between membranes need to be done with tape single sided or double sided, or folded and lathing, or mastic bound. If construction is exposed to wind, it will be applied by the additional lathing recovery;
- The horizontal lathing (technique applied to hold the membrane during the insulation blowing) is not sufficient when the construction is exposed to wind;
- The compatibility between membrane and tape requires verification. Indeed, although the initial bonding seems appropriate, hygrothermal or mechanical (wind) ageing may degrade the airtightness of these collages;
- When the membranes are glued with a flexible stand, folding the two membranes with double-sided tape gives significant performance; However, if the building is exposed to the wind, the protection of such recovery is necessary.

2.5 Walls whose airtightness is ensured by industrialized systems (sandwich panels, architectural concrete panels, ...).

The prefabricated sandwich panel construction requires improved joints between panels design. For these walls, the ageing procedure is a first application of pressure cycles, variations in temperature and finally humidity variations.

- When the links between sandwich panels is provided by a closed cell foam seal and a silicone, if the closed cell seal is compressed by fixing the panel, the continuous seal is ensured. The features of construction must allow the compression of the seal. Continuity at the corners remains difficult. The silicone gasket completes the seal. The ageing will not have any effect on the airtightness of this system
- If the sealing wall is provided by a closed cell foam seal, the gasket must be compressed with panels fixations therefore continuity is sealed. Continuity at the corners remains difficult. However, additional gasket or mastics must ensure the continuity of the seal. The compressed seals have not shown any degradation;

3 CONCLUSIONS

Several techniques can be identified in terms of airtightness sustainability. Some resist more at higher loads; it is then necessary to adapt the construction method according to the stresses. Thus, a construction particularly exposed to the wind requires additional lathing on membrane's connections. It is also important that the manufacturer communicates on the compatibility and performance of their product.

The results of this research will be used in future publications of the Belgian Building Research Institute.

4 ACKNOWLEDGEMENTS

The results presented in this paper were obtained within the DREAM research project supported by the Walloon Region in Belgium.

CONTROL OF INDOOR CLIMATE SYSTEMS IN ACTIVE HOUSES

Peter Holzer[*1], Peter Foldbjerg[2]

1 Institute of Building Research & Innovation
Luftbadgasse 3/8
1090 Vienna, Austria
peter.holzer@building-research.at

2 VELUX A/S
Ådalsvej 99
2970 Hørsholm, Denmark

ABSTRACT

The term of "Active House" recently developed, addressing houses that target a balanced optimization of indoor environmental quality, energy performance and environmental performance. According to the idea of not only being energy efficient and eco-friendly, Active Houses equally focus on indoor environmental qualities, in particular daylight and air. With their tendency towards intensive sun penetration, natural ventilative systems and generally intensive connections to the exterior, Active Houses challenge the balance of technical and individual indoor climate control.

The paper in hand presents and discusses challenges and findings from six model Active House homes in five European countries:
How to design and run heating, cooling control systems in Active Houses.
How to balance hybrid ventilation systems in Active Houses.
How to open the houses to the delight of sunlight but effectively protect against overheating.
Finally how to balance technical control and still encourage spontaneous individual interference.

From 20010 to 2013 six model homes in Active House Standard have been erected and monitored in five European countries, funded by the company of VELUX. The authors have been deeply involved in design, erection and post occupancy performance monitoring of these model homes.

KEYWORDS

Active Houses, Indoor environmental control, daylight, hybrid ventilation

1 INTRODUCTION

The term of "Active House" recently developed, addressing houses that target a balanced optimization of indoor environmental quality, energy performance and environmental performance. A lively international network, the "Active House Alliance", has been founded and numerous pilot projects in a widespread range of nations and climates have been erected.[1]

According to the idea of not only being energy efficient and eco-friendly, Active Houses equally focus on indoor environmental qualities, in particular daylight and air. With their tendency towards intensive sun penetration, natural ventilative systems and generally

[1] For further information see http://www.activehouse.info (22.08.2014).

intensive connections to the exterior, Active Houses challenge the balance of technical and individual indoor climate control.

Funded and very actively accompanied by the Danish roof window manufacturer VELUX A/S there have been Active Houses erected in five different European countries and had them intensively monitored through a period of at least one full year, including energy monitoring, comfort monitoring and social monitoring.[1]

The paper in hand presents and discusses the findings from those six model homes as regards the aspects of technical control such as

- How to design and run heating, cooling control systems in Active Houses.
- How to balance hybrid ventilation systems in Active Houses.
- How to open the houses to the delight of sunlight but effectively protect against overheating.
- Finally how to balance technical control and still encourage spontaneous individual interference.

2 ACTIVE HOUSES

2.1 Goal and Definition

Goal and definition of Active Houses is precisely but still lively and even poetically defined in the Active House Specifications.[2]

The most poetic definition is: *Active Houses are Buildings that give more than they take.*

The lively programmatic definition is, already structured in three fields of equal interest:

Indoor Environmental Quality
An AH creates healthier and more comfortable life.
An AH ensures a generous supply of daylight and fresh air.
Energy
An AH is energy efficient.
All AH's energy needed is supplied by renewable sources.
Environment
An AH interacts positively with the environment
through an optimized relationship with the local context

Precise quantitative benchmarks as regards indoor environmental standards, energy demand key figures and environmental impacts are defined in Active House Specifications, having already been mentioned before.

It's this balanced approach of Indoor Environmental Quality, Energy and Environment that makes the Active House quite unique amongst other building standards and which, indeed, brings along some challenges, not least in technical control systems.

[1] For further information see http://www.velux.com/sustainable_living/demonstration_buildings (22.08.2014)
[2] Active House Specifications, 2nd edition, http://www.activehouse.info/download-specifications (23.08.2014)

2.2 Specific Features and Physical Challenges of Active Houses

Following the balanced three tier approach of Indoor Environmental Quality, Energy and Environment leads to specific features and physical challenges of Active Houses, which can be addressed as benefits as well as threats:

Usually, an Active House is generously daylit, it is generously supplied with fresh air and it connects significantly to the out-of-doors. Thus, seen positively, it's fast and strong reactive to the healing sun, to the warm breeze, to sounds and smells of nature. Seen negatively, it is fast and strong reactive to the burning sun, to heatwaves and storms, to external noise and air pollution.

This paradox of very different possible perceptions of the same physical qualities finds its equivalence in the possible and yet contradicting expectations towards the qualities of technical control systems, between full automation and individual behaviour.

2.3 The VELUX model homes 2020

From 2010 to 2013 six model homes in Active House Standard have been erected and monitored in five European countries, funded by the company of VELUX. Altogether, this "model home 2020" program was the biggest and most practical concerted research action in the field of Active Houses. The authors have been deeply involved in design, erection and post occupancy performance monitoring of these model homes.

Detailed information on the research program of model home 2020 can be found at http://www.velux.com/sustainable_living/demonstration_buildings

Home for Life, Denmark

Green Lighthouse, Denmark

Sunlighthouse, Austria

LichtAktiv Haus, Germany

Carbon Light Homes, Great Britain

Maison Air et Lumière, France

3 DISCUSSION OF INDOOR ENVIRONMENTAL QUALITY TARGETS

3.1 Comfort versus Health

Indoor Environmental Quality, often but improperly referred to as Indoor Comfort, is a major issue of today's building design research. Design decisions towards Indoor Comfort are intrinsically linked to the energy demand of buildings. And, which is overseen too often, Indoor Environmental Quality is not only about comfort but it's very much about health. This correlation is gaining importance rapidly, with today's urban lifestyle leading to 95% of the lifetime being spent indoors. Thus, it is important to discuss Indoor Comfort very much together with Indoor Health.

Indoor Comfort is a state of mind, based on a set of physiological, psychological and physical parameters. You may speak of the physio-psycho-dualism of comfort. Furthermore, Indoor Comfort very much varies with the level of personal expectations and with the level of personal possibilities. And finally, Indoor Comfort gives instantaneous feedback: A person per definition feels uncomfortable in the instant moment it feels uncomfortable.

Indoor Health is expected to be much less psychologically based, but being much more based on physiological settings which themselves having their roots back in evolution of mankind. The basic physiological needs cannot be changed and triggered to the same extent as comfort expectations can be. And unhealthy indoor environments do not give such strong and instantaneous feedback than uncomfortable environments do.

Thus: In building design indoor health is a "must have" issue and indoor comfort is a "nice to have". To have both is perfect and nothing less is the aim of modern sustainable building design.

3.2 Indoor Thermal Comfort & Health Targets

3.2.1 The physiological thermal comfort definition according to Fanger

Traditional 20th century indoor comfort research focused on the physiological aspect only. Fanger et al. developed a physical model for thermal comfort starting from the 1960's, being now worldwide established and forming the basis for HVAC design around the globe.

The model is based on the assumption that thermal comfort derives from an equated body's heat balance together with a limitation of the sweat rate and a limitation of the skin temperature. Within this model, general thermal comfort can be described as a strict function of six physiological and physical parameters: activity level, clothing level, air temperature, radiant temperature, air humidity and air velocity. Additionally four phenomena of local discomfort are described, again on physical level. The Fanger-model was validated by numerous experiments with groups of test persons in climatic chambers and is implemented into international standardization, such as ISO 7730 and ASHRAE 55. [1, 2]

3.2.2 The physio-psychological thermal comfort definition according to Humphreys

As an important improvement, exhaustive empiric comfort research has been carried out by Humphreys et al., investigated thermal comfort not merely as a function of physiological parameters, but as a both physiological and psychological status. This additional approach is known as the Adaptive Principle, leading to the following findings: People are not passive receptors of their thermal environment, but continually interact with it. People become adjusted to the conditions they normally experience. People therefore must be studied in their everyday habitats. Discomfort arises from insufficient adaptive opportunity. The Principles of Adaptive Comfort have been derived from and validated by field studies in all climate zones of the world, with over 200.000 comfort-votes. Very recently they have been even implemented into international standardization, with the limitation of validity to non AC, but free running mode buildings, again in ISO 7730 and ASHRAE 55 and EN 15251.[3]

3.2.3 Indoor Thermal Environments and Health

Against to the comprehensive research outputs on indoor thermal environments and comfort, there's a significant lack of research outputs on indoor thermal environments and health. The new question arising is: Is it healthy to continuously stay in the thermal comfort zone?

First research steps have been made by Wouter van Marken Lichtenbelt (van Marken Lichtenbelt, 2011), and Christel M.C. Jacquot (Jacquot, 2014) indicating a possible positive health aspect of periodically leaving the strict comfort zone.

3.2.4 Personal Thermal Comfort Expectations

It has to be understood that human indoor comfort expectations and indoor comfort targets are intrinsically tied to the possibly achievable level of indoor comfort. Thus, comfort cannot be discussed as an independent variable: The demand level rises together with the supply level.

---

[1] ISO 7730, 2005
[2] AHRAE 55, 2013
[3] EN 15251, 2007

Numerous investigations on comfort sensation in free running mode buildings revealed a systematic connection between the existing outdoor comfort, the personal adaptive options and the indoor comfort expectations: The neutral temperature (i.e. the temperature preferred by the inhabitants) significantly depends on the mean outdoor temperature. This dependency is most significant under 'free-running mode' conditions (no heating, no cooling)

Bridging to modern homes, especially to Active Houses, it's an important question, which elements of a house are perceived "forgiving" in the tradition of free running mode buildings and which ones are perceived as "demanding" in the tradition of HVAC buildings. Existing results, not at last from the model home 2020 program reveal a close correlation of thermal comfort in Active Houses with the adaptive comfort model of Humphreys et al. as set in EN 15251. Thus, the Active House Specifications define indoor thermal comfort according to the EN 15251 benchmarks.

3.3 Indoor Lighting Comfort & Health Targets

As regards lighting there's young but comprehensive research on both comfort and health available and has found its way into standardisation already. Besides the author published his PhD on the topic of Quality and Quantity of indoor daylight supply.[1]

It is known that apart from visual comfort with its characteristic 500 lx minimum illumination level there are significantly higher requirements from the perspective of alertness and circadian rhythm, reaching up to a demand of 1.000 lx up to 10.000 lx at the retina.

Furthermore it is known that health aspects of light do not only correlate with illumination levels but intrinsically with the spectral distribution of the radiative intensity, which tends to be significantly changed by daylight passing glass panes.

As a result, daylight must be regarded as a major indoor comfort and health issue. Daylight design should guarantee levels of daylight factor of 5% and above. Furthermore, the penetration of indoor space by direct sunlight is an important issue. And finally, periodically skin exposition to spectrally unfiltered sunlight is a must have.

3.4 Indoor Air Quality Comfort & Health Targets

As regards indoor air quality it is necessary to differentiate between the aspects of a) building design and b) the aspects of interior design, materials used in interior design and finally usage itself. From the point of view of building design the ventilation rate is the major parameter of air quality, with air quality being measured by the CO2 concentration level or by the VOC concentration level. In the Active House specifications the relative CO2 concentration level above outdoor CO2 concentration is defined as the benchmark.

[1] Holzer, Hammer (2009)

4 BASIC ASPECTS OF INDOOR ENVIRONMENT CONTROL SYSTEMS

4.1 Logic Structure of Control Systems

Technically every control system consists of at least the basic elements illustrated in the flow chart below:

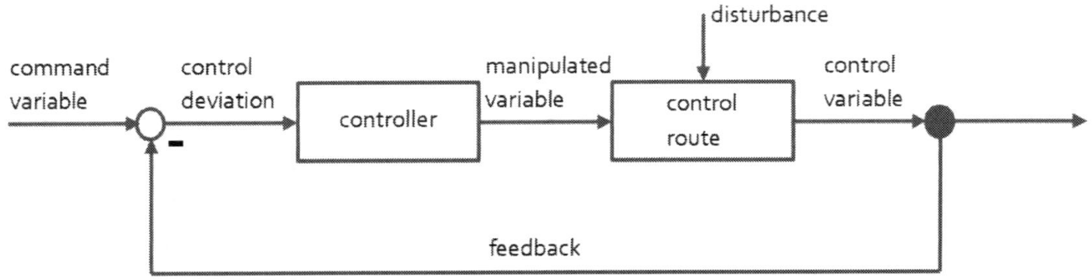

Figure 1: flow chart of an automated technical control

At the beginning of the control system the command variable represents the desired set point of the specific quality that shall be controlled, e.g. the setpoint of the room's air temperature. At the end of the control system the control variable represents the actual value of the specific quality that shall be controlled, e.g. the room's air temperature.
The difference between them is the control deviation, being reported to the controller, being the "brain" of the control system, translating the control deviation into the manipulated variable, e.g. the flow temperature and/or the flow rate of the heating circuit.
The manipulated variable is introduced into the control route, which is the full system of interdepending phenomena that finally influence the control variable, not at least including the disturbances, e.g. the outside temperature. In case of an indoor environment control the control route is nothing less than the house itself, even including its users and their hardly predictable behaviour.

4.2 The range of users' intervention

Practically, an indoor environmental control system may be

1. fully automated without users' intervention
2. automated with users' intervention as regards setpoints
3. users' manual operation
4. automation plus manual operation,
   guided by feedback from the automated control system

5 LEARNINGS AND RECOMMENDATIONS

Based on the experiences of the six VELUX model homes 2020 together with experiences from other post occupancy monitoring, the following chapters sum up the major learnings and recommendations as regards the technical control of indoor climate systems in Active Houses.

5.1 Heating Control Systems

All six model homes have automated, thermostat-controlled heating systems, in the majority of the houses on basis of single room controllers. The controllers work without problems or disturbances and inhabitants never reported problems with this system.

It's our recommendation always automating the heating on a thermostat-controlled basis with single room controllers. The single-room controllers allow specific temperature setpoints for different rooms, what is especially beneficial in case of sleeping rooms. If single room controllers are in operation, the traditional coupling of the heating system's flow temperature to the outside temperature may be replaced by a demandside-controlled flow temperature, which turns out beneficial, since in passive solar driven homes the occurrence of maximum heat load doesn't necessarily come along with the occurrence of minimum outside temperature.

5.2 Ventilative Cooling Control Systems

The six model homes are located in middle/northern Europe, from 48°N to 57° northern latitude. Thus they haven't been equipped with technical cooling, but with Ventilative Cooling by night ventilation.

In the model homes Ventilative Cooling was a major issue in building design and turned out working up to the high expectations in reality – if automatically controlled. The combination of automated window opening and automated sunblind, controlled from room temperature, outside temperature and solar irradiation did the job.

Ventilative Cooling, if automated, did stand the test. Experimental operation without night ventilation during summer heat waves resulted in up to five degrees higher indoor temperatures than with night ventilation under comparable outside climate conditions.

Monitoring proved that when people overrode the building automation as regards Ventilative Cooling and sunblind control they only made it worse. It's our learning and recommendation to automate ventilative cooling together with sunblind operation. Additionally, heat protection could be an issue to user information. See chapter 5.6.

5.3 Natural and Hybrid Comfort Ventilation Systems

All model homes are equipped with automated windows for both Ventilative Cooling and comfort ventilation. Many of the houses are additionally equipped with mechanical ventilation systems.

Apart from Ventilative Cooling the automated windows have been controlled according to indoor VOC levels. This, again, did work without any complaints during the warm seasons. In the winter season good performance and acceptability was recorded from pivot hung, horizontally oriented ventilation openings, positioned in the upper regions of the walls. They proved to be much better adjustable and much less draft risky than side hung windows with vertical orientation.

There have been recordings form occupants noticing the noise from the actuators during the first weeks after moving to the place, but getting used to the sound soon.

It's our learning and recommendation to automate ventilative cooling together with sunblind operation. Additionally, heat protection could be an issue to user information. See chapter 5.6.

As regards hybrid ventilation it was a significant learning to properly and strictly define the setpoint when automated natural ventilation stops and mechanical ventilation with heat recovery takes over. From theoretical analysis and from practical experience a one hour mean value of between 12°C and 14°C outside temperature proved to be a good choice as regards energy optimization as well as regards comfort.

Finally, there turned out a specialty of automating windows, which is also true for automating sunblinds: The automated operation of both windows and sunblinds is instantaneously visible to occupants. Unexpected 'behaviour' of the control system causes instant discomfort and complaints, which is very different from e.g. a heating control, which cannot be recognized in a comparably immediate way.

5.4 Daylight Supply and Sun Protection Systems

It's a major target of Active Houses to extensively supply the rooms with daylight. Daylight Factors of 5% and higher are aimed at, for good reasons. See chapter 3.3.

Thus, Active Houses tend to have windows oriented not only driven by energy design rules, but by daylight design rules, which can be sometimes contradictory and what always leads to significant glazing. As a result, Active Houses are fast reactive towards direct sunlight. An effective and automated system of movable shadings is obligatory for achieving good summer comfort.

Sunblind control performed best when correlated to the outside irradiation, which certainly have to differentiate between the compass orientations.

There turned out a specialty of automating windows, which is also true for automating sunblinds: The automated operation of both windows and sunblinds is instantaneously visible to occupants. Unexpected 'behaviour' of the control system causes instant discomfort and complaints, which is very different from e.g. a heating control, which cannot be recognized in a comparably immediate way.

Therefore, sunblind control must be designed deliberately correct, using the solar irradiation at the specific façade as a trigger for shutting and being equipped with a time lag of a quarter of an hour in case of shutting and half an hour in case of reopening.

5.5 Building Automation together with user information and individual control

The model homes have been equipped with extensive monitoring and control devices. Users have been informed via significantly big, wall-mounted info screens, informing about actual and historic levels of air temperature, VOC, humidity, energy consumption and supply and others.

In the beginning there have been reportings from users feeling overloaded by all the options of the system, but after getting used to it, those complaints disappeared. About half of the users set to not changing any setpoints apart from simple room temperature. The other half actively used the options for personalizing the control system.

Based on the already mentioned learnings on the users' sensitivity towards automation of the 'visible' devices as sunblinds and windows and based on the learnings on users' difficulties to intuitively operate the houses towards heat protection we consider the idea of suggestive control system feedback being a good choice.

The idea behind suggestive control system feedback is to automate the basic functions and additionally inform the user on firstly the reasons behind the automated actions taken and on secondly additional actions the user might take to achieve best comfort.

For example, during a hot summer day, the system would shut the windows and shut the sunblinds, while the info screen would say: "It's a hot and sunny day. In Order to keep the house cool, the sunblinds and ventilation openings have been automatically shut and will stay shut until irradiation and temperature will lower. It's recommended to keep the manually operated windows and doors shut and to avoid unnecessary internal heat loads."

Designing such a suggestive control system feedback could be a challenging and rewarding subject to an engineer-psychologist-cooperation.

6 CONCLUSIONS

The evaluation of the six Active Houses of the model home 2020 program revealed an overall good performance of the indoor environment control systems.

- Heating control caused no problems.
- Ventilative Cooling Control should be automated, with combined operation of automated windows and sunblinds. Individual operation won't work.
- Hybrid comfort ventilation calls for a strict switch point between automated window ventilation and mechanical ventilation.
- Daylight Supply and Sun Protection can be well balanced with effective sunblinds, again automated. Control algorithms have to be 'understandable' to the users.
- Beyond technical automation it's essential offering intuitively manually operable (not only manually telecommanded) devices such as windows, doors, awnings.
- Building Automation together with user information and individual control offers additional chances balancing the gap between automation and individuality.

7 ACKNOWLEDGEMENTS

The paper in hands is based on the research reports of the model home 2020 program, funded and scientifically lead by VELUX A/S.

8 REFERENCES

ISO 7730 (2005). Ergonomics of the thermal environment -- Analytical determination and interpretation of thermal comfort using calculation of the PMV and PPD indices and local thermal comfort criteria.

ASHRAE Standard 55 (2013). Thermal Environmental Conditions for Human Occupancy.

van Marken Lichtenbelt, WD. & Schrauwen, P. (2011). Implications of non-shivering thermogenesis for energy balance regulation in humans. Am J Physiol Regul Integr Comp Physiol 301, R285-296.

Jacquot, C.et al. (2014). Influence of thermophysiology on thermal behavior: the essentials of categorization. Physiology & Behavior 128 (2014) 180–187

EN 15251 (2007). Indoor environmental input parameters for design and assessment of energy performance of buildings addressing indoor air quality, thermal environment, lighting and acoustics.

Holzer, P., Hammer, R., Krec, K. (supervisor) (2009). Quality and Quantity of indoor daylight supply. Potential study specifically taking into account the radiation induced thermal indoor quality and the human photophysiology. Technical University Vienna, 2009.

TIPS FOR IMPROVING REPEATABILITY OF AIR LEAKAGE TESTS TO EN AND ISO STANDARDS

Colin Genge

ABSTRACT

Numerous tests are being performed throughout Europe. While most are or appear to be successful others have high calculated uncertainty values and others don't correlate well when repeated by the same tester with the same equipment or where someone else does the repeated test. Some feel that equipment calibration is the key to consistent results but in most cases that it could be one of the smallest causes for lack of repeatability. We will take a look at how much different factors affect results and how to get the best results.

Repeating the same test on the same building with the same equipment and operators over time has yielded results at 50 Pa that are within +/-2% of each other while results at 4 Pa are within +/-15%. We can assume those errors are all due to wind which was light in all cases of the experiment so far. Since this precludes any building set up error, operator error or equipment error, it is useful to take a look at how EN and ISO Standards should be used to first to reduce the effects of wind and secondly to look at the combined effects of gauge and fan calibration as well as set up.

Building set up accounted for a 15% variation in results at 75 Pa on a large building. The tests were performed by 6 teams of experts on the same building but at different times. Each team was instructed on how to set up the building and all teams took results in the same manner.

Retrotec has performed several rounds of air leakage tests on a fixed enclosure to determine what the variation due to wind. Some were done at the minimum number of 5 points while another set was done at 12 Points. Most often the results were closer to the mean of all tests than the calculated uncertainty indicated but 12% of the tests had uncertainties that were 4 times greater than the calculated value. More testing is being performed to arrive at suitable recommendations.

EN and ISO both require gauge accuracy of +/- 2 Pa from 10 to 100 Pa. when this accuracy range is introduced to results, the potential errors increase by 3% at 50 Pa and 22 % at 4 Pa. Clearly most gauges in current use are typically within 0.2 Pa which eliminates 90% of this error but this is an area that must be controlled if repeated tests after retrofit are to be correlated.

EN and ISO both require flow measurement accuracy of +/- 7% which can realistically be reduced to say 5% but not much lower. Combining gauge and fan accuracy could give a variation of 9% at 50 Pa and 37% at 4 Pa.

Adding up all these error is usually done as the sum of squares. Currently the situation looks like this:

Reference	gauge	fan	wind	set up	sum of squares
4	0.22	0.07	0.15	0.15	0.31
50	0.03	0.07	0.02	0.15	0.17

With some tweaking our potential is:

Reference	gauge	fan	wind	set up	sum of squares
4	0.05	0.05	0.05	0	0.09
50	0.01	0.05	0.02	0	0.05

To accomplish this goal, careful guidelines must be established for performing the test which will be proposed. The Set Up variable must be controlled by local committees who are enforcing their standards. Currently, tests are being performed to the absolute minimum of the Standards will start to cause problems in future when results from one test group and another do not coincide.

Recommendations using the current version of EN and ISO Standards:

1. Perform a field calibration check of your gauge and tubes before each test. Simply connect the yellow tube between Channel A and B. The results should be within 2% and should not fall rapidly indicating a tube leak. A large difference will identify a tube blockage. Check the other color tubes in the same fashion.
2. After the building has been set up and before any automated test is attempted, run the fan up to full speed to ensure you can reach at least 60 Pa on the existing range. If you reach 100 Pa, change to a lower flow range. If you cannot achieve 50 Pa, change to a higher flow range. Then, see how low you can go on that range, let's say 22 Pa. Then set the pressure range in your software to run the fan from just under the maximum to 25 Pa. Before starting your automated test, you must ensure you can reach the maximum and minimum pressures and still get flow readings. Ensure your gauge is set to the range your fan is on and that the gauge will display a flow reading.
3. Although the Standards say the Minimum Test Pressure must be 10 Pa, I believe it really means it must not be lower than 10 Pa. If your Bias pressure is 2 Pa initially and your test pressure is set to 10 Pa on the gauge, later this Baseline will be subtracted so your actual test pressure will be only 8 Pa.
4. Test from the maximum to minimum test pressure to decrease the likelihood that a door or some building feature will open during the test and spoil the result. If you do it the other way around and your arrival criteria is +/- 1 Pa then you system could start taking readings at only 9 Pa if you had set your lowest test pressure to 10 Pa. With a 2 Pa Bias pressure, your true test point may only be 7 Pa which will fall below the Minimum Allowable Test Pressure.
5. The above steps will reduce the chances that you'll have to change ranges in mid test. This slows the test down and will always produce a small notch because calibrations from one range to another are not exact. This notch, if it appears at the top or bottom of your test will cause the result at 4 Pa in particular to vary a lot.

THE INDOOR AIR QUALITY OBSERVATORY - OUTCOMES OF A DECADE OF RESEARCH AND PERSPECTIVES

Séverine Kirchner* and Corinne Mandin

University of Paris-Est, Scientific and Technical Center for Building (CSTB)
Observatory of Indoor Air Quality (OQAI),
84 avenue Jean Jaurès Champs sur Marne Cedex02
Marne-la-Vallée, France
**Corresponding author: severine.kirchner@cstb.fr*

ABSTRACT

For over a decade now, the OQAI — Observatoire de la qualité de l'air intérieur [French observatory for indoor air quality] — has been leading research into indoor air quality and occupant comfort in living spaces: housing, schools, offices, leisure spaces. The objective is to gain insight into patterns of pollution and occupant discomfort in order to root out the causes, evaluate the health risks, and issue policy recommendations.
Further to a national-scale survey on dwellings, OQAI has launched two new national-scale campaigns focused on indoor air quality and comfort in schools and office buildings. A national system for collecting indoor air quality and occupant comfort data on newly-built or retrofitted energy-efficient buildings has been deployed since 2013. The OQAI also continues to press ahead with its work on indoor air quality indexes and participate to the evaluation of the health-impact of emerging pollutants such semi-volatile organic compounds.

KEYWORDS

Indoor air quality, comfort, population exposure, national-scale survey

1 INTRODUCTION

The French observatory for indoor air quality (OQAI) has the mission to gain insight and inventory knowledge on indoor environments in order to evaluate and control the health risks tied to chemical, physical or biological pollutants in indoor living spaces. Through an array of actions and initiatives, the OQAI helps drive improved air quality and occupant comfort in the national building stock as part of a wider energy efficiency policy. Founded in 2001, it is public founded by three French ministries — Housing, Environment and Health — in collaboration with ADEME [French environment and energy management agency] and ANSES [French agency for food, environmental and occupational health and safety]. Operational mobilization and scientific coordination is led by the CSTB [Scientific and Technical Centre for Building].
The governance of OQAI is organized around three main boards: a Supervisory Board defining and coordinating OQAI research strategy, ensuring the OQAI stays independent and immune to lobbying and promoting open accountability; a Scientific Advisory Board

ensuring quality and scientific cogency across all OQAI research; an Advisory Panel whose mission is to canvass opinion and feedback from organizations and institutional stakeholders potentially impacted by OQAI research projects and help identify specific issues or situational contexts. CSTB is the fieldwork operator appointed to handle OQAI programs leading and coordinating a network of scientific and implementing partners needed for OQAI to fulfil its missions.

OQAI currently manages six programs related to housing, living spaces and venues for children, office building, energy efficient buildings, decision support tools and information & communication.

2 THE AIR AT HOME: A NATIONAL INVENTORY

The first nation-scale survey conducted by OQAI on 2003-2005 was focused on the primary residences in mainland France. The aim of this survey was to have a first snapshot of indoor air quality (IAQ) in these occupied buildings and to sketch out a preliminary shortlist of the factors driving this indoor air pollution among building conditions (i.e. environment, building materials, equipment's, ventilation status, etc.) and household activities.

A total of 567 randomly selected primary residences were investigated. The target pollutants were selected from the indoor air pollutants shortlist originally established by the OQAI back in 2002 (Mosqueron et al., 2003; Kirchner, 2011) and revised in response to talks held by the OQAI Scientific Advisory Board, where certain biocontaminants featured high on the agenda: animal allergens, carbon monoxide, volatile organic compounds and aldehydes, particulate matter, radon and gamma radiation. Other parameters were collected such as carbon dioxide, temperature and relative humidity, exhaust airflow rates at air outlets in the houses specifically equipped with ventilation ducts.

This field campaign generated a snapshot giving the very first picture of the kind of air people breathe every day by people living in mainland French towns and cities. The vast dataset compiled (featuring over 650 variables) was distilled to yield the following lessons (Kirchner et al., 2009):

- indoor pollution has its own specific profile, where pollutant concentrations are often higher than outdoors;
- mould, which is a cause of allergies and respiratory disorders, is found in 37% of homes;
- 10% of homes are exposed to multiple pollution; up to 8 compounds were found at high indoor concentrations, confirming how different households face different pollution issues;
- formaldehyde is found in 100% of homes.

In France and across Europe, the pollutant distributions from the OQAI nationwide housing survey have been used to assess the health risk tied to numerous substances and to inform the risk management and prevention measures, such as for contextualizing existing guideline values.

A clear set of pollution factors emerged, including occupancy density, the presence of an attached garage, outdoor air and soil pollution, water damage, smoking, dry-cleaning clothing, and the use of air fresheners, candles or incense sticks. A 2009 report on ventilation levels in French housing stock highlighted that (Deroubaix et al., 2009):

- overall, looking at the entire building stock studied, room air change rate in housing is not dependent on the ventilation systems installed;
- the occupant – through his behaviour and building use, by opening windows for example – actually plays a role in shaping indoor air quality as important as the building's own technical performances and ventilation modes;
- 56% of housing equipped with mechanical ventilation do not meet the minimum air exchange rate requirements for buildings, thus highlighting the need to step up efforts on system scale-up, implementation and maintenance measures.

The research also identified a relationship between pollution and occupant health. There was a higher prevalence of respiratory symptoms (asthma and rhinitis) in homes exposed to pollution by one or more volatile organic compounds (Billionnet et al., 2011). The objective assessment of fungal contamination of the dwellings also extends previous results suggesting positive associations between exposure to molds and asthma and chronic bronchitis (Hulin et al., 2013).

The housing program is currently completed with a focus on household exposure to semi-volatile organic compounds (SVOCs). SVOCs such as phthalates, polybrominated diphenyl ethers, synthetic musks and polychlorinated biphenyls are attracting growing concern due to their ubiquitous presence and their potential adverse health effects, particularly endocrine disruption. The findings, once compiled, will be the first results obtained in France at housing stock scale.

As part of a wider overarching project named ECOS (cumulative indoor exposures to semi-volatile organic compounds in dwellings) co-coordinated by the EHESP and the CSTB, this project concerns data obtained from analyzing samples of indoor airborne particles and floor-settled dust collected over the course of two national-scale surveys: the OQAI-led 2003–2005 national campaign on indoor air quality in primary residences in mainland France; and the CSTB-coordinated 2008–2009 national "Lead in Homes" campaign on contamination by lead and eight other metals in the homes of children aged 6 months to 6 years. Pre-campaign groundwork phases of these studies provide solid assurances of their current feasibility. First, the substances of interest have been ranked and prioritized based on a review of air and dust concentrations reported in the scientific literature, and toxicity reference values (Bonvallot et al., 2010). Next, the EHESP [French school of advanced research on public health issues] developed dedicated analytical methods for the prioritized compounds (Mercier et al., 2014). Finally, laboratory tests on dust samples have made it possible to confirm zero breakdown of the analysed molecules during the freezing-phase step between collection and analysis (Blanchard et al., 2014). The analyses are underway (Mandin et al., 2014a, 2014b) and should be published early 2015.

3 LIVING SPACES AND VENUES FOR CHILDREN

No French study has yet emerged that gives the 'big picture' on indoor air quality and occupant comfort in schools at a national scale (Kirchner et al., 2011). What little information is available is generally limited to some pollutants such as formaldehyde or benzene.

However, the fact is that indoor air pollutant concentrations can sometimes prove higher in schools than in other living spaces and venues, due to a combination of dense rates of occupancy, frequent use of cleaning agents and classroom supplies like felt pens, rubbers, chalk, and so on, and very low air exchange.

The aim of the on-going OQAI nationwide campaign in schools is to gain key insight into air quality and occupant comfort in classrooms and dorms at nursery and primary schools, so as to propose appropriate solutions for improving the school indoor environment. The campaign is coordinated by the CSTB with technical mobilization and on-the-ground logistics support from an array of partner teams, field investigators and laboratories across France.

600 classrooms from 300 schools randomly-selected across mainland France are instrumented to measure a panel of pollutants: volatile or semi-volatile organic compounds, nitrogen dioxide, particles, metals, and allergens. Measurements and occupant questionnaire are also taken to document visual and acoustic comfort. In addition, walkthrough survey allows describing the building condition and occupation.

The results will be analyzed and unveiled in 2016. The new knowledge obtained will be processed to propose specific opportunities for improving school environments so that children get better protection and a better space in which to learn and grow.

4 OFFICES: NATIONAL-SCALE OFFICE BUILDING SURVEY

In France today, there is still a dearth of information on indoor air quality in office buildings and the perceived comfort levels of their occupants, yet studies in other countries show that there could be indoor pollution profiles and features that are specific to office environments. People spend significant amounts of time at the office, and the air they breathe can impact not just their health but also their on-job performance. Therefore, responding to demand from public authorities and environmental health and safety agencies, in 2013 the OQAI launched a national-scale office building survey (Mandin et al., 2012).

The data collection campaign, set to cover all of France, is running until 2016. It focuses on a sample population of 300 buildings, and is designed to representatively cover as many different office-space configurations as possible. The campaign is coordinated by the CSTB with technical mobilization and on-the-ground logistics support from an array of partner teams, field investigators and laboratories deployed across France.

The campaign will serve to provide a state-of-play inventory of office buildings in France in terms of both indoor environment quality and occupant comfort and health. The stakes and issues involved, given the amount of time much of the French population spends at the office, make the OQAI's new "office spaces" campaign a major step forward for occupational health risk assessment and prevention efforts.

Furthermore, with issues tied to reducing energy demand high on the agenda, especially in the building sector, the campaign extends to recording the energy efficiency of the office buildings surveyed.

The campaign is deployed in two phases:

- Phase 1, 300 office buildings, 2013 to 2015 including an all-day survey to: a) collect the data on the building, its envelope, facilities and immediate environment, b) measure various air quality parameters (volatile organic compounds, aldehydes, ultrafine particles, plus temperature, relative humidity and CO_2 concentrations); and c) poll the occupants on their perceptions of in-building health and comfort. A team of 3 investigators is mobilized for this day-long effort.

- Phase 2, starting in 2016, focusing in on a subsample of 50 office buildings, to drill down deeper into the data collected during phase 1. Phase 2 will analyze a new subset of air quality parameters: fungal and bacterial contamination; man-made mineral fibres and asbestos fibres (if found under the phase 1 survey); and cat, dog, and mite allergens. Noise measurements and light levels will also be measured during the week-long field survey.

Phase 1 was launched in June 2013. The results will be analyzed and unveiled in 2016.

In the meantime, the early emerging trends in air quality and occupant comfort in new or recently-retrofitted office buildings will be published in 2014 under the European OFFICAIR project (www.officair-project.eu). The OQAI coordinated and led the French strand of this work, in 21 office buildings (Dijkstra et al., 2014; Mandin et al., 2014c).

5 ENERGY-EFFICIENT BUILDINGS: FULLY COMPREHENSIVE DATASETS ON ENERGY-EFFICIENT BUILDINGS

Since 2008, OQAI has launched a program dedicated to energy efficient building aiming at evaluate the indoor environment and occupants' habits in these new buildings with reinforced airtight envelope.

A first study was carried out in seven energy-efficient newly built houses before and during the houses' first year of occupancy (Derbez et al., 2014a). It provided the indoor concentrations of some pollutants such volatile organic compounds (VOCs), aldehydes and particles and made possible to find some hypotheses concerning the pollutants sources. The monitoring of IAQ and occupant comfort over time was extended by two supplementary years in two of these houses (Derbez et al., 2014b). IEQ was generally acceptable over time except when ventilation systems were shut down. The ventilation systems presented some shortcomings, including the failure to reach the designed exhaust air flow rate and induced occupant dissatisfaction. Regarding the measured pollutants, houses didn't present any specific indoor air pollution. The variability of indoor air quality over time was explained by the high emissions from the new building materials, products, and paints during the first months after completion and then more episodically by human activities during occupancy. Regarding the thermal comfort and even if occupants were globally satisfied, overheating and under heating were observed.

Based on this first experience, the OQAI is heading the deployment of a unique platform in France developed to collect indoor air quality and occupant comfort data in newly-built or freshly-refurbished buildings (Derbez et al., 2014c).

The objective is to populate a dedicated database with all available information on energy-efficient buildings. This is to be achieved by proposing a harmonized protocol targeted at the public-sector agencies and private-sector organizations tasked with measuring air quality in these buildings.

Within the framework of its "Energy-efficient buildings" agenda for action, the OQAI is deploying a data feedback and information sharing platform named "OQAI – ENERGY-EFFICIENT BUILDINGS". This platform is open to all public-sector agencies and private-sector stakeholders with an interest in assessing indoor air quality and occupant comfort in buildings that meet the latest building regulations on thermal performance.

The purpose of this platform is to provide support tools for the deployment of new-build projects and decisions on thermal efficiency retrofit options, to help identify the steps to be taken and optimize the transitional retrofitting of existing building stock.

It is grounded in the principle of steadily building up a common pool of data, obtained through the deployment of a "harmonized measurement, collection and transfer protocol" developed by the OQAI with input from its network of science and technology partners.

This "harmonized protocol" spans the entire data collection and sampling strategy in order to systematically characterize a panel of indicators on indoor air quality and occupant comfort in energy-efficient buildings. The protocol-compliant data thus collected is transferred into a national reference database run by the OQAI, and the output provided to ministries and government agencies.

A hundred-plus operations are already on-going in several regions of France, notably under impetus from the PREBAT programme, thanks to backing from the ADEME [French environment and energy management agency] and local-level co-financers. These first-wave operations are deployed by AASQA [local air quality monitoring authorities], CEREMA [national technical centre for risks, environment, and infrastructure planning], INERIS [national institute for industrial environment and risks], universities, consultants, etc.

6 DECISION SUPPORT TOOLS

The development of decision support tools, such as indoor air quality indicators or predictive indoor air pollution models, is the focus of a dedicated OQAI action programme.

Since 2006, the OQAI has been leading work to produce composite indoor air quality indexes capable of giving an intuitive snapshot of the full range of indoor air pollution geometries, not just to yield a fresh information tool but also to facilitate building/facility management. The first step - compilation of an inventory of the air quality indexes available in France and the international arena in 2006 - showed that existing indexes associated occupant symptoms not just with indoor air quality but also the quality of the wider indoor environment.

In 2007, CSTB conducted a survey to gain insight on the perceived incentives and disincentives of using indoor air quality indexes among facility managers (residential, tertiary-sector and school buildings), institutional clients, and other key people. The conclusions suggested that indexes held potential for sparking awareness of the indoor air quality issue, which at the time was still poorly understood and plagued by misconceptions. Indexes were believed to be a way to gauge the scale of the problem, from both an environmental perspective and a health safety standpoint. However, they were also perceived as a threat carrying an array of risks, not just in terms of health but also on the social, psychological, economic and legal fronts, all of which could incur the liability of the building managers if the indexes are 'negative'.

In 2009, the OQAI developed an indicator to qualify the level of stuffiness of the air in a classroom (schools) or activity room (day-care centre) while children are present: index of air stuffiness in schools, acronymed "ICONE". ICONE takes into account both the frequency and intensity of CO_2, which are measured over a minimum period of one week while children are present (Ramalho et al., 2013). The ICONE rating ranks stuffy air on a scale of zero to five: 0 is non-stuffy air (CO_2 concentration under 1,000 ppm 100% of the time) and 5 represents an extreme stuffy air condition (CO_2 concentration over 1,700 ppm 100% of the time). An indicator device was also developed in order to provide a visual cue at all times to allow management of open doors and windows to achieve the best balance between satisfactory indoor air quality and superfluous energy expenditure. OQAI is currently finalizing a new device to measure and visualize the air stuffiness in housing, after the final test results came in from trial-volunteer occupants.

Beside these developments, a first exploratory study assessing the social–economic cost of indoor air pollution in France has been conducted (Kopp et al., 2014). This research, led jointly with the Université Paris–Sorbonne, ANSES and the CSTB, was conducted on six pollutants: benzene, trichloroethylene, radon, carbon monoxide, particles and environmental tobacco smoke. Based on the method used, the French social cost of indoor pollution, described by the six pollutants, would be of the order of € 19 billion for one year. While these results have to be considered as a first estimation due to assumptions taken into account, it appears that the magnitude of these costs is significant and that the particulate matters make up a major part.

7 KNOWLEDGE FOR OUTREACH

OQAI-acquired knowledge and insight is widely extended through science outreach to industry and the wider public.

An illustrative example is the research used to construct the inventory of French and international data on indoor air quality, which is regularly updated and downloadable directly from the OQAI website (www.oqai.fr).

The OQAI also shares its knowledge through extension and outreach service programmes undertaken jointly with the building sector, the health community, and the wider public. The OQAI workshops offer a forum for various audiences to voice opinion on the issues covering topic areas: socio-economic costs due to IAQ (2014), IAQ in schools (2013), Photocatalytic air purification (2012), the potential of botanical air filtration (2010), Ventilation at home (2008), etc..

Since 2003, OQAI has hosted conference seminars, which serves as a regular forum for dialogue with building air quality sector professionals. The OQAI has followed up each of these conference seminars with papers and reviews keynoting the discussions.

8 CONCLUSIONS

Over the course of a decade of research, the OQAI has acquired a unique body of knowledge and insights into the living spaces and venues where the French spend most of their time: the wealth of information can be harnessed to describe living spaces at building-type scale and hence across the entire occupant population, while integrating the full range of real-world occupancy settings and scenarios. This reference data is used by health risk assessors to inform on exposure levels in populations where representative data is often notably lacking, and by community facility managers to forge public policy such as the mandatory monitoring of indoor air quality in public buildings and labelling of construction products, floorings, wall coverings, paints and varnishes regarding their volatile pollutant emissions.

However the task is far from complete, as there is still only a fragmentary understanding of indoor air quality and its driving factors in a significant proportion of our living spaces. Effective countermeasures to guard against at-risk situations will hinge on gaining a deeper understanding of these spaces.

Every day sees new construction materials and new building products and fittings hit the marketplace, yet experience teaches us that new products sometimes cause unforeseen problems for indoor air quality. The building industry is continually adapting and readapting, especially now that greenhouse gas mitigation policy has the building sector in its sights. The construction sector is set to make a major contribution on rational energy use, bringing with it radical shifts that should be tightly monitored to stay a step ahead of any potential impacts on indoor air quality, occupant comfort and health.

The holistic approach strategy that OQAI adopted from the outset to survey living spaces has given shape to a one-of-its-kind research programme producing unique insights into building technical performances, building equipment and their immediate environment, and unique research output on the occupants, their patterns of behaviour and their perceptions of the spaces they live in. Indeed, OQAI-led research has demonstrated how occupants play a major role in shaping indoor air pollution levels, largely through their behaviour patterns that will either worsen (smoking, using cleaning agents and DIY products, odour neutralizers, etc.) or improve (regularly airing rooms) their indoor environment. If we are to learn to build healthy, comfortable buildings, it is vital to gain a more fine-grained understanding of occupant usage patterns, needs and perceptions of the living space they evolve and interact with.

9 ACKNOWLEDGEMENTS

The French Indoor Air Quality Observatory is funded by the Ministries in charge of Housing, Environment and Health, the Scientific and Technical Building Centre (CSTB), ADEME (French environment and energy management agency) and ANSES (French agency for food, environmental and occupational health and safety).
The project focused on household exposure to semi-volatile organic compounds is co-funded through the national research programme on endocrine disruptors and the ANSES research program.
The authors thank the network of scientific and operational partners involved in the different OQAI programs.

10 REFERENCES

Billionnet C., Gay E., Kirchner S., Leynaert B., Annesi-Maesano I. (2011). Quantitative assessments of indoor air pollution and respiratory health in a population-based sample of French dwellings. *Environmental Research*, 111/3, April 2011, p. 425-434 [doi:10.1016/j.envres.2011.02.008]

Blanchard O., Mercier F., Ramalho O., Mandin C., Le Bot B., Glorennec P. (2014) Measurements of semi-volatile organic compounds in settled dust: influence of storage temperature and duration. *Indoor Air* 2014;24: 125-135.

Bonvallot N., Mandin C., Mercier F., LeBot B., Glorennec P. (2010) Health ranking of ingested semi-volatile organic compounds in house dust: an application to France. *Indoor Air* 2010;20:458–472.

Derbez M., B, Cochet V., Lethrosne M., Pignon C., Riberon J., Kirchner S. (2014a). Indoor air quality and comfort in seven newly built, energy-efficient houses in France. *Building and Environment*, 72, 173-187.

Derbez M., Berthineau B., Cochet V., Mandin C., Pignon C., Riberon J., Wyart G., Kirchner S. (2014b). Longitudinal study of indoor air quality and comfort of two low-energy single-family houses. *Indoor Air 2014, 13th International conference on indoor air quality and climate,* July 7-12, 2014, Hong Kong, CHN.

Derbez M., Lucas J.P., Ramalho O., Riberon J., Mandin C., Kirchner S. (2014c). French national data collection system on indoor air quality and comfort in energy-efficient buildings. *Indoor Air 2014, 13th International conference on indoor air quality and climate,* July 7-12, 2014, Hong Kong, CHN.

Deroubaix P., Lucas J.P., Ramalho O., Ribéron J. & Kirchner S. (2009). Ventilation in French buildings. *Air Information Review*, 30/4, 2009, p. 1-8.

Dijkstra N.E., De Kluizenaar Y., F, Bluyssen P.M., Mandin C., Mihucz V.G., Wolkoff P., Hänninen O., De Oliveira Fernandes E., Silva G.V., Carrer P., Bartzis J. (2014). Modern office related determinants of dry eye complaints: the Officair study. *Indoor Air 2014, 13th International conference on indoor air quality and climate*, July 7-12, 2014, Hong Kong, CHN

Hulin M., Moularat S., Kirchner S., Robine E., Mandin C., Annesi-Maesano I. (2013). Positive associations between respiratory outcomes and fungal index in rural inhabitants of a representative sample of French dwellings. *International Journal of Hygiene and Environmental Health*, 216/2 March [doi:10.1016/j.ijheh.2012.02.011]

Kirchner S., Derbez M., Duboudin C., Elias P., Garrigue J., Grégoire A., Lucas J.P., Pasquier N., Ramalho O., Weiss N.(2009).Indoor air quality in French dwellings
AIVC - International Energy Agency, Contributed report 12, p. 1-30

Kirchner S. (dir.). (2011). Qualité d'air intérieur, qualité de vie : 10 ans de recherche pour mieux respirer Observatoire de la Qualité de l'Air Intérieur. CSTB, September 2011, 208 p. 2011 [ISBN 978-2-86891-505-4].

Kopp P., Boulanger G., Bayeux T., Mandin C., Kirchner S., Vergriette B., Pernelet-Joly V. (2014). Socio-economic costs due to indoor air pollution: a tentative estimation for France. *Indoor Air 2014, 13th International conference on indoor air quality and climate,* July 7-12, 2014, Hong Kong, CHN.

Mandin C., Ramalho O., Riberon J., Kirchner S. (2012). IAQ, ventilation, comfort and health in office buildings: a French nationwide survey Ventilation 2012, *10th International conference on industrial ventilation,* September 17-19, 2012, Paris, FRA.

Mandin C., Mercier F., Lucas J.P., Ramalho O., Blanchard O., Bonvallot N., Raffy G., Gilles E., Glorennec P., Le Bot B. (2014a). ECOS-POUSS: a nationwide survey of semi-volatile organic compounds in home settled dust. *Indoor Air 2014, 13th International conference on indoor air quality and climate,* July 7-12, 2014, Hong Kong, CHN.

Mandin C., Mercier F., Lucas J.P., Ramalho O., Gilles E., Blanchard O., Bonvallot N., Glorennec P., Le Bot B. (2014b). ECOS-PM: a nationwide survey of semi-volatile organic compounds in indoor air. *Indoor Air 2014, 13th International conference on indoor air quality and climate,* July 7-12, 2014, Hong Kong, CHN.

Mandin C., Fossati S., Canha N., Cattaneo A., Cornelissen E., Hänninen O., de Kluizenaar Y., Mihucz V.G., De Oliveira Fernandes E., Peltonen M., Sakellaris Y., Saraga D., Ventura G., Mabilia R., Perreca E., Szigeti T., Carrer P., Bartzis J. (2014c) Indoor air quality in office buildings in Europe: the Officair project. *Indoor Air 2014, 13th International conference on indoor air quality and climate*, July 7-12, 2014, Hong Kong, CHN

Mercier F., Gilles E., Raffy G., Glorennec P., Mandin C., Le Bot B. (2014). A multi-residue method for the simultaneous analysis of several classes of semi-volatile organic compounds in airborne particles. *Indoor Air 2014, 13th International conference on indoor air quality and climate,* July 7-12, 2014, Hong Kong, CHN.

Mosqueron L., Nedellec V., Kirchner S., Gauvin S., Dor F., Cabanes P., Golliot F., Blanchard O., Derbez M., De Blay F., Lieuter-Colas F. (2009). Ranking indoor pollutants according to their potential health effect, for action priorities and costs optimization in the French permanent survey on indoor air quality. *Healthy Buildings*, December 7-11, 2003, Singapore, SGP, Vol. 3, p. 138-143

Ramalho O., Mandin C., Ribéron J., Wyart G. (2013) Air Stuffiness and Air Exchange Rate in French Schools and Day-Care Centres, *International Journal of Ventilation* 2013;12:175-180.

MODEL ERROR DUE TO STEADY WIND IN BUILDING PRESSURIZATION TESTS

François Rémi Carrié[1] and Valérie Leprince[2]

1 ICEE
93 rue Molière
69003 Lyon, France

2 PLEIAQ
84 C Avenue de la libération
69330 Meyzieu, France

ABSTRACT

We have analysed the steady wind model error based on a simplified building model with one leak on the windward side and one on the leeward side of the building. Our model gives an analytical expression of this error that depends on the leakage distribution and pressure coefficients. Using a test pressure of 50 Pa in this model, standard measurement protocol constraints contain the steady wind model error within about 3% and 11% with wind speeds below 6 m s^{-1} and 10 m s^{-1}, respectively. At 10 Pa, the error is in the range of 35% and 60% at 6 m s^{-1} and 10 m s^{-1}, respectively. The results are very sensitive to the leakage distribution at the low pressure point. Averaging pressurization and depressurization test results may be more or less beneficial depending on the leakage distribution, pressure coefficients and test pressure.

KEYWORDS

Airtightness, pressurization test, infiltration, measurement

NOMENCLATURE

C	Air leakage coefficient ($m^3\ s^{-1}\ Pa^{-n}$)
C_p	Pressure coefficient (-)
n	Flow exponent (-)
p	Pressure relative to external pressure (Pa)
q	Volumetric airflow rate ($m^3\ s^{-1}$)
U	Wind speed at the building level (m s^{-1})
x	Dimensionless pressure (-)
y	Dimensionless pressure coefficient (-)
z	Dimensionless leakage distribution (-)
Greek symbols	
□*p*	Pressure difference (Pa)
□□*x*□	Error of *x* (units of *x*)
□	Air density (kg m^{-3})

Subscripts and superscripts

bd	Pertaining to blower door measurement device
down	Downstream (leeward façades)
est	Estimated value
H	High pressure point
i	Interior of building
j	Index of a variable
k	Integer
model	Pertaining to model errors
s	Pressure measurement station
up	Upstream (windward façades)
wind	Pertaining to wind errors
zp	Zero-flow pressure measurement

1 INTRODUCTION

The most common way to perform a building airtightness test is to measure the airflow rate leaking through the building envelope at a given pressure. The test protocol is detailed in several standards (e.g. ISO 9972). Sherman and Palmiter (1994) have analysed uncertainties in those tests due in particular to precision and bias errors of pressure and flow measurement devices, as well as the deviation of the flow exponent. Other authors have estimated the repeatability and reproducibility of those tests (see for example Delmotte and Laverge, 2011).

The error due to wind is known to be a major source of error in building pressurization tests. However, it has rarely been studied in depth. To our knowledge, only recently Walker and collaborators (2013) have investigated the impact of the wind on the uncertainties based on the analysis of large measurement datasets and give interesting practical guidelines to reduce the size of the uncertainties due to wind.

To further understand the impact of wind on the results of pressurization tests, this paper looks more specifically at the governing equations giving the airflow rate through the blower door as a function of wind speed. It proposes an analytical approach to characterize the error due to a steady wind in building pressurization tests with a one-zone model.

2 BUILDING IDEALIZATION AND DIMENSIONAL ANALYSIS

In our analysis, we assume that the building can be represented by a single zone separated from the outside by 2 types of walls: walls on the windward side of the building which are subject to the same upwind pressure; and walls on the leeward side which are subject to the same downwind pressure. We further assume that the airflow rate through the leaks of the envelope is given by a power-law with the same flow exponent. Therefore, the building can be represented by only 2 leaks, one upwind, and one downwind. In this simple case, the true leakage flow coefficient of the building is strictly equal to the sum of the leakage flow coefficients. The leakage airflow rate at p_i is:

$$q_{bd} = C_{up}\left(p_{up} - p_i\right)^n + C_{down}\left(p_{down} - p_i\right)^n \tag{1}$$

The zero-flow pressure may be derived analytically from the mass balance equation:

$$C_{up}\left(p_{up} - p_{zp,i}\right)^n + C_{down}\left(p_{down} - p_{zp,i}\right)^n = 0 \tag{2}$$

where:

$$p_{up} = C_{p,up}\frac{\rho U^2}{2} \qquad p_{down} = C_{p,down}\frac{\rho U^2}{2} \tag{3}$$

Therefore, assuming C_{up} and $C_{p,up}$ are not null:

$$p_{zp,i} = \frac{1 + \left(\frac{C_{down}}{C_{up}}\right)^{1/n}\frac{C_{p,down}}{C_{p,up}}}{1 + \left(\frac{C_{down}}{C_{up}}\right)^{1/n}} C_{p,up}\frac{\rho U^2}{2} \tag{4}$$

It is useful to use dimensionless quantities to reduce the number of parameters.
Assuming $U \neq 0$, let:

$$x_j = \frac{p_{up}}{p_j};\, y = -\frac{C_{p,down}}{C_{p,up}};\quad z = \frac{C_{down}}{C_{up}} \tag{5}$$

Therefore, if $y\, z^{1/n} = 1$ then $p_{zp,i} = 0$; else:

$$x_{zp,i} = \frac{p_{up}}{p_{zp,i}} = \frac{1+z^{1/n}}{1-y\,z^{1/n}} \tag{6}$$

If the pressurization test is based on the leakage airflow rate measurement at a single pressure station $p_i = p_s$, the estimate of the leakage flow coefficient is:

$$C_{est} = \frac{C_{up}(p_{up}-p_s)^n + C_{down}(p_{down}-p_s)^n}{(p_{zp,i}-p_s)^n} \tag{7}$$

and the error on the estimated leakage airflow rate at any reference pressure is:

$$\begin{aligned}\frac{\delta q}{q} &= \frac{q_{est}-q_{nowind}}{q_{nowind}} = \frac{C_{est}-(C_{up}+C_{down})}{(C_{up}+C_{down})} \\ &= \frac{C_{up}(p_{up}-p_s)^n + C_{down}(p_{down}-p_s)^n - (C_{up}+C_{down})(p_{zp,i}-p_s)^n}{(C_{up}+C_{down})(p_{zp,i}-p_s)^n}\end{aligned} \tag{8}$$

In dimensionless quantities, this gives:

$$\left(\frac{\delta q}{q}\right)_{model,wind} = \frac{1}{1+z}\frac{(1-x_s)^n + z\,(1+y x_s)^n}{\left(1-\dfrac{1-y\,z^{1/n}}{1+z^{1/n}}x_s\right)^n} - 1 \tag{9}$$

Standard test protocols implicitly require the test pressure to be much greater than the upstream pressure. Therefore, x_s is small compared to 1 and assuming $y\,z^{1/n} \neq 1$, developing equation (9) in Taylor series truncated at order 2 near x_s gives to the following equation:

$$\begin{aligned}\left(\frac{\delta q}{q}\right)_{model,wind} &= \frac{1}{1+z}\left(n\left(y\,z + \frac{1+z}{x_{zp,i}} - 1\right)x_s\right. \\ &\left. + \left(\frac{n(n-1)}{2}(1+y\,z^2) + (y\,z-1)\frac{n^2}{x_{zp,i}} + (1+z)\frac{n(n+1)}{2\,x_{zp,i}^2}\right)x_s^2\right) + O(x_s^3)\end{aligned} \tag{10}$$

Note that the first order term is null only when $z = 0$ or $z = 1$. This expansion remains true with $x_{zp,i} \to \infty$ which is equivalent to $p_{zp,i} \to 0$ for $U \neq 0$.

3 PRESSURIZATION TEST CONDITIONS

3.1 Standard constraints for airtightness pressurization tests

For a test to be valid according to ISO 9972, the following constraints must be met:

— Constraint (a): $|p_L| \geq 10$ Pa
— Constraint (b): $|p_L| \geq 5\,|p_{zp,i}|$
— Constraint (c): $|p_H| \geq 50$ Pa
— Constraint (d): $|p_{zp,i}| \leq 5$ Pa

Since we are looking at the error at one specific pressure station, constraint (c) is not relevant for us, but we have applied the other constraints. Note that these constraints apply to measured pressure differences, not to the induced pressure difference.

3.2 Range of input parameters

The sensitivity analysis performed further in this paper is restricted to the following ranges of the input parameters (note that constraints apply which may further restrict those ranges):

- The wind velocity U varies between 0 and 10 m s^{-1};
- The pressure measurement stations are initially set to 50 Pa and 10 Pa. Note however that constraint (b) may impose a pressure measurement station higher than those values.
- Table 1 gives 3 pairs of possible values of $C_{p,up}$ and $C_{p,down}$ inspired from Liddament (1996).
- The leakage distribution ratio ($z = \frac{C_{down}}{C_{up}}$) ranges from 0.1 to 160.

3 RESULTS

4.1 Impact of wind velocity and zero-flow pressure

Given that $p_{zp,i}$ is strictly decreasing with z from p_{up} down to p_{down}, limiting the zero-flow pressure to 5 Pa (Constraint (d)) for all possible input values implies $U \leq 3.45$ m s^{-1} for our range of C_p values. Our model confirms that Constraint (d) is unlikely to be met above 6 m s^{-1} as stated in ISO 9972. In fact, it is impossible for values of z greater than 1.5. However, this is not true for values of z smaller than approximately 1.5 (the zero-flow pressure may be much smaller than 5 Pa for wind velocities greater than 6 m s^{-1}). Figure 1 shows that the error (in absolute value) is below about 3% for wind speeds up to 6 m s^{-1} at building height and 11% up to 10 m s^{-1} for a test pressure at 50 Pa. The error is much greater at a test pressure of 10 Pa: it is in the region of 40% and 60% for wind speeds up to 6 m s^{-1} and 10 m s^{-1}, respectively (Figure 2).

4.2 Averaging pressurization and depressurization

Developing equation (9) in Taylor series (for $x_s << 1$) shows that averaging the results of two tests performed at opposite pressures p_s and $-p_s$ eliminates the first order term of the series and thereby yields a smaller error when the first order term is not null. This is confirmed in Figure 1-Figure 2 where the error of the average value is indeed smaller, but the benefit is relatively small.

4.3 Relaxing Constraint (d)

Constraint (d) ($|p_{zp,i}| \leq 5$ Pa) is directly related to the wind pressure (see equation (4)). Therefore, relaxing constraint (d) allows testing at higher wind velocities. We have not found a significant impact on the error range when relaxing constraint (d) (see Figure 1-Figure 2). It remains in the same range in either pressurization or depressurization mode (when averaging or not).

4.4 Restricted leakage distribution

With a restricted leakage distribution to $2 < z < 8$ which is meant to represent a deviation of a factor of 2 from an even distribution[1], we have found a significant effect on the error (Figure 3-Figure 4), in particular at 10 Pa, where the error drops below 3%.

5. CONCLUSIONS AND FUTURE WORK

One key result is that alone, the model error due to the wind on the estimated airflow rate is relatively small for the high pressure point, but it can become very significant with a low pressure point. While the error lies within 12% for wind speeds up to 10 m s^{-1} at 50 Pa, it can reach 60% at the low pressure point (10 Pa). However, these results are very sensitive to the leakage distribution for the low pressure point. Our analysis does not include other errors, in particular those due to precision and bias of the measurement devices used or the deviation of the flow exponent over the range of pressures. Because tests are usually performed at multiple pressure stations including relatively low pressures (close to 10 Pa), it would be useful to extend our model to include these additional source of errors.

6. REFERENCES

[1] ISO 9972. 2006. Thermal performance of buildings – Determination of air permeability of buildings - Fan pressurization method. ISO. 30 pp.

[2] Sherman, M.H., and Palmiter, L. 1994. Uncertainty in Fan Pressurization Measurements. Airflow Performance of Envelopes, Components and Systems STP 1255: 262-283. Also published as LBNL report LBNL-32115.

[3] Delmotte, C. and Laverge, J. 2001. Interlaboratory tests for the determination of repeatability and reproducibility of buildings airtightness measurements. In proceedings of the 32nd AIVC conference. pp. 183-187.

[4] Walker, I.S., Sherman, M.H., Joh, J. and Chan, W.R. 2013. Applying large datasets to developing a better understanding of air leakage measurement in homes. International Journal of Ventilation. ISSN 1473-3315, 111: 323-337.

[5] Liddament, M.W. 1996. A guide to energy efficient ventilation. AIVC Document AIC-TN-VENTGUIDE-1996. ISBN 0946075859. 254 pp.

---

[1] Evenly distributed leaks on the facades of a cube would give $z = 4$ (there are 3 facades on the leeward side plus the roof).

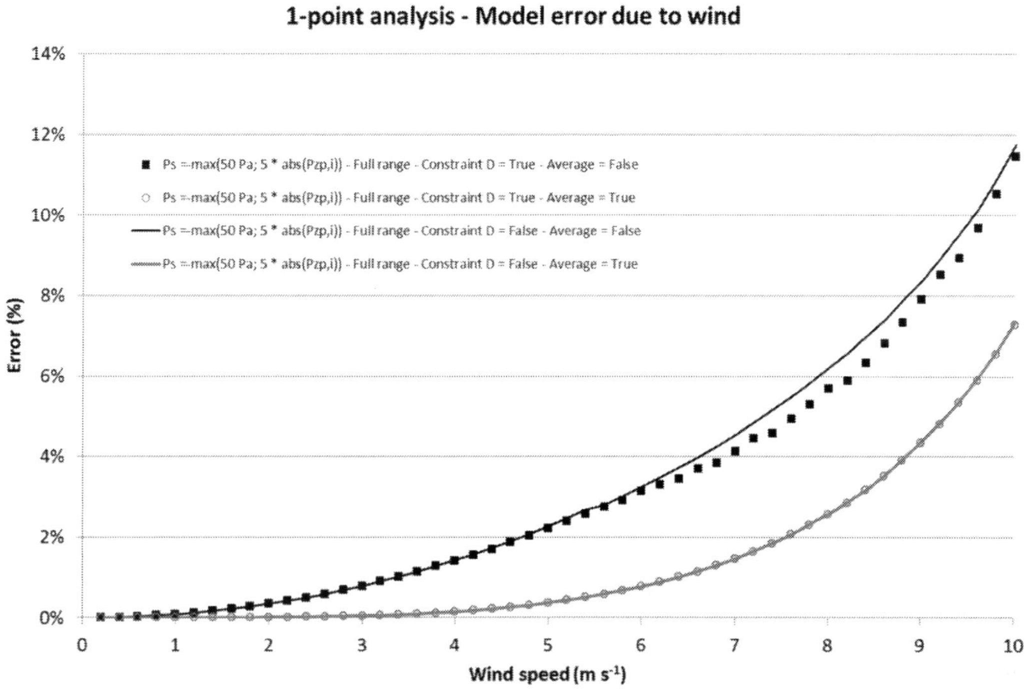

Figure 1 : Error due to wind as a function of wind speed and leakage distribution.

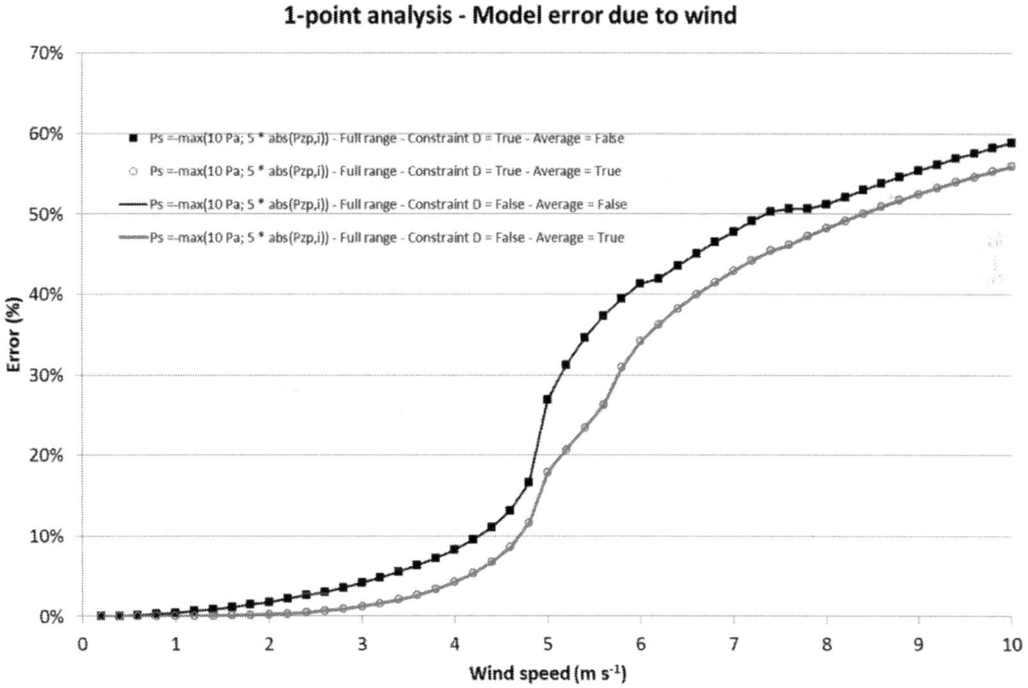

Figure 2 : Error due to wind as a function of wind speed and leakage distribution.

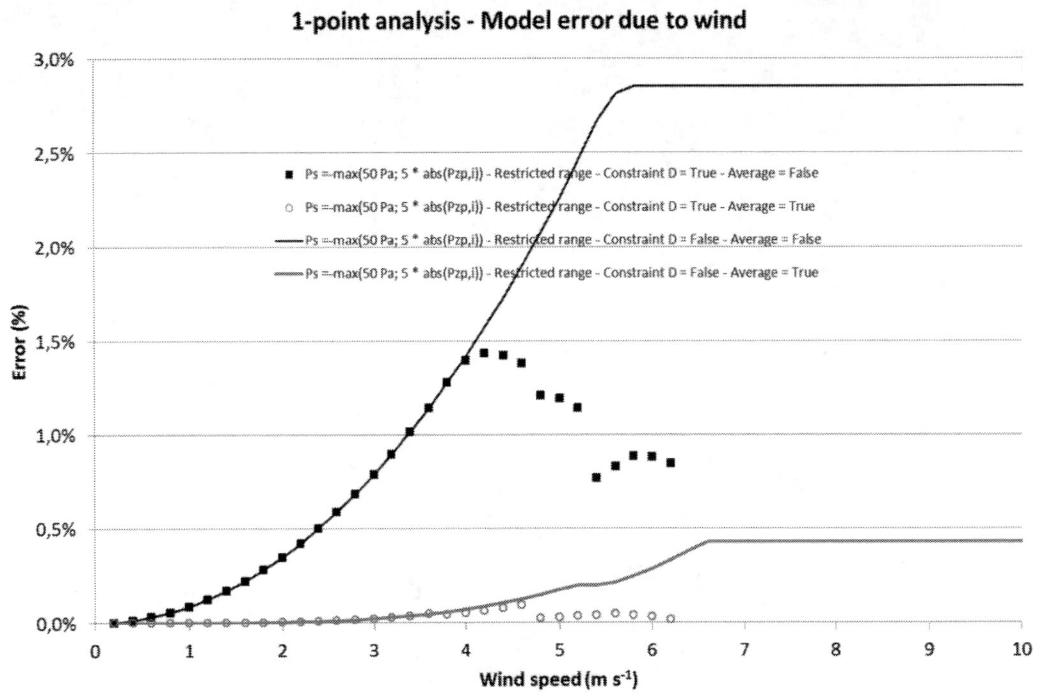

Figure 3 : Error due to wind as a function of wind speed and leakage distribution.

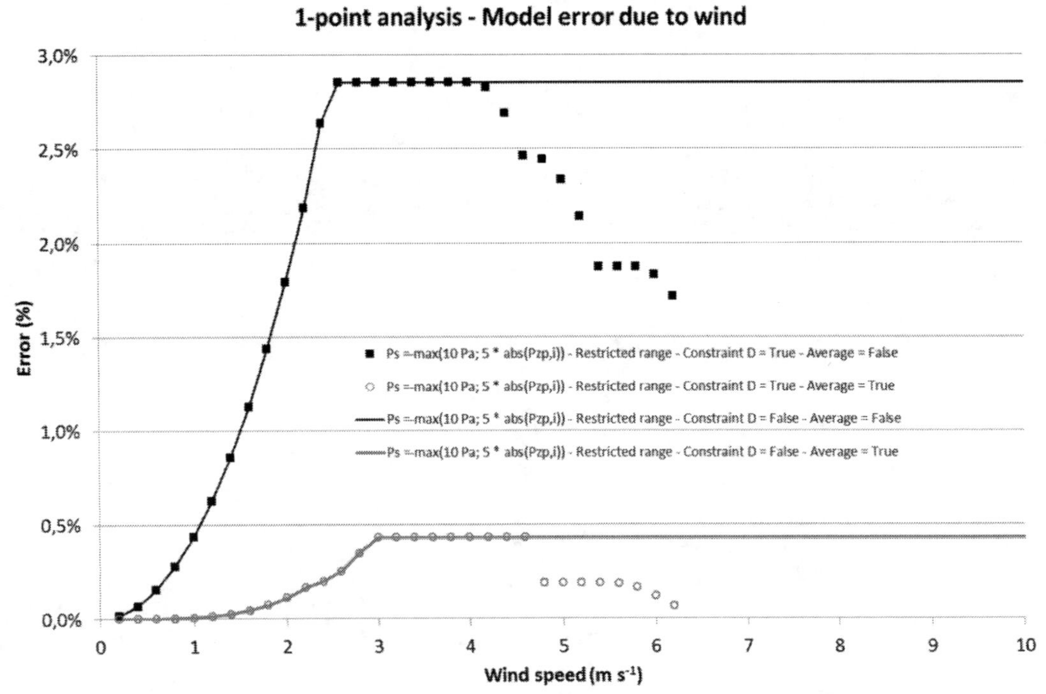

Figure 4 : Error due to wind of average result between pressurization and depressurization tests as a function of wind speed and leakage distribution.

Test case ID	$C_{p,up}$	$C_{p,down}$	$y = -\frac{C_{p,down}}{C_{p,up}}$
1	0.05	-0.30	6.0
2	0.25	-0.50	2.0
3	0.50	-0.70	1.4

Table 1: Values of wind pressure coefficients used in sensitivity analyses.

TEMPERATURE AND PRESSURE CORRECTIONS FOR POWER-LAW COEFFICIENTS OF AIRFLOW THROUGH VENTILATION SYSTEM COMPONENTS AND LEAKS

François Rémi Carrié

ICEE
93 rue Molière
69003 Lyon, France

ABSTRACT

The characterization of power-law coefficients of the airflow through ventilation system components and ductwork or building leaks should include corrections on the airflow rate measurement because of two phenomena: a) the temperature and pressure conditions at the flow measurement device may not be the same as those seen by the test object; b) the temperature and pressure conditions experienced by the object may differ from reference conditions. This paper gives the analytical expression of these corrections depending on the air viscosity, air density and flow exponent. Corrections may be significant in possible test conditions and therefore should be applied systematically when performing such measurements.

KEYWORDS

Airflow measurement, ventilation, building, airtightness, leak

NOMENCLATURE

A	Cross-sectional area (m^2)
C	Airflow rate coefficient (power-law coefficient) (m^3 s^{-1} Pa^{-n})
C_d	Discharge coefficient (-)
k	Airflow rate coefficient (power-law coefficient) ($m^{2-3/n}$ $s^{-2+1/n}$)
L	Length (m)
n	Flow exponent (-)
p	Relative pressure (Pa)
p_{atm}	Absolute pressure (Pa)
q_v	Volumetric airflow rate (m^3 s^{-1})

Greek symbols

α, β, γ	Real numbers (-)
Δp	Pressure difference (Pa)
μ	Dynamic viscosity of the air (kg m^{-1} s^{-1})
ρ	Air density (kg m^{-3})

Subscripts and superscripts

i	Pertaining to inside of ductwork or envelope where test object is located
e	Pertaining to outside of ductwork or envelope where test object is located
max	Maximum value
$meas$	At airflow measurement device
$object$	Pertaining to test object
$test$	Pertaining to conditions near the test object
0	Pertaining to reference temperature and pressure conditions

Physical constants

$T_0 = 293{,}15$ K

$p_{atm;0} = 101325$ Pa

$\rho_0 = 1.204$ kg m^{-3}

1 INTRODUCTION

Ventilation system components and leaks in ductwork or buildings are commonly characterized by a power-law with parameters that may be determined experimentally by creating artificially a series of pressure differences across the components or leaks. There may be differences either:

— between the temperature observed at the flow measurement device and the temperature experienced by the tested object;

— between the temperature experienced by the tested object and the temperature in reference conditions.

Therefore, to be able to compare test results, one shall apply corrections to account for both effects described above. There are a number of standards that propose such corrections, but there are differences between those.

The objective of this paper is to give the common background for these corrections. The analyses shown here apply when mass conservation between the test object and the flow measurement device applies.

2 BACKGROUND

We may divide the types of methods used to characterize ventilation system components or leaks in 4 major categories:

a. Envelope or ductwork depressurisation tests
b. Envelope or ductwork pressurisation tests
c. Ventilation system component tests with measurement device downstream
d. Ventilation system component tests with measurement device upstream

Common application of these methods is represented in Figure 1. This figure shows that the conditions seen by the test object are different depending on the method. Note however that all methods implicitly assume that there is no significant temperature and absolute pressure change across the test object, i.e., that the object can be characterized at the temperature and absolute pressure conditions of the air entering the object.

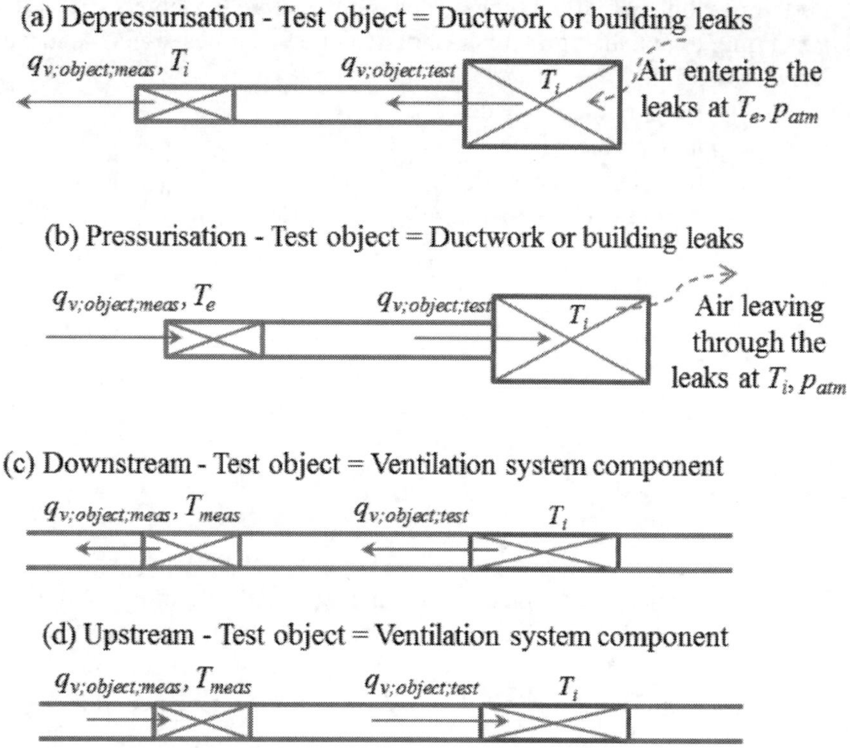

Figure 1: Common application of methods used to characterize ventilation system components or leaks

3 PRESSURE-FLOW RELATIONSHIP

The relationship between flow and pressure through a ventilation system component or through leak(s) is commonly expressed as follows:

$$q_{v;object;test} = C_{object;test} \cdot \Delta p^n \quad (1)$$

Where

$q_{v;object;test}$ is the airflow rate through the test object ($m^3\ s^{-1}$)
$C_{object;test}$ is the flow coefficient that characterizes the test object ($m^3\ s^{-1}\ Pa^{-n}$)
Δp is the pressure across the test object (Pa)
n is the flow exponent ($0.5 \leq n \leq 1.0$)

The coefficient C depends on the only physical characteristics of the system, which are:
— The dynamic viscosity of the air, μ ($kg\ m^{-1}\ s^{-1}$)
— The air density, ρ ($kg\ m^{-3}$)
— The characteristic length of the system, L (m)

Therefore, C is a combination of terms of the form: $\mu^\alpha \cdot \rho^\beta \cdot L^\gamma$ where α, β, γ are unknown exponents. Dimensional homogeneity allows us to write the following system[1]:

Dimension	C		μ	ρ	L
Length (m)	$3+n$	=	$-\alpha$	-3β	γ
Time (s)	$2n-1$	=	$-\alpha$		

(2)

[1] Note that 1 Pa = 1 N m^{-2} = 1 kg $m^{-1}\ s^{-2}$

Mass (kg)	$-n$	=	α	β

Therefore:

$$\begin{aligned} \alpha &= 1-2n \\ \beta &= n-1 \qquad (3) \\ \gamma &= 2n+1 \end{aligned}$$

Leading to:

$$C \propto \mu^{1-2n}\,\rho^{n-1}\,L^{2n+1} \qquad (4)$$

Note that equation (4) is consistent with the correction proposed by Walker (1998) and ASTM E 779-2010.

When $n = 0.5$, we find:

$$C \propto \rho^{-0.5}\,L^{2} \qquad (4)$$

In this particular case ($n = 0.5$), proportionality is consistent with the commonly used relationship for an inertial flow through a perfect orifice:

$$q_v = C_d \cdot A \cdot \left(\frac{2\,\Delta p}{\rho}\right)^{0.5} \qquad (4)$$

Another common expression of the flow-pressure relationship is:

$$\Delta p = k \cdot \rho \cdot q_v^{1/n} \qquad (5)$$

Similarly, we obtain:

$$k \propto \mu^{2-\frac{1}{n}}\,\rho^{\frac{1}{n}-2} \qquad (6)$$

4 MASS CONSERVATION THROUGH THE TEST OBJECT

The temperature and pressure conditions around the test object are not necessarily the same as the ones near the airflow measurement device. Therefore, mass conservation gives:

$$q_{v;object;test} = \frac{\rho_{meas}}{\rho_{test}} q_{v;object;meas} \qquad (7)$$

5 FLOW COEFFICIENT CORRECTION

To allow comparison between tests, we are interested in the flow through the test object under reference conditions ("0") for the same pressure difference, Δp. Therefore:

$$q_{v;object;test} = C_{object;test} \cdot \Delta p^n \tag{8}$$

and

$$q_{v;object;0} = C_{object;0} \cdot \Delta p^n \tag{9}$$

If n is unknown, $C_{object;test}$ and n may be determined together with a regression analysis based on equation (8). (Note that this implicitly assumes that all test points are obtained under similar temperature and pressure conditions.) Then, using equation (4) allows one to obtain the airflow rate coefficient at reference temperature and pressure conditions:

$$\frac{C_{object;0}}{C_{object;test}} = \left(\frac{\mu_{test}}{\mu_0}\right)^{2n-1} \left(\frac{\rho_{test}}{\rho_0}\right)^{1-n} \tag{10}$$

This also yields:

$$q_{v;object;0} = \frac{C_{object;0}}{C_{object;test}} \frac{\rho_{meas}}{\rho_{test}} q_{v;object;meas} \tag{11}$$

and therefore:

$$q_{v;object;0} = \left(\frac{\mu_{test}}{\mu_0}\right)^{2n-1} \left(\frac{\rho_{test}}{\rho_0}\right)^{1-n} \frac{\rho_{meas}}{\rho_{test}} q_{v;object;meas} \tag{12}$$

Note that when $n = 0.5$, the air viscosity vanishes from equation (12).

6 APPLICATION TO PRESSURISATION TESTS WITH UNKNOWN FLOW EXPONENT

This paragraph is relevant to tests performed at multiple pressure stations to obtain both the airflow rate coefficient and the flow exponent. In this case, we proceed with the following steps:

- Apply mass conservation correction equation (7) to obtain the air leakage flow rate through the test object for various pressure differences;
- Extract $C_{object;test}$ and n from the values of $q_{v;object;test}$ obtained;
- Apply equation (10) to characterize the envelope or ductwork airtightness in reference conditions.

In depressurisation mode, the air entering the measurement device is inside air (so at temperature T_i) while the air passing through the leaks is at temperature T_e (see Figure 1). Therefore:

$$q_{v;object;test} = \frac{\rho_i}{\rho_e} q_{v;meas} \tag{13}$$

$$C_{object;0} = \left(\frac{\mu_e}{\mu_0}\right)^{2n-1} \left(\frac{\rho_e}{\rho_0}\right)^{1-n} C_{object;test} \tag{14}$$

In pressurisation mode, we obtain:

$$q_{v;object;test} = \frac{\rho_e}{\rho_i} q_{v;meas} \tag{15}$$

$$C_{object;0} = \left(\frac{\mu_i}{\mu_0}\right)^{2n-1} \left(\frac{\rho_i}{\rho_0}\right)^{1-n} C_{object;test} \tag{16}$$

These equations are identical to those used in ASTM E 779-2010. They differ from ISO 9972 where viscosity dependence is neglected.

7 APPLICATION TO PRESSURISATION TESTS WITH DEFAULT FLOW EXPONENT

This paragraph is relevant either:

- When the measurement is performed at only one pressure station;
- When the leakage airflow rate(s) at one or several pressure stations is (are) compared to a threshold value.

In both cases, the exponent is set to a default value and the underlying criterion is:

$$q_{v;object;0} \leq C_{max} \cdot \Delta p^n \tag{17}$$

Therefore, we will need to derive $q_{v;object;0}$ to check this criterion.

In pressurisation mode:

$$q_{v;object;0} = \left(\frac{\mu_i}{\mu_0}\right)^{2n-1} \left(\frac{\rho_i}{\rho_0}\right)^{1-n} \frac{\rho_e}{\rho_i} q_{v;object;meas} \tag{18}$$

In depressurisation mode:

$$q_{v;object;0} = \left(\frac{\mu_e}{\mu_0}\right)^{2n-1} \left(\frac{\rho_e}{\rho_0}\right)^{1-n} \frac{\rho_i}{\rho_e} q_{v;object;meas} \tag{19}$$

8 APPLICATION TO CHARACTERIZATION OF VENTILATION SYSTEM COMPONENTS

Equation (7) is used to obtain $q_{v;object;test}$. However, unlike in envelope or ductwork pressurisation tests, the temperature of the air flowing through the measurement device is specific to the test apparatus. It may not be assumed in all cases that $T_{meas} = T_i$ or $T_{meas} = T_e$ as we did previously: this may or may not be true depending on where the air flowing through the measurement device comes from.

As for the airflow rate or the airflow coefficient, because the test conditions seen by the object are those inside the ducts in which the object is placed, we may write:

$$q_{v;object;0} = \left(\frac{\mu_i}{\mu_0}\right)^{2n-1} \left(\frac{\rho_i}{\rho_0}\right)^{1-n} \frac{\rho_{meas}}{\rho_i} q_{v;object;meas} \tag{20}$$

9 TEMPERATURE AND PRESSURE DEPENDENCE OF AIR VISCOSITY AND AIR DENSITY

To apply those corrections, it is useful to know the how the air viscosity and air density vary with temperature and absolute pressure. When the humidity can be neglected, the following correlations may be used:

$$\rho = 1.204 \cdot \frac{p_{atm}}{p_{atm;0}} \frac{T_0}{T} \tag{21}$$

and

$$\mu = (17{,}1 + 0{,}048(T - 273{,}15)) \cdot 10^{-6} \tag{22}$$

Giving $\mu_0 = 18.06\ 10^{-6}$ kg m^{-1} s^{-1}

Of course, depending on the accuracy needed and test conditions, other correlations may be used, such as those proposed in ASTM E779 (2010) and CR 14738 (2002).

10 NUMERICAL APPLICATION

To see the magnitude of the changes induced by those corrections, we have calculated the airflow rate correction defined as follows in various test conditions:

For equations (18) and (19): $\text{Airflow correction} = \frac{q_{v;measured} - q_{v;0}}{q_{v;0}}$ (23)

For equation (20): $\text{Airflow correction} = \frac{q_{v;object} - q_{v;0}}{q_{v;0}}$ (24)

The results show relatively high deviations in possible test conditions (see Table 1 and Table 2).

p_{atm} (Pa)	ϑ_i (°C)	ρ_i (kg m^{-3})	μ_i (kg m^{-1} s^{-1})	Airflow correction assuming $\rho_{meas} = \rho_i$ Eq. (20), n = 0,5	Eq. (20), n = 0,65
101325	0	1,29	17,1 $\cdot 10^{-6}$	-3,5%	-0,8%
101325	5	1,27	17,3 $\cdot 10^{-6}$	-2,6%	-0,6%
101325	10	1,25	17,6 $\cdot 10^{-6}$	-1,7%	-0,4%
101325	15	1,22	17,8 $\cdot 10^{-6}$	-0,9%	-0,2%
101325	20	1,20	18,1 $\cdot 10^{-6}$	0,0%	0,0%
101325	25	1,18	18,3 $\cdot 10^{-6}$	0,8%	0,2%
101325	30	1,16	18,5 $\cdot 10^{-6}$	1,7%	0,4%
90000	0	1,15	17,1 $\cdot 10^{-6}$	2,4%	3,4%
90000	5	1,13	17,3 $\cdot 10^{-6}$	3,4%	3,6%
90000	10	1,11	17,6 $\cdot 10^{-6}$	4,3%	3,8%
90000	15	1,09	17,8 $\cdot 10^{-6}$	5,2%	4,0%
90000	20	1,07	18,1 $\cdot 10^{-6}$	6,1%	4,2%
90000	25	1,05	18,3 $\cdot 10^{-6}$	7,0%	4,4%
90000	30	1,03	18,5 $\cdot 10^{-6}$	7,9%	4,6%

Table 1: Airflow correction (in %) derived from equation (20).

		Airflow correction assuming ϑ_e = 10 °C			
p_{atm} (Pa)	ϑ_i (°C)	Eq. (18), n = 0,5	Eq. (18), n = 0,65	Eq. (19), n = 0,5	Eq. (19), n = 0,65
101325	0	0,1%	2,8%	-5,2%	-3,9%
101325	5	-0,8%	1,2%	-3,5%	-2,2%
101325	10	-1,7%	-0,4%	-1,7%	-0,4%
101325	15	-2,6%	-1,9%	0,0%	1,4%
101325	20	-3,4%	-3,4%	1,8%	3,1%
101325	25	-4,2%	-4,8%	3,5%	4,9%
101325	30	-5,0%	-6,2%	5,2%	6,6%
90000	0	6,2%	7,2%	0,6%	0,1%
90000	5	5,2%	5,5%	2,4%	2,0%
90000	10	4,3%	3,8%	4,3%	3,8%
90000	15	3,4%	2,2%	6,1%	5,6%
90000	20	2,5%	0,7%	8,0%	7,5%
90000	25	1,6%	-0,8%	9,8%	9,3%
90000	30	0,8%	-2,3%	11,6%	11,1%

Table 2: Airflow correction derived from equations (18) and (19).

11 CONCLUSION

Our analyses show that because the airflow rate coefficient depends on air viscosity, air density and flow exponent, the corrections that have to be applied to airflow rates through ventilation system components and leaks to characterize these objects depend on these variables as well. However, when the flow is inertial (n = 0.5), these corrections do not depend on air viscosity. Still, the corrections may be significant, which calls for including these effects systematically when such measurements are performed.

12 REFERENCES

1.ASTM E779-10. 2010. Standard Test Method for Determining Air Leakage Rate by Fan Pressurization. ASTM. 10 pp

2.CR 14378. 2002. Ventilation for buildings - Experimental determination of mechanical energy loss coefficients of air handling components. CEN Report 14378.

3.ISO 9972. 2006. Thermal performance of buildings – Determination of air permeability of buildings - Fan pressurization method. ISO. 30 pp.

4.Walker, I.S., Wilson, D.J., Sherman, M.H. 1998. A comparison of power law to quadratic formulations or air infiltration calculations. *Energy and Buildings*, 27: 293-299. Also published as LBNL report LBNL-41447.

TOPICAL SESSION: STATUS OF THE REVISION OF EN VENTILATION STANDARDS SUPPORTING THE EPBD RECAST

Jaap Hogeling[*1], Gerhard Zweifel[2],
Bjarne W. Olesen[3], and François Rémi Carrié[4]

1 ISSO
Kruisplein 25
Rotterdam, The Netherlands
**Corresponding author: j.hogeling@isso.nl*

2 Lucerne University of Applied Sciences and Arts
Technikumstrasse 21
CH-6048 Horw

3 Technical University of Denmark
ICIEE, Department of Civil Engineering
Nils Koppels Elle, Building 402
DK-2800 Kgs, Lyngby, Benmark

4 ICEE
93 rue Molière
69003 Lyon, France

1 SYNOPSIS

This session will discuss the major changes of the ventilation standards supporting the implementation of the Energy Performance of Buildings Directive recast. It is foreseen that some of these standards will serve as basis of future ISO standards as well.

2 SESSION PROGRAMME

Chairpersons: Jaap Hogeling (Netherlands) and Gerhard Zweifel (Switzerland)

The session will start with an introduction giving the overall context of the revision of the EPBD standards. Then, session will be structured around 3 presentations that will outline the major changes in the versions submitted for public enquiry during the summer 2014.

- The second generation CEN-EPBD standards, Jaap Hogeling, Chair of CEN TC 371 "Project Committee - Energy Performance of Buildings"
- Indoor environmental parameters: revision of EN 15251, Bjarne Olesen, Convenor of CEN TC 156 / WG 19 "Joint Working Group between CEN/TC 156 and CEN/TC 371 - Revision of EN 15251;2007"
- Calculations methods: revision of EN 15241-15242-15243, Gerhard Zweifel, Convenor of TC 156 / WG 21 "Revision of calculation standards EN 15241, 15242 and 15243"
- Inspection of ventilation systems and air conditioning systems, François Rémi Carrié, Convenor of TC 156 / WG 23 "Guidelines for inspection of ventilation systems and air conditioning systems"

3 CONTEXT

The Energy Performance of Buildings Directive, first published in 2002 (2002/91/EC), required all EU countries to take measures to improve the energy performance of new and renovated buildings, in particular through energy regulations and energy certification schemes for buildings, as well as inspections of boilers and air-conditioners. To help member states achieve this while allowing fair comparisons of best practices and use of construction products, the European Commission issued a mandate to European standardisation bodies which resulted in the publication of a first EPBD package standards in 2007-2008.

In parallel, the European Concerted Action as well as European projects of the Intelligent Energy Europe such as IEE-CENSE and Harmonac provided feedback on the first EPBD calling for a second mandate to revise the standards to support the EPBD directive recast (2010/31/EU).

According to mandate M480, "the improved set of EPBD standards shall become a systematic, clear and comprehensive package for the benefit of professionals, Member States and relations with third countries." This new mandate work was divided into two phases:

- Phase 1 (2011-2012): This phase included:
  - Preparation of basic principles and rules, together with the member states, and based on the overall requirements & expectations on the standards
  - Development of a modularly structured overarching standard
  - Preparation of technical rules for drafting all standards. To ease the understanding and application of the standards, templates have been developed for the standard texts as well as for checking the consistency of the equations and results provided by the standards.
- Phase 2 (2012-2015): This phase deals with the actual revision of the standards in separate working groups.

With regard to ventilation, there are 7 EPBD standards under revision within CEN Technical Committee 156 (Ventilation for Buildings) (Table 1).

Table 1: Ventilation EPBD standards under revision and scope

	Input parameters for energy calculations	Calculation	Inspection
Standard(s) under revision and scope	EN 15251 – Indoor environmental input parameters EN 13779 — Performance requirements for ventilation, air conditioning and room-conditioning systems (non-residential buildings)	EN 15241, EN 15242, EN 15243 These 3 standards have been re-arranged into a set of 5 standards dealing with calculation methods for energy requirements of ventilation and air-conditioning systems (generation, distribution and emission).	EN 15239, EN 15240 Guidelines for inspection of ventilation and air conditioning systems These 2 standards have been merged into a single standard dealing with ventilation and air conditioning systems, while the two aspects were divided between EN 15239 and EN 15240.

The texts are progressively submitted to TC 156 approval and progression to public enquiry. The enquiry phase should last until spring 2015 and, after addressing comments, the new standards should be published by the end of 2015.

INTEGRATED APPROACH AS A PREREQUISITE FOR NEARLY ZERO ENERGY SCHOOLS IN MEDITERRANEAN REGION-ZEMEDS PROJECT

Niki Gaitani[1*], Clara Ferrer[2], Mat Santamouris[1], Claudia Boude[3], Michael Gerber[4], Roberta Ansuini[5],

1. National & Kapodistrian University of Athens, Group of Building Environmental Research, Building of Physics-5, University Campus, 15784, Athens, Greece

2. Fundació Privada Ascamm - Parc Tecnològic del Vallès, Av. Universitat Autònoma, 23 - 08290 Cerdanyola del Vallès (Barcelona) - Spain

3. Gefosat, 11 ter avenue Lepic – 34070 Montpellier - France

4. Local Energy Agency of Montpellier
2 Place Paul Bec
34 000 Montpellier - France

5. Province of Ancona, via Ruggeri, 5, Ancona, Italy

**Corresponding author: ngaitani@phys.uoa.gr*

ABSTRACT

The present article deals with an integrated nearly zero-energy buildings (nZEB) approach for the retrofit of Mediterranean school buildings, in the framework of ZEMedS project. EU energy policy encourages member states to start converting building stock into nZEB and public authorities to adopt exemplary actions. In Mediterranean regions of Italy, Greece, Spain and France, there are approximately 87.000 schools, consuming in a rough estimation around 2Mtoe/year. The final energy consumption of typical Mediterranean school built in the period 60-80's was estimated, on an average of 100 kWh/m^2/y in final energy for all uses. An holistic approach should combine measures to achieve energy performance and indoor environmental quality (IEQ), as well as strategies to involve the users, particularly the future generations. ZEMedS project covers a complete renovation path, tackling strategies for the envelope, the systems and renewable energy applications as well as the energy management and users' behaviour. In this context, the first results are presented with case studies of school buildings to be analyzed in terms of the energy efficiency and cost optimality so as to define a detailed renovation action plan.

KEYWORDS
NZEB, Schools, Energy-efficient Retrofitting, IEQ, Mediterranean

1 NZEB EU FRAMEWORK

Buildings represent the largest available source of cost effective energy saving and CO_2 reduction potential within Europe. The aim to reduce energy consumption in buildings has led to Zero Energy Building (ZEB) concept. EU energy policy encourages Member States (MS) to start converting building stock into nearly zero-energy buildings (nZEB) and public

authorities to adopt exemplary actions (EPBD recast). The Energy Efficiency Directive aims to accelerate the refurbishment rate of public buildings through a binding target. The commitment to implement energy efficiency in buildings requires efforts from all MS to contribute to energy efficiency in the building sector, through the adoption of suitable regulatory and policy instruments. Energy Efficiency Directive (EED, 2012/27/EU) adopted in October 2012, includes a requirement for Member States to develop long term renovation strategies for their national building stocks. EED was developed in order to help deliver the EU's 20% headline target on energy efficiency by 2020, as well as to pave the way for further improvements thereafter. Alongside EED, the Energy Performance of Buildings Directive (EPBD, 2010/31/EU) recast in 2010, sets out numerous requirements including energy performance certification of buildings, inspection regimes for boilers and air conditioning plants, and requirements for new buildings to be nearly zero energy. EPBD sets minimum energy performance standards for buildings undergoing renovation. According to Article 2.2 of the EPBD recast "'nearly zero-energy building' means a building that has a very high energy performance, as determined in accordance with Annex I. The nearly zero or very low amount of energy required should be covered to a very significant extent by energy from renewable sources, including energy from renewable sources produced on-site or nearby;"
Specifically Annex I states that "The energy performance of a building shall be determined on the basis of the calculated or actual annual energy that is consumed in order to meet the different needs associated with its typical use and shall reflect the heating energy needs and cooling energy needs (energy needed to avoid overheating) to maintain the envisaged temperature conditions of the building, and domestic hot water needs". Each Member State shall ensure that, as from 1 January 2014, 3% of the total floor area of heated and/or cooled buildings owned and occupied by its central government is renovated each year to meet at least the minimum energy performance requirements that it has set in application of Article 4 of Directive 2010/31/EU. Furthermore, the Energy Roadmap 2050, published on the 15 December 2011, goes beyond the 2020 goals and provides an analysis of the long term energy policy orientations: EU is committed to reducing greenhouse gas emissions to 80-95% below 1990 levels by 2050.

In Europe most renovation activity currently achieves only modest energy savings, perhaps 20-30%, but this needs to increase to profound renovations of at least 60% if the full economic potential is to be realized. Buildings Performance Institute Europe (BPIE) has studied impact of different renovation pathways on the resulting energy and carbon savings. The outcome shows scenarios where both the rate and the depth of renovation were substantially increased, rapid decarbonisation of the energy supply system, could be achieved. The qualitative definition and the different approaches worldwide to achieve net zero have led to discussion amongst experts. There have been attempts to tackle a wide spectrum of additional specifications and issues pertaining to terminology and definitions around buildings that consume very low or zero energy (or carbon), including those with net energy production – 'energy positive' (Ferrante, 2012). Torcellini et al (2006) have reported four well-documented definitions based on extensive data from existing low energy buildings: net-zero site energy, net-zero source energy, net-zero energy costs and net-zero energy emissions.

Moreover, many methodologies have been proposed. They deal with different features such as: Metric of the balance: delivered energy, primary energy, CO_2 (equivalent) emissions, energy cost, Period of balance: annual, monthly, Type of energy use: operating energy, total energy and energy use and EE (embodied energy), Type of balance: generation/use, grid in/out, Renewable supply options: footprint, on-site, off-site, Conversion factors (for primary energy and CO_2 emissions). Even with the recent methodologies and projects, the complexity of nZEB concept and the existing national and regional policies have led to this current situation, where only few countries have already set up a nZEB standard. At the end of November 2012, according to the European Commission only 9 Member States (BE, DK,

CY, FI, LT, IE, NL, SE and UK) had reported their nZEB national plans to the Commission as required. As regards the practical definition of nZEBs, only 5 Member States (BE, CY, DK, IE and LT) presented a definition that contains both a numerical target and a share of renewable energy sources.
As the qualitative nature of the nZEB definition leaves some room for interpretation, MS may follow different paths and uphold different standards in order to fulfil the directive.
ZEMedS project focuses on renovating schools to nZEB in the Mediterranean regions, an area which represents 17% of EU-27 population. This document presents the current situation in 4 Mediterranean countries (France, Greece, Italy and Spain) regarding nZEB approach and first developments to tackle school nZEB renovation in a holistic approach.

2 CURRENT SITUATION OF MEDITERRANEAN SCHOOLS

2.1 Spanish School Buildings

Spain is gradually implementing the European Directives related to energy efficiency in buildings. As regards to nearly Zero Energy Buildings, a roadmap is still to be made available and no national voluntary labels exist in this approach or even in positive energy buildings. Concerning best practices, some initiatives are in place to reduce energy consumption in schools, regarding mainly the use and management. However, no school has been identified to be renovated following a holistic approach and reaching low energy consumption. In Spain, the lack of data regarding energy performance of current buildings is an important barrier when it is time to renovate the built stock. In addition, even that the current indoor conditions are not satisfying minimum standards in many schools (low ventilation rates, overheating, glare problems, etc.), indoor environmental quality has not been assessed in scholar buildings up to date. A program to carry out energy and IEQ assessments in schools would constitute a good option to provide data to a “Schools observatory” or even a “Buildings observatory”. Even so, some available information on mean energy consumption in 354 Catalan schools have shown a wide range of values, 68-122 kWh/m^2/year, being the thermal contribution 60-90%. At first, buildings built before first thermal regulations in force are generally supposed to consume more energy than recent built. However, increased comfort and use of new technologies, has led to the situation where often schools built last decades may consume more than that from the 60-70’s.

2.2 Greek School Buildings

In Greece, transposition of the European Directive 2009/28/EC took effect in June 2010 by the national law N.3851/2010 on RES (FEK 85/A/4.6.2010). All public buildings by 2015 and all new buildings by 2020 should cover their primary energy consumption from RES, combined heat and power, district heating or cooling, and energy efficient heat pumps. The latest Greek energy regulation exacts from the 2010/31 EPBD recast was laid out on February of 2013. This law describes a more command and control approach and also encompasses the 2020’s nZEB time-restriction. However, research has to be done to define the Greek roadmap for nZEBs, with numerical indicator for energy demand and the share of renewable energy sources. So far there is not any national law that embodies the 2012/27 EED as far as renovation rates of public buildings are concerned. Greece currently has 15,446 schools of which 4,500 are over 45 years old. The total energy consumption of school buildings is around 270,000 MWh. As of 2011, in order to get a new building permit it is necessary to achieve an annual solar fraction of 60% for sanitary hot water production from solar thermal systems (Greek NREAP, 2010), or demonstrate the technical difficulties that prevented compliance. New buildings and existing buildings undergoing major renovation must be able to obtain a class B energy certificate upon completion and are required to have certain minimum U-values and heat recovery in central air-conditioning units. According to the scientific research and literature, the average energy consumption in Greek Secondary Schools was bill-based estimated at 16 kWh/m²/y for electricity and for space heating with oil at 68 kWh/m²/y (Gaitani, 2010). The mean energy consumption has been categorized per climate zone with a range from 49kWh/m²/year up to 90kWh/m²/year (Dascalaki, 2012). For Greek Schools, the School Building Organisation (SBO) is credited by the national budget for all the expenditures related to infrastructure throughout the country. SBO is undertaking the construction of schools through the alternative finance method of PPPs.

2.3 Italian School Buildings

Italy adopted the EPBD-Recast Directive in August 2013 but the decrees (action plan and definitions) are still missing. This is slowing down the diffusion of the nZEB concept and its application, as technical regulation in force is still the one related to the previous Directive 2002/91/CE – EPBD. Regulation constraints refer mainly to heating consumption, while for cooling consumption only few aspects are considered. More than 60% of Italian school buildings were built without any energy-related regulation in force (before 1976) and less than 10% were built after the adoption of the Law 10:1991 which is the first regulation in Italy introducing clear constraints about energy efficiency. Currently, the lack of appropriate energy management may often conduct to the situation during mid seasons where the heating system is on but school users open the windows. The great majority of schools are public, and as the public bodies facing a lot of economical problems in the last years due to the financial crisis, the possibilities for deep renovations are minor. The status of school buildings is getting worse but there is a vast lack in funding. Furthermore, a lot of school buildings would need a seismic upgrading, that is considered more urgent than energetic upgrading, and however it is a lot more expensive.

2.4 French School Buildings

In France a series of targets have been set by the Environment Round Table and implemented in law as of 2009. The widespread development of new, low consumption buildings has been encouraged, the next step being the development of positive energy buildings by 2020. Extensive renovations of the existing building stock are under way, with the goal being about 400,000 renovations per year (Ecofys, 2013) leading to a 38% reduction in primary energy consumption by 2020. Public buildings are to be renovated to achieve a minimum reduction of 40% in energy and 50% in greenhouse gas emissions within 8 years (French NREAP, 2010). Local authorities are the decision makers when it comes to renovating a school. But they often do not have information about the energy consumption or indicators to assess comfort in their buildings. Thus, the decision to renovate a school is not always dependent on the level of consumption of the building but, in fact, is a political choice. However, when a renovation project is approved, the local authorities set the goals of energy efficiency and comfort level and budget of the operation. They appoint a technical team (architects and consultants) for this project and to connect with companies that carry out the mission. In addition, other induced work is necessary and can have a significant impact on the budget of the operation. Indeed, other regulatory constraints related to the accessibility of the disabled or fire safety must be taken into account. Cities so often encounter funding problems and must set priorities. Currently, thermal regulation for existing buildings does not go far enough and to achieve a NZEB goal, is up to the owners to set the target at the beginning of the project. Although they might lack the expertise in this domain. In France, especially in Languedoc-

Roussillon, the first examples of successful renovation are becoming increasingly known. They were initiated and financially supported by the Languedoc -Roussillon and ADEME through calls for proposals. They gave a good example of how the project should be managed in terms of energy performance and how to take into consideration summer comfort. Nevertheless, the accurate results will be derived in the future and current quantitative indicators are missing.

2.5 MED particularities

The majority of the regions with Mediterranean climates have relatively mild winters and hot summers. Although significant variations can be found among the places that satisfy the Mediterranean climate criteria, the countries bordering the basin share some similarities: in almost all the coastline cities, the minimum yearly average temperature is between 5–10°C and the maximum is between 27–34°C.For the past few decades there has been a sizeable increase in summer cooling demand in the Mediterranean area, especially in urban areas. Global warming is expected to adversely affect both the environment and human activities in the Mediterranean area, with scenarios for average yearly air temperatures predicting an increase between 2.2 and 5.1°C by 2100, or even sooner than that. According to the IPCC (2007), an average temperature rise above 1.5°C is likely to have severe impacts in local environments and ecosystems, while increased temperatures are expected to bring about longer heat waves, decreased precipitation and a longer summer in general. These predictions should be taken seriously into consideration when renovating buildings and, particularly, schools.
School holidays in Mediterranean are mainly in Summer (2 months for teachers, 3 months for students) and this is why the almost the totality of schools has not a cooling system. As a result, schools have a not-too-high consumption, but this causes also a lot of comfort problems (from April to October).
Decision makers in MED regions are facing a complex situation, where needed energy renovations should integrate not only energy savings but also health and comfort requirements, especially in the framework of climate change. Moreover, current economic crisis makes this need more challenging. Overall cost analysis would be a great tool for decision makers to help identify the best solution. Training of decision makers as well as of the designers is necessary to raise awareness of these issues.

3 INTEGRATED APPROACH TOWARDS NZEB

European supported project ZEMedS, standing for Zero Energy Mediterranean Schools, will help overcome two main barriers: insufficient knowledge about existing technologies to renovate schools up to nZEB goals and inefficient tools for decision makers undergo these renovations. ZEMedS includes, among other activities, developing high quality TOOLkit for both building designers and decision makers, studying 10 real schools and developing tendering specifications.

3.1 Integrated Design

Buildings demand energy in their life cycle, directly for their construction, operating energy as well as for the rehabilitation/ demolition. In addition, they demand indirectly through the embodied energy. Integrated design is a concerted methodology which emphasizes the development of a holistic building design. This approach invites all parties into the planning process, and takes into account all the areas of building's expertise plus the users.

In the framework of ZEMedS project, it is proposed to follow a holistic approach. Integrated Design methodology offers taking into account all the related criteria and actors involved. According to MaTrID project, ID steps are the following: 0. Project development; 1. Design basis; 2. Iterative problem solving; 3. On track monitoring; 4. Delivery; 5. In use.
Focusing on school renovation, the project SchoolVentCool proposes a school renovation methodology that considers: technical and structural aspects, ecological aspects, educational and socioeconomic aspects and economic aspects. The methodology starts with having an overview of the entire building portfolio of public bodies, in order to make the investment following a master plan.

Figure 1SchoolVentCool methodology

Finally, integrated approach must consider from the beginning the costs and benefits during all building life cycle. If the focus is on the investment costs, without having the overall picture of the savings and benefits in the energy use, but also in better health learning outcomes, among other, short-term oriented decisions may be taken, compromising the way towards nZEB goals.

3.2 Energy efficiency in MED Schools

Realization of nearly zero energy buildings entail a wide range of measures related to technologies, systems, and solutions with varying degrees of complexity depending upon the location and surrounding environmental conditions. Lighting, heating, cooling, ventilation and various plug loads need to be minimized while behavioural aspects of the users should be analysed to become more environmental conscious.

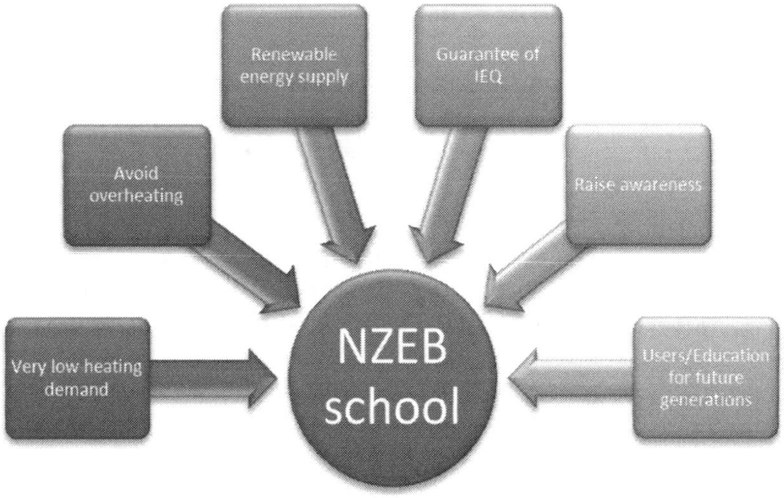

Figure 2Key issues for NZEB renovation in Mediterranean schools

As shown in Fig 1, NZEB renovation of Mediterranean schools not only should consider the energy performance (reduction of energy consumption and renewable energy supply), but should also incorporate criteria which may compromise the objectives of reducing consumption, such as the behaviour of building's user. In Mediterranean climate, summer comfort needs to be seriously addressed. The designers and decision makers of school buildings need to pay particular attention to the quality of indoor environment which will impact on health but also in raising awareness of future generations about these issues.

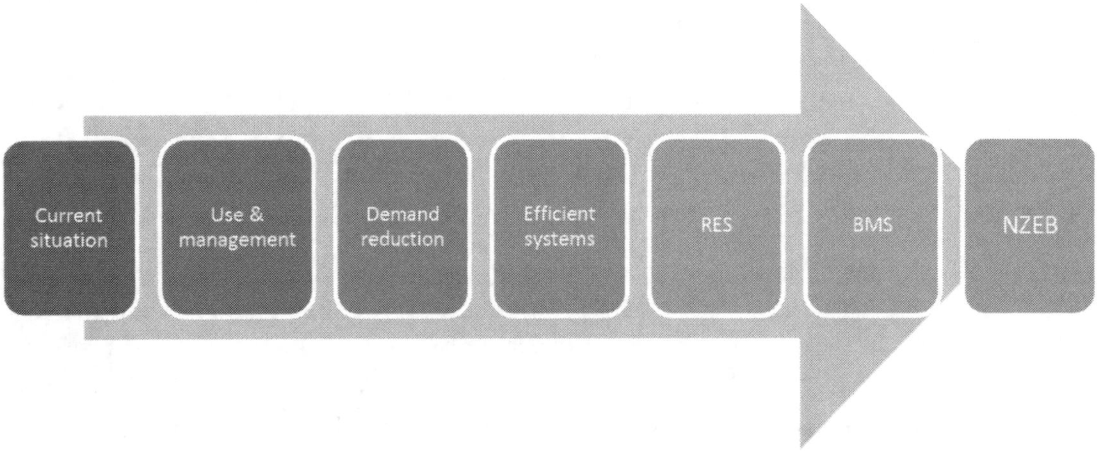

Figure 3Energy steps paving the way to NZEB

ZEMedS project proposes an "energy steps" methodology (fig. 3) in order to achieve the final nZEB goal. Energy strategies to be implemented need to pay particular attention to:

- Users. Energy management should not rely on their sole responsibility. However they are going to play a key role in energy consumption in the school and beyond, when pupils will leave de school. For this reason, users need to be part of the renovation process and take an active role in school every day energy management;
- Heating is the main contributor to the energy demand. Passive heating techniques have been largely applied in colder climates. In MED regions they need to be implemented, taking care of not increasing cooling demand and of cost-effective criteria. For example, heat recovery ventilation may be not cost-effective for a school located on the MED coast.
- Overheating problems appear because of solar radiation and users' heat production (internal gains). In Mediterranean regions, just a range of passive cooling techniques can result to avoid implementing a cooling system. At least, solar shading, cool surfaces and ventilative cooling are to prescribe for all the cases;
- Renewable energy supply. The leading RES potential in whole MED is solar energy. However, other sources may be needed. Biomass from local woods is generally underexploited; however, sustainability and gas emissions need to be studied previously. Wind and geothermal will constitute the other major alternatives.
- Ventilation (See following chapter)

Particularly, a detailed energy and cost analysis will be applied with EnergyPlus software for a year in 10 typical school buildingsfrom 5 regions (Barcelona (ESP) Montpellier (FR), Ancona (IT), Peristeri (GR) , Tuscany (IT)). The main objectives are (1) to assess how an existing building is far from the performances of a nZEB, and (2) is to assess the effort in retrofitting up to NZEB goal.

3.3 Indoor Environmental Quality

Together with the energy demand, the general requirements of indoor climate conditions should be considered, in order to avoid possible adverse consequences. The school buildings in general are characterized by a high density of people per unit area, which is associated with increased concentrations of certain pollutants and therefore with reduced attentiveness of students and less ability to learn. Moreover, students are a vulnerable group necessitating a special attention. Characterized by: (1) Greater respiratory rate compared with adults; (2) Lower weight; (3) Higher levels of physical activity; (4)The human lung continues to develop until adolescence. As it concerns Indoor Environmental Quality (IEQ) in Mediterranean schools, the key issues could be summarized:

- Indoor environment/Ventilation: The priority is to reduce indoor pollutant sources (and outdoors when possible), After that, a ventilation strategy needs to be defined according to many criteria (building characteristics, local climate, current outdoor and indoor pollutants, energy performance, thermal comfort, users, cost). The existing strategy is usually natural ventilation but often results in poor indoor conditions. Setting the ventilation requirements is quite challenging (current standards for schools may vary 1000-1500 ppm limit value for CO_2, while ventilation rates range 5.5-12.5 l/s/person). In addition, according to ANSES, current knowledge in IAQ and health effects is not enough consistent to set limit values that will guarantee health protection, performance and comfort perception. Some key findings (Simoni et al, 2010) point out that a higher frequency of symptoms linked to child asthma could be associated to CO_2 concentrations higher than 1000 ppm. Apart from that, ANSES states that a unique CO_2 indicator is not enough to monitor and guarantee IAQ because of other so important pollutants that may not be always linked (VOC's, particulate matter (PM2.5; PM10...)...). The chosen ventilation strategy should guarantee a healthy air, help preventing overheating and achieve energy performance.

- Daylighting is often easy to be guaranteed as most of the schools do not usually have important depths and are also equipped with large window areas. Nevertheless shading system must be optimized with daylight requirements to avoid glare problems

- Thermal discomfort conditions have to be considered carefully mainly because of the high occupant density in classrooms and because of the negative influences that an unsatisfactory thermal environment has on learning performance of the pupils. Overheating is a problem for MED school building design. Generally, there are no cooling systems in school buildings. As new buildings are built with more thermal insulation and to improved standards of air tightness, concerns are emerging of an increased risk of overheating. Alongside the key issue of indoor air quality, overheating is a risk that needs to be managed carefully as we move further towards the aim of nZEB and now ranks among greatest concerns that need to be addressed as a priority

4 CONCLUDING REMARKS

The main objectives of ZEMedS project includes the assessment of energy conservation potentials in Mediterranean schools in relation to the environmental quality perspectives. School buildings feature indoor air quality problems while their energy consumption and global environmental quality could be improved significantly. Many studies have identified

the lack of date base, knowledge, experience and best-practice examples as barriers in refurbishment projects. School renovation should be focused on energy savings but also on improving indoor environment for better learning outcomes. Setting the priorities on different needs (safety, maintenance, spatial requirements, energy savings, etc.) is profoundly depend on the budget availability and the existing funding channels. Nowadays, the economic crisis in most of Mediterranean countries has led to very reduced self financing capacity, being the budgets allocated to cover only urgent needs and significantly reducing the capacities of municipalities and regional administrations.
ZEMedS actions are mainly addressed to involve two target groups, school decision makers and building designers, by providing technical and financial assistance on nZEB renovation of schools in Mediterranean climates, and giving support in succeeding in implementing school renovation with nZEB goals. An holistic approach will combine cost optimal measures to achieve energy performance and indoor environmental quality (IEQ) in schools. ZEMedS project will pave the way for a complete renovation path, tackling strategies for the envelope, the systems and renewable energy applications as well as the energy management and users' behaviour of Med Schools.

References

1. European Parliament and Council (2010) DIRECTIVE 2010/31/EU of 19 May 2010 on the energy performance of buildings (recast). Official Journal of the European Union, 18.6.2010

2. Directive 2012/27/EU of the European Parliament and of the Council of 25 October 2012 on energy efficiency, amending Directives 2009/125/EC and 2010/30/EU and repealing Directives 2004/8/EC and 2006/32/EC

3. [French NREAP] French Ministry for Ecology, Energy, Sustainable Development and Sea (2010) National Action plan for the promotion of renewable energies 2009-2020

4. [Greek NREAP] Greek Ministry of Environment, Energy and Climate Change (2010) National Renewable Energy Action Plan in the Scope of Directive 2009/28/EC

5. [Italian NREAP] Italian Ministry for Economic Development (2010) Italian National Renewable Energy Action Plan

6. [Spanish NREAP] Spanish Ministry of Industry, Tourism and Commerce (2010) Spain's National Renewable Energy Action Plan 2011-2020

7. IEA ECB&CS Annex 36 - Retrofitting in Educational Buildings - Energy Concept Advisor for Technical Retrofit Measures. Heike Kluttig Evaluation of Questionnaire on the Energy Consumption of Educational Buildings . In: T. Mroz& H. Erhorn, State of the Art Overview: Questionnaire Evaluations. IEA ECB&CS Annex 36

8. ECOFYS 2011. Panorama of the European non-residential construction sector

9. Ferrante, A (2012) 'Zero- and low-energy housing for the Mediterranean climate', Advances in Building Energy Research, 6:1, pp. 81-118

10. Gaitani N. PhD. Energy Rating & Environmental Quality of School Buildings in Greece (2010)

11. Dascalaki, E.G., Balaras, C.A., Gaglia, A.G., Droutsa, K.G., Kontoyiannidis, S. (2012) 'Energy Performance of Buildings – EPBD in Greece', Energy Policy 45, pp. 469–477

12. Ecofys (2013). Towards nearly zero-energy buildings - Definition of common principles under the EPBD - Final report. Available at: http://ec.europa.eu/energy/efficiency/buildings/doc/nZEB_full_report.pdf [Accessed 12th September 2013]

13. Torcellini, P., Pless, S., &Deru, M., Crawley, D. (2006) Zero energy buildings: A critical look at the definition; Preprint. Available at http://www.nrel.gov/docs/fy06osti/39833.pdf [Accessed 12th September 2013]

14. 2013 ANSES (French Agency for Food, Environmental and Occupational Health & Safety). CO2 concentrations in indoor air and health effects (Concentrations de CO2 dans l'air intérieur et effets sur la santé). Scientific edition (in French) Free download: https://www.anses.fr/en/documents/AIR2012sa0093Ra.pdf

15. Integrated Design Guidelines. MaTrID project. http://www.integrateddesign.eu/downloads/MaTrID_Process-Guideline.pdf
16. Brochure SchoolVentCool_The Way Towards Your Cool School. http://www.schoolventcool.eu/

17. HealthVent project. http://www.healthvent.byg.dtu.dk/

18. Simoni M, Annesi-Maesano I, Sigsgaard T, Norbäck D, Wieslander G, Nystad W, Canciani M, Sestini P, Viegi G (2010), School air quality related to dry cough, rhinitis and nasal patency in children, European Respiratory Journal, 35:742-749

19. ZEMedS: http://www.zemeds.eu/

Event sponsors

The conference organizers are grateful to the following companies for their support to this event

International Network for Information on Ventilation and Energy Performance (INIVE)
Lozenberg 7
B-1932 Sint-Stevens-Woluwe
Belgium

ISBN 978-1-5108-3763-8

CURRAN ASSOCIATES INC.
proceedings.com